Genomic Applicat

George J. Netto • Iris Schrijver
Editors

Genomic Applications in Pathology

Springer

Editors
George Jabboure Netto
Departments of Pathology,
Oncology and Urology
Johns Hopkins University
Baltimore, MD, USA

Iris Schrijver
Department of Pathology
Stanford University School of Medicine
Stanford, CA, USA

ISBN 978-1-4939-4647-1 ISBN 978-1-4939-0727-4 (eBook)
DOI 10.1007/978-1-4939-0727-4
Springer New York Heidelberg Dordrecht London

Printed on acid-free paper

Springer is part of Springer Science+Business Media (www.springer.com)

Preface

The pathologist has an increasingly central role in the management of cancer patients in the era of personalized oncology. Molecular diagnostic and genomic applications are rapidly penetrating the daily practice of the pathologist as the list of actionable genetic alterations in solid and hematologic malignancies continues to expand. At the same time, a paradigm shift in the diagnostic approach for inherited genetic diseases, infectious diseases, and pharmacogenetics is unfolding. As a result, a plethora of clinical genomic applications is being rapidly implemented in diagnostic molecular pathology laboratories as we move closer to the anticipated reality of "precision medicine."

This textbook provides a comprehensive resource of genomic applications to practicing molecular pathologists and hematopathologists, general and subspecialized practicing pathologists, as well as pathology trainees. The target audience also includes oncologists, geneticists, and other medical and surgical clinicians. The 33 chapters encompass a state-of-the-art review of the scientific principles underlying current and emerging genomic technologies and the bioinformatics approaches required to effectively analyze the daunting amount of data generated by next-generation sequencing. Implementation roadmaps for various clinical assays including single gene, gene panel, whole exome, and whole genome assays are addressed. Topics related to reporting and the pathologist's and laboratorian's role in the interpretation and clinical integration of genomic test results are discussed. Practice-related considerations including the regulatory framework, reimbursement, legal and ethical issues as related to genomic testing are also included. Importantly, chapters on genomic applications for site-specific solid tumors and hematologic and lymphoid neoplasms provide a review with practical and actionable information regarding the latest advances. Finally, genomic applications in pharmacogenomics, inherited genetic diseases, and infectious diseases are also highlighted.

As this most exciting field continues to evolve rapidly, the information in this textbook provides an up-to-date framework for the transition of next-generation sequencing applications from bench to bedside, for genomic assay development, and for responsible implementation of genome-scale testing. We hope that you will enjoy the keen insights from our 62 expert authors and that this text will prove to be a valuable tool in your practice, as it is to ours.

George J. Netto
Iris Schrijver

Contents

Part I Genomic Technologies

1 **Current Massively Parallel Sequencing Technologies: Platforms and Reporting Considerations** ... 3
John R. ten Bosch, Wayne W. Grody

2 **Emerging Next-Generation Sequencing Technologies** ... 19
Matthew W. Anderson

3 **Transcriptome Sequencing (RNA-Seq)** ... 31
Sugganth Daniel, Alberto Paniz-Mondolfi, Federico A. Monzon

4 **MiRNA Expression Assays** ... 45
Cornelia Braicu, Beatriz M. Maia, Ioana Berindan-Neagoe, George A. Calin

5 **Circulating Tumor Cells: Enrichment and Genomic Applications** ... 71
Marija Balic, Richard J. Cote

6 **Circulating Cell-Free DNA for Molecular Diagnostics and Therapeutic Monitoring** ... 85
Julia A. Beaver, Ben H. Park

Part II Practice Related Aspects of Clinical Genomics

7 **Genomic Pathology: Training for New Technology** ... 101
Richard L. Haspel

8 **Implementation of Genome Sequencing Assays** ... 111
Joshua L. Deignan

9 **Regulatory and Reimbursement Issues in Genomic Testing** ... 119
Jan A. Nowak

10 **Patents and Proprietary Assays** 129
Roger D. Klein

11 **Ethical Issues in Clinical Genetics and Genomics** 141
Henry T. Greely

Part III Clinical Implementation of Diagnostic Genomics

12 **Transitioning Discoveries from Cancer Genomics Research Laboratories into Pathology Practice** 159
Tamara Jamaspishvili, Jeremy A. Squire

13 **Bioinformatics Tools in Clinical Genomics** 177
David K. Crockett, Karl V. Voelkerding

14 **Next Generation Sequencing for Single-Gene Analysis** 197
Hao Ho, Christopher D. Gocke

15 **Next-Generation Sequencing for Gene Panels** 209
Michael O. Dorschner

16 **Implementation of Exome Sequencing Assay** 225
Arunkanth Ankala, Madhuri R. Hegde

17 **Implementation of Genome Sequencing Assays** 241
Tina M. Hambuch, Carri-Lyn Mead

18 **Clinical Information Systems in the Era of Personalized Medicine** 259
Jonathan Nowak, Lynn Bry

19 **Reporting Clinical Genomic Assay Results and the Role of the Pathologist** 277
Janina A. Longtine

Part IV Genomic Applications in Oncology

20 **Genomic Applications in Hematologic Oncology** 297
Kevin E. Fisher, Charles E. Hill

21 **Genomic Applications in Brain Tumors** 321
Matija Snuderl

22 **Genomic Applications in Head and Neck Cancers** 341
Sun M. Ahn, Nishant Agrawal

23 **Genomic Applications in Breast Carcinoma** 359
Leticia De Mattos-Arruda, Britta Weigelt, Jorge S. Reis-Filho

24 **Genomic Applications in Pulmonary Malignancies** 383
Reinhard Büttner, Margarete Odenthal, Sabine Merkelbach-Bruse

25 Genomic Applications in Colorectal and Pancreatic Tumors 415
Andreas Jung, Ralph Hruban

26 Molecular Pathology of Genitourinary Cancers: Translating the Cancer Genome to the Clinic 435
Joshua I. Warrick, Scott A. Tomlins

27 Genomic Applications in Gynecologic Malignancies 465
Sarah Chiang, Luciano G. Martelotto, Britta Weigelt

28 Genomic Applications in Epithelial Ovarian Malignancies 489
James D. Brenton

29 Genomic Applications in Soft Tissue Sarcomas 503
Eva Wardelmann, Wolfgang Hartmann

Part V Genomic Applications of Inherited and Infectious Diseases

30 Genomic Applications in Inherited Genetic Disorders 535
Bryan L. Krock, Rong Mao, D. Hunter Best, Elaine Lyon

31 Genomic Applications in Pharmacogenomics 553
Rehan Qayyum

32 Sequencing Cell Free DNA in the Maternal Circulation to Screen for Down Syndrome and Other Common Aneuploidies 563
Glenn E. Palomaki

33 Genomic Applications in the Clinical Management of Infectious Diseases 581
Martina I. Lefterova, Niaz Banaei, Benjamin A. Pinsky

Index 605

Contributors

Nishant Agrawal Department of Otolaryngology – Head and Neck Surgery, Johns Hopkins Medical Institutions, Baltimore, MD, USA

Sun M. Ahn Department of Otolaryngology – Head and Neck Surgery, Johns Hopkins Medical Institutions, Baltimore, MD, USA

Matthew W. Anderson Diagnostic Laboratories, BloodCenter of Wisconsin, Milwaukee, WI, USA

Arunkanth Ankala Department of Human Genetics, Emory University School of Medicine, Atlanta, GA, USA

Marija Balic Division of Oncology, Department of Internal Medicine, Medical University of Graz, Graz, Austria

Niaz Banaei Department of Pathology, Stanford University School of Medicine, Stanford, CA, USA

Division of Infectious Diseases and Geographic Medicine, Department of Medicine, Stanford University School of Medicine, Stanford, CA, USA

Clinical Microbiology and Virology Laboratories, Stanford University School of Medicine, Stanford, CA, USA

Julia A. Beaver The Sidney Kimmel Comprehensive Cancer Center at Johns Hopkins, Baltimore, MD, USA

Ioana Berindan-Neagoe Research Center for Functional Genomics, Biomedicine and Translational Medicine and The Department of Immunology, Iuliu Hatieganu University of Medicine and Pharmacy, Cluj-Napoca, Romania

Department of Functional Genomics and Experimental Pathology, The Oncological Institute "Prof. Dr. Ion Chiricuta", Cluj-Napoca, Romania

D. Hunter Best ARUP Laboratories, Department of Pathology, University of Utah School of Medicine, Salt Lake City, UT, USA

John R. ten Bosch Molecular Genetics Laboratory, Department of Genetics, Kaiser Permanente Medical Center, San Jose, CA, USA

Cornelia Braicu Research Center for Functional Genomics, Biomedicine and Translational Medicine, Iuliu Hatieganu University of Medicine and Pharmacy, Cluj-Napoca, Romania

James D. Brenton Functional Genomics of Ovarian Cancer Laboratory, Cancer Research UK Cambridge Institute, University of Cambridge, Cambridge, UK

Lynn Bry Department of Pathology, Brigham and Women's Hospital, Harvard Medical School, Boston, MA, USA

Reinhard Büttner Center for Integrated Oncology, Institute of Pathology, University Hospital Cologne, Köln-Bonn, Cologne, Germany

George A. Calin Department of Experimental Therapeutics, The University of Texas MD Anderson Cancer Center, Houston, TX, USA

Center for RNA Interference and Non-Coding RNAs, The University of Texas MD Anderson Cancer Center, Houston, TX, USA

Sarah Chiang Department of Pathology, Memorial Sloan-Kettering Cancer Center, New York, NY, USA

Richard J. Cote Department of Pathology, University of Miami Miller School of Medicine, Miami, FL, USA

David K. Crockett ARUP Laboratories, Department of Pathology, University of Utah School of Medicine, Salt Lake City, UT, USA

Sugganth Daniel Department of Pathology and Immunology, Baylor College of Medicine, Houston, TX, USA

Joshua L. Deignan Department of Pathology and Laboratory Medicine, David Geffen School of Medicine at UCLA, Los Angeles, CA, USA

Michael O. Dorschner Department of Pathology, University of Washington, Seattle, WA, USA

Kevin E. Fisher Molecular Genetic Pathology, Emory University Hospital, Atlanta, GA, USA

Christopher D. Gocke Division of Molecular Pathology, Department of Pathology, Johns Hopkins University School of Medicine, Baltimore, MD, USA

Henry T. Greely Center for Law and the Biosciences, Stanford University, Stanford, CA, USA

Wayne W. Grody Department of Pathology and Laboratory Medicine, UCLA School of Medicine, Los Angeles, CA, USA

Department of Pediatrics and Human Genetics, UCLA School of Medicine, Los Angeles, CA, USA

Molecular Diagnostic Laboratories and Clinical Genomics Center, UCLA Medical Center, Los Angeles, CA, USA

Tina M. Hambuch Illumina Clinical Services Laboratory, San Diego, CA, USA

Wolfgang Hartmann Gerhard-Domagk-Institute of Pathology, University Hospital Münster, Münster, Germany

Richard L. Haspel Beth Israel Deaconess Medical Center and Harvard Medical School, Boston, MA, USA

Madhuri R. Hegde Department of Human Genetics, Emory University School of Medicine, Atlanta, GA, USA

Charles E. Hill Molecular Diagnostics Laboratory, Emory University Hospital, Atlanta, GA, USA

Hao Ho Division of Molecular Pathology, Department of Pathology, Johns Hopkins University School of Medicine, Baltimore, MD, USA

Ralph Hruban Departments of Pathology and Oncology, Johns Hopkins University, Baltimore, MD, USA

Tamara Jamaspishvili Department of Pathology and Molecular Medicine, Queen's University, Kingston, ON, Canada

Andreas Jung Pathologisches Institut, Ludwig-Maximilians Universität München, Munich, Germany

Roger D. Klein Department of Molecular Pathology, Robert J. Tomsich Pathology and Laboratory Medicine Institute, Cleveland Clinic, Cleveland, OH, USA

Bryan L. Krock ARUP Laboratories, Department of Pathology, University of Utah School of Medicine, Salt Lake City, UT, USA

Martina I. Lefterova Department of Pathology, Stanford University School of Medicine, Stanford, CA, USA

Janina A. Longtine Department of Pathology, The Mount Sinai Medical Center, New York, NY, USA

Elaine Lyon ARUP Laboratories, Department of Pathology, University of Utah School of Medicine, Salt Lake City, UT, USA

Beatriz M. Maia Department of Experimental Therapeutics, The University of Texas MD Anderson Cancer Center, Houston, TX, USA

Rong Mao ARUP Laboratories, Department of Pathology, University of Utah School of Medicine, Salt Lake City, UT, USA

Luciano G. Martelotto Department of Pathology, Memorial Sloan-Kettering Cancer Center, New York, NY, USA

Leticia De Mattos-Arruda Department of Medical Oncology, Vall d'Hebron Institute of Oncology, Vall d'Hebron University Hospital, Barcelona, Spain

Universitat Autònoma de Barcelona, Barcelona, Spain

Department of Pathology, Memorial Sloan Kettering Cancer Center, New York, NY, USA

Carri-Lyn Mead Illumina Clinical Services Laboratory, San Diego, CA, USA

Sabine Merkelbach-Bruse Center for Integrated Oncology, Institute of Pathology, University Hospital Cologne, Köln-Bonn, Cologne, Germany

Federico A. Monzon Department of Pathology and Immunology, Baylor College of Medicine, Houston, TX, USA

Department of Molecular and Human Genetics, Baylor College of Medicine, Houston, TX, USA

Jonathan Nowak Department of Pathology, Brigham and Women's Hospital, Harvard Medical School, Boston, MA, USA

Jan A. Nowak Department of Pathology and Laboratory Medicine, NorthShore University HealthSystem, Evanston, IL, USA

Margarete Odenthal Center for Integrated Oncology, Institute of Pathology, University Hospital Cologne, Köln-Bonn, Cologne, Germany

Glenn E. Palomaki Division of Medical Screening and Special Testing, Department of Pathology and Laboratory Medicine, Women and Infants Hospital/Alpert Medical School of Brown University, Providence, RI, USA

Savjani Institute for Health Research, Windham, ME, USA

Alberto Paniz-Mondolfi Department of Pathology and Immunology, Baylor College of Medicine, Houston, TX, USA

Ben H. Park The Sidney Kimmel Comprehensive Cancer Center at Johns Hopkins, Baltimore, MD, USA

Benjamin A. Pinsky Department of Pathology, Stanford University School of Medicine, Stanford, CA, USA

Division of Infectious Diseases and Geographic Medicine, Department of Medicine, Stanford University School of Medicine, Stanford, CA, USA

Clinical Microbiology and Virology Laboratories, Stanford University School of Medicine, Stanford, CA, USA

Rehan Qayyum Division of General Internal, Department of Medicine, Johns Hopkins School of Medicine, Baltimore, MD, USA

Jorge S. Reis-Filho Department of Pathology, Memorial Sloan-Kettering Cancer Center, New York, NY, USA

Matija Snuderl Department of Pathology, NYU Langone Medical Center and Medical School, New York, NY, USA

Jeremy A. Squire Department of Pathology and Molecular Medicine, Queen's University, Kingston, ON, Canada

Scott A. Tomlins Departments of Pathology and Urology, University of Michigan Medical School, Ann Arbor, MI, USA

Michigan Center for Translational Pathology, University of Michigan Medical School, Ann Arbor, MI, USA

Comprehensive Cancer Center, University of Michigan Medical School, Ann Arbor, MI, USA

Karl V. Voelkerding ARUP Laboratories, Department of Pathology, University of Utah School of Medicine, Salt Lake City, UT, USA

Eva Wardelmann Gerhard-Domagk-Institute of Pathology, University Hospital Münster, Münster, Germany

Joshua I. Warrick Department of Pathology, University of Michigan Medical School, Ann Arbor, MI, USA

Britta Weigelt Department of Pathology, Memorial Sloan-Kettering Cancer Center, New York, NY, USA

PART I

GENOMIC TECHNOLOGIES

CHAPTER 1

Current Massively Parallel Sequencing Technologies: Platforms and Reporting Considerations

JOHN R. TEN BOSCH, WAYNE W. GRODY

Introduction 3

Platform Chemistry 4

Rapid Sequencing Revolution 5

Enrichment Techniques 7

Different Tests, Different Outcomes 8

Analysis 9

Interpretation and Reporting 11

Dealing with Incidental Findings 13

Gene Patents 14

Follow-up, Reanalysis, and Duty to Recontact 14

Conclusions 15

Introduction

The Human Genome Project was officially completed in 2003 with the publication of the (near-)complete sequence of 3.3 billion nucleotides in the haploid genome. Launched in 1990, the Project took 13 years and a budget of about $3 billion to sequence the first human genome. Well, at least that was the timeframe of the publicly funded Human Genome Project. A later entrant, the so-called private genome project pursued by the biotechnology company Celera, actually accomplished the same goal in a much shorter time, namely, in about 5 years. This was accomplished by a contrasting sequencing strategy; instead of the organized, targeted, chromosome-by-chromosome, BAC-by-BAC (bacterial artificial chromosome) approach, the Celera group utilized "shotgun" sequencing, a more global sequencing of countless random DNA fragments which were only reassembled at the end into the complete, ordered human genome.

In a similar way, DNA sequencing as performed by both diagnostic and research laboratories all over the world has recently undergone a dramatic transformation in speed and throughput. Instead of the traditional approach of sequencing one small (150–200 bp) DNA region at a time, using a specific pair of complementary primers targeted to just that area of interest, the new sequencing platforms utilize a shotgun approach, randomly shearing the entire genome into over 300 million small fragments, sequencing each of them repeatedly in parallel, and then reconstructing the resulting sequences, using sophisticated computer software, into the complete genome. Just as the entry of the Celera project spurred the total genome sequencing effort to an earlier completion, the advent of this "next-generation" or "massively parallel" DNA sequencing (NGS) technology has truly been a "game-changer", allowing for practical and timely sequencing of large panels of genes, of all the coding regions of the genome (the exome), or of the whole genome itself in

G.J. Netto and I. Schrijver (eds.), *Genomic Applications in Pathology*,
DOI 10.1007/978-1-4939-0727-4_1,

individual research subjects, patients, or non-human samples. And because the approach requires no preexisting knowledge of the target regions (only random/universal primers are used), the technology has opened the way to much new gene discovery and new organism identification (e.g., the microbiome). For the clinical molecular diagnostic laboratory, it has fueled a transition from traditional single-gene testing to a new world of genome-wide sequence analysis in the clinical setting [1].

Platform Chemistry

Early adopters of NGS technology were 454 Life Sciences (acquired by Roche) and Solexa (acquired by Illumina). The 454 platform is a pyrosequencing-based system that produces long sequencing reads currently up to 1 kb in length [2]. As the first commercially available NGS instrument, 454 was a marked improvement to traditional sequencing methods. However, as other competitors such as Illumina and the ABI SOLiD entered the market, 454 struggled to keep pace with the advances in sequencing throughput that led to a sharp decline in sequencing costs over several years (Fig. 1.1). Nevertheless, 454 for years remained a viable alternative in the niche market of groups requiring extra-long reads for the sequencing of complex genomic regions such as the HLA genes [3].

The *S*equencing by *O*ligonucleotide *Li*gation and *D*etection (SOLiD) platform, developed by Life Technologies, sequences DNA fragments via multiple rounds of ligation to fluorescent probes. During the sequencing process each nucleotide is interrogated twice, in separate ligation reactions, to ensure the accuracy of the call. As a result, SOLiD raw reads are highly accurate and require less oversampling than other NGS platforms. Whereas the individual nucleotide calls on these reads were of high quality, the reads were short compared to those produced by Illumina, which had dramatically lengthened reads via chemistry improvements and the adoption of paired-end sequencing. For this reason and a variety of others, the SOLiD platform has fallen out of favor, as evidenced by the acquisition of NGS competitor Ion Torrent by Life Technologies in 2010.

The two NGS platforms most common in diagnostic laboratories today are Illumina and

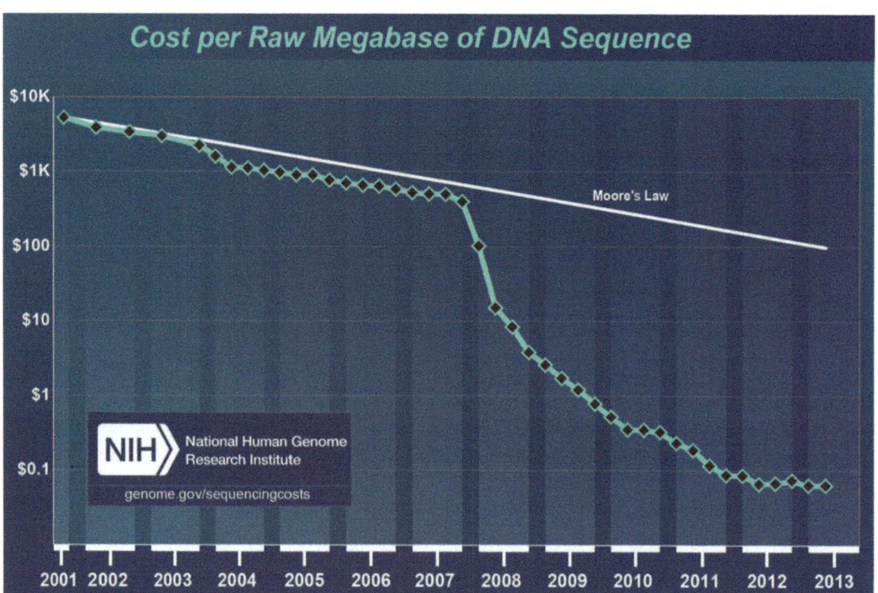

Figure 1-1 Cost of DNA sequencing over time. The cost of a raw megabase of DNA sequence over time is compared to Moore's law. Wikipedia defines Moore's law as the "observation that over the history of computing hardware, the number of transistors on integrated circuits doubles approximately every two years." Since the advent of next-generation sequencing (NGS), the cost of sequencing has outpaced Moore's law by a wide margin. Data and chart provided by the National Human Genome Research Institute (NHGRI)

Ion Torrent. Illumina technology is based on "sequencing-by-synthesis" chemistry, which is enabled by nucleotides containing reversible dye-terminators in the reaction mixture [4]. The dye-terminators halt the extension of growing fragment chains at each of the hundreds of millions of fragment colonies on the flow cell (Fig. 1.2a). The incorporated, dye-conjugated nucleotides are detected via laser excitation and image capture with a high-resolution CCD (charge-coupled device) camera. The dye-terminators are chemically removed before the next sequencing cycle, which begins with the addition of fresh, reversible dye-terminator nucleotides to the flow cell.

Detection of the fluorescence emitted during each sequencing cycle is facilitated by the creation of clonal fragment clusters prior to sequencing. These clonal clusters emit amplified fluorescent signals capable of detection by the instrument's CCD camera. The clusters are formed by first immobilizing a library of fragments onto the flow cell. The fragments are added to the flow cell at a dilute concentration so that, once formed, the clusters seldom overlap. The surface of the flow cell is coated with primers that enable bridge amplification of the immobilized fragments and the formation of the clonal clusters in the presence of polymerase chain reaction (PCR) reagents. Formation of the clonal clusters is the final step before sequencing.

Ion Torrent developed a novel NGS method based on the change in pH that results from the release of a hydrogen ion during nucleotide incorporation [5]. Ion Torrent chips contain millions of microwells with ion-sensitive field-effect transistor (ISFET) sensors. Beads containing emulsion PCR-amplified clonal DNA fragments are deposited onto the chips and loaded into the sensor-containing microwells via centrifugation. Each microwell accommodates a single bead. Unmodified nucleotides are then added to the chip in a stepwise fashion, and the ISFET sensor detects the pH change that results from the addition of a nucleotide to the growing DNA chain (Fig. 1.2b). If more than one nucleotide is incorporated, the change in ion concentration should be proportional. Long homopolymer stretches, however, may not adhere to this rule.

The Ion Torrent instrument completes sequencing runs in a relatively short period of time by industry standards, in part because it does not rely on optics. Because Ion Torrent sequencing chips incorporate the same technology used in the semiconductor industry, any advancements in complementary metal–oxide–semiconductor (CMOS) technology will likely improve the ion chips as well.

As with the Illumina platform, sensitivity in the detection of incorporated nucleotides is facilitated by the amplification of template fragments. The Ion Torrent method uses emulsion PCR to sequester clonal fragments onto beads that are then deposited into individual microwells on the Ion chip. Prior to emulsion PCR, beads are conjugated with unique DNA fragments. The beads are then added to a water–oil mixture containing emulsion droplets infused with PCR reagents. The droplets envelop and isolate individual beads, thereby enabling the clonal amplification of the attached DNA fragments.

Rapid Sequencing Revolution

Once NGS technology was commercialized, subsequent iterations of the original technologies were designed to lower the cost per base of sequencing by generating longer reads and increasing read capacity (Fig. 1.1). Many in the field thought so-called third-generation sequencing technologies, such as single-molecule NGS or nanopore sequencing, would be the next wave of innovations to take hold in the laboratory. Although these third-generation technologies still hold much promise (see the next chapter in this book), improvements in existing NGS run times have made the greatest impact in laboratories that apply this technology. This is especially true in clinical laboratories, where short turnaround time is often paramount [6].

Benchtop sequencers, which derive their name from relatively small laboratory footprints, were the first NGS instruments to offer significant advancements in sequencing run times [7, 8]. These fast run times are, in part, a result of their smaller sequencing capacity, though instrument and chemistry enhancements certainly contributed as well. Benchtop sequencers were, and continue to be, much more affordable and easier to maintain than

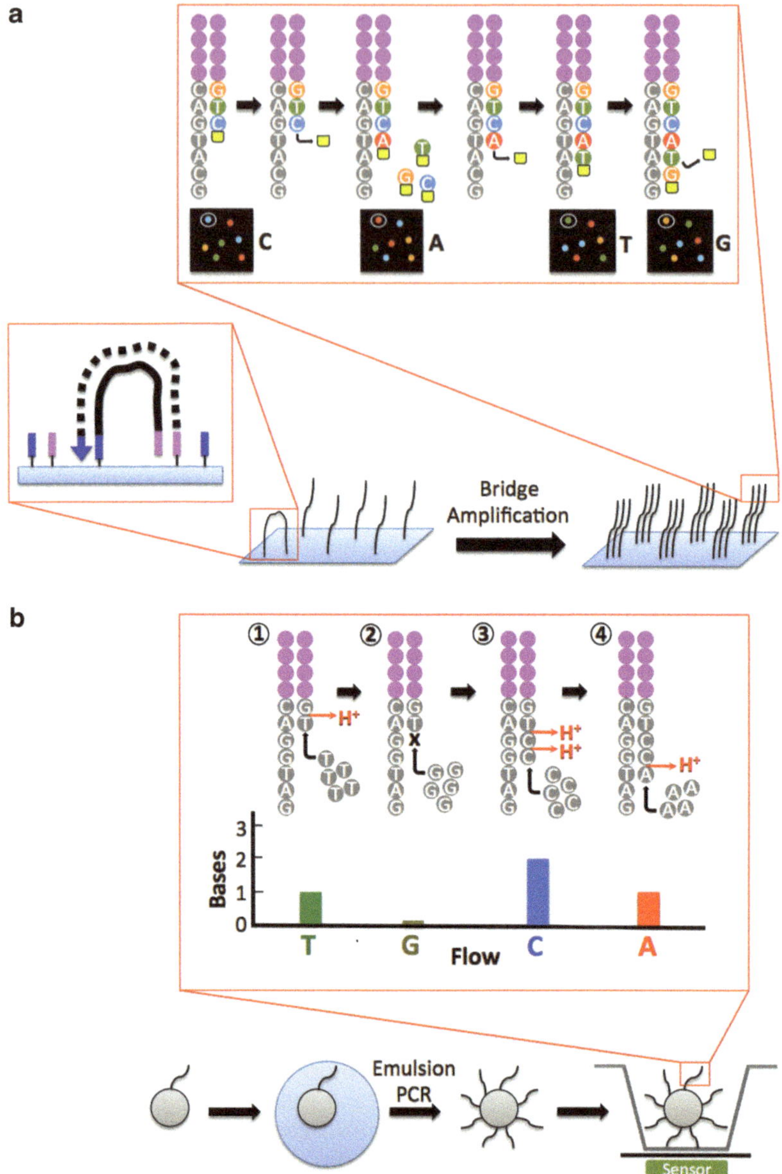

Figure 1-2 NGS Platforms. (**a**) Schematic depiction of the Illumina NGS platform chemistry. DNA fragments are attached to the flow cell and subsequently amplified into clonal clusters. Universal primer ends (*purple* and *blue*) coat the surface of the flow cell and facilitate the formation of the clonal clusters via bridge amplification of individual fragments. Sequencing occurs by the addition of fluorescent nucleotides with reversible terminators (*yellow*). Following laser excitation, images of the flow cell are captured and the incorporated nucleotide at each cluster is identified. The next cycle of sequencing begins by the removal of the reversible terminator and the addition of fresh nucleotides. (**b**) Schematic depiction of the Ion Torrent NGS platform chemistry. Beads containing clonal DNA fragments are produced using emulsion PCR. In this process, a bead containing a single DNA fragment and PCR components is enveloped within an aqueous droplet in an oil mixture. Following amplification, the clonal beads are isolated, enriched, and added to individual microwells on the Ion chip (one bead per well). Unlabeled nucleotides are then added stepwise to the microwells. Incorporated nucleotides are identified by the presence of hydrogen ions released from reaction that produces the phosphodiester bond during polymerization. These hydrogen ions are detected by the ISFET sensor beneath the microwell (*green*). Because the number of hydrogen ions released is proportional to the nucleotides incorporated, the addition of multiple nucleotides in a single cycle can be detected. If the nucleotide present in the microwell is not complementary to the next base in the growing DNA chain, no reaction occurs and no hydrogen ions are released.

their large-capacity predecessors and counterparts. This aspect is why benchtop sequencers remain extremely popular in clinical laboratories with fixed-gene NGS panels. In addition to their affordability, benchtop sequencers typically produce adequate sequence coverage of most gene panels, despite multiplexing, with the added benefit of a quicker turnaround time.

The two most successful NGS benchtop instruments have been the Illumina MiSeq and the Ion Torrent Personal Genome Machine (PGM). The MiSeq instrument leverages the same sequencing-by-synthesis chemistry used by the large-capacity Illumina sequencers, but with shorter run times. Its rapid sequencing runs are the result of a smaller sequencing capacity as well as enhanced fluidics and automated, onboard cluster generation. The popularity of the MiSeq platform prompted Illumina to transfer several of its features to a new large-capacity HiSeq instrument. These improvements confer flexibility to the new HiSeq, which allow the user to choose between a rapid-run or high-output mode. However, even in the "low throughput" rapid-run mode, the HiSeq lives up to its name by producing enough sequence to cover an entire human genome at >30× coverage.

The first commercial instrument sold by Ion Torrent was the PGM, a benchtop sequencer. The PGM is scalable and can accept one of three different capacity Ion sequencing chips, the largest of which produces up to 2 Gb of sequence with 400 bp reads. On-instrument sequencing with the PGM is exceptionally fast because the PGM has no moving parts and no optics, both of which have been rate limiting on other sequencing platforms. More recently, Ion Torrent released its larger capacity Proton instrument. The Ion Torrent Proton generates up to 10 Gb of sequence in a single day with the promise of a Proton chip that will soon enable whole-genome sequencing within a single working day.

Enrichment Techniques

The rapid adoption of NGS technology in diagnostics would not have occurred so rapidly without the introduction of several easy and efficient techniques for isolating regions of the genome. Sequencing even a small gene panel is extremely labor-intensive with traditional PCR techniques and sequencing larger panels is virtually impossible given the time constraints of most clinical tests. Over the past few years, several genomic enrichment techniques have been developed to overcome this bottleneck in targeted NGS. Most of these techniques fall under three general categories of genomic enrichment: bait hybridization, highly multiplexed PCR, and microfluidic technologies. Although there is variability between the systems, a general rule of thumb is that the higher throughput techniques target a smaller number of genes than the techniques that enrich thousands of genes.

Bait hybridization with microarrays was one of the first enrichment techniques developed [9–11]. This method quickly became one of the most popular when a solution-based approach using biotinylated bait probes was developed [12]. Initially offered by Agilent technologies, these solution-based hybridization techniques use long nucleic acids designed to minimize cross-hybridization of undesired sequence as the bait probes. Briefly, genomic DNA is subjected to NGS library preparation. These adapter-ligated DNA fragments are then hybridized to a pool of bait probes complementary to genomic regions of interest. Once hybridized, the target sequences are enriched using streptavidin-coupled magnetic beads and bead washing. The enriched DNA fragments are then amplified in preparation for sequencing.

Traditional PCR has modest multiplexing capability, much less than is required for even the smallest of NGS gene panels. Several different methods have been developed to overcome this limitation of traditional PCR. Circularization of library fragments is one technique that can be used to isolate and enrich for thousands of targets in multiplex [13]. Recently commercialized by Agilent Technologies, this technique, deemed the Haloplex system, is now able to isolate the tens of thousands of targets necessary for an entire human exome. The Illumina TruSeq Amplicon enrichment system uses a technology derived from their SNP GoldenGate genotyping assay to amplify up to 1,536 targets in a single reaction. The TruSeq system, like Agilent Haloplex, isolates regions of interest using oligonucleotide probes with universal priming sites and target-specific ends that facilitate amplification. However, the TruSeq Amplicon system bypasses the need for template circularization by using probes that flank each target. Ion Torrent similarly developed a technique to overcome the barriers of traditional multiplex PCR. Called Ion AmpliSeq, this technology requires only minimal starting DNA material to enrich DNA from 12 to 6,144 targets for sequencing on the Ion Torrent PGM platform.

Microfluidic platforms such as Raindance and Fluidigm take advantage of proprietary instruments that compartmentalize PCR templates and reagents into thousands of PCR minireactors. Despite somewhat similar concepts, the Raindance platform delivers a large set of target sequences from a single sample [14], while the Fluidigm platform isolates a smaller number of targets from multiple samples simultaneously.

Different Tests, Different Outcomes

Choosing an NGS platform can be a difficult task. As a result of rapid NGS innovations, laboratories are not only investing in the current capabilities of a platform but also count on future improvements that will allow that platform to keep pace with the rest of the field. Nevertheless, the half-life of even the most successful NGS instruments, much like the computer infrastructure that supports them, tends to be fairly short. Indeed, starting an NGS laboratory should be considered not a one-time investment, but an ongoing obligation. For this reason, some institutions have chosen to concentrate laboratory resources on NGS analysis and interpretation and to outsource the actual sequencing to another clinical testing laboratory. At least one institution has successfully adopted this approach in order to offer whole-genome sequencing (WGS) to its patients [15].

The principal criterion to consider when deciding on a NGS platform is the purpose for which it will be used. For example, the needs of a laboratory that considers NGS for a carrier screening assay will be much different than the needs of a laboratory that performs WGS—or a laboratory that leverages deep-coverage NGS to identify somatic mutations in a small subset of cancer genes or drug-resistance mutations in subclones of bacterial or viral microorganisms. How many targets must be sequenced per assay and at what read depth? Which kinds of mutations must be detected? What is the expected sample volume and turnaround time for the test? How much DNA will be available for sequencing? These questions highlight some of the most critical parameters to consider before purchasing an NGS instrument.

The requirements of a laboratory performing WGS and a laboratory performing whole-exome sequencing (WES) could be much different given that the protein-coding portions represent only ~1.5 % of the genome. However, the throughput requirements of the two laboratories might be quite similar if the WES laboratory used indexed DNA barcodes to combine multiple samples in each run. In such a scenario, both laboratories would require full-capacity runs of a high-throughput instrument to obtain the necessary amount of sequence in a single instrument run.

WES is an assay that targets all of the approximately 25,000 protein-coding genes in the genome [16]. Diagnostic WES tends to be favored when the phenotype of a patient does not suggest a particular disorder or group of genes. This is in contrast to targeted panels that include genes that, when mutated, contribute to a common, or related set of syndromes (e.g., hearing loss, cardiac abnormalities). Fixed-gene panels cost less and typically guarantee minimum sequence coverage for all of the genes in the panel. However, because a limited number of genes are sequenced in these panels, WES may still be indicated in event of a negative result.

WES targets more genes than panel assays, but its sensitivity is lower [17, 18]. This is because there is more DNA to sequence and because the targeted DNA is difficult to enrich uniformly. Of course, higher coverage for WES is possible, but only at a higher price and with diminishing returns. Most laboratories currently offering clinical exome sequencing do so with recognition that certain regions of the exome might be missed. That being said, with average coverage rates of ~100–200-fold, only a small minority of genes or exons tend to be missed. These low-coverage regions usually have a skewed GC percentage or other sequence-related issues that make enrichment difficult, so coverage problems are often predictable [19, 20].

WGS targets the entire genome, but at increased overall cost and, usually, at lower coverage than WES. However, WGS sample preparation is much easier and is more affordable than the preparation required for WES or even panel sequencing, given that no enrichment step is required. In addition, sequencing data are obtained on intergenic regions, and it is becoming increasingly clear that much intergenic sequence has a biochemical or regulatory function and is not, in fact, "junk DNA" [21].

Mutation detection using NGS is even more complex with cancer samples [22, 23]. This is, in part, due to the fact that cancers are heterogeneous, both at the cellular level with tissue being a mixture of tumor and normal cells, and at the genetic level with different populations of cells harboring different combinations of mutations. In addition, many tumor types have a high rate of genomic rearrangement. Within chromosomal regions of increased ploidy, nucleotide-level variants are diluted even further. Each of these characteristics of tumor samples makes it difficult to detect the potentially mosaic somatic mutations that helped drive cancer transformation. Furthermore, it may be necessary to sequence normal patient tissue in order to distinguish somatic from inherited variants in the patient sample of interest. These challenges have limited most clinical cancer sequencing to relatively small panels of genes that are sequenced to a very high sequencing depth [24, 25].

Analysis

The College of American Pathologists (CAP) laboratory accreditation checklist for molecular pathology defines NGS as two "inextricably linked" processes that must be independently validated and maintained: the analytical or "wet bench" process and the bioinformatics pipeline [26]. Included in the analytical wet bench procedure is library preparation, enrichment, indexing of pooled samples, and the sequencing process itself. Confirmatory testing of reported NGS findings is also included in this section. Bioinformatics includes the pipeline used to support the analysis, interpretation, and reporting of NGS results (Fig. 1.3a). Bioinformatics therefore includes the algorithms used to analyze the results as well as the scripts used to tie together the analysis steps. It also includes any in-house databases used to interpret and store identified variants.

NGS analysis consists of three discrete processes described as primary, secondary, and tertiary stages. In brief, these stages include the conversion of raw NGS data to DNA sequence; the mapping that identifies sequence variants; and the annotation and filtering of variants [27]. Each stage of analysis provides an opportunity for the integration of quality control (QC) measures to avoid potential false calls. NGS bioinformatics pipelines incorporate these QC elements into an automated workflow that ties together the distinct steps of sequence analysis. Most laboratory developed or in-house pipelines use external analysis tools to process the data and internal scripting to facilitate the movement of data through pipeline and file format optimization [28]. Commercial NGS analysis solutions are often composed of proprietary analysis tools that could possibly be strung together to provide an integrated workflow.

Primary NGS data analysis is the process of converting raw data (e.g., images or sensor data) to DNA sequence. Such analysis often occurs on instrument using vendor-provided software. The most common file format output of primary analysis is the FASTQ file [29]. The FASTQ file format is a variant of the well established FASTA format. However, FASTQ files contain both sequence and individual base quality scores. Inclusion of the quality scores allows the trimming and/or removal of poor quality reads prior to mapping. Trimming may be beneficial if a significant decrease in base quality is observed toward the end of a read.

Secondary analysis consists of quality assurance (QA) filtering of raw reads, alignment of reads, and variant calling. NGS platforms operate in a "shotgun" manner, meaning that sequencing reads are obtained by a random sampling of the genomic DNA. For human re-sequencing, individual reads must be mapped to a reference genome in order to determine the locations from which they originated. To permit the mapping of sequence reads that harbor variants, alignment parameters should allow for slight deviations from the reference sequence. Unfortunately, more complex variants, such

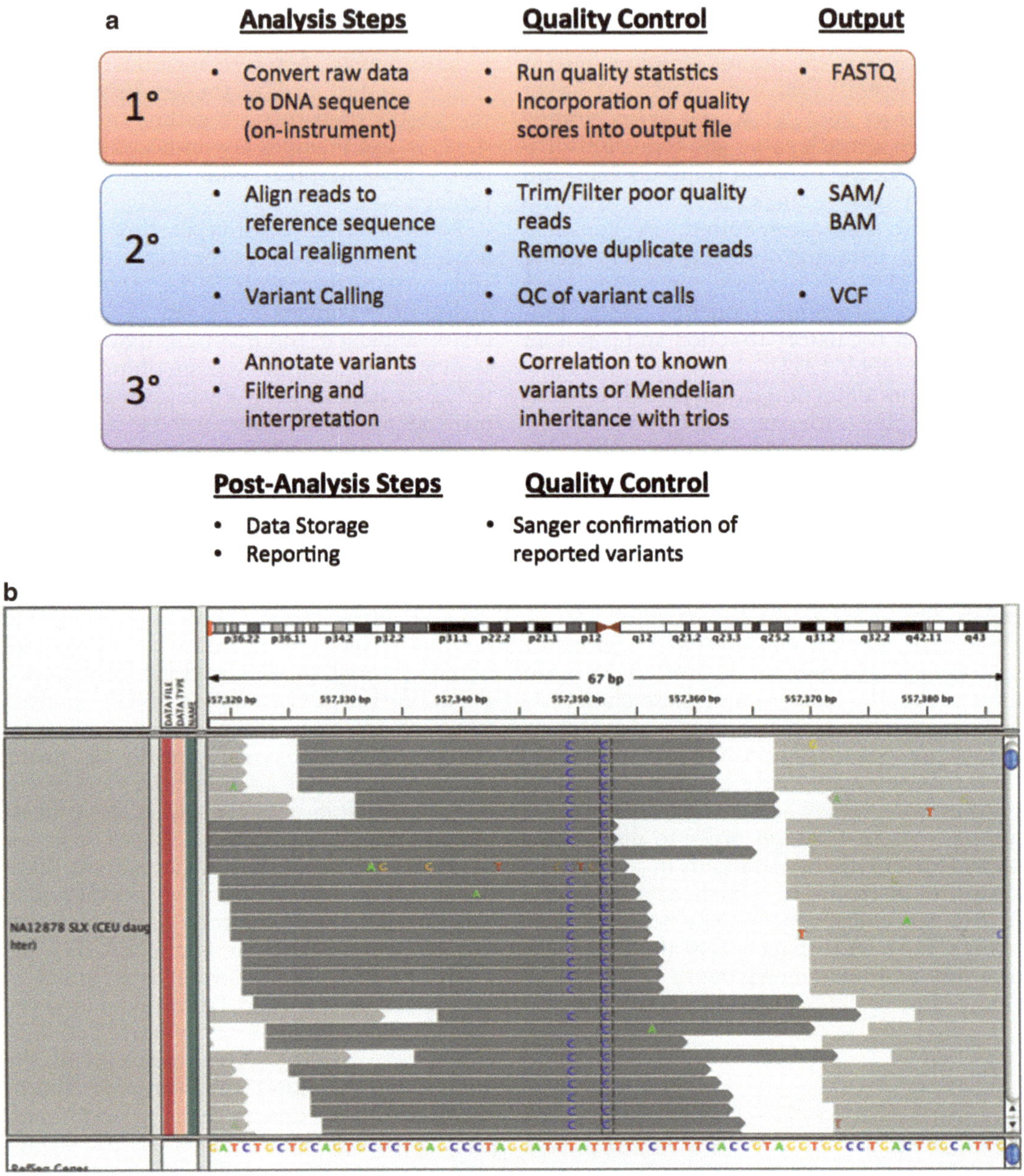

Figure 1-3 NGS analysis. (**a**) Sample NGS analysis pipeline and common post-analysis steps. Analysis steps are split into primary, secondary, and tertiary stages of analysis. (**b**) Visualization of secondary alignment data using the Integrated Genomics Viewer (IGV). IGV is an open-source data visualization viewer created by the Broad Institute [32]. (**c**) Sample tertiary analysis workflow. In this example, the correlation of a phenotype with that of the patient can be performed either at the beginning of the analysis or closer to the end. Where this filter is applied can have profound implications on the variants seen and ultimately reported

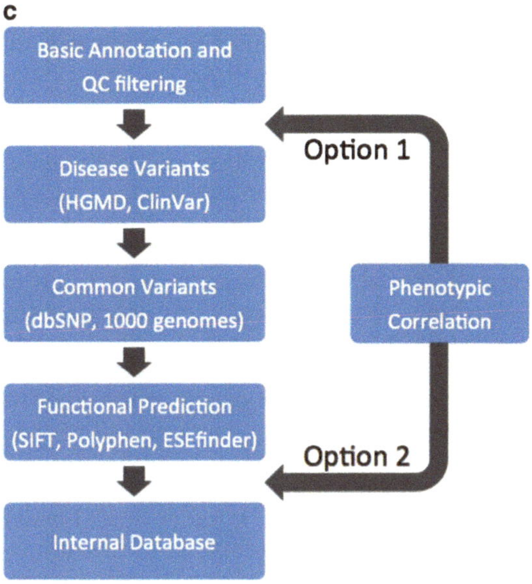

Figure 1-3 (continued)

as indels, are among the most difficult to map precisely because each conflicting nucleotide makes it less likely the read will be recognized as a derivation of the reference sequence [30]. Longer reads help counter this problem by providing additional sequence for comparison to the reference. Longer reads are also useful for alignment to repetitive regions of the genome (such as short tandem repeats and trinucleotide repeat expansions). Reads from pseudogenes are particularly problematic because they can mimic variant-containing reads from the functional gene relative. Another strategy to help avoid misalignment is to employ a local realignment method to fine-tune the results from the initial mapping. When using two rounds of alignment, mapping parameters can be loosened during the first step so that more computationally intensive algorithms can be focused on the reads that require the most attention—those with mismatched nucleotides.

There are several alignment algorithms for human re-sequencing, many of which are freely available as open source software [31]. Each of these mapping algorithms varies in terms of speed, memory requirement, and sequencing platforms supported. The standard output file format for sequence alignment data is the Sequence Alignment/Map (SAM) file and its compressed binary twin, the BAM file [32]. Conversions between these file formats, as well as a variety of other manipulations, can be performed by the open source SAMtools or other similar utilities. In addition, alignment files can be uploaded into applications, such as the open source Integrated Genomics Viewer (IGV), for the visualization of aligned reads and sequence coverage of NGS data [33] (Fig. 1.3b).

Variant calling is the process of identifying the differences between the aligned reads and the reference sequence. The universal format for the variant file is the Variant Call Format (VCF) file, a tab-delimited text file that includes reference and variant nucleotides, chromosome positions, unique variant identifiers, quality scores, and any number of other potential fields the user wishes to add [34]. Tertiary analysis includes annotation of the variant file and the filtering that follows. Populating the supplementary fields of the VCF file typically involves extracting information from databases, both internal and external, to help elucidate the significance of each variant. These databases include dbSNP and the 1000 Genomes Project [35, 36], both of which are used to identify common single nucleotide polymorphisms (SNPs). These SNPs are typically benign variants found in asymptomatic individuals. The Human Gene Mutation Database (HGMD) and the promising public archive, ClinVar, are examples of curated databases that gather information about the significance of variants both benign and pathogenic [37, 38]. Inclusion of phenotypic information associated with a particular gene or variant has proven to be an effective tool for exome and genome sequencing laboratories that deal with thousands and sometimes millions of variants per sample. Another common annotation practice is the inclusion of values from prediction tools that assign variant scores based on the projected pathogenicity of variants [39, 40].

Interpretation and Reporting

Whenever a genomic assay is performed on a patient, there is always a possibility that an incidental or "off-target" result will be found.

This has been true for routine chromosome analysis and also for genomic microarray assays [41–44]. Because exome and genome sequencing are such high-resolution assays capable of detecting an extremely large number of variants, the rates of incidental findings and variants of uncertain clinical significance (VUS) with these tests are the highest yet observed. Indeed, a typical whole-exome sequence produces >20,000 incidental variants, while a whole-genome sequence yields >3 million. This phenomenon not only makes genome-wide test interpretations many orders of magnitude more complex than single-gene or gene-panel tests but also introduces an ethical conundrum as to what should be reported out on particular cases (and when).

Prior to the advent of genome-based diagnostics, the interpretation of molecular assays was, for the most part, performed manually and without the use of software to filter out variants based on a set of assumptions. One of the few exceptions to this practice is the limitation of results from a large mutation panel to a small number of well-characterized mutations, a practice common in laboratories that perform cystic fibrosis screening [45, 46]. These targeted mutation panels are useful in that they identify the most common disease-causing mutations while avoiding VUS. The decision to include or exclude variants from a final list is made only after careful consideration of each candidate mutation [47]. The final mutation panel can then be "locked" so the same loci are accepted or rejected with each run of the assay.

In contrast, the automated filtering of NGS datasets, as is done in exome sequencing, occurs without any preconceived notion of the variants that will pass the bioinformatic pipeline filters [34]. These filters are designed based on a set of assumptions that are put in place to deal with the hundreds or thousands of variants obtained from NGS assays. The use of these filters is meant to ensure that the variants most likely to be disease causing are prioritized. However, any mutations that do not fit the filter assumptions will be missed.

A sample variant annotation and filtering routine is shown in Fig. 1.3c. First, variants are annotated using information from a public genome database. The affected gene is recorded, as is the nature of the mutation, be it an amino acid change, splice site disruption, or one that affects transcript production or stability. Curated clinical-grade disease databases such as HGMD and ClinVar are then queried to determine if any variants have been previously identified as pathogenic. Variants are subsequently checked against dbSNP and 1000 Genome data (and any suitably large internal set of "control" genomes/exomes the laboratory may have accrued) to determine allele frequency and whether the observed changes have been identified or may be inferred (based on their background frequency) to be benign polymorphisms. The remaining variants are then subjected to algorithms that predict whether the variant is a putative mutation based on an amino acid change or splice site interruption. Finally, variants are checked against an internal database to determine if the laboratory has previously observed them and, if so, how they were characterized. Some laboratories may also choose to sequence parental samples, or samples from other relatives, or companion benign tissue from the cancer patient, to aid in the interpretation of variants found in the proband [48].

Pipeline filters that incorporate the phenotypic information associated with the variants detected in these assays will dictate the rate of VUS and incidental variants ultimately reported, as well as the sensitivity of the test [6, 49]. As is shown in Fig. 1.3c, these phenotypic filters are applied either at the beginning of the filtering process or toward the end. When phenotypic filters are applied only after disease-causing mutations are identified, some laboratories will feel obligated to report them even if they are indeed unrelated to the test indications. In contrast, when genes are preselected using keywords related to the patient's phenotype, all variants that occur outside of this virtual gene panel are removed from consideration prior to when analysis begins. In such a scenario, incidental findings would likely not be found, but neither would significant mutations that occur in genes yet to be correlated with the disease phenotype. A somewhat hybrid strategy is to perform multiple rounds of interpretation, which increase the size of the virtual gene panel each time until a significant variant is identified or the entire exome is unmasked.

Dealing with Incidental Findings

As discussed above, it is inevitable that a laboratory performing clinical WES or WGS is going to be confronted with incidental (also sometimes referred to as secondary, off-target, or unexpected) findings. At time of this writing, there is intense debate in the genetics community surrounding how such findings should be handled, a debate which is likely to continue for some time [50, 51]. The classic example, often used as a basis for discussion, is the finding of a clearly pathogenic mutation in a gene associated with one of the adult-onset, dominant cancer syndromes (such as familial breast/ovarian cancer or Lynch syndrome) in a young child or baby undergoing genome-wide sequencing for an unrelated condition such as congenital deafness, seizure disorder, or autism. The medical genetics community has long adhered to an ethical policy of not performing predictive/presymptomatic genetic testing for adult-onset disorders in children, unless there is some medical or surgical intervention that would need to be introduced in childhood in order to prevent or minimize the condition. Since *BRCA-associated* breast cancers, for example, do not occur in childhood, nor would a baby girl ever be a candidate for prophylactic mastectomy or oophorectomy, testing for *BRCA* mutations would never be sanctioned in such a young patient. But what happens if the laboratory happens to stumble upon one incidentally during sequencing for one of the unrelated disorders like congenital hearing loss? Now that it has been seen, should it be reported? Could nondisclosure, per the existing ethical policy, eventually cause harm to the child or her mother (if the mutation were in fact passed down from her)? Could a compromise solution be designed whereby the incidental finding is "flagged" in the electronic medical record to reappear and be reported out when the girl reaches age 18? Or should patients or parents be offered a multitiered consent form prior to testing, in which they get to select which types, if any, of incidental findings they wish or do not wish to receive?

Up until recently, such decisions were left to the discretion of the laboratory director and/or the ordering physician, and dealt with on a case-by-case basis. But we now have in hand a recommended guideline, issued by the American College of Medical Genetics and Genomics (ACMG) [52]. The guideline was the product of over a year of intense deliberation by a diverse committee of clinicians and laboratorians who, in the end, decided quite firmly to plant themselves in the "duty to warn" camp. The major recommendations are:

- Documented mutations in a select list of high-penetrance, potentially lethal but actionable conditions *must be sought and reported*.
- The same rules apply to sequencing of healthy parents in a "trio" or benign companion tissue when doing tumor sequencing.
- No distinction is made between adult and pediatric patients.
- These results are given to the ordering clinician who has responsibility for deciding which, when and how to convey to the patient.
- The *patient cannot opt out* from receiving these incidental findings.

The target list consists of high-penetrance familial cancer syndromes, cardiomyopathies, malignant hyperthermia, and other conditions meeting the criteria. Generating the most subsequent controversy have been the proviso that these targets must be actively sought out (as opposed to incidentally "stumbling" upon them) and the elimination of an "opt out" choice for patients. One of the rationales used to justify these conclusions is the analogy with radiology: a radiologist is obligated to report all abnormal findings seen in a chest X-ray, regardless of the indication or specific suspicion upon which the X-ray had been ordered. However, these points could also be interpreted as infringing on both patient and laboratory autonomy (the latter by establishing an implied responsibility for sufficient coverage and validation of a set of target genes that the laboratory had never claimed to be within its purview).*

Not surprisingly, feelings run strong on both sides of this debate [53, 54]. Like all ACMG guidelines, these are considered "recommendations" rather than mandatory "standard of care". Until the dust settles, it is

*A subsequent policy clarification announced at the March 2014 ACMG annual meeting has loosened the "opt-out" restriction somewhat.

probably most prudent for each laboratory to carefully consider the options and then have a written procedure of its own for handling incidental findings, one that is made clear to clinicians and patients prior to the ordering and performance of the test. If a laboratory chooses to inform the provider of incidental findings, clear policies that dictate the variants to be reported should be in place so there is no confusion on what variants to report. Among the criteria that need to be considered, it should be established whether all incidental findings will be reported or only those for which knowledge of the results is actionable. Will the parents of minors be notified of incidental findings? What about other family members at risk? Whether or not the laboratory intends to report incidental findings of variants that are not well characterized should also be made clear. Of course, laboratories choosing to provide only targeted gene panels will not face this dilemma.

Gene Patents

The issue of intellectual property and restrictive gene patents has been one with which the molecular diagnostic community has had to contend almost since its inception. All of us have examples of molecular tests we have had to remove from our menus after receiving "cease-and-desist" letters from the exclusive gene patent-holder. While we do not like it, we have learned to live with it in the context of setting up (or not) single-gene test offerings. But how can we confront such restrictions at the genomic level, where it is estimated that 30–40 % of *all* genes have some intellectual property tied to them [55]? Won't this kill genome-wide testing before it even leaves the starting-gate? Fortunately, this has now become something of a moot point, with the dramatic Supreme Court ruling in the *Association for Molecular Pathology et al. vs. Myriad Genetics* case, which was brought as a direct challenge to the restrictive intellectual property tied to the *BRCA1* and *BRCA2* genes, allowing only Myriad Genetics to offer the full-gene sequencing test. In brief, the Court ruled in June 2013 that genes represent "products of nature" and therefore cannot be patented, thus invalidating not only the *BRCA* patents but those for all other genes, as well [56]. Curiously, the Court did let stand as patentable subject matter one part of Myriad's claims dealing with specific cDNA probes (and, by implication, PCR primers). However, the days when cDNA probes were required for genetic testing are well behind us, and NGS in particular requires no prior knowledge or specific primers at all for the DNA to be sequenced. Thus, despite all the other challenges in implementing and reporting NGS tests, fear of gene patent infringement should no longer be a hindrance.

Follow-up, Reanalysis, and Duty to Recontact

Unlike long-established analytes in clinical chemistry or microbiology, our knowledge of the clinical implications of genetic and genomic variants is constantly changing as new discoveries are published and additional mutations and polymorphisms are deposited in DNA databases. Thus, the clinical interpretation of a particular nucleotide variant today may not be the same is it might be next year or even next month. This begs the question, long debated in the genetics and oncology communities, about whether or not there is a "duty to recontact" patients as our knowledge of previously tested targets changes, and if so, whether the responsibility for doing so should fall on the testing laboratory or on the ordering clinician. While some laboratories that focus largely on a particular gene or set of genes, such as Myriad Genetics, keep extensive databases of the variants found in those genes and flag recurrent ones for recontact even years later if the interpretation has changed, consensus has emerged that such a requirement would place an untenable burden on the average clinical molecular diagnostics laboratory dealing with many different disorders on a daily basis [57]. Obviously, this challenge would be multiplied by many orders of magnitude when performing sequencing tests on all the genes in the genome, and no one laboratory or director can conceivably be held responsible for keeping up-to-the-minute on all the literature pertaining to 25,000 genes. Given that reality, and the concern that even an implied duty to recontact would entail an open-ended laboratory–physician–patient relationship that could not be met, current thought in the field is that the responsibility for monitoring developments that might result in a revised interpretation of genomic results should fall to the patient and/or their physician, either of whom can, when appropriate, request reanalysis of archived exome or genome sequence data or submit a new specimen for re-sequencing [58].

Conclusions

An outsider viewing the current state of NGS for clinical purposes is likely to be intimidated by the many challenges and hurdles it presents: astronomically expensive instruments, an infinite and constantly changing knowledge base, potential errors in capture, sequencing and alignment, the huge numbers of VUS produced on every case, the need for multidisciplinary interpretations that could take hundreds of hours, uncertainties about costs and reimbursement, and so on. But just as we recently saw a solution to the gene patent problem, we can be certain that these remaining challenges will be met as well in the coming years. Based on trends thus far, we can be sure that genome-level DNA sequencing will continue to improve in accuracy, user-friendliness, speed, and cost-effectiveness. As lower cost continually expands the market for these tests, it is even likely that many of the ethical questions that we find so difficult to answer at present—such as the return of incidental findings—will sort themselves out. Indeed, there may come a time in the not-so-distant future when we may wonder why our predecessors ever agonized so much over these questions, when society might come to accept routine WGS of every newborn as no more controversial than the standard heel-stick for metabolic disease screening that we accept now. When or whether that ever comes to pass is at this point an open question. But there can be no question that NGS will assume an ever-increasing role in molecular diagnostic testing in the years to come, ultimately usurping or replacing Sanger sequencing and other traditional methods.

REFERENCES

1. ten Bosch JR, Grody WW. Keeping up with the next generation: massively parallel sequencing in clinical diagnostics. J Mol Diagn. 2008;10:484–92.
2. Margulies M, Egholm M, Altman WE, Attiya S, Bader JS, Bemben LA, Berka J, Braverman MS, et al. Genome sequencing in microfabricated high-density picolitre reactors. Nature. 2005;437(7057):376–80.
3. Trachtenberg EA, Holcomb CL. Next-generation HLA sequencing using the 454 GS FLX system. Methods Mol Biol. 2013;1034:197–219.
4. Bentley DR, Balasubramanian S, Swerdlow HP, Smith GP, Milton J, Brown CG, Hall KP, et al. Accurate whole human genome sequencing using reversible terminator chemistry. Nature. 2008; 456(7218):53–9.
5. Rothberg JM, Hinz W, Rearick TM, Schultz J, Mileski W, Davey M, Leamon JH, Johnson K, et al. An integrated semiconductor device enabling non-optical genome sequencing. Nature. 2011;475(7356): 348–52.
6. Saunders CJ, Miller NA, Soden SE, Dinwiddie DL, Noll A, Alnadi NA, Andraws N, Patterson ML, Krivohlavek LA, Fellis J, Humphray S, Saffrey P, Kingsbury Z, Weir JC, Betley J, Grocock RJ, Margulies EH, Farrow EG, Artman M, Safina NP, Petrikin JE, Hall KP, Kingsmore SF. Rapid whole-genome sequencing for genetic disease diagnosis in neonatal intensive care units. Sci Transl Med. 2012;4(154): 154ra135.
7. Loman NJ, Misra RV, Dallman TJ, Constantinidou C, Gharbia SE, Wain J, Pallen MJ. Performance comparison of benchtop high-throughput sequencing platforms. Nat Biotechnol. 2012;30(5):434–9.
8. Jünemann S, Sedlazeck FJ, Prior K, Albersmeier A, John U, Kalinowski J, Mellmann A, Goesmann A, von Haeseler A, Stoye J, Harmsen D. Updating benchtop sequencing performance comparison. Nat Biotechnol. 2013;31(4):294–6.
9. Albert TJ, Molla MN, Muzny DM, Nazareth L, Wheeler D, Song X, Richmond TA, Middle CM, Rodesch MJ, Packard CJ, Weinstock GM, Gibbs RA. Direct selection of human genomic loci by microarray hybridization. Nat Methods. 2007;4(11): 903–5.
10. Hodges E, Xuan Z, Balija V, Kramer M, Molla MN, Smith SW, Middle CM, Rodesch MJ, Albert TJ, Hannon GJ, McCombie WR. Genome-wide in situ exon capture for selective resequencing. Nat Genet. 2007;39(12):1522–7.
11. Okou DT, Steinberg KM, Middle C, Cutler DJ, Albert TJ, Zwick ME. Microarray-based genomic selection for high-throughput resequencing. Nat Methods. 2007;4(11):907–9.
12. Gnirke A, Melnikov A, Maguire J, Rogov P, LeProust EM, Brockman W, Fennell T, Giannoukos G, Fisher S, Russ C, Gabriel S, Jaffe DB, Lander ES, Nusbaum C. Solution hybrid selection with ultra-long oligonucleotides for massively parallel targeted sequencing. Nat Biotechnol. 2009;27(2):182–9.
13. Porreca GJ, Zhang K, Li JB, Xie B, Austin D, Vassallo SL, LeProust EM, Peck BJ, Emig CJ, Dahl F, Gao Y, Church GM, Shendure J. Multiplex amplification of large sets of human exons. Nat Methods. 2007;4(11):931–6.

14. Tewhey R, Warner JB, Nakano M, Libby B, Medkova M, David PH, Kotsopoulos SK, Samuels ML, Hutchison JB, Larson JW, Topol EJ, Weiner MP, Harismendy O, Olson J, Link DR, Frazer KA. Microdroplet-based PCR enrichment for large-scale targeted sequencing. Nat Biotechnol. 2009;27(11):1025–31.
15. Bick D, Dimmock D. Whole exome and whole genome sequencing. Curr Opin Pediatr. 2011; 23(6):594–600.
16. Ng SB, Turner EH, Robertson PD, Flygare SD, Bigham AW, Lee C, Shaffer T, Wong M, Bhattacharjee A, Eichler EE, Bamshad M, Nickerson DA, Shendure J. Targeted capture and massively parallel sequencing of 12 human exomes. Nature. 2009;461(7261): 272–6.
17. Chin EL, da Silva C, Hegde M. Assessment of clinical analytical sensitivity and specificity of next-generation sequencing for detection of simple and complex mutations. BMC Genet. 2013;14:6.
18. Clark MJ, Chen R, Lam HY, Karczewski KJ, Chen R, Euskirchen G, Butte AJ, Snyder M. Performance comparison of exome DNA sequencing technologies. Nat Biotechnol. 2011;29(10):908–14.
19. Dohm JC, Lottaz C, Borodina T, Himmelbauer H. Substantial biases in ultra-short read data sets from high-throughput DNA sequencing. Nucleic Acids Res. 2008;36(16):e105.
20. Treangen TJ, Salzberg SL. Repetitive DNA and next-generation sequencing: computational challenges and solutions. Nat Rev Genet. 2011;13(1):36–46.
21. ENCODE Project Consortium. An integrated encyclopedia of DNA elements in the human genome. Nature. 2012;489(7414):57–74.
22. Ding L, Wendl MC, Koboldt DC, Mardis ER. Analysis of next-generation genomic data in cancer: accomplishments and challenges. Hum Mol Genet. 2010;19(R2):R188–96.
23. Gundry M, Vijg J. Direct mutation analysis by high-throughput sequencing: from germline to low-abundant, somatic variants. Mutat Res. 2012; 729(1–2):1–15.
24. Beadling C, Neff TL, Heinrich MC, Rhodes K, Thornton M, Leamon J, Andersen M, Corless CL. Combining highly multiplexed PCR with semiconductor-based sequencing for rapid cancer genotyping. J Mol Diagn. 2013;15(2):171–6.
25. Hadd AG, Houghton J, Choudhary A, Sah S, Chen L, Marko AC, Sanford T, Buddavarapu K, Krosting J, Garmire L, Wylie D, Shinde R, Beaudenon S, Alexander EK, Mambo E, Adai AT, Latham GJ. Targeted, high-depth, next-generation sequencing of cancer genes in formalin-fixed, paraffin-embedded and fine-needle aspiration tumor specimens. J Mol Diagn. 2013;15(2):234–47.
26. College of American Pathologists (CAP). Checklist for molecular pathology laboratories. www.cap.org. Accessed Jun 2013.
27. Altmann A, Weber P, Bader D, Preuss M, Binder EB, Müller-Myhsok B. A beginners guide to SNP calling from high-throughput DNA-sequencing data. Hum Genet. 2012;131(10):1541–54.
28. Torri F, Dinov ID, Zamanyan A, Hobel S, Genco A, Petrosyan P, Clark AP, Liu Z, Eggert P, Pierce J, Knowles JA, Ames J, Kesselman C, Toga AW, Potkin SG, Vawter MP, Macciardi F. Next generation sequence analysis and computational genomics using graphical pipeline workflows. Genes (Basel). 2012;3(3):545–75.
29. Cock PJ, Fields CJ, Goto N, Heuer ML, Rice PM. The Sanger FASTQ file format for sequences with quality scores, and the Solexa/Illumina FASTQ variants. Nucleic Acids Res. 2010;38(6):1767–71.
30. Ruffalo M, LaFramboise T, Koyutürk M. Comparative analysis of algorithms for next-generation sequencing read alignment. Bioinformatics. 2011;27(20): 2790–6.
31. Li H, Homer N. A survey of sequence alignment algorithms for next-generation sequencing. Brief Bioinform. 2010;11(5):473–83.
32. Li H, Handsaker B, Wysoker A, Fennell T, Ruan J, Homer N, Marth G, Abecasis G, Durbin R, 1000 Genome Project Data Processing Subgroup. The sequence alignment/map format and SAMtools. Bioinformatics. 2009;25(16):2078–9.
33. Thorvaldsdóttir H, Robinson JT, Mesirov JP. Integrative genomics viewer (IGV): high-performance genomics data visualization and exploration. Brief Bioinform. 2013;14(2):178–92.
34. Stitziel NO, Kiezun A, Sunyaev S. Computational and statistical approaches to analyzing variants identified by exome sequencing. Genome Biol. 2011; 12(9):227.
35. Sherry ST, Ward MH, Kholodov M, Baker J, Phan L, Smigielski EM, Sirotkin K. dbSNP: the NCBI database of genetic variation. Nucleic Acids Res. 2001;29(1):308–11.
36. Clarke L, Zheng-Bradley X, Smith R, Kulesha E, Xiao C, Toneva I, Vaughan B, Preuss D, Leinonen R, Shumway M, Sherry S, Flicek P, 1000 Genomes Project Consortium. The 1000 Genomes Project: data management and community access. Nat Methods. 2012;9(5):459–62.
37. Stenson PD, Mort M, Ball EV, Howells K, Phillips AD, Thomas NS, Cooper DN. The human gene mutation database: 2008 update. Genome Med. 2009;1(1):13.
38. Riggs ER, Wain KE, Riethmaier D, Savage M, Smith-Packard B, Kaminsky EB, Rehm HL, Martin CL, Ledbetter DH, Faucett WA. Towards a universal clinical genomics database: the 2012 international standards for cytogenomic arrays consortium meeting. Hum Mutat. 2013;34(6):915–9.
39. Adzhubei IA, Schmidt S, Peshkin L, Ramensky VE, Gerasimova A, Bork P, Kondrashov AS, Sunyaev SR. A method and server for predicting damaging missense mutations. Nat Methods. 2010;7(4):248–9.
40. Kumar P, Henikoff S, Ng PC. Predicting the effects of coding non-synonymous variants on protein function using the SIFT algorithm. Nat Protoc. 2009;4(7):1073–81.
41. Pieters JJ, Verhaak CM, Braat DD, van Leeuwen E, Smits AP. Experts' opinions on the benefit of an incidental prenatal diagnosis of sex chromosomal aneuploidy: a qualitative interview survey. Prenat Diagn. 2012;32(12):1151–7.
42. Cerretini R, Acevedo S, Chena C, Belli C, Larripa I, Slavutsky I. Evaluation of constitutional chromosome aberrations in hematologic disorders. Cancer Genet Cytogenet. 2002;134(2):133–7.

43. Bruno DL, Stark Z, Amor DJ, Burgess T, Butler K, Corrie S, Francis D, Ganesamoorthy D, Hills L, James PA, O'Rielly D, Oertel R, Savarirayan R, Prabhakara K, Salce N, Slater HR. Extending the scope of diagnostic chromosome analysis: detection of single gene defects using high-resolution SNP microarrays. Hum Mutat. 2011;32(12):1500–6.
44. Schaaf CP, Scott DA, Wiszniewska J, Beaudet al. Identification of incestuous parental relationships by SNP-based DNA microarrays. Lancet. 2011;377(9765):555–6.
45. Grody WW, Cutting GR, Watson MS. The cystic fibrosis mutation "arms race": when less is more. Genet Med. 2007;9(11):739–44.
46. Johnson MA, Yoshitomi MJ, Richards CS. A comparative study of five technologically diverse CFTR testing platforms. J Mol Diagn. 2007;9(3):401–7.
47. Grody WW, Cutting GR, Klinger KW, Richards CS, Watson MS, Desnick RJ. Laboratory standards and guidelines for population-based cystic fibrosis carrier screening. Genet Med. 2001;3(2):149–54.
48. de Ligt J, Willemsen MH, van Bon BW, Kleefstra T, Yntema HG, Kroes T, Vulto-van Silfhout AT, Koolen DA, de Vries P, Gilissen C, del Rosario M, Hoischen A, Scheffer H, de Vries BB, Brunner HG, Veltman JA, Vissers LE. Diagnostic exome sequencing in persons with severe intellectual disability. N Engl J Med. 2012;367(20):1921–9.
49. Need AC, Shashi V, Hitomi Y, Schoch K, Shianna KV, McDonald MT, Meisler MH, Goldstein DB. Clinical application of exome sequencing in undiagnosed genetic conditions. J Med Genet. 2012;49(6):353–61.
50. Kohane IS, Masys DR, Altman RB. The incidentalome: a threat to genomic medicine. JAMA. 2006;296:212–5.
51. Green RC, Berg JS, Berry GT, Biesecker LG, Dimmock DP, Evans JP, Grody WW, Hegde MR, Kalia S, Korf BR, Krantz I, McGuire AL, Miller DT, Murray MF, Nussbaum RL, Plon SE, Rehm HL, Jacob HJ. Exploring concordance and discordance for return of incidental findings from clinical sequencing. Genet Med. 2012;14(4):405–10.
52. Green RC, Berg JS, Grody WW, et al. ACMG recommendations for reporting of incidental findings in clinical exome and genome sequencing. Genet Med. 2013;15:565–74.
53. Wolf SM, Annas GJ, Elias S. Point-counterpoint. Patient autonomy and incidental findings in clinical genomics. Science. 2013;340:1049–50.
54. Allyse M, Michie M. Not-so-incidental findings: the ACMG recommendations on the reporting of incidental findings in clinical whole genome and whole exome sequencing. Trends Biotechnol. 2013;31(8):439–41.
55. Rosenfeld J, Mason CE. Pervasive sequence patents cover the entire human genome. Genome Med. 2013;5:27.
56. Supreme Court of the United States. Association for Molecular Pathology et al. vs. Myriad Genetics, Inc. 2013. http://www.supremecourt.gov/opinions/12pdf/12-398_1b7d.pdf
57. Hirschhorn K, Fleisher LD, Godmilow L, et al. Duty to re-contact. Genet Med. 1999;1:171–2.
58. Pyeritz RE. The coming explosion in genetic testing—is there a duty to recontact? N Engl J Med. 2011;365:1367–9.

CHAPTER 2

Emerging Next-Generation Sequencing Technologies

MATTHEW W. ANDERSON

Introduction 19
Advantages of Single-Molecule Sequencing 19
Single-Molecule Cycle Sequencing 20
Real-Time Single-Molecule Sequencing with Polymerase 21
Sequencing Through Direct Imaging 23
Sequencing with Protein Nanopores 24
Solid-State Nanopore Sequencing 26
Conclusions 27

Introduction

What would be the ideal sequencing machine? Can a single sequencing technology replace all the tools of molecular pathology and cytogenetics that we currently utilize to interrogate the genome in both health and disease? An ideal sequencing instrument should detect all types of genomic variation including structural [single nucleotide polymorphisms (SNP's), indels, copy number variation, inversions, chromosomal rearrangements], epigenomic, and transcriptional. Long read lengths are required to enable efficient genomic assembly and accurate phasing, and the detection method must produce highly accurate base calls to minimize errors and reduce costly iterative sequencing. Finally, the system should be inexpensive, be easy to maintain and operate, and require short run times. At the time of writing this chapter, such a sequencing machine simply does not exist. However, engineers, physicists, and biologists in both industry and academia are actively working to solve the major technical challenges facing the development of new sequencing technologies. While this chapter focuses on discussing these "emerging" sequencing technologies, the reader is cautioned that the development of new sequencing technologies occurs at a prodigious pace. In addition, many of the cutting-edge advances in sequencing technology are being developed within a commercial environment, where it is difficult for those outside the company to obtain detailed and vetted information about instrument performance. With those caveats in mind, this chapter endeavors to provide a broad overview of emerging new sequencing technologies and some of the potential applications in nucleic acid analysis which will be enabled by these technological advances.

Advantages of Single-Molecule Sequencing

Biomolecular detection (whether for sequencing, chemistry, or immunology) typically requires a signal amplification step for robust and reproducible analyte detection. For both

G.J. Netto and I. Schrijver (eds.), *Genomic Applications in Pathology*,
DOI 10.1007/978-1-4939-0727-4_2, © Springer Science+Business Media New York 2015

Sanger and next-generation sequencing, signal amplification occurs through PCR amplification of the target DNA, ensuring that the fluorescence or luminescence signals generated during the sequencing reaction are sufficiently strong. Although PCR is the mainstay of molecular biology protocols, it is not without its disadvantages. Sequence artifacts can be generated during the PCR reaction to include nucleotide misincorporation events, amplification bias due to GC content, preferential allele amplification, and the formation of chimeric sequences during later PCR cycles, all of which can be reflected in the sequencing results [1–4]. When smaller regions of DNA are amplified, significant contextual information (phase, haplotype, etc.) is also lost.

To solve these issues, most emerging high-throughput sequencing technologies are designed to sequence individual nucleic acid molecules. Single-molecule sequencing offers a number of practical advantages. Sample preparation is greatly simplified, because there is less experimental manipulation required to create sequencing libraries. For example, the ability to directly sequence RNA would eliminate the additional steps typically required to convert RNA into cDNA prior to sequencing. Single-molecule approaches would theoretically reduce the required amount of input DNA, an important consideration for the analysis of rare cellular populations or individual cells. Single-molecule sequencing also enables long templates to be sequenced in phase, preserving long-range structural variation.

Single-Molecule Cycle Sequencing

In 2003, Stephen Quake and colleagues were the first to report single-molecule DNA sequencing through the use of fluorescence microscopy and fluorescence resonance energy transfer (FRET) [5]. Using this technique, the authors were able to detect the incorporation of up to five nucleotides on a single DNA template. Although the strategy was promising, the relatively short molecular distance over which FRET can occur limited the theoretical sequencing read length to approximately 15 basepairs (bp). A year later, Quake cofounded Helicos BioSciences (Cambridge, MA), with the goal of developing a commercial sequencing instrument based on single-molecule sequencing. In the Helicos sequencing strategy, FRET-based detection is replaced by a "sequencing by synthesis" approach, in which fluorescently labeled nucleotides are added sequentially during the sequencing reaction. Similar to other high-throughput sequencing technologies, only one type of nucleotide (A, T, G, or C) is added to the reaction during each cycle of sequencing. Therefore, not every template molecule incorporates a nucleotide during each round of sequencing.

The Helicos sequencing protocol is relatively simple. First, sequencing libraries are prepared by randomly fragmenting genomic DNA to produce short (100–200 bp) fragments. Next, multiple adenosine nucleotides are added to the 3′ end of the DNA fragments to allow the template molecules to hybridize to poly-T oligonucleotide anchors on the surface of the flow cell (Fig. 2.1). The terminal adenosine nucleotide is fluorescently labeled so that each template molecule can be spatially localized on the flow cell surface, prior to the start of sequencing. During the sequencing reaction, DNA polymerase and one of four fluorescently labeled nucleotides are sequentially added to the flow cell. Each nucleotide is modified with a terminator moiety to prevent multiple nucleotide additions during each sequencing cycle. After nucleotide incorporation, the array is imaged and the terminator moiety and fluorescence label are removed to enable subsequent rounds of sequencing.

In 2008, the company shipped its first sequencing instrument and reported the use of the technology to sequence the M13 phage genome [6]. Utilizing data generated from sequencing his own genome with the Helicos instrument [7], Quake and his group reported an average read length of 33 bp, and an error profile composed predominantly of deletions and insertions (approximately 3–5 % overall). Although the raw read error rate was relatively high, the overall consensus accuracy was 99 % for SNP's due to a high depth of coverage (28×). Although the same genomic sequence was later analyzed by a multidisciplinary group at Stanford to create a personal genome-based clinical assessment [8], the

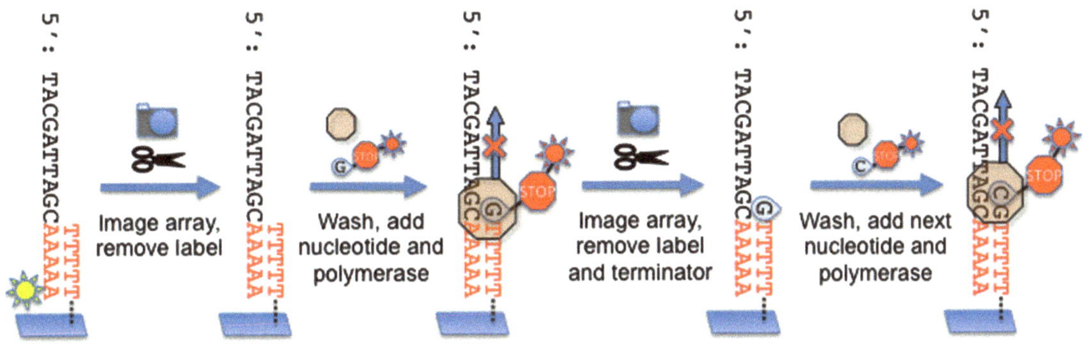

Figure 2-1 Helicos sequencing chemistry. DNA template molecules are modified by the 3′ addition of adenosine nucleotides, and hybridized to poly-T oligonucleotides covalently linked to the surface of the flow cell. The terminal adenosine nucleotide is fluorescently labeled to allow the instrument to record the location of each template molecule on the flow cell surface. Prior to the start of sequencing, the fluorescently labeled 3′ adenosine is cleaved and washed away. During each round of sequencing, polymerase and a single fluorescently labeled nucleotide (A, T, G, or C) is added to the flow cell. The labeled nucleotides are modified with a cleavable terminator residue which prevents multiple base incorporation during each cycle. After nucleotide incorporation, the array is imaged and the fluorescence signal is recorded for each template molecule. Once the images are captured, the fluorescent label and terminator are removed to regenerate a template suitable for the next round of nucleotide addition. Image and legend reproduced with permission [48]

high error rate, short read length, and high cost per base made the Helicos technology impractical for whole-genome sequencing. The Helicos single-molecule sequencing approach has since been shown to be advantageous for other applications including direct RNA sequencing (RNA-seq) [9] and chromatin profiling (CHiP-seq) [10], but the company struggled to maintain commercial operations due in large part to the high cost of the instrument and limitations of the platform for whole-genome sequencing applications.

Real-Time Single-Molecule Sequencing with Polymerase

What if one could directly observe DNA polymerase as it synthesizes DNA? DNA sequence information would be generated in "real-time" at a rate equal to nucleotide incorporation catalyzed by DNA polymerase, with read lengths theoretically limited only by the processivity of the polymerase or the size of the DNA template. By eliminating the iterative sequencing cycles required by current high-throughput sequencing technologies, the overall cost of obtaining a complete genome sequence could also be significantly reduced.

Whereas the potential advantages of real-time single-molecule DNA sequencing are readily apparent, designing a sequencing instrument that can "eavesdrop" on a single DNA polymerase molecule is extremely challenging from an engineering and biophysics perspective. For example, the detection method must be able to accurately detect signals generated from the activity of a single DNA polymerase molecule, and true nucleotide incorporation events must be discerned against a background of high concentrations of unbound labeled nucleotides. Two solutions to this problem have been developed, either through physically confining DNA polymerase to a small observation volume, or by the use of FRET to detect when a labeled nucleotide is in close proximity to the polymerase active site.

In 2003, Levene et al. reported the development of "zero-mode waveguide" (ZMW) technology, a technique that utilizes nanoscale holes in a metal film to restrict incident laser light to a small focused detection volume approximating 10^{-21} l (zeptoliter) [11]. By creating an extremely focused region in which laser light can excite a fluorophore, ZMVs enable single-molecule analysis in the presence of high concentrations of fluorescently labeled ligands. As a demonstration of the applicability of ZMW technology for DNA sequencing applications, the authors immobilized DNA polymerase and M13 phage DNA within the detection volume of ZMWs. After the addition of a fluorescently labeled nucleotide, temporally distinct fluorescent signals were detected within the ZMWs consistent with polymerase-catalyzed nucleotide incorporation events.

In 2004, Pacific Biosciences (Menlo Park, CA) was founded to develop a DNA sequencing instrument using ZMW technology. The company published its first proof-of-concept study describing single-molecule real-time (SMRT®) sequencing in 2009 [12], and released its first commercial DNA sequencing instrument (PacBio RS) in 2010. In the SMRT sequencing method, DNA template libraries are prepared by shearing genomic DNA into 250 bp to 10 kilobase (kb) fragments and ligating hairpin adapters to each end of the molecule to create a circular DNA template. Primed DNA templates lacking the hairpin adapters can also be sequenced, but the number of reads generated is reduced considerably [13]. Individual DNA polymerase molecules bound to DNA template are then localized at the bottom of the ZMW through simple diffusion and biotin/streptavidin interactions [14]. A mixture of nucleotides is subsequently added to the chip, with each nucleotide uniquely labeled with a different fluorophore attached to the base via linkage to the phosphate chain. Unbound nucleotides rapidly diffuse in and out of the ZMV detection volume, far too quickly to be registered as a fluorescence signal by the detector (Fig. 2.2). When a nucleotide enters the active site of the polymerase, its motion is dramatically slowed, allowing time for the laser to excite the fluorophore and generate a fluorescent signal. DNA polymerase cleaves the phosphate chain as the nucleotide is incorporated, freeing the fluorophore to rapidly diffuse out of the detection volume of the ZMW. The reaction reconstitutes a free 3′ hydroxyl group, which can then be used for the next round of nucleotide addition.

SMRT sequencing has a high per base error rate (15–20 %) dominated by insertions, presumably due to non-templated nucleotides binding to the active site of the polymerase. Interestingly, the error rate profile appears to be random, and not context-specific (i.e., homopolymer errors) as opposed to other high-throughput sequencing platforms [15].

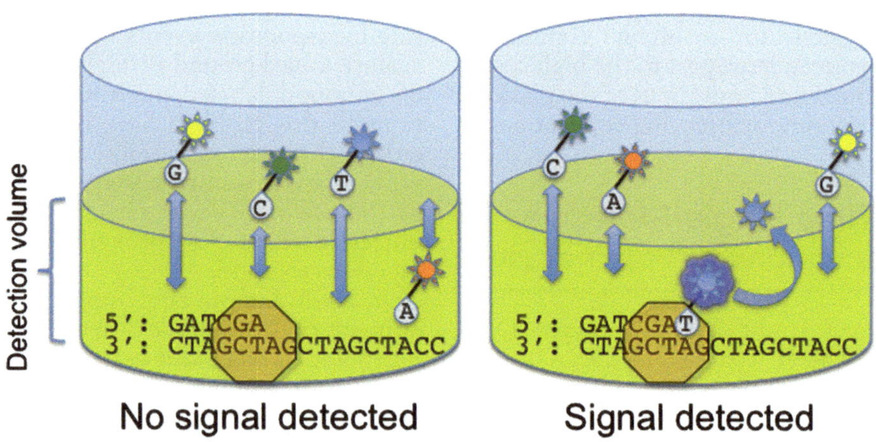

Figure 2-2 Real-time single-molecule sequencing with polymerase. In SMRT chemistry, DNA polymerase and template molecules are immobilized at the bottom of a zero-mode waveguide (ZMW) sequencing well. The ZMW focuses laser energy to create an extremely small detection volume at the bottom of the ZMW where the polymerase and template molecule are localized. A mixture of all four nucleotides is added to the ZMW, each uniquely labeled with a different fluorescence moiety. Unincorporated nucleotides rapidly diffuse in and out of the detection volume at a rate too fast for a fluorescence signal to be recorded. As a nucleotide is incorporated, the fluorescent moiety is held within the detection volume long enough to be excited by the laser and give off a fluorescence signal which can be recorded. During nucleotide addition, the fluorophore is cleaved away as the phosphodiester bond is formed. The liberated fluorophore rapidly diffuses out of the ZMW detection volume, terminating the fluorescence signal for that particular nucleotide incorporation event. Image and legend reproduced with permission [48]

Much of the high per-base error rate can be overcome through iterative sequencing of the circular templates, resulting in high consensus accuracy (99 %) [12]. Because the error rate is independent of sequence context, GC-rich and homopolymer regions of the genome can be sequenced and analyzed including the CGG repeat region of the *FMR1* gene implicated in fragile X syndrome [16].

An attractive feature of SMRT sequencing is the ability to directly detect modified bases such as 5-methylcytosine [17]. Compared to standard high-throughput sequencing techniques which rely on bisulfite treatment of the DNA library to characterize methylation [18], SMRT sequencing can directly detect modified bases during the sequencing reaction as changes in the kinetics of DNA polymerase. Because no prior chemical modification of the DNA library is required, multiple different base modifications on the same DNA template molecule can be detected simultaneously [19].

Despite its numerous advantages, the PacBio RS may not be the platform of choice for all sequencing applications. In its current iteration, the PacBio RS utilizes a chip containing approximately 150,000 ZMWs, half of which are read at one time to generate less than 100,000 reads per run. For counting applications such as RNA expression analysis, higher throughput is necessary to attain the depth of sequencing coverage required to detect lower abundance RNA species. Although low accuracy long reads (>5 kb) can be useful for genome assembly and phasing, higher per base accuracy will be required to successfully analyze highly polymorphic genomic regions, such as the major histocompatibility complex. Finally, the instrument is significantly more expensive than current high-throughput sequencers. If these challenges can be overcome, the PacBio RS system may be useful for routine clinical sequencing applications such as HLA genotyping, microbiology [20], and oncology [21].

Life Technologies (Carlsbad, CA) has developed another approach for single-molecule polymerase-based sequencing that utilizes FRET to detect nucleotide incorporation events. Although there are no published data on this technology as yet, presentations describing the method (termed "Starlight") suggest that the strategy involves generating FRET between quantum dot-labeled DNA polymerase molecules and fluorescently labeled nucleotides [22]. When a fluorescently labeled nucleotide enters the active site of the polymerase, two signals are generated during FRET. The signal from the quantum dot decreases, indicating that a nucleotide is bound to the active site, while the signal from the bound fluorescently labeled nucleotide increases. Theoretically, the presence of two distinct but temporally related signals from the same nucleotide incorporation event could result in highly accurate base calls. Despite the potential of the Starlight technology, plans for commercialization remain unclear as Life Technologies devotes considerable resources to the ongoing development of the Ion Torrent platform.

Sequencing Through Direct Imaging

Direct visualization of biological macromolecules has long been proposed as an approach to determine nucleic and amino acid sequences. The potential benefits of sequencing through direct imaging include extremely long read lengths, fast analysis, and preservation of large-scale structural variation. Optical-based approaches to mapping long DNA fragments have been developed utilizing either restriction enzymes or fluorescent labeling [23], but the resolution of light or fluorescence microscopy is far too low to allow for single-base identification. Recently, scanning transmission electron microscopy (STEM) has been explored as a potential direct imaging sequencing technology. In fact, the use of electron microscopy to image DNA is not new. In the 1970s, STEM was used to generate low resolution images of purified genomic DNA from both *Drosophila* [24] and human samples [25]. Secondary structures of DNA (hairpin loops, etc.) were readily observable, and the images provided support for the presence of inverted repeat sequences in the human genome. However, for STEM to become useful for nucleic acid sequencing, it must demonstrate sufficient resolution to accurately image and identify each nucleotide by its unique chemical structure.

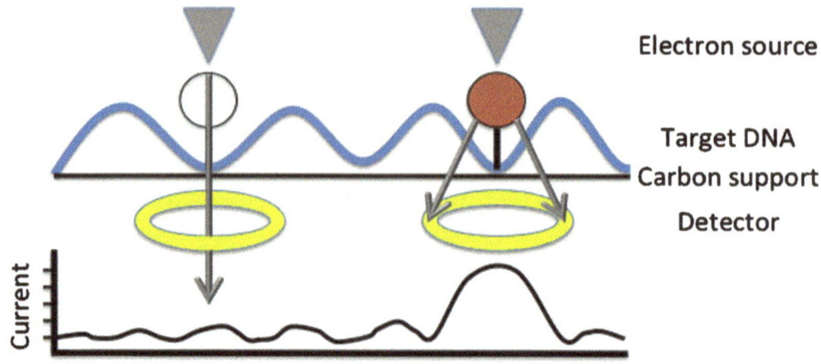

Figure 2-3 Sequencing with electron microscopy. Fragments of DNA are labeled through the incorporation of nucleotides modified with heavy atoms (*red circle*). Labeled DNA template molecules are then stretched and applied to a carbon substrate overlying a circular detector. When the electron beam encounters an unlabeled nucleotide (*open circle*) the path of the beam is undisturbed and the electrons pass through the center of the detector. In contrast, heavy atoms scatter the electron beam, resulting in increased current within the detector. Figure adapted from Bell et al. [26]

In STEM, resolution is directly related to the ability of a target atom to scatter the electron beam. Heavier atoms are more easily visualized as the higher atomic mass results in increased electron scattering. Unfortunately, STEM cannot readily distinguish between each base in a nucleic acid sequence because natural nucleotides differ by only a few atoms with low atomic mass. Therefore, STEM-based approaches to sequence DNA must involve modifications to the target DNA (i.e., heavy atom labeling) to make the nucleotides "visible" to the electron microscope.

Recently, the first report describing STEM to sequence DNA was published [26]. In this approach, target DNA molecules are labeled by performing PCR in the presence of thymine nucleotides modified with a single mercury atom. Mercury-labeled DNA molecules are then purified, linearized, and deposited onto a carbon substrate. The labeled DNA strands are imaged with the electron microscope, and modified nucleotides are detected by an increase in current as the electron beam is scattered by the heavy atom label (Fig. 2.3). For their initial experiments, the authors used M13 phage and a synthetic DNA molecule that contained labeled thymine nucleotides at well-defined positions in the sequence. Although STEM was able to detect labeled thymines in the test DNA molecules, only about half of the labeled thymine residues predicted by the test sequence were identified either due to inefficient incorporation during PCR or loss during processing of the sequencing templates. In addition, there was partial overlap between the signals generated by background current and the current generated by labeled thymine.

Although promising, several technical hurdles must be overcome before STEM becomes a viable approach to DNA sequencing. Improved methods must be devised to uniquely label each nucleotide and ensure a high degree of label incorporation into the sequencing templates. The commercial potential of STEM for direct DNA sequencing is unclear, but mapping of long DNA fragments by either STEM or optical-based approaches may ultimately provide important complementary structural information to guide genome assembly from short-read high-throughput sequencing data [27, 28].

Sequencing with Protein Nanopores

To date, all single-molecule sequencing approaches require labeling (fluorescence, heavy atoms, etc.) of the template molecule or nucleotides. Labeling adds complex preparation steps to the sequencing workflow, increases reagent costs, and can have adverse effects on the sequencing reaction such as inhibiting the action of polymerase. An alternative approach to avoid the use of labels

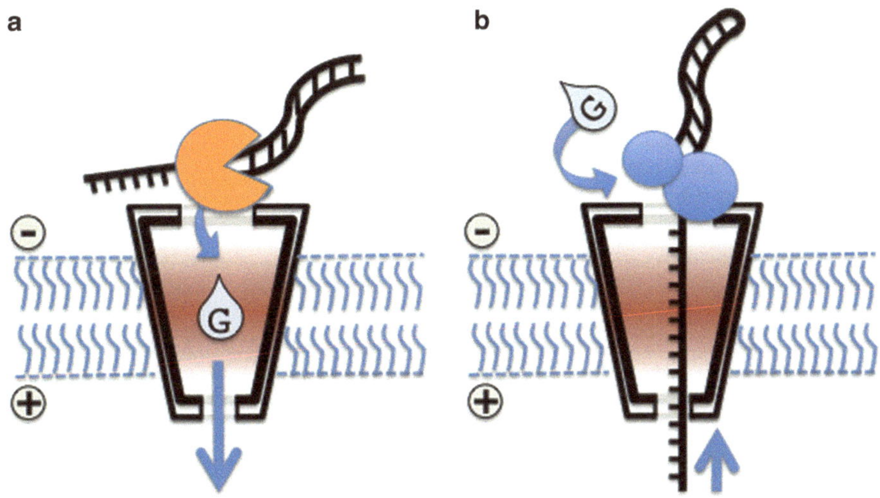

Figure 2-4 Nanopore sequencing. (**a**) Nanopore sequencing with exonuclease. Nucleotides are liberated from the 3′ hydroxyl terminus of DNA through the catalytic action of exonuclease. Free nucleotides enter the nanopore and generate a current signal specific to each base. (**b**) Nanopore sequencing with DNA polymerase. Double-stranded DNA is modified with a hairpin adapter, a blocking oligonucleotide is added, and the complex is bound to DNA polymerase. The blocking oligonucleotide is removed as the DNA interacts with the nanopore (not shown), allowing a segment of single-stranded DNA to enter the nanopore. Free nucleotides are added to the buffer surrounding the entrance of the nanopore, and DNA polymerase incorporates the nucleotides into the DNA strand in a controlled step-wise manner. During each base addition, the single-stranded region of the DNA template is withdrawn from the nanopore a distance equivalent to the length of one nucleotide. Each discrete movement of the template within the nanopore generates a current signal, which can be recorded and analyzed to determine the sequence

would be through electrochemical detection of nucleotides in the DNA sequence. Electrochemical detection would eliminate the use of expensive enzymes and labels, and could represent the fastest and simplest option for DNA sequencing.

In the early 1990s David Deamer, Daniel Branton, and George Church filed a patent application for the use of protein membrane channels (nanopores) as a method for electrochemical nucleic acid sequencing. The authors envisioned that an ionic gradient could be established across a lipid bilayer containing nanopores. Nucleic acids passing through the nanopores would disrupt the flow of ions, resulting in changes in current that could be used to decode the nucleotide sequence. In an initial proof-of-concept study [29], Deamer and coworkers showed that RNA and DNA molecules could transit through α-hemolysin nanopores derived from *Staphylococcus aureus*, and the passage of nucleic acids through the nanopore was reflected by changes in ionic current. Unfortunately, nucleic acid molecules passed through the nanopore far too quickly for each nucleotide in the sequence to generate a unique base-specific change in current. Therefore, new strategies had to be devised to control the transit of nucleic acids through the nanopore structure in order for each nucleotide to be reliably detected for sequencing.

In recent years, several important advances have been made to bring nanopore sequencing closer to reality [30]. Modifications have been made to several naturally occurring nanopores to improve the signals generated as nucleotides or polynucleotides traverse the nanopore [31, 32], and enzymatic approaches have been developed to control the movement of nucleic acids through the nanopore [33–35]. One strategy developed by Hagan Bayley and colleagues is to use exonuclease to cleave individual nucleotides from a nucleic acid polymer, and then detect the free nucleotides as they flow through the nanopore channel (Fig. 2.4a) [33]. The advantage to this approach is that the signals are significantly less complex because only four distinct current signals are generated, each unique to a different nucleotide. However, the accuracy of this technique is critically dependent on close alignment between the exonuclease and the nanopore to ensure that each liberated nucleotide is detected exactly in order according to the nucleotide sequence on the template strand.

A second and potentially more promising approach is to utilize molecular motors to control the movement of single-stranded nucleic acid in the nanopore [36]. In 2012, two groups led by Jens Gundlach and Mark Akeson showed that DNA polymerase (DNAP) could control the rate of nucleic acid translocation through a protein nanopore to generate well-defined ionic currents [34, 35]. In these experiments, DNA template molecules are modified at the 3′ end with a hairpin adaptor to prevent DNAP from extending the template. Next, a blocking oligonucleotide is hybridized to the template molecule, exposing a short single-stranded segment of the template DNA. DNAP then binds to the 3′ end of the blocking oligonucleotide, but extension is inhibited by the presence of abasic residues. Voltage is applied to the lipid bilayer, and the single-stranded segment of the DNAP/template complex enters the nanopore (Fig. 2.4b). The force of the electric current removes the blocking oligonucleotide, exposing a primer with a 3′ hydroxyl group that DNAP can utilize for synthesis of DNA. Free nucleotides are then added to the reaction, DNAP extends the primer, and the single-stranded segment of the template molecule is pulled out of the nanopore in a stepwise fashion. Because the retrograde movement through the nanopore is controlled by nucleotide addition by DNAP, the template molecule moves and traverses the nanopore over the length of one nucleotide at a time. Using this technique, complex patterns of current were observed, suggesting that the signals were influenced not only by the nucleotide sequence, but also by local sequence context and interactions between the template strand and the interior wall of the nanopore. More consistent current tracings were produced with trinucleotide repeat sequences, suggesting that base calling algorithms could be designed to recognize these defined patterns to generate sequence reads.

Much of the commercial development of protein-based nanopore sequencing has been pursued by Oxford Nanopore Technologies (Oxford, UK). Both exonuclease and molecular motor approaches for nanopore sequencing are being developed, and the company has made several technical advances such as enhancing the stability of the lipid bilayer [37]. A prototype instrument has been unveiled, but detailed performance metrics have not been released.

Solid-State Nanopore Sequencing

An alternative approach to the use of engineered protein pores for sequencing would be to construct nanopores from inorganic materials (solid-state nanopores). Leveraging recent advances in materials science, solid-state nanopores have been fabricated from silicon [38], graphene [39], and carbon nanotubes [40]. Theoretically, solid-state nanopores would be highly stable and could be manufactured utilizing existing infrastructure built by the semiconductor industry.

Various approaches have been developed to detect nucleic acids as they pass through a solid-state nanopore. Nabsys (Providence, RI) is developing a solid-state nanopore sequencing system that relies on hybridization of labeled probe oligonucleotides to single-stranded DNA [41]. After hybridization, the remaining template molecule is converted to double-stranded DNA and coated with a DNA-binding protein to increase the resistance as the template moves through the nanopore [42]. The template sequence can then be decoded by determining the relative positions of the labeled probes. Although contiguous stretches of DNA can be sequenced using this method, an added benefit is that the density of the probes can be reduced to enable mapping of large genomes [43]. However, for this strategy to be effective the nanopore must be able to precisely determine the nucleotide distance between the probes.

An added advantage of solid-state nanopores is that they can be modified with electrochemical sensors that can detect and identify nucleotides as they pass through the nanopore. For example, solid-state nanopores have been fitted with nanoelectrodes that transfer tunneling current through passing nucleotides to generate unique current signatures [44]. Chemically modified probes have also been designed that facilitate tunneling current by forming complementary interactions with each base as it passes through the nanopore [45]. In an elegant solution, graphene ribbons have been used to identify nucleotides in a near-planar orientation by taking advantage of natural pi-stacking interactions contributed by the aromatic rings of the nucleobases [46]. Finally, IBM and Roche have reportedly been collaborating on a "DNA transistor" which uses electrical current to control DNA translocation through the nanopore and determine the nucleotide sequence simultaneously [47]. Although solid-state nanopore sequencing is still in the research and development stage, it is clear that these approaches have significant potential to offer fast, reliable, and cost-effective sequencing.

Conclusions

This chapter provides an overview of the evolution of single-molecule sequencing technologies, from fluorescence-based approaches to direct electrochemical detection via solid-state nanopores. Although it is difficult to predict which sequencing technology will become a commercial success, there appears to be an inexorable progression towards the goal of sequencing individual nucleic acid molecules with virtually no sample preparation. Given the complexity of the human genome, it is likely that no one single technology will provide a complete solution for genomic analysis. However, emerging single-molecule sequencing approaches appear poised to revolutionize clinical molecular diagnostics if they can deliver on the promise of fast, cost-effective, and accurate high-throughput sequencing.

REFERENCES

1. Walsh PS, Erlich HA, Higuchi R. Preferential PCR amplification of alleles: mechanisms and solutions. PCR Methods Appl. 1992;1:241–50.
2. Cha RS, Thilly WG. Specificity, efficiency, and fidelity of PCR. PCR Methods Appl. 1993;3:S18–29.
3. Shuldiner AR, Nirula A, Roth J. Hybrid DNA artifact from PCR of closely related target sequences. Nucleic Acids Res. 1989;17:4409.
4. Mutter GL, Boynton KA. PCR bias in amplification of androgen receptor alleles, a trinucleotide repeat marker used in clonality studies. Nucleic Acids Res. 1995;23:1411–8.
5. Braslavsky I, Hebert B, Kartalov E, Quake SR. Sequence information can be obtained from single DNA molecules. Proc Natl Acad Sci U S A. 2003;100:3960–4.
6. Harris TD, Buzby PR, Babcock H, Beer E, Bowers J, Braslavsky I, Causey M, Colonell J, Dimeo J, Efcavitch JW, Giladi E, Gill J, Healy J, Jarosz M, Lapen D, Moulton K, Quake SR, Steinmann K, Thayer E, Tyurina A, Ward R, Weiss H, Xie Z. Single-molecule DNA sequencing of a viral genome. Science. 2008; 320:106–9.
7. Pushkarev D, Neff NF, Quake SR. Single-molecule sequencing of an individual human genome. Nat Biotechnol. 2009;27:847–52.
8. Ashley EA, Butte AJ, Wheeler MT, Chen R, Klein TE, Dewey FE, Dudley JT, Ormond KE, Pavlovic A, Morgan AA, Pushkarev D, Neff NF, Hudgins L, Gong L, Hodges LM, Berlin DS, Thorn CF, Sangkuhl K, Hebert JM, Woon M, Sagreiya H, Whaley R, Knowles JW, Chou MF, Thakuria JV, Rosenbaum AM, Zaranek AW, Church GM, Greely HT, Quake SR, Altman RB. Clinical assessment incorporating a personal genome. Lancet. 2010;375:1525–35.
9. Ozsolak F, Platt AR, Jones DR, Reifenberger JG, Sass LE, McInerney P, Thompson JF, Bowers J, Jarosz M, Milos PM. Direct RNA sequencing. Nature. 2009;461:814–8.
10. Goren A, Ozsolak F, Shoresh N, Ku M, Adli M, Hart C, Gymrek M, Zuk O, Regev A, Milos PM, Bernstein BE. Chromatin profiling by directly sequencing small quantities of immunoprecipitated DNA. Nat Methods. 2010;7:47–9.
11. Levene MJ, Korlach J, Turner SW, Foquet M, Craighead HG, Webb WW. Zero-mode waveguides for single-molecule analysis at high concentrations. Science. 2003;299:682–6.
12. Eid J, Fehr A, Gray J, Luong K, Lyle J, Otto G, Peluso P, Rank D, Baybayan P, Bettman B, Bibillo A, Bjornson K, Chaudhuri B, Christians F, Cicero R, Clark S, Dalal R, Dewinter A, Dixon J, Foquet M, Gaertner A, Hardenbol P, Heiner C, Hester K, Holden D, Kearns G, Kong X, Kuse R, Lacroix Y, Lin S, Lundquist P, Ma C, Marks P, Maxham M, Murphy D, Park I, Pham T, Phillips M, Roy J, Sebra R, Shen G, Sorenson J, Tomaney A, Travers K, Trulson M, Vieceli J, Wegener J, Wu D, Yang A, Zaccarin D, Zhao P, Zhong F, Korlach J, Turner S. Real-time DNA sequencing from single polymerase molecules. Science. 2009;323: 133–8.
13. Coupland P, Chandra T, Quail M, Reik W, Swerdlow H. Direct sequencing of small genomes on the Pacific Biosciences RS without library preparation. Biotechniques. 2012;53:365–72.
14. Korlach J, Marks PJ, Cicero RL, Gray JJ, Murphy DL, Roitman DB, Pham TT, Otto GA, Foquet M, Turner SW. Selective aluminum passivation for targeted immobilization of single DNA polymerase molecules in zero-mode waveguide nanostructures. Proc Natl Acad Sci U S A. 2008;105:1176–81.

15. Carneiro MO, Russ C, Ross MG, Gabriel SB, Nusbaum C, DePristo MA. Pacific biosciences sequencing technology for genotyping and variation discovery in human data. BMC Genomics. 2012;13:375.
16. Loomis EW, Eid JS, Peluso P, Yin J, Hickey L, Rank D, McCalmon S, Hagerman RJ, Tassone F, Hagerman PJ. Sequencing the unsequenceable: expanded CGG-repeat alleles of the fragile X gene. Genome Res. 2013;23(1):121–8.
17. Flusberg BA, Webster DR, Lee JH, Travers KJ, Olivares EC, Clark TA, Korlach J, Turner SW. Direct detection of DNA methylation during single-molecule, real-time sequencing. Nat Methods. 2010;7:461–5.
18. Lister R, Ecker JR. Finding the fifth base: genome-wide sequencing of cytosine methylation. Genome Res. 2009;19:959–66.
19. Fang G, Munera D, Friedman DI, Mandlik A, Chao MC, Banerjee O, Feng Z, Losic B, Mahajan MC, Jabado OJ, Deikus G, Clark TA, Luong K, Murray IA, Davis BM, Keren-Paz A, Chess A, Roberts RJ, Korlach J, Turner SW, Kumar V, Waldor MK, Schadt EE. Genome-wide mapping of methylated adenine residues in pathogenic Escherichia coli using single-molecule real-time sequencing. Nat Biotechnol. 2012;30:1232–9.
20. Rasko DA, Webster DR, Sahl JW, Bashir A, Boisen N, Scheutz F, Paxinos EE, Sebra R, Chin CS, Iliopoulos D, Klammer A, Peluso P, Lee L, Kislyuk AO, Bullard J, Kasarskis A, Wang S, Eid J, Rank D, Redman JC, Steyert SR, Frimodt-Moller J, Struve C, Petersen AM, Krogfelt KA, Nataro JP, Schadt EE, Waldor MK. Origins of the E. coli strain causing an outbreak of hemolytic-uremic syndrome in Germany. N Engl J Med. 2011;365:709–17.
21. Smith CC, Wang Q, Chin CS, Salerno S, Damon LE, Levis MJ, Perl AE, Travers KJ, Wang S, Hunt JP, Zarrinkar PP, Schadt EE, Kasarskis A, Kuriyan J, Shah NP. Validation of ITD mutations in FLT3 as a therapeutic target in human acute myeloid leukaemia. Nature. 2012;485:260–3.
22. Pennisi E. Genomics. Semiconductors inspire new sequencing technologies. Science. 2010;327:1190.
23. Neely RK, Deen J, Hofkens J. Optical mapping of DNA: single-molecule-based methods for mapping genomes. Biopolymers. 2011;95:298–311.
24. Schmid CW, Manning JE, Davidson N. Inverted repeat sequences in the Drosophila genome. Cell. 1975;5:159–72.
25. Deininger PL, Schmid CW. An electron microscope study of the DNA sequence organization of the human genome. J Mol Biol. 1976;106:773–90.
26. Bell DC, Thomas WK, Murtagh KM, Dionne CA, Graham AC, Anderson JE, Glover WR. DNA base identification by electron microscopy. Microsc Microanal. 2012;18:1049–53.
27. Teague B, Waterman MS, Goldstein S, Potamousis K, Zhou S, Reslewic S, Sarkar D, Valouev A, Churas C, Kidd JM, Kohn S, Runnheim R, Lamers C, Forrest D, Newton MA, Eichler EE, Kent-First M, Surti U, Livny M, Schwartz DC. High-resolution human genome structure by single-molecule analysis. Proc Natl Acad Sci U S A. 2010;107:10848–53.
28. Zhou S, Wei F, Nguyen J, Bechner M, Potamousis K, Goldstein S, Pape L, Mehan MR, Churas C, Pasternak S, Forrest DK, Wise R, Ware D, Wing RA, Waterman MS, Livny M, Schwartz DC. A single molecule scaffold for the maize genome. PLoS Genet. 2009;5: e1000711.
29. Kasianowicz JJ, Brandin E, Branton D, Deamer DW. Characterization of individual polynucleotide molecules using a membrane channel. Proc Natl Acad Sci U S A. 1996;93:13770–3.
30. Branton D, Deamer DW, Marziali A, Bayley H, Benner SA, Butler T, Di Ventra M, Garaj S, Hibbs A, Huang X, Jovanovich SB, Krstic PS, Lindsay S, Ling XS, Mastrangelo CH, Meller A, Oliver JS, Pershin YV, Ramsey JM, Riehn R, Soni GV, Tabard-Cossa V, Wanunu M, Wiggin M, Schloss JA. The potential and challenges of nanopore sequencing. Nat Biotechnol. 2008;26:1146–53.
31. Wu HC, Astier Y, Maglia G, Mikhailova E, Bayley H. Protein nanopores with covalently attached molecular adapters. J Am Chem Soc. 2007;129: 16142–8.
32. Derrington IM, Butler TZ, Collins MD, Manrao E, Pavlenok M, Niederweis M, Gundlach JH. Nanopore DNA sequencing with MspA. Proc Natl Acad Sci U S A. 2010;107:16060–5.
33. Clarke J, Wu HC, Jayasinghe L, Patel A, Reid S, Bayley H. Continuous base identification for single-molecule nanopore DNA sequencing. Nat Nanotechnol. 2009;4:265–70.
34. Manrao EA, Derrington IM, Laszlo AH, Langford KW, Hopper MK, Gillgren N, Pavlenok M, Niederweis M, Gundlach JH. Reading DNA at single-nucleotide resolution with a mutant MspA nanopore and phi29 DNA polymerase. Nat Biotechnol. 2012;30:349–53.
35. Cherf GM, Lieberman KR, Rashid H, Lam CE, Karplus K, Akeson M. Automated forward and reverse ratcheting of DNA in a nanopore at 5-A precision. Nat Biotechnol. 2012;30:344–8.
36. Lieberman KR, Cherf GM, Doody MJ, Olasagasti F, Kolodji Y, Akeson M. Processive replication of single DNA molecules in a nanopore catalyzed by phi29 DNA polymerase. J Am Chem Soc. 2010;132: 17961–72.
37. Kang XF, Cheley S, Rice-Ficht AC, Bayley H. A storable encapsulated bilayer chip containing a single protein nanopore. J Am Chem Soc. 2007;129: 4701–5.
38. Rosenstein JK, Wanunu M, Merchant CA, Drndic M, Shepard KL. Integrated nanopore sensing platform with sub-microsecond temporal resolution. Nat Methods. 2012;9:487–92.
39. Garaj S, Hubbard W, Reina A, Kong J, Branton D, Golovchenko JA. Graphene as a subnanometre trans-electrode membrane. Nature. 2010;467: 190–3.
40. Liu H, He J, Tang J, Liu H, Pang P, Cao D, Krstic P, Joseph S, Lindsay S, Nuckolls C. Translocation of single-stranded DNA through single-walled carbon nanotubes. Science. 2010;327:64–7.
41. Balagurusamy VS, Weinger P, Ling XS. Detection of DNA hybridizations using solid-state nanopores. Nanotechnology. 2010;21:335102.
42. Kowalczyk SW, Hall AR, Dekker C. Detection of local protein structures along DNA using solid-state nanopores. Nano Lett. 2010;10:324–8.

43. Thompson JF, Oliver JS. Mapping and sequencing DNA using nanopores and nanodetectors. Electrophoresis. 2012;33:3429–36.
44. Ohshiro T, Matsubara K, Tsutsui M, Furuhashi M, Taniguchi M, Kawai T. Single-molecule electrical random resequencing of DNA and RNA. Sci Rep. 2012;2:501.
45. Ohshiro T, Umezawa Y. Complementary base-pair-facilitated electron tunneling for electrically pinpointing complementary nucleobases. Proc Natl Acad Sci U S A. 2006;103:10–4.
46. Min SK, Kim WY, Cho Y, Kim KS. Fast DNA sequencing with a graphene-based nanochannel device. Nat Nanotechnol. 2011;6:162–5.
47. Harrer S, Ahmed S, Afzali-Ardakani A, Luan B, Waggoner PS, Shao X, Peng H, Goldfarb DL, Martyna GJ, Rossnagel SM, Deligianni L, Stolovitzky GA. Electrochemical characterization of thin film electrodes toward developing a DNA transistor. Langmuir. 2010;26:19191–8.
48. Anderson MW, Schrijver I. Next generation DNA sequencing and the future of genomic medicine. Genes. 2010;1:38–69.

CHAPTER 3

Transcriptome Sequencing (RNA-Seq)

SUGGANTH DANIEL, ALBERTO PANIZ-MONDOLFI, FEDERICO A. MONZON

Introduction 31
Microarrays, Tag Methods, and RNA-Seq 31
Sequencing Platforms 32
RNA Sequencing 33
Transcriptome Assembly 33
Specific RNA-Seq Approaches 35
Clinical Applications of RNA-Seq 35
Inherited Conditions 35
Complex Conditions 36
Cancer 36
Clinical Microbiology Applications 38
Conclusions 39

Introduction

The transcriptome is the entire assembly of RNA transcripts in a given cell type, including protein coding RNA such as messenger RNA (mRNA), and noncoding transcripts like ribosomal RNA (rRNA), transfer RNA (tRNA), micro RNA (miRNA), and other noncoding RNA (ncRNA) [1]. As opposed to the genome, which is shared by all cells in a given organism, the transcriptome is specific to a given tissue or cell type, or even specific to the single cell level. Transcriptome sequencing (RNA-Seq) is a recently developed technology that uses high-throughput sequencing approaches to determine the sequence of all RNA transcripts in a given specimen. This technology has rapidly deepened our understanding of alternative splicing, the functional elements of the genome, as well as fusion transcripts in cancer and has increased our understanding of gene expression profiles of various tissues and cells [2, 3]. The goal of RNA-Seq is to identify and catalogue the myriad forms of known and novel transcripts as well as identifying splicing and/or allelic usage patterns that are present in specific cells or situations. RNA-Seq is regarded as an unbiased technique to assess differential gene expression because it does not depend on the use of probe hybridization and can be used to study gene expression in various stages of development and in benign and malignant disease processes [2–6].

Microarrays, Tag Methods, and RNA-Seq

Gene expression profiling by microarrays (GEM) surfaced at the end of the 1990s and became a well-established research technology in the early 2000s and was introduced a few years later as a clinical tool [7–9]. The microarray is a hybridization-based technology, consisting of oligonucleotide probes complementary to target sequences that are immobilized on a solid substrate; RNA transcripts are fluorescently labeled and are hybridized to the arrays, which are then scanned with a laser. The signal intensity is used to determine the relative abundance of transcripts in a sample, creating what is known as a gene expression profile [7]. The

G.J. Netto and I. Schrijver (eds.), *Genomic Applications in Pathology*,
DOI 10.1007/978-1-4939-0727-4_3, © Springer Science+Business Media New York 2015

increased understanding of the underlying chemistries and kinetics along with quality control by laboratories and organizations such as the MicroArray Quality Control consortium (MAQC) contributed to dramatic improvements in the quality of microarray technology and of gene expression data [10, 11]. This led to an explosion in the number of gene expression profiling studies that expanded our knowledge of transcriptomic differences between different tissue types, benign and malignant diseases and allowed us to identify the existence of molecular subtypes of tumors, such as the intrinsic subtypes of breast cancer and molecular subtypes of leukemias and lymphomas [12–14].

Although now a very robust technology, GEM has some drawbacks. As the conditions for hybridization are constant to all the microarray probes, there is bias towards probes whose hybridization kinetics are favored. Another limitation is the short dynamic range for evaluating relative transcript abundance (around three logs) [15]. For this reason, other approaches were developed, including tag-based techniques such as serial analysis of gene expression (SAGE), cap analysis of gene expression (CAGE), and massively parallel signature sequencing (MPSS). These approaches allowed unbiased transcript detection by sequencing but were limited by their dependence on Sanger sequencing, which made them laborious and difficult to scale up [16]. In addition, GEM and tag-based approaches shared a common problem, namely, the difficulty to differentiate between transcript isoforms (splicing variants), even though specialized microarrays spanning exon junctions to identify spliced isoforms and high-resolution genomic tiling microarrays are available [17].

The recent development of high-throughput massively parallel sequencing (next-generation sequencing or NGS) set the stage for the development of sequencing-based transcriptome evaluation (RNA-Seq). This approach has clear advantages over microarrays and other methods, as the unbiased nature of RNA-Seq allows the detection of known and unknown transcripts, and has enabled the identification of new splice isoforms and gene fusions. RNA-Seq has other advantages such as: nucleotide sequence level resolution (it allows for determination of allele specific expression of transcripts), high dynamic range (due to its digital nature), detection of noncoding transcripts, and the ability to detect posttranscriptional sequence changes (mutations or editing). Comparisons between expression GEM and RNA-Seq show strong concordance among these technologies and RNA-Seq data [18, 19].

Sequencing Platforms

Multiple NGS platforms can be used for RNA-Seq [20–22]. We will briefly describe the salient features of these sequencing instruments but the most commonly used technologies have been described in detail in the preceding chapters.

The Roche 454 Genome Sequencer adopted the pyrosequencing chemistry, the first non-electrophoretic, bioluminescence method of sequencing by synthesis (SBS) and was the first to introduce sequencing in a large-scale parallel manner [23]. In this technology, adapter-ligated DNA fragments are attached to capture beads and amplified individually via emulsion polymerase chain reaction (PCR). Each bead is then packed into a pico-sized well where in nucleotides are flowed in sequential order. Incorporation of nucleotides generates pyrophosphate that is then converted into a light signal by an enzymatic reaction. Light signals are detected and transformed into nucleotide sequences. In a typical run, this platform is able to generate one million reads and an average read length of 400–500 nucleotides [24].

The SOLiD (Sequencing by Oligonucleotide Ligation and Detection) technology utilizes bead clonal amplification. However, it uses ligation, rather than polymerization, as the means of detecting bases to a sequencing template. SOLiD generates short reads, ranging from 25 to 50 nucleotides [25].

The Illumina platform (formerly Solexa) also uses SBS with a surface-based clonal amplification template method, coupled with the four-color nucleotide detection method. It uses reversible termination to ensure that only a single nucleotide is extended each sequencing cycle, and the incorporated nucleotides are detected after each cycle. The read length depends on the number of cycles of

DNA synthesis. The throughput is ten times or greater compared to the 454 but read lengths are typically 100–150 base pairs (bp) [26].

The Ion Torrent technology from Life Technologies uses a non-optical method based on an integrated semiconductor circuit. In this system, clonal amplification is performed on beads that are then distributed in micro-wells (similar to 454 above). Then, nucleotides are supplied in a step-wise fashion to each parallel sequencing reaction when complementary nucleotides are incorporated and hydrolysis of the triphosphate bond liberates a single proton, which causes a shift in pH. This pH change is detected by a sensor and converted to an electric voltage generating a signal [27]. Read lengths with this technology are currently up to 400 bp; throughput depends on the instrument used and on the size of the semiconductor chip.

Third-generation sequencers perform single-molecule sequencing and do not use PCR prior to sequencing [28]. Single-molecule sequencing was first developed by Helicos Biosciences, which has also developed direct RNA sequencing (DRS) which circumvents the need for cDNA synthesis [29]. Pacific Biosciences has developed the first single-molecule real-time (SMRT) sequencing platform, in which single DNA polymerase molecules are attached to the bottom surface of individual detectors, enabling signal capture in real time without the requirement of a termination step. This technology achieves substantially longer reads averaging 2–3 kilobases (kb) and up to 7 kb [30].

RNA Sequencing

Currently, direct sequencing of RNA is not routinely used. RNA-Seq thus always involves the generation of cDNA libraries that can then be sequenced with the approaches outlined above [5]. To generate cDNA, one can use mRNA targeted approaches to select polyA RNA using oligo(dT) selection and random hexamer primed first-strand cDNA synthesis [31]. This technique helps to remove rRNA sequences, which constitute ~80 % of the total RNA. Poly(A) selection is very effective at enriching mRNAs in eukaryotes, but this selection approach will miss ncRNAs and mRNAs that lack a poly(A) tail. If one wants to retain RNAs without a poly(A) tail in the assembled transcriptome, rRNA contamination can instead be removed by hybridization-based depletion methods [32, 33]. The cDNA is then fragmented by DNase I, followed by addition of a deoxyadenine base to the 3′ ends and ligation to adapters. These adapter-ligated cDNA fragments are subsequently amplified and sequenced in a high-throughput manner to obtain short sequence reads with any of the platforms described above [5].

Transcriptome Assembly

Sequence reads obtained from the common NGS platforms are often short and therefore need to be reconstructed into full-length transcripts, with the exception of short sequence length RNA classes such as miRNA [34]. Short reads can be converted to longer reads by the paired-end protocol, in which 75–150 bp are sequenced from both ends of short DNA fragments (100–250 bp), and the overlapping reads are digitally stitched together [35]. Before assembly and mapping, preprocessing of sequence data is performed to remove low quality reads and artifacts. RNA-Seq artifacts such as sequencing adaptors, low-complexity reads and near-identical reads that are derived from PCR amplification are removed with software tools such as SeqTrim and TagDust [36, 37]. Sequencing errors are removed or corrected by using the quality score, a probability function that a specific base in the sequence is correct and/or the k-mer frequency which is the number of times a short oligonucleotide of length k appears in a set of DNA sequences. Very low frequency k-mers usually originate from sequencing errors and reads containing these errors can be removed. However, this carries the danger of removing very rare genuine transcripts [34].

After completion of preprocessing, transcriptome assembly can be achieved by either reference genome based assembly or "de novo" assembly (Fig. 3.1). The reference based method comprises three parts. First,

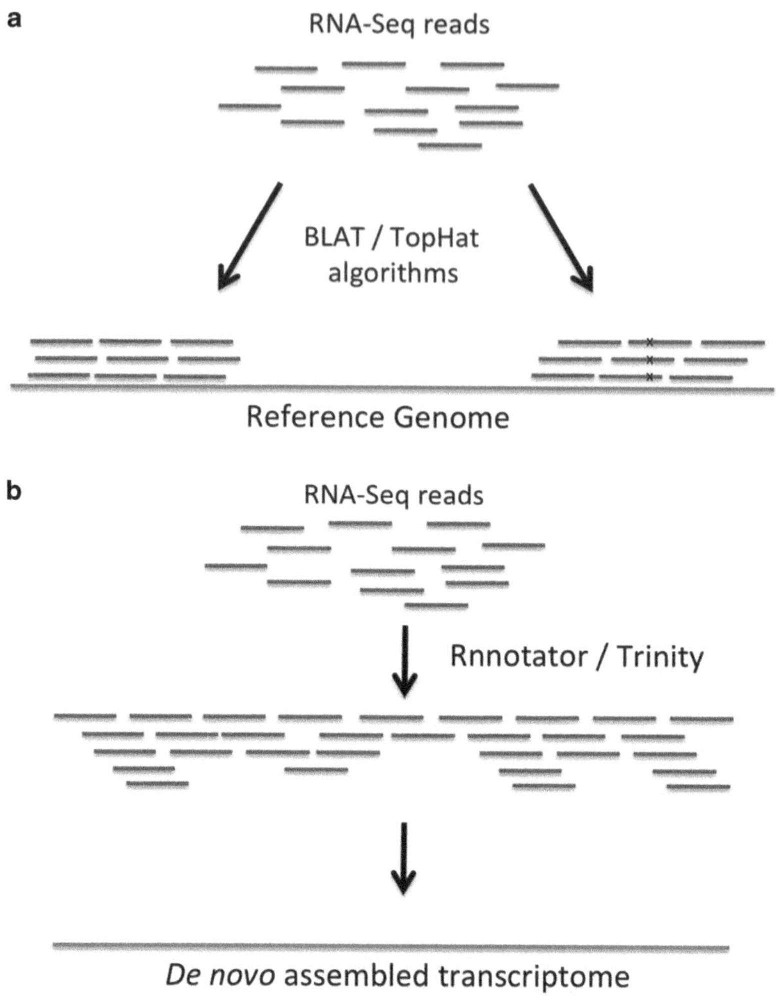

Figure 3-1 Alternatives for transcriptome assembly. (**a**) Transcriptome reads are aligned to a reference genome to determine mapping position and concordance with the reference. (**b**) Overlapping sequences in transcriptome reads are used to infer the transcriptome sequence

alignment of the reads to a reference genome with a "splice-aware aligner" such as BLAT or TopHat [38, 39] is completed, secondly, overlapping reads from the same locus are clustered in a graph to arrive at all possible isoforms, and finally the graph is traversed with programs such as Cufflinks or Scripture to decide on individual isoforms [40]. Reference based methods require less computing power, eliminate some artifacts and errors as these would not align to the reference genome, and are very sensitive to rare transcripts. It is important to note that errors caused by the short read aligners can carry over into assembly and that spliced reads spanning longer introns can be missed. "Multi-reads," where a sequence aligns equally well to several loci in the genome can be excluded, however, this will leave gaps in the final sequence assembly [34].

"De novo" assembly does not a use a reference genome. Instead, it leverages the redundancy in the short reads and uses the overlaps to assemble the transcriptome with software tools such as Rnnotator and Trinity [41]. Trinity was developed specifically for RNA-Seq data and prevents overlapping genes on the same strand being erroneously interpreted as fusion transcripts. It also groups related linear sequences that represent alternative isoforms or paralogous gene families into nonlinear structures containing "bubbles" and alternative ends. In general "de novo" strategies demand lots of computing power [41, 42]. Cloud computing is an alternative and cloud based genome assemblers have been developed [34].

Specific RNA-Seq Approaches

Generation of strand-specific information has been pursued after the realization that antisense transcription events were not just background noise, but were indeed functional in normal and diseased states and quite frequent [43]. Standard RNA-Seq uses double stranded cDNA which erases strand specific information, therefore techniques such as using unique ligation adapters for the first strand of cDNA synthesis, chemical marking of the second strand and direct RNA sequencing of the first strand of cDNA have been adopted [44].

Targeted RNA-Seq has also been developed, based on capture of sequences of interest, to reduce the cost of sequencing. This approach allows the evaluation of a subset of transcripts of interest (e.g., specific fusion transcripts in leukemia) or characterization of rare transcripts. Targeted selection can be performed by using methods such as cDNA hybrid selection with biotinylated oligonucleotide baits on a solid phase or in solution. Targeted RNA-Seq can provide significant savings in cost, time, and computing power and can be used for the development of clinically useful panels [45, 46].

Direct single-molecule RNA sequencing (DRS) has been developed by Helicos Biosciences [47]. This approach frees RNA-Seq from biases introduced by cDNA synthesis, end repair, ligation and amplification procedures, although this method is still limited by read lengths for detection of fusion mRNA transcripts [47]. Although this technique shows promise to enable single cell and circulating nucleic acid applications due to its low RNA input requirements, the high error rate (4–5 %), lack of paired read capability and the high cost of this technology have prevented its widespread use and efforts towards clinical development [48].

Clinical Applications of RNA-Seq

Clinical gene expression profiling developed in the last decade, with applications for cancer diagnosis and prognostication [9, 49–51] and for transplant rejection detection [52], some of which have obtained US Food and Drug Administration (FDA) clearance. However, despite great promise, widespread use of transcriptome profiling for chronic inflammatory, neurological, and infectious diseases, for example, has not become reality [53–55]. Aided by the rapidly decreasing cost of data generation, and the ongoing technical advances of NGS, RNA-Seq has the potential to become a powerful tool in the management and treatment of human disease, although this technology is not yet in clinical use [56, 57]. In the paragraphs below, we discuss examples of RNA-Seq applications being developed for different clinical scenarios.

Inherited Conditions

Since the completion of the human genome there has been a steady increase in the identification of genes responsible for monogenic conditions [58–60]. NGS has been successfully employed for DNA based diagnostic assays such as disease specific panels and whole exome/genome approaches [61–63]. RNA-Seq is being used to study phenotypic variation among individuals affected with genetic diseases, and for diagnosis in cases where DNA-based sequencing and deletion/duplication analyses are unsuccessful. In a recent example of the possible contribution of RNA-Seq to clinical diagnostics, Chandrasekharappa and collaborators showed that the addition of RNA-Seq to NGS-based DNA sequencing and array comparative genomic hybridization (aCGH) allows the detection of more disease alleles in patients with Fanconi Anemia (FA) [64]. The use of

RNA-Seq allowed the identification of exon skipping associated with synonymous, missense and nonsense mutations, as well as intronic pathogenic mutations in FA genes. RNA-Seq has also proved to be an important tool to improve our understanding of complex phenotypes in multigenic disorders, as seen in Down syndrome, for example [65]. In a recent study by Costa et al., RNA-Seq was performed in human trisomic endothelial progenitor cells, revealing differential expression of genes expressed at low levels, novel regions of active transcription outside known loci, identification of non-polyadenylated long and short noncoding RNAs, identification of novel splice isoforms and novel extended untranslated regions for known genes which could represent novel miRNA targets or regulatory sites for gene transcription [65]. This approach could help better understand the mechanisms involved in the generation of Down syndrome phenotypes and the observed individual variability.

Complex Conditions

Genome-wide association studies (GWAS) linking SNPs with specific phenotypes of complex traits and common diseases have shown that only a small fraction of associated SNPs falls within coding regions and that most are intronic or intergenic [66]. This suggests that these nucleotide variants affect gene expression rather than protein function [67]. These variants, therefore, are known as expression quantitative trait loci (eQTL). Given that RNA-Seq allows for the integrative analysis of variation in allele specific transcript sequence, regulatory sites, and expression levels, it is expected that its use will help better understand regulatory variation at single base resolution and this could translate into better insight into the molecular pathogenesis of complex traits [68]. A recent example of success with this approach, although not done with RNA-Seq, is the demonstration that a common noncoding polymorphism at the 1p13 locus, that was found to be associated with changes in plasma low-density lipoprotein cholesterol isoforms and risk for myocardial infarction, affects the expression of a gene that regulates lipoprotein production by the liver [69].

Cancer

RNA-Seq has led to a large number of new discoveries in cancer [56, 67]. As discussed above, RNA-Seq represents a promising tool for revealing and interpreting the complex transcriptomes of human cancers, based on its superior capacity to identify gene fusions, ability to detect bias in expression of mutated alleles, and its improved dynamic range in quantifying gene expression with lower detection limits [56]. The unbiased nature of RNA-Seq also has proved to be an important tool in assessing pseudogene expression in human cancers. In a recent study by Kalyana-Sundaram et al., the authors were able to establish lineage and categorize cancer specific pseudogene expression profiles to the level of disease subtyping, which suggested that pseudogenes might play a role in cellular differentiation and cancer progression [70].

Perhaps one of the most common uses of clinical transcriptome profiling is the identification of tissue origin in cancers of unknown primary site. Several commercial platforms are clinically available for this purpose [9, 71, 72]. Although this application has not yet been migrated to RNA-Seq platforms, it is foreseeable that this will occur in the near future, because this technology enables not only the identification of the site of origin based on the expression pattern but also the detection of expressed mutations and alternative splicing events that could be of utility for therapy selection [73]. An interesting development in this area is the emerging ability to obtain expression profiles from single cells. Recently, Ramsköld et al. have applied this approach to the identification of melanoma circulating tumor cells (CTCs), based on distinct gene expression patterns obtained from just a handful of isolated CTCs [74]. This approach opens the door to the development of noninvasive tests for the determination of tissue of origin and/or for the detection of mutations and fusion transcripts (see below) that can direct targeted therapy.

Breast Cancer (BC) management is another area where transcriptome profiling has been incorporated into routine clinical management, with the use of gene expression profiles that are prognostic of tumor recurrence and that are used for therapeutic management in early stage patients [49]. While the current BC prognostic assays use either quantitative reverse transcriptase PCR (RT-PCR) or microarrays, efforts to translate these panels into RNA-Seq platforms are already underway [75]. This has the potential to expand content on these panels and to incorporate therapeutic biomarkers in these prognostic tests [50, 75].

Because RNA-Seq provides full transcript sequences and is capable of assembling transcripts without relying on preexisting reference sequences, it has been used for the detection of fusion transcripts, both for those previously known to be associated with specific tumors as well as for previously unidentified fusions [56]. Although recurrent gene fusions have been well documented in hematologic malignancies and sarcomas for several decades, the discovery of recurrent gene fusions in epithelial solid tumors is relatively recent. In 2005, Tomlins et al. reported the discovery of fusion transcripts between the *TMPRSS2* and the *ETS* transcription factor genes in prostate cancer [76]. This discovery transformed our understanding of solid tumors and opened the door to an avalanche of studies reporting recurrent fusions in a variety of tumor types [77–83]. Some of these discoveries, such as the identification of *ALK* rearrangements in lung cancer, have already been incorporated into the diagnostic algorithms and management strategies of cancer patients [84, 85]. RNA-Seq based assays are expected to be used to detect these fusion transcripts in some solid tumor types, such as thyroid cancers, among others [46]. The discovery of new fusion transcripts is not limited to solid tumors; novel fusions increasingly are being reported in hematologic malignancies, especially in those that were characterized by a normal conventional karyotype [86, 87]. It is likely that the number of reported fusion transcripts in human cancers will continue to rise as the use of RNA-Seq becomes more prevalent. As in the case of lung cancer, some of these fusion transcripts might be useful for diagnosis, prognosis or selection of targeted therapies. Thus, we can expect that detection of these transcripts (whether by RNA-Seq or not) will become routine in the practice of pathology. A forecast of the upcoming role of RNA-Seq in clinical cancer medicine is the recent report of the use of transcriptome profiling in the identification of *FLT3* overexpression in a patient with acute lymphoblastic leukemia who had recurred after transplant [88]. In this case, reported in the lay press, RNA-Seq was used on a research basis and the finding of high levels of *FLT3* led to the use of Sunitinib, which had been shown to inhibit FLT3 in vitro, with a durable remission achieved. Although this example is just one anecdotal case reported outside of a controlled clinical trial, using high-throughput sequencing for therapeutic guidance has already been incorporated in clinical trials [89].

One of the challenges in the application of RNA-Seq to cancer diagnostics is the fact that metastatic tumors show evolution in their mutational patterns when compared to primary tumors. This has recently been demonstrated by Shah et al., who performed whole genome and transcriptome sequencing of a metastatic invasive lobular breast carcinoma occurring 9 years after its initial diagnosis, and showed significant differences in the mutational landscape between the primary and metastatic tumors [90]. Not only is this problem confined to differences between primary and metastatic tumors, but significant intra-tumoral heterogeneity also exists within primary tumors at the level of DNA sequence variations [91] and at the level of gene expression patterns [92]. These findings highlight an important problem in the development of personalized cancer medicine: ensuring that the specimen tested adequately represents the disease to be treated [93, 94].

In summary, basic and clinical research with RNA-Seq is providing us with a treasure trove of information that should allow us to better understand tumor initiation, progression, and resistance to therapy. However, RNA-Seq cannot answer all remaining questions in oncology. The complexity of cancer has been made evident by recent research efforts and it is now clear that understanding cancer biology and our ability to personalize treatment and to impact outcomes will require the use of all available "omics" technologies, because not all molecular alterations that drive tumor behavior are detected by a single approach [95]. Examples of this are pediatric tumors, in which mutations are less frequent than in adults, and where transcriptome analyses have recently identified disregulated genes that might uncover new targeted therapeutic approaches [96].

Clinical Microbiology Applications

The advent of NGS has made it possible to study and identify a large number of microbial populations in humans and to start defining the normal microbiome as well as microbiome changes associated with abnormal states [97, 98]. The rapid development of microbiome studies mostly relies on the use of 16S rRNA gene sequencing, which is based on DNA sequencing and has become a transforming force in clinical microbiology [99]. However, RNA-Seq technology has created new opportunities for the study of bacterial gene expression [100, 101]. One of the advantages of the RNA-Seq approach is the possibility to study unculturable bacteria or bacteria that cannot be isolated [102]. As such, transcriptome analysis by RNA-Seq has been applied to various clinically relevant microorganisms including: *Staphylococcus aureus*, *Streptococcus pneumoniae*, *Mycoplasma pneumoniae*, *Listeria monocytogenes*, *Helicobacter pylori*, *Salmonella typhi*, *Vibrio cholerae*, *Chlamydia trachomatis* and *Bacillus anthracis*, among others [101]. Whereas RNA-Seq offers several advantages over prior technologies, prokaryote RNA biology poses specific new challenges for this technology [103]. These include the absence of a poly-A-tail (which allows for easy retrieval of coding RNA in eukaryotes) [104], the highly unstable nature of bacterial RNA [105], and the fact that up to 50–80 % of bacterial RNA preparations are composed of ribosomal rRNA and tRNA [106]. However, sequencing-based microbial transcriptome studies have been made possible by removing, at least partially, the rRNA and/or tRNA through a variety of extraction methods including r/tRNA depletion through hybridization with magnetic bead-linked complementary oligonucleotides, or the use of terminator exonucleases involved in specific degradation of transcripts with a 5′ monophosphate group [103, 104, 107, 108]. These bacterial RNA-Seq studies have contributed to a more refined understanding of bacterial gene expression and its impact on microbial ecology [109], physiology and ultimately its potential use in clinical settings [102, 103].

Perhaps one of the most important aspects in the clinical practice of microbiology is the ability to predict or assess microbial virulence and pathogenicity [101, 102]. One of the most important findings in this area is the identification of a larger number of untranslated regions (UTRs) in bacterial transcripts [110]. These UTRs contain riboswitches and binding sites of regulatory small RNAs (sRNAs) and are likely involved in the regulation of gene expression in bacteria, including the expression of genes related to pathogenicity [110]. RNA-Seq experiments have discovered that sRNAs account for up to 20 % of bacterial RNA, including antisense RNAs [109], and these sRNAs appear to have regulatory roles [102, 111]. In a study by Perkins et al. focused on *Salmonella typhi*, strand-specific cDNA sequencing (ssRNA-seq) was used to identify transcriptionally active genes, revealing a large number of previously unknown transcribed regions, including novel noncoding RNAs, some of which might impact the expression of virulence genes [108]. Sharma et al. obtained similar results in transcriptome profiles of *Helicobacter pylori* and were able to establish a correlation between the size of 5′UTRs and cellular function, concluding that UTR size correlated with pathogenicity [112]. In a new twist in the study of understanding host–pathogen interactions, Westermann and colleagues have recently proposed the development of "Dual RNA-Seq", in which both pathogen and host RNA are sequenced together [113].

As detailed above, RNA-Seq is starting to provide an in-depth view of pathogen transcriptomes and this research is predicted to have a direct impact not only on clinical diagnostics and epidemiology but also in the future progress of the field of pathogenomics. Nevertheless, many challenges remain. Single-molecule technology could allow sequencing of full-length polycistronic transcripts, which are commonly found in bacteria, and uncover how alternative transcription origins are utilized and regulated [102]. Probably the biggest hurdle for clinical implementation of RNA-Seq is the lack of reliable clinically based genotype–phenotype correlations that will enable clinical decision making based on bacterial expression profiles [109].

Conclusions

RNA sequencing is making it possible to study transcriptomes at unprecedented resolution and with the ability to detect previously unknown noncoding and fusion transcripts. This technology is currently being applied to the study of inherited, neoplastic, and infectious disorders. Results from these transcriptome analyses are increasing our understanding of normal and disease processes and it is expected that this new knowledge will translate into clinical applications in the near future. Certainly, this technology seems to be moving to real clinical utility fastest in the area of oncology. This is illustrated by the use of RNA sequencing in a patient with acute lymphoblastic leukemia followed by success in the use of specific targeted therapy, based on the transcriptome profile [88]. RNA sequencing now has been incorporated in at least one ongoing clinical trial in patients with advanced cancers [83, 89]. In addition, RNA-Seq has been demonstrated to improve the diagnosis of genetic diseases and has the potential to reveal important clues about bacterial pathogenicity. It is therefore safe to predict that RNA sequencing will become one of the new applications of NGS in clinical and anatomic pathology.

REFERENCES

1. Piétu G, Mariage-Samson R, Fayein NA, Matingou C, Eveno E, Houlgatte R, et al. The Genexpress IMAGE knowledge base of the human brain transcriptome: a prototype integrated resource for functional and computational genomics. Genome Res. 1999;9(2):195–209.
2. Ozsolak F, Milos PM. RNA sequencing: advances, challenges and opportunities. Nat Rev Genet. 2011;12:87–98 (Review on RNA-Seq).
3. Morozova O, Hirst M, Marra MA. Applications of new sequencing technologies for transcriptome analysis. Annu Rev Genomics Hum Genet. 2009;10:135–51.
4. Shah SP, Köbel M, Senz J, Morin RD, Clarke BA, Wiegand KC, Leung G, Zayed A, et al. Mutation of FOXL2 in granulosa-cell tumors of the ovary. N Engl J Med. 2009;360(26):2719–29 (RNA-Seq for fusion transcripts).
5. Wang Z, Gerstein M, Snyder M. RNA-Seq: a revolutionary tool for transcriptomics. Nat Rev Genet. 2009;10(1):57–63 (Review on RNA-Seq).
6. Collisson EA, Cho RJ, Gray JW. What are we learning from the cancer genome? Nat Rev Clin Oncol. 2012;9(11):621–30 (Review on NGS in cancer).
7. Schena M, Shalon D, Davis RW, Brown PO. Quantitative monitoring of gene expression patterns with a complementary DNA microarray. Science. 1995;270:467–70.
8. Golub TR, Slonim DK, Tamayo P, Huard C, Gaasenbeek M, Mesirov JP, et al. Molecular classification of cancer: class discovery and class prediction by gene expression monitoring. Science. 1999;286:531–7.
9. Monzon FA, Lyons-Weiler M, Buturovic LJ, Rigl CT, Henner WD, Sciulli CM, Dumur CI, Medeiros F, Anderson GG. Clinical validation of a gene expression test for identification of tumor tissue of origin. J Clin Oncol. 2009;27(15):2503–8.
10. Shi L, Reid LH, Jones WD, Shippy R, Warrington JA, Baker SC, et al. The MicroArray Quality Control (MAQC) project shows inter- and intraplatform reproducibility of gene expression measurements. Nat Biotechnol. 2006;24:1151–61.
11. Malone JH, Oliver B. Microarrays, deep sequencing and the true measure of the transcriptome. BMC Biol. 2011;9:34 (Review on Microarrays and RNA-Seq).
12. Perou CM, Sørlie T, Eisen MB, van de Rijn M, Jeffrey SS, Rees CA, et al. Molecular portraits of human breast tumours. Nature. 2000;406(6797):747–52.
13. Yeoh EJ, Ross ME, Shurtleff SA, Williams WK, Patel D, Mahfouz R, et al. Classification, subtype discovery, and prediction of outcome in pediatric acute lymphoblastic leukemia by gene expression profiling. Cancer Cell. 2002;1(2):133–43.
14. Alizadeh AA, Eisen MB, Davis RE, Ma C, Lossos IS, Rosenwald A, et al. Distinct types of diffuse large B-cell lymphoma identified by gene expression profiling. Nature. 2000;403(6769):503–11.
15. Sharov V, Kwong KY, Frank B, Chen E, Hasseman J, Gaspard R, Yu Y, Yang I, Quackenbush J. The limits of log-ratios. BMC Biotechnol. 2004;4:3.
16. Harbers M, Carninci P. Tag-based approaches for transcriptome research and genome annotation. Nat Methods. 2005;2:495–502.
17. Bertone P, Stolc V, Royce TE, Rozowsky JS, Urban AE, Zhu X, et al. Global identification of human transcribed sequences with genome tiling arrays. Science. 2004;306:2242–6.
18. Marioni JC, Mason CE, Mane SM, Stephens M, Gilad Y. RNA-seq: an assessment of technical reproducibility and comparison with gene expression arrays. Genome Res. 2008;18:1509–17.
19. Bradford JR, Hey Y, Yates T, Li Y, Pepper SD, Miller CJ. A comparison of massively parallel nucleotide sequencing with oligonucleotide

microarrays for global transcription profiling. BMC Genomics. 2010;11:282.
20. Metzker ML. Sequencing technologies—the next generation. Nat Rev Genet. 2010;11:31–46 (Excellent review on sequencing platforms).
21. Lee H, Tang H. Next generation sequencing technologies and fragment assembly algorithms. Methods Mol Biol. 2012;855:155–74.
22. Glenn TC. Field guide to next-generation DNA sequencers. Mol Ecol Resour. 2011;11(5):759–69.
23. Rothberg JM, Leamon JH. The development and impact of 454 sequencing. Nat Biotechnol. 2008;26(10):1117–24.
24. Leamon JH, Lee WL, Tartaro KR, Lanza JR, Sarkis GJ, deWinter AD, et al. A massively parallel PicoTiterPlate based platform for discrete picoliter-scale polymerase chain reactions. Electrophoresis. 2003;24:3769–77 (Roche 454).
25. Valouev A, Ichikawa J, Tonthat T, Stuart J, Ranade S, Peckham H, et al. A high-resolution, nucleosome position map of C. elegans reveals a lack of universal sequence-dictated positioning. Genome Res. 2008;18:1051–63.
26. Bentley DR, Balasubramanian S, Swerdlow HP, Smith GP, Milton J, Brown CG, et al. Accurate whole human genome sequencing using reversible terminator chemistry. Nature. 2008;456: 53–9.
27. Rothberg JM, Hinz W, Rearick TM, Schultz J, Mileski W, Davey M, et al. An integrated semiconductor device enabling non-optical genome sequencing. Nature. 2011;475(7356):348–52.
28. Sam LT, Lipson D, Raz T, Cao X, Thompson J, Milos PM, et al. A comparison of single molecule and amplification based sequencing of cancer transcriptomes. PLoS One. 2011;6(3):e17305.
29. Harris TD, Buzby PR, Babcock H, Beer E, Bowers J, Braslavsky I, et al. Single-molecule DNA sequencing of a viral genome. Science. 2008;320: 106–9.
30. Eid J, Fehr A, Gray J, Luong K, Lyle J, Otto G, et al. Real-time DNA sequencing from single polymerase molecules. Science. 2009;323: 133–8.
31. Morin R, Bainbridge M, Fejes A, Hirst M, Krzywinski M, Pugh T, et al. Profiling the HeLa S3 transcriptome using randomly primed cDNA and massively parallel short-read sequencing. Biotechniques. 2008;45(1):81–94. doi:10.2144/000112900.
32. Nagalakshmi U, Waern K, Snyder M. RNA-Seq: a method for comprehensive transcriptome analysis. Curr Protoc Mol Biol. 2010;Chapter 4:Unit 4.11.1-13.
33. He S, Wurtzel O, Singh K, Froula JL, Yilmaz S, Tringe SG, et al. Validation of two ribosomal RNA removal methods for microbial metatranscriptomics. Nat Methods. 2010;7:807–12.
34. Martin JA, Wang Z. Next-generation transcriptome assembly. Nat Rev Genet. 2011;12(10): 671–82.
35. Rodrigue S, Materna AC, Timberlake SC, Blackburn MC, Malmstrom RR, Alm EJ, et al. Unlocking short read sequencing for metagenomics. PLoS One. 2010;5:e11840.
36. Falgueras J, Lara AJ, Fernández-Pozo N, Cantón FR, Pérez-Trabado G, Claros MG. SeqTrim: a high-throughput pipeline for pre-processing any type of sequence read. BMC Bioinformatics. 2010;11:38.
37. Lassmann T, Hayashizaki Y, Daub CO. TagDust—a program to eliminate artifacts from next generation sequencing data. Bioinformatics. 2009;25:2839–40.
38. Kent WJ. BLAT-the BLAST-like alignment tool. Genome Res. 2002;12:656–64.
39. Trapnell C, Pachter L, Salzberg SL. TopHat: discovering splice junctions with RNA-seq. Bioinformatics. 2009;25:1105–11.
40. Trapnell C, Williams BA, Pertea G, Mortazavi A, Kwan G, van Baren MJ, et al. Transcript assembly and quantification by RNA-seq reveals unannotated transcripts and isoform switching during cell differentiation. Nat Biotechnol. 2010;28: 511–5.
41. Grabherr MG, Haas BJ, Yassour M, Levin JZ, Thompson DA, Amit I, et al. Full-length transcriptome assembly from RNA-seq data without a reference genome. Nat Biotechnol. 2011;29: 644–52.
42. Iyer MK, Chinnaiyan AM. RNA-Seq unleashed. Nat Biotechnol. 2011;29(7):599–600.
43. Faghihi MA, Wahlestedt C. Regulatory roles of natural antisense transcripts. Nat Rev Mol Cell Biol. 2009;10:637–43.
44. Levin JZ, Yassour M, Adiconis X, Nusbaum C, Thompson DA, Friedman N, et al. Comprehensive comparative analysis of strand-specific RNA sequencing methods. Nat Methods. 2010;7: 709–15.
45. Levin JZ, Berger MF, Adiconis X, Rogov P, Melnikov A, Fennell T, et al. Targeted next-generation sequencing of a cancer transcriptome enhances detection of sequence variants and novel fusion transcripts. Genome Biol. 2009;10: R115.
46. Nikiforov YE, Yip L, Nikiforova MN. New strategies in diagnosing cancer in thyroid nodules: impact of molecular markers. Clin Cancer Res. 2013;19(9):2283–8. 46.
47. Ozsolak F, Platt AR, Jones DR, Reifenberger JG, Sass LE, McInerney P, et al. Direct RNA sequencing. Nature. 2009;461:814–8.
48. Ozsolak F. Third-generation sequencing techniques and applications to drug discovery. Expert Opin Drug Discov. 2012;7(3):231–43.

49. Prat A, Ellis MJ, Perou CM. Practical implications of gene-expression-based assays for breast oncologists. Nat Rev Clin Oncol. 2011;9(1):48–57.
50. Baehner FL, Lee M, Demeure MJ, Bussey KJ, Kiefer JA, Barrett MT. Genomic signatures of cancer: basis for individualized risk assessment, selective staging and therapy. J Surg Oncol. 2011;103(6):563–73.
51. Chibon F. Cancer gene expression signatures—the rise and fall? Eur J Cancer. 2013;49(8):2000–9.
52. Pham MX, Teuteberg JJ, Kfoury AG, et al. Gene-expression profiling for rejection surveillance after cardiac transplantation. N Engl J Med. 2010;20:1890–900.
53. Cooper-Knock J, Kirby J, Ferraiuolo L, Heath PR, Rattray M, Shaw PJ. Gene expression profiling in human neurodegenerative disease. Nat Rev Neurol. 2012;8(9):518–30. doi:10.1038/nrneurol.2012.156. Epub 2012 Aug 14.
54. Drew JE. Cellular defense system gene expression profiling of human whole blood: opportunities to predict health benefits in response to diet. Adv Nutr. 2012;3(4):499–505.
55. Shih B, Watson S, Bayat A. Whole genome and global expression profiling of Dupuytren's disease: systematic review of current findings and future perspectives. Ann Rheum Dis. 2012;71(9):1440–7. doi:10.1136/annrheumdis-2012-201295. Epub 2012 Jul 6.
56. Meldrum C, Doyle MA, Tothill RW. Next-generation sequencing for cancer diagnostics: a practical perspective. Clin Biochem Rev. 2011;32:177–95.
57. Saunders CJ, Miller NA, Soden SE, et al. Rapid whole-genome sequencing for genetic disease diagnosis in neonatal intensive care units. Sci Transl Med. 2012;4:154ra135.
58. Pagon RA, Adam MP, Bird TD, Dolan CR, Fong CT, Stephens K, editors. GeneReviews™ [Internet]. Seattle, WA: University of Washington, Seattle; 1993–2013. http://www.ncbi.nlm.nih.gov/books/NBK1116/
59. Xue Y, Chen Y, Ayub Q, Huang N, Ball EV, Mort M, Phillips AD, Shaw K, Stenson PD, Cooper DN, Tyler-Smith C, the 1000 Genomes Project Consortium. Deleterious- and disease-allele prevalence in healthy individuals: insights from current predictions, mutation databases, and population-scale resequencing. Am J Hum Genet. 2012;91:1022–32.
60. Naidoo N, Pawitan Y, Soong R, Cooper DN, Ku CS. Human genetics and genomics a decade after the release of the draft sequence of the human genome. Hum Genomics. 2011;5(6):577–622.
61. Teekakirikul P, Kelly MA, Rehm HL, Lakdawala NK, Funke BH. Inherited cardiomyopathies: molecular genetics and clinical genetic testing in the postgenomic era. J Mol Diagn. 2013;15(2):158–70.
62. Mayer AN, Dimmock DP, Arca MJ, Bick DP, Verbsky JW, Worthey EA, Jacob HJ, Margolis DA. A timely arrival for genomic medicine. Genet Med. 2011;13(3):195–6.
63. Schrijver I, Aziz N, Farkas DH, et al. Opportunities and challenges associated with clinical diagnostic genome sequencing: a report of the Association for Molecular Pathology. J Mol Diagn. 2012;14(6):525–40.
64. Chandrasekharappa SC, Lach FP, Kimble DC, Kamat A, Teer JK, Donovan FX, NISC Comparative Sequencing Program, et al. Massively parallel sequencing, aCGH, and RNA-Seq technologies provide a comprehensive molecular diagnosis of Fanconi anemia. Blood. 2013;121(22):e138–48.
65. Costa V, Angelini V, D'Apice L, Mutarelli M, Casanassimi A, et al. Massive-scale RNA-Seq analysis of non ribosomal transcriptome in human trisomy 21. PLoS One. 2011;6(4):e18493.
66. Freedman ML, Monteiro AN, Gayther SA, et al. Principles for the post-GWAS functional characterization of cancer risk loci. Nat Genet. 2011;43:513–21.
67. Costa V, Aprile M, Esposito R, Ciccodicola A. RNA-Seq and human complex diseases: recent accomplishments and future perspectives. Eur J Hum Genet. 2013;21(2):134–42.
68. Majewski J, Pastinen T. The study of eQTL variations by RNA-seq: from SNPs to phenotypes. Trends Genet. 2011;27(2):72–9.
69. Musunuru K, Strong A, Frank-Kamenetsky M, Lee NE, Ahfeldt T, Sachs KV, et al. From non-coding variant to phenotype via SORT1 at the 1p13 cholesterol locus. Nature. 2010;466(7307):714–9. doi:10.1038/nature09266.
70. Kalyana-Sundaram S, Kumar-Sinha C, Shankar S, Robinson DR, Wu YM, Cao X, et al. Expressed pseudogenes in the transcriptional landscape of human cancers. Cell. 2012;149(7):1622–34.
71. Monzon FA, Koen TJ. Diagnosis of metastatic neoplasms: molecular approaches for identification of tissue of origin. Arch Pathol Lab Med. 2010;134(2):216–24.
72. Erlander MG, Ma XJ, Kesty NC, Bao L, Salunga R, Schnabel CA. Performance and clinical evaluation of the 92-gene real-time PCR assay for tumor classification. J Mol Diagn. 2011;13(5):493–503. doi:10.1016/j.jmoldx.2011.04.004. Epub 2011 Jun 25.
73. Varadhachary G. New strategies for carcinoma of unknown primary: the role of tissue-of-origin molecular profiling. Clin Cancer Res. 2013;19(15):4027–33.

74. Ramsköld D, Luo S, Wang YC, Li R, Deng Q, Faridani OR, et al. Full-length mRNA-Seq from single-cell levels of RNA and individual circulating tumor cells. Nat Biotechnol. 2012;30(8):777–82.
75. Sinicropi D, Qu K, Collin F, Crager M, Liu ML, Pelham RJ, et al. Whole transcriptome RNA-Seq analysis of breast cancer recurrence risk using formalin-fixed paraffin-embedded tumor tissue. PLoS One. 2012;7(7):e40092.
76. Tomlins SA, Rhodes DR, Perner S, et al. Recurrent fusion of TMPRSS2 and ETS transcription factor genes in prostate cancer. Science. 2005;310:644–8.
77. Santoro M, Melillo RM, Fusco A. RET/PTC activation in papillary thyroid carcinoma: European Journal of Endocrinology Prize Lecture. Eur J Endocrinol. 2006;155:645–53.
78. Soda M, Choi YL, Enomoto M, Takada S, Yamashita Y, Ishikawa S, et al. Identification of the transforming EML4-ALK fusion gene in non-small-cell lung cancer. Nature. 2007;448:561–6.
79. Maher CA, Kumar-Sinha C, Cao X, Kalyana-Sundaram S, Han B, Jing X, Sam L, Barrette T, Palanisamy N, Chinnaiyan AM. Transcriptome sequencing to detect gene fusions in cancer. Nature. 2009;458(7234):97–101.
80. Edgren H, Murumagi A, Kangaspeska S, Nicorici D, Hongisto V, Kleivi K, Rye IH, Nyberg S, Wolf M, Borresen-Dale AL, Kallioniemi O. Identification of fusion genes in breast cancer by paired-end RNA-sequencing. Genome Biol. 2011;12(1):R6.
81. Robinson DR, Kalyana-Sundaram S, Wu YM, Shankar S, Cao X, Ateeq B, et al. Functionally recurrent rearrangements of the MAST kinase and Notch gene families in breast cancer. Nat Med. 2011;17:1646–51.
82. Seshagiri S, Stawiski EW, Durinck S, Modrusan Z, Storm EE, Conboy CB, et al. Recurrent R-spondin fusions in colon cancer. Nature. 2012;488:660–4.
83. Wu YM, Su F, Kalyana-Sundaram S, Khazanov N, Ateeq B, Cao X, et al. Identification of targetable FGFR gene fusions in diverse cancers. Cancer Discov. 2013;3(6):636–47.
84. Lindeman NI, Cagle PT, Beasley MB, Chitale DA, Dacic S, Giaccone G, et al. Molecular testing guideline for selection of lung cancer patients for EGFR and ALK tyrosine kinase inhibitors: guideline from the College of American Pathologists, International Association for the Study of Lung Cancer, and Association for Molecular Pathology. J Mol Diagn. 2013;15(4):415–53.
85. Kwak EL, Bang YJ, Camidge DR, Shaw AT, Solomon B, Maki RG, et al. Anaplastic lymphoma kinase inhibition in non-small-cell lung cancer. N Engl J Med. 2010;363(18):1693–703. doi:10.1056/NEJMoa1006448.
86. Wen H, Li Y, Malek SN, Kim YC, Xu J, Chen P, et al. New fusion transcripts identified in normal karyotype acute myeloid leukemia. PLoS One. 2012;7(12):e51203. doi:10.1371/journal.pone.0051203. Epub 2012 Dec 12.
87. Zhou JB, Zhang T, Wang BF, Gao HZ, Xu X. Identification of a novel gene fusion RNF213-SLC26A11 in chronic myeloid leukemia by RNA-Seq. Mol Med Rep. 2013;7(2):591–7. doi:10.3892/mmr.2012.1183.
88. Kolata G. In leukemia treatment, glimpses of the future. The New York Times. July 8, 2012; A1.
89. Roychowdhury S, Iyer MK, Robinson DR, Lonigro RJ, Wu YM, Cao X, et al. Personalized oncology through integrative high-throughput sequencing: a pilot study. Sci Transl Med. 2011;3:111ra21.
90. Shah P, Morin RD, Khattra J, Prentice L, Pugh T, Burleigh A, Delaney A, Gelmon K, Guliany R, Senz J, et al. Mutational evolution in a lobular breast tumour profiled at single nucleotide resolution. Nature. 2009;461:809–13.
91. Gerlinger M, Rowan AJ, Horswell S, Larkin J, Endesfelder D, Gronroos E, et al. Intratumor heterogeneity and branched evolution revealed by multiregion sequencing. N Engl J Med. 2012;366(10):883–92.
92. Sottoriva A, Spiteri I, Piccirillo SG, Touloumis A, Collins VP, Marioni JC, Curtis C, Watts C, Tavaré S. Intratumor heterogeneity in human glioblastoma reflects cancer evolutionary dynamics. Proc Natl Acad Sci U S A. 2013;110(10):4009–14.
93. Stevens JB, Horne SD, Abdallah BY, Ye CJ, Heng HH. Chromosomal instability and transcriptome dynamics in cancer. Cancer Metastasis Rev. 2013;32(3–4):391–402.
94. Dienstmann R, Rodon J, Barretina J, Tabernero J. Genomic medicine frontier in human solid tumors: prospects and challenges. J Clin Oncol. 2013;31(15):1874–84.
95. Mardis ER. Applying next-generation sequencing to pancreatic cancer treatment. Nat Rev Gastroenterol Hepatol. 2012;9:477–86.
96. Downing JR, Wilson RK, Zhang J, Mardis ER, Pui CH, Ding L, Ley TJ, Evans WE. The pediatric cancer genome project. Nat Genet. 2012;44(6):619–22.
97. Human Microbiome Project Consortium. Structure, function and diversity of the healthy human microbiome. Nature. 2012;486(7402):207–14.
98. Li E, Hamm CM, Gulati AS, Sartor RB, Chen H, Wu X, et al. Inflammatory bowel diseases phenotype, C. difficile and NOD2 genotype are associated with shifts in human ileum associated microbial composition. PLoS One. 2012;7(6):e26284.
99. Woo PC, Lau SK, Teng JL, Tse H, Yuen KY. Then and now: use of 16S rDNA gene sequencing for

bacterial identification and discovery of novel bacteria in clinical microbiology laboratories. Clin Microbiol Infect. 2008;14(10):908–34.
100. Croucher NJ, Thomson NR. Studying bacterial transcriptomes using RNA-seq. Curr Opin Microbiol. 2010;13(5):619–24.
101. Pinto AC, Melo-Barbosa HP, Miyoshi A, Silva A, Azevedo V. Application of RNA-seq to reveal the transcript profile in bacteria. Genet Mol Res. 2011;10(3):1707–18.
102. Güell M, Yus E, Lluch-Senar M, Serrano L. Bacterial transcriptomics: what is beyond the RNA horiz-ome? Nat Rev Microbiol. 2011;9(9):658–69. doi:10.1038/nrmicro2620.
103. van Vliet AHM. Next generation sequencing of microbial transcriptomes: challenges and opportunities. FEMS Microbiol Lett. 2010;302:1–7.
104. Passalacqua KD, Varadarajan A, Ondov BD, Okou DT, Zwick ME, Bergman NH. Structure and complexity of a bacterial transcriptome. J Bacteriol. 2009;191:3203–11.
105. Deutscher MP. Degradation of stable RNA in bacteria. J Biol Chem. 2003;278:45041–4.
106. Condon C. Maturation and degradation of RNA in bacteria. Curr Opin Microbiol. 2007;10:271–8.
107. Yoder-Himes DR, Chain PS, Zhu Y, Wurtzel O, Rubin EM, Tiedje JM, Sorek R. Mapping the Burkholderia cenocepacia niche response via high-throughput sequencing. Proc Natl Acad Sci U S A. 2009;106:3976–81.
108. Perkins TT, Kingsley RA, Fookes MC, et al. A strand-specific RNA-seq analysis of the transcriptome of the typhoid bacillus Salmonella typhi. PLoS Genet. 2009;5:e1000569.
109. Pallen MJ, Loman NJ, Penn CW. High-throughput sequencing and clinical microbiology: progress, opportunities and challenges. Curr Opin Microbiol. 2012;13:625–31.
110. Sorek R, Cossart P. Prokaryotic transcriptomics: a new view on regulation, physiology and pathogenicity. Nat Rev Genet. 2010;11:9–16.
111. Toledo-Arana A, Dussurget O, Nikitas G, Sesto N, Guet-Revillet H, Balestrino D, et al. The Listeria transcriptional landscape from saprophytism to virulence. Nature. 2009;459(7249):950–6. doi:10.1038/nature08080. Epub 2009 May 17.
112. Sharma CM, Hoffmann S, Darfeuille F, Reignier J, et al. The primary transcriptome of the major human pathogen Helicobacter pylori. Nature. 2010;464:250–5.
113. Westermann AJ, Gorski SA, Vogel J. Dual RNA-seq of pathogen and host. Nat Rev Microbiol. 2012;10:618–30.

CHAPTER 4

MiRNA Expression Assays

CORNELIA BRAICU, BEATRIZ M. MAIA, IOANA BERINDAN-NEAGOE, GEORGE A. CALIN

Introduction 45
MiRNA in Human Disease 46
Which Types of Biological Samples Can Be Used for MiRNA Studies? 48
MiRNA Extraction Protocols 50
MiRNA Quantity and Quality Control 52
MiRNA Stability 53
Advanced Techniques for Examining MiRNA Expression 53
Data Processing and Normalization 58
Analysis of MiRNA Function and Gene Interaction 62
MiRNA Mimics and Inhibitors 63
Conclusions 64
Acknowledgements 65

Introduction

The clinical promise of the research application of microRNAs (miRNAs) is of great interest for clinicians as well as pathologists. In this chapter, we describe different assays, methods, and technologies used to identify, characterize, and confirm miRNA expression in human diseases.

MiRNAs are small noncoding RNAs (ncRNAs) that play highly important roles in posttranscriptional regulation of protein-coding gene expression and are increasingly useful for early diagnosis, treatment, and assessment of treatment response of different pathologies. NcRNAs are currently classified into two main categories according to their length: short and long noncoding RNAs [1, 2]. Short ncRNAs are 18–200 nucleotides long and comprise transcript variants such as miRNAs, small interfering RNAs, transfer RNAs, and piwi-interacting RNAs [3]. In 2006, Andrew Fire and Craig Mello received a Nobel Prize for their discovery of RNA interference—gene silencing by double-stranded RNA [4], a proof of the tremendous scientific interest generated by the discovery of such noncoding RNAs. MiRNAs are single-stranded RNA structures that are 18–22 nucleotides long. They regulate the expression of genes by binding to different sites in the 3′ untranslated regions of several target mRNAs, which causes mRNA cleavage (degradation) or translational repression [5].

Different miRNA profiles have proven to be relevant in both physiological and pathological processes [1–3]. Therefore, in the new age of personalized medicine, while researchers demonstrated miRNAs as having multiple cellular functions and connected the complexity of miRNA profiles with several human diseases, including cancer and cardiovascular and autoimmune diseases [1, 2, 5, 6], clinicians aim to use these miRNA expressions as tools for diagnostic approaches and individualized therapeutic options.

Currently, there is a wide range of platforms for miRNA identification, confirmation, and profiling, each having specific strengths and limitations [6, 7]. In this sense,

G.J. Netto and I. Schrijver (eds.), *Genomic Applications in Pathology*,
DOI 10.1007/978-1-4939-0727-4_4, © Springer Science+Business Media New York 2015

focusing on miRNA extraction, quality control, molecular profiling, expression profiling, and sequencing in different types of biological samples (cells, tissues [in vitro and in vivo], and biological fluids) is of paramount importance as these methods can be useful for pathologists as additional methods to classical immunohistochemistry in order to validate specific diagnoses.

In this chapter, we emphasize methods for the assessment of miRNA expression profiles in cells, tissues, and body fluids and highlight the main advantages and disadvantages of these methods in specific biological applications. This complete picture can help pathologists combine new methods and classical ones for diagnostic and therapeutic purposes.

MiRNA in Human Disease

MiRNAs are actively being assessed as possible biomarkers of prognosis, therapy prediction, and target of therapy for multiple diseases [2, 5, 8–11]. Their biological significance was illustrated initially in cancer and recently proved to be implicated in several other pathologies like cardiovascular disease, obesity, diabetes, autoimmune diseases, neurodegenerative diseases, and viral infections [7, 12, 13]. MiRNAs exhibit a tissue-specificity profile and are present in complex regulatory networks of cellular function, tissue differentiation, and maintenance of cell identity and evolution from embryogenesis to adult life. Also, multiple fundamental processes are modulated by miRNAs, such as cell differentiation, apoptosis, tumor initiation, invasion, and metastasis.

Alterations of particular signaling pathways or biological processes are often correlated with a specific pathology. The capacity of miRNAs to modulate key signaling pathways has been extensively documented. One relevant example is the miR-34 family, which includes important components of the p53 pathways. p53 is one of the most studied tumor-suppressor genes and directly transactivates miR-34a and miR-34b/miR-34c followed by modulation of the cell cycle and apoptosis [14, 15]. MiRNAs such as miR-125b [16] and miR-101 [17] were demonstrated to regulate the major controller of cancer growth in the prostate, namely, androgen-receptor signaling. Similarly, miR-106b [18] and miR-23b [19] can target members of the phosphatidylinositol 3-kinase/Akt/PTEN signaling pathway in this type of cancer [20]. MiRNA implication in cancer is also related to the capacity to regulate apoptosis, cellular growth, and proliferation. For example, miR-14 is required for growth control and *let-7* family miRNAs are regulators of the proto-oncogene *RAS*. Interestingly, the miRNA mir-17 cluster is located on human chromosome 13 that is frequently amplified in B-cell lymphomas [21, 22].

Understanding miRNA roles in different disease processes is an ongoing process that remains far from complete, although preclinical and clinical data are so far encouraging [23]. Below we provide a broad overview of miRNAs as biomarkers of prognosis, therapy prediction, and targets of therapy, demonstrating their possible roles in clinical management in the genomic and personalized medicine era.

Cancer

Several miRNAs are located at genomic regions linked to cancer [21]. During the cellular transformation, some miRNAs are specifically deregulated and their altered expression and functions lead to important disease phenotypes. Calin and colleagues provided the first identification of miRNA involvement in cancer more than a decade ago (2002) [24] in chronic lymphocytic leukemias (CLL). In this study, miR-15a and miR-16-1 were shown to be deleted or downregulated in most CLL samples indicating the potential of miRNA regulatory control over target genes [24], including BCL-2 [25]. The same group subsequently described a unique miRNA signature associated with prognostic factors and disease progression in CLL [26]. Additionally, TCL-1, an important oncogene in B-cell CLL responsible for the aggressive form of the disease, was shown to be regulated by miR-29 and miR-181, two miRNAs differentially expressed in CLL [27], and the role of miRNAs and ncRNAs was broadly described in this disease [28]. The expression levels of miR-21 were shown to be significantly higher in patients with poor prognosis and were able to predict the overall survival. A score termed 21FK based on miR-21 evaluation by quantitative reverse transcription-PCR (qRT-PCR), fluorescence in situ hybridization (FISH), and karyotype was proposed to predict patients' survival. Patients with low 21FK score demonstrated significantly better survival [29]. More recently, miR-155 was shown to be overexpressed in B cells from individuals with monoclonal B-cell lymphocytosis (MBL, a premalignant condition) and was successfully identified circulating in microvesicles of both MBL and CLL patients. miR-155 overexpression was also observed, in patients with CLL who did not reach complete responses after therapy, pointing to miR-155 as a powerful biomarker for progression in individuals with MBL and for prediction to therapy in individuals with CLL [30].

Various miRNAs are recognized to control the expression of tumor-suppressor genes and oncogenes, whereas others have predictive value for specific treatment response and survival, including miR-21 and miRNA-221/222, which are usually overexpressed in various types of cancer cells. miR-21 is the top most upregulated miRNA in solid human cancers, as compared with matching noncancerous tissue [6, 7, 11]. miR-21 has been shown as able to promote tumor proliferation and invasion in gastric cancer through the suppression of PTEN expression [31]. Also, increased expression of miR-21 in non-small-cell lung cancer was significantly associated with worse survival and increased risk of lymphoid infiltration [32]. In colorectal cancer patients, serum levels of miR-21 were correlated with recurrence and mortality, making it a potential prognostic marker in this type of tumors [33]. In invasive ductal carcinomas of the breast, high expression of miR-21 was associated with important clinicopathological features such as tumor size, stage, grade, negativity of estrogen receptor (ER), positivity for human epidermal growth factor receptor 2 (HER2), high Ki-67 expression, mastectomy, and lower overall survival. All these findings indicate that miR-21 is a potentially important prognostic factor in breast cancer [34]. Other well-studied miRNAs are the members of the miR-17-92 family, consisting of six miRNAs with the same seed sequence, part of a cluster located on human chromosome 13q31. This region was shown to be frequently amplified in several types of lymphoma [35] and solid tumors [36]. In oral squamous cell carcinomas, miR-17/20a was shown to regulate cell migration inhibition and demonstrated negative correlation with TNM-stage and lymphatic metastasis [36]. High levels of miR-miR-17, miR-20a, and miR-92-1, along with miR-15a and miR-16-1, suggested poor prognosis in multiple myeloma (MM) once it was associated with shorter progression-free survival [37]. In colon cancer, upregulation of this cluster was also correlated with poor prognosis and miR-17 expression was identified as an independent prognostic factor in this tumor [38].

Potential utility of urine miRNAs as noninvasive biomarkers in urologic cancers such as bladder, prostate, and renal cell carcinomas has been suggested [20]. Increased levels of miR-126, miR-182, and miR-199a have been documented in urine of bladder cancer patients indicating the potential of miRNAs as a biomarker of the disease [39].

MiRNAs can also play an important role in drug resistance. In colorectal cancer for example, miR-222 plays a role in the development of multidrug resistance by modulation of ADAM-17 [36]. Similarly, miR-122 and miR-29a were shown to contribute to resistance to adriamycin and docetaxel in breast cancer [40]. Downregulation of miR-29 is capable of increasing resistance to cisplatin in ovarian cancer cells [41]. In this sense, future studies targeting miRNAs related to drug resistance can be of paramount importance in clinical practice.

MiRNAs can be targeted by the epigenetic machinery. Taking this into account, studies on the impact of methylation and acetylation on miRNA expression and cancer have been largely pursued with important clinical implications. MiR-129-2 is frequently methylated in hepatocellular carcinoma cells (HCC) [42] and in CLL, adversely impacting survival in the latter [43].

Not surprisingly, miRNAs were also shown to be capable of targeting the epigenetic machinery. As an example, members of miR-29 family can negatively modulate DNA methyltransferases DNMT3A and DNMT3B enzymes in lung cancer. Lower survival rates were observed in patients with higher levels of DNMT3A in lung cancer [44]. This indicates that interactions between the miRNome and the epigenome can provide new grounds for future studies of therapy-based strategies in cancer.

Adipogenesis

MiRNAs are deregulated in adipose tissue from obese patients. MiRNAs appear to play regulatory roles in many biological processes associated with obesity, including adipocyte differentiation, insulin action, lipid storage processes, and fat metabolism [45, 46]. Several miRNAs were described in adipocytes and appear to have a role in the modulation of adipogenesis, which may impact the targeting of adipogenesis dysfunction by controlled delivery of miRNA structures [45, 47]. During adipogenesis, miRNAs can accelerate or inhibit adipocyte differentiation and hence regulate fat cell development. In addition, miRNAs may regulate adipogenic lineage commitment in multipotent stem cells and hence govern fat cell numbers. Recent findings suggest miR-519d to be associated with human obesity, but larger case–control studies are needed. Few miRNA targets have been experimentally validated in adipocytes. Both miR-27 and miR-519d target PPAR family members, which are well-established regulators of fat cell development [48]. Most of the studies on miRNAs in adipogenesis are based on murine models, reporting the activity of miRNAs in various processes regulating adipogenesis. In mice models, the cluster miR-17-92, miR-200, and miR-103 has a pro-adipogenic role and let-7 and miR-27a/b are anti-adipogenic [45–48].

Cardiovascular Diseases

The latest investigations furnish the evidence that miRNAs modulate a wide range of cardiac functions with developmental, pathophysiological, and also clinical implications [49]. MiRNA expression analysis has led to the discovery of a potential role for miR-1, miR-16, miR-27b, miR-30d, miR-126, miR-133, miR-143, and the let-7 family in mammalian heart development [50]. Recently, deregulated expression of miR-1 and miR-133 was reported in human heart failure [51, 52] and cardiac hypertrophy [53]. MiRNAs have been suggested as biomarkers for cardiovascular diseases: increasing levels of miR-1, miR-133, miR-499, and miR-208 were associated with cardiac injury after acute myocardial infarction; reduced levels of miR-126 were linked to the development of coronary artery disease or diabetes [54].

Autoimmune Diseases

Immune modulatory genes are highly regulated by miRNAs. Toll-like receptors, known mediators in microbial infections, and their ligands induce miR-155 expression, immune cell survival, and cytokine signaling suppression [55]. MiRNA is heavily implicated in the molecular mechanisms that regulate the immune system during the development of autoimmune diseases. Examples of specific types of autoimmune disorders in which miRNAs are involved in immune system response evolution include rheumatoid arthritis, where miR-155 and miR-146a are deregulated; multiple sclerosis where miR-18b and miR-599 are associated with disease relapse; and systemic lupus erythematosus where miR-155 and miR-146a appear to be involved in upregulating and downregulating disease activity, respectively. MiR-941 is potentially involved in the pathogenesis of inflammatory bowel disease [55]. In Sjogren's syndrome, an autoimmune disease affecting salivary and lacrimal glands, miR-146a and miR-155 were found to be deregulated during disease development. MiR-21 has been shown to be activated in T cells in psoriasis, enhancing dermal inflammation [56]. Several miRNAs such as miR-510, miR-191, and miR-342 were found to be deregulated in association with type I diabetes type. In this condition, miR-21 expression was also highly correlated with the severity of the disease [57].

Which Types of Biological Samples Can Be Used for MiRNA Studies?

An important issue in the evaluation of miRNA expression is the quality of biological samples and the RNA isolation method used to obtain an appropriate quantity of total

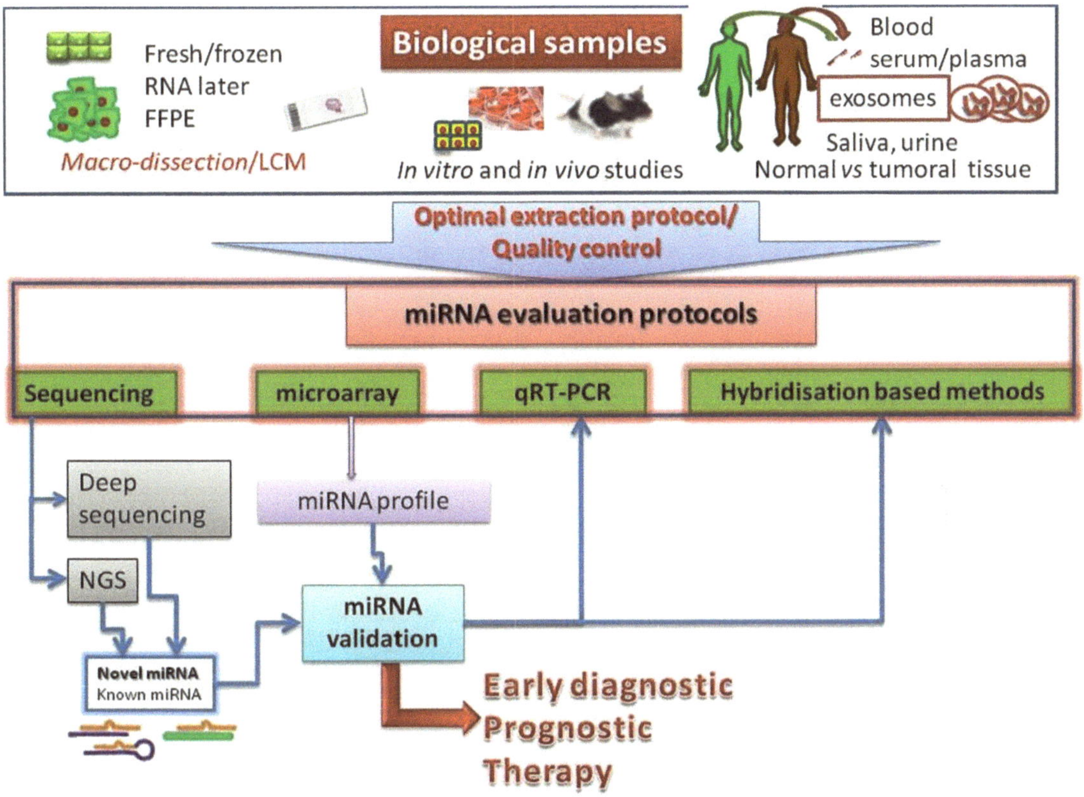

Figure 4-1 Biological samples and methods for microRNA evaluation. The figure illustrates a variety of biological samples and evaluation methods for miRNA assessment, targeting the achievement of routine early diagnosis, prognostication, and therapy for these molecules. *LCM* laser capture microdissection

RNA in order to achieve reproducible, reliable results [8]. Because of their short sequences, miRNAs are known to be relatively stable and homogeneous compared to RNA and DNA. They are well preserved in a wide variety of biological samples, including various body fluids such as blood serum/plasma, saliva, and urine [1, 2, 5, 8, 9, 58]. MiRNAs can also be isolated from fresh tissues, cell cultures, and more recently from formalin-fixed paraffin-embedded (FFPE) tissue blocks and archived materials, using various methods (Fig. 4.1). The extraction of miRNAs from FFPE samples allows pathologists to retrospectively use archival paraffin blocks from previously diagnosed patients with invaluable corresponding clinical follow-up data [10, 11, 59] to assess the relationship with response to treatment and overall survival. The in situ *hybridization* (ISH) technique for miRNA evaluation has also been developed for FFPE samples as a useful method for comparison with immunohistochemistry (IHC).

The major limitation of using FFPE tissue samples in molecular biology applications is the less than optimal quality of RNA extracted from them; specifically, the nucleic acids are degraded to fewer than 300 base pairs in length and are chemically modified during formalin fixation [10, 11]. However, due to the short sequence of miRNAs, they can be successfully extracted from FFPE tissue blocks using several commercially available kits [10, 59]. Additionally, miRNA expression profiles in FFPE tissue are closely similar to those in fresh tissue, proving that, if prepared appropriately, FFPE tissue samples are excellent resources for miRNA expression investigations [10, 11, 59]. Several preparation methods are commonly used prior to miRNA isolation [60, 61]. During extraction and purification of RNA, preventing the loss of small RNA species is the main concern. Therefore, using a robust miRNA isolation technique and ensuring the stability of stored miRNA samples that were isolated using these methods are highly important [12, 60, 61]. MiRNA expression and localization studies have demonstrated the potential for use of these molecules as both screening and early diagnosis markers for various pathologies through their identification in body fluids. Therefore, development of reliable instruments for the assessment and quantification of these miRNAs, which are found at minute concentrations, is important. Fresh tissue samples, immune cells like leukocytes, and cell lines usually furnish large quantities of high-quality miRNAs, which are required for miRNA profiling studies (Fig. 4.1). Several kits are now available for miRNA extraction of good quality and quantity, including from FFPE tissues.

Regarding the need for matched pairs of normal and tumor samples, both normal and tumor tissue samples ideally should be obtained from the same patient, as described by Yanaihara et al. [58]. Access to both normal and tumor cells can be highly insightful for pathologists when evaluating miRNA expression. Nevertheless, tumor cell lines can also be suitable models for studying the effects of downregulated and upregulated miRNA expression, in spite of the fact that these may contain genetic abnormalities never identified in patient tumors. A recommended step at the start of a miRNA-related study is to select clinical samples with different characteristics (e.g., colorectal tumor samples with and without microsatellite instability, blood cells obtained from CLL patients with poor and good prognosis) in relatively similar numbers (more than 30 in each group for performance of various statistical analyses). If the number of normal controls is not sufficient for conferring statistical power, the data can still be analyzed by comparing the two (or more) clinically distinct sets of samples. Because disease-oriented profiling uses few types of samples (e.g., malignant and normal control samples in the simplest example), replicates of each sample are not needed for profiling. In more detailed studies using limited numbers of samples that are very different biologically (e.g., cells transfected with a specific reagent and non-transfected controls), analysis of three replicates of each sample is critical.

Investigators have observed large differences in miRNA expression in serum and plasma samples. The possibility has been raised that plasma and serum might display several differences in their miRNA content. These differences in miRNA concentration may be due to the fact that the coagulation process modifies the pattern of blood miRNA [62]. However, differences can also be caused by platform-dependent variations in measurement of miRNAs [12, 63].

MiRNA Extraction Protocols

Sample preparation and RNA extraction approaches can have direct consequences for miRNA analysis and profiling, especially with samples that are prone to miRNA degradation (Fig. 4.2). All protocols for miRNA extraction have a basic step of cell/tissue lysis with the exception of those for body fluids. For tissue samples, performance of this step using mechanical disruption of the sample is recommended, as researchers have observed higher efficiency in quantity when using a homogenizer. The samples must be processed according to the manufacturer's extraction protocol, and every effort needs to be made in order to eliminate contaminants, such as xylene, chloroform, and TRIzol.

Extraction of miRNAs using TRIzol or TRIReagent is a method initially proposed as an approach to facilitate the elimination of proteins from nucleic acids. This method is recommended primarily for cells and tissues with increased expression of endogenous RNases or when separation of cytoplasmic RNA from nuclear RNA is required [64, 65]. The TRIzol/TRIReagent extraction procedure is a reliable method for isolation of miRNA species, as reported in studies that compared it with other commercially available kits for miRNA extraction. When correctly stored and managed using these procedures, miRNA samples have not exhibited degradation [64–66] (Table 4.1). However, a study called attention to the loss of short structured RNAs with low GC

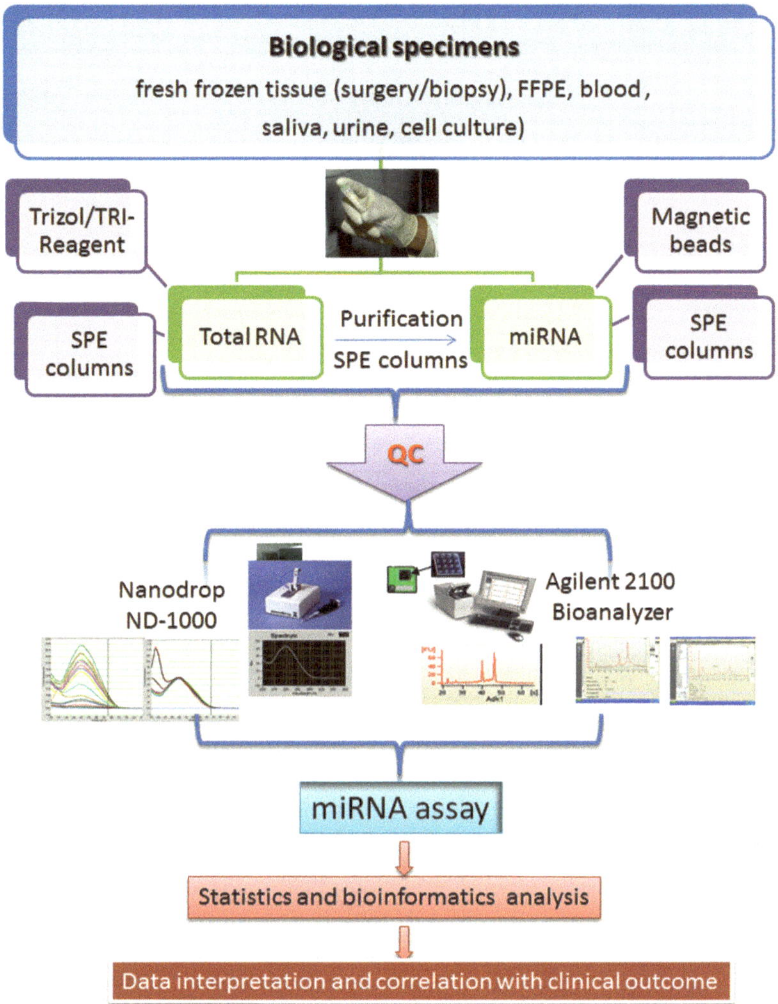

Figure 4-2 MiRNA extraction protocols, quality control, and quantification. The recovery of miRNA from biological samples through Trizol/TRIReagent, SPE columns, and magnetic beads is shown on the *top* of the figure. MiRNA quality control through the use of a NanoDrop or a Bioanalyzer is highly recommended prior to miRNA assays, statistics and bioinformatics analysis, and data correlation with clinical outcome in order to generate accurate results. *SPE* solid-phase extraction

content during extraction with this reagent when using a small number of cells [67].

The development of commercial miRNA extraction kits, such as the mirVana miRNA isolation kit (Ambion), mirPremier microRNA isolation kit (Sigma-Aldrich), miRNeasy Mini Kit (Qiagen), and miRCURY (Exiqon), was based on the principle of solid-phase extraction (SPE), which significantly improves the retention of small RNA species on SPE columns in a selective manner. Consequently, any redundant materials in the extraction procedures, including large RNAs, are eliminated prior to the final step of the protocol, in which small RNAs are eluted. The use of SPE appears to be the most widely applied effective approach to miRNA extraction.

Use of magnetic beads also ensures an easy and rapid workflow for miRNA extraction, thus avoiding the use of hazardous chemicals. Specifically, miRNAs are immobilized on magnetic beads, decanting the contaminants in a solution. MiRNA purification is done by a magnetic field to extract the magnetic beads

Table 4-1 Biological Samples Used for MiRNA Evaluation and Their Utility

Sample type		MiRNA quantity (ng)	Utility	References
Cell lines		≥1,000	Identification of new miRNA; validation of action mechanisms and target pathways; identification of novel target therapies	[11, 15, 16, 18]
Biopsy	Macrodissection	1–100	Identification of novel early diagnostic biomarkers, novel prognostic biomarkers, biomarkers for prediction of the response to treatment or resistance to treatment, clinical investigation, clinical trials, translational research	[13, 15, 16, 18, 19]
	LCM	≤1–10		
Fresh/frozen tissue	Macrodissection	≥1,000		
	LCM	10–200		
FFPE	Macrodissection	10–200		
	LCM	≤1–40		
Serum		1–100		
Plasma		1–100		
Saliva		≤1–20		
Urine		≤1–40		

FFPE formalin fixed paraffin embedded, *LCM* laser capture microdissection

from the solution, after decontamination [68]. This extraction method has many benefits for different samples with different concentrations of miRNAs, such as serum/plasma, saliva, and urine.

MiRNA Quantity and Quality Control

Quantification of nucleic acids, including miRNAs, extracted from different types of samples is essential for quality and quantity control. It is recommended to use the same amounts of miRNA when comparing various biological samples [66]. Recent experimental data demonstrated that the miRNA expression profile is affected by RNA integrity [10, 60, 65]. Samples with low RNA integrity exhibited the highest miRNA concentrations, because when RNA is degraded this results in the formation of small RNA species, leading to an overestimation of the miRNA amount. Consequently, miRNAs should not be considered as individual fractions or integrated parts of the total RNA. However, assessment of RNA integrity must be a routine step in assessing miRNA expression patterns [61, 63]. Importantly, measuring RNA concentrations using different platforms is a challenging process, so comparing miRNA profiles in samples evaluated using different platforms can be very complex.

In practice, miRNAs are quantified using the same techniques as those used to quantify DNA and mRNA (Fig. 4.2). The lack of specificity that arises during miRNA quantification and quality control is the main inconvenience of the available techniques. The most commonly used methods for quantifying miRNAs are spectrophotometric approaches (μ-volume quantification, e.g., by the NanoDrop instrument series), which measure the concentrations of total nucleic acid species but are not able to differentiate miRNA species; microfluidic systems (2100 Bioanalyzer [Agilent Technologies]); Experion automated electrophoresis system (Bio-Rad); and capillary gel electrophoresis (QIAxcel advanced system [Qiagen]).

Staining with fluorescent RNA-binding dyes (e.g., RiboGreenRNA assay kit [Promega, Ambion]) is preferred when assessing nucleic acids at low concentrations. Most of the methods used for integrity evaluation are based on an assessment of 18S and 28S ribosomal RNA, followed by different algorithms for analysis. These methods have different sensitivities, producing a wide range of results, and they therefore require objective comparisons of the experimental data [63, 66].

MiRNA Stability

For all miRNA assay procedures sample processing should be performed in an RNase-free environment in order to prevent degradation of nucleic acids and to produce consistent results. At present, a wide range of RNase-free materials and reagents is commercially available. To prevent degradation of nucleic acids, including miRNA species, protocols for storage indicate that RNAs should be deposited at −80 °C for prolonged periods to enable the generation of reproducible and reliable data. In the case of cDNAs synthesized based on miRNA sequence, the appropriate storage temperature for miRNA expression profile studies is −20 °C. In a recent study, researchers presented the TRIzol/TRIReagent system as the gold standard for miRNA extraction [61]. Rapid degradation of miRNAs and cDNAs appeared to be independent of the extraction method. In order to reduce the degradation processes, proper storage conditions for RNA and cDNA are essential and should be used to ensure the accuracy of the experimental data and to allow correlation of different studies presented in the literature [69].

Advanced Techniques for Examining MiRNA Expression

MiRNA expression profiles are studied to determine the roles of these small molecules in particular biological processes. MiRNA profiles in tumors, obtained by different approaches (blotting, hybridization, qRT-PCR, microarray, or next-generation sequencing [NGS]) have supported their roles as potential biomarkers for diagnosis of and prognosis for cancer [70].

MiRNAs can be evaluated in a wide range of biological samples with variable amounts of quantity and different qualities of miRNA content, including fresh and FFPE tissue samples, in which laser capture microdissection (LCM) is required to obtain homogeneous tissue samples even at a lower yield (<10 ng of miRNA/sample) than in macrodissection. Plasma/serum, saliva, and urine are other types of samples in which low concentrations of miRNA are obtained and where quantification of miRNA is a major challenge.

Once investigators observe alterations of miRNA profiles in patients with different pathologies [71] the quantifying of primary miRNA (pri-miRNA), precursor miRNA (pre-miRNA) transcripts, and mature miRNAs is required to distinguish among different isomiR species, i.e., sequence variants of miRNAs [72], as well as changes in mature form processing. Assessing miRNA expression profiles as a diagnostic, prognostic, or therapeutic tool can also be performed, using molecular and biological methodologies including Northern blotting, polymerase chain reaction (PCR), qRT-PCR, ISH, miRNA microarrays, and NGS [73, 74].

Northern Blotting

Northern blotting was the first technique used to identify miRNAs [75, 76], and up until recently, it was the only standardized and most widely used assay for small RNA research. This technique can reliably detect the expression profiles of miRNAs of interest, determine their sizes, and accurately quantify and identify a predictive population of miRNAs with a specific role in a disease [77].

Despite its frequent use, Northern blotting has several technical limitations that restrict its use as a miRNA expression-profiling tool. These include the relatively large sample quantity requirement and poor sensitivity of routine analysis once mature miRNAs are very short, and their prevalence in total RNA is very low. Furthermore, the close sequence similarity among miRNAs in the same families poses a challenge to the specificity of miRNA detection using Northern blotting [78].

Over the years, new Northern blot versions have been developed in an attempt to increase the sensitivity of the procedure and reduce total assay time. Nevertheless, Northern blotting and its variants are considered to be medium-throughput miRNA-screening techniques.

Quantitative Reverse Transcription-Polymerase Chain Reaction

qRT-PCR is one of the most frequently used approaches to achieve gene-expression quantification, including for mature miRNA and pre-miRNA expression profiling [60, 79]. In this procedure, small quantities of miRNA or total RNA are reverse-transcribed into cDNA, followed by a quantitative PCR analysis in which accumulation of reaction products is observed in real time [7]. qRT-PCR has excellent sensitivity and sequence specificity, and it is the most often used method for expression profiling and confirmation of miRNA findings that were obtained by other methodologies, particularly by microarray assays. Most qRT-PCR-based miRNA-expression quantification approaches are specific to 3′ end sites of targeted miRNAs [60, 79]. In qRT-PCR applications using hydrolysis probes, both mature miRNAs and their precursors must be assayed.

There are several advantages in using hydrolysis probes, such as the capacity to detect specific miRNA precursors [7] and the achievement of higher reaction efficiency for mature rather than miRNA precursors. Other available qRT-PCR protocols for miRNAs use a single nonspecific dye, such as SYBR Green. These methods require treatment with DNase for removing the genomic DNA [65] but are less costly and have a lower detection limit and a higher sequence specificity and accuracy than the hydrolysis probe-based methods [49, 79].

Given that most miRNAs are 21–22 nucleotides long, a classic PCR primer of approximately the same size length imposes limitations on miRNA evaluation, because at least two non-overlapping primers are required for the exponential amplification phase in qRT-PCR [79, 80]. One way to circumvent this limitation is to extend the length of the miRNA, generally via polyadenylation [81, 82]. Because all RNA structures with 3′ ends will be polyadenylated using this procedure, it cannot differentiate among pre-miRNAs and mature miRNAs that are capable of activation of the RISC protein complex for the silencing of genes. Another restriction of this polyadenylation-based technique is its inability to quantify species containing 2′-oxymethyl modifications at their 3′ ends that will block polyadenylation. Companies like Qiagen, Stratagene [83], Agilent Technologies [84], and Invitrogen (NCode) developed commercial qRT-PCR assays for miRNA [85]. Additionally, Exiqon [86] and Eurogentec [87] have developed assays that require proprietary reagents, such as locked nucleic acid (LNA), and complex modification steps that restrict their routine implementation.

In Situ Hybridization

ISH is used for validation of experimental data and evaluation of relative expression levels [88, 89] and can also be applied to discern biological differences between pre-miRNAs and mature miRNAs [90, 91]. Therefore, ISH provides powerful complementary data for confirmation of target miRNA, which enables progression from high-throughput investigations to more focused examination of the roles and localization of individual miRNAs [60, 90].

Despite the technical difficulty associated with adapting ISH methods to miRNA quantification [90, 92], ISH is a robust assay that has been applied to miRNA studies in a large number of organisms from bacteria to various eukaryote cell types [60, 79, 88, 91]. Paralogous miRNAs such as miR-29a and miR-29b differ by unique central nucleotides and a few nucleotides at their 3′ ends, existing in different cellular compartments (nuclear and cytoplasmic, respectively). Such differences in localization can only be illustrated using ISH techniques [60, 79]. ISH techniques also offer the advantage of delineating the cellular distribution of the miRNA(s) and also long noncoding RNAs in tissue samples composed of different cell types (Fig. 4.3).

Microarrays

DNA microarray technology continues to evolve following its initial use for the measurement of differences in the amount of target DNA and RNA sequences among biological samples [74, 93–95]. Recently, DNA microarrays have been shown to be a powerful tool for the evaluation of alterations in an abundance of miRNA species [96, 97].

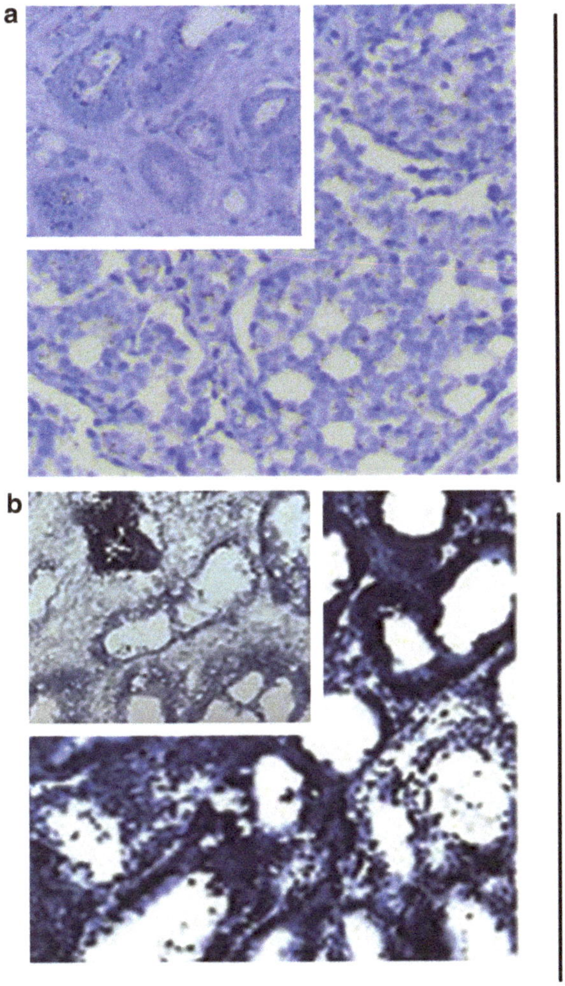

Figure 4-3 In situ hybridization (ISH) for miRNAs and long noncoding RNAs. Examples of ISH are provided. (**a**) miR-224 in prostate cancer. Staining for miR-224 is shown in *brown* in the *small box* (higher magnification—10×) and in the *bigger box* (lower magnification—40×). (**b**) ISH in colorectal cancer. The *small box* illustrates ISH for U6 (normalizer) and the *bigger one* for long noncoding RNA CCAT2

Although microarrays enhance the throughput of miRNA analysis with a high sensitivity and specificity, microarray-derived findings must be confirmed using qRT-PCR assays [93–95]. A comprehensive workflow for microarray data production, analysis, and interpretation is provided in Fig. 4.4.

Despite the large amount of experimental data obtained from DNA microarrays used for the evaluation of miRNA profiles in pathologic conditions, only a few studies have focused on intraplatform and interplatform correlations [60]. The reproducibility of data obtained using different miRNA-detection technologies with a focus on comparing the sensitivity and specificity of microarray platforms has also rarely been explored [74, 93]. Such studies, however, point to a poor correlation of mRNA gene expression among different microarray platforms in terms of reproducibility and equivalence. This is in part due to the lack of optimization of microarray protocols and the deficiency in complete and accurate data annotation of the commercially available platforms, lack of correct probe matching among technologies, discrepancies in data normalization, and interlaboratory or intralaboratory variability in technical expertise [98].

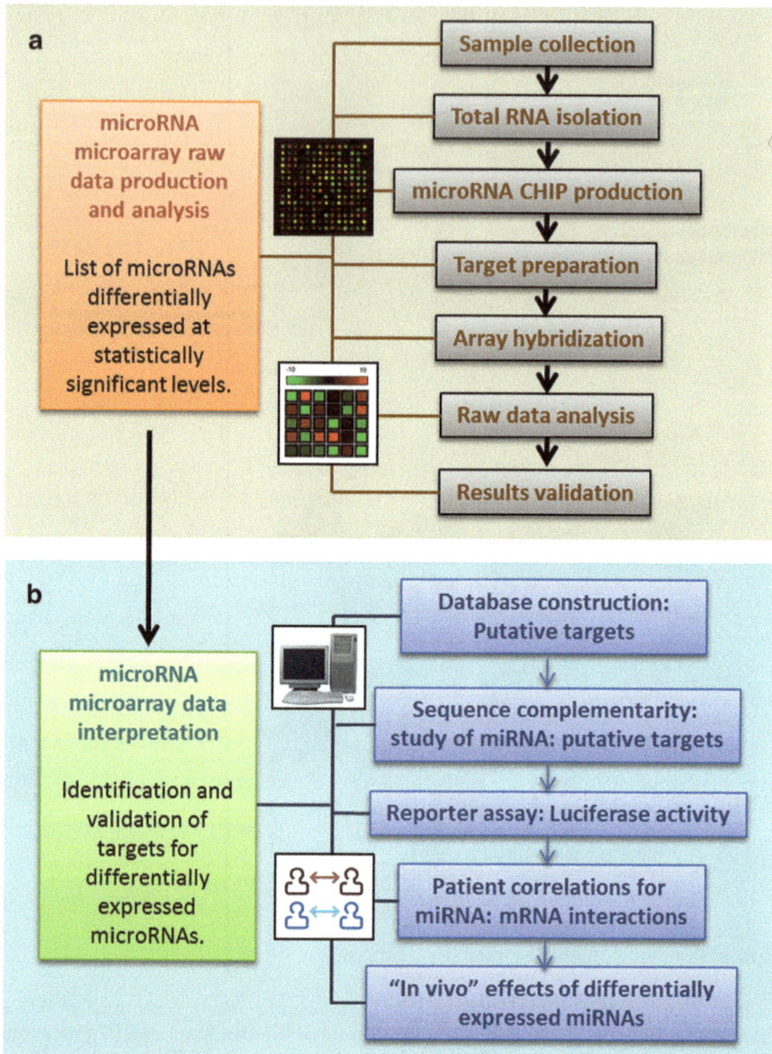

Figure 4-4 Comprehensive workflow for microarray data production, analysis, and interpretation. Obtaining a differentially expressed miRNA list through the use of microarray technology is possible following the workflow shown in (**a**). The identification and validation of targets for these differentially expressed miRNAs follow the workflow shown in (**b**)

The short length of miRNA sequences represents the main drawback for primer design, selection of specifically labeled probes, and, generally, optimization of reactions for the concomitant detection of collective miRNAs by microarray technologies [13, 94]. Also, because miRNAs are short, even small variations in their length and/or the GC nucleotide content have major consequences on their biochemical properties, which are particularly visible at the melting temperature (T_m) of the miRNA in a hybridization reaction [13].

Commercially available microarray platforms for the analysis of miRNA expression are listed in Table 4.2. Selecting a microarray platform is difficult, as each platform has benefits and limitations in printing and surface technology, slide design, labeling and detection chemistry (one or two colors), hybridization conditions, probe design, and cost per sample [93, 94].

Table 4-2 **Microarray Technology Platforms for MiRNA Expression**

No.	Technology	Probe technology	Labeling technology	Detection chemistry	miRbase version/ number of human miRNAs	Quantity (ng)	References
1	Affymetrix GeneChip miRNA 3.0 Array	High-density arrays of 25 oligonucleotide probes	FlashTag™ Biotin RNA Labeling kit (Genisphere)	One color	17/1733 human	130–500	[48, 51]
2	Illumina, BeadChip miRNA arrays v2	Biotin-labeled oligo-dT primer with a universal sequence at the 5′-end	Adaptation of the DASL (cDNA-mediated annealing, selection, extension, and ligation)	One color	12/1146 human and 656 mouse	40–200	[52]
3	Agilent, miRNA Microarray	60-mer SurePrint technology	Agilent Labeling (v1) uses CIP (GE) and T4 RNA ligase	One color	12/866 human and 89 viral	100	[48, 53]
4	Exiqon Human, miRCURY LNA miRNA	LNA-based capture probes	miRCURY LNA microRNA, Hi-Power Labeling Kit, Hy3/Hy5	Two color	19/3100 human, mouse, and rat	30	[48, 110]
5	Invitrogen, the NCode Human miRNA Microarray V3	34–44 complementary oligonucleotides	NCode™ Rapid miRNA Labeling System	Two color	10/380 human	30–1,000	[48]

Affymetrix has developed a microarray that includes miRNAs from several species [95]. This platform and that developed by Illumina contain probes for human and murine species [95, 97]. The Exiqon platform uses LNAs to increase the specificity [95]. Agilent Technologies offers a microarray design with considerable flexibility in analysis [97]. Microarray technologies are especially suitable for high-throughput miRNA expression profiling given the large number of samples that can be analyzed [7].

Next-Generation Sequencing

NGS is used for the assessment of known miRNA sequence patterns and for identifying novel miRNA species, which traditional approaches are incapable of detecting [7, 99, 100]. Sequencing miRNAs provides information not only about the levels of expression but also about single-nucleotide polymorphisms (SNPs), posttranscriptional RNA modifications—particularly in 3′-terminal regions—extension with a single nucleotide, and variations in miRNA length. Posttranscriptional modifications of miRNAs produce multiple mature miRNA variants, which are referred to as isomiRs [7, 101]. The ideal NGS approach should require a low amount of total RNA, should have no or a low sequence bias, and is amenable to performing multiplex assays with easily designed experimental protocols [102].

High-throughput sequencing for miRNA expression profiling is currently performed on several commercial NGS platforms including SOLiD (Applied Biosystems), HiSeq2000 (or Genome Analyzer IIX; Illumina), and GS FLX+454 (Roche) as well as smaller scale NGS platforms, such as Ion Torrent and Ion Proton (Applied Biosystems), MiSeq (Illumina), and GS Junior454 (Roche). Sequencing technologies have also been applied to miRNA analysis but are expensive and not widely accessible. NGS platforms, which use less than 1 μg of total RNA, start with the preparation of a cDNA library [100, 103] from total RNA and miRNA samples [103] followed by massively parallel sequencing of the millions of distinct DNA molecules in the library [104, 105]. They are able to generate millions of short sequence reads in order to provide miRNA profiles or to identify novel miRNAs [4, 99, 101].

NGS platforms for miRNA high-throughput sequencing and novel miRNA characterization have high sensitivity and resolution and can differentiate among highly similar sequences, such as isomiRs. The main limitation of these approaches is that they are not suitable for absolute quantification of miRNA amounts. They also require highly complex bioinformatics analysis of sequence reads. Specialized bioinformatics software programs as miRDeep [106] and miRExpress [99, 107] and Web-based tools such as miRanalyzer [99, 103, 108] and miRCat [109] are available for miRNA-expression profiling as well as discovery [99] making the identification of known and novel miRNAs an easier process.

As a summary, Table 4.3 lists the main principles used in miRNA assays and techniques and Table 4.4 lists the main platforms used for evaluation of miRNA-expression patterns along with their strengths and limitations.

Examination of miRNAs is considerably more complex and technically demanding compared to that of mRNAs due to several factors: short miRNA lengths, difficulty in distinguishing different miRNA types (pre-miRNA, pri-miRNA, and mature miRNA), fluctuating T_ms for primers and probes, different RNA ligase conditions, and high homology between probes among miRNA families.

Data Processing and Normalization

Prior to data normalization, data preprocessing for miRNA pattern analysis is required and comprises a set of corrections of values specific for the platform used: baseline corrections and threshold settings for qRT-PCR approaches, background adjustment for microarrays, and/or screening for small amounts of RNA-sequencing data [110]. Subsequent to these initial steps, data interpretation requires selection of the optimal normalization strategy for proper estimation of the biological variation among samples to avoid systematic and technical errors [111].

Table 4-3 Principles of MiRNA Assays and Technologies

Assay	Technologies	Method principle	Applications	References
miRNA extraction protocols	Trizol/TRIReagent	Single-extraction assay for DNA, RNA, and proteins; chloroform extraction and ethanol precipitation of RNA—recovers small RNA	Fresh frozen tissue, cell cultures	[12, 15, 16, 18]
	Column commercial kits	Solid-phase extraction (SPE)	FFPE samples	
	Magnetic beads	Magnetic microspheres for binding small RNA species, purification in magnetic field	Low concentration of miRNA (serum/plasma, saliva, or urine)	
miRNA purification protocols	Column commercial kits	SPE	Enrich miRNA fraction from biological specimens	[16, 19, 20, 23]
	Magnetic beads	Magnetic microspheres for binding small RNA species, purification in magnetic field		
qRT-PCR RNA $\rightarrow$ cDNA reverse transcription, specific amplification for target miRNA	Stem loop qRT-PCR and Taqman qRT-PCR	Polyadenylation of the miRNA at the 3-end; reverse transcription reactions is based on stem–loop primers specific for 3′ end of microRNA. Then the amplicons are synthetized based on specific forward and revers primer	Validation of microarray and NGS	[6, 33, 34, 36]
	Poly(T) adaptor primer for poly(A) miRNA and SYBRGreen qRT-PCR	Polyadenylated miRNA to cDNA and cDNA detection by qRT-PCR using technology	Discriminates between miRNAs that differ by as few as a single nucleotide	

(continued)

Table 4-3 **(continued)**

Assay		Technologies	Method principle	Applications	References
Microarrays	RNA → cDNA, cDNA labeling, hybridization with solid-phase probes, washing, detection of signal, data analysis	Affymetrix-GeneChip miRNA 3.0 Array	High-density arrays of 25 oligonucleotide probes	Genome-wide miRNA expression pattern for multiple samples can be processed in parallel in a standardized protocol	[15, 50–53, 55]
		Illumina-BeadChip miRNA arrays v2	Biotin-labeled oligo-dT primer with a universal sequence at the 5′-end and adaptation of the DASL (cDNA-mediated annealing, selection, extension, and ligation)		
		Agilent, miRNA Microarray	60-mer SurePrint technology and Agilent Labeling (v1) use CIP (GE) and T4 RNA ligase		
		Exiqon Human-miRCURY LNA miRNA	LNA-based capture probes and miRCURY LNA microRNA, Hi-Power Labeling Kit, Hy3/Hy5		
		Invitrogen—The NCode Human miRNA Microarray V3	34–44 complementary oligonucleotides and NCode™ Rapid miRNA Labeling System		

Table 4-3 **(continued)**

Assay		Technologies	Method principle	Applications	References
NGS—High-throughput sequencing platforms	Library generation (miRNA transcriptomic profile, genome/DNA of interest, fragmentation, adaptor ligation, PCR amplification), simultaneously sample and amplification product sequencing	*SOLiD* (Applied Biosystems)	Ligation-based chemistry with dibase-labeled probes	miRNA expression pattern and novel miRNA to sequence millions of fragments; simultaneously in system multiplex; differentiate among similar sequences like isomiRS	[6, 15, 56–61]
		HiSeq2000 (Illumina)	Sequencing by synthesis, using *HiSeq2000* (or Genome Analyzer IIX from Illumina)		
		GS FLX+454 sequencing (Roche)	Pyrosequencing		
Smaller scale next-generation sequencing platforms		*Ion Torrent* and *Ion Proton* (Applied Biosystems)	Sequencing chemistry is based on the proton release when a nucleotide is incorporated by the polymerase in the DNA molecule, resulting in a detectable local change of pH		
		MiSeq (Illumina)	When a nucleotide is incorporated into a strand of DNA by a polymerase, a hydrogen ion is released as a by-product		
		GS Junior 454 (Roche)	Library preparation with fusion primers than empPCR amplification		

Table 4-4 **Technology Platforms for MiRNA Expression Analysis: Advantages and Disadvantages**

Technology	Advantages	Limitations	References
Northern blot	"Gold standard"	Low sensitivity and reproducibility, requires a large amount of biological material	[29, 31, 32]
Real-time PCR	Low sample amount, high sensitivity and sequence specificity; discriminates between pre- and mature miRNA	High price	[6, 33, 34, 36–38]
In situ hybridization	Cellular localization	Low sensitivity, background effect	[33, 34, 44, 46]
Microarray	High-throughput assay	Low sensitivity; it requires data validation; low price	[6, 48, 50, 54]
Next-generation sequencing	High-throughput assay; discovering novel or rare miRNAs	High price	[6, 56, 57, 59]

The precision of miRNA analysis is dependent on proper data normalization. A common method of normalization of qRT-PCR inputs is the use of constant endogenous controls or reference miRNAs, such as U6, RNU44, and RNU48 [112]. When using an endogenous control, the standardized approach is applying the $2^{-\Delta CT}$ method [112] with the formula $\Delta CT = CT_{miRNA} - CT_{endogenous\ control}$.

Statistical tools are important in any high-throughput technology, including the examination of miRNA species. The statistical analyses needed differ according to the technology used. The analysis should be carried out with consideration of the biochemistry, biological material characteristics [110, 111], and intrinsic limitations of each miRNA profiling platform [111, 112]. Regarding microarrays, to reduce the overall variance in data interpretation, researchers developed different normalization methods for specific miRNA microarray platforms, taking into account the type of sample, method of RNA extraction, dye labeling (one or two colors), hybridization and washing conditions, and efficacy of scanning [110–112].

Raw data analysis, ideally, should be performed by two distinct bioinformaticians using two independent methods of analysis. Using this approach, we were able for example to identify a unique miRNA signature associated with prognostic factors and disease progression in patients with B-cell CLL [26]. Another strategy for analyzing raw data was generated by microarray images using the GenePix Pro software program (Molecular Devices). For a detailed review, we suggest reading Volinia et al. [113] as well as multiple additional publications describing various methods [114–117].

Analysis of MiRNA Function and Gene Interaction

MiRNAs function primarily by inhibiting translation, degrading the target mRNA, or altering mRNA stability. This is facilitated by the partial sequence homology of the miRNA seed sequence with the 3′ untranslated regions of the target mRNAs [118, 119]. Because of this unique feature, an individual miRNA can have multiple targets and regulate a large number of protein-coding genes [120].

As master gene regulators, miRNAs can impact a variety of cellular pathways and functions. Consequently, miRNAs regulate diverse biological processes that are critical to development, cell death, proliferation, and differentiation. Likewise, given their presence in all eukaryotic cells, miRNAs are implicated in the deregulation of multiple pathways, leading to a variety of diseases in humans, animals, and plants [119, 120]. Thus, understanding how miRNA expression is regulated in normal as well as disease-specific cellular processes is critical. For example, a variety of miRNAs play important roles in various aspects of cellular immunity. miR-29 family members play critical roles in determining the molecular bases for innate and adaptive immune responses toward intracellular bacterial infections [119]. Other examples are miR-208a, miR-208b, and miR-499, which, along with MHC genes, have important functions in the formation of a regulatory circuit that controls cardiac hypertrophy and leads to heart failure [121, 122].

Interactions between miRNAs and biologically important targets can be direct (sequence complementarity) or indirect (e.g., via a transcription factor influenced by the miRNA). Therefore, the step of target identification is one of the most important ones for the biological characterization of a miRNA role in the pathogenic mechanisms of a specific disease. The best approach for target identification is to confirm the negative expression correlation using protein samples obtained from the same study subjects who had miRNA expression profiled. If a negative correlation is found between array miRNA expression and Western blot protein expression, it should be confirmed by transfecting the miRNA of interest in at least two cell lines that express the protein that is thought to be a putative target of that miRNA. To date, investigators have mainly found negative correlations between miRNA expression and their targets (miRNA inhibition of transcription or translation). This does not, however, preclude the existence of positive correlations. The finding that miR-122 has a positive influence on replication of hepatitis C virus by interacting with the 5′-noncoding region of the virus is an indication of such positive correlations [123].

MiRNA Mimics and Inhibitors

Manipulation of mRNAs with specific oligonucleotides that mimic miRNAs provides new opportunities for a better understanding of the molecular mechanisms of diseases, opening up new avenues of research aimed at treatment of pathologies. The concept that the cause of a specific pathogenesis is solely based on alteration of the expression of protein-coding genes is no longer tenable with the discovery of miRNAs. Pathogenic miRNA alterations can be reversed by positively or negatively modulating the expression of miRNAs, highlighting the potential of these small molecules as pharmacological targets.

Because of their small size, miRNAs are easier to transfer than other molecules, such as DNA or mRNA [124]. Antagomirs, anti-miRs, or agonist microRNAs are novel chemically designed oligonucleotides that prevent the interaction of miRNAs with their target mRNA molecules. MiRNA mimics have the opposite effect, mimicking the effect of miRNAs in the cells. Both antagomirs and miRNA mimics can be readily introduced into cell lines via cold transfection using electroporation or viral and nonviral vectors [125].

Viruses widely used for miRNA transfer include retroviruses, lentiviruses, adenoviruses, adeno-associated viruses, and herpes simplex viruses [126]. Their favorable rate of cell uptake and gene and/or tissue specificity has led to their wide use as transfer vectors. Their main drawbacks are production difficulties and limited administration capacity due to host-induced acute inflammatory responses [127]. The antitumoral effect achieved with the administration of adeno-associated viral vector scAAV8 expressing miR-26a, for example, in transgenic mice with conditional expression of Myc in the liver illustrates the potential role of miRNA in therapeutics [128]. In another study, overexpression of miR-133 delivered into a mouse model using an adenovirus vector led to a significant reduction in the number of cardiac myocytes in the left ventricle and increased expression of fetal genes pointing to miR-133 as a therapeutic target in cardiac hypertrophy [51].

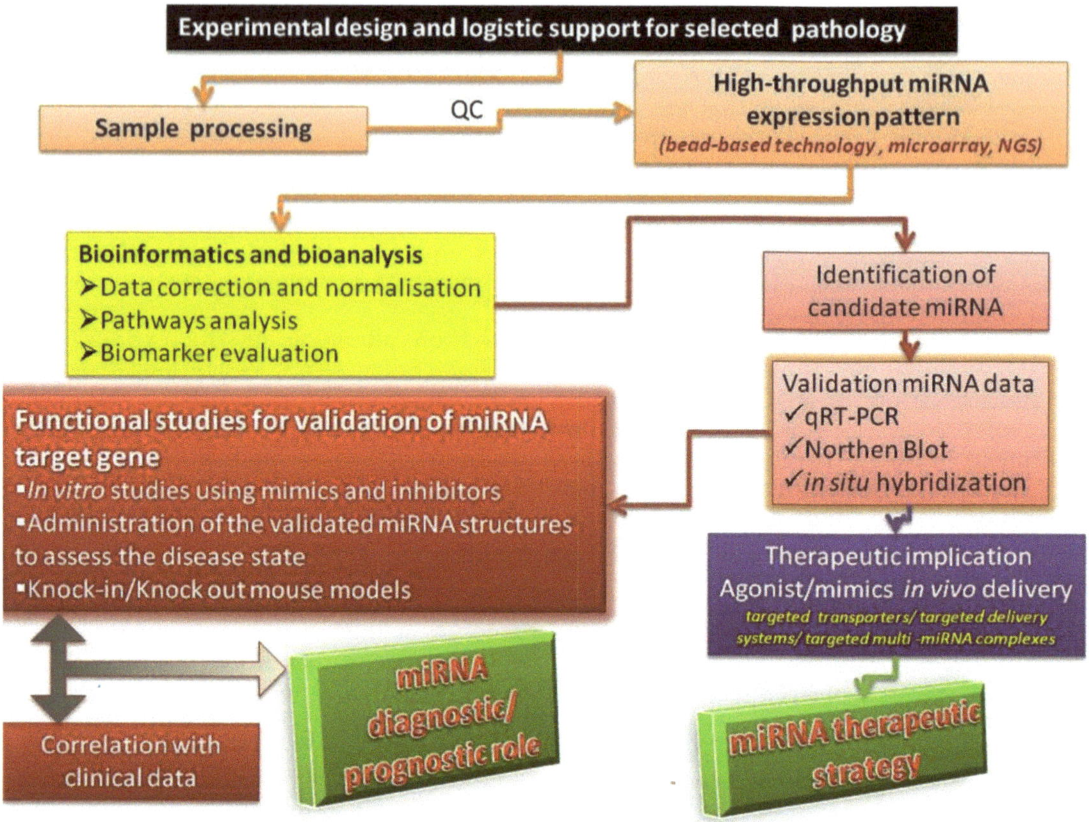

Figure 4-5 Experimental design and logistic support for evaluation of miRNAs in a selected pathology. A comprehensive experimental design and logistic support start with sample processing, quality control, and high-throughput miRNA expression pattern and bioinformatics analysis. Once a candidate miRNA is identified, a validation process takes place and functional studies are aimed at the elucidation of diagnostic, prognostic, or therapeutic roles for the found miRNA

Nonviral miRNA delivery systems are constructed from synthetic or natural compounds. These systems are generally less toxic and less immunogenic than viral systems, and they can be easily secured and repeatedly utilized. Their main disadvantages are reduced efficiency and short expression durations. For example, Rai et al. [129] used a liposomal system carrying miR-7 to block the expression of mutant epidermal growth factor receptor in lung cancer cells.

The main obstacle for clinical applications of such treatment strategies remains the limited efficiency in miRNA delivery. Over the years, researchers have developed many delivery systems, but the focus is still on emerging concepts for the delivery of therapeutic miRNA mimics and antagomirs [130].

As mentioned above, miRNAs inhibit the expression of genes relevant to different biological and pathological processes. Not surprisingly, alteration of the miRNA profile has been linked to multiple human illnesses, including cancer and cardiovascular and autoimmune diseases [131–133]. In this sense, miRNAs continue to have a strong potential as biomarkers for diagnosis [70, 134], prognosis [135, 136], and therapeutic response prediction (Fig. 4.5).

Conclusions

MiRNAs, first discovered over two decades ago, have proven to play a major role in gene regulation and transcription in normal as well as abnormal cells and in many different diseases. Cancer, cardiovascular diseases, immune system deregulation, and adipogenesis, all represent platforms for investigation into miRNA communication and molecular profiling in order to better establish an early diagnostic and a therapeutic option for different patient subgroups. From a pathologists' point of view, miRNA's discovery was a breakthrough for diagnosis, prognostication, and therapy prediction. Bringing these molecules and all their complexity from bench to the bedside is, at present, a big challenge. The investigative frontier in miRNA biology is currently centered on the detection of novel structures of miRNAs, identification of the molecular targets, and potential as yet unknown roles.

Currently, many research applications for miRNAs that could be applied to clinical practice can be found among published data, mainly in the cancer field. Important examples are miR-15a and miR-16-1 downregulation in CLL; miR-155 expression levels and miR-21 panels as powerful biomarkers also in CLL; association of miR-21 with important clinicopathological characteristics in non-small-cell lung cancer, colorectal cancer, and invasive ductal carcinomas of the breast; and association between the miR-17-92 family with many solid tumors, lymphomas, and myelomas.

From the therapeutic perspective, based on clinical trial data, controlling the levels of a single miRNA has resulted in only a limited effect on the expression level for a target gene. The latter could be due to the activation of compensatory pathways or an inefficient delivery.

The changes in miRNA pattern have relevant biological significance and may provide useful information for identification of the early stages of disease, for clinical diagnoses, or for the identification of therapeutic targets with important implications in personalized treatment.

By using ISH, the pathologist is able to use FFPE or fresh tissue samples with high performance quality and reproducible results. This has made it possible to refine and validate the data obtained by microarrays and to localize particular miRNAs in a given tissue. This new technique indicates the remarkable stability of miRNAs in archival human tissues. Additional functional genomic techniques are still necessary to profile the upregulated and downregulated miRNAs in a patient. Microarray technology can be used for the assessment of miRNAs in normal and tumor tissues. As the expression profiles and function of miRNAs during disease development and progression are further elucidated using NGS, their role as biomarkers and therapeutic targets will likely continue to increase.

Acknowledgements

Dr. Calin is the Alan M. Gewirtz Leukemia & Lymphoma Society Scholar. He is supported as a fellow of the University of Texas MD Anderson Cancer Center Research Trust, as the University of Texas System Regents Research Scholar, and by the CLL Global Research Foundation. Work in Dr. Calin's laboratory is supported in part by the National Institutes of Health/National Cancer Institute (CA135444); a Department of Defense Breast Cancer Idea Award; Developmental Research Awards in Breast Cancer, Ovarian Cancer, Brain Cancer, Prostate Cancer, Multiple Myeloma, Leukemia (P50 CA100632), and Head and Neck (P50 CA097007) Specialized Program of Research Excellence grants; Sister Institution Network Fund grants in CLL and colon cancer; the Laura and John Arnold Foundation; the RGK Foundation; and the Estate of C. G. Johnson, Jr. Drs. Berindan-Neagoe and Braicu were financed by a POSCCE grant (709/2010) entitled Clinical and Economical Impact of Proteome and Transcriptome Molecular Profiling in Neoadjuvant Therapy of Triple Negative Breast Cancer (BREASTIMPACT). This research is supported in part by the MD Anderson Cancer Center Support Grant CA016672. We thank Don Norwood in the Department of Scientific Publications at MD Anderson for expert editorial assistance.

REFERENCES

1. Shah MY, Calin GA. The mix of two worlds: non-coding RNAs and hormones. Nucleic Acid Ther. 2012;23(1):2–8.
2. Redis RS, Calin S, Yang Y, et al. Cell-to-cell miRNA transfer: from body homeostasis to therapy. Pharmacol Ther. 2012;136:169–74.

3. Enfield KS, Pikor LA, Martinez VD, Lam WL. Mechanistic roles of noncoding RNAs in lung cancer biology and their clinical implications. Genet Res Int. 2012;2012:737416.
4. Fire A, Xu S, Montgomery MK, et al. Potent and specific genetic interference by double-stranded RNA in Caenorhabditis elegans. Nature. 1998;391:806–11.
5. Bartel DP. MicroRNAs: genomics, biogenesis, mechanism, and function. Cell. 2004;116:281–97.
6. Negrini M, Nicoloso M, Calin G. MicroRNAs and cancer—new paradigms in molecular oncology. Curr Opin Cell Biol. 2009;21:470–9.
7. Pritchard CC, Cheng HH, Tewari M. MicroRNA profiling: approaches and considerations. Nat Rev Genet. 2012;13(5):358–69.
8. Yoshizawa JM, Wong DT. Salivary microRNAs and oral cancer detection. Methods Mol Biol. 2013;936:313–24.
9. Allegra A, Alonci A, Campo S, et al. Circulating microRNAs: new biomarkers in diagnosis, prognosis and treatment of cancer (review). Int J Oncol. 2012;41(6):1897–912. doi:10.3892/ijo.2012.1647.
10. Li J, Smyth P, Flavin R, et al. Comparison of miRNA expression patterns using total RNA extracted from matched samples of formalin-fixed paraffin-embedded (FFPE) cells and snap frozen cells. BMC Biotechnol. 2007;29:36.
11. Calin GA, Croce CM. MicroRNA signatures in human cancers. Nat Rev Cancer. 2006;6(11):857–66.
12. Etheridge A, Lee I, Hood L, et al. Extracellular microRNA: a new source of biomarkers. Mutat Res. 2011;717(1–2):85–90.
13. Krichevsky AM. MicroRNA profiling: from dark matter to white matter, or identifying new players in neurobiology. Scientific World Journal. 2007;7:155–66.
14. Bommer GT, Gerin I, Feng Y, Kaczorowski AJ, Kuick R, Love RE, Zhai Y, Giordano TJ, Qin ZS, Moore BB, et al. p53-mediated activation of miRNA34 candidate tumor-suppressor genes. Curr Biol. 2007;17:1298–307.
15. Mendell J, Olson E. MicroRNAs in stress signaling and human disease. Cell. 2012;148(6):1172–87.
16. Shi XB, Xue L, Yang J, et al. An androgen-regulated miRNA suppresses Bak1 expression and induces androgen-independent growth of prostate cancer cells. Proc Natl Acad Sci U S A. 2007;104:19983–8.
17. Cao P, Deng Z, Wan M, et al. MicroRNA-101 negatively regulates Ezh2 and its expression is modulated by androgen receptor and HIF-1alpha/HIF-1beta. Mol Cancer. 2010;9:108.
18. Poliseno L, Salmena L, Riccardi A, et al. Identification of the miR-106b 25 microRNA cluster as a proto-oncogenic PTEN-targeting intron that cooperates with its host gene MCM7 in transformation. Sci Signal. 2010;3:ra29.
19. Zaman MS, Thamminana S, Shahryari V, et al. Inhibition of PTEN gene expression by oncogenic miR-23b-3p in renal cancer. PLoS One. 2012;7:e50203.
20. Mlcochova H, Hezova R, Stanik M, Slaby O. Urine microRNAs as potential noninvasive biomarkers in urologic cancers. Urol Oncol. 2014;32(1):41.e1–9. pii: S1078-1439(13)00197-X.
21. Calin GA, Sevignani C, Dumitru CD, Hyslop T, Noch E, Yendamuri S, Shimizu M, Rattan S, Bullrich F, Negrini M, et al. Human microRNA genes are frequently located at fragile sites and genomic regions involved in cancers. Proc Natl Acad Sci U S A. 2004;101:2999–3004.
22. Alvarez-Garcia I, Miska EA. MicroRNA functions in animal development and human disease. Development. 2005;132(21):4653–62.
23. Lu M, Zhang Q, Deng M, Miao J, Guo Y, et al. An analysis of human MicroRNA and disease associations. PLoS One. 2008;3(10):e3420.
24. Calin GA, Dumitru CD, Shimizu M, et al. Frequent deletions and down-regulation of micro-RNA genes miR15 and miR16 at 13q14 in chronic lymphocytic leukemia. Proc Natl Acad Sci U S A. 2002;99(24):15524–9.
25. Cimmino A, Calin GA, Fabbri M, et al. miR-15 and miR-16 induce apoptosis by targeting BCL2. Proc Natl Acad Sci U S A. 2005;102(39):13944–9.
26. Calin GA, Ferracin M, Cimmino A, et al. A microRNA signature associated with prognosis and progression in chronic lymphocytic leukemia. N Engl J Med. 2005;353:1793–801.
27. Pekarsky Y, Santanam U, Cimmino A, et al. Tcl1 expression in chronic lymphocytic leukemia is regulated by miR-29 and miR-181. Cancer Res. 2006;66(24):11590–3.
28. Calin GA, Pekarsky Y, Croce CM. The role of microRNA and other non-coding RNA in the pathogenesis of chronic lymphocytic leukemia. Best Pract Res Clin Haematol. 2007;20(3):425–37.
29. Rossi S, Shimizu M, Barbarotto E, et al. microRNA fingerprinting of CLL patients with chromosome 17p deletion identify a miR-21 score that stratifies early survival. Blood. 2010;116(6):945–52.
30. Ferrajoli A, Shanafelt TD, Ivan C, et al. Prognostic value of miR-155 in individuals with monoclonal B-cell lymphocytosis and patients with B chronic lymphocytic leukemia. Blood. 2013;122(11):1891–9.
31. Wang JL, Hu Y, Kong X, et al. Candidate microRNA biomarkers in human gastric cancer: a systematic review and validation study. PLoS One. 2013;8(9):e73683.

32. Wang Y, Li J, Tong L, et al. The prognostic value of miR-21 and miR-155 in non-small-cell lung cancer: a meta-analysis. Jpn J Clin Oncol. 2013;43(8):813–20.
33. Menéndez P, Padilla D, Villarejo P, et al. Prognostic implications of serum microRNA-21 in colorectal cancer. J Surg Oncol. 2013;108(6): 369–73. doi:10.1002/jso.23415.
34. Lee JA, Lee HY, Lee ES, Kim I, Bae JW. Prognostic implications of microRNA-21 overexpression in invasive ductal carcinomas of the breast. J Breast Cancer. 2011;14(4):269–75.
35. Ota A, Tagawa H, Karnan S, et al. Identification and characterization of a novel gene, C13orf25, as a target for 13q31-q32 amplification in malignant lymphoma. Cancer Res. 2004;64(9): 3087–95.
36. Chang CC, Yang YJ, Li YJ, et al. MicroRNA-17/20a functions to inhibit cell migration and can be used a prognostic marker in oral squamous cell carcinoma. Oral Oncol. 2013;49(9): 923–31.
37. Gao X, Zhang R, Qu X, et al. MiR-15a, miR-16-1 and miR-17-92 cluster expression are linked to poor prognosis in multiple myeloma. Leuk Res. 2012;36(12):1505–9.
38. Yu G, Tang JQ, Tian ML, et al. Prognostic values of the miR-17-92 cluster and its paralogs in colon cancer. J Surg Oncol. 2012;106(3):232–7.
39. Hanke M, Hoefig K, Merz H, et al. A robust methodology to study urine microRNA as tumor marker: microRNA-126 and microRNA-182 are related to urinary bladder cancer. Urol Oncol. 2010;28(6):655–61.
40. Zhong S, Li W, Chen Z, Xu J, Zhao J. miR-222 and miR-29a contribute to the drug-resistance of breast cancer cells. Gene. 2013;531(1):8–14. pii: S0378-1119(13)01122-0.
41. Yu PN, Yan MD, Lai HC, et al. Downregulation of miR-29 contributes to cisplatin resistance of ovarian cancer cells. Int J Cancer. 2013;134(3): 542–51.
42. Wong KY, Yim RL, Kwong YL, et al. Epigenetic inactivation of the MIR129-2 in hematological malignancies. J Hematol Oncol. 2013;6:16.
43. Lu CY, Lin KY, Tien MT, et al. Frequent DNA methylation of MiR-129-2 and its potential clinical implication in hepatocellular carcinoma. Genes Chromosomes Cancer. 2013;52:636–43.
44. Fabbri M, Garzon R, Cimmino A, et al. MicroRNA-29 family reverts aberrant methylation in lung cancer by targeting DNA methyltransferases 3A and 3B. Proc Natl Acad Sci U S A. 2007;104:15805–10.
45. Sacco J, Adeli K. MicroRNAs: emerging roles in lipid and lipoprotein metabolism. Curr Opin Lipidol. 2012;23(3):220–5.
46. Romao JM, Jin W, Dodson MV, Hausman GJ, Moore SS, Guan LL. MicroRNA regulation in mammalian adipogenesis. Exp Biol Med (Maywood). 2011;236(9):997–1004.
47. Ortega FJ, Mercader JM, Catalán V, Moreno-Navarrete JM, Pueyo N, Sabater M, Gómez-Ambrosi J, Anglada R, Fernández-Formoso JA, Ricart W, Frühbeck G, Fernández-Real JM. Targeting the circulating microRNA signature of obesity. Clin Chem. 2013;59(5):781–92.
48. McGregor RA, Choi MS. microRNAs in the regulation of adipogenesis and obesity. Curr Mol Med. 2011;11(4):304–16.
49. Ono K, Kuwabara Y, Han J. MicroRNAs and cardiovascular diseases. FEBS J. 2011;278(10):1619–33. doi:10.1111/j.1742-4658.2011.08090.x. Epub 2011 Mar 30.
50. Thum T, Catalucci D, Bauersachs J. MicroRNAs: novel regulators in cardiac development and disease. Cardiovasc Res. 2008;79:562–70.
51. Care A, Catalucci D, Felicetti F, Bonci D, Addario A, Gallo P, et al. MicroRNA-133 controls cardiac hypertrophy. Nat Med. 2007;13:613–8.
52. Yang B, Lin H, Xiao J, Lu Y, Luo X, Li B, et al. The muscle-specific microRNA miR-1 regulates cardiac arrhythmogenic potential by targeting GJA1 and KCNJ2. Nat Med. 2007;13:486–91.
53. Oliveira-Carvalho V, da Silva MM, Guimarães GV, Bacal F, Bocchi EA. MicroRNAs: new players in heart failure. Mol Biol Rep. 2013;40(3):2663–70.
54. Fichtlscherer S, Zeiher AM, Dimmeler S. Circulating microRNAs: biomarkers or mediators of cardiovascular diseases? Arterioscler Thromb Vasc Biol. 2011;31(11):2383–90.
55. Hu R, O'Connell M. MicroRNA control in the development of systemic autoimmunity. Arthritis Res Ther. 2013;15(1):202. doi:10.1186/ar413.
56. Meisgen F, Xu N, Wei T, Janson PC, Obad S, Broom O, Nagy N, Kauppinen S, Kemény L, Ståhle M, Pivarcsi A, Sonkoly E. MiR-21 is up-regulated in psoriasis and suppresses T cell apoptosis. Exp Dermatol. 2012;15:312–4.
57. Mi QS, He HZ, Dong Z, Isales C, Zhou L. microRNA deficiency in pancreatic islet cells exacerbates streptozotocin-induced murine autoimmune diabetes. Cell Cycle. 2010;15: 3127–9.
58. Yanaihara N, Caplen N, Bowman E, et al. Unique microRNA molecular profiles in lung cancer diagnosis and prognosis. Cancer Cell. 2006;9(3):189–98.
59. Liu A, Xu X. MicroRNA isolation from formalin-fixed, paraffin-embedded tissues. Methods Mol Biol. 2011;724:259–67.
60. Nelson PT, Wang WX, Wilfred BR, Tang G. Technical variables in high-throughput miRNA expression profiling: much work remains to be done. Biochim Biophys Acta. 2008;1779(11):758–65.

61. Mraz M, Malinova K, Mayer J, Pospisilova S. MicroRNA isolation and stability in stored RNA samples. Biochem Biophys Res Commun. 2009;390(1):1–4.
62. Wang K, Yuan Y, Cho J-H, McClarty S, Baxter D, et al. Comparing the MicroRNA spectrum between serum and plasma. PLoS One. 2012; 7(7):e41561. doi:10.1371/journal.pone.0041561.
63. Becker C, Hammerle-Fickinger A, Riedmaier I, Pfaffl MW. mRNA and microRNA quality control for RT-qPCR analysis. Methods. 2010;50(4): 237–43.
64. Rio DC, Ares Jr M, Hannon GJ, Nilsen TW. Purification of RNA using TRIzol (TRI reagent). Cold Spring Harb Protoc. 2010; 2010(6):pdb.prot5439.
65. Ach RA, Wang H, Curry B. Measuring microRNAs: comparisons of microarray and quantitative PCR measurements, and of different total RNA prep methods. BMC Biotechnol. 2008;8:69.
66. Bustin SA, Benes V, Garson JA, et al. The MIQE guidelines: minimum information for publication of quantitative real-time PCR experiments. Clin Chem. 2009;55(4):611–22.
67. Kim YK, Yeo J, Kim B, Ha M, Kim VN. Short structured RNAs with low GC content are selectively lost during extraction from a small number of cells. Mol Cell. 2012;46(6):893–5.
68. Yoo CE, Kim G, Kim M, et al. A direct extraction method for microRNAs from exosomes captured by immunoaffinity beads. Anal Biochem. 2012;431(2):96–8.
69. Bravo V, Rosero S, Ricordi C, Pastori RL. Instability of miRNA and cDNAs derivatives in RNA preparations. Biochem Biophys Res Commun. 2007;353(4):1052–5.
70. Vaz C, Ahmad HM, Sharma P, et al. Analysis of microRNA transcriptome by deep sequencing of small RNA libraries of peripheral blood. BMC Genomics. 2010;11:288.
71. Jensen SG, Lamy P, Rasmussen MH, et al. Evaluation of two commercial global miRNA expression profiling platforms for detection of less abundant miRNAs. BMC Genomics. 2011; 12:435.
72. Sablok G, Milev I, Minkov G, et al. isomiRex: web-based identification of microRNAs, isomiR variations and differential expression using next-generation sequencing datasets. FEBS Lett. 2013;587(16):2629–34.
73. Kim SW, Li Z, Moore PS, et al. A sensitive non-radioactive northern blot method to detect small RNAs. Nucleic Acids Res. 2010;38(7):e98.
74. Wang B, Howel P, Bruheim S, et al. Systematic evaluation of three microRNA profiling platforms: microarray, beads array, and quantitative real-time PCR array. PLoS One. 2011;6(2):e17167.
75. Sempere LF, Freemantle S, Pitha-Rowe I, et al. Expression profiling of mammalian microRNAs uncovers a subset of brain-expressed microRNAs with possible roles in murine and human neuronal differentiation. Genome Biol. 2004; 5:R13.
76. Lee RC, Feinbaum RL, Ambros V. The C. elegans heterochronic gene lin-4 encodes small RNAs with antisense complementarity to lin-14. Cell. 1993;75(5):843–54.
77. Setoyama T, Ling H, Natsugoe S, Calin GA. Non-coding RNAs for medical practice in oncology. Keio J Med. 2011;60(4):106–13.
78. de Planell-Saguer M, Rodicio MC. Analytical aspects of microRNA in diagnostics: a review. Anal Chim Acta. 2011;699(2):134–52.
79. Kumar P, Johnston BH, Kazakov SA. miR-ID: a novel, circularization-based platform for detection of microRNAs. RNA. 2011;17(2):365–80.
80. Schmittgen TD, Jiang J, Liu Q, Yang L. A high-throughput method to monitor the expression of microRNAs precursor. Nucleic Acids Res. 2004;32:43–53.
81. Reichenstein I, Aizenberg N, Goshen M, et al. A novel qPCR assay for viral encoded microRNAs. J Virol Methods. 2010;163:323–8.
82. Ro S, Park C, Jin J, et al. A PCR-based method for detection and quantification of small RNAs. Biochem Biophys Res Commun. 2006;351: 756–63.
83. Stratagene® qRT-PCR platforms. http://www.stratagene.com/products/displayProduct.aspx?pid=820. Accessed Dec 2012.
84. Highly-specific miRNA QRT-PCR detection. http://www.genomics.agilent.com/CollectionSubpage.aspx?PageType=Product&SubPageType=ProductData&PageID=308. Accessed Feb 2013.
85. NCode™ miRNA amplification system. http://tools.invitrogen.com/content/sfs/manuals/ncode_rna_amp_man.pdf. Accessed Feb 2013.
86. Raymond CJ, Roberts BS, Garrett-Engele P. Simple, quantitative primer-extension PCR assay for direct monitoring of microRNAs and short-interfering RNAs. RNA. 2005;11: 1737–44.
87. Locked nucleic acids. http://www.eurogentec.com/EGT/files/GOL-LNALEAFLET-1107-V1.pdf. Accessed Feb 2013.
88. Leung AK, Sharp PA. Function and localization of microRNAs in mammalian cells. Cold Spring Harb Symp Quant Biol. 2006;71:29–38.
89. Nass D, Rosenwald S, Meiri E, et al. MiR-92b and miR-9/9* are specifically expressed in brain primary tumors and can be used to differentiate primary from metastatic brain tumors. Brain Pathol. 2009;19(3):375–83.
90. Bustos-Sanmamed P, Laffont C, Frugier F, Lelandais-Brière C, Crespi M. Analyzing small and long RNAs in plant development using non-radioactive in situ hybridization. Methods Mol Biol. 2013;959:303–16.

91. Obernosterer G, Leuschner PJ, Alenius M, Martinez J. Post-transcriptional regulation of microRNA expression. RNA. 2006;12(7):1161–7.
92. Shi Z, Johnson J, Sharon Stack M. Fluorescence in situ hybridization for microRNA detection in archived oral cancer tissues. J Oncol. 2012;2012:903581.
93. Yauk CL, Rowan-Carroll A, Stead JD, Williams A. Cross-platform analysis of global microRNA expression technologies. BMC Genomics. 2010;11:330.
94. Aldridge S, Hadfield J. Introduction to miRNA profiling technologies and cross-platform comparison. Methods Mol Biol. 2012;822:19–31.
95. Duttagupta R, DiRienzo S, Jiang R, et al. Genome-wide maps of circulating miRNA biomarkers for ulcerative colitis. PLoS One. 2012;7(2):e31241.
96. Chen J, April CS, Fan JB. miRNA expression profiling using Illumina Universal BeadChips. Methods Mol Biol. 2012;822:103–16.
97. D'Andrade PN, Fulmer-Smentek S. Agilent microRNA microarray profiling system. Methods Mol Biol. 2012;822:85–102.
98. Benes V, Castoldi M. Expression profiling of microRNA using real-time quantitative PCR, how to use it and what is available. Methods. 2010;50(4):244–9.
99. Liu J, Jenning S, Tong W, Hong H. Next generation sequencing for profiling expression of miRNAs: technical progress and applications in drug development. J Biomed Sci Eng. 2011;4:666–76.
100. Yang Q, Lu J, Wang S, Li H, Ge Q, Lu Z, et al. Application of next-generation sequencing technology to profile the circulating microRNAs in the serum of preeclampsia versus normal pregnant women. Clin Chim Acta. 2011;412:2167–73.
101. Schulte JH, Marschall T, Martin M, et al. Deep sequencing reveals differential expression of microRNAs in favorable versus unfavorable neuroblastoma. Nucleic Acids Res. 2010;38:5919–28.
102. Wang H, Ach RA, Curry B. Direct and sensitive miRNA profiling from low-input total RNA. RNA. 2007;13:151–9.
103. Lu C, Meyers BC, Green PJ. Construction of small RNA cDNA libraries for deep sequencing. Methods. 2007;43:110–7.
104. Buermans HP, Ariyurek Y, van Ommen G, et al. New methods for next generation sequencing based microRNA expression profiling. BMC Genomics. 2010;11:716.
105. Tian G, Yin X, Luo H, et al. Sequencing bias: comparison of different protocols of microRNA library construction. BMC Biotechnol. 2010;10:64.
106. Friedländer MR, Chen W, Adamidi C, et al. Discovering microRNAs from deep sequencing data using miRDeep. Nat Biotechnol. 2008;26:407–15.
107. Wang WC, Lin FM, Chang WC. miRExpress: analyzing high-throughput sequencing data for profiling microRNA expression. BMC Bioinformatics. 2009;10:328.
108. Hackenberg M, Sturm M, Langenberger D. miRanalyzer: a microRNA detection and analysis tool for next-generation sequencing experiments. Nucleic Acids Res. 2009;37:W68–76.
109. Moxon S, Schwach F, Dalmay T. A toolkit for analysing large-scale plant small RNA datasets. Bioinformatics. 2008;24:2252–3.
110. Bargaje R, Hariharan M, Scaria V, Pillai B, et al. Consensus miRNA expression profiles derived from interplatform normalization of microarray data. RNA. 2010;16(1):16–25.
111. Meyer SU, Pfaffl MW, Ulbrich SE. Normalization strategies for microRNA profiling experiments: a 'normal' way to a hidden layer of complexity? Biotechnol Lett. 2010;32(12):1777–88.
112. Deo A, Carlsson J, Lindldof A. How to choose a normalization strategy for miRNA quantitative real-time (qPCR) arrays. J Bioinform Comput Biol. 2011;9:795–812.
113. Volinia S, Calin GA, Liu CG, et al. A microRNA expression signature of human solid tumors defines cancer gene targets. Proc Natl Acad Sci U S A. 2006;103:2257–61.
114. SAM: significance analysis of microarrays. http://www-stat.stanford.edu/~tibs/SAM/index.html. Accessed Feb 2013.
115. PAM: prediction analysis for microarrays. http://www-stat.stanford.edu/~tibs/PAM/index.html. Accessed Feb 2013.
116. Liu CG, Calin GA, Meloon B, et al. An oligonucleotide microchip for genome-wide microRNA profiling in human and mouse tissues. Proc Natl Acad Sci U S A. 2004;101(26):9740–4.
117. Iorio MV, Ferracin M, Liu CG, et al. MicroRNA gene expression deregulation in human breast cancer. Cancer Res. 2005;65(16):7065–70.
118. Williams MD, Mitchell GM. MicroRNA in insulin resistance and obesity. Exp Diabetes Res. 2012;2012:484696.
119. Mo YY. MicroRNA regulatory networks and human disease. Cell Mol Life Sci. 2012;69:3529–31.
120. Salmena L, Poliseno LB, Tay Y, Kats L, Pandolfi PP, et al. A ceRNA hypothesis: the Rosetta Stone of a hidden RAN language? Cell. 2011;146:353–8.
121. Dillman W. Cardiac hypertrophy and thyroid hormone signaling. Heart Fail Rev. 2010;15:125–32.
122. Callis TE, Pandya K, Seok HY, et al. MicroRNA-208a is a regulator of cardiac hypertrophy and

conduction in mice. J Clin Invest. 2009;119: 2772–27786.
123. Jopling CL, Yi M, Lancaster AM, Lemon SM, Sarnow P. Modulation of hepatitis C virus RNA abundance by a liver-specific MicroRNA. Science. 2005;309:1577–81.
124. Yue J. miRNA and vascular cell movement. Adv Drug Deliv Rev. 2011;63(8):616–22.
125. Li W, Szoka Jr FC. Lipid-based nanoparticles for nucleic acid delivery. Pharm Res. 2007;24: 438–49.
126. Walther W, Stain U. Viral vectors for gene transfer: a review of their use in the treatment of human diseases. Drugs. 2000;60:249–71.
127. Al-Dosari MS, Gao X. Nonviral gene delivery: principle, limitations and recent progress. AAPS J. 2009;11(4):671–81.
128. Couto LB, High KA. Viral vector-mediated RNA interference. Curr Opin Pharmacol. 2010;10(5): 534–42.
129. Rai K, Takigawa N, Ito S, et al. Liposomal delivery of microRNA-7-expression plasmid epidermal growth factor receptor tyrosine kinase inhibitor-resistance in lung cancer cells. Mol Cancer Ther. 2011;10(9):1720–7.
130. Gondi CS, Rao JS. Concepts in in vivo siRNA delivery for cancer therapy. J Cell Physiol. 2009;220(2):285–91.
131. Lujambio A, Lowe SW. The microcosmos of cancer. Nature. 2012;482:347–55.
132. Ferracin M, Zagatti B, Rizzotto L, et al. MicroRNAs involvement in fludarabine refractory chronic lymphocytic leukemia. Mol Cancer. 2010;9:123.
133. Ru P, Steele R, Newhall P, et al. MicroRNA-29b suppresses prostate cancer metastasis by regulating epithelial-mesenchymal transition signaling. Mol Cancer Ther. 2012;11:1166–73.
134. Kandalam MM, Beta M, Maheswari UK, et al. Oncogenic microRNA 17-92 cluster is regulated by epithelial cell adhesion molecule and could be a potential therapeutic target in retinoblastoma. Mol Vis. 2012;18:2279–87.
135. Fassina A, Marino F, Siri M, et al. The miR-17-92 microRNA cluster: a novel diagnostic tool in large B-cell malignancies. Lab Invest. 2012; 92(11):1574–82.
136. Lu J, Getz G, Miska EA, et al. MicroRNA expression profiles classify human cancers. Nature. 2005;435(7043):834–8.

CHAPTER 5

Circulating Tumor Cells: Enrichment and Genomic Applications

MARIJA BALIC, RICHARD J. COTE

Introduction 71
Enrichment Techniques for CTC 71
Detection Methods 73
Molecular Characterization of CTC 75
Conclusions 81

Introduction

Worldwide, cancer remains among the most common causes of morbidity and mortality constantly presenting challenges in diagnosis and clinical management. The effective implementation of cancer screening methods and the improvements in treatment strategies have led to a decrease in cancer mortality in the last decades. However, despite the advances in multidisciplinary treatment strategies and the development of more efficacious systemic therapy, patients with metastatic disease remain currently incurable.

Cancer spreads locally and through blood and lymphatic vessels, leading to distant metastases. Although the processes underlying metastasis remain largely unknown, single cells or cell clusters detaching from the primary tumor and invading surrounding vessels are regarded to be the carriers of metastatic spread [1]. Surrogate markers of this spread have been studied for decades [2]. Regional lymphatic spread has long been evaluated in lymph nodes, with a more recent focus on the assessment of sentinel nodes. Whereas the detection of metastatic hematogenous dissemination was initially focused on bone marrow analysis, it has increasingly shifted to peripheral blood, given the invasive nature of the bone marrow aspiration procedure. Occult metastatic spread is not detectable by routine diagnostic and staging methods, and is known in literature as the presence of occult tumor cells, disseminated tumor cells, micrometastases, or, in peripheral blood, circulating tumor cells (CTC).

Enrichment Techniques for CTC

The greatest challenge in the detection of CTC is their rarity in peripheral blood. Very few CTC will be present, even in patients with advanced stages of metastatic disease. These cells have to be detected among white blood cells ($5–10 \times 10^6$/ml), red blood cells ($5–9 \times 10^9$/ml), and platelets ($2.5–4 \times 10^8$/ml). The frequency of CTC is often less than 1 CTC per ml of peripheral blood.

A variety of techniques are currently in use for enrichment and detection of CTC. All these techniques are based on biological and/or physical properties that help distinguish CTC from all the normal blood cells. A brief overview is provided here and the reader is referred to excellent, detailed reviews on the topic [3–5].

G.J. Netto and I. Schrijver (eds.), *Genomic Applications in Pathology*,
DOI 10.1007/978-1-4939-0727-4_5, © Springer Science+Business Media New York 2015

By far, affinity-based enrichment is the most commonly employed strategy. Affinity-based methods take advantage of antigens that are differentially expressed by CTC but not by blood cells (e.g., EpCAM, positive selection) [6, 7], or vice versa (e.g., CD45, negative selection) [8, 9]. A significant drawback of positive enrichment strategies is that they are only effective for CTC that show high expression of the target capture antigen; CTC with low or no expression are not enriched. For example, EpCAM is heterogeneously expressed, even by tumor types that are considered to have high expression levels (such as breast cancer). Furthermore, some epithelial tumors may completely lack EpCAM expression (e.g., renal cell carcinoma). Evidently, the latter is also true for non-epithelial cancers such as melanoma.

The most common method of affinity-based selection is immunomagnetic separation using magnetic beads equipped with antibodies that bind to either CTC or to blood cell antigens [10]. Other forms of immunomagnetic enrichment use columns or cartridges that allow for automation of the process [7, 11]. In recent years, microchip-based affinity methods have been described [6, 12].

Among the oldest approaches for cell enrichment are those that take advantage of differences in cell density (a physical property of the cell). An example of such a method is Ficoll Hypaque separation [13], which separates red blood cells from nucleated cells in the peripheral blood or bone marrow aspirate, including tumor cells. Despite low recovery yield and poor enrichment of the tumor cells by density gradients, it remained the standard approach for many years [14, 15].

An alternative property of tumor cells that has long been considered to be potentially useful for enrichment is cell size [16]. Tumor cells, particularly those derived from epithelial tissues, are larger than most blood cells [17]. An advantage of this approach is that a broader range of tumors is potentially amenable to size-based separation, without dependence on their inherent heterogenous antigen expression. Whereas the use of size to enrich CTC in blood has been considered for almost 50 years [16], it is only recently that size-based separation techniques have become commercially available [18, 19]. We have developed a size-based microfilter for enrichment and detection of CTC [20], which seems to be highly efficient and faster than affinity-based separation techniques. Further, our platform can be used for detailed molecular characterization [21], which is of particular interest for genomic analysis of CTC.

Size-Based Microchips

Several platforms that use size as the capture method have been described [17]. Our group has developed a simple size-based parylene microfilter for isolation of CTC with the potential for integrated downstream RNA, DNA, and multi-marker protein characterization. The ability to fabricate high-density pore filters allowed for enhancement of both the enrichment factor and the recovery rate of CTC. In initial studies, the achieved recovery rate of tumor cells, spiked into peripheral blood, was >90 %. The filtration of a 7.5 ml blood sample could be performed within a few minutes. Using the microfilter, we were able to detect CTC in peripheral blood samples from 51 of 57 metastatic cancer patients that included prostate, colon, breast, and bladder cancer patients, compared to 26 of 57 patients for whom the CellSearch™ method was employed. The mean number of recovered CTC was 5.5 times higher by the microfilter device compared to CellSearch™ [20]. Another advantage of the parylene filter is that parylene is optically transparent making it possible for the capture platform to simultaneously be used as the analysis platform.

Alternative size-based separation platforms have also been described [22]. The main difference between the parylene filter and other sized-based separation platforms is the density and regularity of the pores. Most filters are track-etched, produced by ionizing radiation. This results in an irregular pore distribution, with a low density and significant overlap of pores. Some holes are large enough to allow CTC to pass through. The parylene filter is designed with advanced lithography techniques, resulting in a highly regular and dense pore pattern. The size and shape of the

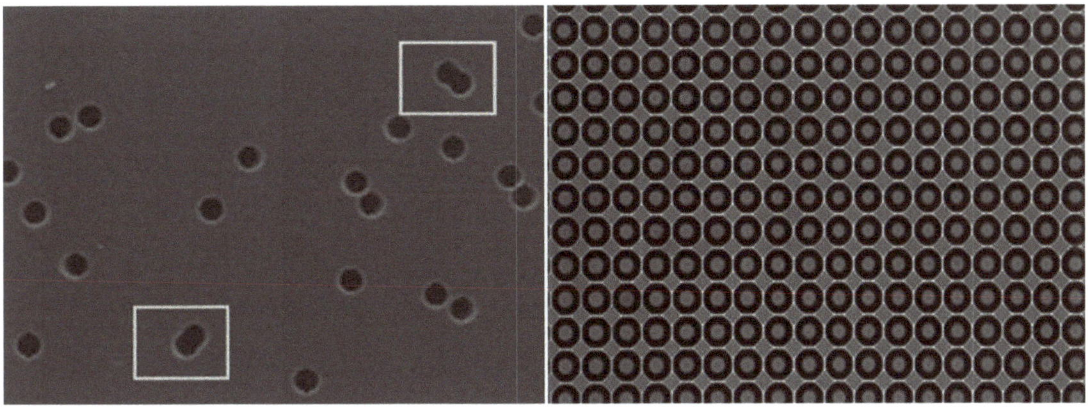

Figure 5-1 Comparison of the design of track-etch filter and a parylene filter. On the *left*, holes large enough to allow CTC to pass through are framed. The parylene filter is designed with advanced lithography techniques, resulting in a highly regular and dense pore pattern (*right* image) [20, 21]

pores can be precisely controlled. The two different designs are depicted in Fig. 5.1.

Affinity-Based Microchips

One of the first reported microchips for CTC enrichment was an affinity-based microfluidic chip with microposts coated with EpCAM antibodies [6, 12]. CTC were successfully isolated in each clinical blood sample tested in all study subjects that included metastatic lung, prostate, pancreatic, breast, and colon cancer patients. Using this method, monitoring of CTC was performed in metastatic non-small-cell lung cancer patients demonstrating correlation of CTC count with tumor response [23]. The affinity-based microfluidic chip was also capable of capturing tumor cells from which DNA could be extracted for *EGFR* mutation analysis [12]. A limitation of such chips is their requirement for a very slow flow of blood for efficient capture of CTC, often taking more than 10 h to process 7.5 ml of blood [6].

Other Microchip Methods

Among existing alternative methods, a microfluidic device that utilizes deterministic hydrodynamic flow and size-based separation is described next [24–26]. Inside a microfluidic chamber the device contains a micropost array, and the diameter of the circular micropost, the distance between the microposts in individual rows, and the row-to-row shift determine its performance. With such a device, separation of plasma from blood cells, different types of blood cells from each other, and DNA fragments of different size can be performed. The time required for relatively large volumes of blood samples is considerably longer.

Some devices have been developed using dielectrophoretic forces applied through the microelectrode arrays onto the field. Based on their electric properties, cells are positioned at a specific distance from electrodes [25, 27]. This technique has demonstrated 100 % separation efficiency when defined numbers of cultured human breast cancer MDA-231 cells were spiked into peripheral blood [28, 29]. A severe limitation of this technique is that only a very small sample volume (30 μl) can be processed. Given the low number of CTC (often less than 1 CTC per ml), this technique would not be expected to be useful for the detection or isolation of CTC in clinical samples.

Detection Methods

CTC enrichment methods allow for the ratio of target CTC to background cells to be significantly enhanced. However, none of the currently available enrichment methods

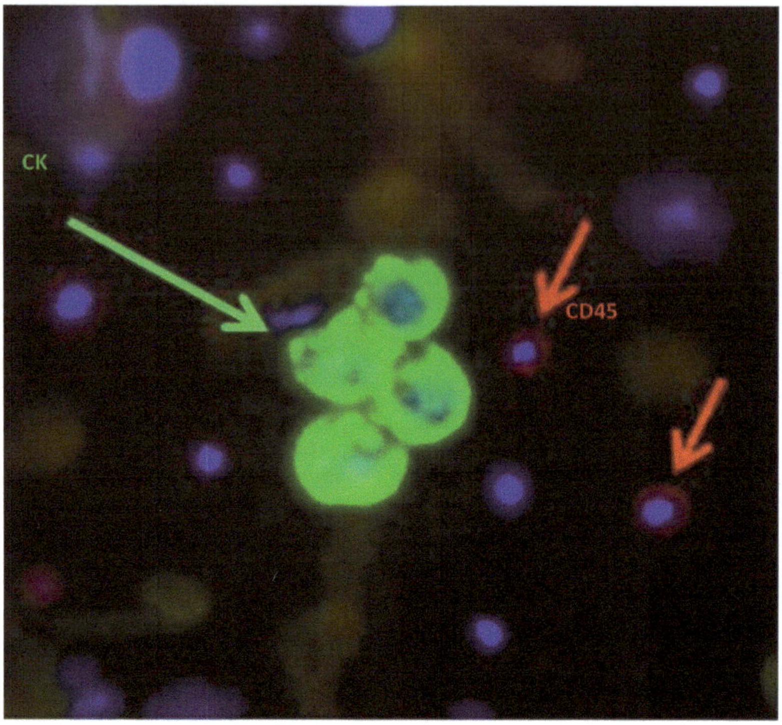

Figure 5-2 Detection of a circulating tumor cell in a background of hematopoietic cells. CTC are identified as cells that are large in size (i.e., 12–25 μm), nucleated (as demonstrated on 4′6-diamidino-2-phenylindole (DAPI) positivity), cytokeratin (CK) positive (*green*), and CD45 negative and display other morphologic features consistent with malignant cells (e.g., a high nuclear to cytoplasmic ratio). In contrast, the lymphocytes display a CK negative/CD45 positive (*red*) profile. *CTC* circulating tumor cells, *CK* cytokeratin

results in a pure population of tumor cells. Following enrichment, all separation techniques require a method to distinguish CTC from the nonspecifically captured cells. Several approaches can be performed to identify tumor cells: cytomorphological characterization of CTC, immunohistochemical/immunofluorescent (IHC/IF) detection of tumor specific antigens, or various real time polymerase chain reaction (RT-PCR) approaches. Cytomorphological characterization relies on classification of tumor cells based on their distinct morphological features [19]. Immunocytochemical detection of CTC relies on antibody based detection of cells using antibodies specific for epithelial cells. Most commonly used antibodies are cytokeratins, including both low- and high-molecular-weight cytokeratins [30]. This method is now often combined with markers such as CD45 that identify the background blood (non-CTC) cells. A representative cytokeratin-positive cell in a background of blood cells is depicted in Fig. 5.2.

Multiplex IHC/IF approaches enable simultaneous visualization of multiple markers on a single cell [31], such as in the representative image for detection of cytokeratins and the putative breast cancer stem cell marker aldehyde dehydrogenase (ALDH) (Fig. 5.3). Detection of CTC by IHC/IF has one major potential drawback, namely, the potential to miss cells that lack the expression of the targeted antigen. The potential for such omission is further suggested by recent literature demonstrating epithelial–mesenchymal transition (EMT) and expression of mesenchymal markers by epithelial CTC [32]. The simultaneous use of multiple cytokeratins of high and low molecular weight, including stem cell associated cytokeratins, has minimized this concern [30, 33].

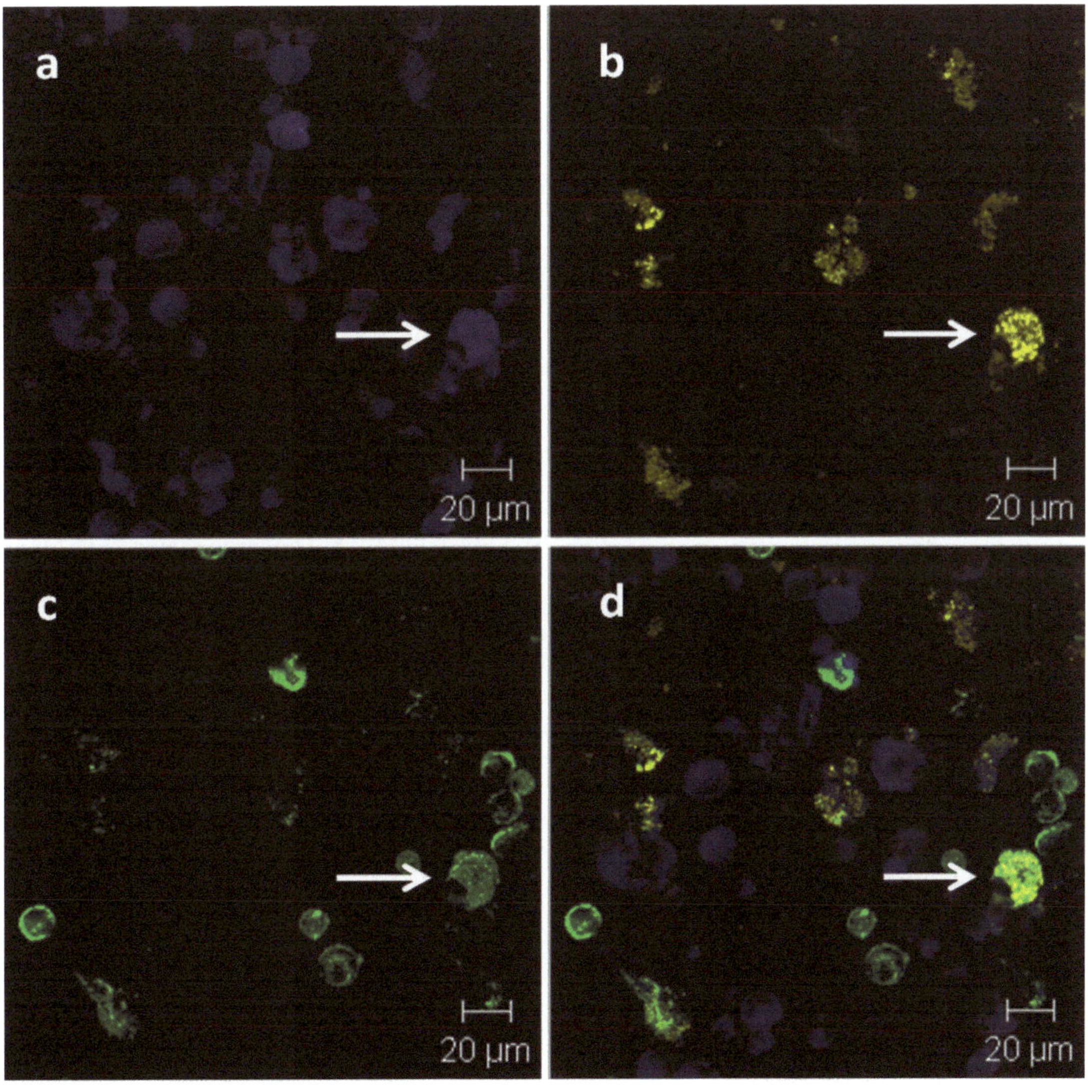

Figure 5-3 Double immunofluorescence for a putative breast cancer stem cell marker ALDH and CK in a Cytospin preparation from a mixture of cancer cell lines (MCF7, SUM159, SUM1315, and HEPG2). The sample was assessed for (**a**) DAPI (*blue*), (**b**) ALDH using a secondary antibody labeled with DyLight 550 (*yellow*), (**c**) CK with a secondary antibody labeled with Alexa Fluor 488 (*green*), and (**d**) all markers merged. The cell indicated with an *arrow* is a representative cell positive for ALDH and CK (*DAPI* 4′6-diamidino-2-phenylindole, *ALDH* aldehyde dehydrogenase, *CK* cytokeratin)

Molecular Characterization of CTC

It is becoming apparent that intratumoral heterogeneity is one of the many factors that could be responsible for therapeutic failure and drug resistance in cancer. Tumor cell heterogeneity is attributed, in part, to one of several key biologic mechanisms of cancer progression. Chief among these are the presence of a perpetual supply of cancer stem cells (CSC) [34], the process of EMT, and the heterogeneity of expression of molecules that determine drug response and resistance, including those that are therapy targets among cancer cells [35–37]. For example, variability in estrogen and progesterone receptor expression and Her2-neu status among mammary cancer cells will result in heterogeneity of response to therapies directed against these targets [38, 39]. Given the likely role of CSC in tumor heterogeneity and resistance [40], molecular characterization of CSC in the setting of metastatic dissemination has increasingly gained interest.

Cancer Stem Cells

Experimental evidence in support of a "cancer stem cell model" in various malignancies is mounting [41]. Such a model assumes the presence of a small proportion of cells on top of a hierarchy of tumor cells with the capability of sustaining tumor formation and growth, self-renewal and differentiation. Several markers of CSC have been identified, including a $CD44^{+}CD24^{-/low}$ phenotype and ALDH expression in breast and other cancers [42, 43], and CD133 expression in non-small-cell lung cancer and brain tumors [44, 45].

In breast cancer, and presumably other cancers, as well, the identification of distinct properties and molecular biomarkers of CSC may help in the development of more effective treatment and novel therapeutic targets [46]. In general, the presence of occult micrometastases is the rationale behind the use of systemic adjuvant chemotherapy following a definitive local treatment of a primary tumor [47]. The dissemination of CSC may be responsible for the failure of adjuvant chemotherapies in a proportion of early-stage breast cancer patients [48]. Several in vitro studies have demonstrated that putative breast CSC are resistant to conventional treatment strategies, including radiation and chemotherapy [49–51]. Consequently, the identification of breast CSC among CTC may be a promising strategy to assess their malignant potential and identify novel therapeutic targets. A major hurdle for such an approach is the hereto limited available knowledge regarding CTC phenotypes and the fact that CSC represent only a proportion of enriched CTC.

CSC in Dissemination

Pooled analysis of data from nine prior breast cancer studies which included a large number of early-stage patients revealed that the presence of disseminated tumor cells (DTC) in bone marrow is associated with a poor prognosis. Surprisingly, a significant proportion of patients with DTC had a favorable survival outcome 10 years or more following the diagnosis [2]. One potential reason for such an outcome could be the ability of DTC to remain dormant in distant organs. Based on the prior observations, we hypothesized that CSC not only exist within the primary tumor but may represent the most potent and virulent cells metastasizing from primary breast cancer to distant locations. In order to test our hypothesis, we performed a study analyzing DTC from breast cancer patients enrolled in the ACOS-OG Z-00010 trial for the putative breast CSC phenotype $CD44^{+}CD24^{-/low}$. The large majority of DTC in examined patients had the putative CSC phenotype [48]. This study provided the first demonstration that DTC/CTC are primarily composed of CSC, in contrast to primary and metastatic tumors in which fewer than 10 % of cells have a CSC phenotype [48, 52]. This finding has significant biologic implications, as it suggests the relative enrichment for breast CSC in the process of metastasis [53]. The fact that the ACOS-OG Z-00010 trial patients were early-stage I and II breast cancer patients in whom only 3 % of BM samples were positive for DTC, made the finding even more significant [54].

Several studies have since confirmed these findings. In a prospective analysis of bone marrow aspirates from high-risk breast cancer patients, using cell sorting by flow cytometry, Reuben et al. [55] were able to show a high percentage of CSC in DTC. Using a similar approach, Theodoropoulos et al. demonstrated the presence of CTC in 67 % of patients with metastatic disease, with 35 % of CTC displaying the $CD44^{+}/CD24^{low}$ CSC phenotype [56]. In another flow cytometry study, evaluating peripheral blood from breast cancer patients at variable stages, Wang et al. showed an increasing percentage of putative CSC in correlation with higher tumor stage [57]. The above findings further emphasize the need for reliable CTC enrichment methods allowing for detailed molecular characterization.

Epithelial–Mesenchymal Transition

The ability of tumor cells to undergo EMT is crucial for local invasion and gaining access to the blood stream through intravasation [58]. The plasticity of tumor cells and their capability to transform and acquire mesenchymal characteristics may be derived from CSC [59]. EMT is associated with a specific set of genetic changes that lead to increased tumor cell motility and an invasive phenotype. These changes are typically characterized by loss of E-Cadherin expression and subsequent translocalization of β-catenin from the cell membrane into the nucleus, increased expression of vimentin, production of matrix metalloproteinase enzymes, and upregulation of various EMT-inducing transcription factors such as Twist, Snail, and Slug [60]. Thus, EMT provides a potential mechanistic basis for how CTC intravasate in primary tumors to reach the circulation, and subsequently extravasate from the circulation to seed tumor implants at distant secondary sites. Several studies have evaluated the expression of EMT associated markers in CTC. In a study involving metastatic breast cancer patients, Aktas et al. revealed at least one of three EMT markers (Akt2, PI3K, and Twist1), assessed by RT-PCR, to be expressed by the CTC population in 62 % of patients harboring CTC. Patients with CTC who were positive for EMT were more likely to fail to respond to palliative chemotherapy, antibody or hormonal therapy [61]. Evaluating CTC expression of EMT markers Twist and vimentin by immunofluorescence, Kallergi et al. found vimentin/Twist expressing CTC in 77 % of early-stage breast cancer patients compared to 100 % of patients with metastatic disease [62]. In a recent study involving 11 breast cancer patients who were serially monitored for CTC phenotype, mesenchymal phenotype CTC were more likely to be associated with disease progression. In one index patient, the authors were able to demonstrate a reversible shift between the epithelial and mesenchymal phenotype corresponding to response to treatment and disease progression, respectively [63].

Additional potential mechanisms that could facilitate tumor cell dissemination include ameboid motility and collective migration of cell clusters [64]. The occurrence of circulating tumor micro-emboli (CTM) in metastatic lung cancer patients was demonstrated by Hou et al. [65]. In this study, the authors showed that single CTC expressed apoptosis related markers at a higher rate than CTM. These findings suggest that collective migration of tumor cells in circulation may offer a survival advantage to the tumor.

Nevertheless, given the recent evidence that at least a portion of CTC are cells transitioning between the epithelial and mesenchymal state [63] that possess stem cell-like properties and the ability of reversible modulation [66], the functional characterization of these processes in CTC is crucial. Development of new technologies that will enhance sensitivity and efficiency of CTC detection will facilitate functional characterization of CTC invasiveness, aggressiveness, plasticity, and tumorigenic potential. Functional characterization will, in turn, help further clarify the mechanisms of tumor cell dissemination.

Recent advances in CTC enrichment have allowed for better genetic and molecular characterization of CTC. The latter is carried out by various strategies that include fluorescent in situ hybridization (FISH), comparative genomic hybridization (CGH), PCR-based techniques and immunofluorescence. These studies have shed light on the oncogenic profile and metastatic potential of CTC and have allowed the comparison of the genetic profile of tumor metastases and CTC to that of their primary tumor counterpart.

Global Genomic Profiling of CTC

One approach to the genetic characterization of CTC is the evaluation of whole-genome copy number alterations in CTC in comparison to their primary tumor origin [67]. The earliest studies profiling single cells of DTC and CTC were performed by Dr. Christoph Klein's group [46, 68, 69] and provided fundamental knowledge on dissemination and metastasis. Using genomic and transcriptomic profiling, the authors confirmed the malignant origin of single cells detected in bone marrow of breast cancer patients. Subsequently, based on comprehensive molecular analysis performed on mice transplanted with single premalignant *HER-2* transgenic glands, the same group was able to define metastasis as an early event [70]. The comparative analyses in these early studies of primary tumors, metastases, and DTC provided the impetus for the molecular characterization of systemic disease rather than solely focusing on primary tumors [71].

Technical limitations of CTC isolation efficiency and the difficulties of performing whole-genome analyses on rare cells have limited the number of CTC genomic profile studies. Magbanua et al. [72] developed a novel approach for the molecular profiling of CTCs utilizing sequential immunomagnetic enrichment steps and flow cytometry sorting to isolate CTC. The latter was followed by whole-genome DNA amplification and array CGH analysis. The authors were thus able to unveil a wide range of copy number alterations in CTC obtained from peripheral blood samples of patients with advanced breast cancer. In order to delineate genomic alterations specific to CTC, the authors performed a comparative analysis between the CTC CGH dataset and a previously published dataset of primary tumor CGH [73]. Several copy number gains were more frequent in CTC compared to primary tumors. A sub-analysis performed in patients with known *HER-2* status of primary tumors revealed focal amplification of *HER-2* in CTC obtained from two patients with *HER-2* positive primary tumors. In contrast, in eight patients with *HER-2* negative primary tumors, the CTC samples showed no *HER-2* copy number gain in six and low-level gains in the remaining two.

In an earlier study, Paris et al. demonstrated that copy number profiles of CTC detected in castration resistant prostate cancer patients were similar to those of their paired solid tumor DNA and distinct from corresponding DNA from the remaining depleted mononuclear blood cells after EpCAM enrichment of CTC [74]. The study demonstrated the usefulness of the collagen adhesion matrix based CTC enrichment method for genetic analysis of CTC in prostate cancer patients. Similarly, Magbanua et al. were able to show the utility of immunomagnetic enrichment followed by fluorescence activated cell sorting for isolating CTC in castration resistant prostate cancer. The isolated CTC were successfully used to perform copy number profiling, evaluate progression and monitor response to therapy [75].

Recently, Heitzer et al. have tested whether tumor-specific copy number alterations can be detected in the peripheral blood of patients with cancer [76]. The authors evaluated the plasma DNA concentration and the fraction of DNA fragments in patients with colorectal and breast cancer and in healthy controls along with CTC detection by CellSearch. The presence of biphasic DNA size distribution was associated with increased CTC counts in cancer patients. Further, plasma DNA was screened for mutations with deep sequencing and an ultrasensitive mutation-detection method. In patients with biphasic DNA size distribution, an elevated concentration of mutated plasma DNA fragments was also detectable. The authors suggested that detection and characterization of plasma DNA in cancer patients may be useful for monitoring the response of cancer patients to ongoing treatment.

The above studies offer a proof of principle of the feasibility of CTC analysis for the assessment of genomic alterations associated with cancer progression and of monitoring response to targeted therapy. Prospective trials will be needed to evaluate the usefulness of such noninvasive approaches to genomic based therapy response prediction assays.

Our group has evaluated the feasibility of IHC/IF labelling of CTC captured by the parylene filter described above, for subsequent microdissection using a precise and contact-free laser microdissection system (PALM) and DNA extraction. Figure 5.4 shows MCF7 breast cancer cells that were captured and microdissected using this method. The captured CTC underwent DNA extraction and whole-genome amplification. The DNA quality was verified by multiplex PCR according to a previously published protocol for the evaluation of the DNA quality prior to CGH analysis [77]. Whole-genome array CGH analysis using a 44 K Agilent array successfully revealed the expected array CGH profile of MCF7 cells.

The greatest advances in the treatment of cancer have been made with combinations of

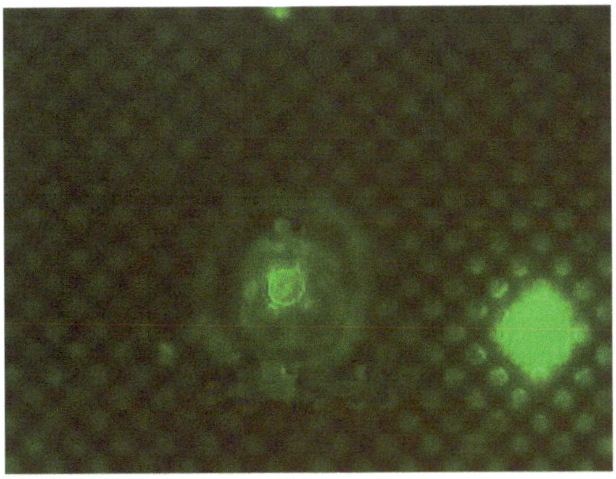

Figure 5-4 Following CK (cytokeratin) immunostaining (*green*), laser microdissection of the parylene filter was performed under a precise and contact-free laser microdissection system (PALM). The microdissection of the filter represents a necessary step for single cell analyses

targeted therapies. The identification of important pathways and their components as well as the characterization of the predictive value of specific molecular changes have led to an improvement in individualization of the treatment of cancer patients. With increasing knowledge of molecular targets and biomarkers, it may become necessary to perform genomic profiling of a large number of genomic changes in cancer tissue, metastatic sites, and CTC prior to the implementation of anticancer therapies. Once multi-targeted drugs or combinations of targeted therapies become clinically applicable, genomic profiling may help to optimize treatment. Such profiling may require next-generation sequencing of a larger number and combination of genes, or profiling of many translational products. Currently, the technologies are limited. For example, the number of markers that can be analyzed is limited, regardless of their structures (i.e., proteins, mRNA, DNA, etc.). Whereas global profiling may add substantial information to the understanding of metastasis, heterogeneity of tumors, and the biology of disease, the current analyses are largely experimental. Analyses of specific changes, for now, may have a more specific clinical role in determining treatment strategies. The development of technologies capable of determining a large number of clinically relevant biomarkers is exigent.

Mutation Analyses of CTC

Currently, mutation analysis of genetic alterations that predict response to targeted therapies in metastatic-stage cancer patients is performed on primary rather than metastatic tumor samples. Examples include assessment of *KRAS* and *BRAF* mutations, and *EGFR* mutations in colorectal cancer and lung adenocarcinoma, respectively [78, 79]. Whereas determinants of response to a given targeted therapy would ideally be assessed in the metastatic tumor being treated, obtaining such samples through an invasive surgical or imaging procedure is associated with significant morbidity and/or expense. Finding a source of tumor that can be noninvasively accessed, represents the most relevant population of tumor cells being treated, and could be serially accessed through the course of therapy, would be ideal. CTC may well represent such a source of tumor cells.

One of the early studies that demonstrated the ability to detect specific mutations in captured CTC to predict response to targeted therapies is the study by Punnoose et al. [80]. The authors were able to show the potential of CTC mutation analysis of *KRAS* along with further molecular analyses to provide real-time information on tumor biomarker status, including EGFR, Her-2neu, or ER on spiked cancer cells from various cancer cell lines. Maheswaran et al. demonstrated the feasibility of detecting *EGFR* mutations in CTC in metastatic lung cancer patients undergoing tyrosine kinase inhibitor treatment [12]. CTC were enriched using a microfluidic device containing microposts coated with antibodies against epithelial cells. *EGFR* mutation analysis was performed on DNA that was recovered using allele-specific PCR.

Dharmasiri et al. demonstrated the technical feasibility of detecting *KRAS* mutations in colon cancer cell lines spiked into peripheral blood [81]. Subsequently, Yang et al. [82] and Mostert et al. [83] were able to detect *KRAS* mutations in CTC in the peripheral blood of colorectal cancer patients. The discrepancies occasionally found between tissue and CTC *KRAS* mutation status were suggested to be due to the limited number of CTC available [83]. Gasch et al. have recently been able to analyze single CTC, obtained by CellSearch based enrichment, for several genetic alterations in 49 metastatic colorectal cancer patients [76]. Considerable heterogeneity among patients and within individually analyzed cells from the same patient was found in regard to *EGFR* expression and genetic alterations in *EGFR*, *KRAS*, and *PIK3* [76].

Androgen receptor mutation is one of the mechanisms leading to castration resistance in advanced prostate cancer [84]. Jiang et al. [85] have established an approach to detect such mutations in CTC from these patients after enrichment of CTC with the CellSearch method. Such an approach may facilitate the development of more effective treatment strategies in advanced prostate cancer.

Technological advances in CTC enrichment and sequencing have made it possible to perform genomic profiling on CTC. Many of the described and available technologies have the potential to be further developed for genomic profiling of CTC pools and single CTC in clinical samples. This will facilitate studies of CTC as a tool for liquid biopsy but also the evaluation of intra-CTC heterogeneity. For downstream analyses, there are many possibilities, depending on the questions to be addressed. Good single cell DNA quality may allow for global array CGH profiling and next-generation sequencing of larger regions or selected genes. Finally, specific PCR protocols on various platforms may be employed to detect particular genetic alterations and mutations.

Transcriptional CTC Profiling

Transcriptional profiling of CTC presents a significant technical challenge. A study by Smirnov et al. was one of the first to attempt global gene expression profiling of CTC in colorectal, prostate, and breast cancer patients [86]. Global gene expression profiles of CTC-enriched and corresponding CTC-depleted portions were generated and a list of CTC-specific genes was obtained. Subsequently, using quantitative RT-PCR, the authors were able to differentiate the expression level of a set of CTC specific genes in patients compared to normal controls. The study illustrated, for the first time, the feasibility of performing global gene expression profiling in CTC.

Barbazan et al. [87] performed whole-transcriptome amplification and gene expression analyses on affinity-enriched CTC from metastatic colorectal cancer patients. A 410 gene CTC signature was identified by hierarchical clustering, which included genes related to cell movement, cell adhesion, cell death, proliferation, cell signalling, and interaction. Confirmation of several genes was performed by quantitative RT-PCR in an independent set of patients. Sieuwerts et al. [88] brought attention to the fact that profiling a low number of CTC may result in discrepant estrogen receptor and HER-2 status profiles compared to primary tumor—a finding that could impact the use of current therapeutic strategies in breast cancer [35, 89].

Gene expression profiling studies, such as those evaluating the expression profiles of EMT related and CSC signatures in CTC [30, 48, 51–54], have enabled a more detailed evaluation of the biologic events associated with CTC and cancer metastasis. These studies provide preliminary support for the utility of CTC genomic assessment as a tool for exploring the biology of metastasis.

Epigenomic and miRNA Characterization of CTC

Epigenetic events are fundamental to normal processes of development and differentiation, and are increasingly found to play a substantial role in carcinogenesis. Aberrant DNA methylation profiles, histone modification and the alterations in micro RNA (miRNA) are examples of epigenetic alterations associated with cancer formation. Therefore, assessment of epigenomic alterations in CTC is crucial to further our understanding of the biology of cancer metastasis. As with all other types of genomic analyses, the rarity of CTC in patient samples presents technical challenges to epigenetic applications.

So far, only few studies have evaluated DNA methylation in CTC, attempting to correlate CTC occurrence with the methylation status of circulating DNA [55–57]. Likewise, few studies have either addressed the association of cancer miRNA alterations and CTC occurrence or the expression of miRNA in CTC. Sieuwerts et al. [88] were able to demonstrate the expression of ten miRNAs in CTC in metastatic breast cancer patients. Such studies are bound to become more frequent as new prognostic and therapeutic applications related to epigenetic alterations in cancer emerge.

Conclusions

The ability to detect and characterize CTC remains a technical challenge. Advancements in CTC enrichment, detection and characterization methods are rapidly being made. In the past decade, molecular assessment of CTC at the single cell level has provided the foundation for improved understanding of the biology of metastatic cancer spread. We have witnessed an era of great technical advancement that has led to the improved sensitivity of CTC detection and a better definition of recently discovered molecular processes related to CTC occurrence. Advances in next-generation sequencing and bioinformatics will no doubt potentiate the field of CTC analyses, and, through a better understanding of the biologic events associated with cancer metastasis, help establish novel strategies for cancer treatment. Definition of single molecular targets, such as mutations detected in CTC, may soon influence the treatment of cancer patients. In the not so distant future, analyses of "liquid biopsies," to define a large number of molecular targets and potential mechanisms of resistance in a given patient, will become reality. Such analyses will dynamically guide the treatment of cancer patients, and parallel their cancer progression status.

REFERENCES

1. Malanchi I, Santamaria-Martinez A, Susanto E, Peng H, Lehr HA, Delaloye JF, et al. Interactions between cancer stem cells and their niche govern metastatic colonization. Nature. 2012;481(7379):85–9.
2. Braun S, Vogl FD, Naume B, Janni W, Osborne MP, Coombes RC, et al. A pooled analysis of bone marrow micrometastasis in breast cancer. N Engl J Med. 2005;353(8):793–802.
3. Balic M, Williams A, Dandachi N, Cote RJ. Micrometastasis: detection methods and clinical importance. Cancer Biomark. 2010;9(1–6):397–419.
4. Alix-Panabieres C, Pantel K. Circulating tumor cells: liquid biopsy of cancer. Clin Chem. 2013;59(1): 110–8.
5. Lin H, Balic M, Zheng S, Datar R, Cote RJ. Disseminated and circulating tumor cells: role in effective cancer management. Crit Rev Oncol Hematol. 2011;77(1):1–11.
6. Nagrath S, Sequist LV, Maheswaran S, Bell DW, Irimia D, Ulkus L, et al. Isolation of rare circulating tumour cells in cancer patients by microchip technology. Nature. 2007;450(7173):1235–9.
7. Cristofanilli M, Budd GT, Ellis MJ, Stopeck A, Matera J, Miller MC, et al. Circulating tumor cells, disease progression, and survival in metastatic breast cancer. N Engl J Med. 2004;351(8):781–91.
8. Liu Z, Fusi A, Klopocki E, Schmittel A, Tinhofer I, Nonnenmacher A, et al. Negative enrichment by immunomagnetic nanobeads for unbiased characterization of circulating tumor cells from peripheral blood of cancer patients. J Transl Med. 2011;9:70.
9. Lustberg M, Jatana KR, Zborowski M, Chalmers JJ. Emerging technologies for CTC detection based on depletion of normal cells. Recent Results Cancer Res. 2012;195:97–110.
10. Naume B, Borgen E, Beiske K, Herstad TK, Ravnas G, Renolen A, et al. Immunomagnetic techniques for the enrichment and detection of isolated breast carcinoma cells in bone marrow and peripheral blood. J Hematother. 1997;6(2):103–14.
11. Iinuma H, Okinaga K, Adachi M, Suda K, Sekine T, Sakagawa K, et al. Detection of tumor cells in blood using CD45 magnetic cell separation followed by nested mutant allele-specific amplification of p53 and K-ras genes in patients with colorectal cancer. Int J Cancer. 2000;89(4):337–44.
12. Maheswaran S, Sequist LV, Nagrath S, Ulkus L, Brannigan B, Collura CV, et al. Detection of mutations in EGFR in circulating lung-cancer cells. N Engl J Med. 2008;359(4):366–77.
13. Rosenberg R, Gertler R, Friederichs J, Fuehrer K, Dahm M, Phelps R, et al. Comparison of two density gradient centrifugation systems for the enrichment of disseminated tumor cells in blood. Cytometry. 2002;49(4):150–8.
14. Cote RJ, Cordon-Cardo C, Reuter VE, Rosen PP. Immunopathology of adrenal and renal cortical tumors. Coordinated change in antigen expression is associated with neoplastic conversion in the adrenal cortex. Am J Pathol. 1990;136(5):1077–84.
15. Cote RJ, Peterson HF, Chaiwun B, Gelber RD, Goldhirsch A, Castiglione-Gertsch M, et al. Role of immunohistochemical detection of lymph-node metastases in management of breast cancer. International Breast Cancer Study Group. Lancet. 1999;354(9182):896–900.
16. Fleischer RL. Cancer filter deja vu. Science. 2007; 318(5858):1864.
17. Vona G, Sabile A, Louha M, Sitruk V, Romana S, Schutze K, et al. Isolation by size of epithelial tumor cells: a new method for the immunomorphological and molecular characterization of circulating tumor cells. Am J Pathol. 2000;156(1):57–63.
18. Paterlini-Brechot P, Benali NL. Circulating tumor cells (CTC) detection: clinical impact and future directions. Cancer Lett. 2007;253(2):180–204.
19. Vona G, Estepa L, Beroud C, Damotte D, Capron F, Nalpas B, et al. Impact of cytomorphological detection of circulating tumor cells in patients with liver cancer. Hepatology. 2004;39(3):792–7.

20. Lin HK, Zheng S, Williams AJ, Balic M, Groshen S, Scher HI, et al. Portable filter-based microdevice for detection and characterization of circulating tumor cells. Clin Cancer Res. 2010;16(20):5011–8.
21. Zheng S, Lin H, Liu JQ, Balic M, Datar R, Cote RJ, et al. Membrane microfilter device for selective capture, electrolysis and genomic analysis of human circulating tumor cells. J Chromatogr A. 2007; 1162(2):154–61.
22. Farace F, Massard C, Vimond N, Drusch F, Jacques N, Billiot F, et al. A direct comparison of Cell Search and ISET for circulating tumour-cell detection in patients with metastatic carcinomas. Br J Cancer. 2011; 105(6):847–53.
23. Sequist LV, Nagrath S, Toner M, Haber DA, Lynch TJ. The CTC-chip: an exciting new tool to detect circulating tumor cells in lung cancer patients. J Thorac Oncol. 2009;4(3):281–3.
24. Davis JA, Inglis DW, Morton KJ, Lawrence DA, Huang LR, Chou SY, et al. Deterministic hydrodynamics: taking blood apart. Proc Natl Acad Sci U S A. 2006;103(40):14779–84.
25. Huang LR, Cox EC, Austin RH, Sturm JC. Continuous particle separation through deterministic lateral displacement. Science. 2004; 304(5673):987–90.
26. Inglis DW, Davis JA, Austin RH, Sturm JC. Critical particle size for fractionation by deterministic lateral displacement. Lab Chip. 2006;6(5):655–8.
27. Wang XB, Yang J, Huang Y, Vykoukal J, Becker FF, Gascoyne PR. Cell separation by dielectrophoretic field-flow-fractionation. Anal Chem. 2000;72(4): 832–9.
28. Becker FF, Wang XB, Huang Y, Pethig R, Vykoukal J, Gascoyne PR. Separation of human breast cancer cells from blood by differential dielectric affinity. Proc Natl Acad Sci U S A. 1995;92(3):860–4.
29. Gascoyne PR, Wang XB, Huang Y, Becker FF. Dielectrophoretic separation of cancer cells from blood. IEEE Trans Ind Appl. 1997;33(3):670–8.
30. Cote RJ, Beattie EJ, Chaiwun B, Shi SR, Harvey J, Chen SC, et al. Detection of occult bone marrow micrometastases in patients with operable lung carcinoma. Ann Surg. 1995;222(4):415–23. discussion 23–5.
31. Balic M, Rapp N, Stanzer S, Lin H, Strutz J, Szkandera J, et al. Novel immunofluorescence protocol for multimarker assessment of putative disseminating breast cancer stem cells. Appl Immunohistochem Mol Morphol. 2011;19(1):33–40.
32. Raimondi C, Gradilone A, Naso G, Vincenzi B, Petracca A, Nicolazzo C, et al. Epithelial-mesenchymal transition and stemness features in circulating tumor cells from breast cancer patients. Breast Cancer Res Treat. 2011;130(2):449–55.
33. Bartkowiak K, Effenberger KE, Harder S, Andreas A, Buck F, Peter-Katalinic J, et al. Discovery of a novel unfolded protein response phenotype of cancer stem/progenitor cells from the bone marrow of breast cancer patients. J Proteome Res. 2010;9(6):3158–68.
34. Lorico A, Rappa G. Phenotypic heterogeneity of breast cancer stem cells. J Oncol. 2011;2011: 135039.
35. Fehm T, Becker S, Duerr-Stoerzer S, Sotlar K, Mueller V, Wallwiener D, et al. Determination of HER2 status using both serum HER2 levels and circulating tumor cells in patients with recurrent breast cancer whose primary tumor was HER2 negative or of unknown HER2 status. Breast Cancer Res. 2007; 9(5):R74.
36. Fehm T, Muller V, Aktas B, Janni W, Schneeweiss A, Stickeler E, et al. HER2 status of circulating tumor cells in patients with metastatic breast cancer: a prospective, multicenter trial. Breast Cancer Res Treat. 2010;124(2):403–12.
37. Tobin LA, Robert C, Nagaria P, Chumsri S, Twaddell W, Ioffe OB, et al. Targeting abnormal DNA repair in therapy-resistant breast cancers. Mol Cancer Res. 2012;10(1):96–107.
38. Liedtke C, Broglio K, Moulder S, Hsu L, Kau SW, Symmans WF, et al. Prognostic impact of discordance between triple-receptor measurements in primary and recurrent breast cancer. Ann Oncol. 2009; 20(12):1953–8.
39. Nogami T, Shien T, Tanaka T, Doihara H, Taira N, Takabatake D, et al. The discordance between primary breast cancer lesions and pulmonary metastatic lesions in expression of aldehyde dehydrogenase 1-positive cancer cells. Breast Cancer. 2014;21(1):58–65.
40. Chakrabarty A, Bhola NE, Sutton C, Ghosh R, Kuba MG, Dave B, et al. Trastuzumab-resistant cells rely on a HER2-PI3K-FoxO-survivin axis and are sensitive to PI3K inhibitors. Cancer Res. 2013;73(3):1190–200.
41. Azizi E, Wicha MS. Cancer stem cells—the evidence accumulates. Clin Chem. 2012;59(1):205–7.
42. Ginestier C, Hur MH, Charafe-Jauffret E, Monville F, Dutcher J, Brown M, et al. ALDH1 is a marker of normal and malignant human mammary stem cells and a predictor of poor clinical outcome. Cell Stem Cell. 2007;1(5):555–67.
43. Al-Hajj M, Wicha MS, Benito-Hernandez A, Morrison SJ, Clarke MF. Prospective identification of tumorigenic breast cancer cells. Proc Natl Acad Sci U S A. 2003;100(7):3983–8.
44. Wang S, Xu ZY, Wang LF, Su W. CD133+ cancer stem cells in lung cancer. Front Biosci (Landmark Ed). 2013;18:447–53.
45. Singh SK, Clarke ID, Hide T, Dirks PB. Cancer stem cells in nervous system tumors. Oncogene. 2004;23(43):7267–73.
46. Schardt JA, Meyer M, Hartmann CH, Schubert F, Schmidt-Kittler O, Fuhrmann C, et al. Genomic analysis of single cytokeratin-positive cells from bone marrow reveals early mutational events in breast cancer. Cancer Cell. 2005;8(3):227–39.
47. Cote RJ. Occult metastases: real harm or false alarm? J Thorac Cardiovasc Surg. 2003;126(2):332–3.
48. Balic M, Lin H, Young L, Hawes D, Giuliano A, McNamara G, et al. Most early disseminated cancer cells detected in bone marrow of breast cancer patients have a putative breast cancer stem cell phenotype. Clin Cancer Res. 2006;12(19):5615–21.
49. Duru N, Fan M, Candas D, Menaa C, Liu HC, Nantajit D, et al. HER2-associated radiation resistance of breast cancer stem cells isolated from HER2-negative breast cancer cells. Clin Cancer Res. 2012;18(24):6634–47.
50. Fillmore CM, Kuperwasser C. Human breast cancer cell lines contain stem-like cells that self-renew, give rise to phenotypically diverse progeny and survive chemotherapy. Breast Cancer Res. 2008;10(2):R25.

51. Kakarala M, Wicha MS. Implications of the cancer stem-cell hypothesis for breast cancer prevention and therapy. J Clin Oncol. 2008;26(17):2813–20.
52. Abraham BK, Fritz P, McClellan M, Hauptvogel P, Athelogou M, Brauch H. Prevalence of CD44+/CD24-/low cells in breast cancer may not be associated with clinical outcome but may favor distant metastasis. Clin Cancer Res. 2005;11(3):1154–9.
53. Wicha MS. Cancer stem cells and metastasis: lethal seeds. Clin Cancer Res. 2006;12(19):5606–7.
54. Giuliano AE, Hawes D, Ballman KV, Whitworth PW, Blumencranz PW, Reintgen DS, et al. Association of occult metastases in sentinel lymph nodes and bone marrow with survival among women with early-stage invasive breast cancer. JAMA. 2011; 306(4):385–93.
55. Reuben JM, Lee BN, Gao H, Cohen EN, Mego M, Giordano A, et al. Primary breast cancer patients with high risk clinicopathologic features have high percentages of bone marrow epithelial cells with ALDH activity and CD44CD24lo cancer stem cell phenotype. Eur J Cancer. 2011;47(10):1527–36.
56. Theodoropoulos PA, Polioudaki H, Agelaki S, Kallergi G, Saridaki Z, Mavroudis D, et al. Circulating tumor cells with a putative stem cell phenotype in peripheral blood of patients with breast cancer. Cancer Lett. 2010;288(1):99–106.
57. Wang N, Shi L, Li H, Hu Y, Du W, Liu W, et al. Detection of circulating tumor cells and tumor stem cells in patients with breast cancer by using flow cytometry: a valuable tool for diagnosis and prognosis evaluation. Tumour Biol. 2012;33(2):561–9.
58. Lim J, Thiery JP. Epithelial-mesenchymal transitions: insights from development. Development. 2012;139(19):3471–86.
59. Scheel C, Weinberg RA. Phenotypic plasticity and epithelial-mesenchymal transitions in cancer and normal stem cells? Int J Cancer. 2011;129(10): 2310–4.
60. Lee K, Nelson CM. New insights into the regulation of epithelial-mesenchymal transition and tissue fibrosis. Int Rev Cell Mol Biol. 2012;294:171–221.
61. Aktas B, Tewes M, Fehm T, Hauch S, Kimmig R, Kasimir-Bauer S. Stem cell and epithelial-mesenchymal transition markers are frequently overexpressed in circulating tumor cells of metastatic breast cancer patients. Breast Cancer Res. 2009;11(4):R46.
62. Ignatiadis M, Kallergi G, Ntoulia M, Perraki M, Apostolaki S, Kafousi M, et al. Prognostic value of the molecular detection of circulating tumor cells using a multimarker reverse transcription-PCR assay for cytokeratin 19, mammaglobin A, and HER2 in early breast cancer. Clin Cancer Res. 2008;14(9): 2593–600.
63. Yu M, Bardia A, Wittner BS, Stott SL, Smas ME, Ting DT, et al. Circulating breast tumor cells exhibit dynamic changes in epithelial and mesenchymal composition. Science. 2013;339(6119):580–4.
64. Friedlein R, von Kieseritzky F, Braun S, Linde C, Osikowicz W, Hellberg J, et al. Solution-processed, highly-oriented supramolecular architectures of functionalized porphyrins with extended electronic states. Chem Commun (Camb). 2005;21(15):1974–6.
65. Hou JM, Krebs M, Ward T, Sloane R, Priest L, Hughes A, et al. Circulating tumor cells as a window on metastasis biology in lung cancer. Am J Pathol. 2011;178(3):989–96.
66. May CD, Sphyris N, Evans KW, Werden SJ, Guo W, Mani SA. Epithelial-mesenchymal transition and cancer stem cells: a dangerously dynamic duo in breast cancer progression. Breast Cancer Res. 2011;13(1):202.
67. Sun YF, Yang XR, Zhou J, Qiu SJ, Fan J, Xu Y. Circulating tumor cells: advances in detection methods, biological issues, and clinical relevance. J Cancer Res Clin Oncol. 2011;137(8):1151–73.
68. Klein CA. Parallel progression of primary tumours and metastases. Nat Rev Cancer. 2009;9(4):302–12.
69. Klein CA, Seidl S, Petat-Dutter K, Offner S, Geigl JB, Schmidt-Kittler O, et al. Combined transcriptome and genome analysis of single micrometastatic cells. Nat Biotechnol. 2002;20(4):387–92.
70. Husemann Y, Geigl JB, Schubert F, Musiani P, Meyer M, Burghart E, et al. Systemic spread is an early step in breast cancer. Cancer Cell. 2008;13(1):58–68.
71. Stoecklein NH, Klein CA. Genetic disparity between primary tumours, disseminated tumour cells, and manifest metastasis. Int J Cancer. 2010;126(3):589–98.
72. Magbanua MJ, Sosa EV, Roy R, Eisenbud LE, Scott JH, Olshen A, et al. Genomic profiling of isolated circulating tumor cells from metastatic breast cancer patients. Cancer Res. 2013;73(1):30–40.
73. Fridlyand J, Snijders AM, Ylstra B, Li H, Olshen A, Segraves R, et al. Breast tumor copy number aberration phenotypes and genomic instability. BMC Cancer. 2006;6:96.
74. Paris PL, Kobayashi Y, Zhao Q, Zeng W, Sridharan S, Fan T, et al. Functional phenotyping and genotyping of circulating tumor cells from patients with castration resistant prostate cancer. Cancer Lett. 2009;277(2):164–73.
75. Magbanua MJ, Sosa EV, Scott JH, Simko J, Collins C, Pinkel D, et al. Isolation and genomic analysis of circulating tumor cells from castration resistant metastatic prostate cancer. BMC Cancer. 2012;12:78.
76. Heitzer E, Auer M, Hoffmann EM, Pichler M, Gasch C, Ulz P, et al. Establishment of tumor-specific copy number alterations from plasma DNA of patients with cancer. Int J Cancer. 2013;133(2):346–56.
77. van Beers EH, Joosse SA, Ligtenberg MJ, Fles R, Hogervorst FB, Verhoef S, et al. A multiplex PCR predictor for aCGH success of FFPE samples. Br J Cancer. 2006;94(2):333–7.
78. Amado RG, Wolf M, Peeters M, Van Cutsem E, Siena S, Freeman DJ, et al. Wild-type KRAS is required for panitumumab efficacy in patients with metastatic colorectal cancer. J Clin Oncol. 2008;26(10):1626–34.
79. Paez JG, Janne PA, Lee JC, Tracy S, Greulich H, Gabriel S, et al. EGFR mutations in lung cancer: correlation with clinical response to gefitinib therapy. Science. 2004;304(5676):1497–500.
80. Punnoose EA, Atwal SK, Spoerke JM, Savage H, Pandita A, Yeh RF, et al. Molecular biomarker analyses using circulating tumor cells. PLoS One. 2010; 5(9):e12517.
81. Dharmasiri U, Njoroge SK, Witek MA, Adebiyi MG, Kamande JW, Hupert ML, et al. High-throughput selection, enumeration, electrokinetic manipulation, and molecular profiling of low-abundance circulating tumor cells using a microfluidic system. Anal Chem. 2011;83(6):2301–9.

82. Yang MJ, Chiu HH, Wang HM, Yen LC, Tsao DA, Hsiao CP, et al. Enhancing detection of circulating tumor cells with activating KRAS oncogene in patients with colorectal cancer by weighted chemiluminescent membrane array method. Ann Surg Oncol. 2010;17(2):624–33.
83. Mostert B, Jiang Y, Sieuwerts AM, Wang H, Bolt-de Vries J, Biermann K, et al. KRAS and BRAF mutation status in circulating colorectal tumor cells and their correlation with primary and metastatic tumor tissue. Int J Cancer. 2013;133(1):130–41.
84. Taplin ME, Rajeshkumar B, Halabi S, Werner CP, Woda BA, Picus J, et al. Androgen receptor mutations in androgen-independent prostate cancer: Cancer and Leukemia Group B Study 9663. J Clin Oncol. 2003;21(14):2673–8.
85. Jiang Y, Palma JF, Agus DB, Wang Y, Gross ME. Detection of androgen receptor mutations in circulating tumor cells in castration-resistant prostate cancer. Clin Chem. 2010;56(9):1492–5.
86. Smirnov DA, Zweitzig DR, Foulk BW, Miller MC, Doyle GV, Pienta KJ, et al. Global gene expression profiling of circulating tumor cells. Cancer Res. 2005;65(12):4993–7.
87. Barbazan J, Alonso-Alconada L, Muinelo-Romay L, Vieito M, Abalo A, Alonso-Nocelo M, et al. Molecular characterization of circulating tumor cells in human metastatic colorectal cancer. PLoS One. 2012;7(7):e40476.
88. Sieuwerts AM, Mostert B, Bolt-de Vries J, Peeters D, de Jongh FE, Stouthard JM, et al. mRNA and microRNA expression profiles in circulating tumor cells and primary tumors of metastatic breast cancer patients. Clin Cancer Res. 2011;17(11):3600–18.
89. Munzone E, Nole F, Goldhirsch A, Botteri E, Esposito A, Zorzino L, et al. Changes of HER2 status in circulating tumor cells compared with the primary tumor during treatment for advanced breast cancer. Clin Breast Cancer. 2010;10(5):392–7.

CHAPTER 6

Circulating Cell-Free DNA for Molecular Diagnostics and Therapeutic Monitoring

JULIA A. BEAVER, BEN H. PARK

Introduction 85
Applications in Maternal–Fetal Medicine 86
Applications in Cancer Diagnostics 86
ccfDNA Detection Technologies 89
Additional Medical Applications of ccfDNA 94
Conclusions 94

Introduction

Presence of circulating cell-free DNA (ccfDNA) in human blood was originally discovered in 1948 by Mandel and Métais [1]. As the name implies, it refers to free-floating, "naked" DNA that is found in the blood and is thought to be derived from normal cells as well as cells in various disease states. Interest in the potential medical utility of ccfDNA resurfaced in the last decade when scientists began to explore a use for it in maternal–fetal medicine and oncology. More recently, techniques to identify and quantitate ccfDNA have been applied to other disease processes such as sepsis, myocardial infarction, stroke, and diabetes [2–5].

It has been postulated that ccfDNA is shed into the circulation by macrophage release of necrotic or apoptotic cellular debris [6]. It has also been suggested that direct secretion of ccfDNA into the plasma is possible [7]. Thus, despite its presence in the circulation, the origin(s) of ccfDNA is (or are) not fully known but could involve multiple not mutually exclusive mechanisms. ccfDNA can be found in many different human fluids such as whole blood, serum, plasma, and urine, and ccfDNA fragments range from 70 to 1,200 base pairs in length [8–10]. Researchers have shown that ccfDNA has a variable half-life on the order of 15 min to a few hours and is quickly cleared by the kidney and the liver [10–13]. Patients with conditions such as metastatic cancer, trauma, myocardial infarction, and sepsis have on average a higher concentration of overall ccfDNA than normal controls [9, 14–18]. Potential uses for ccfDNA based assays in the diagnosis, prognosis, and monitoring of various disease states are currently under investigation. Several different assays are being used for the identification and quantification of ccfDNA but currently there is no standard platform.

G.J. Netto and I. Schrijver (eds.), *Genomic Applications in Pathology*,
DOI 10.1007/978-1-4939-0727-4_6, © Springer Science+Business Media New York 2015

Applications in Maternal–Fetal Medicine

In the late 1970s, fetal cells were first discovered in the maternal circulation and subsequent work demonstrated that a small percentage of ccfDNA originating from the fetus could also be found in the maternal blood [19–21]. The advantages of noninvasive diagnostic testing that using maternal–fetal ccfDNA would afford have fueled significant subsequent interest. The size of fetally derived ccfDNA was determined to be <300 base pairs (bp), whereas ccfDNA fragments derived from maternal cells are >300 bp [22]. It appears that the fetal fraction of ccfDNA accounts for approximately 3–6 % of the total ccfDNA population, although studies differ slightly, and as gestation continues the percentage may rise [23, 24].

Given the small percentage of fetal ccfDNA, different technologies with varying sensitivities have been examined to detect and quantify this minority population of DNA. Work using fetus-specific methylated markers, digital polymerase chain reaction (PCR) based technologies, and assays querying for paternal short tandem repeat sequences, has allowed for the examination of fetal aneuploidies and sex determination in multiple reported studies [20, 25, 26]. Fan et al. used direct shotgun sequencing followed by chromosome mapping to establish the overrepresentation or underrepresentation of chromosomes in maternal plasma ccfDNA, thereby identifying potential aneuploidy [27]. Similarly, other investigators have employed next-generation massively parallel sequencing of maternal plasma DNA to detect increased representations of chromosomes 21, 18 and 13 to identify fetuses harboring these chromosomal abnormalities [28, 29]. In addition, advances in detection of ccfDNA from maternal plasma have allowed for Rhesus D genotyping and the detection of paternally inherited genetic disorders [30, 31]. The feasibility of detecting relatively rare fetal DNA molecules in maternal peripheral blood was established with these earlier studies which set the stage for the applications of these technologies in cancer diagnostics.

Applications in Cancer Diagnostics

Using radioimmunoassays, Leon et al. were the first to show that the average amount of ccfDNA in cancer patients was increased compared to healthy individuals without cancer [32]. However, the range of ccfDNA concentrations in cancer patients was found to vary substantially between 0 and >1,000 ng/ml of blood with normal subjects typically exhibiting ccfDNA concentrations between 0 and 100 ng/ml [12, 33–35]. Given the significant overlap of ccfDNA concentrations in normal and cancer patients, it became apparent that the total quantity of ccfDNA could not be used as a reliable diagnostic tool.

Similar to fetal DNA detection from maternal blood, the promise of ccfDNA in cancer diagnostics and monitoring is based on the ability to detect the small population of cell-free plasma tumor DNA (ptDNA) from the larger population of normal ccfDNA through the identification of tumor-specific (somatic) variations. It should be emphasized that the plasma fraction of blood contains *cell-free* ptDNA and this is distinct from current efforts studying circulating tumor cells (CTCs) as cancer biomarkers. Studies have demonstrated that for a patient with a tumor containing approximately 3×10^{10} cells, tumor DNA comprises 3.3 % of the ccfDNA found in the blood stream daily [9]. In addition, multiple groups have demonstrated that the size of ptDNA molecules is smaller than that of ccfDNA derived from normal cells, and typically ranges from 70 to 200 bp [8, 9, 36].

Cancer DNA harbors numerous somatic changes that include mutations, epigenetic alterations, and amplifications, as well as rearrangements resulting from translocations and deletions or insertions. Naturally, ptDNA would also harbor these genetic and epigenetic changes. Specifically, the ptDNA contains the same mutations and genomic rearrangements in tumor suppressor genes or oncogenes which are driving the development and progression of the cancer. In addition, so-called "passenger" mutations or genetic alterations that are likely the result of tumor genetic instability but not of direct functional consequence, are somatic changes that would be represented in ptDNA, therefore both driver and passenger mutations/alterations could serve as potential cancer markers [37].

The ability to use a patient's blood sample to perform a "liquid biopsy" allows for the identification of residual micrometastatic cancer and provides a noninvasive test to query for specific mutations without surgical intervention. In theory, the liquid biopsy would be a real-time assessment of molecular tumor genotype (qualitative) and existing tumor burden (quantitative). The short half-life of ptDNA lends itself to be a reliable marker of tumor burden and possibly of response to therapies. Therefore, the potential applications for clinical oncology that stem from ptDNA detection are vast. This technology has the capacity to completely change the paradigm of how clinicians make decisions regarding adjuvant systemic therapies as well as therapies for metastatic disease. In the adjuvant setting one could theoretically test each patient post-surgery to determine if there is residual micrometastatic disease in order to make an informed assessment of the need for adjuvant systemic treatment and prevent the administration of toxic systemic therapies when not needed. Additionally, if validated, such a technique could guide the substitution of different forms of therapy (e.g., chemotherapy versus hormonal versus biologic) in the adjuvant setting or when treating metastatic disease by triggering a change of regimen when a decrease in "personalized" DNA markers is not achieved. Furthermore, real-time knowledge of the molecular profile of a tumor without the need for a biopsy would also help drive rational therapies, clinical trial enrollment and create new surrogate endpoints allowing for a more rapid pace of drug approval. Until the last few years, the technology to realize these applications had not existed. However, with the introduction of next-generation sequencing (NGS) technologies and improvements in digital and emulsion PCR, there is now the capacity to identify DNA-based genetic biomarkers that are unique to a patient's cancer and to perform subsequent analysis of the patient's plasma to quantify the amount of residual tumor burden via the measurement of ptDNA.

Mutations in proto-oncogenes and tumor suppressor genes found in tumor tissues can be detected in the patient's plasma using the above mentioned technologies. Mutations in *TP53* for example were found in 42.9 % of the plasma DNA samples from patients harboring *TP53* mutations in their tumor [38]. Similarly, the relatively common p.V600E *BRAF* mutation has been shown to be present in ptDNA and has been used to monitor patient response to *BRAF* directed therapy [39]. Several studies have examined mutant *KRAS* in a primary tumor and identified corresponding *KRAS* mutations in the plasma. However, these studies demonstrated varying sensitivities for ptDNA detection ranging from 27 to 100 % [40–42].

In 2005, investigators used a digital PCR based technique termed BEAMing (which stands for its four primary components: Beads, Emulsion, Amplification, and Magnetics, described below) to identify patients with point mutations in mutant *APC* molecules in both early stage and metastatic colorectal cancer patients [9]. The authors found 100 % concordance between *APC* mutations in the plasma and the known solid tumor *APC* mutations of six metastatic colorectal cancer patients. They also analyzed 16 patients with early stage colorectal cancer with known *APC* mutations and found that 63 % had detectable mutant *APC* DNA in their plasma. On average, they described 11.1 % of the total *APC* gene fragments in the plasma of metastatic patients to be mutant compared to only 0.04–0.9 % in early stage patients, explaining the likely reason for decreased sensitivity of detection in that group. Subsequent work by the same investigators examined the plasma of 18 patients undergoing therapy for colorectal cancer and correlated the amount of ptDNA with tumor burden using BEAMing for four genes (*APC*, *PIK3CA*, *TP53*, and *KRAS*) [10]. This work identified a median percentage of 0.18 % of ptDNA when separate samples were analyzed for different mutated genes, finding similar concentrations of mutant genes in ptDNA. These patients were followed after surgery with subsequent blood draws and assessment of their ptDNA during chemotherapy and surveillance. The findings seemed to correlate with clinical status.

Investigators have also evaluated "hotspot" *PIK3CA* mutations (a gene commonly mutated in breast and other cancers) in metastatic breast cancer patients. In a retrospective study, 49 archival matched tumor and plasma samples were examined for exon 9 and 20 hotspot *PIK3CA* mutations using BEAMing for both tumor tissues and plasma. They found 100 % concordance between the presence and type of *PIK3CA* mutations in the tumor and plasma. However, a subsequent prospective study by the same group identified an approximately 70 % concordance of *PIK3CA* mutational status between tumor tissues and peripheral blood. This seemingly disparate result may have been the effect of tumor heterogeneity and clonal evolution, since the prospective study used archived primary cancer tissues along with blood drawn at the time of study entry as the source for tumor mutational and ptDNA analyses, respectively. Blood samples had been drawn concurrently with the tumor tissue acquisition in the retrospective study. Changes in *PIK3CA* mutational status were only seen in patients whose tumors were harvested more than 3 years prior to the blood draw for ptDNA analysis [43]. These results raise concerns regarding the use of archival specimens when assessing mutation status and performing genetic profiling in cancer patients with metastatic disease, as the mutational and genomic spectrum may differ significantly between primary and metastatic sites of disease [44, 45].

More recently, several groups have used BEAMing to identify point mutations involved with disease progression. For example, Taniguchi and colleagues have recently demonstrated the ability to detect second site p.T790M epidermal growth factor receptor (*EGFR*) mutations in non-small-cell lung cancer patients treated with EGFR kinase inhibitors [46]. Interestingly, they also detected the same mutation in a significant fraction of patients who were not treated with these inhibitors suggesting the existence of a minority population of cancer cells that might lead to the emergence of this clonal population upon EGFR kinase inhibitor therapy. In addition, two separate studies reported the use of BEAMing to detect the emergence of *KRAS* mutations that conferred resistance to antibody mediated EGFR targeted therapies [47, 48]. Taken together, these studies suggest that a powerful potential use of assessing ptDNA status in treated cancer patients is the ability to monitor for the emergence of resistant clones with a particular mutation or genotype.

In hematologic malignancies, there are established techniques available to detect minimal residual disease in the blood exploiting tumor-specific DNA rearrangements. For example, in Chronic Myeloid Leukemia (CML) detection of the *BCR-ABL1* fusion transcript by quantitative real-time PCR (qPCR) using PCR primers specific for the fusion transcript has allowed for real-time monitoring of the disease burden and the ability to follow response to treatment using peripheral blood or bone marrow samples [49]. The capacity to perform such assays in solid tumor malignancies has yet to transition into clinical practice given the relative rarity of known recurrent somatic rearrangements in solid tumor malignancies. Because leukemias are by definition blood based diseases, abundance of leukemic cells in the peripheral blood and/or bone marrow facilitates the use of qPCR of fusion transcripts as a reliable measure of disease burden. In contrast, the ability to identify circulating tumor cells (CTCs) for most solid malignancies is still hindered by low sensitivity though newer methods for improving capture and therefore sensitivity of isolating CTCs have shown promise [50].

With the understanding that tumor cells "shed" DNA as ccfDNA, and the advent of NGS technologies, several groups have now demonstrated the ability to identify tumor-specific genetic rearrangements that are patient-specific. The technology of "mate paired end" sequencing can identify many of the genomic alterations found in cancers including mutations, translocations, amplifications, deletions, etc. [51, 52]. In 2008, Campbell et al. identified the presence of multiple patient-specific somatic rearrangements in cancer using massively parallel sequencing [51]. To date, hundreds of solid tumor samples have been subjected to this form of NGS, with rearrangements found in virtually all samples and the majority of samples containing more than ten rearrangements. Using a technology termed PARE for Personalized Analysis of Rearranged Ends (a technique described later in this chapter), investigators from two groups identified somatic rearrangements in primary tumor tissue, designed unique patient-specific PCR primers and were able to detect and quantify personal markers in the plasma of five cancer patients [53, 54]. The sensitivity for detecting rearranged DNA was calculated to be 1 cancer genome equivalent among 390,000 normal genome equivalents. To minimize false negative results both groups recommended the use of multiple somatic rearrangement markers to increase the reliability of detection. Theoretically these markers should be 100 % specific, since each marker is validated to detect only tumor-specific rearrangements.

More recently, further work has verified that genomic rearrangements can be directly identified from the plasma of metastatic cancer patients using NGS and specific bioinformatics criteria [55]. In their study, Leary et al. expand upon the use of NGS by analyzing the copy number of chromosomes found in the plasma of metastatic cancer patients compared to healthy controls. Similar to efforts in fetal medicine using NGS of maternal plasma to query copy number changes of various chromosomes, the study demonstrated a 0.61- to 1.97-fold copy number increase in the plasma of cancer patients compared to normalized controls. Thus, it is possible to detect patients with metastatic cancer compared to normal controls by assessing copy number alterations present in ccfDNA. This appears to be feasible if the percentage of ptDNA compared to ccfDNA is at least 0.75 %. At this level of ptDNA the assay had a sensitivity of >90 % and a specificity of >99 %. However, it should be noted that the sensitivity and specificity of this technique are dependent upon the depth of sequencing data collected, which reflects the number of DNA molecules that are assayed for each individual.

As alluded to above, the differing levels of sensitivity among ptDNA detection studies may reflect the amount of genome equivalents sampled by the investigators as well as the techniques used. With the knowledge that there is a greater amount of ccfDNA (both tumor and normal derived) in metastatic cancer patients compared to early stage cancer patients, increasing the amount of genome equivalents sampled in early stage cancer patients is likely to improve the sensitivity of these assays [32]. In addition, tumor heterogeneity can result in a low clonal frequency of a given mutation within a solid tumor mass. In this situation, wild-type sequences shed from other tumor cells and normal cells may significantly decrease the amount of ptDNA for the given mutation, and will not be reflective of the overall tumor burden [40, 56, 57]. Similar issues with dilution of ptDNA by total ccfDNA have been hypothesized to cause difficulties in the detection of loss of heterozygosity (LOH) in ptDNA in several studies [12, 58–60]. As discussed above, the use of multiple somatic alterations as markers can mitigate some of these concerns. However, investigators have also considered the possibility of using stool, urine, and increased volumes of plasma to improve the sensitivity of detecting rare mutations within ptDNA [61].

The ratio of long to short DNA fragments (DNA integrity) is also being studied as a possible biomarker of tumor presence and tumor burden. It is technically feasible to detect noncoding repetitive DNA sequences such as ALU sequences in ccfDNA, and the length and ratio of these markers can determine the DNA integrity within ccfDNA. This has led to studies examining whether changes in these markers are prognostic and/or diagnostic in several types of cancers [62, 63]. Testing for DNA integrity could be broadly applicable for many cancer subtypes and therefore could also improve sensitivities of current assays. In addition, studies examining epigenetic alterations in the plasma of cancer patients, specifically detection of promoter hypermethylation by methylation-specific PCR have been performed in various cancer subtypes and hold significant promise as another biomarker of cancer burden [64, 65].

ccfDNA Detection Technologies

Total ccfDNA can be isolated from the blood, plasma, and urine of patients. The majority of ccfDNA studies in maternal–fetal medicine have used column-based extraction methods and/or other automated techniques [66–68]. Once isolated, ccfDNA can be measured by fluorescence-based methods utilizing PicoGreen staining or UV spectrometry, or by qPCR with detection by intercalating dyes such as SYBR green or with dual labeled fluorescent/quencher probes (for example, hydrolysis probe technology). For the purpose of detecting variant molecules within ccfDNA such as fetal DNA or cancer ptDNA, various methods have been developed. BEAMing, Droplet Digital PCR, and PARE are among such methods which allow for the detection and quantification of minority ccfDNA populations such as ptDNA and will be described below.

BEAMing

Incorporating an emulsion digital PCR based technology, the BEAMing technique is able to identify and quantitate rare genetic molecules found in a larger population of normal or wild-type DNA molecules [10, 69–71]. In essence, this technology allows for the rapid, massively, and in parallel assessment of individual DNA molecules to separate rare genetic variants from a "sea" of normal DNA. Thus, BEAMing is a method of digital PCR, which assesses individual DNA molecules after serial dilution and/or separation such that the read-out yields individual reactions with a binary result of either the presence or absence of variant DNA. Because ptDNA represents only a small fraction of total ccfDNA, digital PCR techniques, and BEAMing in particular, were developed to separate this minority population so that mutations in ptDNA could be readily detected. For applications in oncology, BEAMing has successfully been used with high sensitivity and specificity to detect somatic point mutations within ptDNA and was shown to correlate the amount of ptDNA with disease burden as described earlier in this chapter [9, 10, 69, 72].

The technology of BEAMing allows for single molecule PCRs to be performed on magnetic beads in water-in-oil emulsions with genetic variants subsequently quantified by flow cytometry. The complete method of BEAMing consists of six parts as initially described by Dressman et al. (Fig. 6.1) and subsequently updated and improved upon by Diehl et al. [10, 69]. What follows is a brief description of the various steps:

Quantification of human genomic DNA from plasma samples: Plasma is subjected to qPCR in order to quantify the amount of total plasma DNA using a modified version of a human LINE-1 qPCR [72].

Preamplification: Primer sets are designed which correspond to the candidate genes mutated in the original tumor sample. An initial PCR amplification of multiple loci is performed for these regions using the isolated plasma DNA as the template material. Subsequently, in a second amplification, individual amplicons are generated by nested primer pairs to create the DNA template.

Step 1: Coupling oligonucleotides to beads: 10 μM oligonucleotides are attached to streptavidin bound beads by incubation in a bind-and-wash buffer using a magnet. The oligonucleotides are adjusted to have a dual biotin group at the 5′ end and average 41 base pairs in length. After incubation each bead binds approximately 10^5 oligonucleotide molecules.

Step 2: Preparing microemulsions: Microemulsions are prepared using a combination of oil and aqueous phases in a 2 ml round bottom cryogenic vial [70, 73, 74]. PCR reagents, previously amplified DNA template, and oligonucleotide-coupled beads are added to the aqueous phase prior to drop wise combination of the two phases. The droplets should appear similar and all beads should be enclosed in 3–9 μm diameter droplets. In addition 5–35 % of the beads should contain PCR products to optimize the technique.

Step 3: PCR cycling: The water-in-oil emulsions are added to a PCR plate at approximately 80 μl into eight wells. The first few rounds of PCR are facilitated by additional forward primer added to the solution.

Step 4: Magnetic capture of beads: After amplification the emulsions are broken apart with nonionic detergents, the beads are pelleted and the oil and a portion of the aqueous solution are removed. The remaining solution is removed through attachment of the beads to a magnet and subsequent washes. It is estimated that more than 10,000 PCR products are present on each bead.

Step 5: Sequence differentiation: This step employs the hybridization of fluorescein-conjugated or biotin-conjugated oligonucleotides to the PCR amplified bead product for analysis of the variants. One probe is made specific for

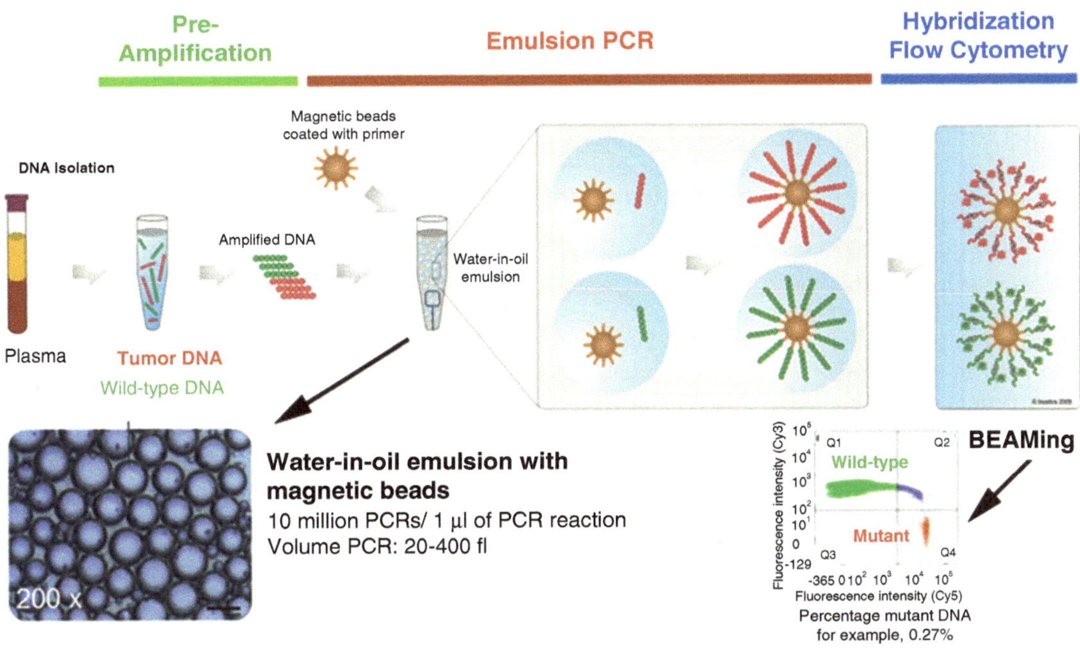

Figure 6-1 BEAMing. The workflow for the BEAMing process is shown. Plasma DNA is extracted and pre-amplified prior to being partitioned into water-in-oil emulsions with magnetic beads. PCR amplification is then carried out on the beads within emulsions, and then emulsions are broken to release amplified DNA on the beads. Fluorescent probes specific for mutant or wild-type DNA are then hybridized to the PCR products and subjected to flow cytometry (from Inostics)

the wild-type allele and another is made specific for the variant allele. The probes are bound to their respective targets through various washing and incubation steps.

Step 6: Flow cytometry: Magnetic flow cytometry and sorting of the beads is performed to quantify the variant and the wild-type population. Multiple platforms have been used including BD Bioscience FACSan, LSR I and IIm FACSCalibur, FACSAria, and fluorescence microscopy.

Whereas BEAMing provides a sensitive and specific platform to quantify a very small fraction of mutant ptDNA fragments, it poses some challenges. Given the fixed number of DNA fragments in each sample, as well as BEAMing's limitation of detection, it is possible that in early stage cancer patients, ptDNA may not be detectable due to a low level of genome equivalents. In addition, should the initial PCR product have errors from the DNA polymerase those errors would persist in the ultimate detection of mutant DNA. This could affect sensitivity due to mismatches between the probe and the target DNA molecule both at the mutation being queried, and also in adjacent nucleotides that could affect annealing properties of a probe. Another potential limitation of BEAMing is the time for developing assays for every mutation. This could be relevant for the use of BEAMing to measure residual tumor burden, because testing of multiple somatic alterations will likely be required due to the inherent genetic instability and therefore heterogeneity of most human cancers. Thus, in such a scenario, BEAMing would require the development and validation of multiple somatic markers with each mutation necessitating optimization, adding further complexity and costs.

Droplet Digital PCR

Droplet digital PCR is another variation of emulsion based digital PCR based technology that, in principle, is similar to BEAMing. Analogous to BEAMing, the basis behind Droplet digital PCR is that the target sample is separated into partitions of single molecules which individually undergo PCR. The major difference is in the partitioning methods. While BEAMing uses microliter emulsions that are generated using mechanical means, droplet technologies use specialized capillary equipment to generate emulsions that are in the nanoliter and even picoliter size range. In theory, this allows for more precise and accurate quantification of nucleic acid variants and may afford higher sensitivities in certain situations. These next-generation PCR platforms are capable of separating DNA samples into tens of thousands to millions of individual droplets with a random distribution. The target alleles in the sample are then amplified and using a dual labeled probe each droplet provides a fluorescently positive or negative signal via hydrolysis from polymerase activity, allowing for quantification of a given target molecule. There are currently several different droplet digital PCR platforms each with advantages and disadvantages with respect to sensitivity, costs, and other factors.

Microfluidic Based Digital PCR

Although less mature than emulsion based digital PCR, the use of microfluidics for digital PCR analysis has been gaining in popularity with advances in technology. Due to its less developed technology relative to emulsion based digital PCR, this chapter will only provide a brief overview of this nascent field. Microfluidic technology is not new, but its application towards digital PCR has only recently been applied compared to other competing methods. Microfluidics involves the separation of fluids into progressively smaller channels and chambers such that it is possible to force single DNA molecules into specified compartments with the use of "one way gates." In essence, the partitioning of DNA molecules that normally occurs with emulsions is instead achieved through the use of microfluidic "chips" or arrays. These arrays require exquisite design and implementation considerations, and heretofore the ability to create such reagents at a resolution to allow for a massive in parallel approach for digital PCR was not feasible. However, similar to emulsion based PCR, improvements in technology along with decreased costs have resulted in viable platforms for microfluidics based digital PCR, with commercial launches imminent.

PARE

PARE (personalized analysis of rearranged ends) (Fig. 6.2) was developed to exploit somatic rearrangements in cancer DNA for biomarker development. The principles of PARE are similar to the use of qPCR analysis for CML using the unique fusion transcript *BCR-ABL1* that is only present in cancer cells but not in normal cells within the affected individual. A prime difference, however, is the use of ccfDNA, rather than cellular derived RNA, for measuring tumor burden. With the advent of next-generation mate paired end sequencing, Leary et al. demonstrated that unique somatic rearrangements resulting from translocations, amplifications, and deletions could be identified in a patient's tumor [54]. Exploiting this knowledge to develop personalized biomarkers, they showed that PCR primers that amplified only these rearrangements could be used in qPCR assays to detect ptDNA from metastatic breast and colon cancer patients. Moreover and similar to BEAMing, quantitative ptDNA measurements could be used to follow responses to systemic therapies. The advantage of this approach is that specificity for tumor DNA comes from the primers used, similar to its *BCR-ABL1* transcript counterpart in CML. This obviates the need to design specific probes and assays for every somatic cancer alteration that is to be used as a biomarker of residual disease and response to therapy. A potential disadvantage to this approach, however, is that the identification of somatic rearrangements presents more challenges in data interpretation and bioinformatics compared to detecting the point mutations that are used for BEAMing. As an example, the costs of performing whole-exome sequencing have dramatically decreased in recent years, and this approach therefore lends itself to rapid and easily verifiable somatic mutation detection that would be needed for BEAMing and other digital PCR approaches. In contrast, PARE does not necessarily rely upon digital PCR technologies (although likely will require digital PCR for low level ptDNA detection), but whole-exome NGS will likely be inadequate to identify somatic rearrangements in cancer due to the limits of this technology, and therefore, whole-genome sequencing will be needed, requiring

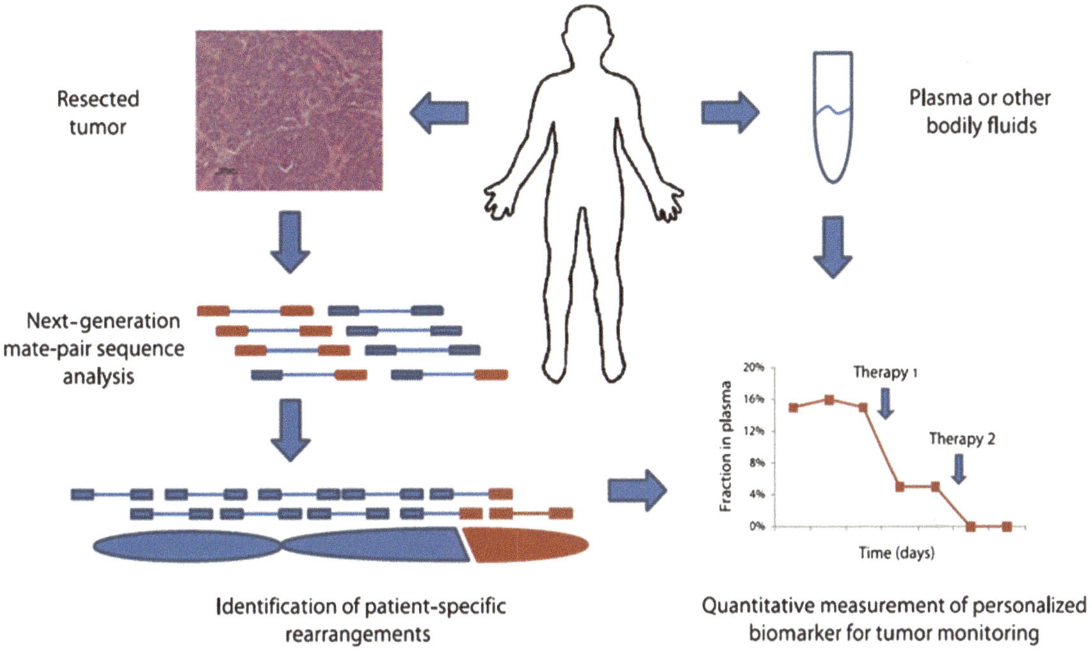

Figure 6-2 PARE for monitoring ptDNA. Cancer DNA is extracted from resected tumor tissues and then used for NGS to identify somatic genomic rearrangements (from translocations, deletions, amplifications). Plasma is also obtained from the patient and ptDNA extracted. Cancer-specific PCR primers that will amplify only structural rearrangements are then used to query ptDNA via qPCR. ptDNA plasma tumor DNA

significantly more time and costs. It should be noted that these two platforms are not necessarily competing, and indeed the use of both somatic rearrangements and mutations could be complementary for the use of ptDNA as personalized cancer biomarkers. The PARE technique is detailed in the supplementary information of McBride et al. [53]. See Fig. 6.2 [54].

What follows is a brief description of the various steps:

Step 1: Identification of tumor-specific genomic rearrangements: Genomic DNA is extracted via standard protocols from the original tumor sample. Appropriate libraries are created from this genomic DNA according to the sequencing platform to be utilized. Massively parallel, paired-end sequencing is then performed according to the technology being used. Campbell et al. described paired-end sequencing of at least 25–35 bp from either end as generating the ideal coverage for detecting appropriate rearrangements [51]. Once sequenced, alignment is performed to the reference human genome. Using bioinformatics and other computational algorithms, putative genomic rearrangements can be identified.

Step 2: Identifying tumor-specific rearrangements: A variety of bioinformatics tools can be utilized to identify putative intrachromosomal and interchromosomal somatic rearrangements. At least two reads spanning the same rearrangement should be considered, for example paired end reads on the same chromosome mapping <100 kb apart or both ends mapping within 50 kb of a change in copy number are indicators of possible rearrangements. Once putative rearrangements are identified, PCR primers on either side of the presumed rearrangements can be designed and sequencing of PCR products in both tumor and normal can be performed to confirm a bona fide somatic rearrangement.

Step 3: Design of primers to detect tumor-specific rearrangements: Each confirmed tumor-specific somatic rearrangement is examined for suitability as a marker by assessing copy number changes, uniqueness of surrounding DNA, and its possible role as a causal gene/noncoding region in cancer and/or drug resistance. Ideally three to four or more somatic rearrangements should be identified per patient as genetic heterogeneity and tumor evolution can result in the loss of a given marker over time [43]. Based on the approach of quantifying *IGH* rearrangements in acute lymphoblastic leukemia, the next step entails the use of nested PCR to amplify the rearrangements [75].

Step 4: Extraction of free DNA from serum and subsequent quantification of somatic rearrangements: Plasma has been shown to be a richer and higher quality source of ccfDNA than serum, but plasma used for ccfDNA analysis must be spun down after collection in EDTA within 2 h of venipuncture. The exact amount of plasma to collect is unclear due to the wide range of ccfDNA found in patients with early versus late stages of cancer. However, in general, the greater the volume of blood in a given sample, the more genome equivalents of ccfDNA are present. Therefore, the potential and sensitivity of detecting ptDNA is directly affected by the number of genome equivalents assayed, and therefore secondarily affected by the amount of blood drawn from an individual patient. After genomic DNA extraction either qPCR or digital PCR can be employed depending on the platform(s) used, to identify and quantitate the amount of ptDNA in the patient at the time of blood draw

In summary, PARE offers great promise in individualized cancer medicine. Genetic biomarkers assayed by the PARE technique can be created to detect disease burden in the plasma of early stage and metastatic cancer patients. However, issues similar to those encountered with BEAMing exist in terms of sensitivity of the technique, and the quantity of ptDNA sampled, if low, may lead to false negative results. Whereas the current cost and speed of NGS could be perceived to be prohibitive for PARE use in clinical applications, the anticipated decreasing costs and improvements in technology will likely greatly enhance the ease and applicability of this technique.

Additional Medical Applications of ccfDNA

In addition to cancer and maternal–fetal medicine, there are other potential uses for ccfDNA in medical diagnostics. However, the majority of these applications involve the simple quantification of total ccfDNA as there is no mutant or minority variant DNA population to analyze. For example, investigators have studied patients who have undergone traumas or burns and correlated the amount of ccfDNA with the severity of injury, patient outcome and length of hospital stay [16, 17, 76]. Others have analyzed the ability of ccfDNA levels to predict the need for ventilator use in an ICU setting with two of the three studies demonstrating increased mortality and ventilator requirements in patients with high levels of ccfDNA [4, 18, 77]. In addition, ccfDNA levels were found to predict occurrence of sepsis. Studies attempting to characterize the change in ccfDNA associated with myocardial infarction for prognostic purposes have also been conducted. For example, one study demonstrated that higher levels of ccfDNA corresponded with increasing levels of myocardial damage and subsequent cardiac outcome [78]. A separate study correlated classic cardiac ischemic markers such as troponin and creatinine kinase with increasing levels of ccfDNA [3]. Similar research has also been carried out in patients with stroke, and it was shown that levels of ccfDNA in these patients have prognostic significance [2]. Finally, ccfDNA levels have been examined to better assess sickle cell crisis. In one study, a higher level of ccfDNA was found in sickle cell patients diagnosed with acute pain crisis [79]. Further research in these areas will undoubtedly lead to new and better methods of assessing ccfDNA for the improved management of human health and disease.

Conclusions

Future directions for the field of digital PCR and nucleic acid detection include improvements in technology, potential new assays to measure circulating nucleic acids from body fluids and tissues, and new applications of these methods to other disease states. From a technical standpoint, the use of additional circulating nucleic acids is being examined as an adjuvant or superior method of detecting residual disease and mutation profiling in cancer and other disease states. Specifically, circulating messenger RNA (mRNA) has been shown to be present in the serum of humans although the cellular origin of circulating mRNA is less clear than that of ccfDNA. Circulating cell free mRNA can be detected using microarray technologies or reverse transcription qPCR [80]. Investigators are currently examining whether circulating mRNA could aid in differentiating between bacterial versus viral infections in critically ill patients. In addition, the ability of donor-specific ccfDNA and circulating mRNA to predict graft rejection in transplant patients has been explored. Interestingly, for these assays, it appears that urinary mRNA may afford a more reliable predictor of rejection [81]. Circulating mRNA is also being investigated in the early diagnosis of diabetic retinopathy and diabetic neuropathy. By screening for organ-specific mRNA in the plasma, investigators are attempting to diagnose these diabetic complications sooner, with the hope that earlier intervention would lead to improved outcomes [5, 82]. Finally, microRNAs (miRNAs), which are noncoding RNA species that can regulate gene expression of coding genes, have been described in the serum of cancer patients with B cell lymphoma and serum miRNA levels have also been shown to correlate with metastases [83, 84].

Although great progress has been made in exploiting ccfDNA for cancer and other disease states, there is much room for further discovery and progress. A current critical barrier is the difficulty in detecting a relatively small percentage of mutant or variant molecules in a majority fraction of associated normal/wild-type ccfDNA. Although this chapter has described some of the current methods for quantification and detection of these small populations of ccfDNA, there is currently neither an industry standard nor widespread clinical acceptance for the use of ccfDNA, nor a uniformly agreed upon platform. Future studies are needed to establish the best techniques to quantify, detect, and monitor ccfDNA, and appropriate criteria for ccfDNA surveillance will need to be developed with prospective clinical trials. Through standardization and improved detection technologies, the future holds great promise for the development and implementation of clinical assays that will enable ccfDNA to help clinicians and their patients make better and more informed therapeutic decisions.

REFERENCES

1. Mandel P, Metais P. Les acides nucleiques du plasma sanguin chez l'homme. C R Acad Sci Paris. 1948;142:241–3.
2. Lam NY, et al. Plasma DNA as a prognostic marker for stroke patients with negative neuroimaging within the first 24 h of symptom onset. Resuscitation. 2006;68(1):71–8.
3. Antonatos D, et al. Cell-free DNA levels as a prognostic marker in acute myocardial infarction. Ann N Y Acad Sci. 2006;1075:278–81.
4. Saukkonen K, et al. Association of cell-free plasma DNA with hospital mortality and organ dysfunction in intensive care unit patients. Intensive Care Med. 2007;33(9):1624–7.
5. Sandhu HS, et al. Measurement of circulating neuron-specific enolase mRNA in diabetes mellitus. Ann N Y Acad Sci. 2008;1137:258–63.
6. Choi JJ, Reich III CF, Pisetsky DS. The role of macrophages in the in vitro generation of extracellular DNA from apoptotic and necrotic cells. Immunology. 2005;115(1):55–62.
7. Stroun M, et al. The origin and mechanism of circulating DNA. Ann N Y Acad Sci. 2000;906:161–8.
8. Jahr S, et al. DNA fragments in the blood plasma of cancer patients: quantitations and evidence for their origin from apoptotic and necrotic cells. Cancer Res. 2001;61(4):1659–65.
9. Diehl F, et al. Detection and quantification of mutations in the plasma of patients with colorectal tumors. Proc Natl Acad Sci U S A. 2005;102(45):16368–73.
10. Diehl F, et al. Circulating mutant DNA to assess tumor dynamics. Nat Med. 2008;14(9):985–90.
11. Lo YM, et al. Rapid clearance of fetal DNA from maternal plasma. Am J Hum Genet. 1999;64(1): 218–24.
12. Fleischhacker M, Schmidt B. Circulating nucleic acids (CNAs) and cancer—a survey. Biochim Biophys Acta. 2007;1775(1):181–232.
13. Emlen W, Mannik M. Effect of DNA size and strandedness on the in vivo clearance and organ localization of DNA. Clin Exp Immunol. 1984;56(1):185–92.
14. Chang CP, et al. Elevated cell-free serum DNA detected in patients with myocardial infarction. Clin Chim Acta. 2003;327(1–2):95–101.
15. Wimberger P, et al. Impact of platinum-based chemotherapy on circulating nucleic acid levels, protease activities in blood and disseminated tumor cells in bone marrow of ovarian cancer patients. Int J Cancer. 2011;128(11):2572–80.
16. Lo YM, et al. Plasma DNA as a prognostic marker in trauma patients. Clin Chem. 2000;46(3):319–23.

17. Chiu TW, et al. Plasma cell-free DNA as an indicator of severity of injury in burn patients. Clin Chem Lab Med. 2006;44(1):13–7.
18. Rhodes A, et al. Plasma DNA concentration as a predictor of mortality and sepsis in critically ill patients. Crit Care. 2006;10(2):R60.
19. Herzenberg LA, et al. Fetal cells in the blood of pregnant women: detection and enrichment by fluorescence-activated cell sorting. Proc Natl Acad Sci U S A. 1979;76(3):1453–5.
20. Lo YM, et al. Presence of fetal DNA in maternal plasma and serum. Lancet. 1997;350(9076):485–7.
21. Dennis Lo YM, Chiu RW. Prenatal diagnosis: progress through plasma nucleic acids. Nat Rev Genet. 2007;8(1):71–7.
22. Li Y, et al. Cell-free DNA in maternal plasma: is it all a question of size? Ann N Y Acad Sci. 2006; 1075:81–7.
23. Lo YM, et al. Quantitative analysis of fetal DNA in maternal plasma and serum: implications for noninvasive prenatal diagnosis. Am J Hum Genet. 1998;62(4):768–75.
24. Lun FM, et al. Microfluidics digital PCR reveals a higher than expected fraction of fetal DNA in maternal plasma. Clin Chem. 2008;54(10):1664–72.
25. Vainer OB, et al. A new Y chromosome marker for noninvasive fetal gender determination. Ann N Y Acad Sci. 2008;1137:157–61.
26. Vecchione G, et al. Fetal sex identification in maternal plasma by means of short tandem repeats on chromosome x. Ann N Y Acad Sci. 2008;1137:148–56.
27. Fan HC, et al. Noninvasive diagnosis of fetal aneuploidy by shotgun sequencing DNA from maternal blood. Proc Natl Acad Sci U S A. 2008;105(42): 16266–71.
28. Chiu RW, et al. Noninvasive prenatal diagnosis of fetal chromosomal aneuploidy by massively parallel genomic sequencing of DNA in maternal plasma. Proc Natl Acad Sci U S A. 2008;105(51):20458–63.
29. Chiu RW, Lo YM. Clinical applications of maternal plasma fetal DNA analysis: translating the fruits of 15 years of research. Clin Chem Lab Med. 2013; 51(1):1–8.
30. Lo YM. Fetal RhD genotyping from maternal plasma. Ann Med. 1999;31(5):308–12.
31. Li Y, et al. Detection of paternally inherited fetal point mutations for beta-thalassemia using size-fractionated cell-free DNA in maternal plasma. JAMA. 2005;293(7):843–9.
32. Leon SA, et al. Free DNA in the serum of cancer patients and the effect of therapy. Cancer Res. 1977;37(3):646–50.
33. Allen D, et al. Role of cell-free plasma DNA as a diagnostic marker for prostate cancer. Ann N Y Acad Sci. 2004;1022:76–80.
34. Chun FK, et al. Circulating tumour-associated plasma DNA represents an independent and informative predictor of prostate cancer. BJU Int. 2006; 98(3):544–8.
35. Schwarzenbach H, et al. Detection and monitoring of cell-free DNA in blood of patients with colorectal cancer. Ann N Y Acad Sci. 2008;1137:190–6.
36. Giacona MB, et al. Cell-free DNA in human blood plasma: length measurements in patients with pancreatic cancer and healthy controls. Pancreas. 1998; 17(1):89–97.
37. Chen X, et al. Detecting tumor-related alterations in plasma or serum DNA of patients diagnosed with breast cancer. Clin Cancer Res. 1999;5(9): 2297–303.
38. Garcia JM, et al. Extracellular tumor DNA in plasma and overall survival in breast cancer patients. Genes Chromosomes Cancer. 2006;45(7):692–701.
39. Shinozaki M, et al. Utility of circulating B-RAF DNA mutation in serum for monitoring melanoma patients receiving biochemotherapy. Clin Cancer Res. 2007;13(7):2068–74.
40. Castells A, et al. K-ras mutations in DNA extracted from the plasma of patients with pancreatic carcinoma: diagnostic utility and prognostic significance. J Clin Oncol. 1999;17(2):578–84.
41. Kopreski MS, et al. Somatic mutation screening: identification of individuals harboring K-ras mutations with the use of plasma DNA. J Natl Cancer Inst. 2000;92(11):918–23.
42. Dianxu F, et al. A prospective study of detection of pancreatic carcinoma by combined plasma K-ras mutations and serum CA19-9 analysis. Pancreas. 2002;25(4):336–41.
43. Higgins MJ, et al. Detection of tumor PIK3CA status in metastatic breast cancer using peripheral blood. Clin Cancer Res. 2012;18(12):3462–9.
44. Gerlinger M, et al. Intratumor heterogeneity and branched evolution revealed by multiregion sequencing. N Engl J Med. 2012;366(10):883–92.
45. Yachida S, et al. Distant metastasis occurs late during the genetic evolution of pancreatic cancer. Nature. 2010;467(7319):1114–7.
46. Taniguchi K, et al. Quantitative detection of EGFR mutations in circulating tumor DNA derived from lung adenocarcinomas. Clin Cancer Res. 2011;17(24): 7808–15.
47. Diaz Jr LA, et al. The molecular evolution of acquired resistance to targeted EGFR blockade in colorectal cancers. Nature. 2012;486(7404):537–40.
48. Misale S, et al. Emergence of KRAS mutations and acquired resistance to anti-EGFR therapy in colorectal cancer. Nature. 2012;486(7404):532–6.
49. Branford S. Chronic myeloid leukemia: molecular monitoring in clinical practice. Hematology Am Soc Hematol Educ Program. 2007;2007:376–83.
50. Yu M, et al. Circulating breast tumor cells exhibit dynamic changes in epithelial and mesenchymal composition. Science. 2013;339(6119):580–4.
51. Campbell PJ, et al. Identification of somatically acquired rearrangements in cancer using genome-wide massively parallel paired-end sequencing. Nat Genet. 2008;40(6):722–9.
52. Stephens PJ, et al. Complex landscapes of somatic rearrangement in human breast cancer genomes. Nature. 2009;462(7276):1005–10.
53. McBride DJ, et al. Use of cancer-specific genomic rearrangements to quantify disease burden in plasma from patients with solid tumors. Genes Chromosomes Cancer. 2010;49(11):1062–9.
54. Leary RJ, et al. Development of personalized tumor biomarkers using massively parallel sequencing. Sci Transl Med. 2010;2(20):20ra14.
55. Leary RJ, et al. Detection of chromosomal alterations in the circulation of cancer patients with whole-genome sequencing. Sci Transl Med. 2012;4(162): 162ra154.

56. Ryan BM, et al. A prospective study of circulating mutant KRAS2 in the serum of patients with colorectal neoplasia: strong prognostic indicator in postoperative follow up. Gut. 2003;52(1):101–8.
57. Wang S, et al. Potential clinical significance of a plasma-based KRAS mutation analysis in patients with advanced non-small cell lung cancer. Clin Cancer Res. 2010;16(4):1324–30.
58. Schwarzenbach H, et al. Cell-free tumor DNA in blood plasma as a marker for circulating tumor cells in prostate cancer. Clin Cancer Res. 2009;15(3):1032–8.
59. Boddy JL, et al. Prospective study of quantitation of plasma DNA levels in the diagnosis of malignant versus benign prostate disease. Clin Cancer Res. 2005;11(4):1394–9.
60. Schwarzenbach H, et al. Comparative evaluation of cell-free tumor DNA in blood and disseminated tumor cells in bone marrow of patients with primary breast cancer. Breast Cancer Res. 2009;11(5):R71.
61. Diehl F, et al. Analysis of mutations in DNA isolated from plasma and stool of colorectal cancer patients. Gastroenterology. 2008;135(2):489–98.
62. Umetani N, et al. Prediction of breast tumor progression by integrity of free circulating DNA in serum. J Clin Oncol. 2006;24(26):4270–6.
63. Umetani N, et al. Increased integrity of free circulating DNA in sera of patients with colorectal or periampullary cancer: direct quantitative PCR for ALU repeats. Clin Chem. 2006;52(6):1062–9.
64. Ellinger J, et al. CpG island hypermethylation in cell-free serum DNA identifies patients with localized prostate cancer. Prostate. 2008;68(1):42–9.
65. Taback B, Saha S, Hoon DS. Comparative analysis of mesenteric and peripheral blood circulating tumor DNA in colorectal cancer patients. Ann N Y Acad Sci. 2006;1075:197–203.
66. Clausen FB, et al. Improvement in fetal DNA extraction from maternal plasma. Evaluation of the NucliSens magnetic extraction system and the QIAamp DSP virus kit in comparison with the QIAamp DNA blood mini kit. Prenat Diagn. 2007;27(1):6–10.
67. Rodriguez de Alba M, et al. Noninvasive prenatal diagnosis of monogenic disorders. Expert Opin Biol Ther. 2012;12 Suppl 1:S171–9.
68. Legler TJ, et al. Fetal DNA: strategies for optimal recovery. Methods Mol Biol. 2008;444:209–18.
69. Dressman D, et al. Transforming single DNA molecules into fluorescent magnetic particles for detection and enumeration of genetic variations. Proc Natl Acad Sci U S A. 2003;100(15):8817–22.
70. Diehl F, et al. BEAMing: single-molecule PCR on microparticles in water-in-oil emulsions. Nat Methods. 2006;3(7):551–9.
71. Li M, et al. BEAMing up for detection and quantification of rare sequence variants. Nat Methods. 2006;3(2):95–7.
72. Rago C, et al. Serial assessment of human tumor burdens in mice by the analysis of circulating DNA. Cancer Res. 2007;67(19):9364–70.
73. Tawfik DS, Griffiths AD. Man-made cell-like compartments for molecular evolution. Nat Biotechnol. 1998;16(7):652–6.
74. Ghadessy FJ, Ong JL, Holliger P. Directed evolution of polymerase function by compartmentalized self-replication. Proc Natl Acad Sci U S A. 2001;98(8):4552–7.
75. Morley AA, et al. Sensitive and specific measurement of minimal residual disease in acute lymphoblastic leukemia. J Mol Diagn. 2009;11(3):201–10.
76. Lam NY, et al. Time course of early and late changes in plasma DNA in trauma patients. Clin Chem. 2003;49(8):1286–91.
77. Wijeratne S, et al. Cell-free plasma DNA as a prognostic marker in intensive treatment unit patients. Ann N Y Acad Sci. 2004;1022:232–8.
78. Rainer TH, et al. Plasma beta-globin DNA as a prognostic marker in chest pain patients. Clin Chim Acta. 2006;368(1–2):110–3.
79. Vasavda N, et al. Circulating DNA: a potential marker of sickle cell crisis. Br J Haematol. 2007;139(2):331–6.
80. O'Driscoll L, et al. Feasibility and relevance of global expression profiling of gene transcripts in serum from breast cancer patients using whole genome microarrays and quantitative RT-PCR. Cancer Genomics Proteomics. 2008;5(2):94–104.
81. Muthukumar T, et al. Messenger RNA for FOXP3 in the urine of renal-allograft recipients. N Engl J Med. 2005;353(22):2342–51.
82. Shalchi Z, et al. Retina-specific mRNA in the assessment of diabetic retinopathy. Ann N Y Acad Sci. 2008;1137:253–7.
83. Roth C, et al. Circulating microRNAs as blood-based markers for patients with primary and metastatic breast cancer. Breast Cancer Res. 2010;12(6):R90.
84. Lawrie CH, et al. Detection of elevated levels of tumour-associated microRNAs in serum of patients with diffuse large B-cell lymphoma. Br J Haematol. 2008;141(5):672–5.

PART II

Practice Related Aspects of Clinical Genomics

CHAPTER 7

Genomic Pathology: Training for New Technology

RICHARD L. HASPEL

Introduction 101

Integration of Genomic Testing into Clinical Care 101

Are Physicians Prepared for the Genomic Era? 103

Current Physician Training in Genomics 103

The Important Role of Pathologists in the Genomic Era 104

Single Program Approaches to Genomic Pathology Training 105

A National Approach to Genomic Pathology Education 106

Future Directions 108

Conclusions 108

Acknowledgement 109

Introduction

Sequencing of the first human genome took over 10 years and cost more than $2 billion [1]. Current massively parallel next-generation methods allow a whole genome to be sequenced in weeks at costs under $10,000 [2]. Pathologists, as the directors of clinical laboratories, have the expertise to effectively translate genomic technology to patient care. To play this important role, pathologists must be trained in genomic methods and result interpretation. This chapter provides evidence demonstrating the need for genomic pathology education, addresses the progress to date of several educational initiatives, and suggests possible ways to improve future training.

Much of molecular pathology involves testing for single gene variants (e.g., *BRCA*). For the purpose of this chapter, genomics refers to analysis of large portions of the genome with a single "test." Aside from the whole genome, only gene-coding regions (exome) or expressed genes (transcriptome) can be sequenced. Chip-based testing, as well as other approaches, can be utilized in the analysis of hundreds of genes, millions of single-nucleotide polymorphisms (SNPs) or copy number variation across the genome.

Integration of Genomic Testing into Clinical Care

Genomic testing is being incorporated into almost all areas of medicine. In oncology, genomic analysis of tumors has already led to personalized chemotherapy, as exemplified by the following cases. In 2007, a patient was diagnosed with an oral adenocarcinoma [3]. Despite excision and adjuvant radiation, the tumor metastasized to the lung. Although treated with an epidermal growth factor receptor (EGFR) tyrosine kinase inhibitor (based on increased EGFR immunohistochemical (IHC) staining), the tumor

G.J. Netto and I. Schrijver (eds.), *Genomic Applications in Pathology*,
DOI 10.1007/978-1-4939-0727-4_7, © Springer Science+Business Media New York 2015

continued to grow. Lacking additional chemotherapeutic options, whole-genome and -transcriptome sequencing was performed on a lung biopsy specimen. This analysis demonstrated upregulation of the *RET* oncogene, and this finding was confirmed with fluorescent in situ hybridization (FISH) analysis as well as IHC staining. Subsequent treatment with a RET tyrosine kinase inhibitor led to stabilization of disease for 4 months. Upon disease progression, sequencing of a new biopsy specimen revealed mutations that could bypass the RET inhibition.

In a second illustrative oncology case, a patient developed what appeared, by morphology, to be acute promyelocytic leukemia (APML) [4]. The corresponding *PML-RAR* rearrangement, however, could not be detected using a standard FISH assay. Using next-generation sequencing (NGS) methods, a cytogenetically cryptic *PML-RAR* fusion was identified and this result was subsequently confirmed by FISH and polymerase chain reaction (PCR) assays. Determination of the genetic basis of the disease led to appropriate treatment with all-trans retinoic acid. The entire diagnostic process was completed in approximately 7 weeks.

In addition to whole-genome sequencing, gene-panels are becoming more commonly used in a variety of cancers. For breast cancer, both a 21-gene and a 70-gene assay performed on tumor samples are commercially available to provide information regarding risk of recurrence and possible need for chemotherapy [5]. A 167-gene assay has also been developed to help determine appropriate management of cytologically indeterminate thyroid nodules [6]. A 13-gene panel of oncogenes has been used to guide pharmacologic management of cancer patients. In a prospective study of salivary duct carcinoma cases, the assay influenced treatment decisions in six of eight patients tested [7].

Genomic technology has also been applied to non-neoplastic diseases. A 15-month-old presented with intractable inflammatory bowel disease (IBD) requiring multiple surgical interventions [8]. Whole-exome sequencing led to the discovery of a variant in the X-linked inhibitor of apoptosis gene (XIAP). Deficiency of this gene can lead to hemophagocytic histolymphiocytosis (HLH). Although not typically associated with colitis, given the prognostic implications of HLH, the patient underwent a bone marrow transplant and the procedure appears to have cured the patient's IBD. Genomic testing has also revealed the genetic cause of other rare diseases [9]. Results may not always have an immediate direct effect on patient care but can lead to insights into the disease pathophysiology.

Genomics applications are not limited to disease state and are increasingly being offered to healthy individuals. Pre-conception risk assessment and prenatal diagnosis are examples of the latter applications. For women planning a pregnancy, a commercially available single test assesses risk for over 100 genetic diseases and is offered at a relatively low cost of approximately $350 [10]. A second NGS-based test to detect fetal trisomy 13, 18, and 21 in maternal peripheral blood is also commercially available [11]. Using a sample of the mother's blood, cell-free DNA is isolated and sequenced and the amount of representation from each chromosome is quantified. For example, an excess of chromosome 21 DNA is consistent with Down Syndrome in the fetus. Recent head-to-head studies have also shown that chromosomal microarrays compare favorably to standard karyotyping in regard to prenatal diagnosis and determining abnormalities associated with stillbirth [12, 13].

Genomic testing has also been performed on healthy individuals outside the setting of pregnancy. In transfusion medicine, high-throughput assays have been developed to determine the blood group antigen genotypes of healthy donors [14]. This genotyping will allow better donor–recipient matching and identification of donors with rare variants for which classic serologic methods are of limited utility. In regard to the potential role of genomics in preventive medicine, a study published in 2010 described the whole-genome sequencing of a blood sample from a 40-year-old male with a family history of sudden cardiac death and coronary artery disease [15]. Over two million SNPs and 752 copy number variants were analyzed with results suggesting an increased risk of myocardial infarction and diabetes. In addition, several variants in genes associated with sudden cardiac death and response to medication were identified. In another study, 20 samples were collected over a 14-month period from a 54 year old male [16]. Whole-genome and transcriptome analysis was performed and over one million gene-associated variants were identified. Some of these variants were associated with increased risk for diseases such as diabetes and basal cell carcinoma. Interestingly, the patient developed indicators of type 2 diabetes during the course of the study.

In addition to sequencing human genomes, NGS methods are being applied to microbiologic testing. During the 2011 *E. coli* outbreak in Europe, the entire sequence of the causative organism was determined in a less than a week [17]. During a recent tuberculosis outbreak in Canada, whole-genome sequencing of 32 isolates led to the determination of the outbreak epidemiology when traditional methods failed [18].

Are Physicians Prepared for the Genomic Era?

The application of genomic testing to patient care will only continue to increase. As such, physicians must be prepared to understand appropriate ordering practice and the interpretation of these new assays. Unfortunately, there is evidence that many physicians do not understand single-gene molecular testing, let alone genomic analysis.

A study from 1997 examined physician practice in testing for the *APC* gene variant associated with familial adenomatous polyposis [19]. In approximately 20 % of cases, an inappropriate strategy for pre-symptomatic testing was used and 32 % of the results were misinterpreted by ordering physicians. Over a decade later, physician ability to appropriately utilize and interpret genetic tests does not appear to have improved. In 2010, genetic counselors at a large reference laboratory examined test ordering practice for 36 molecular tests and corrected inappropriate orders [20]. There were issues with approximately 30 % of ordered test (1,200 orders). In the majority of these cases (68 %), the wrong test was ordered. The most frequent inappropriate test orders were for *NF1* deletion/duplication testing and alpha globin sequencing where 80 % and 64 % of orders were cancelled, respectively.

Physicians are aware of their need for acquiring additional knowledge of genetic testing. In one survey-based study of over 200 internists, while 65 % stated that they have counseled a patient on a genetic issue and 44 % had ordered a genetic test in the past 6 months, 74 % rated their knowledge of genetics as "somewhat poor" or "very poor" and approximately 80 % indicated a need for additional training [21]. In another study involving 401 family physicians, 55 % reported that they had no awareness of the Genetic Information Non-Discrimination Act (GINA) [22].

Current Physician Training in Genomics

Clearly, many of today's practicing physicians have difficulty interpreting single gene testing and are not prepared for the genomic era. The lack of knowledge is perhaps not surprising given the content of medical school courses in genetics. In a 2007 study evaluating 112 of these courses in the USA and Canada, only 11 % included "practical training" in medical genetics [23]. Consistent with the latter finding, a major conclusion of a separate focus-group based study involving family medicine residents was that medical school genetics training "dealt with rare disorders and was not clinically relevant" [24].

Others have recognized this gap in medical education and there have been numerous publications calling for greater training in genomic medicine [25–28]. There are, however, relatively few published initiatives to improve health professional knowledge in this area. Since the above study of medical school genetics courses, there has been limited evidence of widespread integration of genomic medicine education into medical school training [29–31]. There is, however, evidence for innovation such as the "genes-to-society" curriculum developed at Johns Hopkins University [32]. Implemented in 2009, this 15 month course endeavors to incorporate genomics as a "horizontal strand" during the first and second years of medical school.

Beyond medical school, a set of core competencies approved by the European Society of Medical Genetics was published in 2010 [33]. Whereas labeled as addressing the "challenge of genomic medicine," the competencies established for nongenetic specialists are not sufficiently detailed and include only basic goals such as to "identify individuals with or at risk of a genetic condition" and "manage patients with genetic conditions, using accepted guidelines." In the USA, a project that aims at generating interactive tools for genomic medicine education as well as teaching modules related to genomic biology, disease susceptibility, and pharmacogenomics remains in the planning stages [34].

The Important Role of Pathologists in the Genomic Era

Pathologists are in a unique position to assist in translating genomic technology to clinical care. Pathologists already direct the laboratories offering single gene testing and have the expertise in ensuring accurate and precise results. As the authors of the aforementioned study involving detection of a cryptic *PML-RAR* fusion transcript wrote, "to fully use this potentially transformative technology to make informed clinical decisions, standards will have to be developed that allow for CLIA-College of American Pathologists certification of whole-genome sequencing" [4].

Almost all specimens used for genomic testing will pass through the pathology laboratory. In anatomic pathology, a pathologist must first determine that there is a malignant process before sending for assays that determine prognosis or potential chemotherapy regimens in a given neoplastic disease. In addition, the pathologist must also ensure that an appropriate sample is sent. Determining the type of processing (fresh versus frozen or formalin-fixed) and the portion of the specimen to analyze are crucial in providing accurate results [35]. In clinical pathology, whether in the blood bank, microbiology, hematology, or molecular pathology laboratories, pathologists have access to samples for genomic analysis. Furthermore, pathologists are already versed in incorporating genetic data into pathology reports that enable other clinicians to understand the results and act appropriately.

Given the experience and training of pathologists in sample preparation, assay validation and quality control, one can argue that without pathologists overseeing genomic testing, there is the potential for patient harm. As evidence of such potential danger, in 2009, a direct to consumer (DTC) genomic testing company mixed up samples, leading to clients receiving incorrect results suggesting risk for a variety of diseases [36].

Commercially offered DTC genomic tests typically use gene chips to study over a million SNPs. Some SNPs are associated with increased risk of disease based on genome wide association studies (GWAS) [37] that use a case control design to determine SNPs associated with a specific disease or trait. The results from GWAS, however, are meant to be used on a population basis and not to determine individual risk. The previously mentioned whole-genome sequencing studies on healthy individuals for the assessment of sudden cardiac death and diabetes also used GWAS data to determine disease susceptibility [15, 16]. As noted by the authors of one of these studies, when a patient has multiple risk factors for a disease (e.g., diet, smoking status, medication use, genetic variants), currently "no methods exist for statistical integration of such conditionally dependent risks" [15].

This difficulty of determining true pretest probability, as well as which genetic variants are clinically significant, has significant implications for patient care. In a study where identical samples from five individuals were sent to two different DTC genomic testing companies, the results were discordant in 33 % of the time (e.g., an individual received a report of an increased risk for a disease from one company and average or decreased risk for the same disease from the other) [38].

Pathologists are well versed in many of the statistical issues that arise in the setting of genomic testing [39]. For example, a test with very high test specificity (>99 %) may still have a low positive predictive value if the prevalence of the disease in the population tested is very low. This issue is compounded considering that most genomic tests are made up of many individual "tests" (e.g., a multi-gene panel) which increases the risk of false-positive results [40]. In the genomic era, pathologists' familiarity with issues related to statistics, accuracy, precision, and quality control will be vitally important.

Given the above, pathologists need to be centrally involved in translating genomic methods to patient care. As genomic testing will affect all areas of medicine, however, pathologists will need to collaborate with other specialists such as genetic counselors and medical geneticists. In a pilot study, pathologists developed a workflow for tumor analysis consisting of sample processing, sequencing and result validation [2]. While the authors "anticipate that the molecular genetics and pathology communities will move high-throughput sequencing toward CLIA certification, which will ultimately reduce costs and improve turnaround time," they also describe the formation of a "genomic" tumor board consisting of oncologists, medical geneticists, ethicists as well as pathologists.

It is important to note, however, that there are currently less than 3,000 molecular geneticists and genetic counselors in the USA with only approximately 500 being certified each year [34]. In contrast, there are approximately 20,000 board certified pathologists [41]. As such, pathologists not only have the expertise but the workforce needed to translate genomic testing to patient care, even if only a subset of pathologists will specialize in this area.

Single Program Approaches to Genomic Pathology Training

As can be surmised from the above discussion, there is a strong case for training pathologists in genomics. In 2010, a group of representatives from leading pathology organizations, insurance consortiums, industry, the National Institutes of Health (NIH), and the military met at the Banbury Conference Center at Cold Spring Harbor Laboratory to discuss the future of genomic pathology. Recommendations from the meeting listed seven "blue dot" projects to help ensure that pathologists play a significant role in applying genomic technology to patient care [42]. One of these projects (Blue dot project #1) had the goal "to ensure that every Accreditation Council for Graduate Medical Education (ACGME)-approved residency in pathology in North America includes a mandatory curriculum in genomics and personalized medicine." An editorial published in the same issue of the American Journal of Clinical Pathology as the Banbury Conference recommendations stated "although all seven projects certainly have merit and are important to pathologists … project 1 is, without doubt, a 'no-brainer'" and "the need to introduce NGS and whole-genome technology topics into medical student and pathology resident education is mandatory" [43].

Given current medical school training, individuals entering pathology residency would be expected to have a limited background in genetics and genomics. Although pathology residency programs are currently required by The Accreditation Council of Graduate Medical Education (ACGME) to have training in "molecular biology," the information form submitted prior to site inspection only asks programs to document instruction related to FISH, PCR, DNA sequencing and microarray techniques [44]. There are also no genomic pathology-specific requirements for molecular genetic pathology fellowship programs.

In the absence of current ACGME requirements, but recognizing the importance of educating pathologists in genomic methods, several pathology residency programs have established genomic pathology curricula and two programs have published their approach. In 2009, faculty at Beth Israel Deaconess Medical Center (BIDMC) established a mandatory resident curriculum in genomic pathology. Knowledge, affective, and performance-based objectives were included [45]. First, residents attended three lectures. An introductory lecture provided an overview of genomics and the important role a pathologist is expected to play in genomic testing. The second focused on genomic testing methods such as NGS. Recognizing the need for inter-specialty collaboration, the third lecture focused on communicating genetic and genomic test results to patients and was given by BIDMC genetic counselors.

The three lectures provided a strong knowledge-base for the other components of the curriculum. To demonstrate the ability to apply this knowledge, residents were asked to select a paper on a disease of their choice that used genomic methods. With a faculty advisor, the resident reviewed the paper and delivered a 15-min presentation to his/her peers describing the findings. A wide variety of conditions including both malignant (e.g., melanoma) and nonmalignant (e.g., macular degeneration) were discussed. Demonstrating the thoroughness of the literature review, in the first year of administering the genomic pathology curriculum at BIDMC, two residents found an error on a DTC company Web site. For a variant associated with multiple sclerosis risk, the Web site listed the odds ratio as 1.8. When the residents examined the cited paper, the odds ratio was actually 1.37 [46].

The final component of the curriculum, offering residents free-of-charge DTC genomic testing, allowed participants to appreciate affective issues related to genomic testing (i.e., the impact testing has on patients). A company was selected that utilized SNP analysis and GWAS to determine risk for 40 conditions and also provided genetic counselors to help answer questions. The testing was completely voluntary, not required to participate in the curriculum and results were only seen by the resident who ordered the testing. In addition, the curriculum was scheduled over several months (while residents were on other rotations), so participants were able to hear the lectures and have an adequate knowledgebase before deciding on the testing. Each year testing was offered, over 70 % of residents participated. In an anonymous survey, no residents felt coerced into participating and several commented that the testing added to their understanding of genomic pathology. In addition, a key driver of adult learning is relevance and the "need to know" [47]. Several residents used their testing results to decide on the topic for their presentation. While there has been some debate on the utility and ethics of educational DNA testing, self-testing on a smaller scale is not a new concept in clinical pathology training [29, 31, 48, 49]. At some programs, for example, a resident may perform laboratory testing (e.g., a type and screen) on their own blood sample.

The BIDMC curriculum has been published and key components including the lectures and resident presentations are available online [50–52]. The curriculum is currently undergoing revision and there are now plans to integrate genomics training into the month-long molecular pathology and cytogenetics rotation. Exercises related to annotation of genomic data and communication of results to patients are also being created.

Pathology faculty at Stanford University have published the genomic pathology curriculum offered to pathology residents at Stanford. This mandatory series of ten core lectures was started in 2010 and made available online in 2012 [53, 54]. The first lecture provides an introduction to methods for measuring and manipulating nucleic acids and includes a discussion of polymerase chain reaction and sequencing technology. The following three lectures provide a background on types of genetic variation as well microarray and NGS methods. The subsequent five lectures cover specific clinical applications of genomics in areas including inherited disorders, solid tumors, pharmacogenomics, HLA genetics, and hematopoietic cancers. The final lecture addresses ethical, regulatory, and economic issues in genomic pathology. In 2011, the Stanford University Pathology Department also began offering an advanced genomic medicine elective for residents, faculty, and fellows who "plan to work actively with genomic data." This elective is taught in a small-group interactive environment and includes additional instruction in NGS, genetic variation and sequence analysis.

A National Approach to Genomic Pathology Education

In 2010 a survey was distributed to members of the Pathology Residency Program Directors Section (PRODS) of the Association of Pathology Chairs (APC) in order to obtain a better assessment of current national practice [55]. Of 185 programs surveyed, 42 (23 %) responded. While 93 % of programs provided training in molecular pathology, only 31 % had any training in genomic pathology-related topics such as NGS and DTC genetic testing. And, whereas 91 % of programs without training wanted to have a curriculum, lack of faculty expertise (52 %) and time in the resident schedule (76 %) were cited as major barriers. Due to these issues, 74 % of programs did not plan on initiating training in the following year. Respondents rated availability of online modules as the most helpful tool in implementing a new curriculum or for improving an existing curriculum in genomic medicine.

The survey results prompted the creation of a PRODS committee to facilitate integration of genomic pathology training into residency programs. The Training Residents in Genomics (TRIG) Working Group is made up of experts in medical education and molecular genetic pathology as well as members of leading pathology organizations. The American Society for Clinical Pathology (ASCP) provides administrative support. The TRIG Working Group includes three past presidents of the Association for Molecular Pathology (AMP), a past editor-in-chief of *The Journal of Molecular Diagnostics*, and the former chief of the molecular pathology section of the National Cancer Institute. Recognizing the need for collaboration across specialties, the National Society of Genetic Counselors (NSGC), the American College of Medical Genetics and Genomics, and the National Coalition for Health Professional Education in Genetics (NCHPEG) have appointed representatives. The latter is a group of over 50 organizations, including consumer and volunteer groups, government agencies, private industry, managed care organizations, and genetics professional societies, working collaboratively to improve genetics education.

The first major goal of the TRIG Working Group is to develop a national genomic pathology curriculum including teaching tools. The members decided to begin with a structured evaluation of the BIDMC curriculum. More extensive curriculum objectives were developed and presented at a PRODS session at the July 2011 APC meeting. By March 2012, four PowerPoint lectures with lecture notes had been created. The lectures were initially posted with free access on the Intersociety Council for Pathology Information (ICPI) Web site and are now also available on a separate TRIG Working Group Web site [56, 57]. The curriculum includes an introductory lecture followed by lectures on genomic methods, applying genomic technology to clinical care and communicating with the patient.

A second major goal of the TRIG Working Group is to promote training of pathology residents in genomics. Towards this end, members of the working group have given presentations at the annual meetings of leading pathology organizations including the Academy of Clinical Laboratory Physicians and Scientists, ASCP, and the United States and Canadian Academy of Pathology (USCAP). Of particular note, at the 2012 USCAP Annual Meeting, portions of the TRIG Working Group lectures were both presented and distributed in booklet form at a first-ever joint companion meeting of ASCP, AMP, and the American Society for Investigative Pathology. Based on its success, a second companion meeting, including presentation of two TRIG lectures, was held at the 2012 ASCP Annual Meeting. Future sessions incorporating the TRIG lectures with an interactive genomic data analysis component are being planned for the 2013 College of American Pathologists (CAP) Annual Meeting and the 2013 ASCP Annual Meeting.

Several articles have also been published regarding the progress of the TRIG Working Group. These include a peer-reviewed manuscript published in *Personalized Medicine* as well as informational articles published in *Critical Values* and on the NCHPEG Web site [55, 58, 59]. Further demonstrating a cross-specialty collaborative approach, abstracts have been accepted for platform presentations at the annual meetings of NSGC and NCHPEG.

The third major goal of the TRIG Working Group is to assess the degree of resident training and knowledge in genomic pathology. Administered by the ASCP, the yearly pathology resident in-service exam (RISE) is taken by almost all residents in the USA. Scores on the exam allow residents to gauge their progress and have recently been correlated with board exam performance [60]. Beginning in 2012, the TRIG Working Group has contributed both knowledge and survey questions to the RISE. The survey questions can help assess resident attitudes and perceived ability related to genomic pathology as well as directly query whether individual programs offer training. As such, the use of the RISE allows a comprehensive approach to nationally determine the current state of resident training in genomic pathology.

The TRIG Working Group represents a novel approach to teaching new technologies to pathology trainees. Typically, development of a national curriculum involves members of a single pathology organization creating a list of curricular objectives [39, 61]. While useful as an initial step, rarely are actual tools for teaching the objectives provided. From its inception, the TRIG Working Group has been utilitarian and collaborative in its approach and in 2 years not only created a curriculum but developed four free-of-charge PowerPoint lectures with notes to assist programs in teaching genomic pathology. There have already been over 1,500 visits to the lectures on the ICPI Web site.

Finally, published pathology curricula are often not evaluated in regard to outcomes [39, 61]. The RISE survey and knowledge questions created by the TRIG Working Group will provide valuable data on the degree and efficacy of resident training in genomic pathology on a national scale. Rarely has an assessment tool with the scope of the RISE been used to study curricular improvement.

Through a grant awarded in 2012 from the NIH, the TRIG Working Group will be able to further develop a genomic pathology curriculum as well as online modules, resident workshops and assessment tools with the ASCP providing educational design support. The curriculum will be tested at four residency sites and national trends in genomics training will be assessed using the RISE. At the end of the 5 year funding period, the ultimate goal is to ensure genomics training in >90 % of pathology residency programs in the USA.

Future Directions

The efforts of individual residency programs and the TRIG Working Group are important steps forward in genomic pathology education. More, however, can be done to ensure that current and future pathologists assimilate genomic testing and help provide quality patient care. The ACGME should establish standards related to genomic pathology. The new accreditation system (becoming active for pathology in 2014) establishes "milestones" for each resident to achieve during his or her training [62]. Creating milestones related to genomic pathology would require programs to provide instruction in this area. The American Board of Pathology should also include genomics related questions on board certification exams. Needing to know the material to become board-certified would incentivize residents to learn genomic pathology.

Training in genomic pathology needs to begin in medical school and extend beyond residency. There have already been some innovative single institution approaches in undergraduate genomics education but pathologists, because they are already playing a major role in teaching during the first 2 years of medical school, should take an active part in incorporating genomic pathology instruction into coursework. The Undergraduate Medical Educators Section (UMEDS) of the APC has taken the first step towards a structured, national approach by surveying members regarding current incorporation of genomic topics into their pathology courses. Beyond residency, molecular genetic pathology as well as other pathology fellowships need to incorporate genomics training. Continuing medical education programs also need to be established to teach practicing pathologists. Training beyond residency could build on the teaching tools created by individual programs, the TRIG Working Group as well as resources developed for physicians in other specialties. Assessment tools to determine efficacy should be established for these educational programs.

Conclusions

As the director of the National Human Genome Research institute wrote in 2011, "It is time to get serious about genomics education for all healthcare professionals" [28]. Based on published data, few medical schools and residency programs are training physicians in the utility and interpretation of genomic testing. Pathologists, with their access to tissue samples and expertise in laboratory testing, must play a leading role in ensuring the safe application of genomics to patient care.

Further work is needed to educate pathologists in genomics and should build on the resources created by individual programs as well as the TRIG Working Group. The latter provides a collaborative and structured model for curriculum design and assessment, not only for genomics education, but in other novel technologies.

Acknowledgement

This work was supported by the National Institutes of Health (1R25CA168544-01).

REFERENCES

1. National Human Genome Research Institute. The human genome project completion: Frequently asked questions. 2010. http://www.genome.gov/11006943. Accessed 5 Feb 2013.
2. Roychowdhury S, Iyer MK, Robinson DR, et al. Personalized oncology through integrative high-throughput sequencing: a pilot study. Sci Transl Med. 2011;3:111ra121.
3. Jones SJM, Laskin J, Li YY, et al. Evolution of an adenocarcinoma in response to selection by targeted kinase inhibitors. Genome Biol. 2010;11:R82.
4. Welch JS, Westervelt P, Ding L, et al. Use of whole-genome sequencing to diagnose a cryptic fusion oncogene. JAMA. 2011;305:1577–84.
5. Azim HA Jr, Michiels S, Zagouri F, et al. Utility of prognostic genomic tests in breast cancer practice: the IMPAKT 2012 Working Group Consensus Statement. Ann Oncol. 2013. [epub ahead of print].
6. Alexander EK, Kennedy GC, Baloch ZW, et al. Preoperative diagnosis of benign thyroid nodules with indeterminate cytology. N Engl J Med. 2012;367:705–15.
7. Nardi V, Sadow PM, Juric D, et al. Detection of novel actionable genetic changes in salivary duct carcinoma helps direct patient treatment. Clin Cancer Res. 2013;19:480–90.
8. Worthey EA, Mayer AN, Syverson GD, et al. Making a definitive diagnosis: successful clinical application of whole exome sequencing in a child with intractable inflammatory bowel disease. Genet Med. 2011;13:255–62.
9. Heinzen EL, Swoboda KJ, Hitomi Y, et al. De novo mutations in ATP1A3 cause alternating hemiplegia of childhood. Nat Genet. 2012;44:1030–4.
10. Counsyl Web site. https://www.counsyl.com. Accessed 5 Feb 2013.
11. Sequenom Web site. http://www.sequenomcmm.com/Home/Health-Care-Professionals/Trisomy-21. Accessed 5 Feb 2013.
12. Wapner RJ, Martin CL, Levy B, et al. Chromosomal microarray versus karyotyping for prenatal diagnosis. N Engl J Med. 2012;367:2175–84.
13. Reddy UM, Page GP, Saade GR, et al. Karyotype versus microarray testing for genetic abnormalities after stillbirth. N Engl J Med. 2012;367:2185–93.
14. Denomme GA, Johnson ST, Pietz BC. Mass-scale red cell genotyping of blood donors. Transf Apher Sci. 2011;44:93–9.
15. Ashley EA, Butte AJ, Wheeler MT, et al. Clinical assessment incorporating a personal genome. Lancet. 2010;375:1525–35.
16. Chen R, Mias GI, Li-Pook-Than J, et al. Personal omics profiling reveals dynamic molecular and medical phenotypes. Cell. 2012;148:1293–307.
17. Bio-IT World Staff. German teams, BGI and Life Technologies identify deadly European E. Coli strain. 2011. http://www.bio-itworld.com/news/06/02/2011/German-teams-BGI-Life-Technologies-Identify-E-coli-strain.html. Accessed 5 Feb 2012.
18. Gardy JL, Johnston JC, Ho Sui SJ, et al. Whole-genome sequencing and social-network analysis of a tuberculosis outbreak. N Engl J Med. 2011;364: 730–9.
19. Giardiello FM, Brensinger JD, Petersen GM, et al. The use and interpretation of commercial APC gene testing for familial adenomatous polyposis. N Engl J Med. 1997;336:823–7.
20. ARUP Laboratories. Value of genetic counselors in the laboratory. 2011.http://www.aruplab.com/files/resources/genetics/White-paper-1-value-of-GCs-in-lab.pdf. Accessed 5 Feb 2013.
21. Klitzman R, Chung W, Marder K, et al. Attitudes and practices among internists concerning genetic testing. J Genet Couns. 2013;22:90–100.
22. Laedtke AL, O'Neill SM, Rubinstein WS, et al. Family physicians' awareness and knowledge of the Genetic Information Non-Discrimination Act (GINA). J Genet Couns. 2012;21:345–52.
23. Bell MA, Wales PS, Torbeck LJ, et al. The current status of medical genetics instruction in U.S. and Canadian medical schools. Acad Med. 2007;82: 441–5.
24. Telner D, Carroll JC, Talbot Y. Genetics education in medical school: a qualitative study exploring educational experiences and needs. Med Teach. 2008;30: 192–8.
25. Guttmacher AE, Porteous ME, McInerney JD. Educating health-care professionals about genetics and genomics. Nat Rev Genet. 2007;8:151–7.
26. Salari K. The dawning era of personalized medicine exposes a gap in medical education. PLoS Med. 2009;6(8):e1000138.
27. Nelson EA, McGuire AL. The need for medical education reform: genomics and the changing nature of health information. Genome Med. 2010;2(3):18.
28. Feero WG, Green ED. Genomics education for health care professionals in the 21st century. JAMA. 2011;306(9):989–90.

29. Salari K, Pizzo PA, Prober CG. Commentary: to genotype or not to genotype? Addressing the debate through the development of a genomics and personalized medicine curriculum. Acad Med. 2011;86: 925–7.
30. Dhar SU, Alford RL, Nelson EA, et al. Enhancing exposure to genetics and genomics through an innovative medical school curriculum. Genet Med. 2012;14:163–7.
31. Walt DR, Kuhlik A, Epstein SK, et al. Lessons learned from the introduction of personalized genotyping into a medical school curriculum. Genet Med. 2011;13:63–6.
32. Wiener CM, Thomas PA, Goodspeed E, et al. "Genes to society"—the logic and process of the new curriculum for the Johns Hopkins University School of Medicine. Acad Med. 2010;85:498–506.
33. Skirton H, Lewis C, Kent A, et al. Genetic education and the challenge of genomic medicine: development of core competences to support preparation of health professionals in Europe. Eur J Hum Genet. 2010;18:972–7.
34. Patay BA, Topol EJ. The unmet need of education in genomic medicine. Am J Med. 2012;125:5–6.
35. Goetz L, Bethel K, Topol EJ. Rebooting cancer tissue handling in the sequencing era: toward routine use of frozen tumor tissue. JAMA. 2013;309:37–8.
36. MacArthur D. Sample swaps at 23andMe: a cautionary tale. 2010. http://scienceblogs.com/geneticfuture/2010/06/sample_swaps_at_23andme_a_caut.php. Accessed 5 Feb 2013.
37. Pearson TA, Manolio TA. How to interpret a genome-wide association study. JAMA. 2008;299:1335–44.
38. Ng PC, Murray SS, Levy S, Venter JC. An agenda for personalized medicine. Nature. 2009;461:724–6.
39. Smith BR, Wells A, Alexander CB, et al. Curriculum content and evaluation of resident competency in clinical pathology (laboratory medicine): a proposal. Am J Clin Pathol. 2006;125(Suppl):S3–37.
40. Kohane IS, Masys DR, Altman RB. The incidentalome: a threat to genomic medicine. JAMA. 2006;296:212–5.
41. ICPI. Career opportunities in pathology. http://www.pathologytraining.org/trainees/documents/recruit.ppt. Accessed 5 February 2013.
42. Tonellato PJ, Crawford JM, Boguski MS, et al. A national agenda for the future of pathology in personalized medicine: report of the proceedings of a meeting at the Banbury Conference Center on genome-era pathology, precision diagnostics, and preemptive care: a stakeholder summit. Am J Clin Pathol. 2011;135:668–72.
43. Ross JS. Next-generation pathology. Am J Clin Pathol. 2011;135:663–5.
44. ACGME. Hospital based accreditation: pathology. 2008. http://www.acgme.org/acgmeweb/tabid/142/ProgramandInstitutionalGuidelines/Hospital-Based Accreditation/Pathology.aspx. Accessed 5 Feb 2013.
45. Kern DE, Thomas PA, Howard DM, et al. Curriculum development for medical education: a six step approach. Baltimore: The Johns Hopkins University Press; 1998.
46. Elliott R, Jacobs WC. Multiple sclerosis. 2010.http://genomicmedicineinitiative.org/wp-content/uploads/2010/03/Multiple-Sclerosis.pdf. Accessed 5 Feb 2013.
47. Knowles MS, Holton III EF, Swanson RA. The adult learner 6th edition. Burlington: Elsevier; 2005.
48. Callier SL. Swabbing students: should universities be allowed to facilitate educational DNA testing? Am J Bioeth. 2012;12:32–40.
49. Grenzen JR, Krasowski MD. Resident training in clinical chemistry. Clin Lab Med. 2007;27:343–58.
50. Haspel RL, Arnaout R, Briere L, et al. A curriculum in genomics and personalized medicine for pathology residents. Am J Clin Pathol. 2010; 133(6): online supplement. http://ajcp.ascpjournals.org/site/misc/pdf/Haspel_online.pdf Accessed 5 Feb 2013.
51. Haspel RL, Arnaout R, Briere L, et al. A call to action: training pathology residents in genomics and personalized medicine. Am J Clin Pathol. 2010;133:832–4.
52. BIDMC Department of Pathology. Genomic medicine initiative. 2010. genomicmedicineinitiative.org. Accessed 5 Feb 2013.
53. Schrijver I, Natkunam Y, Galli S, et al. Integration of genomic medicine into pathology residency training: the Stanford Open Curriculum. J Mol Diagn. 2013;15:141–8.
54. Stanford University Faculty. Stanford open curriculum in genomic medicine. 2012.http://www.youtube.com/playlist?list=PLfTljtR5bxMcTg8hgQp9sA4YQwicpSAQv. Accessed 5 Feb 2013.
55. Haspel RL, Atkinson JB, Barr FG, et al. TRIG on track: educating pathology residents in genomic medicine. Pers Med. 2012;9:287–93.
56. The TRIG Working Group. Lecture materials from the Training Residents in Genomics (TRIG) Working Group. 2012. http://www.pathologytraining.org/trig_lecture.htm. Accessed 5 Feb 2013.
57. TRIG Working Group Website. 2013. ascp.org/TRIG. Accessed 8 June 2013.
58. Haspel RL. How can we teach genomic medicine to pathology professionals? Crit Values. 2013;6:19.
59. Haspel RL. NCHPEG Member news: Free genomic medicine PowerPoint lectures now available. 2012. http://www.nchpeg.org/index.php?option=com_content&view=article&id=400:nchpeg-member-news-free-genomic-medicine-powerpoint-lectures-now-available-from-the-intersociety-council-for-pathology-information-&catid=92:in-practice-fall-2012&Itemid=271. Accessed 5 Feb 2013.
60. Rinder HM, Grimes MM, Wagner J, et al. Senior pathology resident in-service examination (RISE) scores correlate with outcomes of the American Board of Pathology (ABP) certifying examinations. Am J Clin Pathol. 2011;136:499–506.
61. Talbert ML, Dunn ST, Hunt J, et al. Competency-based education for the molecular genetic pathology fellow: a report of the association for molecular pathology training and education committee. J Mol Diagn. 2009;11:497–507.
62. ACGME. The next accreditation system. http://www.acgme-nas.org/index.html. Accessed 5 Feb 2013.

CHAPTER 8

Implementation of Genome Sequencing Assays

JOSHUA L. DEIGNAN

Introduction 111
Equipment and Reagents 111
What is a CLIA-Certified/CAP-Accredited Laboratory? 112
Personnel 112
Requisition, Clinical Information, and Informed Consent 113
Sample Selection 114
Reporting of Results and Variant Interpretation 114
Bioinformatics Requirements and Data Storage 116
Test Validation 116
Proficiency Testing 116
Conclusions 117

Introduction

Whereas genome sequencing assays started as a tool of the research laboratory, they have now found their way into the clinical laboratory where they are increasingly being adopted for clinical laboratory testing. The requirements for a clinical laboratory test whose results will be communicated back to the ordering physician for patient care are evidently more stringent than those of most research assays, the results of which are not directly used for patient care. Therefore, the task of implementing a genome sequencing assay in the clinical environment poses a great challenge regardless of whether the assay is a gene panel, exome sequencing (ES), or genome sequencing (GS) assay. As more clinical laboratories attempt to incorporate next-generation sequencing (NGS) technology into their molecular diagnostic toolbox, the need for professional NGS standards and guidelines will become increasingly pressing. In fact, several NGS guidelines have already emerged in the literature [1–3]. The purpose of this chapter is to give the reader an overview of the various issues that should be considered when a clinical laboratory director makes the decision to evaluate genome sequencing as a potential platform for clinical testing.

Equipment and Reagents

One of the first decisions a clinical laboratory needs to make is what type of NGS platform to purchase. Currently, there are several different commercially available platforms, some with more widespread use than others and each with its own unique technical characteristics and attributes that a laboratory needs to consider. Some of the ways in which the various platforms may currently differ are cost (ranging from the hundreds of thousands of dollars to nearly a million dollars), length of sequence reads (short versus long), overall total sequencing capacity, instrument size/footprint (small

G.J. Netto and I. Schrijver (eds.), *Genomic Applications in Pathology*,
DOI 10.1007/978-1-4939-0727-4_8, © Springer Science+Business Media New York 2015

versus large), turnaround time (days to weeks), and ease of use. The reader is referred to several existing reviews on this topic, as the technical specifications of the various platforms currently in use have previously been covered in much greater detail than can be discussed here [4, 5]. Laboratories are also encouraged to communicate with colleagues at other institutions who have practical experience using the various types of platforms before making their own financial investment. Furthermore, given that the technical specifications and cost of NGS platforms are continuously changing, it is difficult to weigh the decision to financially invest now or wait for an upgraded or a completely new platform.

As part of deciding on what platform to invest in, laboratories need to decide what type of assay is to be implemented. They may desire to launch a gene panel, covering a known set of clinically relevant genes for a defined condition (e.g., cardiomyopathy or deafness); they may desire to launch an ES test, covering the vast majority of the protein-coding regions in order to assist with the diagnosis of Mendelian disorders; or they may desire to launch a GS test, assessing as much of the entire genomic sequence as technically feasible, in order to also find potentially pathogenic intronic variants and copy number alterations (gains and losses) for phenotypes such as intellectual disability. Small, benchtop sequencers tend to have a lower total sequencing capacity, which could be sufficient for a limited gene panel for which the test can be developed to sequence only a handful of specific regions at a much greater depth than typically achieved for more comprehensive testing, whereas larger sequencers with a higher total sequencing capacity would be more optimal for analyzing variants at a sufficient depth across the exome or the genome. Newer benchtop sequencers, that have the potential to perform ES and GS with a smaller footprint, are also beginning to emerge. Finally, if a laboratory decides to only sequence a set of genes/regions (such as a gene panel or the exome), the laboratory is encouraged to evaluate as many of the various available technologies and chemistries as possible in order to find the one that works best for the intended purpose of the test [4].

What is a CLIA-Certified/CAP-Accredited Laboratory?

Every existing clinical laboratory is familiar with the Clinical Laboratory Improvement Amendments (CLIA), which provide quality standards and guidance on how a clinical laboratory is required to operate. These regulations apply to every laboratory that performs testing on human specimens for clinical purposes (not research laboratories). Such laboratories are required by law to obtain appropriate CLIA certification that allows them to offer diagnostic, prognostic, and other clinical tests. Most clinical laboratories, including molecular diagnostic laboratories, are also accredited by the College of American Pathologists (CAP), which is one of the accrediting organizations that are permitted to enforce CLIA regulations through onsite inspections, proficiency testing, and other educational resources. CLIA regulations pertain to all areas of the clinical laboratory. The CAP issues a "Molecular Pathology" checklist that specifies the set of quality, patient and employee safety, and test performance parameters that a clinical molecular diagnostic laboratory is required to adopt in order to obtain CAP accreditation. The checklist is the basis for onsite CAP inspections. The CAP has only recently published an additional checklist (as part of the "Molecular Pathology" checklist) that is specific for laboratories performing clinical NGS in order to offer guidance on requirements pertaining to such a complex testing environment. In addition to the above specific checklists, like any other section of a clinical laboratory, molecular diagnostic laboratories are required to fulfill a set of requirements listed in the "Laboratory General" checklist which pertains to all areas of a clinical laboratory.

Personnel

Whereas personnel who perform NGS assays in research laboratories are typically at the undergraduate, graduate, or postdoctoral fellow level, current CLIA regulations require that only individuals who are appropriately trained perform clinical laboratory testing. For all clinical molecular diagnostic testing, including NGS, all of the qualification requirements for personnel working in laboratories that perform high-complexity testing apply. Under CLIA regulations, individuals must have at least an associate degree (or the equivalent) with a major in a laboratory science in order to perform high-complexity testing. Though individuals are not required to be licensed, it can be more challenging to find employment without certification. There are only a few mechanisms that can be pursued in order to obtain a license. In most states, documentation of relevant education as well as a satisfactory score on the American Society for Clinical Pathology (ASCP) Molecular Biology examination is sufficient in order to obtain licensure to perform clinical molecular diagnostic testing. Other states, like New York and California, have additional specific requirements. For example, technologists working in clinical molecular diagnostic laboratories in California may obtain one of the two certifications, either a Clinical Laboratory Scientist certification, allowing them to perform any type of clinical laboratory testing, or a Clinical Genetic Molecular Biologist Scientist certification. The latter will only allow the certified personnel to perform clinical molecular diagnostic testing. Both require relevant educational experience, 1 year of full-time training in an approved training program, and the obtaining of a satisfactory score on an exam.

With regard to NGS, many molecular biology certification and training programs do not currently address the complex challenges associated with these types of assays. For example, generalist certification in California tends to focus more time on other areas of the clinical laboratory system, such as chemistry, hematology, and microbiology, than on molecular diagnostic testing. Nevertheless, a technologist who obtained a generalist certification would be legally licensed to perform clinical NGS testing even without much experience. Some of these shortcomings are currently being addressed by various professional groups including the Association for Molecular Pathology (AMP). As NGS assays involve both a "wet lab" and a "dry" bioinformatics component, finding additional appropriately licensed individuals with a strong bioinformatics background also poses a challenge, as most of those individuals are likely to have gained that experience as part of a graduate program and may not desire a future career as a clinical laboratory technologist. However, because the bioinformatics analysis is also a part of the analytical component of any NGS clinical test, those individuals should be appropriately licensed, as well.

Requisition, Clinical Information, and Informed Consent

Like any other clinical laboratory test, NGS assays will need a requisition form for appropriate test ordering. For gene panels, laboratories will need to decide whether to list the analyzed genes directly on the requisition, which may include 50–100 genes, or publish that information elsewhere. Laboratories may also want to be able to offer physicians the option of selecting which genes to analyze, though this would require the laboratory to have a bioinformatics-based mechanism for masking the results from specific genes during the analysis and interpretation steps. Because it would be impossible to list all of the relevant regions covered as part of an exome or a genome sequencing test, that information should not be on a requisition form. One possibility is to refer the ordering physician to a website, where information about the specific gene and exon coverage could be provided in detail. This would also allow the ordering physician to determine whether a given clinical NGS test will be of medical benefit to the particular patient in question before ordering the test.

Depending on the intended purpose of the test, acquiring sufficient clinical information from the ordering physician will be a critical component for proper interpretation of the results. Information such as specific phenotypic keywords, suspected diagnoses, and any information pertaining to family history should be provided, and the laboratory will need to have a process in place to address NGS test requests for which the clinical information was not initially provided.

Clinical samples, such as whole blood, have a limited stability and sample quality will be compromised if stored for an extended period of time while the ordering physician is contacted for additional clinical information. Having a genetic counselor on staff as part of the NGS laboratory operation may be necessary to allow for immediate contact with the ordering physician in order to discuss and clarify cases for which the clinical indication is unclear. Given the high reagent cost for this type of clinical testing, the NGS laboratory is less likely to perform testing if there is uncertainty as to whether this type of testing is appropriate. Due to the complexity of NGS assays, informed consent from each patient should also be obtained by the ordering physician prior to ordering the test. The consent document should convey the purpose of the specific NGS test, its limitations, possible unintended consequences such as unexpected consanguineous familial relationships, and the type of sample to be obtained. If the laboratory intends to use either the remnant sample such as genomic DNA or the patient sequence data for future research after the test has been performed and reported, this requires human subjects' research consent as well as Institutional Review Board (IRB) approval. As part of this consent process, a mechanism by which the patient can opt for his or her sample not to be used in such manner and be discarded following test completion should be provided, as well. Consultation with the appropriate IRB and institutional or laboratory legal counsel is recommended during the process of creating any informed consent documentation.

Sample Selection

Most clinical molecular diagnostic testing is done on whole-blood and tissue samples. Tissue samples are either formalin fixed and paraffin embedded (FFPE) or snap frozen in liquid nitrogen immediately after a clinical procedure and stored at −80 °C until the laboratory performs the test. Laboratories wishing to set up an NGS assay will need to decide which type(s) of sample(s) they are going to accept, based on the intended purpose of the test. For germline analysis to look for variants implicated in Mendelian disorders, whole blood is the preferred specimen type. Laboratories may also wish to accept pre-extracted genomic DNA in order to facilitate international requests. However, in such cases, it is recommended that the final report contain a disclaimer stating that the genomic DNA used for the analysis was extracted outside the clinical laboratory and was tested at the request of the ordering physician, that the accuracy of the identifying information provided with the specimen regarding its patient of origin cannot be independently confirmed by the clinical laboratory, and that the integrity of the specimen could not be verified. This will help protect the clinical laboratory of potential liability, in case there was a sample mix-up in the laboratory where the DNA extraction was performed. Laboratories that elect to accept FFPE tissue for various NGS oncology assays should be aware of inherent limitations associated with FFPE tissue use. Formalin fixation leads to DNA cross-linking that often results in fewer long intact DNA fragments compared to those obtained from fresh frozen tissue samples. Therefore, assays utilizing FFPE tissue-derived DNA need to be designed for amplification of only short genomic sequence fragments (100–200 bp). As previously mentioned, fresh frozen tissue obtained after biopsy will usually result in a higher quality and more intact DNA but requires special arrangements during transportation to the molecular laboratory, including transport on wet or dry ice, storage in a −80 °C freezer, and additional biohazard precautions. Both FFPE and fresh frozen tissue may also be requested by the ordering physician to assess potential mosaicism in typically germline Mendelian conditions. A policy regarding the acceptability of this type of specimen for the latter purpose, based on the nature of the clinical test, needs to be established by the laboratory.

Reporting of Results and Variant Interpretation

For molecular diagnostic tests targeting a single clinically relevant mutation, laboratories typically report findings as positive, indicating that the specified mutation was observed, or negative, indicating that the specified mutation was not observed. For larger mutation panels, such as those indicated for conditions such as cystic fibrosis, laboratories typically report findings as positive when (a) given mutation(s) was/were observed or negative, indicating that no mutations were observed. For full-gene sequencing assays (such as those for *BRCA1/2*-associated breast cancer risk performed using traditional Sanger sequencing methods), laboratories may report pathogenic variants or any variants they observe, whether pathogenic or benign, or they may report that they did not observe any variants in the exon sequences of the gene(s) analyzed. However, for NGS assays, there will likely be too many variants observed to include all of them in the report. Therefore, labs will need to decide on the best mechanism to pursue in order to convey relevant or potentially relevant information back to the ordering physician.

Formal guidelines currently exist on how to compose a clinical NGS report [3]. Similar to any basic laboratory report, laboratories should provide an overview statement at the top of the report regarding whether any clinically relevant finding was identified, especially in relation to the clinical indication for ordering the test. The latter point further illustrates the importance of providing appropriate clinical information by the ordering physician on the test requisition. Reporting issues unique to laboratories performing ES and GS assays emerge from the fact that such tests may also find clinically relevant variants in genes that are unrelated to the primary clinical concerns in the tested patient, usually referred to as "incidental findings," as well as many variants which are known to be benign, such as polymorphisms that are very common in the population [6]. How to appropriately handle these incidental findings will be best addressed by professional guidance documents [7]. According to the existing CAP NGS checklist, the laboratory should have a policy on how it will be reporting (or not reporting) the abovementioned incidental findings. The decision regarding which variants to report will ultimately reside with the laboratory director, but this information should be clearly conveyed to both the ordering physician and the patient prior to performing the test.

In addition to reporting and attempting to interpret clinically relevant variants, another, equally important, component of a clinical report is a statement regarding the limitations in the laboratory's ability to fully assess the clinical question based on the method of testing. For certain tests, this may be as simple as stating that only specific genes are covered in a given gene panel assay and listing the anticipated clinical sensitivity and specificity of only testing for variants in those genes. For others, such as an ES test, the limitation disclaimer statement may require a description of the sequence capture method used indicating that only a certain percentage of the clinically relevant variants are expected to be sufficiently sequenced by this test. Some laboratories may elect to only interpret and report single-nucleotide variants (SNVs), whereas others may also wish to interpret and report small insertions and deletions (indels). Other laboratories may also include larger copy number alterations. While some molecular laboratories may perform confirmation of clinically relevant variants using Sanger sequencing, others may forego confirmation by an alternative methodology altogether [8]. Regardless, the decision as to how to perform a test ultimately resides with the laboratory director, and it is in the best interest of the laboratory to be as clear and upfront as possible with providing such information so that both the ordering physician and the patient know ahead of time what clinical information the test is likely to provide as well as its limitations.

In addition to deciding which variants to report and providing statements regarding the limitations of the assay, the laboratory should utilize as many resources as possible when providing its interpretation of individual variants. Such resources include the American College of Medical Genetics and Genomics (ACMG) sequencing guidelines, among others [9]. In addition to relatively well-curated existing databases, molecular laboratories should closely assess family history and variant population frequency, because many of the variants observed in NGS assays, especially ES and GS, are not likely to have been previously reported in association with a similar phenotype. Rare variants are more likely to be pathogenic than common ones, but the presence of a rare potentially pathogenic variant in a clinically relevant gene does not prove causality.

Bioinformatics Requirements and Data Storage

Most molecular diagnostic tests typically comprise primarily a "wet lab" component with minimal calculations required prior to finalizing a result, e.g., calculation of the size of an *FMR1* CGG allele for fragile X syndrome. For more recent FDA-approved assays, such as commercially available assays for *KRAS* and *BRAF* mutation analyses in colorectal adenocarcinoma and melanoma, respectively, no calculations by the technologists are required. Software programs, which are part of the FDA test submission and approval of such assays, handle any calculations in the background prior to generating a result. In contrast, as previously mentioned, NGS assays call for a large bioinformatics analytical component, which currently is performed manually for the most part.

At present, there are only a handful of commercially available software packages to assist with NGS assay result interpretations. However, most laboratories offering this type of testing on a clinical basis have assembled their own in-house-developed bioinformatics pipeline consisting of various computer programs and variant databases [10, 11]. Ideally, a laboratory should have several qualified bioinformaticists who are familiar with both bioinformatics and molecular biology. As previously mentioned, the fact remains that bioinformaticists who are also trained in the requirements needed in a CLIA-certified clinical laboratory are difficult to find. Given that much of the downstream interpretation and reporting will be based on the results of the bioinformatics analysis, it is crucially important for the laboratory personnel not to treat this part of the assay as a "black box" but instead to see it as one in which there is constant discussion, evaluation, and questioning, so that it can evolve with subsequent improvements in external databases and other sources of clinical information.

Laboratories that plan to offer NGS assays should arrange for a marked increase in data storage requirements. The information produced by NGS platforms far exceeds the capacity of a typical desktop computer hard drive or DVD. Therefore, laboratories should be prepared for a financially significant investment in this aspect of the testing, as well. In fact, the financial investment in bioinformatics can often be equivalent to the financial investment required for acquiring the NGS platform itself.

Test Validation

Most clinical laboratory tests involving NGS are currently considered to be laboratory-developed tests (LDTs) and require a full validation, as opposed to the usually more limited verification required for FDA-cleared or -approved products. The main elements of required assessment as stipulated by CLIA regulations pertain to accuracy (how well do the results match what they should be), precision (would a laboratory get the same result from the same sample any day or time), analytical sensitivity (what percentage of a mutant allele could a laboratory detect among a background of normal alleles), analytical specificity (what is the effect of neighboring mutations or interfering substances on the test results), reference range (what does the laboratory consider to be normal), and reportable range (what is included in the test in terms of genomic regions) [12, 13]. The reader is referred to other recent publications that contain detailed suggestions for validation of NGS-based testing including the AMP- and CAP-issued guidelines on NGS assays [1–3].

Proficiency Testing

The requirement for proficiency testing is defined in the CLIA regulations and applies to all molecular tests. Laboratories must participate in proficiency testing at least semiannually by one of several available mechanisms. Typical proficiency testing surveys, such as those administered by the CAP, consist of three blinded samples sent out to every laboratory performing a given clinical laboratory test (e.g., a cystic fibrosis carrier screening mutation panel) twice per year. Laboratories are expected to treat these samples in the same manner as they would treat real patient samples. Once the clinical testing is performed on the proficiency testing sample, final results and interpretations are submitted back to the organization administering the proficiency test. The laboratory is subsequently graded as "acceptable" or "unacceptable" based on either existing values which are known to be the "true" values or based on consensus findings among the majority of participating laboratories.

As one would expect, proficiency testing for ES/GS assays presents several unique challenges. First, most clinical laboratories would find it to be cost prohibitive to test three additional samples, twice yearly, simply for this purpose at the current cost of a few thousand dollars per sample. Secondly, there are only a handful of samples with known genotypes throughout the genome (such as J. Craig Venter's genomic DNA) [14] that would be ideal for proficiency testing for ES and GS assays. Thirdly, as already described, laboratories vary in terms of the genomic regions they interrogate and which types of variants they interpret, so it is difficult to determine which variants in the genome to use for grading purposes if a sample was submitted to multiple laboratories. Finally, inherent in the complexity of NGS assays is a certain rate of false-positive and -negative results. Therefore, with the exception of some gene panels, where Sanger sequencing is used to fill in areas of low coverage, a laboratory could not be expected to observe 100 % of the clinically relevant variants.

The CAP and other organizations are currently in the process of establishing formal, method-based proficiency testing that is expected to be ready in 2014 or 2015. Until such surveys become available, laboratories are encouraged to establish interlaboratory sample exchange activity with other clinical laboratories performing similar clinical NGS testing as well as to enroll in the two currently available CAP sequencing proficiency surveys (also known as SEC and SEC1). These sequencing surveys can serve to assess a laboratory's ability to analyze and interpret variants of interest using correct Human Genome Variation Society (HGVS) nomenclature using provided or newly created Sanger sequencing traces, respectively [15].

Conclusions

Clinical laboratories need to be aware of the challenges associated with implementing NGS assays before deciding whether or not to make the investment. A laboratory may choose to postpone setting up this type of testing until the instrumentation cost and bioinformatics investment become less financially prohibitive. That being said, this technology is not likely to disappear from the clinical arena, so clinical laboratories, as well as practicing and training physicians, should remain continually aware of the improvements that are taking place in the various platforms, databases, and bioinformatics programs, as these technologies are already impacting the medical management of patients and will likely continue to do so in the foreseeable future.

REFERENCES

1. Schrijver I, Aziz N, Farkas DH, Furtado M, Gonzalez AF, Greiner TC, Grody WW, Hambuch T, Kalman L, Kant JA, Klein RD, Leonard DG, Lubin IM, Mao R, Nagan N, Pratt VM, Sobel ME, Voelkerding KV, Gibson JS. Opportunities and challenges associated with clinical diagnostic genome sequencing: a report of the Association for Molecular Pathology. J Mol Diagn. 2012;14(6):525–40. doi:10.1016/j.jmoldx.2012.04.006. Epub 2012 Aug 20. PubMed PMID: 22918138.
2. Gargis AS, Kalman L, Berry MW, Bick DP, Dimmock DP, Hambuch T, Lu F, Lyon E, Voelkerding KV, Zehnbauer BA, Agarwala R, Bennett SF, Chen B, Chin EL, Compton JG, Das S, Farkas DH, Ferber MJ, Funke BH, Furtado MR, Ganova-Raeva LM, Geigenmüller U, Gunselman SJ, Hegde MR, Johnson PL, Kasarskis A, Kulkarni S, Lenk T, Liu CS, Manion M, Manolio TA, Mardis ER, Merker JD, Rajeevan MS, Reese MG, Rehm HL, Simen BB, Yeakley JM, Zook JM, Lubin IM. Assuring the quality of next-generation sequencing in clinical laboratory practice. Nat Biotechnol. 2012;30(11):1033–6.
3. Rehm HL, Bale SJ, Bayrak-Toydemir P, Berg JS, Brown KK, Deignan JL, Friez MJ, Funke BH, Hegde MR, Lyon E, Working Group of the American College of Medical Genetics and Genomics

Laboratory Quality Assurance Commitee. ACMG clinical laboratory standards for next-generation sequencing. Genet Med. 2013;15(9):733–47.
4. Clark MJ, Chen R, Lam HY, Karczewski KJ, Chen R, Euskirchen G, Butte AJ, Snyder M. Performance comparison of exome DNA sequencing technologies. Nat Biotechnol. 2011;29(10):908–14.
5. Glenn TC. Field guide to next-generation DNA sequencers. Mol Ecol Resour. 2011;11(5):759–69.
6. Kohane IS, Masys DR, Altman RB. The incidentalome: a threat to genomic medicine. JAMA. 2006;296(2):212–5.
7. Green RC, Berg JS, Grody WW, Kalia SS, Korf BR, Martin CL, McGuire AL, Nussbaum RL, O'Daniel JM, Ormond KE, Rehm HL, Watson MS, Williams MS, Biesecker LG, American College of Medical Genetics and Genomics. ACMG recommendations for reporting of incidental findings in clinical exome and genome sequencing. Genet Med. 2013;15(7): 565–74.
8. Strom SP, Lee H, Das K, Vilain E, Nelson SF, Grody WW, Deignan JL. Assessing the necessity of confirmatory testing for exome-sequencing results in a clinical molecular diagnostic laboratory. Genet Med. 2014. doi:10.1038/gim.2013.183.
9. Richards CS, Bale S, Bellissimo DB, Das S, Grody WW, Hegde MR, Lyon E, Ward BE, Molecular Subcommittee of the ACMG Laboratory Quality Assurance Committee. ACMG recommendations for standards for interpretation and reporting of sequence variations: revisions 2007. Genet Med. 2008;10(4):294–300.
10. Ledergerber C, Dessimoz C. Base-calling for next-generation sequencing platforms. Brief Bioinform. 2011;12(5):489–97.
11. Li H, Homer N. A survey of sequence alignment algorithms for next-generation sequencing. Brief Bioinform. 2010;11(5):473–83.
12. Jennings L, van Deerlin VM, Gulley ML, College of American Pathologists Molecular Pathology Resource Committee. Recommended principles and practices for validating clinical molecular pathology tests. Arch Pathol Lab Med. 2009;133(5):743–55.
13. Mattocks CJ, Morris MA, Matthijs G, Swinnen E, Corveleyn A, Dequeker E, Müller CR, Pratt V, Wallace A, EuroGentest Validation Group. A standardized framework for the validation and verification of clinical molecular genetic tests. Eur J Hum Genet. 2010;18(12):1276–88.
14. Levy S, Sutton G, Ng PC, Feuk L, Halpern AL, Walenz BP, Axelrod N, Huang J, Kirkness EF, Denisov G, Lin Y, MacDonald JR, Pang AW, Shago M, Stockwell TB, Tsiamouri A, Bafna V, Bansal V, Kravitz SA, Busam DA, Beeson KY, McIntosh TC, Remington KA, Abril JF, Gill J, Borman J, Rogers YH, Frazier ME, Scherer SW, Strausberg RL, Venter JC. The diploid genome sequence of an individual human. PLoS Biol. 2007;5(10):e254.
15. Richards CS, Palomaki GE, Lacbawan FL, Lyon E, Feldman GL. Three-year experience of a CAP/ ACMG methods-based external proficiency testing program for laboratories offering DNA sequencing for rare inherited disorders. Genet Med. 2013; 16:25–32.

CHAPTER 9

Regulatory and Reimbursement Issues in Genomic Testing

JAN A. NOWAK

Introduction 119
Laboratory Regulations 119
Laboratory Regulations for Genomic Testing 121
Laboratory Developed Tests 121
The FDA and Medical Practice in Clinical Laboratories 122
Reimbursement for Genomic Tests 124
Conclusions 126

Introduction

The development of advanced genetic sequencing technologies and genomic testing services challenges the existing regulatory framework for clinical laboratory testing. These challenges will demand a refinement and adaptation on the part of laboratories, professional and accrediting organizations, vendors and manufacturers, and regulatory agencies of existing standards and practices to accommodate novel genomic technologies and clinical applications. The value of existing concepts of laboratory developed tests and companion diagnostics will need revision in order to accommodate genomic sequencing assays. In some circumstances, these novel technologies will challenge established definitions of disease, and the foundations of medical practice.

Similarly, reimbursement for genomic sequencing assays will demand a reassessment of the traditional understanding of laboratory testing, moving away from simple chemical analyses to syndromically defined genetic inquiries necessitating increasing amounts of professional work and involvement. The existing CPT (current procedural terminology) structure can be modified to accommodate these new realities, but will also demand reevaluation of existing concepts of laboratory testing and professional services related to test interpretation and usage.

Laboratory Regulations

Traditionally, novel technologies have been introduced into medical laboratories and clinical usage through fairly defined routes. Technically superior analytical methodologies or entirely novel tests would first emerge from research laboratories into clinical settings in select centers with specific interests in a particular analyte. These initial experiences would typically be reported and vetted at professional society meetings and in peer reviewed publications where they might attract the attention of other professionals who would incorporate these technical advancements into their own laboratories' offerings and assays. Eventually such advancements would supplant extant methodologies and ultimately become standard of care. Assays, methods, and associated instrumentation which offer investment opportunities

G.J. Netto and I. Schrijver (eds.), *Genomic Applications in Pathology*,
DOI 10.1007/978-1-4939-0727-4_9, © Springer Science+Business Media New York 2015

might be developed commercially and marketed, further disseminating a particular technology and promoting further standardization of methodology and clinical usage. This route of introduction was made possible by the established ability of individual laboratories to modify and develop analytical assays for clinical use as authorized by the Clinical Laboratory Improvement Act (CLIA) of 1988.

Under CLIA, laboratories are authorized to implement their analytic procedures for clinical use if they adhere to these basic requirements of laboratory developed tests (LDTs) and Food and Drug Administration (FDA) cleared or approved tests [1]:

42 CFR 493.1253 Standard: Establishment and Verification of Performance Specifications

(1) Verification of performance specifications. Each laboratory that introduces an unmodified, FDA-cleared or approved test system must do the following things before reporting patient test results:
  (i) Demonstrate that it can obtain performance specifications comparable to those established by the manufacturer for the following performance characteristics:
    (A) Accuracy.
    (B) Precision.
    (C) Reportable range of test results for the test system.
  (ii) Verify that the manufacturer's reference intervals (normal values) are appropriate for the laboratory's patient population.
(2) Establishment of performance specifications. Each laboratory that modifies an FDA-cleared or approved test system, or introduces a test system not subject to FDA clearance or approval (including methods developed in-house and standardized methods such as text book procedures, Gram stain, or potassium hydroxide preparations), or uses a test system in which performance specifications are not provided by the manufacturer must, before reporting patient test results, establish for each test system the performance specifications for the following performance characteristics, as applicable:
  (i) Accuracy.
  (ii) Precision.
  (iii) Analytical sensitivity.
  (iv) Analytical specificity to include interfering substances.
  (v) Reportable range of test results for the test system.
  (vi) Reference intervals (normal values).
  (vii) Any other performance characteristic required for test performance.
(3) Determination of calibration and control procedures. The laboratory must determine the test system's calibration procedures and control procedures based upon the performance specifications verified or established under paragraph (b)(1) or (b)(2) of this section.
(4) Documentation. The laboratory must document all activities specified in this section.

Sec. 493.1254 Standard: Maintenance and Function Checks

(a) Unmodified manufacturer's equipment, instruments, or test systems. The laboratory must perform and document the following:
  (1) Maintenance as defined by the manufacturer and with at least the frequency specified by the manufacturer.
  (2) Function checks as defined by the manufacturer and with at least the frequency specified by the manufacturer. Function checks must be within the manufacturer's established limits before patient testing is conducted.
(b) Equipment, instruments, or test systems developed in-house, commercially available and modified by the laboratory, or maintenance and function check protocols are not provided by the manufacturer. The laboratory must do the following things:
  (1) (i) Establish a maintenance protocol that ensures equipment, instrument, and test system performance that is necessary for accurate and reliable test results and test result reporting.
    (ii) Perform and document the maintenance activities specified in paragraph (b)(1)(i) of this section.
  (2) (i) Define a function check protocol that ensures equipment, instrument, and test system performance that

is necessary for accurate and reliable test results and test result reporting.

(ii) Perform and document the function checks, including background or baseline checks, specified in paragraph (b)(2)(i) of this section. Function checks must be within the laboratory's established limits before patient testing is conducted.

Laboratory Regulations for Genomic Testing

Establishment of performance specifications for a single analyte is often challenging, but addressing these seemingly simple concepts for genomic breadth test procedures is particularly daunting. Genomic sequencing assays are technically complex, and capable of addressing multiple "analytes" on a scale not previously seen. Consequently, the usual methods and standards for addressing the analytical performance characteristics required by CLIA are impractical and alternative approaches to assuring compliance will need to be developed. Furthermore, the technical, chemical, and interpretive complexity of these genomic technologies requires that commercial developments play a much more significant role, and at a much earlier stage in test evolution, in bringing these assays to reality. Finally, the complexity of the information achievable by genomic sequencing assays, not to mention the volume of information, makes clinical validation and clinical utility difficult to define in the traditional sense. While the popular press continues to highlight the coming revolution of genomic medicine, professional societies struggle to conform to traditional practices for assuring quality and uniformity in testing, vendors and manufacturers tread cautiously in a rarified atmosphere of regulatory guidance, with the regulatory agencies themselves uncertain of how to best perform their missions. In the interim, some entrepreneurial laboratories have proceeded to offer genomic testing, with their own interpretations of CLIA requirements and of what constitutes clinical validity. Some laboratories have offered such testing directly to patients, or "consumers," obviating any allegiance to CLIA altogether. The issue being challenged is not simply the correct mechanism of new test evolution, but much more fundamentally what constitutes the practice of clinical laboratory medicine, and to what extent genomic information is a part of that practice.

Although some might argue that a revolutionary approach to regulatory oversight of genomic testing is in order, it is reasonable to expect that the foundation of laboratory medicine practice as defined by CLIA continue, and that we will more likely see an evolution of established practices to accommodate genome based testing. Understanding that, it is worthwhile exploring several current regulatory concepts that will impinge on the clinical adaption of genomic testing methodologies.

Laboratory Developed Tests

The ability of laboratories to develop and implement their own assays (LDTs) has been a cornerstone of medical advancement whereby improvements in laboratory analyses could be introduced into clinical usage in a controlled and regulated fashion. With increasing acceptance, such laboratory developed assays eventually overtake and replace extant methodologies. It is likely that the vast majority of clinical assays in use today had their origins in LDTs. The responsibility for assuring LDT assay credibility rests with those entities authorized by CLIA to inspect such laboratories for compliance with the regulations, as well as with the professional standards of those entrusted with administering those laboratories, typically pathologists, in their roles as laboratory Medical Directors. In recent years, this concept of the LDT or laboratory modified test has been challenged from at least two quarters. The emergence of highly technically complex testing systems has fostered the establishment of large, high volume corporate laboratories often devoted to a single, frequently proprietary, assay. Test complexities, as well as IP restrictions, typically preclude adoption of such tests by other laboratories. Whereas CLIA regulations are generally respected through internal studies and validations, the assurances that accompany widespread use by multiple laboratories and institutions, affecting proficiency testing, peer to peer debate and review, and ultimately quality improvement, would seem to be diminished by this adaptation of the LDT paradigm.

The traditional laboratory developed or modified test has also been challenged by manufacturers who have endeavored to develop specific "companion diagnostic tests," a concept promoted by FDA to ensure that appropriate tests and devices are available to support specific therapeutic claims. The requirement that specific tests be developed, validated, and approved by FDA in concert with specific therapeutic drugs necessitates considerable expense on the part of the developer, costs that are not realized by the enterprising LDT developer. Consequently, the FDA has been petitioned by some test manufacturers to enforce its claimed jurisdiction over LDTs and require that all LDTs be subject to formal review by FDA before being placed into clinical service. Whereas the notion of a companion diagnostic has some legitimacy for the agency in ensuring that validated and approved testing methodologies are available for all approved drugs that require them, extension of the concept to mean that the approved assay methodologies are the best available and are the only tests that can be used in concert with a particular drug fails terribly in practice when one envisages the impracticality that a "one drug—one test—one clinical condition" constraint creates. Furthermore, having ordained a particular assay as the test of choice, FDA effectively disincentivizes any further test improvement (and in that, may be franchising certain "harms" to patients). The concept is also oblivious to already established LDT based tests. For some analytes, LDTs have been the standard of laboratory practice for multiple years, with documented proficiency, often with performance characteristics exceeding those of a more recently approved companion diagnostic. None of this historic information is reviewed or even acknowledged whenever FDA touts the approval of yet another "companion diagnostic" as the latest major step in the advancement of personalized medicine.

A more realistic approach would be to recognize that certain analytes, and not tests, have bearing on the performance of particular drugs. This notion is rooted in our understanding of biology, which is ultimately the basis of molecular medicine. Tests which are designed to evaluate these specific "companion analytes" must demonstrate certain performance characteristics in order to be valid. The necessary performance characteristics are defined in the peer reviewed literature and are constantly subject to review and revision. A specific assay may be assigned "companion analyte test" status if it meets certain performance requirements. Laboratory assays used in clinical trials (clinical trial assays, CTA's) may deserve some special recognition in that they frequently, but not always, help define the necessary performance characteristics for the biological analyte. Formal approval of CTA's as in vitro diagnostic products (IVDs) could substantiate their utility without introducing the negative consequences that accompany "companion diagnostic" designation.

Movement away from the "companion diagnostic" terminology is also desirable in underscoring that laboratory test results, in themselves, are never diagnostic. Every laboratory test result needs to be interpreted and understood in the context of a specific patient by the patient's physician. The laboratory test result, however informative, never dictates the course of treatment. That decision needs to be made individually for each patient. To think otherwise is the ultimate depersonalization of medicine.

The FDA and Medical Practice in Clinical Laboratories

Whereas the practice of laboratory medicine is under the purview of CLIA, FDA also claims significant jurisdiction over clinical testing in that such tests, and the instruments and reagents used to perform them, are medical devices, specifically, IVDs. FDA includes LDTs in this category and considers laboratories which use LDTs to be medical device manufacturers. The history of "enforcement discretion" by FDA in administering to traditional LDTs suggests that there is clear understanding within the agency of the difference between a commercially developed product intended for distribution and sale and a locally developed biochemical assay or modification of an approved assay to better serve a limited population. The notion that a corporate laboratory can offer a menu of proprietary LDTs has presented a challenge to FDA's traditional practice in overseeing LDTs. FDA has been developing a refinement of its LDT oversight policy for several years but has yet to release any guidance.

FDA has proposed a three tier classification system for laboratory tests based on "risk" of potential harm to a patient. The lowest tier includes tests which generate results which are typically interpreted in the context of other clinical and laboratory test values and which, in themselves, are unlikely to cause harm to a patient should they be "incorrect." The high risk category includes tests which FDA holds to be "determinative" for a particular drug or course of therapy. Examples include tests for targeted agents *HER2* fluorescence in situ hybridization (FISH) for Herceptin in the treatment of breast cancer, *EGFR* mutation testing for the use of Tarceva in the treatment of non small cell lung cancer (NSCLC), and *BRAF* mutation testing for the treatment of metastatic malignant melanoma with Vemurafinib. FDA has advocated that any tests in the Tier 3 risk group must be reviewed and approved by FDA prior to clinical implementation, a requirement that is currently not imposed on LDTs. This line of thinking would create a conundrum for many LDT performing laboratories, both in terms of the work required for submission to FDA for test approval, and in terms of practicality because the same analyte may be evaluated for different purposes in different circumstances. The *BRAF* p.Val600Glu (commonly known as p.V600E) mutation may be predictive of sensitivity to vemurafinib therapy in metastatic malignant melanoma, but can be used as a corroborative diagnostic marker in evaluation of FNA (fine needle aspirate) specimens of thyroid nodules, and as a surrogate for *MLH1* promoter hypermethylation in cases of microsatellite unstable colorectal cancers, bearing on the likelihood of Lynch Syndrome. The *BRAF* mutation also has prognostic significance for colorectal cancer. Thus, in different circumstances the same analyte (and the same analytical assay) may be assigned to different risk Tiers. The concept that laboratory tests can be neatly categorized further deteriorates when one considers that many laboratory tests can become "determinative" in some clinical situations, e.g., a single blood glucose measurement, a single serum potassium determination, or an antibiotic sensitivity profile for a bacterial pathogen. Whether performed with an FDA approved IVD or with an LDT, it must be acknowledged that no laboratory test is perfect (or diagnostic, as discussed earlier) and that an understanding of its performance characteristics and its application in any specific clinical scenario demands professional knowledge and judgment. FDA is creating a conundrum in pursuing too literal an interpretation of "risk" in applying this categorization scheme. The implication is that FDA is the ultimate arbiter of how tests should be performed, how their results should be interpreted, and how the results should be used clinically, actions which arguably exceed FDA's mission.

While awaiting FDA promised guidance on LDTs, several professional organizations have issued their own proposals. Most of these have grappled with the risk concept put forward by FDA. The College of American Pathologists (CAP) reserved the highest risk category to tests that utilize a non-transparent algorithm and are not subject to the usual checks and balances afforded by obvious tests performed by many laboratories, whose performance can be assessed through traditional proficiency testing programs and peer inspections. CAP acknowledged that there may be some tests that deserve greater scrutiny, i.e., those that are associated with targeted therapies, but which are more appropriately placed in a lower risk tier. The quality performance of such tests, performed as LDTs, could be assured through a stepped up inspection system that would include peer review of test validation data. The Association for Molecular Pathology (AMP) has issued a similar proposal.

The current complex and convoluted regulatory oversight environment will continue to be challenged by the genomic sequencing assays as will our society's fundamental understanding and concepts of health, disease, and the practice of medicine. Such challenges should be welcome if they strengthen the basic principles and purposes for the rules and agencies that were established to assure high quality clinical laboratory testing. Conversely, established oversight mechanisms will need to be adapted to accommodate the novel characteristics of genomic sequencing assays. It should be clear, too, that oversight of clinical laboratory tests and testing does not reside in any one law or agency. This responsibility is shared by a wide array of government agencies, accrediting organizations, professional societies, payers, clinicians, and medical specialists each of whom focus attention on part of the complex fabric that we consider oversight. Consequently, we should expect the "new" rules for oversight of genomic testing to emerge from a variety of quarters that will address issues of safety, efficacy, quality assurance, analytical and clinical validity, and clinical utility.

Reimbursement for Genomic Tests

Reimbursement for Pathology and Laboratory Services is intrinsically linked to the American Medical Association (AMA)'s CPT system for accurately describing medical, surgical, and diagnostic services. Once defined and codified, specific services are assigned values by the Centers for Medicare and Medicaid Services (CMS) or private payers through a number of different mechanisms, which typically endeavor to accommodate the various technical and professional components necessary for providing those services in specific clinical situations.

The vast majority of laboratory test codes are analyte specific (e.g., 84295 Sodium; serum, plasma, or whole blood) making it readily apparent to payers what test was performed, and with linkage to specific ICD9 (international classification of diseases, ninth revision) codes, in what clinical setting, allowing for some degree of confidence of appropriate test usage. Such transparency was lacking for the molecular pathology codes available through 2012 which were descriptive of test component methodologies (e.g., 83896 nucleic acid probe, each) and demanded the application of multiple codes for any one analyte, further obscuring what service was being provided, and making linkage to specific clinical scenarios impossible. Consequently, in 2009 the AMA CPT Editorial Panel commissioned a working group to develop a revised coding scheme in order to recognize molecular pathology tests in a manner consistent with other laboratory and pathology services. These new molecular codes were first introduced in 2012 and fully implemented in 2013 with the retirement of the older methodology based codes.

Designed with input from a variety of stakeholders including public and private payers, test vendors, laboratories, trade and professional organizations, the new molecular pathology CPT codes are exquisitely analyte specific, frequently with descriptors that suggest the clinical scenario for test usage (e.g., 81241 F5 (coagulation Factor V) (e.g., hereditary hypercoagulability) gene analysis, Leiden variant). Organized into two levels, the Tier 1 codes accommodate the most commonly performed molecular diagnostic tests. The rationale for establishing this grouping of tests was based on an acknowledgement that the methodologies for performing those assays, while perhaps still varied, had matured sufficiently that a single reimbursement value could fairly accommodate all laboratory practice settings. The Tier 2 grouping was intended to accommodate less common tests for which methodologies had not sufficiently evolved to define common practice parameters, either in terms of test design, instrumentation, or professional work. These tests were grouped into levels based on an acknowledged level of test complexity using existing methodologies, with traditional Sanger sequencing as the prototype methodology used by most laboratories. Categorization in Tier 2 is not expected to be permanent. With time, as methodologies become more standardized, a Tier 2 test could eventually be assigned a specific Tier 1 code. The resemblance of the Tier 2 scheme to the surgical pathology code levels (88300–88309) is not happenstance, recognizing the functionality of the surgical pathology codes. A significant difference from the surgical pathology codes is that the Tier 2 codes cannot be self-assigned. Unassigned tests are relegated to the unlisted procedure code (81479 unlisted molecular pathology procedure).

Like the Tier 1 codes, the Tier 2 codes endeavor to be analyte specific, with the vast majority defining specific genetic variations that often time define a specific inherited syndrome. Perusal of the Tier 2 codes shows that the majority are essentially syndromically defined, i.e., this is the molecular inquiry needed to evaluate one specific (genetic) clinical question. This emerging characteristic of the molecular pathology codes distinguishes these tests from traditional laboratory tests. These complex molecular assays move beyond simply analytical detection of a biochemical analyte to a more involved evaluation of genetic complexity related to a specific clinical scenario. In that sense, it is more appropriate to refer to these evaluations as "services" rather than "tests," acknowledging the significant professional knowledge and judgment necessary to appropriately perform, understand, and interpret such tests.

This is the reimbursement setting in which clinical laboratories are beginning to perform assays that utilize next generation sequencing (NGS). Anticipating the issue of how to recognize NGS based testing within CPT, the AMP Economic Affairs Committee initiated discussions in early 2012 that would, hopefully, enlighten the topic. AMP released its proposal coding scheme in March, 2013 [2]. The AMA CPT Editorial Panel, working through its Molecular Pathology Advisory Group (MPAG) has scheduled stakeholder meetings that will open the discussion with an eye to having a functional coding system in place for 2014.

The development of a codification scheme for NGS tests brings to the forefront questions that go far beyond simple reimbursement and call for serious introspection and debate about how and why any test should be used, ethical considerations in reviewing and making available unsolicited genetic information, and even questions regarding the basic principles of medical practice. The challenge in addressing NGS testing is to recognize unique laboratory and pathology services in a manner that is consistent with existing pathology related services and codes. Consideration of the existing Tier 1 and Tier 2 Molecular Pathology codes emphasizes three elements that need to be accommodated: transparency, clinical utility, and professional work. In the context of NGS tests, all three of these elements depart from their traditional meanings in laboratory testing.

Whereas transparency is readily achieved for the simple Tier 1 "analyte specific" codes, the definition blurs with the Tier 2 codes where the analyses are of many genetic alterations in a single gene whose commonality is an association, often to varying degrees, with specific clinical syndromes and scenarios. There may be no one specific "analyte," and the service that is requested is more appropriately described as a syndromically related genetic evaluation. This distinction will take on even greater import with NGS where the genetics evaluation can be readily extended to involve multiple relevant genes for their contribution to specific clinical scenarios. The "analyte" is now more a clinical question than a biochemical entity.

The issue of clinical utility for simple tests, molecular or biochemical, is generally acknowledged in the descriptor, if not by the fact that a CPT code has been assigned to a specific assay. The clinical descriptor takes on more importance in the Tier 2 codes where, in many circumstances, the test itself is essentially defined by the clinical syndrome. We can anticipate that NGS tests will, in many circumstances, be utilized in a similar manner to provide multigene evaluations relevant to specific clinical questions.

The unique capability of NGS methodologies to interrogate multiple target sequences simultaneously challenges the current mode of clinical laboratory testing. Typically, a single analytical test is ordered, the test is performed, a result is reported, a bill is generated, and the laboratory or physician is compensated for the work performed. The technical output of an NGS assay could generate data far in excess of what is needed to address a specific clinical question. The opportunity to "re-query" an existing NGS database defines a new kind of medical service that markedly deviates from traditional laboratory tests. A "re-query" could reflect a series of iterative inquiries to evaluate a complex phenotype, for example, or could be unrelated to the primary clinical question. An example of the latter might an evaluation of drug metabolizing gene variants for the purpose of selecting therapy and dosing some months after primary evaluation of a malignant tumor for diagnosis and classification. The technical work for the "re-query" would be different than for the initial sequencing, although costs of data storage, retrieval, editing, and quality assurance (QA) could still be substantial. The predominant work for the re-query would be professional evaluation of the data and interpretation in the context of the question being addressed.

With these considerations in mind, the AMP NGS proposal sought to categorize potential uses of NGS tests, identifying usages that could complement or replace older methodologies, and highlighting those unique features of NGS tests that herald novel medical services which will need novel CPT codes to adequately recognize the technical and professional services provided. Examples put forward in the AMP proposal are listed in Table 9.1. Some of these usages reflect some technical advantage or efficiency in addressing clinical problems already defined by the Tier 2 codes. Unless the newer technology generates novel clinically useful information, the existing Tier 2 codes would appear sufficient. Other usages of NGS, however, are categorically new tests. For some applications, the clinical question will define the extent of inquiry and the services can be defined in those terms. Undoubtedly, broader genomic inquiries will emerge as NGS becomes more common and feasible. For those inquiries which have demonstrated clinical utility, a CPT coding scheme which follows the aforementioned principles will be able accommodate their introduction into clinical service.

Table 9-1 **Examples of Clinical Genomic Sequencing Assays (Excerpted From [2])**

1. Aneuploidy detection in circulating cell-free fetal DNA (chromosome 21 only or 21, 18, and 13)
2. Disorder-specific multigene evaluations for heritable disorders
3. Identification of rare genetic defects in individual patients
4. Multi-gene evaluation of a neoplasm for diagnostic, prognostic, and/or therapeutic decision-making
5. Clonality assessment in lymphoma
6. Whole-exome and whole-genome analysis of a neoplasm
7. Microbiome evaluations

Conclusions

The implementation of genomic testing in medicine is a major milestone in a journey that had its beginnings more than a half century ago with the discovery that DNA not only serves as the basis of heredity but is also the regulatory medium for cellular differentiation and regulation. It is tempting to regard such a technological advancement as revolutionary, but it is important to remember that the genetic underpinnings and our molecular understanding of disease processes are only one domain in the practice of medicine and our approach to the individual patient. It is likewise important to remember that current regulatory and reimbursement mechanisms have evolved to complement the practice of medicine. Although imperfect in some details, our current oversight mechanisms have functioned well in ensuring the quality and availability of laboratory testing and evaluation, and just as importantly, have served to promote constant innovation and improvement.

Progress on this front will demand a response from all parties including laboratories, professional and accrediting organizations, vendors and manufacturers, and regulatory agencies in order to adapt existing standards and practices to novel genomic technologies and clinical applications.

Reimbursement for genomic sequencing assays will demand a reassessment of the traditional understanding of laboratory testing, moving away from simple chemical analyses to syndromically defined genetic inquiries necessitating increasing amounts of professional work and involvement. Existing CPT structure can be modified to accommodate these new realities, but will demand reevaluation of existing concepts of laboratory testing and professional services related to genomic test interpretation and clinical usage. These challenges offer an opportunity to focus on those elements that have served us well and identify those which have not, and to evolve new systems to accommodate these new technologies in a way that promotes better clinical care of each patient.

REFERENCES

1. 42 CFR 493.1253: Standard—establishment and verification of performance specifications, and 42 CFR 493.1254—standard—maintenance and function checks.
2. Proposal to address CPT coding for genomic sequencing procedures, Association for Molecular Pathology Economic Affairs Committee, Mar 2013.

CHAPTER 10

Patents and Proprietary Assays

ROGER D. KLEIN

Introduction 129
Patent Overview 129
History of Gene Patents 130
Evidence for and against 131
Key Legal Cases 132
Implication of Recent Court Decisions for Genetic Testing 138

Introduction

In the diagnostic realm, patents on relationships between human gene variants and clinical phenotypes, as well as on the underlying genetic sequences themselves, have proven to be extremely controversial. Pathologists, geneticists, other laboratory professionals, as well as some patient and consumer organizations have criticized such patents for increasing test costs, decreasing innovation, reducing patient access, restricting patients' choices of providers and their access to second opinions, inhibiting clinical and basic research, and fostering the development of proprietary databases of medically significant genetic findings [1–3].

Author Michael Crichton joined the chorus of critics in his 2006 novel *Next*, going as far as to include an appendix to the book that exposed the "evils" of gene patents and advocated a ban on them, views he also expressed in a New York Times column [4]. In February 2007, Congressmen Xavier Becarra (D-Calif.) and David Weldon (R-Fla.) introduced "The Genomic Research and Accessibility Act" (HR 977), a bill that would have banned future patents on all nucleic acid sequences.

Conversely, proponents of gene patents have argued that these patents incentivize gene discovery, as well as investments in and commercialization of genetic tests. Gene patents, it has been argued, benefit patients by encouraging discoveries of genetic relationships and the development and introduction of new assays that in the absence of patents would not have been brought to fruition.

This chapter chronicles the history of human gene patents, discusses arguments for and against gene patents, and presents key legal cases that impact on or directly address the validity and permissible scope of such patents. Finally, the implications of these recent legal developments for diagnostic testing are discussed.

Patent Overview

A US utility patent confers upon the patent holder the right to exclude others from making, using, selling, offering to sell, or importing an invention or a product made by a patented process, for 20 years from the filing date [5]. The basis for the US patent system is found in the Constitution, which in Article I, Section 8, Clause 8 states, "The Congress shall have Power...To promote the Progress of Science and useful Arts, by securing for limited Times to Authors and Inventors the exclusive Right to their respective Writings

G.J. Netto and I. Schrijver (eds.), *Genomic Applications in Pathology*,
DOI 10.1007/978-1-4939-0727-4_10,

and Discoveries;..." Patent exclusivity has historically been justified by the incentives it generates for inventors to create, commercialize, and disclose new inventions, the benefits from which will accrue to society at large.

Congress enacted the first US patent laws in 1790. The Patent Act of 1790 was repealed and replaced in 1793, and the patent laws have subsequently been modified on numerous occasions. The basic structure of the current Patent Act was established in 1952, when the patent laws were reenacted in their entirety. Since passage of the 1952 Act, the patent laws have been amended several times, recently undergoing significant revisions by way of the America Invents Act of 2011.

Under US patent law, patentable inventions must be novel, non-obvious, and useful [6]. In addition, under "written description" and "enablement" requirements a patent must describe the patented invention in what is termed its "specification," "in such full, clear, concise, and exact terms as to enable any person skilled in the art to which it pertains, or with which it is most nearly connected, to make and use the same." Moreover, the specification must set forth the "best mode," in the mind of the inventor, of practicing the invention [7]. Within the specification, patent "claims" define the invention's features, establishing the boundaries of what is claimed, much as a real estate deed delineates the boundaries of a plot of land.

Patent applications are submitted to the United States Patent and Trademark Office (USPTO) where they are rejected, or allowed and issued. "Processes, machines, manufactures, and compositions of matter" can be patented [8], but patents may not be obtained on products of nature or, under the "natural phenomenon doctrine," "laws of nature, natural phenomena, and abstract ideas" [9].

Patent infringement, which encompasses the making, using, selling, offering to sell, or importing of a patented product or a product made by a patented process, can occur through: direct infringement of the patent; [10] inducement of others to infringe the patent [11]; or contributing to another's infringement of the patent [12]. For example, prior to recent US Supreme Court decisions, a laboratory could have been found to have directly infringed a gene patent if it tested for mutations in a patented gene, or variants claimed in a patented genotype–phenotype association.

In order to be found liable for inducing another to infringe a patent, a party must have actively, intentionally, and knowingly solicited or assisted another to infringe the patent, with the solicited individual or entity itself having directly infringed the patent. Thus, a laboratory that used educational materials to promote an offered test for a patented genetic association to physicians who then ordered the test, received the results, and thought about the association during management of their patients, could until recently have been found to have induced the direct infringement of the patent by the ordering physicians.

Finally, sale of a material component of a patented invention that has no substantial use other than as a component of the invention denotes contributory infringement. Applying this definition, the laboratory in the preceding example could also have been found liable for contributory infringement for providing testing for the patented genetic association.

History of Gene Patents

The legitimization of gene patents in the USA appears to have been an outgrowth of legal and political changes that were initiated in response to the economic dislocations of the late 1970s and early 1980s. During this period, the country was plagued by high unemployment, high inflation, and a decline in economic confidence. In response, Congress took a number of steps to encourage the growth of domestic technology industries. Among the most significant of these were changes to the US patent system.

To maximize the economic value derived from our substantial federal investments in basic science research, Congress in 1980 passed the Bayh-Dole Act, which encouraged universities to patent, and thereby commercialize, inventions arising out of government sponsored research grants [13]. In the years subsequent to the passage of Bayh-Dole, federal financial commitments dedicated to biomedical research dramatically increased. National Institutes of Health funding of biomedical research ballooned from approximately $5 billion in the late 1970s to $26 billion in 2003 [14]. Because of these governmental actions, the number of patents assigned to universities increased from 264 in 1979 to 3,291 in 2002 [15, 16].

In another important event, in 1980 the US Supreme Court ruled in *Diamond v. Chakrabarty* [17] that man-made, living organisms could be patented. In its decision, the Supreme Court urged a broad interpretation of patent eligibility, holding that "anything under the sun that is made by man," including living organisms, can be patented. Finally, in an effort to provide national uniformity and add greater certainty and expertise to the application of patent law, in 1982 Congress created the Court of Appeals for the Federal Circuit (CAFC), with exclusive appellate jurisdiction for patent cases [18].

Since its inception, Federal Circuit decisions have affected the biotech sector significantly by generally expanding patent-eligible subject matter and strengthening the rights of patent holders relative to potential infringers. Many patents have since been issued on a range of biotech inventions, from transgenic mice and leukemia-derived cell lines to recombinant drugs and vaccines. Thousands of patents have also been awarded on human gene sequences, genetic variants, and more recently, genotype–phenotype correlations [19].

The coalescence of the preceding events set the stage for the enormous growth of the US biotech industry. For example, from 1982 to 2002, US Food and Drug Administration (FDA) approvals for biotech drugs and vaccines grew from 2 to 35. The number of US biotech companies expanded from 225 in 1977 to 1,457 in 2001. Biotech employment mushroomed from 700 in 1980 to 191,000 in 2001. In addition, the industry's growth has created hundreds of thousands of jobs in related industries [20, 21].

It has been argued that in awarding gene patents the US Patent and Trademark Office and the CAFC merely followed the Supreme Court's instruction in *Chakrabarty* to interpret patent eligibility broadly [22]. Importantly, post-Chakrabarty our patent system looked to chemical law precedents as a basis for awarding gene patents, and treated DNA itself as a chemical despite its dual roles as a physical substance and a store of biological information. In *Amgen v. Chugai Pharmaceutical Co*. the CAFC wrote, "A gene is a chemical compound, albeit a complex one" [23].

Prior precedents in chemical law upheld the patenting of isolated, purified compounds such as aspirin, epinephrine, vitamin B12 and prostaglandins [24–27]. The Patent Office applied these legal precedents to isolated DNA sequences. This direct superimposition of chemical law precedents to DNA permitted circumvention of the "product of nature" doctrine's longstanding prohibition against patenting natural substances, and allowed for the issuance of patents on isolated, purified human genes.

Evidence for and against

To practitioners in the field it appears obvious that gene patents have significantly inhibited the provision of genetic testing services [28]. Many providers have discontinued or have been prevented from providing molecular genetic testing for inherited breast and ovarian cancer, Duchenne muscular dystrophy, spinocerebellar ataxias, genes causing Long QT syndrome, as well as the *FLT3* internal tandem duplication in patients with intermediate risk acute myelogenous leukemia (AML), the *JAK2* p.Val617Phe (better known as V617F) variant in myeloproliferative neoplasms, and many others.

Intuitively, one would expect that monopolistic behavior would lead to increased prices and decreased patient access to testing. Although there is some support for this contention, true markets do not exist for health care services in the USA because of the roles of third party insurance and government as major payers. Further, prices are difficult to obtain, which makes comparisons of actual charges difficult [29]. However, for single gene discoveries and their subsequent introduction into clinical testing, the notion that gene patents have been a necessary stimulus seems dubious. In general, rather than encouraging the introduction of new tests, gene patents have tended to cause laboratories to discontinue tests they had already been performing.

Most human genes on which clinical testing has been performed have been discovered by university faculty members. For these professors, publication and solicitation of grants based upon their discoveries is necessary for academic promotion and even professional "survival," rendering patents a superfluous incentive. Inherited diseases are frequently rare, offering very limited market potential. Yet many such genes have been discovered despite an apparent lack of significant commercial or monetary potential because of the research interests of the investigator.

Lastly, it is usually relatively inexpensive to design, develop, validate, and perform genetic tests using justifiably patented tools and techniques. This is in contrast to pharmaceuticals, which require costly, extensive periods of discovery and testing, and must undergo an expensive approval process, features that support the need for robust patent protection [30].

Although the preceding discussion regarding the adverse effects of gene patents on the introduction of new molecular genetic assays holds true for most assays, the relative impact of gene-related patents on some tests based on multi-analyte gene expression profiling seems less clear. A central feature of these assays is a reliance on proprietary mathematical algorithms that proponents claim allow for correlation of the expression patterns of for example multiple mRNAs, sometimes in combination with other parameters, with relevant clinical characteristics such as diagnosis, prognosis, or response to drug therapy. A variety of such tests are oncology oriented.

Implementation of these types of expression assays typically requires prolonged and potentially expensive periods of study in order to establish sufficient clinical validity and utility to justify their use. Arguably, at the time of this writing few such assays have been crossed this threshold and are supported by high level evidence of this nature. In addition, expression profiling tests may in the future require FDA approval or clearance, increasing development costs. Therefore, exclusivity may be necessary to attract sufficient funding to advance those assays that ultimately prove worthy in clinical care. Arguably, some inventive work has occurred in such assays through establishment of the gene "signature." Moreover, although the assays incorporate or rely on natural, biological associations, they also can generally be "invented around," and therefore pose less risk of tying up essential natural phenomena. Interestingly, although patent protection may be essential to bring assays of this type to market, individual gene or nucleic acid patents could otherwise obstruct their development by restricting use of the genes available for inclusion in the test.

Key Legal Cases

Bilski v. Kappos

In *Bilski v. Kappos*, the USPTO rejected a patent application in which the submitted claims covered a process of hedging commodities against price fluctuations. The method involved contracting to purchase commodities at fixed prices from sellers who wanted to hedge against a fall in prices, while contracting to sell commodities at fixed prices to consumers who were hedging against a rise in prices. On appeal the CAFC upheld the USPTO's decision denying a patent [31].

Prior to *Bilski*, the rule at the CAFC was that patentable processes had to produce a "useful, concrete, and tangible result." In *Bilski*, the CAFC articulated a new standard, its "machine or transformation test." The CAFC sitting as the entire court, termed *en banc*, held that patentable processes must be tied to a particular machine or apparatus or must transform a particular article into a different state or thing, and that this transformation must be central to the purpose of the process. Bilski's hedging process, the CAFC ruled, failed to meet the machine or transformation test, and therefore was ineligible to receive a patent [31].

The US Supreme Court affirmed the lack of patent eligibility of the claimed hedging process, but refined the CAFC's reasoning [32]. Although the machine or transformation test may be a "useful and important clue or investigative tool" for deciding whether some processes are patent eligible inventions under 35 U.S.C. section 101 of the Patent Act, the Supreme Court held that it is not the sole test of patent eligibility by which such processes are to be evaluated.

Some gene patent claims that assert ownership over genotype–phenotype associations have framed these natural laws as a series of steps, thus characterizing them as processes. The *Bilski* decision influences the framework under which the patent eligibility of process claims is evaluated. Therefore, although it was narrowly crafted to the specific set of business facts before the Supreme Court in the case, *Bilski* has relevance for the assessment of the patent eligibility of process claims involving human genes.

KSR Int'l Co. v. Teleflex Inc.

In the 2007 case of *KSR Int'l Co. v. Teleflex Inc.*, a unanimous US Supreme Court relaxed the legal standards for determining patent obviousness under section 103 of the Patent Act [33]. KSR added a sensor to one of its previously designed automobile throttle pedals. Teleflex then sued KSR for infringement of a patent that claimed the combination of an adjustable automobile accelerator pedal and an electronic sensor. In response, KSR argued that the patent was invalid because its subject matter was obvious. The district court agreed with KSR, and ruled that the accelerator–sensor combination was obvious. The CAFC reversed the lower court decision.

In upholding the patent, the CAFC applied what was termed its "teaching, suggestion, or motivation" test (TSM test) for obvious determinations. Under this test, a patent claim could only be found obvious if there was "some motivation or suggestion to combine the prior art teachings" present in the previous body of knowledge in the field, the nature of the problem the solution sought to solve, or the knowledge of a person who possessed ordinary skill in the field. That an approach was "obvious to try," the CAFC wrote, had under previous precedents long been irrelevant.

The Supreme Court rejected the CAFC's rigid, formalistic, and narrow process for obviousness determination in favor of a more "expansive and flexible approach," ruling that the throttle pedal-sensor combination had been obvious at the time of the patent application. Importantly, the Supreme Court held that obviousness to try a problem-solving approach can in fact render a patent obvious under circumstances in which there is a demonstrated need for a discovery, and a finite number of identified, predictable solutions to the problem. The Court wrote:

> When there is a design need or market pressure to solve a problem and there are a finite number of identified, predictable solutions, a person of ordinary skill has good reason to pursue the known options within his or her technical grasp. If this leads to the anticipated success, it is likely the product not of innovation but of ordinary skill and common sense. In that instance the fact that a combination was obvious to try might show that it was obvious under §103 [33].

Many patented genes were initially mapped to a chromosomal region prior to their discovery. In addition, many medically important genes are involved in sequential biochemical pathways, in which disease-related perturbations were known before the identification of particular genetic associations. Therefore, it would have been obvious to look for variants in these genes among a finite number of genes during genetic studies of the relevant disorder. Finally, cDNA sequences are directly derived from the exon sequences of native genes, and can also be deduced from the amino acid sequences of the proteins for which they encode, likely rendering significant numbers of patent claims on cDNA obvious. In light of the preceding, the Supreme Court's decision in *KSR* potentially affects the validity of many gene-related patents.

In Re Kubin

In 2009, the case of *In Re Kubin* provided the CAFC with an early opportunity to apply the obviousness paradigm the Supreme Court set forth in *KSR* [34]. In *Kubin*, the USPTO refused to award a patent on the full gene and cDNA sequences of the Natural Killer Cell Activation Inducing Ligand (NAIL), a natural killer (NK) cell surface receptor that plays a role in cellular activation. The Patent Office rejected the application both on obviousness grounds under 35 U.S.C. section 103 and for inadequate written description under 35 U.S.C. section 112.

On appeal, the CAFC affirmed the Patent Office's decision, agreeing that delineation of the NAIL gene sequences were obvious in light of the prior art, which included knowledge of the existence of the NAIL protein, but not its protein sequence. Citing the case of *Graham v. John Deere Co.* [35], the CAFC reviewed the factual inquiries necessary for a legal finding of obviousness. These, the CAFC wrote, include: "(1) the scope and content of the prior art; (2) the differences between the prior art and the claims at issue; (3) the level of ordinary skill in the art at the time the invention was made; and (4) objective evidence of non-obviousness, if any." Under the aforesaid criteria, the CAFC found that the NAIL gene sequences were obvious.

Applying the Supreme Court's *KSR* decision the CAFC reversed one of its previous DNA cases, *In re Deuel*, in which it had held that "obvious to try" an approach was an inappropriate test for obviousness [36]. In *In re Deuel*, the CAFC had reversed the Patent Office's conclusion that the existence of a prior art reference describing a method of gene cloning together with the partial amino-acid sequence of the protein, rendered the underlying cDNA sequence obvious. Instead, the *In re Deuel* Court found that knowledge of the protein sequence was itself insufficient to generate the sequence of the underlying cDNA and, therefore, that the sequence was non-obvious. Further, the CAFC eliminated "obviousness to try" as a potential determinant of obviousness. The Court wrote:

> [T]he existence of a general method of isolating cDNA or DNA molecules is essentially irrelevant to the question whether the specific molecules themselves would have been obvious, in the absence of other prior art that suggests the claimed DNAs... 'Obvious to try' has long been held not to constitute obviousness. A general incentive does not make obvious a particular result, nor does the existence of techniques by which those efforts can be carried out.

In light of the Supreme Court's prior rejection of the CAFC's "obvious to try" doctrine in *KSR*, the CAFC in *Kubin* found that the NAIL cDNA and full gene sequences were obvious to try, and therefore obvious under section 103. The CAFC stated:

> In light of the concrete, specific teachings of Sambrook and Valinente, artisans in this field, as found by the Board in its expertise, had every motivation to seek and every reasonable expectation of success in achieving the sequence of the claimed invention. In that sense, the claimed invention was reasonably expected in light of the prior art and 'obvious to try.'

Mayo v. Prometheus

In *Mayo Collaborative Services v. Prometheus Laboratories, Inc.* [37], Prometheus Labs sued Mayo Clinic in the District Court for the Southern District of California for infringement of two patents covering the post-administration correlation of blood levels of the thiopurine metabolites 6-methyl mercaptopurine and 6-thioguanine with thiopurine efficacy and related side effects. Both patents were written in the form of stepwise processes, the relevant claims of which included the generic steps of: (1) administering the drug; (2) measuring the metabolite levels; and (3) describing the metabolite concentrations above and below which are associated with an increased likelihood of toxicities or lack of efficacy respectively, then informing the ordering physician of the potential need to decrease or increase the drug dose. Thus, the patent in effect claims the reference range for thiopurine drugs. Mayo Clinic had been utilizing Prometheus' test, but in 2004 announced that it was going to offer its own internally developed test for metabolites. Prometheus sued Mayo for patent infringement. Mayo Clinic argued that Prometheus' patents covered unpatentable natural phenomena, and were therefore invalid as a matter of law under section 101 of the Patent Act. The District Court agreed with Mayo, and ruled that Prometheus' patents were invalid. The CAFC reversed the District Court, instead holding that the patents claimed methods of treatment. Moreover, the CAFC held, the in vivo metabolism of thiopurine agents constituted transformations of matter under that Court's "machine or transformation test," a test which was discussed earlier in this chapter in connection with *Bilski v. Kappos*. In *Bilski*, the Supreme Court clarified that although the "machine or transformation test" is an important and useful clue to patent eligibility, it is not a definitive test for it.

Mayo appealed to the Supreme Court, which following its decision in *Bilski*, accepted *Mayo v. Prometheus* and immediately returned it to the CAFC for reconsideration. On remand, the CAFC reaffirmed its earlier decision reversing the District Court's determination that Prometheus' patents were invalid.

Mayo again appealed to the Supreme Court, and the Court accepted the case. In a nine-to-zero decision, the Supreme Court held that the processes claimed in Prometheus' patents were not patent eligible. The Court recognized that an unpatentable biological correlation lay at the center of Prometheus' patents. In order to receive a process patent that purports to claim an application of a natural law, the Court noted, sufficient inventive effort must be added to the natural law so as to ensure that the patent is "significantly more than a patent upon the natural law itself." Moreover, the Court emphasized that the addition of routine steps cannot convert the natural law into a patentable process. As the Court explained, "If a law of nature is not patentable, then neither is a process of reciting a law of nature, unless that process has additional features that provide practical assurance that the process is more than a drafting effort designed to monopolize the law of nature itself." The Court succinctly summarized: "[T]o transform an unpatentable law of nature into a patent-eligible application of such a law, one must do more than simply state the law of nature while adding the words 'apply it'."

The unanimity, clarity, and strength of the Supreme Court's opinion in support of this ruling standing alone implies that analogous patents covering genotype–phenotype associations are also invalid. This conclusion is bolstered by the Court of Appeals' affirmance of the District Court finding of invalidity of Myriad Genetics' sequence comparison claims in *Association for Molecular Pathology v. Myriad Genetics* discussed subsequently [38], and is reinforced and further strengthened by the Supreme Court's decision finding human DNA patent ineligible in the *Myriad* case. *Mayo v. Prometheus* and *AMP v. Myriad* have important implications for genomic analyses performed using next generation sequencing, and for genetic testing as a whole.

Association for Molecular Pathology v. Myriad Genetics, Inc.

Finally, in *Association for Molecular Pathology v. Myriad Genetics, Inc.* [38], a lawsuit sponsored by the American Civil Liberties Union, various medical and professional societies, health care providers, and breast cancer patients sued Myriad Genetics, the University of Utah Research Foundation, and the USPTO seeking to invalidate key composition of matter and process claims of patents covering the wild-type and mutated sequences of the *BRCA1* and *BRCA2* genes, as well as correlations between variants in those sequences and the predisposition to breast and ovarian cancer.

In total, the plaintiffs challenged 15 claims contained in 7 patents. They argued that these patent claims were invalid under section 101 of the Patent Act of 1952, and unconstitutional under Article I, Section 8, Clause 8 and the First and Fourteenth Amendments, because they asserted ownership of natural products, natural laws, natural phenomena, abstract ideas, and basic human knowledge or thought. In response, Myriad argued that its patents claimed DNA sequences that were identical to those in the human body, but because the sequences were isolated from the body they constituted human inventions.

Myriad also asserted that its patented associations between variants in *BRCA1* and *BRCA2* and the hereditary predisposition to breast and ovarian cancers were actually diagnostic methods involving sequence comparisons, not patents on the biological relationships themselves. The District Court distilled the lawsuit into a single fundamental question, "Are isolated human genes and the comparison of their sequences patentable?"

The Judge, Robert W. Sweet, emphasized the centrality of knowledge of molecular biology to the proper disposition of the case, as well as the importance of any potentially relevant additional inventive steps [39]. On page 27 of its opinion, Judge Sweet wrote: "An understanding of the basics of molecular biology is required to resolve the issues presented and to provide the requisite insight into the fundamentals of the genome, that is, the nature which is at the heart of the dispute between the parties…" The Court devoted the next 19 pages of the opinion to a thorough review of generally accepted principles of molecular biology. It concluded the section with the recognition that some inventive work was involved in the initial sequencing of the *BRCA1* and *BRCA2* genes stating: "However, because sequencing requires knowledge of the sequence of a portion of the target sequence, some ingenuity and effort is required for the initial sequencing of a target DNA." Expert declarations by Mark A. Kay, M.D., Ph.D. and this chapter's author helped the Court sort out the extent, significance and relevance of this work to the validity of the claims at issue.

In the pertinent sections of their dueling declarations, Kay attempted to emphasize the inventive aspects of sequencing a newly discovered product, while Klein delineated the breadth of the patents and the natural products and laws they claimed; the routine, insubstantial, and non-transformative steps involved in performing genetic testing, and the relationship of genetic testing to other forms of medical diagnosis. In paragraph 183 of his declaration, Dr. Kay described the steps involved in sequencing a newly identified product:

> To sequence a particular target, at least part of the target sequence must be known to design a suitable primer. The initial sequencing of a target sequence requires ingenuity far beyond the mere application of routine laboratory techniques and usually involves a significant amount of trial and error. A primer is used to initiate the sequencing reaction at the desired location of a target sequence. A primer is an artificial DNA fragment, usually between 15 and 30 nucleotides long, that binds specifically to the target nucleotide sequence. The nucleotide sequence of the primer is complementary to the target sequence such that the bases of the primer and the bases of the target sequence bind to each other.

By contrast, in paragraphs 32–34 Dr. Klein wrote:

> The claims at issue in this case do not cover diagnostic tools or actual methods used in genetic testing. Nor are they analogous to patents on medical instruments. Rather they claim DNA sequences which are themselves the subject of medical inquiry. Further, they incorporate generic steps in an effort to describe the biological relationships between mutations in *BRCA1* and *BRCA2* and the predisposition to cancer in the abstract patent language of a 'process.' However, the key steps in genetic testing, DNA extraction, amplification, and sequencing can now be performed using routine, automated methods. Nevertheless, the defendants claim the exclusive right to read and compare *BRCA1* and *BRCA2* sequences irrespective of the method used, whether that method is in existence now or will be invented in the future. Correlating a patient's gene sequence with the predisposition to disease is simply another form of medical diagnosis, similar to correlating elevations in blood glucose with diabetes, a heart murmur with mitral stenosis, or the patterns on a pathology slide with a particular type of tumor and its optimal therapy. Automated sequencers reveal the sequence of the nucleotides visually in what is called a chromatogram. That chromatogram is then "read" (by software and visual inspection) to determine a patient's gene sequence. DNA extraction and sequencing are not transformative activities. Rather extraction is a routine, non-substantial preparatory step that allows for PCR amplification and sequencing. Sequencing is an automated procedure. DNA extraction, PCR, and sequencing do not involve transformations that are central to the purpose of the process of reading a patient's gene sequence. Unlike "tanning, dyeing, making waterproof cloth, vulcanizing India rubber, or smelting ores," which are performed for the purpose of physically transforming substances so as to create what are essentially new materials for their own sake, the purpose of genetic testing is solely to read the sequence of the DNA, not to transform it into something else. Only in this way can the patient and her physician learn whether a medically relevant mutation is present in her body.

On March 29, 2010 in a landmark decision, the District Court held that the composition of matter claims on the *BRCA1* and *BRCA2* gene sequences and their cDNAs, and the process claims covering the correlations between mutations in *BRCA1* and *BRCA2* and the predisposition to breast cancer and ovarian cancer are invalid as a matter of law. In evaluating the composition of matter claims on the isolated gene sequences, the

Court emphasized the unique informational characteristics contained in the DNA sequence, and the preservation of that native sequence in isolated DNA, stating "Because the claimed isolated DNA is not markedly different from native DNA as it exists in nature it constitutes unpatentable subject matter under 35 U.S.C. section 101." Similarly, the Court found comparison claims of known wild-type and patient sequences for diagnosis, claims that in effect asserted ownership over the biological relationships between *BRCA1* and *BRCA2* mutations and the predisposition to breast cancer, invalid as merely claiming abstract mental processes.

On appeal, the CAFC on July 29, 2011 in a two-to-one decision reversed the District Court, holding that isolated human gene sequences are patent eligible. However, the CAFC upheld the lower court's ruling that Myriad's sequence comparison claims were invalid. The plaintiffs appealed the case to the Supreme Court.

Immediately after deciding *Mayo v. Prometheus*, the Supreme Court accepted *AMP v. Myriad*, threw out the CAFC's decision, and sent the case back to the Court of Appeals for further consideration in light of its decision in *Mayo*. After rebriefing of the case and a second round of oral arguments, the CAFC again held two-to-one that isolated human genes are patent eligible on the grounds that they represent new compositions of matter that do not exist in nature.

As in the CAFC's previous decision in the case, each judge wrote a separate opinion. All three judges agreed that *BRCA1* and *BRCA2* cDNA should be patent eligible, reasoning that cDNA is not naturally occurring and is made by man. The central disagreement among these judges was whether separating human DNA from its chromosome and other cellular constituents renders it a patent eligible invention.

The two judges who determined that human DNA is patent eligible came to the same conclusion using different reasoning. One judge, who authored what was nominally the primary opinion for the Court, opined that because separating a gene from its chromosome involves breaking covalent bonds, a DNA sequence removed from its natural environment is a new chemical. Another judge relied at least in part on the past practice of the USPTO in granting such patents, and the reliance of companies and inventors on that practice. This judge said may have voted differently had the question come before her on a "blank canvas."

The dissenting judge wrote that the breaking of covalent bonds alone did not create a new molecule, and was not determinative of patent eligibility. Rather, he concluded that the genes' DNA sequences are identical whether the genes are within or outside the body, and because of this that these DNAs are fundamentally the same molecule, irrespective of location. For the dissenting judge, the importance of the sequence of nucleotides in the DNA molecules substantially outweighed the importance of any chemical differences between the DNA in the body and DNA isolated from it.

However, the CAFC ultimately chose to disregard the constancy of the gene's most fundamental and relevant property, its coding sequence. On behalf of the Court Judge Alan Lourie wrote: "The isolated DNA molecules before us are not found in nature. They are obtained in the laboratory and are man-made, the product of human ingenuity." Judge Lourie maintained that native and isolated gene sequences have distinct chemical structures and identities because the native genes have been separated from associated proteins and the chromosomes on which they naturally reside, either through the cleaving of covalent bonds or by synthesis. In addition, the CAFC again held that Myriad's sequence comparison claims were invalid. The plaintiffs once more appealed the case to the Supreme Court.

On June 13, 2013, in an historic nine-to-zero decision authored by Justice Clarence Thomas, the Supreme Court held that naturally occurring DNA sequences are "products of nature" that are not patent eligible. The court acknowledged Myriad's contribution to the field, but noted that its discoveries were limited to the identifying the precise location and sequence of the *BRCA1* and *BRCA2* genes. The Court stated: "In this case…Myriad did not create anything. To be sure, it found an important and useful gene, but separating that gene from its surrounding genetic material is not an act of invention." Moreover, the Court referred back to Myriad's patent claims, which themselves confirmed that the fundamental essence of DNA lies in its information content.

> "Myriad's claims," the Court wrote, "are simply not expressed in terms of chemical composition, nor do they rely in any way on the on the chemical changes that result from the isolation of a particular section of DNA. Instead, the claims understandably focus on the genetic information encoded in the *BRCA1* and *BRCA2* genes. If the patents depended upon the creation of a unique molecule, then a would-be infringer could arguably avoid at least Myriad's patent claims on entire genes…by isolating a DNA sequence that included both the *BRCA1* or *BRCA2* gene and one additional nucleotide pair. Such a molecule would not be chemically identical to the molecule 'invented' by Myriad. But Myriad obviously would resist that outcome because its claim is concerned primarily with the information contained in the genetic sequence, not with the specific chemical composition of a particular molecule."

Finally, the Court did rule that cDNA is patent eligible because it is not naturally occurring. However, patent eligibility, as the Court pointed out in a footnote, does not necessarily equate to patentability under other sections of the Patent Act that this decision did not address. Moreover, because cDNA is not essential for the performance of most genetic testing, the ruling that cDNA is patent eligible is unlikely to have a significant impact on molecular genetic testing going forward.

Implication of Recent Court Decisions for Genetic Testing

In two recent decisions relevant to genetic testing, both unanimous, the Supreme Court reaffirmed its longstanding prohibitions on patenting natural laws and products of nature. In *Mayo*, the Court was clear that characterizing a biological association as a process does not, without adding a truly inventive step, convert the association into a patent eligible application of a natural law. *Mayo* was an extremely important decision, which seemingly means that method patents that attempt to claim associations between genetic variants and clinical phenotypes are invalid. In *Association for Molecular Pathology*, the Supreme Court found that naturally occurring human DNA sequences are not patentable, rendering patents on human genes invalid. When read together these two cases appear to have removed the intellectual property barriers associated with testing for genetic mutations and relationships to clinical phenotypes, whether testing is for identification of predisposition to disease, therapeutic responsiveness, medicinal side effects, or tumor behavior. Thus, the Supreme Court has helped facilitate the introduction of large-scale sequencing into clinical practice, and has thereby encouraged the advancement, development, and implementation of personalized medicine.

REFERENCES

1. Klein RD. Gene patents and genetic testing in the United States. Nat Biotechnol. 2007;25:989–90.
2. Klein RD. Legal developments and practical implications of gene patenting on targeted drug discovery and development. Clin Pharmacol Ther. 2010;87: 633–5.
3. Cook-Deegan R, Conley JM, Evans JP, Vorhaus D. The next controversy in genetic testing: clinical data as trade secrets? Eur J Hum Genet. 2013;21:585–8.
4. Michael C. Patenting Life, N.Y. TIMES, 13 Feb 2007, at A23. http://www.nytimes.com/2007/02/13/opinion/13crichton.html.
5. 35 U.S.C. § 154(a) (2012).
6. 35 U.S.C. §§ 101–103 (2012).
7. 35 U.S.C. § 112 (2012).
8. 35 U.S.C. § 101 (2012).
9. Diamond v. Diehr, 450 U.S. 175, 185 (1981).
10. 35 U.S.C. § 271(a) (2012).
11. 35 U.S.C. § 271(b) (2012).
12. 35 U.S.C. § 271(c) (2012).
13. 35 USC §§ 200–212 (2012).
14. Moses H, Dorsey ER, Matheson DHM, Their SO. Financial anatomy of biomedical research. JAMA. 2005;294:1333–42.
15. Rai AK, Eisenberg RS. Bayh-Dole reform and the progress of medicine. Law Contemp Probs. 2003;66:289–314.
16. http://www.uspto.gov/web/offices/ac/ido/oeip/taf/univ/asgn/table_1_2005.htm.
17. Diamond v. Chakrabarty, 447 US 303 (1980).
18. Federal Courts Improvement Act of 1982, 96 Stat. 25, codified as 28 USC 1295 (2012).
19. Caulfield T, Cook-Deegan RM, Kieff FS, Walsh JP. Nat Biotechnol. 2006;24:1091–4.
20. http://www.bio.org/ataglance/bio/.

21. Lee SB, Wolfe LB. Biotechnology industry. In: Encyclopaedia of occupational health and safety. 4th ed. International Labour Organization.
22. Brief for Respondent, Association for Molecular Pathology v. Myriad Genetics, No. 12-398, 569 US: (7 Mar 2013); U.S. patent and trademark office utility guidelines, 66 Fed. Reg. 1092 (5 Jan 2001).
23. Amgen v. Chugai Pharmaceutical Co., 927 F.2d 1200 (1990), cert. denied, 502 US 856 (1991).
24. Kuehmsted v. Farbenfabriken, 179 F. 701 (7th Cir. 1910), cert. denied, 220 US 622 (1911) (acetyl salicylic acid).
25. Parke-Davis & Co. v. H.K. Mulford & Co., 189 F. 95 (SDNY 1911), aff'd, 196 F. 496 (2d Cir. 1912) (epinephrine).
26. Merck & Co. v. Olin Mathieson Chemical Corp., 253 F.2d 156 (4th Cir. 1958) (Vitamin B12).
27. In re Bergstrom, 427 F.2d 1394 (CCPA 1970) (PGE, PGF).
28. Cho MK, Illangasekare S, Weaver MA, Leonard DGB, Merz JF. Effects of patents and licenses on the provision of clinical genetic testing services. J Mol Diagn. 2003;5:3–8.
29. Secretary's Advisory Committee on Genetics, Health, and Society. Gene patents and licensing practices and their impact on patient access to genetic tests. 2010. http://oba.od.nih.gov/oba/SACGHS/reports/SACGHS_patents_report_2010.pdf.
30. Bessen J, Meurer MJ. Patent failure: how judges, bureaucrats, and lawyers Put innovators at risk. Princeton, NJ: Princeton University Press; 2008.
31. Bilski v. Kappos, 545 F.3d 943 (Fed. Cir 2008) (en banc).
32. Bilski v. Kappos, 130 S.Ct. 3218 (2010).
33. KSR Int'l Co. v. Teleflex Inc., 550 US 398 (2007).
34. In re Kubin, 561 F.3d 1351 (Fed. Cir. 2009).
35. Graham v. John Deere Co., 383 U.S. 1 (1966).
36. In Re Deuel, 51 F.3d 1552 (Fed. Cir. 1995).
37. Mayo Collaborative Services v. Prometheus Laboratories, Inc., 132 S.Ct. 1289 (2012).
38. Association for Molecular Pathology v. Myriad Genetics, Inc., No. 12-398, 569 US (2013)
39. Association for Molecular Pathology v. United States Patent and Trademark Office, No. 09-4515 (SDNY filed 12 May 2009).

CHAPTER 11

Ethical Issues in Clinical Genetics and Genomics

HENRY T. GREELY

Introduction 141

"Targeted" Genetic Testing: Ethical and Legal Issues 141

Other Issues 144

Steps Toward Broadband Genetic and Genomic Testing 146

Broadband Genomic Testing: Ethical and Legal Issues 148

Deciding toTest: Medical Appropriateness and Informed Consent 148

Test Accuracy 149

Communicating Results 151

DTC Marketing 152

Other Issues 153

Conclusions 154

Introduction

Genetic testing has been clinically available for over 50 years and has generated ethical and legal questions for at least that long. The methods for genetic testing have changed dramatically over the years and the details of the ethical and legal issues have changed dramatically, as well, but the basic problems have not [13–15]. Today we are on the edge of clinical whole-genome sequencing ("WGS") technologies. As we move from "retail" or targeted tests, few in number and done only a few at a time, to "wholesale" or broadband tests, where hundreds of different important results may appear from one test, the old types of ethical and legal issues will arise, but their implications will change enormously. These implications are just beginning to be discussed [2, 6, 17, 18, 21, 22, 25, 27].

The ethical and legal issues that have arisen with contemporary genetic testing can be grouped into five rough categories: the decision to test, accuracy, communicating results, direct-to-consumer ("DTC") testing, and "other." This chapter will look first at those issues under current targeted testing. After a brief section on the existing beginnings of broadband genomic testing, the chapter will then reexamine those five areas as they are likely to appear in the coming whole-genome sequence world where today's students will spend the vast majority of their professional careers—and their lives as patients, family members, and citizens.

"Targeted" Genetic Testing: Ethical and Legal Issues

Most existing genetic tests have been "targeted." The tests are for one or a handful of genetic or chromosomal variants known to be strongly associated with particular diseases. Sometimes the patients suggest the genetic test to their doctors, but typically the doctors, listening to their patients' concerns and family

G.J. Netto and I. Schrijver (eds.), *Genomic Applications in Pathology*,
DOI 10.1007/978-1-4939-0727-4_11, © Springer Science+Business Media New York 2015

histories and after completion of a clinical evaluation and examination, recommend the test. Either way, concrete risks lead to specific tests for particular genetic variants.

This is still largely the world of clinical genetics. Pregnant couples seek information from obstetricians about Down syndrome, about genetic diseases known to run in their families, or about conditions that carrier screening has shown they might pass down to their children. Parents whose children have various abnormal conditions ask pediatricians or geneticists to confirm tentative diagnoses through genetic tests. Adults with symptoms of genetic diseases, or who learn of strong family genetic risks, seek testing. In every case, the genetic tests examine only one or a handful of genetic or chromosomal variants. These tests give rise to difficult ethical and legal problems, today as in the past five decades.

Deciding to Test: Medical Appropriateness and Informed Consent

The first ethical question is whether the test should be done at all. Medically inappropriate or unconsented tests are generally unethical. In the targeted context, these decisions are made about a specific test, for one or a few genes linked to one to or a handful of conditions.

A test might be inappropriate because it is a good test but not for this patient. Thus, a woman with an extensive family history that does not include breast or ovarian cancer normally should not be offered *BRCA* testing. Sometimes the test might be inappropriate, because it does not work—or, perhaps, has not been proven to work. Normally one would not want to order a genetic test based on, say, one small and unreplicated study showing a weak association between a genetic variant and susceptibility to a disease.

Some argue that tests are different—they are "only" information and do not have the possible harms and costs of "actual" interventions. This is clearly wrong. For the patients, inaccurate test results, either because of a bad test or an inaccurate result from an otherwise good test, will cause unnecessary anxiety or false reassurance, as well as the possibility of a cascade of subsequent tests and interventions. For the medical system, the costs of unnecessary tests, as well as unnecessary or inappropriate follow-ups, lead to the waste of resources that might have been put to better use. Inappropriate tests have *only* risks and costs with no balancing benefits.

The fact that a test is medically appropriate, however, does not guarantee that it should be ordered. Although there are a few exceptions, tests, like other medical interventions usually require the informed consent of the patient.

One public health exception in genetics applies to newborns. Every state provides some neonatal genetic screening (and, where appropriate, follow-up testing); only two states require parental consent [28]. Neonatal screening looks for serious disease where early detection can make a huge difference in the child's life. Phenylketonuria (PKU) is the canonical example. The roughly 1 in 50,000 American children born with this autosomal recessive genetic disorder *will* suffer from severe intellectual disabilities, unless their condition is detected early and they are put on a stringent diet, in which case their development is close to normal. If the issue is saving the brains of infants, the public health imperative trumps the need for informed consent. American states began to require neonatal genetic testing in the 1970s, starting with PKU and a handful of similar diseases. Those requirements expanded dramatically in the mid-2000s and now most states require mandatory screening for 30–50 genetic conditions (not all, perhaps, appropriately) [28].

Apart from these kinds of exceptional cases, genetic testing, like other medical interventions, requires informed consent. In general the law requires that the patient be told enough about the risks and benefits of the proposed intervention—and its reasonable alternatives—to be able to make an informed and intelligent decision. In determining what information suffices, American states take two different approaches. In most states, health care providers are required to provide the information that similarly situated, reasonable providers would give in those circumstances. This physician-centered consent standard is similar to the general standard for medical malpractice.

The other approach centers on patients. It requires providers to give the patient the information that might make a difference to a reasonable patient. Neither of these standards is easy to apply to real cases, although guidelines or consensus statements from professional organizations can be powerful evidence for what a reasonable physician would do.

In genetic testing, the issues of informed consent revolve around information about the accuracy of the test (both analytic validity and clinical validity) as well as the potential benefits and risks to the patient. It is important to remember that, legally and ethically, informed consent is not a signature on a form, but a *process* of informing the patient about the procedure, its benefits and risks; giving the patient a chance to ask questions; and ascertaining the patient's decision. A signature is some evidence that the process took place, but performing an adequate process is crucial.

Test Accuracy

An inaccurate test is both an ethical and legal problem. Regulatory questions around accuracy are addressed in other chapters of this book, but a few points might be usefully made here.

The Federal Food, Drug, and Cosmetics Act generally defines medical tests as medical devices. If a test is required to be shown safe and effective, the FDA looks not just at its analytic validity but also at its clinical validity. FDA, however, has not generally regulated laboratory tests done in licensed clinical laboratories. It takes the position that it has the authority to do so, but that it is exercising its discretion not to regulate these "Laboratory Developed Tests" ("LDTs"). (On June 4, 2013 the American Clinical Laboratory Association filed a "citizen's petition" with the FDA arguing that the agency does not, in fact, have statutory authority to regulate LDTs, which it calls "the practice of medicine." This may lead to a court decision on the question [9]). The FDA says it has not regulated clinical laboratories because they are under physicians' control and are regulated by the states and by the federal Clinical Laboratory Improvements Amendments Act ("CLIA"), with the College of American Pathologists playing substantial role through its accreditation procedures. These regulatory schemes, however, look mainly at the analytic validity of the test—how well it was done?—not its clinical validity—is it clinically accurate? That second, but very important, question is rarely subject to regulatory review. Two exceptions are worth noting. First, the FDA does regulate tests that are sold as "kits," either directly to consumers or to physicians who in turn use the kit. Second, the state regulatory agencies may have the power to examine clinical utility, which at least one state, New York, has asserted.

Communicating Results

In traditional testing the laboratory returns test results to the physician who ordered them. It is then the physician's responsibility to make sure that those results are conveyed to patients in ways that will allow the patient to understand their implications. For practitioners who are medical geneticists, communicating test results for a particular genetic disease is a task well within their expertise. Other physicians may need to make a special study of genetic testing in order to convey results usefully. Some will choose to use genetic counselors in this role. Genetic counselors, as a result of licensure and reimbursement issues, almost always work under the supervision of a physician.

DTC Testing

Traditionally, genetic tests were ordered, processed, and returned like other medical tests—through ordering physicians. Physicians would make the decision that a patient should have a genetic test, would supervise the collection and dispatch of the sample, would receive directly from the laboratory the results, and would communicate those results to the patient. Another approach to genetic tests arose in the last decade. Instead of being ordered by, and returned through, a physician, DTC tests are ordered by the now consumer (not patient) and their results returned directly to the consumer.

At least three firms were reported to have been offering DTC genetic testing as early as

2003, but the real rise to prominence for this field started in 2007, when three highly publicized DTC genetics firms started operations: 23andMe, Navigenics, and deCODEme, an offshoot of a preexisting Icelandic company. All three firms used single nucleotide polymorphism (SNP) chips to provide a wide range of information to their customers and provided SNP chip results directly to consumers for a wide range of genetic susceptibilities. Their focus varied. 23andMe, at least initially, focused on "genetic entertainment," fun facts about ancestry and nonmedical traits, such as dry or wet type earwax. Navigenics and deCODEme, on the other hand, focused on health traits from the beginning, an approach increasingly adopted by 23andMe. Each looked at hundreds of thousands of SNPs to provide information about scores of traits. For prices that initially ranged from just under $1,000 (23andMe) to about $3,000 (Navigenics), the companies would provide their analysis of a customer's genetic susceptibilities. At the same time, other firms began to offer DTC genetic tests for single genes or for particular traits or risks, some medical and some not. DTC testing is advertised, for example, for athletic ability or for romantic compatibility.

For many DTC tests, observers doubt whether DTC genetic test results have any proven value and while a physician might be able to choose useful tests, a consumer may fall prey to inappropriate tests or even quackery. At the other end of the test process, commentators have worried that consumers who directly receive their own genetic results might misinterpret them, either overreacting to apparent bad news by taking inappropriate actions (up to and including suicide) or, perhaps even more worrisome, overreacting to good news by avoiding or ceasing good health practices or medical behaviors, such as regular breast cancer screening [1].

Concerns over the spread of DTC genetic testing came to a head in May 2010. That month another DTC firm, Pathway Genomics, announced that it was going to sell a SNP-chip based DTC genomics product in collaboration with the Walgreens drug store chain. The product would have been very similar to that offered over the Internet by 23andMe, Navigenics, and deCODEme. Instead of ordering online, getting a tube in the mail, spitting in it, and returning it in a mailer to the company, the consumer would buy a kit at Walgreens that included the tube, spit in it, and mail it.

The FDA decided to react, sending warning letters to Pathway Genomics (and Walgreens). In the aftermath, the previous DTC firms, whose long-running marketing had been Internet-based, also received similar letters. In the following 3 years, the FDA has announced its intent to apply some form of regulation to genetic testing and has held several public hearings to get advice on how to regulate. As of this writing, however, no new regulatory scheme has emerged [10].

At about the same time several states took an increased interest in DTC genetic testing. The New York Public Health Department ordered the various DTC companies to stop marketing to people in New York because they had not demonstrated that the offered tests were clinically valid. California issued a similar "show cause" order to some of the companies, but, unlike New York, quickly settled in return for some fairly minor concessions, none requiring any proof of clinical validity.

While the FDA's stalled focus has expanded past the DTC test that sparked its public action, DTC genetic tests remain available. Ironically, however, two of the three original DTC SNP-chip companies, Navigenics and deCODEme, have been acquired by other firms, which have stopped offering DTC services. 23andMe's focus appeared to be shifting from earning money directly from consumers to earning money from pharmaceutical and biotech firms interested in the data it had collected. The firm hit a roadblock in November 2013, when the FDA ordered it to stop providing health information. It now gives customers only genealogical information and access to their raw SNP data, but is in talks with FDA about resuming its earlier practices[31].

Other Issues

Genetic testing, even in its "targeted" version, raises other questions. This section will discuss six specific concerns: confidentiality, discrimination, testing children, family relationships, updating test results, and the relationship of clinical genetic testing to research.

When genetic testing is part of the medical process, the results are health information, protected under state law and the Health Insurance Portability and Accountability Act (HIPAA) regulations just as much, and as little, as any other health information. All health information is subject to both unauthorized and authorized breaches of confidentiality. Unauthorized breaches can come from hackers, misbehaving insiders, or, most frequently, lost laptops or other electronic devices. None of this leakage can be completely prevented. It can only be limited, through appropriate detection and punishment as well as well-designed systems.

Authorized disclosures include disclosures to other health personnel, in many cases to insurers or other payors, to law enforcement, and to those with court orders. Although the law allows these disclosures, patients may not expect them and might be upset about them. For example, law enforcement officials seeking to determine whether a suspect's DNA matches that found at a crime scene might seek genetic information about a suspect from health providers. In one famous case, of the BTK ("bind, torture, and kill") serial killer from Kansas, the police sought and received from a University clinic part of a pap smear from the daughter of a suspect. After it was genotyped for the 13 "CODIS" markers, the short tandem repeats used for forensic identification in the USA, the investigators were able to determine that the crime scene DNA could have come from her father. They arrested the father, took a DNA sample from him that matched the crime scene DNA, and he pleaded guilty. What the daughter felt about this is not known.

The targeted nature of traditional genetic testing affects the incentives around breaching confidentiality in two conflicting ways. If a patient's records only contain the results of a genetic test for one condition, they will not contain much, if any, information about other conditions. Instead of a disclosure leading to information about many genes and many disease risks or traits, it can only lead to information about one. On the other hand, targeted testing is usually only done for people at increased risk of carrying some deleterious genetic variant. Thus, for example, the chance that someone tested specifically for a *BRCA1* mutation actually has one is greater than the chance that a random person has one.

Genetic discrimination has been discussed for almost as long as genetic testing has been used. Concern over this kind of discrimination seemed widespread even though there was very little evidence that insurers or employers had, in fact, used genetic test results to discriminate [13]. Perhaps some of the problem was that physicians and genetic counselors felt compelled to tell patients of the risk of genetic discrimination, whether or not there was strong evidence of its actual existence.

During the 1990s and through the 2000s, the USA adopted more and more protections against genetic discrimination in health insurance and employment. When Congress adopted HIPAA in 1996 it took two important but little-noticed steps concerning genetic discrimination. First, it banned the use of genetic risks as preexisting conditions for most health coverage. Second, it forbade employers or unions from applying any medical underwriting, based on genetics or otherwise, to those with employment-provided coverage. The vast majority of Americans with private health coverage obtain it as a result of employment; this ban on medical underwriting meant that none of their risks, genetic or otherwise, could be considered. Americans who needed individually underwritten coverage increasingly were protected by state laws banning genetic discrimination in health coverage. By 2008, over 45 states had some bans on the use of genetic information in health coverage; some states also banned employment discrimination.

In 2008 Congress passed the Genetic Information Nondiscrimination Act (GINA), banning, as a matter of federal law, discrimination in health insurance or in employment that was based on "genetic information," defined broadly enough to cover not just the results of tests on DNA, RNA, or proteins, but also on family history. GINA applies to employment and health coverage; it does not apply to life insurance, disability insurance, or long-term care insurance, although many states have now passed statutes extending protection to those or other areas.

Clinical genetics had reached a consensus that testing of children should be limited to only those conditions for which the test results would make a medical difference before they become adults [26]. Testing for sickle cell disease can affect early childhood health care; testing for Tay-Sachs disease can affect, sadly, childhood prognosis. Testing children for *BRCA1* mutations cannot change their medical management during childhood, though there is some fear that the results might change, and in negative ways, how their parents treat them. Children therefore have generally not been tested for adult onset diseases (This consensus may be beginning to fray.) [26].

Families share genetic variants, which means family relationships complicate genetic testing, but in two different ways. On the one hand, one family member's test result provides some information about another family member's genome. If a patient has a dominant genetic variant conferring high risk for a disease, unless it is a new mutation, one of the patient's parents must have had that variant and hence that risk. The patient's sibs and children will also have a 50 % chance of carrying the variant. Patients are currently encouraged to inform their possibly affected relatives about these risks; the obligations of health care providers if the patients refuse are still unclear. On the other hand, genetic testing may provide unexpected and unwanted family information. Some nontrivial percentage of men who believe they are a child's father may be the father in every way *except* being the genetic father. If genetic testing is done of the child and both (putative) parents, this "false paternity" is easily detected. If genetic testing is done of two putative siblings, they may be shown to be half-siblings. Who, if anyone, should be told of these genetic family relationships continues to be a largely opaque area in genetic testing.

One good aspect of (appropriate) genetic tests is they only have to done once. Except for tumors, a person's genome does not change substantially over a lifetime. The meaning of that genomic sequence may change, however, as new information is discovered about connections between various genome sequences (or collections of sequences) and disease risks. Ideally, the clinician will be aware of possible changes in the meaning of a patient's tested genetic variants. This is particularly important, of course, if the tested individual has a "variant of unknown significance," as because its significance may change from unknown to known. But even variants that are classified as pathogenic or clinically insignificant may change their meaning with new knowledge, particularly knowledge of other, modifying sequence segments.

Finally, clinical genetic testing puts pressure on the boundary between medical practice and medical research. Especially if a patient's variant is of unknown or unclear significance, either medical information on the patient at the time of testing or subsequent follow-up information may be of great research value. And yet patients may not want their cases used in research, even if their data are "anonymized." Finding the line between using these potentially invaluable data for research and respecting a patient's right not to be used in research is a continuing tension.

Steps Toward Broadband Genetic and Genomic Testing

The division between today's targeted genetic tests and tomorrow's broadband genomic tests is neat, but not entirely accurate. For one thing, "targeted" specific tests for specific DNA variants are likely to continue into the future, at least until highly accurate WGS becomes a nearly universal part of good medical care. More importantly, some "broadband" genetic or genomic tests already exist. Three important examples are multiplex tandem mass spectrometry as a part of neonatal genetic screening, SNP chips, and array-based comparative genomic hybridization (aCGH). For varying reasons, however, none of them has been deeply revolutionary.

The adoption of tandem mass spectrometry for neonatal genetic screening in the mid-2000s was a major change in neonatal screening [16, 28]. Neonatal genetic screening has its roots in the early 1970s with the rise of state-mandated tests for all infants for a handful of serious genetic diseases. Tandem mass spectrometry technology changed this model. This method of looking at proteins (not directly at DNA) allowed simultaneous screening for scores of genetic diseases. The cost of the machine was substantial; the costs of adding extra screening tests were minimal. The result was the rapid expansion of mandatory screening in every American state to include 30–50 different conditions, including many for which the value of early intervention is unclear.

Tandem mass spectrometry allows multiplex testing and hence is a "broadband" form of genetic testing, but in more meaningful ways it remains targeted. Although it tests for scores of diseases, each disease is quite rare. The vast majority of screened babies will have no positive results and almost no babies will be so terribly unlucky as to have positive results for two different diseases, each of which is found in only one baby in, say, 10,000.

SNP chips provide another example of a "sort of" broadband testing technology. These arrays, pioneered in the mid-1990s for research uses by Affymetrix and other firms, allow the rapid categorization of a DNA sample by a set of chosen SNPs, locations in the genome where some people have one nucleotide, while others have a different nucleotide. The original SNP chips could simultaneously detect which nucleotide was found in thousands of different locations in a DNA sample for a few thousand dollars, whereas today's chips can look at several million SNPs for a few hundred dollars.

SNP chips are also a broadband technology—of a sort. They can deliver millions of pieces of genomic data for a small price. The problem is that the data they provide has very little clinical value. Although it is possible to have medically important SNPs, these are unusual. SNPs are almost never medically significant in themselves; their medical significance, if any, comes from the fact that they are usually found along with nearby alleles that are medically significant.

The result is that the medical risk attributable to having an A instead of a G in a particular SNP will usually be quite small in absolute terms, even if it is statistically significant. Statistical significance can make it useful as a research tool because it indicates that something nearby (and hence inherited along with that SNP allele) may be causally related to a disease, but the small risk makes it almost useless for direct medical application. Thus, 23andMe's SNP chip results come with the interpretation of the customer's "genetic risk" of more than 200 diseases. For almost all of those diseases, the absolute change in risk is too small to be useful. For example, one of the strongest disease associations claimed by 23andMe is that a particular set of SNPs will lead to a person having a fourfold higher than normal risk of being diagnosed with Crohn's disease. The underlying risk of Crohn's disease in the general population, however, is 0.7 %. A fourfold increase takes that risk all the way to 2.8 %. Few people will change their lives because their risks of that disease are 1 in 35 instead of 1 in 143.

aCGH is another "partially broadband" technology. It is a useful tool for revealing whether, for any given spot, a sample has too little, too much, or just the right amount of DNA. This is mainly important for recognizing copy number variations, ranging from whole chromosomes (thus seeing, for example, whether a sample has three copies of chromosome 21 or only one copy of the X chromosome) to insertions or deletions of DNA that are several thousand base pairs long. This technology, therefore, is also somewhat broadband: one test will reveal any regions of the sampled genome that have other than the usual two copies (or, in the case of the X and Y chromosomes in males, other than the normal one copy). It has thus become a widely used tool for the testing for aneuploidies, such as trisomy 13, 18, and 21, as well as for aneuploidies of the sex chromosomes – X0, XXY, XXX, XYY, and others. Apart from those major aneuploidies, however, the importance and meaning of copy number variations remain generally unclear. Some have been associated, at least on a research basis, with various conditions, but the number of clinically meaningful associations between copy number variations at the less than whole chromosome level and particular diseases remains, as yet, small.

WGS (and its less comprehensive relative, whole-exome sequencing or WES) seems to provide the best of all these tests, or holds the promise of doing so. Like SNP chips and aCGH, it looks across the whole genome (or, in the case of WES, all the parts of the genome that directly code for protein), but, unlike them, it will provide powerful information about many sites. These characteristics raise all the problems of targeted genetic testing and more.

Broadband Genomic Testing: Ethical and Legal Issues

We have already entered the era of clinical WES and WGS. Thousands of people have now had whole exomes or genomes sequenced from their bodies' tissues—some for curiosity, some in search of a diagnosis for a mysterious childhood syndrome, and some to have tumors sequenced in the hopes of finding a better treatment against their individual cancer variant. (Although in many of the cases so far, particularly those involving cancer, only the exome has been sequenced, the remainder of this chapter will refer to WGS to include both exome and genome, in the assumption that cheap WGS will eventually drive out WES.)

These numbers will grow even faster as the price of WGS (and its analysis) goes down, while its accuracy improves. Soon, people whose doctors think they need a test for a particular genetic risk will be offered WGS instead of testing for just the appropriately targeted genes. Before much longer, we will see neonatal genetic screening for 30–50 diseases replaced by WGS with its power to predict thousands of diseases. Eventually almost all people with access to good health care will probably have their whole-genome sequences in their electronic medical records.

Wide use of WGS, effectively for screening purposes, holds the promise of substantial health benefits, if done wisely. It also holds the certainty of substantial ethical, legal, and practical challenges during its implementation. Those challenges, like the issues confronting traditional medical genetic testing, can be seen in five categories: accuracy, informed consent, return of results, DTC provision, and "other." In each case, the move to WGS or WES complicates the issues, sometimes massively.

Deciding to Test: Medical Appropriateness and Informed Consent

The initial problems for clinical WGS are deciding when to use such a test and how to obtain informed consent.

With a traditional single trait or single gene, one generally asks whether the test is appropriate for the patient. That will still be the case in some uses of WGS. Looking for a genomic cause for a mysterious syndrome or looking for points of attack in a tumor's genome will usually need to be done through WGS because the specific variants of interest could be anywhere in the genome. The use of WGS purely for screening purposes, however, either as neonatal genetic testing or as a routine part of medical care, raises different questions. Screening is not the same as looking for the cause or nature of a known or suspected condition. Decisions to screen require a different, and generally broader, calculation of the individual and societal costs and benefits, rather than purely an emphasis on an individual patient. Someone—whether governments, professional organizations, consensus conferences, or others—will need to decide whether and to what extent population-wide screening uses of WGS will be appropriate. That decision will not normally rest with an individual's physician or perhaps even with the individual.

Between the extremes of necessary WGS and screening WGS lies opportunistic use of WGS. A patient may present good medical reasons for getting genetic testing targeted at a particular gene or trait, but if the price of WGS is near, or below, the price of the targeted testing, it may be tempting—to the doctor, the patient, and whoever is paying for the test—to order WGS. It is conceivable that the WGS could be examined *only* for the targeted genes or traits, in which case the other information would be discarded and in that case it would, in effect, just be another method of doing targeted testing. But, more likely, the WGS information would be used both to answer questions about the trait of interest and to screen this particular patient. In that case, the physician should only recommend, or order, WGS if she is confident that, on average, the information it brings will have some net benefit to the patient. If WGS has not yet been established to have positive (or, at least, nonnegative) expected value as a general screening tool, the physician should resist the temptation—and possibly the urgings of the patient and the payor—to order a medically useless WGS test instead of the targeted test.

WGS makes informed consent much more difficult and, in some ways, frankly impossible. With traditional testing the patient can learn about the advantages and disadvantages of being tested for particular genes or traits and make an informed decision whether to accept that testing. With WGS the patient is being tested, at least potentially, for *every* genetically influenced trait and every stretch of genome. No patient can learn about each one of the thousands of genetic traits before deciding to accept or reject testing on each specific trait. There are too many traits, and too little time.

Instead, patients will need to learn about the kinds of results that WGS could provide them and to decide what kinds of results they want. Are they interested in hearing about genomic variations that are only of reproductive significance? Do they want to learn only about risks above a certain cut-off, for example those that are more than double the average risk and that have an absolute risk of over, say, 10 %? Do they want to learn about risks for which, at least at this point, there are no useful medical interventions? That conversation could, at least in theory, take place after the WGS was performed and before results were returned, but facing these questions might help the patient decide whether to undertake WGS at all. It is thus better done as part of the informed consent process.

In addition to information about the possible results, patients will also need to be told some background facts about WGS. These include not only what it is, but what the realistic chances are that the WGS will produce false positives and false negatives that could affect them. They also need to consider the possible effects on their genetic privacy, as well as the possible implications for their parents, siblings, children, and other genetic relatives. And they need to know that the interpretation of their genome may change, so that WGS will not be a "do it once and forget it" procedure.

Test Accuracy

It will be neither ethically nor legally appropriate to make widespread use of grossly inaccurate sequencing results in clinical decisions, and, at least in the near future, accuracy of WGS will be a major issue, both for its analytic and its clinical validity.

The first question will be: "how well will WGS detect the actual sequence of the genome?" Laboratories, researchers, and clinicians know how well the current testing protocols identify genetic variations, but each sequencing machine and each protocol under which the WGS is done will affect this basic sequence accuracy. How accurate are the sequencers? We do not really know. Some companies have reported accuracy levels; Complete Genomics, for example, reported as early as 2008 that its sequence calls were 99.999 % correct, making 1 error in 100,000 calls. (This would, in a 3.4 billion base pair haploid genome, still mean 34,000 errors.) In 2011, a team headed by Michael Snyder compared two sequences generated by two different firms' sequencers, those of Illumina and of Complete Genomics [20]. They reported that of 3.7 million single nucleotides where the tested individual was known to vary from the reference human genome, two firms agreed in their calls of only 88 % of them. One of the firms appeared to be more accurate on the divergent calls than the other, but in both cases a large fraction of their discordant SNP calls was wrong. Still worse, the team looked at the calls made on short insertions (about 50 base pairs or fewer) and short deletions (about 200 base pairs and fewer). The concordance rate on these "indels" between the two platforms was only about 25 %.

In addition, we know that there are other aspects these sequencing techniques do poorly at this time. These include determining the length of repeating sequences, some of which are involved in serious human diseases, such as Huntington disease. They also do not necessarily "phase" the results by revealing on which chromosome different variants are found. If a genome shows two different deleterious mutations in a gene known to be involved in an autosomal recessive disease, the person would be unaffected but a carrier if both of the mutations are on the same chromosome but should be affected if they are on different chromosomes.

WGS is nowhere near being ready to be used by itself across the genome for clinical purposes, although some centers may offer it. Any clinical use currently must require a protocol that confirms the most important findings using independent methods. The Snyder team suggests a number of strategies, from sequencing samples using multiple platforms, to doing WES to validate findings in the more important parts of the genome, to using more established Sanger sequencing or array capture technologies to validate particularly important findings [20]. The total accuracy of the resulting WGS will depend not just on the raw accuracy of the sequencing machine, but of the accuracy of the whole protocol, taken together. It is not clear whether the FDA will require such accuracy testing; someone, possibly the College of American Pathologists, must.

The clinical side of accuracy is even more daunting. At least there is some "gold standard" of reality to the sequence (analytic validity), but the medical implications of that sequence often will be much less clear. Some genetic variants, including presumably those most commonly found in humans, can be safely considered nonpathogenic; others, with long track records in medical genetics, such as the 185delAG mutation in *BRCA1* or more than 39 CAG repeats in the *HTT* gene, are equally well known to be serious. But WGS will turn up hundreds of thousands of VUS's, sequence variants that are not known to be either clearly safe or dangerous. How will they be called?

Apart from testing and interpretation in academic laboratories, one possibility is that different firms will spring up to provide, as a service, genome sequence interpretation, either with or independently from the actual sequencing work. (At least one such firm, Personalis, already exists, though it limits itself now to research uses only.) In other cases, the firms that do the sequencing may also do the interpretation. If multiple firms provide interpretative services, will their answers be consistent—and will their answers be "right"?

The DTC SNP business has already provided a worrisome example of this issue. In 2009 a team headed by Craig Venter sent samples from the same five individuals to both 23andMe and Navigenics and compared the results [24]. Both firms did well (though not perfectly) in calling the underlying SNPs; they agreed more than 99.7 % of the time on the 500,000–1 million SNPs. But when Venter's group looked at the interpretation of the results with respect to 13 diseases, they found that the two firms disagreed on the relative risk about a third of the time. For four diseases, the firms agreed entirely; for another seven, they agreed half the time or less. In 2010 the US Government Accountability Office did a similar study, with similar results [12].

Now imagine the difficulties of different companies interpreting whole-genome sequences, each with thousands of different genetic risks. If each company makes its own calls based on its own proprietary, nonpublic decision-making algorithms, it will be impossible to compare the bases for the calls. More importantly, it will make it very difficult for anyone—a doctor providing a second opinion in this case or researchers interested in that specific disease—ever to assess who is right.

A similar problem already exists with respect to the "traditional" genetic testing of *BRCA1* and *BRCA2*, where Myriad Genetics has refused since 2004 to share its database of sequence variants and patient outcomes, making informed second opinions nearly impossible. Although the Supreme Court's decision in *Association for Molecular Pathology v. Myriad Genetics, Inc.* [5] has removed Myriad's patent monopoly on testing for *BRCA1* and *BRCA2* mutations, their database of mutations, the fruit of that monopoly, remains their property. (There is an initiative called "Free the Data!" that is trying to create a similar, open database, in part by asking physicians and patients who used Myriad to share their, de-identified, test results.)

This interpretive step is both crucial and difficult. It will become even more difficult without broad and open access to information needed to assess the likely accuracy of any one interpretation as shown by patient outcomes. At the research level, the "Global

Alliance," announced in early June 2013, hopes to avoid some of these problems by providing broad access, with research subject consent, to genomic and health data from sequencing experiments [11, 19]. Something similar is likely to prove necessary for clinical uses.

Communicating Results

In clinical WGS, communicating results will be no different in kind than in traditional genetic testing. The testing laboratory (in this case, perhaps through a separate WGS analysis firm) will return results to the physician who ordered them, who will in turn need to explain the relevant ones to the patient. In degree, however, the difference in quantity with WGS will effectively transform the activity.

Traditional testing provides information usually on just one disease or risk, while WGS will provide information—positive, negative, or indeterminate—on any risk, trait, or disease that is associated with genomic sequences. Even today, that amounts to thousands of diseases and susceptibility risks, as well as hundreds of pharmacogenomic traits, hundreds of SNP chip associations (as WGS necessarily reveals SNPs as well), and scores of physical or behavioral traits not related to disease. Each of those numbers will only grow as our understanding of the relationships between specific genomic sequences, or combinations of sequences, and phenotypes grows. This embarrassment of riches leads to two questions: what information should be returned and how should it be returned?

The first question has already begun to be debated, at least around its edges. In 2013 the American College of Medical Genetics and Genomics (ACMG) issued guidelines recommending that when a patient is offered WGS as part of the investigation of a particular risk or disease, the laboratory must return to the physician, and the physician should discuss with the patient any highly penetrant, serious risks for which there is good medical intervention, whether or not they have anything to do with the reason for seeking genetic testing [17, 18]. The ACMG guidelines list 56 such conditions as of the time of writing. Thus, if someone was tested for long Q-T syndrome risk, but the WGS showed high risks for Lynch syndrome, that second risk would have to be disclosed. The European Society for Human Genetics (ESHG), by contrast, has recommended that only expressly sought-after risks be disclosed [29]. And at least one article has been published decrying the ACMG position for removing from patients the right to remain ignorant of their risks [30], coupled with another article supporting the ACMG position [23] and an ACMG "clarification" [3].

Note that the debate, thus far, is merely about what results physicians *should* return when they are doing the WGS on their patient for a specific indication or indications and the patients have not requested any additional information. It does not deal with screening uses of WGS at all, or, even when it is being used for targeted purposes, with what physicians should do when patients request broader information, or how physicians, before offering WGS, should determine what their patients want. As discussed above, the categories of results that patients want to receive back from WGS should be a major topic of conversation before ordering such a test.

The ESHG position seems both morally and, at least in the USA, legally questionable. A physician has a fiduciary obligation to put his or her patient's interests first. To *not* to tell the patient about known or readily discoverable information concerning a highly penetrant and serious genetic risk, for which there is a good medical intervention, seems a breach of the physician's obligation to the patient [7]. Physicians who, when examining lungs in a CAT scan, see (or read in a radiologist's report) something highly suspicious in the liver will not ignore this unlooked for finding. At least some American case law finds that they can be liable if they do. A patient perhaps should be able to retain a "right to ignorance," but only after being informed of the possibility of such findings and expressly requesting not to be informed. If a patient made that decision after adequate information, and the physician documented the decision, the physician's legal risks should be minimal.

But what about the cases, which ultimately should become the majority, where the testing is being done for screening purposes or where the patient says "tell me everything." In both cases, the patient should be told at least whatever scientifically valid information seems likely to be significant. (The patient who wants to be told "everything" could also be given a flash drive with his whole sequence.) What might be significant to one patient might not be important to another, so the physician and patient will need to have had some discussion about what the patient wants, preferably before the test was ordered, but if not then, before results are returned.

In the absence of that discussion, the physician should return at least the kinds of highly penetrant, serious disease risks for which good medical intervention exists, as listed by the ACMG Guidelines, but that must be viewed as a minimum. Those criteria were created for cases in which the testing was being performed for one specific indication but other things turned up; in screening, the idea is to look generally for problems. What information should be returned, absent an express discussion, should probably include disease risks that are substantial (in absolute risk) and significantly higher than the average person's (in relative risk). Whether they should go beyond those that are medically actionable is a harder question, and one that really needs to be answered directly by the patient.

The second question is of a different nature—it is not a normative question of "what" should be done, but a practical question of "how." In the screening context, especially with a patient who wants to be told "everything" (even if qualified to everything significant), the numbers could quickly become unwieldy.

In 2009 Stanford bioengineer Steve Quake had his genome sequenced on a sequencing machine he had invented. In 2010, 32 authors published a medical analysis of this genome [4]. They concluded about 100 findings should be shared with him. Even 3 min of discussion of each of 100 issues amounts to 5 h of counseling. Few counselors could provide good information on 100 random genetic issues; fewer could talk for 5 h, and even fewer patients could listen for 5 h. And no one would pay for that counseling [25].

Information, including genomic information, is only useful if it is properly understood. Patients who do not understand the significance of their results, as noted above, might overreact or underreact to them in ways that could be risky. Yet how many patients will understand, on their own, the meaning of more than a 100 genetic risks? And how many of their personal care physicians will know themselves, or be able to convey if they do know, the meaning of more than a handful of those risks? Using genetic counselors sounds good, but there are about 3,000 genetic counselors in North America today and they are all busy. Thus, the interpretation by appropriately trained pathologists and their direct involvement in multidisciplinary teams that work together to arrive at an interpretation that takes into account the patient's clinical context and wishes, and that is versed in the delivery of such interpretations, may become increasingly important.

The challenge, perhaps biggest for WGS, will be coming up with economically feasible ways to convey complex, important probabilistic information to hundreds of millions of people, many with very limited knowledge of genetics. The movie and television industries, computer game companies, advertising agencies and others need to be engaged in order to produce systems that allow individuals to find out, probably online, what their WGS means. Face-to-face counseling with trained professionals should, however, remain part of communicating results; face-to-face, real-time communication (perhaps online through some HIPAA-compliant version of Skype or a similar venue) gives a better chance to answer questions and to spot confusion or misunderstandings.

DTC Marketing

The current lack of regulation of DTC for traditional and SNP chip testing also applies to WGS. Only one firm seems to be currently offering DTC WGS: a firm named DNA DTC that is offering WGS for about $7,000 (and WES for about $900) [8]. Interestingly, it is offering only the naked sequence, without any interpretation, calling it a nonmedical service—customers will have to figure out what it means, medically or otherwise, on their own.

Any concerns about DTC for individual tests or SNP chips are magnified for WGS, because of the breadth of information—and misinformation—the whole sequence can convey. Both the decision to order WGS without any professional advice and the return of information from WGS without any professional advice seem reckless. The customers may not know what they are ordering or what they have received.

Other Issues

In discussing traditional testing methods, this chapter looked at six concerns: confidentiality, discrimination, testing children, family relationships, updating test results, and the relationship of clinical genetic testing to research. WGS makes each of those more complicated.

With WGS, breaching confidentiality will become more tempting. With traditional genetic testing a breach of confidentiality leads only to information about one (or a few) variants or traits. WGS information, because of its breadth, is more useful to someone who wants to find out something about the patient, whether a hacker or the police. Police might, for example, seek clinical WGS records for a suspect to compare with crime scene DNA; electronic medical record databases could, in effect, supplement the forensic databases of convicts and arrestees. Some of the confidentiality breaches may be unauthorized and hence illegal, but others, such as those following court orders, will be fully legal.

WGS would also magnify, in some respects, discrimination concerns because one can learn of many different risks in a person. On the other hand, WGS will show that everyone has *some* genetic weaknesses. In the long run, this seems likely to lead to less discrimination, as the perpetrators realize they could themselves become victims and because society sees more tangibly the shared nature of these risks.

If WGS is done on children, either for neonatal screening or as part of a test for a particular disease, it will strain the current consensus on not testing children for conditions for which nothing need be done in childhood. WGS will provide information on both childhood onset and adult onset diseases; if the latter information is not revealed immediately to the parents, then what should be done with it—should it be put in an envelope marked "do not open until your 18th birthday"? Also, WGS for a child might reveal genetic risks for the child's parents or siblings that could be important right away. The ACMG Guidelines prompted more controversy by recommending the immediate return to parents of positive results in children's sequences for any of its 57 listed genes [30]. This controversy will have to be resolved before widespread use of WGS in children.

Comparing WGS makes spotting family relationships very easy. With the number of variants revealed by WGS, even fairly distant relationships between people, such as second or third cousin, should be ascertainable with a high degree of likelihood based on the percentages of variants shared. This increased utility for finding family relationships is another factor that may drive people to try to breach confidentiality—such as frustrated genealogists or state officials trying to find a genetic father to sue for child support.

Updating may be the issue that complicates WGS the most. With tests for a single gene, updating the result may or may not ever be necessary. A clear and definitive "safe" or "high risk" variant seems unlikely to change; a VUS, however, would need to monitored and updated, either from the physician's personal knowledge of the area or from some periodic reanalysis. With WGS, not only will each patient have tens of thousands of VUS's, but the overall interpretation of the human genome will change every day with new discoveries on particular variants. The individual's genome will not change, but the interpretation of any individual's genome is likely to change fairly frequently. Not only will some VUS's be resolved, but also many "known" variants will have their effects refined, with better knowledge of risk, severity, age of onset, or the effects (positive or negative) of particular environmental factors or of variants in other genes.

In effect, the interpretation of a patient's genome will be a test that will have to be repeated regularly, like a blood pressure test or a pap smear. Once optimal accuracy of WGS has been achieved, no office visit or actual test may be required, but the existing genome sequence will need to be run through the analysis protocol every few years. The field should agree on just how often such reinterpretation is appropriate. (Of course, if the cost of WGS becomes low enough, it may be cheaper to resequence the patient every few years than to store the data. This would, however, require the costs of sequencing to continue to plummet at the same time the declining costs of computer memory stalled.)

WGS will increase the pressure to blur the line between research and clinical data. Looking at whole-genome sequences along with a host of clinical, phenotypic information in the electronic medical records of millions of patients will likely be the most useful way to make progress in understanding the effects of different sequence variants, alone and in combination. Yet those patients may not have agreed to have their records used in research. And given the wealth of potentially identifying details contained both in the phenotypes in medical records, as well as in the genotypes, almost any patient could, with some effort, be re-identified from his or her records. In the long run, it would be good for everyone for this information to be as widely available for research as possible. But to do so without the consent of the patients would not only be unethical and, most likely, illegal, but it would risk a political backlash against medical research and genetics generally. Resolving this dilemma will require careful effort.

Conclusions

We have had clinical genetic testing for over 40 years. It has raised some complex ethical and legal questions, but those questions have largely been answered, if not perfectly, at least satisfactorily. New technologies, and particularly WGS, will raise those ethical and legal questions to new and higher levels. The old answers may be useful as guides, but they cannot provide satisfactory solutions by themselves. This change in degree really is a change in kind. Implementing clinical WGS effectively will require serious improvements in sequencing technologies, but even more serious interdisciplinary efforts to deal with the ethical, legal, and, ultimately, very practical problems it raises. The time for such efforts is, quite urgently, now.

REFERENCES

1. Am College Med Genet (2008) ACMG statement on direct-to-consumer genetic testing. https://www.acmg.net/docs/ACMG_Statement_on_DTC_4.07.08.pdf
2. Am College Med Genet (2012) Points to consider in the clinical application of genomic sequencing. http://www.acmg.net/StaticContent/PPG/Clinical_Application_of_Genomic_Sequencing.pdf. Accessed 5 Jul 2013
3. Am College Med Genet (2013) Incidental findings in clinical genomics: a clarification.www.acmg.net/docs/Incidental_Findings_in_Clinical_Genomics_A_Clarification.pdf. Accessed 5 Jul 2013.
4. Ashley EA et al. Clinical assessment incorporating a personal genome. Lancet. 2010;375:1525–35.
5. Association for Molecular Pathology v. Myriad Genetics, Inc. 2013. _U.S. __, 133 S.Ct. 1958.
6. Biesecker LG. Opportunities and challenges for the integration of massively parallel genomic sequencing into clinical practice: lessons from the ClinSeq project. Genet Med. 2012;14:393–8.
7. Clayton EW et al. Managing incidental genomic findings: legal obligations of clinicians. Genet Med. 2013;15:624–9.
8. DNA DTC. Products. 2013. http://www.dnadtc.com/. Accessed 2 Sept 2013.
9. Genome Web. Lab group submits citizen petition against FDA regulation of lab developed tests. 2013. http://www.genomeweb.com/print/1237821?hq_e=el&hq_m=1588054&hq_l=1&hq_v=162223e076. Accessed 24 Aug 2013.
10. Genome Web. Members of Congress Urge Obama Administration to Release FDA Draft Guidance on LDT Regulation. 2013. http://www.genomeweb.com/clinical-genomics/members-congress-urge-obama-administration-release-fda-draft-guidance-ldt-regula. Accessed 24 Aug 2013.
11. Global Alliance. Creating a global alliance to enable responsible sharing of genomic and clinical data. 2013. http://www.broadinstitute.org/news/globalalliance. Accessed 2 Sept 2013.
12. Government Accountability Office. Direct-to-consumer genetic tests: misleading test results are further complicated by deceptive marketing and other questionable practices. Washington, DC: GAO; 2010.
13. Greely HT. "Genotype Discrimination": the complex case for some legislative protection. Univ PA Law Rev. 2001;149:1483–505.
14. Greely HT. Human genomics research: new challenges for research ethics. Perspect Biol Med. 2001;44(2):221–9.
15. Greely HT. The uneasy ethical and legal underpinnings of large-scale genomic biobanks. Annu Rev Genomics Hum Genet. 2007;8:343–64.

16. Greely HT. Testing infant destinies. Nature. 2012;492:192.
17. Green RC et al. ACMG recommendations for reporting of incidental findings in clinical exome and genome sequencing. Genet Med. 2013;15:565–74.
18. Green RC et al. Clinical genome sequencing. In: Ginsburg G, Willard H, editors. Genomic and personalized medicine. 2nd ed. San Diego: Elsevier; 2013. p.102–22.
19. Hayden EC. Geneticists push for global data sharing. Nature. 2013;498:16–7.
20. Lam HY et al. Performance comparison of whole genome sequencing platforms. Nat Biotech. 2012;30:78–82.
21. Manolio TA et al. Implementing genomic medicine in the clinic: the future is here. Genet Med. 2013;15(4):258–67.
22. Mayer AN et al. A timely arrival for genomic medicine. Genet Med. 2011;13:195–6.
23. McGuire AL et al. Ethics and genomic incidental findings. Science. 2013;340:1047–8.
24. Ng PC et al. An agenda for personalized medicine. Nature. 2009;461:724–6.
25. Ormond KE et al. Challenges in the clinical application of whole-genome sequencing. Lancet. 2010;375:1749–51.
26. Ross LF et al. Technical report: ethical and policy issues in genetic testing and screening of children. Genet Med. 2013;15:234–45.
27. Schrijver I et al. Opportunities and challenges associated with clinical diagnostic genome sequencing: a report of the association for molecular pathology. J Mol Diagn. 2012;15:525–40.
28. Timmermans S, Buchbinder M. Saving babies?: the consequences of newborn screening. Chicago: University of Chicago Press; 2012.
29. van El CG et al. Whole-genome sequencing in health care: recommendations of the European Society of human genetics. Euro J Hum Gen. 2013;21:580–4.
30. Wolf SM et al. Patient autonomy and incidental findings in clinical genomics. Science. 2013;340:1049–50.
31. Zettler PJ et al. 23andMe, the Food and Drug Administration, and the future of genetic testing. JAMA. 2014;174(4):493–4. doi:10.1001/jamainternmed.2013.14706.

PART III

Clinical Implementation of Diagnostic Genomics

CHAPTER 12

TRANSITIONING DISCOVERIES FROM CANCER GENOMICS RESEARCH LABORATORIES INTO PATHOLOGY PRACTICE

TAMARA JAMASPISHVILI, JEREMY A. SQUIRE

Introduction 159
Cancer-Associated Biomarker Categories 160
Cancer Genomics: From Research to Pathology Practice 162
Impact of Genetic Biomarkers on Drug Development and Clinical Trial Designs 168
The Translational Research Continuum 169
The Biomarker Development Process 170
Conclusions 173

Introduction

Cancers of an identical primary site can be heterogeneous in molecular pathogenesis, clinical course, and treatment responsiveness, which reflects the existence of multiple cancer subtypes [1]. The differentiation of these subtypes is often based on biomarkers that distinguish important cancer features such as the aggressiveness of the disease (prognostic biomarkers) or the response to treatment (predictive biomarkers). The latter have fueled an increasing interest in biomarkers, given the potential they hold for individualized or personalized medicine. This new field focuses on differences between people and the potential for these differences to influence medical outcomes. With individualized or "precision" medicine, a person's cancer may be subtyped based on an explicit biomarker that is present or absent, or that may have increased or decreased expression levels. This may result in a greater likelihood of receiving treatment that is appropriate and effective for a specific tumor in a particular cancer patient. Individualized medicine contrasts markedly with the traditional "empiric method," which uses a standardized treatment for the whole patient population with an established presentation of disease symptoms, based on long-standing generic descriptions of the average patient (Fig. 12.1).

Nowadays, tumor biomarkers, together with new genomic and proteomic technologies, provide powerful tools for the early identification of cancer patients, recurrent disease and for defining therapeutic responsiveness. In spite of the rapid developments in biotechnology and genomics, the pace of acceptance of new markers in clinical practice is surprisingly low. The slow uptake is due to the substantial reasons presented below and elsewhere [1–3]. In this chapter we (1) summarize importance of personalized medicine and describe some of the biomarkers and genetic tests which are being used in pathology practice now; (2) describe the translational research cycle and draw attention to some of the challenges faced in

G.J. Netto and I. Schrijver (eds.), *Genomic Applications in Pathology*,
DOI 10.1007/978-1-4939-0727-4_12, © Springer Science+Business Media New York 2015

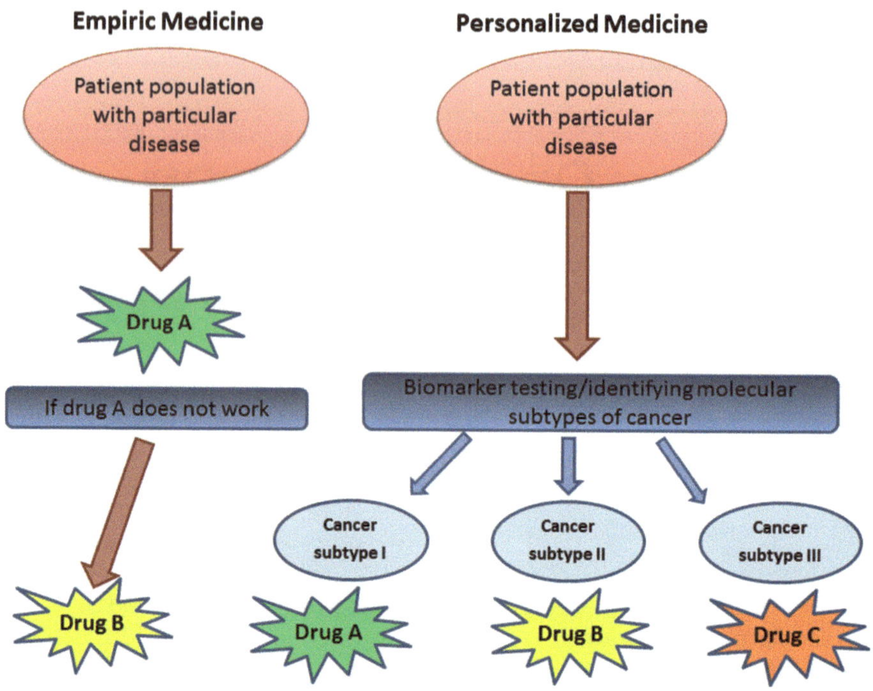

Figure 12-1 Empiric treatment versus patient-oriented treatment. Individualized medicine is contrasted with the traditional "empiric method," which uses a standardized treatment for all patients with a certain disease

delivering practice-changing discoveries; (3) discuss the impact of genomic biomarkers on the design of new clinical trials; and (4) briefly review the guidelines and recommendations for moving successful biomarkers into clinical practice.

Cancer-Associated Biomarker Categories

Personalized, i.e., patient-oriented, research refers to a continuum from initial studies in humans to comparative effectiveness and outcomes research, and the integration of this research into the health care system and clinical practice. The goal of patient-oriented research is to optimize the translation of innovative diagnostic and therapeutic approaches to the point-of-care, as well as to help researchers meet the challenge of contributing to high-quality, cost-effective health care [4]. It involves ensuring that the right patient receives the right clinical intervention at the right time, ultimately leading to better health outcomes [5, 6]. In order to make patient-oriented care effective, there is a great need to discover more promising, reliable cancer-specific biomarkers and translate them successfully into clinical use.

In general, biomarkers are biological measurements that are used to aid clinical practice. The National Cancer Institute defines a biomarker as a "biological molecule found in blood, other body fluids, or tissues that is a sign of a normal or abnormal process, or of a condition or disease" [7]. A biomarker may be used to see how well the body responds to a treatment for a disease or condition [8]. The Biomarkers Consortium (managed by the Foundation of National Institutes of Health) states that "biomarkers are characteristics that are objectively measured and evaluated as indicators of normal biological processes, pathogenic processes, or pharmacologic responses to therapeutic intervention" [9].

There are five different categories of cancer biomarker measurements that and they can be assayed either once at baseline (diagnostic, prognostic, and predictive) or repeatedly (disease screening, disease monitoring, and molecular imaging) during the course of the disease. A marker may belong to a single or to multiple biomarker categories.

A diagnostic biomarker is an indicator measurement that will aid in the detection of malignant disease in an individual. PSA (prostate-specific antigen) is the best-known cancer biomarker for early detection of prostate cancer. Serum PSA has been widely used for almost 25 years in screening for prostate cancer and has brought about a dramatic increase in early detection of the disease. Unfortunately, the low specificity of elevated serum PSA as a cancer biomarker results in a significant number of men who do not actually have prostate cancer, undergoing unnecessary needle core biopsies [10, 11]. To address these concerns the US Preventive Services Task Force recently reconsidered the potential harms and relative benefits of using PSA as a screening biomarker. It was found that there was insufficient evidence to recommend routine use of PSA as a screening test at any age (see section "The Biomarker Development Process"). The PCA3 (prostate cancer antigen 3) RNA biomarker test has been introduced as a simple additional urine assay to address the significant diagnostic dilemma in new cases of prostate cancer [12, 13]. The specificity of this test in prostate cancer is 74 % compared to only 17 % for serum PSA, which at least increases the potential for this type of assay in predicting the likelihood of a positive needle core biopsy [14–16].

Screening biomarkers are an important subclass of biomarkers that have both high specificity and sensitivity for detecting cancer at the population level. These biomarkers are designed to robustly differentiate patients with disease from those without a disease. A perfect screening biomarker should have 100 % sensitivity and 100 % specificity, but at present none of the available biomarkers achieve these ideal performance standards. Another good example of a currently used screening biomarker is the widespread testing for HPV (human papillomavirus) DNA as part of cervical cancer-screening programs. The HPV molecular test is more sensitive with a high negative predictive value, than either conventional cytology (PAP smear) or liquid-based cytology methods. HPV DNA testing is used as an adjunct to conventional methods in the USA and in some European countries, and it can detect the presence of viral subtypes that may provide additional risk stratification in HPV-positive women. Clinical trials are ongoing to clarify the preferable method of technologies for HPV testing to incorporate it into the current population level screening process [17].

Prognostic biomarkers are often defined as measurements made at diagnosis that provide information about patient prognosis. Prognostic biomarkers may predict disease recurrence (disease-free survival) and/or cancer-related death (cancer specific survival) or overall survival for an individual patient in the absence of treatment or in the presence of standard primary treatment. Thus, prognostic markers typically give information about patient outcomes and tumor aggressiveness. For example, estrogen receptor (ER) positive breast cancer patients have longer survival in the absence of systematic therapy than those patients who are ER negative [18]. CA-125, which is present in a subset of ovarian cancers, is not used for detection of early cancers because the serum levels are elevated in only 50 % of patients with stage I disease [19, 20]. This biomarker is usually used to evaluate response to chemotherapy, relapse, and disease progression in ovarian cancer patients. Gupta and Lis performed comprehensive evaluation of the existing literature on the prognostic role of CA125 and suggested that postoperative levels of serum CA125 are also a strong prognostic factor for estimating overall survival and progression free survival in ovarian cancer [21].

Disease monitoring biomarkers are assays that are performed repeatedly over time. A change in disease status during treatment will be reflected by a concomitant change in the biomarker status. Examples of biomarkers used for such monitoring are as follows: PSA in prostate cancer, CA125 in ovarian cancer, CEA in colorectal cancer, CA19-9 in pancreatic cancer, and CA15-3- or CA27.29 in breast cancer.

Predictive biomarkers are used to predict response or resistance to a specific cancer therapy, i.e., they are used to identify the patients who are likely or unlikely to benefit from a specific treatment. For example, in addition to its role as a prognosticator, tumor ER positivity is considered to be a predictive biomarker in breast cancer because such patients are far more likely to benefit from antiestrogen therapy such as Tamoxifen. On the other hand, ER negativity is a predictive biomarker for benefit from conventional cytotoxic chemotherapy. Human epidermal growth factor receptor 2 (Her2/neu) amplification is a predictive marker for benefit from Trastuzumab (Herceptin®), Doxorubicin, and Taxanes [22, 23]. In some situations, predictive biomarkers can be used to identify patients who may not benefit from a particular drug. For example, advanced colorectal cancer patients whose tumors have KRAS mutations are typically poor candidates for treatment with epidermal growth factor receptor (EGFR) antibodies [24, 25].

Cancer Genomics: From Research to Pathology Practice

The completion of the Human Genome Project in 2003 stimulated a shift in emphasis from studying genes and proteins as individual biomarkers to understanding their interaction in pathways of therapeutic importance. Thus, genomics, proteomics, transcriptomics, and metabolomics are now providing excellent opportunities for researchers to learn more about complex diseases like cancer by studying the overall response of cells to a mutation, or to changes in the disease microenvironment. It is important to note that technologies that are used for biomarker discovery are often not exactly the same technologies that will be routinely used in a clinical laboratory. However, it is clear that discoveries made using genomic and proteomic technologies, coupled with advances derived from applied bioinformatics, are showing great promise for simpler and more cost-effective analysis of clinical samples.

Genomic Technologies Used for Biomarker Discovery

GENE EXPRESSION ARRAYS

Gene expression analysis has been one of the first high-throughput molecular profiling technologies with widespread adoption for biomarker discovery. Microarrays enable simultaneous analysis of tens of thousands of genes and thus the rapid identification of new potential biomarkers. Gene expression analysis measures the activity of cellular RNA (mRNA) in a tissue or bodily fluid at a given point in time, and it may provide information about the current status of a disease or the likelihood of future disease. RNA levels are dynamic and change as a result of pathology or environmental signals [26]. Certain patterns of gene activity may be used to diagnose a disease or to predict how an individual will respond to treatment over time. Methods used for gene expression analysis are diverse, ranging from real-time reverse transcription polymerase chain reaction (RT-PCR) to microarray screening technologies, which have been widely used in research, and are now beginning to be applied in clinical settings.

The most significant genomic biomarkers that have emerged in recent years include *BCR-ABL1* for CML (chronic myeloid leukemia) diagnosis and monitoring of treatment responses [27], Her2/neu for diagnosis and prognosis of the breast cancer subtype which benefits from monoclonal antibody (Herceptin®/trastuzumab) treatment [28], and detection of *EGFR* (epidermal growth factor receptor) and *KRAS* mutations for predictive purposes in lung [29] and metastatic colon cancer [30]. Discoveries from molecular profiling of RNA and DNA are now generating many new candidate biomarkers that have potential similar to these successful genomic biomarkers.

The use of DNA expression microarrays has provided one of the most powerful tools to discover subsets of clinically important genes in human cancer [31]. Such expression arrays have been used to obtain major insights into progression, prognosis, and response to therapy on the basis of gene expression profiles (see the section on gene expression tests, below). Typically microarrays have been used to discover subsets of genes whose expression levels can be used to provide a distinct molecular subclassification of disease state. Once such a distinguishing genetic signature with likely clinical relevance has been discovered, custom-made arrays or other molecular biology methods are used to develop preclinical or clinical testing.

GENOME-WIDE ASSOCIATION STUDIES (GWAS)

GWAS is a comprehensive approach that identifies and correlates single nucleotide polymorphisms (SNPs) to complex diseases such as cancers and is predominantly carried out with SNP microarrays specifically designed to interrogate millions of different polymorphisms in the human genome. GWAS is also very helpful as a biomarker discovery tool. Results obtained from GWAS are typically cross-referenced with data from the HapMap Project or the 1000 Genomes Project in a process called imputation that aims to substitute values for missing data [32]. The advantage of GWAS is that it is unbiased and less likely to miss important genes or pathways than methods that use selected genes. Analysis of the large complex datasets generated by GWAS poses several challenges: (1) it requires large sample numbers and advanced bioinformatics to determine statistical significance; (2) there often remains a high likelihood of false positive associations; (3) with such marked biostatistical complexity, small differences may be missed due to stringent biostatistical corrections. Novel integrative genomics approaches are being introduced that combine GWAS information with gene expression data to assess putative functional relationships between genetic variants and their associated biological pathways [33].

A strategy alternative to GWAS is called the candidate-driven or hypothesis-driven approach [34]. In this knowledge-based approach, researchers evaluate which genes to examine based on the scientific literature and they compare the expression levels of those genes and their particular pathways in a group of individuals with disease and in those without disease. The advantage of this method is that it focuses on pathways or genes that have a higher likelihood of being successful candidates based on decades of systematic research. However, in contrast to GWAS, the approach may be biased as it relies heavily on the literature and on existing bioinformatics datasets that are often limited or incomplete, so that candidate approaches run the risk of missing important genes and pathways.

NEXT-GENERATION SEQUENCING (NGS)

The comprehensive screening power of NGS promises to help mine the remaining "unannotated regions" of the genome for novel sequence-based biomarkers that are below the resolution levels for detection by conventional microarray analysis [35]. In NGS all sequence information from a patient sample is aligned to a full-length reference genome to match all sequencing reads to their exact genomic locations [36]. Counting the number of sequencing reads that align to a given genomic location is analogous to microarray intensities for a probe with a specific sequence and this metric can provide an estimate of relative expression levels. With slight modification in the NGS experimental design DNA copy number, expression levels and differential methylation can be determined. Sequencing technologies can further identify variation between samples by identifying genomic locations, whereas reads that do not perfectly match the reference genome may indicate individual genetic variation such as SNPs, loss of heterozygosity (LOH) as well as copy number variation (CNV) [37, 38].

Role of Bioinformatics and Genomic Datasets in the Public Domain

In order to facilitate the biomarker discovery process, it was recognized that there was a need for freely accessible datasets containing comprehensive information associated with DNA and with RNA expression. Most journals now require that investigators make such genomic data publically available in a standardized format for open access in silico analysis. All data must be MIAME (Minimum information about a microarray experiment)-compliant. In other words, MIAME comprises the minimum requirements that should always be included with published microarray datasets, as suggested by the Functional Genomics Data society (http://www.fged.org). The most popular genomic datasets are GEO, ONCOMINE, and ArrayExpress Archive, described below.

GEO (The Gene Expression Omnibus) is the biggest public repository that was designed to utilize features of the most commonly used molecular profiling methods in use today. These include data generated from microarray analyses as well as sequence technologies and include gene expression profiling, noncoding RNA profiling, chromatin immunoprecipitation (ChIP) profiling, genome methylation profiling, SNP genomic variation profiling, array comparative genomic hybridization (aCGH), serial analysis of gene expression (SAGE), and protein arrays (http://www.ncbi.nlm.nih.gov/geo/).

ONCOMINE is a cancer microarray database and Web-based data-mining platform aimed at facilitating discovery from genome-wide expression analyses [39]. Using the ONCOMINE platform, researchers can easily compare gene expression profiles between cancer and normal samples, compare gene expression between different molecular, pathological, and clinical cancer subtypes, and investigate expression of genes in pathways and networks associated with cancer. It is possible to identify pathways, processes, chromosomal regions, and regulatory motifs activated in cancer and also search for genes that distinguish and predict cancer types and subtypes (http://www.oncomine.org).

ArrayExpress Archive/Gene Expression Atlas is a European database that contains functional genomics experiments including gene expression data. Here, researchers can query and download data collected according to MIAME and MINSEQE (Minimum information about a high-throughput nucleotide sequencing experiment) standards. It is also an atlas that can be queried for individual gene expression under different biological conditions across experiments (http://www.ebi.ac.uk/arrayexpress).

Integration Approaches to in Silico Datasets

For in silico analysis, information is extracted from publicly available genomic datasets and then analyzed by the researcher using a computer to look for various patterns associated with particular diseases. In silico analysis can be applied, for example, to determine the location of mutations in a certain tumor suppressor gene, to look for copy number changes for particular genes, and to compare gene/protein expression patterns between cancerous and normal samples. Commercial bioinformatics software (such as Nexus™, BioDiscovery Inc., California, USA; or Partek®, Partek Inc., Saint Louis, USA) enable users to manage, integrate, visualize, and analyze data generated from high-throughput gene expression analysis, aCGH, SNP arrays, and NGS datasets.

The advantages of in silico methods are that they are rapid and avoid the need for expensive experiments to evaluate a biomarker's clinical value. Moreover, bioinformatics permits the investigator to search for a biomarker in one dataset and attempt to validate it in another. However, the utility of in silico analysis depends on the quality of the clinical data collected, as well as the coverage and accuracy of the annotations used to report the genomic data. It can also be difficult to compare results across several datasets because of the differences in genomic methods. For these reasons, in silico analysis in biomarker discovery is often considered an initial step that must be followed by rigorous experimental validation prior to preclinical investigation.

Clinically Applicable Gene-Based Assays

A very important aspect of marker development is to translate it to the clinic, once its usefulness has been established. A potential marker can be tested in different sources, including tumor tissues and body fluids such as serum and urine. The methods used should be of rapid execution, reliable, and ideally not very expensive. As our understanding of complex diseases grows, additional biomarkers are being identified and developed into new and improved diagnostic tools that can analyze multiple biomarkers simultaneously. Often, such biomarker assays establish a complex molecular profile of the disease and provide an estimate of the likelihood of a response to a given treatment. They combine the values of multiple variables to yield a single patient-specific result. Such multigene assays commonly use PCR tests or gene expression microarrays, the results of which are integrated into an algorithm to organize and prioritize individual markers, thereby producing a readily accessible result [40]. The common examples of this modality are discussed below and some are already FDA cleared or approved.

GENE EXPRESSION TESTS

With the discovery in recent years of many potential tumor-biomarkers and the growing notion that a panel of markers rather than one marker alone will predict the most accurate outcome, development of new detection systems has become desirable. In spite of the fact that microarray technologies are costly, gene expression tests are increasingly being implemented in modern clinical practice as an aid to conventional diagnostic, prognostic, and predictive decision tools used in cancer management. Some of the most recently used examples are discussed below.

ColoPrint® (Agendia, Amsterdam, The Netherlands) is a microarray-based gene expression profile to predict the risk of distant recurrence of stage II and III colon cancer. ColoPrint® combines a multigene panel, which includes seven colon cancer-related genes and five reference genes, with a proprietary algorithm for determining risk of recurrence (http://www.agendia.com). ColoPrint uses the same technology, methods and quality control as FDA-cleared assays (i.e., MammaPrint®), though it is not approved by FDA. Similarly, Genomic Health Inc. provided the Oncotype DX® Colon Cancer test for Stage II colon cancer patients by evaluating expression levels of 12 genes. The results of the test are reported as a quantitative Recurrence Score® result, which is a score between 0 and 100 that correlates with the likelihood of a person's chances of having the cancer return [41]. At present this test it is not FDA approved. The assay is only performed by the developers in their Clinical Laboratory Improvement Amendments (CLIA) commercial laboratory. Genomic Health also provides MMR (mismatch repair) testing by immunohistochemistry on colon tumor samples, which, in combination with Oncotype DX® may help the clinician in making treatment decisions (http://www.oncotypedx.com). Stage II colon cancer patients with MMR-deficient (MMR-D) tumors have a much lower risk of recurrence compared to patients with MMR-proficient (MMR-P) tumors [42].

BluePrint® is an 80-gene expression signature which classifies breast cancer into Basal-type, Luminal-type, and ERBB2-type cancers. The BluePrint® Molecular Subtyping Profile, combined with the patient's MammaPrint® (see below) test results, provides a greater level of clinical information to assist in therapeutic decision-making (http://www.agendia.com). BluePrint® does not require FDA clearance because it is considered a Class I, low-risk device under FDA regulations.

MyPRS™/MyPRS Plus™ (my prognostic risk signature) is a tool for guiding treatment in patients with multiple myeloma. It analyzes all of the nearly 25,000 genes in a patient's genome to determine the gene expression profile (GEP) that is associated with a particular patient's condition. The GEP is made up of the 70 most relevant genes (GEP70) which aide in the prediction of the patient's outcome (http://www.signalgenetics.com). Both MyPRS™ and MyPRS Plus were developed by Myeloma Health, LLC who determined performance characteristics in a CLIA-certified laboratory. The FDA has indicated that these tests do not require either clearance or approval at present.

MammaPrint® (Agendia, Amsterdam, The Netherlands) is based on microarray technology using 70 cancer-related and about 1,800 non-cancer-related genes (http://www.agendia.com). The test stratifies patients into two distinct groups: low risk or high risk for distant recurrence, with no intermediate-risk patients. With low-risk patients, hormonal therapy (e.g., Tamoxifen) might be sufficient avoiding the necessity of aggressive treatment such as chemotherapy. The test was cleared by the FDA as a class II device in 2007. However, the FDA did not evaluate treatment outcomes as a result of use of this "prognostic" device. In addition, the EWG (The EGAPP working group) (Evaluation of Genomic Applications in Practice and Prevention) found that "data were adequate to support an association between the MammaPrint Index and 5 or 10 year metastasis rates, but the relative efficacy of testing in ER-positive and ER-negative women is not clear." Also, study subjects were European, and how characteristics of other demographic populations might affect test performance is not known [43]. The MINDACT (Microarray In Node-negative Disease may Avoid Chemotherapy) trial is designed to compare the effectiveness of MammaPrint test results versus clinical evaluation in predicting 15-year disease-free survival and overall survival (EORTC-European Organization for Research and Treatment of Cancer, MINDACT 2008). This trial will compare clinical response to endocrine therapy alone versus endocrine therapy combined with chemotherapy regimens (anthracycline-based, docetaxel–capecitabine, Letrozole).

The Oncotype DX® Breast Cancer test (Genomic Health Inc., Redwood City, CA) uses RT-PCR to study gene expression profiles in formalin-fixed, paraffin-embedded (FFPE) breast cancer tissues. Oncotype DX analyzes expression of 21 genes, 16 cancer related, and five normative [43]. The test is intended for Stage I or II, lymph node negative and ER-positive breast cancer patients, who will be treated with Tamoxifen. Results are reported as a Recurrence Score™ (RS; scale of 0–100). Patients are divided into low, intermediate, and high-risk categories. Oncotype DX® claims to provide information beyond conventional risk assessment tools, including how likely the woman is to benefit from chemotherapy in addition to Tamoxifen therapy (http://www.genomichealth.com). The TAILORx (Trial Assigning Individualized Options for Treatment) trial is primarily designed to determine the benefit of chemotherapy for women with intermediate risk. Oncotype DX results will be issued in 2013. The test is not FDA cleared but is available at the Genomic Health Inc. CLIA-certified laboratory. Oncotype DX® was initially developed just for women with early-stage invasive, estrogen receptor-positive (ER+), node-negative breast cancer patients. However, recent clinical trials have demonstrated both prognostic (the capability of predicting distant recurrence) and predictive (to assess the potential benefit of additional adjuvant chemotherapy) significance in node-positive, ER+women. The current guidelines suggest that if the test is proven to have prognostic as well as predictive significance in node-positive patients, the test may be subject to additional regulation and therefore pre-market approval by the FDA [44].

The most extensively studied tests among those listed above are Oncotype DX® Breast Cancer and MammaPrint®. In many developed countries these new tests are already offered for clinical use, but there remains a need for more comprehensive long-term studies to assess whether test outcomes lead to clear beneficial effects for patients and are cost-effective.

PROTEIN CHIPS

Similar to using DNA chips for identification of gene expression profiles in particular tumors, the advent of "protein chips," which enables the analysis of thousands of proteins expressed by a single tumor sample at the same time, has helped researchers to better understand the molecular basis of disease, including disease susceptibility, diagnosis, progression, and potential points of therapeutic interference. The basic format of most protein chips is similar to that of DNA chips, such as the use of glass or plastic printed with an array of molecules (e.g., antibodies) that can capture proteins. Ideally, a protein chip containing a panel of molecules such as antibodies would be able to predict a cancer state by a simple serum or urine test. This technology is likely to see considerable additional development and application in the coming years [45].

FLUORESCENCE IN SITU HYBRIDIZATION (FISH)

Quantification of multiple mRNA levels in tumors is expensive, technically demanding and not readily available in a routine clinical setting. FISH provides an alternative way to diagnose and identify predictive or prognostically important genetic alterations. The method is simple, fast, and reliable and therefore has been widely accepted for clinical use in human cancer. It is used to assess various genetic alterations (amplifications, deletions, translocations). FISH can detect genomic anomalies over a much greater dynamic size range than other techniques. In the past decade, the technique has been developed to include multicolor FISH assays so it is now possible to assess complex genomic alterations [46]. Recent improvements have been made to FISH in the form of chromogenic in situ hybridization (CISH) and silver-enhanced in situ hybridization (SISH). These techniques use peroxidase enzyme-labeled probes whose signals do not decay over time and allow the specimen to be viewed using bright-field microscopy. CISH and SISH have been used to assess Her2/neu gene status [47].

Assessment of Her2/neu amplifications in breast cancer, to assess prognosis and to predict treatment outcome, is the most common example of FISH use in clinical settings [48]. Other examples include the recently developed commercialized test eXagenBC. The latter promises to provide a tailored prognosis in node-positive and node-negative breast cancer patients and is based on assessment of DNA copy numbers of three genomic regions (around *CYP24*, *PDCD6IP*, and *BIRC5*) for ER-positive and Progesteron (PR)-positive tumors and three different genes (*NR1D1*, *SMARCE1*, *and BIRC5*) for ER-negative, PR-negative tumors in both node-negative and node-positive patients. The eXagenBC test uses a prognostic index (PI) from an algorithm to integrate the information from the three genes and predict recurrence rates. This test may greater accuracy compared to other criteria for recurrence risk assessment and therefore has been suggested for routine clinical use [49].

Additional promising prognosticators are fusion genes such as *TMPRSS2-ERG* translocations and *PTEN* deletions in prostate cancer which show great promise for identification of aggressive prostate cancers. *PTEN* deletions have been associated with earlier biochemical relapse following radical prostatectomy. Prostate cancers showing homozygous *PTEN* deletions, termed "*PTEN* null," have been strongly associated with metastasis and androgen independent progression, i.e., castration resistant prostate cancers (CRPC) [50–53]. One important new FISH biomarker is the echinoderm microtubule-associated protein-like 4-anaplastic lymphoma kinase (*EML4-ALK*) fusion gene, present in a small subset of non-small-cell lung cancers (NSCLC). Such tumors are particularly sensitive to ALK inhibitors such as Crizotinib which has been recently approved by FDA (2011) for the treatment of locally advanced or metastatic non-small-cell lung cancers that are ALK-positive. The FDA also approved the Vysis ALK Break-Apart FISH Probe Kit (Abbott Molecular, Inc.) that is a diagnostic test designed to detect rearrangements of the *ALK* gene in NSCLC [54, 55].

POLYMERASE CHAIN REACTION (PCR)

Clinical diagnostic applications of real-time PCR or real-time quantitative PCR (qPCR) have been widely implemented by hospital-based clinical laboratories [56]. In translational research, qPCR is simple and one of the fastest, most reliable and cheapest molecular techniques for the validation of a newly discovered biomarker. A qPCR assay can be used to identify gene amplifications, deletions, fusions, overexpression, and mutations down to single base changes, and therefore, these very sensitive and specific molecular tests are among the most widely used methods to translate recent discoveries in cancer research into clinical practice.

Examples of clinically applicable qPCR assays in cancer diagnostics and prognostics include the detection of *BCR-ABL1* transcripts in patients with chronic myeloid leukemia (CML) who are then subjected to tyrosine kinase inhibitor (Imatinib/Gleevec®) treatment as a first-line therapy and to quantification of minimal residual disease (MRD) by qPCR [57]. Recently highly sophisticated methods have been developed using DNA-based and RNA-based PCR assays for the detection of *BCR-ABL1* transcripts that were previously not detectable by conventional PCR methods [58, 59]. Thyroid cancer is another example where qPCR assays play an important role: in this case they have a diagnostic and predictive role. Real-time PCR can be used to diagnose papillary thyroid carcinomas (PTCs) harboring a point mutation in *BRAF* or *RAS*, or a *RET-PTC* rearrangement (>70 %), and they can help diagnose follicular thyroid carcinomas (FTCs) that harbor either *RAS* mutations or *PAX8/PPARγ* rearrangements [60]. *RAS* mutations may also be found in benign thyroid lesions. In addition, sporadic and hereditary medullary thyroid carcinomas (MTCs) are both associated with point mutations in the *RET* gene. Thus, molecular testing is now an important component of thyroid cancer diagnosis and management [60, 61].

Assays that simultaneously amplify (or detect) two or more target fragments (or detect sequence changes within target fragments) are termed duplex and multiplex real-time PCRs, respectively. It is noteworthy that the multiplexing of biomarkers has many advantages over single biomarker measurements, especially when trying to identify the best diagnostic or prognostic models for various human cancers (prostate cancer, as an example, is discussed below) [62]. One commercially available real-time PCR assay (HemaVision, DNA technology, Aarhus, Denmark), is widely used in clinical laboratories to simultaneously detect 28 fusion genes and more than 80 breakpoints and splice variants in patients with acute myelogenous leukemia (AML) and acute lymphoid leukemia (ALL) ([63]; http://www.biocompare.com).

Classical cytogenetic methods (e.g., conventional karyotyping) continue to provide well-established diagnostic findings to clinicians. However, the detection of certain genetic abnormalities (translocations or fusion genes) that often have been missed by conventional cytogenetics is now feasible with high reliability using newer molecular techniques that have advantages over traditional methods. These may include a shorter turnaround time, automated analyses, and a lack of the prior requirement of dividing cells [64].

Impact of Genetic Biomarkers on Drug Development and Clinical Trial Designs

Genetic biomarkers now have tremendous impact in every phase of drug development, from drug discovery to preclinical evaluations through each phase of clinical trials and into routine clinical use [65]. In the early phases of drug development, biomarkers are used to evaluate the activity of small molecule therapeutics in animal models, to investigate mechanisms of action and to provide essential preclinical data needed for the various later stages of clinical trials. If the preclinical phase of drug development is successful, then it is followed by an application to the FDA as an investigational new drug (IND). The purpose of an IND is "to ensure that subjects will not face undue risk of harm" in a clinical investigation that involves the use of a drug. The IND is the mechanism by which the investigator, or pharmaceutical sponsor, provides the requisite information to obtain authorization to administer an investigational agent to human subjects. By doing so, the compound can be tested for dose response, efficacy and toxicity [66]. After an IND is approved, the next steps are clinical phases 1, 2, and 3. Phase 1 trials determine safety and dosage and identify side effects (patient number: 20–80); Phase 2 trials are used to obtain an initial assessment of efficacy and to further explore safety of the drug or treatment in a larger number of patients (100–300); In Phase 3 trials the treatment is given to large groups of patients (>1,000) to confirm effectiveness, monitor side effects, compare efficacy to established treatments, and collect information that will allow it to be used safely.

In clinical trials which are designed to validate and assess the usefulness of a prognostic biomarker or one that is predictive for the usefulness of a specific therapy, the major issues are to obtain sufficient statistical evidence of treatment benefit in patients who are positive for the predictive or prognostic biomarker, and then to examine the biological relationships associated with the biomarker's expression and the molecular pathways targeted by the therapeutic agent. Often, such studies utilize a retrospective analysis of a biomarker in available tissues from patients with known response who have been treated similarly [67]. Before initiating studies to confirm the clinical utility of a novel biomarker, it is necessary to conduct validation trials in which several criteria must be met. First, specific testable hypotheses must be proposed based on scientific evidence of the predictive properties of the putative biomarker relative to the existing (standard) treatment. In addition, any prognostic benefit is assessed as well. A novel biomarker is considered promising for clinical utility when it demonstrates the following features in the validation study: (1) the marker is independently associated with clinical outcome; (2) its biological effects are specific for the cancer of interest as opposed to normal tissues, other disease states, or other cancers; (3) the marker's prevalence in the target population is high; and (4) the methods of marker measurement are feasible and reproducible.

In the next phase of the evaluation of clinical utility of the predictive or prognostic biomarker, two major issues have to be considered: the selection of an appropriate patient population and the choice of the most appropriate end point. For example, when evaluating predictive markers of therapeutic efficacy in the adjuvant setting, the primary end point usually is overall, disease-free, or recurrence-free survival. Possible primary end points for metastatic disease trials would include response rate, time to progression, survival, or risks of toxicity [67].

With respect to clinical trial designs for new drugs or treatment options and companion biomarkers, randomized controlled trials (RCT) are the most popular, because they limit the potential for bias by randomly assigning one arm to an intervention and the other arm to non-intervention (or placebo). This minimizes the chance that the incidence of confounding (particularly unknown confounding) variables will differ between the two groups. Currently, some Phase 2 and most Phase 3 drug trials (see below) are randomized, double-blind, and placebo-controlled. Traditional RCT designs are not always well suited for drugs with molecular targets and associated biomarkers. For example, a standard randomized approach in a clinical trial for Trastuzumab would not be very effective without the use of an enrichment design, because the drug has little effect on Her2/neu negative patients. Because almost 75 % of patients are Her2/neu negative, a standard design would require a large sample size to detect the treatment effect of Trastuzumab on Her2/neu positive patients [68]. An enrichment clinical trial design is used to evaluate a treatment or a drug in which the effect can be readily demonstrated on a specific subset of the study population. Often such a subset is identified by a biomarker test that is used to select those patients who are likely to respond well to the treatment. Efficiency of the study thus depends on the prevalence of test-positive patients and on the relative effectiveness of the new treatment in test-negative patients [69]. In the enrichment designs the number of randomized patients is often substantially smaller than for a standard design. Wang et al. [70, 71] proposed modified designs for enrichment studies, accruing either both test-positive and test-negative patients or only marker-positive patients during the first stage, and then accruing test-negative patients if the results are promising for the marker-positive patients at the end of the first stage. Freidlin and Simon took a similar approach and introduced "adaptive threshold designs" for situations in which the proportion of patients sensitive to the new drug is low [72]. When a new treatment is broadly effective, an adaptive design has the power to detect the overall effect similar to a RCT traditional design. Moreover, adaptive designs substantially reduce the chance of false rejection of effective new treatments. Overall, the design of a biomarker-informed clinical trial must provide rapid and robust laboratory results for the molecular target in a manner that minimizes the overall false-positive rate. The availability of timely and standardized biomarker assay results is particularly important because new patients cannot be assigned into a biomarker directed study arm until the laboratory findings are available to the center that coordinates the trial.

The Translational Research Continuum

Despite the rapid pace of biomarker discovery in recent years, there are still very few validated genetic biomarkers of proven and robust clinical utility [73]. This poor performance reflects that the clinical development of new biomarkers is just as difficult as the development and approval of a new drug. Here we will outline the bench to bedside pipeline and discuss how best to facilitate the successful development of biomarkers and molecular targeted treatments, respectively. Throughout the cancer research process, many challenges are faced during the transition of a new discovery from the "research bench" through the phases of laboratory and clinical validations. Unfortunately the majority of "exciting discoveries" never succeed in overcoming the rigorous evaluations and are not accepted as part of routine clinical practice or used for laboratory testing by pathologists (Fig. 12.2).

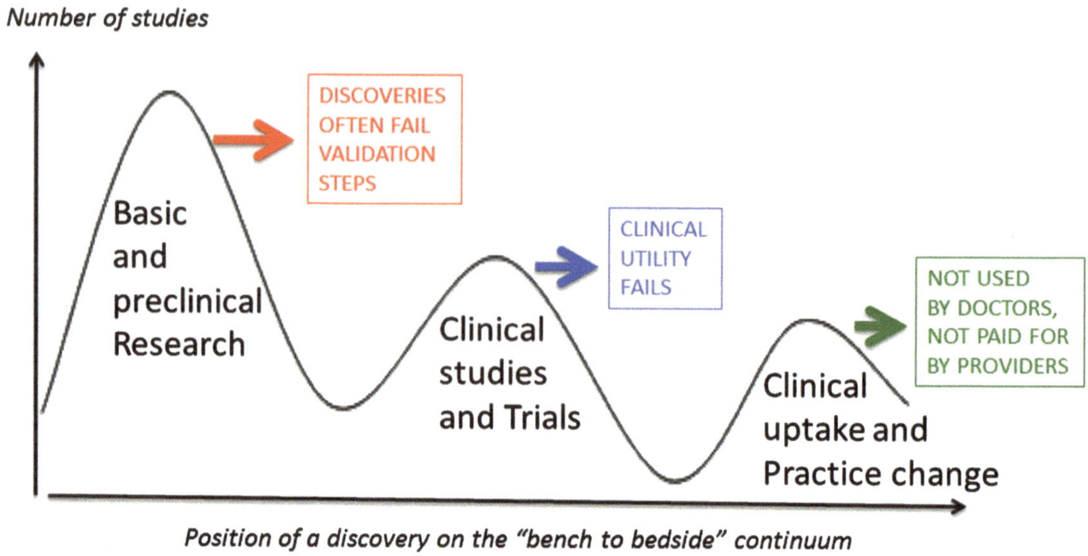

Figure 12-2 The translational research continuum. This graph schematically depicts the three major obstacles that impede an exciting research discovery (*leftmost* peak) moving though the validation phase from preclinical research into clinical trials (*middle* peak) and onto clinical or laboratory practice (small peak on *right*). The graph illustrates the continuing gap between basic biomedical research and clinical research and knowledge. This gap limits the capacity to translate the results of provocative discoveries generated by basic biomedical laboratory research to the bedside, as well as to successfully engage and educate health care providers in the benefits of the discoveries (see Canadian Institute of Health Research. "Canada's Strategy for Patient-Oriented Research." August 2011)

Challenges in Preclinical and Clinical Research

A major factor contributing to the lack of use of genetic biomarkers in clinical trials is the poor quality of published preclinical data. This has been the focus of a recent commentary by Begley and Ellis [3]. IND trials rely heavily on the literature and on having a comprehensive understanding of the agent's target, its associated biomarker and the various downstream consequences of the drug. Very often, however, the biological hypothesis around a new agent and its companion biomarker is uncertain or questionable. The lack of reproducibility of preclinical "research assays" when applied to patient samples may prevent the application of novel biomarkers in a clinical setting. Some of the issues that are considered to be associated with poor uptake of research biomarkers by trialists and clinical laboratories are summarized in Table 12.1.

The Biomarker Development Process

The biomarker development process requires multiple collaborative mechanisms, knowledge networks and consortia to facilitate biomarker fruition in clinical practice. The critical limitation in biomarker development is the lack of a proper structure in the biomarker discovery process as is present in the process of testing a new drug. After proving, among other things, the clinical validity and clinical utility of a newly discovered biomarker (see below), a biomarker is not considered "validated" and cannot be recommended for use in clinical practice until independent research groups at multiple sites have demonstrated concordant results in separate trials. The challenge is firstly to determine which data are required to perform these studies, and secondly, to obtain, share and pool these data together,

Table 12-1 Challenges in Preclinical and Clinical Research

Challenges in preclinical research:

- Research staff does not use SOPs (Standard Operating Procedure) or operate following GLP (Good Laboratory Practice) standards
- Biased comparison groups in the study (case versus controls)
- Statistically underpowered study size, inappropriate statistical analyses, including over-fitting of data

Challenges in clinical research:

- Independent groups are unable to generate concordant results due to the lack of coordination between biomarker research laboratories/lack of standardized protocols across laboratories
- Lack of "good quality samples." So-called "convenience samples" (from local bio-repository) may be too homogeneous to provide evidence for clinical relevance of biomarker to the whole population of the patients
- Clinical heterogeneity often leads to wrong conclusions
- New testing technologies lack appreciation of inter-laboratory performance, standardization, quality control, and cost-effectiveness and cannot be used widely by clinical laboratories (e.g., mass spectrometric protein profiling)
- Lack of pre-analytical studies
- Lack of funding for translational research

and to provide adequate support to analyze the pooled datasets. A solution would be to apply uniform standards, which should facilitate effective translation of newly discovered biomarkers to the clinical setting. Therefore, numerous collaborative mechanisms, knowledge networks, and consortia have emerged in order to facilitate biomarker discovery and enhance the delivery process to the clinic. Examples of such mechanisms such as the Early Detection Research Network (EDRN) and The Biomarkers Consortium (TBC) demonstrate the value of a national coordinated approach [74, 75].

Guidelines (known as the Standards for Reporting of Diagnostic Accuracy, or STARD statement) have been developed for diagnostic studies and were inspired by CRGs (Cochrane Review Groups) in 1999. For prognostic studies, guidelines known as REMARK criteria, were developed by NCI-EORTC (National Cancer Institute-European Organisation for Research and Treatment of Cancer) [76–78]. The STARD initiative serves to improve the reporting quality diagnostic accuracy in publications. The statement consists of a checklist of 25 items and the decision to include items in the checklist was based on evidence linking these items to either bias, variability in results, or limitations of the applicability of results to other settings [76]. The checklist can be used to verify that all essential elements are included in the report of a research study.

REMARK (REporting recommendations for tumor MARKer prognostic studies) guidelines were developed primarily for studies of prognostic markers, especially those evaluating a single tumor marker while possibly adjusting for other known prognostic factors. The guidelines suggest relevant information that should be provided about the study design, preplanned hypotheses, patient and specimen characteristics, assay methods, and statistical analysis methods [77].

While some biomarkers have already been approved by the FDA, the use of others has been recommended in clinical guidelines by various cancer societies [5]. A recent example of this is a test for epidermal growth factor receptor (*EGFR*) mutation in patients with advanced NSCLC, which determines whether or not first-line EGFR tyrosine kinase inhibitor therapy is indicated [5, 79]. The introduction of biomarkers into routine clinical practice is considered in the framework-Tumor Marker Utility Grading System (TMUGS) was designed to evaluate the clinical utility of tumor markers and to propose a hierarchy of "levels of evidence" that might be used to determine if available data support the use of a marker or not [80]. TMUGS provides guidelines to determine the clinical utility of known and future tumor markers, as well as guidance on biomarker assay design, interpretation, and use in clinical practice. This evidence scale has been widely cited and used for deciding whether to recommend the use of a tumor marker in clinical practice and for design and conduct of tumor marker studies [81, 82]. This evidence scale has recently been revised to distinguish data generated from prospective clinical trials, in which the marker is the primary objective of the study, from those in which archived specimens are used [1, 67, 83]. Starting in 2000, the Office of Public Health Genomics (OPHG) at the Centers for Disease Control and Prevention (CDC) established the analytic framework ACCE Model Project based on four main criteria for evaluating a genetic tests: (1) *Analytic validity* is a component of clinical validity (see below) describing how accurately and reliably the test measures the genotype of interest. Analytic validity assesses technical test performance and includes analytic sensitivity (detection rate), analytic specificity (false positive rate), reliability (repeatability of test results), and assay robustness (resistance to small changes in pre-analytic or analytic variables); (2) *Clinical validity* describes the accuracy with which a test predicts a particular clinical outcome and clearly separates two subgroups of patients with different outcomes within a large population. When a test is used diagnostically, clinical validity measures the association of the test with the disorder [84] and when used predictively it measures the probability that a positive test will result in the appearance of the disorder within a stated time period; (3) *Clinical utility* is a balance of benefits and harms when the test is used to influence patient management, i.e., the evidence that the use of the marker improves outcomes compared to not using it. Evaluation of clinical utility factors in available information about the effectiveness of the interventions for people who test positive and the consequences for individuals with false positive or false negative results; (4) *Ethical, Legal and Social Implications* (ELSI) refer to other implications which may arise in the context of using the test and cut across clinical validity and clinical utility criteria. In 2004, a new initiative, termed EGAPP™ (Evaluation of Genomic Applications in Practice and Prevention) was created by OPHG at the CDC "to better organize and support a rigorous, evidence-based process for evaluating genetic tests and other genomic applications that are in transition from research to clinical and public health practice in the U.S." [43, 85].

Table 12-2 Stratification of Evidence by Quality (proposed by USPSTF)

Level 1:	Evidence obtained from at least one properly designed randomized controlled trial
Level 2-a:	Evidence obtained from well-designed controlled trials without randomization
Level 2-b:	Evidence obtained from well-designed cohort or case–control analytic studies, preferably from more than one center or research group
Level 2-c:	Evidence obtained from multiple time series with or without the intervention. Dramatic results in uncontrolled trials might also be regarded as this type of evidence
Level 3:	Opinions of respected authorities, based on clinical experience, descriptive studies, or reports of expert committees

Source: U.S. Preventive Services Task Force (August 1989). Guide to clinical preventive services: report of the U.S. Preventive Services Task Force. Publishing. pp. 24–. ISBN 978-1-56806-297-6

The US Preventive Services Task Force (USPSTF) is an independent panel of non-Federal experts in prevention and evidence-based medicine and is composed of primary care providers. The USPSTF strives "to make accurate, up-to-date, and relevant recommendations about preventive services in primary care. It conducts scientific evidence reviews of a broad range of clinical preventive health care services (such as screening, counseling, and preventive medications) and develops recommendations for primary care clinicians and health systems" (http://www.uspreventiveservicestaskforce.org). These recommendations are published in the form of "Recommendation Statements." Also, the USPSTF stratifies the evidence by quality about the effectiveness of treatments or screening by three different levels (Table 12.2). For example, in 2002, USPSTF deemed the evidence to be insufficient to recommend routine use of PSA as a screening test among men younger than age 75. The recommendation, however, does not include the use of PSA test for surveillance after diagnosis or treatment of prostate cancer.

The USPSTF reviewed the available evidence again in 2011 and in a draft report concluded that population benefit from PSA screening was inconclusive, recommending against PSA-based prostate cancer screening at any age [86, 87]. The USPSTF makes evidence-based recommendations about clinical preventive services such as screenings, counseling services, or preventive medications. Currently the majority of USPSTF recommendations are not in favor of widespread use of cancer screening using biomarkers. However, as more DNA based biomarkers are developed from sequencing projects it seems likely that the benefits of screening may outweigh the risks for some of the diseases where early intervention can prevent disease progression (http://www.uspreventiveservicestaskforce.org/uspstopics.htm#AZ).

Conclusions

Various consortia, grading systems and collaborative initiatives discussed in this chapter are basically founded and developed in North America and are part of the goal to provide evidence-based medicine, which seeks to assess the strength of the evidence of risks and benefits of treatments, diagnostic tests, and biomarkers. Similar systems exist in Europe though they are not discussed here. Networking infrastructures throughout the world developed to date have a goal of sharing and pooling analyzed data to complete the biomarker discovery → development → validation continuum. Increased collaboration between such consortia may, under the right conditions, accelerate biomarker development. Global harmonization of guidelines in the years ahead will likely underpin the success of biomarker translation from bench to bedside.

REFERENCES

1. Simon RM, Paik S, Hayes DF. Use of archived specimens in evaluation of prognostic and predictive biomarkers. J Natl Cancer Inst. 2009;101:1446–52.
2. Brooks JD. Translational genomics: the challenge of developing cancer biomarkers. Genome Res. 2012; 22:183–7.
3. Begley CG, Ellis LM. Drug development: raise standards for preclinical cancer research. Nature. 2012;483:531–3.
4. Strategy for Patient-Oriented Research-CIHR. http://www.cihr-irsc.gc.ca/e/41204.html. Accessed 10 Feb 2013.
5. Ziegler A, Koch A, Krockenberger K, et al. Personalized medicine using DNA biomarkers: a review. Human Genetics. 2012;131:1627–38.
6. Henney AM. The promise and challenge of personalized medicine: aging populations, complex diseases, and unmet medical need. Croat Med J. 2012;53: 207–10.
7. Madu CO, Lu Y. Novel diagnostic biomarkers for prostate cancer. J Cancer. 2010;1:150–77.
8. NCI dictionary of cancer terms-National Cancer Institute. http://www.cancer.gov/dictionary?cdrid=45618. Accessed 10 Feb 2013.
9. Foundation for the National Institutes of Health. http://www.fnih.org/work/key-initiatives/biomarkers-consortium. Accessed 10 Feb 2013.
10. Rao AR, Motiwala HG, Karim OM. The discovery of prostate-specific antigen. BJU Int. 2008;101:5–10. doi:10.1111/j.1464-410X.2007.07138.x.
11. Barry MJ. PSA testing for early diagnosis of prostate cancer. N Engl J Med. 2001;344:1373–7.
12. Haese A, de la Taille A, van Poppel H, et al. Clinical utility of the PCA3 urine assay in European men scheduled for repeat biopsy. Eur Urol. 2008;54: 1081–8.
13. Kirby RS, Fitzpatrick JM, Irani J. Prostate cancer diagnosis in the new millennium: strengths and weaknesses of prostate specific antigen and the discovery and clinical evaluation of prostate cancer gene 3 (PCA3). BJU Int. 2009;103:441–5.
14. Day JR, Jost A, Reynolds MA. PCA3: from basic molecular science to the clinical lab. Cancer Lett. 2011;301:1–6.
15. Laxman B, Morris DS, Yu J, et al. A first-generation multiplex biomarker analysis of urine for the early detection of prostate cancer. Cancer Res. 2008;68: 645–9.
16. Jamaspishvili T, Kral M, Khomeriki I, et al. Quadriplex model enhances urine-based detection of prostate cancer. Prostate Cancer Prostatic Dis. 2011;14:354–60.
17. Canadian Task Force on Preventive Health Care, Pollock S, Dunfield L, et al. CMAJ. 2013;185:35–45.
18. Kiba T, Inamoto T, Nishimura T, et al. The reversal of recurrence hazard rate between ER positive and negative breast cancer patients with axillary lymph node dissection (pathological stage I-III) 3 years after surgery. BMC Cancer. 2008;8:323.
19. Nossov V, Amneus M, Su F, et al. The early detection of ovarian cancer: from traditional methods to proteomics. Can we really do better than serum CA-125? Am J Obstet Gynecol. 2008;199:215–23.
20. Nustad K, Bast Jr RC, Brien TJ, et al. Specificity and affinity of 26 monoclonal antibodies against the CA 125 antigen: first report from the ISOBM TD-1 workshop. Int Soc Oncodev Biol Med. 1996;17:196–219.
21. Gupta D, Lis CG. Role of CA125 in predicting ovarian cancer survival – a review of the epidemiological literature. J Ovarian Res. 2009;2:13.
22. Gennari A, Sormani MP, Pronzato P, et al. HER2 status and efficacy of adjuvant anthracyclines in early

breast cancer: a pooled analysis of randomized clinical trials. J Natl Cancer Inst. 2008;100:14–20.
23. Hayes DF, Thor AD, Dressler LG, et al. HER2 and response to paclitaxel in node-positive breast cancer. N Engl J Med. 2007;357:1496–506.
24. Amado RG, Wolf M, Peeters M, et al. Wild-type KRAS is required for panitumumab efficacy in patients with metastatic colorectal cancer. J Clin Oncol. 2008;26:1626–34.
25. Simon R. Clinical trial designs for evaluating the medical utility of prognostic and predictive biomarkers in oncology. Per Med. 2010;7:33–47.
26. UnitedHealthcare®. Medical Policy 2012; PDF. Accessed 18 Feb 2013.
27. Tohami T, Nagler A, Amariglio N. Laboratory tools for diagnosis and monitoring response in patients with chronic myeloid leukemia. Isr Med Assoc J. 2012;14:501–7.
28. Leong AS, Zhuang Z. The changing role of pathology in breast cancer diagnosis and treatment. Pathobiology. 2011;78:99–114.
29. Kim ST, Sung JS, Jo UH, et al. Can mutations of EGFR and KRAS in serum be predictive and prognostic markers in patients with advanced non-small cell lung cancer (NSCLC)? Med Oncol. 2013;30:328.
30. Laszlo L. Predictive and prognostic factors in the complex treatment of patients with colorectal cancer. Magy Onkol. 2010;54:383–94.
31. Trevino V, Falciani F, Barrera-Saldana HA. DNA microarrays: a powerful genomic tool for biomedical and clinical research. Mol Med. 2007;13:527–41.
32. Howie B, Fuchsberger C, Stephens M, et al. Fast and accurate genotype imputation in genome-wide association studies through pre-phasing. Nat Genet. 2012;44:955–9.
33. Hicks C, Asfour R, Pannuti A, et al. An integrative genomics approach to biomarker discovery in breast cancer. Cancer Inform. 2011;10:185–204.
34. Le Marchand L, Wilkens LR. Design considerations for genomic association studies: importance of gene-environment interactions. Cancer Epidemiol Biomarkers Prev. 2008;17:263–7.
35. Metzker ML. Sequencing technologies – the next generation. Nat Rev Genet. 2010;11:31–46.
36. Trapnell C, Salzberg SL. How to map billions of short reads onto genomes. Nat Biotechnol. 2009;27:455–7.
37. Fertig EJ, Slebos R, Chung CH. Application of genomic and proteomic technologies in biomarker discovery. Am Soc Clin Oncol Educ Book. 2012;32: 377–382.
38. Schrijver I, Aziz N, Farkas DH, et al. Opportunities and challenges associated with clinical diagnostic genome sequencing: a report of the Association for Molecular Pathology. J Mol Diagn. 2012;14:525–40.
39. Rhodes DR, Yu J, Shanker K, et al. ONCOMINE: a cancer microarray database and integrated data-mining platform. Neoplasia. 2004;6:1–6.
40. College of American Pathologists In Vitro Diagnostic Multivariate Index Assays (IVDMIA). http://www.cap.org/apps/cap.portal?_nfpb=true&cntvwrPtlt_actionOverride=%2Fportlets%2FcontentViewer%2Fshow&cntvwrPtlt%7BactionForm.contentReference%7D=committees%2Ftechnology%2Fivdmia.html&_pageLabel=cntvwr. Accessed 18 Feb 2013.
41. Oncotype DX official website. http://www.oncotypedx.com/en-US/Colon/HealthcareProfessionals/RecurrenceRisk/ScoreResult. Accessed 18 Feb 2013.
42. Venook AP, Niedzwiecki D, Lopatin M. Validation of a 12-gene colon cancer recurrence score (RS) in patients (pts) with stage II colon cancer (CC) from CALGB 9581. Abstract # 3518. ASCO annual meeting. Chicago, IL, June 2011.
43. Evaluation of Genomic Applications in Practice and Prevention (EGAPP) Working Group. Recommendations from the EGAPP Working Group: can tumor gene expression profiling improve outcomes in patients with breast cancer? Genet Med. 2009;11:66–73.
44. Health Service Executive Website. Oncotype DX to be made available in all designated cancer centres. http://hse.ie/eng/services/list/5/nccp/news/Oncotype%20DX%20to%20be%20made%20available%20in%20all%20designated%20cancer%20centres.html.
45. Massie CE, Mills IG. Mapping protein-DNA interactions using ChIP-sequencing. Methods Mol Biol. 2012;809:157–73.
46. Bayani J, Squire JA. Application and interpretation of FISH in biomarker studies. Cancer Lett. 2007;249: 97–109.
47. Penault-Llorca F, Bilous M, Dowsett M, et al. Emerging technologies for assessing Her2 amplification. Am J Clin Pathol. 2009;132:539–48.
48. Mass RD, Press MF, Anderson S, et al. Evaluation of clinical outcomes according to Her2 detection by fluorescence in situ hybridization in women with metastatic breast cancer treated with trastuzumab. Clin Breast Cancer. 2005;6:240–6.
49. Davis LM, Harris C, Tang L, et al. Amplification patterns of three genomic regions predict distant recurrence in breast carcinoma. J Mol Diagn. 2007;9:327–36.
50. Yoshimoto M, Cutz JC, Nuin PA, et al. Interphase FISH analysis of PTEN in histologic sections shows genomic deletions in 68 % of primary prostate cancer and 23 % of highgrade prostatic intra-epithelial neoplasias. Cancer Genet Cytogenet. 2006;169: 128–37.
51. Yoshimoto M, Cunha IW, Coudry RA, et al. FISH analysis of 107 prostate cancers shows that PTEN genomic deletion is associated with poor clinical outcome. Br J Cancer. 2007;97:678–85.
52. Squire JA. TMPRSS2-ERG and PTEN loss in prostate cancer. Nat Genet. 2009;41:509–10.
53. Bismar TA, Yoshimoto M, Vollmer RT, et al. PTEN genomic deletion is an early event associated with ERG gene rearrangements in prostate cancer. BJU Int. 2011;107:477–85.
54. Sholl LM, Weremowicz S, Gray SW, et al. Combined use of ALK immunohistochemistry and FISH for optimal detection of ALK-rearranged lung adenocarcinomas. J Thorac Oncol. 2013;8:322–8.
55. Guideline from the College of American Pathologists (CAP), International Association for the Study of Lung Cancer (IASLC), Association for Molecular Pathology (AMP). Molecular testing guideline for selection of lung cancer patients for EGFR and ALK tyrosine kinase inhibitors. https://www.iaslc.org/sites/default/files/wysiwyg-assets/cap_iaslc_amp_webcast_powerpointv9_1.pdf.
56. Murphy J, Bustin SA. Reliability of real-time reverse-transcription PCR in clinical diagnostics: gold standard or substandard? Expert Rev Mol Diagn. 2009;9:187–97.

57. Bartley PA, Ross DM, Latham S. Sensitive detection and quantification of minimal residual disease in chronic myeloid leukaemia using nested quantitative PCR for BCR-ABL DNA. Int J Lab Hematol. 2010;32:222–8.
58. Ross DM, Branford S, Seymour JF, et al. Patients with chronic myeloid leukemia who maintain a complete molecular response after stopping imatinib treatment have evidence of persistent leukemia by DNA PCR. Leukemia. 2010;24:1719–24.
59. Goh HG, Lin M, Fukushima T, et al. Sensitive quantitation of minimal residual disease in chronic myeloid leukemia using nanofluidic digital polymerase chain reaction assay. Leuk Lymphoma. 2011;52:896–904.
60. Nikiforov YE. Molecular diagnostics of thyroid tumors. Arch Pathol Lab Med. 2011;135:569–77.
61. Cantara S, Cappezzone M, Marchisotta S, et al. Impact of proto-oncogene mutation defect in cytological specimens from thyroid nodules improves the diagnostic accuracy of cytology. J Clin Endocrinol Metab. 2010;95:1365–9.
62. Fu Q, Zhu J, Van Eyk JE. Comparison of multiplex immunoassay platforms. Clin Chem. 2010;56:314–8.
63. Choi HJ, Kim HR, Shin MG. The author response: diagnostic standardization of leukemia fusion gene detection system using multiplex reverse transcriptase-polymerase chain reaction in Korea. J Korean Med Sci. 2011;26:1401.
64. King RL, Naghashpour M, Watt CD, et al. A comparative analysis of molecular genetic and conventional cytogenetic detection of diagnostically important translocations in more than 400 cases of acute leukemia, highlighting the frequency of false-negative conventional cytogenetics. Am J Clin Pathol. 2011;135:921–8.
65. Dancey JE, Bedard PL, Onetto N, et al. The genetic basis for cancer treatment decisions. Cell. 2012;148: 409–20.
66. Holbein ME. Understanding FDA regulatory requirements for investigational new drug applications for sponsor-investigators. J Investig Med. 2009;57: 688–94.
67. Sargent DJ, Conley BA, Allegra C, et al. Clinical trial designs for predictive marker validation in cancer treatment trials. J Clin Oncol. 2005;23:2020–7.
68. Pusztai L, Broglio K, Andre F, et al. Effect of molecular disease subsets on disease-free survival in randomized adjuvant chemotherapy trials for estrogen-receptor positive breast cancer. J Clin Oncol. 2008;26:4679–83.
69. Simon R. Advances in clinical trial designs for predictive biomarker discovery and validation. Curr Breast Cancer Rep. 2009;1:216–21.
70. Wang SJ, O'Neill RT, Hung HM. Approaches to evaluation of treatment effect in randomized clinical trials with genomic subset. Pharm Stat. 2007;6:227–44.
71. Liu A, Li Q, Yu KF, et al. A threshold sample-enrichment approach in a clinical trial with heterogeneous subpopulations. Clin Trials. 2010;7:537–45.
72. Freidlin B, Simon R. Adaptive signature design: an adaptive clinical trial design for generating and prospectively testing a gene expression signature for sensitive patients. Clin Cancer Res. 2005;11:7872–8.
73. McDermott U, Downing JR, Stratton MR. Genomics and the continuum of cancer care. N Engl J Med. 2011;364:340–50.
74. Srivastava SS, Kramer BS. Early detection cancer research network. Lab Invest. 2000;80:1147–8.
75. Zerhouni EA, Sanders CA, von Eschenbach AC. The Biomarkers Consortium: public and private sectors working in partnership to improve the public health. Oncologist. 2007;12:250–2.
76. Bossuyt PM, Reitsma JB, Bruns DE, et al. The STARD statement for reporting studies of diagnostic accuracy: explanation and elaboration. Clin Chem. 2003;49:1–6.
77. McShane LM, Altman DG, Sauerbrei W, et al. Reporting recommendations for tumor marker prognostic studies (REMARK). J Natl Cancer Inst. 2006;97:1180–4.
78. Hayes D, Ethier S, Lippman M. New guidelines for reporting tumor marker studies in breast cancer research and treatment: REMARK. Breast Cancer Res Treat. 2006;100:237–8.
79. Keedy VL, Temin S, Somerfield MR, et al. American Society of Clinical Oncology provisional clinical opinion: epidermal growth factor receptor (EGFR) Mutation testing for patients with advanced non-small-cell lung cancer considering first-line EGFR tyrosine kinase inhibitor therapy. J Clin Oncol. 2011;29:2121–7.
80. Hayes DF, Bast RC, Desch CE. Tumor marker utility grading system: a framework to evaluate clinical utility of tumor markers. J Natl Cancer Inst. 1996;88:1456–66.
81. Locker GY, Hamilton S, Harris J, et al. ASCO 2006 update of recommendations for the use of tumor markers in gastrointestinal cancer. J Clin Oncol. 2006;24:5313–27.
82. Harris L, Fritsche H, Mennel R, et al. American Society of Clinical Oncology 2007 update of recommendations for the use of tumor markers in breast cancer. J Clin Oncol. 2007;25:5287–310.
83. Freidlin B, McShane LM, Korn EL. Randomized clinical trials with biomarkers: design issues. J Natl Cancer Inst. 2010;102:152–60.
84. Burke W, Atkins D, Gwinn M, et al. Genetic test evaluation: information needs of clinicians, policy makers, and the public. Am J Epidemiol. 2002;156: 311–8.
85. Teutsch SM, Bradley LA, Palomaki GE, et al. The evaluation of genomic applications in practice and prevention (EGAPP) initiative: methods of the EGAPP Working Group. Genet Med. 2009;11: 3–14.
86. Chou R, Croswell JM, Dana T, et al. Screening for prostate cancer: a review of the evidence for the U.S. Preventive Services Task Force. Ann Intern Med. 2011;155:762–71.
87. Prensner JR, Rubin MA, Wei JT, et al. Chinnaiyan. Beyond PSA: the next generation of prostate cancer biomarkers. Sci Transl Med. 2012;4:127rv3.

CHAPTER 13

Bioinformatics Tools in Clinical Genomics

DAVID K. CROCKETT, KARL V. VOELKERDING

Abbreviations 177
Introduction 177
Methods 178
Conclusions 193

Abbreviations

BAM	Binary alignment mapping file format
BWA	Burrows Wheeler aligner
BWT	Burrows Wheeler transform
GATK	Genome analysis tool kit
HGMD	Human gene mutation database
IGV	Integrative genomics viewer
NGS	Next-generation sequencing
OMIM	Online Mendelian inheritance in man
TVC	Torrent variant caller
VCF	Variant call file format
VUS	Variant of uncertain significance
WGS	Whole-genome sequencing

Introduction

Whether one works on the bench or at the bedside, we are increasingly interfacing with the current avalanche of large datasets. Recent advances in high throughput analysis platforms such as next-generation sequencing (NGS) beg the question of how these data will impact and be utilized by the field of clinical pathology. Historically, large research projects such as the Human Genome Project did not realize their full utility without the subsequent bioinformatics analysis and data interpretation. Because genomic information is increasingly being used in the practice of medicine, bioinformatics is becoming an essential component in medical research and in the clinical diagnostic laboratory setting. As the cost and labor required to sequence human DNA continues to drop, this trend will continue [1]. It is important to emphasize that without expert computational analysis the sequencing results themselves are, in essence, just a very large file of A's, T's, C's, and G's. Thus, there is the need to better understand the field of bioinformatics and how it may affect clinical pathology in the near term.

It is interesting to note that the actual term *bioinformatics* appeared well before the current "genomics revolution." In 1978, a Dutch theoretical biologist (Paulien Hogeweg) first coined this term in reference to the study of information processes in biotic systems [2]. One common definition of bioinformatics is: "Research, development, or application of computational tools and approaches for expanding the use of biological, medical, behavioral or health data" [3]. This includes any computational methods to acquire, store, organize, archive, analyze, or visualize such data. To others, bioinformatics is simply a grammatical contraction of "biological informatics" and may call to mind the computer science disciplines of information science or

G.J. Netto and I. Schrijver (eds.), *Genomic Applications in Pathology*,
DOI 10.1007/978-1-4939-0727-4_13,

information technology, underscoring the large amounts of data to be analyzed and managed [4]. It is also quite notable that the majority of this computational work is not in the Microsoft Windows® environment. A more typical setting is command line parsing, scripting and making queries on Unix/Linux hardware using programming languages and tools such as Perl, Python, Java, R and SQL, among others.

The unprecedented data volumes and the qualitative and uniquely quantitative nature of NGS data have driven a renaissance in bioinformatics research and development resulting in the proliferation of a diversity of open-source and commercial algorithms and software to support the computational processing, analysis, and interpretation of NGS results [5]. These efforts have facilitated a broad dissemination of NGS into every facet of biomedical research and more recently into a multiplicity of clinical diagnostic applications from multi-gene panels to exome and whole-genome sequencing (WGS).

Every laboratory adopting NGS undergoes a learning curve with respect to analyzing NGS data.[1] This has proven to be a significant bottleneck due to the specialized nature of bioinformatics knowledge and lack of personnel trained in the discipline [6]. In this context, the goal of the current chapter is to introduce basic concepts and principles of bioinformatics required for the analysis of NGS data. The spectrum of NGS data generation, processing and alignment, variant calling and interpretation is discussed. The Illumina and Ion Torrent sequencing technologies and associated data analyses are emphasized due to their current dominant roles in the NGS landscape. A subsection is devoted to computational approaches for the identification of candidate genes from exome and WGS studies [7]. The chapter concludes with a discussion of in silico predictors and test reporting strategies.

[1] Pathologists can strive to become more aware of this rapidly moving field by familiarizing themselves with resources such as Web sites like SeqAnswers [67], BioStar [68], and Bio-IT World [69], software and server tutorials such as Galaxy [70] or GATK [16], continuing education or certificates such as NGS short courses [71] and AMIA 10 × 10 [72] offerings.

Methods

Next Generation Sequencing Bioinformatics

The bioinformatics processing of NGS data can be operationally divided into three major steps:

1. Generation of a sequence read file containing linear nucleotide sequence (e.g., ACTGGCA) accomplished using instrument specific software.
2. Mapping and aligning sequence reads to a reference sequence and identifying differences (variants) between sequence reads and reference.
3. Annotation and interpretation of variants with respect to phenotype.

Steps 2 and 3 use either open source or commercial algorithms and software, as well as variant databases. Each of these steps is next presented in greater detail. We first discuss bioinformatics pertinent to the analysis of Illumina sequence data followed by Ion Torrent.

Illumina: Bioinformatics Overview

NGS raw sequence base data from Illumina are comprised of four color (ATCG) fluorescent images optically recorded after each successive sequencing cycle. The images captured reflect single nucleotide base incorporation into individual sequencing clusters, with each cluster representing a clonal amplicon as seen in Fig. 13.1. Illumina utilizes a quality control measure termed the "chastity" filter for acceptance or rejection of individual clusters and this filter is applied after the first 25 cycles of a sequencing run. Specifically, during the first 25 cycles the highest fluorescent intensity base incorporated into a cluster is recorded and its intensity is compared to the next highest fluorescent base recorded for the cluster. This information is used to calculate the chastity filter ratio which is derived by taking the fluorescence of the highest fluorescent intensity base and dividing it by the fluorescence of the same highest fluorescent intensity base plus the fluorescence of the next highest fluorescent intensity base. A ratio of 0.6 or greater is considered a "passing" ratio.

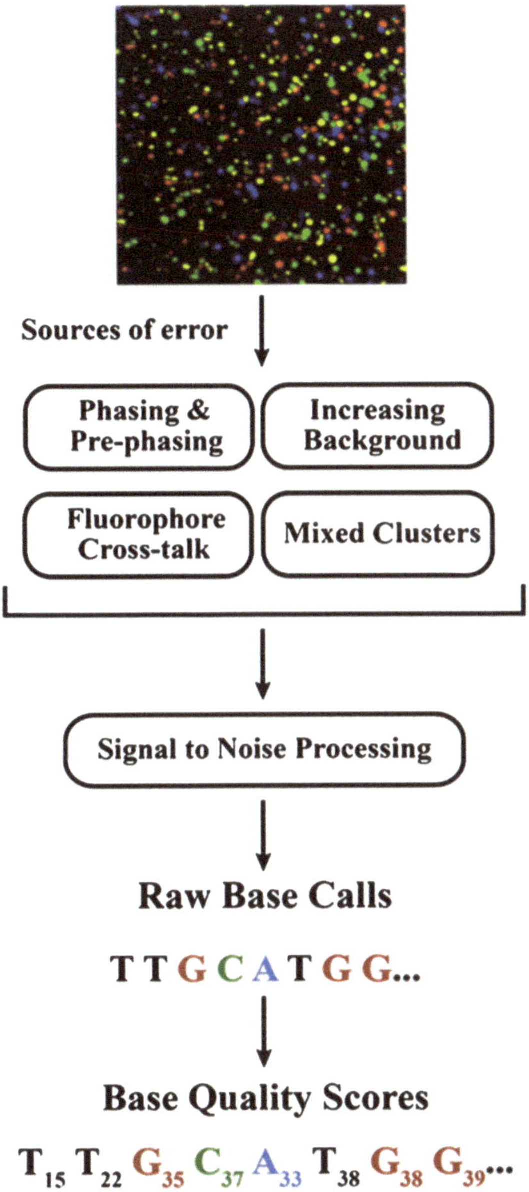

Figure 13-1 Processing of signal to noise and generation of base calls for Illumina sequence data. A flow cell image generated by the Illumina Genome Analyzer is shown at the *top*. This represents an overlay of the four unique reversible dye terminators (fluorophores) in *red, yellow, blue,* and *green*. Respective fluorophores are incorporated into individual clonal clusters during a sequencing cycle. Process steps for generating raw base calls and sources of error are indicated. Base calls and associated quality scores from actual sequence are shown at the *bottom*. *A* adenine, *C* cytosine, *G* guanine, *T* thymine. Reprinted with permission from Coonrod et al. [7]

A cluster "failure" is defined when two or more base incorporation events have chastity ratio values less than 0.6 in the first 25 cycles of a sequencing run. An important reason for a cluster to fail is when cluster densities are too high resulting in cluster overlap and mixed sequencing signals. The dominant error mode for Illumina sequencing is the category of single nucleotide substitutions.

An additional factor contributing to the overall error rate is the phenomenon termed "phasing." Each Illumina cluster is comprised of ~1,000 clonal amplicon strands and each strand is sequenced in parallel, providing a sum total signal for the individual cluster. As each base is incorporated individually within a growing DNA strand, base incorporation can become out of phase within a clonal cluster if one base is skipped (phasing) or multiple bases are incorporated in a single cycle (pre-phasing), resulting in nonuniform fluorescence within a clonal cluster [8]. Additionally, background fluorescence on the flow cell increases during the analytical run, which results in a decreased signal to noise ratio.[2]

Several software packages are available for Illumina base calling. The chastity filter that comes with the Illumina platform removes clusters of low uniformity. The Illumina application Bustard corrects or filters base calls for cross talk, phasing, and pre-phasing, then assigns the base with the highest signal to noise intensity ratio as the base call and calculates a base quality (Q) score [$Q = -10 \times \log10(e)$]. The Q score is logarithmically related to error probability (e) and is conceptually analogous to the Phred quality score used in Sanger sequencing [9, 10]. For example, bases with Q20 and Q30 quality scores have a 1:100 and 1:1,000 probability of being called incorrectly, respectively.

[2]As noted above, errors can occur by virtue of mixed signals in a cluster composed of more than one unique, overlapping clonal amplicon, especially if they have not been eliminated by the chastity filter. In addition, overlap in the emission spectra of each of the four fluorophores can make it difficult to determine which base was incorporated (fluorophores cross talk) when clonal clusters are physically close to each other.

The Q score is calculated for each base along the sequence read and is used as a standard quality control metric for downstream data analysis.

For Illumina sequencing, a key metric is the number of bases with quality scores equal to or greater than 30. As an example, in a representative WGS dataset comprised of 100 base length pair end reads, approximately 88 % of bases had Q scores of 30 or greater. After assignment of the Q scores, the Illumina sequence files are converted into a text-based file format termed FASTQ. The FASTQ file contains sequence reads that have passed filter metrics and each read is associated with an identifier that indicates its location on the flow cell (e.g., lane and tile). The linear sequence is displayed and each base is assigned a base quality score designated using ASCII coding. The FASTQ file format, also employed by Ion Torrent and other NGS technologies, has become the *de facto* information exchange currency in NGS.

Illumina: Sequence Mapping, Alignment, and Variant Calling

Sequence reads in FASTQ files are used for two main computational purposes: assembly and alignment. While the majority of diagnostic applications employ alignment to a reference sequence, assemblies are performed when no reference genome exists for the sequenced DNA (e.g., uncharacterized or novel bacteria and viruses). Algorithms used for assembly seek and join overlapping sections of sequence reads to generate longer length "contigs." The length of contigs can be increased by using longer and paired end sequence reads to yield a genomic scaffold onto which subsequent alignments can be performed.

Mapping and aligning are the processes of determining the best match between the sequencing reads and the reference sequence. Due to the large number of sequence reads in NGS datasets, NGS alignment algorithms use approaches to decrease computational time and the two major approaches are a sophisticated data compression method termed the Burrows Wheeler Transform (BWT) and a method based on a Hash Table. The open source Burrows Wheeler Aligner (BWA) algorithm [11] that has become one of the standards for sequence alignment uses the BWT method [12, 13]. An example of a popular commercial aligner that utilizes a Hash Table method is Novoalign [14]. With Hash Table alignment, either the reference sequence or the sequence reads are first converted into a population of shorter length sequences with each sequence given an identifier for computational tracking. The use of shorter length sequences (also known as "seeds"), allows mapping and alignment to proceed more rapidly [15]. When using either BWT or Hash Table based aligners, parameters for initial mapping and alignment need to be established, including number of nucleotide mismatches permitted across a given read or seed length and whether gaps in alignment are allowed to accommodate insertions and deletions (indels). The operator may elect default settings or set more or less restrictive parameters. The output of most alignment algorithms is a file format termed SAM (sequence alignment map) which contains read position information and orientation in relationship to the reference sequence and a confidence value for the alignment. A reduced size, binary version of SAM is the BAM format. Using initial alignment criteria, which are typically more permissive than secondary algorithms, the output is a dataset known to contain inaccuracies.

After initial alignment, SAM/BAM files are used as inputs into secondary algorithms to refine and increase alignment accuracy prior to identifying variants (differences between the sequencing reads and the reference sequence). Two popular open source software for refining alignments and calling variants are the Genome Analysis Tool Kit (GATK) [16] and SAMtools [17]. As displayed in Fig. 13.2, three major refinement steps are used in GATK: (1) local realignment to improve accuracy of indel calls; (2) removal of reads with the same start and end points, referred to as PCR duplicates; and (3) recalibration of base quality scores. PCR duplicates originate by sequencing identical fragments generated by PCR during library preparation. Nucleotide errors introduced during PCR can be propagated and appear in variant files. To mitigate this, PCR duplicate

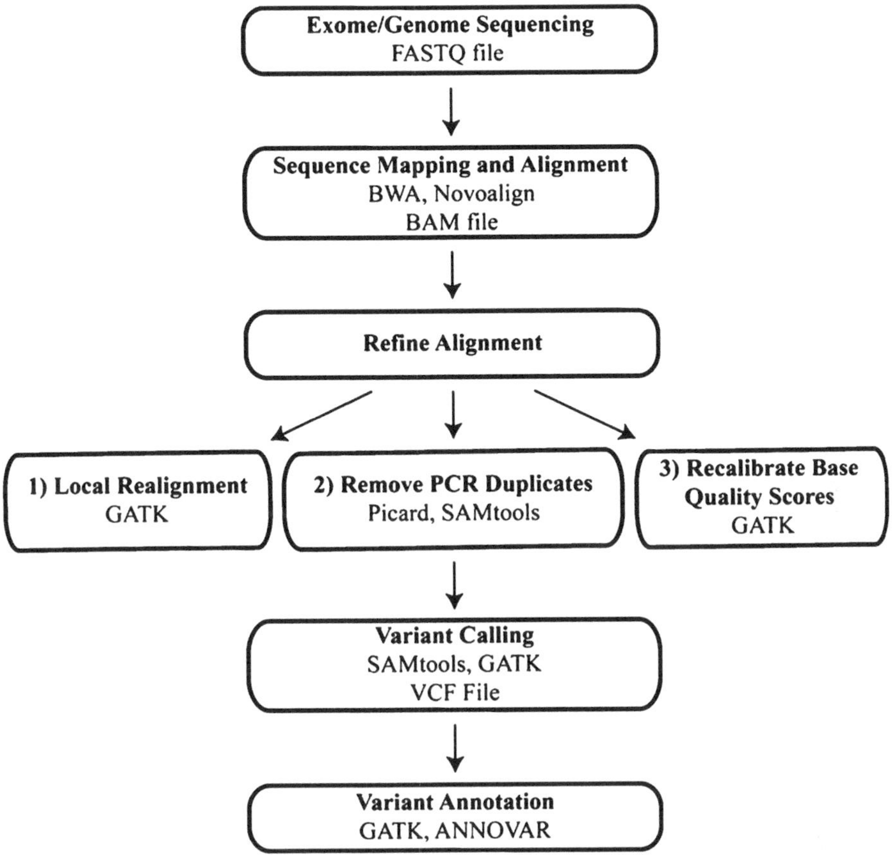

Figure 13-2 A representative bioinformatics workflow for analysis of Illumina sequence data. Steps required to generate a final annotated variant list from raw sequencing data are also indicated. Where applicable, open-source programs are listed along with the file type generated. *BWA* Burrows-Wheeler Aligner, *GATK* Genome Analysis Toolkit, *PCR* polymerase chain reaction, *VCF* Variant Call Format. Reprinted with permission from Coonrod et al. [7]

removal is performed leaving only a single read with overall highest base qualities. The impact of local realignment on indel accuracy in detecting a 3 bp deletion in the FOXP3 gene is illustrated in Fig. 13.3 with aligned reads visualized in the open-source Integrative Genomics Viewer (IGV) [18, 19]. Recalibrating base quality scores is done to adjust Phred-like quality scores generated by the Illumina platform, which have been shown to deviate from the true error rate. After the initial and refined alignments, variants are tabulated in a Variant Call File format (VCF) that contains several parameters including the chromosomal position of the variant, reference base, and the alternative base(s) (e.g., single nucleotide variant or SNV, indel).

Illumina: Coverage and Variant Calling

For many applications, NGS libraries are comprised of randomly overlapping fragments, exceptions being certain targeted enrichment approaches that employ PCR. As such, multiple reads align to the reference in a staggered or non-staggered fashion. This multiplicity can be quantified by enumerating the number of reads aligned to the reference sequence at a given location and is termed the "read coverage depth." Bidirectional sequencing yields forward and reverse strand sequences, and under ideal conditions their percentages would be approximately equal. The percentage of reads containing a variant is referred to as the

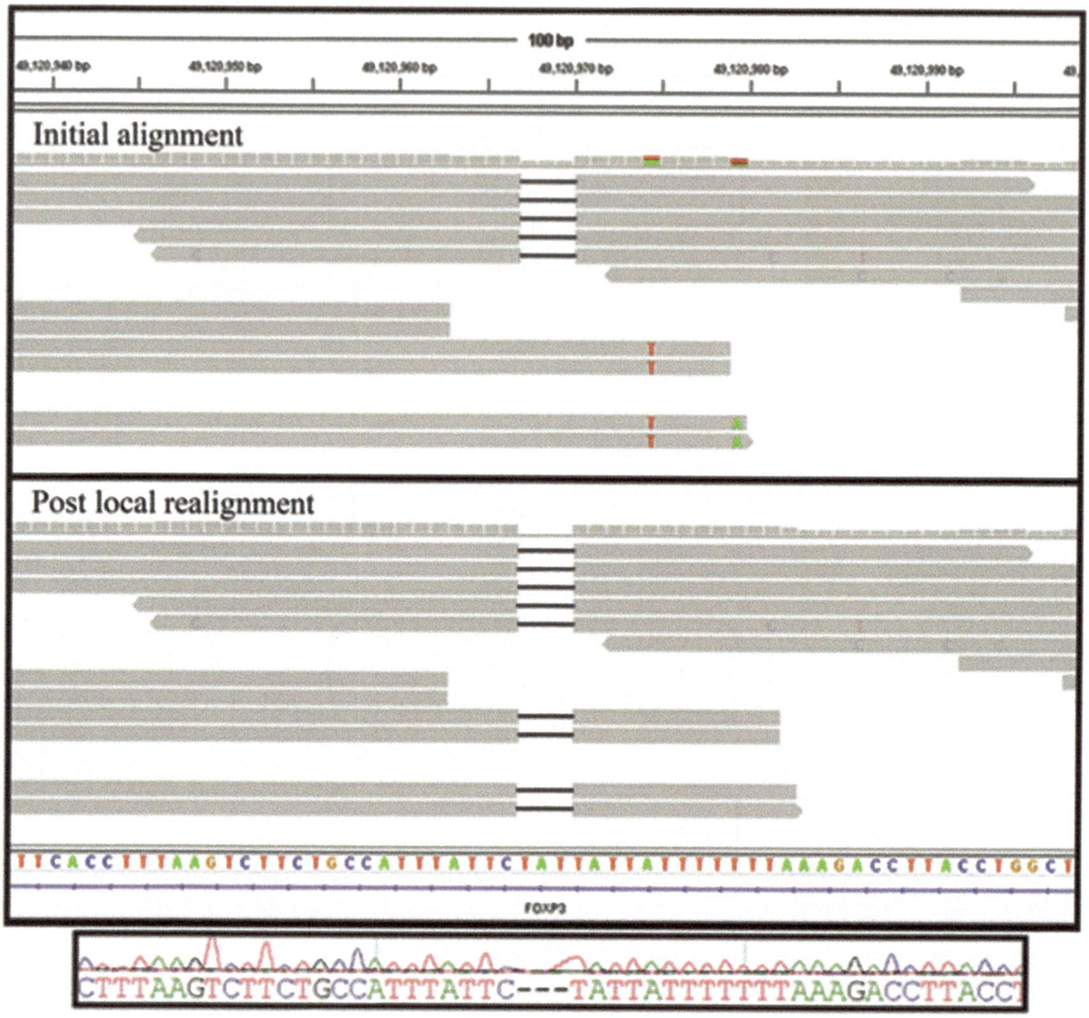

Figure 13-3 Generating refined alignments may improve local mapping around insertions and/or deletions. The local realignment of an indel visualized in the Integrative Genomics Viewer enhances results from initial mapping with an apparent 3-bp deletion in the FOXP3 gene on the X chromosome of a male. Initially, four reads contain the deletion (depicted by black bars within the read) and four reads do not contain the deletion. Important to note is that in the four reads that do not contain the deletion, six flanking single-nucleotide variants (SNVs) are present (a T variant in *red* and an A variant in *green*). The initial alignment suggests heterozygosity for the deletion on the X chromosome, but this is unlikely, given that the sequence reads are derived from a male. The *lower panel* (post local realignment) shows all reads in agreement to contain the 3-bp deletion. In addition, the flanking, false positive SNVs are no longer present. The Sanger sequencing trace shown below confirms the deletion and zygosity of the g.49120967_49120971delTAT deletion. Reprinted with permission from Coonrod et al. [7]

"allelic read percentage." Figure 13.4 depicts a heterozygous single nucleotide variant (SNV) and the overall concepts of coverage and allelic read percentage. Germline variant calling accuracy is greater when there is a consensus among aligned reads consistent with either heterozygosity or homozygosity. Ideally, a sample with a heterozygous SNV would have approximately a 50/50 ratio of reads containing variant and reference nucleotides, and forward and reverse strand reads would be equally represented in both variant- and reference-containing reads. A homozygous variant would be expected to be present in 100 % of aligned sequences with equal representation of forward and reverse

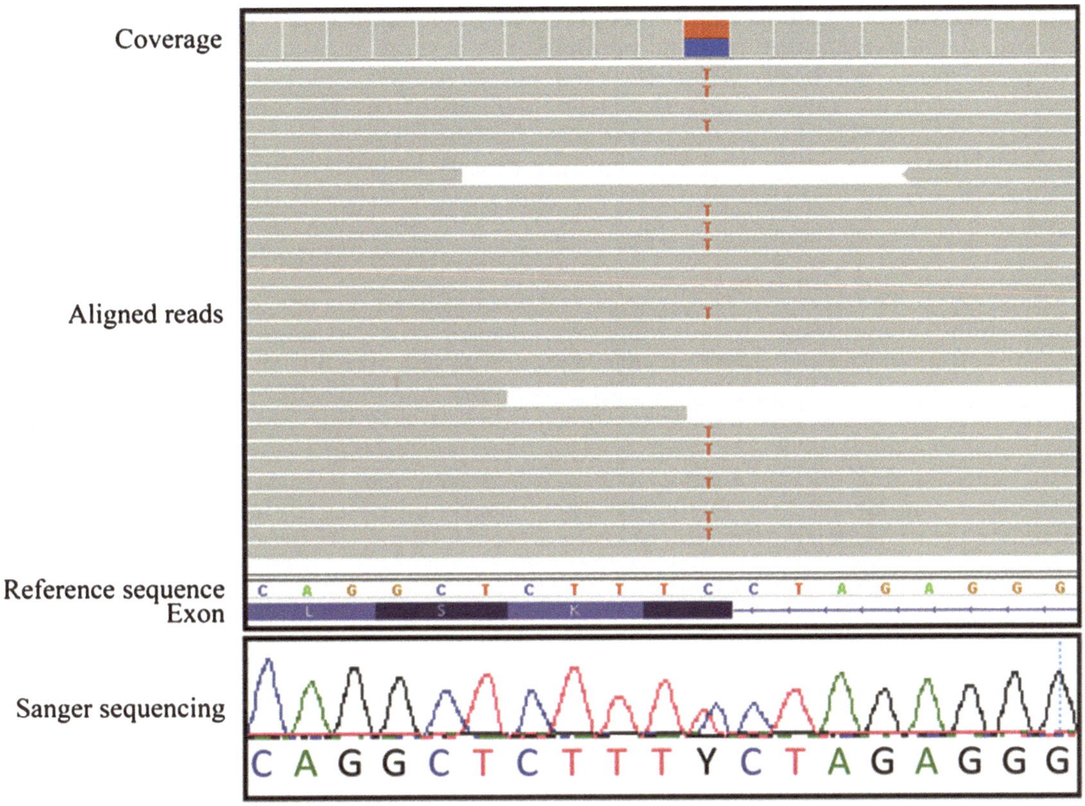

Figure 13-4 Illumina sequence reads and alignment as seen in a popular genome browser (Integrative Genomics Viewer, IGV). *Gray boxes* across the *top* represent read depth (coverage). Mapped and aligned reads are shaded *gray*, with variants from the reference highlighted by a unique color. In this example, a cytosine to thymine (C > T) variant change is present, with variant T's highlighted in *red* in the aligned reads. The heterozygous change is also indicated above (in Coverage) and, in this case, because the variant is heterozygous, the *box* is split into two colors, *red* for the variant nucleotide (T) and blue for the reference nucleotide (C). The reference nucleotide sequence (reference sequence) is shown below the aligned reads along with the location of the exon and amino acid translation of the sequence (exon). This region was also Sanger sequenced to confirm the heterozygous (C > T) variant. The Sanger trace from this sequence is shown at *bottom*. Reprinted with permission from Coonrod et al. [7]

strands. Empirically, a wider range of allelic read percentages and forward and reverse strand percentages (manifesting as "read strand bias") is observed.

Read strand bias arises from several sources including differential PCR amplification of library fragments, sequencing errors in difficult to sequence regions, and misalignment of related sequences (e.g., pseudogenes or highly homologous genes). The minimum number of reads required for variant calling needs to be empirically determined per application. For germline variant detection, a coverage of approximately 30-fold has been found to offer a balance between sensitivity, specificity, and sequencing costs in the research setting, whereas clinical laboratories typically aim for higher coverage depths in an effort to increase variant detection and improve variant call confidence [20–22]. Coverage depths across sequenced regions are variable due to factors such as differential ligation of adapters to fragments during library preparation and differential amplification during clonal expansion, thus necessitating that sufficient sequencing is performed to meet clinically required coverage.

Finally, whereas the above discussed 50/50 and 100 % variant ratios are relevant to identifying heterozygous and homozygous germline variants, respectively, they do not apply when identifying variants in heterogeneous scenarios such as somatic variants in cancer samples comprised of a mixture of tumor and normal cell populations. To identify low allelic read percentage somatic variants, increased read coverage in the several hundred up to thousand-fold range are being used for clinical testing [23, 24].

Illumina: Variant Annotation

Annotation is accomplished by introducing a variant call file into a program that ascribes additional features to variants. Examples of open source software that contain annotation functions are ANNOVAR [25, 26], GATK [16] and snpEff [27], among others. Annotation outputs include many features including chromosomal location of base change from reference, whether the variant is in a gene and its respective location (e.g., exon, intron, splice site), and the consequence of the change to a codon (e.g., synonymous versus nonsynonymous, missense versus frameshift), and zygosity (e.g., homozygous or heterozygous). Often incorporated into annotation software programs are algorithms that predict the functional impact of variants on proteins such as Sorting Intolerant from Tolerant (SIFT) [28–30], Polyphen2 [31, 32] and Mutation Taster [33, 34].

Ion Torrent: Bioinformatics Overview

To generate DNA sequence information, the Ion Torrent™ technology relies on the relatively simple biochemical componentry of DNA polymerase and natural nucleotides. The nucleotide incorporation detection process monitors hydrogen ion release as known nucleotides are incorporated into growing DNA strands. Individual reaction well hydrogen ion signals are detected by a proprietary ion sensing technology that utilizes field effect transistors scaled in a massively parallel configuration using semiconductor technology. Analogous to Illumina technology, signal to noise ratios are algorithmically converted into nucleotide base calls with associated quality scores. The linear sequence file output is converted into the FASTQ format which can be put into a variety of open source and commercial software for subsequent mapping and alignment and variant calling. In practice, most groups use software developed by Ion Torrent which has been optimized for Ion Torrent sequence read data and which is discussed next in greater detail.

Ion Torrent: Data Flow

Each ion sequencing chip contains a high-density array of micro-machined wells that are placed over an ion-sensitive layer and a proprietary ion sensor. During each nucleotide flow over the chip, when a nucleotide is incorporated into a growing strand of DNA, a hydrogen ion is released. The ion release changes the pH of the solution and is detected by the chip's ion sensor. The raw pH value from each well is converted into a voltage and captured as a digital representation of that voltage. On the other hand, if the nucleotide that flows over the chip is not complementary, no incorporation occurs and thus no change in pH or voltage is recorded. In this way, analysis of these data can reveal the base incorporated during the nucleotide flow. This process transforms the chemical information to digital information in a conceptually very simple and direct manner.

The Ion Sequencer outputs raw sequencing data in the form of Data Acquisition (DAT) files. These DAT files reflect the conversion of raw pH values in each well into a digital representation of the change in voltage. The raw DAT data files are then transferred to the Torrent Server for analysis pipeline processing. On the Torrent Server, the raw signal measurements are converted into incorporation measurements and then into base calls for each read. Figure 13.5 shows the steps in the Torrent pipeline from the point of view of the data files that are generated. As shown, each data file output by one step is input into the next step in the pipeline.

Ion Torrent: Sequence Generation

The sequence generation step is also called base calling. This step determines the actual sequence of individual nucleotide bases in each sample. The Torrent algorithm, BaseCaller, runs automatically during the

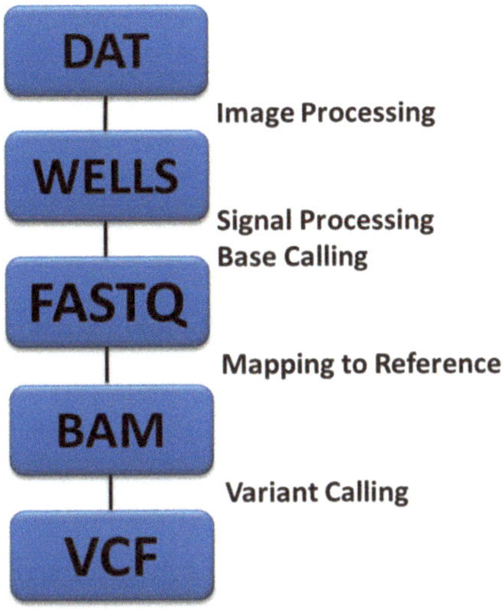

Figure 13-5 The bioinformatics pipeline from Torrent Suite is as follows: DAT (Data Acquisition) contains raw voltage measurements from the chip. WELLS contains the nucleotide incorporation signals for the flow for each well. FASTQ contains the nucleotide calls (sequence calls), and associated quality values. BAM (Binary Alignment Map) contains mapped reads with their alignment to the reference genome. VCF (Variant Call Format) contains variants called on the input DNA sample. Each variant call details how a given DNA position (sequence) found in the sample differs from the reference genome, for instance, by an insertion of bases, deletion of bases, or change in a base

Torrent Suite pipeline and is optimized for Torrent data. During an optimized sequencing run, the majority of wells in the sequencing chip contain a clonal DNA template. The DNA sample has one template (fixed) strand and a synthesized strand. The sample's synthesized strand grows in length whenever a base from a nucleotide flow is incorporated. In order to be incorporated into the growing strand, the nucleotide base in each flow cycle must be complementary to the next base on the template's fixed strand. The bases contained in each nucleotide flow solution are known beforehand.

This simple process is base calling for one position in the sequence. However, nucleotide incorporation happens in each well of the chip. Millions of wells per chip and hundreds of flows per run make base calling a massively parallel operation. These pH measurements over the entire chip will occur several times per second as the sequencing process takes advantage of semiconductor technology to increase throughput. It is important to note that the Torrent Suite base calling algorithm is optimized for Ion Torrent data. Although the base calling module uses fairly stringent filters that are designed to increase the accuracy of results, filters can be adjusted if a given sequencing application requires maximizing the number of reads. After sequence data are generated, the Torrent Server automatically performs sequence alignment on those data.

Ion Torrent: Sequence Alignment

During sequence alignment (also known as mapping), base calls generated by the Torrent Suite analysis are aligned to a reference genome in the BAM file format. Several alignment metrics are also produced at this time. The Torrent Suite utilizes the Torrent Mapping Alignment Program (TMAP). This is a sequence alignment software program that is developed specifically to meet Ion Torrent data mapping challenges.[3]

TMAP incorporates three common alignment algorithms (BWA-short, BWA-long and Sequence Search and Alignment by Hashing Algorithm, SSAHA). The main indexing structure in TMAP uses a compressed suffix array (FM-index) based on the Burrows–Wheeler transform (block-sorting compression). The initial alignment approach uses all three algorithms to quickly produce a list of candidate mapping locations. These candidate locations are then refined using the more accurate Smith-Waterman algorithm. Resulting alignments are aggregated to identify the optimized mapping location. User-defined parameters then determine if all alignments, a subset of alignments, or a random best alignment is reported. TMAP employs a two-stage mapping strategy to maintain sensitivity and specificity while significantly reducing runtime. In this two-stage mapping, reads that do not align

[3] Ion Torrent™ data's particular qualities require special consideration during the alignment process for several reasons, including: (a) Reads generated by Ion Sequencers are variable in length and are expected to increase as the technology matures. (b) The principal error mode associated with Ion data relates to miscalling homopolymer lengths and results in insertion or deletion errors during alignment and post-processing.

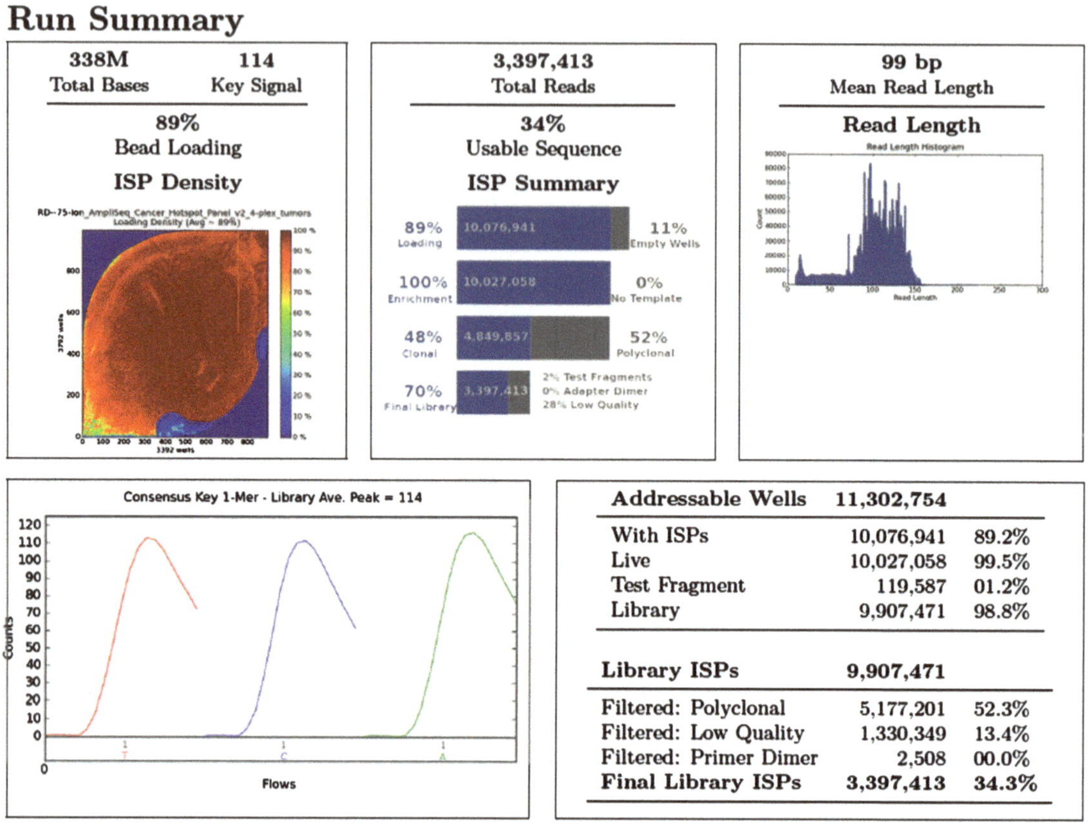

Figure 13-6 Example view of the metrics available through the Torrent Suite Browser's Reports summary, which includes chip loading quality estimates, ion sphere particles (ISP) efficiency, read length, alignment estimates, and variant statistics (not shown). Courtesy of Life Technologies

during the first pass are given to the second stage with a new set of algorithms and/or parameters. The output of sequence alignment is a BAM file containing mapped reads. Each BAM file can be analyzed to obtain various metrics, including quality estimates and read length estimates. Various chip loading and alignment metrics can be viewed in the Torrent Browser's Reports summary pages as shown in Fig. 13.6.

TMAP has key advantages over other alignment tools. To deal with varied length reads and error profiles that are specific to Ion Torrent™ data, the re-implemented versions of the three popular alignment algorithms have been optimized. In this way, TMAP results are expected to perform significantly better when compared against the original algorithms alone. Since TMAP amalgamates the candidate locations for all three methods and identifies the best alignment, final accuracy and specificity can benefit from the advantages of each individual algorithm. TMAP has also been optimized for computational performance and with the TMAP index that is shared between alignment algorithms. The overall performance is comparable to other alignment software in terms of CPU load and RAM utilization. It is notable that the combined performance of multiple algorithms is equal to or better than the total performance of running each algorithm separately. However, some technical issues to avoid when using TMAP are described.[4]

[4]TMAP is recommended for alignment, as its algorithms are tuned to handle Ion Torrent data in an optimal manner. Some common issues to avoid when using TMAP are: (a) TMAP is not an assembler. TMAP is an alignment tool that gives the location where a particular read from the sequencing instrument aligns to the reference genome, but TMAP

In terms of realignment, the Torrent Browser supports redoing sequence alignment through two different interfaces. The first is by way of a Plugin, where the Alignment plugin runs the TMAP alignment module and optionally supports aligning against a different reference genome. Second is Reanalysis, where the Run Report reanalyze feature supports rerunning TMAP and also supports changing TMAP parameters for the new alignment.

Ion Torrent: Variant Calling

Generating sequence reads from the Ion Personal Genome Machine® (PGM™) Sequencer and Ion Proton™ Sequencer is only the first step in understanding the biological meaning of that sequence. The detection of single nucleotide polymorphisms (SNPs) and insertions or deletions (indels) in the generated sequencing data is essential to nearly all sequencing applications. Variant calling is the key step for finding genetic differences between samples and for understanding resulting biological associations. To perform variant detection, sequencing reads are first mapped to a reference genome to generate a read pileup. This read pileup is compared to the reference sequence, and SNPs and indel variants are identified. Ion's variant calling algorithms make calls based on the consensus accuracy, independently of the variants identified in raw reads.

Several software strategies exist for calling variants using Ion Torrent sequence data. These approaches were built to support different sequencing applications, such as targeted sequencing using Ion AmpliSeq™ or Ion TargetSeq™ selection technologies, as well as traditional genome or exome sequencing experiments. These software options are listed here with each software approach and their respective workflows described in more detail below.

1. Torrent Variant Caller plugin. This is a SNP and indel calling analysis module that is part of the Torrent Suite Software and is accessible through the Torrent Browser. It is designed to be initiated automatically after sequencing data have been generated and bases called. This plugin can also be initiated manually to process previously generated datasets.
2. Ion Reporter™ Software. This is a cloud-based software service that provides both variant calling and annotations. It incorporates log and traceability features that are essential to researchers performing routine sequencing assays.
3. Third-party software. Several commercial partners have provided software tools for detecting variants and visualization.

The Torrent Variant Caller (TVC) plugin is a secondary analysis software tool designed to call SNPs and indel variants. It is a built-in component of the larger Torrent Suite Software package. This plugin accepts the aligned reads (.bam file) generated by Torrent Suite Software as input. The plugin produces an output file (.vcf) of an annotated list of SNPs and indel variants called in each sample. By simply configuring the run plan before running the chip on the Ion PGM™ Sequencer, a user can set the TVC plugin to run automatically upon completion of the primary analysis of the chip. The TVC plugin can also be run (or rerun)

---

does not create a consensus file for the reference. (b) TMAP currently does not support RNA-Seq data. (c) Alignment programs must be optimized for the sequencing platform. The user can optionally use a different alignment program (outside of the Torrent Browser and the Torrent Suite Software). To align Torrent data with other alignment programs (beside TMAP), the user must ensure that the program's filters are set for Torrent data, not for data from other platforms. Incorrect results may occur if the data are not aligned with the correct parameter settings. (d) TMAP performs best when a restrictive error tolerance (such as maximum threshold of five mismatches) is not specified. (e) Typically, quality scores in repetitive sequences are lower than in nonrepetitive sequences. (f) It is not recommended to install other utilities named tmap on one's Torrent Server. For example, The European Molecular Biology Open Software Suite (EMBOSS) also includes a utility named tmap. If one installs EMBOSS on the Torrent Server, it will be likely to see name conflicts with the two different tmap programs. (g) Running TMAP using a partial or incomplete reference sequence may cause reads originating from homologous regions to be incorrectly mapped to target regions, which in turn may cause a downstream variant calling application to produce false positive variant calls.

Variant Calls

Chrom	Position	Gene Sym	Target ID	Type	Zygosi...	Ref	Variant	Var Freq	P-value	Cov	Ref Cov	Var Cov	HotSpot ID
chr3	178936091	PIK3CA	AMPL35831	SNP	Het	G	A	21.9%	1.00e-10	3610	2819	791	COSM763;
chr4	1807894	FGFR3	AMPL4116...	SNP	Hom	G	A	99.7%	7.94e-9	737	2	735	---
chr4	55141055	PDGFRA	AMPL43181	SNP	Hom	A	G	99.7%	1.00e-10	3749	10	3739	---
chr4	55593464	KIT	AMPL46166	SNP	Het	A	C	71.0%	7.94e-6	8169	2349	5802	COSM28026;
chr5	112175770	APC	AMPL59934	SNP	Het	G	A	96.0%	6.31e-6	4478	170	4298	
chr7	55249063	EGFR	AMPL493...	SNP	Hom	G	A	100.0%	1.58e-9	1060	0	1060	---
chr7	116339672	MET	AMPL61096	SNP	Het	C	T	38.0%	1.00e-10	5339	3309	2030	---
chr10	43613843	RET	AMPL78961	SNP	Hom	G	T	99.8%	1.26e-10	1823	0	1820	---
chr12	25398284	KRAS	AMPL553...	SNP	Het	C	A	20.9%	1.00e-10	4355	3435	912	COSM520;

Showing 9 rows

Figure 13-7 The Torrent Suite Variant Calls summary table provides details about each variant called, including chromosome, position, and sequence coverage. For each variant called, the genomic position listed in the "Position" column is also hyperlinked to open the Integrative Genomics Viewer (IGV) and displays all reads pertaining to that variant. Courtesy of Life Technologies

manually at any time after the primary analysis has completed.[5]

When completed, TVC outputs five primary report tables. The most inclusive is the Variant Calls table. This table provides details about each variant called, including chromosome, position, and sequence coverage. For each variant called, the genomic position listed in the "Position" column is hyperlinked to open the Integrative Genomics Viewer (IGV) and displays all reads pertaining to that variant as seen in Fig. 13.7. The TVC plugin's results page also contains a File Links section, which lists the output files generated by the Torrent Variant Caller plugin. This allows any output files to be conveniently available for loading into IGV or other third-party tools for further visualization or analysis.

Release of Torrent Suite Software v3.4 (from February 2013) offered several improvements, including faster signal processing and improved variant calling. Another feature was improved well characterization ("beadfinding") for more accurate background model signal processing, and better phase parameter estimation. Additional filter options designed to increase the accuracy of the results were also been incorporated. Importantly, users may wish to reanalyze and optimize data based on their sequencing application. For example, data quality may weigh most importantly for detection of rare gene variants. On the other hand, maximizing read depth for counting applications such as gene expression may be most important. By simple software interface, filtering can also be returned to the less stringent filtering of earlier versions or additional, more stringent filters can be added to provide the most accurate output [35].

Ion Reporter™ Software is a suite of bioinformatics tools meant to streamline and simplify analysis, reporting, and archiving of sequencing data. The initial design is for

[5]The TVC plugin offers several advantages. It has been optimized for Ion Torrent data, it is an included component of the Torrent Suite Software (automatically updated with each new release), and is supported by Ion Torrent. To initiate an analysis with the TVC plugin, users set up three key configurations: (a) Workflow—Users can select from a set of preconfigured workflows, according to the expected variant frequency in the sample (germline—all variants greater than 20 %, or somatic—at least one variant less than 20 %) and the library type. (b) Scope of analysis—Users can provide a reference sequence for alignment that is larger than the regions sequenced (e.g., the whole human genome) and then restrict the variant calling analysis to a specified region by uploading a target regions file in BED format. The regions file reduces the run time of the TVC plugin. (c) Scope of reporting—The user can also require the scope of the variant report to include a specified region, even if a variant has not been specifically identified in that region, by uploading a hotspots region file in BED format.

Also note that typically, germline variant frequency refers to a relatively pure population, whereas somatic variant frequency is found in a mixed population.

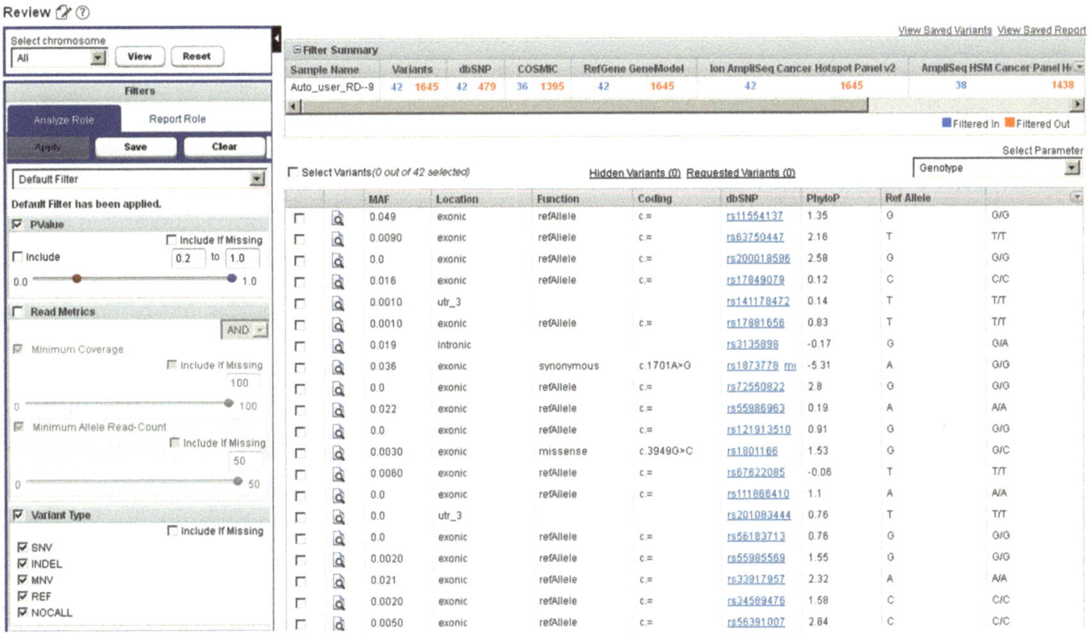

Figure 13-8 A representative view of the Ion Reporter's workflow for gene variant analysis including filtering, annotation, evidence review, internal comments, and report generation. Courtesy of Life Technologies

researchers performing repeated analysis of sequencing assays. Ion Reporter™ Software also integrates comprehensive public annotations to reduce the bioinformatics work needed to understand the impact of detected variants. Although the annotation and interpretation scope is much wider than that of the TVC plugin, the underlying variant calling algorithms are the same as those of the TVC plugin. Some advantages of using Ion Reporter™ Software are that it automatically adds annotations to variants, gives preconfigured workflows (to compare pairs and trios of samples), and provides an audit trail and version control. It also scales to efficiently utilize the computing power of cloud resources.

Currently, the Ion Reporter™ Software can perform mapping, variant calling, and annotation, starting with the input of an unaligned BAM file. Alternatively, it can perform only variant annotation starting with a variant file in VCF format produced by the TVC plugin. To manually or automatically move sequencing data from the Torrent Server to Ion Reporter™ Software, users launch the Ion Reporter™ Uploader plugin. If preconfigured prior to starting a sequencing run, the Ion Reporter™ Uploader plugin uploads the unaligned read BAM file to Ion Reporter and automatically performs mapping, variant calling, and annotation. Ion Reporter™ Software data upload, analysis, and storage are packaged for each chip and purchased as an addition to the other consumables. Ion Reporter™ Software's variant review, interpretation and report generation screen is shown in Fig. 13.8.

In terms of third-party software, Ion Torrent has partnered with commercial software providers including DNASTAR (SeqMan) and SoftGenetics (NextGENe), Partek and Avadis NGS. The intent is to provide alternative solutions for end-to-end workflows focusing on variant calling. Each of these software products provides a comprehensive solution, which allows users to identify variants, annotate those variants, and perform multi-sample comparisons. In addition, these products allow users to integrate additional genomics data into a single viewer. Workflows specific to Ion Torrent must be selected within third-party software solutions for optimal results.[6]

All of the initial data analysis steps for NGS reads as described above must be in place prior to the additional filtering, annotation and interpretation steps leading to gene discovery or clinical diagnostics.

Exome and Genome Sequencing for Causal and Candidate Gene Discovery

The use of exome and genome sequencing for causal and candidate gene identification requires additional bioinformatics strategies beyond variant calling. Genomes contain approximately 3–3.5 million noncoding and coding variants, whereas exomes contain 15,000–20,000 variants in coding regions. Whether starting with genome or exome datasets, the primary approach to identifying causal or candidate genes is to:

1. Focus on variants that are in coding or splice site regions.
2. Exclude higher population frequency variants by assuming their non-pathogenicity.
3. Determine if remaining variants have known or predicted deleterious impact on gene function.
4. Rank variants on the basis of deleterious impact and presence within genes of biological relevance to patient phenotype, and co-segregation with that phenotype if a family study is being performed.

Efficiently accomplishing these steps requires the establishment of an integrated process that draws upon variants databases, [e.g., 1,000 Genomes, dbSNP, Online Mendelian Inheritance in Man (OMIM), Human Gene Mutation Database (HGMD)] and variant impact prediction tools, [e.g., Sorting Intolerant From Tolerant (SIFT), Polymorphism Phenotyping (PolyPhen), Genomic Evolutionary Rate Profiling (GERP)]. The above approach is referred to as a logic tree or heuristic filtering method, and in practice, most clinical laboratories performing causal and candidate gene identification have established in house custom approaches to accomplish these process steps. An emerging trend is the development of several commercial software to accomplish heuristic filtering, yet one limitation of heuristic methods is that they do not provide any measure of statistical uncertainty for a given variant or candidate gene. In this context, new causal and candidate gene discovery prediction algorithms are being developed such as the Variant Annotation, Analysis, and Selection Tool (VAAST) that compares allele frequencies between cases, controls, and background datasets in conjunction with modeling variant severity by amino acid substitution analysis to provide a list of variants, each associated with a VAAST ranking score and a p-value [36, 37]. The p-value is a measure of the probability that a variant is statistically significant in a case, as compared to the control dataset.

Another recently reported approach to predicting causative variants describes a statistical method using a weighted sum approach, which takes into account "background" variation in genes to avoid having large or highly variable genes in the population rank high on the candidate list, can accommodate related or unrelated datasets, can incorporate linkage or functional data, and uses a computational approach to generate a measure of statistical certainty (p-value) for individual genes [38]. One could combine both heuristic and probabilistic approaches as shown in Fig. 13.9, and the outputs of each approach are then cross-compared to generate causal and candidate gene lists. One commercial software that incorporates both heuristic and probabilistic (i.e., VAAST) approaches has been developed by Omicia [39]. In terms of final variant interpretation

[6] Issues to note include avoiding the use of the TVC plugin or Ion Reporter™ Software on results from primary analysis methods (base calling and aligning) not optimized for Ion Torrent data. Running the TVC plugin or Ion Reporter™ Software over a long reference genome without defining a target region with a BED file causes run times to be significantly longer than with a BED file. One should avoid deriving conclusions about variants based only on raw accuracy or consensus accuracy. The TVC plugin uses consensus accuracy for candidate calls and also uses the raw reads to model errors and true variants. Therefore, the variant calls made by the TVC plugin cannot be verified solely by viewing the reads in an alignment browser, because the browser does not illustrate the power of the variant caller algorithm across multiple reads at the same position.

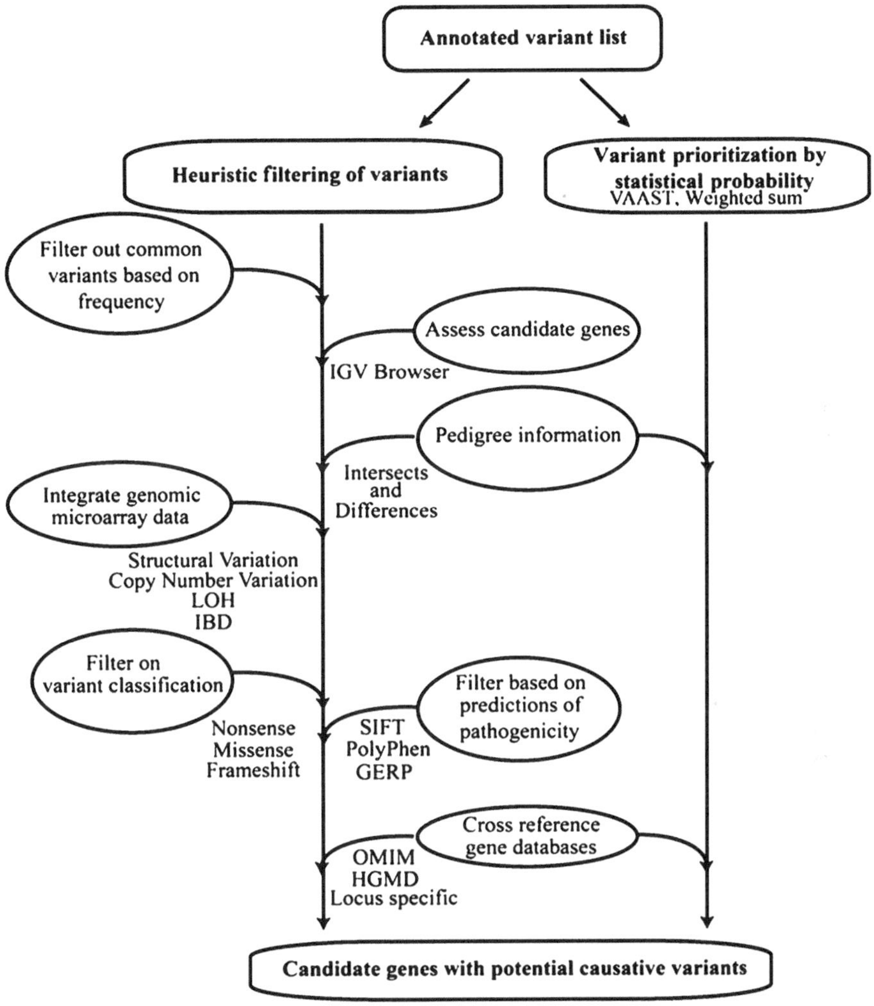

Figure 13-9 Diagram of approaches for candidate gene discovery from exome and genome sequencing data. Annotated variant lists can be analyzed with heuristic filtering approaches, statistical probability approaches, or a combination of both, to generate candidate gene lists. Multiple process steps are involved in heuristic filtering as depicted. Steps that incorporate pedigree information and cross-referencing of gene databases are critical components of both heuristic and statistical probability approaches. *GERP* Genomic Evolutionary Rate Profiling, *HGMD* Human Gene Mutation Database, *IBD* identity by descent, *IGV* Integrative Genomics Viewer, *LOH* loss of heterozygosity, *OMIM* Online Mendelian Inheritance in Man, *PolyPhen* Polymorphism Phenotyping, *SIFT* Sorting Intolerant from Tolerant, *VAAST* Variant Annotation, Analysis and Selection Tool. Reprinted with permission from Coonrod et al. [7]

for clinical use, other notable commercial efforts include approaches such as Knome knoSOFT [40], Ingenuity Variant Analysis [41], SV Bio Genome to Mutation (G2M) [42], and the Clinical Genomicist Workstation from Washington University in St. Louis [43].

Finally, heuristic or probabilistic approaches can also be complemented by including information from high density microarrays. For example, array data in family studies can be mined to identify linkage regions and regions of identity by descent that co-segregate with the phenotype of affected individuals in a pedigree. Focusing on these regions may reduce and prioritize specific genomic regions for evaluation.

In Silico Predictors

Medical genetics involves diagnosis, management, and determining risk of hereditary disorders, and genotype to phenotype correlation of gene variants in disease is a major focus [44, 45]. In monogenic diseases, gene mutations are typically curated as either pathogenic or benign. However, many gene variants must be classified as "unknown" or of "uncertain" significance because they have not been clearly associated with a clinical phenotype. The expense of time and labor to validate disease association of a given variant of uncertain significance (VUS) may be cost prohibitive [46, 47]. To help bridge this genotype:phenotype gap, the use of prediction algorithms to narrow the uncertain "grey area" between pathogenic and benign sequence variants is often utilized [48–51].

There are several established methods for predicting mutation severity, many of which have been available online for years. Prediction tools such as PolyPhen [52] and SIFT [30] are primarily based on multiple alignment and amino acid substitution penalties. More recently introduced, MutPred [50] calculates probability of deleterious mutations by disrupted molecular mechanism. Additionally, PMut [53] is a neural net, based and trained on human mutations. However, prediction algorithms are not always in agreement with curated data or each other and are, as yet, primarily research tools [54–56]. A brief description of a representative sample of these online prediction tools may serve to improve our understanding.

SIFT was first published in 2003 by Ng and Heinikoff from work done at the Fred Hutchinson Cancer Research Center in Seattle [30]. The algorithm predicts whether an amino acid substitution will affect the function of a protein based on both sequence homology to various orthologs and physical properties of amino acids. SIFT is a multistep procedure that: (1) searches for and chooses similar sequences, (2) makes an alignment of these sequences, and (3) calculates scores based on the amino acids appearing at each position in the alignment. It was initially developed and trained on nsSNP datasets from LacI, Lysozyme, and HIV protease [57]. This algorithm works especially well when adequate numbers of sequence homologs are available for multiple alignment. Conversely, poor performance is seen when multiple alignment is not reliable or completely unavailable.

PolyPhen is an EMBL based tool from 2002 from Ramensky et al. [52]. It was developed to predict the possible impact of an amino acid substitution on the structure and function of a human protein using physical and comparative considerations. It was originally developed from a set of disease-causing mutations in human proteins with known structures extracted from the SWISS-PROT database, and correlated to the OMIM database [31]. Because the algorithm relies on predicted structural disruption, it works especially well when protein structure is known and less reliably when a solved protein structure is not available.

PMut was first published in 2005 by the Molecular Modeling Unit at the Institut de Recerca Biomédica, Parc Científic de Barcelona, Spain [53]. It is based on a two layer neural network and was trained using human mutation data. It allows for either prediction of single point amino acid mutations or scanning of mutational hot spots. Results are obtained by alanine scanning and by identifying massive mutations and genetically accessible mutations. A graphical interface for Protein Data Bank (PDB) structures, when available, and a database containing hot spot profiles for all nonredundant PDB structures are also accessible from the PMut server.

MutPred is a recently developed prediction algorithm by Li, Mooney, and Radivojac [50]. It builds on the established SIFT method but offers improved classification accuracy based upon protein sequence, and models changes of structural features and functional sites between wild-type and mutant sequences with output of probabilities of gain or loss of structure and function. It was trained on a set of disease-associated SNPs from cancer and the OMIM disease archive. This predicted disruption of molecular function again works especially well for well-studied proteins, for which homolog and solved structure is available.

Gene variants are currently being identified at a tremendous pace. Recent endeavors such as the NCBI Genetic Testing Registry, MutaDATABASE, 1,000 Genomes and the Human Variome Project draw attention to this growing interest in gene variant annotation and clinical interpretation in human disease [58–61]. Furthermore, accurate prediction of phenotypic severity for novel mutations and uncertain gene variants as relating to disease function is of great importance to medicine and biology. Informatics tools for predicting disease severity of uncertain gene variants may assist in the improvement of genetically informed patient care.

Currently, there is no widely accepted computational predictor in clinical use for evaluating gene variants of uncertain clinical significance. Furthermore, the lack of a standardized framework and quantitative metrics for evaluation of disease association of novel variants and VUS remains an obstacle to widespread implementation of proposed guidelines and definitions in gene test reporting. Recently, advantages of gene specific algorithms have been reported, where algorithms are trained and tested on a well-defined disease setting with known genotype-phenotype outcomes [62, 63]. Figure 13.10 displays the concept of gene "reference intervals" analogous to traditional laboratory testing. Additionally, approaches that combine algorithm results such as Condel or Consensus have been reported, where benchmarking or ranking agreement of predicted phenotype severity across several complimentary algorithms may provide research priority for novel variants and VUS [64, 65]. Figure 13.11 shows the utility of combining several computational predictors into a single weighted scoring metric.

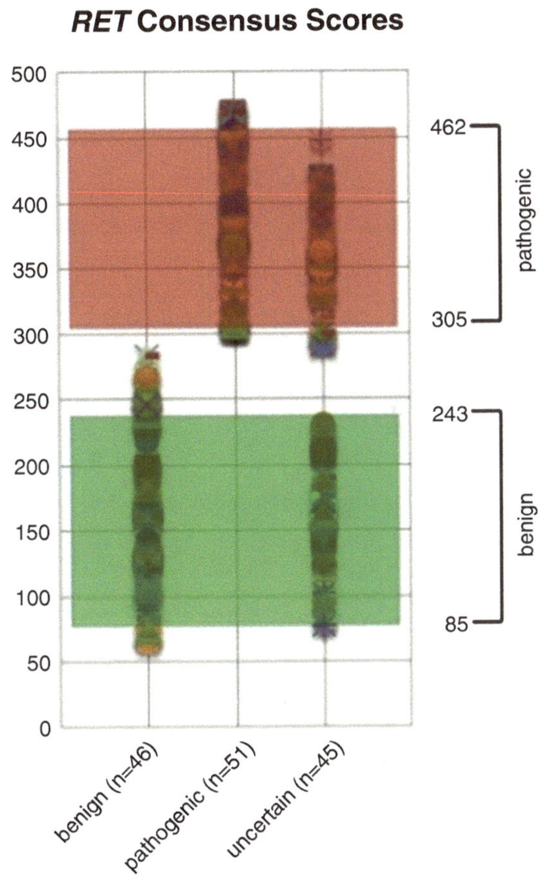

Figure 13-10 Visualization of Consensus scores for *RET* gene variants and disease association including known benign, known pathogenic, and VUS. This demonstrates the principle of gene-specific "reference interval" metrics to improve the utility of in silico predictors. Reprinted with permission from Crockett et al. [65]

Conclusions

As bioinformatics moves into the mainstream of clinical laboratory workflow, several caveats may be appropriate [66] based on key historical lessons from the genomics revolution. These include:

1. *Do not confuse more data with insight*: it can be difficult to extract clinically relevant conclusions from ever increasing amounts of data in a reliable fashion.
2. *Do not confuse insight with value*: while many solid scientific findings may be interesting, they may do little to improve existing laboratory practices or to significantly improve current clinical outcomes.
3. *Do not overestimate one's ability to interpret the data*: even the best data often afford only limited insight into clinical health outcomes.
4. *Do not underestimate the implementation challenges*: leveraging large datasets successfully requires a clinical laboratory system prepared to embrace and effectively handle new methodologies, requiring significant investment of time and capital, and the alignment of economic interests.

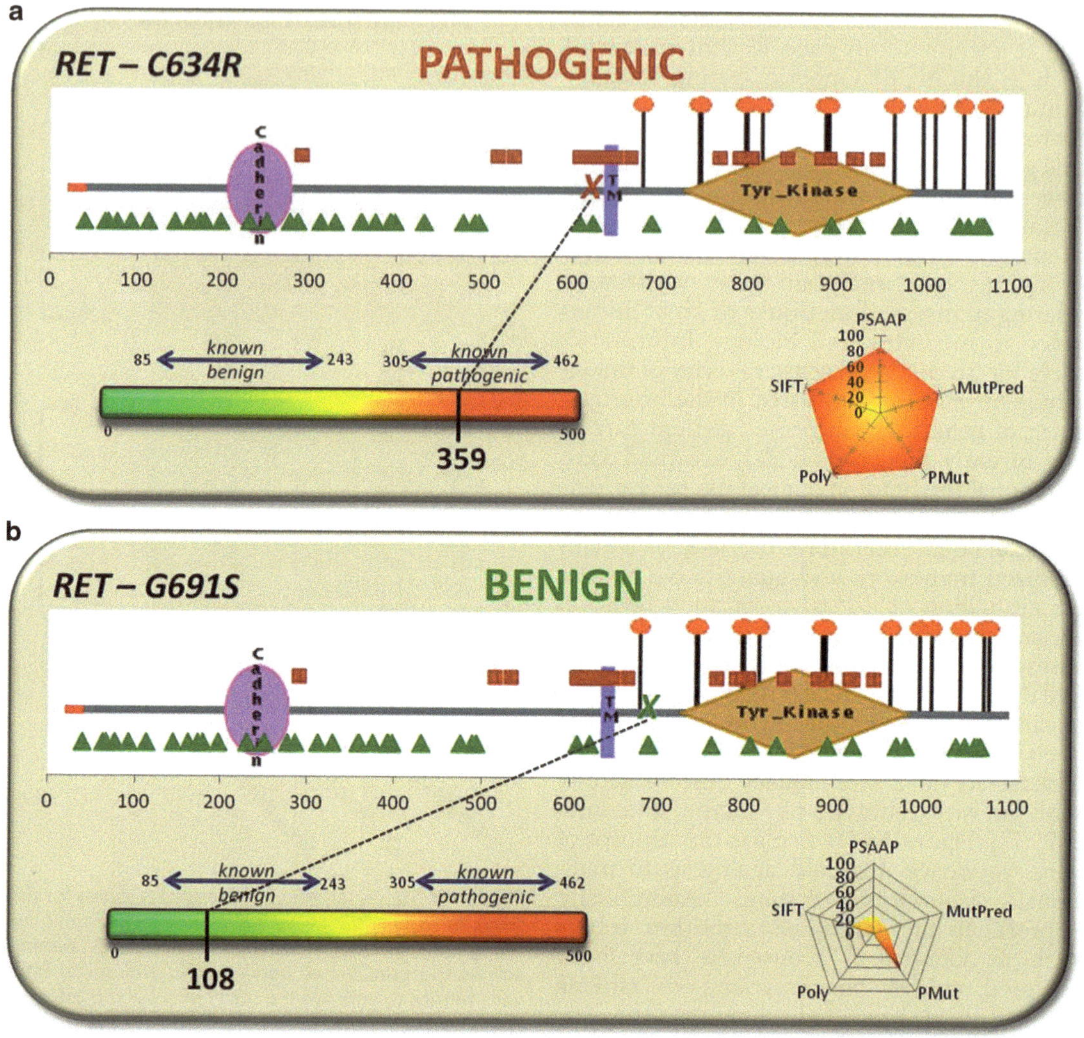

Figure 13-11 Visualization of Consensus scoring using known gene variants plotted on the RET_HUMAN (UniProt #P07949) protein (image courtesy of HPRD.org) and using a weighted algorithm output with radar plots to summarize predictor evidence for (**a**) pathogenic gene variant C634R scoring 359 and (**b**) benign variant G691S with a Consensus score of 108. Reprinted with permission from Crockett et al. [65]

REFERENCES

1. Data from the NHGRI Genome Sequencing Program (GSP). http://www.genome.gov/sequencingcosts/. Accessed 20 Feb 2013.
2. Hogeweg P. The roots of bioinformatics in theoretical biology. PLoS Comput Biol. 2011;7(3):e1002021. doi:10.1371/journal.pcbi.1002021.
3. Huerta M. NIH working definition of bioinformatics and computational biology. 2000. http://www.bisti.nih.gov/docs/CompuBioDef.pdf. Accessed 6 May 2014.
4. Bioinformatics.org wiki. http://www.bioinformatics.org/wiki/Bioinformatics. Accessed 24 Feb 2013.
5. Moorthie S, Hall A, Wright CF. Informatics and clinical genome sequencing: opening the black box. Genet Med. 2013;15(3):165–71. doi:10.1038/gim.2012.116.
6. Mardis ER. A decade's perspective on DNA sequencing technology. Nature. 2011;470(7333):198–203. doi:10.1038/nature09796.
7. Coonrod EM, Durtschi JD, Margraf RL, Voelkerding KV. Developing genome and exome sequencing for candidate gene identification in inherited disorders: an integrated technical and bioinformatics approach. Arch Pathol Lab Med. 2013;137(3):415–33. doi:10.5858/arpa.2012-0107-RA.
8. Ledergerber C, Dessimoz C. Base-calling for next-generation sequencing platforms. Brief Bioinform. 2011;12(5):489–97. doi:10.1093/bib/bbq077.
9. Ewing B, Hillier L, Wendl MC, Green P. Base-calling of automated sequencer traces using phred. I. Accuracy assessment. Genome Res. 1998;8(3):175–85.

10. Ewing B, Green P. Base-calling of automated sequencer traces using phred. II. Error probabilities. Genome Res. 1998;8(3):186–94.
11. Burrows-Wheeler Aligner (BWA). http://bio-bwa.sourceforge.net. Accessed 15 Jan 2013.
12. Li H, Durbin R. Fast and accurate short read alignment with Burrows-Wheeler transform. Bioinformatics. 2009;25(14):1754–60. doi: btp324 [pii] 10.1093/bioinformatics/btp324.
13. Li H, Durbin R. Fast and accurate long-read alignment with Burrows-Wheeler transform. Bioinformatics. 2010;26(5):589–95. doi:btp698 [pii] 10.1093/bioinformatics/btp698.
14. Novoalign. http://www.novocraft.com/main/index.php. Accessed 4 Mar 2013.
15. Li H, Homer N. A survey of sequence alignment algorithms for next-generation sequencing. Brief Bioinform. 2010;11(5):473–83. doi:bbq015 [pii] 10.1093/bib/bbq015.
16. Genome Analysis Toolkit (GATK). http://www.broadinstitute.org/gatk/. Accessed 4 Mar 2013.
17. SAMtools. http://samtools.sourceforge.net. Accessed 4 Mar 2013.
18. Interated Genomics Viewier (IGV). http://www.broadinstitute.org/igv/. Accessed 15 Jan 2013.
19. Robinson JT, Thorvaldsdottir H, Winckler W, Guttman M, Lander ES, Getz G, Mesirov JP. Integrative genomics viewer. Nat Biotechnol. 2011;29(1):24–6. doi:nbt.1754 [pii] 10.1038/nbt.1754.
20. Brockman W, Alvarez P, Young S, Garber M, Giannoukos G, Lee WL, Russ C, Lander ES, Nusbaum C, Jaffe DB. Quality scores and SNP detection in sequencing-by-synthesis systems. Genome Res. 2008;18(5):763–70. doi: gr.070227.107 [pii] 10.1101/gr.070227.107.
21. Dohm JC, Lottaz C, Borodina T, Himmelbauer H. Substantial biases in ultra-short read data sets from high-throughput DNA sequencing. Nucleic Acids Res. 2008;36(16):e105. doi: gkn425 [pii] 10.1093/nar/gkn425.
22. Ajay SS, Parker SC, Abaan HO, Fajardo KV, Margulies EH. Accurate and comprehensive sequencing of personal genomes. Genome Res. 2011;21(9):1498–505. doi: gr.123638.111 [pii] 10.1101/gr.123638.111.
23. Marchetti A, Del Grammastro M, Filice G, Felicioni L, Rossi G, Graziano P, Sartori G, Leone A, Malatesta S, Iacono M, Guetti L, Viola P, Mucilli F, Cuccurullo F, Buttitta F. Complex mutations & subpopulations of deletions at exon 19 of EGFR in NSCLC revealed by next generation sequencing: potential clinical implications. PLoS One. 2012;7(7):e42164. doi: 10.1371/journal.pone.0042164.
24. Wagle N, Berger MF, Davis MJ, Blumenstiel B, Defelice M, Pochanard P, Ducar M, Van Hummelen P, Macconaill LE, Hahn WC, Meyerson M, Gabriel SB, Garraway LA (2012) High-throughput detection of actionable genomic alterations in clinical tumor samples by targeted, massively parallel sequencing. Cancer Discov 2 (1):82-93. doi:2159-8290.CD-11-0184 [pii] 10.1158/2159-8290.CD-11-0184.
25. Annovar. http://www.openbioinformatics.org/annovar/. Accessed 4 Mar 2013.
26. Wang K, Li M, Hakonarson H. ANNOVAR: functional annotation of genetic variants from high-throughput sequencing data. Nucleic Acids Res. 2010;38(16):e164. doi: gkq603 [pii] 10.1093/nar/gkq603.
27. SnpEff Annotation. http://www.broadinstitute.org/gatk/guide/tagged?tag=snpeff. Accessed 21 Nov 2012.
28. Sorting Intolerant from Tolerant (SIFT). http://sift.jcvi.org/. Accessed 4 Mar 2013.
29. Ng PC, Henikoff S. Predicting deleterious amino acid substitutions. Genome Res. 2001;11(5):863–74. doi:10.1101/gr.176601.
30. Ng PC, Henikoff S. SIFT: predicting amino acid changes that affect protein function. Nucleic Acids Res. 2003;31(13):3812–4.
31. Sunyaev S, Ramensky V, Koch I, Lathe 3rd W, Kondrashov AS, Bork P. Prediction of deleterious human alleles. Hum Mol Genet. 2001;10(6):591–7.
32. PolyPhen2. http://genetics.bwh.harvard.edu/pph2/. Accessed 25 Feb 2013.
33. Mutation Taster. http://www.mutationtaster.org/index.html. Accessed 6 Dec 2012.
34. Schwarz JM, Rodelsperger C, Schuelke M, Seelow D. MutationTaster evaluates disease-causing potential of sequence alterations. Nat Methods. 2010;7(8):575–76. doi: nmeth0810-575 [pii] 10.1038/nmeth0810-575.
35. Torrent Suite Documentation. http://ioncommunity.lifetechnologies.com/docs/DOC-3343. Accessed 17 Jan 2013.
36. Yandell M, Huff C, Hu H, Singleton M, Moore B, Xing J, Jorde LB, Reese MG. A probabilistic disease-gene finder for personal genomes. Genome Res. 2011; doi:gr.123158.111 [pii] 10.1101/gr.123158.111.
37. Rope AF, Wang K, Evjenth R, Xing J, Johnston JJ, Swensen JJ, Johnson WE, Moore B, Huff CD, Bird LM, Carey JC, Opitz JM, Stevens CA, Jiang T, Schank C, Fain HD, Robison R, Dalley B, Chin S, South ST, Pysher TJ, Jorde LB, Hakonarson H, Lillehaug JR, Biesecker LG, Yandell M, Arnesen T, Lyon GJ. Using VAAST to identify an X-linked disorder resulting in lethality in male infants due to N-terminal acetyltransferase deficiency. Am J Hum Genet. 2011; doi: S0002-9297(11)00210-2 [pii] 10.1016/j.ajhg.2011.05.017.
38. Ionita-Laza I, Makarov V, Yoon S, Raby B, Buxbaum J, Nicolae DL, Lin X. Finding disease variants in Mendelian disorders by using sequence data: methods and applications. Am J Hum Genet. 2011;89(6):701–12. doi: S0002-9297(11)00481-2 [pii] 10.1016/j.ajhg.2011.11.003.
39. Omicia. http://www.omicia.com/. Accessed 27 Feb 2013.
40. Knome, Inc. http://www.knome.com/. Accessed 12 Mar 2013.
41. Ingenuity Systems. http://www.ingenuity.com/. Accessed 11 Mar 2013.
42. SV Bio. http://www.svbio.com/. Accessed 26 Feb 2013.
43. GPS@WUSTL. http://gps.wustl.edu/. Accessed 14 Feb 2013.
44. Weinstein ND. What does it mean to understand a risk? Evaluating risk comprehension. J Natl Cancer Inst Monogr. 1999;25:15–20.
45. Ensenauer RE, Michels VV, Reinke SS. Genetic testing: practical, ethical, and counseling considerations. Mayo Clin Proc. 2005;80(1):63–73.

46. Nowak R. Genetic testing set for takeoff. Science. 1994;265(5171):464–7.
47. Machens A, Gimm O, Hinze R, Hoppner W, Boehm BO, Dralle H. Genotype-phenotype correlations in hereditary medullary thyroid carcinoma: oncological features and biochemical properties. J Clin Endocrinol Metab. 2001;86(3):1104–9.
48. Wei Q, Wang L, Wang Q, Kruger WD, Dunbrack Jr RL. Testing computational prediction of missense mutation phenotypes: functional characterization of 204 mutations of human cystathionine beta synthase. Proteins. 2010;78(9):2058–74.
49. Kumar P, Henikoff S, Ng PC. Predicting the effects of coding non-synonymous variants on protein function using the SIFT algorithm. Nat Protoc. 2009;4(7): 1073–81.
50. Li B, Krishnan VG, Mort ME, Xin F, Kamati KK, Cooper DN, Mooney SD, Radivojac P. Automated inference of molecular mechanisms of disease from amino acid substitutions. Bioinformatics. 2009; 25(21):2744–50.
51. Dorfman R, Nalpathamkalam T, Taylor C, Gonska T, Keenan K, Yuan XW, Corey M, Tsui LC, Zielenski J, Durie P. Do common in silico tools predict the clinical consequences of amino-acid substitutions in the CFTR gene? Clin Genet. 2010;77(5):464–73.
52. Ramensky V, Bork P, Sunyaev S. Human non-synonymous SNPs: server and survey. Nucleic Acids Res. 2002;30(17):3894–900.
53. Ferrer-Costa C, Gelpi JL, Zamakola L, Parraga I, de la Cruz X, Orozco M. PMUT: a web-based tool for the annotation of pathological mutations on proteins. Bioinformatics. 2005;21(14):3176–8.
54. Spencer DS, Stites WE. The M32L substitution of staphylococcal nuclease: disagreement between theoretical prediction and experimental protein stability. J Mol Biol. 1996;257(3):497–9.
55. Kang HH, Williams R, Leary J, Ringland C, Kirk J, Ward R. Evaluation of models to predict BRCA germline mutations. Br J Cancer. 2006;95(7):914–20.
56. Engelhardt BE, Jordan MI, Muratore KE, Brenner SE. Protein molecular function prediction by Bayesian phylogenomics. PLoS Comput Biol. 2005; 1(5):e45.
57. Ng PC, Henikoff S. Accounting for human polymorphisms predicted to affect protein function. Genome Res. 2002;12(3):436–46.
58. Javitt G, Katsanis S, Scott J, Hudson K. Developing the blueprint for a genetic testing registry. Public Health Genomics. 2010;13(2):95–105.
59. Bale S, Devisscher M, Van Criekinge W, Rehm HL, Decouttere F, Nussbaum R, Dunnen JT, Willems P. MutaDATABASE: a centralized and standardized DNA variation database. Nat Biotechnol. 2011;29(2): 117–8.
60. Durbin RM, Abecasis GR, Altshuler DL, Auton A, Brooks LD, Durbin RM, Gibbs RA, Hurles ME, McVean GA. A map of human genome variation from population-scale sequencing. Nature. 2010;467(7319):1061–73.
61. Cotton RG, Al Aqeel AI, Al-Mulla F, Carrera P, Claustres M, Ekong R, Hyland VJ, Macrae FA, Marafie MJ, Paalman MH, Patrinos GP, Qi M, Ramesar RS, Scott RJ, Sijmons RH, Sobrido MJ, Vihinen M. Capturing all disease-causing mutations for clinical and research use: toward an effortless system for the Human Variome Project. Genet Med. 2009;11(12):843–9.
62. Crockett DK, Piccolo SR, Ridge PG, Margraf RL, Lyon E, Williams MS, Mitchell JA. Predicting phenotypic severity of uncertain gene variants in the RET proto-oncogene. PLoS One. 2011;6(3):e18380. doi:10.1371/journal.pone.0018380.
63. Thompson BA, Greenblatt MS, Vallee MP, Herkert JC, Tessereau C, Young EL, Adzhubey IA, Li B, Bell R, Feng B, Mooney SD, Radivojac P, Sunyaev SR, Frebourg T, Hofstra RM, Sijmons RH, Boucher K, Thomas A, Goldgar DE, Spurdle AB, Tavtigian SV. Calibration of multiple in silico tools for predicting pathogenicity of mismatch repair gene missense substitutions. Hum Mutat. 2013;34(1):255–65. doi:10.1002/humu.22214.
64. Gonzalez-Perez A, Lopez-Bigas N. Improving the assessment of the outcome of nonsynonymous SNVs with a consensus deleteriousness score, Condel. Am J Hum Genet. 2011;88(4):440–9. doi:10.1016/j.ajhg.2011.03.004.
65. Crockett DK, Ridge PG, Wilson AR, Lyon E, Williams MS, Narus SP, Facelli JC, Mitchell JA. Consensus: a framework for evaluation of uncertain gene variants in laboratory test reporting. Genome Med. 2012;4(5):48. doi:10.1186/gm347.
66. Turning Information into Impact. http://www.forbes.com/sites/davidshaywitz/2012/12/30/turning-information-into-impact-digital-healths-long-road-ahead/. Accessed 26 Feb 2013.
67. SEQanswers. http://seqanswers.com/. Accessed 12 Feb 2013.
68. Biostar - Bioinformatics Explained. http://www.biostars.org/. Accessed 12 Feb 2013.
69. BioIT-World. http://www.bio-itworld.com/. Accessed 12 Feb 2013.
70. Galaxy NGS server. https://main.g2.bx.psu.edu/. Accessed 12 Feb 2013.
71. NHGRI Short Course on NGS data. http://www.soph.uab.edu/ssg/courses/nhgri_r25. Accessed 12 Feb 2013.
72. Training Health Care Professionals to Serve as Informatics Leaders. Current 10x10 Virtual Offerings. http://www.amia.org/education/10x10-courses. Accessed 12 Feb 2013.

CHAPTER 14

Next Generation Sequencing for Single-Gene Analysis

HAO HO, CHRISTOPHER D. GOCKE

Introduction 197

Applications of Single-Gene NGS 198

Next Generation Sequencing Platforms 202

Next Generation Sequencing Technologies 202

Cost of Next Generation Sequencing 203

Conclusions 204

Introduction

The first generation of sequencing technologies was developed in the 1970s by Sanger [1, 2] and Maxam and Gilbert [3]. Frederick Sanger's sequencing method is based on DNA synthesis making use of dideoxynucleotide analogues (radiolabelled or fluorescently labeled) to cause chain termination. By contrast, Allan Maxam and Walter Gilbert performed DNA sequencing through chemical degradation in which terminally labeled DNA fragments were chemically cleaved at specific bases and analyzed by gel electrophoresis. Since Maxam and Gilbert's method was more technically challenging and less amenable to being scaled up, Sanger sequencing ultimately prevailed and became the "gold standard" for decoding DNA sequence in the past three decades.

The first automated DNA sequencer, which performed partial automation of DNA sequence analysis through fluorescence detection of DNA fragments, was invented at Caltech in 1986 [4]. Subsequent improvement of the technology led to the introduction of the first commercial DNA sequencer by ABI in 1996, using slab gel electrophoresis (ABI Prism 310). It was then improved upon 2 years later by the ABI Prism 3700 which utilized automated reloading of up to 96 capillaries with polymer matrix. While fully automated, the chief limitations of these capillary instruments are low throughput and high cost, resulting in it taking 13 years and nearly three billion dollars to complete the human genome sequencing project [5].

Due to the limitations of automated Sanger sequencing, new and improved technologies for sequencing large amounts of DNA have been developed in recent years, collectively referred to as Next Generation Sequencing (NGS). NGS, also known as massively parallel sequencing, relies on miniaturization of individual sequencing reactions by immobilizing spatially separated templates to a solid surface or support. This allows thousands to billions of individual sequencing reactions to be performed in parallel and be distinctly detected by digital imaging or electrical sensing, easily overcoming the limited scalability of "first generation sequencing" by eliminating the electrophoresis step for sequence separation. The end results are vastly improved throughput at a fraction of the original cost, reducing the price from

G.J. Netto and I. Schrijver (eds.), *Genomic Applications in Pathology*,
DOI 10.1007/978-1-4939-0727-4_14, © Springer Science+Business Media New York 2015

>$1,000 down to 10 cents or less per megabase, and putting the dream of whole-genome sequencing for $100–$1,000 within reach. This, of course, only includes the reagent cost. The dramatic shifts in cost and accessibility brought by NGS have the potential not only to revolutionize the field of genomics but also to open up a new world of medical diagnostics in which application of NGS technology will have a high impact in the foreseeable future [6–12].

Applications of Single-Gene NGS

Minimal Residual Disease Detection

For certain hematopoietic malignancies such as acute lymphoblastic leukemia (ALL), there is a strong correlation between the presence of post-treatment minimal residual disease (MRD) and adverse clinical outcomes [13–15]. As a result, accurate assessment of MRD is critical in risk stratification (standard, intermediate or high risk) of ALL patients to predict clinical outcome and thus to provide guidance for proper management of the disease [16]. Currently, multiparameter flow cytometry (mpFC) and quantitative polymerase chain reaction (qPCR) are the main strategies for the assessment of MRD, but each method has its own drawbacks [17]. Multiparameter flow cytometry, which relies on the detection of a unique collection of antigens specific for leukemia, has sensitivity on the order of 1 cell in 10^4. However, data interpretation can be challenging and is operator- and/or laboratory-dependent, and not infrequently is confounded by variable expression of leukemic antigens in the post-therapy setting. On the other hand, although PCR amplification of immunoglobulin heavy chain (*IGH*) or T-cell receptor (TCR) genes, or oncogenic fusion transcripts for MRD detection of leukemia may achieve higher sensitivity (1 cell in 10^5 or better), it often requires the use of patient-specific primers to assess the genetic rearrangement or translocation unique to each individual patient's disease—a task that can be expensive and labor intensive with difficulty in achieving uniformity [18–20].

Theoretically, most, if not all, of the possible IgH or TCR rearrangement configurations can be sequenced by NGS using a set of consensus primers that are able to amplify all existing *IGH* or TCR segments. Automation of the procedure can not only eliminate operator-dependency of data interpretation (as in mpFC) but also obviate the need to develop patient-specific reagents (for qPCR). In fact, Wu et al. [20] have shown that with targeted sequencing of *TCRB* and *TCRG* using an Illumina HiSeq NGS platform, they were able to detect MRD of T-ALL that was 10- to 100-fold lower than the limit of detection (LOD) of mpFC. However, the study was limited to the subtypes of T-ALL that had undergone TCR rearrangements (35 of 43 cases), and was not applicable to those at a more primitive stage (e.g., early thymic precursor immunophenotype). Additionally, Gaward et al. [21] showed that similar strategies can be applied in MRD monitoring of B-ALL by sequencing the *IGH* locus. Interestingly, their findings also provided new insight into the molecular mechanisms by which clonal evolution occurs in B-ALL patients. In summary, although few studies have yet been performed, the advent of NGS may present a more rapid, sensitive, informative, and cost-effective method for MRD testing in the future.

Oncologic Testing

In an era where target-based therapies are becoming a norm in the management of oncologic malignancies, pretreatment screening for predictive biomarkers is crucial in identification of cancer-specific genetic alteration(s) that are susceptible to available therapeutic modalities. Currently, this kind of "personalized" therapy is well established in the management of patients diagnosed with lung and colorectal cancers (e.g., EGFR and BRAF inhibitors), melanoma (e.g., BRAF inhibitors), and certain hematological malignancies (e.g., tyrosine kinase inhibitors in CML), but is expanding rapidly to other tumors as well. Traditional methods of mutational analysis such as Sanger sequencing, pyrosequencing, and allele-specific PCR have been widely used for this purpose. However, due to limited bandwidth and throughput of these older technologies, the depth of analysis has been confined to certain known mutational "hotspots" of individual genes—a practice that potentially can miss other significant genetic aberrations elsewhere. In order to supply the ever-growing breadth of information required to deliver truly personalized therapeutic interventions, increasingly the NGS approach has been utilized for cancer genomics analysis because it has the ability to simultaneously detect various genetic alterations in thousands of different genes in a single run [22–29].

Using the Roche GS Junior 454 NGS platform, a small pilot experiment by Borras et al. analyzing FFPE samples from colorectal and lung cancers showed that the approach is efficient and accurate in detecting all existing *KRAS* mutations [23]. Shindoh et al. used the SOLiD 4 platform to perform cDNA screening on primary specimens and cell lines derived from lung cancer, breast cancer and melanoma, and found that the system can efficiently identify various genetic alterations in *EGFR*, *KRAS*, *NRAS*, and *ERBB2* genes [24]. With targeted sequencing of *BRCA1* and *BRCA2* genes using two NGS platforms (SOLiD 4 and Ion Torrent PGM), Chan et al. reported that both systems are highly sensitive and specific for single nucleotide polymorphisms (SNPs), though the PGM platform lacks specificity in insertion/deletion (indel) calling [26]. It is clear that the single-gene approach is being built on by manufacturers and diagnostic laboratories, resulting in panels of various sizes directed at cancer-related genes broadly or at tumor-specific or druggable genes in particular.

Besides solid tumors, NGS is also being applied in screening of individual genetic alterations and monitoring of disease progression in hematologic malignancies such as chronic myelomonocytic leukemia (e.g., testing the *TET2*, *CBL*, *RAS*, and *RUNX1* genes) [30, 31], myelodysplastic syndrome (e.g., *TP53*) [30], and myeloproliferative neoplasms (e.g., *JAK2*) [29]. Moreover, Grossmann et al. proved that NGS can be used successfully in the assessment of GC-rich genes such as *CEBPA* (in AML patients) and found it to be highly sensitive for mutation analysis of this gene [28].

The great depth of coverage inherent in NGS provides both benefits and drawbacks to single-gene or small panel testing. For example, most NGS tests have the ability to easily identify mutations in a target gene when they represent as little as 5 % of the total. While this is useful in analyzing challenging clinical samples with a low tumor burden, it makes initial validation or confirmation by some other methods with lower limits of detection (particularly Sanger sequencing) difficult or impossible. The detection of small numbers of mutated molecules also challenges existing paradigms. For example, many patients are found to have low levels of the drug-resistant mutant T790M in the *EGFR* gene prior to therapy with small molecule inhibitors of the gene; the clinical significance of this finding, particularly with regard to therapy, remains unclear [32]. Tumor heterogeneity is being characterized by sensitive methods in a variety of other genes. Will the detection of a minor subclone with a sensitizing mutation have the same import as the mutation in the bulk of the tumor? How hard will diagnostic laboratories have to look for such subclones? Studies with highly sensitive assays will be needed to answer these and similar questions.

Infectious Diseases

Infectious disease diagnosis and screening are additional areas in which NGS can have major impacts, gradually replacing the traditional molecular tests that are based mainly on Sanger sequencing. In the diagnosis and genotyping of Hepatitis C (HCV) and Human immunodeficiency viral (HIV) infections, NGS allows the combination of both steps in a single reaction [33, 34]. Once a diagnosis of infection is established, assessment of intrapatient viral genetic variation becomes crucial for evaluation of viral evolutionary dynamics and identifying emerging resistant strains not only in order to provide guidance for optimal antiviral therapy but also to serve as a valuable source of information for designing effective vaccines. In this regard, NGS has revolutionized the field by simplifying the once time-consuming and expensive assessment of intrahost viral genetic diversity of HCV and HIV into a cost-effective procedure at an unprecedented resolution [33, 35]. Moreover, similar to the oncologic target-based therapy, NGS is being applied to assess the co-receptor tropism of HIV-1 prior to treatment with CCR5 antagonist [36].

Not surprisingly, the use of NGS is being extended to the detection and classification of other known viruses (e.g., HPV genotyping) [37] and screening for unknown disease-causing microorganisms in pathology samples [38]. With its ability to simultaneously detect multiple infectious agents, NGS has been proven invaluable in the metagenomic analysis of infectious diseases during local outbreaks (e.g., norovirus), pandemics (e.g., avian influenza), or global epidemics (e.g., seasonal influenza virus) [39–41]. The fields of biodefense against bioterrorism will also benefit from NGS as this new technology can not only rapidly detect the presence of specific pathogenic agent(s) but also perform subtyping/subclassification and drug resistance profiling at the same time, which can expedite implementation of counterterrorism measures [42–44].

Inherited Diseases

Candidate genes responsible for inherited disorders have traditionally been identified through linkage studies [45]. Currently, a known genetic cause has been assigned to more than 3,500 Mendelian disorders (http://www.ncbi.nlm.nih.gov/omim). Although classical genome-wide linkage studies are effective at elucidating causal variants for some inherited diseases, those that are sporadic, extremely rare or occur de novo are usually not amenable to this method. With the advent of NGS, whole-exome/genome sequencing became feasible and served as a powerful tool for probing the genetic defects of those rare syndromes or complex diseases that had remained elusive in their etiologies [46, 47].

As mentioned above, the major contribution of NGS in medical genetics thus far lies not in screening of known single-gene mutations, but rather in the discovery of allelic variants or novel genetic pathways associated with rare inherited syndromes that are beyond the reach of traditional linkage analysis. For instance, since its original description in 1981, the underlying cause of Kabuki syndrome—a rare, sporadic disorder with multiple congenital anomalies—had remained intractable to conventional approaches of gene discovery. Through massively parallel sequencing of the exomes of ten unrelated probands, Ng et al. were able to demonstrate that Kabuki syndrome is due to mutations in the *MLL2* gene [48]. Using similar strategies, the same group of scientists was also able to uncover the underlying cause of Miller syndrome, another rare inherited disease [49]. Likewise, Hoischen et al. were able to characterize Schinzel–Giedion syndrome as an entity secondary to de novo mutations in the *SETBP1* gene by analyzing the exomes from only four affected individuals [50]. The mysteries of other as-yet-unexplained disorders are also being gradually solved through whole-exome/genome sequencing by various groups [46, 51].

As in other fields of study, the ability to decode multiple genes in parallel (e.g., in gene panels) allows NGS to be applied in the screening and monitoring of complex diseases such as inherited retinal degeneration (IRD) for which genetic testing has become increasingly important for proper diagnosis, prognosis, and development of personalized therapy [52].

Human Leukocyte Antigen Typing

The human leukocyte antigen (HLA) class I and class II gene loci consist of more than 7,000 alleles, giving rise to >4,600 distinct HLA proteins, and thus are the most polymorphic genes in the human genome known to date [53]. The *HLA-B7* gene [54] was the first HLA gene to be cloned (in 1980) and was subsequently used as a probe in Southern blot analyses to mark the advent of restriction fragment length polymorphism (RFLP) in the study of HLA genomic polymorphism [55]. This cumbersome method was later replaced by the sequence specific oligonucleotide approach in combination with PCR [56–58], and the use of sequence-based typing for procedures requiring high resolution HLA typing, such as hematopoietic stem cell transplantation [59]. Besides being labor-intensive, time-consuming, and expensive (due to the highly polymorphic nature of the HLA loci), typing ambiguity remains a critical challenge for the current methods secondary to their limitations in genomic coverage and the difficulties in determining the cis-trans relationships between variants [53].

Although still at an experimental stage, a massively parallel sequencing approach can help to overcome HLA typing ambiguity by virtue of its ability to perform deep sequencing with high coverage of the entire HLA region, combined with clonal amplification to provide in-phase sequencing of linked polymorphisms [53, 60]. Lind et al. sequenced six known samples using NGS and obtained 100 % concordance in all analyzed HLA loci [60]. In a double blind study that enrolled eight independent laboratories to genotype the same 20 samples for multiple HLA loci, using the 454 GS FLX platform coupled with CONEXIO ATF software, Holcomb et al. were able to achieve an overall concordance of 97.2 % with the known genotypes [61], pointing to the inter-laboratory reliability of this approach in high resolution HLA genotyping. Moreover, Erlich et al. developed a novel NGS protocol for HLA class I typing and showed the superiority of this method relative to the current sequence-specific oligonucleotide-based gold standard in terms of typing accuracy while maximizing throughput and minimizing cost, providing concrete support for NGS as a reliable, efficient, and scalable approach for HLA typing [62].

Archeological and Mitochondrial Studies

The study of ancient DNA began in the early 1980s with amplification of small DNA sequences using bacterial cloning followed by sequencing [63, 64]. This inefficient and labor-intensive technique became obsolete with the development of PCR a few years later [65, 66], which combined with cloning and Sanger sequencing to form the classical methodology in molecular archeology [67]. Traditionally, mitochondrial DNA (mtDNA) is used as a target primarily because it is present in many copies per cell in contrast to the two copies of nuclear DNA. Moreover, the much smaller mitochrondrial genome size and lack of mtDNA recombination also simplify data analysis. Even so, only targeted regions of the mtDNA are used in most studies due to technical limitations (low throughput and high cost) of the classical methodology [68].

Development of NGS has opened up new possibilities in the field of archeology. The new technology can not only sequence the complete mtDNA genome with relative ease but also render the previously unthinkable whole-nuclear-genome sequencing of an extinct species a distinct possibility. In fact, using massively parallel sequencing, the complete genomes of three long extinct hominid groups (Neanderthals, Denisovans, and Palaeo-Eskimo) were decoded recently [69–71]. By uncovering the genetic diversity and composition of our ancient ancestors through whole-genome sequencing, NGS has helped to overcome a major restriction confronted by the classical methodology in human evolutionary study. In summary, the high sensitivity and efficiency of NGS have markedly enhanced our ability to generate vast amounts of high-quality data from ancient DNA in a relatively short time—a feat that in turn will help to unfold the evolutionary history of human and other species with unprecedented resolution and rapidity.

Mitochondrial disorders are also amenable to identification by NGS. These include neurologic and neuromuscular disorders such as MELAS (mitochondrial encephalomyopathy, lactic acidosis, and stroke-like episodes) and MERRF (myoclonic epilepsy with ragged red fibers). These conditions are often heterogeneous in their phenotype, in part because of the relative distributions of mutant and wild-type molecules in the affected tissues due to the cytoplasm mode of transmission [72]. The small size of the mitochondrial genome (about 16,000 bases, smaller than the span of most nuclear genes) permits easy targeting and the quantitative nature of NGS permits accurate genotyping. In theory, scientists could also employ NGS to study mitochondrial genomes for forensic purposes, but this methodology will need further validation to meet the required standards [73].

Other Clinical Applications

NGS has emerging applications in areas such as forensic studies [74–76], post-bone marrow transplant engraftment testing [77], monitoring of transplanted organ rejection [78], and prenatal screening [79, 80]. With maturation and continual improvement, there is no reason to doubt that NGS will eventually find its way into many other different fields of study and clinical application, as well.

Next Generation Sequencing Platforms

Due to continual advancement of current technologies, new and improved NGS platforms are being introduced at a breakneck pace. We will only briefly touch upon the few currently commercially available NGS platforms here as more details regarding these instruments are available in a separate chapter.

Roche GS FLX 454 Sequencing System

The Roche GS FLX Genome Sequencer was introduced in 2004 as the first commercial NGS platform. The instrument uses the pyrosequencing technology (see below) to detect light emitted (after a series of enzymatic reactions) from pyrophosphate release during nucleotide incorporation. The sequencer is able to produce an average read length of 400 bp, although the company claims that the updated version, the GS FLX + System, can have read lengths up to 1,000 bp with output of 700 Mbp. A smaller desktop version using the same principle is also in use, known as the GS Junior System.

Illumina/Solexa Genome Analyzer

The Illumina/Solexa Genome Analyzer systems employ the cyclic reversible termination method (also based on a sequencing-by-synthesis principle) with fluorescently labeled nucleotides to produce read lengths up to 2× 100 bp (paired-end reads) and output of up to 600 Gbp per run. The HiSeq series consists of four platforms (Hiseq 1,000, 1,500, 2,000, and 2,500), differing in sequencing speed and output capacity. A small benchtop version known as the MiSeq is also available.

Life Technologies SOLiD Genetic Analyzer

The SOLiD system performs sequencing by synthesis using a unique color space mechanism. Instead of using DNA polymerase, it uses DNA ligase to link specific fluorescently labeled octamers to the DNA fragment being read, followed by fluorescence detection and another round of ligation. The end result is that each nucleotide is read twice and thus markedly reduces the error rate of sequencing—from 1 % in general to <0.1 %. The latest SOLiD 5500xl W system can produce throughput of up to 320 Gbp/run with read lengths of up to 75 bp.

Ion Torrent Personal Genome Machine

Similar to the Roche 454 system, the Ion Torrent Personal Genome Machine (PGM) uses a sequencing-by-synthesis technology. However, instead of capturing emitted light, it uses an ion sensor to directly detect pH change resulting from H^+ release upon nucleotide incorporation. By abrogating the multiple enzymatic and imaging steps, the PGM essentially eliminates the lag time between nucleotide incorporation and the readout, making it the fastest benchtop sequencer currently available. Throughput of the PGM is scalable, ranging from 10 Mbp (using Ion 314 chip) to 100 Mbp (Ion 316 chip) to 1 Gbp (Ion 318 chip) with read lengths of up to 400 bp. A more advanced version of the instrument, the Ion Proton System is also available. Using the same sequencing principle but with higher density chips, the Ion Proton System reportedly is able to sequence the entire human genome on a single chip in just a few hours.

Next Generation Sequencing Technologies

A variety of new technologies that may show promise for massively parallel sequencing (e.g., Nanopore sequencing technology and Single-molecule Real-time sequencing technology) are under development [9, 11, 81–84].

Given that NGS technologies are discussed in greater detail in other chapters, we will only briefly summarize a few mature and commercially available technologies.

Pyrosequencing

Pyrosequencing was developed based on the sequencing-by-synthesis principle, but instead of using dideoxynucleotides for chain termination as in Sanger sequencing, it relies on light detection by a CCD camera after a series of chemical reactions upon nucleotide incorporation. Four enzymes are required, namely, the Klenow DNA polymerase I, ATP sulfurylase, luciferase, and apyrase. In addition, two enzyme substrates are also needed, the adenosine phosphosulfate (APS) and D-LUCIFERIN. The first reaction occurs when the four nucleotides are added, one at a time, to give rise to pyrophosphate (PPi) when a complimentary nucleotide is incorporated into the newly synthesized DNA. The released PPi then serves as a substrate for ATP sulfurylase to produce ATP, which is converted to light by luciferase in the presence of its substrate, D-luciferin. Light detection by the CCD camera signifies the incorporation of the specific nucleotide just added. The unincorporated nucleotides and excess ATP between each addition of different bases are removed by the enzyme apyrase [85].

Cyclic Reversible Termination

The cyclic reversible termination (CRT) sequencing method is similar to Sanger sequencing in that it is based on the sequencing-by-synthesis principle with chain termination, but unlike Sanger sequencing, which uses dideoxynucleotides that lead to irreversible chain termination, the CRT employs fluorescently modified nucleotides attached to a removable terminating/inhibiting group to allow reversible termination of DNA synthesis. Unlike pyrosequencing, all four nucleotides, each labeled with a different fluorescent dye, are added simultaneously to the reaction, and imaging is then performed after a washing step to clear away any unincorporated nucleotides. A cleavage step then ensues to remove the fluorescent label and the terminating/inhibiting group on the newly incorporated nucleotide(s) to allow the next cycle of nucleotide addition [11].

Sequencing by Ligation

Also based on the sequencing-by-synthesis principle, sequencing by ligation is another cyclic method that uses DNA ligase instead of DNA polymerase for strand extension. A mixture of differently fluorescently labeled octameric probes are added to the primed template, and DNA ligase is then used to link the complementary probe to the synthesized DNA strand. The next step is to wash away the non-ligated probes, followed by fluorescence imaging to identify the ligated probe. The cycle can be repeated multiple times by cleavage of the fluorescent dye and regeneration of the 5′-phosphate group after each cycle to generate reads of up to 50 bp. A different anchor primer 1 bp shorter ($n-1$) than the previous one is then used for round 2 of the sequencing; and up to a total of 5 rounds of sequencing can be performed (using $n-2$, $n-3$, and $n-4$ ladder primer sets). The redundant nature of this sequencing system results in double interrogation at each nucleotide position to significantly enhance its base calling accuracy to >99.9 % [11].

Cost of Next Generation Sequencing

The cost of DNA sequencing has taken an average of a four-order-of-magnitude plunge since the advent of NGS in 2005. According to data from National Human Genome Research Institute (www.genome.gov), NGS has helped to reduce the raw cost of sequencing one megabase of DNA from $1,000 in 2005 to a mere 10 cents or less in 2012. In 2005, sequencing a whole genome cost approximately 17 million dollars, while by 2012 it had dropped to around $6,000. It is important to note that raw sequencing costs do not include annotation and interpretation, which are now the most expensive part of the process. Joking reference to the "$1,000 genome/$1,000,000 interpretation" makes the point that the bioinformatic tools and medical infrastructure needed to place the data in context have not yet matured. The cost of DNA sequencing varies among the different NGS platforms, ranging approximately from as low as 7 cents (Illumina Hiseq 2000) to $10 per megabase (Roche 454 GS FLX) [12]. In a more realistic scenario within a regular DNA diagnostic lab in which smaller scale DNA sequencing is the norm, the cost savings between NGS and Sanger sequencing are less dramatic. While comparing the cost of *BRCA* mutation screening using two NGS platforms (SOLiD 4 and Ion Torrent PGM), Chan et al. reported that NGS systems can afford more than twofold (Ion Torrent PGM) to more than fourfold (SOLiD 4) cost savings relative to Sanger sequencing. Moreover, turnaround time was reduced dramatically relative to Sanger sequencing [26]. This, however, very much depends on the exact application and will be quite different for single-gene applications compared to, for example, whole-genome sequencing.

Conclusions

NGS has helped to unwind, at an unprecedented pace, the mysteries embedded in the complicated genomes of human and other organisms. The efficiency, scalability, and affordability of NGS technologies will also turn whole-exome or whole-genome sequencing into a routine assay in clinical labs in the near future—a feat that was unthinkable just a few years ago with Sanger sequencing. Although promising, NGS is still in its infancy in the realm of clinical molecular diagnostics. With its impressive and ever-expanding range of applications, there is no doubt that NGS will have a tremendous impact on the future of personalized medicine.

REFERENCES

1. Sanger F, Coulson AR. A rapid method for determining sequences in DNA by primed synthesis with DNA polymerase. J Mol Biol. 1975;94:441–8.
2. Sanger F, Nicklen S, Coulson AR. 1992 DNA sequencing with chain-terminating inhibitors. Biotechnology. 1977;24:104–8.
3. Maxam AM, Gilbert W. A new method for sequencing DNA. Proc Natl Acad Sci U S A. 1977;74:560–4.
4. Smith LM, Sanders JZ, Kaiser RJ, Hughes P, Dodd C, Connell CR, Heiner C, Kent SB, Hood LE. Fluorescence detection in automated DNA sequence analysis. Nature. 1986;321:674–9.
5. International Human Genome Sequencing Consortium. Finishing the euchromatic sequence of the human genome. Nature. 2004;431:931–45.
6. Shendure J, Aiden EL. The expanding scope of DNA sequencing. Nat Biotechnol. 2012;30:1084–94.
7. Natrajan R, Reis-Filho JS. Next generation sequencing applied to molecular diagnostics. Expert Rev Mol Diagn. 2011;11:425–44.
8. Schadt EE, Turner S, Kasarskis A. A window into third-generation sequencing. Hum Mol Genet. 2010;19:R227–40.
9. Pareek CS, Smoczynski R, Tretyn A. Sequencing technologies and genome sequencing. J Appl Genet. 2011;52:413–35.
10. Su Z, Ning B, Fang H, Hong H, Perkins R, Tong W, Shi L. Next generation sequencing and its applications in molecular diagnostics. Expert Rev Mol Diagn. 2011;11:333–43.
11. Metzker ML. Sequencing technologies—the next generation. Nat Rev Genet. 2010;11:31–46.
12. Liu L, Li Y, Li S, Hu N, He Y, Pong R, Lin D, Lu L, Law M. Comparison of next generation sequencing systems. J Biomed Biotechnol. 2012;2012:251364.
13. Flohr T, Schrauder A, Cazzaniga G, Panzer-Grumayer R, van der Velden V, Fischer S, Stanulla M, Basso G, Niggli FK, Schafer BW, Sutton R, Koehler R, Zimmermann M, Valsecchi MG, Gadner H, Masera G, Schrappe M, van Dongen JJ, Biondi A, Bartram CR. Minimal residual disease-directed risk stratification using real-time quantitative PCR analysis of immunoglobulin and T-cell receptor gene rearrangements in the international multicenter trial AIEOP-BFM ALL 2000 for childhood acute lymphoblastic leukemia. Leukemia. 2008;22:771–82.
14. Schrappe M, Valsecchi MG, Bartram CR, Schrauder A, Panzer-Grumayer R, Moricke A, Parasole R, Zimmermann M, Dworzak M, Buldini B, Reiter A, Basso G, Klingebiel T, Messina C, Ratei R, Cazzaniga G, Koehler R, Locatelli F, Schafer BW, Arico M, Welte K, van Dongen JJ, Gadner H, Biondi A, Conter V. Late MRD response determines relapse risk overall and in subsets of childhood T-cell ALL: results of the AIEOP-BFM-ALL 2000 study. Blood. 2011;118:2077–84.
15. van Dongen JJ, Seriu T, Panzer-Grumayer ER, Biondi A, Pongers-Willemse MJ, Corral L, Stolz F, Schrappe M, Masera G, Kamps WA, Gadner H, van Wering ER, Ludwig WD, Basso G, de Bruijn MA, Cazzaniga G, Hettinger K, van der Does-van den Berg A, Hop WC, Riehm H, Bartram CR. Prognostic value of minimal residual disease in acute lymphoblastic leukaemia in childhood. Lancet. 1998;352:1731–8.
16. Willemse MJ, Seriu T, Hettinger K, d'Aniello E, Hop WC, Panzer-Grumayer ER, Biondi A, Schrappe M, Kamps WA, Masera G, Gadner H, Riehm H, Bartram CR, van Dongen JJ. Detection of minimal residual disease identifies differences in treatment response between T-ALL and precursor B-ALL. Blood. 2002;99:4386–93.

17. Campana D. Progress of minimal residual disease studies in childhood acute leukemia. Curr Hematol Malig Rep. 2010;5:169–76.
18. Thol F, Kolking B, Damm F, Reinhardt K, Klusmann JH, Reinhardt D, von Neuhoff N, Brugman MH, Schlegelberger B, Suerbaum S, Krauter J, Ganser A, Heuser M. Next generation sequencing for minimal residual disease monitoring in acute myeloid leukemia patients with FLT3-ITD or NPM1 mutations. Genes Chromosomes Cancer. 2012;51:689–95.
19. Campana D. Minimal residual disease monitoring in childhood acute lymphoblastic leukemia. Curr Opin Hematol. 2012;19:313–8.
20. Wu D, Sherwood A, Fromm JR, Winter SS, Dunsmore KP, Loh ML, Greisman HA, Sabath DE, Wood BL, Robins H. High-throughput sequencing detects minimal residual disease in acute T lymphoblastic leukemia. Sci Transl Med. 2012;4:134ra63.
21. Gawad C, Pepin F, Carlton VE, Klinger M, Logan AC, Miklos DB, Faham M, Dahl G, Lacayo N. Massive evolution of the immunoglobulin heavy chain locus in children with B precursor acute lymphoblastic leukemia. Blood. 2012;120:4407–17.
22. Cronin M, Ross JS. Comprehensive next generation cancer genome sequencing in the era of targeted therapy and personalized oncology. Biomark Med. 2011;5:293–305.
23. Borras E, Jurado I, Hernan I, Gamundi MJ, Dias M, Marti I, Mane B, Arcusa A, Agundez JA, Blanca M, Carballo M. Clinical pharmacogenomic testing of KRAS, BRAF and EGFR mutations by high resolution melting analysis and ultra-deep pyrosequencing. BMC Cancer. 2011;11:406.
24. Shindoh N, Yoda A, Yoda Y, Sullivan TJ, Weigert O, Lane AA, Kopp N, Bird L, Rodig SJ, Fox EA, Weinstock DM. Next generation cDNA screening for oncogene and resistance phenotypes. PLoS One. 2012;7:e49201.
25. Devonshire AS, Sanders R, Wilkes TM, Taylor MS, Foy CA, Huggett JF. Application of next generation qPCR and sequencing platforms to mRNA biomarker analysis. Methods. 2012;59:89–100.
26. Chan M, Ji SM, Yeo ZX, Gan L, Yap E, Yap YS, Ng R, Tan PH, Ho GH, Ang P, Lee AS. Development of a next generation sequencing method for BRCA mutation screening: a comparison between a high-throughput and a benchtop platform. J Mol Diagn. 2012;14:602–12.
27. Normanno N, Rachiglio AM, Roma C, Fenizia F, Esposito C, Pasquale R, La Porta ML, Iannaccone A, Micheli F, Santangelo M, Bergantino F, Costantini S, De Luca A. Molecular diagnostics and personalized medicine in oncology: Challenges and opportunities. J Cell Biochem. 2012;114:514–24.
28. Grossmann V, Schnittger S, Schindela S, Klein HU, Eder C, Dugas M, Kern W, Haferlach T, Haferlach C, Kohlmann A. Strategy for robust detection of insertions, deletions, and point mutations in CEBPA, a GC-rich content gene, using 454 next generation deep-sequencing technology. J Mol Diagn. 2011;13:129–36.
29. Kohlmann A, Grossmann V, Haferlach T. Integration of next generation sequencing into clinical practice: are we there yet? Semin Oncol. 2012;39:26–36.
30. Kohlmann A, Grossmann V, Klein HU, Schindela S, Weiss T, Kazak B, Dicker F, Schnittger S, Dugas M, Kern W, Haferlach C, Haferlach T. Next generation sequencing technology reveals a characteristic pattern of molecular mutations in 72.8 % of chronic myelomonocytic leukemia by detecting frequent alterations in TET2, CBL, RAS, and RUNX1. J Clin Oncol. 2010;28:3858–65.
31. Delhommeau F, Dupont S, Della VV, James C, Trannoy S, Masse A, Kosmider O, Le Couedic JP, Robert F, Alberdi A, Lecluse Y, Plo I, Dreyfus FJ, Marzac C, Casadevall N, Lacombe C, Romana SP, Dessen P, Soulier J, Viguie F, Fontenay M, Vainchenker W, Bernard OA. Mutation in TET2 in myeloid cancers. N Engl J Med. 2009;360:2289–301.
32. Gazdar AF. Activating and resistance mutations of EGFR in non-small-cell lung cancer: role in clinical response to EGFR tyrosine kinase inhibitors. Oncogene. 2009;28 Suppl 1:S24–31.
33. Escobar-Gutierrez A, Vazquez-Pichardo M, Cruz-Rivera M, Rivera-Osorio P, Carpio-Pedroza JC, Ruiz-Pacheco JA, Ruiz-Tovar K, Vaughan G. Identification of hepatitis C virus transmission using a next generation sequencing approach. J Clin Microbiol. 2012;50:1461–3.
34. Redd AD, Collinson-Streng A, Martens C, Ricklefs S, Mullis CE, Manucci J, Tobian AA, Selig EJ, Laeyendecker O, Sewankambo N, Gray RH, Serwadda D, Wawer MJ, Porcella SF, Quinn TC. Identification of HIV superinfection in seroconcordant couples in Rakai, Uganda, by use of next generation deep sequencing. J Clin Microbiol. 2011;49:2859–67.
35. Beerenwinkel N, Gunthard HF, Roth V, Metzner KJ. Challenges and opportunities in estimating viral genetic diversity from next generation sequencing data. Front Microbiol. 2012;3:329.
36. Archer J, Weber J, Henry K, Winner D, Gibson R, Lee L, Paxinos E, Arts EJ, Robertson DL, Mimms L, Quinones-Mateu ME. Use of four next generation sequencing platforms to determine HIV-1 coreceptor tropism. PLoS One. 2012;7:e49602.
37. Meiring TL, Salimo AT, Coetzee B, Maree HJ, Moodley J, Hitzeroth II, Freeborough MJ, Rybicki EP, Williamson AL. Next generation sequencing of cervical DNA detects human papillomavirus types not detected by commercial kits. Virol J. 2012;9:164.
38. Katano H, Sato S, Sekizuka T, Kinumaki A, Fukumoto H, Sato Y, Hasegawa H, Morikawa S, Saijo M, Mizutani T, Kuroda M. Pathogenic characterization of a cervical lymph node derived from a patient with Kawasaki disease. Int J Clin Exp Pathol. 2012;5:814–23.
39. Nakamura S, Yang CS, Sakon N, Ueda M, Tougan T, Yamashita A, Goto N, Takahashi K, Yasunaga T, Ikuta K, Mizutani T, Okamoto Y, Tagami M, Morita R, Maeda N, Kawai J, Hayashizaki Y, Nagai Y, Horii T, Iida T, Nakaya T. Direct metagenomic detection of viral pathogens in nasal and fecal specimens using an unbiased high-throughput sequencing approach. PLoS One. 2009;4:e4219.
40. Nakamura S, Nakaya T, Iida T. Metagenomic analysis of bacterial infections by means of high-throughput DNA sequencing. Exp Biol Med (Maywood). 2011;236:968–71.
41. Mokili JL, Rohwer F, Dutilh BE. Metagenomics and future perspectives in virus discovery. Curr Opin Virol. 2012;2:63–77.

42. Serizawa M, Sekizuka T, Okutani A, Banno S, Sata T, Inoue S, Kuroda M. Genomewide screening for novel genetic variations associated with ciprofloxacin resistance in Bacillus anthracis. Antimicrob Agents Chemother. 2010;54:2787–92.
43. Wright AM, Beres SB, Consamus EN, Long SW, Flores AR, Barrios R, Richter GS, Oh SY, Garufi G, Maier H, Drews AL, Stockbauer KE, Cernoch P, Schneewind O, Olsen RJ, Musser JM. Rapidly progressive, fatal, inhalation anthrax-like infection in a human: case report, pathogen genome sequencing, pathology, and coordinated response. Arch Pathol Lab Med. 2011;135:1447–59.
44. Kuroda M, Sekizuka T, Shinya F, Takeuchi F, Kanno T, Sata T, Asano S. Detection of a possible bioterrorism agent, Francisella sp., in a clinical specimen by use of next generation direct DNA sequencing. J Clin Microbiol. 2012;50:1810–2.
45. Botstein D, Risch N. Discovering genotypes underlying human phenotypes: past successes for mendelian disease, future approaches for complex disease. Nat Genet. 2003;33(Suppl):228–37.
46. Coonrod EM, Durtschi JD, Margraf RL, Voelkerding KV. Developing genome and exome sequencing for candidate gene identification in inherited disorders. Arch Pathol Lab Med. 2012;137:415–33.
47. Li Y, Vinckenbosch N, Tian G, Huerta-Sanchez E, Jiang T, Jiang H, Albrechtsen A, Andersen G, Cao H, Korneliussen T, Grarup N, Guo Y, Hellman I, Jin X, Li Q, Liu J, Liu X, Sparso T, Tang M, Wu H, Wu R, Yu C, Zheng H, Astrup A, Bolund L, Holmkvist J, Jorgensen T, Kristiansen K, Schmitz O, Schwartz TW, Zhang X, Li R, Yang H, Wang J, Hansen T, Pedersen O, Nielsen R, Wang J. Resequencing of 200 human exomes identifies an excess of low-frequency non-synonymous coding variants. Nat Genet. 2010;42: 969–72.
48. Ng SB, Bigham AW, Buckingham KJ, Hannibal MC, McMillin MJ, Gildersleeve HI, Beck AE, Tabor HK, Cooper GM, Mefford HC, Lee C, Turner EH, Smith JD, Rieder MJ, Yoshiura K, Matsumoto N, Ohta T, Niikawa N, Nickerson DA, Bamshad MJ, Shendure J. Exome sequencing identifies MLL2 mutations as a cause of Kabuki syndrome. Nat Genet. 2010; 42:790–3.
49. Ng SB, Buckingham KJ, Lee C, Bigham AW, Tabor HK, Dent KM, Huff CD, Shannon PT, Jabs EW, Nickerson DA, Shendure J, Bamshad MJ. Exome sequencing identifies the cause of a mendelian disorder. Nat Genet. 2010;42:30–5.
50. Hoischen A, van Bon BW, Gilissen C, Arts P, van Lier B, Steehouwer M, de Vries P, de Reuver R, Wieskamp N, Mortier G, Devriendt K, Amorim MZ, Revencu N, Kidd A, Barbosa M, Turner A, Smith J, Oley C, Henderson A, Hayes IM, Thompson EM, Brunner HG, de Vries BB, Veltman JA. De novo mutations of SETBP1 cause Schinzel-Giedion syndrome. Nat Genet. 2010;42:483–5.
51. Ku CS, Naidoo N, Pawitan Y. Revisiting Mendelian disorders through exome sequencing. Hum Genet. 2011;129:351–70.
52. Shanks ME, Downes SM, Copley RR, Lise S, Broxholme J, Hudspith KA, Kwasniewska A, Davies WI, Hankins MW, Packham ER, Clouston P, Seller A, Wilkie AO, Taylor JC, Ragoussis J, Nemeth AH. Next generation sequencing (NGS) as a diagnostic tool for retinal degeneration reveals a much higher detection rate in early-onset disease. Eur J Hum Genet. 2012;21:274–80.
53. Erlich H. HLA DNA typing: past, present, and future. Tissue Antigens. 2012;80:1–11.
54. Sood AK, Pereira D, Weissman SM. Isolation and partial nucleotide sequence of a cDNA clone for human histocompatibility antigen HLA-B by use of an oligodeoxynucleotide primer. Proc Natl Acad Sci U S A. 1981;78:616–20.
55. Erlich HA, Stetler D, Sheng-Dong R, Ness D, Grumet C. Segregation and mapping analysis of polymorphic HLA class I restriction fragments: detection of a novel fragment. Science. 1983;222:72–4.
56. Saiki RK, Bugawan TL, Horn GT, Mullis KB, Erlich HA. Analysis of enzymatically amplified beta-globin and HLA-DQ alpha DNA with allele-specific oligonucleotide probes. Nature. 1986;324:163–6.
57. Saiki RK, Walsh PS, Levenson CH, Erlich HA. Genetic analysis of amplified DNA with immobilized sequence-specific oligonucleotide probes. Proc Natl Acad Sci U S A. 1989;86:6230–4.
58. Blake E, Mihalovich J, Higuchi R, Walsh PS, Erlich H. Polymerase chain reaction (PCR) amplification and human leukocyte antigen (HLA)-DQ alpha oligonucleotide typing on biological evidence samples: casework experience. J Forensic Sci. 1992;37:700–26.
59. Santamaria P, Lindstrom AL, Boyce-Jacino MT, Myster SH, Barbosa JJ, Faras AJ, Rich SS. HLA class I sequence-based typing. Hum Immunol. 1993;37: 39–50.
60. Lind C, Ferriola D, Mackiewicz K, Heron S, Rogers M, Slavich L, Walker R, Hsiao T, McLaughlin L, D'Arcy M, Gai X, Goodridge D, Sayer D, Monos D. Next generation sequencing: the solution for high-resolution, unambiguous human leukocyte antigen typing. Hum Immunol. 2010;71:1033–42.
61. Holcomb CL, Hoglund B, Anderson MW, Blake LA, Bohme I, Egholm M, Ferriola D, Gabriel C, Gelber SE, Goodridge D, Hawbecker S, Klein R, Ladner M, Lind C, Monos D, Pando MJ, Proll J, Sayer DC, Schmitz-Agheguian G, Simen BB, Thiele B, Trachtenberg EA, Tyan DB, Wassmuth R, White S, Erlich HA. A multi-site study using high-resolution HLA genotyping by next generation sequencing. Tissue Antigens. 2011;77:206–17.
62. Erlich RL, Jia X, Anderson S, Banks E, Gao X, Carrington M, Gupta N, DePristo MA, Henn MR, Lennon NJ, de Bakker PI. Next generation sequencing for HLA typing of class I loci. BMC Genomics. 2011;12:42.
63. Higuchi R, Bowman B, Freiberger M, Ryder OA, Wilson AC. DNA sequences from the quagga, an extinct member of the horse family. Nature. 1984; 312:282–4.
64. Paabo S. Molecular cloning of ancient Egyptian mummy DNA. Nature. 1985;314:644–5.
65. Paabo S. Ancient DNA: extraction, characterization, molecular cloning, and enzymatic amplification. Proc Natl Acad Sci U S A. 1989;86:1939–43.
66. Paabo S, Higuchi RG, Wilson AC. Ancient DNA and the polymerase chain reaction. The emerging field of molecular archaeology. J Biol Chem. 1989;264: 9709–12.
67. Cooper A, Poinar HN. Ancient DNA: do it right or not at all. Science. 2000;289:1139.

68. Rizzi E, Lari M, Gigli E, De Bellis G, Caramelli D. Ancient DNA studies: new perspectives on old samples. Genet Sel Evol. 2012;44:21.
69. Green RE, Krause J, Briggs AW, Maricic T, Stenzel U, Kircher M, Patterson N, Li H, Zhai W, Fritz MH, Hansen NF, Durand EY, Malaspinas AS, Jensen JD, Marques-Bonet T, Alkan C, Prufer K, Meyer M, Burbano HA, Good JM, Schultz R, Aximu-Petri A, Butthof A, Hober B, Hoffner B, Siegemund M, Weihmann A, Nusbaum C, Lander ES, Russ C, Novod N, Affourtit J, Egholm M, Verna C, Rudan P, Brajkovic D, Kucan Z, Gusic I, Doronichev VB, Golovanova LV, Lalueza-Fox C, de la RM, Fortea J, Rosas A, Schmitz RW, Johnson PL, Eichler EE, Falush D, Birney E, Mullikin JC, Slatkin M, Nielsen R, Kelso J, Lachmann M, Reich D, Paabo S. A draft sequence of the Neandertal genome. Science. 2010;328:710–22.
70. Reich D, Green RE, Kircher M, Krause J, Patterson N, Durand EY, Viola B, Briggs AW, Stenzel U, Johnson PL, Maricic T, Good JM, Marques-Bonet T, Alkan C, Fu Q, Mallick S, Li H, Meyer M, Eichler EE, Stoneking M, Richards M, Talamo S, Shunkov MV, Derevianko AP, Hublin JJ, Kelso J, Slatkin M, Paabo S. Genetic history of an archaic hominin group from Denisova Cave in Siberia. Nature. 2010;468:1053–60.
71. Rasmussen M, Li Y, Lindgreen S, Pedersen JS, Albrechtsen A, Moltke I, Metspalu M, Metspalu E, Kivisild T, Gupta R, Bertalan M, Nielsen K, Gilbert MT, Wang Y, Raghavan M, Campos PF, Kamp HM, Wilson AS, Gledhill A, Tridico S, Bunce M, Lorenzen ED, Binladen J, Guo X, Zhao J, Zhang X, Zhang H, Li Z, Chen M, Orlando L, Kristiansen K, Bak M, Tommerup N, Bendixen C, Pierre TL, Gronnow B, Meldgaard M, Andreasen C, Fedorova SA, Osipova LP, Higham TF, Ramsey CB, Hansen TV, Nielsen FC, Crawford MH, Brunak S, Sicheritz-Ponten T, Villems R, Nielsen R, Krogh A, Wang J, Willerslev E. Ancient human genome sequence of an extinct Palaeo-Eskimo. Nature. 2010;463:757–62.
72. DiMauro S, Hirano M. MELAS. In: Pagon RA, Adam MP, Bird TD, Dolan CR, Fong CT, Stephens K, editors. GeneReviews. Seattle, WA: University of Washington; 2010.
73. Bandelt HJ, Salas A. Current next generation sequencing technology may not meet forensic standards. Forensic Sci Int Genet. 2012;6:143–5.
74. Zietkiewicz E, Witt M, Daca P, Zebracka-Gala J, Goniewicz M, Jarzab B, Witt M. Current genetic methodologies in the identification of disaster victims and in forensic analysis. J Appl Genet. 2012;53:41–60.
75. Berglund EC, Kiialainen A, Syvanen AC. Next generation sequencing technologies and applications for human genetic history and forensics. Investig Genet. 2011;2:23.
76. Alvarez-Cubero MJ, Saiz M, Martinez-Gonzalez LJ, Alvarez JC, Eisenberg AJ, Budowle B, Lorente JA. Genetic identification of missing persons: DNA analysis of human remains and compromised samples. Pathobiology. 2012;79:228–38.
77. Tiercy JM. Immunogenetics of hematopoietic stem cell transplantation: the contribution of microsatellite polymorphism studies. Int J Immunogenet. 2011;38:365–72.
78. Snyder TM, Khush KK, Valantine HA, Quake SR. Universal noninvasive detection of solid organ transplant rejection. Proc Natl Acad Sci U S A. 2011; 108:6229–34.
79. Palomaki GE, Kloza EM, Lambert-Messerlian GM, Haddow JE, Neveux LM, Ehrich M, van den BD, Bombard AT, Deciu C, Grody WW, Nelson SF, Canick JA. DNA sequencing of maternal plasma to detect down syndrome: an international clinical validation study. Genet Med. 2011;13:913–20.
80. Palomaki GE, Deciu C, Kloza EM, Lambert-Messerlian GM, Haddow JE, Neveux LM, Ehrich M, van den BD, Bombard AT, Grody WW, Nelson SF, Canick JA. DNA sequencing of maternal plasma reliably identifies trisomy 18 and trisomy 13 as well as Down syndrome: an international collaborative study. Genet Med. 2012;14:296–305.
81. Rizzo JM, Buck MJ. Key principles and clinical applications of "next generation" DNA sequencing. Cancer Prev Res (Phila). 2012;5:887–900.
82. Timp W, Mirsaidov UM, Wang D, Comer J, Aksimentiev A, Timp G. Nanopore sequencing: electrical measurements of the code of life. IEEE Trans Nanotechnol. 2010;9:281–94.
83. Korlach J, Bjornson KP, Chaudhuri BP, Cicero RL, Flusberg BA, Gray JJ, Holden D, Saxena R, Wegener J, Turner SW. Real-time DNA sequencing from single polymerase molecules. Methods Enzymol. 2010;472: 431–55.
84. Fang G, Munera D, Friedman DI, Mandlik A, Chao MC, Banerjee O, Feng Z, Losic B, Mahajan MC, Jabado OJ, Deikus G, Clark TA, Luong K, Murray IA, Davis BM, Keren-Paz A, Chess A, Roberts RJ, Korlach J, Turner SW, Kumar V, Waldor MK, Schadt EE. Genome-wide mapping of methylated adenine residues in pathogenic Escherichia coli using single-molecule real-time sequencing. Nat Biotechnol. 2012;30:1232–9.
85. Ahmadian A, Ehn M, Hober S. Pyrosequencing: history, biochemistry and future. Clin Chim Acta. 2006;363:83–94.

CHAPTER 15

Next-Generation Sequencing for Gene Panels

MICHAEL O. DORSCHNER

Introduction 209
Assay Design and Considerations for Developing an NGS Gene Panel 210
Example Procedure 212
Reagent and Oligonucleotide Preparation 214
Procedure 215
Data Analysis 222
Conclusions 222

Introduction

Next-generation sequencing (NGS) technologies have revolutionized molecular diagnostics over the past several years. These rapidly evolving platforms are moving quickly from the research bench to the clinical laboratory. At the same time, the enormous outputs of these massively parallel technologies have dramatically reduced the cost of DNA sequencing [1]. A single sequencing run can generate from a few gigabases to over half a terabase of data. Analysis and storage of data on this scale are no small undertaking, requiring specialized bioinformaticists, computational biologists, and substantial information technology (IT) infrastructure. Over the past few years, molecular diagnostic laboratories have begun to convert conventional Sanger sequencing-based assays to NGS-based gene panels.

Molecular diagnostic laboratories often begin implementing NGS with indication or disease-targeted multi-gene panels [2]. Panel-based approaches have been applied to the diagnosis of heterogeneous disorders with overlapping, difficult-to-distinguish phenotypes. Multi-gene panels for cardiomyopathies [3], cancer predisposition [4], and X-linked intellectual disability [5] were among the first NGS assays to be launched by clinical laboratories (Table 15.1). With disease-targeted panels, all of the clinically relevant genes, from 10 or more to well over 100, can be examined concurrently, putting an end to the serial gene-by-gene diagnostic odysseys imposed by traditional Sanger-based approaches. NGS gene panels offer several advantages by (1) reducing the time a clinician spends on test selection, (2) reducing the collective turnaround time from test initiation to the reporting of results to patients, and (3) limiting testing to only those genes with proven clinical utility for a given phenotype.

Some academic and commercial laboratories have begun offering clinical exome and genome sequencing. Exome and genome sequencing come with a variety of issues to consider as follows: (1) the likelihood of incidental findings, not relevant to the indication for which the test was ordered and whether the lab is responsible for reporting these variants; (2) a lack of sequence coverage for disease-relevant loci, which are likely to be better targeted with panel testing; and (3) the increased likelihood of identifying and reporting more variants of unknown significance, as laboratories are typically not genome-wide experts.

G.J. Netto and I. Schrijver (eds.), *Genomic Applications in Pathology*,
DOI 10.1007/978-1-4939-0727-4_15, © Springer Science+Business Media New York 2015

Table 15-1 Next-generation Sequencing Gene Panels Offered by Clinical Laboratories

Next-generation sequencing gene panel	Number of genes
Hereditary cancer	10–50
Somatic/tumor cancer	10–200
Cardiomyopathies (dilated, hypertrophic)	50–70
Arrhythmias	10–30
Hearing loss (syndromic, non-syndromic)	23–72
Neurodegenerative (dementia, Parkinson's, ALS, dystonia)	4–75
X-linked mental retardation	30–150
RASoptahies	~10
Mitochondrial disorders	35–400

ALS amyotrophic lateral sclerosis

Laboratories that offer disease-targeted panels often have expertise relevant to the interpretation of the variants detected by the assay. The American College of Medical Genetics and Genomics currently recommends exome or genome sequencing only for cases in which a disease-targeted panel is likely to produce a negative result or for disorders for which a targeted test is not available [6].

Along with the commercialization of several NGS platforms came technologies for performing target enrichment [7]. Polymerase chain reaction (PCR), the mainstay of Sanger-based diagnostics, is not practical for processing multi-gene panels at an appreciable scale. Technologies based on highly multiplexed PCR, such as TruSeq Custom Amplicon Assays™ (Illumina), AmpliSeq™ (Life Technologies), and droplet PCR (Raindance Technologies), have recently been launched. However, these assays remain more expensive than hybridization-based enrichment strategies. Customizable, in-solution hybridization-based technologies include SureSelect™ (Agilent Technologies), SeqCap EZ™ (NimbleGen), and TargetSeq™ (Life Technologies). These platforms, based on hybridization of biotinylated RNA or DNA probes to sample fragment libraries, are used to capture from 1 kilobase sequences to genomic regions of more than 24 megabases [8–10] (Fig. 15.1). Users can choose to target coding segments or entire transcribed regions, within the limitations of the capture and sequencing technologies. In-solution, hybridization-based approaches are widely used among molecular diagnostic laboratories. Many clinical laboratories have tested multiple platforms, ultimately focusing on one enrichment strategy, in an effort to streamline laboratory workflows.

In this chapter, an example of a procedure for NGS-based gene panel testing is provided. This is only one of many possible examples, using equipment and reagents that could be exchanged for others, but it provides insight into the conceptual workflow of NGS-based testing of gene panels in a laboratory-developed assay setting. The example procedure was developed using the Agilent SureSelect in-solution sequence capture system for target enrichment, followed by sequencing with an Illumina HiSeq or MiSeq system. The procedure commences with the construction of sample-specific fragment libraries with platform-specific indexed adapters, followed by enrichment of the libraries for sequence targets of interest, sequencing of those targets, and analysis (Fig. 15.2).

Assay Design and Considerations for Developing an NGS Gene Panel

After selecting a panel of genes for assay development, the user of our example wet-bench procedure will need to design enrichment probes. Custom probe designs can be easily generated using, for example, the online Agilent SureDesign software (https://earray.chem.agilent.com/suredesign). Depending on the target of interest, probes can be designed to capture only exons or entire genomic regions. If copy number variants (CNVs) will need to be ascertained, it is advised to design probes across the entire genomic region, starting 10 kilobases upstream of the first exon and ending 10 kilobases downstream of the final exon. This should provide sufficient read depth to call

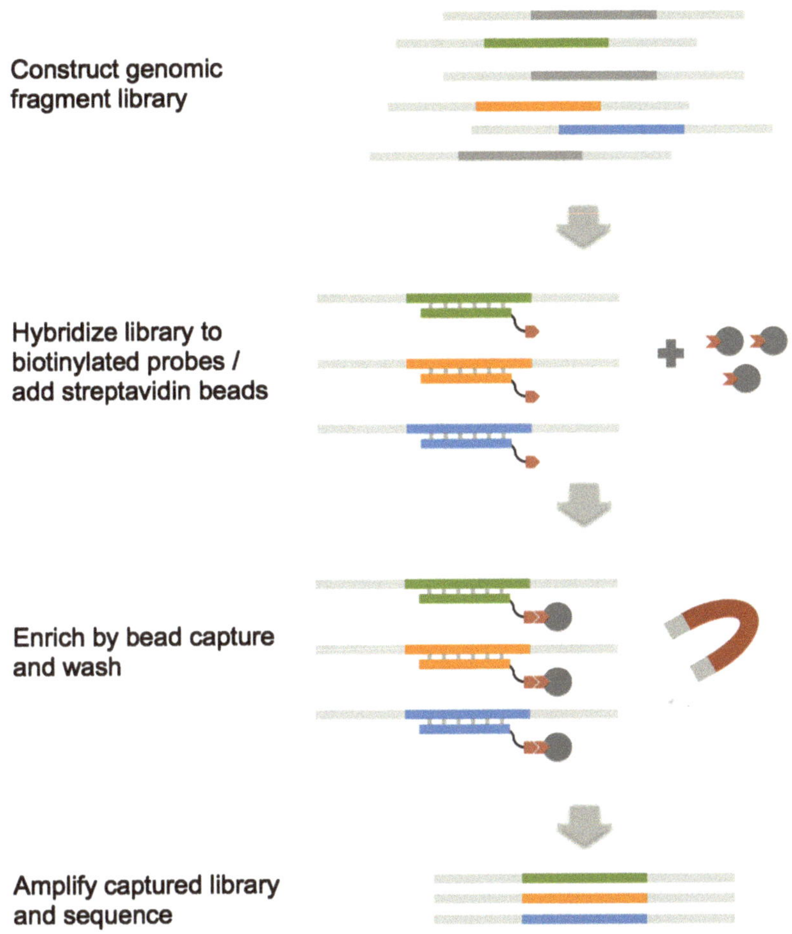

Figure 15-1 In-solution sequence capture for target enrichment. After the construction of a genomic fragment library, the library is hybridized to capture probes and enriched. Subsequently, the captured library can be amplified and sequenced

small insertions/deletions using some of the CNV detection algorithms [11].

In addition to selection of the target regions, the user must also consider the existence of repetitive regions, such as segmental duplications or paralogs and highly repetitive elements, including short interspersed elements (SINEs) and long interspersed elements (LINEs). Depending on the sequence similarity of these loci, they can be difficult to enrich and sequence. It can also be challenging to align reads accurately to the reference. With the Illumina platform, the current maximum read length on a HiSeq is 100 bp and 250 bp on a MiSeq. Longer paired-end reads can dramatically improve the accuracy of read mapping, but this will depend on the length of identity between or among paralogous regions of the genome.

GC content can influence the ability to capture and sequence targets. The first exons of many genes are GC rich. Despite the inclusion of probes to enrich libraries for these regions, read coverage is often low and sometimes absent over these segments. The user may need to include additional probes over these regions to enhance sequence capture.

After the design is completed, the user can review the results by downloading what is called a BED file. The BED file can be uploaded to the UCSC Genome Browser (http://genome.ucsc.edu/) to view probe positions and target coverage. If the desired targets are not adequately covered by probes,

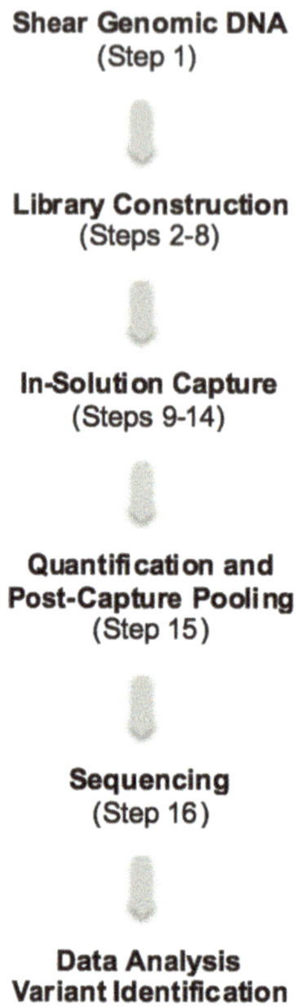

Figure 15-2 Step-by-step workflow. This example of a laboratory procedure for an NGS-based gene panel assay begins with the construction of sample-specific fragment libraries with platform-specific indexed adapters, followed by enrichment of the libraries for sequence targets of interest, sequencing of those targets, and data analysis

the user can redesign the probe set by modifying design parameters in the SureDesign software. When the custom design has been finalized, the user can request an online quote and consider proceeding with placing an order.

Example Procedure

Materials

1. Qubit dsDNA HS Quantification kit (Life Technologies, P/N [part number] Q32851)
2. P20, P200, P1000 filtered pipet tips (Rainin, P/N RT-L10F, RT-L200F, and RT-L1000)
3. Covaris microTUBE AFA Fiber Screw-Cap 6×16 mm (P/N 520096) for the M220 series, Covaris microTUBE AFA Fiber Pre-Slit Snap-Cap 6×16 mm (P/N 520045) for the S220 series, *or* Covaris 96 microTUBE Plate (P/N 520078) for the E220 series
4. Ultrapure Distilled Deionized water (Life Technologies, P/N 10977-015)
5. Eppendorf LoBind Microcentrifuge Tubes 1.5 mL (Fisher, P/N 13-6987-91)
6. 0.2 mL PCR Tube Strips, 8-Tube, full-height with domed caps (Bio-Rad, P/N TBC-0201)
7. 0.2 mL PCR Tube Strips, 8-Tube, low-profile (Bio-Rad, P/N TLS-0801)
8. Optical Flat 8-Cap Strips (Bio-Rad, P/N TCS-0803)
9. Domed 8-Cap Strips (Bio-Rad, P/N TCS-0801)
10. Minicentrifuge, with tube strip rotor (Fisher, P/N 05-090-100)
11. NEB End repair kit (NEB, P/N E6050L)
12. NEB Next A-tailing Kit (NEB, P/N E6053L)
13. Agencourt AMPure XP Beads, 60 mL (Beckman Coulter, P/N A63881)
14. 70 % Ethanol (see section "Reagent and Oligonucleotide Preparation")
15. 20 % PEG/2.5 M NaCl solution (see section "Reagent and Oligonucleotide Preparation")
16. T4 Ultrapure Ligase and 10× T4 Ultrapure Buffer (Enzymatics Inc., P/N L603-HC-L)
17. KAPA Library Amplification Readymix (KAPA Biosystems, P/N KK2612)
18. DNA 1000 kit (Agilent, P/N 5067-1504)
19. SureSelect XT Custom Library (Agilent, part numbers vary depending on the size of the target region)
20. Dynabeads M-270 Streptavidin (Life Technologies, P/N 65305)
21. Library Quantification Kit/Illumina/Universal (Kapa Biosystems, P/N KK4824)
22. 1 M Tris–HCl, pH 7.5 (Fisher, P/N BP1758-100)
23. Tween 20 (Fisher, P/N BP337-100)
24. Oligonucleotides (Integrated DNA Technologies or other commercial vendors) (Tables 15.2 and 15.3)

Equipment

1. P10, P20, P200, and P1000 pipets (Rainin, P/N L-10XLS, L-20XLS, L-200XLS, and L-1000XLS)

Table 15-2 **Oligonucleotide Sequences**

Oligo name	Sequence (5′–3′)
Universal	AATGATACGGCGACCACCGAGATCTACACTCTTTCCCTACACGACGCTCTTCCGATC*T
Index	/5P/GATCGGAAGAGCACACGTCTGAACTCCAGTCAC [idx] ATCTCGTATGCCGTCTTCTGCTTG
Block-U	AATGATACGGCGACCACCGAGATCTACACTCTTTCCCTACACGACGCTCTTCCGATCT
Block-U-RC	AGATCGGAAGAGCGTCGTGTAGGGAAAGAGTGTAGATCTCGGTGGTCGCCGTATCATT
Block-I	CAAGCAGAAGACGGCATACGAGAT [idx] GTGACTGGAGTTCAGACGTGTGCTCTTCCGATC
Block-I-RC	GATCGGAAGAGCACACGTCTGAACTCCAGTCAC [idx] ATCTCGTATGCCGTCTTCTGCTTG
TS-PCR-1	AATGATACGGCGACCACCGAGA
TS-PCR-2	CAAGCAGAAGACGGCATACGAG

U universal, *RC* reverse complement, *idx* index, *5P* 5′ phosphorylation, *asterisk* phosphorothioate bond, *I* index-containing, *TS* TruSeq™, *PCR* polymerase chain reaction

Table 15-3 **Index Sequences**

ID	Sequence
1	ATCACG
2	CGATGT
3	TTAGGC
4	TGACCA
5	ACAGTG
6	GCCAAT
7	CAGATC
8	ACTTGA
9	GATCAG
10	TAGCTT
11	GGCTAC
12	CTTGTA
13	AGTCAA
14	AGTTCC
15	ATGTCA
16	CCGTCC
18	GTCCGC
19	GTGAAA
20	GTGGCC
21	GTTTCG
22	CGTACG
23	GAGTGG
25	ACTGAT
27	ATTCCT

ID identification number

2. Covaris® M220, S220, or E220 Focused-Ultrasonicator, connected to a computer with SonoLab™ software
3. Microcentrifuge (Eppendorf 5430)
4. One 24 position Tube IsoRack with 0 °C IsoPack (Eppendorf, P/N 22510053)
5. Dyna-Mag 2 Magnetic Stand (Life Technologies, P/N 12321D)
6. 2100 Bioanalyzer (Agilent Technologies)
7. Speed vacuum with tube or plate adaptor (Eppendorf Vacufuge Plus)
8. HiSeq 1000/2000, HiSeq 1500/2500, or MiSeq system (Illumina)
9. CFX96 Touch Real-Time PCR Detection System (Bio-Rad, P/N 185-5484)
10. T100 Thermal Cycler (Bio-Rad, P/N 186-1096)
11. Qubit 2.0 Fluorometer (Life Technologies, P/N Q32866)

Reagent and Oligonucleotide Preparation

1. 100 mL 20 % PEG/NaCl solution
   In a beaker, add:
   (a) 50 mL 5 M NaCl (Life Technologies, P/N AM9759)
   (b) 20 g PEG 8000 (Fisher, P/N BP233-100)
   (c) Water, up to 100 mL total volume
   Autoclave for 15 min. Allow the solution to cool because separation may occur upon heating.
2. 100 mL 70 % ethanol:
   In a graduated cylinder, measure 70 mL 100 % ethanol. Add 30 mL of molecular grade water, and mix.
3. Generation of 50 μM adapter stocks:
   (a) Resuspend the Universal adapter oligo and Indexed oligos at 100 μM in 10 mM Tris pH 8.0 and 50 mM NaCl.
   (b) For each indexed adaptor combine 25 μL of Index oligo and 25 μL of Universal oligo in a clean 0.2 mL PCR tube.
   (c) Heat to 95 °C for 5 min in a thermal cycler.
   (d) Remove the tubes from the thermal cycler, and briefly spin to ensure that contents are at the bottoms of the tubes. Allow to cool at room temperature for 30 min.
4. Prepare index-specific hybridization blockers (ISHB):
   (a) Resuspend the Block-U, Block-U-RC, Block-I, and Block-I-RC oligos at 1,000 μM in 10 mM Tris pH 8.0 and 50 mM NaCl.
   (b) For each index used combine 0.5 μL of each oligo Block-U, Block-U-RC, Block-I, and Block-I-RC. 0.6 μL of this stock will be used during the hybridization reaction setup in Step 9. This blocker solution can be used in place of the Agilent-provided blockers.
5. Prepare library amplification oligonucleotides:
   (a) Resuspend TS-PCR-1 and TS-PCR-2 oligos at 100 μM in 10 mM Tris pH 8.0.

Procedure

This procedure provides step-by-step instructions for generating genomic fragment libraries and subsequent target enrichment, using the Agilent SureSelect in-solution capture system. With minor modification, the libraries generated by this protocol could be enriched with any other hybridization-based approach. This protocol is designed for processing batches of eight samples. Volumes and quantities can be scaled according to the needs of the user.

Step 1: Shear DNA

Covaris provides a shearing guide for each instrument model with recommended settings to generate specific fragment sizes. The user should test these settings prior to shearing valuable samples. The optimal insert size for 2×100 bp paired-end sequencing is 250 bp (225–275 bp). The average insert size will be larger than the total of the paired-end reads to avoid generating overlapping sequence at the 3′ ends.

1. Quantify DNA using the Qubit dsDNA quantification kit according to the manufacturer's instructions.
2. Prepare the Covaris instrument. Ensure that the reservoir is filled with deionized water and the water temperature has equilibrated and degassed, prior to use.
3. Dilute 1 µg of high-quality gDNA with 1× low TE buffer in a 1.5 mL LoBind tube to a total volume of 130 µL.
4. Transfer the 130 µL of DNA sample to the proper Covaris microTube, making sure not to introduce bubbles.
5. Secure the microTube in the tube holder, and shear the DNA using the appropriate settings to generate fragments of 150–200 bp (or other fragment size ranges, depending on specific needs).
6. Optional: Repeat DNA quantification prior to proceeding to Step 2.

Step 2: End Repair

Shearing will create double-stranded fragments with overhangs. A combination of T4 polynucleotide kinase and T4 DNA polymerase will convert these to 5′-phosphorylated, blunt ends.

Table 15-4 End-repair Master Mix Components

Reagent	Per reaction (µL)	For 8 reactions (µL)
10× End-repair buffer	6	52.8
End-repair enzyme mix	5	44
Total	11	96.8

1. Prepare a master mix from the components listed in Table 15.4 in a clean 1.5 mL LoBind tube on ice or a cooling rack.
2. Dispense 11 µL of master mix into each 0.2 mL PCR tube.
3. Dispense 50 µL of sheared DNA to each master mix-containing PCR tube. Mix carefully and thoroughly by pipetting up and down ten times.
4. Briefly spin to bring contents to the bottoms of the tubes.
5. Incubate tubes in a thermal cycler for 30 min at 20 °C.
6. Remove AMPure XP beads from the refrigerator, and gently shake to resuspend beads. The beads must be at room temperature prior to use.
7. After the 20 °C incubation, shake the AMPure XP beads to ensure complete resuspension, and transfer 110 µL of bead solution to each end-repaired DNA.
8. Mix and incubate for 5 min at room temperature.
9. Place tubes on a strip/plate magnet for 5 min to separate beads from solution.
10. Visually confirm that the beads have moved to the side of the tube and the solution is clear. Aspirate 171 µL of clear solution from each tube without disturbing the beads, and discard.
11. Dispense 180 µL of freshly prepared 70 % ethanol to each tube and incubate for 30 s at room temperature. Aspirate ethanol, and repeat for a total of two washes. Let beads dry for 5 min. Under- or over-drying beads will reduce yield.
12. Remove tubes from the magnet, add 43 µL of nuclease-free water, and mix.

Table 15-5 Adenylation Master Mix Components

Reagent	Per reaction (μL)	For 8 reactions (μL)
10× dA-tailing buffer	5	44
Klenow fragment	3	26.4
Total	8	70.4

Table 15-6 Ligation Master Mix Components

Reagent	Per reaction (μL)	For 8 reactions (μL)
10× T4 ultrapure buffer	5	44
Ultrapure ligase	5	44
Total	10	88

Do not elute DNA off of the beads. The beads and DNA will be carried forward into subsequent reactions.

13. Samples can be stored at −20 °C if not proceeding to the next step.

Step 3: Adenylate End-Repaired DNA

1. Prepare a master mix from the components listed in Table 15.5 in a clean 1.5 mL LoBind tube on ice or cooling rack.
2. Dispense 8 μL of adenylation master mix into each of the 0.2 mL PCR tubes.
3. Add 42 μL of end-repaired DNA (including beads) to each master mix-containing PCR tube. Mix carefully and thoroughly by pipetting up and down ten times.
4. Briefly spin to bring contents to the bottoms of the tubes.
5. Incubate in a thermal cycler for 30 min at 37 °C.
6. Vortex the 20 % PEG/2.5 M NaCl solution. The solution must be at room temperature before use.
7. Transfer 90 μL of PEG solution to each tube containing the A-tailed DNA, and mix 20+ times. Incubate for 5 min at room temperature.
8. Place the tube on a magnet for 2 min to separate the beads from the solution.
9. Visually confirm that the beads have moved to the side of the tube and the solution is clear. Aspirate 140 μL of clear solution from each reaction tube, and discard.
10. Dispense 200 μL of freshly prepared 70 % ethanol to each tube and incubate for 30 s at room temperature. Aspirate the ethanol, and discard. Repeat for a total of two washes. Let the beads dry for 5 min. Under- or over-drying beads will reduce yield.
11. Remove the tubes from the magnet, add 38 μL of nuclease-free water, and mix.
12. Place the tube back on the magnet. Wait for the solution to clear, and remove 1 μL of solution for quantification. Remove the tube from the magnet, and mix briefly. Do not elute DNA off of the beads.
13. Samples can be stored at −20 °C if not proceeding to the next step.

Step 4: Quantification of Adenylated DNA

1. Using the 1 μL of solution retrieved from step 12 in section "Step 3: Adenylate End-Repaired DNA", quantify the DNA using the Qubit dsDNA HS Quantification kit.
2. Note any samples that are less than 10 ng/μL. Samples at lower concentrations may exhibit reduced library complexity, potentially impacting variant detection.

Step 5: Ligate Indexed Adapters to Adenylated DNA

Prior to performing the ligation, determine how the samples will be sequenced. If the samples will be pooled, each will require a specific indexed adaptor so that reads can be demultiplexed after sequencing.

1. Remove T4 Ultrapure Ligase, 10× T4 Ultrapure Buffer, and indexed adapters (50 μM) from the freezer. Prepare ligation master mix by combining the reagents listed in Table 15.6.
2. Remove Agencourt AMPure XP Beads and 20 % PEG solution from the

refrigerator and equilibrate to room temperature for at least 30 min. The beads and PEG solution must be at room temperature before use. Vortex very well to resuspend all beads.

3. Vortex and centrifuge the adenylated DNA from *Step 3* and place on ice or cold block.
4. Prepare ligation master mixes by combining the following for each sample:
   (a) Dilute the indexed 50 μM adapters 1:3 with ultrapure distilled water into clean 1.5 mL microfuge tubes to make a 16.7 μM solution.
   (b) Add 5 μL of diluted adaptor to each sample.
5. Add 10 μL of ligation master mix to each sample tube, and mix thoroughly by gently pipetting up and down ten times. Avoid bubble formation.
6. Briefly spin to bring contents to the bottoms of the tubes.
7. Incubate at room temperature (25 °C) for 15 min.
8. Briefly spin to bring contents to the bottoms of the tubes, and proceed with bead cleanup.
9. Vortex the 20 % PEG/2.5 M NaCl solution.
10. Transfer 90 μL of PEG solution to the tube containing the ligated DNA.
11. Mix 20 times, and incubate for 5 min at room temperature.
12. Place the tube on a strip/plate magnet for 2 min to separate the beads from solution.
13. Visually confirm that the beads have moved to the side of the tube and the solution is clear.
14. Aspirate 140 μL of clear solution from the reaction tube, and discard.
15. Dispense 180 μL of freshly prepared 70 % ethanol to each well of the tube and incubate for 30 s at room temperature. Aspirate the ethanol, and discard. Repeat for a total of two washes. Let the beads dry for 5 min. Under- or over-drying beads will reduce yield.
16. Remove the tube from the magnet, add 50 μL of nuclease-free water, and mix.
17. Place tube on magnet for 1 min to separate beads.
18. Aspirate and transfer 50 μL eluent to a new labeled tube.

Table 15-7 **Pre-capture Library Amplification Components**

Reagent	Per reaction (μL)	For 8 reactions (μL)
TS-PCR-1 (20 μM)	2.5	44
TS-PCR-2 (20 μM)	2.5	44
KAPA Library Amplification Mix	50	880
Total	55	968

Step 6: Amplify Pre-capture Library

1. Thaw PCR primers (TS-PCR-1 and TS-PCR-2) at room temperature, and keep KAPA Library Amplification Mix on ice.
2. Label two 0.2 mL PCR strip tubes per sample to prepare for two 50 μL PCRs each.
3. Spin down TS-PCR-1 and TS-PCR-2 (20 μM) primer tubes and the indexed libraries.
4. Divide each library prepared in the previous step into the two labeled PCR tubes (22.5 μL/tube).
5. Prepare 50 μL of master mix for each library/sample by combining the reagents listed in Table 15.7.
6. Add 27.5 μL of amplification master mix to each library-containing tube.
7. Seal with PCR tube caps, and centrifuge the PCR tubes briefly and place in a thermal cycler.
8. Amplify libraries with the thermal cycling profile in Table 15.8.
9. Store at 4 °C or continue with Step 7.

Step 7: Purify Amplified Pre-capture Libraries

1. Remove AMPure XP beads from 4 °C and keep at room temperature for at least 30 min. Vortex generously to resuspend all beads.
2. Pool the 2 × 50 μL PCR reactions (per library) into a 1.5 mL tube.
3. Shake the Agencourt AMPure XP beads to resuspend.

Table 15-8 **Pre-capture Library Amplification Thermal Cycling Protocol**

Step	Temperature/duration	Cycles
Activation/denaturation	98 °C, 30 s	1
Amplification	98 °C, 10 s 60 °C, 30 s 72 °C, 30 s	8
Final extension	72 °C, 5 min	1
Hold	16 °C	

4. Dispense 180 μL of beads into each pooled PCR reaction.
5. Mix well, and incubate for 5 min at room temperature.
6. Place the reaction tube onto a Dyna-Mag 2 rack for 5 min to separate beads from the solution.
7. Visually confirm that the beads have moved to the side of the tube and the solution is clear.
8. Aspirate 280 μL of clear solution from the tubes, and discard.
9. Dispense 200 μL of 70 % ethanol to each tube and incubate for 30 s at room temperature. Aspirate the ethanol, and discard. Repeat for a total of two washes. Let the beads dry for 5 min. Under- and over-drying beads will reduce yield.
10. Take the tubes off the magnet, add 50 μL water to each reaction tube, and mix. Place the reaction plate onto a magnet for 1 min to separate the beads.
11. Transfer 50 μL eluent to a new labeled tube.

Step 8: Pre-capture Library Assessment

1. Assess the fragment size distribution and concentration by running a sample of each library on Bioanalyzer 2100 (DNA 1000 Chip).
2. When assessing the fragment size distribution, keep in mind that the average fragment size should be the size of the sheared DNA from Step 1 plus 100 bp to account for the ligation of the adapters. If your target insert size is 200 bp, you should expect to see an average post-capture fragment size of 300 bp (±25 bp). If the fragment size distribution is significantly above or below the expected size, i.e., 100 bp smaller or larger, you should consider repeating the procedure.

Step 9: Hybridization with Sequence Capture Probes

Blockers are added to inhibit the adapters from cross hybridizing and "daisy-chaining" during the hybridization. Full-length index-specific blockers are more efficient at inhibiting cross-hybridization than blockers targeted to the common portions of the adapters. The Agilent-provided blockers are replaced with the blockers described in section "Reagent and Oligonucleotide Preparation".

1. Pre-capture indexed library must be at a concentration of at least 147–221 ng/μL. Use a vacuum concentrator to concentrate the samples, if needed. Do not heat above 45 °C.
2. Combine the reagents listed in Table 15.9 to generate the hybridization buffer.
3. Warm the hybridization buffer to 65 °C if a precipitate forms.
4. Label three clean 8 PCR tube strips: *A*, *B*, and C.
5. Dispense 40 μL of hybridization buffer into each tube of strip *A*.
6. Prepare a SureSelect RNase Block dilution according to Table 15.10. Note: The amount of RNAse block solution needed will depend on the size of the capture target.
7. Combine the appropriate amount of diluted RNase block solution and SureSelect capture library according to Table 15.11 in strip C. Mix by pipetting.

Table 15-9 Hybridization Buffer Components

Reagent	Per reaction (μL)	For 8 reactions (μL)
SureSelect Hyb #1	25	200
SureSelect Hyb #2 (red cap)	1	8
SureSelect Hyb #3 (yellow cap)	10	80
SureSelect Hyb #4	13	104
Total	49 (40 μL/ rxn)	392 (40 μL/ rxn)

Rxn reaction

Table 15-10 RNase Block Dilution

Capture size	Per reaction (μL)		For 8 reactions (μL)	
	RNase	Water	RNase	Water
<3.0 Mb	1	9	8	72
≥3.0 Mb	1	3	8	24

Table 15-11 Combining RNase Block Dilution With the SureSelect Library

Capture size	Per reaction (μL)		For 8 reactions (μL)	
	RNase	Library	RNase	Library
<3.0 Mb	5	2	40	16
≥3.0 Mb	2	5	16	40

Each tube should contain 7 μL of RNase block/capture library mixture.

8. Prepare SureSelect Block Mix by combining the components listed in Table 15.12. The SureSelect Index Block #3 is replaced with the ISHB assembled in section "Reagent and Oligonucleotide Preparation". An individual SureSelect Block Mix will be prepared for each index used.
9. In strip *B*, prepare the pre-capture indexed libraries for target enrichment:
   (a) Add 3.4 μL of 147–221 ng/μL of indexed library to each tube.
   (b) Add 5.6 μL of the corresponding SureSelect Block Mix (containing ISHB).

Table 15-12 SureSelect Adapter Block Mix

Reagent	Per reaction (μL)
SureSelect indexing block #1	2.5
SureSelect block #2	2.5
Index-specific block	0.6
Total	5.6

   (c) Mix thoroughly by pipetting up and down.
   (d) Cap the tubes and place in the thermal cycler.
   (e) Heat the library/adapter block mix to 95 °C for 5 min; then hold at 65 °C. Use a heated lid set at 105 °C.
10. Equilibrate hybridization buffer-containing strip *A* to 65 °C for 5 min before proceeding to the next step.
11. Place the strip into the thermal cycler and incubate at 65 °C for 2 min.
12. Maintain all three strips at 65 °C while transferring 13 μL of the hybridization buffer from each tube of strip *A* to the SureSelect Capture Library tubes in strip C.
13. Transfer 9 μL of indexed library/adapter block mix from each tube of strip *B* to the corresponding tube of strip C.
14. Seal strip C, now containing 29 μL of solution, tightly with a new strip cap.
15. Incubate the hybridization strip at 65 °C for 24 h.

Step 10: Bead Capture and Post-hybridization Washes

Prior to beginning the post-hybridization washes, prepare a clean, RNase-free workspace. Examine the volume of each hybridization reaction. If greater than 5 μL has evaporated, repeat the hybridization. Prewarm SureSelect wash buffer 2 to 65 °C.

1. Bead preparation:
   (a) Resuspend the DynaBeads by vortexing.
   (b) To each of 8 clean 1.5 mL LoBind tubes, add 50 μL DynaBeads.
   (c) To each tube of beads:

- Add 200 µL of SureSelect binding buffer, and vortex for 5 s.
- Place tubes on a magnetic stand until the solution clears (1–2 min).
- Aspirate and discard the solution. Do not disturb beads.
- Repeat for a total of three washes.

(d) Resuspend beads in 200 µL of SureSelect binding buffer per hybridization.

2. Capture hybrid library:
   (a) Add the contents of each hybridization reaction to a tube of prepared beads.
   (b) Mix beads and hybrid library by pipetting up and down slowly ten times.
   (c) Place each tube on a tube rotator and mix for 30 min at room temperature. Make sure that the solution is being adequately mixed.
   (d) Centrifuge briefly and place tubes on magnetic stand. After the solution has cleared (1–2 min), aspirate the solution and discard.
   (e) Resuspend the beads in 500 µL of SureSelect buffer #1 by briefly vortexing for 5 s.
   (f) Centrifuge briefly and place tubes on magnetic separator. After the solution has cleared (1–2 min), aspirate the solution and discard.
3. Stringency washes:
   (a) Resuspend beads in 500 µL of prewarmed SureSelect wash buffer #2.
   (b) Vortex briefly for 5 s.
   (c) Incubate tubes at 65 °C for 10 min, mixing periodically.
   (d) Centrifuge briefly and place the tubes on the magnetic separator. After the solution has cleared (1–2 min), aspirate the solution and discard.
   (e) Repeat (a)–(d) for a total of three washes.
   (f) Make sure that all wash buffer has been removed.
4. Add 50 µL of SureSelect elution buffer, and vortex for 5 s to resuspend the beads.
5. Incubate tubes for 10 min at room temperature. Mix periodically.
6. Centrifuge briefly and place the tubes on the magnetic separator. After the solution has cleared (1–2 min), aspirate the solution and transfer it to clean 1.5 mL LoBind tubes.
7. Add 50 µL SureSelect neutralization buffer to the captured DNA. Mix briefly.

Step 11: Post-hybridization Library Cleanup

1. Remove AMPure XP Beads from refrigerator, and gently shake to resuspend the beads. The beads must be at room temperature before use.
2. Resuspend and transfer 180 µL of the beads to each 100 µL tube of captured DNA.
3. Mix, and incubate for 5 min at room temperature.
4. Place the tubes on a magnetic stand for 5 min to separate the beads from the solution.
5. Visually confirm that the beads have moved to the side of the tube and that the solution is clear.
6. Aspirate ~280 µL of clear solution from the tube without disturbing the beads, and discard.
7. Dispense 500 µL of freshly prepared 70 % ethanol to each tube and incubate for 30 s at room temperature. Aspirate ethanol, and repeat for a total of two washes. Let beads dry for 5 min. Under- and over-drying beads will reduce yield.
8. Add 15 µL of molecular biology-grade water, and incubate at room temperature for 2 min.
9. Place the tube on a magnetic stand. After the solution clears (2–3 min), aspirate 15 µL of water and transfer to a clean 1.5 mL LoBind tube.
10. Repeat Steps 8 and 9 above, adding the second eluent to the first for a total of 40 µL of captured DNA.

Step 12: Amplification of Post-capture Library

The goal of this step is to generate enough material for sequencing. Only the minimum number of cycles to generate sufficient material should be performed. Minimize the number of cycles to ensure maintenance of library complexity.

1. Prepare post-capture amplification mix by combining the reagents listed in Table 15.13.

Table 15–13 Post-capture Amplification Mix

Reagent	Per reaction (×4) (μL)	For 8 reactions (×4) (μL)
PCR-grade water	50	420
TS-PCR-1 (20 μM)	5.0	42
TS-PCR-1 (20 μM)	5.0	42
KAPA Library Amplification Mix	100	840
Total	160	1,344 (160 μL/rxn)

Rxn reaction

Table 15-14 Post-capture Library Amplification Thermal Cycling Potocol

Step	Temperature/duration	Cycles
Activation/denaturation	98 °C, 30 s	1
Amplification	98 °C, 10 s 60 °C, 30 s 72 °C, 30 s	6
Final extension	72 °C, 5 min	1
Hold	16 °C	

2. Prepare four PCR reaction tubes for each library. Dispense 40 μL of amplification master mix into each tube; add 10 μL of library to each tube for a total of 50 μL per reaction.
3. Place sealed reactions in a thermal cycler. Amplify the captured DNA using the thermal cycling protocol in Table 15.14.
4. When amplifications are complete, proceed with Step 13 or store at 4 °C for up to 72 h.

Step 13: Purify Amplified Post-capture Libraries

1. Pool the 4 × 50 μL PCR reactions into a single clean 1.5 mL microcentrifuge tube.
2. Gently shake the AMPure XP bottle to resuspend any beads that may have settled. The beads must be at room temperature before use.
3. Transfer 360 μL of beads to each tube containing pooled reactions.
4. Mix well, and incubate for 5 min at room temperature.
5. Place the tubes onto a Dyna-Mag 2 rack for 5 min to separate the beads from solution.
6. Visually confirm that the beads have moved to the side of the tube and the solution is clear.
7. Aspirate 560 μL of clear solution from each tube, and discard.
8. Dispense 300 μL of 70 % ethanol to each tube and incubate for 30 s at room temperature. Aspirate the ethanol, and discard. Repeat for a total of two washes. Dry the beads for 5 min. Under- or over-drying of the beads will reduce yield.
9. Take the tubes off the magnet, add 20 μL of water to each tube, and mix.
10. Place the tubes on the Dyna-Mag 2 rack for 2 min to separate the beads.
11. Transfer 20 μL eluent to a new tube.
12. Add 2 μL of 1 % Tween (final 0.1 %). The samples are now ready for Q/C.

Step 14: Quality Control Analysis of Captured Libraries

1. Assess the fragment size distribution and concentration by running a sample of each library on Bioanalyzer 2100 (DNA 1000 Chip) as performed in Step 8.
2. Record the average fragment size for each library. The proper size will be important for accurately calculating the library concentrations in Step 15.

Step 15: Library Quantification by Real-Time Quantitative PCR

1. Quantify each library using the KAPA Biosystems Library Quantification Kit according to the manufacturer's specification. Run 1:50,000 and 1:100,000 dilutions for each library in triplicate.
2. Make sure to take into account the difference in size between the library fragment

size and the standards included in the kit. Use the average fragment size determined in Step 14. This will have an impact on the final molar concentration.

3. Samples should be adjusted to a final concentration of 10 nM and pooled (if desired).
4. *Equimolar library pooling*. Precise pooling of libraries can be challenging. The reproducibility of the qPCR-based assay described in Step 15 is very high but is very dependent on the pipetting skills of the operator. Make sure to run qPCR reactions in triplicate. One should expect to see variances between libraries (as measured by sequence read output) of less than 0.5×.

Step 16: Illumina Sequencing

1. Sequence libraries according to the manufacturer's instructions.
2. *Minimal read depth for accurate variant calling*. As a general rule of thumb, a minimum of 20–30× coverage over each target base will ensure that both alleles are detected, if the individual is heterozygous at a given nucleotide. Clinical laboratories will typically set the coverage threshold at 100× or greater for targeted gene panels, as it is not difficult (or expensive) to obtain coverages at or above this level.
3. *Sequence run evaluation*. Accuracy of NGS has improved greatly over the past several years. For 2 × 100 bp sequencing run on a HiSeq instrument, greater than 85 % of bases should yield Q30 or greater. For clinical work, it may be advisable to sacrifice read length in favor of accuracy. Some laboratories still perform 2 × 50 bp or 2 × 75 bp—as the first portion of the read is the most accurate on a per nucleotide basis.
4. *Library complexity*. If your library does not possess adequate complexity, your results may not reveal the full complement of expected variants, as the template has undergone a bottleneck during the procedure. As you become more familiar with the process, you can add various Q/C cutoffs or metrics to reduce the likelihood of low-complexity data. Low complexity will also lower the sensitivity of the assay.

Data Analysis

A wide variety of software tools are available for analysis of next-generation sequence data. Many laboratories use the software pipeline provided by Illumina or other commonly used tools, such as the Burrows–Wheeler Aligner [12] and Genome Analysis Tool Kit (GATK) [13, 14]. Each laboratory will need to evaluate analytical tools and determine how to set software parameters needed for their specific application(s). Most software tools are optimized for the detection of germline variants. If users need to identify somatic mutations found in tumors, they will need to optimize the specific parameters of each application to ensure robust detection. Specific software applications have become available, such as Mutect [15], SomaticSniper [16], or SNVmix [17, 18], just for this purpose. One would need to test their analysis pipeline empirically with a rigorous validation protocol, before applying their use to clinical diagnostic work.

In some cases, it may be useful to verify alignments and variant detection visually. The Integrative Genomics Viewer [19] can be used to examine a wide array of sequence data, including read alignments, variants, and coverage data.

Conclusions

NGS gene panels provide comprehensive, rapid, and cost-effective technology for clinical genetic testing. Single-gene testing only identifies the causative variant in 10–20 % of clinically diagnosed complex genetic diseases. By combining all of the known genes for a given phenotype, testing can be performed in a more effective manner, thereby reducing the time required to make a molecular diagnosis. Currently, the optimal platform for development and implementation of NGS gene panels is an in-solution, hybridization-based enrichment system as described in the example provided in this chapter. These assays are automatable, reproducible, and highly sensitive for the detection of single-nucleotide variants, small insertions/deletions, and copy number variants. NGS gene panels will likely continue to be the preferred testing modality for many applications in the foreseeable future, even as the costs of whole-exome and whole-genome sequencing fall.

REFERENCES

1. Metzker ML. Sequencing technologies—the next generation. Nat Rev Genet. 2010;11:31–46.
2. Rehm HL. Disease-targeted sequencing: a cornerstone in the clinic. Nat Rev Genet. 2013;14:295–300. doi:10.1038/nrg3463.
3. Teekakirikul P, Kelly MA, Rehm HL, Lakdawala NK, Funke BH. Inherited cardiomyopathies: molecular genetics and clinical genetic testing in the postgenomic era. J Mol Diagn. 2013;15(2):158–70.
4. Pritchard CC, Smith C, Salipante SJ, Lee MK, Thornton AM, Nord AS, Gulden C, Kupfer SS, Swisher EM, Bennett RL, Novetsky AP, Jarvik GP, Olopade OI, Goodfellow PJ, King MC, Tait JF, Walsh T. ColoSeq provides comprehensive lynch and polyposis syndrome mutational analysis using massively parallel sequencing. J Mol Diagn. 2012;14(4):357–66.
5. Gecz J, Shoubridge C, Corbett M. The genetic landscape of intellectual disability arising from chromosome X. Trends Genet. 2009;25(7):308–16.
6. ACMG Board of Directors. Points to consider in the clinical application of genomic sequencing. Genet Med. 2012;14:759–61.
7. Mamanova L, Coffey AJ, Scott CE, Kozarewa I, Turner EH, Kumar A, Howard E, Shendure J, Turner DJ. Target-enrichment strategies for next-generation sequencing. Nat Methods. 2010;7(2):111–8.
8. Gnirke A, Melnikov A, Maguire J, Rogov P, LeProust EM, Brockman W, Fennell T, Giannoukos G, Fisher S, Russ C, Gabriel S, Jaffe DB, Lander ES, Nusbaum C. Solution hybrid selection with ultra-long oligonucleotides for massively parallel targeted sequencing. Nat Biotechnol. 2009;27(2):182–9.
9. Fisher S, Barry A, Abreu J, Minie B, Nolan J, Delorey TM, Young G, Fennell TJ, Allen A, Ambrogio L, Berlin AM, Blumenstiel B, Cibulskis K, Friedrich D, Johnson R, Juhn F, Reilly B, Shammas R, Stalker J, Sykes SM, Thompson J, Walsh J, Zimmer A, Zwirko Z, Gabriel S, Nicol R, Nusbaum C. A scalable, fully automated process for construction of sequence-ready human exome targeted capture libraries. Genome Biol. 2011;12(1):R1.
10. Shearer AE, Hildebrand MS, Smith RJ. Solution-based targeted genomic enrichment for precious DNA samples. BMC Biotechnol. 2012;12:20.
11. Nord AS, Lee M, King MC, Walsh T. Accurate and exact CNV identification from targeted high-throughput sequence data. BMC Genomics. 2011; 12:184.
12. Li H, Durbin R. Fast and accurate short read alignment with Burrows-Wheeler transform. Bioinformatics. 2009; 25:1754–60.
13. McKenna A, Hanna M, Banks E, Sivachenko A, Cibulskis K, Kernytsky A, Garimella K, Altshuler D, Gabriel S, Daly M, DePristo MA. The Genome Analysis Toolkit: a MapReduce framework for analyzing next-generation DNA sequencing data. Genome Res. 2010;20:1297–303.
14. DePristo M, Banks E, Poplin R, Garimella K, Maguire J, Hartl C, Philippakis A, del Angel G, Rivas MA, Hanna M, McKenna A, Fennell T, Kernytsky A, Sivachenko A, Cibulskis K, Gabriel S, Altshuler D, Daly M. A framework for variation discovery and genotyping using next-generation DNA sequencing data. Nat Genet. 2011;43:491–8.
15. Cibulskis K, Lawrence MS, Carter SL, Sivachenko A, Jaffe D, Sougnez C, Gabriel S, Meyerson M, Lander ES, Getz G. Sensitive detection of somatic point mutations in impure and heterogeneous cancer samples. Nat Biotechnol. 2013;31(3):213–9.
16. Larson DE, Harris CC, Chen K, Koboldt DC, Abbott TE, Dooling DJ, Ley TJ, Mardis ER, Wilson RK, Ding L. SomaticSniper: identification of somatic point mutations in whole genome sequencing data. Bioinformatics. 2012;28(3):311–7.
17. Goya R, Sun MG, Morin RD, Leung G, Ha G, Wiegand KC, Senz J, Crisan A, Marra MA, Hirst M, Huntsman D, Murphy KP, Aparicio S, Shah SP. SNVMix: predicting single nucleotide variants from next-generation sequencing of tumors. Bioinformatics. 2010;26(6):730–6.
18. Roth A, Ding J, Morin R, Crisan A, Ha G, Giuliany R, Bashashati A, Hirst M, Turashvili G, Oloumi A, Marra MA, Aparicio S, Shah SP. JointSNVMix: a probabilistic model for accurate detection of somatic mutations in normal/tumour paired next-generation sequencing data. Bioinformatics. 2012;28(7):907–13.
19. Thorvaldsdóttir H, Robinson JT, Mesirov JP. Integrative Genomics Viewer (IGV): high-performance genomics data visualization and exploration. Brief Bioinform. 2012;14:178–92. doi:10.1093/bib/bbs017.

CHAPTER 16

Implementation of Exome Sequencing Assay

ARUNKANTH ANKALA, MADHURI R. HEGDE

Introduction 225
Whole-Exome Sequencing: Methodology 226
Exome Sequencing: A Transformative Technology 228
Implementation of Exome Sequencing in Clinical Medicine 230
Limitations and Challenges of Implementing Exome Sequencing Assays 233
A Look to the Future of Whole-Exome Assays 234
Conclusions 234

Introduction

The paradigm shift from genetics to genomics was put in motion by a revolutionary study that described sequencing of the entire genome of *Mycoplasma genitalium* in a single run on a Roche 454 instrument [58]. The study revealed a highly parallel sequencing-by-synthesis method capable of sequencing 25 million bases, at 99 % or more accuracy, in a single 4 h run. Subsequently, several high-throughput flow cell-based sequencing methods became commercially available from Illumina (San Diego, California), Roche (454 Life Sciences Corporation, Branford, Connecticut), and Life Technologies (Carlsbad, California). These developments marked the beginning of a new era based on next-generation sequencing (NGS). Simultaneously, several sequence capture or target enrichment methods were evolving to improve the throughput and specificity of sequencing technology [1, 6, 30, 37, 55, 74]. With the rapid development of these advanced sequencing technologies, per-base sequencing costs are declining drastically, to a level at which almost complete resequencing of the human genome is becoming affordable, even in clinical settings [3, 57, 64, 87]. Nevertheless, the infrastructure requirements, analysis burden, and turnaround time requirements involved in clinically interpreting entire patient genomes for mutation detection bear significant issues. Whole-exome sequencing (WES), in contrast, which interrogates the roughly 1 % of the human genome that represents the entire coding region and harbors 85 % or more of causative mutations, is quite feasible and much more affordable in a clinical setting.

The successful implementation of NGS technology in clinical laboratories for diagnostic purposes began with gene panels designed to specifically target and sequence multiple genes related to a particular disorder. Soon several disease specific or phenotype specific gene panels became clinically available [31, 40, 48, 51, 70, 98, 99, 101]. These included highly heterogeneous disorders, such as congenital disorders of glycosylation (CDG), congenital muscular dystrophies (CMD), limb girdle muscular dystrophies, dilated cardiomyopathy, and mitochondrial

G.J. Netto and I. Schrijver (eds.), *Genomic Applications in Pathology*,
DOI 10.1007/978-1-4939-0727-4_16,

disorders, each with several subtypes of overlapping phenotypes and associated with a large number of causative genes [2, 98]. Traditional molecular diagnostic approaches for such diseases followed a sequential, Sanger sequencing-based gene-by-gene analysis of known disease-associated genes. However, with the advent of NGS technologies and the decline in per-base sequencing cost, the NGS panel approach has become a significantly cheaper and quicker option, available as a single test. Subsequently, with the availability of better sequence chemistries and easier workflows, NGS technology moved into other clinical arenas, including cancer diagnosis [66], human leukocyte antigen locus characterization [39, 81], and pathogen genome sequencing for the purpose of evaluating resistance [85]. Rapid identification of novel disease genes and the revealing locus and allelic heterogeneity of inherited genetic disorders, both Mendelian and complex, has established WES as a comprehensive clinical test.

In this chapter, we discuss the various roles of WES in clinical medicine and provide an overview of how WES has transformed the diagnostic outlook on genetic disorders. We highlight the major successes and challenges of implementing WES assays in clinical genetics, concluding with a note on the future of whole-exome assays.

Whole-Exome Sequencing: Methodology

Exome Capture and Next-Generation Sequencing

WES refers to sequencing of the entire protein-coding region of the human genome. This is achieved by parallel sequencing of all targeted regions (exons) using NGS technologies. Irrespective of the manufacturer and sequencing platform, the basic methodology or principles involved in WES are similar (Fig. 16.1). First, genomic DNA is fragmented either by optimized sonication or by restriction digestion to generate uniform libraries of DNA strands. This fragmented DNA is then enriched for protein-coding regions of the genome (exons), using unique adapter ligation chemistry that is proprietary to each individual commercial manufacturer [18]. Adapter-ligated DNA fragments are captured and amplified either on a solid surface (bridge amplification on a glass slide) or in solution (emulsion PCR on micro-beads). Finally, different massively parallel sequencing technologies are used to sequence all target DNA regions and produce what are called sequence reads, of different lengths, depending on the technology used. Sequence reads are computationally aligned to a reference exome and analyzed for sequence variations. The experimental design allows for each nucleotide to be represented in a large number of reads, which is referred to as "read depth" or "coverage." Variant annotation using analytical pipelines helps filter false positives and non-contributive calls to identify causal mutations. WES therefore serves as a comprehensive method for rapid identification of exonic mutations, such as missense, nonsense, splice site, and small deletion and insertion mutations (indels); however, detection of copy number variations (CNVs) and structural variations (SVs) is still an issue.

Sequence Analysis and Variant Detection

Massively parallel sequencing of the entire exome generates terabytes of information. Sorting through and making sense of such massive volumes of data to identify causative genes and mutations requires multistep bioinformatics analysis. Upon initial generation of sequence base call files, they are converted into the more commonly used FASTQ file format for storage and later analysis [18]. Several open-source and in-house-developed software programs can be used to align sequence reads to a best-match location of a reference sequence and stored in what is called the BAM (binary alignment) file format [49]. These aligned reads are then processed to call out sequence variants depending on the presence and zygosity of variants. Information from this analysis, which includes inferred single nucleotide polymorphisms (SNPs) and insertions and deletions (indels) along with base coverage, quality, and score, is stored in a different file format termed variant call format (VCF) [22, 50]. Finally, each single call in the entire set of variants is annotated with a

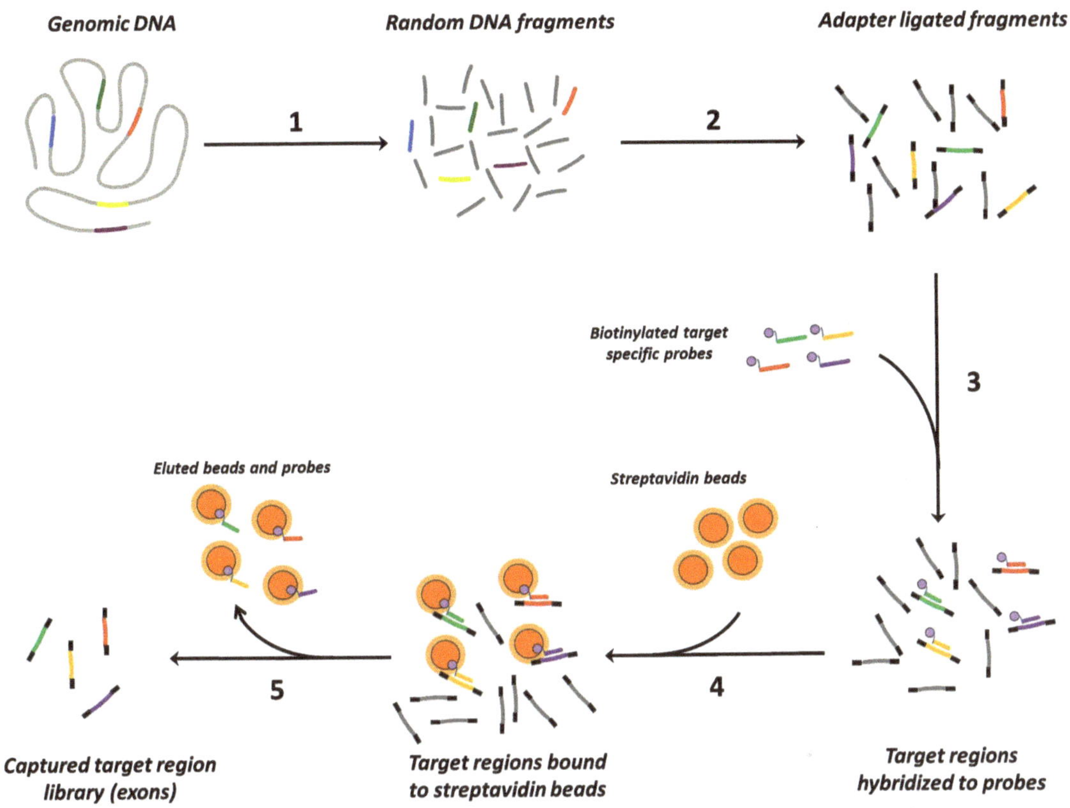

Figure 16-1 Basic methodology of exome capture or target enrichment for whole-exome sequencing. The various steps involved are indicated by numbers. In step 1, genomic DNA is randomly fragmented into more or less uniform shorter segments, either by ultrasonication or restriction digestion with enzymes. In step 2, adapters with sequencing motifs and indices are ligated to the fragments. In step 3, biotinylated probes that are specific for target regions (exons) are added and allowed to hybridize. Step 4 involves addition of streptavidin beads to selectively capture all target regions by binding biotin. While the streptavidin beads (with bound target regions) are held by a magnet, unbound nonspecific DNA fragments are separated and washed away. Finally, in step 5, target regions are eluted by denaturation from the biotinylated probes. Although alternative methods for adapter ligation may be available, the basic concept for target (exon) capture is similar

variety of customizable information, including gene name, genomic and cDNA coordinates, amino acid change, and functional classification, to help with the interpretation of causative variants [104].

Variant Analysis and Molecular Diagnosis

Analysis of the variants and identification of the disease causative gene and mutations in WES are daunting tasks compared to the traditional single-gene sequencing approach. Several predictive algorithms are being developed and made commercially available, but their reliability and interpretative ability is not well established. Most of the clinical laboratories that offer WES assays currently include various parameters, such as the functional effect of the observed variant, relevance of the gene to the clinical presentation, and mode of inheritance, to filter variant calls through in-house-validated pipelines and algorithms (Fig. 16.2). Finally, short-listed candidate variants are confirmed by the gold standard Sanger sequencing. Confirmed variants may fall into different categories based on previous association and functional effects of the variant (Table 16.1). In the event a new disease gene is identified, disease association requires further evidence. In silico analysis by prediction algorithms based on evolutionary conservation of the amino acid or nucleotide

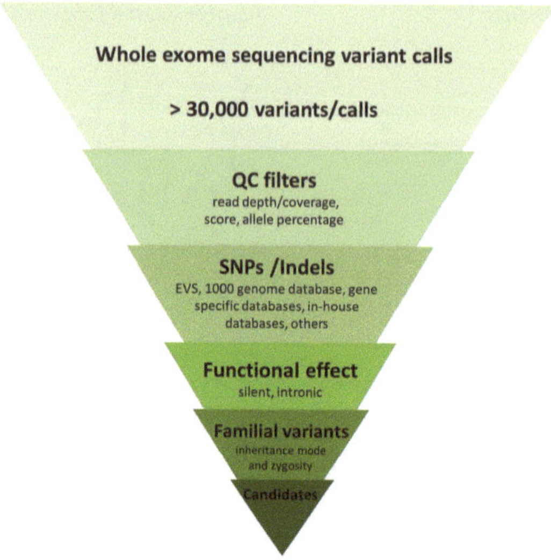

Figure 16-2 Basic pipeline for variant filtration in whole-exome sequencing analysis. Various parameters are included in WES algorithms to filter and remove nonpathogenic and false-positive variants from whole-exome variant data to create a manageable dataset (150–250 variants) that includes the candidate causative mutations. As indicated in the data filtration funnel, variants that do not meet QC metrics, such as those with poor coverage (<20×), are considered less likely to be real, treated as false positives, and therefore filtered. Variants with a minor allele frequency of >0.01 are polymorphisms by definition and less likely to be pathogenic. Silent changes and intronic variants beyond the consensus splice donor/acceptor sequences are less likely to be pathogenic and are often filtered in initial rounds of analysis. Familial variants may also be carefully filtered based on zygosity and segregation pattern. Though the basic parameters followed are common to all commercial and laboratory-developed algorithms, the thresholds and ranges for acceptability may vary. EVS = Exome Variant Server (NHLBI Exome Sequencing Project)

may increase confidence in an association, but is not definitive [18]. Segregation of mutations in the gene with presence of disease among family members may also provide additional evidence, but does not necessarily or fully associate the gene with disease. Functional studies are best, when available, because they may not only establish disease association but also provide insight into disease pathogenesis and treatment options. Alternatively, identification of mutations in the novel gene in unrelated individuals with similar phenotypes by rapid targeted single-gene testing may establish disease association, as well. While diagnostic laboratories focus on finding a pathogenic change in a known disease-associated gene, research testing of exomes is driven by the additional goal of new gene discovery and may include extensive functional analysis to establish disease association with the gene.

Exome Sequencing: A Transformative Technology

NGS approaches, and especially WES, have created hope for patients who may have already undergone a diagnostic odyssey of invasive approaches and clinical tests, and yet remain in the dark as to the underlying genetic cause of their condition. The potential of WES to provide molecular diagnoses by screening nearly all human exons for mutations was recognized early on, and attempts to explore its diagnostic potential were soon underway, heralding a new era in clinical and medical genetics.

WES as a Diagnostic Assay-Proven Potential

WES has facilitated characterization of several recessive as well as dominant diseases, revealing associations with new disease genes. Recessive traits, which are more commonly highlighted in consanguineous families, are comparatively easier to diagnose and implicate through WES because affected individuals within the family carry causative mutations in segments that are homozygous by descent. For example, in the case of first cousin mating, these regions account for approximately 10 % of the entire exome, thereby restricting the search to this small region. For dominant traits, however, the process is less straightforward. Molecular characterization of dominant traits is complicated by several factors, including reduced penetrance for certain genes, locus heterogeneity, and alleles that affect reproductive fitness. In such scenarios, the finding of independent *de novo* variants in the same gene among multiple unrelated affected individuals provides considerable evidence for disease association irrespective

Table 16-1 **Predictive Value and Significance of Confirmed Whole-Exome Sequencing Variants**

Variant category	Clinical diagnostic value	Functional value	Further action
Previously reported mutation	Establish diagnosis and disease subtype	Understand clinical spectrum	–
Novel mutation in known disease-associated gene	Establish familial mutation	Expand allelic and phenotypic heterogeneity	–
Mutation in known disease-unrelated gene	Establish disease diagnosis	Expand locus heterogeneity of disease	Functional studies, characterize disease subtype
Potential pathogenic variant/mutation in previously unknown gene	Not predictable or actionable	Hypothesize new disease/gene	Characterize disease type, functional studies

of allelic heterogeneity. The first successful demonstration of the potency of WES for rare variant identification and disease diagnosis came from an unexpected diagnosis of a patient referred for possible Bartter syndrome [16]. Due to an inconclusive clinical presentation, WES was performed for this individual, and informed variant analysis led to the identification of a homozygous mutation in the *SLC26A3* gene. This study provided the first proof of concept of the application of WES for genetic disease diagnosis. Even though the gene was previously known to be disease causing (congenital chloride-losing diarrhea, CLD), the clinical overlap of the patient's phenotype with that of Bartter syndrome [36] obviated suspicion of the gene. Substantial family information, including that of consanguinity, inheritance mode of the disease, and regions of excessive homozygosity due to identity by descent, helped with the molecular characterization. Moreover, reevaluation of additional study subjects with a presumptive diagnosis of Bartter syndrome identified mutations in *SLC26A3*. These findings not only established the diagnostic ability of WES but also expanded the phenotypic variability of *SLC26A3*-associated CLD.

Whole-Exome Sequencing Facilitates Gene Discovery

Traditional gene mapping tools, such as homozygosity mapping, linkage analysis, karyotyping, and copy number variation (CNV) analysis, have led to the identification of new disease genes [41, 44, 45, 102]; however, these methods require analysis of a cohort of multiple unrelated affected individuals to narrow down genomic regions of interest, before finally zeroing in on the candidate gene. In contrast, WES of a single family or a parent–proband trio can result in rapid gene identification. This was first reported approximately 5 years after the launch of the technology in 2005 [72]. Using WES, two potentially pathogenic variants were identified in a novel candidate gene, *DHODH*, thus implicating the gene in the autosomal recessive Miller syndrome. This condition is characterized by severe micrognathia, cleft lip or palate, limb defects, coloboma, and supernumerary nipples [65]. Even though the disease had been described several decades ago, not much about the causal gene or mode of inheritance was known until this study. Despite little understanding of how *DHODH* mutations cause Miller syndrome, the subsequent identification of mutations in additional patients by targeted gene sequencing confirmed disease association without functional analysis. Shortly thereafter, another novel disease gene association was reported by the same group, which identified *MLL2* (*KMT2D*) to be the causative gene for Kabuki syndrome [71]. These findings strongly suggested that exome sequencing of a small number of affected individuals from unrelated kindred, or of multiple individuals from a single affected family, could be a powerful and efficient strategy for the identification of rare disease genes.

From Medical Genetics to Medical Genomics: A Shift in Paradigm

Beginning in early 2008, the NIH's Undiagnosed Diseases Program (UDP) began offering clinical WES as a pilot program, with initial funds totaling $280,000 [60]. UDP's explicit objectives were to provide molecular diagnosis to patients who remained undiagnosed despite thorough workup and to discover novel disease genes and disorders to gain insight into the pathogenesis of the clinical manifestations. After receiving several thousand applications from prospective participants, 160 individuals were enrolled, and the huge task of deciphering the underlying genetic causes began. Included was a healthy Colombian couple with two sons affected with an uncharacterized neurological illness, presenting with seizures, tremors, and several other complications. When one of the sons succumbed to the disease, the second son of the family was enrolled in the above-mentioned multi-institute initiative in hopes of identifying the underlying cause. After collaborative efforts for more than a year, a definitive diagnosis came from WES analysis. Furthermore, the molecular diagnosis was also established for almost 25 % (39/160) of the enrolled individuals overall. Novel disease genes, including *NT5E*, associated with arterial calcification disorder [90], and *HINT3*, an aprataxin-related gene causative of a familial distal myopathy [28], were identified, as well. Most of the diagnoses made, however, included known rare (≤1 in 10,000) or ultra-rare (<60 cases reported) diseases in individuals who had previously undergone multiple molecular and/or biochemical genetic tests. UDP's experience suggested that, with comprehensive phenotypic information, accurate bioinformatics tools, and a methodological approach, WES can be an economical single test for disease diagnosis.

Implementation of Exome Sequencing in Clinical Medicine

Whereas the suitability of WES for clinical medicine was initially debated, the emerging consensus is that the future of diagnostic exome sequencing has already begun [54, 63]. As new genes and diseases are identified through clinical WES, the test is gaining popularity. Expected reductions in cost and improved reimbursement are also likely to mean wider implementation of WES in clinical medicine.

Mendelian Disorders and Exome Sequencing

The conventional approach, still widely in practice, for molecular diagnosis of single-gene Mendelian disorders follows serial interrogation of all exons and exon–intron boundaries of known disease-specific genes via traditional polymerase chain reaction (PCR) amplification and the gold standard Sanger sequencing. Unlike complex traits and disorders such as autism and intellectual disability, which can involve several causative genes and variants, Mendelian disorders are generally associated with mutations in a single gene. With the utilization of clinical genetics and molecular diagnosis, however, locus heterogeneity and overlapping disease phenotypes have shown that, even for Mendelian disorders, making a molecular diagnosis is less straightforward than previously thought. This notion favored the application of multi-gene panels in which all common disease-related genes are interrogated simultaneously through NGS. Consequently, there are now several individual disease gene panels available [2, 40, 98]. Even though the panel approach has reduced the diagnostic odyssey for patients and boosted diagnostic capacity, a substantial fraction of patients still remain without a molecular diagnosis. This can be attributed, in part, to the inability to detect mutations in regulatory and intronic regions. Nevertheless, most such cases are believed to be due to the involvement of previously unknown disease genes. One important feature in support of this is the occurrence of more than 85 % of causative mutations for Mendelian disorders in exonic regions of the genome [12]. This percentage, together with the growing potential of WES as a diagnostic tool, makes it a preferred approach for rare Mendelian disorders with genetic and phenotypic heterogeneity. Notably, however, causative variants detectable by a combination of conventional methodologies, including homozygosity mapping and candidate gene selection, may be missed by WES [10, 69]. Bloch-Zupan et al. [10] report a case of homozygous mutations in the *SMOC2* gene, responsible for dental developmental defects, which were initially missed by WES due to poor coverage [10]. Overall, however, whereas homozygosity mapping or linkage analysis may be preferred for consanguineous and large pedigrees, WES is proving to be the most informative of these diagnostic tests [13, 14, 26, 59]. In some cases, WES has provided an accurate molecular diagnosis in patients previously diagnosed with a different disease, further cementing the value of this assay in clinically heterogeneous Mendelian disorders [47]. Besides establishing a molecular diagnosis in patients and providing carrier testing opportunities for family members, the identification of causative mutations in Mendelian diseases also guides patient management and family counseling [4], and opens up opportunities for therapeutic intervention and participation in clinical studies [75]. Finally, the identification of new disease genes and causative mutations contributes to our understanding of disease phenotype, pathogenesis, and gene function [77].

Complex Disorders and Exome Sequencing

Common complex diseases constitute a major part of overall disease burden in the general population. Most common diseases are complex, with extensive genetic heterogeneity resulting in clinically indistinguishable phenotypes. This includes conditions such as autism, intellectual disability, cardiac disease, and diabetes. X-chromosome-linked intellectual disability alone has been associated with more than 100 different genes. Similarly, autism spectrum disorders are linked to multiple genes, with no single gene accounting for more than 1 % of cases [9]. It is obvious that, even more so than for single-gene Mendelian disorders, the WES approach is advantageous for multifactorial and multigenic complex disease characterization. Recently, one single WES study investigating the genetic etiology of autosomal recessive forms of intellectual disability identified 50 novel candidate genes [68]. These include genes encoding proteins involved in transcription, translation, cell-cycle control, and fatty acid and energy metabolism critical for normal brain development and function. The discovery of such novel disease-associated genes not only improves our understanding of the underlying cause of disease manifestations but can also suggest novel targets for therapy and management.

Unlike most Mendelian disorders, diseases with complex genetic etiologies involve coding variants that present as risk factors rather than direct causes of disease. Such risk factors found by traditional methods to date include an *APOE* genotype that plays a role in late-onset alzheimer's disease, complement factor H polymorphism in age-related macular degeneration, and an *LRRK2* risk variant in Parkinson's disease [19, 43, 92]. The application of WES to complex disease diagnosis will enable the identification of similar common protein-coding risk alleles, as well as rare risk alleles. Genome-wide association studies (GWAS) have been revolutionary in terms of uncovering common variants associated with complex disorders, but have not satisfactorily explained the heritability of these traits [17, 56, 62, 83]. With the advent of WES, the focus of complex trait genetics has shifted towards low-frequency and rare variants [79, 97], and the link between variants and complex traits is on its way to becoming clearer [11, 23, 27, 73, 80]. The routine use of WES in clinical laboratories will most likely identify more and more rare variants that have a strong causative effect on phenotype, unlike the common variants that, individually, contribute only minimally [24, 42].

Application of WES to Neoplastic Diseases

Historically, pathologists have relied on histomorphology to classify and diagnose neoplasms [8, 21]. Recent progress in cancer genomics, however, has pointed towards the utility of a more granular approach through the identification of genetic alterations common to morphologically diverse tumor types and through the discrimination of subgroups within what was thought to be a single tumor type [7]. Consequently, WES has been applied to tumor diagnostics to obtain a comprehensive picture of copy number alterations (CNAs) and of pathogenic mutations [52]. The potential of WES to detect somatic CNAs in cancer syndromes has been explored, as well [52, 82]. In a study involving 17 matched tumor and normal tissues from patients with metastatic castrate-resistant prostate cancer, targeted WES analysis successfully identified various common CNAs, such as androgen receptor (*AR*) gain and *PTEN* loss [52]. This study and others suggest that somatic CNAs that involve the amplification of oncogenes or deletion of tumor suppressors and are significant contributors to cancer etiology can now be monitored more comprehensively using WES than array-based technologies [15]. Unlike germline mutations, somatic mutation and CNA detection in cancer are performed by simultaneous exome sequencing of normal and tumor tissue from the same individual, followed by a comparison of copy number ratios of exonic regions in the two sample types [52]. This approach of analyzing the relative coverage (of tumor versus normal sample) distinguishes a true chromosomal deletion from a lack of coverage due to technical limitations. WES thus offers the combined efficiency of both array comparative genomic hybridization (aCGH), which detects CNAs by relative probe frequency [78], and single nucleotide polymorphism (SNP) array, which detects loss of heterozygosity (LOH) and absence of heterozygosity (AOH) by zygosity changes at known SNP loci [61]. Whereas the prohibitive cost and analysis burden of whole-genome sequencing (WGS) have limited its clinical application thus far, successful detection of somatic *DNMT3A* mutations in acute monocytic leukemia [105], *PBRM1* mutations in renal carcinoma [100], *BAP1* mutations in metastasizing uveal melanomas [34], and *AR*, *NCOA2*, *PTEN*, *RB1*, and *TP53* CNAs in prostate cancer [94] by WES are confirming it as a cancer diagnostic and monitoring assay option.

There are several advantages to using WES for cancer genomics. First, it provides an exon-level resolution of CNAs. Second, the vast data available through comprehensive sequencing projects such as The Cancer Genome Atlas (TCGA) can be leveraged because whole-exome data for thousands of cancer cases from multiple studies are publicly available [95]. This makes integrative cancer detection strategies possible and drives personalized medicine approaches. Genotype-directed therapies are transforming cancer care, as seen with several drugs and target inhibitors in various cancer types, including chronic myeloid leukemia, colorectal adenocarcinoma, and melanoma [25, 53, 76]. The role of coexisting or co-occurring passenger mutations, separate from the driver mutations that actually cause the clonal expansion of cancer cells, is also being investigated so the two can be distinguished [5, 33]. Comparison of WES data across multiple patients is expected to contribute to the teasing out of the two, which could in turn translate into new drug targets. Despite these advantages, WES still has some limitations. These are primarily pertaining to coverage of certain exons and of genes with complex sequence context, as a result of which some mutations and CNAs may be missed. Additionally, CNAs involving gene-poor regions may not be detected due to assay design. Gene fusion events or chimeric gene products unique to cancer etiology and the more frequent large chromosomal aberration events, such as translocations, large deletions, or inversions, are not detected by WES. A comprehensive approach of various NGS technologies including WES, WGS, and transcriptome analysis is being explored, but clinical applicability is still rudimentary [67, 86, 93, 94].

From Diagnosis to Therapy: Advances in Clinical Care

Despite the proven potential of WES for clinical diagnostic purposes, one common criticism of the technology is the lack of evidence for its clinical usefulness. Pharmacogenomics is one area in which WES is expected to play a major role, especially by identifying variants that contribute to genotype-specific responses to drugs. One such example is related to the substitution of glutamic acid for valine at position 600 (p. V600E) in the *BRAF* gene in individuals with malignant melanoma [20]. This specific mutation acts by conferring a constant flux through the mitogen-activated protein kinase (MAPK) pathway, thereby promoting malignancy. The genotype-specific drug vemurafenib (PLX4032), recently approved by the FDA, is used for targeted intervention of metastatic melanoma [46, 106]. Eventually, however, tumor cells were found to develop resistance to the drug over time, but in a cohort of 20 melanoma patients treated with vemurafenib, WES identified the underlying cause for the development of drug resistance: a gain in copy number (by 2–13 times) of the mutant p.V600E *BRAF* allele [88].

Several other targeted therapies, such as imatinib for chronic myeloid leukemia, trastuzumab for breast cancer, irinotecan and panitumab for colorectal cancer, and erlotinib for lung cancer, may all be monitored for their treatment effect and resistance development using WES. Implementation of WES in the context of personalized medicine is highlighted by a recent study reporting a novel genetic risk factor linked to the VACTERL association [89]. A heterozygous mutation in the *CPSI* gene, identified by WES in monozygotic twins, is suspected of being the risk factor associated with the severe pulmonary artery hypertension observed post-surgery in the twin who underwent surgery. Generally, homozygous or compound heterozygous mutations in *CPSI* are associated with a rare urea cycle disorder; however, through WES analysis the authors clarified that there were no discordant *de novo* mutations between the two twins and that the observed complication must have been due to the combination of the observed heterozygous variant and an environmental trigger: in this case, surgery.

Limitations and Challenges of Implementing Exome Sequencing Assays

Despite being quite comprehensive, WES has yet to overcome several technical and analytic challenges before it can replace the current gold standard of Sanger sequencing, or even targeted NGS panels. These challenges are summarized here. The first and foremost technical challenge is the inefficiency to capture and sequence all target exons. Contrary to what is suggested by its name, WES currently misses around 5–8 % of the human exome because of low or no coverage [16]. Most of this is explained by sequence context, such as with high or low GC content or the presence of highly homologous pseudogenes [38]. Capture of all target exons is, of course, essential to avoid false-negative interpretations due to the presence of potentially causative mutations in missed exons. Highly repetitive sequences, which include interspersed repeats and tandem repeats, constitute more than half of the human genome. These highly homologous regions are co-enriched and co-sequenced along with the target regions [96]. This challenge may be countered by increasing the sequence read size, which is still limited with current NGS technologies. However, several alternative approaches, such as paired end sequencing and correlation of average read depth differences to detect repeat regions, are being explored [96]. A second challenge is storage and management of the vast amount of sequencing data generated by the technology. This demands a large investment in infrastructure and technology, which is a major strain for diagnostic laboratories. A third limitation is the variant detection capability of WES. With high coverage and read depth, point mutations and small indels in exonic regions can be detected with high efficiency, but those in regulatory regions are not. In addition, larger multi-exon or multi-gene deletions and duplications, which contribute to a significant proportion of the mutation spectrum for several genes, as well as gene-fusion or chimeric events common in cancer, are not efficiently detected. Besides variant detection capability, another major challenge of the test involves assessment of the clinical implications of variants identified. Most of the observed variants may not be clinically predictable or actionable due to lack of sufficient evidence. However, with the routine practice of WES and accumulation of relevant information, this concern would gradually be reduced. The fifth challenge to implementing WES assays in clinical care is the requirement of additional training for physicians to help them interpret test results and reports. With a more comprehensive set of variants available for consideration in the patient's clinical context, clinicians who see the patient, if trained in this area, would be able to make the optimal interpretation as to the causative gene. Alternatively or ideally simultaneously, extensive phenotypic information may be collected beforehand and made available to the pathologists and laboratorians interpreting the data. Finally, a considerable challenge facing the clinics and laboratories that offer these tests is the constantly changing technology. Recently, members of the Standardization of Clinical Testing workgroup (Nex-StoCT) have laid out guidelines for the validation and implementation of NGS-based tests [29]. With NGS technology changing all the time, however, these aspects also change and can become a hurdle to implementation.

Despite the challenges and limitations, WES and WGS have stirred tremendous interest, with the future of clinical care promising expedited diagnosis and more personalized medicine. Moreover, implementation of WES in medical practice will potentially aid the advancement of our understanding of human biology and pathogenesis.

A Look to the Future of Whole-Exome Assays

Current commercially available NGS technologies have already revolutionized the diagnostic capacity of modern clinical genetics. Nevertheless, advanced so-called "third-generation" sequencing technologies, such as Helicos Heliscope (Helicos Biosciences Corporation, Cambridge, MA), PacBio SMRT (Pacific Biosciences, California), and Nanopore sequencers (Oxford Nanopore Technologies, Oxford, UK), are being actively developed to further improve genomic sequencing applications [32]. These third-generation sequencing platforms differ from the current technologies in that the initial target capture and enrichment step, which involves DNA amplification, is no longer required. The input patient DNA is sequenced and analyzed at the single-molecule level with the help of engineered protein polymerases [32]. This will not only cut cost and turnaround time but also have the added advantage of avoiding any in vitro amplification bias. Upon thorough validation and optimization of their diagnostic ability, these future technologies promise to move today's medical practice to the anticipated next level of care.

Currently, even more so than the sequencing technology and needed coverage improvements, the progress in data analysis tools and candidate variant filtration is of major concern. WES alone, which interrogates about 1 % of the human genome, returns a list of about 20,000 variant calls [91]. Family information, such as the mode of inheritance within a family, linkage analysis or variant data, i.e., the WES profile of unaffected family members, helps eliminate familial normal variations and track down disease-causing mutations [60], but performing additional tests including WES on multiple family members increases diagnostic costs and is not ideal for a variety of reasons. As more and more exomes are analyzed and sequence variants reported in publicly available databases, however, variant analysis and disease diagnosis by WES will certainly become easier and faster.

Meanwhile, with the implementation of WES and NGS technologies in clinical pathology becoming more common, the need for trained pathologists capable of interpreting the data and assessing the potential impact on an individual's health is growing. The training of future pathologists is now under discussion, and teaching curricula in genomics and personalized medicine are being actively developed for residents [35, 84]. A national committee of Pathology Program Directors and other experts has also recently formed to develop model curricula and promote their widespread implementation [35, 103]. The implementation of WES and WGS in clinical practice has, therefore, added a new dimension to the already multifaceted roles of pathologists.

Conclusions

With more than 85 % of causative mutations harbored in as little as 1 % of the entire human genome, the use of WES as the most efficient strategy for disease diagnosis seems well justified. Even though WGS has the potential to identify CNVs and point mutations in exons, as well as in regulatory regions of the introns, the cost, time, and the analysis burden currently involved has meant WGS is on hold for clinical implementation, at least for now. Substantial proof-of-principle studies and evidence of diagnostic capability, affordability, and feasibility in the clinical setting have supported the use of WES. Currently, it is offered for clinical diagnosis by multiple major clinical laboratories across the USA, and as the technology improves and becomes less expensive, more laboratories are beginning to develop the test.

Clinicians who contemplate ordering a WES assay should first consider other available tests, such as relatively comprehensive gene panels. Gene panels, which interrogate only a limited number of genes, each more or less associated with the patient's clinical presentation, more completely retain the integrity of the individual's genetic information. Appropriate ethical guidelines and data-masking features during data analysis will likely overcome this difference eventually and make WES widely acceptable for rare diseases, cancer, and prenatal and infectious disease diagnosis. Finally, reductions in cost, more robust technologies, and improved data storage processes will soon make clinical WGS feasible, as well. The future of medical care can be envisioned as an integrated approach, with pathologists, geneticists, and other physicians all contributing to make informed decisions about patient management and treatment.

REFERENCES

1. Albert TJ, Molla MN, Muzny DM, Nazareth L, Wheeler D, Song X, Richmond TA, Middle CM, Rodesch MJ, Packard CJ, Weinstock GM, Gibbs RA. Direct selection of human genomic loci by microarray hybridization. Nat Methods. 2007; 4:903–5.
2. Ankala A, Kohn JN, Dastur R, Gaitonde P, Khadilkar SV, Hegde M. Ancestral founder mutations in Calpain-3 in indian agarwal community—historical, clinical and molecular perspective. Muscle Nerve. 2013;47:931–7.
3. Ansorge WJ. Next-generation DNA sequencing techniques. N Biotechnol. 2009;25:195–203.
4. Antonarakis SE, Beckmann JS. Mendelian disorders deserve more attention. Nat Rev Genet. 2006;7:277–82.
5. Barton NH. Genetic hitchhiking. Philos Trans R Soc Lond B Biol Sci. 2000;355:1553–62.
6. Bashiardes S, Veile R, Helms C, Mardis ER, Bowcock AM, Lovett M. Direct genomic selection. Nat Methods. 2005;2:63–9.
7. Berman J. Modern classification of neoplasms: reconciling differences between morphologic and molecular approaches. BMC Cancer. 2005; 5:100.
8. Berman JJ. Tumor taxonomy for the developmental lineage classification of neoplasms. BMC Cancer. 2004;4:88.
9. Betancur C. Etiological heterogeneity in autism spectrum disorders: more than 100 genetic and genomic disorders and still counting. Brain Res. 2011;1380:42–77.
10. Bloch-Zupan A, Jamet X, Etard C, Laugel V, Muller J, Geoffroy V, Strauss JP, Pelletier V, Marion V, Poch O, Strahle U, Stoetzel C, Dollfus H. Homozygosity mapping and candidate prioritization identify mutations, missed by whole-exome sequencing, in SMOC2, causing major dental developmental defects. Am J Hum Genet. 2011;89:773–81.
11. Bogardus C. Missing heritability and GWAS utility. Obesity (Silver Spring). 2009;17:209.
12. Botstein D, Risch N. Discovering genotypes underlying human phenotypes: past successes for mendelian disease, future approaches for complex disease. Nat Genet. 2003;33(Suppl):228–37.
13. Cabral RM, Kurban M, Wajid M, Shimomura Y, Petukhova L, Christiano AM. Whole-exome sequencing in a single proband reveals a mutation in the CHST8 gene in autosomal recessive peeling skin syndrome. Genomics. 2012;99:202–8.
14. Campeau PM, Lu JT, Sule G, Jiang MM, Bae Y, Madan S, Hogler W, Shaw NJ, Mumm S, Gibbs RA, Whyte MP, Lee BH. Whole-exome sequencing identifies mutations in the nucleoside transporter gene SLC29A3 in dysosteosclerosis, a form of osteopetrosis. Hum Mol Genet. 2012;21: 4904–9.
15. Chang H, Jackson DG, Kayne PS, Ross-Macdonald PB, Ryseck RP, Siemers NO. Exome sequencing reveals comprehensive genomic alterations across eight cancer cell lines. PLoS One. 2011;6:e21097.
16. Choi M, Scholl UI, Ji W, Liu T, Tikhonova IR, Zumbo P, Nayir A, Bakkaloglu A, Ozen S, Sanjad S, Nelson-Williams C, Farhi A, Mane S, Lifton RP. Genetic diagnosis by whole exome capture and massively parallel DNA sequencing. Proc Natl Acad Sci U S A. 2009;106:19096–101.
17. Cirulli ET, Goldstein DB. Uncovering the roles of rare variants in common disease through whole-genome sequencing. Nat Rev Genet. 2010;11:415–25.
18. Coonrod EM, Durtschi JD, Margraf RL, Voelkerding KV. Developing genome and exome sequencing for candidate gene identification in inherited disorders. Arch Pathol Lab Med. 2012; 137:415–33.
19. Corder EH, Saunders AM, Strittmatter WJ, Schmechel DE, Gaskell PC, Small GW, Roses AD, Haines JL, Pericak-Vance MA. Gene dose of apolipoprotein E type 4 allele and the risk of

Alzheimer's disease in late onset families. Science. 1993;261:921–3.
20. Davies H, Bignell GR, Cox C, Stephens P, Edkins S, Clegg S, Teague J, Woffendin H, Garnett MJ, Bottomley W, Davis N, Dicks E, Ewing R, Floyd Y, Gray K, Hall S, Hawes R, Hughes J, Kosmidou V, Menzies A, Mould C, Parker A, Stevens C, Watt S, Hooper S, Wilson R, Jayatilake H, Gusterson BA, Cooper C, Shipley J, Hargrave D, Pritchard-Jones K, Maitland N, Chenevix-Trench G, Riggins GJ, Bigner DD, Palmieri G, Cossu A, Flanagan A, Nicholson A, Ho JW, Leung SY, Yuen ST, Weber BL, Seigler HF, Darrow TL, Paterson H, Marais R, Marshall CJ, Wooster R, Stratton MR, Futreal PA. Mutations of the BRAF gene in human cancer. Nature. 2002;417:949–54.
21. de Coronado S, Haber MW, Sioutos N, Tuttle MS, Wright LW. NCI thesaurus: using science-based terminology to integrate cancer research results. Stud Health Technol Inform. 2004;107: 33–7.
22. DePristo MA, Banks E, Poplin R, Garimella KV, Maguire JR, Hartl C, Philippakis AA, del Angel G, Rivas MA, Hanna M, McKenna A, Fennell TJ, Kernytsky AM, Sivachenko AY, Cibulskis K, Gabriel SB, Altshuler D, Daly MJ. A framework for variation discovery and genotyping using next-generation DNA sequencing data. Nat Genet. 2011;43:491–8.
23. Dickson SP, Wang K, Krantz I, Hakonarson H, Goldstein DB. Rare variants create synthetic genome-wide associations. PLoS Biol. 2010;8: e1000294.
24. Do R, Kathiresan S, Abecasis GR. Exome sequencing and complex disease: practical aspects of rare variant association studies. Hum Mol Genet. 2012;21:R1–9.
25. Druker BJ, Talpaz M, Resta DJ, Peng B, Buchdunger E, Ford JM, Lydon NB, Kantarjian H, Capdeville R, Ohno-Jones S, Sawyers CL. Efficacy and safety of a specific inhibitor of the BCR-ABL tyrosine kinase in chronic myeloid leukemia. N Engl J Med. 2001;344:1031–7.
26. Dundar H, Ozgul RK, Yalnizoglu D, Erdem S, Oguz KK, Tuncel D, Temucin CM, Dursun A. Identification of a novel twinkle mutation in a family with infantile onset spinocerebellar ataxia by whole exome sequencing. Pediatr Neurol. 2012;46:172–7.
27. Eichler EE, Flint J, Gibson G, Kong A, Leal SM, Moore JH, Nadeau JH. Missing heritability and strategies for finding the underlying causes of complex disease. Nat Rev Genet. 2010;11: 446–50.
28. Gahl WA, Markello TC, Toro C, Fajardo KF, Sincan M, Gill F, Carlson-Donohoe H, Gropman A, Pierson TM, Golas G, Wolfe L, Groden C, Godfrey R, Nehrebecky M, Wahl C, Landis DM, Yang S, Madeo A, Mullikin JC, Boerkoel CF, Tifft CJ, Adams D. The National Institutes of Health Undiagnosed Diseases Program: insights into rare diseases. Genet Med. 2012;14:51–9.
29. Gargis AS, Kalman L, Berry MW, Bick DP, Dimmock DP, Hambuch T, Lu F, Lyon E, Voelkerding KV, Zehnbauer BA, Agarwala R, Bennett SF, Chen B, Chin EL, Compton JG, Das S, Farkas DH, Ferber MJ, Funke BH, Furtado MR, Ganova-Raeva LM, Geigenmuller U, Gunselman SJ, Hegde MR, Johnson PL, Kasarskis A, Kulkarni S, Lenk T, Liu CS, Manion M, Manolio TA, Mardis ER, Merker JD, Rajeevan MS, Reese MG, Rehm HL, Simen BB, Yeakley JM, Zook JM, Lubin IM. Assuring the quality of next-generation sequencing in clinical laboratory practice. Nat Biotechnol. 2012;30:1033–6.
30. Gnirke A, Melnikov A, Maguire J, Rogov P, LeProust EM, Brockman W, Fennell T, Giannoukos G, Fisher S, Russ C, Gabriel S, Jaffe DB, Lander ES, Nusbaum C. Solution hybrid selection with ultra-long oligonucleotides for massively parallel targeted sequencing. Nat Biotechnol. 2009;27:182–9.
31. Gowrisankar S, Lerner-Ellis JP, Cox S, White ET, Manion M, LeVan K, Liu J, Farwell LM, Iartchouk O, Rehm HL, Funke BH. Evaluation of second-generation sequencing of 19 dilated cardiomyopathy genes for clinical applications. J Mol Diagn. 2010;12:818–27.
32. Gullapalli RR, Desai KV, Santana-Santos L, Kant JA, Becich MJ. Next generation sequencing in clinical medicine: challenges and lessons for pathology and biomedical informatics. J Pathol Inform. 2012;3:40.
33. Hanahan D, Weinberg RA. The hallmarks of cancer. Cell. 2000;100:57–70.
34. Harbour JW, Onken MD, Roberson ED, Duan S, Cao L, Worley LA, Council ML, Matatall KA, Helms C, Bowcock AM. Frequent mutation of BAP1 in metastasizing uveal melanomas. Science. 2010;330:1410–3.
35. Haspel RL, Arnaout R, Briere L, Kantarci S, Marchand K, Tonellato P, Connolly J, Boguski MS, Saffitz JE. A call to action: training pathology residents in genomics and personalized medicine. Am J Clin Pathol. 2010;133:832–4.
36. Hebert SC. Bartter syndrome. Curr Opin Nephrol Hypertens. 2003;12:527–32.
37. Hodges E, Xuan Z, Balija V, Kramer M, Molla MN, Smith SW, Middle CM, Rodesch MJ, Albert TJ, Hannon GJ, McCombie WR. Genome-wide in situ exon capture for selective resequencing. Nat Genet. 2007;39:1522–7.
38. Hoischen A, Gilissen C, Arts P, Wieskamp N, van der Vliet W, Vermeer S, Steehouwer M, de Vries P, Meijer R, Seiqueros J, Knoers NV, Buckley MF, Scheffer H, Veltman JA. Massively parallel sequencing of ataxia genes after array-based enrichment. Hum Mutat. 2010;31:494–9.

39. Holcomb CL, Hoglund B, Anderson MW, Blake LA, Bohme I, Egholm M, Ferriola D, Gabriel C, Gelber SE, Goodridge D, Hawbecker S, Klein R, Ladner M, Lind C, Monos D, Pando MJ, Proll J, Sayer DC, Schmitz-Agheguian G, Simen BB, Thiele B, Trachtenberg EA, Tyan DB, Wassmuth R, White S, Erlich HA. A multi-site study using high-resolution HLA genotyping by next generation sequencing. Tissue Antigens. 2011;77: 206–17.
40. Jones MA, Bhide S, Chin E, Ng BG, Rhodenizer D, Zhang VW, Sun JJ, Tanner A, Freeze HH, Hegde MR. Targeted polymerase chain reaction-based enrichment and next generation sequencing for diagnostic testing of congenital disorders of glycosylation. Genet Med. 2011;13:921–32.
41. Kerem B, Rommens JM, Buchanan JA, Markiewicz D, Cox TK, Chakravarti A, Buchwald M, Tsui LC. Identification of the cystic fibrosis gene: genetic analysis. Science. 1989;245: 1073–80.
42. Kiezun A, Garimella K, Do R, Stitziel NO, Neale BM, McLaren PJ, Gupta N, Sklar P, Sullivan PF, Moran JL, Hultman CM, Lichtenstein P, Magnusson P, Lehner T, Shugart YY, Price AL, de Bakker PI, Purcell SM, Sunyaev SR. Exome sequencing and the genetic basis of complex traits. Nat Genet. 2012;44:623–30.
43. Klein RJ, Zeiss C, Chew EY, Tsai JY, Sackler RS, Haynes C, Henning AK, SanGiovanni JP, Mane SM, Mayne ST, Bracken MB, Ferris FL, Ott J, Barnstable C, Hoh J. Complement factor H polymorphism in age-related macular degeneration. Science. 2005;308:385–9.
44. Kurotaki N, Imaizumi K, Harada N, Masuno M, Kondoh T, Nagai T, Ohashi H, Naritomi K, Tsukahara M, Makita Y, Sugimoto T, Sonoda T, Hasegawa T, Chinen Y, Tomita Ha HA, Kinoshita A, Mizuguchi T, Yoshiura Ki K, Ohta T, Kishino T, Fukushima Y, Niikawa N, Matsumoto N. Haploinsufficiency of NSD1 causes Sotos syndrome. Nat Genet. 2002;30:365–6.
45. Lander ES, Botstein D. Homozygosity mapping: a way to map human recessive traits with the DNA of inbred children. Science. 1987;236: 1567–70.
46. Lee JT, Li L, Brafford PA, Van Den Eijnden M, Halloran MB, Sproesser K, Haass NK, Smalley KSM, Tsai J, Bollag G. PLX4032, a potent inhibitor of the B-Raf V600E oncogene, selectively inhibits V600E-positive melanomas. Pigment Cell Melanoma Res. 2010;23:820–7.
47. Leidenroth A, Sorte HS, Gilfillan G, Ehrlich M, Lyle R, Hewitt JE. Diagnosis by sequencing: correction of misdiagnosis from FSHD2 to LGMD2A by whole-exome analysis. Eur J Hum Genet. 2012;20:999–1003.
48. Lemke JR, Riesch E, Scheurenbrand T, Schubach M, Wilhelm C, Steiner I, Hansen J, Courage C, Gallati S, Bürki S. Targeted next generation sequencing as a diagnostic tool in epileptic disorders. Epilepsia. 2012;53:1387–98.
49. Li H, Durbin R. Fast and accurate short read alignment with Burrows-Wheeler transform. Bioinformatics. 2009;25:1754–60.
50. Li H, Handsaker B, Wysoker A, Fennell T, Ruan J, Homer N, Marth G, Abecasis G, Durbin R. The sequence alignment/map format and SAMtools. Bioinformatics. 2009;25:2078–9.
51. Lin X, Tang W, Ahmad S, Lu J, Colby CC, Zhu J, Yu Q. Applications of targeted gene capture and next-generation sequencing technologies in studies of human deafness and other genetic disabilities. Hear Res. 2012;288:67–76.
52. Lonigro RJ, Grasso CS, Robinson DR, Jing X, Wu YM, Cao X, Quist MJ, Tomlins SA, Pienta KJ, Chinnaiyan AM. Detection of somatic copy number alterations in cancer using targeted exome capture sequencing. Neoplasia. 2011;13: 1019–25.
53. Lynch TJ, Bell DW, Sordella R, Gurubhagavatula S, Okimoto RA, Brannigan BW, Harris PL, Haserlat SM, Supko JG, Haluska FG, Louis DN, Christiani DC, Settleman J, Haber DA. Activating mutations in the epidermal growth factor receptor underlying responsiveness of non-small-cell lung cancer to gefitinib. N Engl J Med. 2004;350:2129–39.
54. Majewski J, Rosenblatt DS. Exome and whole-genome sequencing for gene discovery: the future is now! Hum Mutat. 2012;33:591–2.
55. Mamanova L, Coffey AJ, Scott CE, Kozarewa I, Turner EH, Kumar A, Howard E, Shendure J, Turner DJ. Target-enrichment strategies for next-generation sequencing. Nat Methods. 2010; 7:111–8.
56. Manolio TA, Collins FS, Cox NJ, Goldstein DB, Hindorff LA, Hunter DJ, McCarthy MI, Ramos EM, Cardon LR, Chakravarti A, Cho JH, Guttmacher AE, Kong A, Kruglyak L, Mardis E, Rotimi CN, Slatkin M, Valle D, Whittemore AS, Boehnke M, Clark AG, Eichler EE, Gibson G, Haines JL, Mackay TF, McCarroll SA, Visscher PM. Finding the missing heritability of complex diseases. Nature. 2009;461:747–53.
57. Mardis ER. Next-generation DNA sequencing methods. Annu Rev Genomics Hum Genet. 2008;9:387–402.
58. Margulies M, Egholm M, Altman WE, Attiya S, Bader JS, Bemben LA, Berka J, Braverman MS, Chen YJ, Chen Z, Dewell SB, Du L, Fierro JM, Gomes XV, Godwin BC, He W, Helgesen S, Ho CH, Irzyk GP, Jando SC, Alenquer ML, Jarvie TP, Jirage KB, Kim JB, Knight JR, Lanza JR, Leamon JH, Lefkowitz SM, Lei M, Li J, Lohman KL, Lu H, Makhijani VB, McDade KE, McKenna MP, Myers EW, Nickerson E, Nobile JR, Plant R, Puc BP, Ronan MT, Roth GT, Sarkis GJ, Simons JF,

Simpson JW, Srinivasan M, Tartaro KR, Tomasz A, Vogt KA, Volkmer GA, Wang SH, Wang Y, Weiner MP, Yu P, Begley RF, Rothberg JM. Genome sequencing in microfabricated high-density picolitre reactors. Nature. 2005; 437:376–80.
59. Martinez FJ, Lee JH, Lee JE, Blanco S, Nickerson E, Gabriel S, Frye M, Al-Gazali L, Gleeson JG. Whole exome sequencing identifies a splicing mutation in NSUN2 as a cause of a Dubowitz-like syndrome. J Med Genet. 2012;49: 380–5.
60. Maxmen A. Exome sequencing deciphers rare diseases. Cell. 2011;144:635–7.
61. McCarroll SA, Kuruvilla FG, Korn JM, Cawley S, Nemesh J, Wysoker A, Shapero MH, de Bakker PI, Maller JB, Kirby A, Elliott AL, Parkin M, Hubbell E, Webster T, Mei R, Veitch J, Collins PJ, Handsaker R, Lincoln S, Nizzari M, Blume J, Jones KW, Rava R, Daly MJ, Gabriel SB, Altshuler D. Integrated detection and population-genetic analysis of SNPs and copy number variation. Nat Genet. 2008;40:1166–74.
62. McClellan J, King MC. Genetic heterogeneity in human disease. Cell. 2010;141:210–7.
63. Mefford HC. Diagnostic exome sequencing: are we there yet? N Engl J Med. 2012;367:1951–3.
64. Metzker ML. Sequencing technologies: the next generation. Nat Rev Genet. 2009;11:31–46.
65. Miller M, Fineman R, Smith DW. Postaxial acrofacial dysostosis syndrome. J Pediatr. 1979;95: 970–5.
66. Mitra AP, Pagliarulo V, Yang D, Waldman FM, Datar RH, Skinner DG, Groshen S, Cote RJ. Generation of a concise gene panel for outcome prediction in urinary bladder cancer. J Clin Oncol. 2009;27:3929–37.
67. Morin RD, Johnson NA, Severson TM, Mungall AJ, An J, Goya R, Paul JE, Boyle M, Woolcock BW, Kuchenbauer F, Yap D, Humphries RK, Griffith OL, Shah S, Zhu H, Kimbara M, Shashkin P, Charlot JF, Tcherpakov M, Corbett R, Tam A, Varhol R, Smailus D, Moksa M, Zhao Y, Delaney A, Qian H, Birol I, Schein J, Moore R, Holt R, Horsman DE, Connors JM, Jones S, Aparicio S, Hirst M, Gascoyne RD, Marra MA. Somatic mutations altering EZH2 (Tyr641) in follicular and diffuse large B-cell lymphomas of germinal-center origin. Nat Genet. 2010; 42:181–5.
68. Najmabadi H, Hu H, Garshasbi M, Zemojtel T, Abedini SS, Chen W, Hosseini M, Behjati F, Haas S, Jamali P, Zecha A, Mohseni M, Puttmann L, Vahid LN, Jensen C, Moheb LA, Bienek M, Larti F, Mueller I, Weissmann R, Darvish H, Wrogemann K, Hadavi V, Lipkowitz B, Esmaeeli-Nieh S, Wieczorek D, Kariminejad R, Firouzabadi SG, Cohen M, Fattahi Z, Rost I, Mojahedi F, Hertzberg C, Dehghan A, Rajab A, Banavandi MJ, Hoffer J, Falah M, Musante L, Kalscheuer V, Ullmann R, Kuss AW, Tzschach A, Kahrizi K, Ropers HH. Deep sequencing reveals 50 novel genes for recessive cognitive disorders. Nature. 2011;478:57–63.
69. Need AC, Shashi V, Hitomi Y, Schoch K, Shianna KV, McDonald MT, Meisler MH, Goldstein DB. Clinical application of exome sequencing in undiagnosed genetic conditions. J Med Genet. 2012;49:353–61.
70. Neveling K, Collin RW, Gilissen C, van Huet RA, Visser L, Kwint MP, Gijsen SJ, Zonneveld MN, Wieskamp N, de Ligt J. Next-generation genetic testing for retinitis pigmentosa. Hum Mutat. 2012;33:963–72.
71. Ng SB, Bigham AW, Buckingham KJ, Hannibal MC, McMillin MJ, Gildersleeve HI, Beck AE, Tabor HK, Cooper GM, Mefford HC, Lee C, Turner EH, Smith JD, Rieder MJ, Yoshiura K, Matsumoto N, Ohta T, Niikawa N, Nickerson DA, Bamshad MJ, Shendure J. Exome sequencing identifies MLL2 mutations as a cause of Kabuki syndrome. Nat Genet. 2010;42:790–3.
72. Ng SB, Buckingham KJ, Lee C, Bigham AW, Tabor HK, Dent KM, Huff CD, Shannon PT, Jabs EW, Nickerson DA, Shendure J, Bamshad MJ. Exome sequencing identifies the cause of a mendelian disorder. Nat Genet. 2010;42:30–5.
73. Nielsen R. Genomics: in search of rare human variants. Nature. 2010;467:1050–1.
74. Okou DT, Steinberg KM, Middle C, Cutler DJ, Albert TJ, Zwick ME. Microarray-based genomic selection for high-throughput resequencing. Nat Methods. 2007;4:907–9.
75. Oti M, Brunner HG. The modular nature of genetic diseases. Clin Genet. 2007;71:1–11.
76. Paez JG, Janne PA, Lee JC, Tracy S, Greulich H, Gabriel S, Herman P, Kaye FJ, Lindeman N, Boggon TJ, Naoki K, Sasaki H, Fujii Y, Eck MJ, Sellers WR, Johnson BE, Meyerson M. EGFR mutations in lung cancer: correlation with clinical response to gefitinib therapy. Science. 2004;304:1497–500.
77. Peltonen L, Perola M, Naukkarinen J, Palotie A. Lessons from studying monogenic disease for common disease. Hum Mol Genet. 2006;15 Spec No 1, R67–74.
78. Pinkel D, Segraves R, Sudar D, Clark S, Poole I, Kowbel D, Collins C, Kuo WL, Chen C, Zhai Y, Dairkee SH, Ljung BM, Gray JW, Albertson DG. High resolution analysis of DNA copy number variation using comparative genomic hybridization to microarrays. Nat Genet. 1998;20: 207–11.
79. Prasad A, Merico D, Thiruvahindrapuram B, Wei J, Lionel AC, Sato D, Rickaby J, Lu C, Szatmari P, Roberts W, Fernandez BA, Marshall CR, Hatchwell E, Eis PS, Scherer SW. A discovery resource of rare copy number variations in

individuals with autism spectrum disorder. G3 (Bethesda). 2012;2:1665–85.
80. Pritchard JK. Are rare variants responsible for susceptibility to complex diseases? Am J Hum Genet. 2001;69:124–37.
81. Proll J, Danzer M, Stabentheiner S, Niklas N, Hackl C, Hofer K, Atzmuller S, Hufnagl P, Gully C, Hauser H, Krieger O, Gabriel C. Sequence capture and next generation resequencing of the MHC region highlights potential transplantation determinants in HLA identical haematopoietic stem cell transplantation. DNA Res. 2011;18: 201–10.
82. Sathirapongsasuti JF, Lee H, Horst BAJ, Brunner G, Cochran AJ, Binder S, Quackenbush J, Nelson SF. Exome sequencing-based copy-number variation and loss of heterozygosity detection: ExomeCNV. Bioinformatics. 2011;27:2648–54.
83. Schork NJ, Murray SS, Frazer KA, Topol EJ. Common vs. rare allele hypotheses for complex diseases. Curr Opin Genet Dev. 2009;19: 212–9.
84. Schrijver I, Natkunam Y, Galli S, Boyd SD. Integration of genomic medicine into pathology residency training: the stanford open curriculum. J Mol Diagn. 2013;15:141–8.
85. Serizawa M, Sekizuka T, Okutani A, Banno S, Sata T, Inoue S, Kuroda M. Genomewide screening for novel genetic variations associated with ciprofloxacin resistance in Bacillus anthracis. Antimicrob Agents Chemother. 2010;54: 2787–92.
86. Shah SP, Morin RD, Khattra J, Prentice L, Pugh T, Burleigh A, Delaney A, Gelmon K, Guliany R, Senz J, Steidl C, Holt RA, Jones S, Sun M, Leung G, Moore R, Severson T, Taylor GA, Teschendorff AE, Tse K, Turashvili G, Varhol R, Warren RL, Watson P, Zhao Y, Caldas C, Huntsman D, Hirst M, Marra MA, Aparicio S. Mutational evolution in a lobular breast tumour profiled at single nucleotide resolution. Nature. 2009;461: 809–13.
87. Shendure J, Ji H. Next-generation DNA sequencing. Nat Biotechnol. 2008;26:1135–45.
88. Shi H, Moriceau G, Kong X, Lee MK, Lee H, Koya RC, Ng C, Chodon T, Scolyer RA, Dahlman KB, Sosman JA, Kefford RF, Long GV, Nelson SF, Ribas A, Lo RS. Melanoma whole-exome sequencing identifies (V600E)B-RAF amplification-mediated acquired B-RAF inhibitor resistance. Nat Commun. 2012;3:724.
89. Solomon BD, Pineda-Alvarez DE, Hadley DW, Program NCS, Teer JK, Cherukuri PF, Hansen NF, Cruz P, Young AC, Blakesley RW, Lanpher B, Mayfield Gibson S, Sincan M, Chandrasekharappa SC, Mullikin JC. Personalized genomic medicine: lessons from the exome. Mol Genet Metab. 2011;104:189–91.
90. St. Hilaire C, Ziegler SG, Markello TC, Brusco A, Groden C, Gill F, Carlson-Donohoe H, Lederman RJ, Chen MY, Yang D. NT5E mutations and arterial calcifications. N Engl J Med. 2011;364: 432–42.
91. Stower H. The exome factor. Genome Biol. 2011;12:407.
92. Tan EK. Identification of a common genetic risk variant (LRRK2 Gly2385Arg) in Parkinson's disease. Ann Acad Med Singapore. 2006;35:840–2.
93. Taylor BS, Ladanyi M. Clinical cancer genomics: how soon is now? J Pathol. 2011;223:318–26.
94. Taylor BS, Schultz N, Hieronymus H, Gopalan A, Xiao Y, Carver BS, Arora VK, Kaushik P, Cerami E, Reva B, Antipin Y, Mitsiades N, Landers T, Dolgalev I, Major JE, Wilson M, Socci ND, Lash AE, Heguy A, Eastham JA, Scher HI, Reuter VE, Scardino PT, Sander C, Sawyers CL, Gerald WL. Integrative genomic profiling of human prostate cancer. Cancer Cell. 2010;18:11–22.
95. TCGA. Integrated genomic analyses of ovarian carcinoma. Nature. 2011;474:609–15.
96. Treangen TJ, Salzberg SL. Repetitive DNA and next-generation sequencing: computational challenges and solutions. Nat Rev Genet. 2012; 13:36–46.
97. Vaags AK, Lionel AC, Sato D, Goodenberger M, Stein QP, Curran S, Ogilvie C, Ahn JW, Drmic I, Senman L, Chrysler C, Thompson A, Russell C, Prasad A, Walker S, Pinto D, Marshall CR, Stavropoulos DJ, Zwaigenbaum L, Fernandez BA, Fombonne E, Bolton PF, Collier DA, Hodge JC, Roberts W, Szatmari P, Scherer SW. Rare deletions at the neurexin 3 locus in autism spectrum disorder. Am J Hum Genet. 2012;90:133–41.
98. Valencia CA, Ankala A, Rhodenizer D, Bhide S, Littlejohn MR, Keong LM, Rutkowski A, Sparks S, Bonnemann C, Hegde M. Comprehensive mutation analysis for congenital muscular dystrophy: a clinical PCR-based enrichment and next-generation sequencing panel. PLoS One. 2013;8:e53083.
99. Valencia CA, Rhodenizer D, Bhide S, Chin E, Littlejohn MR, Keong LM, Rutkowski A, Bonnemann C, Hegde M. Assessment of target enrichment platforms using massively parallel sequencing for the mutation detection for congenital muscular dystrophy. J Mol Diagn. 2012;14:233–46.
100. Varela I, Tarpey P, Raine K, Huang D, Ong CK, Stephens P, Davies H, Jones D, Lin ML, Teague J, Bignell G, Butler A, Cho J, Dalgliesh GL, Galappaththige D, Greenman C, Hardy C, Jia M, Latimer C, Lau KW, Marshall J, McLaren S, Menzies A, Mudie L, Stebbings L, Largaespada DA, Wessels LF, Richard S, Kahnoski RJ, Anema J, Tuveson DA, Perez-Mancera PA, Mustonen V, Fischer A, Adams DJ, Rust A, Chan-on W,

Subimerb C, Dykema K, Furge K, Campbell PJ, Teh BT, Stratton MR, Futreal PA. Exome sequencing identifies frequent mutation of the SWI/SNF complex gene PBRM1 in renal carcinoma. Nature. 2011;469:539–42.

101. Vasta V, Ng SB, Turner EH, Shendure J, Hahn SH. Next generation sequence analysis for mitochondrial disorders. Genome Med. 2009;1:100.

102. Vissers LE, van Ravenswaaij CM, Admiraal R, Hurst JA, de Vries BB, Janssen IM, van der Vliet WA, Huys EH, de Jong PJ, Hamel BC, Schoenmakers EF, Brunner HG, Veltman JA, van Kessel AG. Mutations in a new member of the chromodomain gene family cause CHARGE syndrome. Nat Genet. 2004;36:955–7.

103. Wall DP, Tonellato PJ. The future of genomics in pathology. F1000 Med Rep. 2012;4:14.

104. Wang K, Li M, Hakonarson H. ANNOVAR: functional annotation of genetic variants from high-throughput sequencing data. Nucleic Acids Res. 2010;38:e164.

105. Yan XJ, Xu J, Gu ZH, Pan CM, Lu G, Shen Y, Shi JY, Zhu YM, Tang L, Zhang XW, Liang WX, Mi JQ, Song HD, Li KQ, Chen Z, Chen SJ. Exome sequencing identifies somatic mutations of DNA methyltransferase gene DNMT3A in acute monocytic leukemia. Nat Genet. 2011;43:309–15.

106. Yang H, Higgins B, Kolinsky K, Packman K, Go Z, Iyer R, Kolis S, Zhao S, Lee R, Grippo JF, Schostack K, Simcox ME, Heimbrook D, Bollag G, Su F. RG7204 (PLX4032), a selective BRAFV600E inhibitor, displays potent antitumor activity in preclinical melanoma models. Cancer Res. 2010;70:5518–27.

CHAPTER 17

IMPLEMENTATION OF GENOME SEQUENCING ASSAYS

TINA M. HAMBUCH, CARRI-LYN MEAD

Introduction 241
Pre-analytical Considerations: Test Definition, Physician Support, and Process Development 243
Analytical Considerations: Analytical and Bioinformatics Validations and Quality Control 244
Post-analytical Considerations: Interpretation and Reporting 246
Communication and Support 253
Infrastructure Considerations 254
Ongoing Quality Assessment and Control 255
Conclusions 256
Glossary 256

Introduction

The rapid adoption and use of genetic sequencing in clinical laboratories has been largely driven by the evolution of faster, better, and cheaper sequencing methods. The inflection point for clinical application was likely the introduction of automated capillary sequencing methods in the 1990s. With the introduction of next-generation sequencing (NGS) into the clinical space and the some early significant successes, the growth of genetic sequencing in the clinical laboratory is likely to accelerate even more. Today, genetic testing is available for more than 2,000 genes by clinical laboratories around the world (http://www.genetests.org) [1], and this number increases annually. The rapid increase in availability of genetic sequence information has also enabled clinical discovery, which then forms the basis of new clinical tests. As our knowledge of disease biology and genetics increases, the reach and utility of clinical genetic testing will only continue to expand and improve. The implementation of NGS will undoubtedly further accelerate both discovery and testing. In this chapter, we focus on the implementation of whole-genome sequencing (WGS) as a clinical laboratory test. This chapter is organized according to the workflow, and sections are arranged in terms of pre-analytic, analytic, and post-analytic considerations (Fig. 17.1).

Whereas WGS may appear to be a single test, it has many possible indications for use, and each requires different handling throughout the process. Therefore, we discuss the possible clinical indications for testing and the pre-analytical, analytical, and post-analytical requirements for each of these applications. These issues are addressed with regard to current professional and regulatory best practices, guidelines, and resources [2, 3]. However, this field is evolving rapidly, and whereas the principles in this chapter are likely to remain consistent, many details such as specific resources or databases that are discussed are likely to change; therefore, it is strongly recommended that additional resources be consulted when implementing WGS in a clinical laboratory. It

G.J. Netto and I. Schrijver (eds.), *Genomic Applications in Pathology*,
DOI 10.1007/978-1-4939-0727-4_17, © Springer Science+Business Media New York 2015

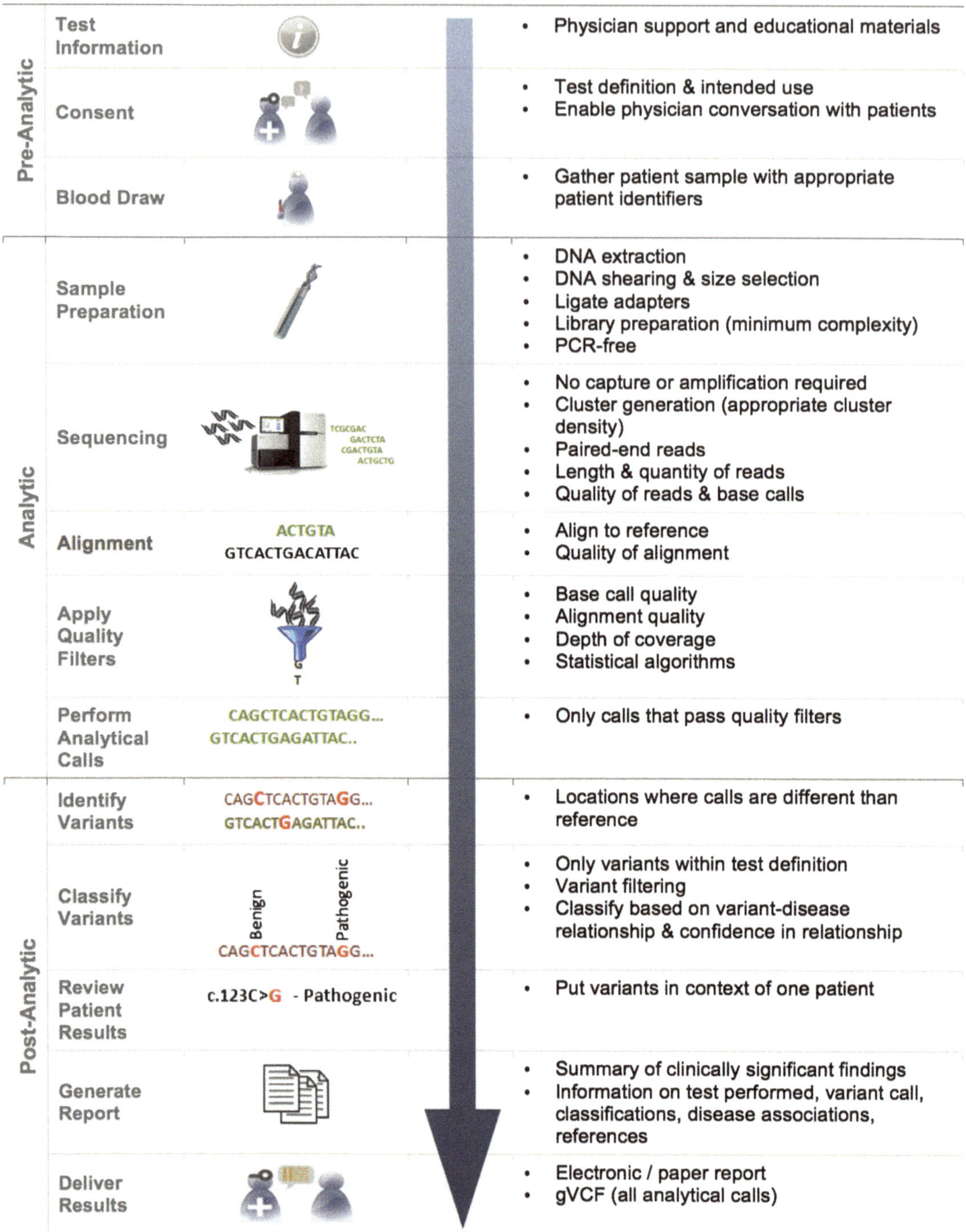

Figure 17-1 Process and workflow for genome sequencing. This figure depicts the major steps in the processing of a genome through next-generation sequencing. The pre-analytic section illustrates the important steps in establishing the test and good communication between the ordering physician and the laboratory. The analytic section shows the processing of the physical sample in the laboratory and the calling of the data using bioinformatics processes. The post-analytic section depicts the steps involved in aggregating information about the results, interpreting those results, and generating a report that can be returned to the ordering physician. The sections in this chapter provide detailed descriptions of these steps

is an exciting time to be involved in clinical genetic testing, as there is an opportunity to help drive important advances in medical care. However, WGS is also a nascent test type and as such the uses, potential, challenges, and concerns have not been entirely characterized yet. Good communication between laboratory and physician, careful analytical and bioinformatics processes, and thoughtful policy development are necessary to offer WGS as a clinical test.

Pre-analytical Considerations: Test Definition, Physician Support, and Process Development

The pre-analytical phase encompasses all steps taken prior to the actual testing of the sample. The introduction of any new test in a clinical laboratory requires several considerations prior to the physical launch of this clinical test. Several guidelines have been published to aid clinical laboratorians with the evaluation of when, how, and why to implement a new test ([4]; CLSI publications (multiple); CAP checklists). These guidelines include discussions of assessing clinical and regulatory concerns as well as financial and workflow considerations. Additional guidelines and recommendations specific to the implementation and offering of genomic sequencing testing have also recently been published [5–8]. The principles established in previous guidelines and best practice recommendations are very much still applicable and certainly should be included in the planning process. However, when the test involves a relatively new methodology that can be applied in a number of different ways, these multiple considerations must be refined and developed by the individual laboratories offering the testing. Newer guidelines, such as the ACMG (American College of Medical Genetics and Genomics) clinical laboratory standards for NGS [9], are particularly useful for considering the additional complexities that this new technology may introduce.

In the case of WGS, as it is commonly referred to, it is important to begin the initial assessment with a test definition and intended use statement to clarify the capabilities and expectations. First, it must be clear that WGS, and in particular clinical WGS, is not representative of every base position of the entire genome, nor can it detect all the types of sequence variants that might be present in a whole genome. For example, all sequencing methodologies tend to be error prone in regions with large nucleotide repeat expansions, such as the CGG repeat expansion associated with Fragile X disease; WGS using NGS is no exception to the rule. Additionally, while WGS is potentially able to detect many types of variants, including single nucleotide variants, copy number variants, insertions, deletions, and translocation events, it is not able to detect all of these different types of variants with the same levels of sensitivity and specificity. In particular, for clinical WGS, thresholds or statistical algorithms can be used to determine whether each variant call meets strict quality metrics that are used to ensure that when calls are made in a clinical context, they meet a minimum threshold of accuracy. This will be discussed further in the analytical portion of this chapter, but is called out here to emphasize that clinical WGS requires additional rigor that might, in some cases, reduce the WGS representation. Some enthusiastic doctors may consider ordering this test for a patient, without realizing that WGS might require supplementary testing in order to prove useful, and, in some cases, may not be the most appropriate test. Therefore, the first consideration when deciding to offer clinical WGS in one's laboratory should be to consider what the test can and cannot be used for, and the degree to which the clinical laboratory is able to support the wide range of potential clinical questions for which WGS might be employed.

In defining the intended use statement for a WGS test, important components to consider include the following considerations:

1. Is WGS to be used as a preliminary screen, confirmatory test, test in aid of diagnosis, or as a test to make prognostic or management decisions after a diagnosis has been made?
2. Is it intended to address conditions caused by inherited or somatic genetic variants?

3. What types of variants are clinically relevant for the population being assessed, and how well does the technology detect these different types of variants?
   (a) Will multiple analyses or methods be combined?
4. What are the technical requirements for the condition(s) being assessed?
5. Who are the ordering physicians and what level of support will they need?
   (a) Are genetic counselors available to support questions from physicians?
   (b) What marketing materials, clear instructions, and definitions of terms will be needed? Will supplementary educational materials be needed?
6. Consent and information return policies for the laboratory
   (a) Who owns or has access to results, and for how long?
   (b) Do results constitute only the clinical report, or could the data be reanalyzed to address different questions at other dates?

When the clinical laboratory answers these questions, it rapidly becomes clear that the same whole-genome sequence could serve to support multiple different test definitions, and might require different support staff and educational materials, as well as multiple processing and reporting policies. A thorough evaluation of the laboratory, the population it serves, and the abilities and needs of both parties will be critical to defining how WGS is offered. It may be beneficial for a laboratory to perform a thorough analysis of the community it serves and to identify the most important needs of that population before defining and offering a test.

Today, the most common use of WGS is in the assessment of rare disease of suspected genetic etiology where symptoms may be overlapping or nonspecific and first tier testing has been inconclusive [10–12] or where WGS presents the fastest possible aid for differential diagnostic evaluation [13]. Inherent in this approach is the expectation that the disease is caused by variants in a single gene (sometimes called monogenic or Mendelian conditions). The primary intention of clinical testing is not gene discovery; however, as with microarray testing, variants may be identified in genes for which the gene function is not yet established, but only suspected or perhaps completely undefined. In such cases, if those variants are thought to be likely causative, additional testing may be required to establish clinical validity and ideally clinical laboratories should have plans for how to make such recommendations to physicians who have ordered the test.

Analytical Considerations: Analytical and Bioinformatics Validations and Quality Control

The analytic phase of the testing begins after the pre-analytic phase and involves all of the processes that enable the actual testing of the sample to produce the analytical result. For NGS, this includes DNA extraction, DNA shearing and size selection, adapter ligation, library preparation, cluster generation, sequencing, alignment, variant calling, and all of the quality metrics associated with every stage of these processes. The process of DNA extraction depends on the type of sample being received, which may differ between different types of WGS tests. Diagnostic testing for Mendelian conditions is typically performed using DNA extracted from peripheral blood, whereas other types of testing may accept other types of samples, for example saliva. Preparation of the sample would therefore include the extraction of DNA from the specific sample type being tested. As part of this process, the evaluation of the quality and quantity of the DNA must occur prior to testing and meet all established quality parameters.

The subsequent process of DNA shearing and size selection, adapter ligation, library preparation, cluster generation, and sequencing are generally the same for all WGS testing, with potential differences being associated with factors such as read length and targeted depth of coverage. Before offering a test clinically, the clinical laboratory must validate the test for the specific performance metrics that were established in the test definition. For example, if one intends to detect substitutions, that must be validated, but a validation performed on substitutions does not assess the ability to accurately identify copy number variation (CNV) or insertion or deletion (indel) events. Additionally, regions that are variable across the genome, such as high and low GC regions, should be evaluated to understand the consistency of base calling quality. Validations are intended to evaluate the analytical sensitivity and specificity, limits of detection, and reportable range. During the process of validation, quality metrics and filters should be established that can then be used during ongoing quality control and assessment.

When considering an entire genome and the resulting number of data points that must be considered in that evaluation, multiple tiered validation approaches may be appropriate. One method of validation is to test the performance with a "truth set" of DNA. Many samples are available that have been sequenced using orthologous technology and contain specific, well-characterized and clinically valid mutations that are known to be causative for diseases (repositories such as Coriell, Hospital for Sick Kids, etc.). Many of these available samples include parent–child pedigrees, so in addition to confirming variant detection, subtraction and filters can be tested using these known relationships. The analysis should account for background conflicts that can be attributed to de novo mutations in every generation (<100/genome). The number of conflicts observed that exceed this background rate is dependent on the choice of aligner and variant caller and the settings that have been used to align reads and make genotype calls. Another approach involves deep sequencing of targeted regions that are relevant to the WGS; in such a case, it is recommended that multiple samples representing various regions of the genome, various GC and other regional genomic characteristics, and various types and complexities of variants are included in the analysis. These samples, if amplified by PCR and sequenced, typically result in a pool of sequence data that is quite deep, for example several hundreds of thousands or even millions of independently sequenced fragments. Re-sampling (bootstrapping) analyses can then be used to evaluate the depth and quality filters that yield high quality sequence. If done across multiple regions and using multiple samples, this experiment can be very useful in establishing the confidence in specific types of calls and in assessing how they calibrate to quality metrics. Additionally, confidence in different types of calls made and for different genomic regions (e.g., percent GC) can be established. Validation of WGS should be updated in the event of any processing changes, regardless of whether the chemistry or platform changes.

The quality of an NGS sequence relies on both the sequencing platform itself and the methods used to analyze the resulting data. For that reason, validations must be designed to establish both the sequencing and the pipeline used for analysis. Specific methods to evaluate the bioinformatics pipeline separate from the platform can be performed using datasets that are rapidly becoming available through efforts such as the National Institute of Standards and Technology (NIST) and Genetic Testing Reference Materials Coordination Program (GetRM). Synthetically generated data can also be used to test specific challenges to calling algorithms.

Transformation of signals produced during NGS into genetic calls of DNA bases involves a highly complex process that utilizes sophisticated bioinformatic analyses. Generally, there are three steps in the analysis—(1) preprocessing of reads, (2) alignment, and (3) variant calling. Preprocessing involves filtering out raw sequence data that do not meet certain quality criteria. The process of alignment involves mapping of reads to the reference human genome sequence, which may be obtained from the National Center for Biotechnology Information (NCBI), University of California Santa Cruz (UCSC) Genome Browser, or Ensembl. There are many tools that employ different algorithms to align reads; each offers trade-offs on speed and accuracy [14, 15]. Mapping is complicated by the fact that the reference genome is incomplete and because humans have some regions that may be individually variable. Because of this, approximately 5–10 % of reads will fail to be aligned. Mapping quality is measured and the confidence score assigned with each read placement. One of the community-accepted standards to represent alignments is in Binary Alignment Map (BAM) file format [16], which captures the above-mentioned data, allows efficient compression and enables random access of reads (when sorted) that align to a particular segment of the genome. Once the alignment procedure is complete, the BAM file serves as input to the next step in the bioinformatics pipeline—variant calling, where genetic variants are identified. Depending on the intended use of the test, a variety of variant calling tools, each one specializing in detecting small (single nucleotide variants (SNVs) and indels) or large genomic alterations (structural variants (SVs) and CNVs), might be employed. In some cases, several tools might be used in conjunction to identify different types of variants.

Variant calling algorithms are typically based on two main paradigms—the first one involves relying on base counting and allelic fraction to distinguish between a heterozygous and homozygous genotype call, the second involves probabilistic methods (Bayes' theorem) to calculate posterior probability given the observed read data and a genomic prior probability [17]. The latter method accounts for noise in the data and helps provide a measure of statistical uncertainty associated with each genotype call in the form of a score. The score is usually a representation of the confidence in the genotype call. Although many algorithms report on variant positions, it is important to consider that the reference genome may contain a non-wild type allele, and to monitor the quality of the positions called as homozygous to the reference; no calls and poor quality homozygous reference calls should be considered in the downstream interpretation effort.

During the validation process, a clinical laboratory that is implementing WGS should be aware of and test for potential artifacts in processing. For example, the reference genome is not necessarily wild type. Perhaps the most prominent example of this is that the reference genome carries the Factor V Leiden mutation. Therefore, if a laboratory is only considering the variants that are called against the reference, such mutations may be missed in an individual who also carries this genotype. Assessment of reference allele frequency based on the 1,000 Genomes Project data shows that there are approximately 63,000 positions in the genome where the reference genome carries an allele that is present in populations at less than 1 % allele frequency. Additionally, for regions of the genome such as Human Leukocyte Antigen (HLA) locus, there is not necessarily a "wild type" per se, and additional information such as phasing may be necessary to confidently evaluate the variants found. While similar challenges exist for many types of clinical tests, laboratories should be aware of and prepared to manage such issues.

One challenge to the implementation of WGS in a clinical laboratory is that the analytical validity may not be the same for all regions of the genome, nor is it for all types of variants that may be of interest. The specific weaknesses and strengths of WGS must be considered when launching a test, and then communicated effectively and evaluated, potentially on a case-by-case basis, for appropriateness given the needs of the test in that specific situation.

Post-analytical Considerations: Interpretation and Reporting

The post-analytic process occurs after the analytic phase and includes the interpretation and reporting of the analytical calls produced in the analytic phase. As with the previous phases, the type of testing being performed has a significant impact on the post-analytic process.

A genome is approximately 3.1 billion data points, and contains around three to four million variable positions, including on average 9,600 amino acid changing positions and 73 premature termination positions (internal data). Given such a large amount of information, a thoughtful plan must exist for how to identify and evaluate the information that is most likely to be relevant and informative to the clinical questions being considered.

After achieving confidence in the quality of the genetic calls that have been made and defining the regions for which calling can be done with confidence, the clinical implications of the call should be assessed. This process can be divided into annotation, in which information and meta-data are gathered about the variant calls, interpretation, in which all the information is evaluated in the clinical context, and reporting, in which the information is communicated back to the ordering physician.

Historically, the assessment of clinical validity, or the strength of the relationship between a variant (or call) and a disease has been recommended but not required in genetic reporting. This is changing, and recent College of American Pathologists (CAP) guidelines now address how clinical laboratories should support the assessment of the clinical implications of a call. For instance, in cases with a single gene, this typically consists of an expert or panel of experts within the laboratory who evaluate each variant based on peer-reviewed publications and other factual evidence and categorize them for inclusion in the report. This process has become significantly more sophisticated in recent years, as there are now several databases and online tools that can aid in the assessment of clinical implications of variants.

The process of information gathering can be automated and is commonly referred to as annotation. As WGS is implemented in a laboratory, using a series of automated tools for annotation becomes necessary to support the large number of variants that are detected and require downstream evaluation. Tools are available online, such as the Variant Effects Predictor (VEP, part of the Ensembl suite of resources) that will gather information from a variety of databases as well as predict characteristics such as amino acid change based on transcript. Recommendations for types of information that should be gathered can be found in official publications by CAP and ACMG; these include information about which gene and transcript a specific variant is found in, the position of the variant in the genome, the DNA and amino acid change produced by the variant (using Human Genome Variation Society (HGVS) nomenclature), the consequence of the variant (e.g., intronic, upstream, missense, stop gained, synonymous), characteristics such as frequency of the variant and conservation of that position. Additionally, in silico structure or function prediction software such as SIFT [18] or PolyPhen [19] may provide additional information. When implementing the annotation process, it is very important to assess the annotation software suites that will be used, and confirm that the variant is being searched correctly and the information gathered is being downloaded and displayed properly. It is also important to keep in mind the unproven nature of many of the prediction software tools; while these may be useful in assessment, they are not yet reliable.

Having annotated the positions, interpretation of the variants for reporting can begin. The evaluation of evidence around what the clinical implications of variants might be is a critical process that is guided by both professional expertise [20, 21] and a pipeline that can support such evaluations (Fig. 17.2). Several biological and clinical characteristics of a variant should be considered. Biological characteristics include the type of variant, where it occurs in the gene, the frequency of the variant and possibly in silico evaluations of the variant. Clinical characteristics include whether the variant has been reported to be associated with a condition or phenotype and can take the form of case study reports, case–control studies and functional evaluations of the effect of the mutation in vitro or in vivo. These peer-reviewed publications in which a clinical phenotype or functional effect has been measured in individuals who carry the variant that is being reviewed are often the most compelling evidence in variant assessment. Careful literature searches, or searches of appropriate databases may be helpful in identifying the full body of literature that exists. It is important to remember that these databases may or may not be updated regularly, and may or may not be complete with regard to the actual publications that exist. Furthermore, many variants have been characterized in databases based on old information, and therefore, if a database reports a variant as pathogenic or of uncertain significance, it is important that the clinical laboratory perform an updated and independent assessment to ensure that this information is still valid. The gene in which the variant was detected must also be considered, and the degree to which the relationship of the gene relative to the disease is relevant. This includes a familiarity with phenomena such as whether certain types of mutations (e.g., activating) or regions of genes (specific exons) are known to be more or less likely to be associated with a particular disease. With regard to the disease, specifics of the mode of inheritance, prevalence of disease, and age of onset are important considerations. For example, if a disease has a prevalence of 1/100,000 and is autosomal recessive, then, using Hardy–

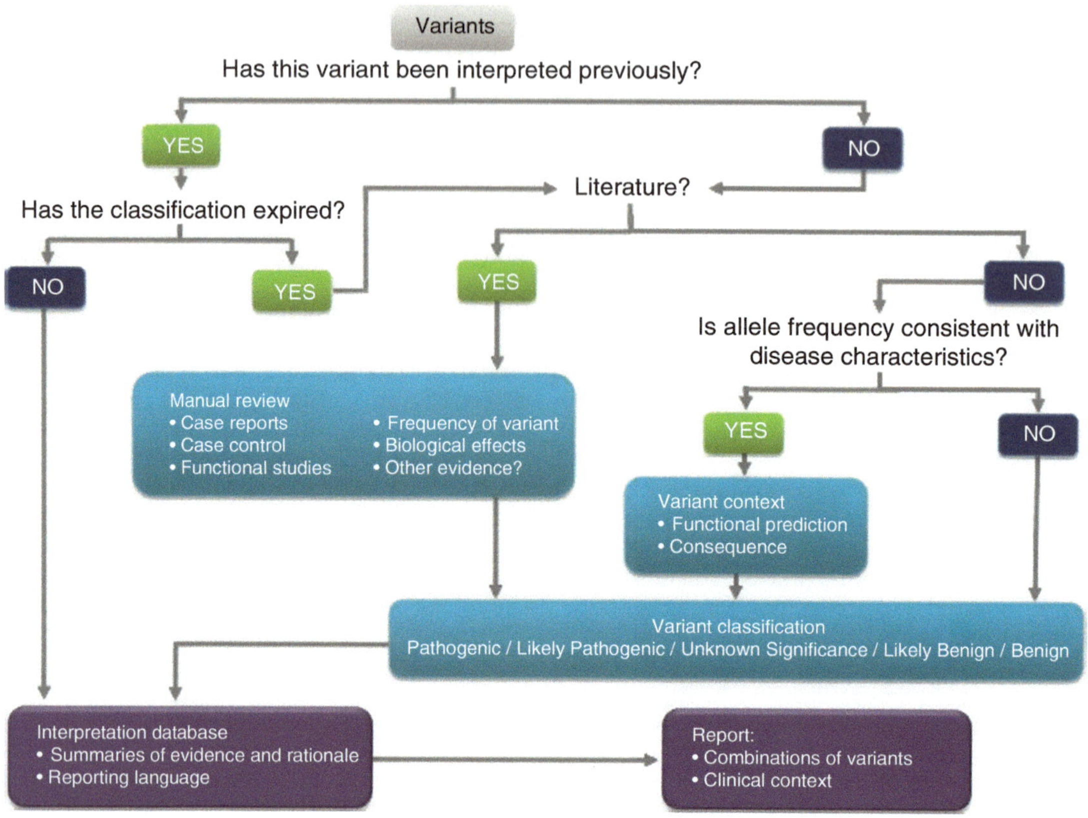

Figure 17-2 Decision tree for the evaluation of clinical implications associated with sequencing calls. The process shown is the one that the Illumina Clinical Services Laboratory uses for the evaluation of evidence that links a particular allele to a clinical condition

Weinberg principles, a variant with a frequency higher than 1 % is unlikely to be causing that disease. Likewise, if mutations known to cause disease are exclusively gain-of-function, then a stop mutation or silent mutation is less likely to be considered pathogenic. Finally, when reporting the results, the clinical questions and context of the patient must be considered: is this a diagnostic evaluation or a carrier screen? What other tests have already been performed? Is there any additional phenotypic information that might be relevant for the results and how they should be considered? This is a complex set of considerations and requires knowledge of clinical and technical genetics.

Literature has historically been an important source of information regarding the clinical associations of a genetic variant to a disease. The ability to publish case reports has been critical to help identify genes and variants that are suspicious for possibly causing a disease. However, these are often just first steps and subsequent studies in which cases are evaluated against controls, either in pedigrees or populations, and additional functional evaluations may be critical for providing evidence that a variant is likely causative for a disease. Very often an initial report will appear in which a variant is found in a gene that is known to be associated with disease, and the variant may seem a compelling explanation for the disease, but further studies show that the variant is also found in unaffected individuals or has no effect on the protein function. It is therefore imperative to review all the literature associated with a variant, as well as to evaluate how strong the data in the papers are before determining one's confidence level regarding a variant's pathogenicity or lack thereof. Some articles are more robust than others, and should be weighted accordingly. This process is often where clinical laboratories have real specialties because they typically have a well-trained, clinically oriented staff of MDs, PhDs, and Genetic Counselors who review the evidence presented in these papers and bring all of the considerations listed in Fig. 17.2 into the final reporting language. This process is challenging and time-consuming at a single gene level, but represents what is likely the biggest challenge at the level of genome sequencing.

There are many tools and approaches that have been or can be developed to help manage this burden on the clinical laboratory. Clearly defining how filtering tools will be applied based on the case at hand and indications for use will significantly reduce the interpretive burden at this point. Other approaches, such as ruling out variants with high frequencies or those that are synonymous, before having to do additional downstream investigation, are commonly practiced in clinical laboratories. When choosing to do this, however, it is important to consider aspects such as how common a disease might be in a particular ethnic population and incomplete penetrance, which might lead a laboratory to incorrectly ruling the variant out. Natural language processing tools have also been suggested to assist with the burden of reading published literature. Collection of the literature associated with variants requires a standardized set of terms, such as that used by the HGVS [22, 23] that search tools can use. The extent to which natural language software and other software approaches can automate the evaluation of variants is still highly debated. It is clear that these tools are invaluable for collating the information. One challenge is simply reading the papers. Individuals must be able to read through a paper, evaluate the strength of evidence regardless of author's conclusions, and document this. This currently requires professionals spending a significant amount of time sifting through that information. Every clinical laboratory faced with the numbers of variants to be assessed in WGS will be challenged to hire a qualified staff large enough to support such efforts. For this reason, databases in which such information is available and could be shared become extremely valuable. At the same time, each laboratory that builds up these databases incurs a huge expense in this effort. How the laboratories can create community access that will benefit other laboratories and ultimately the patients while still paying for the effort required of them to create this information is an interesting and active area of exploration.

Designing the Post-analytic Process for Monogenic Conditions

Given the daunting number of variants to consider, approaches must be developed to apply filters so that only variants of potential relevance are identified and evaluated. Both biological and clinical features can be used to help refine the search for genomic information. In cases where parental samples are available, a geneticist or genetic counselor should begin with taking a family history to identify whether the current condition is most likely to be autosomal recessive, autosomal dominant (possibly with reduced penetrance), or de novo inheritance. If both or even one parent sample can be sequenced along with the affected individual, then subtraction can be performed across the entire genome in order to evaluate variants that meet the biological hypothesis of the following conditions:

- Autosomal recessive, in which one would expect to find at least two variants within a single gene, one inherited from each parent. To perform this search, all three samples are sequenced and the child's variants are filtered to match the expectation of two variants in a gene, one from each parent. This can significantly reduce the number of variants that must be considered. After this subset of gene/variants is identified, the genes and specific variants can be filtered further. For example, common variants with allele frequencies above 5 % might be excluded from consideration; when making such decisions, patient ethnicity, prevalence of the condition in that ethnic group, penetrance, and modes of inheritance should be considered because sometimes common variants are pathogenic. Through the use of these types of filters, the resulting subset of variants should be of a tractable number that can be individually evaluated by qualified clinical laboratory staff.
- Autosomal dominant, in which one would expect to find only a single causative variant

within a gene. This model is more difficult because there are significantly more possible variants to evaluate; however, if there is a family history (even with reduced penetrance), one can subtract variants from the unaffected side of the family and look for matches to the presumed carrier parent (who may or may not be affected). Again, additional filters to remove high frequency variants can be applied and the resulting variants can be considered.

- De novo, in which the causative variant arose within the proband. In this case, all variants inherited from both parents can be subtracted and only those variants that arose in the affected individual can be considered.

For all of the above methods, the process may also include the evaluation of the resulting variant set in the context of the clinical phenotype or a defined set of genes that are set out by the physician/medical geneticist. This could take the form of a filtering tool that enables the list of variants to those in genes known to be associated with the phenotype, or simply as part of the context that the clinical laboratory staff uses during the evaluation process.

These types of approaches based on filtering by modes of inheritance and parental genomes are currently the most popular way of WGS testing. However, this does require the added expense of sequencing multiple genomes in order to identify potentially causative variant(s). Sometimes, the parental samples may not be available, or the additional cost may be prohibitive. In such cases, a clinical phenotype approach can be used on its own. This is not likely to be as effective as a biological approach, but it has been used successfully in several cases. For this approach, one requires access to thorough clinical phenotype information, such as all presenting features or previous testing results (e.g., no increased creatine kinase). This information can then be used to search through phenotype-to-gene information that is available in various databases (e.g., Online Mendelian Inheritance in Man (OMIM) [24]), or accessible within phenotype software tools, to identify and rank order genes that might be involved with the symptoms affecting the proband. Then, all variants within that subset of genes can be considered, with additional filters applied to remove variants that are too common to be likely involved with disease.

Each of these approaches is labor intensive and requires a clinical laboratory staff trained in the evaluation of genetic disease, preferably formally trained and certified through the American Board of Medical Genetics (ABMG), American Board of Pathology (ABP) or the American Board of Genetic Counseling (ABGC). In the case where the first assessment is found inconclusive, multiple different approaches might need to be performed. The clinical laboratory team performing the filtering and variant assessment should expect to spend several hours per genome evaluating the resulting variants, and this type of effort should be budgeted for in the planning for this type of testing. For laboratories that have implemented these types of approaches, diagnostic yields ranging from 30 to 40 % have been reported (personal communication); keeping in mind that these are often patients for whom all other testing has failed. The cost and time investment of the WGS test must be considered against the potential costs and consequences to the affected individuals of undiagnosed genetic disease.

Designing the Post-analytic Process for Oncology Applications

Another possible use for WGS is in the assessment of the molecular profile of tumors in patients who have already been diagnosed with cancer. This type of testing can be useful in determining candidate therapeutic treatments when standard of care approaches have been exhausted. In these cases, the tumor sample and a normal sample of DNA are procured from the patient. Variants found in the normal sample are subtracted from the tumor sample, so that only variants that have arisen somatically can be identified. In a somewhat unique manner, most laboratories that perform clinical oncological testing will have a tumor board associated with the laboratory that reviews the findings and contributes to the interpretation. Results from this type of testing, the clinical laboratory and the associated tumor board may be able to identify the most promising chemotherapeutic options based on the presentation of the molecular profile. Of particular interest for oncological applications are large chromosomal rearrangements, insertion or deletion events, and copy number variants that can be identified. Anecdotal reports of these approaches have been very encouraging [25, 26].

The analyses required for detection of tumor variants are significantly more complex than those described above for Mendelian conditions. In the analytic phase, special consideration about the sample type should be given based on the type of cancer being tested. For example, blood samples in leukemic patients would likely be more representative of the tumor rather than the normal signal and the type of tissue most appropriate for the normal sample should be thoughtful considered. Beyond that, the analytic process for the normal sample is essentially the same as what would be done for the monogenic conditions (described above). Tumor samples, however, require special additional processing and handling. To begin with, the DNA isolated from a tumor may be from fresh, fresh frozen, or more commonly, from formalin-fixed paraffin-embedded (FFPE) tissue. The different tissues may require significantly different extraction techniques and evaluations of the quality of that DNA. Laboratories must evaluate their abilities to support each of these extraction techniques and subsequent evaluations of appropriate DNA quality and quantity. The downstream informatics processing of tumor samples also has some unique requirements. Tumor samples are often contaminated with some amount of normal cells. Quantifying this fraction is difficult and imprecise and has implications for downstream informatics processing that must be incorporated into the process. Additionally, NGS methods sequence individual molecules separately, and therefore, in a diploid situation a heterozygote would be expected to have approximately half of the sequences showing one variant and half with the other. The algorithms that have been developed for NGS typically have been developed to optimize for this scenario, and general recommendations regarding the required number of independent sampling events are also usually made with this expectation. However, a tumor does not represent a diploid scenario. Therefore, one must establish at what frequency one wishes to detect somatic variants; this might be 20 %, 5 %, 1 %, or less. Depending on what the laboratory decides, sequencing must be done to a depth that ensures likely detection of variants. The depth required to attain the required sensitivity can be estimated using a sampling statistic:

$$P(x,p,N)=\sum_{K=X}^{N-X}\frac{N!}{(X!)(N-X)!}p^{X}q^{(N-X)}$$

Empirical validation will be discussed in the validation section. However, in addition to different processing requirements, the bioinformatics algorithms used to detect variants may also need to be optimized, and additional or alternative algorithms may be needed. In some cases, different algorithms may be called for to detect different types of variants, for example copy number or structural variants (chromosomal rearrangements). Laboratories planning to launch tumor–normal WGS analyses should be prepared to evaluate these needs and plan appropriately for implementation. This can be an arduous process and a team may be needed to identify the requirements and evaluate the appropriate set of tools for implementation.

Cancer is not the only disease type that is associated with the occurrence of somatic variants; certain genetic conditions (often associated with hemi-hypertrophy or skin lesions and increased likelihood of developing cancer later in life) may also demonstrate these and be of interest for a clinical molecular lab. Additionally, in testing for mitochondrial diseases, it may be critical to enable detection of mitochondrial heteroplasmy. All of these applications involve the challenges described above for tumor scenarios, and may require the same or similar planning and evaluations before implementation.

Designing the Post-analytic Process for Screening for Fetal Aneuploidies

WGS can also be used for various forms of screening tests. Screening involves identifying genetic variants with potential clinical implications, typically before there is any clinical presentation, and often that would be confirmed by additional testing before any medical action is taken. Currently, the most common and popular screen involving WGS is for aneuploidy in prenatal settings. Commonly called noninvasive prenatal screening or testing (NIPS or NIPT), this involves performing deep sequencing of either targeted regions or the whole genome in an effort to identify chromosomal regions that are present at non-diploid copy numbers. These kinds of screens have only been available in the last few years, but their sensitivity and specificity is greatly improved over serum screening paradigms and therefore is being rapidly adopted, particularly for high-risk pregnancies. These tests are performed from a maternal blood sample where the DNA for the testing is fetal DNA circulating in the maternal blood stream, and thus considered noninvasive from the perspective of the fetus. Because this testing requires isolation and enhancement of the fetal DNA, specific planning should be given to additional techniques that might be necessary for implementation, such as for DNA isolation and quality evaluation to ensure that the appropriate quality and quantity of DNA is present to perform testing. This test also requires quantification of genomic regions that are present at non-diploid quantities, and the subsequent analyses.

Designing the Post-analytic Process for Predisposition and Carrier Screening

Finally, WGS can also be used for more traditional screening of genetic variants for which individuals may be carriers or at risk. While this type of testing is currently more likely to be performed using targeted panels, it is possible to employ WGS for this purpose. The post-analytic process for this type of testing is heavily dependent on the test definition provided in the pre-analytic phase. Typically, this would have identified a set of genes that would be included in the test, and this set would have established clinical utility of testing for a specified set of diseases. In the case of WGS, this can be a many-to-many relationship where there may be many genes tested that are providing information about predisposition or carrier status for one disease, but also any one gene could have multiple diseases clinically associated with it. The test definition would also define the regions within those genes that are included in the test (e.g., exonic regions, parts of intronic regions directly adjacent to the exons). Therefore, the set of variants requiring interpretation from the analytical stage would be filtered to those included in the established test definition. Most laboratories performing this testing restrict reporting to those variants assessed as clinically significant; however, ancillary documentation of all assessed variants and their classifications are included in some cases.

Once the clinical implications of a particular individual's variants have been decided, the information must be put into the clinical context for which the test was ordered. Incidental findings can potentially be quite numerous, and additionally, a single answer might not be found. There could be three or four variants in two or three genes that plausibly lead to a patient's symptoms and have equally inconclusive or conclusive evidence supporting them. Indeed, one discovery is that in at least a few cases, patients have been discovered to be suffering from more than one genetic disease [10] explaining the perplexing clinical presentation. Reports must be flexible enough to enable the benefit of a personalized survey of the genome, but standardized enough to enable clear communication of results. A searchable electronic report might be the best solution; this could provide links to disease descriptions and additional evidence that practitioners could then have access to as needed. The goal is to provide a succinct answer to the major question of the moment, but also to enable both the physician and patient to benefit from the additional information that may be present and of concern. One challenge with whole-genome evaluation is that our understanding of genetics and biology is not perfect or complete; most variants that will be detected will be of uncertain significance. This is also an area in which it would be of benefit for clinical laboratories to communicate more effectively with physicians and genetic counselors. Whereas it will require an upfront time commitment as well as tools that enable communication, it might well be worthwhile for laboratories to ensure that doctors and genetic counselors have access to the following information before they receive their reports:

1. The standards that a laboratory uses in order to make calls

2. How laboratories classify variants into the standard bins of Pathogenic, Likely Pathogenic, variants of uncertain significance (VUS), Likely Benign, Benign, or other
3. How much confidence a practitioner should have in that call
4. What the weaknesses of the test are, and any recommendations regarding additional testing that could supplement these weaknesses

Communication tools might be readily located on clinical laboratory Web sites, where quick, 5-min podcast type communications might provide both doctors and genetic counselors with information that can significantly increase the power and confidence they have when using a test.

An ongoing challenge will still be the large number of variants about which people are uncertain. While a large number of VUS is a point of concern, this is not new to the field. The International Standards for Cytogenomic Arrays Consortium (http://www.iscaconsortium.org) [27] has demonstrated approaches to dealing with the large number of novel and uncertain variants that are detected in individuals when genomic evaluations are standardly performed. In less than a decade the cytogenetics community has made huge strides in understanding the nature and degree of variation at the cytogenetic level. Similar approaches could be used in the field of sequencing to better understand the nature of human genetic variation, which will aid significantly in improving and refining interpretation in the future. Meanwhile, clinical laboratories can make every effort to communicate a priori that this is an anticipated outcome of these tests, and help to prepare physicians and genetic counselors for managing the information.

Communication and Support

Once a test has been defined and the performance specifications and abilities established, it is critical to develop support materials. The laboratory should also be staffed with trained genetic support specialists. These specialists should be available to help physicians decide if WGS is the best test for the presenting situation, and also to help plan for alternative or supplemental testing that might be necessary. It is of particular importance, but also particularly challenging to communicate this when the very title "whole-genome sequencing" might imply all things to everyone. It is helpful to provide information through a Web site that can help individuals evaluate what the test supports and what it does not.

Depending on the breadth of WGS services that a laboratory intends to offer, it may be helpful to develop an overview section that clarifies which tests offer what, and are likely to be most appropriate. As information such as analytical validity, limitations of detection, and reportable regions need to be included in test definitions, and because these will be variable depending on the application of WGS, it is likely that it will be necessary to create multiple test definitions and descriptions. Including general educational materials will help physicians and patients navigate the options and choose most appropriately. Importantly, information should be readily available to help physicians understand the limits of detection, such as an ability to detect variants present in the sample at, for example, 10 % but not 5 % in tumor samples, or the ability to detect deletions within certain size ranges. Laboratories should be prepared to monitor and track their capabilities to make calls of any type throughout the genome. As tests are ordered, laboratory staff evaluate the test requisitions and evaluate the laboratory's ability to support the request. If there are concerns about whether WGS is appropriate for the sample being ordered, the laboratory should contact the physician and discuss the options before the testing is initiated.

Genetic counseling is a best practice recommendation for genetic tests in which the results may have direct medical indications for immediate family members or in which the results might be predictive. WGS produces information that meets those criteria, not only for the specific indication of the testing, but also for secondary findings. The ACMG has issued a series of recommendations for clinical genomic testing, counseling, and consent [28]. The ACMG has stated that genome or exome sequencing is appropriate in a series of circumstances that include strong reason to suspect a genetic etiology, symptoms associated with multiple genetic conditions for which simultaneous evaluation of multiple genes can be practical, inconclusive previous tests, and, in special cases, prenatal diagnosis. WGS is not advised at this time for prenatal or newborn screening. The recommendations specifically advise that the following elements be addressed in counseling and consent sessions: (1) pretest counseling including written documentation, (2) discussion of potential for incidental findings, (3) discussion of expected outcomes as well as incidental findings to be returned to physician, (4) potential benefits, risks and limitations of testing and if there are alternatives, (5) distinction between clinical testing and research, (6) potential for results to be identifiable in databases, and (7) policies for updating information. It is also recommended that such testing only be performed on minors in cases where the testing can lead to diagnosis for conditions in which interventions might be possible, and under institutional review board (IRB) approved research. Additionally, the ACMG has recommended that everyone who has access to WGS, regardless of indications, should have results reported for a set of 56 conditions. These conditions represent highly penetrant genetic conditions for which there are potentially life-saving interventions available. Although these recommendations have been controversial, it is indicative of the medical community's rapid adoption and preparations to manage this information in regular clinical practice.

After WGS analysis and interpretation has been performed, additional communication with the ordering physician is likely to be necessary. While inconclusive test results are not uncommon for physicians, findings may require additional communication, particularly with regard to the management or further testing of VUS.

Infrastructure Considerations

After identifying what the WGS test will be used for, the clinical laboratory should consider the current infrastructure and any possible additional needs that would require additional build out. Depending on what resources and infrastructure a laboratory has, an assessment of necessary components includes the following:

- Facility
  - NGS sequencers are not usually very bulky, but they require space that is stable, climate controlled and has both power and Internet support. Specific requirements include uninterruptible power supply (UPS) and e-power setup, with heat, ventilation, and air conditioning (HVAC), temperature and humidity control at around 68–72 °F and 70 % relative humidity. Laboratories are required to practice space separation between pre- and post-amplification activities, and ideally would have negative pressure control on rooms that could have contamination, use a pressure-controlled hood. Additional safety precautions may also be necessary depending on specific requirements.
- Staff
  - NGS is considered to be high complexity testing and involves many steps. A well-trained staff is critical for this. Typically, a staff to support WGS will require people with expertise in high complexity molecular assays, genetics analyses, bioinformatics, and genetic counseling.
- Workflow process
  - WGS may be among the easier of the NGS assays to perform in that there are no capture or amplification steps (Fig. 17.1). Nonetheless, there are still several manual steps required and each of these can potentially introduce a contaminant or sample swap. In order to avoid such complications, a good workflow process and ideally a laboratory information management system (LIMS) to track and document a sample's process through the assay steps should be implemented. Assessment of steps in the process that can be error-prone is critical to designing a workflow in the laboratory that is robust, and consideration of appropriate controls, performance metrics, and tracking systems is prudent. In particular, positive sample controls are recommended because pre-analytical sample swapping is one of the

most common errors introduced into clinical testing.

- Computing and bioinformatics infrastructure
  - A high-performance storage and computing cluster (a set of connected computers that work together as a single system) is necessary to perform whole-genome sequence analyses in high volumes. These analyses can be performed on a computing cluster consisting of many multi-core computers. An evaluation of these needs should be based on predicted volumes and specific analytical requirements for the test(s) that will be supported. Additionally, a tracking system for recording quality metrics across and within each sequencing run, lane, and sample is extremely useful for catching runs that go poorly and not wasting time and money on failed runs. These types of tracking system can also enable users to identify when additional sequencing will be necessary. Finally, bioinformaticians who are skilled in these analyses are important members of the NGS clinical team.
  - A data management system for storage of genomic information should be planned for before implementing WGS in the clinical laboratory. Various guidelines suggest that sequencing results that could be used in evaluation of hereditary conditions should be stored for multiple years [3, 4]. The recently released CAP NGS checklist requires that data be stored for a minimum of 2 years to enable reanalysis of NGS results. This is in addition to other requirements around storage of actual clinical deliverables. What will be stored, and how it will be stored requires thorough consideration.
  - Many software tools are available to support the multiple steps involved in WGS analysis. An evaluation of which tools should be used based on the intended use of the test should be performed. Once the right set of tools is identified, users may need to create a workflow using custom scripts that enable the usage of several tools, keeping in mind that input and output abilities and requirements may be variable among these tools. The software tools used in the analytical calling and downstream analysis and classification of variants are among the most variable aspects of clinical WGS being performed today. It is critical that laboratories understand the caveats and limitations associated with any of the software tools being used in their data analysis pipeline.
- Security
  - It is likely that WGS will be considered impossible to make anonymous. Privacy concerns around how these data are stored, when and how they are updated, who should have access, and what should go into the medical record are currently not well addressed by policies. However, laboratories are thinking about how this is likely to change and what safeguards and options they will be able to offer the doctors and patients who are interested in ordering WGS.

Ongoing Quality Assessment and Control

After validations have been performed, quality filters and metrics established, mechanisms are developed to monitor ongoing performance during testing of clinical samples. The process of genome sequencing can be divided up into three stages: wet-lab processing, bioinformatic analysis, and interpretation and report generation. The wet-lab component encompasses DNA extraction, DNA shearing and size selection, ligation of oligonucleotide adaptors to create a size-selected library and physical isolation of the library fragments during amplification and sequencing. Each step of the process should be considered for the implications of a failure or contamination event; accordingly, the quality monitoring should be designed to detect the most likely or significant possible failures. Specifically, DNA extraction, library preparation, cluster generation, and the sequencing run should be assayed for quality. There are many ways in which quality can be monitored, and these include establishing run metrics at various steps, performing quality assessment steps (such as quantitative PCR (qPCR), DNA quantification and purity measures, run metric measures). Robotics and automation are valuable additions that can be made to a protocol to minimize the possibility of human error. Future advances to further combine the sequencing laboratory steps with automation will increasingly assure a reduction in potential errors. Controls can also be useful in the assessment of run quality. External controls, such as lambda DNA fragments, can be spiked into samples to measure the success of the run. Alternatively, orthologous assays such as microarrays can be utilized to measure sequencing accuracy at a very high level by comparing the concordance of calls from a genomic level microarray to the sequencing calls.

Proficiency testing is one method that is used as part of ongoing quality assessment. The molecular pathology on-site inspections by the CAP occur every 2 years, but ongoing proficiency testing with both intra- and inter-laboratory analysis improves testing procedures and helps to prevent errors (reviewed in [4]). As several clinical laboratories are currently offering genomic level sequencing, alternative proficiency testing programs are used to enable laboratories offering exome and genome sequencing to compare their calls. In a recent exchange between the Illumina Clinical Services Laboratory and the University of California, Los Angeles (UCLA) molecular pathology laboratory comparing two samples that had been run and reported in both laboratories, both laboratories made calls for 3,573,631 sites, of which 19,340 represented variants from the reference. Across all the calls made, 16 positions were called discordantly between the two laboratories. Investigation of such discordantly called sites, along with relative quality metrics from each run and the types of variants these sites represented (e.g., high GC regions or repeat regions) will help participating laboratories improve quality.

Conclusions

The implementation of clinical WGS is not trivial, and the suggestions made in this chapter highlight the need for well-trained teams that bring diverse expertise to the clinical laboratory. One challenge that is often raised is the lack of experts available; this is a legitimate concern and for that reason community efforts for establishing guidelines, and promoting education and best practices are critically needed. Ongoing training and certification, active participation in societies and meetings, and regular review of recent guidelines and publications will be necessary particularly during the early phases when the learning curve will be steep and policies are likely to evolve. That said, this is also a great opportunity for clinical laboratorians to work closely with their medical practitioner colleagues, as well as with experts in diverse fields such as bioinformatics, population genetics, and information technology to create a new approach to evaluating, diagnosing, and managing genetic disease using entire genomes of information.

Glossary

Proband Affected individual on whom testing is being performed.

Mendelian condition A condition that is caused by variants within a single gene and that can be passed to offspring in an autosomal dominant or autosomal recessive pattern.

Disease prevalence Proportion of a population to have a condition.

Allele frequency Proportion of a particular allele among all alleles for a gene.

REFERENCES

1. Gene tests. http://www.genetests.org. October 2013.
2. Clinical Laboratory Standards Institute (CLSI). Multiple standards publications relating to laboratory process and sequencing. http://www.clsi.org/standards/.
3. College of American Pathologists (CAP). Molecular pathology checklist. 2013:1–67.
4. Chen B, Gagnon MC, Shahangian S, Anderson NL, Howerton DA, Boone DJ. Good laboratory practices for molecular genetic testing for heritable diseases and conditions: Department of Health and Human Services, Public Health Service, Centers for Disease Control and Prevention. MMWR Recomm Rep. 2009;58(No. RR-6):1–37.
5. Gargis AS, Kalman L, Berry MW, Bick DP, Dimmock DP, Hambuch T, et al. Assuring the quality of next-

generation sequencing in clinical laboratory practice. Nat Biotechnol. 2012;30(11):1033–6.
6. Green RC, Berg JS, Grody WW, Kalia SS, Korf BR, Martin CL, et al. ACMG recommendations for reporting of incidental findings in clinical exome and genome sequencing. Genet Med. 2013;15:565–74.
7. Group AMPWGAW. The Association for Molecular Pathology's approach to supporting a global agenda to embrace personalized genomic medicine. J Mol Diagn. 2011;13(3):249–51.
8. Schrijver I, Aziz N, Farkas DH, Furtado M, Gonzalez AF, Greiner TC, et al. Opportunities and challenges associated with clinical diagnostic genome sequencing: a report of the Association for Molecular Pathology. J Mol Diagn. 2012;14(6):525–40.
9. Rehm HL, Bale SJ, Bayrak-Toydemir P, Berg JS, Brown KK, Deignan JL, Friez MJ, Funke BH, Hegde MR, Lyon E, Working Group of the American College of Medical Genetics and Genomics Laboratory Quality Assurance Committee. 2013:1–15.
10. Ng SB, Buckingham KJ, Lee C, Bigham AW, Tabor HK, Dent KM, et al. Exome sequencing identifies the cause of a Mendelian disorder. Nat Genet. 2010; 42(1):30–5.
11. Dinwiddie DL, Kingsmore SF, Caracciolo S, Rossi G, Moratto D, Mazza C, et al. Combined DOCK8 and CLEC7A mutations causing immunodeficiency in 3 brothers with diarrhea, eczema, and infections. J Allergy Clin Immunol. 2013;131(2):594–7.
12. Worthey EA, Mayer AN, Syverson GD, Helbling D, Bonacci BB, Decker B, et al. Making a definitive diagnosis: successful clinical application of whole exome sequencing in a child with intractable inflammatory bowel disease. Genet Med. 2011; 13(3):255–62.
13. Saunders CJ, Miller NA, Soden SE, Dinwiddie DL, Noll A, Alnadi NA, et al. Rapid whole-genome sequencing for genetic disease diagnosis in neonatal intensive care units. Sci Transl Med. 2012;4:154.
14. Flicek P, Birney E. Sense from sequence reads: methods for alignment and assembly. Nat Methods. 2009;6(11 Suppl):S6–12.
15. Li H, Homer N. A survey of sequence alignment algorithms for next-generation sequencing. Brief Bioinform. 2010;11:473–83.
16. Li H, Handsaker B, Wysoker A, Fennell T, Ruan J, Homer N, et al. The Sequence Alignment/Map format and SAMtools. Bioinformatics. 2009;25: 2078–9.
17. Nielsen R, Paul JS, Albrechtsen A, Song YS. Genotype and SNP calling from next-generation sequencing data. Nat Rev Genet. 2011;12(6):443–51.
18. Kumar P, Henikoff S, Ng PC. Predicting the effects of coding non-synonymous variants on protein function using the SIFT algorithm. Nat Protoc. 2009;4(7): 1073–81.
19. Adzhubei IA, Schmidt S, Peshkin L, Ramensky VE, Gerasimova A, Bork P, Aondrashov AS, Sunyaev SR. A method and server for predicting damaging missense mutations. Nat Methods 2010; 7:248–9.
20. Maddalena A, Bale S, Das S, Grody W, Richards S, Committee ALQA. Technical standards and guidelines: molecular genetic testing for ultra-rare disorders. Genet Med. 2005;7:571–83.
21. ALPCW Group. ACMG recommendations for standards for interpretation of sequence variations. Genet Med. 2000;2:302–3.
22. Antonarakis SE, The Nomenclature Working Group. Recommendations for a nomenclature system for human gene mutations. Hum Mutat. 1998;11:1–3.
23. den Dunnen JT, Antonarakis SE. Mutation nomenclature extensions and suggestions to describe complex mutations: a discussion. Hum Mutat. 2000;15:7–12.
24. Online Mendelian Inheritance in Man, OMIM®. Baltimore, MD: McKusick-Nathans Institute of Genetic Medicine, Johns Hopkins University; 2013. http://omim.org/.
25. Jones SJ, Laskin J, Li YY, Griffith OL, An J, Bilenky M, et al. Evolution of an adenocarcinoma in response to selection by targeted kinase inhibitors. Genome Biol. 2010;11(8):R82.
26. Mwenifumbo JC, Marra MA. Cancer genome-sequencing study design. Nat Rev Genet. 2013;14(5): 321–32.
27. International Standards for Cytogenomic Arrays Consortium. 2013. http://www.iscaconsortium.org.
28. Grody WW, Thompson BH, Gregg AR, Bean LH, Monahan KG, Schneider A, Leb RV. ACMG position statement on prenatal/preconception expanded carrier screening. Genet Med. 2013;15:482–3.

CHAPTER 18

Clinical Information Systems in the Era of Personalized Medicine

JONATHAN NOWAK, LYNN BRY

Introduction 259
Clinical Systems Support of Genomic Testing 260
Genomic Standards for Clinical Systems and Data Interoperability 264
Factors to Consider within the CLIA Laboratory that Performs Genomic Testing 266
Factors That Involve but, by Necessity, Extend Beyond the CLIA Laboratory, Including Data Warehousing, Integrated Reporting Across Diagnostic Specialties, Decision Support Tools, and Effective Warehousing of Genomic Information 269
Conclusions 274

Introduction

Scientific and technical advances continue to further our understanding of how genetic alterations affect human health and the development of disease. Integrating genomic findings in the delivery of patient care represents an exciting area of medicine. The capacity to interpret and leverage this new source of information, however, and to do so in a broad and high-throughput manner via clinical information systems remains a key challenge.

In spite of the challenges, institutions and testing CLIA laboratories should recognize that existing clinical systems, operating procedures, and standards to support interoperability across systems do provide important resources to enable genomic analyses in patient care. Beyond individual patient testing, strategies need to also focus on delivery of genomic content to healthcare providers, and on the means to warehouse this information, both to assist in ongoing research and development (R&D) activities to support CLIA testing, and to evaluate outcomes from the use of genomic data in patient diagnosis, prognosis, and management.

Clinical genomics brings many new concepts to CLIA laboratories and healthcare institutions. Some factors affect processes within the testing lab, whereas others require additional institutional input to solve. In this chapter, we focus on four common areas that influence effective use and development of clinical information systems to support the use of genomic data in health care:

1. Developing clinical systems to support genomic testing.
2. Genomic standards for clinical systems and data interoperability.
3. Factors to consider within the CLIA laboratory that performs genomic testing.
4. Factors that involve but, by necessity, extend beyond the CLIA laboratory, including data warehousing, integrated

G.J. Netto and I. Schrijver (eds.), *Genomic Applications in Pathology*,
DOI 10.1007/978-1-4939-0727-4_18, © Springer Science+Business Media New York 2015

reporting across diagnostic specialties, decision support tools, and effective warehousing of genomic information.

Clinical Systems Support of Genomic Testing

Developing Clinical Infrastructure to Support Genomic Testing

As yet, the lack of end-to-end solutions to support data handling across technical, bioinformatics, and interpretive workflows requires that laboratories and institutions undertake projects of substantive complexity to implement genomic testing for clinical purposes. As platforms and vendor solutions improve, the efforts and costs required should drop. However, given the current complexities inherent in implementing genomic testing, particularly at the level of multi-gene panels and exome sequencing, laboratories and institutions need to develop a cohesive plan that defines the testing to be undertaken and the resources needed to support it. Broadly, analyses should include a business plan, institutional initiatives to be supported, as well as clearly specifying clinically actionable contributions to patient care. Standard methods for project management and integration of information systems [1] can assist in developing a robust plan. At a high level, these methods commonly incorporate the steps discussed below.

Development of Use Cases for Clinical Genomic Testing

What are the cases for genomic testing? Define the reasons and evidence to support testing, including clinical utility and support of clinical trials or translational research programs. Evaluate what types of testing will be performed, from multiplex panels to high-density arrays, targeted amplicon sequencing, exome or genome sequencing, as each has different needs in terms of information systems support within and external to the testing laboratory. Laboratories often do best to play to local and institutional strengths. To ensure that one-off processes are not developed in technical and IT plans, laboratories should select at least two areas of focus that may relate to disease and type of testing (germline, somatic, infectious disease), but no more than three to four to ensure adequate focus and development of infrastructure that can handle testing within a reasonable time frame.

Requirements Gathering

Given the use cases, what resources and infrastructure are needed to support them, from the point of ordering genomic tests to reporting results back to the ordering physician and associated electronic health records (EHR)? In addition, it is important to consider population-based aspects needed to interpret and improve testing, such as data warehousing for population-based analyses in evaluating significance of new variants and for the demonstration of outcomes per availability of genomic results. Other factors to evaluate include re-analysis of genomic data with respect to clinical triggering events, such as future patient visits or the need to "push" new and medically actionable findings to clinicians as clinical evidence regarding the significance of particular variants improves. Thorough gathering of requirements will touch not only upon the clinical information systems, but includes an evaluation of operational, logistic, and other resources needed to support end-to-end processes.

Validation of Requirements

The validation of requirements is a necessary "sanity check" in the process, to evaluate the requirements and their capacity to be supported with available institutional resources and budgets. It is not uncommon for CLIA laboratories and institutions to revise the initial plans and scope to be in line with what existing resources and funding can facilitate.

Gap Analysis

Given the requirements, what systems and/or resources exist, and which need to be developed? Evaluate costs and resources associated with each, including costs associated with the purchase of new systems, as well as upgrades to existing clinical information systems. After an initial gap analysis, reevaluate use cases and requirements, and iterate as needed to develop a final plan that incorporates areas in which genomic testing can be accomplished in a manner that fits within institutional needs and available budgets and resources.

Functional Specifications

Functional specifications that include IT requirements in terms of software, hardware, and systems integrations need to be created (Fig. 18.1). The specifications will be part of the overall business, financial, and operational plan. The IT components should also include needs for supporting personnel including project managers, clinical systems analysts, database administrators, system administrators, bioinformaticians, and additional supporting computational staff and statisticians. Support of technical platforms, software, and hardware must also be incorporated. Included in these analyses should be an understanding of requirements for attaining a break-even point and determining the return on investment derived from support of clinical research programs within or across institutions.

Timeline and Plan

It is important to generate a timeline for development and integration of resources, showing key milestones to be met and dependencies across clinical, technical, and informatics needs.

After undertaking all these activities, laboratories and institutions may realize that starting with more complex forms of testing such as exome and/or genome analysis can be quite challenging, particularly if local expertise and infrastructure do not already exist. Re-tooling of plans to focus on targeted areas, particularly where vendor kits and informatics solutions may be leveraged to provide a more "black-box" platform, can be helpful, along with a goal to strategize for later plans to incorporate more complex forms of testing. In this manner, more focus can be placed on ensuring that the needed IT systems and infrastructure are in place to support the initial forms of testing as well as the ones to be brought online at later dates.

LIS Versus LIMS: Understanding the Setting in Which CLIA Genomic Testing will Occur

Many institutions face the following challenges when implementing a plan for clinical genomic testing: (1) how to effectively leverage expertise and resources that may exist in a research core performing complex genetic testing, but that is not certified as a CLIA laboratory [2], and (2) how to leverage expertise and resources from the CLIA laboratory for testing that contains many items that are new and, in part, frankly foreign to many clinical laboratories. On this latter point, evaluations for infrastructure to support diagnostic testing will include understanding contributions from existing clinical laboratory information systems (LIS) versus non-CLIA laboratory information management systems (LIMS) that may be encountered in research environments.

Whereas clinical LIS that specialize in supporting complex genomic testing are now becoming commercially available [3, 4], these products remain external to the standard, vendor-based clinical LIS [5] that support high volume testing in clinical laboratories and anatomic pathology services. In implementing programs for clinical genomic testing, CLIA laboratories thus need to assess whether to internally develop needed components, or consider purchasing a dedicated, "best in breed" LIS to support testing, and then focusing on systems integration with the main LIS. Section 18.3 below goes into more detail regarding specific areas to evaluate. As shown in Fig. 18.2, the interactions between the CLIA laboratory's LIS and the infrastructure supporting clinical genomic testing can take a variety of forms:

1. The clinical LIS remains the system of record for data structures and processes that support the essential business process for CLIA testing, such as those regarding clients ordering tests, patient information, insurers, sample and test dictionaries, and fee codes/schedules. Under this scenario, the genomics LIS may operate fairly independently of the clinical LIS but requires means to retrieve and update data structures maintained in the clinical LIS. In this manner, both systems share common data

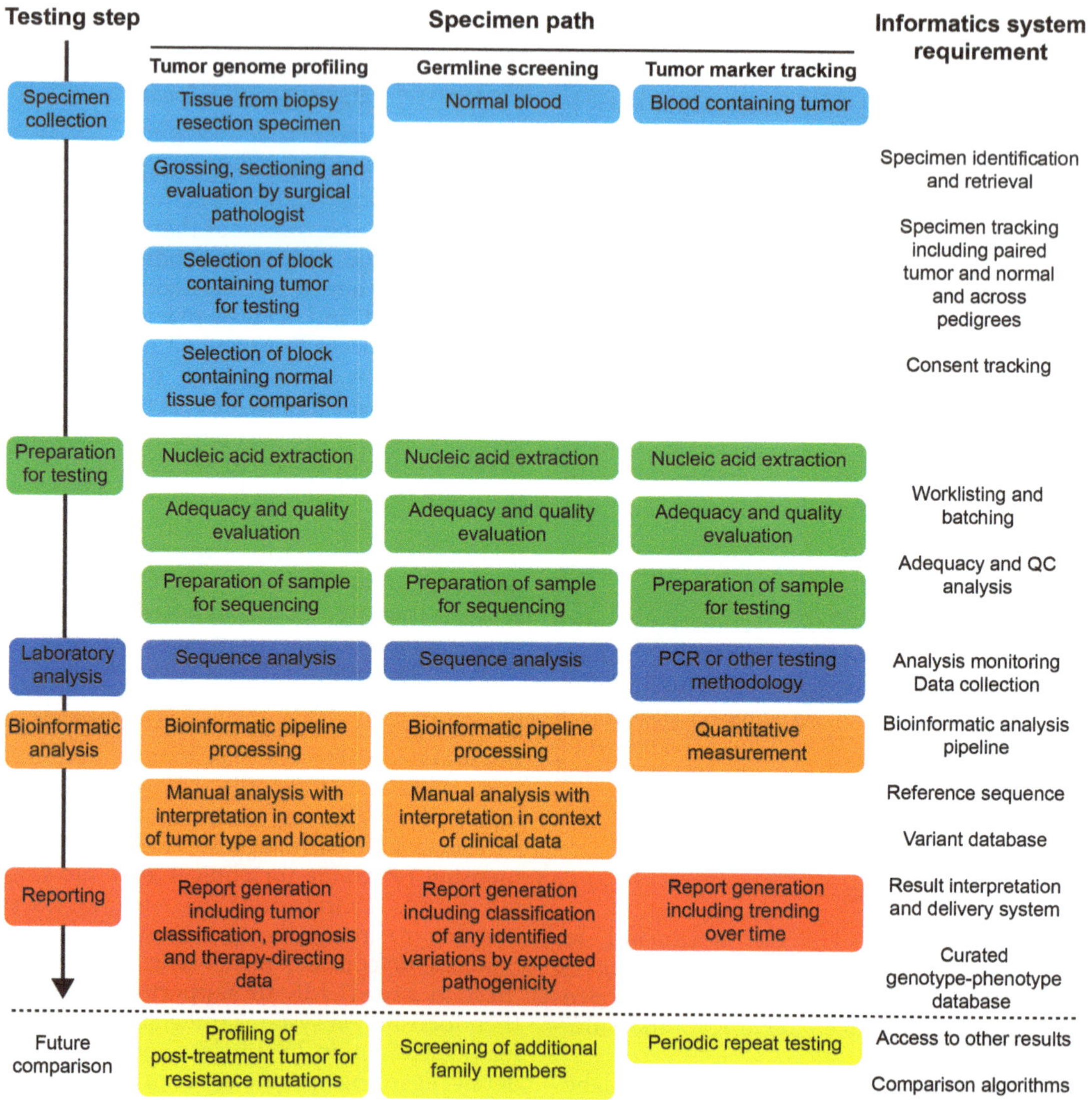

Figure 18-1 Testing pathways and informatics system requirements for representative types of molecular testing

structures and vocabularies, or ontologies, for ordering, testing, reporting, and billing.

2. The clinical LIS handles the up-front business process for test ordering and may handle additional steps including accessioning and work listing. Thereafter, orders are communicated to the genomics LIS, optimally via messaging standards such as those developed by Health Level 7 (HL7) (www.hl7.org) [6]. Receipt into the genomic LIS may require a separate accessioning process upon receipt of the patient order, but subsequent downstream steps are handled within the genomics LIMS, including communication of final results and reports.
3. The clinical LIS handles the initial and final end points of genomic testing, centered around order receipt and return of a final report to the ordering clinician and the EHR. Systems integration with the genomics LIS or LIMS defines operational and IT components needed to facilitate forwarding of needed sample and patient data for CLIA testing to occur, and return

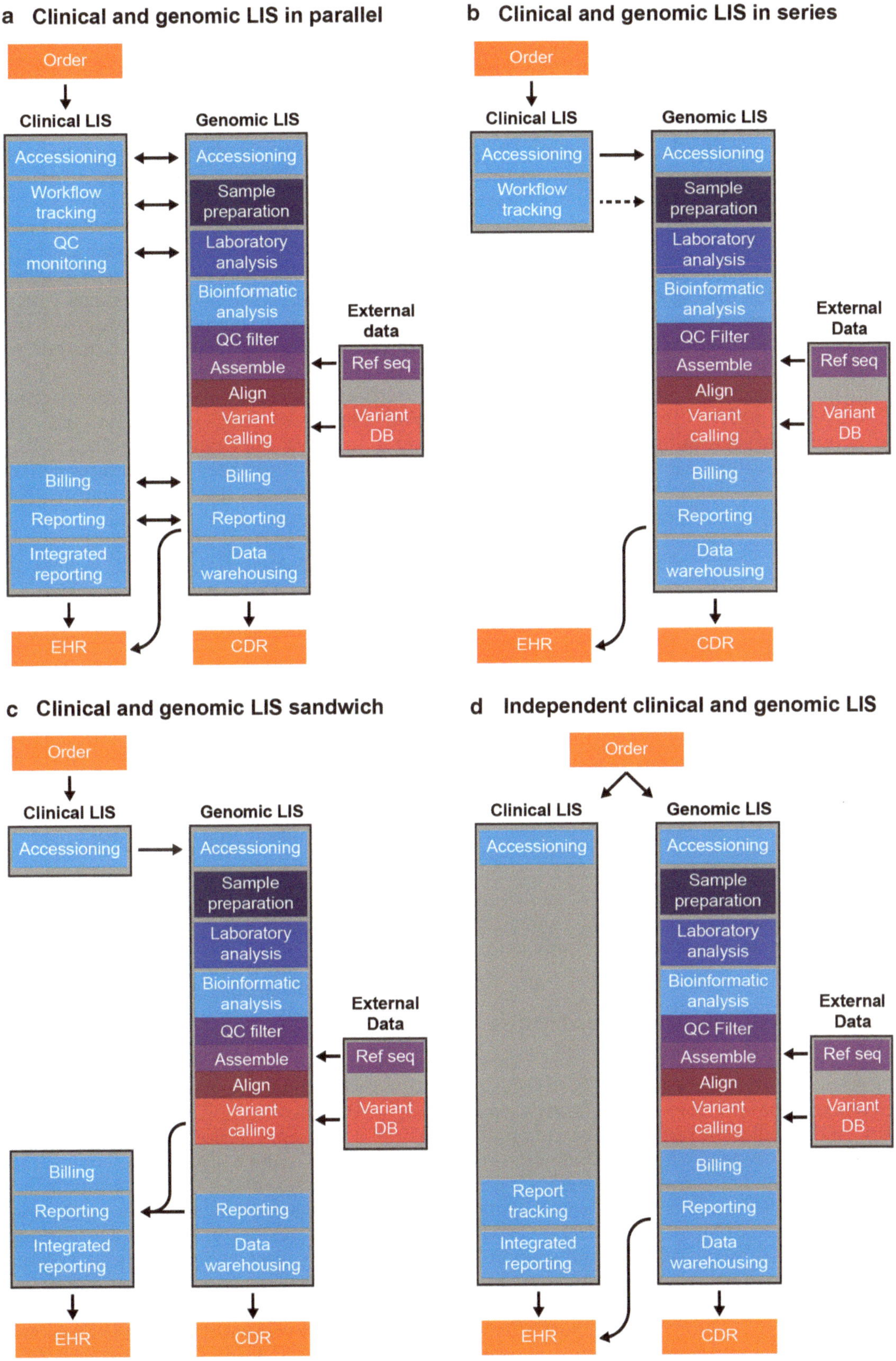

Figure 18-2 Possible configurations for integrating the clinical LIS and the genomic LIS

of results from the genomic analyses, which may include steps from the call of variants to return of a structured report that will be forwarded to the client.

4. Both the clinical LIS and genomics LIS are completely separate, which can occur within a single institution, and is also the structure if leveraging genomic testing from an outside CLIA reference laboratory. In this situation, systems integration will focus on means to communicate orders and receive results from the testing lab.

Of note, laboratories and institutions facing a need to get LIMS and other non-CLIA resources to perform to CLIA specifications in support of clinical genomic testing can refer to the Next Generation Sequencing (NGS) section of the Molecular Pathology checklists [7] developed by the College of American Pathologists. These documents provide standards and quality parameters to be followed in validating clinical LIS and for implementing NGS in a CLIA environment.

Genomic Standards for Clinical Systems and Data Interoperability

While clinical genomics may still be in its infancy, international efforts have developed standards to support data and systems interoperability. Though still new and evolving, the following resources provide means to send, receive, and warehouse genomic data.

Gene-Level Calls and Coordinates

Efforts by many groups including the Human Genome Variation Society (HGVS; [8]), NCBI (www.ncbi.nlm.nih.gov), EMBL (www.embl.de), and medical associations including the American College of Medical Genetics and Genomics (ACMG; www.acmg.net), College of American Pathologists (CAP; www.cap.org), and Association of Molecular Pathology (AMP; www.amp.org) have developed and supported use of common nomenclatures for describing gene variants. At their simplest, these systems first specify the location of a variant relative to its genomic chromosomal position, location within a coding sequence, and/or location with the resulting polypeptide chain of a protein. They next utilize standardized nomenclature to specify gene rearrangements, alternatively spliced transcripts, haploid phasing of variants, and reporting of copy number variants (CNVs). These baseline formats are leveraged in subsequent structures including the variant call format and in HL7 messages to communicate clinical data across systems. However, laboratories should be aware that discrepancies may exist when considering nomenclature systems that focus on cytogenetic versus sequence or transcript-based positions.

Although the HGVS recommendations cover the broad range of common genetic alterations, new applications require continual expansion of the nomenclature. For example, whereas the nomenclature for describing translocations detected by karyotypic or FISH analysis is well defined by the International System for Human Cytogenetic Nomenclature, there is not yet a broadly accepted way for reporting translocations detected by NGS. A second necessary component for standardized reporting is broadly agreed upon reference materials and databases of known variants. Although the sequence of the human genome was declared complete in 2003, analysis and annotation of the sequence are still ongoing, with a reference annotation only completed in 2012, and one that is routinely updated [9, 10].

Genomic File Formats

A variety of standard file formats are utilized during genomic testing, from the .fastq and .bam file formats used to store sequence data in early stages of bioinformatics analyses, to the variant call format (VCF; [11]) that provides a commonly used format for the structure and reporting of variants identified against a reference genome. By storing only the variants identified against a reference, the VCF file greatly reduces the amount of information that needs to be stored or communicated. It has thus become a standard means for communicating variants, whether from targeted sequencing, exome- or genome-level analyses.

Used extensively within the 1000 Genomes Project the VCF format includes metadata elements to store information regarding the gene sequence and specific identified variants. Variants are identified by their genomic coordinates relative to a defined reference genome. Version 4.0 of the format also includes quality information associated with the call of each variant, and filtering information if an external system or algorithm assigned specific information regarding the state or quality of the variant. While the format does not provide defined structures or methods for documenting the pipelines used to perform analyses, this information may be captured in metadata fields or in the file header. As adoption of clinical genomic testing increases, we anticipate that the VCF and underlying supporting structures will become more developed to support widespread clinical testing.

Health Level 7

Health Level 7 (www.hl7.org) is a nonprofit organization that develops standards to support interoperability across healthcare systems. HL7's Clinical Genomics working group has devised standards for communicating pedigree data [12] and a structured Genetic Test Report (GTR; [6]). Both projects contain detailed specifications and implementation guides that may be downloaded from their website. While both are still relatively new and continue to evolve, they provide an internationally developed standard to communicate complex genomic information across systems.

The pedigree model provides a data standard to capture and communicate family relationships for a given patient, including diseases and genetic risk factors. A working example of the model has been implemented for the "My Family Health Portrait" website managed by the U.S. Surgeon General (https://familyhistory.hhs.gov/fhh-web/home.action). While broader adoption within commercial EHRs is being considered, such may require substantive alterations to data structures storing patient information, as well as addressing patient privacy and protection concerns under HIPAA. This latter concern largely relates to linking individuals within a medical database, if one or both parties have not explicitly given consent to do so [13, 14].

HL7's GTR supports reporting of sequence-based variants, cytogenetics, and gene expression studies. Message structures include standard components for communicating the ordering institution, clinician or practice, and patient demographic data. The "Test Details" section includes data structures to communicate reasons for testing (including diagnostic codified data), specimens sent for testing, as well as elements used by the testing laboratory in reporting variants or other findings, interpretations, and additional supporting information to accompany reports.

Version 2 of the GTR, released in January 2013, provides the capacity to link Logical Observation Identifiers Names and Codes (LOINC; [15]) to genetic tests and reports, and has been piloted at various sites worldwide [16, 17].

Of note, the capacity for the Portable Document Format (PDF) to render healthcare data from embedded Extensible Markup Language (XML) [18] offers potential opportunity to store HL7 messages in these file formats. The Healthcare PDF standard may thus provide a means for institutions to store messages in a format that can also generate a human-readable report.

As with any standard, HL7 alone does not provide the application layer needed to perform core functions once data are sent or received, but provides an essential component for defining methods to communicate data across sites. In practice, most large CLIA laboratories and institutions have invested in teams and supporting IT infrastructure to implement and manage HL7 messages. As such, the availability of communication standards for genomic testing has the potential to reduce the time and effort required to otherwise develop and maintain de novo processes. These standards also generally provide an improved capacity to scale as clinical testing and associated volumes of data to be communicated increase.

Standard Data Sources and Content

Several global projects aim to develop standards and content for the clinical interpretation of genomic variants. Whereas projects such as the Online Mendelian Inheritance in Man database (OMIM; www.omim.org), the Kyoto Encyclopedia of Genes and Genomes (KEGG; www.genome.jp/kegg/), and the

Catalog of Somatic Mutations in Cancer (COSMIC; http://cancer.sanger.ac.uk/cancergenome/projects/cosmic/) arose from research activities, these databases often provide content to CLIA laboratories that evaluate the significance of genomic findings. New initiatives, including ClinVar (http://www.ncbi.nlm.nih.gov/clinvar/) and My Cancer Genome (http://mycancergenome.org), aim to provide additional curation and a "CLIA-grade" tool through which laboratories may communicate variants identified as well as contribute supporting evidence regarding the interpretation of variants. As the field progresses, vendor solutions that aggregate or license defined content will also become more widely available.

Factors to Consider within the CLIA Laboratory that Performs Genomic Testing

Various features of the standard LIS may be leveraged to support genomic testing. Box 18.1 highlights many of these components. In contrast, genomic testing typically requires a number of items that are new to the technical and IT staff within a CLIA laboratory. These latter areas, detailed in Box 18.2, are where infrastructure, resources, and personnel need to be developed to support the associated activities.

Box 18-1 LIS FEATURES TO BE LEVERAGED IN GENOMIC TESTING

- Data structure to support business processes for diagnostic testing.
- Support for order entry and interfacing with order entry functions in an electronic health record (EHR).
- Sample accessioning and tracking.
- Test ordering and worklisting.
- Management of quality control (QC) and quality assurance (QA) processes.
- Receipt of results, including interpretation.
- Billing triggers to assist in billing for testing.

Box 18-2 NEW INFORMATION SYSTEM REQUIREMENTS TO CONSIDER IN GENOMIC TESTING

- Decision support to guide clinicians in the ordering of genomic tests.
- Capturing of additional data at the point of ordering genomic tests, such as additional patient consent, pedigree information, or other factors such as tumor cellularity.
- Managing complex technical and quality control steps including library preparation, barcoding, and multiplexing of samples.
- Linking to and managing bioinformatics pipelines, including version control, and monitoring of pipeline performance for individual patient cases and across sequencing runs.
- Developing and managing CLIA-grade content for interpreting and reporting genomic results.
- Integrated reporting of genomic information with other phenotypic analyses including histopathologic and/or clinical laboratory biomarkers.
- Developing data storage resources for genomic information, both for ready retrieval of information when needed, as well as to meet any medico-legal and associated state or local laws regarding the storage of clinical data.
- Leveraging genomic data in clinical decision support.
- Potential need to reevaluate genomic datasets on a periodic basis, relative to defined clinical triggers such as a patient visit or push of new information regarding variants that are medically actionable.
- Data warehousing of genomic results to support evaluation of unknown variants and improve test panels.

Clinical Systems Supporting Order Entry of Genomic Tests

Whereas orders for genomic tests share many aspects with orders routinely placed for other forms of patient testing, several properties merit special attention. A first consideration regards determining when testing is warranted. Particularly in the case of germline testing, the decision to test relies upon integrating data from the patient's medical history, clinical examination, and laboratory findings with pedigree information. Without a centralized mechanism for routinely entering and communicating these data in a structured manner, opportunities to make a genetic diagnosis may be missed. In the case of cancer testing for somatic variants, analyses may be conducted under a research protocol, or complex genomic analyses may only be considered after initial screening tests that use phenotypic markers or focused molecular diagnostic tests. These factors need to be communicated to the ordering clinician, and appropriate pre-existing information needs to be relayed back to the laboratory to direct testing, especially when multistep algorithms are in place.

In addition, the mechanism by which a given gene or genetic region can be tested can also influence how the test may be ordered. Unlike most clinical lab tests, in which the specific technique is a clear component of the test, genomic testing may require, as an example, sequencing multiple regions of the genome while being cognizant of the intrinsic limitations of the assay's technology, such as the inability to detect structural variations or disambiguate sequence from pseudogenes. While many of these processes may remain internal to the laboratory as testing for specific patient cases progresses, testing of certain genomic regions may also require that the ordering physician and supporting personnel be informed of such aspects at the time of ordering.

To fully address these issues, order entry systems are an essential part of the clinical workflow for genomic testing. Genetic test ordering benefits from order entry systems that provide the means to search and compare available assays and link into decision support tools to aid with the selection and ordering of appropriate tests that are supported by medical evidence [19–21]. Ideally, this process could even be automated. For example, patients exhibiting abnormal responses to pharmacologic therapy could be automatically flagged for evaluation of drug metabolism enzymes. Most importantly, such systems free clinicians from the burden of maintaining detailed knowledge about indications for both common and rare genetic tests, while providing ready access to resources that allow them to tailor possible testing to a patient's individual scenario [22, 23]. Notably, although such clinical decision support systems are not yet widely implemented, they are among the most requested EHR functions related to genomic medicine with the ultimate aim of improving personalized health care [21].

Specimen Identification and Tracking for Genetic Tests

Although most clinical LIS are well equipped for tracking a wide variety of specimen types, including those routinely used for genetic testing, several accommodations need to be made for genetic testing. One critical requirement is tracking of the patient materials used for testing, particularly in the case of cancer diagnosis where multiple samples may be sent for molecular and phenotypic analyses. For solid tissue specimens, testing is routinely performed on a subset of the available material, typically a portion of a single paraffin block. Testing may also include solid tissue and fluid samples, such as for B cell clonality assays that could be performed on blood, cerebral spinal fluid (CSF), and tissue, where comparison of results across sites may be critical in guiding therapeutic decisions. Therefore, reports for such cases need to include an unambiguous statement about what material was used for testing. For cancer-based testing, the adoption of automated, whole-slide imaging systems can facilitate the documentation of material used for testing by creating a permanent, high resolution record of the exact material that was used, even if the material on the original slide is consumed to accomplish the testing.

In addition to tracking input material for testing, systems for tracking genetic test material need to have flexible and robust capabilities for handling samples that fail testing or are judged to be technically inadequate for analysis [19]. In each case, the laboratory's information system must be able to identify when cases have not passed quality control checks and divert them for appropriate handling. A final consideration for material tracking is the archiving and storage of samples after testing. Although not all samples are necessarily retained by the laboratory, some forms of testing including chimerism and clonality analyses can rely critically upon the ability to re-test previously analyzed samples. Furthermore, testing of paternity and of extended familial pedigrees often warrant storing tested samples beyond a specified time after reporting the final results. Management of such long-term storage is frequently not incorporated effectively in clinical LIS. Although numerous commercial software packages can handle many of these tasks individually, there is limited integration between these systems and LIS. However, as the volume and complexity of genetic testing grow, the capabilities of these programs should increase and their ability to link to other LIS packages should strengthen.

LIS Tracking of Consents and Results Reporting

Several additional types of information unique to genetic testing often need to be managed within clinical laboratory systems [24]. Genetic testing may require additional consents beyond those obtained for routine clinical testing [25]. In addition to simply tracking patient consents, the LIS may also be called upon to track multiple types or levels of consent for a given test. As testing platforms based on NGS become more prevalent, results frequently include findings of unknown medical significance. Some tests may also identify incidental findings that are not directly related to the initial disease in question but which nevertheless may be medically informative for the patient. Different patients may have distinct preferences about being informed of such results. To address this possibility, recently proposed recommendations for informed consent prior to performing whole-genome sequencing have advocated a category-based model for disclosing different classes of findings [26], though the degree to which CLIA laboratories implement these levels depends upon local and institutional views of genomic testing and use of results. However, the LIS may have to track which results should be released to the patient based on information documented in the consent forms [26, 27]. In addition, as genetic data from a patient may be periodically reevaluated by more sophisticated algorithms drawing from updated knowledge bases, the amount and nature of new information may be quite different from that for which informed consent was initially obtained. Not only will consents need to be designed broadly enough to account for new information from periodic reevaluation, but the clinical infrastructure in CLIA laboratories may be called upon to alert clinicians that new results are available for their patient [26].

Proper interpretation of genetic tests may also require information about multiple individuals from the family pedigree to be associated with the individual being tested, particularly from the parents and siblings of a patient [28]. This information may include the approval to provide the results of testing to other family members. In these situations, multiple specimens from different individuals may need to be linked within the information system so that they are tested and analyzed together before the release of a single report to the patient's EHR. The necessity of linking multiple patients to a single patient record is a rather unique requirement of genetic testing and is conceptually different from the standard one-to-one relationship between patients, specimens, and results that underlines traditional LIS design. In fact, commercial LIS do not readily handle receipt of supporting samples under the individual who provided them. Rather, in most CLIA laboratories additional samples may be accessioned under the primary patient, with additional fields added to uniquely identify the individual and their relationship with the patient being tested.

Standardized Report Formats for Genetic Tests

Traditionally, the LIS reports results in a highly structured and standardized format, though free-text elements in a narrative format may occur. Although genetic results typically contain a mixture of structured data, such as the precise genomic location of an identified mutation, and unstructured data, such as a text-based interpretation of the results, this information is typically reported into the EHR in an unstructured format. However, as the amount and complexity of genetic tests increase, it will be necessary to adopt a standardized and structured template for reporting results [21, 29]. Structured data will not only facilitate comparing and transmitting results among provider systems, but are also essential to enable downstream algorithms and tools to provide decision support to clinicians and patients [30]. Additionally, structured reporting also facilitates the warehousing of genomic information, to create knowledge bases for developing content as well as to enhance laboratory quality control programs that monitor new and previously encountered variants. As described above, several standards are being adopted to facilitate structured and standardized reporting. Efforts to identify and categorize normal variants and disease causing alterations are still evolving rapidly [31]. Therefore, it is essential for molecular reports to include detailed information about the reference material used, as changes to the reference sequence or annotated functional information could potentially alter test interpretation. Finally, it will be important for structured reports to indicate the method used for analysis, because multiple methods may be applicable when testing a given region of the genome [29]. Although Current Procedural Terminology (CPT) codes exist for many molecular diagnostics assays and although these have been recently updated to better reflect current testing practices, they frequently lag behind the introduction of new technologies for genetic testing. Additionally, as they are primarily designed for billing needs, they may not be able to capture the necessary details about how a test was performed, particularly as bioinformatics and computational analyses play an increasing role in the reporting of molecular results. Additional medical nomenclature systems such as the Systematized Nomenclature of Medicine-Clinical Terms (SNOMED-CT), Logical Observation Identifiers Names and Codes (LOINC), and the Unified Medical Language System (UMLS) may be necessary to succinctly and unambiguously communicate testing and analysis methodologies.

Factors That Involve but, by Necessity, Extend Beyond the CLIA Laboratory, Including Data Warehousing, Integrated Reporting Across Diagnostic Specialties, Decision Support Tools, and Effective Warehousing of Genomic Information

Integrated Reporting with Other Anatomic Pathology and Clinical Laboratory Data

Although some molecular assays represent independent laboratory studies, many begin with pre-existing specimens, that have simultaneously undergone non-genetic testing in anatomic or clinical pathology laboratories. As such, the clinical value of the molecular data only becomes apparent when interpreted in the context of the other nonmolecular laboratory data for the specimen. Beyond the realm of certain complex anatomic pathology cases, typically hematopathology and soft tissue pathology, the ordering clinician has traditionally borne the burden of integrating disparate and potentially asynchronously provided results (Fig. 18.3a). However, as the breadth of molecular testing grows, and its interdependence with other laboratory results increases, the capacity to integrate molecular and nonmolecular findings becomes increasingly important and implementation will fall to the clinical diagnostic laboratory.

Cases involving simple and routine data integration have the potential to be handled within the LIS, particularly to integrate genomic findings with pertinent phenotypic markers that have also been performed in the clinical molecular diagnostic laboratory. For example, patients with HIV who receive highly active anti-retroviral therapy (HAART) undergo routine monitoring of their viral load and CD4+ T-cell counts. They may also undergo periodic HIV genotyping, per defined changes in clinical status, to assess development of anti-retroviral drug resistance in the underlying population of HIV virions. These frequently repeated and highly standardized results could be integrated within a standard report format in the LIS to better facilitate longitudinal assessment of response to therapy and continued management. This approach integrates the molecular data with important phenotypic markers, providing an improved context in which to assess the meaning and validity of findings, and their integration in the overall clinical status of the patient.

As another example, a similar approach could be employed for monitoring glucose control in patients with type 2 diabetes by leveraging pharmacogenomic markers predictive of drug responsiveness with phenotypic markers of glycemic control. Pharmacogenomic studies have recently identified polymorphisms in genes affected by sulfonylureas that can predict an individual's response to treatment with these drugs [32]. Beyond guiding initial therapy selection, the patient's genetically predicted sensitivity profile to different drug classes could be retrieved within the LIS and integrated with periodic glucose and hemoglobin A1c data to predict an expected response to sulfonylurea treatment as a comparison with the patient's actual response, and as a measure of treatment compliance [33, 34]. Such an integrated report would be especially helpful in tracking the progression of type 2 diabetes and would have the capacity to assist in the selection of personalized therapy per the patient's underlying genetic background and current phenotypic presentation of the disease.

In more complex cases, molecular data will need to be structured and managed in the LIS in a manner that facilitates integration with other types of pathology information. Such will often be the case with somatic mutation analyses of tumors, where the dataset from diagnostic testing in pathology laboratories often includes histopathological assessment, phenotypic biomarkers, SNPs, and more complex molecular tests. In the past, molecular results for tumor specimens have typically been reported in an isolated fashion either as addenda to already finalized surgical pathology reports or as completely separate reports released into the medical record [35]. However, two related trends in tumor biology are driving the need for integration of anatomic pathology data. First, molecular alterations are increasingly defining tumors and tumor subtypes, as well as aid in the selection of therapies. For example, recent guidelines from the College of American Pathologists for reporting ancillary biomarker studies for lung and colorectal adenocarcinomas underscore the importance of molecular data in the standard characterization for these tumor types [36, 37]. Indeed, information such as the key driver mutation for a lung adenocarcinoma may be one of the single most important results that an oncologist wishes to obtain from a surgical pathology specimen, given the evidence supporting its use in selecting pharmacologic therapy. Therefore, this information will need to be clearly contained within an integrated report for the specimen in much the same way that data necessary for cancer staging are routinely included in surgical pathology reports.

A second aspect driving integrated reporting is the recognition that many types of molecular data cannot be interpreted meaningfully in the absence of additional pathology data. The significance of the same mutation in a given gene may vary widely depending upon the type of tumor in which it is detected [38]. This context dependence is extremely important for many NGS assays on tumors, for which the final interpretation is closely linked to the original tumor type. Simply reporting that a tumor has a mutation in *KRAS* provides limited information without the interpretative context of the associated tumor type. Finally, the sheer amount of data derived from evaluating many tumors with complex assays begins to exceed the point at which manual review and synthesis of findings can be supported in any scalable capacity. The pathologist therefore plays an essential role in providing medical direction and supervision regarding needed data integration and reporting (Fig. 18.3b).

Data Warehousing

In generating genomic results for the individual patient, testing laboratories and institutions should plan to warehouse the information aggregated across cases and populations tested [22]. Key reasons include means to mine the information when evaluating new variants, to

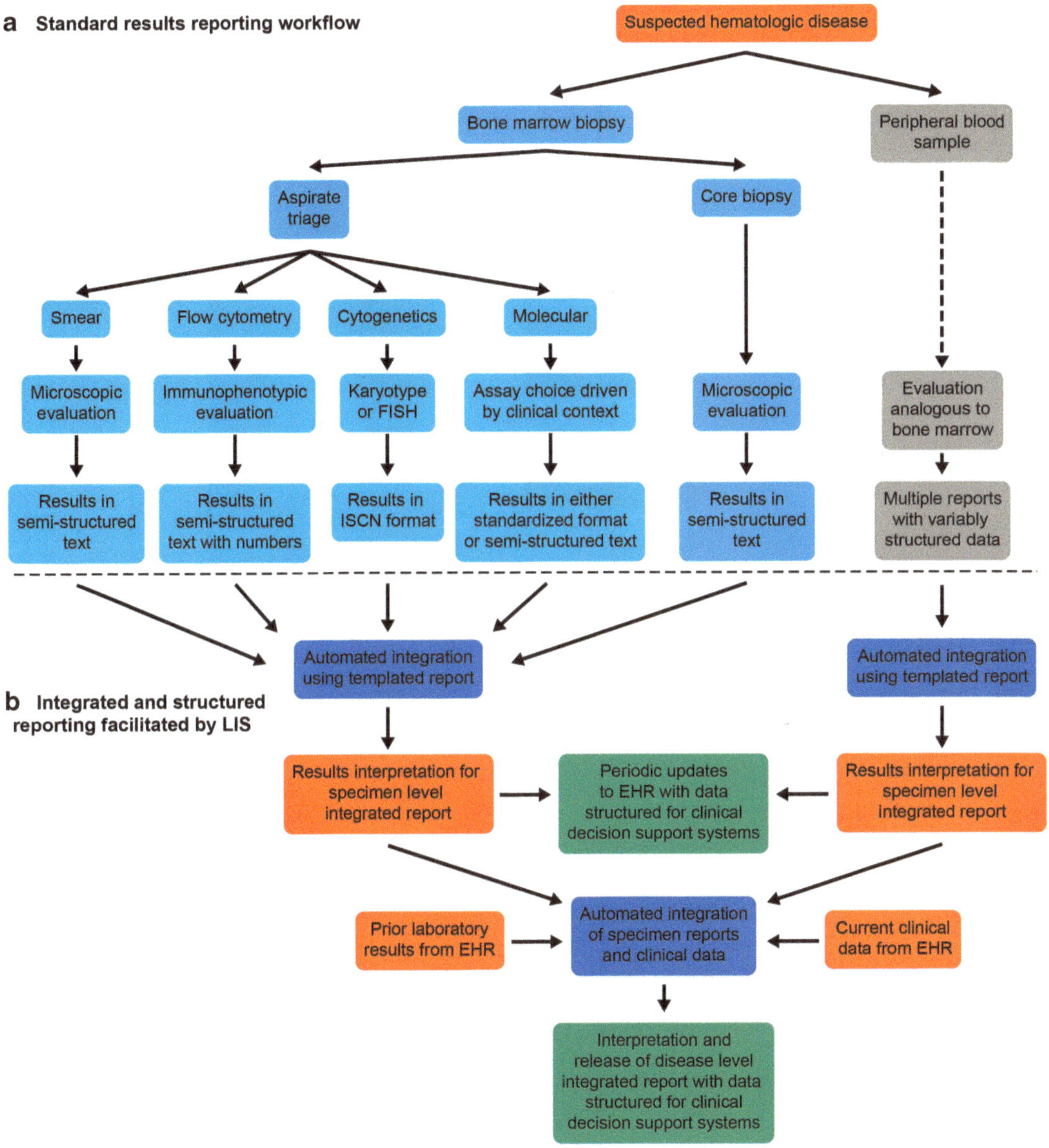

Figure 18-3 The standard pathology reporting workflow and a proposed mechanism for implementing integrated reporting in the LIS. EHR electronic health records

identify their prevalence in certain populations, and/or to assess clinical outcomes. The retrospective datasets can also be leveraged in ongoing CLIA laboratory quality assurance (QA) and quality improvement (QI) activities. The warehoused information also provides an invaluable resource to support active research programs, including translational activities needed to assess new or unknown variants and develop the medical evidence regarding their use in patient care [39–41].

Whereas many open-source and commercial applications have been developed to warehouse genomic information [41, 42], the evaluation of what system, or systems, to use needs to consider existing institutional resources and expertise, underlying funding, and continuing support to maintain the infrastructure.

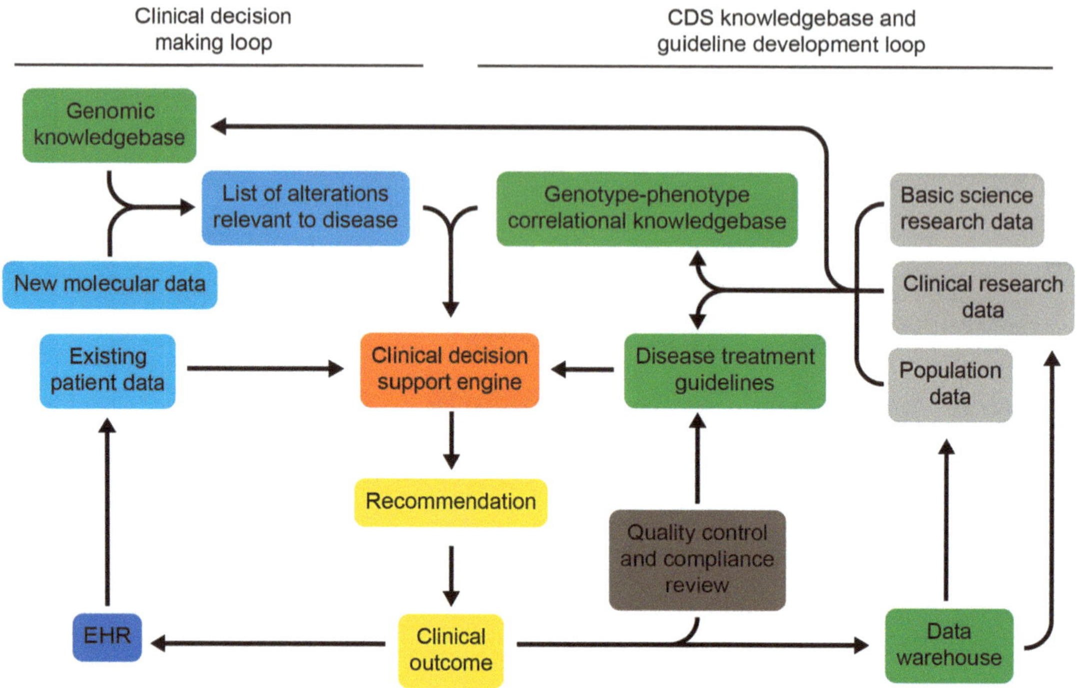

Figure 18-4 Mechanism for incorporating research data with clinical outcomes to improve clinical decision support systems. CDS clinical decision support, EHR electronic health records

In most healthcare institutions, pathology information systems contribute more than half of all data transactions into an EHR. Considering the volume of clinical laboratory and other high-throughput forms of testing, this dataset provides a rich source of phenotypic information, and is commonly the most structured and codified in healthcare systems. In spite of the amount and richness of pathology data, the means to store and effectively warehouse genomic information within the EHR frequently requires resources outside of the pathology department or Clinical Laboratory, in part due to the fact that current commercial LIS are not optimized to generate and manage genomic data. Thus, at an institutional level, it is important that pathologists actively participate in the planning and development of resources to warehouse genomic information, including the tools used to leverage it for basic, translational and clinical activities (Fig. 18.4).

Decision Support

Until recently, and due largely to the non-multiplex nature of many early molecular assays, most genetic test results were interpreted in a manner comparable to single analyte results. While this strategy can work for highly penetrant genetic variants with defined phenotypes and associated medical evidence supporting their use in clinical care, this method proves suboptimal when evaluating multi-gene interactions and the need to present complex information to clinicians [43]. To meet this need, clinical decision support systems (CDSS) are being adapted to incorporate genetic results. CDSS are uniquely suited to analyzing genomic medicine information because of the absolute amount of information generated, the highly structured nature of the genetic results, and the rapidity with which our knowledge and interpretation of genetic variants is increasing [20, 44]. These systems commonly lever-

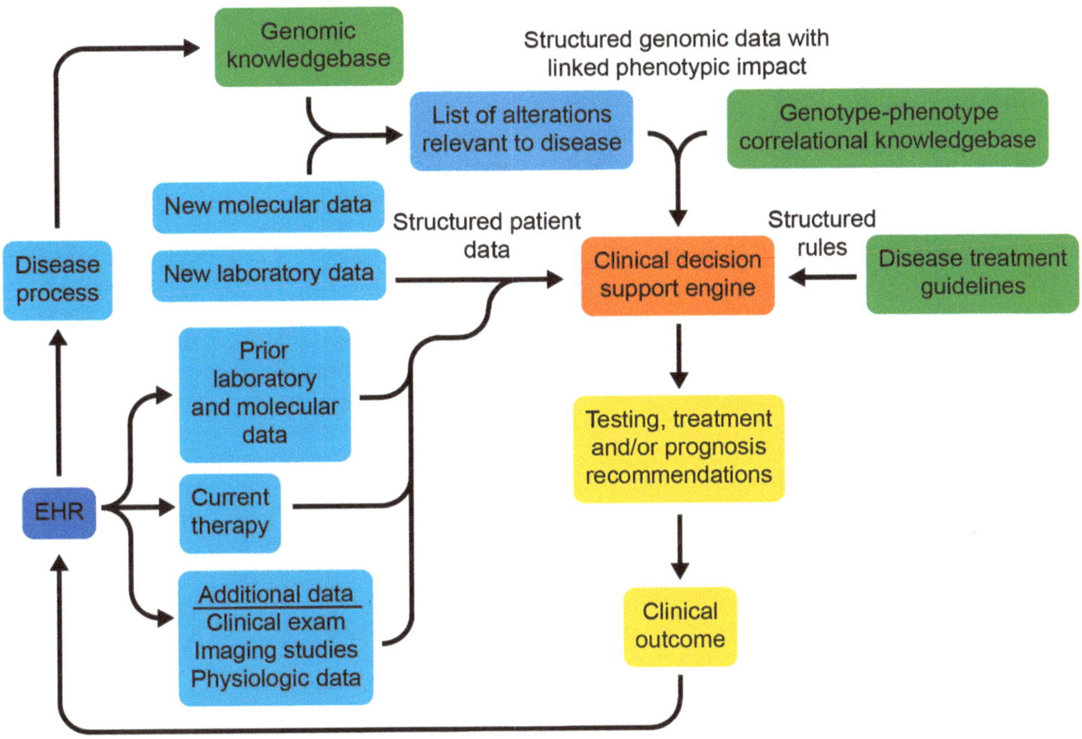

Figure 18-5 Clinical decision support system function and required inputs in ongoing practice

age population-based knowledge bases to provide prevalences and prior information regarding genotype–phenotype relationships for a given disease, with sets of clinical rules or criteria to generate recommendations for clinical action (Fig. 18.5).

Notable examples where CDSS have proven useful in linking genomic information with clinical outcomes are in the areas of pharmacogenomics and cancer management [43]. In pharmacogenomics, many genomic variants have been identified that can predict the likelihood of overall drug effectiveness and the potential for adverse drug reactions. Such systems are now increasingly used to optimize dosage of drugs with narrow therapeutic ranges, minimizing adverse drug reactions, as well as selecting optimal therapy based on the patient's genetic background [45–47]. Though early in their development, a number of CDSS systems have also been developed to aid in cancer risk reduction and cancer management [43, 44] using genetic information. Finally, whereas many of these currently available CDSS tools are widely applicable and show substantial benefit to patients, most operate from a relatively limited knowledge base and set of rules. More sophisticated systems are emerging that draw from broader and more deeply curated knowledge bases, to enable highly complex analyses and interpretations. For example, IBM's Watson health system, developed in collaboration with Memorial Sloan-Kettering Cancer Center, is designed to provide diagnostic and treatment recommendation for cancer patients by merging knowledge from clinical experts with molecular and genomic data, along with outcomes from cancer case histories (http://www-03.ibm.com/innovation/us/watson/watson_in_healthcare.shtml). While the system is designed to enhance the dissemination of practice-changing research to nonexperts, a process which can frequently take more than a decade, it will also function as a much broader platform for guiding decisions in oncology, providing clinicians with a continuously updated set of treatment recommendations that are shaped by genomic data streams and refinements in clinical guidelines.

Regardless of the form that CDSS take, they share common informatics requirements when interacting with LIS. First, genomic information needs to be reported in a machine-readable format in a defined location, and not just released as human-viewable free text results in the EHR [40]. Utilization of a system such as the HL7 Clinical Genomic messaging standard or Clinical Bioinformatic Ontology (CBO) will likely be necessary to unambiguously communicate genetic data between reporting systems and the decision support engine [30]. Analogously, additional nongenetic information will also need to be accessed by the decision support engine to provide context for the evaluation of the genetic result [22]. Whereas existing standards may be adequate to convey a subset of this information, it will also be necessary to use a controlled vocabulary to define patient phenotypic data so that they can be uniformly accessed and understood by the clinical decision engine. Programs such as the Electronic Medical Records and Genomics (eMERGE) Network and PhenX (Consensus Measure for Phenotypes and Exposures) have begun to standardize the collection and annotation of phenotypic information for use in genome-wide association studies, but could also provide a phenotypic reporting system that would easily be adapted to CDSS use [48, 49].

CDSS implementation also requires access to knowledge bases that document and link genotype and phenotype relationships for the disease or medical process of interest. Although the knowledge bases utilized by CDSS thus far have typically been purpose-built, databases such as ClinVar and efforts such as the Clinical Pharmacogenomics Implementation Consortium (http://www.pharmgkb.org/page/cpic) may eventually evolve to act as integrated repositories containing structured genotype–phenotype data that can support automated decision engines. A separate, but closely related resource also required for CDSS is a rule set for generating a recommendation based on the patient data and genotype–phenotype knowledgebase. Although rule sets may be based upon accepted standards for treatment of different conditions, these standards will need to be translated and stored in machine readable structures. Additionally, they will need to be customized and validated at a hospital level to ensure compatibility with established institutional workflows. Common frameworks for representing CDSS rule sets are not yet available, but collaborations such as Health eDecisions (www.healthedecisions.org) are being developed to provide a common syntax for CDSS rules. Finally, clinical decision support systems will need to be structured in a way that permits the information contained within them to be rapidly updated and validated as new genetic data accumulate and therapeutic options and prognostic data evolve. Even though the implementation of clinical decision support systems requires several new bioinformatics and computational tools, many of these resources may be reused, furthering the adoption of CDSS once standards are in place.

Conclusions

Genomic datasets present new challenges to clinical laboratories, pathology departments, and healthcare institutions, particularly in providing a wealth of data for which evidence is often lacking regarding their application to clinical care. Clinical LIS provide an essential set of systems to facilitate ordering, testing, and communication of medically relevant information, yet also need to provide mechanisms by which findings of unknown significance can undergo future evaluation and be warehoused to aid in population-based analyses of findings. At an institutional level, major systems integration together with the development of new systems is frequently needed to enable clinical decision support that adequately utilizes genomic data. Broader adoption of electronic health records and incorporation of new technologies that leverage new computational models and means to store and transmit data will improve our capacity to harness this information, as well as handle anticipated large datasets from other forms of diagnostic testing. Regardless of the testing modality to be considered, standard methods for developing pathways to implement complex plans, with a focus on using robust standards when available, can assist with providing needed systems integration and can facilitate appropriate utilization of such resources.

REFERENCES

1. Irani Z, Love P. Evaluating information systems. Burlington, MA: Elsevier Ltd.; 2008.
2. Centers for Medicare & Medicaid Services. Clinical Laboratory Improvement Amendments. Retrieved from http://www.cms.gov; 2014.
3. GenoLogics—A LIMS for the Next-Gen Omics lab. Retrieved from http://www.genologics.com; 2014.
4. Sapio Sciences—The most configurable and flexible LIMS software available. Retrieved from http://www.sapiosciences.com; 2014.
5. Gale K. Laboratory and the art of enterprise integration. Healthc Financ Manage. 2009;63(10):36–8.
6. (HL7), H.L.S.I. Clinical genomics; 2013.
7. College of American Pathologists Molecular Pathology Checklist. Retrieved from http://www.cap.org; 2014.
8. (HGVS), H.G.V.S. Standards—definitions, symbols, nucleotides, codons, amino acids (v2.0); 2013.
9. International Human Genome Sequencing Consortium. Finishing the euchromatic sequence of the human genome. Nature. 2004;431(7011):931–45.
10. Harrow J, et al. GENCODE: the reference human genome annotation for The ENCODE Project. Genome Res. 2012;22(9):1760–74.
11. Genome Project. VCF (Variant Call Format) version 4.0; 2011.
12. (HL7), H.L.S.I. HL7 Version 3 Standard: Clinical Genomics; Pedigree, Release 1; 2013.
13. Lebo RV, Grody WW. Testing and reporting ACMG cystic fibrosis mutation panel results. Genet Test. 2007;11(1):11–31.
14. IRB Guidebook: Chapter V Biomedical and Behavioral Research: An Overview. Retrieved from http://www.hhs.gov; 2014.
15. Regenstrief Institute. Logical Observation Identifiers Names and Codes (LOINC); 2013.
16. Bosca D, Marco L, Burriel V, Jaijo T, Millán JM, Levin A, Pastor O, Robles M, Maldonado JA. Genetic testing information standardization in HL7 CDA and ISO13606. Stud Health Technol Inform. 2013;192:338–42.
17. Chute CG, Kohane IS. Genomic medicine, health information technology, and patient care. JAMA. 2013;309(14):1467–8.
18. ASTM. Portable Document Format-Healthcare (PDF) a best practices guide. ASTM AIIMASTM—BP-01-08; 2013.
19. Hoffman MA. The genome-enabled electronic medical record. J Biomed Inform. 2007;40(1):44–6.
20. Garg AX, et al. Effects of computerized clinical decision support systems on practitioner performance and patient outcomes: a systematic review. JAMA. 2005;293(10):1223–38.
21. Scheuner MT, et al. Are electronic health records ready for genomic medicine? Genet Med. 2009;11(7):510–7.
22. Downing GJ, et al. Information management to enable personalized medicine: stakeholder roles in building clinical decision support. BMC Med Inform Decis Mak. 2009;9:44.
23. Scheuner MT, Sieverding P, Shekelle PG. Delivery of genomic medicine for common chronic adult diseases: a systematic review. JAMA. 2008;299(11):1320–34.
24. Ronquillo JG. How the electronic health record will change the future of health care. Yale J Biol Med. 2012;85(3):379–86.
25. Lucassen A, Hall A. Consent and confidentiality in clinical genetic practice: guidance on genetic testing and sharing genetic information. Clin Med. 2012;12(1):5–6.
26. Ayuso C, et al. Informed consent for whole-genome sequencing studies in the clinical setting. Proposed recommendations on essential content and process. Eur J Hum Genet. 2013;21(10):1054–9.
27. Bunnik EM, et al. The new genetics and informed consent: differentiating choice to preserve autonomy. Bioethics. 2013;27(6):348–55.
28. Hinton Jr RB. The family history: reemergence of an established tool. Crit Care Nurs Clin North Am. 2008;20(2):149–58. v.
29. Gulley ML, et al. Clinical laboratory reports in molecular pathology. Arch Pathol Lab Med. 2007;131(6):852–63.
30. Ullman-Cullere MH, Mathew JP. Emerging landscape of genomics in the Electronic Health Record for personalized medicine. Hum Mutat. 2011;32(5):512–6.
31. Naidoo N, et al. Human genetics and genomics a decade after the release of the draft sequence of the human genome. Hum Genomics. 2011;5(6):577–622.
32. Aquilante CL. Sulfonylurea pharmacogenomics in Type 2 diabetes: the influence of drug target and diabetes risk polymorphisms. Expert Rev Cardiovasc Ther. 2010;8(3):359–72.
33. Huang C, Florez JC. Pharmacogenetics in type 2 diabetes: potential implications for clinical practice. Genome Med. 2011;3(11):76.
34. Wilcox AR, Neri PM, Volk LA, Newmark LP, Clark EH, Babb LJ, Varugheese M, Aronson SJ, Rehm HL, Bates DW. A novel clinician interface to improve clinician access to up-to-date genetic results. J Am Med Inform Assoc. 2014;21(e1):e117–21. doi:10.1136/amiajnl-2013-001965.
35. Wilkins BS, Clark DM. Making the most of bone marrow trephine biopsy. Histopathology. 2009;55(6):631–40.
36. Cagle PT, et al. Template for reporting results of biomarker testing of specimens from patients with non-small cell carcinoma of the lung. Arch Pathol Lab Med. 2014;138(2):171–4.
37. Bartley AN, et al. Template for reporting results of biomarker testing of specimens from patients with carcinoma of the colon and rectum. Arch Pathol Lab Med. 2014;138(2):166–70.
38. Richards CS, et al. ACMG recommendations for standards for interpretation and reporting of sequence variations: revisions 2007. Genet Med. 2008;10(4):294–300.

39. Kohane IS, Churchill SE, Murphy SN. A translational engine at the national scale: informatics for integrating biology and the bedside. J Am Med Inform Assoc. 2012;19(2):181–5.
40. Masys DR, et al. Technical desiderata for the integration of genomic data into Electronic Health Records. J Biomed Inform. 2012;45(3):419–22.
41. Murphy S, Churchill S, Bry L, Chueh H, Weiss S, Lazarus R, Zeng Q, Dubey A, Gainer V, Mendis M, Glaser J, Kohane I. Instrumenting the health care enterprise for discovery research in the genomic era. Genome Res. 2009;19(9):1675–81.
42. Aronson SJ, Clark EH, Babb LJ, Baxter S, Farwell LM, Funke BH, Hernandez AL, Joshi VA, Lyon E, Parthum AR, Russell FJ, Varugheese M, Venman TC, Rehm HL. The GeneInsight Suite: a platform to support laboratory and provider use of DNA-based genetic testing. Hum Mutat. 2011;32(5):532–6.
43. Welch BM, Kawamoto K. Clinical decision support for genetically guided personalized medicine: a systematic review. J Am Med Inform Assoc. 2013; 20(2):388–400.
44. Belle A, Kon MA, Najarian K. Biomedical informatics for computer-aided decision support systems: a survey. ScientificWorldJournal. 2013;2013:769639.
45. Peterson JF, et al. Electronic health record design and implementation for pharmacogenomics: a local perspective. Genet Med. 2013;15(10):833–41.
46. Pulley JM, Denny JC, Peterson JF, Bernard GR, Vnencak-Jones CL, Ramirez AH, Delaney JT, Bowton E, Brothers K, Johnson K, Crawford DC, Schildcrout J, Masys DR, Dilks HH, Wilke RA, Clayton EW, Shultz E, Laposata M, McPherson J, Jirjis JN, Roden DM. Operational implementation of prospective genotyping for personalized medicine: the design of the Vanderbilt PREDICT project. Clin Pharmacol Ther. 2012;92(1):87–95.
47. Tural C, et al. Clinical utility of HIV-1 genotyping and expert advice: the Havana trial. AIDS. 2002; 16(2):209–18.
48. Pathak J, et al. Evaluating phenotypic data elements for genetics and epidemiological research: experiences from the eMERGE and PhenX network projects. AMIA Summits Transl Sci Proc. 2011;2011: 41–5.
49. Pathak J, et al. Mapping clinical phenotype data elements to standardized metadata repositories and controlled terminologies: the eMERGE Network experience. J Am Med Inform Assoc. 2011;18(4): 376–86.

CHAPTER 19

Reporting Clinical Genomic Assay Results and the Role of the Pathologist

JANINA A. LONGTINE

Introduction 277
Reporting of Single Gene or Gene Panel Results 278
Reporting of Whole-Exome and Whole-Genome Sequencing Results 279
Reporting of Cancer Test Results 280
Communicating Through Reports 281
The Role of the Pathologist 290
Conclusions 291

Introduction

Over the past decade we have seen a rapid rise in the number of clinically relevant molecular diagnostic assays accompanied by increasingly sophisticated technologies and complexity of generated data. This development has been driven in part by the discovery of the molecular underpinnings of disease. Relatively simple genotyping assays designed to detect a single allelic variant in one gene, such as *F5* c.1601G>A (p.Arg534Gln) or Factor V Leiden, advanced to genotyping panels within one gene. The American College of Medical Genetics and Genomics (ACMG)/American Congress of Obstetricians and Gynecologists-recommended panel of 23 mutations for cystic fibrosis screening is an example of the latter [1]. Further knowledge led to assays involving multiple mutations in multiple genes, such as the pathogenic variants in sarcomere proteins associated with hypertrophic cardiomyopathy [2] or the molecular stratification of lung adenocarcinoma used to predict response to targeted therapies [3, 4]. We are now entering an era of even greater complexity (and uncertainty) with the clinical application of exome or genome sequencing. Throughout this time, molecular pathologists and molecular geneticists have developed and implemented diagnostic assays following guidelines for quality assurance and test reporting issued by the College of American Pathologists (CAP), the ACMG, and the Association for Molecular Pathology (AMP). In this new era of high-throughput sequencing ("next-generation sequencing"), there are yet no universal standards for clinical reporting. This chapter reviews the challenges inherent in generating and delivering rational, informative genomic clinical reports and highlights emerging solutions.

G.J. Netto and I. Schrijver (eds.), *Genomic Applications in Pathology*,
DOI 10.1007/978-1-4939-0727-4_19, © Springer Science+Business Media New York 2015

Reporting of Single Gene or Gene Panel Results

Genotyping or targeted sequencing assays are designed to interrogate single-nucleotide variants (SNVs) or small indels with *known* phenotype–genotype correlation. In these assays, the clinical report should follow recommended guidelines of reporting and include laboratory, patient, and sample identifiers, the results indicating that the tested mutation is detected or not using standardized gene nomenclature and provide analytical and clinical interpretations with appropriate documentation from the medical literature [5]. Proceeding to single-gene sequencing or gene panel sequencing created the Pandora's box of variants of unknown significance (VUS), which are DNA variants that have not been reliably characterized as benign or pathogenic. Conventional genetic approaches, using segregation of the variant with disease in large family studies with affected individuals, are effective for assessing the significance of a VUS. This is particularly powerful for high-penetrance, rare variants. Unfortunately, this method is not applicable in the evaluation of most VUS. Some VUS have been characterized as pathogenic using a combination of clinical data and in vitro or animal model experiments that were conducted to prove biologic relevance, but this approach is difficult and not readily applied. Alternatively, for VUS in protein-coding exons, the pathogenicity may be predicted by using bioinformatic tools to assess evolutionary conservation and variants' effects on protein structure [6–8]. The predictive power of these tools is quite variable and may not correlate with clinical disease in humans. There is also a risk of over-interpretation of pathogenicity due to limited understanding of contextual information, such as biologic modifiers [9]. In addition, it can be just as difficult to prove that a variant is benign as it is to prove that it is pathogenic. Common variants in minority populations have not yet been well defined, further confounding interpretation because variants annotated as pathogenic may be SNVs in minority populations. This issue is being addressed through the 1000 Genomes Project (http://www.1000genomes.org last accessed April 13, 2013). For clinical reporting, ACMG has recommended six levels for interpretation of sequence variants which communicate the certainty of the clinical significance of the variant [7] (Table 19.1).

Table 19-1 Interpretive Categories of Germline Sequence Variations

Category	ACMG recommendations	Description
1	Sequence variation is previously reported and is a recognized cause of the disorder.	Pathogenic
2	Sequence variation is previously unreported and is of the type which is expected to cause the disorder.	Likely pathogenic
3	Sequence variation is previously unreported and is of the type which may or may not be causative of the disease.	Unknown significance
4	Sequence variation is previously unreported and is probably not causative of the disease.	Likely benign
5	Sequence variation is previously reported and is a recognized neutral variant.	Benign
6	Sequence variation is not known or expected to be causative of disease, but is found to be associated with a clinical presentation.	

Several databases are available to assist in interpreting VUS (e.g., UCSC Genome Bioinformatics Database and tools (http://www.genome.ucsc.edu last accessed April 15, 2013), Ensembl Genome Browser (http://useast.ensembl.org/index.html last accessed April 15, 2013), and 200 Exomes [10]). However, most public databases are not of clinical grade and contain errors. NCBI dbSNP Short Genetic Variations (http://www.ncbi.nlm.nih.gov/projects/SNP/ last accessed April 27, 2013) is a highly utilized public database of DNA sequence variation that is used to remove common variants in high-throughput sequencing data analysis. However, many entries lack population frequency information or are based on studies with few individuals [11]. dbSNP is not curated and is known to contain pathogenic variants. In 2008, Won et al. showed that 8 % of dbSNP (build 126) sequence variants were also present in the Human Gene Mutation Database (HGMD, http://www.hgmd.cf.ac.uk/ac/index.php, last accessed May 15, 2013) [11–14] highlighting the need for caution when using public databases for clinical annotations. Tong et al. have shown that a high frequency of inaccurate variant annotations were associated with variants discovered prior to the first human genome map and standardized coordinates [15]. This problem will be resolved with full adoption of nomenclature standards, specific reference to the genome build versions in annotation reports, and efforts to standardize clinical annotations of the genome [16].

Reporting of Whole-Exome and Whole-Genome Sequencing Results

With the decrease in cost of "whole" exome (WES) and "whole" genome sequencing (WGS), there has been a push to move from multi-gene panels to these more cost-effective approaches for clinical care. With gene or gene panel testing, the genes are selected because they are known to be implicated in a disease or influence therapeutic options. The test is ordered to address a specific clinical question. The comprehensive data generated by WES or WGS inherently includes a discovery process because only a small proportion of the data can be rationally associated with disease using current knowledge. Therefore, the challenge is to determine how much of the data will be reported in a clinical setting. There are several different approaches. One is that bioinformatic data analysis identifies all variants, but only those known to be associated with the disease in question are fully analyzed, interpreted, and reported. Alternatively, in addition to gene variants known to be relevant to the patient's disease, all gene variants known to be associated with human disease that are medically actionable and analytically verified are reported. For the latter, it is critical to set the bar high to minimize reporting of variants as pathogenic, which may later turn out to be benign. A third approach is to view the data as a resource that could be interrogated over the life of the patient as different medical needs and conditions develop [17].

Another challenge raised by WES/WGS is deciding what to do with the data after reporting. The data can be stored and available for future interrogation as new information emerges. In keeping all of the data, should the laboratory be responsible for reviewing and submitting updated reports on all archived cases when new variant–phenotype correlations are discovered and vetted? This would be most difficult to consistently achieve without an appropriate information technology infrastructure [18]. In addition, there is no clear mechanism for payment of reanalysis at this time. Due to the expense of data storage and the rapid change in technology, it may be more cost effective and easier to repeat the testing from the start rather than store data.

It is important to understand the technical limitations of the utilized assay prior to generating a report. The high-throughput short sequence reads are aligned to a reference genome, a best estimate of the gene sequence is determined (base calling), and variants are identified. The variants are filtered bioinformatically in order to generate a candidate list of pathogenic variants. To critically analyze the candidate list, one must be fully cognizant of how the data were filtered to know what type of variants may or may not have been detected. In addition, one must understand the sequencing methodology (e.g., capture design) and quality metrics of the sequence reads (how well individual regions are sequenced) to generate an informative report. The patient (and clinician) must also understand the limits of the testing and its interpretation through the consenting process.

Reporting of Cancer Test Results

The classification of cancer has been transformed by the discovery of specific cytogenetic and molecular aberrations that identify biologic subgroups of neoplasms within those previously grouped and classified according to histologic type. Recurrent genetic translocations that define subgroups of acute myeloid leukemias (AML) in the WHO classification are prognostic and frequently predictive. They are also often associated with characteristic morphologic and immunophenotypic features [19]. For karyotype normal AMLs, mutations in *NPM1*, *FLT3*, and *CEBPA* provide important prognostic information to guide therapy [20]. The advent of targeted therapy has further advanced cancer molecular diagnostics by identifying "driver" mutations inherent in the pathogenesis of the specific cancer types that are also sensitive to inhibitory therapy. The *BCR–ABL1* fusion gene encoded by t(9:22)(q34;q11.2) is the driver of chronic myelogenous leukemia. Its protein is the target of the tyrosine kinase inhibitor imatinib, and its chimeric mRNA is a sensitive tool for monitoring response to treatment and identifying drug resistance. The discovery of constitutively activating mutations in the *EGFR* gene in lung adenocarcinoma and the therapeutic efficacy of targeted small molecular inhibitors, such as gefitinib and erlotinib, heralded the importance of molecular diagnostics in solid tumor taxonomy [21–23]. Histologic classification is insufficient, and molecular testing is required to identify *EGFR*-mutant, responsive cancers. The molecular stratification of lung adenocarcinoma has continued to evolve, identifying multiple mutually exclusive driver mutations associated with different targeted treatments, such as *ALK* and *ROS1* mutations, and other mutations that predict a lack of response to targeted therapy, such as those affecting the *KRAS* gene. As such, pathologists now must seamlessly integrate molecular and cytogenetic/FISH testing into routine care and incorporate the mutation profile lexicon into their diagnostic armamentarium.

Clinically relevant testing algorithms have been developed in molecular pathology laboratories to sequentially identify hotspot mutations, based on the prevalence in different cancer types. As multiple hotspots in multiple genes became clinically relevant and as multiplex technologies evolved, many laboratories moved to gene mutation profiling to identify "actionable" mutations in a timely and cost-effective manner [24–27]. This evolution raises several important points related to informative cancer mutation reports. Most of the actionable mutations are heterozygous, diluting the targeted mutant alleles (1:1) with non-mutated, wild-type alleles. In addition, clinical specimens are heterogeneous with a mix of normal and tumor cells. Both factors contribute to the potential reduction of the mutant alleles within the cancer DNA specimen and require the development of sensitive genotyping tests with defined limits of detection and rigorous quality controls. In addition, a skilled pathologist is needed to estimate the percent of cancer cells within the specimen to determine specimen adequacy. Both anatomic and clinical pathology training and expertise are helpful. The reporting pathologist must understand the technology and the assay limitations before issuing a negative report.

Some clinical laboratories are now moving to exome panels for cancer mutation profiling. A challenge that emerged with broad genotyping panels and which holds with exome panels or WES/WGS is the need for well-curated cancer mutation knowledge bases. For example, *EGFR* exon 19 deletions and exon 21 p.Leu858Arg point mutation in lung adenocarcinomas confer sensitivity to tyrosine kinase inhibitors, whereas exon 20 insertion mutations are associated with primary resistance to these drugs [28]. Approximately 10-15 % of *EGFR*-mutant lung cancers have less common *EGFR* mutations that were not included in many clinical trials, making it difficult to predict response to targeted therapy [29]. *BRAF* mutation p.Val600Glu (commonly known as p.V600E) has different therapeutic implications for melanoma and colorectal carcinoma. *BRAF* p.V600E mutations lead to constitutive activation of the MAPK signaling pathway. *BRAF* p.V600E-positive metastatic melanomas have a dramatic response (60–80 %) to the selective *BRAF* inhibitor, vemurafenib [30]. However, colon cancer with *BRAF* p.V600E mutation infrequently (<5 %) responds to vemurafenib [31] due to EGFR-mediated MAPK pathway reactivation, leading to vemurafenib resistance [32, 33]. It is a challenge for each laboratory and its staff to be fully informed of the current literature as well as all available clinical trials. Furthermore, exome panels applied across all cancer types may reveal mutations with limited clinical evidence of utility in a particular cancer subtype or reveal more common mutations in an uncommon or an unexpected tumor type. These necessitate time-consuming literature investigations that may yield only small studies or case reports that preclude a truly informative report.

As cancer high-throughput sequencing expands, the need for accurate, curated knowledge databases will increase for pathologists and other laboratory professionals. This is important for the generation of informative reports so that physicians and their patients may better understand the clinical implications of a rendered report. Catalogue of Somatic Mutations in Cancer (COSMIC) has an extensive compilation of mutations found in cancer, but has no clinical data to associate therapeutic response (http://cancer.sanger.ac.uk/cancergenome/projects/cosmic/ last accessioned April 14, 2013). Vanderbilt-Ingram Cancer Center, Nashville, Tennessee has created a freely available, curated online knowledge base for specific mutations in different cancer types, My Cancer Genome (www.mycancergenome.org last accessed April 13, 2013). The site includes an overview of mutations for different cancer types as well as the frequency and clinical significance of each mutation with supporting literature references and information about related clinical trials. To further address the problems associated with less common mutations, they have created a database, DNA-Mutation Inventory to Refine and Enhance Cancer Treatment (DIRECT), intended to serve as a comprehensive electronic catalog of cancer gene response outcomes from individual patients culled from a meta-analysis of published literature [34]. The first entry represents EGFR mutations in non-small-cell lung carcinoma and can be queried for clinically relevant data associated with 188 mutations and response to erlotinib or gefitinib. DIRECT is accessible on the "My Cancer Genome" website.

To assist in communicating cancer-associated predictive mutations, it may be helpful to group mutations into tiers for each tumor type, such that tier 1 are mutations known to be associated with response to a specific treatment in a particular cancer subtype; tier 2 are mutations currently being evaluated in clinical trials; and tier 3 are mutations whose clinical relevance is unknown in a particular cancer subtype. As with germline testing, the encumbrance of the laboratory to update previously issued reports as new information appears is not settled.

Communicating Through Reports

Although it is early in the clinical implementation of high-throughput sequencing, several professional organizations have issued initial guidelines for reporting, including the CAP (through the 2012 Molecular Pathology Checklist), the ACMG [35, 36], and the New York State Department of Health (http://www.wadsworth.org/labcert/TestApproval/forms/NextGenSeq_ONCO_Guidelines.pdf). Figure 19.1 is an example of a report for high-throughput sequencing for germline mutations associated with autism spectrum disorder using an exome panel of 30 genes. The report illustrates key features to be included to enable clear communication of results. Specifically, the report incorporates:

1. Contact information for the laboratory with a link to the laboratory website providing additional information about the test, including an information sheet on the specific genes
2. Three patient identifiers, indication for testing, the test performed, specimen type, dates of specimen collection, and receipt and date of report
3. Abnormal results using Human Genome Variation Society (HGVS) nomenclature (www.hgvs.org/mutnomen) and ACMG-recommended interpretive categories
4. A listing of the genes tested as well as any test limitations, such as inconsistency in sequence quality due to high GC content (as seen in, e.g., *SHANK3*, *ARX*) or areas that lack sufficient coverage to confidently determine mutational status

MOUNT SINAI GENETIC TESTING LABORATORY

Icahn School of Medicine at Mount Sinai

Department of Genetics and Genomic Sciences
Icahn School of Medicine at Mount Sinai

One Gustave L. Levy Place, Box 1497
New York, NY 10029-6574
Tel: 212-241-7518 Fax: 212-241-0139
www.mssm.edu/genetictesting
CLIA # 33D0653419

Page 1 of 3

MOLECULAR GENETICS

Patient Name: **Test Sample**
Date of Birth: **4/3/2010**
Reference #: **N/A**
Indication: **Autism spectrum disorder**
Test Type: **Autism NGS Sequencing Panel**
Specimen Type: **Blood**
Lab #: **1300000AU**
Date Collected: **3/4/2013**
Date Received: **3/4/2013**

Final Report: **4/3/2013**

Referring:

Fax:

RESULTS AND INTERPRETATION

Abnormal Result:
***NSD1* likely pathogenic heterozygous mutation detected: c.4591dupA, p.Met1531Asnfs*4**
***SHANK2* homozygous variant of unknown significance detected: c.1670G>A, p.Ser557Asn**
***TSC2* likely benign heterozygous variant detected: c.5312C>T, p.Pro1771Leu**

GENE*	RESULTS
NRXN1 NM_001135659.1, NM_004801.4, NM_138735.2	No mutations detected
***NSD1* NM_022455.4, NM_172349.2**	**Mutation detected: p.Met1531Asnfs*4**
AHI1 NM_001134830.1, NM_001134831.1, NM_001134832.1, NM_017651.4	No mutations detected
CNTNAP2 NM_014141.5	No mutations detected
TSC1 NM_000368.4, NM_001162426.1, NM_001162427.1	No mutations detected
PTEN NM_000314.4 *plus* a portion of the 5' UTR (chr10:89623220-89623484)	No mutations detected
***SHANK2* NM_012309.3**	**Variant detected: p.Ser557Asn**
DHCR7 NM_001163817.1, NM_001360.2	No mutations detected
CACNA1C only exon 8 (chr12:2613597-2613710 and chr12:2614003-2614116)	No mutations detected
UBE3A NM_000462.3, NM_130838.1, NM_130839.2	No mutations detected
***TSC2* NM_000548.3, NM_001077183.1, NM_001114382.1**	**Variant detected : p.Pro1771Leu**
SHANK3 NM_033517.1 *except* for two exons plus splice sites (chr22:51113470-51113684 and chr22:51135986-51136148)**	No mutations detected
NLGN4X NM_020742.2, NM_181332.1	No mutations detected
AP1S2 NM_003916.3	No mutations detected
CDKL5 NM_001037343.1, NM_003159.2	No mutations detected
PTCHD1 NM_173495.2	No mutations detected
ARX NM_139058.2 *except* for a portion of exon 2 (chrX:25031504-25031920)***	No mutations detected
IL1RAPL1 NM_014271.3	No mutations detected
OTC NM_000531.5	No mutations detected
KDM5C NM_001146702.1, NM_004187.3	No mutations detected
OPHN1 NM_002547.2	No mutations detected
PCDH19 NM_001105243.1, NM_020766.2, NM_001184880.1	No mutations detected
UPF3B NM_023010.3, NM_080632.2	No mutations detected
GRIA3 NM_000828.4, NM_007325.4	No mutations detected
GPC3 NM_001164617.1, NM_001164618.1, NM_001164619.1, NM_004484.3	No mutations detected
SLC9A6 NM_001042537.1, NM_001177651.1, NM_006359.2	No mutations detected
FMR1 NM_001185075.1, NM_001185076.1, NM_001185081.1, NM_001185082.1, NM_002024.5	No mutations detected
SLC6A8 NM_001142805.1, NM_001142806.1, NM_005629.3	No mutations detected
MECP2 NM_001110792.1,NM_004992.3 *plus* a portion of the 3' UTR (chrX:153295704-153295748)	No mutations detected
RAB39B NM_171998.2	No mutations detected

* All coding exons of the genes corresponding to the transcripts listed plus the flanking 5 base pair splice sites are sequenced relative to the hg19 assembly. Exceptions are noted.

Figure 19-1 Example of a report for high-throughput sequencing for germline mutations associated with autism spectrum disorder using an exome panel of 30 genes. Courtesy of the Mount Sinai Genetic Testing Laboratory, Department of Genetics and Genomics, Icahn School of Medicine at Mount Sinai, New York, NY

** Only a single variant has been reported as pathogenic in either of these two excluded exonsin "The Human Gene Mutation Database (HGMD)" (http://www.hgmd.cf.ac.uk/ac/index.php, December 2012, Gauthier *et al.*). Please note that this region is excluded due to inconsistency in sequencing result quality due to high GC content.
***Twelve percent of all *ARX* sequence mutations reported in "The Human Gene Mutation Database (HGMD)" (http://www.hgmd.cf.ac.uk/ac/index.php, December 2012) are located within this region. Therefore, some mutations may be missed. Please note that this region is excluded due to inconsistency in sequencing result quality due to high GC content.

Interpretation: Next generation sequencing of a panel of 30 genes that have been associated with Autism Spectrum Disorders (ASDs) was performed on DNA extracted from the peripheral blood specimen of this patient. A *de novo*, heterozygous, frameshift mutation which consists of a duplication of an A residue at nucleotide 4591 was detected in coding exon 11 of the longest transcript of *NSD1*NM_022455.4: c.4591dupA, p.Met1531Asnfs*4 (HG19 chr5:176675276). This results in a change in the coding sequence at codon position 1531 and causes a premature termination codon four amino acids downstream. This mutation has not been previously reported but is predicted to be pathogenic based on the deleterious nature of the mutation. Mutations in *NSD1* cause Sotos syndrome, which is an autosomal dominant overgrowth condition characterized by a typical facial appearance, learning disabilities, and, in some patients, ASD and additional congenital anomalies.

In addition, a homozygous, missense variant of unknown significance was detected in coding exon 10 of the *SHANK2* gene NM_012309.3: c.1670G>A, p.Ser557Asn (HG19, chr11:70644655). This variant results in an amino acid change from serine to asparagine at position 557 in the major transcript product of *SHANK2* and has been previously reported in dbSNP (rs141184740) and the Exome Variant Server. Both databases show this change in the heterozygous state only, with a frequency of approximately 1%. Genotypes present in dbSNP with a frequency of ≥1% are generally considered benign. To our knowledge this variant has not been reported in the homozygous state. For this reason, it is unclear whether this homozygous change could influence the phenotype of this patient and, therefore, the clinical significance is uncertain.

Furthermore, a likely-benign, maternally-inherited,heterozygous, missense variant was detected in the last exon of the *TSC2* gene NM_000548.3: c.5312C>T, p.Pro1771Leu (HG19, chr16:2138499). This variant results in an amino acid change from proline to leucine at position 1771 and has been previously reported in dbSNP (rs137854214) as rare (<1% heterozygous). Mutations in *TSC2* cause tuberous sclerosis, an autosomal dominant disorder that includes non-malignant tumors of the brain, kidneys, heart, lungs, eyes and skin. In addition, seizures, developmental delay and autism may also be observed in patients with *TSC2* mutations.This change is predicted to be "benign" by the Poly-Phen2 HumVar in silico analysis and "damaging" by the SIFT in silico analysis. Correlation with maternal phenotype is recommended.

Genetic counseling and correlation with the clinical phenotype of this patient is recommended. Please note that parental DNA was analyzed solely for the presence of the p.Met1531Asnfs*4 mutation in *NSD1*, the p.Ser557Asn variant in *SHANK2*, and the p.Pro1771Leu variant in *TSC2*.

This technology may not detect all small insertion/deletions and is not diagnostic for large duplications/deletions and structural genomic variation. In addition, a mutation(s) in a gene not included on the panel could be present in this patient. Although each of the genes on the panel is a rare cause of ASD, this panel is expected to detect 5-10% of mutations present in ASD patients. The sensitivity of this panel is estimated at 99% for single base substitutions and 97% overall. All potentially pathogenic variants were subjected to Sanger sequencing for confirmation of the result. Any benign polymorphisms identified during this analysis were not reported.

Comments:
Please note this test was developed and its performance characteristics were determined by The Mount Sinai Genetic Testing Laboratory and were considered acceptable for patient testing. It has not been cleared or approved by the FDA. The FDA has determined that such clearance or approval is not necessary.

This type of mutation analysis generally provides highly accurate genotype information for point mutations and single nucleotide polymorphisms. Despite this level of accuracy, it should be kept in mind that there are many potential sources of diagnostic error, including misidentification of samples, polymorphisms, or other rare genetic variants that interfere with analysis. In addition, families should understand the limitations of the testing and that rare diagnostic errors may occur for the reasons described.

References:
1. BS Abrahams and DH Geschwind. Advances in autism genetics: on the threshold of a new neurobiology. Nature. 9(5):341-355, 2008.
2. Addington AM et al. A novel frameshift mutation in UPF3B identified in brothers affected with childhood onset schizophrenia and autism spectrum

Figure 19-1 (continued)

Test Sample born 4/3/2012
1300000AU 3/3

disorders. Mol Psychiatry. 16(3):238-9, 2011.
3. Adegbola A et al. A novel mutation in JARID1C/SMCX in a patient with autism spectrum disorder (ASD). Am J Med Genet A. 146A(4):505-11, 2008.
4. Archer HL et al. CDKL5 mutations cause infantile spasms, early onset seizures, and severe mental retardation in female patients. J Med Genet. 43(9):729-34, 2006.
5. Bakkaloglu B et al. Molecular cytogenetic analysis and resequencing of contactin associated protein-like 2 in autism spectrum disorders. Am J Hum Genet. 82(1):165-73, 2008.
6. Berkel S, et al. Mutations in the SHANK2 synaptic scaffolding gene in autism spectrum disorder and mental retardation. Nat Genet. 42(6):489-91, 2010.
7. Borck G et al. Clinical, cellular, and neuropathological consequences of AP1S2 mutations: further delineation of a recognizable X-linked mental retardation syndrome. Hum Mutat. 29(7):966-74,2008.
8. Buxbaum JD et al. Mutation screening of the PTEN gene in patients with autism spectrum disorders and macrocephaly. Am J Genet B Neuropsychiatr Genet. 144B(4):484-91, 2007.
9. Chabrol B et al. Delineation of the clinical phenotype associated with OPHN1 mutations based on the clinical and neuropsychological evaluation of three families. Am J Med Genet A. 138(4):314-7, 2005.
10. Cohen D et al. Specific genetic disorders and autism: clinical contribution towards their identification. J Autism Dev Disord. 35(1):103-6, 2005.
11. Dibbens LM et al. X-linked protocadherin19 mutations cause female-limited epilepsy and cognitive impairment. Nat Genet. 40(6):776-81, 2008.
12. Durand CM et al. Mutations in the gene encoding the synaptic scaffolding protein SHANK3 are associated with autism spectrum disorders. Nat Genet. 39(1)25-7, 2007.
13. Gauthier J et al. Novel de novo SHANK3 mutation in autistic patients. Am J Med Genet B Neuropsychiatr Genet. 150B(3):421-4, 2009.
14. Giannandrea M et al. Mutations in the small GTPase gene RAB39B are responsible for X-linked mental retardation associated with autism, epilepsy, and macrocephaly. Am J Hum Genet. 86(2):185-95, 2010.
15. Guilmatre A et al. Recurrent rearrangements in synaptic and neurodevelopmental genes and shared biologic pathways in schizophrenia, autism, and mental retardation. Arch Gen Psychiatry. 66(9):957-56, 2009.
16. Herman GE et al. Increasing knowledge of PTEN germline mutations: Two additional patients with autism and macrocephaly. Am J Med Genet A. 143(6):589-93, 2007.
17. Jamal SM et al. Novel de novo PCDH19 mutations in three unrelated females with epilepsy female restricted mental retardation syndrome. Am J Med Genet A. 152A(10):2475-81, 2010.
18. Kielinen M, Rantala H, Timonen E, Linna SL, Moilanen I. Associated medical disorders and disabilities in children withautistic disorder: a population-based study. Autism. 8:49-60, 2004.
19. Kim HG et al. Disruption of neurexin 1 associated with autism spectrum disorder. Am J Hum Genet. 82(1):199-207, 2008.
20. Laumonnier F et al. Mutations of the UPF3B gene, which encodes a protein widely expressed in neurons, are associated with nonspecific mental retardation with or without autism. Mol Psychiatry. 15(7):767-776, 2010.
21. Laumonnier, F et al. X-linked mental retardation and autism are associated with a mutation in the NLGN4 gene, a member of the neuroligin family. Am J Hum Genet. 74(3):552-7, 2004.
22. Liao P and Soong TW. CaV1.2 channelopathies: from arrhythmias to autism, bipolar disorder, and immunodeficiency. Pflugers Arch. 460(2):353-9, 2010.
23. Marco EJ and Skuse DH. Autism-lessons from the X chromosome. Soc Cogn Affect Neurosci. 1(3):183-93, 2006.
24. Marshall CR et al. Structural variation of chromosomes in autism spectrum disorder. Am J Hum Genet. 82(2):477-88, 2008.
25. Metzker, ML. Sequencing technologies -the next generation. Nature Reviews, Genetics. 11(1): 31-46, 2010.
26. Miles JH, Hillman RE. Value of a clinical morphology examination in autism. Am J Med Genet. 91:245-253, 2000.
27. Morrow JD, Whitman BY, and Accardo PJ. Autistic disorder in Sotos syndrome: a case report. Eur J Pediatr. 149(8):567-9, 1990.
28. Nawara M et al. The ARX mutations: a frequent cause of X-linked mental retardation. Am J Med Genet A. 140(7):727-32, 2006.
29. Neri G et al. Clinical and molecular aspects of the Simpson-Golabi-Behmel syndrome. Am J Med Genet A. 79(4):279-83, 1998.
30. Noor A et al. Disruption at the PTCHD1 Locus on Xp22.11 in Autism spectrum disorder and intellectual disability. Sci Transl Med. 2(49):49-68, 2010.
31. Ozonoff S, Williams BJ, Gale S, Miller JN. Autismand autistic behavior in Joubert syndrome. J Child Neurol. 14(10):636-41, 1999.
32. Pinto D et al. Functional impact of global rare copy number variation in autism spectrum disorders. Nature. 466(7304):368-72, 2010.
33. Piton A et al. Mutations in the calcium-related gene IL1RAPL1 are associated with autism. Hum Mol Genet. 17(24):3965-74,2008.
34. Piton A et al. Systematic resequencing of X-chromosome synaptic genes in autism spectrum disorder and schizophrenia. Mol Psychiatry. 16(8):867-80, 2011.
35. Schroer RJ et al. Natural history of Christianson syndrome. Am J Med Genet A. 152A(11):2775-83, 2010.
36. Serrano M et al. Neuropsychiatric manifestations in late-onset urea cycle disorder patients. J Child Neurol. 25(3):352-8, 2010.
37. Straus KA eta al. Recessive symptomatic focal epilepsy and mutant contactin-associated protein-like 2. N Engl J Med. 354(13):1370-7, 2006.
38. Stromme P et al. Mutations in the human ortholog of Aristalesscause X-linked mental retardation and epilepsy. Nat Genet. 30(4):441-5, 2002.
39. Takahashi Y et al. A loss-of-function mutation in the SLC9A6 gene causes X-linked mental retardation resembling Angelman syndrome. Am J Med Genet B Neuropsychiatr Genet. 156B(7)799-807, 2011.
40. Tarpey PS et al. Mutations in the gene encoding the Sigma 2 subunit of the adaptor protein 1 complex, AP1S2, cause X-linked mental retardation. Am J Hum Genet. 79(6):1119-24,2006.
41. Tzschach A et al. Novel JARID1C/SMCX mutations inpatients with X-linked mental retardation. Hum Mutat. 27(4):389, 2006.
42. Vissers LE et al. A de novo paradigm for mental retardation. Nat Genet. 42(12):1109-12, 2010.
43. Weaving LS et al. Mutations of CDKL5 cause a severe neurodevelopmental disorder with infantile spasms and mental retardation. Am J Hum Genet. 75(6):1079-93, 2004.
44. Wu Y et al. Mutations in ionotropic AMPA receptor 3 alter channel properties and are associated with moderate cognitive impairment in humans. Proc Natl Acad Sci USA. 104(46)18163-8, 2007.
45. Yan J et al. Analysis of the neuroligin 3 and 4 genes in autism and other neuropsychiatric patients. Mol Psychiatry. 10(4):329-32, 2005.
46. Zahir FR et al. A patient with vertebral, cognitive and behavioural abnormalities and a de novo deletion of NRXN1alpha. J Med Genet. 45(4):239-43, 2008.
47. Zanni G et al. Oligophrenin 1 mutations frequently cause X-linked mental retardation with cerebellar hypoplasia. Neurology. 65(9):1364-9, 2005.
48. Exome Variant Server, NHLBI GO Exome Sequencing Project (ESP), Seattle, WA (URL: http://evs.gs.washington.edu/EVS/) [March, 2013].

Figure 19-1 (continued)

5. The version of the reference genome used (e.g., hg19 (NCBI build 37))
6. Result interpretation including an explanation of supporting evidence and clinical implications as well as the recommendation for genetic counseling
7. Limitations of the technology (e.g., regarding detection of large deletions/duplications)
8. A statement that all pathogenic mutations were confirmed by an alternative methodology (e.g., Sanger sequencing)
9. A FDA disclaimer
10. References supporting the test panel and the interpretation

Some laboratories may choose to include a more detailed description of the methods utilized in the report.

Although similar features should be incorporated into cancer reports, the variant classification for somatic mutations, compared to that of germline variants, requires different levels of interpretation. In cancer genotyping where one tests for somatic variants known to have diagnostic, prognostic, or predictive significance, the interpretive levels can be quite simple, such as the previously mentioned tiers 1, 2, and 3. With cancer exome panels, the breadth of variants identified requires more complex interpretation levels. For example, Genomics and Pathology Services at Washington University School of Medicine in St. Louis use an eight-level variant classifier (Table 19.2) for their cancer exome panel. Figure 19.2 is an example of a report for cancer whole-genome sequence analysis in a normal karyotype acute myeloid leukemia specimen. This report contains all identified protein-altering somatic mutations (nonsynonymous/splice site SNVs and coding indels) and further subcategorizes them into prognostic mutations, recurrent mutations in this cancer type (acute myeloid leukemia), recurrent cancer mutations based on the COSMIC database, and all remaining mutations. It is notable that this report also comments on pertinent negatives, which affect the prognostic risk profile. The germline variants are reported separately. There is a report for germline variants in 43 known cancer susceptibility syndromes and another for all other germline coding variants.

Because WES and WGS of cancer require the patient's germline DNA to be analyzed simultaneously to properly identify the unique cancer somatic mutations, clinical laboratories need to develop policies about identifying and reporting clinically important germline mutations. This holds true for laboratories that perform high-throughput sequencing for only germline analysis because similar "secondary" findings, i.e., those not related to the clinical condition that led to the high-throughput sequencing, will emerge. A 2012 ACMG Board of Directors policy statement on the clinical application of genomic sequencing states that "laboratories and clinics using WGS/WES should have clear policies in place related to disclosure of secondary findings. Patients should be informed of those policies and the types of secondary findings that will be reported back to them and under what circumstances. Patients should be given the option of *not* receiving certain or secondary findings" [35]. Subsequently a Working Group charged by the ACMG Board of Directors developed a minimum list of incidental secondary findings composed of pathogenic variants associated with monogenic disorders to be reported to

Table 19–2 **Somatic Variant Classification Description, Genomics and Pathology Services, Washington University School of Medicine in St. Louis**

Level	Description
1	Clinically actionable for the patient's cancer
2	Clinically prognostic for the patient's cancer
3	Clinically actionable in a cancer other than the patient's cancer
4	Biologic evidence suggesting an alteration of function of the normal protein
5	Previously identified in other patients having the same cancer, but with no known clinical relevance
6	Previously identified in other patients having some other cancer or other diseases, but with no known clinical relevance
7	Novel variants, not previously documented as a polymorphism
8	Previously documented as a known polymorphism

Washington University in St. Louis
SCHOOL OF MEDICINE

Clinical and Molecular Cytogenetics		Name:	**UPN-XXX, XXX**
Laboratory and Genomic Medicine		Culture Number:	**G11-673**
Department of Pathology		Gender:	**U**
St. Louis, MO 63110		Date of Birth/Age:	
Phone: 314-454-8101			
Fax: 314-454-5192			
CLIA 26D0698685			
CAP # 27556-03			

WHOLE GENOME SEQUENCE ANALYSIS

Tissue:	**Bone Marrow**	MRN:	**1111111**	Date Collected:	**2/26/2011**
Physician(s):	**Peter Westervelt, M.D.**	Hospital #:	**111111**	Date Accessioned:	**2/26/2011**
	Timothy J. Ley, M.D.	Facility:	**BJH**	Date of Culture:	**2/26/2011**

Processing: Whole Genome Sequencing

Indication: Newly diagnosed AML, need prognostic classification for treatment and transplantation
Specimen Quality Comment:

WHOLE GENOME SEQUENCE ANALYSIS

***Preliminary/Amended ***

Karyotype:

46,XX
Diagnosis:

CHROMOSOME ANALYSIS:	**NO EVIDENCE OF CLONAL ABERRATIONS**
MOLECULAR ABERRATIONS:	**PATHOGENIC SEQUENCE VARIANTS DETECTED (SEE BELOW FOR DETAILS)**

Interpretation:

TEST PERFORMED

Paired-end whole genome sequencing (WGS) of DNA derived from bone marrow (102.8 Gbp at 28.1X fold coverage) and matched normal counterpart (skin, 98.6 Gbp at 26.2X fold coverage).

CLINICAL HISTORY AND INDICATION FOR TEST

This is a previously 57 Y/O healthy Caucasian female who presented with sudden onset of fatigue and bruisability, and was found to have a peripheral white blood count of 105,000 cells per micro litre, with 85% myeloblasts. Bone marrow (current specimen) revealed 100% myeloblasts with morphological features and cell surface markers consistent with FAB M1 AML.

VARIANT CLASSIFICATION DESCRIPTION: "Level 1" variants are all protein-altering somatic mutations (nonsynonymous/splice site SNVs and coding indels). Within Level 1, mutations are further classified into the following four categories: **(1a)** prognostic mutations, **(1b)** recurrent AML mutations, **(1c)** recurrent cancer mutations [based on data from COSMIC (Catalogue of Somatic Mutations in Cancer database at Sanger Institute, UK], and **(1d)** remaining mutations.

Page 1 of 4

Figure 19-2 Example of a report for cancer whole-genome sequence analysis in a normal karyotype acute myeloid leukemia specimen, courtesy of Dr. Shashikant Kulkarni, Genomics and Pathology Services, Washington University in St. Louis, School of Medicine, St. Louis, MO

UPN-XXX, XXX WHOLE GENOME SEQUENCE ANALYSIS (2/28/2011) G11-673

The "Level 2" report will contain all germline (inherited) variants in 43 known cancer susceptibility syndromes. The "Level 3" report will contain all other germline coding variants. When these variants are available, an addendum to this report, reflecting these variants will be issued.

DNA VARIANTS

Gene affected	Chromosomal location	DNA change	AA change	Mutation type	Variant Classification
FLT3	13q12	insAAGTACTCATT ATCTGAGAA ATCAACGTAG	p.594insSDNEYFYVDF	Internal tandem duplication (ITD)	**Level 1a**
NPM1	5q35	insCATG	p.594insSDNEYFYVDF	Insertion	**Level 1a**
DNMT3A	2p23	delA	p.L723fs	Deletion	**Level 1a**
TTN	2q31	C>T	p.E14263K	Base substitution	**Level 1b**
SMC3	10q25	G>T	p.G662C	Base substitution	**Level 1b**
PTPRT	20q12-q13	G>A	p.P1232L	Base substitution	**Level 1c**
CALHM3	10q24.33	C>T	p.S15	Base substitution	**Level 1d**
CDH24	14q11.2	G>T	P.Y590X	Base substitution	**Level 1d**
GPR123	10q26	C>T	p.T381	Base substitution	**Level 1d**
GPR183	13q32.3	G>A	p.A338V	Base substitution	**Level 1d**
PCDH24	5q35.2	C>T	p.P1004L	Base substitution	**Level 1d**
SLC151A	13q33-q34	C>T	p.W77X	Base substitution	**Level 1d**
DSCAM	21q22.2-q22.3	A>G	p.L1847	Base substitution	**Level 1d**
PDXDC1	16p13.11	G>A	p.E421K	Base substitution	**Level 1d**
LOC728896	17q21.32	G>T	p.F46L	Base substitution	**Level 1d**

INTERPRETATION

1a. Mutations with prognostic significance

***FLT3* (Internal tandem duplication, ITD)**

FLT3 is a member of the class III receptor tyrosine kinase family; *FLT3* and its ligand play an important role in proliferation, survival and differentiation of hematopoietic progenitor cells. *FLT3*-ITD are found in about 20% of all AML cases, and in 28%–34% of cytogenetically normal AML (*Cancer Cell 12,501-513,2007*).

The variant described above has been previously described in the literature with well-established pathogenicity. This variant is associated with prognostic classification of intermediate risk (*Blood, 21, 2010,vol 115, No.3*).

***NPM1* (four base pair deletion in exon 12)**

NPM1 encodes a phosphoprotein with pleiotropic functions. *NPM1* mutations are found in 25%–35% of all adult AML cases, in 45%–64% of karyotypically normal-AML, in 35%–40% of AML with 9q deletion, and in about 15% of AML with trisomy 8. *NPM1* mutations are associated with *FLT3*-ITD (~40%) and *FLT3* TKD mutations. *NPM1* is also associated with higher BM blast counts and serum LDH levels, myelomonocytic or monocytic morphology, and high CD33 and absent CD34 expression.

NPM1 mutation with *FLT3* ITD in cytogenetically normal individuals is generally associated with intermediate risk (*Blood, 21 J 2010 ,vol 115, No.3*).

DNMT3A deletion

DNMT3A mutations are highly recurrent in patients with de novo AML with an intermediate-risk cytogenetic profile and may be independently associated with a poor outcome (*N Engl J Med 2010; 363:2424-2433, December 16, 2010*). Although the impact on outcome has not yet been verified by other groups, this variant may confer adverse risk to AML patients.

Of note, mutations in *CEBPA* were not detected in this genome, nor were cryptic fusions involving *PML, RARA, RUNX1, RUNX1T1, CBFB,* or *MYH11.* These mutations would have reclassified the risk profile of this patient

Page 2 of 4

Figure 19-2 (continued)

UPN-XXX, XXX WHOLE GENOME SEQUENCE ANALYSIS (2/28/2011) G11-673

had they been present.

1b. Mutations that are recurrent in AML, with unknown significance for prognosis.

***TTN* E14263K**

Mutations in *TTN* have been detected in many tumor types including AML, perhaps because of the enormous size of this gene. Their significance is currently unknown (WU AML database).

***SMC3* G662C**

SMC3 is a component of the cohesin complex, which contributes to the coordination of chromosome segregation, the DNA damage response and chromatin regulation by epigenetic mechanisms (*Nature Reviews Cancer 11:199, 2011*). Its role in cancer pathogenesis is unknown, but mutations in this gene and other components of the cohesin complex have been found in other AML patients (WU AML database).

1c. Variants that are found in cancer, but not previously described in AML; unknown significance for prognosis.

***PTPRT* P1232L**

Mutations in this phosphatase gene are common in patients with colon cancer (Science 304:1164, 2004), but have not previously been described in AML. This mutation is predicted to disrupt phosphatase function, and may be relevant for pathogenesis.

1d. Somatic mutations of unknown significance

CALHM3 S15 (silent)
CDH24 Y590X
GPR123 T38I
GPR183 338V
PCDH24 P1004L
SLC151A1 W77X
DSCAM L1847
PDXDC1 E421K
LOC728896 F46L

These sequence variants are of unknown significance based on the current knowledge. Their presence is noted for future reference. An addendum will be issued when an update or change in interpretation for any of the reported variants is noted. An alert will be issued to this individual's management team.

Gene affected	Chromosomal location	DNA change	AA change	Mutation type	Variant Classification
FLT3	13q12	insAAGTACTCATTATCTGAGA A ATCAACGTAG	p.594insSDNEYFYVDF	Internal tandem duplication (ITD)	**Level 1a**
NPM1	5q35	insCATG	p.594insSDNEYFYVDF	Insertion	**Level 1a**
DNMT3A	2p23	delA	p.L723fs	Deletion	**Level 1a**
TTN	2q31	C>T	p.E14263K	Base substitution	**Level 1b**
SMC3	10q25	G>T	p.G662C	Base substitution	**Level 1b**
PTPRT	20q12-q13	G>A	p.P1232L	Base substitution	**Level 1c**
CALHM3	10q24.33	C>T	p.S15	Base substitution	**Level 1d**
CDH24	14q11.2	G>T	P.Y590X	Base substitution	**Level 1d**
GPR123	10q26	C>T	p.T381	Base substitution	**Level 1d**

Figure 19-2 (continued)

UPN-XXX, XXX WHOLE GENOME SEQUENCE ANALYSIS (2/28/2011) G11-673

GPR183	13q32.3	G>A	p.A338V	Base substitution	**Level 1d**
PCDH24	5q35.2	C>T	p.P1004L	Base substitution	**Level 1d**
SLC151A	13q33-q34	C>T	p.W77X	Base substitution	**Level 1d**
DSCAM	21q22.2-q22.3	A>G	p.L1847	Base substitution	**Level 1d**
PDXDC1	16p13.11	G>A	p.E421K	Base substitution	**Level 1d**
LOC728896	17q21.32	G>T	p.F46L	Base substitution	**Level 1d**

Report Electronically Reviewed and Signed Out By

Shashikant Kulkarni, M.S., Ph.D., FACMG Date Reported: 2/28/2011
Medical Director, Cytogenomics & Molecular Pathology and Genomics and Pathology Services
Associate Professor of Pediatrics, Genetics, Pathology and Immunology

Amendments for WHOLE GENOME SEQUENCE ANALYSIS (2/28/2011)

Amended Date: 2/28/2011
Reason: This case was amended to correct text formatting. Diagnosis is unchanged.

Previous Signout Date: 2/28/2011

Amended Date: 4/29/2011
Reason: Finalization of provisional report following the performance of additional pathologic testing

Previous Signout Date: 2/28/2011

Page 4 of 4

END OF REPORT

Figure 19-2 (continued)

the ordering clinician along with the results relevant to the diagnostic indication that prompted the sequencing test [36]. They recommended that all testing laboratories seek and report these variants (in >50 genes) for all clinical germline exome and genome sequencing, including the normal of cancer-normal paired samples, regardless of the indication for testing. The recommendations state that it is the responsibility of the ordering clinician to provide comprehensive pre- and post-testing counseling to the patient. Laboratory policies must clearly state how their reports will communicate results, i.e., those variants thought to be related to the disease under investigation as well as incidental, secondary variants. The laboratory consent forms and processes must also clearly communicate the adopted reporting policy to both the patient and ordering clinician.

The static reports illustrated in this chapter are only examples and necessarily have limitations. Static reports only reflect the knowledge available at the time the report was generated and do not provide an easy mechanism to update and notify the clinician and patient as new information becomes available. There is also no payment structure in place for laboratories to develop and execute this responsibility. In addition, as more and more information becomes available through high-throughput sequencing, both the clinician and the patient will need more time to "digest" the laboratory results. It is likely that a more dynamic model incorporating interactive electronic reporting with one-on-one counseling will evolve.

The Role of the Pathologist

These are exciting and unsettled times as pathology navigates how best to incorporate genomics into clinical practice. Pathologists have always been at the interface of collecting, analyzing, interpreting, and integrating data to effectively communicate clinically relevant results to clinicians and their patients. Pathologists are the direct link between medical data and clinical care. High-throughput sequencing is a disruptive technology for the practice of pathology, especially molecular pathology. A comprehensive test with big data sets will replace a series or a panel of targeted tests and, yet, will still require quality assurance, analysis, interpretation, and effective communication. Historically, pathology has impacted clinical care through the adoption of new technologies that reveal relevant clinical correlates, from the light microscope to the electron microscope through to immunohistochemistry, autoanalyzers, flow cytometry, polymerase chain reaction, mass spectroscopy, and more. High-throughput sequencing should not change this primary role, but we must prepare ourselves for this potential sea change. It will be years before patients routinely obtain high-throughput sequencing tests, but now is the time for us to embrace genomics and firmly engage so that we properly guide and lead its incorporation into clinical care. As a corollary, if technology and promise are the drivers, we must be responsible, cautious, and ethical in reporting results in the context of each patient.

Pathologists play a key role in cancer diagnosis using morphology integrated with ancillary techniques (protein expression, cytogenetics, FISH, molecular diagnostics). This role will continue with high-throughput sequencing. Pathologists understand the importance of the pre-analytical variables of tissue collection and processing on nucleic acid extraction and sequencing accuracy and can optimize this workflow process. Pathologist morphologic skills are essential for selecting the best area of tumor tissue to ensure specimen adequacy and sufficient percentage of tumor cells to detect mutant alleles in heterogeneous samples mixed with benign tissue elements. To fully communicate high-throughput sequencing results, pathologists will need to not only learn the evolving molecular classification of cancer but also know related biologic pathways, targets for drug therapy, and changes related to drug resistance.

There are several ways in which pathologists can overcome challenges and barriers to readily implementing high-throughput sequencing reporting [37].

1. Integrate bioinformaticists, programmers, and statisticians into the reporting team:

   Most high-throughput sequencing data analysis algorithms in use need some programming expertise together with specialized servers to handle and store all of the

data. We need standardized bioinformatics pipelines for refinement in base calling and annotation of identified variants. There is still a lack of physician-friendly computational data analysis tools, so the reporting physicians must work alongside the bioinformaticists to understand the analysis pipeline and be cognizant of potential pitfalls. The development of smaller, lower throughput sequencing instruments, which are more compatible with the clinical lab laboratory, will also provide a better entry for the pathology team to develop confidence and expertise in the details of the pipeline. Clinical decision support systems will also need to be incorporated as the amount of annotated data expands.
2. Create interdisciplinary teams of pathologists, geneticists, oncologists, and translational researchers to build consensus on patient-specific variant interpretation and reporting.
3. Advocate for certified clinical grade annotated variant databases, ideally including population frequencies and clinical relevance for each sequence variant.
4. Develop and utilize structured training programs and curricula to provide pathologists and pathology trainees with genomic literacy and skills in interpreting and reporting high-throughput sequencing data.

Conclusions

Medicine has entered the genomic era in which vast numbers of genetic variants will be incorporated into the criteria for the diagnosis, prognosis, and treatment of disease. Successful implementation requires truly informative and accessible clinical reports to guide physicians and their patients. To achieve that goal, the paradigm for clinical reporting will change. Pathologists should be at the forefront of this transformation.

REFERENCES

1. Watson MS, Cutting GR, Desnick RJ, Driscoll DA, Klinger K, Mennuti M, Palomaki GE, Popovich BW, Pratt VM, Rohlfs EM, Strom CM, Richards CS, Witt DR, Grody WW. Cystic fibrosis population carrier screening: 2004 revision of American College of Medical Genetics mutation panel. Genet Med. 2004;6(6):387–91.
2. Sudmant PH, Kitzman JO, Antonacci F, Alkan C, Malig M, Tsalenko A, Sampas Teekakirikul P, Kelly MA, Rehm HL, Lakdawala NK, Funke BH. Inherited cardiomyopathies: Molecular genetics and clinical genetic testing in the postgenomic era. J Mol Diagn. 2013;15(2):1–13. http://dx.doi.org/10.1016/j.jmoldx.2012.09.002.
3. Cardarella S, Ortiz TM, Joshi VA, Butaney M, Jackman DM, Kwiatkowski DJ, Yeap BY, Janne PA, Lindeman NI, Johnson BA. The introduction of systemic genomic testing for patients with non-small-cell lung cancer. J Thorac Oncol. 2012;7(12):1767–74. doi:10.1097/JTO.0b013e3182745bcb.
4. Sequist LV, Heist RS, Shaw AT, Fidias P, Rosovsky R, Temel JS, Lennes IT, Digumarthy S, Waltman BA, Bast E, Tammireddy S, Morrissey L, Muzikansky A, Goldberg SB, Gainor J, Channick CL, Wain JC, Gaissert H, Donahue DM, Muniappan A, Wright C, Willers H, Mathisen DJ, Ellisen LW, Mino-Kenudson M, Lanuti M, Borger DR, Iafrate AJ, Engelman JA, Dias-Santagata D. Implementing multiplexed genotyping of non-small cell lung cancers into routine clinical practice. Ann Oncol. 2011;22(12):2616–24. doi:10.1093/annonc/mdr489. Epub 2011 Nov 9.
5. Gulley ML, Braziel RM, Halling HC, Hsi ED, Kant JA, Nikiforova MN, Nowak JA, Ogino S, Oliveira A, Polesky HF, Silverman L, Tubbs RR, Van Deerlin VM, Vance GH, Versalovic J, Molecular Pathology Resource Committee, College of American Pathologists. Clinical laboratory reports in molecular pathology. Arch Pathol Lab Med. 2007;131(6):852–63.
6. Adzhubei IA, Schmidt S, Peshkin L, Ramensky VE, Gerasimova A, Bork P, Kondrashov AS, Sunyaev SR. A method and server for predicting damaging missense mutations. Nat Methods. 2010;7(4):248–9. doi:10.1038/nmeth0410-248.
7. Richards CS, Bale S, Bellissimo DB, Das S, Grody WW, Hedge MR, Lyon E, Ward BE, Molecular Subcommittee of the ACMG Laboratory Quality Assurance Committee. ACMG recommendations for standards for interpretation and reporting of sequence variations: Revisions 2007. Genet Med. 2008;10(4):294–300. doi:10.1097/GIM.0b013e31816b5cae.
8. Sim NL, Kumar P, Hu J, Henikoff S, Schneider G, Ng PC. SIFT web server: predicting effects of amino acid substitutions on proteins. Nucleic Acids Res. 2012;40(Web Server issue):W452–257. doi:10.1093/nar/gks539. Epub 2012 Jun 11 Accessed February 10, 2013.
9. Drmanac R. The advent of personal genomic sequencing. Genet Med. 2011;13(3):188–90. doi:10.1097/GIM.0b013e31820f16e6.
10. Li Y, Vinckenbosch N, Tian G, Huerta-Sanchez E, Jiang T, Jiang H, Albrechtsen A, Andersen G, Cao H, Korneliussen T, Grarup N, Guo Y, Hellman I, Jin X, Li Q, Liu J, Liu X, Sparsø T, Tang M, Wu H, Wu R, Yu C, Zheng H, Astrup A, Bolund L, Holmkvist J, Jørgensen T, Kristiansen K, Schmitz O, Schwartz TW, Zhang X, Li R, Yang H, Wang J, Hansen T, Pedersen O, Nielsen R, Wang J. Resequencing of 200 human exomes identifies an excess of low-frequency non-synonymous coding variants. Nat Genet. 2010;42(11):969–72. doi:10.1038/ng.680. Epub 2010 Oct 3.

11. Adams DR, Sincan M, Fuentes Fajardo K, Mullikin JC, Pierson TM, Toro C, Boerkoel CF, Tifft CJ, Gahl WA, Markello TC. Analysis of DNA sequence variants detected by high-throughput sequencing. Hum Mutat. 2012;33(4):599–608. doi:10.1002/humu.22035. Epub 2012 Feb 28.
12. Stenson PD, Mort M, Ball EV, Howells K, Phillips AD, Thomas NS, Cooper DN. The human gene mutation database: 2008 update. Genet Med. 2009;1(1):13. doi:10.1186/gm13.
13. Stenson PD, Ball EV, Mort M, Phillips AD, Shaw K, Cooper DN. The human gene mutation database (HGMD) and its exploitation in the fields of personalized genomics and molecular evolution. Curr Protoc Bioinformatics. 2012;Chapter 1:Unit1.13. doi:10.1002/0471250953.bi0113s39.
14. Won HH, Kim HJ, Lee KA, Kim JW. Cataloging coding sequence variation in human genome databases. PLoS One. 2008;3(10):e3575. doi:10.1371/journal.pone.0003575. Epub 2008 Oct 30.
15. Tong MY, Cassa CA, Kohane IS. Automated validation of genetic variants from large databases: ensuring that variant references refer to the same genomic locations. Bioinformatics. 2011;27(6):891–3. doi:10.1093/bioinformatics/btr029. Epub 2011 Jan 22.
16. Howard HJ, Horaitis O, Cotton RG, Vhinen M, Dalgleish R, Robinson P, Brooks AJ, Axton M, Hoffmann R, Tuffery-Giraud S. The Human variome project (HVP) 2009 Forum "Towards Establishing Standards". Hum Mutat. 2010;31(3):366–7. doi:10.1002/humu.21175.
17. Biesecker LG. Opportunities and challenges for the integration of massively parallel genomic sequencing into clinical practice: lessons from the ClinSeq project. Genet Med. 2012;14(4):393–8. doi:10.1038/gim.2011.78. Epub 2012 Feb 16.
18. Aronson SJ, Clark EH, Varugheese M, Baxter S, Babb LJ, Rehm HJ. Communicating new knowledge on previously reported genetic variants. Genet Med. 2012;14(8):713–9. doi:10.1038/gim.2012.19.
19. Swerdlow SH, Campo E, Harris NL, Jaffe ES, Pileri SA, Stein H, Thiele J, Vardiman J, editors. WHO classification of tumours of haematopoietic and lymphoid tissues. Lyon: International Agency for Research on Cancer; 2008.
20. Estey EH. Acute Myeloid leukemia: 2013 update on risk-stratification and management. Am J Hematol. 2013;88(4):318–27. doi:10.1002/ajh.23404.
21. Lynch TJ, Bell DS, Sordella R, Gurubhagavatula S, Okimoto RA, Brannigan BW, Harris PL, Haserlat SM, Supko JG, Haluska FG, Louis DN, Christiani DC, Settleman J, Haber DA. Activating mutations in the epidermal growth factor receptor underlying responsiveness of non-small-cell lung cancer to gefitinib. N Engl J Med. 2004;350(21):2129–39. Epub 2004 Apr 29.
22. Paez JG, Jänne PA, Lee JC, Tracy S, Greulich H, Gabriel S, Herman P, Kaye FJ, Lindeman N, Boggon TJ, Naoki K, Sasaki H, Fujii Y, Eck MJ, Sellers WR, Johnson BE, Meyerson M. EGFR mutations in lung cancer: correlation with clinical response to gefitinib therapy. Science. 2004;304(5676):1497–500. Epub 2004 Apr 29.
23. Pao W, Miller V, Zakowski M, Doherty J, Politi K, Sarkaria I, Singh B, Heelan R, Rusch V, Fulton L, Mardis E, Kupfer D, Wilson R, Kris M, Varmus H. EGF receptor gene mutations are common in lung cancers from "never smokers" and are associated with sensitivity of tumors to gefitinib and erlotinib. Proc Natl Acad Sci U S A. 2004;101(36):13306–11. Epub 2004 Aug 25.
24. Dias-Santagata D, Akhavanfard S, David SS, Vernovsky K, Kuhlmann G, Boisvert SL, Stubbs H, McDermott U, Settleman J, Kwak EL, Clark JW, Isakoff SJ, Sequist LV, Engelman JA, Lynch TJ, Haber DA, Louis DN, Ellisen LW, Borger DR, Iafrate AJ. Rapid targeted mutational analysis of human tumours: a clinical platform to guide personalized cancer medicine. EMBO Mol Med. 2010;2(5):146–58. doi:10.1002/emmm.201000070.
25. MacConaill LE, Campbell CD, Kehoe SM, Bass AJ, Hatton C, Niu L, Davis M, Yao K, Hanna M, Mondal C, Luongo L, Emery CM, Baker AC, Philips J, Goff DJ, Fiorentino M, Rubin MA, Polyak K, Chan J, Wang Y, Fletcher JA, Santagata S, Corso G, Roviello F, Shivdasani R, Kieran MW, Ligon KL, Stiles CD, Hahn WC, Meyerson ML, Garraway LA. Profiling critical cancer gene mutations in clinical tumor samples. PLoS One. 2009;4(11):e7887. doi:10.1371/journal.pone.0007887.
26. Pao W, Kris MG, Iafrate AJ, Ladanyi M, Jänne PA, Wistuba II, Miake-Lye R, Herbst RS, Carbone DP, Johnson BE, Lynch TJ. Integration of molecular profiling into the lung cancer clinic. Clin Cancer Res. 2009;15(17):5317–22. doi:10.1158/1078-0432.CCR-09-0913. Epub 2009 Aug 25.
27. Su Z, Dias-Santagata D, Duke M, Hutchinson K, Lin YL, Borger DR, Chung CH, Massion PP, Vnencak-Jones CL, Iafrate AJ, Pao W. A platform for rapid detection of multiple oncogenic mutations with relevance to targeted therapy in non-small-cell lung cancer. J Mol Diagn. 2011;13(1):74–84. doi:10.1016/j.jmoldx.2010.11.010. Epub 2010 Dec 23.
28. Cheng L, Alexander RE, MacLennan GT, Cummings OW, Montironi R, Lopez-Beltran A, Cramer HM, Davidson DD, Zhang S. Molecular pathology of lung cancer: key to personalized medicine. Mod Pathol. 2012;25(3):347–69.doi:10.1038/modpathol.2011.215. Epub 2012 Jan 27. Review.
29. Oxnard GR, Janne P. Power in numbers: meta-analysis to identify inhibitor-sensitive tumor genotypes. Clin Cancer Res. 2013;19(7):1634–6. doi:10.1158/1078-0432.CCR-13-0169. Epub 2013 Feb 12.
30. Flaherty KT, Puzanov I, Kim KB, Ribas A, McArthur GA, Sosman JA, O'Dwyer PJ, Lee RJ, Grippo JF, Nolop K, Chapman PB. Inhibition of mutated, activated BRAF in metastatic melanoma. N Engl J Med. 2010;363(9):809–19.doi:10.1056/NEJMoa1002011.
31. Kopetz S, Desai J, Chan E, Hecht P, O'Dwyer J, Lee RJ, Nolop KB, Saltz L. PLX4032 in metastatic colorectal cancer patients with mutant BRAF tumors. J Clin Oncol (meeting abstracts). 2010;28:15_suppl 3534.
32. Corcoran RB, Ebi H, Turke AB, Coffee EM, Nishino M, Cogdill AP, Brown RD, Della Pelle P, Dias-Santagata D, Hung KE, Flaherty KT, Piris A, Wargo JA, Settleman J, Mino-Kenudson M, Engelman JA. EGFR-mediated re-activation of MAPK signaling contributes to insensitivity of BRAF mutant colorectal cancers to RAF inhibition with vemurafenib. Cancer Discov. 2012;2(3):227–35. doi:10.1158/2159-8290.CD-11-0341. Epub 2012 Jan 16.

33. Prahallad A, Sun C, Huang S, Di Nicolantonio F, Salazar R, Zecchin D, Beijersbergen RL, Bardelli A, Bernards R. Unresponsiveness of colon cancer to BRAF (V600E) inhibition through feedback activation of EGFR. Nature. 2012;483(7387):100–3. doi:10.1038/nature10868.
34. Yeh P, Chen H, Andrews J, Naser R, Pao W, Horn L. DNA-mutation Inventory to refine and enhance cancer treatment (DIRECT): a catalogue of clinically relevant cancer mutations to enable genome-directed cancer therapy. Clin Cancer Res. 2013;19:1894–901.
35. ACMG Board of Directors. Points to consider in the clinical application of genomic sequencing. Genet Med. 2012;14(8):759–61. doi:10.1038/gim.2012.74.
36. Green RC, Berg JS, Grody WW, Kalia SS, Korf BR, Martin CL, McGuire A, Nussbaum RL, O'Daniel JM, Ormand KE, Rehm HL, Watson MS, Williams MS, Biesecker LG. ACMG recommendations for reporting of incidental findings in clinical exome and genome sequencing. http://www.acmg.net/docs/ACMG_Releases_Highly-Anticipated_Recommendations_on_Incidental_Findings_in_Clinical_Exome_and_Genome_Sequencing.pdf (2013). Accessed 15 Apr 2013.
37. Schrijver I, Nazneen A, Farkas DH, Furtado M, Ferreira Gonzalez A, Grenier T, Grody W, Hambuch T, Kalman L, Kant JA, Klein RD, Leonard DGB, Lubin IM, Mao R, Nagan N, Pratt VM, Sobel ME, Voelkerding K, Gibson JS. Opportunities and challenges associated with clinical diagnostic genome sequencing. A report of the Association for Molecular Pathology. J Mol Diagn. 2012;14:525–40. doi: 10.1016/j.jmoldx.2012.04.006. Epub 2012 Aug 20.

PART IV

GENOMIC APPLICATIONS IN ONCOLOGY

CHAPTER 20

Genomic Applications in Hematologic Oncology

KEVIN E. FISHER, CHARLES E. HILL

Introduction to Single Gene Assays 297

DNA-Based Single Gene Assays 297

RNA-Based Single Gene Assays 303

Detection of BCR-ABL1 mRNA Transcript 303

Chromosome Assays 305

Conventional Cytogenetics 306

Fluorescence In Situ Hybridization (FISH) 306

Array Comparative Genomic Hybridization (aCGH) 309

Single Nucleotide Polymorphism (SNP) Arrays 311

Whole Genome Assays 314

Next Generation Sequencing (NGS) 314

Conclusions 316

Introduction to Single Gene Assays

Single gene assays used in the diagnosis and monitoring of hematologic malignancies can be broadly divided into DNA-based and RNA-based assays. Fundamentally, each of these types of assays uses similar analytical steps to extract molecular information from a patient sample. These steps include a standard biochemical nucleic acid extraction, template amplification by the polymerase chain reaction (PCR) or PCR-equivalent, and detection of the sequence of interest. DNA-based assays commonly encountered in molecular hematologic oncology that highlight different molecular approaches to DNA analysis include B-lymphocyte and T-lymphocyte receptor gene rearrangement assays, and exon-specific Janus kinase 2 (*JAK2*) and FMS-like tyrosine kinase 3 (*FLT3*) mutation analyses.

DNA-Based Single Gene Assays

B-Cell Immunoglobulin Gene Rearrangement

Both B- and T-lymphocytes (B-cells and T-cells) generate specific immune responses to diverse antigenic stimuli via a series of highly regulated somatic recombination events [1]. B-cells utilize three genes that encode secreted immunoglobulins: *IGH* located on chromosome 14q32, and either *IGκ* located on chromosome 2p12 or *IGλ* located on chromosome 22q11. The *IGH* gene contains approximately 87 variable (V_H), 30 diversity (D_H), and 6 joining (J_H) segments that randomly recombine in a process called VDJ recombination [2].

G.J. Netto and I. Schrijver (eds.), *Genomic Applications in Pathology*,
DOI 10.1007/978-1-4939-0727-4_20, © Springer Science+Business Media New York 2015

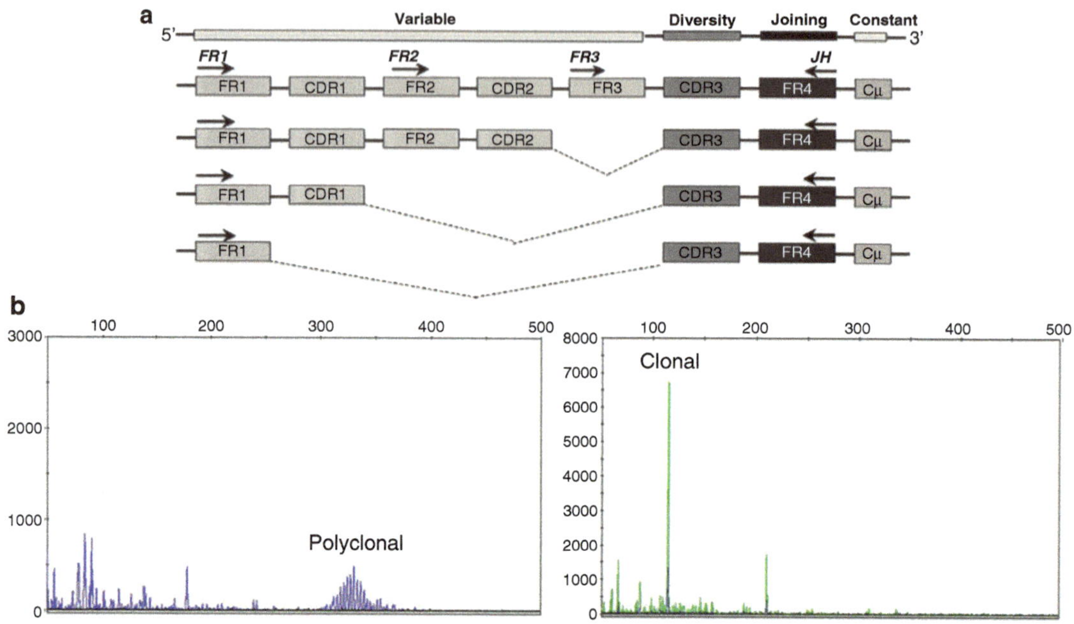

Figure 20-1 *IGH* gene rearrangements. (**a**) Schematic representation of the *IGH* gene rearrangements. There are four relatively conserved framework regions (FR1, FR2, FR3, and FR4) and complementary determining regions (CDR1, CDR2, and CDR3). Detection of VDJ rearrangements is achieved using an FR1, FR2, or FR3 primer (forward) and a JH primer (reverse) located at the FR4 region. (**b**) Capillary electrophoresis tracing of PCR products. The FR1-JH combination of primers generates PCR products ranging from 290 to 360 base pairs (bp), the FR2-JH generates up to 280 bp, and FR3-JH generates the shortest PCR products (70–170 bp) and provides the most reliable amplification. The *left panel* shows a polyclonal background using primers targeting FR1. The *right panel* shows a clonal population of cells (peak) with a rearrangement detected by the FR3 primer set (reprinted from Nikiforova MN et al. Detection of Clonal *IGH* Gene Rearrangements: Summary of Molecular Oncology Surveys of the College of American Pathologists. *Arch Pathol Lab Med* 2007;131:185–189 with permission from *Archives Pathology and Laboratory Medicine*. Copyright 2007 College of American Pathologists)

The constant (C_H) regions are then joined with the VDJ complexes via RNA splicing, and the C_H segment determines the antibody class (e.g., IgG, IgA, others) and allows for class switching. Once VDJ-C recombination occurs, B-cells encounter antigen within the germinal center of lymph nodes. This antigen presentation triggers somatic hypermutation within the complementary determining regions (CDR) of the V region and produces a virtually inexhaustible array of genetic diversity. This process allows for selection of a B-cell clone that codes for an immunoglobulin (Ig) with high specificity for the presented antigen. IgG molecules contain two identical heavy chain and two identical kappa or lambda light chains. *IGH* gene rearrangements occur first, followed by kappa and then lambda rearrangements. The *IGκ* and *IGλ* light chain loci lack D regions and therefore undergo VJ recombination only.

Molecular diagnostics typically utilizes *IGH* and *IGκ* gene rearrangements to identify clonal B-cell populations. In the *IGH* gene, CDRs are flanked by 15–30 amino acid framework regions (FRs) that rarely undergo mutation, thus are amenable to PCR primer targeting [3]. The three major FRs present within the V-region (FR1, FR2, and FR3) serve as the 5′ forward primer binding sites for amplification of the intervening DNA and all reactions share the same 3′ reverse primer binding site (Fig. 20.1). If a clonal proliferation has occurred, the gene rearrangement present in the clone will be overrepresented in the total population of rearrangements, allowing the identification of the cellular proliferation as clonal.

Assessing for B-cell clonality is useful in the diagnostic process for a variety of lymphomas and leukemias, and in providing possible targets for monitoring disease. The three

FR primers cover approximately 95 % of all possible IGH rearrangements, and assessing IGκ analysis is particularly useful for the diagnosis of marginal zone, follicular, and Hodgkin lymphoma in paraffin-embedded tissue [4, 5]. Most laboratories use capillary electrophoresis and fluorescence to detect the amplicons; each of the forward primers is conjugated with a different color fluorescent dye and the amplicons of different sizes (FR1: 290–360 bp, FR2: 235–295 bp, FR3: 69–129 bp) are separated by their charge/mass ratio (Fig. 20.1).

T-Cell Receptor Gene Rearrangement

As with immunoglobulins, multiple gene segments rearrange during T-cell development to encode the T-cell receptor (TCR). TCR-alpha (*TCRα*) and TCR-gamma (*TCRγ*) genes rearrange the V, J, and C loci whereas TCR-beta (*TCRβ*) and TCR-delta (*TCRδ*) genes rearrange the V, D, J, and C loci. Each T-cell possesses a gene that codes for a single TCR subunit gene that is unique in both sequence and length. Each subunit heterodimerizes to form the final TCR, either TCR-αβ or TCR-γδ. However, when assessing for T-cell clonality, *TCRγ* is used because it is rearranged at an early stage of T-cell development, and unlike *TCRδ*, it is not deleted in TCR-αβ cells. Thus, clonal rearrangements of *TCRγ* can be detected in both TCR-αβ and TCR-γδ clonal proliferations [6]. Clonal T-cell rearrangements are seen in greater than 90 % of T-cell leukemias [e.g., pre-T-cell acute lymphoblastic leukemia (pre-T-ALL), T-cell prolymphocytic leukemia, and T-cell large granular lymphocytic leukemia] and 50–75 % of T-cell lymphomas (e.g., peripheral T-cell lymphoma, mycosis fungoides, and anaplastic large cell lymphoma) [7].

Similar to B-cell clonality testing, PCR primers target conserved regions in the V and J exons that flank variable regions within the *TCRγ* locus [8]. Capillary electrophoresis and fluorescence are used to detect the fluorescently tagged amplicons. There is not one shared 3′ reverse primer but rather two primers that target the J exon, thus amplicons of interest are present between 55–85 bp, 155–185 bp, 200–235 bp, and 235–270 bp. There are no consensus criteria for interpretation of these results and each laboratory is responsible for test interpretation. Most laboratories correlate findings with morphologic and other laboratory data.

Diagnostic Challenges in Clonality Testing

A low quantity of B- or T-cells in a sample can lead to "false positive" results or pseudoclonality. When there are very few B- or T-cells in a sample, there is a limited repertoire to amplify. If one of these is slightly more abundant than others or if there is a minimal difference in amplification efficiency, this rearrangement may be preferentially amplified and therefore be of significantly higher peak amplitude compared to the background. This "peak" can be erroneously interpreted as a clonal proliferation.

It is possible for gene rearrangements to occur that yield products outside of the typical range of amplicon sizes. Usually, the product size will be just outside of the predicted ranges. It is important to remember that changes such as somatic hypermutation or other mutational processes can lead to this scenario. Therefore, it is possible to have a "false negative" result due to the rearrangement producing a clonal product that is outside of the predicted size range. When amplification products fit these characteristics, it is important to interpret the results with caution. Sequence analysis can help distinguish a spurious peak from a clonal rearrangement.

Even when the presence of a monoclonal B- or T-cell proliferation is unequivocal, one must remember that the presence of clonality does not confirm malignancy. The results of molecular clonality assays ultimately must always be interpreted in the context of the clinical, morphologic, and immunophenotypic data of the clinicopathologic entity in question. Many benign dermatologic, inflammatory, and infectious disorders may demonstrate clonality in B- or T-cell rearrangement assays [9]. Furthermore, the presence of B-cell clonality or T-cell clonality does not necessarily imply a B-cell or a T-cell malignancy, respectively. For example, approximately 60 % of pre-B-acute lymphoblastic leukemias and 10 % of acute myeloid leukemias (AMLs) can harbor T-cell gene rearrangements, and IGH rearrangements can be seen in 25–30 % of angioimmunoblastic T-cell lymphomas as a result of expanded EBV-positive B-cells [7]. Therefore, the presence of a B-cell or T-cell rearrangement should be viewed independently as evidence of clonality, rather than malignancy or lineage specificity.

JAK2 Mutation Analysis

Whereas B-cell and T-cell gene rearrangement assays investigate discrete regions of multiple genetic loci, some single gene assays target single point mutations that are diagnostically or prognostically relevant. Janus kinase 2 (JAK2) is a tyrosine kinase that mediates signaling downstream of cytokine receptors, such as erythropoietin, thrombopoietin, and granulocyte colony-stimulating factor. JAK-mediated phosphorylation of signal transducers of activated transcription (STAT) proteins mediates target gene expression in the nucleus [10]. Gain of function mutations in the JAK/STAT pathway are principally involved in the development of myeloproliferative disorders.

One such mutation is a single somatic G>T nucleotide change in exon 14 of *JAK2* at position 1849. This mutation codes for a valine to phenylalanine conversion at codon 617 in the JAK2 pseudokinase domain and is seen in more than 95 % of patients with polycythemia vera (PV), and in approximately 50 % of patients with essential thrombocythemia or primary myelofibrosis, 20 % of patients with refractory anemia with ring sideroblasts and thrombocytosis, and 5 % of patients with AML or myelodysplastic syndrome (MDS). When *JAK2* mutational testing is extended to exons 12 and 13, virtually all PV patients will have *JAK2* mutations, thus fulfilling one of the major 2008 WHO diagnostic criteria for PV [7, 11].

A widely utilized testing method for *JAK2* p.V617F mutations employs real-time quantitative PCR (RQ-PCR) using a sequence specific [mutant versus wild-type (WT)] forward primer tagged with a specific probe for detection [12]. This mutant specific or WT primer allows for single base pair discrimination because transcription elongation occurs at a very low rate with a 3′ mismatched base pair (Fig. 20.2). When detection of a product occurs, it is quantified by comparison to a WT and mutant standard curve and calculation of the percent of mutant and WT allele is performed The results are reported both qualitatively (detected/not detected) and quantitatively (as a percentage of mutated allele and percentage of WT allele). Qualitative reporting is useful for initial diagnosis while quantitative reporting allows for assessment of allele burden and "zygosity," which yields additional prognostic information [13], and for monitoring molecular response to therapy [14].

In all RQ-PCR assays, the diagnostic limit of detection must be clearly defined. Eventually, a base pair/primer mismatch will proceed with amplification. Due to this "escape amplification," cycle thresholds are needed to mitigate false-positive amplifications. Cycle thresholds are determined in each laboratory and are analogous to the "limit of blank" in the chemistry laboratory that can be used to distinguish true positives from false positives. Any amplification detected after this threshold is considered to be a result of nonspecific primer annealing and amplification (Fig. 20.2).

A diagnostic dilemma occurs when the mutated allele is present at low detectable levels (e.g., 0.2 % mutant allele) and falls below the defined accepted cutoff level for a positive result (e.g., 1.0 % mutant allele). In these problematic cases, the assay could be rerun with the same sample. If the result still is equivocal, the result should be discussed with the clinician and additional sample could be requested, if clinically indicated.

FLT3 Mutation Analysis

FLT3 is a receptor tyrosine kinase that is normally expressed on hematopoietic stem cells and is lost as hematopoietic cells differentiate. *FLT3* mutations are the most common somatic alterations in AML and occur in approximately 25 % of patients. Two main *FLT3* mutations result in constitutive activation of FLT3 signaling: internal tandem duplication (ITD) mutations in exons 14 and 15 and point mutations in exon 20 that alter the aspartic acid at codon 835 (known as D835). Studies have shown that *FLT3* ITD mutations portend a poor prognosis but the prognostic

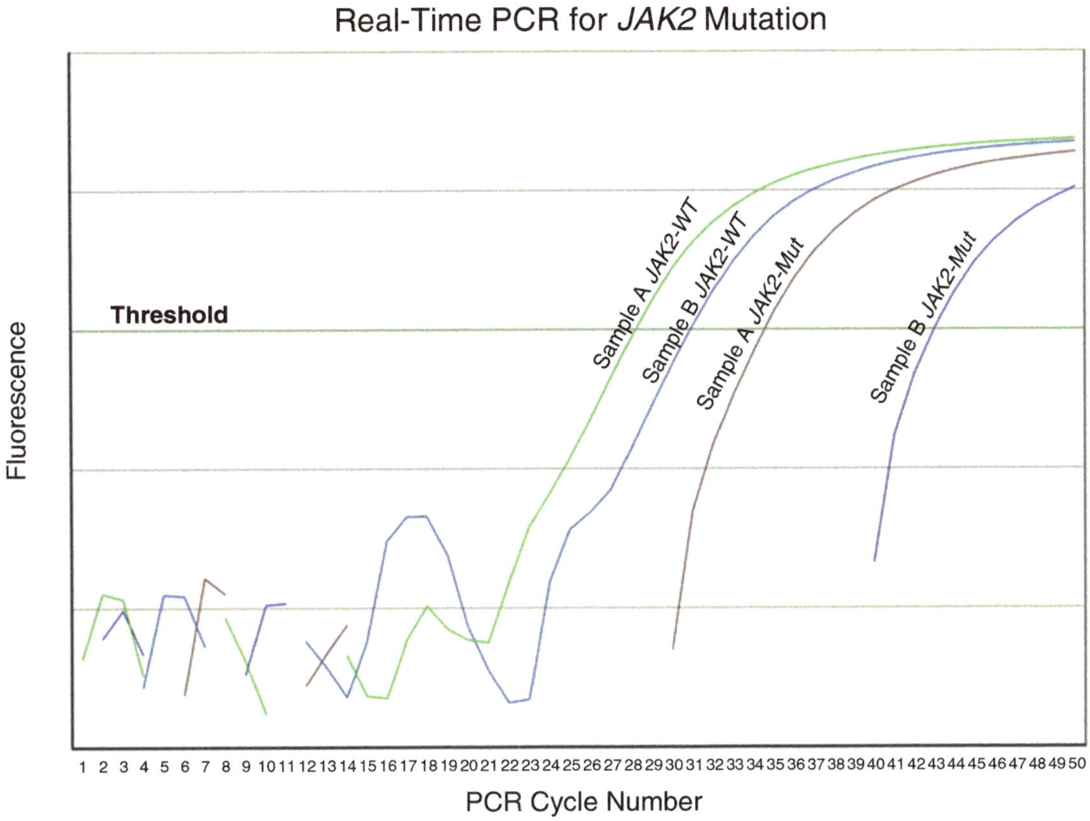

Figure 20-2 Example of real-time PCR testing for the *JAK2* mutation from two patient samples. Sample A demonstrates a small population of neoplastic cells that harbor a mutated *JAK2* allele. The *green line* labeled "Sample A *JAK-WT*" shows the amplification of the abundant wild-type allele (designated by the amplification at an earlier cycle number) while the *brown line* labeled "Sample A *JAK-mut*" shows the amplification of the less abundant mutated allele indicating the presence of the *JAK2* p.V617F mutation. Sample B is an example of nonspecific "escape" amplification. The *blue line* labeled "Sample B *JAK-WT*" is again showing amplification of the abundant wild-type allele. The *purple line* labeled "Sample B *JAK-mut*" crosses the *y*-axis threshold for positive fluorescence intensity at around cycle 43. This is beyond the designated cutoff for the assay which in this case is predetermined by the laboratory as cycle 41. This result would be reported as "negative for the *JAK2* p.V617F mutation"

significance of D835 point mutants is less well understood [15].

The methodology to assess for *FLT3*-ITD mutations is fairly straightforward. PCR can be performed with fluorescently tagged forward and reverse primers targeting exons 14 and 15. In this assay, a result consists of a PCR product that is 330 base pairs (bp) in length. Due to the presence of inserted triplet repeats, ITD mutants yield PCR products that are longer than WT peaks—333 bp or greater in size (Fig. 20.3).

Testing for the D835 mutation can be performed by another method that assesses the presence of point mutations: restriction enzyme digestion. Restriction enzymes are bacterially derived enzymes that cleave double-stranded DNA at specific palindromic sequences. In our assay example, a portion of the *FLT3* TK2 domain is amplified with a forward PCR primer that is fluorescently tagged at the 5′ end, and with a reverse unlabeled PCR primer. This amplification produces a PCR product of 150 bp. The 150 bp PCR product is then subjected to an EcoRV restriction digestion. The restriction enzyme recognizes and cleaves a specific palindromic sequence (5′-GATATC-3′/3′-CTATAG-5′) that is normally present in the WT *FLT3* exon that codes for aspartic acid at

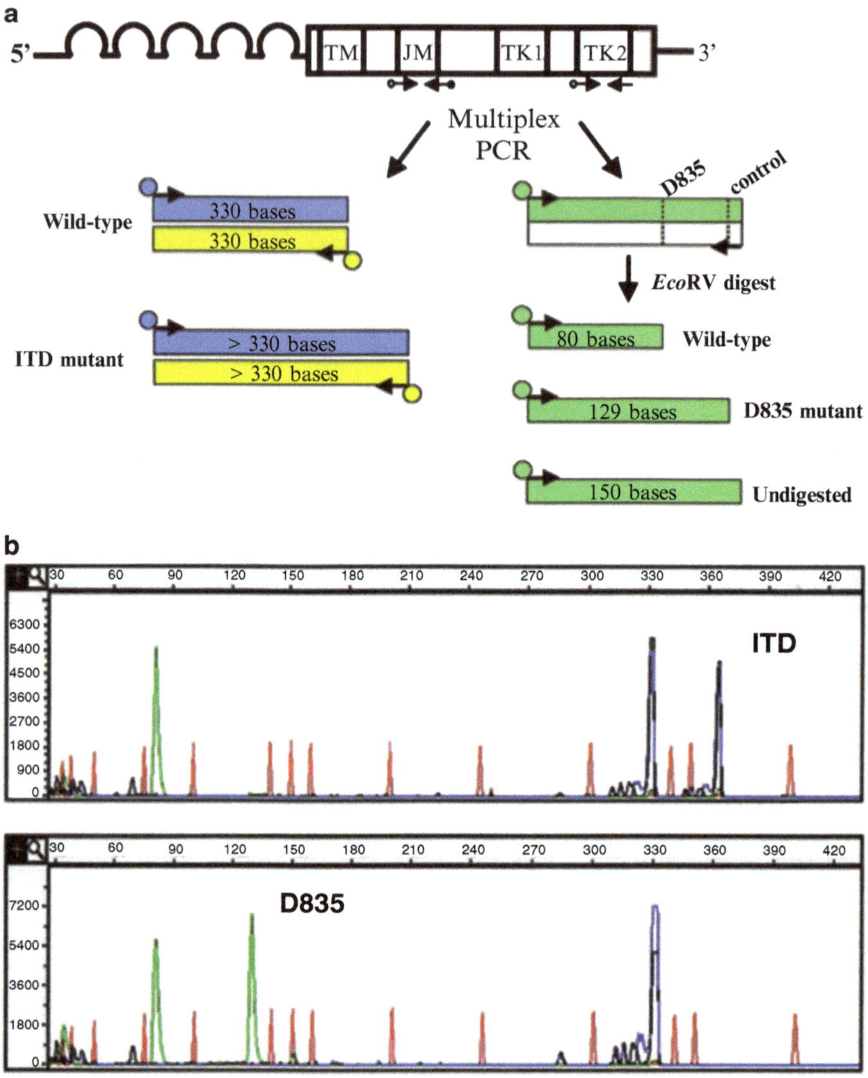

Figure 20-3 A *FLT3* assay. (**a**) Diagram of the *FLT3* assay design. The *FLT3* gene consists of five extracellular immunoglobulin-like domains, a transmembrane domain (TM), a JM domain, and an interrupted kinase domain (TK1 and TK2). PCR primers flanking the JM domain (forward primer labeled with *blue*, reverse primer labeled with *yellow*) and primers specific for the TK2 domain (forward labeled with TET *green*, reverse unlabeled) are multiplexed into a single PCR reaction. After amplification, the PCR products are digested with the restriction enzyme *Eco*RV. The *dotted lines* in the TK2 PCR product represent the *Eco*RV cut sites, with the recognition sequence (GATATC). The JM portion of the PCR yields a wild-type PCR product of 330 bases labeled with both blue and yellow dyes. *FLT3* internal tandem duplication (ITD) mutations result in PCR products that are longer than wild type (>330 bp), also labeled with both *blue* and *yellow*. After digestion, the D835 portion of the assay yields wild-type products sizing at 80 bases that are labeled in *green*. D835 mutant *green-labeled* products size at 129 bases, and undigested *green-labeled* products size at 150 bases. (**b**) Examples of results of the *FLT3* assay. Capillary electrophoresis pherograms; the *x* axis represents size of the PCR products in bases, and the *y* axis represents relative fluorescence intensity. *Red peaks* represent the internal size standard. *Green* PCR product peaks result from the D835 portion of the assay. *Blue* and *black peaks* result from the ITD portion of the assay. (*Top pherogram*): Example of a *FLT3* ITD mutant result. (*Bottom pherogram*): Example of a D835 mutant result (adapted from Murphy KM et al. Detection of FLT3 internal tandem duplication and D835 mutations by a multiplex polymerase chain reaction and capillary electrophoresis assay. *J Mol Diagn* 2003;5(2):96–102 with permission from *Journal of Molecular Diagnostics*. Copyright 2003 Elsevier B.V.)

codon 835. When a mutation occurs, the specific palindromic sequence is altered, and EcoRV no longer recognizes it as a cleavage site. Thus, EcoRV digestion of the WT type D835 PCR product results in an 80 bp fragment that can be detected by capillary electrophoresis. Since D835 mutations eliminate the EcoRV wild-type digestion site, the product is not cleaved, and there is an intact 130 bp fluorescently labeled fragment (Fig. 20.3). Twenty base pairs are lost because there is an EcoRV restriction site in the reverse primer. Because the reverse primer is not tagged with a fluorescent label, the 20 bp product remains undetected. This serves as a useful "digestion control" because the presence of abundant 150 bp amplicon would indicate that the restriction digest failed, or was incomplete [16].

RNA-Based Single Gene Assays

A fundamental difference between DNA-based and RNA-based assays is the need to incorporate a reverse transcriptase step in RNA-based assays in order to convert RNA templates into DNA templates. This is so the DNA template can be easily manipulated by conventional PCR-based methods. The prototypical RNA-based assay from a methodological standpoint is *BCR-ABL1* transcript testing. The presence of *BCR-ABL1* and its underlying chromosomal translocation (the Philadelphia chromosome [Ph]) can be detected in numerous ways, but the assay discussed here is for molecular monitoring of the mRNA fusion gene product.

Detection of BCR-ABL1 mRNA Transcript

BCR-ABL1 fusion genes occur when there is a balanced translocation between a portion of the *BCR* gene on chromosome 22 with the *ABL1* gene on chromosome 9 [t(9;22)(q34;q11.2)]. Depending on which *BCR* and *ABL1* exons fuse with one another, mRNA transcripts of varying sizes are transcribed. Three major mRNA transcripts are described: m-bcr encodes a 190 kDa fusion protein (p190), M-bcr encodes a 210 kDa fusion protein (p210), and μ-bcr encodes a 230 kDa fusion protein (p230) (Fig. 20.4a). The p210 transcript/protein is predominately involved in chronic myelogenous leukemia (CML) [17], the p190 transcript/protein is most frequently associated with *BCR-ABL1*-positive ALL, and patients with the p230 transcript/protein often demonstrate CML with prominent neutrophilic maturation and/or conspicuous thrombocytosis [7]. All *BCR-ABL1* fusion proteins are permissive for oncogenesis.

CML is a myeloproliferative neoplasm that is consistently associated with *BCR-ABL1* translocations and the detection of the translocation is required for the diagnosis. Because normal cells do not harbor t(9;22)(q34;q11) translocations, it is implicit that they should not express fusion transcripts. In CML specifically, monitoring of transcript level during therapy provides important prognostic information. The rationale for accurate quantitative molecular testing in CML arose from the International Randomized Study of Interferon and STI571 (IRIS) trial [18]. This trial demonstrated the superiority of imatinib (a tyrosine kinase inhibitor now used as first-line treatment for CML) over cytarabine and interferon-alpha, and concluded that patients who demonstrated complete cytogenetic remission had a better prognosis than those who did not. It was determined later that patients with a reduction in *BCR-ABL1* transcript levels of at least three $\log_{10}$ by 12 months on imatinib therapy had a negligible risk for disease progression during the subsequent 12 months [19].

Quantitation of *BCR-ABL1* transcripts requires reverse transcriptase PCR in combination with RQ-PCR to provide simultaneous detection and quantitation of *BCR-ABL1* fusion transcripts. Nucleic acid is extracted from leukocytes and the DNA is degraded with DNAse. The remaining RNA is transcribed into cDNA using a reverse transcriptase step. The cDNA is then subjected to RQ-PCR. Forward primers targeting BCR exons e13, e14, and e1 and a reverse primer targeting *ABL1* exon a2 are used. Primers targeting *ABL1* are used for a housekeeping gene endogenous control (Fig. 20.4a).

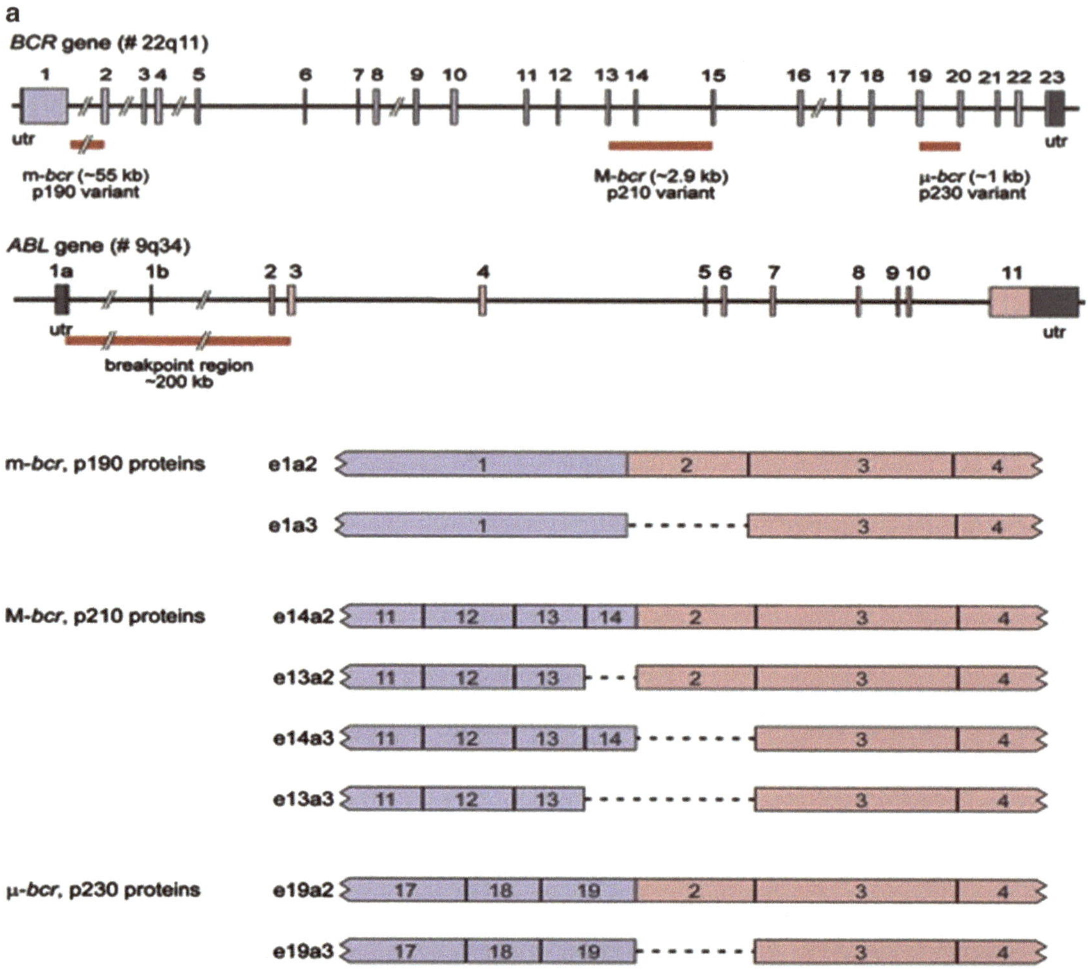

Figure 20-4 Structure of the *BCR* and *ABL1* genes with breakpoint regions and corresponding fusion gene transcripts. (**a**) The *ABL1* gene contains one large breakpoint region (~200 kb), whereas three breakpoint regions have been found in the *BCR* gene: m-*bcr*, M-*bcr*, and μ-*bcr*, which are associated with the p190, p210, and p230 *BCR-ABL1* fusion proteins, respectively. The three well-defined breakpoint regions in the *BCR* gene can produce at least eight different fusion transcripts, because of alternative splicing in the *ABL* gene (splicing to exon 2 or exon 3) and, because the M-*bcr* consists of two intronic regions (intron 13 and intron 14). (**b**) Sample A shows amplification of the intact *ABL1* gene product (*purple line* labeled "Sample A *ABL1*") with no detection of the fusion transcript. Sample B demonstrates amplification of the *BCR-ABL1* fusion transcript. The *green line* labeled "Sample B *ABL1*" shows the amplification of the intact *ABL1* gene product while the *second green line* labeled "*BCR-ABL1*" shows the amplification of the *BCR-ABL1* fusion transcript (adapted from Dekking E et al. Detection of fusion genes at the protein level in leukemia patients via the flow cytometric immunobead assay. *Best Pract Res Clin Haematol* 2010;(3):333–45 with permission from *Best Practice & Research Clinical Haematology.* Copyright 2010 Elsevier B.V.)

The real-time detection of the amplicons occurs with fluorescent probes that target the *BCR-ABL1* transcript. With increasing amplification of the *BCR-ABL1* transcript, the amount of reporter dye excitation increases exponentially and this is proportional to the amount of transcript present in the sample (Fig. 20.4b).

Initially, one of the major drawbacks to quantitative *BCR-ABL1* testing was the lack of a consensus reference standard. Therefore, quantitative results determined in one laboratory could not be reliably reproduced in a separate laboratory. Recently four fixed *BCR-ABL1*-control gene values were established as the first World Health Organization International Standards (WHO IS)

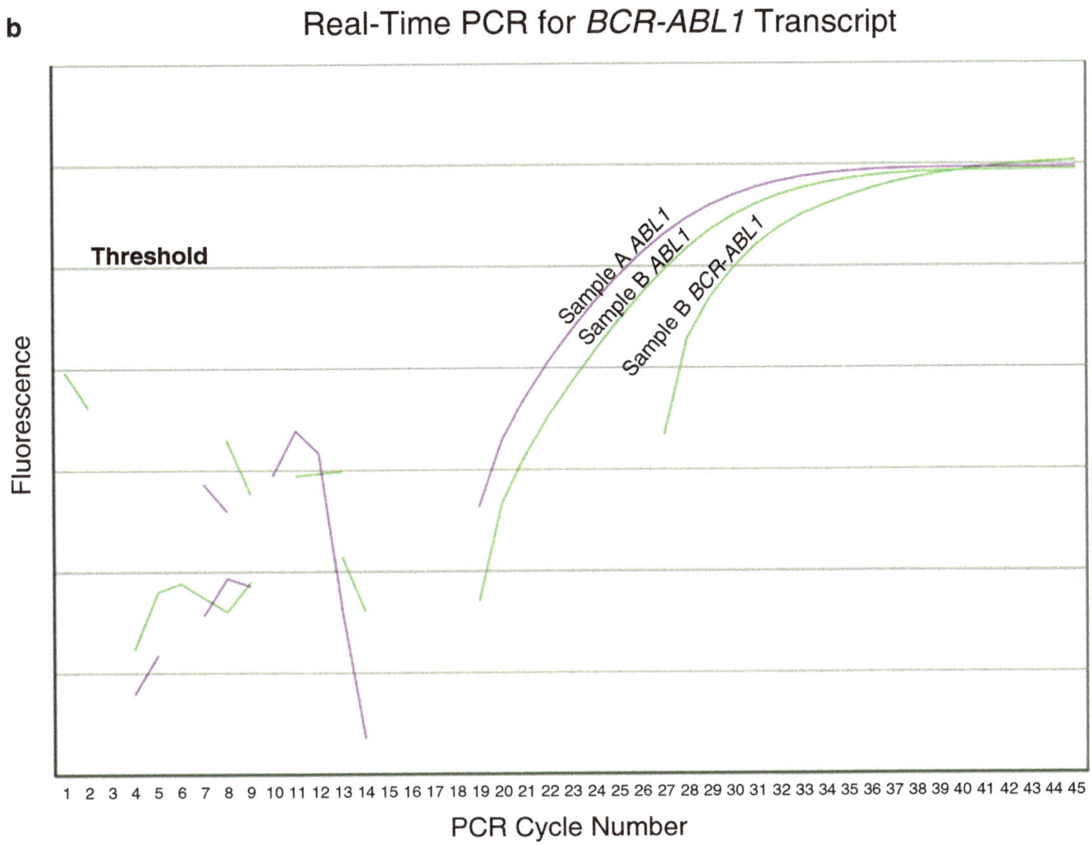

Figure 20-4 (continued)

for quantitation of *BCR-ABL1* [20] and these are now commercially available. The four reference standards are run in conjunction with patient samples and a calculation of a correction parameter (CP) value between the patient values of a laboratory (in copies/mL) and the WHO IS is performed. This design allows that the final value is reported as an IS % ratio that can be compared to values obtained in other laboratories.

Whereas quantitative monitoring of *BCR-ABL1* has become standard of care, there are other factors in the care of these patients that can best be served by more comprehensive molecular analyses. The advent of new pharmaceutical treatments has led to a need to test for *ABL1* kinase domain mutations to assess drug resistance. In addition, there are other chromosomal abnormalities that provide prognostic or therapeutic information. These issues are best addressed by more comprehensive molecular analyses.

Chromosome Assays

Chromosome assays have been utilized in hematologic oncology for almost four decades since the inception of Giemsa staining of replicating chromosomes (commonly known as G banding). The visualization of chromosomes has allowed scientists to detect and decipher aberrations in chromosome structure and define disease entities. Many hematologic cancers require the demonstration of a particular chromosomal abnormality for diagnosis, and the presence of certain chromosomal abnormalities provides prognostic and therapeutic information. Standard methodologies [conventional cytogenetics and fluorescence in situ hybridization (FISH)] and two newer methodologies, single nucleotide polymorphism (SNP) arrays and array comparative genomic hybridization (aCGH), will be discussed.

Conventional Cytogenetics

Conventional cytogenetics techniques are the most commonly used modalities in clinical laboratories to assess chromosomal composition for the diagnosis and therapeutic evaluation of hematologic malignancies [21]. Routine cytogenetic analysis (karyotyping) can achieve resolutions sufficient to detect alterations of a few megabases (Mb) [22]. This range of detection is useful to assess both gains and losses of large regions of the genome, as well as rearrangements within and among chromosomes. Cytogenetic analysis can identify numerous diagnostic chromosomal abnormalities in leukemia and lymphoma such as the Ph chromosome in CML, 8q24 MYC translocations in Burkitt lymphoma, recurrent genetic abnormalities in AML, and deletions of 5q in MDS (Fig. 20.5).

To perform routine karyotypic analysis, live cells are cultured and stimulated to divide, in order to promote mitoses. The cells are arrested in metaphase with a pharmacologic microtubule inhibitor (e.g., colchicine) and treated with trypsin followed by a counterstain (Giemsa or equivalent). This produces differential staining of chromosomal regions leading to light (replicating) and dark (condensed chromatin) bands, each corresponding to a specific area of chromosomal DNA. A cytotechnologist then interprets the chromosomes and their respective bands for any abnormalities, deletions, or duplications. There are, however, some major limitations to routine cytogenetic analysis. In order to perform the assay, cells must be received in a non-fixed, fresh state in order to stimulate mitoses and to perform successful staining. Also, karyotyping is not suited for the detection of microdeletions, cryptic translocations, or small genetic alterations due to the limit of detection which is, at best, a few Mb. More sensitive methodologies such as FISH are available to detect abnormalities not seen on routine cytogenetic analysis.

Fluorescence In Situ Hybridization (FISH)

The introduction of FISH in routine clinical diagnostics has enabled molecular cytogenetics and is now a widely used tool for the diagnosis and monitoring of patients with hematologic malignancies. FISH has aided the identification of structural chromosome rearrangements and is frequently used as a supplemental test to karyotypic analysis [20]. There are a few distinct advantages to FISH compared to conventional chromosome analysis. FISH can be performed on formalin-fixed paraffin-embedded (FFPE) tissue and, therefore, can visualize both metaphase and interphase chromosomes. FISH can visualize DNA segments between 100 kilobases (kb) and 1.5 Mb, versus 2–5 Mb. Also, FISH has a relatively short turnaround time (as little as 16 h in some cases) compared to days for routine cytogenetic analysis. However, in order for FISH analysis to have any utility, one must first select FISH probes that interrogate the chromosome or region of interest.

FISH probes are fluorescently labeled DNA probes that bind to complementary regions of the chromosome of interest. The DNA content of the target probes is made of either locus specific indicators (LSI) or centromere enumeration probes (CEP). LSI probes are specific to unique DNA sequences (target-specific loci) and are composed of 200–600 bp segments that span the 100 kb–1.5 Mb locus of interest. CEP probes are composed of highly repetitive DNA sequences that target common regions of all centromeres or are directed to specific centromeres of particular chromosomes. Each probe is tagged directly or indirectly with a fluorophore. The probes and the target DNA are denatured to yield single-stranded DNA, which allows for the annealing of complementary DNA sequences. The signal probes bind the target of interest and the attached fluorophores are evaluated by fluorescence microscopy. For hematologic malignancies, several types of FISH probes are used, and each of these probes has utility in the appropriate diagnostic context (Table 20.1).

Enumeration (centromere) probes are utilized to examine whether neoplastic cells show a loss or gain of chromosomal number and serve as internal controls for deletion or amplification probes. The loss or gain of chromosome copy number or deletions of certain regions of the chromosome provide diagnostic and prognostic information. For example, characteristic hypo-lobated or non-lobated megakaryocytes are visualized in bone marrow preparations of patients with MDS

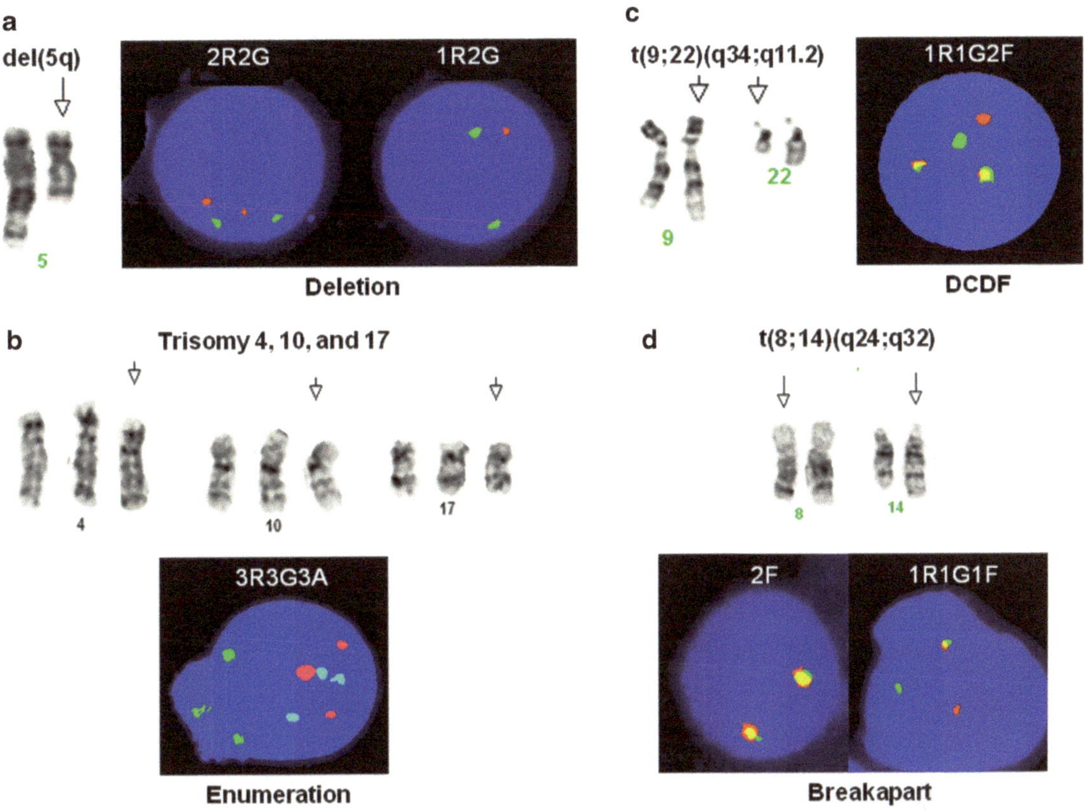

Figure 20-5 Examples of classic cytogenetic abnormalities in hematologic malignancies visualized by conventional karyotyping or fluorescence in situ hybridization. (**a**) Deletion of the short arm of chromosome 5q [del(5q)]. (*Left*) A conventional karyotype shows the absence of chromosomal material from the long arm of chromosome 5. (*Right*) The expected 2R2G (two *red*, two *green*) signal indicating two copies of the long arm of chromosome 5q (R = red) and two centromeres for chromosome 5 (G = green) is shown in the nucleus (*blue circle*) on the *left*. Chromosomal material from the long arm of chromosome 5 is lost from the nucleus on the *right*. This results in an aberrant 1R2G signal. (**b**) Detection of the Philadelphia chromosome [t(9;22)(q34;q11.2)]. (*Left*) A conventional karyotype showing the translocation of genetic material from chromosome 22 (*BCR*) to chromosome 9 (*ABL1*). (*Right*) An aberrant one *red*, one *green*, two fusion (1R1G2F) signal is shown. The appearance of the *yellow* fusion color indicates juxtaposition of chromosomal material from chromosomes 9 and 22. The non-rearranged chromosome is shown with the expected 1R1G signal. (**c**) Trisomy 4, 10, and 17 in a patient with acute lymphoblastic leukemia (ALL). (*Left*) The karyotype revealed an extra chromosome 4, 10, and 17. This combination of cytogenetic abnormalities is prognostically favorable in some ALLs. (*Right*) Enumeration FISH probes with different fluorophores highlight the three copies of each chromosome [R = red, chromosome 7; G = green, chromosome 10; A = aqua, chromosome 17]. (**d**) Cytogenetic and FISH results from a patient with Burkitt lymphoma showing the translocation between *MYC* on chromosome 8 and *IGH* on chromosome 14. (*Left*) The characteristic translocation seen in Burkitt lymphoma involving the *MYC* locus on the long arm of chromosome 8 and the *IGH* locus on the long arm of chromosome 14 is shown [t(8;14)(q24;q32)]. (*Right*) An example of a FISH breakapart probe. The expected 2F (two fusion) signal indicates an intact *MYC* locus (chromosome 8, shown on the *left*). In normal cases, the contiguous configuration of the *MYC* locus juxtaposes the *red* and *green* probes to produce a fusion signal. The loss of the fusion signal (*right*) results in an aberrant one *red*, one *green*, one fusion (1R1G1F) signal. The loss of a fusion signal denotes a "break" in the genetic material from one set of chromosome 8. The appearance of both a *red* and *green* signal indicates the genetic material from chromosome 8 is still detectable, but it is no longer in its contiguous configuration (Images courtesy of Dr. Debra F. Saxe, PhD, Emory University School of Medicine Department of Pathology and Laboratory Medicine)

Table 20-1 Commonly Used FISH Probes for Hematologic Malignancies

Probe type	Locus	Identifies	Abnormal signal	Disease
Enumeration (Centromere)	Telomeres or centromeres of chromosomes 3, 4, 7, 10, 12, and 17	Gain or loss of chromosome number	nR, nG, nB where $n \neq 2$	ALL, MM, CLL, and AMLs with complex karyotypes
Deletion	5p15.2 (G) 5q31 *EGR1* (R)	Loss 5, del(5q)	RG, RGG	MDS, AML
	7p11.1-q11.1 (G) 7q31 (R)	Loss 7, del(7q)		
	13q34 *LAMP1*(G) 3q14 (R)	Loss 13, del(13q)	RG, RGG	MM, CLL
	17p11.1-q11.1 (G) 17p13.1 *TP53* (R)	Loss 17, del(17p)		
Dual color Dual fusion	9q34 *ABL1* (R) 22q11.2 *BCR* (G)	t(9;22) BCR/ABL1	Any F	CML, Ph+ ALL
	15q22 *PML* (R) 17q21.1 *RARA* (G)	t(15;17) PML/RARA	Any F	APL
	11q13 C*CND1* (R) 14q32 *IGH* (G)	t(11;14) IGH/CCND1	Any F	MCL, MM
Breakapart	11q23 *MLL* (F)	*MLL* rearrangements	Any R or G	AML, MDS, mixed-lineage lymphomas
	16q22 *CBFB* (F)	inv(16), t(16;16), del(16q)	Any R or G	AML with inv(16) or t(16;16)
	8q24 C*MYC* (F)	*MYC* rearrangements	Any R or G	Lymphoma (Burkitt)

R red, *G* green, *B* blue, *F* fusion (yellow), *MM* multiple myeloma, *CLL* chronic lymphocytic leukemia/small lymphocytic lymphoma, *AML* acute myeloid leukemia, *MDS* myelodysplastic syndrome, *CML* chronic myelogenous leukemia, *Ph+* Philadelphia chromosome, *ALL* acute lymphoblastic leukemia, *APL* acute promyelocytic leukemia, *MCL* mantle cell lymphoma, *MM* multiple myeloma

associated with isolated del(5q) (Fig. 20.5a), whereas triple trisomy 4, 10, and 17 is considered a prognostically favorable genetic event in some ALLs (Fig. 20.5b) [23].

Some FISH probes are designed such that different fluorescent signals juxtapose to emit a single "fusion" color. The presence or absence of the fusion color can be detected in dual color dual fusion (DCDF) or breakapart probes, respectively. DCDF probes can distinguish balanced translocations such as t(9;22)(q34;q11) involving *BCR* and *ABL1*. The red (R) fluorophore attached to the probe that targets the *ABL1* gene juxtaposes with the green (G) fluorophore attached to the probe that targets the *BCR* gene. This juxtaposition causes the fluorescent signal to fluoresce yellow. The presence of the yellow fusion (F) color is scored as a positive rearrangement (Fig. 20.5b). In contrast, breakapart probes start as a fusion signal and when the locus is disrupted to give a single G or single R signal, it implies that a portion of the locus has rearranged to another chromosome. Breakapart probes are useful for detecting gene rearrangements in loci that rearrange with multiple chromosomes such as the 11q23 MLL locus in ALL and AML or the *MYC* locus in Burkitt lymphoma (Fig. 20.5d) [24].

Though FISH is a highly effective diagnostic tool, there are some limitations. Technically, the scoring of FISH probes and determining cutoffs for positivity can be problematic. A cytogenetic technologist, geneticist, or pathologist counts anywhere from 100 to 200 cells and cutoffs must be well defined to avoid making inappropriate calls. This is particularly true for deletion probes. In addition, like any nucleic acid-based test, sample integrity is very important and can lead to poor hybridization. Fusion signals may be small, weak, or absent due to biologic and technical reasons. Also, quality control is critical to determine that the FISH process is working appropriately.

By its nature, FISH analysis can only interrogate preselected chromosomal regions. By focusing on a specific region of the genome, other significant prognostic or biologic information in other regions of the same or other chromosomes may remain undetected. Theoretically, FISH probes targeting all regions of all chromosomes could be employed, but this is not a practical alternative for most clinical laboratories. For this reason, aCGH or SNP arrays that combine comprehensive coverage of the entire genome with the ability to assess specific regions of interest are attractive new methods for whole chromosome analysis.

Array Comparative Genomic Hybridization (aCGH)

Array CGH, also called molecular karyotyping, is a technique that uses competitive hybridization of fragmented tumor and control DNA to comprehensively interrogate hundreds of discrete genomic loci for DNA copy number gains and losses [25, 26]. The resolution of aCGH depends on the spacing and length of the interrogating DNA probes on the microarray. Whole-genome aCGH platforms exist that utilize equally spaced probes to interrogate the genome at 6–70 kb intervals, although most of the clinically available aCGH platforms use targeted arrays that detect previously characterized aneuploidies or chromosomal abnormalities/rearrangements [27].

In aCGH, equal amounts of fluorescently labeled sample DNA (using Cy3, a green dye, for example) and control DNA (using Cy5, a red dye, for example) are cohybridized to an array containing complementary DNA targets. When the sample or control DNA anneals, the labeling intensities or "spots" of fluorescence are measured. The resulting ratio of the fluorescence intensities is proportional to the ratio of the copy numbers of DNA sequences in the sample and control genomes. If the intensities of the fluorescent red and green dyes are equal, that region of the patient's sample genome is interpreted as having a quantity of DNA equal to the control sample. If there is an altered green:red ratio, this indicates a loss or a gain of the sample DNA at that specific genomic region (Fig. 20.6) [26].

One of the most commonly reported applications of aCGH in hematologic oncology is the detection of chromosomal abnormalities in "cytogenetically normal" malignancies [28]. That is, detecting copy number alterations (CNAs) or chromosomal abnormalities that are not detected by conventional cytogenetics. Because aCGH can reach a genomic resolution of approximately 6 kb, aCGH can detect chromosomal alterations that G-banding or FISH probes lack the resolution to detect. In one study, aCGH detected new cytogenetic abnormalities not seen by karyotype or FISH analyses in 80 % of MDS patients [29]. Other studies have also demonstrated genomic imbalances, cryptic CNAs, or karyotypic alterations in cytogenetically normal MDS patients [30–32]. Similar results have been described in pediatric ALL [33], blastic plasmacytoid dendritic cell neoplasms [34], chronic lymphocytic leukemia [35], and AML [36]. These studies suggest that aCGH is a useful adjunct to conventional chromosomal analyses to assess for both diagnostic and prognostic cytogenetic abnormalities in hematologic malignancies.

Despite its broad and sensitive cadre of applications, aCGH is limited as a comprehensive diagnostic tool because of an inability to detect balanced translocations, copy neutral loss of heterozygosity (CN-LOH, a process whereby a lost portion of the chromosome is reduplicated from the sister chromatid), and uniparental disomy (UPD, a process similar to CN-LOH that involves an entire chromosome). The results of aCGH are also influenced by the amount of tumor sample present, the presence of tumor subclones, and

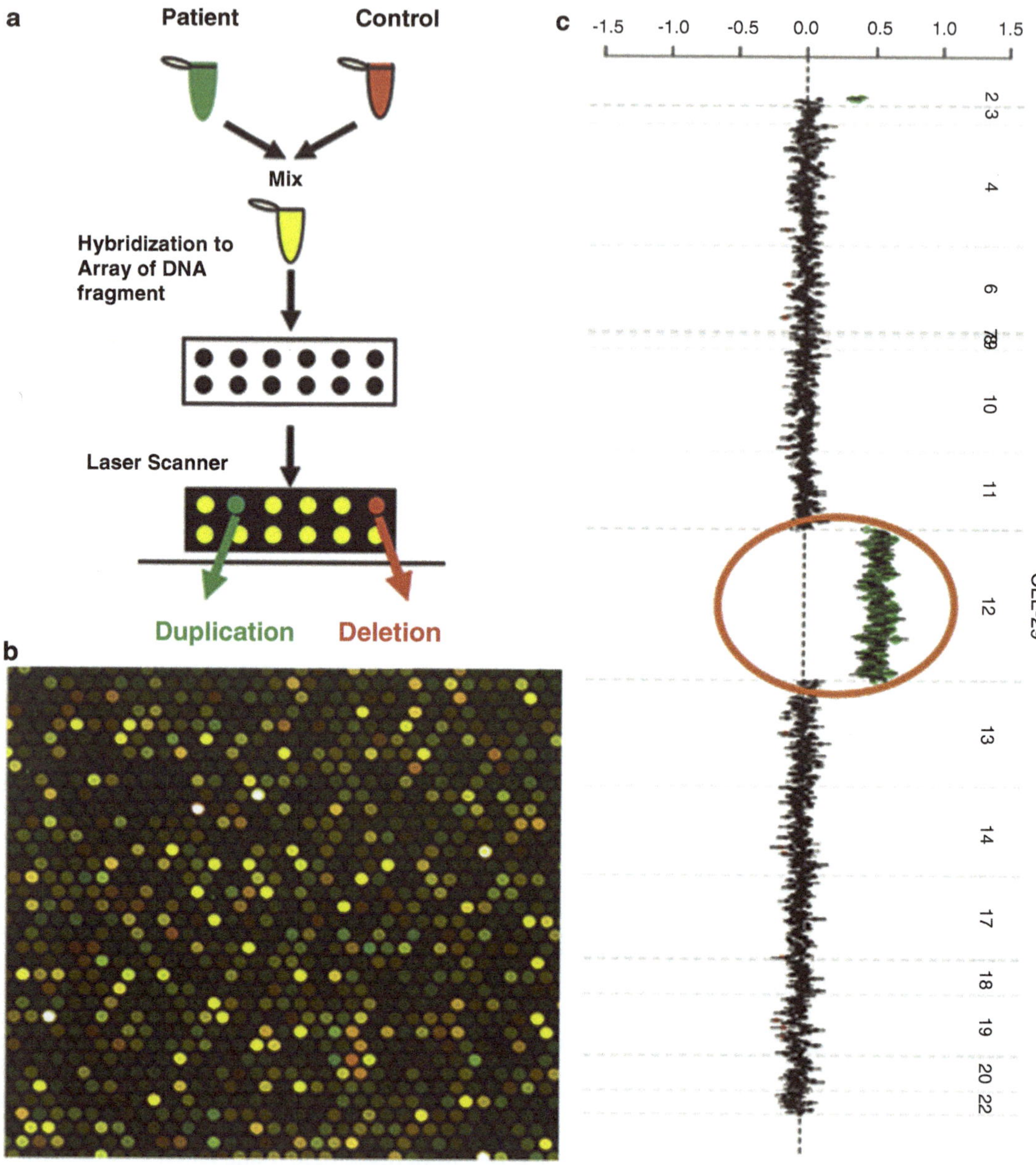

Figure 20-6 An example of array comparative genetic hybridization with gain of chromosome 12 in chronic lymphocytic leukemia. (**a**) Sample DNA is labeled with a green fluorescent dye (Cy3) and reference DNA is labeled with red (Cy5). The two are mixed and competitively cohybridized to an array containing genomic DNA targets that have been fixed to a glass slide. The areas on the slide that appear *green* indicate extra chromosomal material (duplication) in the test sample at that particular region. Areas on the slide that appear *red* indicate relatively less test DNA (deletion) in the sample at that specific spot. *Yellow areas* indicate equal amounts of sample and reference DNA. (**b**) The slides are scanned into image file and an output of scanning depicts hundreds of spots with different ratios of the fluorescence intensities. (**c**) Microarray image files are quantified using software that detects the fluorescent signals and maps them to specific regions of the chromosome. The signals are converted to the data output format shown here. A gain of genetic material from chromosome 12 from a CLL patient sample is indicated in the *red circle* (adapted from Shinawa M and Cheung SW. The array CGH and its clinical applications. *Drug Discov Today* 2008 Sep;13(17–18):760–70 with permission from *Drug Discovery Today*. Copyright 2010 Elsevier B.V.)

the resolution of the microarray platform [25]. Not surprisingly, aCGH platforms that target the entire genome are more expensive and are likely to detect genomic imbalances of unclear significance [26].

Single Nucleotide Polymorphism (SNP) Arrays

SNPs are single base pair changes in genomic DNA that occur (on average) every 1,000–2,000 bases [37]. Because SNPs have a low rate of recurrent mutation, studies have "mapped out" the location of common SNPs along the human genome. These SNP maps serve as reference sequences to allow comparison between DNA of interest (sample) and normal DNA (reference DNA or uninvolved tissue) at the single nucleotide level. The utility of SNP analysis is finding favor in virtually every facet of medicine: pharmacogenetics, neuropsychiatric disorders, and forensics, to name a few [38–40].

For hematologic malignancies, SNP array karyotyping takes advantage of very large numbers of allele-specific probes synthesized on microarrays to detect genome-wide CNAs and allelic imbalances. SNP array karyotyping represents the only platform currently available for genome-scale detection of CN-LOH or UPD. However, much like aCGH, SNP arrays are not designed to detect balanced translocations which, as noted previously, are commonly found in hematopoietic malignancies [41]. That said, SNP array karyotyping is a tool for the diagnosis and monitoring of hematopoietic neoplasms.

SNP arrays interrogate genomic loci to determine the DNA copy number and the genotype. The most common SNP array platforms include Illumina and Affymetrix arrays, which utilize bead [42, 43] or chip technology [44], respectively. In the bead-based Illumina platform, whole genome amplification and fragmentation steps are followed by hybridization to an oligonucleotide bead array (Fig. 20.7). In the Affymetrix technology, genomic DNA is digested by restriction endonucleases, amplified and labeled and hybridized to oligonucleotides on a microarray chip (Fig. 20.8) [45]. SNP arrays offer superior resolution to conventional karyotypic analysis, can detect genetic lesions less than 100,000 bp in size, and nullify the need for mitotically active cells. They can also detect genes involved in unbalanced copy number changes and determine genetic targets of amplifications and deletions [41].

During the past decade, this resolution sensitivity has yielded a wealth of information regarding genomic alterations in hematologic malignancies. For example, SNP arrays identified recurrent abnormalities in the *EBF1* and *PAX5* genes in childhood ALL [46, 47]. CLL and plasma cell myeloma are amenable to SNP array analysis, because there sometimes is difficulty obtaining metaphase chromosomes for conventional karyotyping [41]. SNP arrays identified 24 large (>10 Mb) copy-neutral regions with LOH in some cases of CLL that were not detectable by alternative methods [48]. Also in CLL, investigators found novel mutations in MAX pathway genes that are involved in regulatory mechanisms for cell proliferation, differentiation, and apoptosis [49]. In plasma cell myeloma, SNP arrays consistently recapitulate the findings of FISH analysis and provide further information regarding the particular genes deleted within chromosomal regions associated with poor prognoses [e.g., *CYLD* and *WWOX* in del(16q)] [50]. Perhaps the most significant application for SNP array analysis in hematologic malignancies has been the discovery of prognostically relevant genomic alterations in MDS and AML without recurrent or defined chromosomal translocations [51–53].

SNPs exist in a binary fashion (either SNP A or SNP B). Therefore, a single bead or chip oligonucleotide corresponds to a single allele at the SNP locus. Each SNP allele produces a specific color indicator (red for SNP A and blue for SNP B, for example) when sample DNA is bound. When genomic DNA binds, the relative intensities of red and blue signal at individual SNP loci are evaluated to determine three possible genotypes: homozygous A/A or B/B, or heterozygous A/B [54]. Importantly, the hybridization signals of tumor DNA across more than a million SNP loci are compared to normal diploid DNA (usually buccal mucosa DNA) at the individual probe sites [55]. The comparison of tumor sample to matched normal DNA allows for

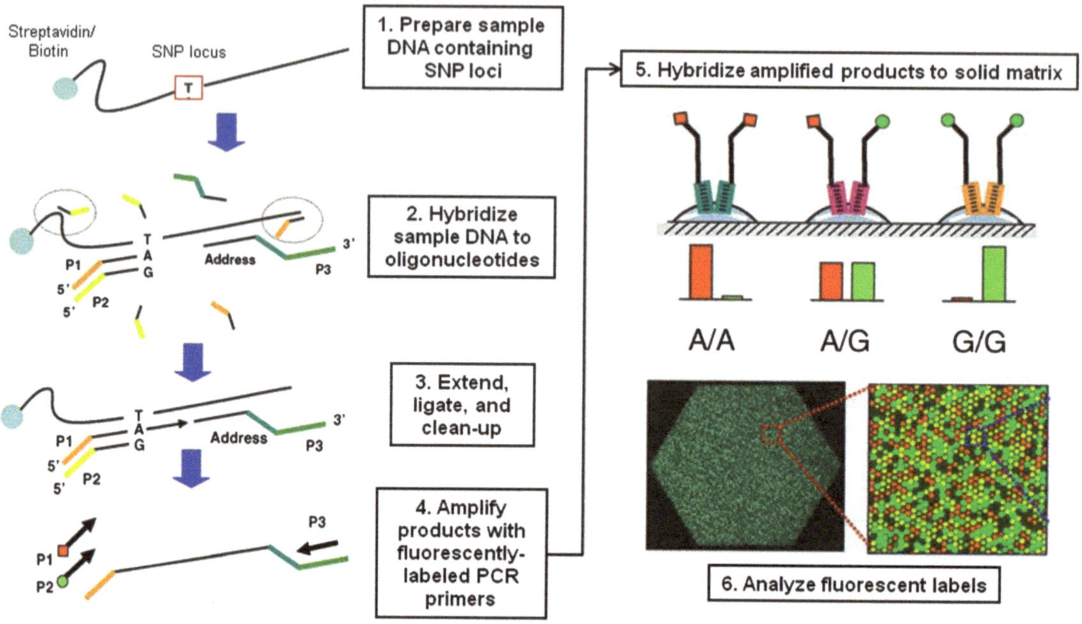

Figure 20-7 Principles of a bead-based (Illumina) SNP array. (*1*) Genomic sample DNA is first "activated" by biotinylation. (*2*) Oligonucleotides that correspond to specific SNPs are then combined with the activated DNA in the oligonucleotide/target annealing step, in which the query oligonucleotides hybridize to the genomic DNA that binds to the paramagnetic particles (via the biotin). Two allele-specific oligonucleotides (ASOs designated *orange* or *yellow*) and one locus-specific oligonucleotide (LSO designated *green*) are designed for each SNP. The ASO contains a 5′ universal sequence that serves as a universal primer for all beads. The LSO contains a unique sequence complementary to a particular bead type designated as the "address." (*3*) DNA polymerase with high specificity for a perfectly matched target sequence at the SNP adds nucleotides between the ASO and the LSO. DNA ligase is used to seal the nick between the extended ASO and the LSO to form PCR templates that can be amplified with universal PCR primers (Step 4). (*4*) PCR amplification is performed with three universal PCR primers (P1, P2, and P3) labeled with Cy3 (*green*), Cy5 (*red*), and biotin, respectively. (*5*) Double-stranded dye-labeled PCR products are converted to single-stranded DNA (ssDNA) on paramagnetic beads (again via binding with biotin). These ssDNA are removed and hybridized to their complement bead type via their unique "address." (*6*) The bound DNA is containing Cy3 and/or Cy5. The dyes are excited by lasers at different wavelengths. Based on the intensities detected from the two channels for the two respective alleles of each SNP, genotypes are designated using computer software (adapted from Shen R et al. High-throughput SNP genotyping on universal bead arrays. *Mutat Res*. 2005 Jun 3;573(1–2):70–82 with permission from *Mutation Research*. Copyright 2005 Elsevier B.V.)

evaluation of chromosomal abnormalities. The abundance of one color (SNP allele) in the tumor sample could represent LOH, secondary to either chromosomal deletion or CN-LOH (Fig. 20.8d).

The source of reference DNA is important in SNP analysis. When an SNP array sample is compared to reference DNA instead of paired normal sample from the same individual, there is a greater risk of a "miscall." In the most common scenario, an individual harbors an inherited copy number variation (CNV, a region of the genome with polymorphous gene segments). CNVs appear frequently in the genome, but the same CNVs are usually not seen among many persons. This has complicated the validation of these regions in reference databases such as the Database of Genomic Variants (DGV) [56]. Virtually all CNV and SNP databases are incomplete and lack comprehensive validation. Thus, it is recommended to simultaneously analyze normal, non-neoplastic DNA from the same patient [41, 54, 57]. Buccal mucosal epithelial cell swabs are routinely used as a source of normal DNA, but buccal mucosa can be

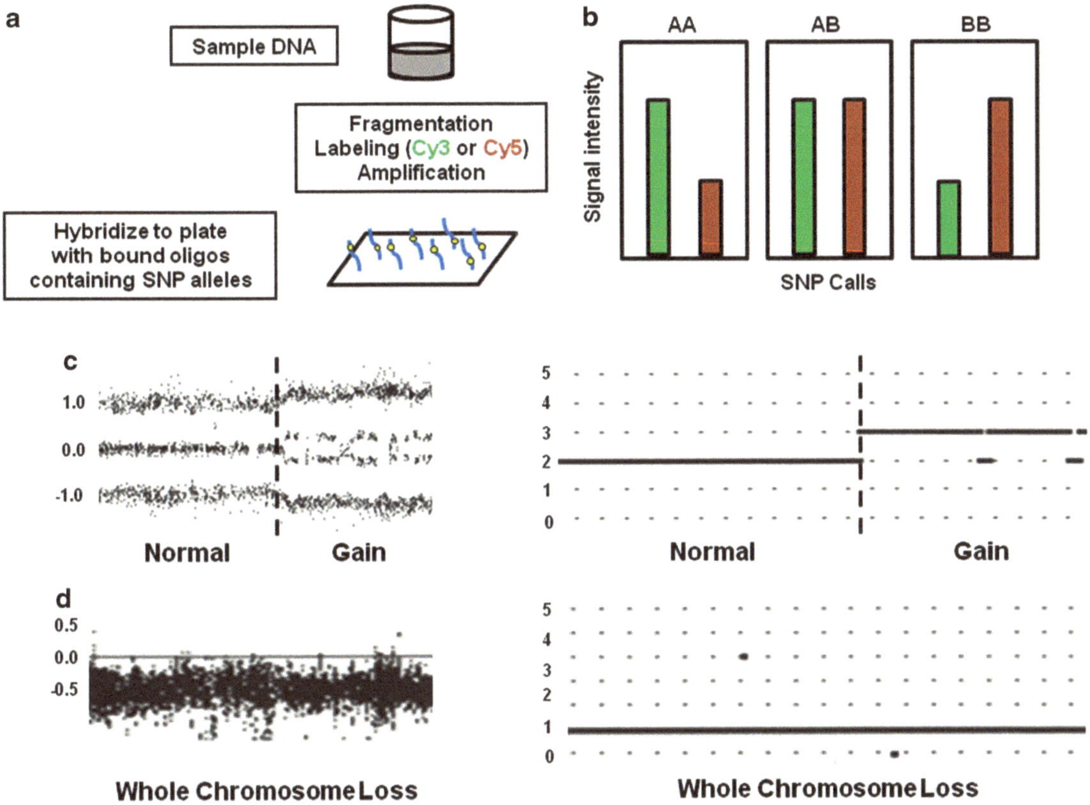

Figure 20-8 Principles of a chip-based (Affymetrix) SNP array. (**a**) Sample DNA is fragmented and labeled with fluorescent dye (e.g., Cy3 and Cy5). DNA is hybridized to oligonucleotide probes (*blue*) corresponding to the individual SNP alleles (*yellow circles*). (**b**) Amplified and labeled DNA is hybridized to probes corresponding to alleles for each SNP locus. This results in a genotyping pattern allowing for determination of the heterozygosity or homozygosity for each allele. At the same time, intensity of the hybridization signals allows for determination of copy number changes. Various software packages allow for generation of karyotyping maps. (**c**) Example data from an Affymetrix 6.0 SNP array showing segmental gain of chromosomal material. The *left panel* is an example of an "allele peak" view. On the *left side* of the *dotted line* there is a mixture of signal intensities of the arbitrarily designed "B" allele. In the normal state (*left* of the *dotted line*) there are equal mixtures of loci homozygous for the B SNP allele (+1), loci heterozygous for the AB allele (0), and homozygous for the A SNP allele (−1). A segmental gain of chromosomal material is shown to the *right* of the *dotted line*. The "allele peak" view now shows the presence of three alleles: BBB (*top line*, +1.5), BBA (*second line from top*, +0.75), BAA (*third line from top*, −0.75), and AAA (*bottom line*, −1.5). The "copy number state" is shown in the *right panel*. This assists in the identification of three alleles. (**d**) Example data from an Affymetrix 6.0 SNP array showing complete loss of chromosomal material from 5q. The *left panel* is an example of a "weighted log2" view indicating overall loss of fluorescent signal. The "copy number state" is shown in the *right panel*. This shows one copy of the allele and is interpreted as a complete loss of genetic material

contaminated with blood, thus introducing tumor DNA into the supposed "normal sample." A skin biopsy may be a source of optimal normal DNA but is a more invasive procedure than a simple mouth swab [54].

Although SNP arrays interrogate a significant portion of the genome in terms of SNP distribution, they still cover less than 0.1 % of the whole genome. With current technology, SNP arrays cannot be designed to distinguish every single base pair change throughout the genome [58]. Thus, they cannot be used to evaluate point mutations in genes in which these are common, such as *FLT3*, *CKIT*, or *PDGFR*. SNP arrays also are unable to detect balanced chromosomal translocations. To resolve this, SNP array analysis is applied subsequent to tests that readily identify translocations and point mutations such as routine cytogenetic analysis/FISH or single gene assays, respectively.

The ability to interrogate all base pairs in hematologic malignancies would provide a comprehensive assessment of the neoplastic genotype and ensure that all clinically relevant information would be obtained from a patient sample. Assays that interrogate the whole genome at the base pair level have recently been introduced to assess this information.

Whole Genome Assays

As the name implies, whole genome assays (at least theoretically) interrogate all nucleotides present in the entire genome or, in the case of whole exome sequencing (WES), interrogate all the base pairs in the coding regions of the genome. Next-generation sequencing (NGS) platforms can sequence the entire genome in a single experimental run, because modern computers possess the massive computing power necessary to manipulate billions of data points simultaneously. The time and cost required to sequence the entire genome have dropped precipitously. In 2001, it took a staggering 10 years and 2.7 billion U.S. dollars to sequence the entire genome. Approximately a decade later, it takes an equally staggering few weeks and a few thousand U.S. dollars to perform the same task [59]. By coupling enormous clinical potential with legitimate affordability and reasonable turnaround time, NGS techniques are projected to become a mainstay for the diagnosis and treatment of hematologic malignancies in the imminent future.

Next Generation Sequencing (NGS)

Three benchtop NGS platforms, The 454 GS Junior (Roche), MiSeq (Illumina), and Ion Torrent PGM (Life Technologies), have emerged as the industry leaders and are used for hematolymphoid malignancy testing. Regardless of the platform, intact genomic DNA must be partitioned into workable amplifiable fragments such that template DNA can undergo massively parallel DNA sequencing. Current methods generally involve randomly breaking genomic DNA into smaller sizes (either by sonication or restriction digestion), amplifying the template DNA (e.g., creating a DNA library), and anchoring the fragments to solid phase components. The solid phase anchored fragmented DNA is partitioned in such a way that simultaneous sequencing reactions can occur. These sequencing reactions are performed on amplified DNA fragments (of the same template sequence) because most imaging systems cannot detect single template fluorescent or luminescent events [60, 61]. The ability to detect the sequence of millions of individual partitioned fragments of genomic DNA simultaneously is the sine qua non of NGS [60–65].

Each benchtop NGS platform performs massive parallel sequencing using a different methodology and as a result, each of the three has differential performance characteristics. At least for bacterial genomes, the MiSeq tends to generate the highest throughput per run with the lowest number of errors but delivers short reads. The 454 GS Junior delivers the longest read length but suffers from more errors than the MiSeq, and the Ion Torrent PGM produces the shortest reads with the most errors but with the fastest throughput and shortest run-time [66, 67]. All three NGS platforms are currently in use in clinical laboratories, but it is unclear and likely too early to determine whether one platform (or another platform in development) will emerge as the "gold standard" for clinical use. Still, for the purposes of this chapter, the application of NGS to hematologic malignancies is the same, regardless of which platform is used.

Applications of NGS to Hematologic Malignancies

The goal of NGS in hematologic malignancies is to identify and interpret genetic variation between the neoplastic population and the matched germline DNA [68]. When optimized, NGS can detect point mutations, insertions, deletions, and chromosomal rearrangements making it an ideal platform to assess known clinically relevant mutations and for the discovery of new mutations or mutations previously undetected by standard methodologies [69].

One of the first documented applications of NGS to hematologic malignancies was performed in 2008 when investigators used NGS to sequence "cytogenetically normal" (by routine karyotyping and FISH analysis) AMLs. They discovered ten somatic mutations in exon coding regions. Of the ten mutations, two mutations, *FLT3* ITDs and *NPM1* exon 12 insertions, were known to be mutated in AML and the other eight novel mutations and the associated genes are currently under intense investigation to determine their contribution to the pathogenesis of AML [70]. This landmark paper served as a "proof of principle" study: one could feasibly use NGS to identify mutations in hematologic malignancies that other methods had failed to detect. A few years later, novel recurrent mutations in *DNMT3A* were discovered in approximately 20 % of AML cases and found to be associated with poor overall survival, but predictive of improved survival in patients less than 60 years of age treated with high-dose induction chemotherapy [68, 71]. In MDS, NGS identified mutations in *TP53*, *EZH2*, *ETV6*, *RUNX1*, and *ASXL1* that were found to be predictors of poor overall survival [72]. In multiple myeloma, investigators discovered previously unknown point mutations in *KRAS*, *BRAF*, and *NRAS* [73]. New mutations were also discovered in CLL and NGS has been investigated as a tool to molecularly monitor clonal evolution in pediatric acute leukemia patients [74–76].

NGS has been reported to recapitulate the results generated by traditional single gene assays, such as identifying immunoglobulin or T-cell receptor rearrangements, *FLT3* mutations, and *JAK2* p.V617F mutations [70, 77, 78]. There are also reports that NGS can identify balanced translocations such as BCR-ABL1 t(9;22)(q34;q11.2) and cryptic translocations not identified by routine cytogenetics [69, 79].

The use of NGS platforms has also found favor in monitoring residual disease and disease recurrence. NGS platforms are able to "look for every aberration" in contrast to conventional methods such as cytogenetics, FISH, or PCR that are designed to assess for predetermined chromosomal abnormalities or mutations. This unique property allows for the detection of tumor subclones that were either present at the initiation of therapy or are evolving in the presence of therapy. Examples of this "escape clone monitoring" were recently described in AML. Investigators found two major clonal evolution patterns during AML relapse: a primary tumor clone acquired additional mutations that evolved into the relapse clone, or a subclone of the primary tumor clone survived initial therapy, gained additional mutations, and then expanded at relapse. These data also suggested that AML cells routinely acquire additional mutations at relapse, and some of these mutations may contribute to clonal selection and chemotherapy resistance [80]. Clonal evolution monitoring with NGS is also described in drug-resistant *BCR-ABL1* mutants [81] and chronic lymphocytic leukemia [82]. As NGS studies accumulate, it is postulated that more sophisticated monitoring and treatment protocols will arise.

The comprehensive data attained from NGS are obviously attractive for use as a clinical tool. However, the trade-off for generating many parallel short templates is loss of sequencing accuracy. NGS platforms have approximately tenfold higher error rates in base pair reads (1 in 1,000 bases at 20-fold coverage) versus Sanger sequencing (1 in 10,000 bases). The depth of sequencing coverage (the number of times a single fragment is amplified) may be insufficient to identify single point mutations in limited sample sizes; neoplastic tissue that comprises 25 % of the input sample that is sequenced at 30-fold coverage still produces an error rate of 5 % [83]. Certain highly repetitive regions of the genome are difficult to examine accurately, partly due to the algorithms used to align the sequencing data [68]. WES allows greater depth of coverage to more accurately detect point mutations in the 1–2 % of the genome that constitutes the protein coding regions. However, certain portions of the exome are also subject to the same sequencing difficulties and pertinent mutations in intronic DNA will be missed. To address the technological challenges inherent to NGS, consensus guidelines for test validation, quality control, proficiency testing, and quality control for NGS testing in clinical laboratories are available and continue to evolve [84].

Both WGS and WES generate tremendous amounts of data that pose unprecedented informatics challenges to analyze, interpret, retrieve, and store, particularly in a Health Insurance Portability and Accountability Act (HIPAA)-compliant manner. A recent report highlighted some of the data analysis challenges that NGS sequencing poses. For the detection of *FLT3* internal tandem duplication, only one of seven software analysis packages (Pindel) reliably detected the aberration with 100 % sensitivity and specificity. Some of the software programs did not detect the duplication in any sample [85]. These data raise questions regarding the "in silico" interpretation of NGS data that are beyond the scope of this chapter.

Suffice it to say, guidelines and recommendations governing the broad clinical application of WGS or WES to hematologic malignancies will require input from pathologists, clinicians, and informatics specialists, among others, to meet the rigorous quality demands required for clinical laboratory testing.

Conclusions

The genomic applications to hematologic malignancies are diverse but are essential for both diagnosis and clinical management, and each testing methodology has utility in the appropriate clinical context. Choosing the most fitting test requires a fund of knowledge for both the disease entity and the testing methods. Single gene testing, routine cytogenetic karyotyping, FISH, aCGH, SNP array, WES, and WGS analyses all possess innate utility and limitations when the clinical question is clearly delineated. As the discoveries and applications of innovative technologies in genomic medicine will continue to evolve, it is the medical professional's responsibility to become familiar with all genomic testing methodologies, in order to request the most suitable test for the diagnosis and management of patients with hematologic malignancies.

REFERENCES

1. Jung D, Giallourakis C, Mostoslavsky R, Alt FW. Mechanism and control of V(D)J recombination at the immunoglobulin heavy chain locus. Annu Rev Immunol. 2006;24:541–70.
2. Bruns DE, Ashwood ER, Burtis CA. Fundamentals of molecular diagnostics. St. Louis, MO: Saunders Elsevier; 2007.
3. Nikiforova MN, Hsi ED, Braziel RM, Gulley ML, Leonard DGB, Nowak JA, et al. Detection of clonal IGH gene rearrangements: summary of molecular oncology surveys of the College of American Pathologists. Arch Pathol Lab Med. 2007;131(2):185–9.
4. Mannu C, Gazzola A, Bacci F, Sabattini E, Sagramoso C, Roncolato F, et al. Use of IGK gene rearrangement analysis for clonality assessment of lymphoid malignancies: a single center experience. Am J Blood Res. 2011;1(2):167–74.
5. Tapia G, Sanz C, Mate JL, Munoz-Marmol AM, Ariza A. Improved clonality detection in Hodgkin lymphoma using the BIOMED-2-based heavy and kappa chain assay: a paraffin-embedded tissue study. Histopathology. 2012;60(5):768–73.
6. Jevremovic D, Viswanatha DS. Molecular diagnosis of hematopoietic and lymphoid neoplasms. Hematol Oncol Clin North Am. 2009;23(4):903–33.
7. Swerdlow SH, International Agency for Research on Cancer, World Health Organization. WHO classification of tumours of haematopoietic and lymphoid tissues. Lyon, France: International Agency for Research on Cancer; 2008.
8. van Dongen JJ, Langerak AW, Bruggemann M, Evans PA, Hummel M, Lavender FL, et al. Design and standardization of PCR primers and protocols for detection of clonal immunoglobulin and T-cell receptor gene recombinations in suspect lymphoproliferations: report of the BIOMED-2 Concerted Action BMH4-CT98-3936. Leukemia. 2003;17(12):2257–317.
9. Leonard DGB. Diagnostic molecular pathology. Philadelphia, PA: W.B. Saunders; 2003.
10. Patnaik MM, Tefferi A. Molecular diagnosis of myeloproliferative neoplasms. Expert Rev Mol Diagn. 2009;9(5):481–92.
11. Tefferi A, Skoda R, Vardiman JW. Myeloproliferative neoplasms: contemporary diagnosis using histology and genetics. Nat Rev Clin Oncol. 2009;6(11):627–37.
12. Hammond E, Shaw K, Carnley B, P'ng S, James I, Herrmann R. Quantitative determination of JAK2 V617F by TaqMan: an absolute measure of averaged copies per cell that may be associated with the different types of myeloproliferative disorders. J Mol Diagn. 2007;9(2):242–8.
13. Lippert E, Girodon F, Hammond E, Jelinek J, Reading NS, Fehse B, et al. Concordance of assays designed for the quantification of JAK2V617F: a multicenter study. Haematologica. 2009;94(1):38–45.
14. Barosi G, Birgegard G, Finazzi G, Griesshammer M, Harrison C, Hasselbalch HC, et al. Response criteria for essential thrombocythemia and polycythemia vera: result of a European LeukemiaNet consensus conference. Blood. 2009;113(20):4829–33.

15. Frohling S, Scholl C, Gilliland DG, Levine RL. Genetics of myeloid malignancies: pathogenetic and clinical implications. J Clin Oncol. 2005; 23(26):6285–95.
16. Murphy KM, Levis M, Hafez MJ, Geiger T, Cooper LC, Smith BD, et al. Detection of FLT3 internal tandem duplication and D835 mutations by a multiplex polymerase chain reaction and capillary electrophoresis assay. J Mol Diagn. 2003;5(2):96–102.
17. Dekking E, van der Velden VHJ, Böttcher S, Brüggemann M, Sonneveld E, Koning-Goedheer A, et al. Detection of fusion genes at the protein level in leukemia patients via the flow cytometric immunobead assay. Best Pract Res Clin Haematol. 2010;23(3): 333–45.
18. O'Brien SG, Guilhot F, Larson RA, Gathmann I, Baccarani M, Cervantes F, et al. Imatinib compared with interferon and low-dose cytarabine for newly diagnosed chronic-phase chronic myeloid leukemia. N Engl J Med. 2003;348(11):994–1004.
19. Druker BJ, Guilhot F, O'Brien SG, Gathmann I, Kantarjian H, Gattermann N, et al. Five-year follow-up of patients receiving imatinib for chronic myeloid leukemia. N Engl J Med. 2006;355(23):2408–17.
20. White HE, Matejtschuk P, Rigsby P, Gabert J, Lin F, Lynn Wang Y, et al. Establishment of the first World Health Organization International Genetic Reference Panel for quantitation of BCR-ABL mRNA. Blood. 2010;116(22):e111–7.
21. Wang N. Methodologies in cancer cytogenetics and molecular cytogenetics. Am J Med Genet. 2002; 115(3):118–24.
22. Bejjani BA, Saleki R, Ballif BC, Rorem EA, Sundin K, Theisen A, et al. Use of targeted array-based CGH for the clinical diagnosis of chromosomal imbalance: is less more? Am J Med Genet A. 2005;134(3): 259–67.
23. Sharathkumar A, DeCamillo D, Bhambhani K, Cushing B, Thomas R, Mohamed AN, et al. Children with hyperdiploid but not triple trisomy (+4,+10,+17) acute lymphoblastic leukemia have an increased incidence of extramedullary relapse on current therapies: a single institution experience. Am J Hematol. 2008;83(1):34–40.
24. Marschalek R. Mechanisms of leukemogenesis by MLL fusion proteins. Br J Haematol. 2011;152(2): 141–54.
25. Jacoby MA, Walter MJ. Detection of copy number alterations in acute myeloid leukemia and myelodysplastic syndromes. Expert Rev Mol Diagn. 2012; 12(3):253–64.
26. Shinawi M, Cheung SW. The array CGH and its clinical applications. Drug Discov Today. 2008; 13(17–18):760–70.
27. Shaikh TH. Oligonucleotide arrays for high-resolution analysis of copy number alteration in mental retardation/multiple congenital anomalies. Genet Med. 2007;9(9):617–25.
28. Shao L, Kang S-HL, Li J, Hixson P, Taylor J, Yatsenko SA, et al. Array comparative genomic hybridization detects chromosomal abnormalities in hematological cancers that are not detected by conventional cytogenetics. J Mol Diagn. 2010;12(5):670–9.
29. Kolquist KA, Schultz RA, Furrow A, Brown TC, Han J-Y, Campbell LJ, et al. Microarray-based comparative genomic hybridization of cancer targets reveals novel, recurrent genetic aberrations in the myelodysplastic syndromes. Cancer Genet. 2011; 204(11):603–28.
30. Paulsson K, Heidenblad M, Strombeck B, Staaf J, Jonsson G, Borg A, et al. High-resolution genome-wide array-based comparative genome hybridization reveals cryptic chromosome changes in AML and MDS cases with trisomy 8 as the sole cytogenetic aberration. Leukemia. 2006;20(5):840–6.
31. Thiel A, Beier M, Ingenhag D, Servan K, Hein M, Moeller V, et al. Comprehensive array CGH of normal karyotype myelodysplastic syndromes reveals hidden recurrent and individual genomic copy number alterations with prognostic relevance. Leukemia. 2011;25(3):387–99.
32. Vercauteren SM, Sung S, Starczynowski DT, Lam WL, Bruyere H, Horsman DE, et al. Array comparative genomic hybridization of peripheral blood granulocytes of patients with myelodysplastic syndrome detects karyotypic abnormalities. Am J Clin Pathol. 2010;134(1):119–26.
33. Dawson AJ, Yanofsky R, Vallente R, Bal S, Schroedter I, Liang L, et al. Array comparative genomic hybridization and cytogenetic analysis in pediatric acute leukemias. Curr Oncol. 2011;18(5):e210–7.
34. Lucioni M, Novara F, Fiandrino G, Riboni R, Fanoni D, Arra M, et al. Twenty-one cases of blastic plasmacytoid dendritic cell neoplasm: focus on biallelic locus 9p21.3 deletion. Blood. 2011;118(17): 4591–4.
35. Higgins RA, Gunn SR, Robetorye RS. Clinical application of array-based comparative genomic hybridization for the identification of prognostically important genetic alterations in chronic lymphocytic leukemia. Mol Diagn Ther. 2008;12(5):271–80.
36. Okada M, Suto Y, Hirai M, Shiseki M, Usami A, Okajima K, et al. Microarray CGH analyses of chromosomal 20q deletions in patients with hematopoietic malignancies. Cancer Genet. 2012;205(1–2): 18–24.
37. Sachidanandam R, Weissman D, Schmidt SC, Kakol JM, Stein LD, Marth G, et al. A map of human genome sequence variation containing 1.42 million single nucleotide polymorphisms. Nature. 2001;409(6822):928–33.
38. Ersland KM, Christoforou A, Stansberg C, Espeseth T, Mattheisen M, Mattingsdal M, et al. Gene-based analysis of regionally enriched cortical genes in GWAS data sets of cognitive traits and psychiatric disorders. PLoS One. 2012;7(2):e31687.
39. Kidd KK, Pakstis AJ, Speed WC, Grigorenko EL, Kajuna SL, Karoma NJ, et al. Developing a SNP panel for forensic identification of individuals. Forensic Sci Int. 2006;164(1):20–32.
40. Tzvetkov M, von Ahsen N. Pharmacogenetic screening for drug therapy: from single gene markers to decision making in the next generation sequencing era. Pathology. 2012;44(2):166–80.
41. Sato-Otsubo A, Sanada M, Ogawa S. Single-nucleotide polymorphism array karyotyping in clinical practice: where, when, and how? Semin Oncol. 2012;39(1):13–25.
42. Fan JB, Gunderson KL, Bibikova M, Yeakley JM, Chen J, Wickham Garcia E, et al. Illumina universal

bead arrays. In: Brian O, Alan K, editors. Methods in enzymology. San Diego, CA: Academic; 2006. p. 57–73.
43. Murray SS, Oliphant A, Shen R, McBride C, Steeke RJ, Shannon SG, et al. A highly informative SNP linkage panel for human genetic studies. Nat Methods. 2004;1(2):113–7. doi:10.1038/nmeth712.
44. Maciejewski JP, Tiu RV, O'Keefe C. Application of array-based whole genome scanning technologies as a cytogenetic tool in haematological malignancies. Br J Haematol. 2009;146(5):479–88.
45. Maciejewski JP, Mufti GJ. Whole genome scanning as a cytogenetic tool in hematologic malignancies. Blood. 2008;112(4):965–74.
46. Kawamata N, Ogawa S, Zimmermann M, Niebuhr B, Stocking C, Sanada M, et al. Cloning of genes involved in chromosomal translocations by high-resolution single nucleotide polymorphism genomic microarray. Proc Natl Acad Sci U S A. 2008;105(33):11921–6.
47. Mullighan CG, Goorha S, Radtke I, Miller CB, Coustan-Smith E, Dalton JD, et al. Genome-wide analysis of genetic alterations in acute lymphoblastic leukaemia. Nature. 2007;446(7137):758–64.
48. Pfeifer D, Pantic M, Skatulla I, Rawluk J, Kreutz C, Martens UM, et al. Genome-wide analysis of DNA copy number changes and LOH in CLL using high-density SNP arrays. Blood. 2007;109(3):1202–10.
49. Edelmann J, Holzmann K, Miller F, Winkler D, Buhler A, Zenz T, et al. High-resolution genomic profiling of chronic lymphocytic leukemia reveals new recurrent genomic alterations. Blood. 2012;120(24):4783–94.
50. Jenner MW, Leone PE, Walker BA, Ross FM, Johnson DC, Gonzalez D, et al. Gene mapping and expression analysis of 16q loss of heterozygosity identifies WWOX and CYLD as being important in determining clinical outcome in multiple myeloma. Blood. 2007;110(9):3291–300.
51. Parkin B, Erba H, Ouillette P, Roulston D, Purkayastha A, Karp J, et al. Acquired genomic copy number aberrations and survival in adult acute myelogenous leukemia. Blood. 2010;116(23):4958–67.
52. Tiu RV, Gondek LP, O'Keefe CL, Elson P, Huh J, Mohamedali A, et al. Prognostic impact of SNP array karyotyping in myelodysplastic syndromes and related myeloid malignancies. Blood. 2011;117(17):4552–60.
53. Yi JH, Huh J, Kim HJ, Kim SH, Kim YK, Sohn SK, et al. Adverse prognostic impact of abnormal lesions detected by genome-wide single nucleotide polymorphism array-based karyotyping analysis in acute myeloid leukemia with normal karyotype. J Clin Oncol. 2011;29(35):4702–8.
54. Heinrichs S, Li C, Look AT. SNP array analysis in hematologic malignancies: avoiding false discoveries. Blood. 2010;115(21):4157–61.
55. Ishkanian AS, Malloff CA, Watson SK, DeLeeuw RJ, Chi B, Coe BP, et al. A tiling resolution DNA microarray with complete coverage of the human genome. Nat Genet. 2004;36(3):299–303.
56. Zhang J, Feuk L, Duggan GE, Khaja R, Scherer SW. Development of bioinformatics resources for display and analysis of copy number and other structural variants in the human genome. Cytogenet Genome Res. 2006;115(3–4):205–14.
57. Le Scouarnec S, Gribble SM. Characterising chromosome rearrangements: recent technical advances in molecular cytogenetics. Heredity (Edinb). 2012;108(1):75–85.
58. LaFramboise T. Single nucleotide polymorphism arrays: a decade of biological, computational and technological advances. Nucleic Acids Res. 2009;37(13):4181–93.
59. Venter JC. Multiple personal genomes await. Nature. 2010;464(7289):676–7. doi:10.1038/464676a.
60. Metzker ML. Sequencing technologies—the next generation. Nat Rev Genet. 2010;11(1):31–46. doi:10.1038/nrg2626.
61. Rothberg JM, Hinz W, Rearick TM, Schultz J, Mileski W, Davey M, et al. An integrated semiconductor device enabling non-optical genome sequencing. Nature. 2011;475(7356):348–52. doi:10.1038/nature10242.
62. Bentley DR, Balasubramanian S, Swerdlow HP, Smith GP, Milton J, Brown CG, et al. Accurate whole human genome sequencing using reversible terminator chemistry. Nature. 2008;456(7218):53–9.
63. Dressman D, Yan H, Traverso G, Kinzler KW, Vogelstein B. Transforming single DNA molecules into fluorescent magnetic particles for detection and enumeration of genetic variations. Proc Natl Acad Sci U S A. 2003;100(15):8817–22.
64. Leamon JH, Lee WL, Tartaro KR, Lanza JR, Sarkis GJ, deWinter AD, et al. A massively parallel PicoTiterPlate based platform for discrete picoliter-scale polymerase chain reactions. Electrophoresis. 2003;24(21):3769–77.
65. Ronaghi M, Uhlén M, Nyrén P. A sequencing method based on real-time pyrophosphate. Science. 1998;281(5375):363–5.
66. Liu L, Li Y, Li S, Hu N, He Y, Pong R, et al. Comparison of next-generation sequencing systems. J Biomed Biotechnol. 2012;2012:251364.
67. Loman NJ, Misra RV, Dallman TJ, Constantinidou C, Gharbia SE, Wain J, et al. Performance comparison of benchtop high-throughput sequencing platforms. Nat Biotechnol. 2012;30(5):434–9. doi:10.1038/nbt.2198.
68. Merker JD, Valouev A, Gotlib J. Next-generation sequencing in hematologic malignancies: what will be the dividends? Ther Adv Hematol. 2012;3(6):333–9.
69. Welch JS, Westervelt P, Ding L, et al. Use of whole-genome sequencing to diagnose a cryptic fusion oncogene. JAMA. 2011;305(15):1577–84.
70. Ley TJ, Mardis ER, Ding L, Fulton B, McLellan MD, Chen K, et al. DNA sequencing of a cytogenetically normal acute myeloid leukaemia genome. Nature. 2008;456(7218):66–72.
71. Patel JP, Gonen M, Figueroa ME, Fernandez H, Sun Z, Racevskis J, et al. Prognostic relevance of integrated genetic profiling in acute myeloid leukemia. N Engl J Med. 2012;366(12):1079–89.
72. Bejar R, Stevenson K, Abdel-Wahab O, Galili N, Nilsson B, Garcia-Manero G, et al. Clinical effect of point mutations in myelodysplastic syndromes. N Engl J Med. 2011;364(26):2496–506.

73. Chapman MA, Lawrence MS, Keats JJ, Cibulskis K, Sougnez C, Schinzel AC, et al. Initial genome sequencing and analysis of multiple myeloma. Nature. 2011;471(7339):467–72. doi:10.1038/nature09837.
74. Campana D. Minimal residual disease monitoring in childhood acute lymphoblastic leukemia. Curr Opin Hematol. 2012;19(4):313–8.
75. Gawad C, Pepin F, Carlton VE, Klinger M, Logan AC, Miklos DB, et al. Massive evolution of the immunoglobulin heavy chain locus in children with B precursor acute lymphoblastic leukemia. Blood. 2012;120(22):4407–17.
76. Ramsay AJ, Martinez-Trillos A, Jares P, Rodriguez D, Kwarciak A, Quesada V. Next-generation sequencing reveals the secrets of the chronic lymphocytic leukemia genome. Clin Transl Oncol. 2012;15(1):3–8.
77. Benichou J, Ben-Hamo R, Louzoun Y, Efroni S. Rep-Seq: uncovering the immunological repertoire through next-generation sequencing. Immunology. 2012;135(3):183–91.
78. Kohlmann A, Grossmann V, Haferlach T. Integration of next-generation sequencing into clinical practice: are we there yet? Semin Oncol. 2012;39(1):26–36.
79. Shaffer LG, Schultz RA, Ballif BC. The use of new technologies in the detection of balanced translocations in hematologic disorders. Curr Opin Genet Dev. 2012;22(3):264–71.
80. Ding L, Ley TJ, Larson DE, Miller CA, Koboldt DC, Welch JS, et al. Clonal evolution in relapsed acute myeloid leukaemia revealed by whole-genome sequencing. Nature. 2012;481(7382):506–10. doi:10.1038/nature10738.
81. Soverini S, De Benedittis C, Machova Polakova K, Brouckova A, Horner D, Iacono M, et al. Unraveling the complexity of tyrosine kinase inhibitor-resistant populations by ultra-deep sequencing of the BCR-ABL kinase domain. Blood. 2013;21:2013.
82. Landau Dan A, Carter Scott L, Stojanov P, McKenna A, Stevenson K, Lawrence Michael S, et al. Evolution and impact of subclonal mutations in chronic lymphocytic leukemia. Cell. 2013;152(4):714–26.
83. Gullapalli RR, Lyons-Weiler M, Petrosko P, Dhir R, Becich MJ, LaFramboise WA. Clinical integration of next-generation sequencing technology. Clin Lab Med. 2012;32(4):585–99.
84. Gargis AS, Kalman L, Berry MW, Bick DP, Dimmock DP, Hambuch T, et al. Assuring the quality of next-generation sequencing in clinical laboratory practice. Nat Biotechnol. 2012;30(11):1033–6. doi:10.1038/nbt.2403.
85. Spencer DH, Abel HJ, Lockwood CM, Payton JE, Szankasi P, Kelley TW, et al. Detection of FLT3 internal tandem duplication in targeted, short-read-length, next-generation sequencing data. J Mol Diagn. 2013;15(1):81–93.

CHAPTER 21

Genomic Applications in Brain Tumors

MATIJA SNUDERL

Introduction 321
Targeted Genomic Assays Used in the Clinical Evaluation of Brain Tumors 322
Whole-Genome Evaluation of Brain Tumors 329
Gliomas 333
Conclusions 337

Introduction

The field of neuro-oncology has evolved significantly over the past two decades. Because brain tumors are difficult to culture, discoveries of genomic rearrangements made by classic cytogenetics in other fields of oncology have not been possible. A new era has started with the implementation of molecular techniques, which led to discoveries of novel diagnostic, prognostic and predictive molecular markers. Whereas histological classification of tumors relies on the morphological features, targeted and genome-wide approaches led to deeper understanding of brain tumor biology and molecular subclassification of morphologic entities. Recent advances in genome-wide techniques have also discovered genes underlying previously well known aberrations. For example, a co-deletion of 1p/19q has long been accepted in the field as a predictive marker in oligodendroglioma. Whole-genome sequencing discovered the underlying role of the *CIC* gene in oligodendroglioma biology. While many methods are still for research use only, there is no doubt that many will be used clinically in the foreseeable future. A variety of assays have been designed to analyze chromosomal rearrangements, copy number changes, point mutations, and epigenetic changes. This is particularly important because most malignant brain tumors have largely resisted standard chemotherapy and radiation therapy and will require more targeted approaches based on the specific biology of the tumors. Genome, transcriptome, and epigenome analyses will likely become a focus for diagnostics and for identifying therapeutic targets.

Gliomas are the most common tumors of the central nervous system (CNS) and often require additional molecular workup, either for diagnosis or for clinical management. In clinical practice, a single gene or target region is usually evaluated. The most commonly used assays include analyses of 1p/19q, *MGMT* methylation and *IDH1* mutation status [1–3]. From a technical point of view they include fluorescence in situ hybridization (FISH), the polymerase chain reaction (PCR), a variety of methylation-specific assays, and sequencing or immunohistochemistry (IHC). These targeted approaches can provide important diagnostic and prognostic or predictive information, particularly in diffuse gliomas. Other common CNS neoplasms such as meningiomas, ependymomas, and medulloblastomas seem to show changes that are far too complex to be detectable by a

G.J. Netto and I. Schrijver (eds.), *Genomic Applications in Pathology*,
DOI 10.1007/978-1-4939-0727-4_21,

single-target approach. Genome-based analysis of the expression profile of medulloblastomas, supported later by copy number and mutation analysis, has pioneered subclassification of a single disease based on molecular characteristics. With a growing number of targets, a whole-genome approach and assay multiplexing are becoming more feasible and cost-effective solutions. With the costs of whole-genome analyses decreasing, one can expect that a large number of specific assays designed for particular targets will be replaced by a panel that is able to evaluate numerous genes of interest.

Targeted Genomic Assays Used in the Clinical Evaluation of Brain Tumors

1p/19q

Loss of chromosomal arms 1p and 19q is the hallmark of oligodendroglial neoplasms. Although frequencies in reports vary, loss of 1p and 19q is detected in up to 80 % of oligodendrogliomas of World Health Organization (WHO) grade II, 60 % of anaplastic oligodendrogliomas WHO grade III, 30–50 % of oligoastrocytomas WHO Grade II, 20–30 % of anaplastic oligoastrocytomas, and approximately 10 % of diffuse astrocytic gliomas.

Numerous studies have confirmed an association between 1p/19q co-deletion and a favorable response to chemotherapy, initially to procarbazine, lomustine, and vincristine, and later to temozolomide, as well as to radiotherapy. Therefore, testing for 1p/19q loss is considered the standard of care. Most neuro-oncologists will use 1p/19q status to make therapeutic decisions, although the clinical approach is not uniform. Many will withhold radiation therapy upfront, even in case of a small residual tumor after surgery, in order to avoid the risk of long-term toxicity and choose chemotherapy with temozolomide or even careful monitoring alone. Radiation, therefore, remains as an option in case of progression. From a diagnostic point of view, 1p/19q loss can help to distinguish oligodendrogliomas from morphologically similar neoplasms such as neurocytomas, clear cell ependymomas and meningiomas, dysembryoplastic neuroepithelial tumors (DNETs), or small cell variant of glioblastoma. The loss of 1p and 19q is mediated by formation of a balanced whole arm translocation involving chromosomes 1 and 19, with subsequent loss of the derivative chromosome der(1;19)(p10;q10) and maintenance of the der(1;19)(q10;p10). The genes responsible for tumorigenesis of oligodendroglioma were enigmatic until recently, when several whole-genome sequencing studies have identified *CIC* (Fig. 21.1) and *FUBP1* gene mutations [4, 5].

In laboratory practice, PCR-based loss of heterozygosity studies and FISH are the most commonly used methods to detect 1p/19q loss. Other less common methods would include arrayed comparative genomic hybridization (aCGH) and multiplex ligation dependent probe amplification (MLPA). MLPA only requires standard PCR instrumentation and capillary gel electrophoreses; however, it does not require the patient's normal DNA sample. Both FISH and PCR methods are technically straightforward but have some advantages and disadvantages. Loss of heterozygosity analysis is a PCR-based method, the major disadvantage of which is the necessity of obtaining a normal blood sample. This can be complicated if blood is not collected at the time of surgery and the patient is discharged when the diagnostic dilemma arises. FISH scoring can be time-consuming, but can provide additional prognostic information. By FISH, tumor cells with 1p/19q present would show two signals for 1p and two for 19q and two control signals of 1q and 19p, respectively (Fig. 21.2a). A typical co-deletion pattern would have nuclei with two signals for 1q and one for 1p, and nuclei with two signals for 19p and only one for 19q (Fig. 21.2b). This is called an absolute deletion [6]. However, some tumors are characterized by polysomy, i.e., gains of either chromosome 1 or 19 or both, with concurrent loss of 1p/19q [7]. These are sometimes referred to as relative deletions. Nuclei will have four or more 1q signals and two or more 1p signals (Fig. 21.2c) or four or more 19p signals and two or more 19q signals. Several studies have confirmed that the additional data about polysomy and relative deletions

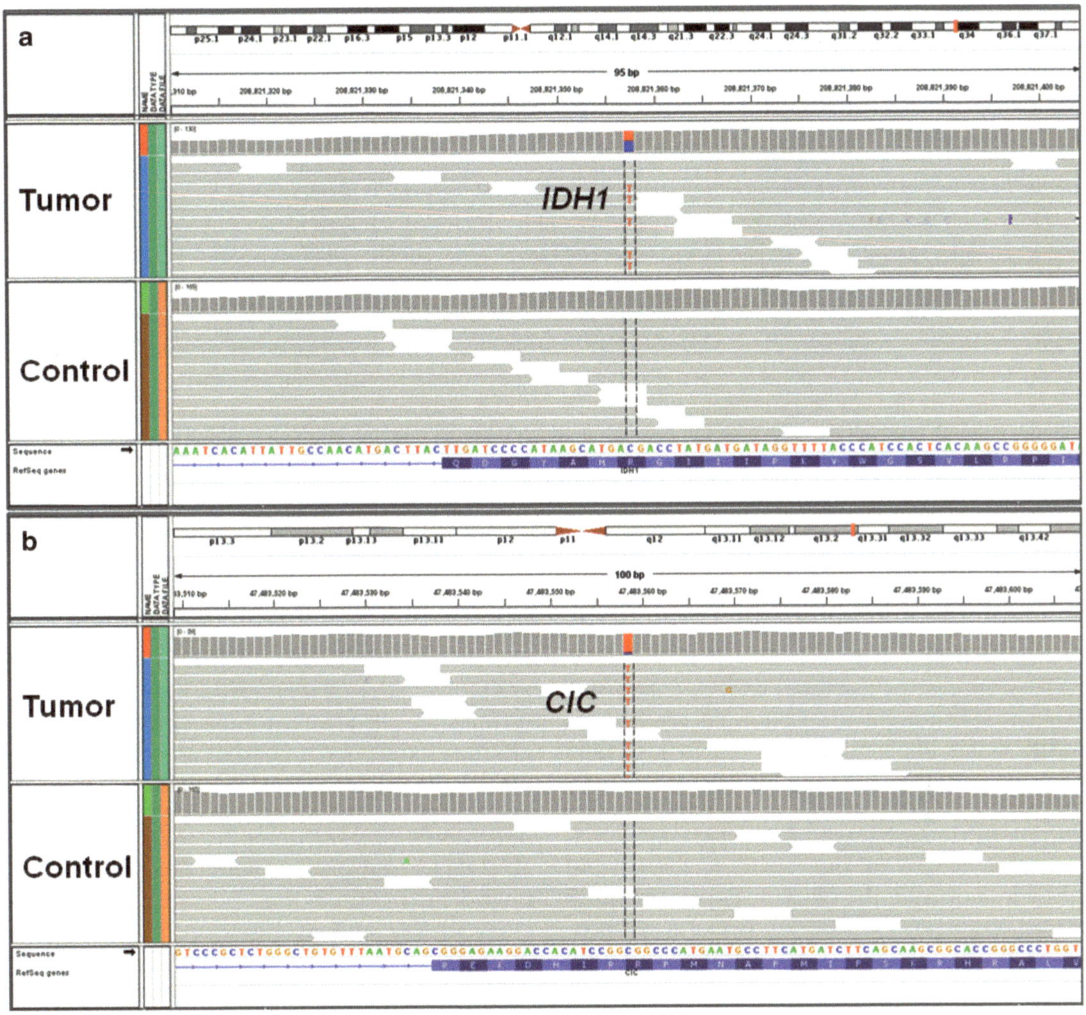

Figure 21-1 Identification of *IDH1* and *CIC* mutations in oligodendroglioma. In oligodendroglioma, whole-genome sequencing (Illumina platform) identifies concurrent mutations in the *IDH1* (c. 395C>T, p.R132H) and *CIC* (c.604C>T, p.R202W) genes. The majority of *IDH1* mutations in gliomas are p.R132H. The majority of mutations in oligodendrogliomas with 1p19q loss and *IDH1* or *IDH2* mutations occur within exons 5 and 20 of the *CIC* gene. The example shown, in the form of the Integrative Genomics Viewer (IGV, Broad Institute) browser view, is from exon 5, which is a highly conserved DNA- interacting HMG domain. Novel non-synonymous mutations can be identified by filtering against the normal sequence pileup and by comparison with the dbSNP database. Figure courtesy of Dr. Stephen Yip, BC Cancer Agency

seen by FISH but not by loss of heterozygosity analysis provide important prognostic information. Concurrent loss of 1p/19q and polysomy predicts early recurrence and poor survival [7, 8]. Therefore, for the clinical assessment of 1p/19q loss, FISH provides better predictive value compared to loss of heterozygosity.

Importantly, although tumors with a classic oligodendroglial morphology are much more likely to have 1p/19q co-deletion, in mixed tumors the molecular pattern is the same in both the oligodendroglial and the astrocytic component and therefore FISH can be performed anywhere in the neoplasm. Interestingly, association between a brain site

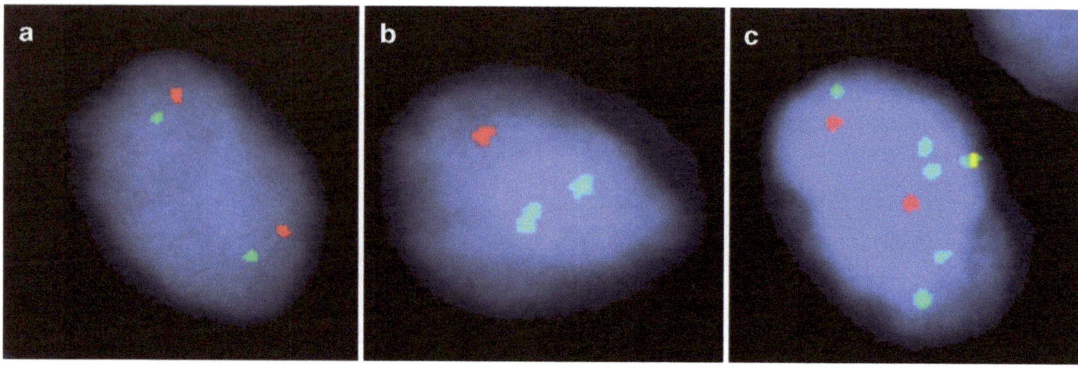

Figure 21-2 Testing for 1p19q in oligodendroglioma by FISH. 1p19q testing is the standard of care for tumors with a suspected oligodendroglial component and serves as a diagnostic and predictive marker. FISH analysis includes control probes for 1q or 19p (*green signal*) and probes of interest for 1p or 19q (*red signal*). *Panel* (**a**) reflects the presence of 1p with two *red* and two *green* signals. In *Panel* (**b**), a classic absolute deletion with loss of one copy of 1p is demonstrated, while two 1q signals remain. *Panel* (**c**) illustrates the so-called relative deletion, also known as superloss, with numerous hybridized control probes of 1q indicating polysomy and loss of ~50 % of the 1p signals. This tumor, by PCR, would appear to have loss of heterozygosity (LOH) similar to the tumor with the absolute deletion illustrated in *Panel* (**b**). However, tumors with superloss tend to be more aggressive

and 1p/19q status has been well documented. Frontal lobe oligodendrogliomas are significantly more likely to carry 1p/19q loss than gliomas from the temporal lobe with the same morphology. The biological reasons for that are currently unknown. Also, 1p/19q loss is exceedingly rare in the pediatric population and does not have the same diagnostic or prognostic importance. For these reasons, molecular results have to be interpreted together with histological and clinical data, and cannot replace a morphological diagnosis. 1p/19q status, WHO grade, morphological diagnosis (oligodendroglioma versus astrocytoma versus oligoastrocytoma), patient age, and performance score are independent statistically significant prognostic variables. Therefore, combining molecular and histopathological data to arrive at a diagnosis has a stronger predictive value than either of them alone.

IDH1

IDH1 mutations are associated with younger patients and with the diagnosis of secondary glioblastoma (sGBM), which develops from a preexisting low-grade glioma (LGG). The majority of diffuse astrocytomas of WHO grade II and anaplastic astrocytomas of WHO grade III carry *IDH1* mutations [9]. *IDH1* mutations are also common in oligodendrogliomas and particularly associated with 1p/19q loss. This suggests that both low-grade astrocytomas and oligodendrogliomas might develop from a common precursor. The overall *IDH1*mutation frequency in astrocytomas, oligodendrogliomas, and oligoastrocytomas is between 50 and 80 %, with no difference based on the WHO grade, including sGBM. In contrast, *IDH1* mutation is exceedingly rare in primary GBMs, which arise without a known low-grade precursor lesion. Because the frequency in other tumors is low, *IDH1* mutation is a very useful diagnostic marker. In the vast majority of cases *IDH1* mutation affects codon 132 and appears heterozygous, with the other gene copy remaining wild type. In gliomas, the most common mutation is p.R132H, which represents about ~90 % of mutations (Fig. 21.1), followed by p.R132C (4 %), p.R132S, and p.R132G in approximately 1.5 % of cases, each. Mutations in the *IDH2* gene are present in approximately 3 % of gliomas. Gliomas with an *IDH1* mutation have a significantly better outcome than wild-type tumors, independent on type and grade. In addition to strong prognostic value, *IDH1* is also useful for diagnosis, where it can help in several ways. First, it distinguishes oligodendrogliomas from other similar looking neoplasms such as neurocytoma, clear cell ependymoma, pilocytic astrocytoma with a prominent oligodendroglial-like component, and dysembryoplastic neuroepithelial tumor (DNT), which all lack *IDH1* mutations. Second, it is very useful in distinguishing between diffuse glioma and reactive gliosis. The existence of a robust antibody to the mutant IDH1 p.R132H protein allows the establishment of this diagnosis even in very small samples and in samples with few infiltrating tumor cells [10]. Although the antibody is specific for only one type of mutated protein, it can identify about 90 % of mutated cases due to the predominance of the p.R132H mutation. Sequencing can be performed to evaluate the remaining cases if clinically warranted.

EGFR and Other Receptor Tyrosine Kinases

Amplifications of receptor tyrosine kinase (RTK) genes play a crucial role in tumorigenesis of malignant glioma and are considered major drivers of tumor growth. Up to 50 % of high grade gliomas have amplification of a RTK gene, most commonly *EGFR, KIT, VEGFR2, PDGFRA*, and *MET* [11–14]. Compared to *TP53* mutation, *IDH1* mutation and 1p/19q loss, the high-level amplification of RTK genes is a relatively late event in the tumorigenesis of glioma. Typically, only one RTK will have a high level amplification in any given tumor. *EGFR* is the most commonly amplified RTK in adult GBMs (~40 %), while in children *PDGFRA* amplification seems to be the most common with frequencies ranging from 5 to 12 %. Up to 30 % of diffuse intrinsic pontine gliomas in children have *PDGFRA* amplification. *EGFR* amplification is a hallmark of primary GBM and is more common in older patients, whereas secondary GBMs that develop from lower-grade gliomas are much less likely to have *EGFR* amplification. In addition to amplification, however, ~50 % of all *EGFR* amplified GBMs include a truncated mutant variant, *EGFRvIII*, with constitutively upregulated tyrosine kinase activity. Not surprisingly, given the prevalence of *EGFR* alterations, significant, but thus far unsuccessful efforts, have been focused on development of therapies targeting *EGFR*. In addition, amplification of *EGFR* has unclear significance as a prognostic marker. Some studies found no association with survival, others reported a negative impact, and some suggested a favorable impact on patient survival. Interestingly, it has been reported that the level of EGFR amplification influences response to therapy [15].

There are several reasons to perform *EGFR* testing in clinical practice [6]. From a molecular diagnostic perspective, *EGFR* amplification is pathognomonic of GBM when seen in the context of a brain tumor and due to the high level of amplification cells can be easily identified, even in samples with low cellularity. *EGFR* amplified cells have been shown to be associated with the invasive edge of gliomas. *EGFR* amplification is also almost mutually exclusive with 1p/19q deletion and *IDH1/2* mutations and is therefore helpful in the differential diagnosis of anaplastic oligodendroglioma versus the small cell variant of GBM, because there is significant morphological overlap between these clinically very different glioma subtypes. *EGFR* amplification is encountered in most small cell GBMs and tumors lack 1p/19q co-deletion, while anaplastic oligodendrogliomas often show 1p19q loss but never *EGFR* amplification [6]. Despite numerous studies, it remains primarily a diagnostic marker, as this alteration does not seem to provide independent prognostic information. Also, despite numerous clinical trials, it does not seem to predict response to EGFR inhibitors or antibodies. The role of other RTKs in clinical diagnostics is less clear. Although associated with distinct molecular subtypes of GBM, *PDGFRA* amplifications can be seen also in lower grade gliomas and *MET* amplifications are present in a small subset of GBMs and LGGs. For clinical practice, one can consider performing FISH or using a DNA array such as aCGH (Fig. 21.3). The advantage of aCGH would be to evaluate multiple targets with one assay. However, whole-genome arrays might not be able to detect minor clones with amplifications and might be less successful with samples in which tumor cell density is low.

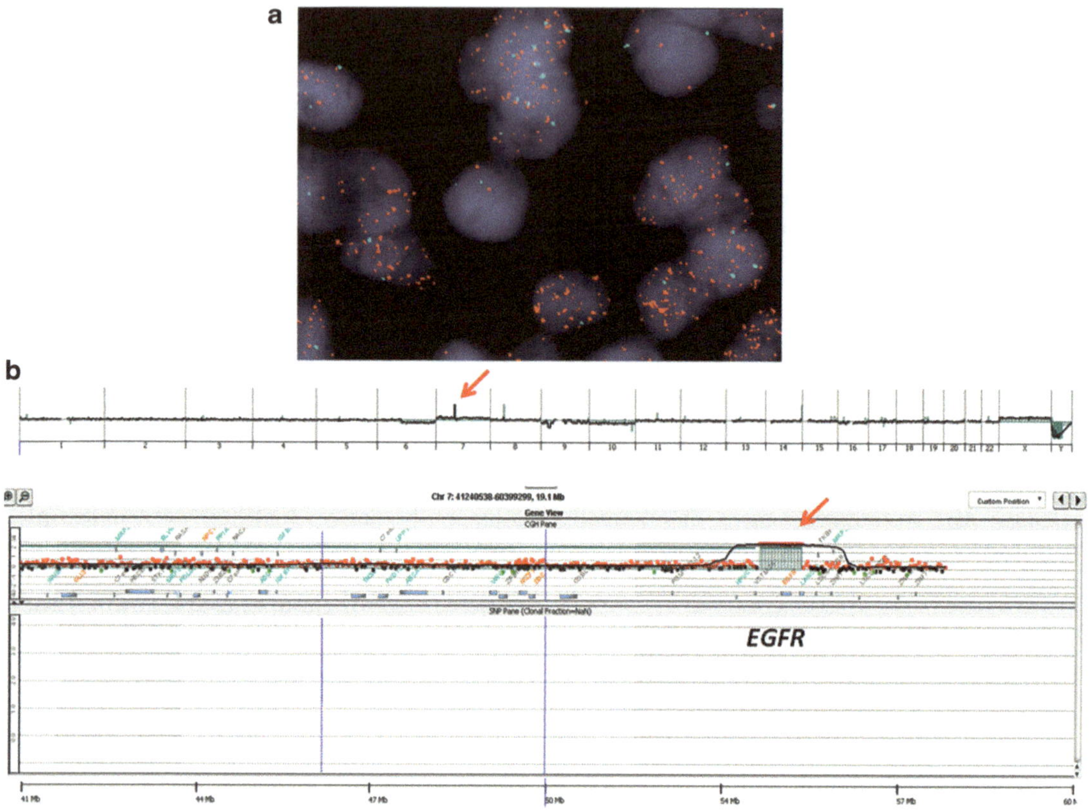

Figure 21-3 *EGFR* amplification in GBM. Amplification of RTKs is one the most common driver mutations in tumorigenesis of GBM. *EGFR* is the most commonly amplified gene in adults. High level amplification can be easily detected by FISH (*Panel* (**a**), *red*: *EGFR*, *blue*: control probe for chromosome 7, nuclei stained by DAPI) or by aCGH (**b**). Whereas the overall genome is relatively stable, chromosome 7 shows a distinct peak of focal amplification (*Panel* (**b**), *top*, *red arrow*). A gain of signal from probes covering *EGFR* gene is seen in the magnified view of the amplified region on chromosome 7 (*Panel* (**b**), *bottom*, *red arrow*)

BRAF

BRAF is a member of the serine/threonine protein kinases family. RAF kinases are part of the MAPK cascade, which regulates cell proliferation, differentiation, and apoptosis and therefore plays a wide variety of biological roles in a tumor cell. Activating mutations in *BRAF* are common in variety of solid tumors including papillary thyroid carcinoma, melanoma, adenocarcinoma of the colon, and pediatric LGGs. In addition to point mutations, pilocytic astrocytomas have a unique mechanism of *BRAF* activation via single-copy gain of *BRAF*, which results from a tandem repeat leading to fusion product *BRAF-KIAA1549* [6]. This is a molecular hallmark of pilocytic astrocytoma and is identified in approximately 80 % of pilocytic tumors in the cerebellum. Several other breakpoint variants have been described such as *KIAA1549* exon16–*BRAF* exon 9, *KIAA1549* exon 15–*BRAF* exon 9, *KIAA1549* exon19–*BRAF* exon 9, *KIAA1549* exon18–*BRAF* exon 10, and *KIAA1549* exon 16–*BRAF* exon 11. The resulting fusion proteins are highly tumorigenic. However, the constitutive activation of *BRAF* can also lead to senescence, particularly in slow-growing neoplasms.

Another mechanism of MAPK pathway activation is a tandem duplication at 3p25 leading to an in-frame fusion between *SRGAP3* and *RAF1*, which has high sequence homology with *BRAF*. The fusion gene retains the activation segment and kinase domain but lacks the inhibitory domain, resulting in constitutive upregulation. Constitutive activation of the MAPK pathway is seen in *NF1* and non-*NF1* associated pilocytic and pilomyxoid astrocytomas arising in the cerebellum. Duplication of the *BRAF* locus at 7q34 was identified in more than half of these tumors. In contrast, *BRAF* gene fusion is unusual in diffuse gliomas, which often contain the *BRAF* p.V600E point mutation, present in 25 % of pediatric astrocytomas. Testing for *BRAF* should therefore include testing for both point mutations and gene fusions, based on the location of the tumor [16]. Overall, the data are unclear about the value for diagnosis and for use as a clinical biomarker. One possibility would be to use it together with *IDH1/2* mutation analyses. *BRAF* rearrangements are absent in gangliogliomas and infiltrative gliomas and uncommon in supratentorial pilocytic astrocytoma. Unfortunately, *BRAF* rearrangement does not distinguish between a pilocytic astrocytoma and a more aggressive pilomyxoid astrocytoma and does not seem to correlate with outcome. Inhibition of *BRAF* activity using targeted therapy might be a possibility in unresectable or disseminated tumors, if these inhibitors can penetrate the blood–brain barrier.

MGMT

The O6-methylguanine-DNA methyltransferase (*MGMT*) gene has been one of the most commonly tested molecular markers in neuropathology and neuro-oncology because of the clear association between *MGMT* promoter hypermethylation and an increased response to alkylating agents [17]. GBM patients with methylated *MGMT* promoter have a significant survival benefit with temozolomide treatment and *MGMT* promoter hypermethylation is one of the strongest prognostic factors for patients with newly diagnosed GBM, including elderly patients [18, 19]. Patients with a hypermethylated *MGMT* promoter who are treated with concomitant and adjuvant temozolomide and radiotherapy had survival rates of ~50 and 15 % at 2 and 5 years, respectively. However, 2- and 5-year survival rates in patients treated with radiotherapy alone were only ~25 and ~5 %, respectively. In patients with GBM lacking *MGMT* promoter hypermethylation 2- and 5-year survival rates were 15 and 8 % when treated with combined radiochemotherapy, but 2 and 0 % when treated with radiotherapy alone. Although the response to temozolomide is best in the methylation-positive group, one might argue that there appears to be some benefit in patients with non-methylated tumors. In the pediatric GBM population, the data are less clear.

The frequency of *MGMT* promoter hypermethylation in the glioma literature varies widely, ranging from ~30 to 70 % in GBM. This is due to technical aspects of the testing, but also to tumor heterogeneity, necrosis and normal tissue contamination. Overall, *MGMT* promoter hypermethylation was observed in ~50–80 % of anaplastic gliomas WHO grade III, and 40–90 % of the diffuse gliomas WHO grade II. The *MGMT* gene is located on 10q26 has a CpG-rich region of 763 bp with 98 CpG sites within the first exon. A promoter and an enhancer region are also located within the CpG island. CpG sites are not methylated in the normal tissue. In tumors, however, the cytosine in CpG sites can be methylated, which leads to altered chromatin structure and prevents binding of transcription factors. The result of this is silencing of the gene expression. *MGMT* is a DNA repair protein that in normal tissue catalyzes the transfer of a methyl group from the O6-position of a guanine DNA nucleotide to a cytosine residue. This is a one-way process, and alkylated MGMT is degraded. In tumors for which alkylating chemotherapy is used, such as temozolomide in malignant gliomas, this process leads to the binding of an alkyl group to the O6-position of guanine, which induces DNA mismatching and DNA-double-strand breakage, resulting in apoptosis. A normally functioning MGMT protein neutralizes the lethal effects of alkylating agents by repairing DNA damage. When *MGMT* is silenced by hypermethylation of the promoter, however, reduced *MGMT* expression is thought to result in tumor cells not being able to repair DNA damage. This enhances the cytotoxic effects of temozolomide. Interestingly, patients with a hypermethylated *MGMT* promoter exhibited a survival benefit even when treated with radiotherapy alone. Therefore, it is possible that MGMT also plays a role in radiotherapy induced DNA damage repair. Another possibility is that *MGMT* methylation is an overall marker of genome methylation status in the tumor and that other DNA repair genes are silenced by promoter hypermethylation as well. Because other genetic alterations associated with a favorable prognosis, such as 1p/19q loss and *IDH1* mutation, often coexist with *MGMT* promoter hypermethylation, the contribution of each remains to be determined. Given the observed effect of temozolomide, even in the nonmethylated subgroup, and the lack of other options, it remains the first drug of choice regardless of *MGMT* status. *MGMT* methylation testing is frequently requested in clinical practice, but the impact of this testing on clinical management is unclear because the therapy remains similar, regardless of the result. *MGMT* testing, however, plays an important role in clinical trials to properly stratify patients. There are several assays that can be used for testing. The most frequently utilized is methylation specific PCR and real-time methylation specific PCR [20]. Other possibilities include methylation specific pyrosequencing and methylation specific MLPA. Contra-intuitively, expression of the protein by IHC does not correlate well with the DNA results and cannot be recommended for clinical practice [20].

INI1

Atypical teratoid/rhabdoid tumor (AT/RT) is characterized by a combination of the presence of a primitive embryonal component and mesenchymal and epithelial components. Rhabdoid cells are not always detectable at the time of diagnosis and the tumor can closely mimic medulloblastoma. The hallmark of AT/RT is a loss of chromosome arm 22q, which carries the *SMARCB1* gene, also known as *INI1* or *hSNF5*, at 22q11. Altered by deletion or mutation, loss of *INI1* is a defining molecular event in this tumor. Families with germ-line mutations of the *SMARCB1* gene have an inherited disposition to rhabdoid tumors everywhere in the body, including the brain AT/RT. Therefore, family members of children with these tumors should be tested for mutations to assess the potential risk. Presence of a reliable antibody for IHC of SMARCB1/INI1 is widely used in clinical diagnostics and has dramatically decreased the number of misdiagnosed tumors [21]. The current standard of care includes testing for INI1 by IHC in all medulloblastomas, primitive neuroectodermal tumors (PNETs) and choroid plexus carcinomas to avoid misdiagnosis. Patients with AT/RTs have an extremely poor outcome, although regimens using high-dose chemotherapy suggest potential benefit. In addition to AT/RT, mosaic loss of *INI1* was described in neurofibromatosis type 2 (NF2) associated schwannomas and in schwannomatosis associated schwannomas, but is rarely seen in sporadic schwannomas. This suggests a role for *INI1* in syndromes associated with multiple schwannomas. IHC is available and is the test method of choice. Sequencing can be performed in some cases or for genetic testing of family members, if a familial syndrome is suspected.

PTEN

PTEN is a tumor suppressor and its loss is common in gliomagenesis. Inactivation of *PTEN* either by mutation or deletion is a frequent feature in many high grade gliomas and leads to upregulation of the AKT pathway. Up to 80 % of all GBMs show a loss of 10q23 containing *PTEN* and up to 40 % of primary GBMs will carry *PTEN* mutations. In the small cell variant of GBM, 10q is almost always lost and this, together with *EGFR* amplification and 1p/19q preservation, comprises a useful molecular panel to diagnose GBM and distinguish it from an anaplastic oligodendroglioma. *PTEN* loss is present in both primary and secondary GBMs and is associated with shorter survival in the pediatric population, making it an interesting diagnostic as well as prognostic marker. It does not, however, seem to be a prognostic marker in adult GBM. *PTEN* loss is most often hemizygous, and testing can be easily performed using FISH [6] or aCGH.

CDKN2A

CDKN2A is located on 9p21 and encodes the p16 protein, which is a key inhibitor of the cell cycle in Rb pathway signaling.

This pathway is one of the most commonly affected pathways in cancers, including gliomas. In gliomas, oligodendrogliomas, and astrocytomas, *CDKN2A* is usually lost via homozygous deletion and associated with high grade tumors and decreased survival. Interestingly, a small subset of pilocytic astrocytomas also demonstrates loss of *CDKN2A*. However, this finding is uncommon in pediatric LGGs overall, as discussed in the whole-genome section below. Similar to *PTEN*, testing can be performed using FISH [6] or aCGH.

Whole-Genome Evaluation of Brain Tumors

Medulloblastoma

Medulloblastoma is the most common malignant brain tumors of childhood. Medulloblastomas, by definition, arise in the posterior fossa, while similarly looking tumors are called PNETs elsewhere in the brain. The WHO classification recognizes several subtypes of medulloblastoma based on morphology: Classic, desmoplastic/nodular, medulloblastoma with extensive nodularity, and large cell/anaplastic medulloblastoma. These subtypes can be associated with age (such as medulloblastoma with extensive nodularity in infants) or with better versus poor prognosis (such as desmoplastic medulloblastoma and large cell/anaplastic medulloblastoma, respectively). Although several of the morphologic subtypes are no longer regarded as separate entities by the WHO, the molecular classification is playing an increasingly important role in the classification of this disease.

Medulloblastoma is a prototypic brain tumor in which molecular tools have provided better understanding of the disease biology by classifying a relatively uniform appearing neoplasm into distinct biological entities and identifying potential therapeutic targets. Association of medulloblastoma with rare entities such as Turcot and Gorlin syndromes not only suggested the role of Wnt and Hedgehog signaling but also suggested that medulloblastomas can have a relatively simple "one pathway" oncogenesis. The real extent of the molecular diversity in medulloblastoma was first revealed using expression profiling. Expression profile studies first identified medulloblastoma, PNET and AT/RT as distinct molecular entities and later studies divided medulloblastomas into 4–6 distinct subgroups, depending on the study. There are several reports of distinct molecular pathways and distinct groups of medulloblastoma [22–25]. Whereas all groups identified the Sonic hedgehog (Shh) and Wnt subgroups as relatively distinct, separation of non-Shh non-Wnt tumors varied based on classifications schemes. The current consensus molecular classification distinguishes four types: Wnt, Shh, Group 3, and Group 4 (Fig. 21.4) [26, 27].

Sequencing of the genes involved in the oncogenic pathways identified Wnt activation in sporadic medulloblastomas via mutations in *CTNNB1*, *AXIN*, and *APC* and mutations activating the Shh pathway including *PTCH1*, *SUFU*, and *SMO*. Whole-genome sequencing studies have revealed novel genes that are mutated in medulloblastoma including *MLL2* in Shh and Wnt groups, *MLL3* in Group 3 and 4, *SMARCA4* in Wnt and Group 3, *DDX3X* in Wnt, *LDB1* and *BCOR* in Shh, and many others. Among the most frequently mutated genes in medulloblastoma are, not surprisingly, *CTNNB1* and *PTCH1*, but also *DDX3X*, *MLL2*, *SMARCA4*, and *KDM6A*. In addition to mutations, chromosomal and smaller copy number changes were identified early on. Isochromosome 17q, formed as a result of loss of chromosome 17p and gain of 17q, was found in approximately 30 % of medulloblastomas and large chromosomal or small copy number changes have been associated with certain subtypes. Examples are loss of chromosome 6 in the Wnt group, loss of long arm of chromosome 9 in Shh group, and amplifications of *MYC* or *MYCN* loci associated with large cell/anaplastic medulloblastomas subtypes and poor outcome. Later, gains and amplifications of *OTX2* and *PVT2* were revealed in pathogenesis of Group 3, *SNCAIP* and *CDK6* in Group 4, and *GLI2* in Shh medulloblastoma. The amount of genomic data is significantly larger than the amount of functional data that would confirm the role of many of these changes; however, the consensus about the four main groups of medulloblastoma remains a practical framework for further studies.

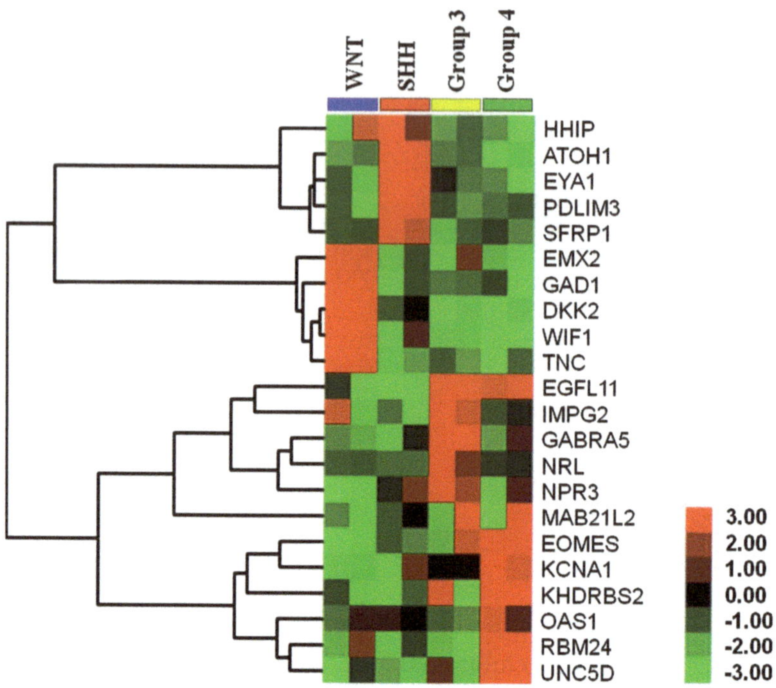

Figure 21-4 Expression profiling of medulloblastoma. Whole-genome expression profiling pioneered the molecular classification of medulloblastoma. Focused panels of selected genes specific for each subgroup can be created to classify tumors [47]. Each tumor profile is represented by a column and each gene expression level by a line. Color coding represents the level of increased (*red*) or decreased (*green*) expression compared to control genes. Two tumors of each subgroup are shown to cluster together. Shh and Wnt tumors are very distinct, but there is some overlap between Group 3 and 4 expression, even with this selected group of genes. Nevertheless, tumors can still be easily categorized. Immunohistochemistry (IHC) for SFRP1, DKK2, NPR3, and KCNA1 has been proposed as a practical panel for classification in clinical practice. Figure courtesy of Joanna Triscott and Dr. Sandra E. Dunn, University of British Columbia

Wnt Group

The importance of recognizing the Wnt group is the very good long-term prognosis in comparison to other groups. Whereas medulloblastoma is more common in males, in Wnt medulloblastomas the sex ratio is approximately 1:1. Wnt medulloblastomas occur at all ages, but are rare in infants and are usually not disseminated at the time of diagnosis. Germ-line mutations of the *APC* gene in the Wnt pathway are associated with Turcot syndrome. Unfortunately, this subtype is also the least common and represents only ~10 % of all medulloblastomas. With long-term survival rates exceeding 90 %, many patients suffer from the long-term therapy associated complications including cognitive decline, endocrine insufficiencies, growth problems and secondary neoplasms rather than from medulloblastoma recurrence. While in other subtypes molecular studies might identify new therapeutic targets, in Wnt patients, the goal instead might be to first modify unnecessary toxic treatments to decrease later morbidities and mortality associated with the treatment. The Wnt medulloblastoma genome is relatively stable, with only a few changes other than monosomy 6. Having said that, tumors with a clear Wnt expression profile and without monosomy 6 also have been described. Furthermore, overexpression of genes in the Wnt pathway has been detected in Shh and Group 3 medulloblastomas, as well. Although most Wnt tumors show classic morphology, a Wnt transcriptional signature is associated with excellent prognosis even in tumors with anaplastic/large cell features.

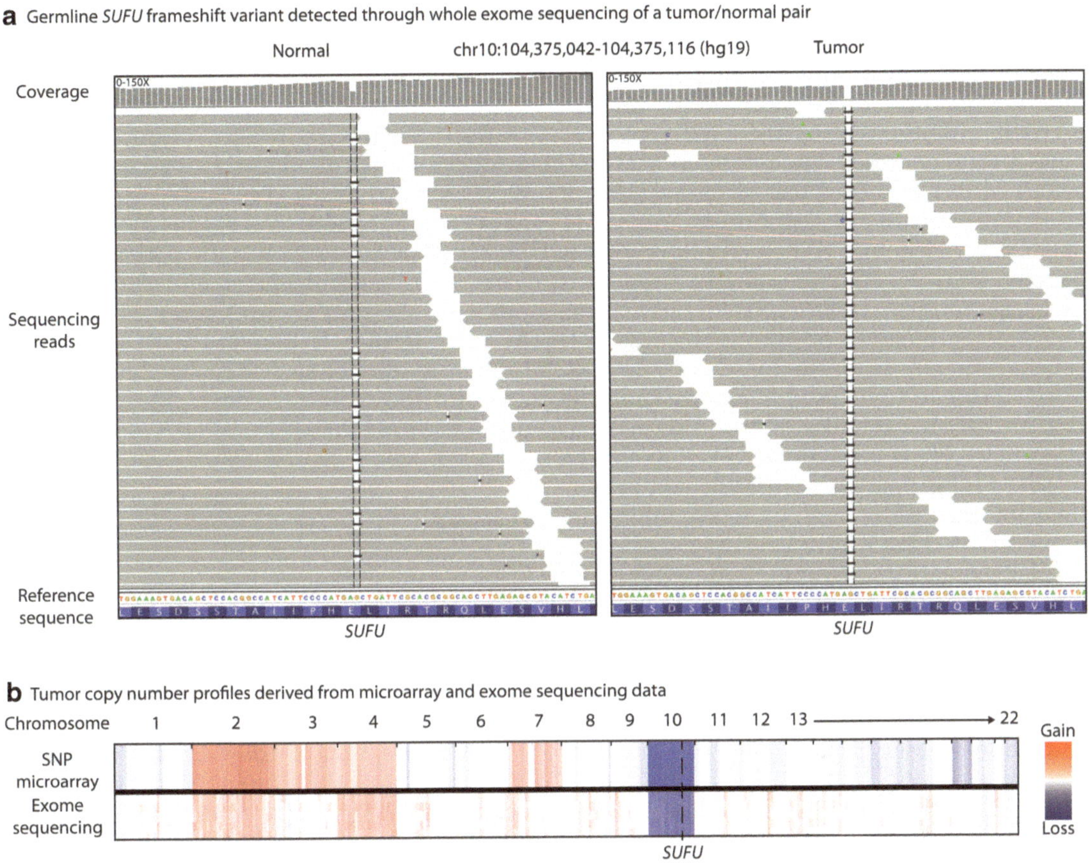

Figure 21-5 Complex genomic features in medulloblastomas with germ-line mutation and somatic loss. (**a**) A patient with a germ-line heterozygous loss-of-function *SUFU* mutation (in the specimen labeled Normal, shown as *empty spots* in some reads) developed a tumor via a somatic loss of chromosome 10, which resulted in complete loss of function of *SUFU* (completely empty column in all reads of the Tumor specimen). The resulting medulloblastoma is therefore classified to be of Shh subtype. (**b**) The overlay of SNP microarray and exome sequencing data reveals the copy number profile of the tumor with loss of chromosome 10. Of note, both a SNP array and exome sequencing can provide copy number data in addition to information about point mutations and indels. Figure courtesy of Dr. Trevor J. Pugh, Broad Institute of MIT and Harvard

Such tumors, however, are rare in the Wnt subgroup. Several methods have been proposed to diagnose Wnt medulloblastoma. It remains to be validated whether IHC for the DKK1 or CTNNB1 proteins, cytogenetic testing for monosomy 6, or a transcriptional signature should be used for clinical testing.

Shh Group

The Shh medulloblastomas are characterized by aberrations involving the Shh signaling pathway, which drive tumor initiation and progression. Germ-line mutations in the Shh receptor *PTCH1* lead to Gorlin syndrome, which includes in its phenotype a predisposition to medulloblastoma. Infantile medulloblastoma is associated with germ-line mutations of the Shh inhibitor *SUFU* (Fig. 21.5). Shh medulloblastomas have a bimodal age distribution with high frequency in infants 0–3 years old and in adults. There is no gender predominance in Shh medulloblastomas. Nodular/desmoplastic medulloblastomas almost invariably belong to the Shh subgroup. However, up to 50 % of Shh subgroup medulloblastomas are not nodular/desmoplastic and therefore histology alone is not sufficient to identify them. Overall, the prognosis of Shh medulloblastoma is similar to Group 4. Patients have a better prognosis than those with Group 3 medulloblastoma, but worse than those with Wnt tumors. Somatic mutations along the entire pathway have been identified, including in *PTCH1*, *SMO*, and *SUFU*, as well as amplifications of their downstream transcription factors *GLI1* and *GLI2*. Although Shh medulloblastomas have been successfully identified on the basis of transcriptional profiling, some have proposed a combination of immunohistochemical stains such as those for the SFRP1 or GAB1 proteins to be used in clinical practice. On the DNA level, loss of chromosome 9q, where *PTCH1* is located (9q22), is exclusive for the Shh group. These tumors can potentially be targeted by small molecule inhibitors against *SMO* [28]. Unfortunately the effect is short-lived and tumors rapidly develop resistance mutations [29]. Furthermore, Shh medulloblastomas that carry aberrations at the more downstream parts of the pathway, such as amplifications of *GLI1*, *GLI2*, and *MYCN* genes would be inherently resistant to such inhibitors.

Group 3

Overall, Group 3 and 4 medulloblastomas are characterized by more overlapping features and are less distinct than Wnt and Shh medulloblastomas. Also, both groups exhibit a higher number of genetic changes, particularly more complex DNA rearrangements. Group 3 tumors occur more commonly in males than females, and are found in infants and children, but are almost never observed in adults. They have a high incidence of large cell/anaplastic histology although many of them are classic medulloblastomas. In addition, they are very frequently metastatic at the time of diagnosis. The Group 3 transcriptional profile is Photoreceptor/GABAergic. A true hallmark of Group 3 tumors, however, is *MYC* overexpression/amplification, to the point that some have proposed to rename them *MYC* Group, instead. Immunohistochemical positivity for NPR3 has been suggested as a Group 3 marker and has been associated with aggressive disease. Group 3 tumors often show gains of chromosomes 1q, 7, and 17q and/or loss of chromosomes 11, 17p, 5q, 10q, and 16q. Gains of 18q are seen in both Groups 3 and 4. Amplification and overexpression of the medulloblastoma oncogene *OTX2* appears to be restricted to Group 3 and Group 4 tumors. Group 3 can be further stratified into Group 3a, which includes all medulloblastomas with *MYC* amplifications, and contains most of the high risk patients. Group 3b patients do not have *MYC* amplifications, and their clinical outcome is similar to Group 4 patients. Because Group 3 patients have the worst prognosis, they have, arguably, the highest need for the most aggressive therapy and for novel targeted therapies.

Group 4

Group 4 medulloblastomas have classic histology and are more prominent in males. Presence of the isochromosome 17q is a hallmark of Group 4, although it can be also seen in a minority of Group 3 medulloblastomas. Furthermore, isolated 17p deletion is seen in both Groups 3 and Group 4, but not in Wnt or Shh medulloblastomas. Recently it has been found that 17p harbors *CTDNEP1*, which is a novel candidate gene in the pathogenesis of medulloblastoma. The other interesting cytogenetic change is common loss of the X chromosome in females with Group 4 medulloblastoma, which is seen in ~80 % of tumors. This is particularly interesting considering that the male–female ratio in Group 4 medulloblastoma is almost 3:1 and therefore suggests that the X chromosome contains potential tumor suppressor genes such as *KDM6A*, as discussed below. *MYCN* and *CDK6* are often amplified in Group 4 medulloblastomas, while they are usually not affected in Group 3 tumors. Tumors are frequently metastatic at presentation and the prognosis is intermediate, similar to non-infant Shh medulloblastomas. Compared to other groups, their pathogenesis is the least understood and there is no consensus about the driver mutation.

Group 4 medulloblastomas are reliably identifiable via their transcriptional profile and have a Neuronal differentiation/Glutamatergic profile. KCNA1 has been suggested as an immunohistochemical marker for this group. However, recent whole-genome studies identified several mutated genes in medulloblastoma that normally participate in histone modifications. *KDM6A* is an interesting example because it is altered by a nonsense mutation and can be considered a novel tumor suppressor gene. *KDM6A*, *BCOR*, *DDX3X*, and other genes mutated in medulloblastoma are located on Xp, which is commonly lost in females with this type of medulloblastoma. In addition to *KDM6A*, other chromatin remodeling genes such as *ZMYM3* and *CHD7* can be mutated in Group 4, suggesting that this group might be defined by mutations in genes responsible for epigenetic modifications [30].

Molecular methods thus far do not play a significant role in the diagnosis of medulloblastoma and the main role of these studies would be prognostic and, hopefully, someday predictive. Whereas the overall biology indicated above is suggestive of a high level of complexity, the actual list of prognostic molecular factors associated with poor or improved survival in medulloblastoma patients is surprisingly short. Markers of good prognosis include increased TrkC mRNA expression and nuclear staining of beta-catenin or positivity for *DKK1* indicating Wnt pathway activation. Markers indicating worse outcomes include *MYC* amplification and/or overexpression, 17p loss and i17q formation, as well as strong p53 immunoreactivity suggestive of underlying *TP53* mutation. A diagnostic panel for clinical practice has not yet been established. Although a combination of immunohistochemical stains has been proposed, validation comparing it with the expression profile is necessary. IHC can be combined with a few FISH targets, most importantly *MYC*, but possibly also *MYCN* and 9q22 for *PTCH1*. A relatively simple pathology stratification for clinical practice can combine a patient's age, medulloblastoma morphology, and a combination of a few immunohistochemical stains with FISH to identify Wnt and Shh pathway status, and *MYC* anomalies. Another possibility is to combine IHC with aCGH or to replace IHC altogether and perform a targeted expression profiling assay using a subset of genes. Lastly, one can perform a focused expression profile array to distinguish these tumors (Fig. 21.4). To date, however, these remain research tools.

Gliomas

Glioblastoma

Glioblastoma (GBM) is the most common malignant brain tumor of adults. It typically develops from astrocytes and mostly arises de novo without a previous low-grade precursor (primary GBM, pGBM). Approximately 10 % of GBM arise from a preexisting LGG and are termed secondary GBM (sGBM). Survival of patients with sGBM is much longer than that of patients with primary GBM. These two types arise along different molecular pathways and have different expression profiles.

A variety of studies have attempted to identify individual genes as well as signaling pathways by combining expression profiling and structural DNA data to identify prognostic and possibly predictive markers. The most commonly affected genes and pathways in GBM include *EGFR* and other receptor tyrosine kinases, the *PI3K/PTEN/AKT* pathway and the *TP53/MDM2/p14* pathway. The most common focal DNA changes are amplification of *EGFR*, amplification of 4q12 which contains *PDGFRA*, *KIT* and *VEGFR2*, and deletion of the *CDKN2A* gene. The large chromosomal variants include loss of 10q, 19q, 22q, and 1p. Pediatric GBMs also commonly show microsatellite instability due to DNA mismatch repair defects, which is uncommon in adult tumors, and have a different spectrum of copy number changes such as gain of 1q, 3q, and 16p as well as loss of 8q and 17p. Based on expression profiling, gliomas were classified into three main groups: Proneural, Mesenchymal, and Proliferative, based on analysis of gene ontology. This classification demonstrated prognostic value and has been confirmed by several studies, including The Cancer Genome Atlas (TCGA) project. Discovery of novel mutations, such as those in the *IDH1*gene, in high grade gliomas and a combination of expression profiling classification studies and DNA alterations led to subsequent identification of additional subtypes of GBM and a more recent classification into Proneural, Neural, Mesenchymal, and Classical subtypes [31]. Each subtype is defined by a combination of expression parameters as well as DNA aberrations such as *EGFR* amplification, *NF1* loss and *PDGFRA/IDH1* alterations.

Classical GBM

High-level *EGFR* amplification, often accompanied by *EGFRvIII* mutation and paired with *EGFR* overexpression, is a hallmark molecular change in Classical GBMs. Another typical finding is loss of chromosome 10. Classical GBMs also lack mutations in *TP53*, which is one of the most commonly mutated genes in GBM. Homozygous deletion of the 9p21.3 locus containing the *CDKN2A* gene that encodes p16INK4A and p14ARF is another frequent event in Classical GBM. Loss of *CDKN2A* is mutually exclusive with loss of other RB pathway genes, such as *RB1*, *CDK4*, or CCDN2, suggesting that the RB pathway is almost exclusively affected through homozygous deletion of *CDKN2A*. Expression profiling identified Notch and Shh signaling pathways to be overexpressed in the Classical group.

Mesenchymal GBM

The molecular hallmark of the Mesenchymal subgroup is a heterozygous deletion of the *NF1* containing region 17q11.2. The majority of these tumors have decreased *NF1* expression. In addition to the heterozygous deletion, *NF1* mutations are also common in this subgroup. Expression profiling indicated upregulation of mesenchymal markers, including *YKL40* and *MET*, as well as genes in the tumor necrosis factor super family pathway and in the NF-kB pathway, both possibly due to high levels of necrosis and inflammation in these tumors.

Proneural GBM

The Proneural GBMs are characterized by aberrations of *PDGFRA*, either by amplification or point mutations, and by mutations in *IDH1*. Similar to *EGFR* amplifications in the Classical group, focal amplifications of the locus at 4q12 that harbors *PDGFRA*, *VEGFR2*, and *KIT* are present in all subtypes of GBM but are observed much more frequently in the Proneural group. Concurrent *PDGFRA* amplification accompanied by high levels of *PDGFRA* expression is almost exclusive for Proneural GBMs. *IDH1* mutations are present in Proneural GBMs which lack *PDGFRA* aberrations. Another common genetic event in this group is loss of *TP53* function. As evidenced by expression profiling, overexpression of oligodendrocytic developmental genes such as *SOX*, *DCX*, *DLL3*, *ASCL1*, *NKX2-2*, and *OLIG2*, and decreased *CDKN1A* expression can be present. High expression of *OLIG2* was previously shown to downregulate the tumor suppressor *CDKN1A*, leading to increased proliferation. Similar to *IDH1* mutations, mutations in *PIK3CA/PIK3R1* were also identified in Proneural tumors without *PDGFRA* abnormalities.

Neural GBM

The Neural subtype seems to have the least identifiable molecular features including aberrations of *PTEN*, *TP53*, *EGFR*, *NF1*, and *ERBB2*, as well as homozygous deletions of *CDKN2A*. None of these, however, dominate. Expression profiling showed upregulation of neuronal markers, including *SLC12A5*, *NEFL*, *GABRA1*, and *SYT1*.

The subclasses seem to differ in their response to aggressive therapy. The greatest response is typical among Classic and Mesenchymal GBMs and no response is common in the Proneural group. Due to the lack of effective targeted therapy against either group it is unclear whether classification has added prognostic or predictive value to currently performed clinical tests. There also seems to be an association between glioma grade and molecular subtype. While GBMs are composed of a mix of subtypes, grade II and grade III diffuse gliomas are almost exclusively Proneural. Furthermore, sGBM are Proneural. These tumors are diagnosed at a younger age, have a high rate of *IDH1* and *TP53* mutations and lower rates of *EGFR* amplification and chromosome 10 loss. Pediatric GBMs are characterized by two distinct mutations of histone H3.3, each defining an epigenetic subgroup of GBM with a distinct global methylation pattern. These mutations were also mutually exclusive with IDH1 mutations, which characterizes a third mutation-defined subgroup of pediatric GBM [32].

Low-Grade Gliomas and Diffuse Intrinsic Pontine Gliomas in Children

LGGs are the most common brain tumors of childhood and in children seem to display different aberrations than the LGGs that are precursors of GBM in adults [33]. Diffuse intrinsic pontine gliomas (DIPG) are a distinct subtype of gliomas, and arise almost exclusively in children [34]. They are typified by aggressive growth and are almost entirely resistant to current therapies. Whole-genome studies have shown that LGG have relatively stable genomes, while DIPG can carry large-scale imbalances with common gains of 2q, 8q, and 9q and losses of 16q, 17p, and 20p. The chromosomal imbalance profile seems to distinguish DIPG from both pediatric and adult GBM. DIPG do not show the loss of *CDKN2A* that is common in adult and pediatric high grade gliomas. The most common focal gain is amplification of *PDGFRA*, followed by *MET*, *IGF1R*, *ERBB4*, and *EGFR*, in contrast with adult GBMs in which *EGFR* amplification is the most common focal gain. By FISH it was also observed that some tumors contained mutually exclusive subclones, with amplifications of *PDGFRA* or *MET*. Whole-genome sequencing identified recurrent mutations in *BRAF*, *RAF1*, *histone H3*, *ATRX*, rearrangements of *MYB* or *MYBL1*, and mutations and duplications of *FGFR1*, all of which seem to be mutually exclusive on the cellular level [33–35]. The number of non-silent mutations and rearrangements is very low with the median number of one mutation per tumor, which suggests that very few alterations are necessary for tumorigenesis. Somatic *histone H.H3* mutations seem to be particularly typical for DIPG, whereas diffuse LGG carry *FGFR1*, *MYB*, and *MYBL1* alterations [36]. As discussed in the section about *BRAF*, mutations and duplications are characteristic for low-grade pilocytic astrocytomas and pleomorphic xanthoastrocytomas.

Genetic Mosaicism and Intratumoral Heterogeneity in Gliomas

Classification studies divided high grade gliomas into distinct subgroups. However, several studies identified a genetic mosaicism of receptor tyrosine kinase (RTK) amplifications in GBMs, which leads to the question of how well defined these groups in fact are. RTKs are commonly amplified in GBMs, the most common being *EGFR* amplification, which is present in ~40 % of cases, and *PDGFRA* amplification in 10–15 % of cases. *PDGFRA* amplification is often accompanied by amplifications of *VEGFR2* and *KIT*, which reside in the same region. The third most common is amplification of *MET*. Each of these RTKs is associated with a particular molecular subtype of GBM as described above, *EGFR* with Classic, *PDGFRA* with Proneural and *MET* with the Mesenchymal subtype.

Studies using FISH illustrated that some GBMs contain a mix of up to three intermingled subpopulations of GBM cells with mutually exclusive amplifications of *EGFR*, *PDGFRA* and *MET*, which arose from the same precursor (Fig. 21.6) [37–39]. GBM cells in which *EGFR* and *PDGFRA* amplification was present within the same tumor cell have also been observed [37, 38]. Different clones tend to inhabit different microenvironments and might play a different role in the growth and progression of GBM [39]. Mosaic amplification has also been described in LGGs [40] and pediatric gliomas [34], further hampering the idea that gliomas reliably can be stratified into distinct subtypes. The authors of one of the studies, which quantified subclones with different amplifications, reported that ratios of subclones are highly variable and minor subclones might not be picked up by whole-genome approaches. One can assume that the heterogeneity observed on the level of RTK can be also present on the level of point mutations, which would make the level of intratumoral heterogeneity even higher [41]. Finally, the presence of different subclones within the tumor raises several issues with regard to sampling of the brain tumors and testing at the time of diagnosis versus later, as well as issues pertaining to the development of therapies that are to be specific for certain molecular subtypes.

Meningioma

Meningiomas represent approximately 30 % of primary CNS tumors of adults. Although most are benign, these tumors tend to recur and require multiple resections. Hence, they

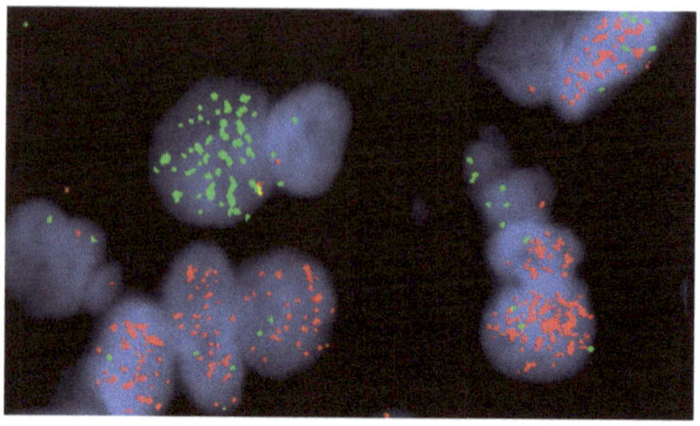

Figure 21-6 Mosaic amplification of RTKs in GBM. Intermingled subpopulations of GBM cells carry high level amplification of *EGFR* (*red*) or *PDGFRA* (*green*). Amplification of an RTK is a strong driver mutation and *EGFR* and *PDGFRA* amplification are considered mutually exclusive on the cellular level. Furthermore, *EGFR* amplification is strongly associated with the Classical subtype and amplification of *PDGFRA* with the Proneural subtype of GBM, two supposedly distinctly different molecular categories of GBM. In this tumor, however, these two populations arose from the same precursor, sharing the same early mutation events such as *TP53* mutation and homozygous deletion of *CDKN2A*. Subclones later developed different, usually mutually exclusive RTK amplifications. This type of heterogeneity can be identified by in situ methods, whereas a whole-genome approach would likely not be able to distinguish whether *EGFR* and *PDGFRA* amplifications occur in the same cell or in different subclones. Furthermore, it might not identify minor subclones

are associated with significant morbidity and mortality. Currently, the prognostic criteria for tumor behavior are largely based on histology, but molecular studies helped identify several potential markers of aggressive behavior.

Meningiomas have a complex karyotype. Not surprisingly, the loss of 22q is the most frequent change seen in meningiomas. The long arm of chromosome 22 harbors the *NF2* gene and meningiomas are commonly associated with neurofibromatosis type 2. *NF2* loss is also present in sporadic meningiomas. Although this can theoretically be useful for diagnosis, it is rarely used due to the sufficient material for histopathologic, immunohistochemical and ultrastructural (electron microscopy) analysis. Loss of 22q in combination with 1p and 14q loss can be used to distinguish meningioma from other dural based tumors, such as hemangiopericytomas and solitary fibrous tumors. Loss of 1p, gain of 1q, and loss of 14q have also been associated with shorter progression free survival in adults. In pediatric cohorts, loss of 22q is commonly observed due to association with neurofibromatosis type 2 and tumors often show loss of 1p and 14q, as well, although the correlation with survival in this population is not clear. Although in gliomas the loss of *CDKN2A* is an early event of tumorigenesis, in meningiomas it is associated with higher grade and short survival.

Other genes associated with meningioma have long been elusive. Whereas loss of 17p has been observed, meningiomas rarely carry *TP53* mutations, which seem to be restricted to anaplastic tumors. The *INI1* gene that resides near *NF2* on 22q has been another candidate gene; however, it is rarely altered in meningiomas. Rare alterations of *PTEN* and *PTCH1* have been described. Whole-genome studies recently succeeded in identifying driver mutations in non-NF2 meningiomas. Meningiomas without *NF2* aberrations can carry mutations in *TRAF7*, *KLF4*, *AKT1*, and *SMO* [42] Mutations in these genes appear mutually exclusive with *NF2* aberrations. There is also a striking spatial distribution: meningiomas associated with *NF2* loss are found in the hemispheres, cerebellum and spinal cord, whereas other tumors originate in the skull base. Meningiomas with *SMO* mutations are frequently present around the skull base midline, which is particularly interesting considering the role of the Shh pathway in midline brain development and its failure resulting in holoprosencephaly. Finally, there is also a striking association between histological type and mutation status. Secretory meningiomas have been defined by concurrent mutations of *TRAF7* and *KLF4* [43]. *NF2* aberrant meningiomas seem to be associated with more aggressive behavior. Considering how many distinct histological variants of meningioma exist, it will be interesting to see whether future studies will associate other morphological subtypes with specific mutations, as well.

Ependymoma

Ependymoma is the second-most common malignant brain tumor of childhood and the most common spinal cord tumor of adults. Histological grading of ependymomas is notoriously unreliable and, given their potential for recurrences after many years of disease free survival and resistance to current therapies, ependymomas represent an ideal target for molecular studies [44]. Many ependymomas have complex genomes with large chromosomal gains and losses, but clear diagnostic, prognostic or predictive markers have not been yet identified. Common genetic abnormalities in ependymoma involve losses of chromosomes 1p, 3, 6q, 9p, 10q, 13q, 16p, 17, 21, and 22q and gains of 1q, 4q, 5, 7, 8, 9, 12q, and 20. Chromosome 22 loss is probably the most frequent overall genetic abnormality in sporadic ependymoma and in *NF2* associated ependymomas. Patients with neurofibromatosis type 2 develop a variety of central nervous system malignancies including ependymomas and meningiomas, which both show loss of 22q. Interestingly, *NF2* mutations have been identified in spinal ependymomas, but are not common in cranial ones and are rare in pediatric intracranial ependymomas, suggesting the presence of another candidate tumor suppressor gene at this locus. Gain of 1q is associated with the posterior fossa location, and, similar to meningiomas, it is associated with high grade features and a marker of poor outcome. In contrast, 6q25.3 deletion is associated with a significantly better outcome. Other recurrent chromosomal abnormalities include monosomy 17 with the 17p arm being more frequently lost, gain of 7, which is typical for spinal tumors, and loss of 13q and 16q. Loss of 10q was found in adult and pediatric ependymoma suggesting that *PTEN* might be a potential tumor suppressor gene in ependymomas. Mutations of *PTEN* were not identified, however. In addition to tumors with very complex karyotypes, a considerable proportion of ependymomas have an almost diploid balanced genome. These often have a worse outcome, suggesting underlying but as yet unidentified mutations. These tumors are mostly intracranial and present in young children. aCGH analyses revealed that ependymomas may be clinically stratified based on the number and type of chromosomal losses and gains. Ependymomas with moderately complex DNA changes, including gains of chromosomes 9, 15q and 18, and loss of chromosome 6 have very good progression free survival and almost 100 % overall survival. The second group has an almost normal diploid genome but does worse than the first group with progression free survival and overall survival at approximately 70 %. The last group of ependymomas, with 1q gain and/or homozygous deletion of 9p21, has the worst prognosis with progression free survival near 0 % and overall survival less than 30 %. Whether this should be tested in clinical practice is currently not clear because first 1q gain and 9p21 loss have to be confirmed as independent prognostic markers. Tumors with overexpression of metalloproteinases and changes of growth to a more infiltrative phenotype resembling high grade gliomas have poor outcome despite gross total resection [45]. Whole-genome studies did not identify recurrent mutations, gains or losses in ependymomas, but there seem to be distinct epigenetic subgroups of ependymomas. Transcriptional profiling of posterior fossa ependymomas identified two distinct subgroups [46]. Group A patients are younger, with laterally located tumors and a balanced genome. Group A tumors are also more invasive and metastatic and are more likely to recur. Group B ependymomas often arise in the posterior fossa of adults and have an unbalanced genome and grow in the midline with minimal invasion, rare metastasis, and good survival.

Conclusions

Current molecular neuropathology provides several tests that help with diagnosis and clinical management of patients with brain tumors. Molecular tests for 1p19q and *MGMT* and testing for *IDH1/2* by IHC and sequencing are well established and are incorporated in clinical practice as well as in clinical trials. Genetics research significantly improved our knowledge about different molecular subtypes within the tumor types, which were previously defined solely based on morphology. Reflecting this heterogeneity will help to better design clinical trials and towards the development of targeted therapies. The current amount of data clearly surpasses our understanding of it and functional studies are needed to identify potential targets. Implementation of new diagnostic technologies in clinical laboratories will play a crucial role in identifying molecular subtypes and correct therapeutic targets.

REFERENCES

1. Riemenschneider MJ et al. Molecular diagnostics of gliomas: state of the art. Acta Neuropathol. 2010; 120(5):567–84.
2. Tabatabai G et al. Molecular diagnostics of gliomas: the clinical perspective. Acta Neuropathol. 2010; 120(5):585–92.
3. von Deimling A, Korshunov A, Hartmann C. The next generation of glioma biomarkers: MGMT methylation, BRAF fusions and IDH1 mutations. Brain Pathol. 2011;21(1):74–87.
4. Bettegowda C et al. Mutations in CIC and FUBP1 contribute to human oligodendroglioma. Science. 2011;333(6048):1453–5.
5. Yip S et al. Concurrent CIC mutations, IDH mutations, and 1p/19q loss distinguish oligodendrogliomas from other cancers. J Pathol. 2012;226(1):7–16.
6. Horbinski C, Miller CR, Perry A. Gone FISHing: clinical lessons learned in brain tumor molecular diagnostics over the last decade. Brain Pathol. 2011; 21(1):57–73.
7. Snuderl M et al. Polysomy for chromosomes 1 and 19 predicts earlier recurrence in anaplastic oligodendrogliomas with concurrent 1p/19q loss. Clin Cancer Res. 2009;15(20):6430–7.
8. Wiens AL et al. Polysomy of chromosomes 1 and/or 19 is common and associated with less favorable clinical outcome in oligodendrogliomas: fluorescent in situ hybridization analysis of 84 consecutive cases. J Neuropathol Exp Neurol. 2012;71(7):618–24.
9. Yan H et al. IDH1 and IDH2 mutations in gliomas. N Engl J Med. 2009;360(8):765–73.
10. Horbinski C et al. Diagnostic use of IDH1/2 mutation analysis in routine clinical testing of formalin-fixed, paraffin-embedded glioma tissues. J Neuropathol Exp Neurol. 2009;68(12):1319–25.
11. Joensuu H et al. Amplification of genes encoding KIT, PDGFRalpha and VEGFR2 receptor tyrosine kinases is frequent in glioblastoma multiforme. J Pathol. 2005;207(2):224–31.
12. Puputti M et al. Amplification of KIT, PDGFRA, VEGFR2, and EGFR in gliomas. Mol Cancer Res. 2006;4(12):927–34.
13. Schlegel J et al. Amplification of the epidermal-growth-factor-receptor gene correlates with different growth behaviour in human glioblastoma. Int J Cancer. 1994;56(1):72–7.
14. Pierscianek D et al. MET gain in diffuse astrocytomas is associated with poorer outcome. Brain Pathol. 2013;23(1):13–8.
15. Hobbs J et al. Paradoxical relationship between the degree of EGFR amplification and outcome in glioblastomas. Am J Surg Pathol. 2012;36(8):1186–93.
16. Horbinski C. To BRAF or not to BRAF: is that even a question anymore? J Neuropathol Exp Neurol. 2013; 72(1):2–7.
17. Paz MF et al. CpG island hypermethylation of the DNA repair enzyme methyltransferase predicts response to temozolomide in primary gliomas. Clin Cancer Res. 2004;10(15):4933–8.
18. Hegi ME et al. MGMT gene silencing and benefit from temozolomide in glioblastoma. N Engl J Med. 2005;352(10):997–1003.
19. Hegi ME et al. Clinical trial substantiates the predictive value of O-6-methylguanine-DNA methyltransferase promoter methylation in glioblastoma patients treated with temozolomide. Clin Cancer Res. 2004; 10(6):1871–4.
20. Cankovic M et al. The role of MGMT testing in clinical practice: a report of the association for molecular pathology. J Mol Diagn. 2013;15:539–55.
21. Eberhart CG. Molecular diagnostics in embryonal brain tumors. Brain Pathol. 2011;21(1):96–104.
22. Pugh TJ et al. Medulloblastoma exome sequencing uncovers subtype-specific somatic mutations. Nature. 2012;488(7409):106–10.
23. Northcott PA et al. Medulloblastomics: the end of the beginning. Nat Rev Cancer. 2012;12(12):818–34.
24. Northcott PA et al. Medulloblastoma comprises four distinct molecular variants. J Clin Oncol. 2011; 29(11):1408–14.
25. Northcott PA et al. Subgroup-specific structural variation across 1,000 medulloblastoma genomes. Nature. 2012;488(7409):49–56.
26. Taylor MD et al. Molecular subgroups of medulloblastoma: the current consensus. Acta Neuropathol. 2012;123(4):465–72.
27. Kool M et al. Molecular subgroups of medulloblastoma: an international meta-analysis of transcriptome, genetic aberrations, and clinical data of WNT, SHH, Group 3, and Group 4 medulloblastomas. Acta Neuropathol. 2012;123(4):473–84.
28. Rudin CM et al. Treatment of medulloblastoma with hedgehog pathway inhibitor GDC-0449. N Engl J Med. 2009;361(12):1173–8.
29. Yauch RL et al. Smoothened mutation confers resistance to a Hedgehog pathway inhibitor in medulloblastoma. Science. 2009;326(5952):572–4.
30. Jones DT et al. Dissecting the genomic complexity underlying medulloblastoma. Nature. 2012;488(7409): 100–5.
31. Verhaak RG et al. Integrated genomic analysis identifies clinically relevant subtypes of glioblastoma characterized by abnormalities in PDGFRA, IDH1, EGFR, and NF1. Cancer Cell. 2010;17(1):98–110.
32. Sturm D et al. Hotspot mutations in H3F3A and IDH1 define distinct epigenetic and biological subgroups of glioblastoma. Cancer Cell. 2012;22(4): 425–37.

33. Zhang J et al. Whole-genome sequencing identifies genetic alterations in pediatric low-grade gliomas. Nat Genet. 2013;45:602–12.
34. Paugh BS et al. Genome-wide analyses identify recurrent amplifications of receptor tyrosine kinases and cell-cycle regulatory genes in diffuse intrinsic pontine glioma. J Clin Oncol. 2011;29(30):3999–4006.
35. Khuong-Quang DA et al. K27M mutation in histone H3.3 defines clinically and biologically distinct subgroups of pediatric diffuse intrinsic pontine gliomas. Acta Neuropathol. 2012;124(3):439–47.
36. Wu G et al. Somatic histone H3 alterations in pediatric diffuse intrinsic pontine gliomas and non-brainstem glioblastomas. Nat Genet. 2012;44(3):251–3.
37. Snuderl M et al. Mosaic amplification of multiple receptor tyrosine kinase genes in glioblastoma. Cancer Cell. 2011;20(6):810–7.
38. Szerlip NJ et al. Intratumoral heterogeneity of receptor tyrosine kinases EGFR and PDGFRA amplification in glioblastoma defines subpopulations with distinct growth factor response. Proc Natl Acad Sci U S A. 2012;109(8):3041–6.
39. Little SE et al. Receptor tyrosine kinase genes amplified in glioblastoma exhibit a mutual exclusivity in variable proportions reflective of individual tumor heterogeneity. Cancer Res. 2012;72(7):1614–20.
40. Motomura K et al. PDGFRA gain in low-grade diffuse gliomas. J Neuropathol Exp Neurol. 2013; 72(1):61–6.
41. Marusyk A, Polyak K. Tumor heterogeneity: causes and consequences. Biochim Biophys Acta. 2010; 1805(1):105–17.
42. Clark VE et al. Genomic analysis of non-NF2 meningiomas reveals mutations in TRAF7, KLF4, AKT1, and SMO. Science. 2013;339(6123):1077–80.
43. Reuss DE et al. Secretory meningiomas are defined by combined KLF4 K409Q and TRAF7 mutations. Acta Neuropathol. 2013;125(3):351–8.
44. Korshunov A et al. Molecular staging of intracranial ependymoma in children and adults. J Clin Oncol. 2010;28(19):3182–90.
45. Snuderl M et al. Prognostic value of tumor microinvasion and metalloproteinases expression in intracranial pediatric ependymomas. J Neuropathol Exp Neurol. 2008;67(9):911–20.
46. Witt H et al. Delineation of two clinically and molecularly distinct subgroups of posterior fossa ependymoma. Cancer Cell. 2011;20(2):143–57.
47. Northcott PA et al. Rapid, reliable, and reproducible molecular sub-grouping of clinical medulloblastoma samples. Acta Neuropathol. 2012;123(4): 615–26.

CHAPTER 22

Genomic Applications in Head and Neck Cancers

SUN M. AHN, NISHANT AGRAWAL

Introduction 341
Risk Factors for HNSCC 342
Molecular Biology of Head and Neck Cancers 343
Clinical Implications of Molecular Alterations in Head and Neck Cancers 346
Conclusions 352

Introduction

Head and neck cancer is the sixth most common cancer worldwide, with more than 40,000 new cases diagnosed every year in the United States and 500,000 new cases globally. The most common type of cancer of the head and neck is squamous cell carcinoma arising from the epithelial layer of the upper aerodigestive tract. Most patients present with locoregionally advanced disease with a less than 50 % 5-year survival rate. Furthermore, those who present with early-stage disease are at a high risk of recurrence or development of a second primary tumor. Patients with recurrent or metastatic disease endure a significantly worse prognosis with a dismal overall survival of approximately 6 months. The primary modes of therapy include surgical resection or radiation therapy for early-stage disease and combination therapy with surgery, radiotherapy, and/or chemotherapy for advanced-stage disease. Surgical extirpation and chemoradiation protocols for the treatment of head and neck cancer often lead to severe functional deficits and cosmetic deformities. Despite all the advances in cancer therapy, the overall survival of patients with head and neck squamous cell carcinoma (HNSCC) has remained unchanged for the past 30 years. Most recently, with the identification of a human papillomavirus (HPV)-related subset of oropharyngeal SCC, the prognosis in this specific patient population has been significantly improved with a greater than 80 % survival rate in contrast to the continuing poor survival with HPV-negative oropharyngeal SCC [1].

Advancements in molecular and genetic research techniques and bioinformatics have led to an explosion of new discoveries in the molecular biology and genetic alterations behind the pathogenesis of HNSCC. Gaining further insights into the mechanisms underlying tumorigenesis and treatment response as well as advancements in screening, diagnosis, and treatment of HNSCC will ultimately lead to improved clinical outcome.

This chapter reviews the specific genetic alterations in HNSCC and their clinical implications and provides a preview of the future applications of this knowledge that are in development.

G.J. Netto and I. Schrijver (eds.), *Genomic Applications in Pathology*,
DOI 10.1007/978-1-4939-0727-4_22, © Springer Science+Business Media New York 2015

Risk Factors for HNSCC

The general principles underlying mechanisms behind tumorigenesis in HNSCC are thought to be similar to those in other solid tumors. Cancer arises from progressive accumulation of genetic or epigenetic alterations that lead to the development of malignant phenotypes. Prolonged, cumulative exposure to certain carcinogens is thought to be the leading cause of specific alterations acquired during tumor progression. Tobacco and alcohol represent the two predominant carcinogens that are synergistically responsible for HNSCC development [2, 3]. In contrast, oral HPV infection, the main cause of the cancer of the oropharynx, is believed to act independently.

Tobacco and Alcohol

Numerous epidemiological studies conducted in different regions of the world have demonstrated a compelling association between alcohol consumption and tobacco use and the risk of HNSCC [4–7]. Several large case–control studies have shown neoplastic effects of tobacco use and alcohol consumption with a linear correlation with both duration and amount with odds ratios of 2- to 6-fold for alcohol and 7- to 20-fold for tobacco. When combined, these carcinogens had multiplicative or even supra-multiplicative effects with a greater than 35- to 200-fold risk for individuals who consume more than two or more packs of cigarettes and more than four drinks per day [4, 6, 8]. One explanation for the synergistic effect of alcohol and tobacco is that alcohol possibly acts as a solvent for penetration of other carcinogens through the mucosa of the upper aerodigestive tract [9].

According to the World Health Organization, tobacco use is the single most preventable risk factor for cancer deaths worldwide, responsible for 22 % of all cancer mortality. There are over 4,800 chemicals in processed tobacco, of which at least 250 are known to be harmful and more than 50 are known carcinogens. These include polycyclic aromatic hydrocarbons, *N*-nitrosamines, aromatic amines, aldehydes, volatile hydrocarbons, and metals [10]. Cigarette smoke also contains free radicals, nitric oxide, and other unstable oxidants that induce oxidative DNA damage [11, 12]. Our body responds to carcinogens by detoxifying and excreting them through a series of enzymatic processes. However, metabolites resulting from detoxification can also be reactive and cause DNA damage through the formation of DNA adducts [13, 14]. Therefore, it has been proposed that genetic susceptibility or polymorphisms in detoxifying enzymes, such as cytochrome P-450 and glutathione S-transferase, can lead to the development of cancer by either failing to deactivate carcinogens or to activate pro-carcinogenic intermediates [15, 16].

The mechanisms by which alcohol exerts its carcinogenic effects have been linked to alcohol metabolism, DNA damage, and DNA methylation. Acetaldehyde, the primary metabolite of ethanol, can form adducts with DNA and thus result in DNA damage [17]. There is also growing evidence that genetic polymorphisms in enzymes for oxidation of ethanol into acetaldehyde modulates alcohol-related cancer risks, which further supports the mechanistic role of acetaldehyde [18]. In addition, heavy alcohol intake leads to nutritional deficiencies, including that of vitamin B12, B6, and A and folate. This may also result in changes in DNA methylation and transcription patterns that promote tumorigenesis [17].

Human Papillomavirus

HPV infection is the most common sexually transmitted infection in the Unites States. There are over 100 HPV subtypes, and these are categorized into low-risk and high-risk subtypes. Among the high-risk group, HPV-16 and HPV-18 are the two leading subtypes responsible for cancer development [19]. In a recent large cross-sectional study conducted as part of the National Health and Nutrition Examination Survey (NHANES), the prevalence of oral HPV infection in the general population was determined to be about 6.9 %, with a prevalence rate of 3.7 % for high-risk HPV infection. The most prevalent HPV subtype detected was HPV-16 with a prevalence of 1.0 % [20].

High-risk HPV has long been known to cause cervical cancer in women, penile cancer in men, and anal cancer in both men and women. Over the past 10 years, there has been overwhelming evidence that implicates HPV as a causative factor in a subset of HNSCC, mainly of the oropharynx, where up to 50–70 % of the cases are associated with high-risk HPV. In contrast to cervical cancer, where HPV-16 and HPV-18 are together known to cause 70 % of the cases, HPV-related HNSCC is exclusively caused by the HPV-16 subtype, with up to 90 % of the cases being HPV-16 positive. While the overall incidence of head and neck cancer is on the decline over the last two decades, the incidence of oropharyngeal cancer is on the rise [21]. This steady incline in the incidence of oropharyngeal cancer is mirrored by the increase in the incidence of HPV-positive oropharyngeal cancer while the rates of HPV-negative oropharyngeal cancer have been decreasing over the same period of time [22].

Molecular Biology of Head and Neck Cancers

The advances in our understanding of cancer genomics have further elucidated the biological complexity of HNSCC. The disease represents a heterogeneous collection of tumors in which multiple genes and pathways are altered (Table 22.1). In-depth understanding of the pathways implicated in HNSCC tumorigenesis is critical for the identification of new "personalized" therapeutic strategies.

TP53

The role of *TP53*, a tumor-suppressor gene on chromosome 17p12, in HNSCC carcinogenesis is well established in the literature. *TP53* is the most commonly mutated gene in HNSCC, with approximately half of all HNSCC tumors having a *TP53* mutation [23–25]. In normal cells, *TP53* plays a critical role in regulating the cell cycle in response to DNA damage. *TP53* is activated by exposure to cellular stress such as DNA damage, which results in the accumulation of active TP53 protein in the nucleus. Through transcriptional induction of downstream signaling pathways, it induces viable cell growth arrest or apoptosis. Therefore, this growth inhibitory effect of *TP53* is vital in preventing the proliferation of cells harboring damaged DNA or of cells with the potential for neoplastic transformation [26]. Functional loss of *TP53* is one of the most common genetic alterations in many types of human cancer, and mutations in this gene play a critical role in malignant transformation [27]. In HNSCC, alterations in *TP53* occur early in the premalignant squamous epithelium, before invasive transformation. For instance, premalignant oral lesions have been shown to harbor *TP53* mutations in up to 35 % of cases [28, 29]. In fact, the incidence of mutations increases with histological progression from severe dysplasia to invasive carcinoma [30]. Furthermore, the frequency of *TP53* genetic alterations in patients with a history of tobacco and alcohol use is almost double that of those without such history [23, 31, 32].

NOTCH

In two recently published whole-exome next-generation sequencing (NGS) studies of HNSCC, *NOTCH1* was identified as a new cancer gene implicated in HNSCC development. In fact, *NOTCH1* has been identified as the second most commonly mutated gene in HNSCC, with a mutation frequency of 14–15 % [24, 25]. Previously, only a few functional studies described a role for *NOTCH1* in squamous cell oncogenesis, specifically of the skin [33, 34]. However, given the large size of the *NOTCH1* gene which comprises 34 coding exons, robust mutational data analysis was required to comprehensively detect these mutations.

There are four *NOTCH* family receptors in humans, *NOTCH1* to *NOTCH4*. *NOTCH1* encodes a transmembrane receptor that functions in regulating normal cell differentiation, lineage commitment, and embryonic development. After ligand binding, the *NOTCH1* intracellular domain (NICD) is cleaved, and the translocation of the NICD to the nucleus is necessary for transcriptional activation of downstream signaling. The *NOTCH1* ligands include Jagged 1 and 2 and Delta-like ligand 1, 3, and 4. After receptor activation through ligand binding, the release of NICD requires a two-step cleavage process. First, the extracellular portion of the protein is released by proteases

Table 22-1 **Common Genetic Aterations in Primary Head and Neck Squamous Cell Carcinoma**

Gene symbol	Gene name	Location	Frequency	Function	Clinical applications
Tumor-suppressor gene					
TP53	Tumor protein p53	17p13.1	47–60 %	Assists in cell cycle arrest to allow DNA repair, apoptosis, or cell senescence	Biomarker for poor prognosis—decreased survival and therapy resistance; for analysis of margin status; adenoviral gene therapy
CDKN2A/p16	Cyclin-dependent kinase inhibitor 2A	9p21.3	9–12 %	Regulates of G1-to-S phase transition in cell cycle and cell senescence	IHC for p16 as a surrogate marker for HPV-related tumor
Oncogene					
EGFR	Epidermal growth factor receptor	7p12	[a]	Activates critical signaling pathways in proliferation, migration, invasion, angiogenesis, and apoptosis	Fluorescent bioconjugated anti-EGFR molecules for intraoperative optical imaging; anti-EGFR-targeted therapies
HRAS	Harvey rat sarcoma viral oncogene homolog	11p15.5	4–5 %	Promotes cell proliferation, differentiation, morphology, and survival	
PIK3CA	Phosphoinositide-3-kinase catalytic alpha polypeptide	3q26.32	6–8 %	Promotes cell growth, survival, and cytoskeleton organization	
Both (tissue dependent)					
NOTCH1	Notch1	9p34.3	14–15 %	Regulates of cell differentiation, lineage commitment, and embryonic development	Therapeutic inhibition or activation of NOTCH1 pathway

IHC immunohistochemistry, *HPV* human papillomavirus
[a]Overexpression in 80–90 % of HNSCC

TNF-alpha-converting enzyme (TACE) and a disintegrin and metalloprotease (ADAM). A second cleavage by gamma-secretase complex liberates the NICD from the membrane [35]. In the nucleus, NICD interacts with transcriptional regulators and activates downstream target genes, such as the *HRT* and *HES* family of genes, which are crucial for cell differentiation and normal embryonic development.

The role of *NOTCH1* in cancer has been recently described, with *NOTCH1* signaling having both oncogenic and tumor-suppressive roles depending on the cellular context. For instance, activating truncation mutations in *NOTCH1* have been reported in acute lymphoblastic leukemia and chronic lymphocytic leukemia, implicating *NOTCH1* as an oncogene in these hematopoietic cancers [36, 37]. In contrast, the initial findings of inactivating mutations in HNSCC and the observation that loss of *NOTCH1* in murine models led to skin carcinogenesis indicated that *NOTCH1* may also act as a tumor-suppressor gene [33, 34]. The data thus far are conflicting with regard to the exact role of *NOTCH1* in HNSCC. Most *NOTCH1* mutations observed in HNSCC affect the epidermal growth factor (EGF)-like ligand-binding domain and are thought to lead to loss of function, suggesting the role of *NOTCH1* as a tumor suppressor [24]. Contrary to the genetic evidence, there is evidence that *NOTCH1* protein levels are elevated in HNSCC, and tumors expressing higher levels of *NOTCH1* protein are associated with reduced survival as well as with chemoresistance [38–41]. In support of the latter, it was demonstrated that inhibition of the *NOTCH1* pathway using gamma-secretase inhibitors (GSIs) prevented the growth of HNSCC cell lines [42]. These findings suggest that activated *NOTCH1* could function as an oncogene. Additional functional studies in vitro and in vivo are required to elucidate the exact role of *NOTCH1* in HNSCC.

CDKN2A/P16

Cyclin-dependent kinase inhibitor 2A (*CDKN2A*) located on chromosome 9p21 is a known tumor-suppressor gene frequently disrupted in HNSCC. It is frequently inactivated through deletion, point mutations, and epigenetic promoter methylation. Loss of heterozygosity on the short arm of chromosome 9 has been reported frequently, and this change has been recognized as an early event in the progression of premalignant lesions to HNSCC [43, 44]. In the abovementioned two large whole-exome NGS projects, *CDKN2A* mutations were identified in 9 % of all tumors [24, 25]. When considering genetic and epigenetic alterations, *p16* inactivation has been detected in as much as 80 % of HNSCC [45].

The *CDKN2A* gene encodes the protein product *p16* that plays an important role in regulating the G1 phase of the cell cycle. The p16 protein binds to cyclin-dependent kinase 4 (CDK4) and CDK6, inhibiting their association with cyclin D1. This inhibition of cyclin D1/CDK4/6 complex activity prevents pRb phosphorylation and the release of E2F transcription factor, leading to the inhibition of the G1-to-S phase transition and thus leading to cell senescence [46, 47]. Therefore, any genetic abnormalities inactivating the *p16* pathway may confer growth advantages in cells, contributing to the tumorigenic process.

EGFR

Malignant transformation of HNSCC is also driven by alterations in growth factor signaling pathways. Epidermal growth factor receptor (EGFR), also known as HER1 or ErbB-1, is a tyrosine kinase receptor that is highly expressed in normal epithelial cells. EGFR is activated by several ligands, which induces receptor dimerization and autophosphorylation, resulting in activation of downstream signaling pathways [48]. These downstream pathways include *MAPK*, *PI3K/AKT*, *ERK*, and *JAK/STAT* genes that are critical for the regulation of cellular proliferation, apoptosis, angiogenesis, migration, and invasion [49]. The *EGFR* gene is overexpressed in 80–90 % of HNSCC via gene amplification and transcriptional activation [50, 51]. In addition to overexpression, a mutant form of *EGFR* known as *EGFRvIII* has been implicated in resistance to anti-*EGFR*-targeted therapies [52]. This mutant form is characterized by a deletion in exons 2–7, leading to a truncated ligand-binding domain, rendering it constitutively active. Over-activation of EGFR signaling via overexpression or activating mutations enables cells to take on a malignant phenotype.

RAS

The *RAS* gene family consists of three genes that function as small GTPase molecules: *KRAS*, *HRAS*, and *NRAS*. The *RAS* genes play a critical role in cell signaling as part of the *RAS–RAF–MEK–MAPK* pathway. This pathway is involved in the regulation of cell proliferation, differentiation, morphology, and survival. The *RAS* gene family mutations have been implicated in approximately one-third of all human cancers, with *KRAS* being the most common and *HRAS* the least common [53]. However, in HNSCC, *KRAS* mutations are virtually absent while *HRAS* mutations have been described at a low frequency of approximately 4–5 % [24, 25].

PIK3CA

The *PI3K–PTEN–AKT* pathway is another critical pathway in HNSCC carcinogenesis. The *PIK3CA* gene is located on chromosome 3q26 and functions to convert phosphatidylinositol (4,5) biphosphate (P4,5P2) into phosphatidylinositol (3,4,5) triphosphate (PIP3), in turn activating Akt/PKB kinases. This results in the promotion of cell growth, survival, and cytoskeleton reorganization [54]. *PIK3A* is downstream of receptor tyrosine kinases such as EGFR, Met, and vascular endothelial growth factor receptor (VEGFR), which are known oncogenes in HNSCC. The prevalence of *PIK3CA* mutations was estimated to be approximately 6–8 % in the above-referenced two recent whole-exome sequencing HNSCC studies [24, 25]. The overactivation of this pathway occurs through both amplification and mutations in *PIK3CA* as well as through *PTEN* loss [54]. PTEN (phosphatase and tensin homologue deleted on chromosome ten) is a key regulator of PI3K function. PTEN reverses the action of PI3K by removing the 3′ phosphate and thus preventing the activation of downstream molecules such as Akt [55]. A study conducted by Pedrero et al. found evidence for *PIK3CA* amplification in 37 % of primary HNSCC tumors and in 39 % of premalignant lesions, indicating that *PIK3CA* amplification could be an early event in HNSCC oncogenesis [56].

Human Papillomavirus

HPV is a non-enveloped small double-stranded, circular DNA virus that infects epithelial cells [57]. The majority of HPV subtypes cause epithelial lesions with low malignant potential, such as warts or papillomas. However, there is a subset of high-risk HPV that leads to precancerous lesions. Interestingly, only a small fraction of people infected with high-risk HPV will eventually develop cancer, often decades after the original infection.

The molecular mechanism behind HPV-driven carcinogenesis has been extensively studied in cervical cancer. The integration of high-risk HPV DNA into the host genome results in the expression of oncogenes E6 and E7 in the host cell. The E6 oncogene binds to tumor suppressor *TP53*, which causes the degradation of *TP53* via ubiquitin-mediated processes. The degradation of *TP53* prevents the host cell from engaging in cell cycle checkpoints and enduring an apoptotic response [58]. The E7 oncogene is the most important driver of cell cycle deregulation through the binding and destabilizing of the tumor suppressor retinoblastoma (pRb). This binding of pRb results in the release of E2F transcription factors, leading to the transcription of genes involved in proliferation and cell cycle progression [59]. One of the main molecular pathways amplified through E7 is the *CDKN2A/p16* gene pathway, which results in the overexpression of p16 protein. Whereas in tobacco-induced HNSCC the abrogation of TP53 and pRb pathways occurs via mutation and epigenetic alterations, in HPV-related HNSCC wild-type TP53 and pRb are functionally inactivated by the viral oncogenes. E7 also induces cellular proliferation by disrupting the activity of cyclin-dependent kinase inhibitors p21 and p27 [59]. E5 is another viral protein that modulates the *EGFR* signaling pathway by delaying the downregulation of *EGFR* and increasing the level of EGFR [60]. In summary, HPV infection induces failures in cell cycle checkpoints, which causes genetic instability and, over time, progression of premalignant lesions to invasive squamous cell carcinoma.

Clinical Implications of Molecular Alterations in Head and Neck Cancers

The ultimate goal of understanding the molecular biology of HNSCC is to help improve patient outcomes. Improving clinical outcomes can be achieved through three sets of clinical applications: (1) applications that help achieve more accurate and earlier detection of disease, improved therapeutic monitoring, and better surveillance of recurrence; (2) accurate markers of predicting prognosis and therapeutic outcome to identify patients that will require aggressive treatment strategies; and (3) identification of novel therapeutic targets tailored to patient's tumor profile. Some of the molecular biomarkers under investigation in each of the above applications are discussed next.

Diagnostic Applications in Head and Neck Cancers

HUMAN PAPILLOMAVIRUS

The two main methods of identifying HPV-related HNSCC in the clinical arena are in situ hybridization (ISH) of HPV DNA and p16 immunohistochemistry (IHC) as a surrogate marker. ISH can be performed using either fluorescently labeled or chromogenic HPV type-specific probes in formalin-fixed and paraffin-embedded sections. The signals originating within the nuclei of tumor cells usually indicate HPV genome integration. Although this method has lower specificity and sensitivity than Southern blot hybridization or polymerase chain reaction (PCR) amplification, it has the advantage of detecting genomic integration and being technically relatively less demanding [61]. With the introduction of signal amplification techniques, the sensitivity of this method has increased significantly, even to the point of detection of one viral copy per cell [62].

P16 overexpression may be used as a surrogate marker for HPV-positive cancers due to the inactivation of the pRb protein by the HPV protein E7 [63]. In many centers and laboratories, IHC of p16 in formalin-fixed, paraffin-embedded tissues is the main method of detecting HPV-16 positivity. In comparison to other techniques, IHC does not require specialized equipment or tissue handling. The concordance rate between HPV-16 ISH and p16 IHC has been shown to be approximately 92–93 % [61, 64]. Despite the strong correlation, several studies have shown that not all p16-positive cancers are due to HPV [61]. The hypothesis is that the discordance is due to the presence of HPV subtypes other than HPV-16. Therefore, the question of whether p16 should be used as proxy for HPV-16 status in the management of HNSCC remains unanswered, and some advocate using a combination of HPV-16 ISH as well as p16 IHC for detection of HPV-16-related HNSCC [61].

Several additional approaches are under investigation as a potential tool for the detection of HPV-related HNSCC. One such approach is the use of PCR to amplify DNA isolated from a subject sample. On the one hand this approach is highly sensitive and may detect even a single copy of the target DNA in a given sample. On the other hand, the high sensitivity can lead to false-positive results due to contamination and may detect HPV genome that is present but may not be causing the malignancy. Currently, PCR-based HPV detection is being investigated as a tool for early detection and surveillance of disease using saliva or serum [65, 66]. PCR-based methods are also being utilized in large epidemiologic studies conducted to determine the general incidence of HPV oral infection [20]. HPV16 serology can detect antibodies produced in response to HPV infection or HPV immunization. However due to the high prevalence rates in the general population, the exact role of this detection tool in the management of HNSCC is yet to be determined. Finally, NGS high-throughput technology may have a role in HPV detection. Several recent studies have demonstrated proof-of-concept results that indicate the ability of NGS to assess HPV status as well as viral load and genomic copies in tumors [25, 67, 68].

For pathologists, HPV testing can play a critical role in specific clinical situations. For instance, detection of HPV in regional or distant metastatic foci can suggest the tumor origin to be likely from the oropharynx [69, 70]. This is especially important given that about 13 % of patients with HNSCC present with a neck mass as their first and only clinical manifestation, and 3–9 % of these patients fail to have their primary site detected upon clinical and radiologic evaluation [71]. In another scenario, detection of HPV in cystic neck lesions can provide compelling evidence of a metastatic malignant process rather than a benign process such as a branchial cleft cyst.

EGFR FOR OPTICAL IMAGING

The presence of positive margins following tumor resection is a known poor prognostic indicator that results from the infiltrative nature of head and neck cancers. Currently, objective intraoperative means of defining tumor margins, other than conventional crude methods of macroscopic and microscopic visual inspection and palpation, are lacking. Therefore, a novel way to reliably identify tumor margins using intraoperative real-time imaging would potentially have a significant impact on decreasing the rate of postoperative positive margins while sparing uninvolved surrounding tissues. Systemic administration of fluorescently labeled antibodies targeting cancer-specific molecules is under investigation in several European clinical trials in multiple types of cancers [72, 73]. In HNSCC, a promising intraoperative imaging system under clinical investigation is one using panitumumab, a monoclonal anti-EGFR antibody, conjugated with indocyanine green dye (panitumumab-IRDye800). In a recent in vivo study using orthotopic HNSCC xenografts, the tumor was clearly delineated from normal tissue on fluorescence guidance as confirmed by histology. The researchers were also able to detect subclinical microscopic residual disease as well as lymph node metastases measuring <1.0 mm [74]. Fluorescent bioconjugated anti-EGFR nanoparticles or peptides are also being investigated in various solid tumors such as esophageal cancer, glioblastoma, and epidermoid tumors [75–77]. These nanoparticles or peptides may prove to be more efficient when compared to fluorescent anti-EGFR antibodies due to a shorter half-life and superior tissue penetration and distribution [78]. Further studies are required to elucidate the potential value of these innovative optical molecular imaging techniques in improving surgical outcome and ultimately patient survival.

Prognostic Applications in Head and Neck Cancers

TP53

Mutations in tumor-suppressor gene *TP53* have been associated with poor survival as well as decreased response to treatment in HNSCC. In a large multicenter prospective study, the presence of any *TP53* mutation was associated with decreased overall survival with a hazard ratio of 1.4, and the presence of *TP53* alterations that disrupt the DNA-binding domain was found to be more significantly associated with decreased survival with a hazard ratio of 1.7 [79]. Furthermore, alterations in the *TP53* gene have been implicated in poor tumor response to chemoradiation. In one study, a 95 % overall incidence of *TP53* inactivation via mutation or deletion was encountered in patients with recurrent HNSCC refractory to radiotherapy [80]. One possible mechanism of radioresistance is through the inhibition of radiation-induced senescence [81]. The risk of locoregional treatment failure following primary radiation treatment or postoperative adjuvant radiation therapy is shown to be significantly greater in patients whose tumor contained mutant *TP53* genes [82, 83]. Finally, *TP53* mutation status has been found to be an independent negative predictor of response to induction and neoadjuvant chemotherapy in both retrospective and prospective studies [84, 85].

The high incidence of *TP53* alterations and the well-established prognostic value of *TP53* attest to the importance of developing a clinically robust tool to detect *TP53* mutations. One specific clinical application that has been extensively studied is *TP53* mutational status in surgical margins. It has been implied that the *TP53* mutational status at histologically tumor-free surgical margins may be critical in predicting locoregional failure, especially since the genetic alterations in *TP53* precede histologically identifiable changes at the tissue level. In fact, the detection of *TP53* mutations via molecular analysis in histologically "negative" margins has been shown to be a reliable prognostic marker of locoregional tumor recurrence [86–88]. Currently, the main method of detecting *TP53* mutations is through IHC, which cannot detect all types of mutations and has limited sensitivity in application to precancerous lesions [89]. Hence, the IHC should be complemented by genetic analysis, via PCR methods or oligonucleotide probe array technique, to increase the sensitivity and specificity of the detection of altered *TP53*.

HUMAN PAPILLOMAVIRUS

Cumulative data from a large number of retrospective and prospective studies have consistently demonstrated a superior outcome in individuals with HPV-positive oropharyngeal SCC (OPSCC) compared to those with HPV-negative tumors [1, 90–92]. In a recent large meta-analysis of 37 studies that addressed the impact of HPV infection on survival outcome, site-specific analysis showed that patients with HPV-positive OPSCC had a 28 % reduced risk of death for overall survival in comparison to patients with HPV-negative oropharyngeal tumors [90]. Similar observations were made for disease-free survival with a hazard ratio of 0.51. Interestingly, there was no difference in the overall survival between HPV-positive and -negative non-oropharyngeal patients. The authors therefore concluded that the observed improved survival benefit for HPV-positive HNSCC patients is specific to the oropharynx subsite. In the first prospective clinical trial to demonstrate survival benefit in HPV-positive HNSCC, Fakhry et al. reported that patients with HPV-positive tumors had a higher response rate after induction chemotherapy (82 % versus 55 %) and after chemoradiation treatment (84 % versus 57 %) [91]. With a median follow-up of 39.1 months, patients with HPV-positive tumors also had improved overall survival (95 % versus 62 %) and decreased risks of progression (with a hazard ratio of 0.27) and risk of death from any cause (hazard ratio of 0.36) than those with HPV-negative tumors [91]. It is also important to note that the positive prognostic benefit of HPV in OPSCC patients is often mitigated by the negative prognostic effects of smoking [1]. In a large-scale retrospective study, patients with OPSCC were able to be stratified into three prognostic groups: low-, intermediate-, and high-risk groups based on HPV status, smoking, and nodal and primary tumor staging. Patients in the high-risk category had a 3-year overall survival of only 46.2 % versus 93.0 % for low-risk patients. Therefore, when determining the best treatment option, it is important to realize that there is a subgroup of HPV-positive patients who may remain in need of more aggressive therapy.

DETOXIFICATION ENZYMES

Detoxification enzymes, such as glutathione S-transferase (GST) and cytochrome P450, oxidize carcinogens into reactive metabolites that can lead to DNA damage and eventual development of cancer. Several studies have evaluated the role of this group of enzymes as prognostic markers in head and neck cancer. One study examining a subtype of GST, GSTT1, found that patients with the functional genotype were three times more likely to die from HNSCC after adjusting for age, primary therapy, and stage of disease [93]. Alternatively, these enzymes may also be a marker for chemotherapy resistance to cisplatin by inactivating reactive oxygen species induced by cisplatin to kill the offending tumor cells [94]. Several studies have demonstrated that patients with high levels of GSTpi had worse overall survival following treatment with chemotherapy [95], and the survival was worst in the group of patients who were treated with chemoradiotherapy and had elevated levels of both GSTpi and TP53 [96]. Finally, one recent study in a Hungarian HNSCC cohort demonstrated that carriers of specific allelic polymorphisms of cytochrome P4501A1 (CYP1A1) and uridine-diphosphate-glucuronosyltransferase 1A1 (UGT1A1) had the worst prognosis [97].

Therapeutic Applications in Head and Neck Cancers

DE-INTENSIFICATION FOR HUMAN PAPILLOMAVIRUS-RELATED HNSCC

The recognition of HPV-associated HNSCC in the younger, nonsmoker, nondrinker population with improved overall prognosis has led some authors to consider revisiting the standard treatment paradigm in this group of patients [98]. The concept of de-intensification for HPV-positive OPSCC has gained attention, with the ultimate goal of achieving acceptable cure rates while minimizing long-term morbidity. Multiple clinical trials to address this question are under way although no evidence-based de-intensification protocol is currently being utilized in the clinical setting. The Eastern Cooperation Oncology Group (ECOG) 1308 study is a phase II clinical trial for patients with stage III/IV resectable HPV-positive OPSCC in which patients receive dose-reduced intensity-modulated radiation therapy (IMRT) or full-dose IMRT depending on their initial response to induction chemotherapy. Oncologic and functional outcome data of minimally invasive surgical approaches, specifically the transoral robotic surgery (TORS), are also slowly emerging with so far promising results [99–103].

HUMAN PAPILLOMAVIRUS VACCINES AND IMMUNOTHERAPY

There are two prophylactic vaccines currently available that have already been shown to be highly effective at preventing cervix infections by high-risk HPV subtypes [104] as well as associated cervical neoplasia [105, 106]. Gardasil(®) is a quadrivalent vaccine containing viruslike particles (VLPs) of subtypes 6, 11, 16, and 18, and Cervarix(®) is a bivalent vaccine containing VLPs of subtypes 16 and 18. These preventive HPV vaccines focus on generating neutralizing antibodies through the humoral immune response for the prevention of future infections. This involves the interaction between cell surface HPV capsid antigen (L1 and/or L2) and an antigen-specific B-cell receptor, which eventually results in the proliferation of HPV antigen-specific B cells via CD4+ helper T cells. Upon subsequent exposures to HPV, memory and plasma B cells produce HPV-specific antibodies that bind to the virus and prevent its entry into host cells. The prophylactic vaccines are anticipated to reduce the incidence of cervical cancer, although their long-term success is yet to be determined. This question will not be fully addressed for decades, until sufficient time has passed post-vaccination when subjects are expected to develop cervical neoplasms. Furthermore, the vaccine's impact on HPV-associated HNSCC is unknown, as none of the studies performed thus far evaluated the effect on the incidence of oral HPV infection or oral immunity to HPV. Finally, this preventive vaccination strategy is not effective for the treatment of existing infections or established HPV-related lesions.

Treatment of established HPV disease requires cell-mediated immunity that recognizes and eliminates virus-infected cells. Therefore, therapeutic HPV vaccination strategies focus on generating cell-mediated immunity for the clearance of infected cells including HPV-associated tumor cells by using intracellular viral protein as the antigen. The HPV E6 and E7 oncoproteins are ideal tumor antigens since they are "foreign" viral proteins that are uniquely expressed by every virus-related cancer cell. The HPV E6 and E7 antigens are utilized to prime naïve T cells to become effector T cells, namely, CD8+ cytotoxic T lymphocytes and CD4+ T helper cells. These effector T cells mediate antigen-specific killing of both infected cells and tumor cells. There are over 30 different phase I and II clinical trials for therapeutic HPV vaccines under way, mostly in cervical cancer but also in head and neck oropharyngeal cancer [107, 108]. One of the challenges with therapeutic vaccination is generating a robust T-cell response specific to the target antigen. Currently multiple strategies are being evaluated to increase the immunogenicity of the therapeutic vaccines [109].

ADENOVIRUS

Viral vector-mediated gene transfer has been investigated as a new experimental strategy to treat advanced and recurrent HNSCC. With this method, once a portion of the viral genome is replaced with the desired genetic sequence, the virus is injected into the tumor and allowed to infect the host cells. This results in propagation of the desired genetic sequence among the tumor cells. Such treatment technique is attractive for the delivery of tumor-suppressor genes to restore those that have been inactivated. It is also a promising option because HNSCC tumors are often accessible for direct injection of gene therapy. As discussed, the *TP53* tumor-suppressor gene is the most commonly mutated gene in HNSCC, and because wild-type TP53 protein functions to arrest aberrant cellular growth, *TP53* adenoviral gene therapy has been studied extensively. In in vitro models, transfection of wild-type *TP53* using adenoviral vectors resulted in high-efficiency expression of normal *TP53* protein as well as growth inhibition in tumor cell lines with homozygous deletion of *TP53* [110]. Additionally, preclinical in vivo studies using adenovirus containing wild-type *TP53* vector (Ad-p53) have shown successful induction of cancer cell apoptosis as well as enhanced response to chemoradiation treatment [111, 112]. A few phase II trials have demonstrated treatment with *TP53* vector adenovirus to be feasible and safe with some evidence of durable, albeit modest, activity in patients with HNSCC [80, 113]. In fact, ONYX-015, an adenovirus engineered to specifically target cells lacking *TP53* function, is approved for the management of early-stage HNSCC in China [114].

EGFR INHIBITORS

The fact that *EGFR* is overexpressed in 80–90 % of HNSCC and plays an important role in its pathogenesis offers a rationale for the development of *EGFR*-targeted therapy. Multiple monoclonal antibodies and small-molecule tyrosine kinase inhibitors (TKIs) have been developed and are under investigation. Cetuximab (™Erbitux) is a chimeric monoclonal antibody directed against EGFR and is the only FDA-approved targeted agent for use against HNSCC. Panitumumab (™Vectibix) is a "fully human" monoclonal antibody that is FDA approved in colorectal cancer but not in HNSCC. It is under investigation for HNSCC in multiple clinical trials in combination with standard chemoradiation therapy. Extensive clinical studies using cetuximab have demonstrated this agent to be particularly useful as an adjuvant to radiotherapy. In the multinational, randomized phase III trial that led to FDA approval of this agent, cetuximab combined with radiotherapy improved locoregional control and reduced mortality without increasing toxicity in patients with locoregionally advanced HNSCC [115]. More recent studies are showing increased benefit in progression-free survival as well as overall survival when adding cetuximab to platinum-based chemotherapy alone [116]. Despite the fact that *EGFR* is overexpressed in 80–90 % of HNSCC, the cumulative data have only shown a marginal survival benefit with *EGFR*-targeted therapies, with treatment efficacy in a mere 20 % of patients. At present, the mechanisms underlying the resistance to *EGFR*-targeted therapies are under investigation in hopes of improving clinical efficacy of this treatment strategy in HNSCC.

While monoclonal antibodies recognize a precise region in the extracellular ligand-binding domain of EGFR and therefore represent more specific EGFR targeting, small-molecule TKIs may cross-react with other kinases and lack specificity for EGFR. However, they have the advantage of being able to target multiple pathways involved in tumorigenesis as well as being conveniently dosed orally. Currently, all investigations using TKIs in HNSCC are limited to phase I and II clinical trials, with one reported phase III trial. Erlotinib and gefitinib are two of the most studied TKIs in HNSCC, but there is a lack of evidence to support their utility in HNSCC. In one randomized phase II study in patients with locally advanced HNSCC being treated with cisplatin and radiotherapy with or without erlotinib, the authors concluded that although erlotinib did not increase toxicity, there was no significant improvement in complete response rate or progression-free survival [117]. Likewise, the only randomized, placebo-controlled phase III trial of TKI in HNSCC evaluating the use of docetaxel with or without gefitinib failed to reveal improvement in outcomes in patients with recurrent or metastatic disease [118]. Another randomized, placebo-controlled phase II trial of gefitinib with chemoradiation therapy had similar disappointing findings [119]. Therefore, despite the convenient oral dosing of TKIs, the lack of positive phase III data limits their incorporation into the standard care for patients with HNSCC.

NOTCH PATHWAY INHIBITORS

The appropriate therapeutic targeting of *NOTCH* will differ depending on whether the tumor contains *NOTCH* gain- or loss-of-function alterations. For tumors harboring activating *NOTCH* mutations which would lead it to function as an oncogene, a variety of GSIs are being investigated as a possible targeted strategy to inactive NOTCH signaling [35]. GSIs act by preventing NICD cleavage and nuclear translocation [120]. GSIs have shown promise in in vitro and in vivo studies of many solid tumors, including breast, lung, colorectal, and pancreatic cancers as well as melanoma and sarcoma [121–125]. Currently, there are several ongoing phase I and II clinical trials of GSIs in advanced solid tumors [35, 126, 127].

In tumor systems where *NOTCH* acts as a tumor-suppressor gene and therefore is inactivated during oncogenesis, the appropriate strategy would be to activate the *NOTCH* signaling pathway. The *NOTCH* pathway is frequently silenced by epigenetic changes, and histone deacetylase (HDAC) inhibitors are under investigation to restore *NOTCH* signaling in cancers. Valproic acid, an HDAC inhibitor, is in clinical development and being studied in many ongoing phase I and II clinical trials in solid tumors [35]. Of note, it is important to recognize that administrating a systemic therapy that inhibits general *NOTCH* signaling may be complicated by loss of *NOTCH* tumor suppressor function in non-tumorous sites, thus potentially inducing secondary malignancy. Likewise, activating *NOTCH* signaling systemically to target a given tumor lineage may lead to activation of *NOTCH* signaling in a normal cell where *NOTCH* may act as an oncogene. Therefore, a better understanding of the exact *NOTCH* signaling pathway alterations within a cancer-specific context is necessary to develop appropriate *NOTCH*-targeting therapeutics.

Conclusions

In this chapter, we reviewed some of the current understanding of the molecular biology of HNSCC and discussed some of the diagnostic, prognostic, and therapeutic clinical implications. The advent of genomic technologies has greatly advanced our knowledge of the molecular changes underlying HNSCC, and the knowledge gained offers new promise for the treatment of this cancer. The hope is that the novel approaches will ultimately result in improved patient outcome through the development of new diagnostic and prognostic indicators as well as new targeted therapies for HNSCC patients.

REFERENCES

1. Ang KK, Harris J, Wheeler R, Weber R, Rosenthal DI, Nguyen-Tan PF, et al. Human papillomavirus and survival of patients with oropharyngeal cancer. N Engl J Med. 2010;363(1):24–35. PubMed PMID: 20530316.
2. Pelucchi C, Gallus S, Garavello W, Bosetti C, La Vecchia C. Alcohol and tobacco use, and cancer risk for upper aerodigestive tract and liver. Eur J Cancer Prev. 2008;17(4):340–4. PubMed PMID: 18562959. Epub 2008/06/20. eng.
3. Hashibe M, Brennan P, Chuang SC, Boccia S, Castellsague X, Chen C, et al. Interaction between tobacco and alcohol use and the risk of head and neck cancer: pooled analysis in the International Head and Neck Cancer Epidemiology Consortium. Cancer Epidemiol Biomarkers Prev. 2009;18(2):541–50. PubMed PMID: 19190158. Pubmed Central PMCID: 3051410. Epub 2009/02/05. eng.
4. Franceschi S, Talamini R, Barra S, Baron AE, Negri E, Bidoli E, et al. Smoking and drinking in relation to cancers of the oral cavity, pharynx, larynx, and esophagus in northern Italy. Cancer Res. 1990;50(20):6502–7. PubMed PMID: 2208109. Epub 1990/10/15. eng.
5. Zheng TZ, Boyle P, Hu HF, Duan J, Jiang PJ, Ma DQ, et al. Tobacco smoking, alcohol consumption, and risk of oral cancer: a case–control study in Beijing, People's Republic of China. Cancer Causes Control. 1990;1(2):173–9. PubMed PMID: 2102288. Epub 1990/09/01. eng.
6. Blot WJ, McLaughlin JK, Winn DM, Austin DF, Greenberg RS, Preston-Martin S, et al. Smoking and drinking in relation to oral and pharyngeal cancer. Cancer Res. 1988;48(11):3282–7. PubMed PMID: 3365707. Epub 1988/06/01. eng.
7. Talamini R, La Vecchia C, Levi F, Conti E, Favero A, Franceschi S. Cancer of the oral cavity and pharynx in nonsmokers who drink alcohol and in nondrinkers who smoke tobacco. J Natl Cancer Inst. 1998;90(24):1901–3. PubMed PMID: 9862628. Epub 1998/12/23. eng.
8. Talamini R, Bosetti C, La Vecchia C, Dal Maso L, Levi F, Bidoli E, et al. Combined effect of tobacco and alcohol on laryngeal cancer risk: a case–control study. Cancer Causes Control. 2002;13(10): 957–64. PubMed PMID: 12588092. Epub 2003/02/18. eng.
9. Wight AJ, Ogden GR. Possible mechanisms by which alcohol may influence the development of oral cancer–a review. Oral Oncol. 1998; 34(6):441–7. PubMed PMID: 9930353. Epub 1999/02/04. eng.
10. Hoffmann D, Hoffmann I, El-Bayoumy K. The less harmful cigarette: a controversial issue. a tribute to Ernst L. Wynder. Chem Res Toxicol. 2001;14(7):767–90. PubMed PMID: 11453723. Epub 2001/07/17. eng.
11. Arora A, Willhite CA, Liebler DC. Interactions of beta-carotene and cigarette smoke in human bronchial epithelial cells. Carcinogenesis. 2001;22(8):1173–8. PubMed PMID: 11470745. Epub 2001/07/27. eng.
12. Hecht SS. Tobacco smoke carcinogens and lung cancer. J Natl Cancer Inst. 1999;91(14):1194–210. PubMed PMID: 10413421. Epub 1999/07/21. eng.
13. Miller JA. Recent studies on the metabolic activation of chemical carcinogens. Cancer Res.

1994;54(7 Suppl):1879s–81. PubMed PMID: 8137303. Epub 1994/04/01. eng.
14. Pfeifer GP, Denissenko MF, Olivier M, Tretyakova N, Hecht SS, Hainaut P. Tobacco smoke carcinogens, DNA damage and p53 mutations in smoking-associated cancers. Oncogene. 2002; 21(48):7435–51. PubMed PMID: 12379884. Epub 2002/10/16. eng.
15. Feng J, Li L, Zhao YS, Tang SQ, Yang HB, Liu SX. Interaction between CYP 2C19*3 polymorphism and smoking in relation to laryngeal carcinoma in the Chinese Han population. Genet Mol Res. 2011;10(4):3331–7. PubMed PMID: 22183943. Epub 2011/12/21. eng.
16. Hashibe M, Brennan P, Strange RC, Bhisey R, Cascorbi I, Lazarus P, et al. Meta- and pooled analyses of GSTM1, GSTT1, GSTP1, and CYP1A1 genotypes and risk of head and neck cancer. Cancer Epidemiol Biomarkers Prev. 2003;12(12):1509–17. PubMed PMID: 14693745. Epub 2003/12/25. eng.
17. Boffetta P, Hashibe M. Alcohol and cancer. Lancet Oncol. 2006;7(2):149–56. PubMed PMID: 16455479. Epub 2006/02/04. eng.
18. Matsuda T, Yabushita H, Kanaly RA, Shibutani S, Yokoyama A. Increased DNA damage in ALDH2-deficient alcoholics. Chem Res Toxicol. 2006;19(10):1374–8. PubMed PMID: 17040107. Epub 2006/10/17. eng.
19. Munoz N, Bosch FX, de Sanjose S, Herrero R, Castellsague X, Shah KV, et al. Epidemiologic classification of human papillomavirus types associated with cervical cancer. N Engl J Med. 2003;348(6):518–27. PubMed PMID: 12571259. Epub 2003/02/07. eng.
20. Gillison ML, Broutian T, Pickard RK, Tong ZY, Xiao W, Kahle L, et al. Prevalence of oral HPV infection in the United States, 2009-2010. JAMA. 2012;307(7):693–703. PubMed PMID: 22282321. Epub 2012/01/28. eng.
21. Society AC. Cancer fact & figures Atlanta: American Cancer Society; 2012. http://www.cancer.org/research/cancerfactsfigures/cancerfactsfigures/cancer-facts-figures-2012.
22. Chaturvedi AK, Engels EA, Pfeiffer RM, Hernandez BY, Xiao W, Kim E, et al. Human papillomavirus and rising oropharyngeal cancer incidence in the United States. J Clin Oncol. 2011;29(32):4294–301. PubMed PMID: 21969503. Pubmed Central PMCID: 3221528. Epub 2011/10/05. eng.
23. Brennan JA, Boyle JO, Koch WM, Goodman SN, Hruban RH, Eby YJ, et al. Association between cigarette smoking and mutation of the p53 gene in squamous-cell carcinoma of the head and neck. N Engl J Med. 1995;332(11):712–7. PubMed PMID: 7854378.
24. Agrawal N, Frederick MJ, Pickering CR, Bettegowda C, Chang K, Li RJ, et al. Exome sequencing of head and neck squamous cell carcinoma reveals inactivating mutations in NOTCH1. Science. 2011;333(6046):1154–7. PubMed PMID: 21798897. Pubmed Central PMCID: 3162986. Epub 2011/07/30. eng.
25. Stransky N, Egloff AM, Tward AD, Kostic AD, Cibulskis K, Sivachenko A, et al. The mutational landscape of head and neck squamous cell carcinoma. Science. 2011;333(6046):1157–60. PubMed PMID: 21798893. Epub 2011/07/30. eng.
26. Haupt S, Berger M, Goldberg Z, Haupt Y. Apoptosis - the p53 network. J Cell Sci. 2003;116(Pt 20):4077–85. PubMed PMID: 12972501.
27. Muller PA, Vousden KH. p53 mutations in cancer. Nat Cell Biol. 2013;15(1):2–8. PubMed PMID: 23263379. Epub 2012/12/25. eng.
28. el-Naggar AK, Lai S, Luna MA, Zhou XD, Weber RS, Goepfert H, et al. Sequential p53 mutation analysis of pre-invasive and invasive head and neck squamous carcinoma. Int J Cancer. 1995;64(3):196–201. PubMed PMID: 7622308.
29. Ogmundsdottir HM, Hilmarsdottir H, Astvaldsdottir A, Johannsson JH, Holbrook WP. Oral lichen planus has a high rate of TP53 mutations. A study of oral mucosa in Iceland. Eur J Oral Sci. 2002;110(3):192–8. PubMed PMID: 12120703. Epub 2002/07/18. eng.
30. Boyle JO, Hakim J, Koch W, van der Riet P, Hruban RH, Roa RA, et al. The incidence of p53 mutations increases with progression of head and neck cancer. Cancer Res. 1993;53(19):4477–80. PubMed PMID: 8402617.
31. Field JK, Zoumpourlis V, Spandidos DA, Jones AS. p53 expression and mutations in squamous cell carcinoma of the head and neck: expression correlates with the patients' use of tobacco and alcohol. Cancer Detect Prev. 1994;18(3):197–208. PubMed PMID: 8076382. Epub 1994/01/01. eng.
32. Hsieh LL, Wang PF, Chen IH, Liao CT, Wang HM, Chen MC, et al. Characteristics of mutations in the p53 gene in oral squamous cell carcinoma associated with betel quid chewing and cigarette smoking in Taiwanese. Carcinogenesis. 2001;22(9):1497–503. PubMed PMID: 11532872. Epub 2001/09/05. eng.
33. Nicolas M, Wolfer A, Raj K, Kummer JA, Mill P, van Noort M, et al. Notch1 functions as a tumor suppressor in mouse skin. Nat Genet. 2003; 33(3):416–21. PubMed PMID: 12590261. Epub 2003/02/19. eng.
34. Proweller A, Tu L, Lepore JJ, Cheng L, Lu MM, Seykora J, et al. Impaired notch signaling promotes de novo squamous cell carcinoma formation. Cancer Res. 2006;66(15):7438–44. PubMed PMID: 16885339. Epub 2006/08/04. eng.
35. Egloff AM, Grandis JR. Molecular pathways: context-dependent approaches to Notch target-

ing as cancer therapy. Clin Cancer Res. 2012; 18(19):5188–95. PubMed PMID: 22773520. Pubmed Central PMCID: 3463728. Epub 2012/07/10. eng.
36. Weng AP, Ferrando AA, Lee W, Morris JP, Silverman LB, Sanchez-Irizarry C, et al. Activating mutations of NOTCH1 in human T cell acute lymphoblastic leukemia. Science. 2004;306(5694): 269–71. Epub 2004/10/09. eng.
37. Puente XS, Pinyol M, Quesada V, Conde L, Ordonez GR, Villamor N, et al. Whole-genome sequencing identifies recurrent mutations in chronic lymphocytic leukaemia. Nature. 2011;475(7354):101–5. PubMed PMID: 21642962. Pubmed Central PMCID: 3322590. Epub 2011/06/07. eng.
38. Zhang ZP, Sun YL, Fu L, Gu F, Zhang L, Hao XS. Correlation of Notch1 expression and activation to cisplatin-sensitivity of head and neck squamous cell carcinoma. Ai Zheng. 2009;28(2): 100–3.
39. Lin JT, Chen MK, Yeh KT, Chang CS, Chang TH, Lin CY, et al. Association of high levels of Jagged-1 and Notch-1 expression with poor prognosis in head and neck cancer. Ann Surg Oncol. 2010;17(11):2976–83. PubMed PMID: 20517681. Epub 2010/06/03. eng.
40. Gu F, Ma Y, Zhang Z, Zhao J, Kobayashi H, Zhang L, et al. Expression of Stat3 and Notch1 is associated with cisplatin resistance in head and neck squamous cell carcinoma. Oncol Rep. 2010;23(3):671–6. PubMed PMID: 20127005. Epub 2010/02/04. eng.
41. Zhang TH, Liu HC, Zhu LJ, Chu M, Liang YJ, Liang LZ, et al. Activation of Notch signaling in human tongue carcinoma. J Oral Pathol Med. 2011;40(1):37–45. PubMed PMID: 20819128. Epub 2010/09/08. eng.
42. Hijioka H, Setoguchi T, Miyawaki A, Gao H, Ishida T, Komiya S, et al. Upregulation of Notch pathway molecules in oral squamous cell carcinoma. Int J Oncol. 2010;36(4):817–22. PubMed PMID: 20198324. Epub 2010/03/04. eng.
43. Califano J, van der Riet P, Westra W, Nawroz H, Clayman G, Piantadosi S, et al. Genetic progression model for head and neck cancer: implications for field cancerization. Cancer Res. 1996; 56(11):2488–92. PubMed PMID: 8653682. Epub 1996/06/01. eng.
44. van der Riet P, Nawroz H, Hruban RH, Corio R, Tokino K, Koch W, et al. Frequent loss of chromosome 9p21-22 early in head and neck cancer progression. Cancer Res. 1994;54(5):1156–8. PubMed PMID: 8118798. Epub 1994/03/01. eng.
45. Miracca EC, Kowalski LP, Nagai MA. High prevalence of p16 genetic alterations in head and neck tumours. Br J Cancer. 1999;81(4):677–83. PubMed PMID: 10574255. Pubmed Central PMCID: 2362902. Epub 1999/11/26. eng.
46. Chen PL, Scully P, Shew JY, Wang JY, Lee WH. Phosphorylation of the retinoblastoma gene product is modulated during the cell cycle and cellular differentiation. Cell. 1989;58(6): 1193–8. PubMed PMID: 2673546.
47. Lukas J, Parry D, Aagaard L, Mann DJ, Bartkova J, Strauss M, et al. Retinoblastoma-protein-dependent cell-cycle inhibition by the tumour suppressor p16. Nature. 1995;375(6531):503–6. PubMed PMID: 7777060. Epub 1995/06/08. eng.
48. Kalyankrishna S, Grandis JR. Epidermal growth factor receptor biology in head and neck cancer. J Clin Oncol. 2006;24(17):2666–72. PubMed PMID: 16763281. Epub 2006/06/10. eng.
49. Klein JD, Grandis JR. The molecular pathogenesis of head and neck cancer. Cancer Biol Ther. 2010;9(1):1–7. PubMed PMID: 20038820. Pubmed Central PMCID: 3138532. Epub 2009/12/30. eng.
50. Rodrigo JP, Ramos S, Lazo PS, Alvarez I, Suarez C. Amplification of ERBB oncogenes in squamous cell carcinomas of the head and neck. Eur J Cancer. 1996;32A(11):2004–10. PubMed PMID: 8943688. Epub 1996/10/01. eng.
51. Ibrahim SO, Vasstrand EN, Liavaag PG, Johannessen AC, Lillehaug JR. Expression of c-erbB proto-oncogene family members in squamous cell carcinoma of the head and neck. Anticancer Res. 1997;17(6D):4539–46. PubMed PMID: 9494565. Epub 1998/03/12. eng.
52. Sok JC, Coppelli FM, Thomas SM, Lango MN, Xi S, Hunt JL, et al. Mutant epidermal growth factor receptor (EGFRvIII) contributes to head and neck cancer growth and resistance to EGFR targeting. Clin Cancer Res. 2006;12(17):5064–73. PubMed PMID: 16951222. Epub 2006/09/05. eng.
53. Hynes NE, Lane HA. ERBB receptors and cancer: the complexity of targeted inhibitors. Nat Rev Cancer. 2005;5(5):341–54. PubMed PMID: 15864276. Epub 2005/05/03. eng.
54. Rogers SJ, Harrington KJ, Rhys-Evans P, O-Charoenrat P, Eccles SA. Biological significance of c-erbB family oncogenes in head and neck cancer. Cancer Metastasis Rev. 2005;24(1):47–69. PubMed PMID: 15785872. Epub 2005/03/24. eng.
55. Leslie NR, Downes CP. PTEN function: how normal cells control it and tumour cells lose it. Biochem J. 2004;382(Pt 1):1–11. PubMed PMID: 15193142. Pubmed Central PMCID: 1133909. Epub 2004/06/15. eng.
56. Pedrero JM, Carracedo DG, Pinto CM, Zapatero AH, Rodrigo JP, Nieto CS, et al. Frequent genetic and biochemical alterations of the PI 3-K/AKT/PTEN pathway in head and neck squamous cell carcinoma. Int J Cancer. 2005;114(2):242–8. PubMed PMID: 15543611. Epub 2004/11/16. eng.
57. Rautava J, Syrjanen S. Biology of human papillomavirus infections in head and neck carcinogen-

esis. Head Neck Pathol. 2012;6 Suppl 1:S3–15. PubMed PMID: 22782219. Pubmed Central PMCID: 3394166. Epub 2012/07/13. eng.
58. Scheffner M, Werness BA, Huibregtse JM, Levine AJ, Howley PM. The E6 oncoprotein encoded by human papillomavirus types 16 and 18 promotes the degradation of p53. Cell. 1990;63(6):1129–36.
59. Münger K, Baldwin A, Edwards KM, Hayakawa H, Nguyen CL, Owens M, et al. Mechanisms of human papillomavirus-induced oncogenesis. J Virol. 2004;78(21):11451–60.
60. Straight SW, Hinkle PM, Jewers RJ, McCance DJ. The E5 oncoprotein of human papillomavirus type 16 transforms fibroblasts and effects the downregulation of the epidermal growth factor receptor in keratinocytes. J Virol. 1993;67(8): 4521–32. PubMed PMID: 8392596. Pubmed Central PMCID: 237836. Epub 1993/08/01. eng.
61. Singhi AD, Westra WH. Comparison of human papillomavirus in situ hybridization and p16 immunohistochemistry in the detection of human papillomavirus-associated head and neck cancer based on a prospective clinical experience. Cancer. 2010;116(9):2166–73. PubMed PMID: 20186832. Epub 2010/02/27. eng.
62. Huang CC, Qiu JT, Kashima ML, Kurman RJ, Wu TC. Generation of type-specific probes for the detection of single-copy human papillomavirus by a novel in situ hybridization method. Mod Pathol. 1998;11(10):971–7. PubMed PMID: 9796725. Epub 1998/10/31. eng.
63. Wittekindt C, Gultekin E, Weissenborn SJ, Dienes HP, Pfister HJ, Klussmann JP. Expression of p16 protein is associated with human papillomavirus status in tonsillar carcinomas and has implications on survival. Adv Otorhinolaryngol. 2005;62:72–80. PubMed PMID: 15608419. Epub 2004/12/21. eng.
64. Shi W, Kato H, Perez-Ordonez B, Pintilie M, Huang S, Hui A, et al. Comparative prognostic value of HPV16 E6 mRNA compared with in situ hybridization for human oropharyngeal squamous carcinoma. J Clin Oncol. 2009;27(36): 6213–21. PubMed PMID: 19884544. Epub 2009/11/04. eng.
65. Capone RB, Pai SI, Koch WM, Gillison ML, Danish HN, Westra WH, et al. Detection and quantitation of human papillomavirus (HPV) DNA in the sera of patients with HPV-associated head and neck squamous cell carcinoma. Clin Cancer Res. 2000;6(11):4171–5. PubMed PMID: 11106228. Epub 2000/12/06. eng.
66. Chuang AY, Chuang TC, Chang S, Zhou S, Begum S, Westra WH, et al. Presence of HPV DNA in convalescent salivary rinses is an adverse prognostic marker in head and neck squamous cell carcinoma. Oral Oncol. 2008;44(10):915–9. PubMed PMID: 18329326. Pubmed Central PMCID: 3215237. Epub 2008/03/11. eng.
67. Conway C, Chalkley R, High A, Maclennan K, Berri S, Chengot P, et al. Next-generation sequencing for simultaneous determination of human papillomavirus load, subtype, and associated genomic copy number changes in tumors. J Mol Diagn. 2012;14(2):104–11.
68. Barzon L, Militello V, Lavezzo E, Franchin E, Peta E, Squarzon L, et al. Human papillomavirus genotyping by 454 next generation sequencing technology. J Clin Virol. 2011;52(2):93–7. PubMed PMID: 21802982. Epub 2011/08/02. eng.
69. Begum S, Gillison ML, Nicol TL, Westra WH. Detection of human papillomavirus-16 in fine-needle aspirates to determine tumor origin in patients with metastatic squamous cell carcinoma of the head and neck. Clin Cancer Res. 2007;13(4):1186–91. PubMed PMID: 17317828. Epub 2007/02/24. eng.
70. Umudum H, Rezanko T, Dag F, Dogruluk T. Human papillomavirus genome detection by in situ hybridization in fine-needle aspirates of metastatic lesions from head and neck squamous cell carcinomas. Cancer. 2005;105(3):171–7. PubMed PMID: 15822131. Epub 2005/04/12. eng.
71. de Braud F, al-Sarraf M. Diagnosis and management of squamous cell carcinoma of unknown primary tumor site of the neck. Semin Oncol. 1993;20(3):273–8.
72. Terwisscha van Scheltinga AG, van Dam GM, Nagengast WB, Ntziachristos V, Hollema H, Herek JL, et al. Intraoperative near-infrared fluorescence tumor imaging with vascular endothelial growth factor and human epidermal growth factor receptor 2 targeting antibodies. J Nucl Med. 2011;52(11):1778–85. PubMed PMID: 21990576. Epub 2011/10/13. eng.
73. van Dam GM, Themelis G, Crane LM, Harlaar NJ, Pleijhuis RG, Kelder W, et al. Intraoperative tumor-specific fluorescence imaging in ovarian cancer by folate receptor-alpha targeting: first in-human results. Nat Med. 2011;17(10):1315–9. PubMed PMID: 21926976. Epub 2011/09/20. eng.
74. Heath CH, Deep NL, Sweeny L, Zinn KR, Rosenthal EL. Use of panitumumab-IRDye800 to image microscopic head and neck cancer in an orthotopic surgical model. Ann Surg Oncol. 2012;19(12):3879–87. PubMed PMID: 22669455. Pubmed Central PMCID: 3595117. Epub 2012/06/07. eng.
75. Qi S, Miao Z, Liu H, Xu Y, Feng Y, Cheng Z. Evaluation of four affibody-based near-infrared fluorescent probes for optical imaging of epidermal growth factor receptor positive tumors. Bioconjug Chem. 2012;23:1149–56. PubMed PMID: 22621238. Epub 2012/05/25. Eng.
76. Agnes RS, Broome AM, Wang J, Verma A, Lavik K, Basilion JP. An optical probe for noninvasive molecular imaging of orthotopic brain tumors overexpressing epidermal growth factor receptor.

Mol Cancer Ther. 2012;11(10):2202–11. PubMed PMID: 22807580. Epub 2012/07/19. eng.
77. Chan LW, Wang YN, Lin LY, Upton MP, Hwang JH, Pun SH. Synthesis and characterization of anti-EGFR fluorescent nanoparticles for optical molecular imaging. Bioconjug Chem. 2013; 24(2):167–75. PubMed PMID: 23273065. Pubmed Central PMCID: 3578128. Epub 2013/01/01. eng.
78. Oliveira S, van Dongen GA, Stigter-van Walsum M. Roovers RC. Mali W, et al. Rapid Visualization of Human Tumor Xenografts through Optical Imaging with a Near-infrared Fluorescent Anti-Epidermal Growth Factor Receptor Nanobody. Molecular imaging: Stam JC; 2011. PubMed PMID: 21635822. Epub 2011/06/04. Eng.
79. Poeta ML, Manola J, Goldwasser MA, Forastiere A, Benoit N, Califano JA, et al. TP53 mutations and survival in squamous-cell carcinoma of the head and neck. N Engl J Med. 2007;357(25):2552–61. PubMed PMID: 18094376.
80. Ganly I, Kirn D, Eckhardt G, Rodriguez GI, Soutar DS, Otto R, et al. A phase I study of Onyx-015, an E1B attenuated adenovirus, administered intratumorally to patients with recurrent head and neck cancer. Clin Cancer Res. 2000;6(3):798–806. PubMed PMID: 10741699.
81. Skinner HD, Sandulache VC, Ow TJ, Meyn RE, Yordy JS, Beadle BM, et al. TP53 disruptive mutations lead to head and neck cancer treatment failure through inhibition of radiation-induced senescence. Clin Cancer Res. 2012;18(1):290–300. PubMed PMID: 22090360. Pubmed Central PMCID: 3251726. Epub 2011/11/18. eng.
82. Koch WM, Brennan JA, Zahurak M, Goodman SN, Westra WH, Schwab D, et al. p53 mutation and locoregional treatment failure in head and neck squamous cell carcinoma. J Natl Cancer Inst. 1996;88(21):1580–6. PubMed PMID: 8901856. Epub 1996/11/06. eng.
83. Alsner J, Sorensen SB, Overgaard J. TP53 mutation is related to poor prognosis after radiotherapy, but not surgery, in squamous cell carcinoma of the head and neck. Radiother Oncol. 2001;59(2):179–85. PubMed PMID: 11325447.
84. Temam S, Flahault A, Perie S, Monceaux G, Coulet F, Callard P, et al. p53 gene status as a predictor of tumor response to induction chemotherapy of patients with locoregionally advanced squamous cell carcinomas of the head and neck. J Clin Oncol. 2000;18(2):385–94. PubMed PMID: 10637254.
85. Cabelguenne A, Blons H, de Waziers I, Carnot F, Houllier AM, Soussi T, et al. p53 alterations predict tumor response to neoadjuvant chemotherapy in head and neck squamous cell carcinoma: a prospective series. J Clin Oncol. 2000;18(7): 1465–73. PubMed PMID: 10735894. Epub 2000/03/29. eng.
86. Brennan JA, Mao L, Hruban RH, Boyle JO, Eby YJ, Koch WM, et al. Molecular assessment of histopathological staging in squamous-cell carcinoma of the head and neck. N Engl J Med. 1995;332(7):429–35.
87. van Houten VM, Leemans CR, Kummer JA, Dijkstra J, Kuik DJ, van den Brekel MW, et al. Molecular diagnosis of surgical margins and local recurrence in head and neck cancer patients: a prospective study. Clin Cancer Res. 2004;10(11): 3614–20. PubMed PMID: 15173067. Epub 2004/06/03. eng.
88. Pena Murillo C, Huang X, Hills A, McGurk M, Lyons A, Jeannon JP, et al. The utility of molecular diagnostics to predict recurrence of head and neck carcinoma. Br J Cancer. 2012;107(7):1138–43. PubMed PMID: 22918395. Pubmed Central PMCID: 3461148. Epub 2012/08/25. eng.
89. Lopez-Martinez M, Anzola M, Cuevas N, Aguirre JM, De-Pancorbo M. Clinical applications of the diagnosis of p53 alterations in squamous cell carcinoma of the head and neck. Med Oral. 2002;7(2):108–20. PubMed PMID: 11887018. Epub 2002/03/12. eng.spa.
90. Ragin CC, Taioli E. Survival of squamous cell carcinoma of the head and neck in relation to human papillomavirus infection: review and meta-analysis. Int J Cancer. 2007;121(8):1813–20. PubMed PMID: 17546592. Epub 2007/06/05. eng.
91. Fakhry C, Westra WH, Li S, Cmelak A, Ridge JA, Pinto H, et al. Improved survival of patients with human papillomavirus-positive head and neck squamous cell carcinoma in a prospective clinical trial. J Natl Cancer Inst. 2008;100(4):261–9. PubMed PMID: 18270337. Epub 2008/02/14. eng.
92. Sedaghat AR, Zhang Z, Begum S, Palermo R, Best S, Ulmer KM, et al. Prognostic significance of human papillomavirus in oropharyngeal squamous cell carcinomas. Laryngoscope. 2009;119(8):1542–9. PubMed PMID: 19522004. Epub 2009/06/13. eng.
93. Geisler SA, Olshan AF, Cai J, Weissler M, Smith J, Bell D. Glutathione S-transferase polymorphisms and survival from head and neck cancer. Head Neck. 2005;27(3):232–42. PubMed PMID: 15668931. Epub 2005/01/26. eng.
94. Nishimura T, Newkirk K, Sessions RB, Andrews PA, Trock BJ, Rasmussen AA, et al. Immunohistochemical staining for glutathione S-transferase predicts response to platinum-based chemotherapy in head and neck cancer. Clin Cancer Res. 1996;2(11):1859–65. PubMed PMID: 9816141. Epub 1996/11/01. eng.
95. Shiga H, Heath EI, Rasmussen AA, Trock B, Johnston PG, Forastiere AA, et al. Prognostic value of p53, glutathione S-transferase pi, and thymidylate synthase for neoadjuvant cisplatin-based chemotherapy in head and neck cancer. Clin Cancer Res. 1999;5(12):4097–104. PubMed PMID: 10632346. Epub 2000/01/13. eng.

96. Schumaker L, Nikitakis N, Goloubeva O, Tan M, Taylor R, Cullen KJ. Elevated expression of glutathione S-transferase pi and p53 confers poor prognosis in head and neck cancer patients treated with chemoradiotherapy but not radiotherapy alone. Clin Cancer Res. 2008;14(18):5877–83. PubMed PMID: 18794100. Epub 2008/09/17. eng.
97. Szanyi I, Rath G, Moricz P, Somogyvari K, Revesz P, Gerlinger I, et al. Effects of cytochrome P450 1A1 and uridine-diphosphate-glucuronosyltransferase 1A1 allelic polymorphisms on the risk of development and the prognosis of head and neck cancers. Eur J Cancer Prev. 2012;21(6):560–8. PubMed PMID: 22273851. Epub 2012/01/26. eng.
98. Sturgis EM, Ang KK. The epidemic of HPV-associated oropharyngeal cancer is here: is it time to change our treatment paradigms? J Natl Compr Canc Netw. 2011;9(6):665–73. PubMed PMID: 21636538. Epub 2011/06/04. eng.
99. Moore EJ, Olsen KD, Kasperbauer JL. Transoral robotic surgery for oropharyngeal squamous cell carcinoma: a prospective study of feasibility and functional outcomes. Laryngoscope. 2009;119(11):2156–64. PubMed PMID: 19824067. Epub 2009/10/14. eng.
100. Weinstein GS, Quon H, O'Malley Jr BW, Kim GG, Cohen MA. Selective neck dissection and deintensified postoperative radiation and chemotherapy for oropharyngeal cancer: a subset analysis of the University of Pennsylvania transoral robotic surgery trial. Laryngoscope. 2010; 120(9):1749–55. PubMed PMID: 20717944. Epub 2010/08/19. eng.
101. White HN, Moore EJ, Rosenthal EL, Carroll WR, Olsen KD, Desmond RA, et al. Transoral robotic-assisted surgery for head and neck squamous cell carcinoma: one- and 2-year survival analysis. Arch Otolaryngol Head Neck Surg. 2010;136(12):1248–52. PubMed PMID: 21173375. Epub 2010/12/22. eng.
102. Cohen MA, Weinstein GS, O'Malley Jr BW, Feldman M, Quon H. Transoral robotic surgery and human papillomavirus status: Oncologic results. Head Neck. 2011;33(4):573–80. PubMed PMID: 21425382. Epub 2011/03/23. eng.
103. Moore EJ, Olsen SM, Laborde RR, Garcia JJ, Walsh FJ, Price DL, et al. Long-term functional and oncologic results of transoral robotic surgery for oropharyngeal squamous cell carcinoma. Mayo Clin Proc. 2012;87(3):219–25. PubMed PMID: 22386176. Pubmed Central PMCID: 3538408. Epub 2012/03/06. eng.
104. Schiller JT, Castellsague X, Garland SM. A review of clinical trials of human papillomavirus prophylactic vaccines. Vaccine. 2012;30 Suppl 5:F123–38. PubMed PMID: 23199956. Epub 2012/12/05. eng.
105. Joura EA, Garland SM, Paavonen J, Ferris DG, Perez G, Ault KA, et al. Effect of the human papillomavirus (HPV) quadrivalent vaccine in a subgroup of women with cervical and vulvar disease: retrospective pooled analysis of trial data. BMJ. 2012;344:e1401. PubMed PMID: 22454089. Pubmed Central PMCID: 3314184. Epub 2012/03/29. eng.
106. Dillner J, Kjaer SK, Wheeler CM, Sigurdsson K, Iversen OE, Hernandez-Avila M, et al. Four year efficacy of prophylactic human papillomavirus quadrivalent vaccine against low grade cervical, vulvar, and vaginal intraepithelial neoplasia and anogenital warts: randomised controlled trial. BMJ. 2010;341:c3493. PubMed PMID: 20647284. Pubmed Central PMCID: 2907480. Epub 2010/07/22. eng.
107. Ma B, Maraj B, Tran NP, Knoff J, Chen A, Alvarez RD, et al. Emerging human papillomavirus vaccines. Expert Opin Emerg Drugs. 2012;17(4):469–92. PubMed PMID: 23163511. Epub 2012/11/21. eng.
108. Center TSKCC. Baltimore: Johns Hopkins Medicine; 2013. http://www.hopkinsmedicine.org/kimmel_cancer_center/research_clinical_trials/clinical_trials/search_results/J11129.
109. Best SR, Niparko KJ, Pai SI. Biology of human papillomavirus infection and immune therapy for HPV-related head and neck cancers. Otolaryngol Clin North Am. 2012;45(4):807–22. PubMed PMID: 22793854. Pubmed Central PMCID: 3398423. Epub 2012/07/17. eng.
110. Zhang WW, Fang X, Mazur W, French BA, Georges RN, Roth JA. High-efficiency gene transfer and high-level expression of wild-type p53 in human lung cancer cells mediated by recombinant adenovirus. Cancer Gene Ther. 1994;1(1):5–13. PubMed PMID: 7621238. Epub 1994/03/01. eng.
111. Liu TJ, Zhang WW, Taylor DL, Roth JA, Goepfert H, Clayman GL. Growth suppression of human head and neck cancer cells by the introduction of a wild-type p53 gene via a recombinant adenovirus. Cancer Res. 1994;54(14):3662–7.
112. Pirollo KF, Hao Z, Rait A, Jang YJ, Fee Jr WE, Ryan P, et al. p53 mediated sensitization of squamous cell carcinoma of the head and neck to radiotherapy. Oncogene. 1997;14(14):1735–46. PubMed PMID: 9135075. Epub 1997/04/10. eng.
113. Nemunaitis J, Clayman G, Agarwala SS, Hrushesky W, Wells JR, Moore C, et al. Biomarkers Predict p53 Gene Therapy Efficacy in Recurrent Squamous Cell Carcinoma of the Head and Neck. Clin Cancer Res. 2009;15(24):7719–25. PubMed PMID: 19996201.
114. Nemunaitis J, Nemunaitis J. Head and neck cancer: response to p53-based therapeutics. Head Neck. 2011;33(1):131–4. PubMed PMID: 20222046.

115. Bonner JA, Harari PM, Giralt J, Azarnia N, Shin DM, Cohen RB, et al. Radiotherapy plus cetuximab for squamous-cell carcinoma of the head and neck. N Engl J Med. 2006;354(6):567–78. PubMed PMID: 16467544. Epub 2006/02/10. eng.
116. Vermorken JB, Mesia R, Rivera F, Remenar E, Kawecki A, Rottey S, et al. Platinum-based chemotherapy plus cetuximab in head and neck cancer. N Engl J Med. 2008;359(11):1116–27. PubMed PMID: 18784101. Epub 2008/09/12. eng.
117. Martins RG, Parvathaneni U, Bauman JE, Sharma AK, Raez LE, Papagikos MA, et al. Cisplatin and radiotherapy with or without erlotinib in locally advanced squamous cell carcinoma of the head and neck: a randomized phase II trial. J Clin Oncol. 2013;31(11):1415–21. PubMed PMID: 23460709. Epub 2013/03/06. eng.
118. Argiris A, Ghebremichael M, Gilbert J, Lee JW, Sachidanandam K, Kolesar JM, et al. Phase III randomized, placebo-controlled trial of docetaxel with or without gefitinib in recurrent or metastatic head and neck cancer: an eastern cooperative oncology group trial. J Clin Oncol. 2013;31(11):1405–14. PubMed PMID: 23460714. Pubmed Central PMCID: 3612594. Epub 2013/03/06. eng.
119. Gregoire V, Hamoir M, Chen C, Kane M, Kawecki A, Julka PK, et al. Gefitinib plus cisplatin and radiotherapy in previously untreated head and neck squamous cell carcinoma: a phase II, randomized, double-blind, placebo-controlled study. Radiother Oncol. 2011;100(1):62–9. PubMed PMID: 21821303. Epub 2011/08/09. eng.
120. Fortini ME. Gamma-secretase-mediated proteolysis in cell-surface-receptor signalling. Nat Rev Mol Cell Biol. 2002;3(9):673–84. PubMed PMID: 12209127.
121. Curry CL, Reed LL, Golde TE, Miele L, Nickoloff BJ, Foreman KE. Gamma secretase inhibitor blocks Notch activation and induces apoptosis in Kaposi's sarcoma tumor cells. Oncogene. 2005;24(42):6333–44. PubMed PMID: 15940249. Epub 2005/06/09. eng.
122. Huynh C, Poliseno L, Segura MF, Medicherla R, Haimovic A, Menendez S, et al. The novel gamma secretase inhibitor RO4929097 reduces the tumor initiating potential of melanoma. PLoS One. 2011;6(9):e25264. PubMed PMID: 21980408. Pubmed Central PMCID: 3182998. Epub 2011/10/08. eng.
123. Maraver A, Fernandez-Marcos PJ, Herranz D, Canamero M, Munoz-Martin M, Gomez-Lopez G, et al. Therapeutic effect of gamma-secretase inhibition in KrasG12V-driven non-small cell lung carcinoma by derepression of DUSP1 and inhibition of ERK. Cancer Cell. 2012;22(2):222–34. PubMed PMID: 22897852. Epub 2012/08/18. eng.
124. Portanova P, Notaro A, Pellerito O, Sabella S, Giuliano M, Calvaruso G. Notch inhibition restores TRAIL-mediated apoptosis via AP1-dependent upregulation of DR4 and DR5 TRAIL receptors in MDA-MB-231 breast cancer cells. Int J Oncol. 2013;43(1):121–30. PubMed PMID: 23686163. Epub 2013/05/21. eng.
125. Yabuuchi S, Pai SG, Campbell NR, de Wilde RF, De Oliveira E, Korangath P, et al. Notch signaling pathway targeted therapy suppresses tumor progression and metastatic spread in pancreatic cancer. Cancer Lett. 2013;335(1):41–51. PubMed PMID: 23402814. Pubmed Central PMCID: 3665739. Epub 2013/02/14. eng.
126. Tolcher AW, Messersmith WA, Mikulski SM, Papadopoulos KP, Kwak EL, Gibbon DG, et al. Phase I study of RO4929097, a gamma secretase inhibitor of Notch signaling, in patients with refractory metastatic or locally advanced solid tumors. J Clin Oncol. 2012;30(19):2348–53. PubMed PMID: 22529266. Epub 2012/04/25. eng.
127. Richter S, Bedard PL, Chen EX, Clarke BA, Tran B, Hotte SJ, et al. A phase I study of the oral gamma secretase inhibitor R04929097 in combination with gemcitabine in patients with advanced solid tumors (PHL-078/CTEP 8575). Invest New Drugs. 2013. PubMed PMID: 23645447. Epub 2013/05/07. Eng.

CHAPTER 23

GENOMIC APPLICATIONS IN BREAST CARCINOMA

LETICIA DE MATTOS-ARRUDA, BRITTA WEIGELT, JORGE S. REIS-FILHO

Introduction 359
The Molecular Classification of Breast Cancer 360
Gene-Expression Prognostic Signatures 364
First-Generation Prognostic Gene Signatures 364
Second-Generation Prognostic Gene Signatures 369
Gene Expression Predictive Signatures 371
Massively Parallel Sequencing and the Impact in Intra-tumor Genetic Heterogeneity 371
Conclusions 373

Introduction

Breast cancer represents the worldwide leading cause of cancer and mortality among women [1]. The realization that breast cancer is a complex and heterogeneous disease in which each entity has a distinct genetic profile, clinical behavior, and response to therapy has created new opportunities for precision medicine [2].

The management of breast cancer is currently based on clinicopathological characteristics derived from the histopathological analysis of the primary tumor and/or its metastasis, which are known to be prognostic per se. They include tumor size, histological grade and type, lymph vascular invasion, and lymph node involvement [3, 4]. In addition to these prognostic markers, only three predictive markers have been incorporated to the armamentarium of breast pathologists. In fact, critical decisions about the repertoire of systemic therapies offered to breast cancer patients are made on the basis of the assessment of the estrogen receptor (ER), progesterone receptor (PR), and the human epidermal growth factor receptor 2 (HER2). ER and PR assessment is based on immunohistochemical analysis, whereas the HER2 status is currently determined on the basis of a combination of immunohistochemistry and in situ hybridization [5, 6]. Given the importance of the results obtained from these tests and the reported limited inter-laboratory agreement, over the last few years, coordinated efforts to ensure high levels of intra- and inter-laboratory and observer concordance have been made, and recent updates of the American Society of Clinical Oncologists (ASCO)/College of American Pathologists (CAP) guidelines have been published. Discussing the details of these guidelines is beyond the scope of this chapter and readers are referred to the published guidelines [5, 6].

Based on combinations of several clinicopathological prognostic factors and predictive markers, some algorithms (e.g., Nottingham Prognosis Index, Adjuvant! Online) have been created to forecast the outcome of

G.J. Netto and I. Schrijver (eds.), *Genomic Applications in Pathology*,
DOI 10.1007/978-1-4939-0727-4_23, © Springer Science+Business Media New York 2015

breast cancer patients and the subsequent benefit of adjuvant systemic therapy [7, 8]. For example, the Internet-based software program Adjuvant! Online, which was tailored for clinical decision-making in early breast cancer, provides a 10-year survival and relapse probability of an individual patient [8, 9]. Despite being widely used in the clinic and validated in different series, these algorithms have not been found to be sufficiently accurate to individualize therapy for patients with early breast cancer [10, 11].

Advances in gene expression analysis during the last decade have contributed to the notion that breast cancer comprises multiple diseases with different risk factors, clinicopathological features, natural history, and response to therapy [2, 12–15]. Microarray-based gene expression profiling has led to the development of a classification system solely based on the expression of the so-called “intrinsic” genes [2] and to the development of prognostic gene signatures, some of which have now been incorporated into clinical practice [16]. On the other hand, microarray-based signatures predictive of response to specific therapeutic agents have proven more challenging to develop, and currently, none of the signatures available has been shown to be of clinical utility.

With the advent of massively parallel sequencing technologies and the ability to determine the entire constellation of somatic and germline genetic aberrations in a cancer, new opportunities for the realization of the potential of precision medicine have emerged [17]. Not only the complexity of breast cancers at the genetic level has been unraveled, but also the fact that breast cancer from specific subtypes may be underpinned by a different repertoire of somatic mutations. Furthermore, activating mutations affecting the *ERBB2* (*HER2*) [18] and *ESR1* [19–21] genes have been recently described; these findings will likely have an impact on the way ER and HER2 testing is performed for breast cancer patients.

Here, we discuss the current role of gene expression profiling as a means of classification, prognostication, and therapy prediction in breast cancer, and the impact of sequencing analyses on breast cancer classification and therapy decisions for breast cancer patients.

The Molecular Classification of Breast Cancer

Microarray-based gene expression profiling has resulted in the development of a breast cancer classification comprising five “intrinsic” molecular subtypes, namely, luminal A, luminal B, HER2 (also known as HER2-enriched), basal-like, and normal breast-like [2, 12, 13]. The term “intrinsic subtypes” stems from the fact that this classification was developed based on hierarchical cluster analysis of breast cancer using the “intrinsic” genes, which are defined as genes that vary more between breast cancers than between repeated samples of the same tumors. The “intrinsic” molecular classification has made it evident that ER-positive and ER-negative breast cancer are essentially different diseases at the transcriptomic level [2, 12, 13, 15, 22]. Furthermore, studies investigating the clinicopathological features of these cancers revealed that if luminal A and basal-like breast cancers are compared, they differ in terms of risk factors, clinicopathological presentation, histopathological features, response to therapy, and outcomes [15].

In-depth analyses of the transcriptomic profiles of luminal A, luminal B, HER2-enriched, basal-like, and normal breast-like revealed important characteristics of these molecular subtypes. Luminal tumors are characterized by the expression of the ER gene (*ESR1*) and ER-related genes, and are subclassified into luminal A and luminal B subtypes based on the level of expression of proliferation-related genes, whereby luminal A tumors display low levels of expression of proliferation-related genes, whereas luminal B cancers display higher levels [23–26]. HER2 or HER2-enriched cancers are characterized by expression of the *HER2* gene (*ERBB2*) and of genes found in the *HER2* amplicon. It should be noted, however, that not all HER2 enriched breast cancers display *HER2* gene amplification, and not all cases diagnosed as HER2-positive according to the ASCO/CAP guidelines are classified as HER2-enriched by microarray analysis [6, 27]. The basal-like subtype was so named because the transcriptomic profiles of these cancers comprise genes that are usually expressed by normal breast epithelial/ basal cells. Normal breast-like cancers, on the other hand, have proven more controversial. There are several lines of evidence to suggest that this subtype is a mere artifact of gene expression profiling, being the result of "intrinsic" subtyping of samples with a disproportionately high content of normal breast epithelial cells and/ or stromal cells [15, 23, 27, 28].

The terms luminal A, luminal B, HER2, and basal-like have been incorporated to the lexicon of oncologists, surgeons, pathologists, and basic scientists, and the "intrinsic" classification has gained widespread use. It should be noted, however, that the analytical validity, clinical validity and, most importantly, clinical utility of the subtypes remains to be determined. Recently, the IMPAKT 2012 Working Group has stated that the molecular subtype classification based on the PAM50 test and the immunohistochemistry analysis of ER, HER2, and Ki67 (14 % cut-off value) does not satisfactorily present information to modify systemic treatment decisions [29]. As for the identification of clinically relevant subtypes of breast cancers, the IMPAKT 2012 Working Group recommended the use of immunohistochemistry for ER and HER2.

Due to limitations of hierarchical clustering analysis for the classification of single breast cancer samples in a prospective manner [30], single sample predictors have been developed [13]. They allow for gene expression-based subtyping of individual tumors based on microarray gene expression profiling. Microarray-based single sample predictors, however, seem to have limited reproducibility and require extensive and rather complex processing of the microarray data to be applied for the classification of individual samples [26, 31]. To overcome these limitations and to allow for the use of archival material, the PAM50 assay has been developed. This is an nCounter-based assay based on the expression of 50 genes, and classifies breast cancers into the four major intrinsic subtypes (i.e., luminal A, luminal B, HER2-enriched, and basal-like; please note that the normal-like subtype was removed as it is currently perceived as a likely artifact of having a high percentage of normal cell contamination) [27]. Importantly, immunohistochemistry surrogate definitions have gained widespread use in the last few years due to their similarities with breast cancer molecular subtypes as defined by gene expression profiling. Indeed, based on the recognition of "intrinsic" breast cancer subtypes, this immunohistochemistry surrogate classification was accept by the 12th St. Gallen International Breast Cancer Conference Expert Panel as a new approach for therapeutic purposes [32]. Nevertheless, it has been recognized that disagreement between the PAM50 assay and immunohistochemistry may lead to different treatment decisions (Table 23.1) [33].

In addition to the "intrinsic" subtypes, microarray-based class discovery studies have resulted in the identification of additional molecular subtypes, which are predominantly of ER-negative phenotype. The molecular apocrine subtype of breast cancer has been identified by independent investigators [34–36], and is characterized by low or no expression of ER, and expression of androgen receptor (AR) and AR-related genes [34–36]. These tumors have been shown to have an aggressive clinical outcome [36] and to display some molecular and histopathological features consistent with apocrine differentiation. Through an analysis of conditional mouse models, breast cancer cell lines, and primary breast cancers, the claudin-low subtype has been identified [28, 37]. These tumors are characterized by low levels of expression of the tight junction proteins claudin 3, 4, and 7 and other adhesion molecules, including E-cadherin, and with transcriptomic features similar to those of breast cancer initiating cells and epithelial-to-mesenchymal transition. In comparison with other intrinsic subtypes, claudin-low tumors display low levels of expression of ER and ER-related genes and intermediate levels of expression of proliferation-related genes. Although initially perceived as a variant of triple-negative breast cancers (TNBCs), up to 33 and 22 % of claudin-low cancers may be ER and HER2 positive by immunohistochemical analysis [28]. From an immunohistochemical standpoint, it should be emphasized that up to 41 and 55 % of tumors classified as claudin-low by gene expression profiling express claudin 3 and E-cadherin, respectively [28].

Table 23-1 **Molecular Subtypes of Breast Cancer According to Gene Expression Profiling and Immunohistochemistry**

Intrinsic subtype as per gene expression profiling	Luminal A	Luminal B	Luminal B	HER2-enriched	Basal-like
2011 St. Gallen as per IHC subtype [32]	ER and/or PR positive, HER2 negative, Ki67 < 14 %	ER and/or PR positive, HER2 negative, Ki-67 ≥14 %	ER and/or PR positive, HER2 positive, Ki-67: any	ER and/or PR negative, HER2 positive	ER and/or PR negative, HER2 negative
Agreement	73–100 %	73–100 %		41–70 %	80 %
Recommendation of adjuvant systemic therapy	Endocrine therapy	Endocrine therapy ± chemotherapy	Chemotherapy + anti-HER2 + endocrine therapy	Chemotherapy + anti-HER2 therapy	Chemotherapy

IHC = immunohistochemistry, ER = estrogen receptor, PR = progesterone receptor

A detailed class discovery study [38] focusing solely on triple-negative breast cancers has revealed the existence of six molecular subtypes, namely, the basal-like A, basal-like B, mesenchymal, mesenchymal stem-like, immunomodulatory, and luminal androgen receptor (LAR) subtypes. The transcriptomic profile of basal-like 1 subtype is characterized by the expression of genes related to cell cycle and cell division components and pathways (cell cycle, DNA replication reactome, G2 cell-cycle pathway, RNA polymerase, and G1 to S cell cycle), whereas basal-like 2 cancers are enriched in genes related to growth factor signaling (epidermal growth factor pathway, nerve growth factor pathway, MET pathway, Wnt/β-catenin, and IGF1R pathway), glycolysis and gluconeogenesis. The mesenchymal type is transcriptomically characterized by the expression of genes related to cell motility (regulation of actin by Rho), extracellular matrix, and cell differentiation pathways [Wnt pathway, anaplastic lymphoma kinase (ALK) pathway, and TGF-β signaling], whereas the mesenchymal stem-like subtype has a transcriptome similar to that of mesenchymal tumors, but is also enriched for genes related to growth factor signaling pathways including inositol phosphate metabolism, epidermal growth factor receptor, platelet-derived growth factor receptor, calcium signaling, G-protein coupled receptor, ERK1/2 signaling, ABC transporter, and adipocytokine signaling. The immunomodulatory subtype displays transcriptomic features consistent with immune response, including immune cell signaling, cytokine signaling, antigen processing and presentation, and NFκB, TNF, and JAK/STAT signaling. The LAR subtype has the most distinctive transcriptomic profile of all subtypes of TNBCs, being characterized by the expression of AR and AR-related genes, steroid biosynthesis, and porphyrin metabolism [38].

This TNBC classification system has provided a framework for the development of specific therapies for subsets of TNBC patients, given that preclinical studies carried out by Lehmann et al. have demonstrated that breast cancer cell lines representative of the different subtypes of TNBCs are sensitive to specific therapeutic agents when grown as xenografts. While basal-like cell lines have been shown to be sensitive to cisplatin, mesenchymal stem-like and LAR cell lines were shown to be sensitive to a dual PI3K and mTOR inhibitor (BEZ235) and an anti-androgen (bicalutamide), respectively [38]. Furthermore, a recent study has demonstrated that the sub-stratification of primary TNBCs into these molecular subtypes is associated with chemotherapy response, as defined by the rate of pathological complete response following neoadjuvant chemotherapy [39]. The majority (52 %) of tumors classified as basal-like 1 underwent pathological complete response, whereas none of the basal-like 2 and only 10 % of the LAR tumors did. Although these results provide evidence to suggest that this classification system may have therapeutic implications, it should be noted that the stability of the subtypes, in particular of the basal-like 2 subtype, remains to be fully established.

Given the multiple microarray-based classifications currently available, it is unclear as to how the molecular subtypes defined in different classification schemes relate to one another. It is plausible that the molecular apocrine subtype is similar to LAR. Masuda et al. compared the results of PAM50 and the six molecular subtypes of TNBCs, and observed that approximately 38 % of tumors classified by PAM50 as basal-like are classified as either basal-like 1 or 2 using the six subtype classification system; the remaining cases are classified into mesenchymal, mesenchymal stem-like, and immunomodulatory. On the other hand, TNBCs classified as a subtype other than basal-like by PAM50 are preferentially of LAR subtype (59 %), the remaining cases being classified as of mesenchymal stem-like or mesenchymal subtype [39].

The group of investigators of the METABRIC study (Molecular Taxonomy of Breast Cancer International Consortium) analyzed approximately 2,000 breast cancers with both microarray-based gene expression arrays and single nucleotide polymorphism (SNP) arrays. Using an integrative approach based on the combination of gene expression features and gene copy number alterations, the authors have demonstrated that the most parsimonious number of molecular subtypes of breast cancer is ten. The molecular subtypes identified by this integrative approach have been reported to be associated with distinct outcomes, and show a limited correlation with the "intrinsic" subtypes [40]. This classification system, although potentially more representative of the heterogeneity of breast cancers at both the transcriptomic and genomic levels, is based on rather elaborate bioinformatic models which currently require results of both gene expression and SNP arrays. The development of simpler methods or surrogates for the identification of the ten molecular subtypes is eagerly awaited.

Gene-Expression Prognostic Signatures

With the advent of microarrays and the ability to carry out de novo discovery of gene expression changes associated with clinicopathological characteristics of breast cancers, numerous attempts were made to identify gene expression patterns associated with distinct phenotypes of breast cancer (e.g., ER status, ductal versus lobular histology, low versus high histological grade), risk of metastatic dissemination, overall survival of breast cancer patients and response to chemotherapy agents and combinatorial therapies [15, 16, 41]. These studies have demonstrated that microarray-based gene expression profiling is a powerful tool to predict simple phenotypes (e.g., the ER status of a cancer, prognosis of patients with ER-positive disease and high histological grade) [41]. On the other hand, the accuracy of microarrays to predict complex phenotypes (e.g., pathological complete response following neoadjuvant chemotherapy) has proven to be more limited [16, 41].

With the development of analysis methods that allowed for the identification of a set of genes predictive of a given phenotype or outcome, one of the questions that received the most attention was the prognostication of breast cancer patients. The interest in prognostication stemmed from the fact that in the early 2000s approximately 70 % of patients with early stage breast cancer used to receive adjuvant chemotherapy. It should be noted that although only 4–12 % of these patients benefit from chemotherapy, all of them were at risk of chemotherapy-induced toxicity and side effects [15, 24]. Several groups endeavored to define microarray signatures that would identify breast cancer patients with a good outcome based on gene expression profiling. A variety of prognostic gene expression signatures have been developed (Table 23.2) [15, 16, 42–46]. These signatures, however, were developed without taking into account the molecular diversity of breast cancer, and the existence of distinct transcriptomic subtypes that have been shown to have distinct clinical behavior.

First-Generation Prognostic Gene Signatures

The so-called "first-generation" signatures, which basically identify the same set of patients as having poor disease prognosis [47, 48] have been shown to be useful only for ER-positive breast cancer patients and have limited if any prognostic power in patients with ER-negative disease. The prognostic information derived from such signatures has been shown to be determined by the expression levels of proliferation-related genes. In fact, although the agreement between different signatures is by no means perfect, several meta-analyses [23, 47, 49] have demonstrated that they identify as of poor prognosis those patients whose tumors have high levels of expression of proliferation-related genes, which have been shown to constitute one of the strongest prognostic factors in ER-positive disease [49, 50]. These first-generation prognostic signatures have negligible discriminatory power in ER-negative disease because the levels of expression of proliferation-related genes are uniformly high in these tumors (Fig. 23.1).

These first-generation prognostic gene signatures also have been shown to provide independent information to that of provided by tumors size and lymph node status [25]. This information, however, has been shown to be time-dependent, with high accuracy for the prediction of early recurrences but limited ability to predict late distant relapses (i.e., a substantially higher accuracy in predictions made at 5 years after breast cancer diagnosis than at 10 years after) [51].

Table 23-2 **Main Characteristics of Commercially Available Gene Expression Signatures in Breast Cancer**

	Gene expression signature						
Characteristic	**MammaPrint®**	**MapQuant DX™**	**Veridex 76-gene**	**Breast cancer index**	**Oncotype DX®**	**PAM50**	**EndoPredict**
Tissue source	Fresh frozen	Fresh frozen/FFPE	Fresh frozen	FFPE	FFPE	Fresh frozen/FFPE/ nCounter	FFPE
Platform	Microarray	Microarray/ qRT-PCR	Microarray	qRT-PCR	qRT-PCR	Microarray/qRT-PCR	qRT-PCR
Gene signature	70 genes	97 genes (gene signature)/8 genes (PCR)	76 genes	2 gene-ratio HOXB13:IL17R/5-gene molecular grade index	21 genes	55 genes (5 genes used for normalization)	8 cancer related genes and 3 reference genes
Main application	Prognosis	Prognosis	Prognosis	Prognostic	Prognostic	Subtype predictor/ Prognostic (ROR score)	Prognostic
Breast cancer population	N0, stage I-II, T1-2, <61 years	ER+, histological grade II	N0	ER+, receiving tamoxifen	ER+ and N0 disease treated with tamoxifen		ER+ postmenopausal treated with tamoxifen only
Outcome	5-year distant metastasis	Grade 1 (good) vs. grade 3 (poor prognosis)	5-year distant metastasis	RFS and OS	10-year DFR	RFS	Distant recurrence
Company provider	Agendia (Amsterdam, Netherlands)	Ipsogen (Marseille, France)	N/A	BioTheranostics (San Diego, CA, USA)	Genomic Health (Redwood City, CA, USA)	ARUP (Salt Lake City, UT)	N/A
FDA approval/ clearance	Yes, prognostication	No	No	No	No	ROR approved for prognostication	No

ER = estrogen receptor, DFR = disease-free relapse, FFPE = formalin-fixed paraffin-embedded, N = lymph nodes, N/A = not available, OS = overall survival, RFS = relapse-free survival, ROR = risk of relapse, qRT-PCR = quantitative reverse transcriptase polymerase chain reaction

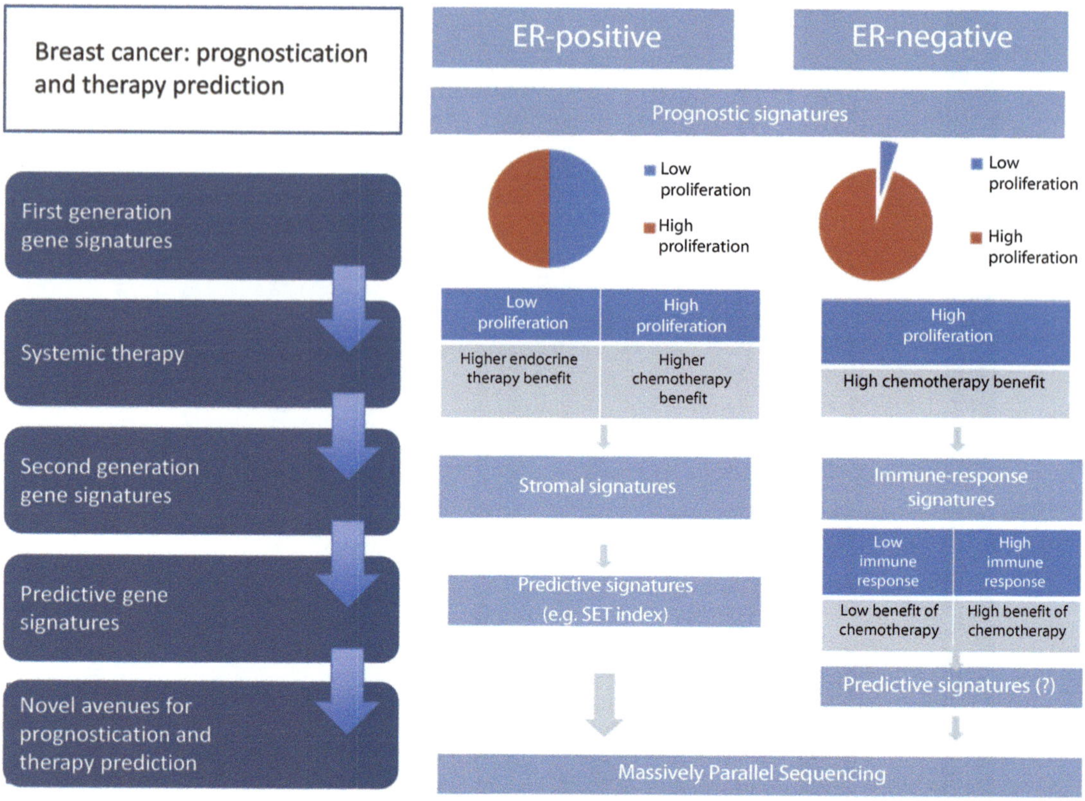

Figure 23-1 Schematic representation of gene expression signatures and their prognostic and predictive value for estrogen receptor (ER)-positive and ER-negative breast cancer. First-generation prognostic gene expression signatures are clinically useful for ER-positive disease and classify patients into good and poor prognoses. Second-generation signatures, which are underpinned by the prognostic value conferred by the expression of immune response-related genes, may play a role in the prognostication of patients with ER-negative breast cancer. The stromal gene signatures and endocrine predictive signatures (such as the SET index) also have the potential to help personalize the therapy for patients with ER-positive disease. New genomic platforms for discovering and validating prognostic and predictive biomarkers (e.g., massively parallel sequencing) are expected to have a dramatic impact on systemic therapy decision making for patients with breast cancer

Owing to the fact that the prognostic power of these first-generation signatures largely stems from the information provided by proliferation-related genes, the classification of breast cancers according to these signatures correlates with response to conventional chemotherapy agents [52–54]. This is not surprising, given that chemotherapy preferentially targets cells that are cycling/proliferating. An important observation, however, that most of the low-risk/good prognosis groups identified by first-generation prognostic signatures may potentially benefit from specific chemotherapy agents (e.g., taxanes) [55, 56].

MammaPrint®

The 70-gene assay (MammaPrint®, Agendia, Netherlands) is a widely used breast cancer multigene classifier assay, and the first US Food and Drug Administration (FDA) cleared assay. MammaPrint® is a microarray-based gene expression profiling assay that assesses the prognosis, as defined by the development of distant metastasis within 5 years of follow-up, for patients with node-negative, stage I–II invasive breast cancers. It initially required RNA extracted from fresh-frozen tumors specimens; however, more recent versions of the assay that can be performed using RNA extracted from archival formalin-fixed, paraffin-embedded (FFPE) samples have been developed.

This gene signature was based on a microarray-based platform that analyzed data from 78 patients with node-negative stage I–II breast cancer who did not receive adjuvant systemic therapy [57]. A prognostic score that categorizes patients into "good" (i.e., no distant metastasis within 5 years of follow-up) and "poor" (i.e., distant metastasis within 5 years of follow-up) outcome groups was generated after supervised analysis of 25,000 genes, which resulted in a list of 70 genes. Although this prognostic signature consists of genes that are to some extent associated with proliferation, invasion, metastasis and angiogenesis, its prognostic power seems to stem mainly from the expression levels of proliferation-related genes alone [47].

This signature was further validated in various cohorts of breast cancer patients (e.g., node-negative, node-positive, HER2-positive), and was shown to provide prognostic information in addition to that provided by standard clinicopathological variables [57–62]. Furthermore, the prognostic groups identified by MammaPrint® seem to correlate with response to chemotherapy; MammaPrint®-defined good prognosis tumors have been reported to derive minimal benefit from chemotherapy, whereas a subset of tumors classified as of poor prognosis have higher rates of chemotherapy response [63].

Prospective validation of MammaPrint® is currently ongoing. The randomized, multicenter trial Microarray In Node negative and 1–3 positive lymph node Disease may Avoid ChemoTherapy (MINDACT) trial has enrolled >6,700 patients, and is comparing MammaPrint® with a tool based on clinicopathological parameters (i.e., Adjuvant! Online) for the selection of patients with negative or 1–3 positive nodes for adjuvant chemotherapy in breast cancer. Its primary objective is to determine whether breast cancer patients with MammaPrint®-defined good prognosis tumors but high-risk clinicopathological features can be safely spared chemotherapy without affecting distant metastases-free survival. In brief, patients classified as of poor prognosis by both clinicopathological features and MammaPrint® will receive chemotherapy, whereas those classified as of good prognosis by both clinicopathological features and MammaPrint® will receive hormone therapy alone, if the tumor is ER-positive. Patients whose results of MammaPrint® and clinicopathological features are discordant will be randomized to receive either chemotherapy or hormone therapy. The results of this trial are eagerly awaited and should be available in the next few years [60].

MapQuant Dx™

Histologic grade, as defined by the Nottingham Grading System [64], has been shown to be a strong prognostic factor; its limited interobserver agreement notwithstanding [65]. The Genomic Grade Index (GGI) Map*Quant* Dx™ (Ipsogen, Marseille, France) was developed as a means to define histologic grade at the transcriptomic level [47, 66, 67]. This signature has been shown to be prognostic in ER-positive disease and, in a way akin to other gene signatures, to correlate with response to chemotherapy. GGI, as a 97-gene measurement of histologic grade, separates classic histological grade into low (GGI-low grade) and high (GGI-high grade) risk groups, as a replacement for grades 1, 2, and 3. In fact, grade 2 reflects the intermediate risk of tumor recurrence, which is not informative for clinical decision-making; the GGI assay reclassifies grade 2 cancers into those with low and high-risk of recurrence [66]. It was primarily developed as a microarray-based assay and required fresh or frozen tissue; however, a more recent version, which recapitulates the performance of the microarray-based version of GGI, based on quantitative reverse transcription (qRT)-PCR, can be applied to RNA extracted from frozen or FFPE samples [68].

Veridex 76-Gene Signature

This 76-gene signature consists of a microarray-based gene signature for prediction of distant metastasis in lymph node-negative breast cancer patients [69]. Akin to MammaPrint®, it requires fresh and frozen samples, generates its prognostic information basically from proliferating genes, and does not reliably predict prognosis for ER-negative disease [47, 51].

The 76-gene signature was developed and validated in single center [69] and multicenter studies [51, 70]. Notably, ER-positive and ER-negative breast cancer subtypes generated separated prognostic gene lists in a training set that analyzed 115 tumors. In these lists, either 60 or 16 genes, respectively, were found to predict 5-year distant metastasis for ER-positive and ER-negative breast cancer subtypes. In an independent testing set of 171 lymph node-negative patients that had not received adjuvant systemic treatment, this 76-gene signature was later shown to be a strong prognostic factor.

A subsequent study showed that the 76-gene signature predicted survival and benefit from adjuvant tamoxifen in lymph node-negative ER-positive patients. A group of patients among the poor prognosis group treated with tamoxifen had 12.3 % of absolute benefit in 10-year distant metastasis when compared to the group of patients who were not treated with tamoxifen [70]. This signature, albeit developed using a series of ER-negative breast cancers, has limited if any prognostic value for patients with ER-negative disease, and has been proven to be time-dependent [15, 16, 51].

Oncotype Dx®

The 21-gene assay (Oncotype DX®, Genomic Health, Redwood City, CA, USA) is one of the most widely used multigene classifier assays. It consists of a quantitative RT-PCR (qRT-PCR)-based signature in which RNA is extracted from formalin-fixed paraffin-embedded tissue samples [71, 72]. The signature measures the expression of 21 genes, of which 16 are cancer-related genes and 5 are reference genes. An algorithm defined to calculate a "recurrence score" (RS) based on that 21-gene list varies from 0 to 100, and classifies patients into three risk groups: low risk (RS <18), intermediate risk (RS from 18 to <31), and high risk (RS ≥31). The RS has been shown to predict the 10-year risk of distant relapse for ER-positive node-negative breast cancer patients, based on the analyses of samples from the National Surgical Adjuvant Breast and Bowel Project (NSABP) B-20 clinical trial [73]. Then, the RS was validated in a large cohort of ER-positive, node-negative tamoxifen-treated patients from the NSABP B-14 trial which resulted in level I evidence to support its prognostic value [65]. In addition, RS has also been shown to be associated with benefit from chemotherapy in patients with ER-positive disease. Chemotherapy benefit is observed in patients whose tumors have a high RS, whereas the benefit from chemotherapy is negligible in patients with low-RS cancers [74]. Currently, the optimal therapeutic strategy for patients with intermediate-RS (i.e., 18 to <31) remains uncertain. The phase 3 Trial Assigning Individualized Options for Treatment (TAILORx), which integrates the Oncotype DX® into the clinical decision-making process, will provide answers for the intermediate risk group. It should be noted that in the TAILORx trial RS from 11 to 25 was selected (rather than RS from 18 to <31). Based on these studies, Oncotype Dx® has been incorporated into clinical guidelines and is recommended by expert panels; furthermore, it has received support from the American Society of Clinical Oncology for its use in early ER-positive node-negative breast cancer [32, 75, 76].

Oncotype DX® has been shown to provide prognostic information in addition to that by standard clinicopathological prognostic parameters [77], offering information above and beyond that of histologic grade and tumor size [78, 79]. This assay has also been revealed to be a useful prognostic test in other scenarios such as (a) ER-positive node-positive patients treated with tamoxifen, (b) ER-positive patients treated with aromatase inhibitors, (c) ER-positive node-negative patients receiving no adjuvant therapy, and (d) node-positive patients treated with doxorubicin-containing chemotherapy [78, 80, 81].

PAM50

As mentioned above, the PAM50 assay (ARUP laboratories, Salt Lake City, UT, USA) was initially conceived as a means to identify the breast cancer "intrinsic" gene subtypes using RNA extracted from FFPE samples; in addition, this assay also provides a risk of relapse prognostic score (referred to as ROR). The ROR score, in the training dataset, predicted the probability of cancer recurrence over 10 years for patients with node-negative tumors who did not receive adjuvant systemic therapy [27]. The prognostic value of ROR score has been further validated for 786 patients with ER-positive breast cancer treated with tamoxifen, showing that the PAM50 and tumor size might give more prognostic information than other clinicopathological variables [82]. Notably, an 11-gene proliferation signature, which is related to cell cycle function, was derived from the 50 genes of the PAM50 assay. The 11-gene signature was found to improve the original model as it was found to have more prognostic value than the expression of Ki67 [82]. The PAM50 assay has shown promise but needs further independent validation. A recent study comparing the prognostic information provided by Oncotype Dx® and PAM50 using over 1,000 samples from the Arimidex, Tamoxifen, Alone or in Combination (ATAC) trial revealed that the PAM50 ROR score yielded significantly more prognostic information than the Oncotype Dx® RS, and that the PAM50 ROR provides independent prognostic information above and beyond that offered by nodal status, tumor size, histopathologic grade, age, and type of endocrine treatment [83].

Breast Cancer Index^SM (BCI)

The Breast Cancer Index (BCI) molecular assay (BCI; BioTheranostics, San Diego, CA, USA) was developed to assess the risk of distant recurrence in ER-positive, node-negative breast cancer patients [84–86]. It is a prognostic assay which combines two gene expression signatures: the HOXB13:IL17BR (H:I) two-gene ratio, which predicts distant recurrence in patients with ER-positive breast cancer treated with tamoxifen [84], and a proliferation-related five-gene molecular grade index (MGI) [85], that distinguishes grade 1 from grade 3 cancers. This dichotomous index (MGI together with *HOXB13:IL17BR*) is based on qRT-PCR using RNA from FFPE tissues and provides more accurate prognosis than either biomarker alone. Furthermore, the BCI, as a continuous risk model that enables prediction of distant recurrence risk, was significantly associated with distant recurrence and breast cancer death [86].

The BCI assay, 21-gene recurrence score, and an immunohistochemical prognostic model (IHC4) were prospectively compared for both early (0–5 years) and late (5–10 years) recurrence in ER-positive, node-negative patients in the TransATAC study [i.e., patients enrolled in the Arimidex, Tamoxifen, Alone or in Combination (ATAC) clinical trial] [87]. The BCI has been shown to be a significant prognostic test for risk of both early and late distant recurrence and could assist in the identification of high-risk patients who would derive benefit from extended endocrine therapy or additional therapy.

EndoPredict Test

EndoPredict is an RNA-based multigene assay that comprises proliferation and ER signaling-related genes for assessing the probability of distant recurrence in patients with ER-positive, HER2-negative breast cancer treated with adjuvant endocrine therapy [42, 44–46]. The EndoPredict test is based on the quantification of mRNA levels of eight cancer genes plus three reference genes in FFPE specimens by qRT-PCR and was shown to provide additional prognostic information, which is independent from clinicopathological parameters (i.e., Adjuvant!Online and Ki67 labeling index) [43]. In two validation cohorts, the EndoPredict test was combined with clinical risk factors (i.e., nodal status and tumor size) into a comprehensive risk score called EPclin, which has been shown to identify a subgroup of "very low" risk patients who may be satisfactorily treated with adjuvant endocrine therapy only [42]. Furthermore, EPclin has been shown to improve the prognostic classification of ER-positive breast cancer patients. The better stratification of patients opens the way to reduce the indication of chemotherapy to those patients who would derive small benefit.

Second-Generation Prognostic Gene Signatures

Many efforts to refine gene expression-based signatures have stood out in the last decade. It is becoming evident that breast cancer behavior is influenced not only by molecular characteristics of the tumor cells but also by the tumor microenvironment, which comprises a network of extracellular and cellular components that interact with tumor cells [88, 89]. The development and invasiveness of breast tumors are affected by the presence of a number of stromal cells (i.e., fibroblasts, myoepithelial cells, endothelial cells, lymphocytes, macrophages) and extracellular matrix molecules [90]. There is growing evidence that immune- and tumor microenvironment-related genes may be of prognostic [91, 92], and to a lesser degree, predictive relevance for ER-negative and highly proliferative ER-positive breast cancers [93–97].

Stromal Gene Signatures

It has been suggested in several studies that stromal gene signatures may add information to that of estrogen expression and tumor proliferation and may determine clinical outcomes in breast cancer [94–96, 98–101]. Although these prognostic studies provided relevant information on the biology of breast carcinomas [95, 96, 98], they did not consider the different breast cancer subtypes or the stroma composition as per tumor subtype. In addition, many gene expression analyses have been performed without prior microdissection of tissue; some studies have not clarified whether stromal gene signatures were associated with a stromal gene itself or to stroma admixed with tumor cells [99]. Stromal metagenes were identified in matching samples containing scarce stroma with other samples rich in stroma [93]. The stromal metagenes tested in a cohort of 684 lymph node-negative breast cancer patients without adjuvant therapy and 259 patients that received tamoxifen highlighted the importance of stromal biology in tumor progression [93]. Moreover, it has been demonstrated that a great proportion of some stromal metagenes differed significantly between ER-positive and ER-negative breast cancers [93]. Notably, a B-cell/plasma cell metagene was found to be an independent prognostic marker for ER-positive highly proliferative tumors, revealing that high expression of the B-cell/plasma cell metagene was related to a favorable prognosis.

In primary breast cancer, a stromal gene expression signature was associated with resistance to preoperative chemotherapy containing anthracyclines [94]. After breast tumor samples have been microdissected, the genes were shown to be expressed in reactive stroma. Agents targeting the stroma are anticipated to provide a new route to overcome resistance to chemotherapy in breast cancer and some drug compounds are currently under clinical development (e.g., NCT01484080).

Immune-Related Signatures

There is burgeoning evidence to suggest that the immune system plays a fundamental role in prognostication of early breast cancer patients. Immune-related gene signatures have gained increased attention in the last decade for ER-positive and ER-negative breast cancer [50, 102–106]. The extent of the impact of these immune-related gene signatures, however, remains to be fully elucidated due to the heterogeneous breast cancer sets that have been tested so far and the variety of techniques that have been used. Interestingly, lymphocyte infiltration, as defined by simple histological and immunohistochemical methods, has been shown to provide strong prognostic information in particular for patients with triple-negative disease [97, 104, 105, 107–109], whereas in other studies the results have been contradictory [110, 111]. Recently, some studies have demonstrated an association between tumor-infiltrating CD8 (+) T lymphocytes and better prognosis in ER-negative tumors (mainly basal-like/triple-negative subtype) [112, 113].

In the neoadjuvant setting, large pooled analysis revealed that patients whose tumors had higher levels of genes relating to the immune response were more likely to have pathological complete response with the administration of chemotherapy based on anthracyclines with or without taxanes [114]. An analysis of the quantity and location of lymphocytic infiltrate at diagnosis was performed using the material from the phase III trial BIG-2-98 (NCT00174655), which compared anthracycline-only chemotherapy versus chemotherapy combining doxorubicin and docetaxel. This analysis has revealed that the amount of lymphocytic infiltrate correlates with the outcome of patients with HER2-positive and ER-negative/HER2-negative disease [115]. In fact, this study revealed the existence of a subset of ER-negative/HER2-negative breast cancers characterized by >50 % of lymphocytes within tumor clusters or tumor stroma (i.e., lymphocyte-predominant cancers) that have an excellent outcome when managed with the chemotherapy regimens employed in the study, with a 5-year disease-free survival of 92 % [115]. These observations provide level II evidence to support the clinical validity of lymphocyte quantification as a prognostic factor for patients with ER-negative/HER2-negative breast cancers treated with chemotherapy [115]. Germane to the translation of these studies into clinically useful biomarkers is the development of robust methods for the quantification of immune response and lymphocytic infiltration in breast cancers.

Gene Expression Predictive Signatures

Predictive gene signatures aim to define the therapeutic response to chemotherapy, endocrine therapy or other target agents [15, 16, 116–120]. Akin to the prognostic gene expression signatures, ER and proliferation have been shown to be the major determinants of response to combinatorial chemotherapy. Thus far, the clinical value of gene expression signatures predictive of response to single chemotherapy agents remains controversial for breast cancer. In fact, there is no robust available gene signature capable of predicting responses to a specific therapeutic agent. Several hypotheses have been advanced to explain the limited success in developing and validating predictive signatures. First, resistance to chemotherapy can be caused by functional alterations in few or single genes, and it is plausible that microarray-based gene expression profiling would not be sufficiently sensitive to identify such genes [116]. Second, intra-tumor genetic heterogeneity plays an important role in determining the emergence of drug resistance. Breast tumors often comprise heterogeneous collections of cancer cells that encompass rare clonal subpopulations, which have different genetic and epigenetic aberrations [121, 122]. Some genetic aberrations, which may be found in single clones of tumors, may drive therapeutic resistance [123]. In fact, because microarrays give an average of the expression profile of the tumor, this technique would not be reliable to identify those rare resistant clones. Finally, multiple other molecular factors at different genetic and epigenetic levels, and also drug resistant mechanisms not related to the tumor itself (e.g., tissue microenvironment, patient metabolism) may determine resistance to therapy [16]. Although some predictive gene expression signatures appear to have predictive value in validation studies (e.g., SET index) [124], their accuracy to determine the response of individual patients may be limited [16].

Massively Parallel Sequencing and the Impact in Intra-tumor Genetic Heterogeneity

The advent of massively parallel sequencing has enabled the analysis of the entire constellation of genetic alterations in cancers to be defined in a matter of days at reasonable costs. Several large scale massively parallel sequencing-based studies of breast cancer have now been completed and demonstrated that (a) the collection of genetic aberrations found in breast cancers is complex with a limited number of genes that are frequently mutated in a substantial proportion of unselected cases [40, 121, 125, 126]; (b) that the number of genes mutated in small minorities of breast cancers is vast; (c) that the repertoire of mutations in luminal and basal-like breast cancers is rather different; and (d) that despite these differences, there is no gene or mutation that defines a subtype of breast cancers [125–128].

These studies, however, have led to the identification of novel driver genes in breast cancer, and, to the realization that the gene that encodes ER-alpha (*ESR1*) [19–21] and HER2 (*HER2, ERBB2*) [18, 129] can be targeted by activating mutations. *ESR1* gene mutations affect the DNA binding domain and some of these mutations have been shown to result in the activation of ER-dependent genes even in the absence of E2, and to require higher doses of tamoxifen and fulvestrant for the inhibition of ER activity [19–21].

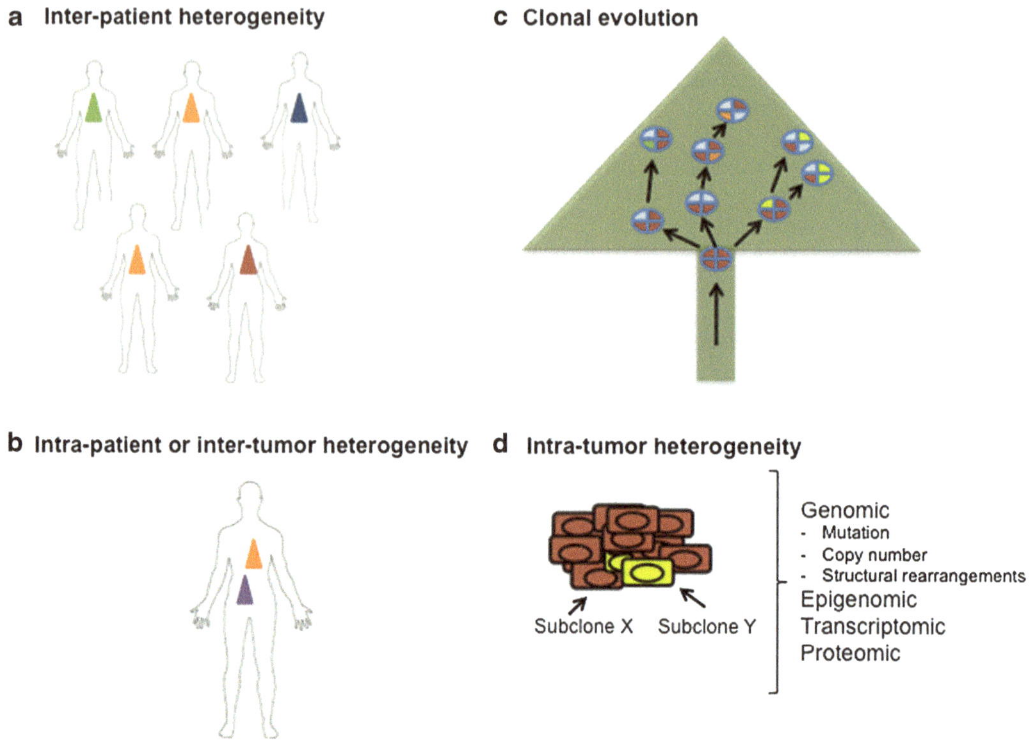

Figure 23-2 Tumor heterogeneity. (**a**) Inter-patient heterogeneity; (**b**) Intra-patient or inter-tumor heterogeneity; (**c**) Clonal evolution and the tree model: mutations shared by all tumor cells proceed from the founder clone which is depicted as the trunk of the tree. The branches are composed by tumor cells that acquire mutations present only in a subset of the tumor cells; (**d**) intra-tumor genetic heterogeneity and the approaches for the characterization of the molecular aberrations in breast cancers

Interestingly, these mutations have been shown to be rare in primary ER-positive breast cancers, but to be present in up to 32 % of ER-positive tumors from patients treated with aromatase inhibitors or subjected to estrogen deprivation [19–21]. *HER2* gene mutations have been shown to be present in up to 1.6 % of breast cancers, and to be preferentially, but not exclusively, found in cases that lack *HER2* amplification or protein overexpression [18]. *HER2* mutations affect the extracellular or tyrosine kinase domains of the gene, and not all mutations have been shown to result in activation of the pathways downstream of HER2. Interestingly, among the mutations that resulted in a measurable phenotype using in vitro and in vivo assays, some mutations have been shown to be bona fide activating mutations, others have been shown to be neomorphic, and finally, one recurrent mutation (p.L755S) has been shown to have limited impact on proliferation, survival or tumorigenicity of human cells, but to cause resistance to lapatinib [18]. Interestingly, cancer cells harboring *HER2* mutations have been shown to be sensitive to HER2 irreversible inhibitors [18] and clinical trials are ongoing to determine whether patients with *HER2* mutant breast cancers would benefit from neratinib (NCT01670877), a HER2 irreversible inhibitor.

Genomic analyses of human cancers have provided direct evidence of spatial [130–132] and temporal [130, 133, 134] intra-tumor genetic heterogeneity, and have shown that a substantial proportion of cancers are composed of mosaics of tumor cells at the time of diagnosis [121, 132], where subclones of cells harbor private mutations in addition to the founder genetic events. Although intra-tumor genetic heterogeneity has been recognized for many years [135], it has been explored in primary breast cancers using massively parallel sequencing approaches in a limited number of studies (Fig. 23.2) [121, 122, 136]. The impact intra-tumor genetic heterogeneity on the biology and, consequently, on treatment design of breast cancer remains to be fully understood. However, the genomic-based analysis of two pairs of matched primary tumors and distant metastatic relapses after adjuvant treatment revealed differences in their mutational makeup [133, 134] and suggested that clonal selection during the metastatic process is likely to occur.

The spatial and temporal intra-tumor genetic heterogeneity observed in solid cancers constitutes a significant challenge for the realization of the potentials of precision medicine, given that genetic biomarker analyses performed in single biopsies for treatment decision-making may differ according to the area of the tumor sampled [130], between the primary tumor and its distant metastases, or even between different metastatic sites [130, 137]. This multiregional separation of molecular aberrations can lead to sampling bias, potentially impairing the interpretation of genomics results derived from individual biopsies. Therefore, approaches to provide a global assessment of the repertoire of somatic genetic aberrations in a cancer are important for the accurate selection of targeted therapies for individual patients.

Deciphering intra-tumor heterogeneity using massively parallel sequencing approaches has important implications that may refine our understanding of breast cancer biology, its genetic diversity and the mechanisms that lead to therapeutic resistance [137–141]. Much effort has been made in this direction, including massively parallel sequencing of single cells [132] and circulating biomarkers [142–145].

Conclusions

Gene expression profiling has provided significant advances in the molecular classification and prognostication of breast cancer, and has given new insights regarding therapeutic prediction. Microarray-based gene expression studies have changed the way breast cancer is perceived and have highlighted that breast cancer comprises a heterogeneous collection of different diseases with distinct molecular characteristics and outcomes. The clinical management of patients is still based on the assessment of ER and HER2 based on immunohistochemistry and immunohistochemistry plus in situ hybridization methods for HER2. It is likely, however, that with the identification of activating mutations of *ESR1* and *ERBB2*, even these tests will have to change [129].

Gene expression profiling studies have guided the identification of prognostic and predictive gene signatures. Within the ER-positive breast cancer subtype, it is now clear that it is possible to identify patients with low-risk and very good prognosis, who may not need chemotherapy. On the other hand, the role of the first-generation gene expression signatures is negligible in ER-negative disease. Despite providing a vast quantity of information, the incorporation of these expression-based assays into clinical decision-making is still limited. Although Oncotype Dx® has been incorporated into the management of patients with ER-positive breast cancers in North America, its use in other countries is still limited. The prospective validation of first-generation prognostic signatures and the definition of the clinical significance of intermediate-RS scores in ongoing clinical trials are eagerly awaited.

Immune response- and tumor microenvironment-related signatures have emerged in the last decade. Their prognostic relevance for ER-negative and highly proliferative ER-positive breast cancers is of great scientific interest but the clinical utility of these signatures has yet to be demonstrated. Furthermore, robust signatures that predict chemotherapy responses in specific subtypes of breast cancer, such as triple-negative, have yet to be developed.

New avenues for discovering and validating prognostic and predictive biomarkers are being developed through massively parallel sequencing approaches, which allow for the characterization of the repertoire of DNA and RNA aberrations in breast cancer. The development of these new technologies will have a dramatic impact on the way prognostication and prediction for breast cancer patients is performed. It is likely that in the next few years, therapy decisions will become increasingly more reliant on the genetic makeup of breast cancers than on anatomic parameters. This paradigm shift will inevitably have a significant impact on the way pathology is practiced, and the role of the pathologist will have to expand above and beyond the histological diagnosis of breast cancers. Germane to the role of pathologists in this era of precision medicine is the recognition that understanding the basic principles of genetic analysis and bioinformatics will be essential for the next generation of diagnostic pathologists.

REFERENCES

1. Jemal A, Bray F, Center MM, Ferlay J, Ward E, Forman D. Global cancer statistics. CA Cancer J Clin. 2011;61(2):69–90.
2. Perou CM, Sorlie T, Eisen MB, van de Rijn M, Jeffrey SS, Rees CA, Pollack JR, Ross DT, Johnsen H, Akslen LA, Fluge O, Pergamenschikov A, Williams C, Zhu SX, Lonning PE, Borresen-Dale AL, Brown PO, Botstein D. Molecular portraits of human breast tumours. Nature. 2000; 406(6797):747–52.
3. Weigelt B, Reis-Filho JS. Histological and molecular types of breast cancer: is there a unifying taxonomy? Nat Rev Clin Oncol. 2009;6(12): 718–30.
4. Weigelt B, Geyer FC, Reis-Filho JS. Histological types of breast cancer: how special are they? Mol Oncol. 2010;4(3):192–208.
5. Hammond ME, Hayes DF, Dowsett M, Allred DC, Hagerty KL, Badve S, Fitzgibbons PL, Francis G, Goldstein NS, Hayes M, Hicks DG, Lester S, Love R, Mangu PB, McShane L, Miller K, Osborne CK, Paik S, Perlmutter J, Rhodes A, Sasano H, Schwartz JN, Sweep FC, Taube S, Torlakovic EE, Valenstein P, Viale G, Visscher D, Wheeler T, Williams RB, Wittliff JL, Wolff AC. American Society of Clinical Oncology/College of American Pathologists guideline recommendations for immunohistochemical testing of estrogen and progesterone receptors in breast cancer. J Clin Oncol. 2010;28(16):2784–95.
6. Wolff AC, Hammond ME, Hicks DG, Dowsett M, McShane LM, Allison KH, Allred DC, Bartlett JM, Bilous M, Fitzgibbons P, Hanna W, Jenkins RB, Mangu PB, Paik S, Perez EA, Press MF, Spears PA, Vance GH, Viale G, Hayes DF. Recommendations for human epidermal growth factor receptor 2 testing in breast cancer: American Society of Clinical Oncology/College of American Pathologists clinical practice guideline update. J Clin Oncol. 2013;31(31): 3997–4013.
7. Todd JH, Dowle C, Williams MR, Elston CW, Ellis IO, Hinton CP, Blamey RW, Haybittle JL. Confirmation of a prognostic index in primary breast cancer. Br J Cancer. 1987;56(4): 489–92.
8. Ravdin PM, Siminoff LA, Davis GJ, Mercer MB, Hewlett J, Gerson N, Parker HL. Computer program to assist in making decisions about adjuvant therapy for women with early breast cancer. J Clin Oncol. 2001;19(4):980–91.
9. Hajage D, de Rycke Y, Bollet M, Savignoni A, Caly M, Pierga JY, Horlings HM, Van de Vijver MJ, Vincent-Salomon A, Sigal-Zafrani B, Senechal C, Asselain B, Sastre X, Reyal F. External validation of Adjuvant! Online breast cancer prognosis tool. Prioritising recommendations for improvement. PLoS One. 2011;6(11):e27446.
10. Mook S, Schmidt MK, Rutgers EJ, van de Velde AO, Visser O, Rutgers SM, Armstrong N, van't Veer LJ, Ravdin PM. Calibration and discriminatory accuracy of prognosis calculation for breast cancer with the online Adjuvant! program: a hospital-based retrospective cohort study. Lancet Oncol. 2009;10(11):1070–6.
11. Campbell HE, Taylor MA, Harris AL, Gray AM. An investigation into the performance of the Adjuvant! Online prognostic programme in early breast cancer for a cohort of patients in the United Kingdom. Br J Cancer. 2009;101(7):1074–84.
12. Sorlie T, Perou CM, Tibshirani R, Aas T, Geisler S, Johnsen H, Hastie T, Eisen MB, van de Rijn M, Jeffrey SS, Thorsen T, Quist H, Matese JC, Brown PO, Botstein D, Lonning PE, Borresen-Dale AL. Gene expression patterns of breast carcinomas distinguish tumor subclasses with clinical implications. Proc Natl Acad Sci U S A. 2001; 98(19):10869–74.
13. Sorlie T, Tibshirani R, Parker J, Hastie T, Marron JS, Nobel A, Deng S, Johnsen H, Pesich R, Geisler S, Demeter J, Perou CM, Lonning PE, Brown PO, Borresen-Dale AL, Botstein D. Repeated observation of breast tumor subtypes in independent gene expression data sets. Proc Natl Acad Sci U S A. 2003;100(14):8418–23.
14. Jeffrey SS, Lonning PE, Hillner BE. Genomics-based prognosis and therapeutic prediction in breast cancer. J Natl Compr Canc Netw. 2005; 3(3):291–300.
15. Reis-Filho JS, Pusztai L. Gene expression profiling in breast cancer: classification, prognostication, and prediction. Lancet. 2011; 378(9805): 1812–23.
16. Weigelt B, Pusztai L, Ashworth A, Reis-Filho JS. Challenges translating breast cancer gene signatures into the clinic. Nat Rev Clin Oncol. 2012;9(1):58–64.
17. Garraway LA, Lander ES. Lessons from the cancer genome. Cell. 2013;153(1):17–37.
18. Bose R, Kavuri SM, Searleman AC, Shen W, Shen D, Koboldt DC, Monsey J, Goel N, Aronson AB, Li S, Ma CX, Ding L, Mardis ER, Ellis MJ. Activating HER2 mutations in HER2 gene amplification negative breast cancer. Cancer Discov. 2013;3(2):224–37.
19. Toy W, Shen Y, Won H, Green B, Sakr RA, Will M, Li Z, Gala K, Fanning S, King TA, Hudis C,

Chen D, Taran T, Hortobagyi G, Greene G, Berger M, Baselga J, Chandarlapaty S. ESR1 ligand-binding domain mutations in hormone-resistant breast cancer. Nat Genet. 2013; 45(12):1439–45.
20. Robinson DR, Wu YM, Vats P, Su F, Lonigro RJ, Cao X, Kalyana-Sundaram S, Wang R, Ning Y, Hodges L, Gursky A, Siddiqui J, Tomlins SA, Roychowdhury S, Pienta KJ, Kim SY, Roberts JS, Rae JM, Van Poznak CH, Hayes DF, Chugh R, Kunju LP, Talpaz M, Schott AF, Chinnaiyan AM. Activating ESR1 mutations in hormone-resistant metastatic breast cancer. Nat Genet. 2013;45(12):1446–51.
21. Li S, Shen D, Shao J, Crowder R, Liu W, Prat A, He X, Liu S, Hoog J, Lu C, Ding L, Griffith OL, Miller C, Larson D, Fulton RS, Harrison M, Mooney T, McMichael JF, Luo J, Tao Y, Goncalves R, Schlosberg C, Hiken JF, Saied L, Sanchez C, Giuntoli T, Bumb C, Cooper C, Kitchens RT, Lin A, Phommaly C, Davies SR, Zhang J, Kavuri MS, McEachern D, Dong YY, Ma C, Pluard T, Naughton M, Bose R, Suresh R, McDowell R, Michel L, Aft R, Gillanders W, DeSchryver K, Wilson RK, Wang S, Mills GB, Gonzalez-Angulo A, Edwards JR, Maher C, Perou CM, Mardis ER, Ellis MJ. Endocrine-therapy-resistant ESR1 variants revealed by genomic characterization of breast-cancer-derived xenografts. Cell Rep. 2013;4(6):1116–30.
22. Gruvberger S, Ringner M, Chen Y, Panavally S, Saal LH, Borg A, Ferno M, Peterson C, Meltzer PS. Estrogen receptor status in breast cancer is associated with remarkably distinct gene expression patterns. Cancer Res. 2001;61(16): 5979–84.
23. Hu Z, Fan C, Oh DS, Marron JS, He X, Qaqish BF, Livasy C, Carey LA, Reynolds E, Dressler L, Nobel A, Parker J, Ewend MG, Sawyer LR, Wu J, Liu Y, Nanda R, Tretiakova M, Ruiz Orrico A, Dreher D, Palazzo JP, Perreard L, Nelson E, Mone M, Hansen H, Mullins M, Quackenbush JF, Ellis MJ, Olopade OI, Bernard PS, Perou CM. The molecular portraits of breast tumors are conserved across microarray platforms. BMC Genomics. 2006;7:96.
24. Weigelt B, Baehner FL, Reis-Filho JS. The contribution of gene expression profiling to breast cancer classification, prognostication and prediction: a retrospective of the last decade. J Pathol. 2010;220(2):263–80.
25. Sotiriou C, Pusztai L. Gene-expression signatures in breast cancer. N Engl J Med. 2009; 360(8):790–800.
26. Weigelt B, Mackay A, A'Hern R, Natrajan R, Tan DS, Dowsett M, Ashworth A, Reis-Filho JS. Breast cancer molecular profiling with single sample predictors: a retrospective analysis. Lancet Oncol. 2010;11(4):339–49.
27. Parker JS, Mullins M, Cheang MC, Leung S, Voduc D, Vickery T, Davies S, Fauron C, He X, Hu Z, Quackenbush JF, Stijleman IJ, Palazzo J, Marron JS, Nobel AB, Mardis E, Nielsen TO, Ellis MJ, Perou CM, Bernard PS. Supervised risk predictor of breast cancer based on intrinsic subtypes. J Clin Oncol. 2009;27(8):1160–7.
28. Prat A, Parker JS, Karginova O, Fan C, Livasy C, Herschkowitz JI, He X, Perou CM. Phenotypic and molecular characterization of the claudin-low intrinsic subtype of breast cancer. Breast Cancer Res. 2010;12(5):R68.
29. Guiu S, Michiels S, Andre F, Cortes J, Denkert C, Di Leo A, Hennessy BT, Sorlie T, Sotiriou C, Turner N, Van de Vijver M, Viale G, Loi S, Reis-Filho JS. Molecular subclasses of breast cancer: how do we define them? The IMPAKT 2012 Working Group Statement. Ann Oncol. 2012;23(12):2997–3006.
30. Mackay A, Weigelt B, Grigoriadis A, Kreike B, Natrajan R, A'Hern R, Tan DS, Dowsett M, Ashworth A, Reis-Filho JS. Microarray-based class discovery for molecular classification of breast cancer: analysis of interobserver agreement. J Natl Cancer Inst. 2011;103(8):662–73.
31. Haibe-Kains B, Desmedt C, Loi S, Culhane AC, Bontempi G, Quackenbush J, Sotiriou C. A three-gene model to robustly identify breast cancer molecular subtypes. J Natl Cancer Inst. 2012;104(4):311–25.
32. Goldhirsch A, Wood WC, Coates AS, Gelber RD, Thurlimann B, Senn HJ. Strategies for subtypes–dealing with the diversity of breast cancer: highlights of the St. Gallen International Expert Consensus on the Primary Therapy of Early Breast Cancer 2011. Ann Oncol. 2011;22(8):1736–47.
33. Bastien RR, Rodriguez-Lescure A, Ebbert MT, Prat A, Munarriz B, Rowe L, Miller P, Ruiz-Borrego M, Anderson D, Lyons B, Alvarez I, Dowell T, Wall D, Segui MA, Barley L, Boucher KM, Alba E, Pappas L, Davis CA, Aranda I, Fauron C, Stijleman IJ, Palacios J, Anton A, Carrasco E, Caballero R, Ellis MJ, Nielsen TO, Perou CM, Astill M, Bernard PS, Martin M. PAM50 breast cancer subtyping by RT-qPCR and concordance with standard clinical molecular markers. BMC Med Genomics. 2012;5:44.
34. Farmer P, Bonnefoi H, Becette V, Tubiana-Hulin M, Fumoleau P, Larsimont D, Macgrogan G, Bergh J, Cameron D, Goldstein D, Duss S, Nicoulaz AL, Brisken C, Fiche M, Delorenzi M, Iggo R. Identification of molecular apocrine breast tumours by microarray analysis. Oncogene. 2005;24(29):4660–71.
35. Doane AS, Danso M, Lal P, Donaton M, Zhang L, Hudis C, Gerald WL. An estrogen receptor-negative breast cancer subset characterized by a hormonally regulated transcriptional program

and response to androgen. Oncogene. 2006; 25(28):3994–4008.

36. Guedj M, Marisa L, de Reynies A, Orsetti B, Schiappa R, Bibeau F, MacGrogan G, Lerebours F, Finetti P, Longy M, Bertheau P, Bertrand F, Bonnet F, Martin AL, Feugeas JP, Bieche I, Lehmann-Che J, Lidereau R, Birnbaum D, Bertucci F, de The H, Theillet C. A refined molecular taxonomy of breast cancer. Oncogene. 2012;31(9):1196–206.
37. Herschkowitz JI, Simin K, Weigman VJ, Mikaelian I, Usary J, Hu Z, Rasmussen KE, Jones LP, Assefnia S, Chandrasekharan S, Backlund MG, Yin Y, Khramtsov AI, Bastein R, Quackenbush J, Glazer RI, Brown PH, Green JE, Kopelovich L, Furth PA, Palazzo JP, Olopade OI, Bernard PS, Churchill GA, Van Dyke T, Perou CM. Identification of conserved gene expression features between murine mammary carcinoma models and human breast tumors. Genome Biol. 2007;8(5):R76.
38. Lehmann BD, Bauer JA, Chen X, Sanders ME, Chakravarthy AB, Shyr Y, Pietenpol JA. Identification of human triple-negative breast cancer subtypes and preclinical models for selection of targeted therapies. J Clin Invest. 2011;121(7):2750–67.
39. Masuda H, Baggerly KA, Wang Y, Zhang Y, Gonzalez-Angulo AM, Meric-Bernstam F, Valero V, Lehmann BD, Pietenpol JA, Hortobagyi GN, Symmans WF, Ueno NT. Differential response to neoadjuvant chemotherapy among 7 triple-negative breast cancer molecular subtypes. Clin Cancer Res. 2013;19(19):5533–40.
40. Curtis C, Shah SP, Chin SF, Turashvili G, Rueda OM, Dunning MJ, Speed D, Lynch AG, Samarajiwa S, Yuan Y, Graf S, Ha G, Haffari G, Bashashati A, Russell R, McKinney S, Langerod A, Green A, Provenzano E, Wishart G, Pinder S, Watson P, Markowetz F, Murphy L, Ellis I, Purushotham A, Borresen-Dale AL, Brenton JD, Tavare S, Caldas C, Aparicio S. The genomic and transcriptomic architecture of 2,000 breast tumours reveals novel subgroups. Nature. 2012; 486(7403):346–52.
41. Shi L, Campbell G, Jones WD, Campagne F, Wen Z, Walker SJ, Su Z, Chu TM, Goodsaid FM, Pusztai L, Shaughnessy Jr JD, Oberthuer A, Thomas RS, Paules RS, Fielden M, Barlogie B, Chen W, Du P, Fischer M, Furlanello C, Gallas BD, Ge X, Megherbi DB, Symmans WF, Wang MD, Zhang J, Bitter H, Brors B, Bushel PR, Bylesjo M, Chen M, Cheng J, Cheng J, Chou J, Davison TS, Delorenzi M, Deng Y, Devanarayan V, Dix DJ, Dopazo J, Dorff KC, Elloumi F, Fan J, Fan S, Fan X, Fang H, Gonzaludo N, Hess KR, Hong H, Huan J, Irizarry RA, Judson R, Juraeva D, Lababidi S, Lambert CG, Li L, Li Y, Li Z, Lin SM, Liu G, Lobenhofer EK, Luo J, Luo W, McCall MN, Nikolsky Y, Pennello GA, Perkins RG, Philip R, Popovici V, Price ND, Qian F, Scherer A, Shi T, Shi W, Sung J, Thierry-Mieg D, Thierry-Mieg J, Thodima V, Trygg J, Vishnuvajjala L, Wang SJ, Wu J, Wu Y, Xie Q, Yousef WA, Zhang L, Zhang X, Zhong S, Zhou Y, Zhu S, Arasappan D, Bao W, Lucas AB, Berthold F, Brennan RJ, Buness A, Catalano JG, Chang C, Chen R, Cheng Y, Cui J, Czika W, Demichelis F, Deng X, Dosymbekov D, Eils R, Feng Y, Fostel J, Fulmer-Smentek S, Fuscoe JC, Gatto L, Ge W, Goldstein DR, Guo L, Halbert DN, Han J, Harris SC, Hatzis C, Herman D, Huang J, Jensen RV, Jiang R, Johnson CD, Jurman G, Kahlert Y, Khuder SA, Kohl M, Li J, Li L, Li M, Li QZ, Li S, Li Z, Liu J, Liu Y, Liu Z, Meng L, Madera M, Martinez-Murillo F, Medina I, Meehan J, Miclaus K, Moffitt RA, Montaner D, Mukherjee P, Mulligan GJ, Neville P, Nikolskaya T, Ning B, Page GP, Parker J, Parry RM, Peng X, Peterson RL, Phan JH, Quanz B, Ren Y, Riccadonna S, Roter AH, Samuelson FW, Schumacher MM, Shambaugh JD, Shi Q, Shippy R, Si S, Smalter A, Sotiriou C, Soukup M, Staedtler F, Steiner G, Stokes TH, Sun Q, Tan PY, Tang R, Tezak Z, Thorn B, Tsyganova M, Turpaz Y, Vega SC, Visintainer R, von Frese J, Wang C, Wang E, Wang J, Wang W, Westermann F, Willey JC, Woods M, Wu S, Xiao N, Xu J, Xu L, Yang L, Zeng X, Zhang J, Zhang L, Zhang M, Zhao C, Puri RK, Scherf U, Tong W, Wolfinger RD, MAQC Consortium. The MicroArray Quality Control (MAQC)-II study of common practices for the development and validation of microarray-based predictive models. Nat Biotechnol. 2010;28(8):827–38.
42. Filipits M, Rudas M, Jakesz R, Dubsky P, Fitzal F, Singer CF, Dietze O, Greil R, Jelen A, Sevelda P, Freibauer C, Muller V, Janicke F, Schmidt M, Kolbl H, Rody A, Kaufmann M, Schroth W, Brauch H, Schwab M, Fritz P, Weber KE, Feder IS, Hennig G, Kronenwett R, Gehrmann M, Gnant M. A new molecular predictor of distant recurrence in ER-positive, HER2-negative breast cancer adds independent information to conventional clinical risk factors. Clin Cancer Res. 2011;17(18):6012–20.
43. Dubsky P, Filipits M, Jakesz R, Rudas M, Singer CF, Greil R, Dietze O, Luisser I, Klug E, Sedivy R, Bachner M, Mayr D, Schmidt M, Gehrmann MC, Petry C, Weber KE, Kronenwett R, Brase JC, Gnant M. EndoPredict improves the prognostic classification derived from common clinical guidelines in ER-positive, HER2-negative early breast cancer. Ann Oncol. 2013;24(3):640–7.
44. Kronenwett R, Bohmann K, Prinzler J, Sinn BV, Haufe F, Roth C, Averdick M, Ropers T, Windbergs C, Brase JC, Weber KE, Fisch K, Muller BM, Schmidt M, Filipits M, Dubsky P, Petry C, Dietel M, Denkert C. Decentral gene expression analysis: analytical validation of the

Endopredict genomic multianalyte breast cancer prognosis test. BMC Cancer. 2012;12:456.
45. Azim Jr HA, Michiels S, Zagouri F, Delaloge S, Filipits M, Namer M, Neven P, Symmans WF, Thompson A, Andre F, Loi S, Swanton C. Utility of prognostic genomic tests in breast cancer practice: the IMPAKT 2012 Working Group Consensus Statement. Ann Oncol. 2013;24(3): 647–54.
46. Dubsky P, Brase JC, Jakesz R, Rudas M, Singer CF, Greil R, Dietze O, Luisser I, Klug E, Sedivy R, Bachner M, Mayr D, Schmidt M, Gehrmann MC, Petry C, Weber KE, Fisch K, Kronenwett R, Gnant M, Filipits M. The EndoPredict score provides prognostic information on late distant metastases in ER+/HER2- breast cancer patients. Br J Cancer. 2013;109(12):2959–64.
47. Wirapati P, Sotiriou C, Kunkel S, Farmer P, Pradervand S, Haibe-Kains B, Desmedt C, Ignatiadis M, Sengstag T, Schutz F, Goldstein DR, Piccart M, Delorenzi M. Meta-analysis of gene expression profiles in breast cancer: toward a unified understanding of breast cancer subtyping and prognosis signatures. Breast Cancer Res. 2008;10(4):R65.
48. Fan C, Oh DS, Wessels L, Weigelt B, Nuyten DS, Nobel AB, van't Veer LJ, Perou CM. Concordance among gene-expression-based predictors for breast cancer. N Engl J Med. 2006;355(6): 560–9.
49. Reyal F, van Vliet MH, Armstrong NJ, Horlings HM, de Visser KE, Kok M, Teschendorff AE, Mook S, van't Veer L, Caldas C, Salmon RJ, van de Vijver MJ, Wessels LF. A comprehensive analysis of prognostic signatures reveals the high predictive capacity of the proliferation, immune response and RNA splicing modules in breast cancer. Breast Cancer Res. 2008;10(6):R93.
50. Desmedt C, Haibe-Kains B, Wirapati P, Buyse M, Larsimont D, Bontempi G, Delorenzi M, Piccart M, Sotiriou C. Biological processes associated with breast cancer clinical outcome depend on the molecular subtypes. Clin Cancer Res. 2008;14(16):5158–65.
51. Desmedt C, Piette F, Loi S, Wang Y, Lallemand F, Haibe-Kains B, Viale G, Delorenzi M, Zhang Y, d'Assignies MS, Bergh J, Lidereau R, Ellis P, Harris AL, Klijn JG, Foekens JA, Cardoso F, Piccart MJ, Buyse M, Sotiriou C. Strong time dependence of the 76-gene prognostic signature for node-negative breast cancer patients in the TRANSBIG multicenter independent validation series. Clin Cancer Res. 2007;13(11):3207–14.
52. Iwamoto T, Bianchini G, Booser D, Qi Y, Coutant C, Ya-Hui Shiang C, Santarpia L, Matsuoka J, Hortobagyi GN, Symmans WF, Holmes FA, O'Shaughnessy J, Hellerstedt B, Pippen J, Andre F, Simon R, Pusztai L. Gene pathways associated with prognosis and chemotherapy sensitivity in molecular subtypes of breast cancer. J Natl Cancer Inst. J Natl Cancer Inst. 2011;103(3): 264–72.
53. Iwamoto T, Pusztai L. Predicting prognosis of breast cancer with gene signatures: are we lost in a sea of data? Genome Med. 2010;2(11):81.
54. Reis-Filho JS, Weigelt B, Fumagalli D, Sotiriou C. Molecular profiling: moving away from tumor philately. Sci Transl Med. 2010;2(47):47ps43.
55. A'Hern RP, Jamal-Hanjani M, Szasz AM, Johnston SR, Reis-Filho JS, Roylance R, Swanton C. Taxane benefit in breast cancer–a role for grade and chromosomal stability. Nat Rev Clin Oncol. 2013;10(6):357–64.
56. Martin M, Prat A, Rodriguez-Lescure A, Caballero R, Ebbert MT, Munarriz B, Ruiz-Borrego M, Bastien RR, Crespo C, Davis C, Rodriguez CA, Lopez-Vega JM, Furio V, Garcia AM, Casas M, Ellis MJ, Berry DA, Pitcher BN, Harris L, Ruiz A, Winer E, Hudis C, Stijleman IJ, Tuck DP, Carrasco E, Perou CM, Bernard PS. PAM50 proliferation score as a predictor of weekly paclitaxel benefit in breast cancer. Breast Cancer Res Treat. 2013;138(2):457–66.
57. van't Veer LJ, Dai H, van de Vijver MJ, He YD, Hart AA, Mao M, Peterse HL, van der Kooy K, Marton MJ, Witteveen AT, Schreiber GJ, Kerkhoven RM, Roberts C, Linsley PS, Bernards R, Friend SH. Gene expression profiling predicts clinical outcome of breast cancer. Nature. 2002;415(6871):530–6.
58. Mook S, Schmidt MK, Viale G, Pruneri G, Eekhout I, Floore A, Glas AM, Bogaerts J, Cardoso F, Piccart-Gebhart MJ, Rutgers ET, Van't Veer LJ. The 70-gene prognosis-signature predicts disease outcome in breast cancer patients with 1-3 positive lymph nodes in an independent validation study. Breast Cancer Res Treat. 2009;116(2):295–302.
59. Buyse M, Loi S, van't Veer L, Viale G, Delorenzi M, Glas AM, d'Assignies MS, Bergh J, Lidereau R, Ellis P, Harris A, Bogaerts J, Therasse P, Floore A, Amakrane M, Piette F, Rutgers E, Sotiriou C, Cardoso F, Piccart MJ. Validation and clinical utility of a 70-gene prognostic signature for women with node-negative breast cancer. J Natl Cancer Inst. 2006;98(17):1183–92.
60. Cardoso F, Van't Veer L, Rutgers E, Loi S, Mook S, Piccart-Gebhart MJ. Clinical application of the 70-gene profile: the MINDACT trial. J Clin Oncol. 2008;26(5):729–35.
61. Knauer M, Cardoso F, Wesseling J, Bedard PL, Linn SC, Rutgers EJ, van't Veer LJ. Identification of a low-risk subgroup of HER-2-positive breast cancer by the 70-gene prognosis signature. Br J Cancer. 2010;103(12):1788–93.
62. Mook S, Knauer M, Bueno-de-Mesquita JM, Retel VP, Wesseling J, Linn SC, Van't Veer LJ, Rutgers EJ. Metastatic potential of T1 breast cancer can be predicted by the 70-gene

MammaPrint signature. Ann Surg Oncol. 2010; 17(5):1406–13.
63. Knauer M, Mook S, Rutgers EJ, Bender RA, Hauptmann M, van de Vijver MJ, Koornstra RH, Bueno-de-Mesquita JM, Linn SC, van't Veer LJ. The predictive value of the 70-gene signature for adjuvant chemotherapy in early breast cancer. Breast Cancer Res Treat. 2010;120(3):655–61.
64. Rakha EA, Reis-Filho JS, Baehner F, Dabbs DJ, Decker T, Eusebi V, Fox SB, Ichihara S, Jacquemier J, Lakhani SR, Palacios J, Richardson AL, Schnitt SJ, Schmitt FC, Tan PH, Tse GM, Badve S, Ellis IO. Breast cancer prognostic classification in the molecular era: the role of histological grade. Breast Cancer Res. 2010;12(4):207.
65. Paik S, Shak S, Tang G, Kim C, Baker J, Cronin M, Baehner FL, Walker MG, Watson D, Park T, Hiller W, Fisher ER, Wickerham DL, Bryant J, Wolmark N. A multigene assay to predict recurrence of tamoxifen-treated, node-negative breast cancer. N Engl J Med. 2004;351(27):2817–26.
66. Sotiriou C, Wirapati P, Loi S, Harris A, Fox S, Smeds J, Nordgren H, Farmer P, Praz V, Haibe-Kains B, Desmedt C, Larsimont D, Cardoso F, Peterse H, Nuyten D, Buyse M, Van de Vijver MJ, Bergh J, Piccart M, Delorenzi M. Gene expression profiling in breast cancer: understanding the molecular basis of histologic grade to improve prognosis. J Natl Cancer Inst. 2006;98(4):262–72.
67. Ivshina AV, George J, Senko O, Mow B, Putti TC, Smeds J, Lindahl T, Pawitan Y, Hall P, Nordgren H, Wong JE, Liu ET, Bergh J, Kuznetsov VA, Miller LD. Genetic reclassification of histologic grade delineates new clinical subtypes of breast cancer. Cancer Res. 2006;66(21):10292–301.
68. Toussaint J, Sieuwerts AM, Haibe-Kains B, Desmedt C, Rouas G, Harris AL, Larsimont D, Piccart M, Foekens JA, Durbecq V, Sotiriou C. Improvement of the clinical applicability of the Genomic Grade Index through a qRT-PCR test performed on frozen and formalin-fixed paraffin-embedded tissues. BMC Genomics. 2009;10:424.
69. Wang Y, Klijn JG, Zhang Y, Sieuwerts AM, Look MP, Yang F, Talantov D, Timmermans M, Meijer-van Gelder ME, Yu J, Jatkoe T, Berns EM, Atkins D, Foekens JA. Gene-expression profiles to predict distant metastasis of lymph-node-negative primary breast cancer. Lancet. 2005;365(9460): 671–9.
70. Foekens JA, Atkins D, Zhang Y, Sweep FC, Harbeck N, Paradiso A, Cufer T, Sieuwerts AM, Talantov D, Span PN, Tjan-Heijnen VC, Zito AF, Specht K, Hoefler H, Golouh R, Schittulli F, Schmitt M, Beex LV, Klijn JG, Wang Y. Multicenter validation of a gene expression-based prognostic signature in lymph node-negative primary breast cancer. J Clin Oncol. 2006;24(11):1665–71.
71. Sparano JA, Paik S. Development of the 21-gene assay and its application in clinical practice and clinical trials. J Clin Oncol. 2008;26(5):721–8.
72. Kim C, Paik S. Gene-expression-based prognostic assays for breast cancer. Nat Rev Clin Oncol. 2010;7(6):340–7.
73. Fisher B, Dignam J, Wolmark N, DeCillis A, Emir B, Wickerham DL, Bryant J, Dimitrov NV, Abramson N, Atkins JN, Shibata H, Deschenes L, Margolese RG. Tamoxifen and chemotherapy for lymph node-negative, estrogen receptor-positive breast cancer. J Natl Cancer Inst. 1997;89(22):1673–82.
74. Paik S, Tang G, Shak S, Kim C, Baker J, Kim W, Cronin M, Baehner FL, Watson D, Bryant J, Costantino JP, Geyer Jr CE, Wickerham DL, Wolmark N. Gene expression and benefit of chemotherapy in women with node-negative, estrogen receptor-positive breast cancer. J Clin Oncol. 2006;24(23):3726–34.
75. NCCN Clinical Practicce Guidelines in Oncology™. Breast Cancer. 2012. http://www.nccn.org/professionals/physician_gls/f_guidelines.asp
76. Harris L, Fritsche H, Mennel R, Norton L, Ravdin P, Taube S, Somerfield MR, Hayes DF, Bast Jr RC. American Society of Clinical Oncology 2007 update of recommendations for the use of tumor markers in breast cancer. J Clin Oncol. 2007;25(33):5287–312.
77. Goldstein LJ, Gray R, Badve S, Childs BH, Yoshizawa C, Rowley S, Shak S, Baehner FL, Ravdin PM, Davidson NE, Sledge Jr GW, Perez EA, Shulman LN, Martino S, Sparano JA. Prognostic utility of the 21-gene assay in hormone receptor-positive operable breast cancer compared with classical clinicopathologic features. J Clin Oncol. 2008;26(25):4063–71.
78. Habel LA, Shak S, Jacobs MK, Capra A, Alexander C, Pho M, Baker J, Walker M, Watson D, Hackett J, Blick NT, Greenberg D, Fehrenbacher L, Langholz B, Quesenberry CP. A population-based study of tumor gene expression and risk of breast cancer death among lymph node-negative patients. Breast Cancer Res. 2006;8(3):R25.
79. Tang CG, Lin AY. Molecular prediction of recurrence of breast cancer. N Engl J Med. 2005;352(15):1605–7. author reply 1605–7.
80. Albain KS, Barlow WE, Shak S, Hortobagyi GN, Livingston RB, Yeh IT, Ravdin P, Bugarini R, Baehner FL, Davidson NE, Sledge GW, Winer EP, Hudis C, Ingle JN, Perez EA, Pritchard KI, Shepherd L, Gralow JR, Yoshizawa C, Allred DC, Osborne CK, Hayes DF. Prognostic and predictive value of the 21-gene recurrence score assay in postmenopausal women with node-positive, oestrogen-receptor-positive breast cancer on chemotherapy: a retrospective analysis of a randomised trial. Lancet Oncol. 2010;11(1):55–65.

81. Dowsett M, Cuzick J, Wale C, Forbes J, Mallon EA, Salter J, Quinn E, Dunbier A, Baum M, Buzdar A, Howell A, Bugarini R, Baehner FL, Shak S. Prediction of risk of distant recurrence using the 21-gene recurrence score in node-negative and node-positive postmenopausal patients with breast cancer treated with anastrozole or tamoxifen: a TransATAC study. J Clin Oncol. 2010;28(11):1829–34.
82. Nielsen TO, Parker JS, Leung S, Voduc D, Ebbert M, Vickery T, Davies SR, Snider J, Stijleman IJ, Reed J, Cheang MC, Mardis ER, Perou CM, Bernard PS, Ellis MJ. A comparison of PAM50 intrinsic subtyping with immunohistochemistry and clinical prognostic factors in tamoxifen-treated estrogen receptor-positive breast cancer. Clin Cancer Res. 2010;16(21):5222–32.
83. Dowsett M, Sestak I, Lopez-Knowles E, Sidhu K, Dunbier AK, Cowens JW, Ferree S, Storhoff J, Schaper C, Cuzick J. Comparison of PAM50 risk of recurrence score with oncotype DX and IHC4 for predicting risk of distant recurrence after endocrine therapy. J Clin Oncol. 2013;31(22):2783–90.
84. Ma XJ, Wang Z, Ryan PD, Isakoff SJ, Barmettler A, Fuller A, Muir B, Mohapatra G, Salunga R, Tuggle JT, Tran Y, Tran D, Tassin A, Amon P, Wang W, Enright E, Stecker K, Estepa-Sabal E, Smith B, Younger J, Balis U, Michaelson J, Bhan A, Habin K, Baer TM, Brugge J, Haber DA, Erlander MG, Sgroi DC. A two-gene expression ratio predicts clinical outcome in breast cancer patients treated with tamoxifen. Cancer Cell. 2004;5(6):607–16.
85. Ma XJ, Salunga R, Dahiya S, Wang W, Carney E, Durbecq V, Harris A, Goss P, Sotiriou C, Erlander M, Sgroi D. A five-gene molecular grade index and HOXB13:IL17BR are complementary prognostic factors in early stage breast cancer. Clin Cancer Res. 2008;14(9):2601–8.
86. Jerevall PL, Ma XJ, Li H, Salunga R, Kesty NC, Erlander MG, Sgroi DC, Holmlund B, Skoog L, Fornander T, Nordenskjold B, Stal O. Prognostic utility of HOXB13:IL17BR and molecular grade index in early-stage breast cancer patients from the Stockholm trial. Br J Cancer. 2011;104(11):1762–9.
87. Sgroi DC, Sestak I, Cuzick J, Zhang Y, Schnabel CA, Schroeder B, Erlander MG, Dunbier A, Sidhu K, Lopez-Knowles E, Goss PE, Dowsett M. Prediction of late distant recurrence in patients with oestrogen-receptor-positive breast cancer: a prospective comparison of the breast-cancer index (BCI) assay, 21-gene recurrence score, and IHC4 in the TransATAC study population. Lancet Oncol. 2013;14(11):1067–76.
88. DeCosse JJ, Gossens C, Kuzma JF, Unsworth BR. Embryonic inductive tissues that cause histologic differentiation of murine mammary carcinoma in vitro. J Natl Cancer Inst. 1975;54(4):913–22.
89. Allinen M, Beroukhim R, Cai L, Brennan C, Lahti-Domenici J, Huang H, Porter D, Hu M, Chin L, Richardson A, Schnitt S, Sellers WR, Polyak K. Molecular characterization of the tumor microenvironment in breast cancer. Cancer Cell. 2004;6(1):17–32.
90. Polyak K, Kalluri R. The role of the microenvironment in mammary gland development and cancer. Cold Spring Harb Perspect Biol. 2010;2(11):a003244.
91. Junttila MR, de Sauvage FJ. Influence of tumour micro-environment heterogeneity on therapeutic response. Nature. 2013;501(7467):346–54.
92. Fridman WH, Pages F, Sautes-Fridman C, Galon J. The immune contexture in human tumours: impact on clinical outcome. Nat Rev Cancer. 2012;12(4):298–306.
93. Bianchini G, Qi Y, Alvarez RH, Iwamoto T, Coutant C, Ibrahim NK, Valero V, Cristofanilli M, Green MC, Radvanyi L, Hatzis C, Hortobagyi GN, Andre F, Gianni L, Symmans WF, Pusztai L. Molecular anatomy of breast cancer stroma and its prognostic value in estrogen receptor-positive and -negative cancers. J Clin Oncol. 2010;28(28):4316–23.
94. Farmer P, Bonnefoi H, Anderle P, Cameron D, Wirapati P, Becette V, Andre S, Piccart M, Campone M, Brain E, Macgrogan G, Petit T, Jassem J, Bibeau F, Blot E, Bogaerts J, Aguet M, Bergh J, Iggo R, Delorenzi M. A stroma-related gene signature predicts resistance to neoadjuvant chemotherapy in breast cancer. Nat Med. 2009;15(1):68–74.
95. Finak G, Bertos N, Pepin F, Sadekova S, Souleimanova M, Zhao H, Chen H, Omeroglu G, Meterissian S, Omeroglu A, Hallett M, Park M. Stromal gene expression predicts clinical outcome in breast cancer. Nat Med. 2008;14(5):518–27.
96. Bergamaschi A, Tagliabue E, Sorlie T, Naume B, Triulzi T, Orlandi R, Russnes HG, Nesland JM, Tammi R, Auvinen P, Kosma VM, Menard S, Borresen-Dale AL. Extracellular matrix signature identifies breast cancer subgroups with different clinical outcome. J Pathol. 2008;214(3):357–67.
97. Mahmoud SM, Paish EC, Powe DG, Macmillan RD, Grainge MJ, Lee AH, Ellis IO, Green AR. Tumor-infiltrating CD8+ lymphocytes predict clinical outcome in breast cancer. J Clin Oncol. 2011;29(15):1949–55.
98. Beck AH, Espinosa I, Gilks CB, van de Rijn M, West RB. The fibromatosis signature defines a robust stromal response in breast carcinoma. Lab Invest. 2008;88(6):591–601.
99. Yang F, Foekens JA, Yu J, Sieuwerts AM, Timmermans M, Klijn JG, Atkins D, Wang Y, Jiang

Y. Laser microdissection and microarray analysis of breast tumors reveal ER-alpha related genes and pathways. Oncogene. 2006;25(9):1413–9.
100. Myhre S, Mohammed H, Tramm T, Alsner J, Finak G, Park M, Overgaard J, Borresen-Dale AL, Frigessi A, Sorlie T. In silico ascription of gene expression differences to tumor and stromal cells in a model to study impact on breast cancer outcome. PLoS One. 2010;5(11):e14002.
101. Mefford D, Mefford J. Stromal genes add prognostic information to proliferation and histoclinical markers: a basis for the next generation of breast cancer gene signatures. PLoS One. 2012;7(6):e37646.
102. Schmidt M, Bohm D, von Torne C, Steiner E, Puhl A, Pilch H, Lehr HA, Hengstler JG, Kolbl H, Gehrmann M. The humoral immune system has a key prognostic impact in node-negative breast cancer. Cancer Res. 2008;68(13):5405–13.
103. Teschendorff AE, Miremadi A, Pinder SE, Ellis IO, Caldas C. An immune response gene expression module identifies a good prognosis subtype in estrogen receptor negative breast cancer. Genome Biol. 2007;8(8):R157.
104. Rody A, Holtrich U, Pusztai L, Liedtke C, Gaetje R, Ruckhaeberle E, Solbach C, Hanker L, Ahr A, Metzler D, Engels K, Karn T, Kaufmann M. T-cell metagene predicts a favorable prognosis in estrogen receptor-negative and HER2-positive breast cancers. Breast Cancer Res. 2009;11(2):R15.
105. Alexe G, Dalgin GS, Scanfeld D, Tamayo P, Mesirov JP, DeLisi C, Harris L, Barnard N, Martel M, Levine AJ, Ganesan S, Bhanot G. High expression of lymphocyte-associated genes in node-negative HER2+ breast cancers correlates with lower recurrence rates. Cancer Res. 2007;67(22):10669–76.
106. Calabro A, Beissbarth T, Kuner R, Stojanov M, Benner A, Asslaber M, Ploner F, Zatloukal K, Samonigg H, Poustka A, Sultmann H. Effects of infiltrating lymphocytes and estrogen receptor on gene expression and prognosis in breast cancer. Breast Cancer Res Treat. 2009;116(1):69–77.
107. Aaltomaa S, Lipponen P, Eskelinen M, Kosma VM, Marin S, Alhava E, Syrjanen K. Lymphocyte infiltrates as a prognostic variable in female breast cancer. Eur J Cancer. 1992;28A(4–5): 859–64.
108. Kotlan B, Gruel N, Zafrani B, Furedi G, Foldi J, Petranyi GG, Fridman WH, Teillaud JL. Immunoglobulin variable regions usage by B-lymphocytes infiltrating a human breast medullary carcinoma. Immunol Lett. 1999;65(3): 143–51.
109. Menard S, Tomasic G, Casalini P, Balsari A, Pilotti S, Cascinelli N, Salvadori B, Colnaghi MI, Rilke F. Lymphoid infiltration as a prognostic variable for early-onset breast carcinomas. Clin Cancer Res. 1997;3(5):817–9.
110. Treilleux I, Blay JY, Bendriss-Vermare N, Ray-Coquard I, Bachelot T, Guastalla JP, Bremond A, Goddard S, Pin JJ, Barthelemy-Dubois C, Lebecque S. Dendritic cell infiltration and prognosis of early stage breast cancer. Clin Cancer Res. 2004;10(22):7466–74.
111. Griffith CD, Ellis IO, Bell J, Burns K, Blamey RW. Density of lymphocytic infiltration of primary breast cancer does not affect short-term disease-free interval or survival. J R Coll Surg Edinb. 1990;35(5):289–92.
112. Liu S, Lachapelle J, Leung S, Gao D, Foulkes WD, Nielsen TO. CD8+ lymphocyte infiltration is an independent favorable prognostic indicator in basal-like breast cancer. Breast Cancer Res. 2012;14(2):R48.
113. Mahmoud S, Lee A, Ellis I, Green A. CD8(+) T lymphocytes infiltrating breast cancer: a promising new prognostic marker? Oncoimmunology. 2012;1(3):364–5.
114. Ignatiadis M, Singhal SK, Desmedt C, Haibe-Kains B, Criscitiello C, Andre F, Loi S, Piccart M, Michiels S, Sotiriou C. Gene modules and response to neoadjuvant chemotherapy in breast cancer subtypes: a pooled analysis. J Clin Oncol. 2012;30(16):1996–2004.
115. Loi S, Sirtaine N, Piette F, Salgado R, Viale G, Van Eenoo F, Rouas G, Francis P, Crown JP, Hitre E, de Azambuja E, Quinaux E, Di Leo A, Michiels S, Piccart MJ, Sotiriou C. Prognostic and predictive value of tumor-infiltrating lymphocytes in a phase III randomized adjuvant breast cancer trial in node-positive breast cancer comparing the addition of docetaxel to doxorubicin with doxorubicin-based chemotherapy: BIG 02-98. J Clin Oncol. 2013;31(7):860–7.
116. Borst P, Wessels L. Do predictive signatures really predict response to cancer chemotherapy? Cell Cycle. 2010;9(24):4836–40.
117. Jansen MP, Foekens JA, van Staveren IL, Dirkzwager-Kiel MM, Ritstier K, Look MP, Meijer-van Gelder ME, Sieuwerts AM, Portengen H, Dorssers LC, Klijn JG, Berns EM. Molecular classification of tamoxifen-resistant breast carcinomas by gene expression profiling. J Clin Oncol. 2005;23(4):732–40.
118. Ayers M, Symmans WF, Stec J, Damokosh AI, Clark E, Hess K, Lecocke M, Metivier J, Booser D, Ibrahim N, Valero V, Royce M, Arun B, Whitman G, Ross J, Sneige N, Hortobagyi GN, Pusztai L. Gene expression profiles predict complete pathologic response to neoadjuvant paclitaxel and fluorouracil, doxorubicin, and cyclophosphamide chemotherapy in breast cancer. J Clin Oncol. 2004;22(12):2284–93.
119. Chang JC, Wooten EC, Tsimelzon A, Hilsenbeck SG, Gutierrez MC, Elledge R, Mohsin S, Osborne CK, Chamness GC, Allred DC, O'Connell P. Gene expression profiling for the prediction of

therapeutic response to docetaxel in patients with breast cancer. Lancet. 2003;362(9381):362–9.
120. Desmedt C, Di Leo A, de Azambuja E, Larsimont D, Haibe-Kains B, Selleslags J, Delaloge S, Duhem C, Kains JP, Carly B, Maerevoet M, Vindevoghel A, Rouas G, Lallemand F, Durbecq V, Cardoso F, Salgado R, Rovere R, Bontempi G, Michiels S, Buyse M, Nogaret JM, Qi Y, Symmans F, Pusztai L, D'Hondt V, Piccart-Gebhart M, Sotiriou C. Multifactorial approach to predicting resistance to anthracyclines. J Clin Oncol. 2011;29(12):1578–86.
121. Shah SP, Roth A, Goya R, Oloumi A, Ha G, Zhao Y, Turashvili G, Ding J, Tse K, Haffari G, Bashashati A, Prentice LM, Khattra J, Burleigh A, Yap D, Bernard V, McPherson A, Shumansky K, Crisan A, Giuliany R, Heravi-Moussavi A, Rosner J, Lai D, Birol I, Varhol R, Tam A, Dhalla N, Zeng T, Ma K, Chan SK, Griffith M, Moradian A, Cheng SW, Morin GB, Watson P, Gelmon K, Chia S, Chin SF, Curtis C, Rueda OM, Pharoah PD, Damaraju S, Mackey J, Hoon K, Harkins T, Tadigotla V, Sigaroudinia M, Gascard P, Tlsty T, Costello JF, Meyer IM, Eaves CJ, Wasserman WW, Jones S, Huntsman D, Hirst M, Caldas C, Marra MA, Aparicio S. The clonal and mutational evolution spectrum of primary triple-negative breast cancers. Nature. 2012;486(7403):395–9.
122. Nik-Zainal S, Van Loo P, Wedge DC, Alexandrov LB, Greenman CD, Lau KW, Raine K, Jones D, Marshall J, Ramakrishna M, Shlien A, Cooke SL, Hinton J, Menzies A, Stebbings LA, Leroy C, Jia M, Rance R, Mudie LJ, Gamble SJ, Stephens PJ, McLaren S, Tarpey PS, Papaemmanuil E, Davies HR, Varela I, McBride DJ, Bignell GR, Leung K, Butler AP, Teague JW, Martin S, Jonsson G, Mariani O, Boyault S, Miron P, Fatima A, Langerod A, Aparicio SA, Tutt A, Sieuwerts AM, Borg A, Thomas G, Salomon AV, Richardson AL, Borresen-Dale AL, Futreal PA, Stratton MR, Campbell PJ. The life history of 21 breast cancers. Cell. 2012;149(5):994–1007.
123. Collisson EA, Cho RJ, Gray JW. What are we learning from the cancer genome? Nat Rev Clin Oncol. 2012;9(11):621–30.
124. Symmans WF, Hatzis C, Sotiriou C, Andre F, Peintinger F, Regitnig P, Daxenbichler G, Desmedt C, Domont J, Marth C, Delaloge S, Bauernhofer T, Valero V, Booser DJ, Hortobagyi GN, Pusztai L. Genomic index of sensitivity to endocrine therapy for breast cancer. J Clin Oncol. 2010;28(27):4111–9.
125. Banerji S, Cibulskis K, Rangel-Escareno C, Brown KK, Carter SL, Frederick AM, Lawrence MS, Sivachenko AY, Sougnez C, Zou L, Cortes ML, Fernandez-Lopez JC, Peng S, Ardlie KG, Auclair D, Bautista-Pina V, Duke F, Francis J, Jung J, Maffuz-Aziz A, Onofrio RC, Parkin M, Pho NH, Quintanar-Jurado V, Ramos AH, Rebollar-Vega R, Rodriguez-Cuevas S, Romero-Cordoba SL, Schumacher SE, Stransky N, Thompson KM, Uribe-Figueroa L, Baselga J, Beroukhim R, Polyak K, Sgroi DC, Richardson AL, Jimenez-Sanchez G, Lander ES, Gabriel SB, Garraway LA, Golub TR, Melendez-Zajgla J, Toker A, Getz G, Hidalgo-Miranda A, Meyerson M. Sequence analysis of mutations and translocations across breast cancer subtypes. Nature. 2012;486(7403):405–9.
126. Ellis MJ, Ding L, Shen D, Luo J, Suman VJ, Wallis JW, Van Tine BA, Hoog J, Goiffon RJ, Goldstein TC, Ng S, Lin L, Crowder R, Snider J, Ballman K, Weber J, Chen K, Koboldt DC, Kandoth C, Schierding WS, McMichael JF, Miller CA, Lu C, Harris CC, McLellan MD, Wendl MC, DeSchryver K, Allred DC, Esserman L, Unzeitig G, Margenthaler J, Babiera GV, Marcom PK, Guenther JM, Leitch M, Hunt K, Olson J, Tao Y, Maher CA, Fulton LL, Fulton RS, Harrison M, Oberkfell B, Du F, Demeter R, Vickery TL, Elhammali A, Piwnica-Worms H, McDonald S, Watson M, Dooling DJ, Ota D, Chang LW, Bose R, Ley TJ, Piwnica-Worms D, Stuart JM, Wilson RK, Mardis ER. Whole-genome analysis informs breast cancer response to aromatase inhibition. Nature. 2012;486(7403):353–60.
127. Cancer Genome Atlas Network. Comprehensive molecular portraits of human breast tumours. Nature. 2012;490(7418):61–70.
128. Stephens PJ, Tarpey PS, Davies H, Van Loo P, Greenman C, Wedge DC, Nik-Zainal S, Martin S, Varela I, Bignell GR, Yates LR, Papaemmanuil E, Beare D, Butler A, Cheverton A, Gamble J, Hinton J, Jia M, Jayakumar A, Jones D, Latimer C, Lau KW, McLaren S, McBride DJ, Menzies A, Mudie L, Raine K, Rad R, Chapman MS, Teague J, Easton D, Langerod A, Oslo Breast Cancer Consortium, Lee MT, Shen CY, Tee BT, Huimin BW, Broeks A, Vargas AC, Turashvili G, Martens J, Fatima A, Miron P, Chin SF, Thomas G, Boyault S, Mariani O, Lakhani SR, van de Vijver M, van't Veer L, Foekens J, Desmedt C, Sotiriou C, Tutt A, Caldas C, Reis-Filho JS, Aparicio SA, Salomon AV, Borresen-Dale AL, Richardson AL, Campbell PJ, Futreal PA, Stratton MR. The landscape of cancer genes and mutational processes in breast cancer. Nature. 2012;486(7403):400–4.
129. Weigelt B, Reis-Filho JS. Activating mutations in HER2: neu opportunities and neu challenges. Cancer Discov. 2013;3(2):145–7.
130. Gerlinger M, Rowan AJ, Horswell S, Larkin J, Endesfelder D, Gronroos E, Martinez P, Matthews N, Stewart A, Tarpey P, Varela I, Phillimore B, Begum S, McDonald NQ, Butler A, Jones D, Raine K, Latimer C, Santos CR, Nohadani M, Eklund AC, Spencer-Dene B, Clark G, Pickering L, Stamp G, Gore M, Szallasi Z, Downward J, Futreal PA, Swanton C. Intratumor

heterogeneity and branched evolution revealed by multiregion sequencing. N Engl J Med. 2012;366(10):883–92.
131. Geyer FC, Weigelt B, Natrajan R, Lambros MB, de Biase D, Vatcheva R, Savage K, Mackay A, Ashworth A, Reis-Filho JS. Molecular analysis reveals a genetic basis for the phenotypic diversity of metaplastic breast carcinomas. J Pathol. 2010;220(5):562–73.
132. Navin N, Kendall J, Troge J, Andrews P, Rodgers L, McIndoo J, Cook K, Stepansky A, Levy D, Esposito D, Muthuswamy L, Krasnitz A, McCombie WR, Hicks J, Wigler M. Tumour evolution inferred by single-cell sequencing. Nature. 2011;472(7341):90–4.
133. Shah SP, Morin RD, Khattra J, Prentice L, Pugh T, Burleigh A, Delaney A, Gelmon K, Guliany R, Senz J, Steidl C, Holt RA, Jones S, Sun M, Leung G, Moore R, Severson T, Taylor GA, Teschendorff AE, Tse K, Turashvili G, Varhol R, Warren RL, Watson P, Zhao Y, Caldas C, Huntsman D, Hirst M, Marra MA, Aparicio S. Mutational evolution in a lobular breast tumour profiled at single nucleotide resolution. Nature. 2009;461(7265):809–13.
134. Ding L, Ellis MJ, Li S, Larson DE, Chen K, Wallis JW, Harris CC, McLellan MD, Fulton RS, Fulton LL, Abbott RM, Hoog J, Dooling DJ, Koboldt DC, Schmidt H, Kalicki J, Zhang Q, Chen L, Lin L, Wendl MC, McMichael JF, Magrini VJ, Cook L, McGrath SD, Vickery TL, Appelbaum E, Deschryver K, Davies S, Guintoli T, Crowder R, Tao Y, Snider JE, Smith SM, Dukes AF, Sanderson GE, Pohl CS, Delehaunty KD, Fronick CC, Pape KA, Reed JS, Robinson JS, Hodges JS, Schierding W, Dees ND, Shen D, Locke DP, Wiechert ME, Eldred JM, Peck JB, Oberkfell BJ, Lolofie JT, Du F, Hawkins AE, O'Laughlin MD, Bernard KE, Cunningham M, Elliott G, Mason MD, Thompson Jr DM, Ivanovich JL, Goodfellow PJ, Perou CM, Weinstock GM, Aft R, Watson M, Ley TJ, Wilson RK, Mardis ER. Genome remodelling in a basal-like breast cancer metastasis and xenograft. Nature. 2010;464(7291):999–1005.
135. Fidler IJ. Tumor heterogeneity and the biology of cancer invasion and metastasis. Cancer Res. 1978;38(9):2651–60.
136. Nik-Zainal S, Alexandrov LB, Wedge DC, Van Loo P, Greenman CD, Raine K, Jones D, Hinton J, Marshall J, Stebbings LA, Menzies A, Martin S, Leung K, Chen L, Leroy C, Ramakrishna M, Rance R, Lau KW, Mudie LJ, Varela I, McBride DJ, Bignell GR, Cooke SL, Shlien A, Gamble J, Whitmore I, Maddison M, Tarpey PS, Davies HR, Papaemmanuil E, Stephens PJ, McLaren S, Butler AP, Teague JW, Jonsson G, Garber JE, Silver D, Miron P, Fatima A, Boyault S, Langerod A, Tutt A, Martens JW, Aparicio SA, Borg A, Salomon AV, Thomas G, Borresen-Dale AL, Richardson AL, Neuberger MS, Futreal PA, Campbell PJ, Stratton MR. Mutational processes molding the genomes of 21 breast cancers. Cell. 2012;149(5):979–93.
137. Swanton C. Intratumor heterogeneity: evolution through space and time. Cancer Res. 2012; 72(19):4875–82.
138. Marusyk A, Almendro V, Polyak K. Intra-tumour heterogeneity: a looking glass for cancer? Nat Rev Cancer. 2012;12(5):323–34.
139. Turner NC, Reis-Filho JS. Genetic heterogeneity and cancer drug resistance. Lancet Oncol. 2012;13(4):e178–85.
140. Yap TA, Gerlinger M, Futreal PA, Pusztai L, Swanton C. Intratumor heterogeneity: seeing the wood for the trees. Sci Transl Med. 2012;4(127): 127ps110.
141. Stephens PJ, Tarpey PS, Davies H, Van Loo P, Greenman C, Wedge DC, Nik-Zainal S, Martin S, Varela I, Bignell GR, Yates LR, Papaemmanuil E, Beare D, Butler A, Cheverton A, Gamble J, Hinton J, Jia M, Jayakumar A, Jones D, Latimer C, Lau KW, McLaren S, McBride DJ, Menzies A, Mudie L, Raine K, Rad R, Chapman MS, Teague J, Easton D, Langerod A, Lee MT, Shen CY, Tee BT, Huimin BW, Broeks A, Vargas AC, Turashvili G, Martens J, Fatima A, Miron P, Chin SF, Thomas G, Boyault S, Mariani O, Lakhani SR, van de Vijver M, van't Veer L, Foekens J, Desmedt C, Sotiriou C, Tutt A, Caldas C, Reis-Filho JS, Aparicio SA, Salomon AV, Borresen-Dale AL, Richardson AL, Campbell PJ, Futreal PA, Stratton MR. The landscape of cancer genes and mutational processes in breast cancer. Nature. 2012;486(7403):400–4.
142. Powell AA, Talasaz AH, Zhang H, Coram MA, Reddy A, Deng G, Telli ML, Advani RH, Carlson RW, Mollick JA, Sheth S, Kurian AW, Ford JM, Stockdale FE, Quake SR, Pease RF, Mindrinos MN, Bhanot G, Dairkee SH, Davis RW, Jeffrey SS. Single cell profiling of circulating tumor cells: transcriptional heterogeneity and diversity from breast cancer cell lines. PLoS One. 2012;7(5): e33788.
143. Forshew T, Murtaza M, Parkinson C, Gale D, Tsui DW, Kaper F, Dawson SJ, Piskorz AM, Jimenez-Linan M, Bentley D, Hadfield J, May AP, Caldas C, Brenton JD, Rosenfeld N. Noninvasive identification and monitoring of cancer mutations by targeted deep sequencing of plasma DNA. Sci Transl Med. 2012;4(136): 136ra168.
144. De Mattos-Arruda L, Cortes J, Santarpia L, Vivancos A, Tabernero J, Reis-Filho JS, Seoane J. Circulating tumour cells and cell-free DNA as tools for managing breast cancer. Nat Rev Clin Oncol. 2013;10(7):377–89.
145. Dawson SJ, Rosenfeld N, Caldas C. Circulating tumor DNA to monitor metastatic breast cancer. N Engl J Med. 2013;369(1):93–4.

CHAPTER 24

Genomic Applications in Pulmonary Malignancies

REINHARD BÜTTNER, MARGARETE ODENTHAL, SABINE MERKELBACH-BRUSE

Introduction 383

Driver Genetic Alterations in Lung Cancer 383

Application of Next-Generation Sequencing (NGS) to Molecular Lung Cancer Pathology 390

Conclusions 401

Notes 401

Introduction

As a result of recent insights into cancer genomes, it has been appreciated that human malignancies arise from a limited set of somatic genetic aberrations driving oncogenic signalling networks. Hence, pathologically altered genes causally driving and maintaining the cancer phenotype are being referred to as "oncogenic drivers." Lung cancer in particular may serve as a paradigm for Personalized Cancer Medicine (PCM) because diagnosing and interfering with the individual set of pathogenic driver mutations results in highly effective, personalized, and frequently less toxic treatment regimens. This has led to the approval of selective small molecules targeting pathologically activated EGFR and ALK receptors in lung cancer and further holds great promise for the majority of other tumor types with prominent oncogenic driver mutations that represent therapeutic targets [1]. However, there are strict requirements for discovery and preclinical validation of oncogenic targets as well as for understanding and responding to mechanisms of resistance [2]. Biomarkers *sensu stricto* define companion diagnostics rather than classical, correlative biomarkers. Lung cancer may also serve as a paradigm for a comprehensive reclassification of tumor entities combining both morphologic and genomic data. Finally, specific changes in the design of clinical studies and approaches to rapidly transfer new therapies into clinical application are paradigmatically exemplified by the experiences from lung cancer (Fig. 24.1).

Driver Genetic Alterations in Lung Cancer

EGFR: Response and Resistance to Targeted Therapy

EGFR (HER1) is the prototypic member of the ERBB family of transmembrane tyrosine kinases also comprising HER2, HER3, and HER4 [3]. It is composed of an extracellular growth-factor-binding domain, a transmembrane segment and an intracellular protein-tyrosine kinase catalytic domain. As a result of ligand binding, the inactive receptor monomers undergo conformational change and receptor dimerization. Receptor activation leads to autophosphorylation as well as

G.J. Netto and I. Schrijver (eds.), *Genomic Applications in Pathology*,
DOI 10.1007/978-1-4939-0727-4_24, © Springer Science+Business Media New York 2015

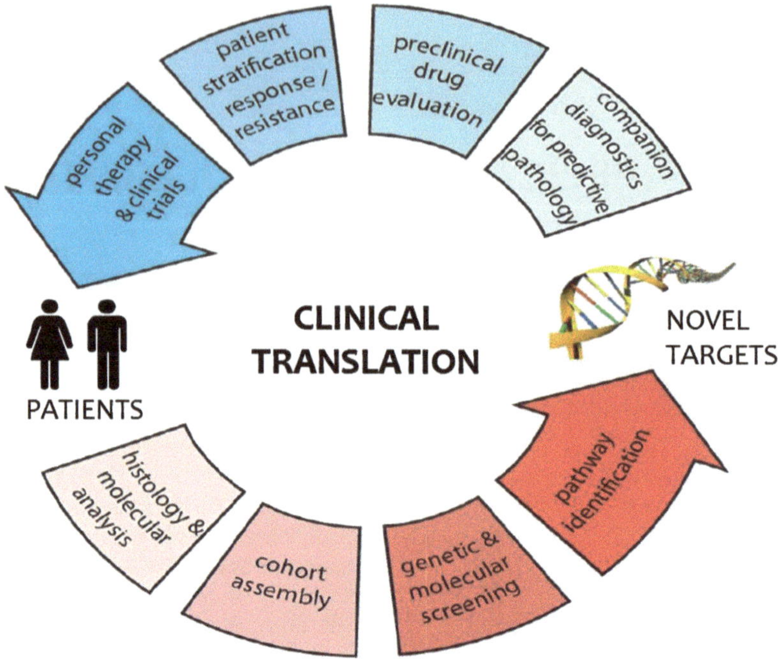

Figure 24-1 Translation of findings from biomarker evaluation to clinical practice. A predictive biomarker defines a subset of patients with an activated oncogenic target that is causally involved in tumor pathogenesis. Hence, a predictive biomarker has been validated in a clinical study for its suitability to select a patient cohort with high probability of responding to a specific therapy. Markers that describe or predict diseases only in a correlative manner do not fulfill the strict criteria of predictive biomarkers

phosphorylation of tyrosine residues of adaptor or signalling molecules. Thereby, two key oncogenic pathways, the RAS/RAF/MAPK pathway and the PI3K/AKT pathway, are activated, promoting cellular proliferation and survival [4]. Due to its function of providing oncogene dependency in NSCLC [5], EGFR has become a therapeutic target using tyrosine kinase inhibitors (TKIs) and/or inhibitory monoclonal antibodies [6].

Since 2004, several investigators have suggested the *EGFR* mutation status but not the *EGFR* gene copy number or expression as the optimal predictor of clinical benefit from EGFR inhibitors in NSCLC [7, 8]. Activating mutations usually occur in the region that encodes the intracellular tyrosine kinase domain and abolish autoinhibition, which keeps the wild-type receptor silent in the absence of ligand [9]. The growth of *EGFR*-mutant NSCLC cells is dependent on aberrant kinase activation. Additionally, the mutant receptor has a higher affinity for the competitive tyrosine kinase inhibitors than for ATP [10].

The impact of *EGFR* mutations on the prediction of therapeutic response was confirmed in the phase III IRESSA Pan-Asia (IPASS) study [11], which revealed significant benefit in progression-free survival for single-agent monotherapy with the tyrosine kinase inhibitor gefitinib versus conventional combination chemotherapy (carboplatin/paclitaxel) in Asian patients with *EGFR*-mutated lung adenocarcinomas. Similar results were found by the Spanish Lung Cancer Group, which compared erlotinib with standard chemotherapy for first-line treatment of European patients with advanced *EGFR*-mutation positive NSCLC [12]. Importantly, both smokers and past-smokers as well as never-smokers benefited from therapy with TKIs in the presence of an activating *EGFR* mutation. Molecular analysis of the *EGFR* mutation status has therefore become a prerequisite for treatment with EGFR inhibitors and is recommended in all NSCLC with the exception of squamous cell carcinoma.

Activating mutations can be found in exons 18–21 of the *EGFR* gene. In-frame deletions of exon 19 which account for about 55 % of mutated cases comprise almost always the amino acid residues p.Leu747 to p.Ala750. The second most prevalent mutations are substitutions in exon 21 which account for approximately 32 %. Nearly 75 % of these substitutions are exchanges from leucine to arginine at codon 858 (p.Leu858Arg) and approximately 10 % from leucine to glutamine at codon 861 (p.Leu861Gln). Mutations in exon 18 account for 12 % of all *EGFR* mutated cases and are predominantly point mutations in codon 719 (experiential data from genotyping >2,000 cases).

EGFR exon 20 insertions comprise approximately 4 % of all mutations in NSCLC and most frequently occur between amino-acid residues 767 and 774 [13]. A rare mutation type recently described is an in-frame exon 19 insertion of six amino acids accounting for approximately 1 % of all *EGFR*-mutant NSCLC [14]. Other rare mutations, distributed throughout all four exons, are supposed to account for 18 % of all *EGFR* mutations [15].

The most prevalent mutations in exon 19 and exon 21 as well as exon 19 insertions confer strong sensitivity to TKIs [14, 16]. Exon 20 insertion mutations, however, are associated with lower sensitivity to inhibitor-based therapy or with primary resistance [13]. A comparative analysis of the outcome of rare *EGFR* mutants across four clinical trials has shown the impossibility to build up clear categories of response for these rare mutations [17]. Findings of the very recent LUX-Lung 3 study comparing the irreversible panHER TKI afatinib versus pemetrexed and cisplatin in first line treatment indicate that the rare mutations may also benefit from TKI treatment, but to a lesser degree [18].

A wide spectrum of techniques has been employed for the detection of *EGFR* mutations. After preparing DNA lysates from sometimes manually microdissected tissue samples, the relevant parts of the *EGFR* gene are amplified by PCR followed by different techniques including Sanger sequencing, pyrosequencing, allele specific PCR, fragment length analyses, or next-generation massively parallel sequencing techniques (see below).

Most patients who initially respond to tyrosine kinase inhibitor therapies will develop secondary resistance. In 50 % of cases acquired resistance is due to the occurrence of p.Thr790Met mutations [19, 20]. This mutation increases the affinity of the binding pocket for ATP, thus interfering with drug binding. Other resistance mechanisms include MET and *HER2* amplification, which lead to the activation of parallel signalling pathways [21, 22]. *MET* amplification drives HER3-dependent activation of PI3K [23]. It is important to identify these pathway resistance mechanisms, because they will allow adoption of second- or third-line therapies, i.e., by appropriate novel TKIs in combination with inhibitory antibodies or MET inhibitors. Rarely, secondary mutations in downstream effectors such as *BRAF* and *PIK3CA* were identified [21, 24]. Two pathway independent mechanisms, namely, transformation to small-cell lung cancer and epithelial-to-mesenchymal transition have also been described [21].

KRAS: Response to TKI Treatment and Therapeutic Options

The *KRAS* gene encodes a member of the family of membrane-bound GTP binding proteins that regulate proliferation, differentiation, and apoptosis through the MAPK, STAT, and PI3K signalling pathways [25]. Activating point mutations occurring in the GTPase domain of *KRAS* are found in 30 % of NSCLC [2]. More than 95 % of mutations are localized in codons 12 and 13 (exon 2), the predominant mutation being p.Gly12Cys (42.3 % of all *KRAS* mutations). These mutations are mutually exclusive with those in *EGFR*. *KRAS* mutations were proposed to be a negative prognostic factor in NSCLC [26]. Additionally they are associated with resistance to EGFR inhibitor based therapy as confirmed by two meta-analyses that found *KRAS* mutations to be a negative predictor of response to single-agent EGFR TKIs in advanced NSCLC [27, 28].

Inhibition of KRAS is difficult as the mutated RAS protein harbors reduced GTPase activity. A more promising approach than targeting the mutant *KRAS* itself is the inhibition of signalling pathways downstream of *KRAS*. Preclinical studies revealed the efficacy of combined inhibition of the PI3K and MAPK pathways [29, 30]. Recently, a prospective study has demonstrated clinical benefit for patients with *KRAS*-mutated NSCLC treated with a MEK1/2 inhibitor in combination with docetaxel versus docetaxel with placebo [31].

BRAF: Activating and Inactivating Mutations

BRAF is a serine–threonine kinase that mediates the RAS family members' activation of downstream proteins in the MAPK pathway [32]. *BRAF* mutations are found in approximately 3 % of Caucasian NSCLC patients. In contrast to melanomas, NSCLC often harbor mutations outside codon p.Val600 [33]. Besides the kinase domain (exon 15), the G loop of the activation domain encoded by exon 11 may be mutated [34, 35].

Whether NSCLC tumors, or other solid tumors harboring p.Val600 mutations clinically respond to treatment with specific BRAF inhibitors is currently under investigation in phase II clinical trials (Center for Integrated Oncology (2012) University Hospitals Cologne and Bonn. http://www.cio-koeln-bonn.de/mediziner/klinische-studien/ Accessed 30 August, 2013). Recently, a single case of a *BRAF* mutated NSCLC responding to therapy with the BRAF inhibitor vemurafenib has been reported [36]. NSCLC cell lines with both p.Val600Glu *BRAF* mutations and mutations other than p.Val600Glu are also sensitive to MEK inhibitors [37], but due to the need of patient selection clinical evidence is still lacking [38, 39].

The SRC protein family inhibitor dasatinib has clinically significant activity unrelated to *EGFR* mutation status and SRC activation, in a small number of patients [40]. In one patient, a novel *BRAF* mutation, p. Tyr472Cys, was detected. This mutation, like other mutations in exons 11 and 15, leads to an impairment of BRAF kinase activity, thus transactivating CRAF, MEK, and ERK [41]. In summary, the small group of *BRAF-mutated* NSCLC patients may be offered one of three different therapy regimens with vemurafenib, MEK-inhibitors or dasatinib, depending on their specific mutational status.

The common point mutations occurring in limited regions of the *KRAS*, *BRAF*, and *PIK3CA* genes can be easily analyzed with sensitive methods like allele-specific PCR or primer extension assays. Special attention has to be paid to inactivating *BRAF* mutations which can occur throughout the entire exons 11 and 15.

HER2: Amplification and Mutation

HER2 is the sole member of the ERBB receptor family without an identified ligand, and hence, it is activated through dimerization with other members of the ERBB family. Activation of the HER2 tyrosine kinase domain turns on signalling via the RAS/RAF/MAPK and the PI3K/AKT pathways.

HER2 gene amplification assessed by fluorescence in situ hybridization (FISH) is demonstrable in only 2 % of NSCLC, whereas 20 % of cases test positive by immunohistochemistry. The authors attribute this to the polysomy of chromosome 17 which occurs in 81 % of NSCLC [42]. Clinical trials with trastuzumab, an antibody to HER2, reported possible benefit for a group of patients with strong HER2 overexpression and FISH positivity, but not for patients solely selected by immunohistochemistry [43, 44]. Our own data suggest that HER2 is the most prominent client protein of the HSP90 chaperon, and a dual inhibitor trial with trastuzumab and AUY922 is currently under way at the Network Genomic Medicine in Cologne (Center for Integrated Oncology (2012) University Hospitals Cologne and Bonn) (http://www.cio-koeln-bonn.de/mediziner/klinische-studien/ Accessed 30 August 2013)

HER2 mutations are present in 2–4 % of NSCLC and occur mostly in exon 20 [45]. In the majority of cases mutations are in-frame insertions causing duplication of amino acids TyrValMetAla (YVMA) at codon 775 which are mutually exclusive with mutations in *KRAS*, *EGFR*, *BRAF*, and *PIK3CA*, as well as with *ALK* rearrangements [46]. *HER2* insertions lead to constitutive receptor activation [47]. Preclinical data suggest that *HER2* mutated tumors are sensitive to HER2 targeted antibodies and tyrosine kinase inhibitors like trastuzumab and lapatinib, but insensitive to those targeting EGFR alone [48]. A partial response to the irreversible tyrosine kinase inhibitor afatinib was described in a small study for three out of three patients harboring *HER2* exon 20 mutations [49]. A single NSCLC case with an exon 20 point mutation was described showing response to treatment with trastuzumab in combination with paclitaxel [50].

Recently, an extracellular domain mutation of *HER2* has been shown to be oncogenic in cellular transformation assays. Oncogenic activity turned out to be sensitive to treatment with HER2 inhibitors [51].

PIK3CA: Alterations and Signalling via AKT and PTEN

Phosphatidyl 3-kinases (PI3K) are heterodimeric lipid kinases, composed of a catalytic and a regulatory subunit, and involved in a wide range of vital cellular processes including proliferation and differentiation. The *PIK3CA* gene encoding the catalytic subunit is frequently mutated in human cancers [52]. Somatic mutations are found in 1–3 % of NSCLC [53, 54] and occur within two hotspot regions, the helical binding domain encoded by exon 9 (p.Glu542Lys and p.Glu545Lys) and the catalytic subunit encoded by exon 20 (p.His1047Arg or Leu). Mutations seem to be more common in tumors with squamous cell histology than in adenocarcinoma (6.5 % versus 1.5 %) and are not mutually exclusive with *EGFR*, *KRAS*, or *BRAF* mutations [55]. In up to 70 % of cases coexisting mutations were found, leading to doubt regarding the role of *PIK3CA* lesions as driver mutations conferring oncogene-dependency [53, 54]. The incidence of *PIK3CA* amplification in NSCLC was found to be 31 % with the overwhelming majority (93.3 %) occurring in squamous cell carcinoma and the remainder in adenocarcinoma [56].

Multiple drugs targeting the PI3K pathway are currently under development. Preclinical studies in other tumor entities show activity of such agents mainly in tumors with *PIK3CA* mutations [57]. Recently, a phase I trial assessing a PI3K inhibitor in solid tumors including lung and breast has provided clear evidence of target inhibition and preliminary antitumor activity [58]. A dual PIK3CA/mTOR inhibitor displayed antitumor activity in gefitinib-resistant NSCLC cell lines that was enhanced in the presence of a *PIK3CA* mutation [59].

As mentioned above, activated PIK3CA may play a role in resistance to EGFR inhibitor therapy. This is not only restricted to acquired resistance, given that *PIK3CA* mutations can co-occur with *EGFR* mutations already present in the primary tumor. In vitro evidence further supported that mutant PIK3CA confers gefitinib resistance [60]. Moreover, loss of PTEN expression in *EGFR* mutant lung cancer cells was identified as contributing to erlotinib resistance [61, 62].

In addition to PIK3CA, other members of the pathway may be altered in certain cancers [63]. This includes deletion or mutation of the pathway inhibitor *PTEN* or activating mutations of *AKT1*. Activated AKT1 phosphorylates downstream effector molecules and thereby activates cell proliferation and survival. Somatic mutations in *AKT1* were found in approximately 1 % of NSCLC, both in adenocarcinoma as well as squamous cell carcinoma [64–66]. According to the COSMIC database (http://www.sanger.ac.uk/genetics/CGP/cosmic/, Accessed 30 August 2013) 84.3 % of *AKT1* mutations reveal a p.Glu17Lys exchange. AKT inhibitors are in clinical development; however, currently there are no clinical data pointing to a role for *AKT1* mutations in selecting patients for targeted treatment of lung cancer.

PTEN downregulates the PI3K/AKT signalling pathway by dephosphorylating PIP3 and thereby inhibiting activation of AKT. Loss of *PTEN* or inactivating *PTEN* mutations result in activation of the PI3K signalling cascade [67]. In NSCLC, *PTEN* mutations have been observed in 4.5–8 % of cases [68, 69]. They are encountered more commonly in squamous cell carcinoma than in adenocarcinoma [69, 70]. Mutations occur throughout the entire open reading frame and often lead to protein truncation. Both homozygous and heterozygous deletions of *PTEN* are observed.

Inhibitors of the PI3K pathway may be active not only in the presence of a *PIK3CA* mutation but also in the case of pathway dysregulation. Ihle et al. [71] showed that besides mutation of *PIK3CA*, loss of PTEN activity was a sufficient predictor of sensitivity to the antitumor activity of the PI3K inhibitor PX-866 in the presence of wild-type *RAS*. Mutant oncogenic *RAS* was proven to be a dominant determinant of resistance, even in tumors with coexisting mutations in *PIK3CA*.

Cell lines harboring different pathway alterations were sensitive to the PI3K inhibitor GDC-0941 [72]. The combination of GDC-0941 with paclitaxel, erlotinib, or a mitogen-activated protein-extracellular signal-regulated kinase inhibitor had greater effects on cell viability than PI3K inhibition alone.

MET: Activation in Untreated NSCLC

The receptor tyrosine kinase hepatocyte growth factor receptor (HGFR) is encoded by the *MET* gene located on chromosome 7 [73]. *MET* point mutations in the semaphorin and juxtamembrane domains occur with a low frequency (1–3 %) in NSCLC [74–76]. *MET* mutations are mutually exclusive with mutations in *EGFR*, *KRAS* and *HER2* [76]. Some of the previously reported *MET* mutations seem to represent SNPs, thus their clinical importance is highly questionable [77]. A mutation affecting the splice site between exons 13 and 14 which leads to a deletion of exon 14 is associated with enhanced ligand-mediated proliferation and tumor growth [74, 76].

In NSCLC not previously treated with EGFR-specific tyrosine kinase inhibitors, high level *MET* amplification is detected in approximately 2–3 % and is associated with poor prognosis [75, 76, 78]. Whether the so-called low-level *MET* amplification resulting, in part, from polysomy confers oncogene-dependency remains to be shown. The efficiency of MET TKIs and monoclonal antibodies is currently under investigation in preclinical and clinical studies. Response to crizotinib, an inhibitor of MET and ALK, has been reported in lung cancer cells with *MET* amplification but not with certain *MET* mutations (p.Asn375Ser and exon 14 deletions) [79, 80]. In accordance with these results, Ou [80] described response to crizotinib therapy in a NSCLC patient with *MET* amplification.

ALK, *RET*, and *ROS1*: Chromosomal Inversions and Translocations

The *ALK* gene encodes a receptor tyrosine kinase, referred to as anaplastic lymphoma kinase. This designation derives from anaplastic large-cell lymphomas, in which a gene fusion between *ALK* and *NPM* (*nucleophosmin*) was first detected [81]. In 2007, fusion of *ALK* with the upstream partner *EML4* was found in NSCLC [82]. The fusion is the result of inversion in chromosome 2.

Approximately 3–5 % of lung adenocarcinomas harbor *ALK* rearrangements [82–84]. Different fusion variants have been reported, all comprising the entire tyrosine kinase domain of *ALK* and varying portions of the *EML4* gene [85]. The fusion results in protein dimerization and therefore constitutive activation of the kinase function [82, 83]. Most of the reported variants start with exon 20 of *ALK* as the first exon of the 3′ part, but rarely, variants starting with exon 19 are described [86, 87]. *EML4-ALK* fusions are usually found in tumors with *EGFR* and *KRAS* wild-type sequences and positivity was found to be associated with resistance to EGFR targeted inhibitors [88, 89]. In a phase I clinical trial, 57 % of patients with an *EML4-ALK* fusion showed an overall response to the dual ALK/MET inhibitor crizotinib [90]. Meanwhile, the compound received regulatory approval for clinical use in the USA and the EU.

The presence of several fusion variants described above, besides the poor RNA quality from FFPE tissue makes it difficult to detect the fusion by RT-PCR. Therefore, the standard method for testing chromosomal rearrangements in FFPE tissue currently is fluorescence in situ hybridization (FISH). Typical signal patterns for the detection of *ALK* rearrangement are shown in Fig. 24.2.

The success of crizotinib therapy is limited by the development of acquired drug resistance [91–93]. Most patients show secondary mutations in the tyrosine kinase domain of *ALK* leading to resistance, but also *ALK* copy number gain, emergence of other oncogenic driver mutations, e.g., in *EGFR* or *KRAS*, or amplification of *KIT* has been detected [92–94]. Recently, the EGF-dependent activation of HER family proteins has been described to be associated with crizotinib resistance [95].

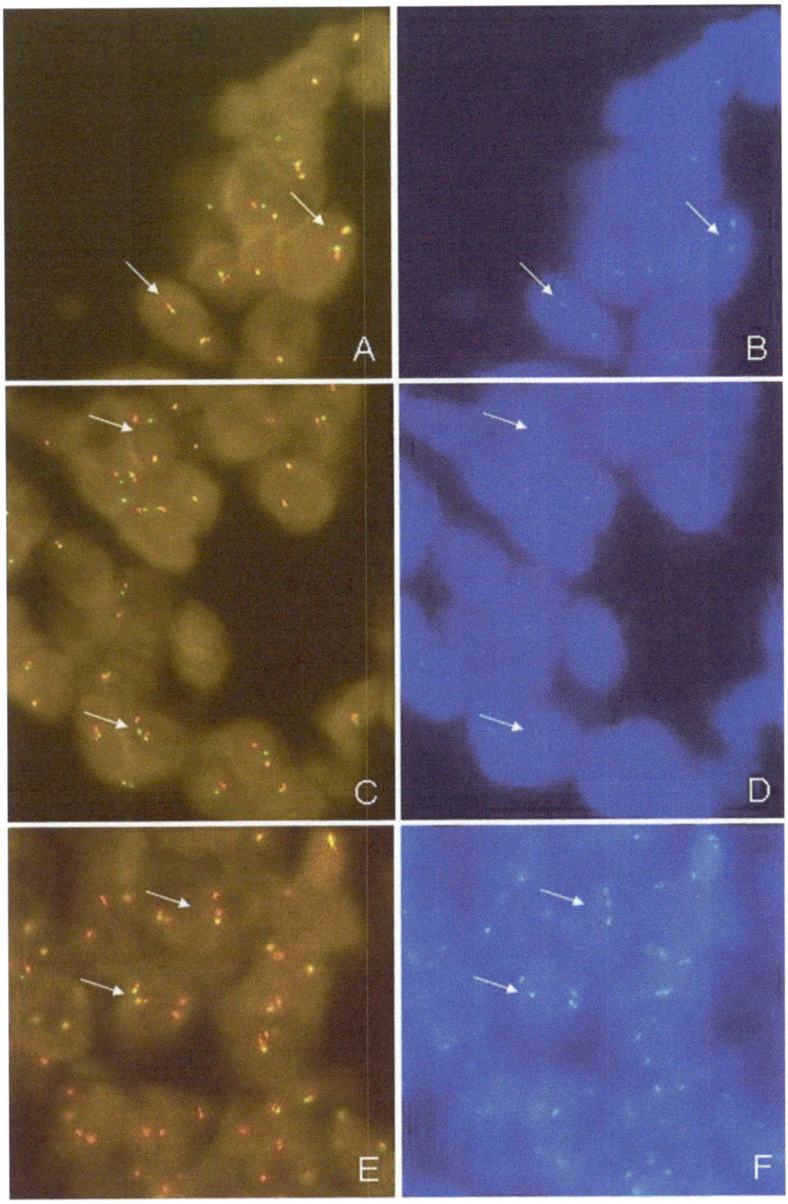

Figure 24-2 Detection of *ALK* rearrangement. The *ALK* rearrangement is detected with the Zyto*Light*® SPEC ALK/EML4 TriCheck™ Assay (Zytovision, Bremerhaven, Germany): Orange and green fluorescent signals for the *ALK* break-apart probes (**a**, **c**, **e**) and blue fluorescent signals for the *EML4* probe (**b**, **d**, **f**) are depicted in three pulmonary adenocarcinoma examples. The first case (**a**, **b**) includes nuclei with two fusion signals (**a**) and two *EML4* signals (**b**) and is considered negative for *ALK* rearrangement. The second case (**c**, **d**) shows break-apart signals for *ALK* (**c**) and three *EML4* signals (**d**) representing the pattern of *ALK-EML4* inversions. In the third case (**e**, **f**) the 5′ part of *ALK* is deleted during inversion, so single red signals (**e**), co-localized with the corresponding blue signals, can be detected

Strategies to overcome resistance are currently tested in preclinical models, including second generation ALK and HSP90 inhibitors as well as combinations with EGFR pathway inhibitors [96, 97].

In addition to the fusion variants describe above, fusion partners other than *EML4* have been identified, including *KIF5B*, which is a microtubule-based motor protein involved in organelle transport. The translocation t(2;10)

(p23;p11) results in the fusion of the first domains of *KIF5B* including the motor domain and the coiled-coil-domain with the tyrosine kinase domain of *ALK* [98]. Another rarely occurring fusion partner is *TFG* (TRK-fused gene) [99].

Another aberration, affecting the *KIF5B* gene, is inv(10)(p11.22q11.2). This inversion was detected in adenocarcinomas of the lung and results in the fusion of *KIF5B* with *RET*, which encodes a receptor tyrosine kinase. The fusion transcript comprises the coiled-coil-domain of *KIF5B* and the tyrosine kinase domain of *RET*. It can be detected in 1–2 % of lung adenocarcinomas [100, 101] and its occurrence is mutually exclusive with other driver mutations in *EGFR*, *KRAS*, *BRAF* and *HER2* [102]. Cells expressing the *KIF5B-RET* fusion transcript are sensitive to multikinase inhibitors [103]. Whether targeting *KIF5B-RET* might provide a new therapeutic strategy is currently tested in a phase II clinical study with the multikinase inhibitor cabozantinib. Another fusion partner of *RET*, CCDC6 (coiled-coil-domain containing 6) was described by Takeuchi [104]. Accordingly, a lung adenocarcinoma cell line transfected with the fused *RET-CCDC6* showed sensitivity to vandetanib [105].

The *ROS1* gene is located on chromosome 6 and encodes a receptor tyrosine kinase of the insulin receptor family. The chromosomal rearrangement of *ROS1* as driver mutation in lung cancer was first described by Rikova [99]. Rearrangements of *ROS1* by translocation or interstitial deletion leading to a fusion with several different partners have been detected in 0.8–1.7 % of NSCLC [104, 106, 107]. The fusion products contain the intact tyrosine kinase domain of *ROS1* and a truncated fraction of the fusion partner, e.g., *TPM3*, *SDC4*, *SLC4A2*, *CD74*, *EZR*, *FIG*, or *LRIG3* [107–109]. In preclinical studies, *ROS1* fusions are associated with sensitivity to the multikinase inhibitor crizotinib [107, 110, 111]. *RET* as well as *ROS1* rearrangements can be detected by FISH using dual color break-apart probes.

FGFR1 and *DDR2*: Alterations in Squamous Cell Carcinoma

The *FGFR1* gene encodes a member of the fibroblast growth factor receptor (FGFR) family [112]. FGFR tyrosine kinases interact with fibroblast growth factors, which regulate cell growth and differentiation. Amplification of *FGFR1* was detected in small-cell lung cancer [113, 114] and even more frequently in squamous cell carcinoma [115, 116]. The region on chromosome 8 spanning the *FGFR1* locus is amplified in about 20 % of lung cancer patients and associated with smoking. The standard method for the detection of gene copy number changes in FFPE tissue is FISH. The patterns of amplification are diverse, mainly due to the incidence of different degrees of polysomy in the tumor cells [117]. However, determination of clinically relevant thresholds await correlation with clinical responses in current phase I trials with FGFR1 inhibitors. Useful clinicodemographic features correlating with *FGFR1* amplification are, as yet, lacking [118].

In preclinical studies specific FGFR inhibitors show activity in lung cancer cell lines that harbor *FGFR1* amplifications [115, 116, 119]. In a phase I "first in man" dose-escalation study, inhibition of the FGFR pathway seemed effective in patients with FGFR dependent lung cancer [120].

Discoidin domain receptors (DDR) 1 and 2 were identified as targets of imatinib, nilotinib, and dasatinib [121]. Recently, gain-of-function mutations in *DDR2* have been described in about 4 % of squamous cell carcinoma of the lung with no hotspots in their distribution [122, 123]. A squamous cell lung cancer patient with the point mutation p. Ser768Arg responded to treatment with dasatinib and erlotinib in an early phase clinical trial [122]. Further clinical trials with dasatinib that specifically target *DDR2* mutations are currently under way.

Application of Next-Generation Sequencing (NGS) to Molecular Lung Cancer Pathology

In order to investigate the different molecular parameters that might individually affect the therapy of lung cancer in a time and cost saving manner, multiplex approaches analyzing many genes simultaneously are needed. Thus, methods which allow multiple target analyses and high-throughput processing are ideally suited for a fast and detailed diagnosis of the mutational status of each lung cancer patient. Approaches of NGS technology provide many advantages in genetic tumor characterization such as high sensitivity in detection of somatic mutations, extremely high capacities allowing deep sequence analyses, and the high potential of sample and target multiplexing.

Parallel Sequencing in Molecular Pathology

The superiority of NGS technologies over conventional methods is due to the generation of amplicon clones or clusters from a single molecule which are then sequenced in parallel. Different platforms on the market have implemented the principle by using different technologies.

One of the first NGS methods was the 454 sequencing approach which was established by 454 Life Science Corporation [124] (Table 24.1). Hereby, DNA is amplified by an emulsion PCR. Each emulsion droplet contains a single adapter-linked DNA template, hybridized to a primer-coated bead that then generates a clonal amplicon colony. Subsequent sequencing takes place in a picoliter-well plate with wells each containing only one single bead with the amplified clone [124]. The DNA clones are sequenced in parallel by the pyrosequencing method.

The technology used for the ion semiconductor sequencing system (Table 24.1) is based on ultrasensitive measurements of pH changes. In order to detect the proton that is released during the nucleotide incorporation, sequencing of the clonal DNA templates is performed in picoliter cartridges of a MOSFET (metal–oxide–semiconductor field-effect transistor) flow cell [125]. Prior to semiconductor sequencing, the template clones are also prepared by emulsion PCR as described for the 454 technology.

In contrast to these technologies, single DNA templates are immobilized on the sequencing flow cell of the Illumina platform and clonal DNA clusters are generated by bridge PCR amplification. Sequencing is performed by DNA strand synthesis using fluorochrome-labelled nucleotides and nucleotide coupling and decapping steps proceeding in tandem [126] (Table 24.1).

The establishment of scaled-down platforms (personal gene analyzers) adapted the different technologies to the throughput of diagnostic laboratories [127–129]. The various NGS systems have different advantages and disadvantages and molecular pathology laboratories have to decide which method represents best their demands on the diagnostic spectrum, the diagnostic variety, or throughput. Table 24.1 summarizes the essential features of currently available NGS approaches with a critical evaluation of their applicability to the routine diagnostic program in molecular pathology.

METHODS

Formalin-fixed and paraffin-embedded tissue (FFPE) remaining after conventional histological and immunohistochemical stainings is widely used for molecular analysis because fresh frozen tumor tissue is only rarely available. The integrity and stability of DNA in FFPE is a limiting factor for the reliability of mutation testing [130]. At the stage of tissue fixation, DNA quality is affected mainly by degradation of target DNA due to the reaction of the phosphodiester backbone with formalin. The main factors affecting the degree of DNA degradation in FFPE are the duration of specimen archiving, the type of fixative used and the duration of fixation prior to paraffin embedding [131, 132]. The small size of the biopsy presents another challenge to molecular analysis, especially in lung cancer as fine needle aspiration biopsy is a commonly used procedure. In practice, however, these parameters are highly variable, particularly in a reference lab setting where the material is received from different diagnostic centers.

During the fixation process, formalin causes deamination of cytosine and adenine, resulting in uracil or hypoxanthine residues in the template DNA [133]. During subsequent PCR, these changes result in C→T/G→A or A→G/T→C transitions. Given that PCR starts from few templates, especially in biopsy samples with small amounts of extracted genomic DNA and thereby low copy numbers of the desired fragment, these artifacts may be amplified and detected as false mutations. C→T/G→A transitions can be prevented by using uracil-*N*-glycosylase prior to PCR.

Table 24-1 **Summary of NGS Platforms (Adapted from Vollbrecht et al. [149])**

	Illumina	**454/Roche**	**Ion torrent (Life Technologies)**	**Solid (Life Technologies)**	**Helicos Heliscope**	**Pacific Biosciences**
Principle	Synthesis based terminator sequencing	Pyrosequencing	Semiconductor sequencing	Sequential dinucleotide ligation	Single-molecule sequencing	Single-molecule sequencing
Signal detection	Four-color fluorescence	Luminescence	Measuring of pH value	Four-color fluorescence	Four-color fluorescence	Four-color fluorescence
Capacity	HiSeq 2000: 600 Gb **MiSeq: 2–8 Gb[a]**	GS Flex: 700 Mb **GS Junior: 35–50 Mb**	Proton: 10 Gb **PGM: 1 Gb**	350 Gb		
Running time	HiSeq 2000: 11–14-d **MiSeq: 18–39 h[a]**	GS Flex: 24 h **GS Junior: 8 h**	Proton: 5 h **PGM: 2–5 h**	7 d	8 d	1 d
Generation of templates	Bridge PCR	Emulsion PCR	Emulsion PCR	Emulsion PCR		
Read length	75–200 bases	600–800 bases	100–400 bases	75 bases	35 bases	6,000 bases
Cost per base	Low	High	Low			
Advantages	– Suited for high throughput – Low costs	– Long read lengths – Short running time	– High flexibility in throughput – Low costs – Short running time	– Low error rate		– Long read lengths – Short running time
Disadvantages	– Reading errors in GC rich regions – Long running time	– High costs – Reading errors in homopolymer stretches	– High number of reading errors in homopolymer stretches	– Long running time – Laborious preparation – Short read lengths	– Long running time – Short read lengths	– High number of reading errors

b=bases, d=days, h=hours
[a]Depending on read length
Bold=platforms more readily suited for the use in routinely performed molecular diagnostics due to costs, throughput, and capacity

Thereby, uracil is removed from the DNA strands and a strand break is created [134]. The occurrence of artifacts can be circumvented by using higher amounts of template DNA wherever available. If that is not possible, we strongly recommend performing multiple independent PCR amplifications from the same sample.

Our group and others have suggested to use 4 % buffered formalin for biopsy specimens for at least 6 h and overnight for larger specimens for optimal fixation [135, 136]. An important aim of the extraction protocol is to preserve the quality of the DNA as well as possible by applying gentle extraction procedures [137]. For subsequent analyses, it is required to purify raw DNA extracts from potential PCR inhibitors such as hemoglobin or melanin [138]. In our experience, automated systems are favored over manual systems due to better standardization of handling steps and the use of magnetic beads for DNA binding instead of spin columns.

During all steps of the analysis process, the prevention of carryover contaminations is an important issue. Respective precautionary measures are described below together with the individual analysis steps and the corresponding notes.

A comprehensive description of the whole workflow can be found in the "Molecular Testing Guideline for Selection of Lung Cancer Patients for EGFR and ALK Tyrosine Kinase Inhibitors" published by the Association for Molecular Pathology [139] and available online (http://www.amp.org/documents/LungBiomarker-AMP-2013-proof.pdf)

Isolation of DNA

Prior to DNA extraction, an experienced pathologist has to evaluate a hematoxylin and eosin (H and E) stained tissue section to define the most appropriate tissue area. Furthermore, the pathologist ideally would mark on the H and E stained section the tumor area for appropriate manual microdissection in order to reduce the amount of non-tumorous tissue (see Note 1). Laser-capture microdissection is also possible in principle; however, it is labor intensive and difficult to apply on a high-throughput level.

For manual microdissection two different methods can be used: First, the marking of the H and E stained slide is transferred to the tissue block before cutting one to six 10 μm thick section rolls in a reaction tube. This method impairs the quality of tissue blocks for further investigations. The alternative protocol is to prepare tissue sections first and to perform the manual microdissection on glass slides after deparaffinization. Depending on the method chosen for manual microdissection two different protocols for deparaffinization can be used (see Note 2).

(a) Deparaffinization of tissue section rolls cut into a reaction tube
   1. Add 1 mL xylol to the reaction tube, vortex, and incubate at room temperature for 10 min.
   2. Centrifuge for 5 min at 13,000 rpm; discard supernatant.
   3. Repeat the first two steps.
   4. Resuspend pellet in 1 mL ethanol, vortex, and incubate at room temperature for 10 min.
   5. Centrifuge for 5 min at 13,000 rpm; discard supernatant.
   6. Repeat the last two steps.
   7. Dry pellet in the open reaction tube for 10 min at 37 °C, resuspend the pellet in suitable buffer (depending on purification kit used) and continue with purification protocol.

(b) Deparaffinization of slide mounted tissue sections prior to macrodissection
   1. Incubate the slides for at least 30 min at 60 °C before deparaffinization.
   2. Incubate slides for 10 min in xylol and ethanol, respectively; repeat the incubation, follow by rehydration in 100, 96, 80, and 70 % ethanol for 5 min each.
   3. Scrape areas marked on the H and E stained section from the deparaffinized slides and transfer into a reaction tube.
   4. Centrifuge for 5 min at 13,000 rpm; discard supernatant.
   5. Resuspend pellet in the appropriate buffer (depending on purification kit used) and continue with purification protocol.

(c) Purification of DNA
   1. Add proteinase K (2 μg/μL) to the appropriate lysis buffer, mix, and incubate overnight at 56 °C.
   2. Purify DNA manually or in an automated way by use of spin columns or magnetic beads according to the manufacturer's instructions.

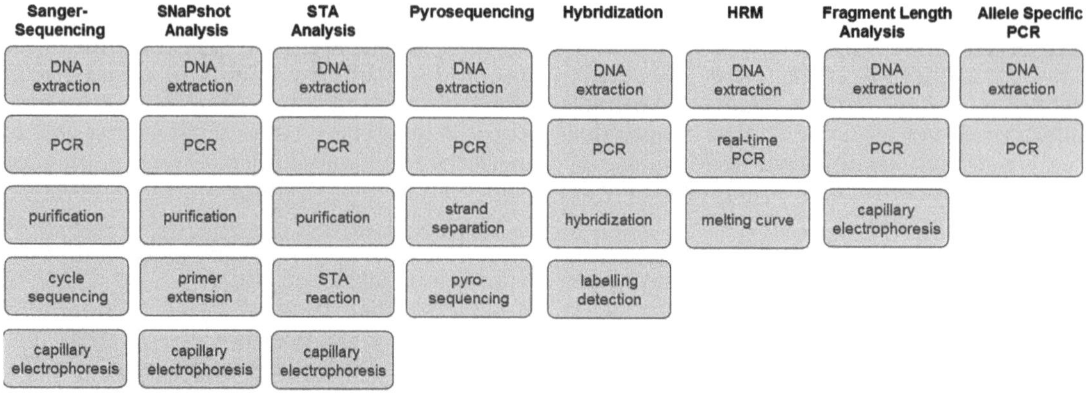

Figure 24-3 Flowchart outlining the protocol steps for different techniques of mutation analysis. The methods included illustrate the example methods described in this chapter. *SnaPshot* single nucleotide probe extension assay, *STA* shifted termination assay, *HRM* high-resolution melting

3. Semi-quantitatively estimate the amount and quality of extracted DNA by agarose-gel electrophoresis or spectrophotometrically using, for example, the NanoDrop ND 1000 (Peqlab, Erlangen, Germany), see Note 3.

Protocols for Mutation Analyses

Several methods are currently used to determine genomic variations, including pyrosequencing, Sanger sequencing, real-time PCR based analysis, fragment length analysis, high-resolution melting (HRM), single nucleotide probe extension assays (such as SNaPshot), reversed hybridization assays, or shifted termination assays (STA). All of the currently available methods require PCR amplification after DNA extraction (Fig. 24.3).

Each individual method has its advantages and disadvantages concerning factors such as mutation spectrum, costs, sensitivity and time required for analysis. Pyrosequencing provides a sensitive method for detecting mutations with 5–10 % allele frequency and further allows the detection of a wide variation of mutations. However, it may not be economical in all settings due to expensive equipment and reagents. Although Sanger sequencing remains the gold standard in many laboratories, it is time-consuming and relatively less sensitive because it is only able to detect mutations with an allele frequency above 15–20 %. On the other hand, it is easy to implement and allows detection of previously unknown mutations.

Several methods which are also commercially available are well suited for the detection of described mutations, e.g., real-time PCR based analyses, single nucleotide probe extension assays (SNaPshot), reversed hybridization assays, or shifted termination assays (STA). They provide high sensitivity and throughput but may fail to detect additional mutations around hot spot regions.

Two methods that allow sensitive screening without obtaining specific sequence information are HRM and fragment length analysis. HRM can distinguish between wild-type and mutated sequences. It is therefore suited for questions where a high rate of wild type is expected because the wild-type cases may be rapidly excluded from further analyses. Fragment length analysis can only detect mutations changing the length of PCR product, e.g., deletions and insertions.

As several methods are commercially available, the following Web-based resources can be used as a basis for laboratory protocols:

http://www.qiagen.com/products/bytechnology/pyrosequencing/pyrosequencing_tutorial.aspx (Pyrosequencing, Qiagen)

http://products.invitrogen.com/ivgn/product/4323163 (SNaPshot, Invitrogen)

http://molecular.roche.com/assays/Pages/cobasEGFRMutationTest.aspx (real-time PCR, Roche)

http://www.zytomed-systems.de/index.php?page=shop.browse&category_id=26&option=com_virtuemart&Itemid=58 (hybridization, Zytomed/Chipron) http://www.trimgen.com/sta-m1.asp (shifted termination assay, STA, TrimGene)

Standard protocols for Sanger Sequencing, HRM, and fragment length analysis are outlined below.

Sanger Sequencing

The dideoxy sequencing method according to Sanger includes several analysis steps. First, when working with DNA from FFPE tissue, the relevant fragments of the DNA have to be amplified. The sequences of primers used for amplification of *EGFR* exons 18–21 by the members of the German Panel for Mutation Testing in NSCLC and a complete compilation of the applied technical procedures is available online (http://www.dgp-berlin.de/downloads/public/protocols/EGFR_Mutations_Protocols_engl.pdf) [140].

Unbound primers and excess nucleotides have to be separated from the PCR fragments prior to cycle sequencing. The cycle sequencing should always be performed as bidirectional sequencing and may be done with different sequencing reagents. As an example, the cycle sequencing protocol for working with the BigDye Terminator Cycle Sequencing Ready Reaction Kit (Applied Biosystems, Darmstadt, Germany) is described here.

The cycle sequencing products have then to be separated from reaction components such as salt ions, dNTPs, and unincorporated dye terminators. The BigDye XTerminator Purification Kit (Applied Biosystems, Darmstadt, Germany) provides a convenient purification tool.

1. Set up standard PCR with the following components (see Note 4).
2. Amplify DNA with standard conditions.

Template:	1–20 μL (according to quantification)
Nucleotide:	100 μM
Reaction buffer:	5 μL
Forward primer:	0.4 μM
Reverse primer:	0.4 μM
Polymerase:	1 U
Desalted water:	up to 50 μL

3. Purify PCR fragments with exonuclease I (Exo I) and FastAP™ thermosensitive alkaline phosphatase (Thermo Scientific, Waltham, MA, USA).
4. Pipet 5 μL PCR product and 0.5 μL Exo I and 1.0 μL FastAP in a PCR tube and centrifuge.
5. Incubate for 15 min at 37 °C, followed by 15 min at 85 °C.
6. Purified product can be used directly or stored at 8 °C.
7. Set up the cycle sequencing reactions.

Template:	1–8 μL (according to electrophoresis)
Primer (forward OR reverse):	10 pmol
Terminator ready reaction mix:	1 μL
Buffer	2 μL
Distilled water:	add 20 μL

8. For each reaction pipet 110 μL of master mix (90 μL Sam Solution and 20 μL XTerminator solution, Applied Biosystems, Darmstadt, Germany) in a well of a 96-well plate.
9. Add the cycle sequencing products, seal plate with the appropriate septum and incubate for 30 min at room temperature with shaking.
10. Centrifuge plate at 1,000 × *g* for 2 min.
11. Run capillary electrophoresis according to manufacturer's instructions.

High-Resolution Melting (HRM)

HRM is a molecular technique for high-throughput screening for mutations in a bounded region. Mutation determination using HRM is based on the dissociation of DNA, when exposed to an increasing temperature in the presence of fluorescent dyes intercalating in double-stranded DNA. The presence of a mutation leads to the formation of DNA heteroduplexes followed by a change in melting behavior (Fig. 24.4) [141]. Some types of mutations are difficult to detect with this method. Because HRM depends on heteroduplex formation the rarely occurring hemizygous mutations where the wild-type

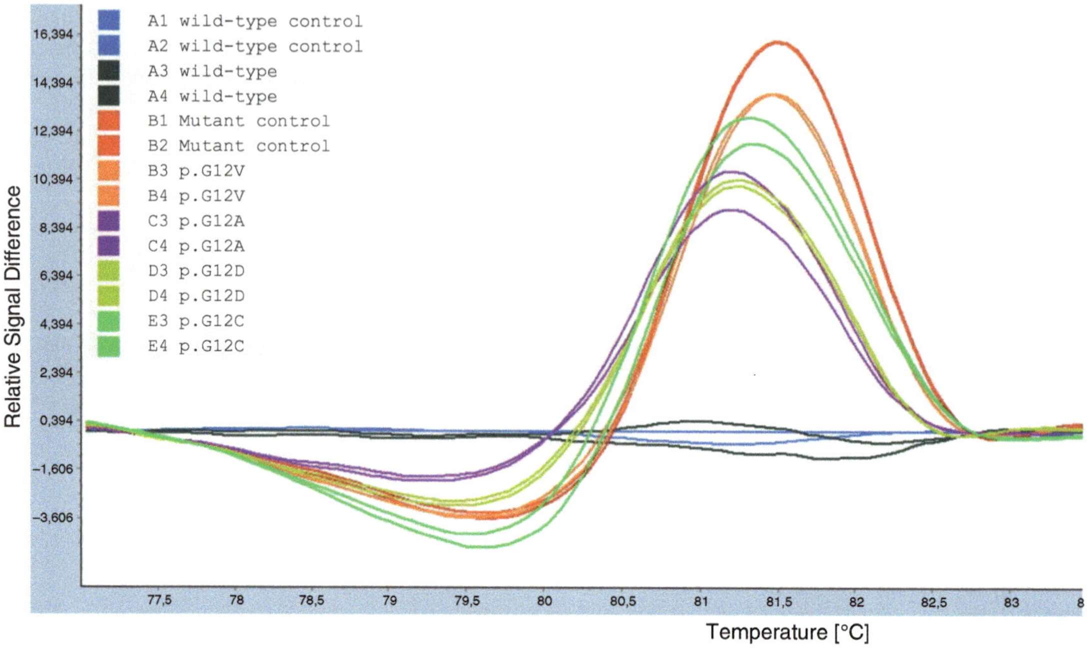

Figure 24-4 HRM for the detection of point mutations in the *KRAS* gene. DNA from wild-type and mutant control cell lines and from five different FFPE samples was used. One of the samples is wild type, while the others are mutated for p.Gly12Val, p.Gly12Ala, p.Gly12Asp, and p.Gly12Cys. Results of HRM are shown as a normalized difference plot

allele is deleted due to chromosome loss might not be detected.

The protocol described here was implemented on a LightCycler 480 (Roche Diagnostics, Mannheim, Germany), but can be easily transferred to other real-time PCR systems. Here, the LightCycler LC 480 High Resolution Melting Master (Roche Diagnostics, Mannheim, Germany) included the use of a generic intercalating dye for double-stranded DNA detection (ResoLight HRM dye) and a hot start PCR enzyme. Primer sequences, amplification conditions and analysis parameters for the relevant exons of *KRAS*, *BRAF*, *PIK3CA* and *AKT1* were described in detail by Ney et al. [141].

1. Dilute template DNA to a concentration of 2 ng/μL.
2. Set up real-time PCR with the following components (see Note 5).

   A master mix can be prepared from Melting Master, both primers, and $MgCl_2$.

Template:	5 μL of the 2 ng/μL dilution
Melting Master:	10 μL
Forward primer:	200 nM
Reverse primer:	200 nM
$MgCl_2$:	depending on assay
Distilled water:	up to 20 μL

3. Pipet water and master mix in 0.2 mL tubes, add template DNA, vortex, and pipet reaction mixture onto the microtiter plate.
4. Seal plate and centrifuge at 1,600 rpm for 2 min.
5. Perform real-time PCR with subsequent melting curve according to manufacturer's instructions, use the detection channel for SYBR Green I assays.
6. Amplification curves can be viewed in the evaluation mode "abs. quant/2nd derivative max".
7. Do not evaluate samples with Cp (crossing point) value >40 cycles.
8. Analyze melting curves with the gene scanning software (Roche Diagnostics, Mannheim, Germany).
9. Set normalization and temperature shifting parameters (depending on assay) and display results as difference plot using one of the wild-type values as the basic value.

10. Samples with a positive deflection from the wild-type curve are considered mutated; for the exact localization of the mutation subsequent Sanger sequencing can be used.

Fragment Length Analysis

Fragment length analysis is a suitable method for the detection of mutations which change the length of amplification products, e.g., insertions and deletions. The most common mutations in exon 19 of *EGFR* are deletions of 15 bp although other do occur. Point mutations occur very rarely, and thus, fragment length analysis is an appropriate method to screen for mutations in this exon.

Amplification of exon 19 is best performed by nested PCR with the reverse primer of the second round labelled with the fluorescent dye FAM. Thereby, PCR products can be easily separated and detected by capillary electrophoresis. To determine the exact mutation status, PCR products of the first round can be analyzed by Sanger sequencing after purification.

The protocol described here is adapted with modifications from Molina-Vila et al. [142]. The primers used are as follows:

1. PCR:
   EGFR Exon 19 forward-outer: 5′-TGGGCAGCATGTGGCACCATC-3′
   EGFR Exon 19 reverse-outer: 5′-AGGTGGGCCTGAGGTTCAG-3′
2. PCR:
   EGFR Exon 19 forward-inner: 5′-ACT CTG GAT CCC AGA AGG TGA G-3′
   EGFR Exon 19 reverse-inner: 6-FAM-5′-CCA CAC AGC AAA GCA GAA ACT C-3′

1. Set up standard PCR with the following components

Template:	see below*	
Nucleotide:	1. PCR: 0.2 mM	2. PCR: 0.25 mM
Reaction buffer:	5 μL	
Forward primer:	1. PCR: 200 nM	2. PCR: 500 nM
Reverse primer:	1. PCR: 200 nM	2. PCR: 500 nM
$MgCl_2$:	2 mM	
Polymerase enzyme:	1 U	
Distilled water:	up to 50 μL	

*template: 1–15 μL (according to quantification) for first PCR; 1 μL of the first PCR is used for the second round.

A master mix can be prepared from nucleotides, reaction buffer, primers, $MgCl_2$ and enzyme.

2. Pipet water and master mix in 0.2 mL tubes, add respective amount of template DNA, vortex, and start PCR reaction with standard conditions.
3. Dilute product of second PCR 1:200; mix 1 μL of dilution with 12 μL HiDi formamide (Applied Biosystems) and 0.25 μL GeneScan-600 LIZ Size Standard (Applied Biosystems).
4. Close tube, vortex, and denature sample for 3 min at 90 °C.
5. Cool sample on ice for at least 10 min, centrifuge briefly and start capillary electrophoresis according to the manufacturer's instructions.
6. Evaluate the samples with the appropriate software, e.g., GeneMapper (Applied Biosystems), see Note 6.

Fluorescence In Situ Hybridization (FISH)

The method of fluorescence in situ hybridization (FISH) can be used for the detection of chromosomal translocations/inversions or amplifications. The labelled probe is directly hybridized onto the tissue slide.

To detect amplifications, two differently labelled probes are usually used: the locus specific probe hybridizes to the gene of interest and a centromere specific probe binds to the corresponding chromosome. For the detection of a translocation, most commonly so-called break-apart assays are used: Two differently labelled probes hybridize adjacent to each other to one of the two translocation partners. Currently, the detection of *FGFR1*, *HER2*, and *MET* amplification and *ALK*, *ROS* and *RET* translocations or inversions is most commonly done using FISH analysis.

Pretreatment of slides may be performed with the half-automated VP2000 processor system (Abbott Molecular, Wiesbaden, Germany) or manually using Abbott's pretreatment reagents. Here, we provide an example protocol working with a wide spectrum of probes from different manufacturers.

1. Prewarm all solutions needed.
2. Mount 3–4 μm tissue sections on sialinized slides, dry overnight at 37 °C, and mark tumor areas with an alcohol-resistant felt pen.

3. Deparaffinize the sections in xylene (3× 10 min) and rehydrate in a graded alcohol series.
4. Dry the sections at 37 °C.
5. Incubate the sections in 0.2 M HCl for 20 min, followed by 3 min in distilled water.
6. Wash the sections for 3 min in 2× SSC wash buffer.
7. Incubate the sections in pretreatment solution (Abbott Molecular) for 30 min, followed by 1 min in distilled water.
8. Wash the sections twice for 5 min in 2× SSC wash buffer.
9. Incubate the sections for 1.5 h at 37 °C in protease solution (0.5 mg/mL protease buffer, Abbott Molecular).
10. Wash the sections twice for 5 min in 2× SSC wash buffer.
11. Incubate the sections for 10 min in 4 % buffered formalin.
12. Wash the sections twice for 5 min in 2× SSC wash buffer.
13. Dry the sections at 37 °C.
14. Add the appropriate amount of probe mix, place a coverslip on the section, and seal with a removable glue.
15. Denature the sections in the presence of probe for 5–10 min at 75 °C and hybridize overnight at 37 °C.
16. Remove the coverslip and perform post-hybridization SSC washes at 72 °C for 2 min.
17. Rinse the sections briefly in 2× SSC wash buffer and counterstain with DAPI (4′,6-Diamidin-2′ phenylindoldihydro chlorid).
18. Scan tumor tissue with a fluorescent microscope with the appropriate filter set by using the 40× or 63× objectives.
19. Evaluate only samples and areas with sharp borders of nuclei, no signs of over-digestion, non-overlapping nuclei, bright and specific green and orange signals in internal control tissue and in the tumor area.
20. For details on evaluation of *ALK*, *RET*, and *ROS* translocations, see Note 7.
21. For details on evaluation of *FGFR1* amplification, see Note 8.

NGS Operation Workflow in Molecular Pathology Diagnostics

Independently from the specific NGS technology, first a template library has to be prepared. DNA templates, selected according to diagnostic requirements can be targeted by hybridization capture procedures [143, 144] or by template-specific multiplex PCR (Fig. 24.5). Customized DNA or RNA capture probe panels representing the genes of interest are commercially available and include the manual for the hybridization and purification process (see protocols a). Up to now at least 200 ng of DNA has to be applied to these technologies [126, 127, 145, 146]. For low available DNA amounts, a whole-genome amplification step has to be carried out before capture hybridization [147] or multiplex PCR approaches are recommended allowing a DNA input as low as 10 ng (see protocols b and c).

To generate single DNA strands and for later sequencing, first 3′A-overhangs are generated and then priming adaptors are ligated (see protocols c). Adaptors might include an individual DNA sequence of eight to ten nucleotides. These barcodes, also called multiple identifier (MID), allow for the handling of different DNA samples simultaneously, each tracked during the sequencing process by a different barcode. Thus, not only multiple targets but also up to 96 different DNA tumor samples may be processed at the same time. In a last step, templates carrying the adapter sequences are selectively amplified by a low cycle adapter-specific PCR, purified and then quantified by either universal adapter-specific Real-Time PCR (see protocol d) or Picogreen measurements (see protocols e). The DNA purification and fragment size selection in between of the respective steps are often automatically performed using magnetic bead based procedures (see protocols f). The workflow of the template specific library preparation using multiplex PCR is summarized in Fig. 24.6. In each working step, the protocol of the reagent providing manufacturer can be followed (see protocols b–f).

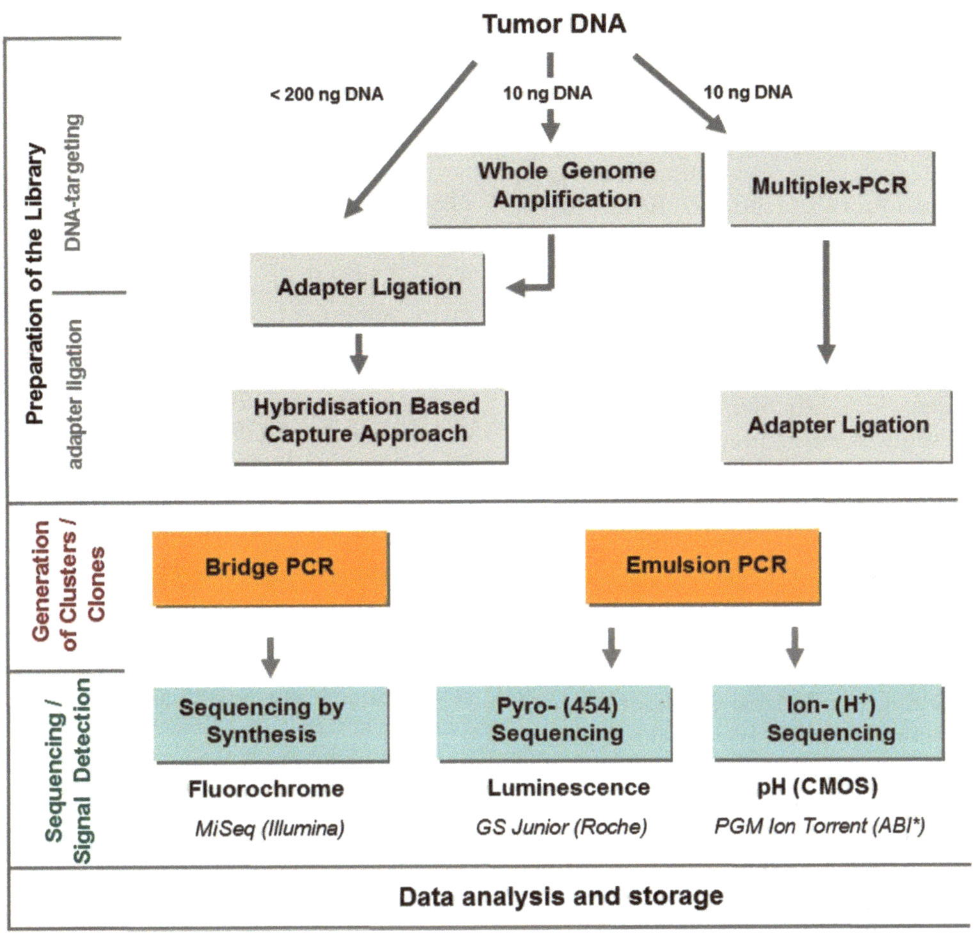

Figure 24-5 Operation workflow of NGS in molecular pathology. Different technical approaches for targeted DNA template library preparation and parallel sequencing are available. The most common approaches for template enrichment and sequencing platforms adapted to the diagnostic requirements are summarized. The scheme is modified according to Vollbrecht et al. [149]

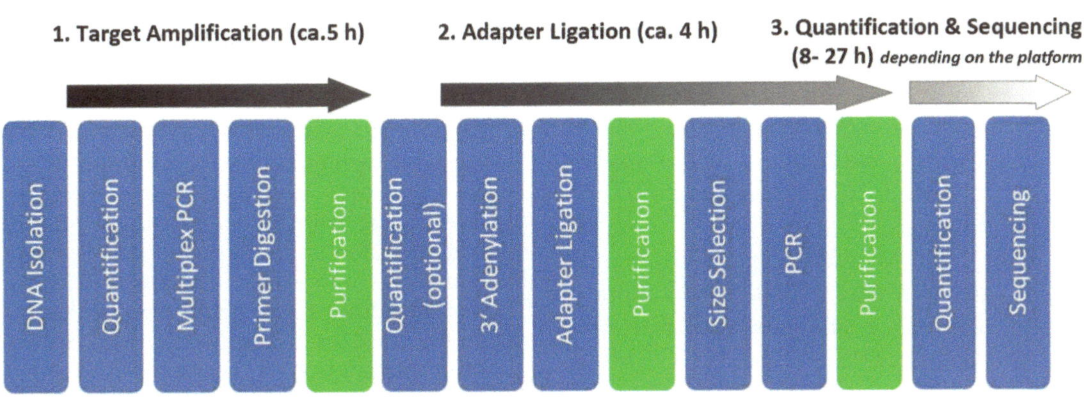

Figure 24-6 An example workflow of library preparation and NGS

Selected Protocols

(a) Capture protocols:
SureSelect: http://www.genomics.agilent.com/article.jsp?pageId=308
Nimblegen: http://www.nimblegen.com/products/seqcap/ez/choice/index.html
Haloplex: http://www.genomics.agilent.com/en/HaloPlex-DNA/HaloPlexPanels/?cid=cat100006&tabId=prod110012
Truseq: http://support.illumina.com/sequencing/sequencing_kits/truseq_dna_sample_prep_kit_v2.ilmn

(b) Multiplex PCR:
Ampliseq: http://tools.invitrogen.com/content/sfs/brochures/Ion-AmpliSeq-Cancer-Hotspot-Panel-Flyer.pdf
Qiagen Multiplex: http://www.sabiosciences.com/NGS.php
New England Biolabs: Overview: https://www.neb.com/applications/library-preparation-for-next-generation-sequencing

(c) Adenylation, ligation:
New England Biolabs: https://www.neb.com/protocols/1/01/01/nebnext-quick-ligation-module-protocol-e6056
Lifetechnologies: http://share.pdfonline.com/6b34e23bb8264f4fb7baa642b7470e0f/MAN0006735_IonAmpliSeqLibraryKit_UG%20Rev4%2015Feb2013.htm

(d) Real time quantification
Qiagen Quantification: http://www.sabiosciences.com/NGS-DNAseqLibQuantArray.php

(e) Picogreen determination
Picogreen quantification: http://probes.invitrogen.com/media/pis/mp07581.pdf
Qbit User Manual: http://www.invitrogen.com/etc./medialib/en/filelibrary/cell_tissue_analysis/Qubit-all-file-types.Par.0519.File.dat/Qubit-2-Fluorometer-User-Manual.pdf

(f) Purification and size selection:
https://www.beckmancoulter.com/wsr-portal/bibliography?docname=Protocol_000387v001.pdf

Needs and Perspectives of NGS in Lung Cancer Diagnostics

In particular in lung cancer, the rapid progress in development of therapeutic strategies requires a continuously increasing spectrum of diagnostic targets. At present, lung cancer hotspot panels, used for multiplex PCR target enrichment and NGS are developed. Although NGS opens up tremendous new perspectives for time, cost, and DNA material saving diagnostics, novel laboratory requirements and quality controls have to be addressed. Thus, experience in data interpretation, threshold settings etc. has to be gathered and the issue of long-time data storage has to be solved [129].

Furthermore, NGS quality controls are not yet defined. According to our experience, the quantity and quality of the input DNA material should be evaluated carefully. Protocols for DNA extraction have to be optimized for high DNA concentration and low salt content of the purified extracts.

As for all other PCR based methods, vertical contaminations from a one run to the next may occur. Because during NGS processing amplicon clones are handled, the high NGS sensitivity can result in detection of amplicons remaining from the preceding run. Therefore, careful cleaning of working places and equipment should be performed. In addition to the conventional control reactions, the following controls are established in our diagnostic laboratory to ensure the detection of vertical contaminations:

- Change of the barcode sets used from run to run.
- Screening for reads with barcode sequences that were applied in the preceding run.

An important perspective in molecular diagnostics, lacking the need of these quality controls, but demanding novel controls, is the NGS analysis of single template molecules. The measurements of single molecules by magnetic tweezer manipulation and subsequent determination of structural DNA fragment alterations is able to cover different approaches in molecular pathology such as sequencing and hybridization analyses [148]. Therefore, this third generation of sequencing strategies might be very promising to be applied in future diagnostics.

Conclusions

In the last few years, lung cancer has become the prime example for the success of personalized therapy. Patients with *EGFR*-mutant adenocarcinomas treated with EGFR inhibitors show a significant survival benefit compared to standard chemotherapy. Also, NSCLC patients bearing *ALK* or *ROS* rearrangements can be treated with specific inhibitors. Several additional therapeutic regimens targeting other consequences of genetic changes as for example the rearrangement of *RET*, are currently investigated in preclinical and clinical studies. Similarly, in squamous cell carcinomas genetic alterations affecting kinases have been found and are currently evaluated clinically, e.g., *FGFR1* amplifications and *DDR2* mutations.

Unfortunately, tumors of patients receiving inhibitor therapies will develop different resistance mechanisms. Because these mechanisms can mostly be explained on the molecular level, it is possible to develop therapeutic approaches overcoming the resistance mechanisms, for example so-called second-generation inhibitors. Known resistance mechanisms include the p.Thr790Met mutation of *EGFR* itself, amplification of *MET*, mutation of *PIK3CA*, as well as transformation to small-cell lung cancer and epithelial-to-mesenchymal transition.

A major challenge now is to implement high-quality molecular diagnostics and personalized treatment strategies in routine clinical practice. With conventional molecular methods represented by sequencing technologies only sequential analysis of the different markers is possible. PCR-based technologies like HRM or allele specific PCR allow multiplexing but are also limited by the detection systems available. Hence, the development of sensitive methods that provide sensitive, accurate, and simultaneous detection of the mutation status of many samples and gene loci is of major interest. Massive parallel sequencing by NGS approaches becomes more important in molecular diagnostic pathology and will replace the conventional technologies within the next few years.

Notes

Note 1. *Isolating DNA from FFPE tissue*
The amount of tumor cells in a FFPE tissue block may be highly variable. Therefore, an experienced pathologist should review each tumor block, indicate the area of highest proportion of tumor cells, and record his estimate of the percentage of tumor cells prior to analysis. Macrodissection in cases with a larger amount of non-tumorous tissue (more than 20 % of the overall area) may be of use to reduce the amount of wild-type DNA.

Note 2. *Avoiding contamination during DNA isolation and amplification*
In general, because PCR amplification is a highly sensitive method and susceptible to carryover contamination, the following fundamental rules should be the gold standard in every diagnostic molecular pathology laboratory. The working area should be divided in pre- and post-PCR sections, optimally three independent rooms for extracting DNA, preparing the PCR, and performing the post-PCR analysis. Each section should have specifically assigned equipment and reagents. Plugged pipette tips have to be used throughout and gloves to be changed between the working sections. Also, during cutting the paraffin blocks by microtome, some rules have to be observed to avoid contamination between blocks:

1. Blocks that are proposed for the same analysis should not be cut consecutively.
2. The microtome and the working area should be cleaned from paraffin material after each individual block.
3. Depending on the type of microtome, the knives should be removed or relocated after each block.
4. If the sections are mounted on slides for manual microdissection, the water bath should be cleaned regularly with filter paper and the water should be changed as often as possible.

Note 3. *Quantity and quality of isolated DNA*
Because the PCR result depends on the number of amplifiable fragments variable amounts of DNA should be used as PCR template, depending on both DNA quantity and extend of DNA fragmentation. It is strongly recommended not to analyze samples with poor DNA quality. In such cases, additional material (e.g., fresh frozen tissue, if available, or another paraffin block) should be used.

Note 4. *Control reactions*
For each amplification experiment, positive and negative controls should be carried along. A sample with water instead of DNA serves as negative control; a positive control may be DNA extracted from FFPE tissue which was amplified successfully in a previous analysis. The control reactions should be checked by agarose gel electrophoresis. The number of PCR cycles should not exceed 40 cycles. If amplification failed twice, even after sample purification, the analysis may be stopped at this point. The analysis can be retried with another paraffin block.

Note 5. *Setting up HRM PCR*
The amplicon size (the shorter the better), exclusion of primer dimers, salt concentration, specific melting products with only one single melting domain, and standardized genomic DNA isolation protocols are important points for implementation of highly sensitive HRM assays.

Each run should include mutated and wild-type DNA as controls. Primers can be designed with the LightCycler Probe Design 2.0 software (Roche Diagnostics) and should be checked for specificity by using the basic local alignment software tool (BLAST) from the National Center for Biotechnology Information (NCBI) (http://blast.ncbi.nlm.nih.gov/Blast.cgi). Performing at least duplicates is necessary to minimize temperature differences on the microtiter plate.

Note 6. *Evaluation of Fragment Length Analysis*
DNA extracted from the following cell lines can be used as positive controls:

- Human adenocarcinoma cell line PC9, carrying the mutation p.E746_A750 del.
- Human adenocarcinoma cell line NCI-H1975 (ATCC# CRL-5908), wild type for *EGFR* exon 19

If wild ty pe, PCR fragments should be 217 bp in length for the first round and 117 bp for the second round.

Note 7. *Evaluation of ALK, RET, and ROS translocations*
Chromosomal rearrangements with *ALK* comprise the inversion on chromosome 2 leading to a fusion with *EML4* and the translocation affecting *ALK*, but not *EML4*, such as *ALK-TGF* or *ALK-KIF5B*. Therefore, the use of a break-apart assay is recommended.

Evaluation is described here for the Zyto*Light*® SPEC ALK/EML4 TriCheck™ (Zytovision, Bremerhaven, Germany). This probe system is designed to discriminate between *ALK* inversions and translocations.

The assay consists of three differently labelled probes, where two probes hybridize distal and proximal to the *ALK* gene breakpoint region, respectively, and the third probe binds to the *EML4* gene. In an interphase nucleus of a normal cell, two orange/green fusion signals and two blue signals are expected. The *EML4-ALK* inversion is indicated by one separate green signal, one separate orange signal, and an additional blue signal. The separate green and orange signals each co-localize with a blue signal. During inversion, the 5′ part of *ALK* can be deleted, so the separate green signal is lost. A signal pattern consisting of one orange/green fusion signal, one orange signal, and a separate green signal as well as two blue signals indicate an *ALK* translocation without involvement of *EML4*.

Break-apart signals have to be detected in at least 15 % of nuclei. Hundred Nuclei are counted to detect the rearrangement. Only nuclei with non-overlapping signals and with the expected number of signals are evaluated. To count signals as two, they have to be separated by at least two signal diameters.

ROS has several rearrangement partners, and an interstitial deletion as well as a translocation may occur. Therefore, the use of a break-apart assay, which is commercially available, is recommended. Typical break-apart signal patterns as described above for *ALK* are expected.

RET inversion on chromosome 10 leads to a fusion between *KIF5B* and *RET*. A break-apart assay for both genes can be used but is not yet commercially available for *RET*. If a *KIF5B* rearrangement is detected, involvement of

ALK has to be ruled out using the *ALK* probe set described above.

Note 8. *Evaluation of FGFR1 amplification*
The evaluation of *FGFR1* amplification is done according to Schildhaus et al. [117], and is described here with the usage of the ZytoLight SPEC FGFR1/CEN 8 Dual Color Probe (ZytoVision, Bremerhaven, Germany). The probe specific for the centromeric region of chromosome 8 is labelled with an orange fluorochrome, the probe specific for the gene region of *FGFR1* is labelled in green.

Some general points have to be considered when counting fluorescent signals [117]:

- Scan the entire tumor area for hot spots of increased *FGFR1* copy numbers.
- Count 20 tumor cell nuclei in three areas, either in three hot spots or in three random areas in case of homogeneous signal distribution. Count cohesive tumor cells; do not selectively consider isolated amplified tumor cells from different areas.
- Count only clearly distinct signals as two separate signals. Count *FGFR1* signal doublets and triplets as one signal. In cases of signal clusters give cluster estimation in steps of five signals, for example, 15, 20, or 25 *FGFR1* signals. Count micro-clusters as five signals.

Green *FGFR1* and orange centromere 8 signals are counted separately. The *FGFR1*/CEN8 ratio, the number of cells with ≥5 and ≥15 *FGFR1* signals and the average *FGFR1* copy number per cell are calculated. Cases are considered as *FGFR1* positive ("amplified") under one of the following conditions:

(1) The *FGFR1*/CEN8 ratio is ≥2.0.
(2) The average number of *FGFR1* signals per tumor cell nucleus is ≥6.
(3) The percentage of tumor cells containing ≥15 *FGFR1* signals or large clusters is ≥10 %.
(4) The percentage of tumor cells containing ≥5 *FGFR1* signals is ≥50 %, with (1–3) representing a high-level and (4) a low-level amplification.

REFERENCES

1. Zander T, Heukamp LC, Bos M, Fassunke J, Mattonet C, Merkelbach-Bruse S, Ko YD, Schlesinger A, Brockmann M, Serke M, Gerigk U, Hekmat K, Engel-Riedel W, Stoelben E, Reiser M, Schulz H, Schmitz SH, Frank K, Buettner R, Wolf J. Regional screening network for characterization of the molecular epidemiology of non-small cell lung cancer (NSCLC) and implementation of personalized treatment. J Clin Oncol. 2012;30(Suppl):abstr CRA10529.
2. Buettner R, Wolf J, Thomas RK. Lessons to learn from lung cancer genomics: the emerging concept of personalized diagnostics and treatment. J Clin Oncol. 2013;31(15):1858–65.
3. Roskoski Jr R. The ErbB/HER receptor protein-tyrosine kinases and cancer. Biochem Biophys Res Commun. 2004;319(1):1–11. doi:10.1016/j.bbrc.2004.04.150.
4. Akca H, Tani M, Hishida T, Matsumoto S, Yokota J. Activation of the AKT and STAT3 pathways and prolonged survival by a mutant EGFR in human lung cancer cells. Lung Cancer. 2006;54(1):25–33. doi:10.1016/j.lungcan.2006.06.007.
5. Onn A, Correa AM, Gilcrease M, Isobe T, Massarelli E, Bucana CD, O'Reilly MS, Hong WK, Fidler IJ, Putnam JB, Herbst RS. Synchronous overexpression of epidermal growth factor receptor and HER2-neu protein is a predictor of poor outcome in patients with stage I non-small cell lung cancer. Clin Cancer Res. 2004; 10(1 Pt 1):136–43.
6. Harari PM. Epidermal growth factor receptor inhibition strategies in oncology. Endocr Relat Cancer. 2004;11(4):689–708. doi:10.1677/erc.1.00600.
7. Lynch TJ, Bell DW, Sordella R, Gurubhagavatula S, Okimoto RA, Brannigan BW, Harris PL, Haserlat SM, Supko JG, Haluska FG, Louis DN, Christiani DC, Settleman J, Haber DA. Activating mutations in the epidermal growth factor receptor underlying responsiveness of non-small-cell lung cancer to gefitinib. N Engl J Med. 2004;350(21):2129–39. doi:10.1056/NEJMoa040938.
8. Paez JG, Janne PA, Lee JC, Tracy S, Greulich H, Gabriel S, Herman P, Kaye FJ, Lindeman N, Boggon TJ, Naoki K, Sasaki H, Fujii Y, Eck MJ, Sellers WR, Johnson BE, Meyerson M. EGFR mutations in lung cancer: correlation with clinical response to gefitinib therapy. Science. 2004;304(5676):1497–500. doi:10.1126/science.1099314.
9. Zhang X, Gureasko J, Shen K, Cole PA, Kuriyan J. An allosteric mechanism for activation of the kinase domain of epidermal growth factor receptor. Cell. 2006;125(6):1137–49. doi:10.1016/j.cell.2006.05.013.
10. Carey KD, Garton AJ, Romero MS, Kahler J, Thomson S, Ross S, Park F, Haley JD, Gibson N, Sliwkowski MX. Kinetic analysis of epidermal growth factor receptor somatic mutant proteins shows increased sensitivity to the epidermal growth factor receptor tyrosine kinase inhibitor, erlotinib. Cancer Res. 2006;66(16):8163–71. doi:10.1158/0008-5472.CAN-06-0453.

11. Mok TS, Wu YL, Thongprasert S, Yang CH, Chu DT, Saijo N, Sunpaweravong P, Han B, Margono B, Ichinose Y, Nishiwaki Y, Ohe Y, Yang JJ, Chewaskulyong B, Jiang H, Duffield EL, Watkins CL, Armour AA, Fukuoka M. Gefitinib or carboplatin-paclitaxel in pulmonary adenocarcinoma. N Engl J Med. 2009;361(10):947–57. doi:10.1056/NEJMoa0810699.
12. Rosell R, Carcereny E, Gervais R, Vergnenegre A, Massuti B, Felip E, Palmero R, Garcia-Gomez R, Pallares C, Sanchez JM, Porta R, Cobo M, Garrido P, Longo F, Moran T, Insa A, De Marinis F, Corre R, Bover I, Illiano A, Dansin E, de Castro J, Milella M, Reguart N, Altavilla G, Jimenez U, Provencio M, Moreno MA, Terrasa J, Munoz-Langa J, Valdivia J, Isla D, Domine M, Molinier O, Mazieres J, Baize N, Garcia-Campelo R, Robinet G, Rodriguez-Abreu D, Lopez-Vivanco G, Gebbia V, Ferrera-Delgado L, Bombaron P, Bernabe R, Bearz A, Artal A, Cortesi E, Rolfo C, Sanchez-Ronco M, Drozdowskyj A, Queralt C, de Aguirre I, Ramirez JL, Sanchez JJ, Molina MA, Taron M, Paz-Ares L. Erlotinib versus standard chemotherapy as first-line treatment for European patients with advanced EGFR mutation-positive non-small-cell lung cancer (EURTAC): a multicentre, open-label, randomised phase 3 trial. Lancet Oncol. 2012;13(3):239–46. doi:10.1016/S1470-2045(11)70393-X.
13. Yasuda H, Kobayashi S, Costa DB. EGFR exon 20 insertion mutations in non-small-cell lung cancer: preclinical data and clinical implications. Lancet Oncol. 2012;13(1):e23–31. doi:10.1016/S1470-2045(11)70129-2.
14. He M, Capelletti M, Nafa K, Yun CH, Arcila ME, Miller VA, Ginsberg MS, Zhao B, Kris MG, Eck MJ, Janne PA, Ladanyi M, Oxnard GR. EGFR exon 19 insertions: a new family of sensitizing EGFR mutations in lung adenocarcinoma. Clin Cancer Res. 2012;18(6):1790–7. doi:10.1158/1078-0432.CCR-11-2361.
15. De Pas T, Toffalorio F, Manzotti M, Fumagalli C, Spitaleri G, Catania C, Delmonte A, Giovannini M, Spaggiari L, de Braud F, Barberis M. Activity of epidermal growth factor receptor-tyrosine kinase inhibitors in patients with non-small cell lung cancer harboring rare epidermal growth factor receptor mutations. J Thorac Oncol. 2011;6(11):1895–901. doi:10.1097/JTO.0b013e318227e8c6.
16. Fukuoka M, Wu YL, Thongprasert S, Sunpaweravong P, Leong SS, Sriuranpong V, Chao TY, Nakagawa K, Chu DT, Saijo N, Duffield EL, Rukazenkov Y, Speake G, Jiang H, Armour AA, To KF, Yang JC, Mok TS. Biomarker analyses and final overall survival results from a phase III, randomized, open-label, first-line study of gefitinib versus carboplatin/paclitaxel in clinically selected patients with advanced non-small-cell lung cancer in Asia (IPASS). J Clin Oncol. 2011;29(21):2866–74. doi:10.1200/JCO.2010.33.4235.
17. Klughammer B, Spleiss O. "Other" (non-activating, non-T790M) EGFR mutations and their clinical implications collected from various Tarceva trials in NSCLC. Eur J Cancer. 2011;47 Suppl 4:S20–1.
18. Yang JC, Schuler M, Yamamoto N, O'Byrne K, Hirsh V, Mok T, Geater SL, Orlov SV, Tsai CM, Boyer MJ, Su WC, Bennouna J, Kato T, Gorbunova V, Lee KH, Shah RNH, Massey D, Lorence RM, Shahidi M, Sequist LV. LUX-Lung 3: a randomized, open-label, phase III study of afatinib versus pemetrexed and cisplatin as first-line treatment for patients with advanced adenocarcinoma of the lung harboring EGFR-activating mutations. J Clin Oncol. 2012;30(Suppl):abstr LBA7500.
19. Kobayashi S, Boggon TJ, Dayaram T, Janne PA, Kocher O, Meyerson M, Johnson BE, Eck MJ, Tenen DG, Halmos B. EGFR mutation and resistance of non-small-cell lung cancer to gefitinib. N Engl J Med. 2005;352(8):786–92. doi:10.1056/NEJMoa044238.
20. Pao W, Miller VA, Politi KA, Riely GJ, Somwar R, Zakowski MF, Kris MG, Varmus H. Acquired resistance of lung adenocarcinomas to gefitinib or erlotinib is associated with a second mutation in the EGFR kinase domain. PLoS Med. 2005;2(3):e73. doi:10.1371/journal.pmed.0020073.
21. Sequist LV, Waltman BA, Dias-Santagata D, Digumarthy S, Turke AB, Fidias P, Bergethon K, Shaw AT, Gettinger S, Cosper AK, Akhavanfard S, Heist RS, Temel J, Christensen JG, Wain JC, Lynch TJ, Vernovsky K, Mark EJ, Lanuti M, Iafrate AJ, Mino-Kenudson M, Engelman JA. Genotypic and histological evolution of lung cancers acquiring resistance to EGFR inhibitors. Sci Transl Med. 2011;3(75):75ra26. doi:10.1126/scitranslmed.3002003.
22. Takezawa K, Pirazzoli V, Arcila ME, Nebhan CA, Song X, de Stanchina E, Ohashi K, Janjigian YY, Spitzler PJ, Melnick MA, Riely GJ, Kris MG, Miller VA, Ladanyi M, Politi K, Pao W. HER2 amplification: a potential mechanism of acquired resistance to EGFR inhibition in EGFR-mutant lung cancers that lack the second-site EGFRT790M mutation. Cancer Discov. 2012;2(10):922–33. doi:10.1158/2159-8290.CD-12-0108.
23. Engelman JA, Zejnullahu K, Mitsudomi T, Song Y, Hyland C, Park JO, Lindeman N, Gale CM, Zhao X, Christensen J, Kosaka T, Holmes AJ, Rogers AM, Cappuzzo F, Mok T, Lee C, Johnson BE, Cantley LC, Janne PA. MET amplification leads to gefitinib resistance in lung cancer by activating ERBB3 signaling. Science. 2007;316(5827):1039–43. doi:10.1126/science.1141478.
24. Ohashi K, Sequist LV, Arcila ME, Moran T, Chmielecki J, Lin YL, Pan Y, Wang L, de Stanchina E, Shien K, Aoe K, Toyooka S, Kiura K, Fernandez-Cuesta L, Fidias P, Yang JC, Miller VA, Riely GJ,

Kris MG, Engelman JA, Vnencak-Jones CL, Dias-Santagata D, Ladanyi M, Pao W. Lung cancers with acquired resistance to EGFR inhibitors occasionally harbor BRAF gene mutations but lack mutations in KRAS, NRAS, or MEK1. Proc Natl Acad Sci U S A. 2012;109(31):E2127–33. doi:10.1073/pnas.1203530109.
25. Riely GJ, Marks J, Pao W. KRAS mutations in non-small cell lung cancer. Proc Am Thorac Soc.2009;6(2):201–5.doi:10.1513/pats.200809-107LC.
26. Karachaliou N, Mayo C, Costa C, Magri I, Gimenez-Capitan A, Molina-Vila MA, Rosell R. KRAS mutations in lung cancer. Clin Lung Cancer. 2012. doi:10.1016/j.cllc.2012.09.007.
27. Linardou H, Dahabreh IJ, Kanaloupiti D, Siannis F, Bafaloukos D, Kosmidis P, Papadimitriou CA, Murray S. Assessment of somatic k-RAS mutations as a mechanism associated with resistance to EGFR-targeted agents: a systematic review and meta-analysis of studies in advanced non-small-cell lung cancer and metastatic colorectal cancer. Lancet Oncol. 2008;9(10):962–72. doi:10.1016/S1470-2045(08)70206-7.
28. Mao C, Qiu LX, Liao RY, Du FB, Ding H, Yang WC, Li J, Chen Q. KRAS mutations and resistance to EGFR-TKIs treatment in patients with non-small cell lung cancer: a meta-analysis of 22 studies. Lung Cancer. 2010;69(3):272–8. doi:10.1016/j.lungcan.2009.11.020.
29. Sos ML, Fischer S, Ullrich R, Peifer M, Heuckmann JM, Koker M, Heynck S, Stuckrath I, Weiss J, Fischer F, Michel K, Goel A, Regales L, Politi KA, Perera S, Getlik M, Heukamp LC, Ansen S, Zander T, Beroukhim R, Kashkar H, Shokat KM, Sellers WR, Rauh D, Orr C, Hoeflich KP, Friedman L, Wong KK, Pao W, Thomas RK. Identifying genotype-dependent efficacy of single and combined PI3K- and MAPK-pathway inhibition in cancer. Proc Natl Acad Sci U S A. 2009;106(43):18351–6. doi:10.1073/pnas.0907325106.
30. Holt SV, Logie A, Davies BR, Alferez D, Runswick S, Fenton S, Chresta CM, Gu Y, Zhang J, Wu YL, Wilkinson RW, Guichard SM, Smith PD. Enhanced apoptosis and tumor growth suppression elicited by combination of MEK (selumetinib) and mTOR kinase inhibitors (AZD8055). Cancer Res. 2012;72(7):1804–13. doi:10.1158/0008-5472.CAN-11-1780.
31. Janne PA, Shaw AT, Pereira JR, Jeannin G, Vansteenkiste J, Barrios C, Franke FA, Grinsted L, Zazulina V, Smith P, Smith I, Crino L. Selumetinib plus docetaxel for KRAS-mutant advanced non-small-cell lung cancer: a randomised, multicentre, placebo-controlled, phase 2 study. Lancet Oncol. 2012. doi:10.1016/S1470-2045(12)70489-8.
32. Leicht DT, Balan V, Kaplun A, Singh-Gupta V, Kaplun L, Dobson M, Tzivion G. Raf kinases: function, regulation and role in human cancer. Biochim Biophys Acta. 2007;1773(8):1196–212. doi:10.1016/j.bbamcr.2007.05.001.
33. Marchetti A, Felicioni L, Malatesta S, Grazia Sciarrotta M, Guetti L, Chella A, Viola P, Pullara C, Mucilli F, Buttitta F. Clinical features and outcome of patients with non-small-cell lung cancer harboring BRAF mutations. J Clin Oncol. 2011;29(26):3574–9. doi:10.1200/JCO.2011.35.9638.
34. Naoki K, Chen TH, Richards WG, Sugarbaker DJ, Meyerson M. Missense mutations of the BRAF gene in human lung adenocarcinoma. Cancer Res. 2002;62(23):7001–3.
35. Brose MS, Volpe P, Feldman M, Kumar M, Rishi I, Gerrero R, Einhorn E, Herlyn M, Minna J, Nicholson A, Roth JA, Albelda SM, Davies H, Cox C, Brignell G, Stephens P, Futreal PA, Wooster R, Stratton MR, Weber BL. BRAF and RAS mutations in human lung cancer and melanoma. Cancer Res. 2002;62(23):6997–7000.
36. Gautschi O, Pauli C, Strobel K, Hirschmann A, Printzen G, Aebi S, Diebold J. A patient with BRAF V600E lung adenocarcinoma responding to vemurafenib. J Thorac Oncol. 2012;7(10):e23–4. doi:10.1097/JTO.0b013e3182629903.
37. Pratilas CA, Hanrahan AJ, Halilovic E, Persaud Y, Soh J, Chitale D, Shigematsu H, Yamamoto H, Sawai A, Janakiraman M, Taylor BS, Pao W, Toyooka S, Ladanyi M, Gazdar A, Rosen N, Solit DB. Genetic predictors of MEK dependence in non-small cell lung cancer. Cancer Res. 2008;68(22):9375–83. doi:10.1158/0008-5472.CAN-08-2223.
38. Hainsworth JD, Cebotaru CL, Kanarev V, Ciuleanu TE, Damyanov D, Stella P, Ganchev H, Pover G, Morris C, Tzekova V. A phase II, open-label, randomized study to assess the efficacy and safety of AZD6244 (ARRY-142886) versus pemetrexed in patients with non-small cell lung cancer who have failed one or two prior chemotherapeutic regimens. J Thorac Oncol. 2010;5(10):1630–6. doi:10.1097/JTO.0b013e3181e8b3a3.
39. Haura EB, Ricart AD, Larson TG, Stella PJ, Bazhenova L, Miller VA, Cohen RB, Eisenberg PD, Selaru P, Wilner KD, Gadgeel SM. A phase II study of PD-0325901, an oral MEK inhibitor, in previously treated patients with advanced non-small cell lung cancer. Clin Cancer Res. 2010;16(8):2450–7. doi:10.1158/1078-0432.CCR-09-1920.
40. Johnson FM, Bekele BN, Feng L, Wistuba I, Tang XM, Tran HT, Erasmus JJ, Hwang LL, Takebe N, Blumenschein GR, Lippman SM, Stewart DJ. Phase II study of dasatinib in patients with advanced non-small-cell lung cancer. J Clin

Oncol. 2010;28(30):4609–15. doi:10.1200/JCO.2010.30.5474.
41. Sen B, Peng S, Tang X, Erickson HS, Galindo H, Mazumdar T, Stewart DJ, Wistuba I, Johnson FM. Kinase-impaired BRAF mutations in lung cancer confer sensitivity to dasatinib. Sci Transl Med. 2012;4(136):136ra170. doi:10.1126/scitranslmed.3003513.
42. Heinmoller P, Gross C, Beyser K, Schmidtgen C, Maass G, Pedrocchi M, Ruschoff J. HER2 status in non-small cell lung cancer: results from patient screening for enrollment to a phase II study of herceptin. Clin Cancer Res. 2003;9(14):5238–43.
43. Gatzemeier U, Groth G, Butts C, Van Zandwijk N, Shepherd F, Ardizzoni A, Barton C, Ghahramani P, Hirsh V. Randomized phase II trial of gemcitabine-cisplatin with or without trastuzumab in HER2-positive non-small-cell lung cancer. Ann Oncol. 2004;15(1):19–27.
44. Krug LM, Miller VA, Patel J, Crapanzano J, Azzoli CG, Gomez J, Kris MG, Heelan RT, Pizzo B, Tyson L, Sheehan C, Ross JS, Venkatraman E. Randomized phase II study of weekly docetaxel plus trastuzumab versus weekly paclitaxel plus trastuzumab in patients with previously untreated advanced nonsmall cell lung carcinoma. Cancer. 2005;104(10):2149–55. doi:10.1002/cncr.21428.
45. Shigematsu H, Takahashi T, Nomura M, Majmudar K, Suzuki M, Lee H, Wistuba II, Fong KM, Toyooka S, Shimizu N, Fujisawa T, Minna JD, Gazdar AF. Somatic mutations of the HER2 kinase domain in lung adenocarcinomas. Cancer Res. 2005;65(5):1642–6. doi:10.1158/0008-5472.CAN-04-4235.
46. Arcila ME, Chaft JE, Nafa K, Roy-Chowdhuri S, Lau C, Zaidinski M, Paik PK, Zakowski MF, Kris MG, Ladanyi M. Prevalence, clinicopathologic associations, and molecular spectrum of ERBB2 (HER2) tyrosine kinase mutations in lung adenocarcinomas. Clin Cancer Res. 2012;18(18):4910–8. doi:10.1158/1078-0432.CCR-12-0912.
47. Shimamura T, Ji H, Minami Y, Thomas RK, Lowell AM, Shah K, Greulich H, Glatt KA, Meyerson M, Shapiro GI, Wong KK. Non-small-cell lung cancer and Ba/F3 transformed cells harboring the ERBB2 G776insV_G/C mutation are sensitive to the dual-specific epidermal growth factor receptor and ERBB2 inhibitor HKI-272. Cancer Res. 2006;66(13):6487–91. doi:10.1158/0008-5472.CAN-06-0971.
48. Wang SE, Narasanna A, Perez-Torres M, Xiang B, Wu FY, Yang S, Carpenter G, Gazdar AF, Muthuswamy SK, Arteaga CL. HER2 kinase domain mutation results in constitutive phosphorylation and activation of HER2 and EGFR and resistance to EGFR tyrosine kinase inhibitors. Cancer Cell. 2006;10(1):25–38. doi:10.1016/j.ccr.2006.05.023.
49. De Greve J, Teugels E, Geers C, Decoster L, Galdermans D, De Mey J, Everaert H, Umelo I, In't Veld P, Schallier D. Clinical activity of afatinib (BIBW 2992) in patients with lung adenocarcinoma with mutations in the kinase domain of HER2/neu. Lung Cancer. 2012;76(1):123–7. doi:10.1016/j.lungcan.2012.01.008.
50. Cappuzzo F, Bemis L, Varella-Garcia M. HER2 mutation and response to trastuzumab therapy in non-small-cell lung cancer. N Engl J Med. 2006;354(24):2619–21. doi:10.1056/NEJMc060020.
51. Greulich H, Kaplan B, Mertins P, Chen TH, Tanaka KE, Yun CH, Zhang X, Lee SH, Cho J, Ambrogio L, Liao R, Imielinski M, Banerji S, Berger AH, Lawrence MS, Zhang J, Pho NH, Walker SR, Winckler W, Getz G, Frank D, Hahn WC, Eck MJ, Mani DR, Jaffe JD, Carr SA, Wong KK, Meyerson M. Functional analysis of receptor tyrosine kinase mutations in lung cancer identifies oncogenic extracellular domain mutations of ERBB2. Proc Natl Acad Sci U S A. 2012;109(36):14476–81. doi:10.1073/pnas.1203201109.
52. Karakas B, Bachman KE, Park BH. Mutation of the PIK3CA oncogene in human cancers. Br J Cancer. 2006;94(4):455–9. doi:10.1038/sj.bjc.6602970.
53. Yamamoto H, Shigematsu H, Nomura M, Lockwood WW, Sato M, Okumura N, Soh J, Suzuki M, Wistuba II, Fong KM, Lee H, Toyooka S, Date H, Lam WL, Minna JD, Gazdar AF. PIK3CA mutations and copy number gains in human lung cancers. Cancer Res. 2008;68(17):6913–21. doi:10.1158/0008-5472.CAN-07-5084.
54. Chaft JE, Arcila ME, Paik PK, Lau C, Riely GJ, Pietanza MC, Zakowski MF, Rusch V, Sima CS, Ladanyi M, Kris MG. Coexistence of PIK3CA and other oncogene mutations in lung adenocarcinoma-rationale for comprehensive mutation profiling. Mol Cancer Ther. 2012;11(2):485–91. doi:10.1158/1535-7163.MCT-11-0692.
55. Kawano O, Sasaki H, Endo K, Suzuki E, Haneda H, Yukiue H, Kobayashi Y, Yano M, Fujii Y. PIK3CA mutation status in Japanese lung cancer patients. Lung Cancer. 2006;54(2):209–15. doi:10.1016/j.lungcan.2006.07.006.
56. Ji M, Guan H, Gao C, Shi B, Hou P. Highly frequent promoter methylation and PIK3CA amplification in non-small cell lung cancer (NSCLC). BMC Cancer. 2011;11:147. doi:10.1186/1471-2407-11-147.
57. O'Brien C, Wallin JJ, Sampath D, GuhaThakurta D, Savage H, Punnoose EA, Guan J, Berry L, Prior WW, Amler LC, Belvin M, Friedman LS, Lackner MR. Predictive biomarkers of sensitivity

to the phosphatidylinositol 3′ kinase inhibitor GDC-0941 in breast cancer preclinical models. Clin Cancer Res. 2010;16(14):3670–83. doi:10.1158/1078-0432.CCR-09-2828.
58. Bendell JC, Rodon J, Burris HA, de Jonge M, Verweij J, Birle D, Demanse D, De Buck SS, Ru QC, Peters M, Goldbrunner M, Baselga J. Phase I, dose-escalation study of BKM120, an oral pan-Class I PI3K inhibitor, in patients with advanced solid tumors. J Clin Oncol. 2012;30(3):282–90. doi:10.1200/JCO.2011.36.1360.
59. Zou ZQ, Zhang XH, Wang F, Shen QJ, Xu J, Zhang LN, Xing WH, Zhuo RJ, Li D. A novel dual PI3Kalpha/mTOR inhibitor PI-103 with high antitumor activity in non-small cell lung cancer cells. Int J Mol Med. 2009;24(1):97–101.
60. Engelman JA, Mukohara T, Zejnullahu K, Lifshits E, Borras AM, Gale CM, Naumov GN, Yeap BY, Jarrell E, Sun J, Tracy S, Zhao X, Heymach JV, Johnson BE, Cantley LC, Janne PA. Allelic dilution obscures detection of a biologically significant resistance mutation in EGFR-amplified lung cancer. J Clin Invest. 2006;116(10):2695–706. doi:10.1172/JCI28656.
61. Sos ML, Koker M, Weir BA, Heynck S, Rabinovsky R, Zander T, Seeger JM, Weiss J, Fischer F, Frommolt P, Michel K, Peifer M, Mermel C, Girard L, Peyton M, Gazdar AF, Minna JD, Garraway LA, Kashkar H, Pao W, Meyerson M, Thomas RK. PTEN loss contributes to erlotinib resistance in EGFR-mutant lung cancer by activation of Akt and EGFR. Cancer Res. 2009;69(8):3256–61. doi:10.1158/0008-5472.CAN-08-4055.
62. Gandhi J, Zhang J, Xie Y, Soh J, Shigematsu H, Zhang W, Yamamoto H, Peyton M, Girard L, Lockwood WW, Lam WL, Varella-Garcia M, Minna JD, Gazdar AF. Alterations in genes of the EGFR signaling pathway and their relationship to EGFR tyrosine kinase inhibitor sensitivity in lung cancer cell lines. PLoS One. 2009;4(2):e4576. doi:10.1371/journal.pone.0004576.
63. Vivanco I, Sawyers CL. The phosphatidylinositol 3-kinase AKT pathway in human cancer. Nat Rev Cancer. 2002;2(7):489–501. doi:10.1038/nrc839.
64. Ding L, Getz G, Wheeler DA, Mardis ER, McLellan MD, Cibulskis K, Sougnez C, Greulich H, Muzny DM, Morgan MB, Fulton L, Fulton RS, Zhang Q, Wendl MC, Lawrence MS, Larson DE, Chen K, Dooling DJ, Sabo A, Hawes AC, Shen H, Jhangiani SN, Lewis LR, Hall O, Zhu Y, Mathew T, Ren Y, Yao J, Scherer SE, Clerc K, Metcalf GA, Ng B, Milosavljevic A, Gonzalez-Garay ML, Osborne JR, Meyer R, Shi X, Tang Y, Koboldt DC, Lin L, Abbott R, Miner TL, Pohl C, Fewell G, Haipek C, Schmidt H, Dunford-Shore BH, Kraja A, Crosby SD, Sawyer CS, Vickery T, Sander S, Robinson J, Winckler W, Baldwin J, Chirieac LR, Dutt A, Fennell T, Hanna M, Johnson BE, Onofrio RC, Thomas RK, Tonon G, Weir BA, Zhao X, Ziaugra L, Zody MC, Giordano T, Orringer MB, Roth JA, Spitz MR, Wistuba II, Ozenberger B, Good PJ, Chang AC, Beer DG, Watson MA, Ladanyi M, Broderick S, Yoshizawa A, Travis WD, Pao W, Province MA, Weinstock GM, Varmus HE, Gabriel SB, Lander ES, Gibbs RA, Meyerson M, Wilson RK. Somatic mutations affect key pathways in lung adenocarcinoma. Nature. 2008;455(7216):1069–75. doi:10.1038/nature07423.
65. Do H, Solomon B, Mitchell PL, Fox SB, Dobrovic A. Detection of the transforming AKT1 mutation E17K in non-small cell lung cancer by high resolution melting. BMC Res Notes. 2008;1:14. doi:10.1186/1756-0500-1-14.
66. Malanga D, Scrima M, De Marco C, Fabiani F, De Rosa N, De Gisi S, Malara N, Savino R, Rocco G, Chiappetta G, Franco R, Tirino V, Pirozzi G, Viglietto G. Activating E17K mutation in the gene encoding the protein kinase AKT1 in a subset of squamous cell carcinoma of the lung. Cell Cycle. 2008;7(5):665–9.
67. Carnero A, Blanco-Aparicio C, Renner O, Link W, Leal JF. The PTEN/PI3K/AKT signalling pathway in cancer, therapeutic implications. Curr Cancer Drug Targets. 2008;8(3):187–98.
68. Kohno T, Takahashi M, Manda R, Yokota J. Inactivation of the PTEN/MMAC1/TEP1 gene in human lung cancers. Genes Chromosomes Cancer. 1998;22(2):152–6.
69. Jin G, Kim MJ, Jeon HS, Choi JE, Kim DS, Lee EB, Cha SI, Yoon GS, Kim CH, Jung TH, Park JY. PTEN mutations and relationship to EGFR, ERBB2, KRAS, and TP53 mutations in non-small cell lung cancers. Lung Cancer. 2010;69(3):279–83. doi:10.1016/j.lungcan.2009.11.012.
70. Lee SY, Kim MJ, Jin G, Yoo SS, Park JY, Choi JE, Jeon HS, Cho S, Lee EB, Cha SI, Park TI, Kim CH, Jung TH. Somatic mutations in epidermal growth factor receptor signaling pathway genes in non-small cell lung cancers. J Thorac Oncol. 2010;5(11):1734–40. doi:10.1097/JTO.0b013e3181f0beca.
71. Ihle NT, Lemos Jr R, Wipf P, Yacoub A, Mitchell C, Siwak D, Mills GB, Dent P, Kirkpatrick DL, Powis G. Mutations in the phosphatidylinositol-3-kinase pathway predict for antitumor activity of the inhibitor PX-866 whereas oncogenic Ras is a dominant predictor for resistance. Cancer Res. 2009;69(1):143–50. doi:10.1158/0008-5472.CAN-07-6656.
72. Spoerke JM, O'Brien C, Huw L, Koeppen H, Fridlyand J, Brachmann RK, Haverty PM, Pandita A, Mohan S, Sampath D, Friedman LS, Ross L, Hampton GM, Amler LC, Shames DS, Lackner MR. Phosphoinositide 3-kinase (PI3K) pathway alterations are associated with histologic subtypes and are predictive of sensitivity to PI3K inhibitors in lung cancer preclinical models.

Clin Cancer Res. 2012. doi:10.1158/1078-0432.CCR-12-2347.
73. Seki T, Hagiya M, Shimonishi M, Nakamura T, Shimizu S. Organization of the human hepatocyte growth factor-encoding gene. Gene. 1991;102(2):213–9.
74. Kong-Beltran M, Seshagiri S, Zha J, Zhu W, Bhawe K, Mendoza N, Holcomb T, Pujara K, Stinson J, Fu L, Severin C, Rangell L, Schwall R, Amler L, Wickramasinghe D, Yauch R. Somatic mutations lead to an oncogenic deletion of met in lung cancer. Cancer Res. 2006;66(1):283–9. doi:10.1158/0008-5472.CAN-05-2749.
75. Okuda K, Sasaki H, Yukiue H, Yano M, Fujii Y. Met gene copy number predicts the prognosis for completely resected non-small cell lung cancer. Cancer Sci. 2008;99(11):2280–5. doi:10.1111/j.1349-7006.2008.00916.x.
76. Onozato R, Kosaka T, Kuwano H, Sekido Y, Yatabe Y, Mitsudomi T. Activation of MET by gene amplification or by splice mutations deleting the juxtamembrane domain in primary resected lung cancers. J Thorac Oncol. 2009;4(1):5–11. doi:10.1097/JTO.0b013e3181913e0e.
77. Krishnaswamy S, Kanteti R, Duke-Cohan JS, Loganathan S, Liu W, Ma PC, Sattler M, Singleton PA, Ramnath N, Innocenti F, Nicolae DL, Ouyang Z, Liang J, Minna J, Kozloff MF, Ferguson MK, Natarajan V, Wang YC, Garcia JG, Vokes EE, Salgia R. Ethnic differences and functional analysis of MET mutations in lung cancer. Clin Cancer Res. 2009;15(18):5714–23. doi:10.1158/1078-0432.CCR-09-0070.
78. Cappuzzo F, Janne PA, Skokan M, Finocchiaro G, Rossi E, Ligorio C, Zucali PA, Terracciano L, Toschi L, Roncalli M, Destro A, Incarbone M, Alloisio M, Santoro A, Varella-Garcia M. MET increased gene copy number and primary resistance to gefitinib therapy in non-small-cell lung cancer patients. Ann Oncol. 2009;20(2):298–304. doi:10.1093/annonc/mdn635.
79. Tanizaki J, Okamoto I, Okamoto K, Takezawa K, Kuwata K, Yamaguchi H, Nakagawa K. MET tyrosine kinase inhibitor crizotinib (PF-02341066) shows differential antitumor effects in non-small cell lung cancer according to MEt alterations. J Thorac Oncol. 2011;6(10):1624–31. doi:10.1097/JTO.0b013e31822591e9.
80. Ou SH, Kwak EL, Siwak-Tapp C, Dy J, Bergethon K, Clark JW, Camidge DR, Solomon BJ, Maki RG, Bang YJ, Kim DW, Christensen J, Tan W, Wilner KD, Salgia R, Iafrate AJ. Activity of crizotinib (PF02341066), a dual mesenchymal-epithelial transition (MET) and anaplastic lymphoma kinase (ALK) inhibitor, in a non-small cell lung cancer patient with de novo MET amplification. J Thorac Oncol. 2011;6(5):942–6. doi:10.1097/JTO.0b013e31821528d3.
81. Morris SW, Kirstein MN, Valentine MB, Dittmer KG, Shapiro DN, Saltman DL, Look AT. Fusion of a kinase gene, ALK, to a nucleolar protein gene, NPM, in non-Hodgkin's lymphoma. Science. 1994;263(5151):1281–4.
82. Soda M, Choi YL, Enomoto M, Takada S, Yamashita Y, Ishikawa S, Fujiwara S, Watanabe H, Kurashina K, Hatanaka H, Bando M, Ohno S, Ishikawa Y, Aburatani H, Niki T, Sohara Y, Sugiyama Y, Mano H. Identification of the transforming EML4-ALK fusion gene in non-small-cell lung cancer. Nature. 2007;448(7153):561–6. doi:10.1038/nature05945.
83. Koivunen JP, Mermel C, Zejnullahu K, Murphy C, Lifshits E, Holmes AJ, Choi HG, Kim J, Chiang D, Thomas R, Lee J, Richards WG, Sugarbaker DJ, Ducko C, Lindeman N, Marcoux JP, Engelman JA, Gray NS, Lee C, Meyerson M, Janne PA. EML4-ALK fusion gene and efficacy of an ALK kinase inhibitor in lung cancer. Clin Cancer Res. 2008;14(13):4275–83. doi:10.1158/1078-0432.CCR-08-0168.
84. Martelli MP, Sozzi G, Hernandez L, Pettirossi V, Navarro A, Conte D, Gasparini P, Perrone F, Modena P, Pastorino U, Carbone A, Fabbri A, Sidoni A, Nakamura S, Gambacorta M, Fernandez PL, Ramirez J, Chan JK, Grigioni WF, Campo E, Pileri SA, Falini B. EML4-ALK rearrangement in non-small cell lung cancer and non-tumor lung tissues. Am J Pathol. 2009;174(2):661–70. doi:10.2353/ajpath.2009.080755.
85. Sasaki T, Rodig SJ, Chirieac LR, Janne PA. The biology and treatment of EML4-ALK non-small cell lung cancer. Eur J Cancer. 2010;46(10):1773–80. doi:10.1016/j.ejca.2010.04.002.
86. Penzel R, Schirmacher P, Warth A. Oncogenic fusions involving exon 19 of ALK. J Thorac Oncol. 2012;7(12):e44. doi:10.1097/JTO.0b013e3182743ad7.
87. Le AT, Varella-Garcia M, Doebele RC. Oncogenic fusions involving exon 19 of ALK. J Thorac Oncol. 2012;7(12):e44. doi:10.1097/JTO.0b013e31826bb94d.
88. Shaw AT, Yeap BY, Mino-Kenudson M, Digumarthy SR, Costa DB, Heist RS, Solomon B, Stubbs H, Admane S, McDermott U, Settleman J, Kobayashi S, Mark EJ, Rodig SJ, Chirieac LR, Kwak EL, Lynch TJ, Iafrate AJ. Clinical features and outcome of patients with non-small-cell lung cancer who harbor EML4-ALK. J Clin Oncol. 2009;27(26):4247–53. doi:10.1200/JCO.2009.22.6993.
89. Kim HR, Shim HS, Chung JH, Lee YJ, Hong YK, Rha SY, Kim SH, Ha SJ, Kim SK, Chung KY, Soo R, Kim JH, Cho BC. Distinct clinical features and outcomes in never-smokers with nonsmall cell lung cancer who harbor EGFR or KRAS mutations or ALK rearrangement. Cancer. 2012;118(3):729–39. doi:10.1002/cncr.26311.

90. Kwak EL, Bang YJ, Camidge DR, Shaw AT, Solomon B, Maki RG, Ou SH, Dezube BJ, Janne PA, Costa DB, Varella-Garcia M, Kim WH, Lynch TJ, Fidias P, Stubbs H, Engelman JA, Sequist LV, Tan W, Gandhi L, Mino-Kenudson M, Wei GC, Shreeve SM, Ratain MJ, Settleman J, Christensen JG, Haber DA, Wilner K, Salgia R, Shapiro GI, Clark JW, Iafrate AJ. Anaplastic lymphoma kinase inhibition in non-small-cell lung cancer. N Engl J Med. 2010;363(18):1693–703. doi:10.1056/NEJMoa1006448.
91. Choi YL, Soda M, Yamashita Y, Ueno T, Takashima J, Nakajima T, Yatabe Y, Takeuchi K, Hamada T, Haruta H, Ishikawa Y, Kimura H, Mitsudomi T, Tanio Y, Mano H. EML4-ALK mutations in lung cancer that confer resistance to ALK inhibitors. N Engl J Med. 2010;363(18):1734–9. doi:10.1056/NEJMoa1007478.
92. Sasaki T, Koivunen J, Ogino A, Yanagita M, Nikiforow S, Zheng W, Lathan C, Marcoux JP, Du J, Okuda K, Capelletti M, Shimamura T, Ercan D, Stumpfova M, Xiao Y, Weremowicz S, Butaney M, Heon S, Wilner K, Christensen JG, Eck MJ, Wong KK, Lindeman N, Gray NS, Rodig SJ, Janne PA. A novel ALK secondary mutation and EGFR signaling cause resistance to ALK kinase inhibitors. Cancer Res. 2011;71(18):6051–60. doi:10.1158/0008-5472.CAN-11-1340.
93. Doebele RC, Pilling AB, Aisner DL, Kutateladze TG, Le AT, Weickhardt AJ, Kondo KL, Linderman DJ, Heasley LE, Franklin WA, Varella-Garcia M, Camidge DR. Mechanisms of resistance to crizotinib in patients with ALK gene rearranged non-small cell lung cancer. Clin Cancer Res. 2012;18(5):1472–82. doi:10.1158/1078-0432.CCR-11-2906.
94. Katayama R, Shaw AT, Khan TM, Mino-Kenudson M, Solomon BJ, Halmos B, Jessop NA, Wain JC, Yeo AT, Benes C, Drew L, Saeh JC, Crosby K, Sequist LV, Iafrate AJ, Engelman JA. Mechanisms of acquired crizotinib resistance in ALK-rearranged lung cancers. Sci Transl Med. 2012;4(120):120ra117. doi:10.1126/scitranslmed.3003316.
95. Tanizaki J, Okamoto I, Okabe T, Sakai K, Tanaka K, Hayashi H, Kaneda H, Takezawa K, Kuwata K, Yamaguchi H, Hatashita E, Nishio K, Nakagawa K. Activation of HER family signaling as a mechanism of acquired resistance to ALK inhibitors in EML4-ALK-positive non-small cell lung cancer. Clin Cancer Res. 2012;18(22):6219–26. doi:10.1158/1078-0432.CCR-12-0392.
96. Katayama R, Khan TM, Benes C, Lifshits E, Ebi H, Rivera VM, Shakespeare WC, Iafrate AJ, Engelman JA, Shaw AT. Therapeutic strategies to overcome crizotinib resistance in non-small cell lung cancers harboring the fusion oncogene EML4-ALK. Proc Natl Acad Sci U S A. 2011;108(18):7535–40. doi:10.1073/pnas.1019559108.
97. Heukamp LC, Thor T, Schramm A, De Preter K, Kumps C, De Wilde B, Odersky A, Peifer M, Lindner S, Spruessel A, Pattyn F, Mestdagh P, Menten B, Kuhfittig-Kulle S, Kunkele A, Konig K, Meder L, Chatterjee S, Ullrich RT, Schulte S, Vandesompele J, Speleman F, Buttner R, Eggert A, Schulte JH. Targeted expression of mutated ALK induces neuroblastoma in transgenic mice. Sci Transl Med. 2012;4(141):141ra191. doi:10.1126/scitranslmed.3003967.
98. Takeuchi K, Choi YL, Togashi Y, Soda M, Hatano S, Inamura K, Takada S, Ueno T, Yamashita Y, Satoh Y, Okumura S, Nakagawa K, Ishikawa Y, Mano H. KIF5B-ALK, a novel fusion oncokinase identified by an immunohistochemistry-based diagnostic system for ALK-positive lung cancer. Clin Cancer Res. 2009;15(9):3143–9. doi:10.1158/1078-0432.CCR-08-3248.
99. Rikova K, Guo A, Zeng Q, Possemato A, Yu J, Haack H, Nardone J, Lee K, Reeves C, Li Y, Hu Y, Tan Z, Stokes M, Sullivan L, Mitchell J, Wetzel R, Macneill J, Ren JM, Yuan J, Bakalarski CE, Villen J, Kornhauser JM, Smith B, Li D, Zhou X, Gygi SP, Gu TL, Polakiewicz RD, Rush J, Comb MJ. Global survey of phosphotyrosine signaling identifies oncogenic kinases in lung cancer. Cell. 2007;131(6):1190–203. doi:10.1016/j.cell.2007.11.025.
100. Ju YS, Lee WC, Shin JY, Lee S, Bleazard T, Won JK, Kim YT, Kim JI, Kang JH, Seo JS. A transforming KIF5B and RET gene fusion in lung adenocarcinoma revealed from whole-genome and transcriptome sequencing. Genome Res. 2012;22(3):436–45. doi:10.1101/gr.133645.111.
101. Kohno T, Ichikawa H, Totoki Y, Yasuda K, Hiramoto M, Nammo T, Sakamoto H, Tsuta K, Furuta K, Shimada Y, Iwakawa R, Ogiwara H, Oike T, Enari M, Schetter AJ, Okayama H, Haugen A, Skaug V, Chiku S, Yamanaka I, Arai Y, Watanabe S, Sekine I, Ogawa S, Harris CC, Tsuda H, Yoshida T, Yokota J, Shibata T. KIF5B-RET fusions in lung adenocarcinoma. Nat Med. 2012;18(3):375–7. doi:10.1038/nm.2644.
102. Wang R, Hu H, Pan Y, Li Y, Ye T, Li C, Luo X, Wang L, Li H, Zhang Y, Li F, Lu Y, Lu Q, Xu J, Garfield D, Shen L, Ji H, Pao W, Sun Y, Chen H. RET fusions define a unique molecular and clinicopathologic subtype of non-small-cell lung cancer. J Clin Oncol. 2012;30(35):4352–9. doi:10.1200/JCO.2012.44.1477.
103. Lipson D, Capelletti M, Yelensky R, Otto G, Parker A, Jarosz M, Curran JA, Balasubramanian S, Bloom T, Brennan KW, Donahue A, Downing SR, Frampton GM, Garcia L, Juhn F, Mitchell KC, White E, White J, Zwirko Z, Peretz T, Nechushtan H, Soussan-Gutman L, Kim J, Sasaki H, Kim HR, Park SI, Ercan D, Sheehan CE, Ross JS, Cronin MT, Janne PA, Stephens PJ. Identification of new

ALK and RET gene fusions from colorectal and lung cancer biopsies. Nat Med. 2012;18(3):382–4. doi:10.1038/nm.2673.
104. Takeuchi K, Soda M, Togashi Y, Suzuki R, Sakata S, Hatano S, Asaka R, Hamanaka W, Ninomiya H, Uehara H, Lim Choi Y, Satoh Y, Okumura S, Nakagawa K, Mano H, Ishikawa Y. RET, ROS1 and ALK fusions in lung cancer. Nat Med. 2012;18(3):378–81. doi:10.1038/nm.2658.
105. Matsubara D, Kanai Y, Ishikawa S, Ohara S, Yoshimoto T, Sakatani T, Oguni S, Tamura T, Kataoka H, Endo S, Murakami Y, Aburatani H, Fukayama M, Niki T. Identification of CCDC6-RET fusion in the human lung adenocarcinoma cell line, LC-2/ad. J Thorac Oncol. 2012;7(12): 1872–6. doi:10.1097/JTO.0b013e3182721ed1.
106. Li C, Fang R, Sun Y, Han X, Li F, Gao B, Iafrate AJ, Liu XY, Pao W, Chen H, Ji H. Spectrum of oncogenic driver mutations in lung adenocarcinomas from East Asian never smokers. PLoS One. 2011;6(11):e28204. doi:10.1371/journal.pone.0028204.
107. Bergethon K, Shaw AT, Ou SH, Katayama R, Lovly CM, McDonald NT, Massion PP, Siwak-Tapp C, Gonzalez A, Fang R, Mark EJ, Batten JM, Chen H, Wilner KD, Kwak EL, Clark JW, Carbone DP, Ji H, Engelman JA, Mino-Kenudson M, Pao W, Iafrate AJ. ROS1 rearrangements define a unique molecular class of lung cancers. J Clin Oncol. 2012;30(8):863–70. doi:10.1200/JCO.2011.35.6345.
108. Jun HJ, Johnson H, Bronson RT, de Feraudy S, White F, Charest A. The oncogenic lung cancer fusion kinase CD74-ROS activates a novel invasiveness pathway through E-Syt1 phosphorylation. Cancer Res. 2012;72(15):3764–74. doi:10.1158/0008-5472.CAN-11-3990.
109. Rimkunas VM, Crosby KE, Li D, Hu Y, Kelly ME, Gu TL, Mack JS, Silver MR, Zhou X, Haack H. Analysis of receptor tyrosine kinase ROS1-positive tumors in non-small cell lung cancer: identification of a FIG-ROS1 fusion. Clin Cancer Res. 2012;18(16):4449–57. doi:10.1158/1078-0432.CCR-11-3351.
110. Davies KD, Le AT, Theodoro MF, Skokan MC, Aisner DL, Berge EM, Terracciano LM, Cappuzzo F, Incarbone M, Roncalli M, Alloisio M, Santoro A, Camidge DR, Varella-Garcia M, Doebele RC. Identifying and targeting ROS1 gene fusions in non-small cell lung cancer. Clin Cancer Res. 2012;18(17):4570–9. doi:10.1158/1078-0432.CCR-12-0550.
111. Yasuda H, de Figueiredo-Pontes LL, Kobayashi S, Costa DB. Preclinical rationale for use of the clinically available multitargeted tyrosine kinase inhibitor crizotinib in ROS1-translocated lung cancer. J Thorac Oncol. 2012;7(7):1086–90. doi:10.1097/JTO.0b013e3182570919.
112. Coutts JC, Gallagher JT. Receptors for fibroblast growth factors. Immunol Cell Biol. 1995;73(6):584–9. doi:10.1038/icb.1995.92.
113. Zhao X, Weir BA, LaFramboise T, Lin M, Beroukhim R, Garraway L, Beheshti J, Lee JC, Naoki K, Richards WG, Sugarbaker D, Chen F, Rubin MA, Janne PA, Girard L, Minna J, Christiani D, Li C, Sellers WR, Meyerson M. Homozygous deletions and chromosome amplifications in human lung carcinomas revealed by single nucleotide polymorphism array analysis. Cancer Res. 2005;65(13):5561–70. doi:10.1158/0008-5472.CAN-04-4603.
114. Schultheis AM, Bos M, Schmitz K, Wilsberg L, Binot E, Wolf J, Buttner R, Schildhaus HU. Fibroblast growth factor receptor 1 (FGFR1) amplification is a potential therapeutic target in small-cell lung cancer. Mod Pathol. 2013. doi:10.1038/modpathol.2013.141.
115. Weiss J, Sos ML, Seidel D, Peifer M, Zander T, Heuckmann JM, Ullrich RT, Menon R, Maier S, Soltermann A, Moch H, Wagener P, Fischer F, Heynck S, Koker M, Schottle J, Leenders F, Gabler F, Dabow I, Querings S, Heukamp LC, Balke-Want H, Ansen S, Rauh D, Baessmann I, Altmuller J, Wainer Z, Conron M, Wright G, Russell P, Solomon B, Brambilla E, Brambilla C, Lorimier P, Sollberg S, Brustugun OT, Engel-Riedel W, Ludwig C, Petersen I, Sanger J, Clement J, Groen H, Timens W, Sietsma H, Thunnissen E, Smit E, Heideman D, Cappuzzo F, Ligorio C, Damiani S, Hallek M, Beroukhim R, Pao W, Klebl B, Baumann M, Buettner R, Ernestus K, Stoelben E, Wolf J, Nurnberg P, Perner S, Thomas RK. Frequent and focal FGFR1 amplification associates with therapeutically tractable FGFR1 dependency in squamous cell lung cancer. Sci Transl Med. 2010;2(62):62ra93. doi:10.1126/scitranslmed.3001451.
116. Dutt A, Ramos AH, Hammerman PS, Mermel C, Cho J, Sharifnia T, Chande A, Tanaka KE, Stransky N, Greulich H, Gray NS, Meyerson M. Inhibitor-sensitive FGFR1 amplification in human non-small cell lung cancer. PLoS One. 2011;6(6):e20351. doi:10.1371/journal.pone.0020351.
117. Schildhaus HU, Heukamp LC, Merkelbach-Bruse S, Riesner K, Schmitz K, Binot E, Paggen E, Albus K, Schulte W, Ko YD, Schlesinger A, Ansen S, Engel-Riedel W, Brockmann M, Serke M, Gerigk U, Huss S, Goke F, Perner S, Hekmat K, Frank KF, Reiser M, Schnell R, Bos M, Mattonet C, Sos M, Stoelben E, Wolf J, Zander T, Buettner R. Definition of a fluorescence in-situ hybridization score identifies high- and low-level FGFR1 amplification types in squamous cell lung cancer. Mod Pathol. 2012;25(11):1473–80. doi:10.1038/modpathol.2012.102.

118. Heist RS, Mino-Kenudson M, Sequist LV, Tammireddy S, Morrissey L, Christiani DC, Engelman JA, Iafrate AJ. FGFR1 amplification in squamous cell carcinoma of the lung. J Thorac Oncol. 2012;7(12):1775–80. doi:10.1097/JTO.0b013e31826aed28.
119. Zhang J, Zhang L, Su X, Li M, Xie L, Malchers F, Fan S, Yin X, Xu Y, Liu K, Dong Z, Zhu G, Qian Z, Tang L, Zhan P, Ji Q, Kilgour E, Smith PD, Brooks AN, Thomas RK, Gavine PR. Translating the therapeutic potential of AZD4547 in FGFR1-amplified non-small cell lung cancer through the use of patient-derived tumor xenograft models. Clin Cancer Res. 2012;18(24):6658–67. doi:10.1158/1078-0432.CCR-12-2694.
120. Wolf J, Lorusso P, Camidge DR, Perez JM, Tabernero J, Hidalgo M, Schuler M, Tian GG, Soria JC, Delord JP, Campone M, Bachelot T, van der Noll R, Ringeisen FP, Nogova L, Sequist LV, Schellens JHM. A phase I dose escalation study of NVP-BGJ398, a selective pan FGFR inhibitor in genetically preselected advanced solid tumors. Cancer Res. 2012;72(8 Suppl 1):Abstract nr. LB-122. doi:1538-7445.AM2012-LB-122.
121. Day E, Waters B, Spiegel K, Alnadaf T, Manley PW, Buchdunger E, Walker C, Jarai G. Inhibition of collagen-induced discoidin domain receptor 1 and 2 activation by imatinib, nilotinib and dasatinib. Eur J Pharmacol. 2008;599(1–3):44–53. doi:10.1016/j.ejphar.2008.10.014.
122. Hammerman PS, Sos ML, Ramos AH, Xu C, Dutt A, Zhou W, Brace LE, Woods BA, Lin W, Zhang J, Deng X, Lim SM, Heynck S, Peifer M, Simard JR, Lawrence MS, Onofrio RC, Salvesen HB, Seidel D, Zander T, Heuckmann JM, Soltermann A, Moch H, Koker M, Leenders F, Gabler F, Querings S, Ansen S, Brambilla E, Brambilla C, Lorimier P, Brustugun OT, Helland A, Petersen I, Clement JH, Groen H, Timens W, Sietsma H, Stoelben E, Wolf J, Beer DG, Tsao MS, Hanna M, Hatton C, Eck MJ, Janne PA, Johnson BE, Winckler W, Greulich H, Bass AJ, Cho J, Rauh D, Gray NS, Wong KK, Haura EB, Thomas RK, Meyerson M. Mutations in the DDR2 kinase gene identify a novel therapeutic target in squamous cell lung cancer. Cancer Discov. 2011;1(1):78–89. doi:10.1158/2159-8274.CD-11-0005.
123. An SJ, Chen ZH, Su J, Zhang XC, Zhong WZ, Yang JJ, Zhou Q, Yang XN, Huang L, Guan JL, Nie Q, Yan HH, Mok TS, Wu YL. Identification of enriched driver gene alterations in subgroups of non-small cell lung cancer patients based on histology and smoking status. PLoS One. 2012;7(6):e40109. doi:10.1371/journal.pone.0040109.
124. Margulies M, Egholm M, Altman WE, Attiya S, Bader JS, Bemben LA, Berka J, Braverman MS, Chen YJ, Chen Z, Dewell SB, Du L, Fierro JM, Gomes XV, Godwin BC, He W, Helgesen S, Ho CH, Irzyk GP, Jando SC, Alenquer ML, Jarvie TP, Jirage KB, Kim JB, Knight JR, Lanza JR, Leamon JH, Lefkowitz SM, Lei M, Li J, Lohman KL, Lu H, Makhijani VB, McDade KE, McKenna MP, Myers EW, Nickerson E, Nobile JR, Plant R, Puc BP, Ronan MT, Roth GT, Sarkis GJ, Simons JF, Simpson JW, Srinivasan M, Tartaro KR, Tomasz A, Vogt KA, Volkmer GA, Wang SH, Wang Y, Weiner MP, Yu P, Begley RF, Rothberg JM. Genome sequencing in microfabricated high-density picolitre reactors. Nature. 2005;437(7057):376–80. doi:10.1038/nature03959.
125. Rothberg JM, Hinz W, Rearick TM, Schultz J, Mileski W, Davey M, Leamon JH, Johnson K, Milgrew MJ, Edwards M, Hoon J, Simons JF, Marran D, Myers JW, Davidson JF, Branting A, Nobile JR, Puc BP, Light D, Clark TA, Huber M, Branciforte JT, Stoner IB, Cawley SE, Lyons M, Fu Y, Homer N, Sedova M, Miao X, Reed B, Sabina J, Feierstein E, Schorn M, Alanjary M, Dimalanta E, Dressman D, Kasinskas R, Sokolsky T, Fidanza JA, Namsaraev E, McKernan KJ, Williams A, Roth GT, Bustillo J. An integrated semiconductor device enabling non-optical genome sequencing. Nature. 2011;475(7356):348–52. doi:10.1038/nature10242.
126. Bentley DR, Balasubramanian S, Swerdlow HP, Smith GP, Milton J, Brown CG, Hall KP, Evers DJ, Barnes CL, Bignell HR, Boutell JM, Bryant J, Carter RJ, Keira Cheetham R, Cox AJ, Ellis DJ, Flatbush MR, Gormley NA, Humphray SJ, Irving LJ, Karbelashvili MS, Kirk SM, Li H, Liu X, Maisinger KS, Murray LJ, Obradovic B, Ost T, Parkinson ML, Pratt MR, Rasolonjatovo IM, Reed MT, Rigatti R, Rodighiero C, Ross MT, Sabot A, Sankar SV, Scally A, Schroth GP, Smith ME, Smith VP, Spiridou A, Torrance PE, Tzonev SS, Vermaas EH, Walter K, Wu X, Zhang L, Alam MD, Anastasi C, Aniebo IC, Bailey DM, Bancarz IR, Banerjee S, Barbour SG, Baybayan PA, Benoit VA, Benson KF, Bevis C, Black PJ, Boodhun A, Brennan JS, Bridgham JA, Brown RC, Brown AA, Buermann DH, Bundu AA, Burrows JC, Carter NP, Castillo N, Chiara ECM, Chang S, Neil Cooley R, Crake NR, Dada OO, Diakoumakos KD, Dominguez-Fernandez B, Earnshaw DJ, Egbujor UC, Elmore DW, Etchin SS, Ewan MR, Fedurco M, Fraser LJ, Fuentes Fajardo KV, Scott Furey W, George D, Gietzen KJ, Goddard CP, Golda GS, Granieri PA, Green DE, Gustafson DL, Hansen NF, Harnish K, Haudenschild CD, Heyer NI, Hims MM, Ho JT, Horgan AM, Hoschler K, Hurwitz S, Ivanov DV, Johnson MQ, James T, Huw Jones TA, Kang GD, Kerelska TH, Kersey AD, Khrebtukova I, Kindwall AP, Kingsbury Z, Kokko-Gonzales PI, Kumar A, Laurent MA, Lawley CT, Lee SE, Lee X, Liao AK, Loch JA, Lok M, Luo S, Mammen RM, Martin JW, McCauley PG, McNitt P, Mehta P, Moon KW, Mullens JW, Newington T, Ning Z, Ling Ng B, Novo SM, O'Neill MJ, Osborne MA, Osnowski A, Ostadan O, Paraschos LL, Pickering

L, Pike AC, Chris Pinkard D, Pliskin DP, Podhasky J, Quijano VJ, Raczy C, Rae VH, Rawlings SR, Chiva Rodriguez A, Roe PM, Rogers J, Rogert Bacigalupo MC, Romanov N, Romieu A, Roth RK, Rourke NJ, Ruediger ST, Rusman E, Sanches-Kuiper RM, Schenker MR, Seoane JM, Shaw RJ, Shiver MK, Short SW, Sizto NL, Sluis JP, Smith MA, Ernest Sohna Sohna J, Spence EJ, Stevens K, Sutton N, Szajkowski L, Tregidgo CL, Turcatti G, Vandevondele S, Verhovsky Y, Virk SM, Wakelin S, Walcott GC, Wang J, Worsley GJ, Yan J, Yau L, Zuerlein M, Mullikin JC, Hurles ME, McCooke NJ, West JS, Oaks FL, Lundberg PL, Klenerman D, Durbin R, Smith AJ. Accurate whole human genome sequencing using reversible terminator chemistry. Nature. 2008;456(7218):53–9.

127. Sikkema-Raddatz B, Johansson LF, de Boer EN, Almomani R, Boven LG, van den Berg MP, van Spaendonck-Zwarts KY, van Tintelen JP, Sijmons RH, Jongbloed JD, Sinke RJ. Targeted next-generation sequencing can replace Sanger sequencing in clinical diagnostics. Hum Mutat. 2013;34(7):1035–42. doi:10.1002/humu.22332.
128. Becker K, Vollbrecht C, Koitzsch U, Koenig K, Fassunke J, Huss S, Nuernberg P, Heukamp LC, Buettner R, Odenthal M, Altmueller J, Merkelbach-Bruse S. Deep ion sequencing of amplicon adapter ligated libraries: a novel tool in molecular diagnostics of formalin fixed and paraffin embedded tissues. J Clin Pathol. 2013;66(9):803–6. doi:10.1136/jclinpath-2013-201549.
129. Desai AN, Jere A. Next-generation sequencing: ready for the clinics? Clin Genet. 2012;81(6): 503–10. doi:10.1111/j.1399-0004.2012.01865.x.
130. Srinivasan M, Sedmak D, Jewell S. Effect of fixatives and tissue processing on the content and integrity of nucleic acids. Am J Pathol. 2002; 161(6):1961–71. doi:10.1016/S0002-9440(10)64472-0.
131. Greer CE, Wheeler CM, Manos MM. Sample preparation and PCR amplification from paraffin-embedded tissues. PCR Methods Appl. 1994;3(6):S113–22.
132. Dubeau L, Chandler LA, Gralow JR, Nichols PW, Jones PA. Southern blot analysis of DNA extracted from formalin-fixed pathology specimens. Cancer Res. 1986;46(6):2964–9.
133. Marchetti A, Felicioni L, Buttitta F. Assessing EGFR mutations. N Engl J Med. 2006;354(5): 526–8. doi:10.1056/NEJMc052564. author reply 526–528.
134. Do H, Dobrovic A. Dramatic reduction of sequence artefacts from DNA isolated from formalin-fixed cancer biopsies by treatment with uracil-DNA glycosylase. Oncotarget. 2012;3(5): 546–58.
135. Zsikla V, Baumann M, Cathomas G. Effect of buffered formalin on amplification of DNA from paraffin wax embedded small biopsies using real-time PCR. J Clin Pathol. 2004;57(6):654–6.
136. Guillou L, Coindre J, Gallagher G, Terrier P, Gebhard S, de Saint Aubain Somerhausen N, Michels J, Jundt G, Vince DR, Collin F, Trassard M, Le Doussal V, Benhattar J. Detection of the synovial sarcoma translocation t(X;18) (SYT;SSX) in paraffin-embedded tissues using reverse transcriptase-polymerase chain reaction: a reliable and powerful diagnostic tool for pathologists. A molecular analysis of 221 mesenchymal tumors fixed in different fixatives. Hum Pathol. 2001;32(1):105–12.
137. Merkelbach S, Gehlen J, Handt S, Fuzesi L. Novel enzyme immunoassay and optimized DNA extraction for the detection of polymerase-chain-reaction-amplified viral DNA from paraffin-embedded tissue. Am J Pathol. 1997; 150(5):1537–46.
138. de Franchis R, Cross NC, Foulkes NS, Cox TM. A potent inhibitor of Taq polymerase copurifies with human genomic DNA. Nucleic Acids Res. 1988;16(21):10355.
139. Lindeman NI, Cagle PT, Beasley MB, Chitale DA, Dacic S, Giaccone G, Jenkins RB, Kwiatkowski DJ, Saldivar JS, Squire J, Thunnissen E, Ladanyi M. Molecular testing guideline for selection of lung cancer patients for EGFR and ALK tyrosine kinase inhibitors: guideline from the College of American Pathologists, International Association for the Study of Lung Cancer, and Association for Molecular Pathology. J Mol Diagn. 2013;15(4):415–53. doi:10.1016/j.jmoldx.2013.03.001.
140. Penzel R, Sers C, Chen Y, Lehmann-Muhlenhoff U, Merkelbach-Bruse S, Jung A, Kirchner T, Buttner R, Kreipe HH, Petersen I, Dietel M, Schirmacher P. EGFR mutation detection in NSCLC—assessment of diagnostic application and recommendations of the German Panel for Mutation Testing in NSCLC. Virchows Arch. 2011;458(1):95–8. doi:10.1007/s00428-010-1000-y.
141. Ney JT, Froehner S, Roesler A, Buettner R, Merkelbach-Bruse S. High-resolution melting analysis as a sensitive prescreening diagnostic tool to detect KRAS, BRAF, PIK3CA, and AKT1 mutations in formalin-fixed, paraffin-embedded tissues. Arch Pathol Lab Med. 2012;136(9): 983–92. doi:10.5858/arpa.2011-0176-OA.
142. Molina-Vila MA, Bertran-Alamillo J, Reguart N, Taron M, Castella E, Llatjos M, Costa C, Mayo C, Pradas A, Queralt C, Botia M, Perez-Cano M, Carrasco E, Tomas M, Mate JL, Moran T, Rosell R. A sensitive method for detecting EGFR mutations in non-small cell lung cancer samples with few tumor cells. J Thorac Oncol. 2008;3(11): 1224–35. doi:10.1097/JTO.0b013e318189f579.
143. Choi M, Scholl UI, Ji W, Liu T, Tikhonova IR, Zumbo P, Nayir A, Bakkaloglu A, Ozen S, Sanjad S, Nelson-Williams C, Farhi A, Mane S, Lifton RP. Genetic diagnosis by whole exome capture and massively parallel DNA sequencing. Proc

Natl Acad Sci U S A. 2009;106(45):19096–101. doi:10.1073/pnas.0910672106.

144. Gnirke A, Melnikov A, Maguire J, Rogov P, LeProust EM, Brockman W, Fennell T, Giannoukos G, Fisher S, Russ C, Gabriel S, Jaffe DB, Lander ES, Nusbaum C. Solution hybrid selection with ultra-long oligonucleotides for massively parallel targeted sequencing. Nat Biotechnol. 2009;27(2):182–9. doi:10.1038/nbt.1523.

145. Spencer DH, Sehn JK, Abel HJ, Watson MA, Pfeifer JD, Duncavage EJ. Comparison of clinical targeted next-generation sequence data from formalin-fixed and fresh-frozen tissue specimens. J Mol Diagn. 2013;15(5):623–33. doi:10.1016/j.jmoldx.2013.05.004.

146. Seyed F, Grunenewald H, Caruccio N. Optimized library preparation method for next-generation-sequencing. Nat Methods. 2009;6:I–II.

147. ElSharawy A, Warner J, Olson J, Forster M, Schilhabel MB, Link DR, Rose-John S, Schreiber S, Rosenstiel P, Brayer J, Franke A. Accurate variant detection across non-amplified and whole genome amplified DNA using targeted next generation sequencing. BMC Genomics. 2012;13:500. doi:10.1186/1471-2164-13-500.

148. Ding F, Manosas M, Spiering MM, Benkovic SJ, Bensimon D, Allemand JF, Croquette V. Single-molecule mechanical identification and sequencing. Nat Methods. 2012;9(4):367–72. doi:10.1038/nmeth.1925.

149. Vollbrecht C, Konig K, Heukamp L, Buttner R, Odenthal M. Molecular pathology of the lungs. New perspectives by next generation sequencing. Pathologe. 2013;34(1):16–24. doi:10.1007/s00292-012-1704-7.

CHAPTER 25

Genomic Applications in Colorectal and Pancreatic Tumors

ANDREAS JUNG, RALPH HRUBAN

Introduction 415
Molecular Classification of Colorectal Cancer 416
Current Molecular Genetic and Genomic Tests in Colorectal Cancer 423
Molecular Classification of Pancreatic Cancer 427
Conclusions 428
Conflict of Interest 429
Glossary 429

Introduction

Classification can be defined as an arrangement of entities in a hierarchical series of nested classes, in which similar or related classes at one hierarchical level are combined comprehensively into more inclusive classes at the next higher level. A class is defined as a collection of similar entities [1]. Classifications help to make complex systems easier to understand and are therefore a very convenient tool in complex biological systems. In tumor biology or oncology the UICC (Union for International Cancer Control) classification, introduced by the French surgeon Pierre Denoix in the 1940s, was the basis for oncologic therapies in practice nowadays. However, the prognostic value of the UICC classification system is still unsatisfactory as approximately 30 % of patients do not benefit from clinical decisions based upon the recommendations of the UICC classification. One of the reasons may be that the UICC classification is mainly based on the descriptive TNM (tumor, nodes, metastasis) classification [2] and does not account for the variations in tumor biology which are caused by many molecular alterations that occur in cancer cells. These can be summarized into some essential pathways [3] which are well known from developmental systems [4]. From this point of view a molecular classification of cancers might be a more genuine approach for the classification of human tumors ultimately leading to more effective treatments.

Tumors are caused by genetic [5] and or epigenetic alterations [6]. These alterations affect oncogenes (gain of function) or tumor suppressor genes (loss of function) which were shown in a plethora of experimental approaches to constitute the driving force of tumor development. In a *simple model* (Table 25.1), alterations in both groups of genes in cancer cells alone are sufficient to explain the biologic behavior of a tumor based on the fact that these alterations change the functional spectrum of the cancer cells or their capabilities either actively (what the cancer cell can do) or passively (how cancer cells react to signals from the tumor environment). In such a model, it is sufficient to analyze genomes and/or

G.J. Netto and I. Schrijver (eds.), *Genomic Applications in Pathology*,
DOI 10.1007/978-1-4939-0727-4_25, © Springer Science+Business Media New York 2015

Table 25-1 **Approaches for the molecular classification of tumors.** A simple model assumes that tumors are driven by tumor intrinsic alterations. Due to (epi)-genetic alterations tumor cells gain the capability to be regulated by stroma cells. In this view it is sufficient to analyze tumor-specific alterations which can be found by screening for mutations as well as the epigenetic landmarks in the genome of the tumor. A more complex model assumes that there is an active interaction of cancer cells with other cells in the environment of the tumor as well as the extracellular matrix (ECM) resulting in the alteration of the functional behavior of the cancer cells. (Epi)-genome-wide analyses are, in this model, only part of the complete picture but are insufficient by themselves as they do not consider the stromal component. Therefore, in this approach, transcriptomes (mRNA, miRNA, noncoding RNA) or the proteome including especially posttranslational modifications are additional tools in the molecular classification of tumors

	Molecular classification	
Model	*Simple*	*Complex*
Driving force	Tumor intrinsic	Tumor stroma interaction
Concept	Cancer is a disease of the genome	Cancer is a complex disorder of genetic alterations and dysregulation of cell-to-cell interactions
Investigation	Genome	
	Epigenome	
	Transcriptome	
	mRNA	
	miRNAome	
	Noncoding RNA	
	Proteome	
	Including: posttranslational modifications	

epigenomes of cancer cells to get a reliable molecular classification. In a *complex model* (Table 25.1), the (epi)-genetic alterations of cancerous cells are only part of a complex dysregulated network which results in the gain of the hallmarks of cancer that are essential for tumor development [7, 8]. These hallmarks are defined by functional alterations in core signaling pathways of living cells. Consequently, a more complex molecular classification is required in this model. The analysis of transcriptomes (mRNA, miRNA, ncRNA) as well as the proteome, taking into account the posttranslational modifications, are additional important tools in this approach [9].

Molecular Classification of Colorectal Cancer

Colorectal cancer is one of the most common tumors in the Western world. Colorectal cancers comprise inherited familial forms accounting for approximately 15 % of cases and a sporadic form. The genetic syndromes are known as:

- FAP (familial adenomatous polyposis),
- HNPCC (hereditary nonpolyposis colorectal cancer, the manifestation of Lynch syndrome in the colorectum),
- MUTYH (MYH) syndrome [5].

Functional/Biological Classification of Colorectal Cancer

“CANCER STEMNESS” COLORECTAL CANCER

Vogelstein and his group almost single handedly unraveled the genetic underpinnings of FAP syndrome associated colorectal cancer [10]. Their work led to the delineation of the multistep carcinogenesis model of colorectal cancer (Fig. 25.1) [11]. This model brought to the forefront several important principles:

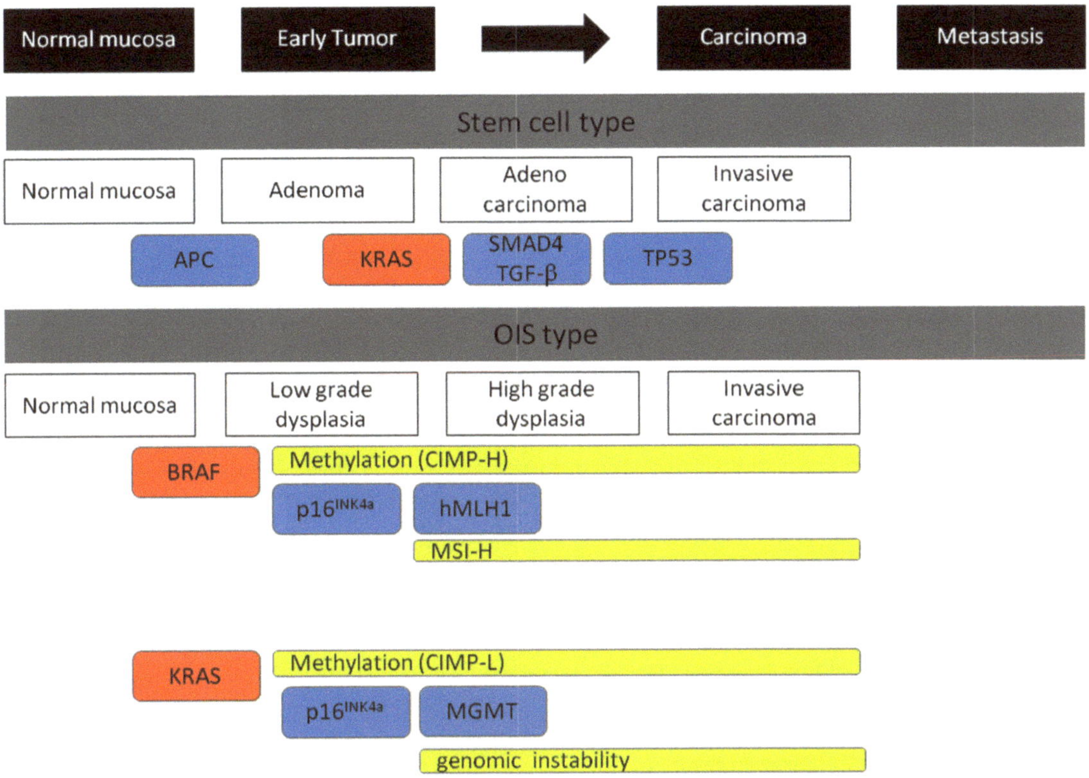

Figure 25-1 Carcinogenesis models of the stem cell- and OIS types of human colorectal cancer. During the progression from normal mucosa to invasive carcinoma, early tumors accumulate mutations that change the capability of cancer cells. Colorectal carcinoma might be separated into at least two different diseases which differ with respect to their way into carcinogenesis, prognosis, and response to 5-FU. The *stem cell type* arises from mutations in the gatekeeper gene *APC*. This is classically followed by mutations in the *KRAS* oncogene, elements of the TGF-β pathway and *TP53* (the guardian of the genome). The mutation in the *APC* gene leads to a Wnt signal independent and thus dysregulated transcriptional activity of β-catenin which shifts cancer cells into a state of cancer stemness. Alternatively, other genes can be involved resulting in an equivalent activation of the signaling pathway. The *OIS type* arises from mutations in an oncogene, most frequently *BRAF* and occasionally *KRAS*. As a consequence the cell cycle is arrested by the upregulation of *CDKN2A* (which encodesp16^{INK4a}) inducing a state of senescence. In parallel the DNA-methyltransferases (DNMTs) are upregulated resulting in a more (CIMP-H) or less (CIMP-L) heavy methylation of the genome. When *CDKN2A* is hit the cell cycle arrest is released and cancer cells start to proliferate. In parallel, the expression of caretaker genes like *MLH1* or *MGMT* is also affected resulting in a genome-wide instability of these tumors. As secondary events, signal pathways that are shared with the stem cell type are [e.g., the Wnt signaling pathway (APC)] are also affected in the OIS type. However, the OIS type does not switch into the stem cell type of colorectal cancer indicating that the type of gatekeeper mutation as well as the temporal acquisition of mutations are important elements in the tumorigenesis process. *Blue* tumor-suppressor gene, *red* oncogene, *yellow* affected function

1. Cancer is a disease of the genome,
2. The accumulation of mutations in oncogenes like *KRAS* and tumorsuppressor genes such as *APC*, *SMAD4* (chromosome 18 related genes) and *p53* (the guardian of the genome) can be associated with distinct progression steps in the tumorigenesis of colorectal cancer,
3. The same mutations are also present in sporadic forms of colorectal cancer.

Here, the tumorsuppressor gene *APC* (adenomatous polyposis coli)—which is the mutated inherited gene in the FAP syndrome—has a pivotal role and was thus named the gatekeeper of colorectal cancer [10]. The functional role of APC was elucidated by the investigation of genetically engineered mouse (GEM) and cell culture models. When inactivating the function of APC in the intestinal tract of adult mice by homologous

Table 25-2 **Molecular classification of colorectal cancers.** On the basis of the initial event (gatekeeper) two different pathways of colorectal carcinogenesis can be distinguished. Alterations in the tumor suppressor gene *APC* result in the induction of a cancer stem cell-like phenotype which is driven by the activity of the transcriptional factor β-catenin resulting in a change of the genetic profile of cancer cells as well as phenotypic and functional changes. Activation of the *BRAF* and, to a lesser extent, the *KRAS oncogenes result in a strong or weak activation of DNA methylation respectively which leads to the development of CIMP which, in turn, results in the loss of gene expression. Here, mainly tumor suppressor genes like the cell cycle inhibitor CDKN2A and/or caretaker genes like MLH1 or MGMT are affected, leading to genomic instability*

Type	Stemness	Oncogene-induced senescence	
Initial event (*gatekeeper gene alteration*)	APC	BRAF	KRAS
Mechanism	β-catenin-dependent gene expression	Methylation CIMP-H	CIMP-L
Impact on	*Proliferation*: cMyc [17], cyclin D_1 [18, 19], $p16^{INK4a}$ [20]	$p16^{INK4a\text{-}off}$ → proliferation	
	Infinite replication: TERT [21, 22] *EMT*: Snail [23], Twist [24], ZEB1 [25], fibronectin [26], vimentin [27] *Migration, invasion*: MMP7 [28], laminin-5γ2 [29], MT-MMP-1 [24], uPA [30], uPAR [31], tenascin C [32] *Cancer stem cell marker*: CD44 [33], CD133 [34], CD166 [34], LGR5 [35]	$MLH1^{off}$ MSH-H	$MGMT^{off}$ DNA instability
Clinical behavior	Bad prognosis	Good prognosis	

recombination of the *APC* gene, affected animals developed adenomas in their intestine [12]. Moreover, as proposed by the multistep carcinogenesis model, the combination of mutations in more than one driver gene like *APC* and *KRAS* [13] or *APC* and *TP53* [14] resulted in more aggressive forms of colorectal cancers.

The APC protein is the central regulatory element of the canonical WNT/β-catenin signaling pathway which regulates the cellular stability of β-catenin [15]. β-catenin is an ambivalent protein [16, 131]. On the one hand it supports, as an integral component of the zonula adherens together with E-cadherin, the epithelial phenotype. On the other hand β-catenin acts in the nucleus as a transcription factor (Table 25.2). In this context, β-catenin induces and maintains stemness of adult stem cells in the intestinal crypts [15] as well as the stemness of cancer stem cells [36, 37]. β-catenin is also responsible for the induction of EMT (epithelial-mesenchymal transition) [38, 132] which is known to induce cancer stemness [39]. Thus, nuclear β-catenin is also an essential prerequisite for metastasis [36, 40] (Table 25.2). As expected, colorectal cancers which are characterized by markers of stemness as well as nuclear β-catenin develop distant metastases at a very high rate [41].

"ONCOGENE-INDUCED SENESCENCE (OIS)" COLORECTAL CANCER

Colorectal cancers arising in the setting of HNPCC (Lynch syndrome) are characterized by the presence of genetic alterations in one of the caretaker genes of the DNA mismatch repair system (MMR); primarily *MLH1*, *MSH2*, *MSH6*, or *PMS2* [5]. Cancer cells with such alterations are no longer able to repair defects in short repetitive sequences known as microsatellites. Microsatellite defects frequently occur during DNA replication and during genome maintenance. As a result, colorectal cancers in Lynch syndrome patients are characterized by accumulation of microsatellite mutations and are designated as microsatellite instability-high (MSI-H) in contrast to colorectal tumors that are characterized by microsatellite stability such as tumors of the stemness type which are termed microsatellite stable (MSS) [42]. Because microsatellites are also found in the coding regions of genes [43] this defect results in a hypermutated phenotype [3, 9, 44].

In contrast to the parallelism of *APC* gene mutations in inherited FAP and sporadic cases, mutations in the DNA mismatch repair genes have not been detected in sporadic forms of MSI-H cancers. Instead in sporadic MSI-H colorectal cancers, methylation of the promoter/exon 1 region of the *MLH1* gene rather than mutation of the gene is responsible for the downregulation of its transcriptional activity [45]. This difference can be used to differentiate sporadic MSI-H colorectal cancers from HNPCCs [45]. Besides methylation of the *MLH1* gene many other genes are also found to be methylated in their regulatory promoter/exon 1 regions which are enriched for CG nucleotides (CpG islands). As the mammalian DNA-methyltransferases (DNMT-1, DNMT-3a, DNMT-3b) methylate the cytosine base in CG dinucleotides, CpG islands are frequently methylated in MSI-H cancers. This phenotype was therefore named CIMP (CpG island methylator phenotype).

The underlying molecular mechanisms of the CIMP were unlocked using GEM models. Activating mutation of the protooncogene *BRAF* in the adult mouse intestine initially was found to be associated with a significant increase in the proliferation rate and growth of the normal mucosa. Affected cells appeared to subsequently counteract this oncogenic activation by the induction of senescence, a process known as oncogene-induced senescence (OIS), that is usually mediated by the upregulation of the cell cycle inhibitor p16^{INK4a} encoded by the *CDKN2A* gene [46]. In the GEMs the resulting OIS was spontaneously followed by the neoplastic transformation of the cancer cells leading to the development of the invasive carcinoma phenotype [47, 48]. The latter neoplastic transformation is accompanied by upregulation of *DNMT* transcriptional activity which results in the methylation of the promoter/exon 1 region of the *CDKN2A* gene and many other genes (CIMP). The epigenetic silencing and loss of the *CDKN2A* expression release the blockade of cell cycle progression [47]. Interestingly, the above complex process was not induced when *KRAS* oncogene activation, rather than *BRAF*, was introduced in the mouse intestinal model [49]. These findings are in line with the observation that *BRAF* mutations are common (46 %) in sporadic hypermutated colorectal cancers of the MSI-H type whereas *KRAS* mutations are mostly encountered in MSS colorectal cancers (43 %) [3]. Interestingly, colorectal tumors that arise via the OIS molecular mechanism display a very characteristic early growth pattern were the glands acquire a serrated architectural pattern. Thus, this type of colorectal cancer has been referred to as "serrated" [50, 51].

Clinically, sporadic MSI-H tumors also differ from their MSS relatives. When comparing the overall survival of patients with MSI-H and MSS colorectal cancers it became evident that patients with MSI-H tumors enjoy a significantly longer survival. When comparing the distribution of MSI-H and MSS cancers with regards to pathologic stage, MSI-H tumors are underrepresented in the stage IV category (27.9 % MSS, 7.9 % MSI-H) [41, 52]. Likewise, only a minority of MSI-H are associated with liver metastases (less than 2.7 %) [53]. This tendency of MSI-H colon cancers to be of earlier stage has argued for adopting a less-intensive therapy protocol. Furthermore, MSI-H colorectal cancers are less likely to be responsive to 5-fluorouracil (5′ FU) monotherapy as, for example, shown in the prospective trial by Jover et al. [54] (Table 25.2). This clinical resistance to 5′ FU therapy has been substantiated at the experimental level. Colorectal cell lines with a defect in the mismatch repair system due to loss of expression of *MLH1* show resistance against 5FU treatment. This resistance is abrogated upon recovering *MLH1* gene expression by transfection strategy [55].

The loss of expression of the *MLH1* gene is due to methylation of its promoter/exon 1 region. Thus, there is an overlap between MSI-H and the CIMP [45] (Fig. 25.1). Given the random nature of DNA methylation throughout different genes, some CIMP colorectal cancers do not necessarily undergo methylation of the *MLH-1* gene and therefore are not MSI-H. The latter accounts for the incomplete overlap of CIMP and MSI-H colorectal cancers [50, 56] and led to the generation of several CIMP/MSI subgroups in the Jass classification system [50]. Additionally, not all MSI-H colorectal cancers show a mutation of *BRAF* [56] indicating that other oncogenes besides *BRAF* might be responsible for the induction of OIS.

MUTYH/MYH COLON CANCER

Germline mutations in the *MUTYH* gene are associated with a recessive form of familial adenomatosis in which *APC* germline mutation is lacking. These patients develop significantly fewer polyps compared to the classical FAP syndrome. Mutations in the *MUTYH* gene result in the occurrence of transversions (C-G→A-T) which eventually also affects the *APC* gene itself. The *MYTYH* gene encodes a DNA glycosylase that excises adenine bases from the DNA backbone as an integral part of the DNA base excision repair system [57, 58].

OTHER TYPES OF COLORECTAL CANCER

Mechanisms other than methylation of the *MLH-1* or *MGMT* genes have been unveiled that are responsible for a mutator genotype in a subset of colorectal cancers. This is exemplified by the occurrence of *POL-E* (DNA polymerase ε) gene mutations [3] which might lead to errors during DNA replication and subsequent accumulation of mutations. These tumors show a hypermethylator phenotype as their mutation rate is up to 50 fold higher (100–500 mutations per million bp) than in MSI-H tumors (10-100 mutations per million bp) [3]. Given that this type of mutation has only been identified recently, the clinical and pathological characteristics including morphologic features, survival and response to therapies are not well defined at this time.

Genotype-Based Classification of Colorectal Cancer: Next-Generation Sequencing

The development of next-generation sequencing (NGS) platforms have made it possible to achieve a more detailed genotyping of a larger number of colorectal tumors [3, 44, 59–61]. Large-scale whole exome sequencing (WES) studies have not only allowed for the discovery of common patterns in colon cancer genetic alterations but also elucidated the significant variations in the set of genetic mutations underlying a given colon cancer on an individual patient level. A genetic landscape of colon cancer has thus emerged where repeated alterations can be depicted as hills and mountains based on their frequency of occurrence [9, 61]. Mountains were represented by mutations that are found in the majority of investigated tumors whereas hills represented low frequency mutations found in fewer tumors. A reassuring finding of WES studies in colon cancer is the fact that all four of the originally identified genes implicated in the above described multistep carcinogenesis model (*APC, KRAS, SMAD4, TP53*) represented salient mountains in the colorectal cancer genetic landscape. On average, an individual colorectal cancer contained approximately 50–70 events. Evidently, not all of these mutations are equally crucial for cancer development. Some represented polymorphisms or were found in regions of the genome without known attributable function. Such mutations were referred to as "passenger mutations," in contrast to those alterations that involved genes which were expected to be essential for the tumor progression and were thus named "driver mutations" [9]. It is these driver mutations that are usually represented in the mountains and hills of the landscape. In colorectal cancer, seven to eight such driver mutations are operational in a given tumor [9, 61]. Furthermore, the driver mutations could be assigned to 12 signaling pathways which regulate the three

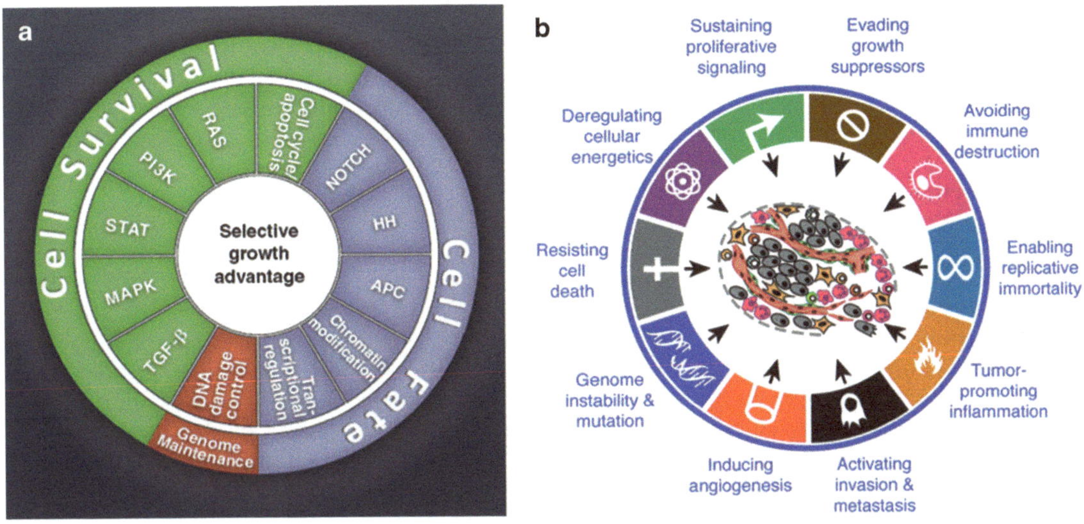

Figure 25-2 Core cellular processes and hallmarks of cancer. (**a**) The three core cellular processes (1) cell survival, (2) cell fate, and (3) genome maintenance (outer ring) are regulated by 12 signaling pathways. (**b**) For the full transformation of a tumor each of the indicated 10 hallmarks of cancer must be activated. For some of them, such as replicative immortality there is only a limited number of mechanisms present (telomerase activation or ALT—alternative lengthening of telomeres) whereas others such as sustaining proliferative signaling can be activated by a multitude of different genes and thus mechanisms. (**a**) from [9], (**b**) from [8]

core cellular processes: cell fate (*APC*), cell survival (*KRAS*, *SMAD4*), and genome maintenance (*TP53*) (Fig. 25.2) [9].

Recent findings from the unsupervised clustering of the mutational data of 276 colorectal tumors [3] revealed two subgroups of colorectal cancers which differ with respect to their mutation rates. The larger group (85 %) displayed a mutation rate of less than 8.24 per 10^6 bp (median: 54 mutations per tumor) whereas the other group (15 %) was characterized by more than 12 mutations/10^6 bp (median: 728 mutations per tumor). The latter group was defined by silencing of the *MLH-1* or *POLE* genes. The two groups were named non-hypermutated and hypermutated, respectively. The most frequently mutated genes in the non-hypermutated group were *APC*, *p53*, *KRAS*, *PI3KCA*, *FBXW7*, *SMAD4*, *TCF7L2*, *NRAS*, *CTNNB1* (*β-Catenin*), *SMAD2*, *WTX* (*FAM123B*), and *SOX9*. The mutations in these genes resulted in either a change in an amino acid or a nonsense mutation. These tumors also harbor chromosomal and subchromosomal alterations like gains in the chromosome arms 1q, 7p, 7q, 8p, 8q, 12q, 13q, 19q, 20p, and 20q as well as deletions in 1p, 4q, 5q, 14q, 15q, 17p, 17q, 18p, 18q, 10, and 22q [62]. The subchromosomal alterations affected *FHIT*, *RBFOX1*, *SMAD3*, *SMAD4*, *APC*, *PTEN*, *WWOA*, and *TCF7L2* by deletion and *IGF2*, *TH* (*tyrosine hydroxylase*), *ASCL2*, *WHSC1L1*, *MYC*, *ERBB2*, and *mir-483* by amplification. Translocations were found for *R-Spondin*, a gene involved in the Wnt-signaling pathway [44]. A different spectrum of mutations was present in the hypermutated group of colorectal cancers. Here, *ACVR2A*, *APC*, *TGFBR2*, *BRAF*, *MSH3*, *MSH6*, *MYO1B*, *TCF7L2*, *CASP8*, and *SLC9A9* were mutated in the majority of cases.

The above findings points to the fact that colorectal tumors in the two groups develop through different sequences of genetic events. One of the major challenges is now to identify experimental and/or bioinformatic algorithms that will elucidate the role of this multitude of genetic alterations. This is especially the case for mutations in the hypermutated cancer group where it remains difficult to categorize a given mutation as occurring by chance (passenger) or as being a driver mutation [63], as was shown for the *TCF7L2* gene [64]. At this point, classifying the found genetic alterations per their biologic pathways could be of help. Thereby mutations found in one tumor are assigned to one or more pathways where the gene product is known to have a fundamental role. Adopting such an approach, it became evident that the Wnt, MAPK and PI3K, TGFβ and p53 pathways were the leading pathways affected in colorectal cancers independent of their hypermutated or non-hypermutated subgroup assignment. An intriguing finding in the majority of tumors was the occurrence of *MYC* activation, a known target of the Wnt-signaling [17] pathway that plays a pivotal role not only in colorectal cancer but in most human tumors [65].

As sequencing technologies continue to evolve in efficiency and accuracy and become lower in cost, the genomic landscape of colon cancer will be further brought into focus. The 1000 genome project and other large-scale initiatives might help shed more light on the genomic alterations of colon cancer and help further classify tumors based on alterations in signaling pathway activities. Genome-wide analysis of colorectal cancer will provide a plethora of data that can be mined in relation to clinicopathologic correlates such as response to specific therapies, or presence/absence of metastases, and can help address pressing management issues such as the need for adjuvant therapy in stage II disease.

Classification of Colorectal Cancer on the Basis of Expression Data

In the late 1990s Robert Weinberg introduced a novel concept for understanding carcinogenesis from a functional perspective. By interrogating the steps required for a normal cell to become neoplastically transformed, Weiberg and his group delineated core signaling pathways involved [66]. To further account for the fact that tumors are complex structures of cancer cells, intratumoral host cells, and the extracellular matrix (ECM) the core signaling pathways were modified to a system of functional classification that includes cancer cell intrinsic functions as well as their reciprocal interactions with the environment. The classes of this classification system are known as the hallmarks of cancer [7, 8]. This concept introduced a paradigm shift from an organ-specific classification system for cancer to a functionally based classification. The model anticipated that tumors with comparable genetic signaling alterations can be treated by the same class of targeted therapeutics. A vivid example of such approach is the treatment of mammary as well as gastric carcinomas with the anti-Her2-targeted antibody trastuzumab (Herceptin®) based on amplification of Her2 in both tumor types [67, 68]. Because the hallmark model is based on functional classes, data from expression profiling (transcriptome, proteome) can also be used as it is possible to assign changes in the expression levels of genes to pathways, or the hallmarks, independent of the knowledge of the (epi)-genetic status of the gene.

Approximately 100 transcriptome analyses of colorectal cancer have been completed with results publicly available in the Genome Omnibus (GEO; http://www.ncbi.nlm.nih.gov/geo/). Eight of these involved more than 100 cases each resulting in a total of 1,181 cases analyzed where annotated clinicopathological correlates were also available. Like other expression profile studies, these studies suffer from the major limitation inherent in transcriptome and proteome analysis, namely the lack of robust reproducibility. Different sets of differentially expressed genes have been found by different groups, which most likely is due to technical factors affecting the reproducibility of gene expression data. In contrast to relatively stable epigenetic and genetic alterations, variations in decay rate of RNA and protein or posttranslational modifications pose major technical hurdles to expression analysis. These have to be considered during probe sampling and in the interpretation of aggregate data [69]. Another important factor is the plasticity of tumors and cancer cells [37, 40, 70, 71]. Thus, different areas of the same tumor are differentially influenced by the environment and as a consequence may give different expression result output [72]. All these limitations render gene expression data sets in colon cancer of limited value at this time.

Current Molecular Genetic and Genomic Tests in Colorectal Cancer

Evaluation of Mismatch Repair Defective Colorectal Cancer and Identification of Lynch Syndrome

As discussed above, loss-of-function defects in MMR genes occur in approximately 15 % of all colorectal cancers. These tumors display characteristically high levels of microsatellite instability that can be exploited in the identification of this molecular subset of colorectal tumors. Most MMR-deficient (dMMR) colorectal cancers are sporadic and identifying them carries a prognostic (i.e., better prognosis stage by stage) and therapeutic (i.e., resistance to 5-fluorouracil chemotherapy) significance. The grave implications of identifying the less prevalent but equally important subset (15–20 %) of familial dMMR colorectal cancers that occur as part of inherited Lynch [hereditary non-polyposis colon cancer (HNPCC)] or Muir-Torre syndromes are self-evident. High penetrance of colorectal cancer in carriers of a germline MMR gene mutation further underlines the importance of accurately recognizing index tumors. Several approaches have been proposed to facilitate such identification, including family-based (Amsterdam II criteria) [73], patient/family-based (Revised Bethesda criteria) [74], morphology-based [75], microsatellite-based, and immunohistochemistry (IHC)-based screening strategies [76]. None of these individually detect all germline mutation carriers [45]. Thus, algorithms including multiple concurrent test strategies have to be used and screening should be applied to all patients with newly diagnosed colorectal cancers if proper management of dMMR cancers and comprehensive screening of Lynch syndrome is to be achieved. Such an approach would assure increased detection of inherited and de novo germline mutations to guide family screening. A recent white paper issued by the Association for Molecular Pathology (AMP) outlines such recommendations and recommends that all new colorectal cancers should be classified into sporadic MMR-proficient, sporadic dMMR, or Lynch dMMR subgroups [45].

DETECTION OF SPORADIC dMMR COLORECTAL CARCINOMA

Sporadic dMMR colorectal adenocarcinoma usually presents in the proximal colon and cecal region. It is more common in older patients and has a female predilection. Morphologically these cancers are characterized by an expansive growth pattern, mucinous differentiation, presence of tumor-infiltrating lymphocytes, and absence of intraglandular "dirty" necrosis. The presence of three of the above features will identify sporadic dMMR cancers with a sensitivity of 98 % and a specificity of 48 %. Similarly, MSI-H colorectal cancers can be predicted by the presence of any one of seven clinicopathologic features including: advanced age, proximal location, and the histomorphologic features described above.

Immunohistochemical evaluation of MMR proteins and MSI evaluation are helpful approaches in identifying sporadic dMMR colorectal cancers. Loss of MLH1 protein is found in 96 % of such tumors whereas MSI-H is encountered in 99 % of cases [77, 78]. Most importantly, the two methods complement each other: all colorectal cancers with promotor hypermethylated *MLH1* could be identified by combining the two methods [79, 80]. Therefore, for accurate detection of sporadic dMMR colorectal cancer, both MSI and MMR protein immunohistochemical evaluation should be adopted in combination. Finally, a *BRAF* c.1799T>A mutation is expected in 46 % of sporadic dMMR cancers [3]. Therefore, the presence of this mutation virtually excludes Lynch syndrome and supports the diagnosis of sporadic dMMR. It is important to remember, however, that the absence of the *BRAF* c.1799T>A mutation only increases the likelihood of Lynch syndrome, a diagnosis that will require supporting clinical features and family history in addition to confirmatory sequencing and/or deletion germline testing [45].

DETECTION OF FAMILIAL dMMR COLORECTAL CARCINOMA (LYNCH SYNDROME)

MSI testing is estimated to detect Lynch syndrome with a sensitivity of 89 % for *MLH1* mutation, 90 % for *MSH2* mutation, and 76 % for *MSH6* mutation. Colorectal Family Registry data revealed that only 0.3 % of dMMR show MMR protein loss without associated MSI-H. Assuming that most of the individuals in this data set are Lynch syndrome patients, MSI criteria alone would miss the diagnosis in 0.3–10 % of new Lynch syndrome patients. In contrast, immunohistochemical assessment of MMR proteins expression is estimated to detect Lynch syndrome with a sensitivity of 81 % for *MLH1* mutation, 88 % for *MSH2* mutation and 76 % for *MSH6*. Based on the data from the Colorectal Family Registry where 11 % of dMMR cancers showed MSI-H without demonstrable associated immunohistochemical loss of MMR protein and with the assumption that most of these pertained to Lynch syndrome patients, immunostains alone would miss 11–12 % of new Lynch syndrome patients [76, 81, 82]. This is, in part, based on the fact that some mutations present in the examined genes do not abolish protein expression, and are therefore not detected by this method.

Neither MSI testing alone nor immunoexpression analysis for MMR proteins alone has a sensitivity of 100 % for detection of germline MMR gene mutations. However, the two assays used in combination would be able to identify all Lynch syndrome cases, as shown in cumulative data from a total of 3,369 patients [76, 77, 79, 80].

Whereas sequencing of germline DNA to detect germline mutations in MMR genes is the most sensitive and specific single strategy to identify Lynch syndrome patients with an estimated sensitivity of 99.5 % and a specificity of 99.96 % for the sequencing method, the required cost and effort associated with sequencing is best reserved for cases where an algorithm like the one suggested below would fail to resolve the diagnosis. In addition, a substantial fraction of mutations cannot be detected by direct sequencing because they are larger rearrangements that require deletion/duplication analysis by MLPA (multiplex ligation-dependent probe amplification) or other methods.

ALGORITHMIC STRATEGIES TO DETECT AND CATEGORIZE dMMR COLORECTAL CANCERS

The incomplete sensitivity of any single test alone to detect Lynch syndrome patients, as described above, lends support to the adoption of a multitest screening algorithm strategy. Almost all dMMR CRC will be detected by using a combination of MSI and IHC. Once a tumor is defined as dMMR, demonstrating loss of protein expression of MSH2/MSH6, MSH6 alone, or PMS2 alone increases the likelihood of Lynch syndrome. In contrast, the concomitant demonstration of high CIMP and *MLH1* promotor hypermethylation will point to a diagnosis of sporadic dMMR cancer. Detection of the *BRAF* c.1799T>A mutation excludes the diagnosis of Lynch syndrome. As outlined in Fig. 25.3, the concurrent use of MSI testing, MMR protein IHC, and *BRAF* c.1799T>A mutation analysis would detect almost all dMMR CRCs, would classify 94 % of all new CRCs into three relevant MMR subgroups (i.e., sporadic dMMR, MMR-proficient, and Lynch syndrome), and would guide secondary molecular testing of the remainder of the cases for more complex hypermethylation, sequencing, and deletion testing [45].

Predictive Markers of Response to Targeted Therapeutics

The recent development of therapeutic antibodies and small molecules agents that are based on the accumulated knowledge of the composition and function of signaling pathways in colorectal carcinoma has brought the therapy of colorectal cancer into the new age of personalized therapy. The new therapeutic strategies differ from the classical ones in that: (1) novel therapeutic agents act with a specific mechanism by targeting a tumor-specific structure and (2) it is theoretically possible to predict the individual tumor response to a given agent based on the analysis of the signaling pathway of the targeted molecule.

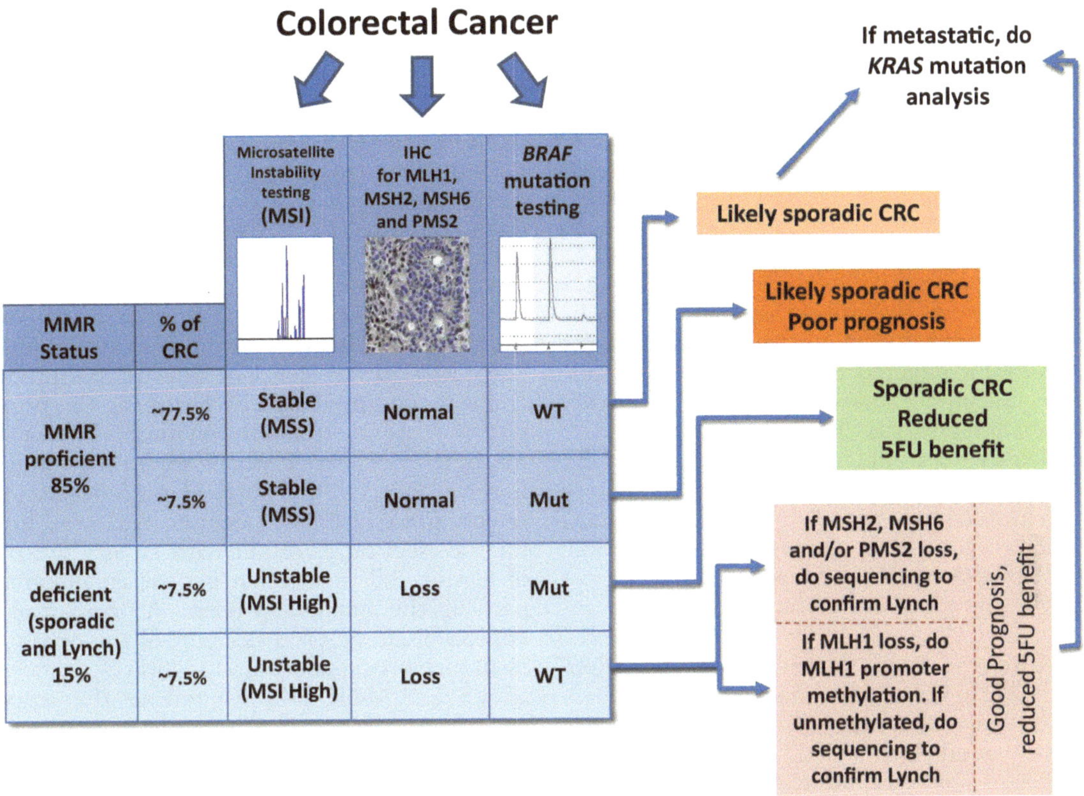

Figure 25-3 *Colorectal cancer characteristics.* Association for Molecular Pathology (AMP) proposed testing strategy and possible test outcomes, downstream additional testing, subgroup assignment, prognosis, and prediction of therapeutic response [45]

TARGETED ANTI-EGFR TREATMENT: TESTING FOR MUTATIONS IN THE KRAS AND NRAS GENES

The efficacy of therapeutic antibodies targeting the epidermal growth factor receptor (EGFR) in treating colorectal cancers is now established as a standard of care as indicated in the National Comprehensive Cancer Network (NCCN) guidelines in the USA, and in others. In one of the first clinical studies addressing the identification of predictive biologic markers of response to anti EGFR treatment in colon cancer, Lievre et al. postulated a potential role for *KRAS* status assessment [83]. This was based on the fact that in colorectal cancer, EGFR mainly activates the intracellular RAS/RAF/MAPK pathway as one of its downstream signaling transduction pathways. Therefore, it seemed rational that anti-EGFR-directed antibodies would not be effective in patients harboring tumors with a mutated (MUT) *KRAS* oncogene (that will be constitutively activated) as found in approximately 40 % of cases. This hypothesis was subsequently proven in several clinical studies addressing different anti-EGFR classes of antibodies including cetuximab (erbitux®) and panitumumab (vectibix®), either in monotherapy or in combination with standard chemotherapeutic backbone therapies like FOLFOX (folic acid, 5-fluorouracil, oxaliplatin) or FOLFIRI (folic acid, 5-fluorouracil, irinotecan) [84–87]. These studies revealed that patients with MUT *KRAS* affecting codons 12, 13, and 61 of exon 2 did not respond and, in fact, could potentially be harmed by anti-EGFR-targeted therapies. Conserved missense mutations in codons 12, 13, and 61 result in prolonged binding of guanosine triphosphate (GTP) and constitutive activation of the KRAS protein. Thus, MUT *KRAS* status became a negative predictive biomarker which is nowadays routinely tested when planning for such therapeutic strategy in patients with colorectal cancer. Unexpectedly, about 40 % of the patients with wild-type (WT) *KRAS* also failed to respond to anti-EGFR-targeted therapies [84]. More recently, it became evident that other mutations in exons 3 and 4 of the *KRAS* gene as well as in exons 2–4 of the *NRAS* oncogene are partially responsible for the lack of response in that subset and should also be tested, as negative predictive biomarkers for anti-EGFR-targeted therapies [88]. Together, the currently known mutations identify approximately 50 % of colorectal cancer patients who should not be offered anti-EGFR therapies.

The *BRAF* (B-type *RAS* associated factor) specific activating transversion mutation, c.1799T>A (p.V600E) accounts for most *BRAF* mutations associated with human cancers, including 10–46 % of colon cancers. Counterintuitively, mutations in the *BRAF* and *PI3K-CA* (phosphatidyl-inositol-3-kinase catalytic subunit alpha) genes did not show predictive value for anti-EGFR response. Instead, mutations in these genes are indicators of a significantly worse prognosis in the setting of metastastatic colorectal cancer [87, 88]. The fact that *BRAF* mutations are encountered in approximately half of MSI-H colorectal cancers [3] and is an indicator of good prognosis [52] in that setting further highlights the reality that the interactions of the signaling pathways in colorectal oncogenesis are yet to be completely understood and that the context and the timing of a given mutation could be paramount for its actual role.

A multitude of technical approaches have been successfully used for the evaluation of *RAS* and *BRAF* mutational status in colorectal carcinoma. These include laboratory developed tests (LDTs) as well as several commercially available/FDA cleared or approved assays [89]. Different sequencing approaches have been used including Sanger sequencing, pyrosequencing, and more recently NGS. Other methods include ARMS®-PCR (amplification refractory mutation system, also known as allele-specific PCR), hybridization, high resolution melting (HRM) analysis, SSCP (single strand conformational polymorphism), RPCR (restriction-based PCR) and FPCR (flip PCR) [90–95]. These techniques differ by their intrinsic technical sensitivity [96]. Sanger sequencing, a technique that is considered the gold standard, is one that requires a relatively high mutant allele percentage to be present within the tested tumor (15–20 %). Superior sensitivity is expected with methods based on allele-specific PCR where 1 % or less in mutant alleles could be detected, an advantage that allows for identification of target mutations in biopsy or pathologic specimens with low tumor burden. Sensitivity limits could be circumvented by enriching the portion of tumor cells in a tested sample by macroscopic- or laser capture microdissection. Involvement by the surgical pathologist in selecting an optimal specimen and area for microdissection, together with an estimation of tumor cell proportion is now an integral part of the analysis process. A "ring trial" approach can be used as a part of the quality assurance process during validation of a newly adopted method [97]. To reduce the costs and efforts required for the analysis, consideration should be given to first assessing mutations in exon 2 of the *KRAS* gene and limiting subsequent assessment of additional *KRAS* and *NRAS* exons (exons 3, 4 and exons 2–4, respectively) to tumors where wild-type (WT) *KRAS* exon 2 is initially found (60 % of tumors). NGS platforms have been making their transition into the diagnostic realm for evaluating somatic mutations in large panels; however, these platforms are often as yet not easily accessible or cost-effective for single gene assays and require sophisticated bioinformatics for analysis.

OTHER "TARGETED" THERAPEUTIC STRATEGIES

The protective effect of NSAIDs (Nonsteroidal Anti-Inflammatory Drugs) use on the development of cancer in the intestinal tract has long been revealed. The underlying molecular mechanism is thought to be that prostaglandin E2 (PGLE2) increases the activity of the Wnt-signaling pathway [98] and therefore exerts tumor promoting effects in the bowel. Consequently, inhibition of cyclooxygenase enzymes (COX-1 and -2) was contemplated as a potential therapy approach for colorectal tumors because both enzymes play a key role in prostaglandin biosynthesis. However, due to their significant side effects, COX-2 inhibitors were not implemented into the therapeutic schemes [99]. Given that the PI3K/AKT/mTOR signaling pathway was shown to activate the COX-system, more recently, colorectal tumors with activation of this pathway by activating mutations in the PI3KCA gene (19 %) [100] have been suggested to be a subset that might profit from the application of NSAIDs like aspirin, which was subsequently shown in an epidemiologic study [101]. This intriguing approach will, of course, have to be prospectively validated in a randomized clinical study.

Finally, in addition to targeting of the EGFR, the VEGFR (vascular endothelial growth factor receptor) has been targeted in colorectal cancer by agents like bevacizumab (Avastin®). Unfortunately no predictive biomarker is currently available for evaluating bevacizumab response [102–104].

Molecular Classification of Pancreatic Cancer

As a better understanding of the genetic alterations in pancreatic cancer emerges, it is becoming clear that a new molecular-based classification is possible [105, 106]. This classification does not discard the well-tested morphologic classification system that has been developed and tested over many decades by expert pathologists; instead, this classification system integrates morphology and molecular findings into a cohesive system with prognostic and therapeutic implications. Here we present a few examples that highlight the value of adding molecular findings to the existing morphologic classification of pancreatic cancer.

Undifferentiated Carcinomas of the Pancreas

In the past, carcinomas of the pancreas with no clear direction of differentiation were lumped together under the designation "undifferentiated carcinoma," and patients with one of these neoplasms were given an extremely poor prognosis [107]. It is now recognized that several molecularly distinct neoplasms can have an undifferentiated morphology, and that these molecular subtypes have prognostic and therapeutic implications.

Pancreatic carcinomas with a medullary histologic appearance (characterized by poor differentiation, a syncytial growth pattern, and pushing borders) often have a distinct molecular profile [108–111]. Unlike most other pancreatic cancers, they are often microsatellite unstable [108–111]. Remarkably, even though these cancers are poorly differentiated, they have a better prognosis than do the usual ductal adenocarcinomas of the pancreas. In addition, patients with a medullary carcinoma of the pancreas are more likely to have a family history of cancer, and medullary cancers may be less sensitive to 5-fluorouracil than ductal adenocarcinomas [108–111].

Anaplastic carcinomas of the pancreas are also neoplasms with no clear direction of differentiation, but unlike medullary carcinomas, anaplastic carcinomas are microsatellite stable and instead are characterized by loss of e-cadherin expression [112]. As a result, the neoplastic cells of truly anaplastic carcinomas are poorly cohesive, very infiltrative and associated with an extremely poor prognosis.

Although not as advanced as it is for some of the other cancer types, there are several opportunities for individualized "targeted" therapy for patients with pancreatic carcinoma. For example, pancreatic cancers that overexpress the nucleoside transporter *ENT1* appear to respond better to gemcitabine-based chemotherapy than pancreatic cancers that do not [113]. In addition, pancreatic cancers in which one of the Fanconi anemia genes has been inactivated, such as *BRCA2*, *BRCA1*, or *PALB2*, appear to be exquisitely sensitive to DNA cross-linking agents such as mitomycin C and to poly-(ADP-ribose) polymerase (PARP) inhibitors [114, 115].

Targeted therapies can also take advantage of the unique morphologic features of ductal adenocarcinoma of the pancreas. For example, pancreatic cancers elicit an intense desmoplastic reaction and albumin-bound paclitaxel (nab-paclitaxel), which is believed to bind to SPARC in this desmoplastic stroma helping to localize the anti-neoplastic agent to the tumor, has been shown to be effective in treating patients with pancreatic cancer [116, 117].

Thus, molecular analyses have helped separate carcinomas of the pancreas into distinct groups with discrete biologies, separate prognoses, and different susceptibilities to various therapies. This molecular classification has clinical implications.

Neuroendocrine Neoplasms

The current morphologic classification system for neuroendocrine neoplasms of the pancreas lumps together small cell carcinomas of the pancreas and neuroendocrine neoplasms with classic well-differentiated neuroendocrine morphology ("salt and pepper nuclei") and a Ki-67 labeling index of >20 % (or >20 mitoses per 10 high power fields) under the designation of neuroendocrine carcinoma [118]. Recent sequencing has, however, demonstrated that these two neoplasms are, in fact, genetically completely distinct. Whole exome sequencing has identified three pathways that are commonly targeted in well-differentiated pancreatic neuroendocrine tumors (PanNETs). These include the *MEN1* gene, the *DAXX* and *ATRX* genes, and genes coding for members of the mammalian target of rapamycin (mTOR) pathway [119, 120]. By contrast, small cell carcinomas of the pancreas lack these signature mutations. Instead, the *TP53* and *RB* genes are targeted in small cell carcinomas [121]. These molecular analyses make it clear that small cell carcinomas and PanNETs with classic neuroendocrine morphology and a Ki-67 labeling index of >20 % should not be grouped together [121].

Molecular analyses also suggest an additional way to classify grade 1 and grade 2 PanNETs. Approximately one in six PanNETs has a mutation in a gene coding for a member of the mTOR pathway [113]. Whereas morphologically indistinguishable from PanNETs without an mTOR pathway mutation, PanNETs with a mutation in an mTOR pathway gene are important to recognize because they are predicted to be sensitive to mTOR pathway inhibitors such as everolimus [122, 123].

Pancreatic Cystic Lesions

In the case of cystic neoplasms of the pancreas, the molecular findings beautifully complement the existing morphologic classification system [124–126]. There are four main types of neoplastic cysts of the pancreas: intraductal papillary mucinous neoplasm (IPMN), mucinous cystic neoplasm (MCN), serous cystic neoplasm, and solid-pseudopapillary neoplasm (SPN). Each of these neoplasms has been well-described morphologically and each has well-characterized clinical features. The exomes of all four of these cystic neoplasms have been sequenced, and a distinct set of genes appears to be altered in each. Virtually all SPNs have a beta-catenin gene mutations, the *VHL* gene is targeted in serous cystadenomas, the *RNF43*, *GNAS*, *KRAS*, *TPS3*, *CDKN2A*, and *SMAD4* genes are targeted in IPMNs, and the *RNF43*, *KRAS*, *TPS3*, *CDKN2A*, and *SMAD4* genes are targeted in MCNs [124–126]. The molecular classification therefore almost perfectly matches the morphologic. This not only validates the classification system, but also has immediate clinical implications because it suggests that the type of cystic neoplasm in the pancreas can be determined simply by sequencing cyst fluid samples obtained endoscopically [127].

Conclusions

Colorectal cancer is a heterogeneous entity which consists of at least of two different types of diseases at the genetic level: (1) a stem cell type and (2) an OIS type. These differences account for the now recognized diversity in clinical behavior among colon cancers as well as in response to chemotherapy. Data from expression profiling are difficult to integrate at this time due to their high heterogeneity. The transition of NGS to the realm of clinical studies and diagnostics promises that a more robust molecular classification of colorectal cancer will be soon within reach. This will help discern genetic differences among (1) colon and rectal cancer [3], (2) left- and right-sided MSS or MSI-H colorectal cancers, respectively, (3) colorectal cancers with or without distant metastases [128], (4) colorectal cancers with solitary and multiple disseminated metastases, (5) and differences in response to chemotherapy [129, 130] and to targeted therapeutics like cetuximab (Erbitux®) or panitumumab (Vectibix®) [130] and others.

The exomes of all of the major types of neoplasms of the pancreas have been sequenced, and neoplasms of the pancreas are now among the best characterized of all neoplasms. An integration of this molecular understanding with the existing morphology-based classification system has helped define new tumor subtypes, discover new markers of cyst type, and has defined therapeutic targets.

Conflict of Interest

Ralph Hruban receives royalty payments from Myriad Genetics for the PALB2 invention.

Andreas Jung is member of Advisory Boards for AMGEN and Merck Serono and received honoraria for presentations.

Glossary

5-FU	5-Fluorouracil
CIMP	CpG island methylator phenotype
CIMP-H	High-grade CIMP
CIMP-L	Low-grade CIMP
ECM	Extracellular matrix
EGFR	Epidermal growth factor
EMT	Epithelio-mesenchymal transition
GEM	Genetically engineered mouse
Her2	EGFR 2
MLH1	MUT L homologue 1
MSH	MUT S homologue
TERT	Telomerase RT component
INK4a	Inhibitor of kinase 4
LEF-1	Lymphocyte enhancing factor-1
MGMT	O^6-Methyl guanosine methyl transferase
miRNA	micro RNA
MMP	Matrix metalloproteinase 7
MSI-H	High-grade microsatellite instability
MT-MMP	Membrane-type MMP
ncRNA	Noncoding RNA
OIS	Oncogene-induced senescence
PMS	Postmitotic segregation
POL E	DNA polymerase ε
TCF	T-cell factor
uPA	Urokinase plasminogen activator
uPAR	uPA receptor

REFERENCES

1. Mayr E, Bock W. Classifications and other ordering systems. J Zool Syst Evol Res. 2002;40: 169–94.
2. Gao P, Song YX, Wang ZN, Xu YY, Tong LL, Sun JX, et al. Is the prediction of prognosis not improved by the seventh edition of the TNM classification for colorectal cancer? Analysis of the surveilla006Ece, epidemiology, and end results (SEER) database. BMC Cancer. 2013; 13:123.
3. The Cancer Genome Atlas. Comprehensive molecular characterization of human colon and rectal cancer. Nature. 2012;487:330–7.
4. Wolpert L. In: Principles of development. 2. revised ed. Oxford University Press; 2001.
5. de la Chapelle A. Genetic predisposition to colorectal cancer. Nat Rev Cancer. 2004; 4:769–80.
6. Baylin SB, Jones PA. A decade of exploring the cancer epigenome—biological and translational implications. Nat Rev Cancer. 2011;11:726–34.
7. Hanahan D, Weinberg RA. The hallmarks of cancer. Cell. 2000;100:57–70.
8. Hanahan D, Weinberg RA. Hallmarks of cancer: the next generation. Cell. 2011;144:646–74.
9. Vogelstein B, Papadopoulos N, Velculescu VE, Zhou S, Diaz Jr LA, Kinzler KW. Cancer genome landscapes. Science. 2013;339:1546–58.
10. Kinzler KW, Vogelstein B. Lessons from hereditary colorectal cancer. Cell. 1996;87:159–70.
11. Fearon ER, Vogelstein B. A genetic model for colorectal tumorigenesis. Cell. 1990;61:759–67.
12. Fodde R, Kuipers J, Rosenberg C, Smits R, Kielman M, Gaspar C, et al. Mutations in the APC tumour suppressor gene cause chromosomal instability. Nat Cell Biol. 2001;3:433–8.
13. Janssen KP, el-Marjou F, Pinto D, Sastre X, Rouillard D, Fouquet C, et al. Targeted expression of oncogenic K-ras in intestinal epithelium causes spontaneous tumorigenesis in mice. Gastroenterology. 2002;123:492–504.
14. Halberg RB, Katzung DS, Hoff PD, Moser AR, Cole CE, Lubet RA, et al. Tumorigenesis in the multiple intestinal neoplasia mouse: redundancy of negative regulators and specificity of modifiers. Proc Natl Acad Sci U S A. 2000; 97:3461–6.

15. Clevers H, Nusse R. Wnt/beta-catenin signaling and disease. Cell. 2012;149:1192–205.
16. Harris H. Tumour suppression: putting on the brakes. Nature. 2004;427:201.
17. He TC, Sparks AB, Rago C, Hermeking H, Zawel L, da Costa LT, et al. Identification of c-MYC as a target of the APC pathway. Science. 1998; 281:1509–12.
18. Shtutman M, Zhurinsky J, Simcha I, Albanese C, D'Amico M, Pestell R, et al. The cyclin D1 gene is a target of the beta-catenin/LEF-1 pathway. Proc Natl Acad Sci U S A. 1999;96:5522–7.
19. Tetsu O, McCormick F. Beta-catenin regulates expression of cyclin D1 in colon carcinoma cells. Nature. 1999;398:422–6.
20. Wassermann S, Scheel SK, Hiendlmeyer E, Palmqvist R, Horst D, Hlubek F, et al. p16INK4a is a beta-catenin target gene and indicates low survival in human colorectal tumors. Gastroenterology. 2009;136:196–205.e2.
21. Hoffmeyer K, Raggioli A, Rudloff S, Anton R, Hierholzer A, Del Valle I, et al. Wnt/beta-catenin signaling regulates telomerase in stem cells and cancer cells. Science. 2012;336:1549–54.
22. Jaitner S, Reiche JA, Schaffauer AJ, Hiendlmeyer E, Herbst H, Brabletz T, et al. Human telomerase reverse transcriptase (hTERT) is a target gene of beta-catenin in human colorectal tumors. Cell Cycle. 2012;11.
23. Conacci-Sorrell M, Simcha I, Ben-Yedidia T, Blechman J, Savagner P, Ben-Ze'ev A. Autoregulation of E-cadherin expression by cadherin-cadherin interactions: the roles of beta-catenin signaling, Slug, and MAPK. J Cell Biol. 2003;163:847–57.
24. Hlubek F, Spaderna S, Jung A, Kirchner T, Brabletz T. Beta-catenin activates a coordinated expression of the proinvasive factors laminin-5 gamma2 chain and MT1-MMP in colorectal carcinomas. Int J Cancer. 2004;108:321–6.
25. Spaderna S, Schmalhofer O, Wahlbuhl M, Dimmler A, Bauer K, Sultan A, et al. The transcriptional repressor ZEB1 promotes metastasis and loss of cell polarity in cancer. Cancer Res. 2008;68:537–44.
26. Gradl D, Kuhl M, Wedlich D. The Wnt/Wg signal transducer beta-catenin controls fibronectin expression. Mol Cell Biol. 1999;19:5576–87.
27. Gilles C, Polette M, Mestdagt M, Nawrocki-Raby B, Ruggeri P, Birembaut P, et al. Transactivation of vimentin by beta-catenin in human breast cancer cells. Cancer Res. 2003;63: 2658–64.
28. Brabletz T, Jung A, Dag S, Hlubek F, Kirchner T. beta-catenin regulates the expression of the matrix metalloproteinase-7 in human colorectal cancer. Am J Pathol. 1999;155:1033–8.
29. Hlubek F, Jung A, Kotzor N, Kirchner T, Brabletz T. Expression of the invasion factor laminin gamma2 in colorectal carcinomas is regulated by beta-catenin. Cancer Res. 2001;61:8089–93.
30. Hiendlmeyer E, Regus S, Wassermann S, Hlubek F, Haynl A, Dimmler A, et al. Beta-catenin up-regulates the expression of the urokinase plasminogen activator in human colorectal tumors. Cancer Res. 2004;64:1209–14.
31. Mann B, Gelos M, Siedow A, Hanski ML, Gratchev A, Ilyas M, et al. Target genes of beta-catenin-T cell-factor/lymphoid-enhancer-factor signaling in human colorectal carcinomas. Proc Natl Acad Sci U S A. 1999;96:1603–8.
32. Beiter K, Hiendlmeyer E, Brabletz T, Hlubek F, Haynl A, Knoll C, et al. beta-Catenin regulates the expression of tenascin-C in human colorectal tumors. Oncogene. 2005;24:8200–4.
33. Wielenga VJ, Smits R, Korinek V, Smit L, Kielman M, Fodde R, et al. Expression of CD44 in Apc and Tcf mutant mice implies regulation by the WNT pathway. Am J Pathol. 1999;154:515–23.
34. Hatzis P, van der Flier LG, van Driel MA, Guryev V, Nielsen F, Denissov S, et al. Genome-wide pattern of TCF7L2/TCF4 chromatin occupancy in colorectal cancer cells. Mol Cell Biol. 2008; 28:2732–44.
35. Barker N, van Es JH, Kuipers J, Kujala P, van den Born M, Cozijnsen M, et al. Identification of stem cells in small intestine and colon by marker gene Lgr5. Nature. 2007;449:1003–7.
36. Pardal R, Clarke MF, Morrison SJ. Applying the principles of stem-cell biology to cancer. Nat Rev Cancer. 2003;3:895–902.
37. Vermeulen L, De Sousa EMF, van der Heijden M, Cameron K, de Jong JH, Borovski T, et al. Wnt activity defines colon cancer stem cells and is regulated by the microenvironment. Nat Cell Biol. 2010;12:468–76.
38. Brembeck FH, Schwarz-Romond T, Bakkers J, Wilhelm S, Hammerschmidt M, Birchmeier W. Essential role of BCL9-2 in the switch between beta-catenin's adhesive and transcriptional functions. Genes Dev. 2004;18: 2225–30.
39. Mani SA, Guo W, Liao MJ, Eaton EN, Ayyanan A, Zhou AY, et al. The epithelial-mesenchymal transition generates cells with properties of stem cells. Cell. 2008;133:704–15.
40. Brabletz T. To differentiate or not—routes towards metastasis. Nat Rev Cancer. 2012;12: 425–36.
41. Neumann J, Horst D, Kriegl L, Maatz S, Engel J, Jung A, et al. A simple immunohistochemical algorithm predicts the risk of distant metastases in right-sided colon cancer. Histopathology. 2012;60:416–26.
42. Fodde R, Smits R, Clevers H. APC, signal transduction and genetic instability in colorectal cancer. Nat Rev Cancer. 2001;1:55–67.

43. Linnebacher M, Gebert J, Rudy W, Woerner S, Yuan YP, Bork P, et al. Frameshift peptide-derived T-cell epitopes: a source of novel tumor-specific antigens. Int J Cancer. 2001;93:6–11.
44. Seshagiri S, Stawiski EW, Durinck S, Modrusan Z, Storm EE, Conboy CB, et al. Recurrent R-spondin fusions in colon cancer. Nature. 2012; 488:660–4.
45. Funkhouser Jr WK, Lubin IM, Monzon FA, Zehnbauer BA, Evans JP, Ogino S, et al. Relevance, pathogenesis, and testing algorithm for mismatch repair-defective colorectal carcinomas: a report of the association for molecular pathology. J Mol Diagn. 2012;14:91–103.
46. Serrano M, Lee H, Chin L, Cordon-Cardo C, Beach D, DePinho RA. Role of the INK4a locus in tumor suppression and cell mortality. Cell. 1996;85:27–37.
47. Carragher LA, Snell KR, Giblett SM, Aldridge VS, Patel B, Cook SJ, et al. V600EBraf induces gastrointestinal crypt senescence and promotes tumour progression through enhanced CpG methylation of p16INK4a. EMBO Mol Med. 2010;2:458–71.
48. Rad R, Cadinanos J, Rad L, Varela I, Strong A, Kriegl L, et al. A genetic progression model of Braf(V600E)-induced intestinal tumorigenesis reveals targets for therapeutic intervention. Cancer Cell. 2013;24:15–29.
49. Bennecke M, Kriegl L, Bajbouj M, Retzlaff K, Robine S, Jung A, et al. Ink4a/Arf and oncogene-induced senescence prevent tumor progression during alternative colorectal tumorigenesis. Cancer Cell. 2010;18:135–46.
50. Jass JR. Classification of colorectal cancer based on correlation of clinical, morphological and molecular features. Histopathology. 2007;50: 113–30.
51. Rex DK, Ahnen DJ, Baron JA, Batts KP, Burke CA, Burt RW, et al. Serrated lesions of the colorectum: review and recommendations from an expert panel. Am J Gastroenterol. 2012;107: 1315–29. quiz 4, 30.
52. Malesci A, Laghi L, Bianchi P, Delconte G, Randolph A, Torri V, et al. Reduced likelihood of metastases in patients with microsatellite-unstable colorectal cancer. Clin Cancer Res. 2007;13:3831–9.
53. Rosty C, Chazal M, Etienne MC, Letoublon C, Bourgeon A, Delpero JR, et al. Determination of microsatellite instability, p53 and K-RAS mutations in hepatic metastases from patients with colorectal cancer: relationship with response to 5-fluorouracil and survival. Int J Cancer. 2001; 95:162–7.
54. Jover R, Zapater P, Castells A, Llor X, Andreu M, Cubiella J, et al. Mismatch repair status in the prediction of benefit from adjuvant fluorouracil chemotherapy in colorectal cancer. Gut. 2006; 55:848–55.
55. Carethers JM, Chauhan DP, Fink D, Nebel S, Bresalier RS, Howell SB, et al. Mismatch repair proficiency and in vitro response to 5-fluorouracil. Gastroenterology. 1999;117:123–31.
56. Weisenberger DJ, Siegmund KD, Campan M, Young J, Long TI, Faasse MA, et al. CpG island methylator phenotype underlies sporadic microsatellite instability and is tightly associated with BRAF mutation in colorectal cancer. Nat Genet. 2006;38:787–93.
57. Sampson JR, Dolwani S, Jones S, Eccles D, Ellis A, Evans DG, et al. Autosomal recessive colorectal adenomatous polyposis due to inherited mutations of MYH. Lancet. 2003;362:39–41.
58. Sieber OM, Lipton L, Crabtree M, Heinimann K, Fidalgo P, Phillips RK, et al. Multiple colorectal adenomas, classic adenomatous polyposis, and germ-line mutations in MYH. N Engl J Med. 2003;348:791–9.
59. Jones S, Zhang X, Parsons DW, Lin JC, Leary RJ, Angenendt P, et al. Core signaling pathways in human pancreatic cancers revealed by global genomic analyses. Science. 2008;321:1801–6.
60. Sjoblom T, Jones S, Wood LD, Parsons DW, Lin J, Barber TD, et al. The consensus coding sequences of human breast and colorectal cancers. Science. 2006;314:268–74.
61. Wood HM, Belvedere O, Conway C, Daly C, Chalkley R, Bickerdike M, et al. Using next-generation sequencing for high resolution multiplex analysis of copy number variation from nanogram quantities of DNA from formalin-fixed paraffin-embedded specimens. Nucleic Acids Res. 2010;38:e151.
62. Wood LD, Parsons DW, Jones S, Lin J, Sjoblom T, Leary RJ, et al. The genomic landscapes of human breast and colorectal cancers. Science. 2007;318:1108–13.
63. Zhang L, Yu J, Willson JK, Markowitz SD, Kinzler KW, Vogelstein B. Short mononucleotide repeat sequence variability in mismatch repair-deficient cancers. Cancer Res. 2001;61: 3801–5.
64. Ruckert S, Hiendlmeyer E, Brueckl WM, Oswald U, Beyser K, Dietmaier W, et al. T-cell factor-4 frameshift mutations occur frequently in human microsatellite instability-high colorectal carcinomas but do not contribute to carcinogenesis. Cancer Res. 2002;62:3009–13.
65. Meyer N, Penn LZ. Reflecting on 25 years with MYC. Nat Rev Cancer. 2008;8:976–90.
66. Hahn WC, Weinberg RA. Modelling the molecular circuitry of cancer. Nat Rev Cancer. 2002;2: 331–41.
67. Bang YJ, Van Cutsem E, Feyereislova A, Chung HC, Shen L, Sawaki A, et al. Trastuzumab in combination with chemotherapy versus chemotherapy alone for treatment of HER2-positive advanced gastric or gastro-oesophageal junction cancer (ToGA): a phase 3, open-label, randomised

controlled trial. Lancet. 2010;376: 687–97.
68. Piccart-Gebhart MJ, Procter M, Leyland-Jones B, Goldhirsch A, Untch M, Smith I, et al. Trastuzumab after adjuvant chemotherapy in HER2-positive breast cancer. N Engl J Med. 2005;353:1659–72.
69. Spruessel A, Steimann G, Jung M, Lee SA, Carr T, Fentz AK, et al. Tissue ischemia time affects gene and protein expression patterns within minutes following surgical tumor excision. Biotechniques. 2004;36:1030–7.
70. Brabletz T, Jung A, Reu S, Porzner M, Hlubek F, Kunz-Schughart LA, et al. Variable beta-catenin expression in colorectal cancers indicates tumor progression driven by the tumor environment. Proc Natl Acad Sci U S A. 2001;98:10356–61.
71. Brabletz T, Jung A, Spaderna S, Hlubek F, Kirchner T. Opinion: migrating cancer stem cells—an integrated concept of malignant tumour progression. Nat Rev Cancer. 2005; 5:744–9.
72. Hlubek F, Brabletz T, Budczies J, Pfeiffer S, Jung A, Kirchner T. Heterogeneous expression of Wnt/beta-catenin target genes within colorectal cancer. Int J Cancer. 2007;121:1941–8.
73. Vasen HF, Watson P, Mecklin JP, Lynch HT. New clinical criteria for hereditary nonpolyposis colorectal cancer (HNPCC Lynch syndrome) proposed by the International Collaborative Group on HNPCC. Gastroenterology. 1999;116: 1453–6.
74. Umar A, Boland CR, Terdiman JP, Syngal S, de la Chapelle A, Ruschoff J, Fishel R, Lindor NM, Burgart LJ, Hamelin R, Hamilton SR, Hiatt RA, Jass J, Lindblom A, Lynch HT, Peltomaki P, Ramsey SD, Rodriguez-Bigas MA, Vasen HF, Hawk ET, Barrett JC, Freedman AN, Srivastava S. Revised Bethesda Guidelines for hereditary nonpolyposis colorectal cancer (Lynch syndrome) and microsatellite instability. J Natl Cancer Inst. 2004;96:261–8.
75. Greenson JK, Huang SC, Herron C, Moreno V, Bonner JD, Tomsho LP, Ben-Izhak O, Cohen HI, Trougouboff P, Bejhar J, Sova Y, Pinchev M, Rennert G, Gruber SB. Pathologic predictors of microsatellite instability in colorectal cancer. Am J Surg Pathol. 2009;33:126–33.
76. Cicek M, Lindor N, Gallinger S, Bapat B, Hopper J, Jenkins M, Young J, Buchanan D, Walsh M, Le Marchand L, Burnett T, Newcomb P, Grady W, Haile R, Casey G, Plummer S, Krumroy L, Baron J, Thibodeau S. Quality assessment and correlation of MSI and IHC markers among population- and clinic-based colorectal tumors. J Mol Diagn. 2011;13(3):271–81.
77. Hampel H, Frankel WL, Martin E, Arnold M, Khanduja K, Kuebler P, Nakagawa H, Sotamaa K, Prior TW, Westman J, Panescu J, Fix D, Lockman J, Comeras I, de la Chapelle A. Screening for the Lynch syndrome (hereditary nonpolyposis colorectal cancer). N Engl J Med. 2005;352:1851–60.
78. Cunningham JM, Kim CY, Christensen ER, Tester DJ, Parc Y, Burgart LJ, Halling KC, McDonnell SK, Schaid DJ, Walsh Vockley C, Kubly V, Nelson H, Michels VV, Thibodeau SN. The frequency of hereditary defective mismatch repair in a prospective series of unselected colorectal carcinomas. Am J Hum Genet. 2001; 69:780–90.
79. Debniak T, Kurzawski G, Gorski B, Kladny J, Domagala W, Lubinski J. Value of pedigree/clinical data, immunohistochemistry and microsatellite instability analyses in reducing the cost of determining hMLH1 and hMSH2 gene mutations in patients with colorectal cancer. Eur J Cancer. 2000;36:49–54.
80. Pinol V, Castells A, Andreu M, Castellvi-Bel S, Alenda C, Llor X, Xicola RM, Rodriguez-Moranta F, Paya A, Jover R, Bessa X. Accuracy of revised Bethesda guidelines, microsatellite instability, and immunohistochemistry for the identification of patients with hereditary nonpolyposis colorectal cancer. JAMA. 2005;293: 1986–94.
81. Palomaki GE, McClain MR, Melillo S, Hampel HL, Thibodeau SN. EGAPP supplementary evidence review: DNA testing strategies aimed at reducing morbidity and mortality from Lynch syndrome. Genet Med. 2009;11:42–65.
82. Barnetson RA, Tenesa A, Farrington SM, Nicholl ID, Cetnarskyj R, Porteous ME, Campbell H, Dunlop MG. Identification and survival of carriers of mutations in DNA mismatch-repair genes in colon cancer. N Engl J Med. 2006;354:2751–63.
83. Lièvre A, Bachet JB, Le Corre D, Boige V, Landi B, Emile JF, Côté JF, Tomasic G, Penna C, Ducreux M, Rougier P, Penault-Llorca F, Laurent-Puig P. KRAS mutation status is predictive of response to cetuximab therapy in colorectal cancer. Cancer Res. 2006;66(8):3992–5.
84. Amado RG, Wolf M, Peeters M, Van Cutsem E, Siena S, Freeman DJ, et al. Wild-type KRAS is required for panitumumab efficacy in patients with metastatic colorectal cancer. J Clin Oncol. 2008;26:1626–34.
85. Bokemeyer C, Bondarenko I, Makhson A, Hartmann JT, Aparicio J, de Braud F, et al. Fluorouracil, leucovorin, and oxaliplatin with and without cetuximab in the first-line treatment of metastatic colorectal cancer. J Clin Oncol. 2009;27:663–71.
86. Peeters M, Price TJ, Cervantes A, Sobrero AF, Ducreux M, Hotko Y, et al. Randomized phase III study of panitumumab with fluorouracil, leucovorin, and irinotecan (FOLFIRI) compared with FOLFIRI alone as second-line treatment in

patients with metastatic colorectal cancer. J Clin Oncol. 2010;28:4706–13.
87. Van Cutsem E, Kohne CH, Lang I, Folprecht G, Nowacki MP, Cascinu S, et al. Cetuximab plus irinotecan, fluorouracil, and leucovorin as first-line treatment for metastatic colorectal cancer: updated analysis of overall survival according to tumor KRAS and BRAF mutation status. J Clin Oncol. 2011;29:2011–9.
88. Douillard J-Y, Oliner KS, Siena S, Tabernero J, Burkes R, Barugel M, et al. Panitumumab-FOLFOX4 Treatment and RAS Mutations in Colorectal Cancer. N Engl J Med. 2013;369:1023.
89. Oliner K, Juan T, Suggs S, Wolf M, Sarosi I, Freeman DJ, et al. A comparability study of 5 commercial KRAS tests. Diagn Pathol. 2010;5:23.
90. Tol J, Dijkstra JR, Vink-Borger ME, Nagtegaal ID, Punt CJ, van Krieken JH, Ligtenberg MJ. High sensitivity of both sequencing and real-time PCR analysis of KRAS mutations in colorectal cancer tissue. J Cell Mol Med. 2010; 14:2122–31.
91. Spittle C, Ward MR, Nathanson KL, Gimotty PA, Rappaport E, Brose MS, Medina A, Letrero R, Herlyn M, Edwards RH. Application of a BRAF pyrosequencing assay for mutation detection and copy number analysis in malignant melanoma. J Mol Diagn. 2007;9:464–71.
92. Dufort S, Richard MJ, de Fraipont F. Pyrosequencing method to detect KRAS mutation in formalin-fixed and paraffin-embedded tumor tissues. Anal Biochem. 2009; 391:166–8.
93. Pichler M, Balic M, Stadelmeyer E, Ausch C, Wild M, Guelly C, Bauernhofer T, Samonigg H, Hoefler G, Dandachi N. Evaluation of high-resolution melting analysis as a diagnostic tool to detect the BRAF V600E mutation in colorectal tumors. J Mol Diagn. 2009;11:40–147.
94. Do H, Krypuy M, Mitchell PL, Fox SB, Dobrovic A. High resolution melting analysis for rapid and sensitive EGFR and KRAS mutation detection in formalin fixed paraffin embedded biopsies. BMC Cancer. 2008;8:142.
95. Lang AH, Drexel H, Geller-Rhomberg S, et al. Optimized allele-specific real-time PCR assays for the detection of common mutations in KRAS and BRAF. J Mol Diagn. 2011;13:23–8.
96. Querings S, Altmuller J, Ansen S, Zander T, Seidel D, Gabler F, et al. Benchmarking of mutation diagnostics in clinical lung cancer specimens. PLoS One. 2011;6:e19601.
97. van Krieken JH, Siebers AG, Normanno N, Quality Assurance for Molecular Pathology group. European consensus conference for external quality assessment in molecular pathology. Ann Oncol. 2013;24:1958–63.
98. Castellone MD, Teramoto H, Williams BO, Druey KM, Gutkind JS. Prostaglandin E2 promotes colon cancer cell growth through a Gs-axin-beta-catenin signaling axis. Science. 2005;310:1504–10.
99. Gupta RA, Dubois RN. Colorectal cancer prevention and treatment by inhibition of cyclooxygenase-2. Nat Rev Cancer. 2001;1:11–21.
100. Neumann J, Wehweck L, Maatz S, Engel J, Kirchner T, Jung A. Alterations in the EGFR pathway coincide in colorectal cancer and impact on prognosis. Virchows Arch. 2013; 463:509–23.
101. Liao X, Lochhead P, Nishihara R, Morikawa T, Kuchiba A, Yamauchi M, et al. Aspirin use, tumor PIK3CA mutation, and colorectal-cancer survival. N Engl J Med. 2012;367:1596–606.
102. Mizushima T, Ide Y, Murata K, Ohashi I, Yasumasa K, Fukunaga M, Takemoto H, Tamagawa H, Hasegawa J, Hata T, Takemasa I, Ikeda M, Yamamoto H, Sekimoto M, Nezu R, Doki Y, Mori M. A phase II study of combined chemotherapy with 5-week cycles of S-1 and CPT-11 plus bevacizumab in patients with metastatic colon cancer. Oncology. 2013;85(6):317–22. [Epub ahead of print] PubMed PMID: 24247419.
103. Vargas GM, Sheffield KM, Parmar AD, Han Y, Gajjar A, Brown KM, Riall TS. Trends in treatment and survival in older patients resenting with stage IV colorectal cancer. Journal of gastrointestinal surgery: official journal of the Society for Surgery of the Alimentary Tract 2014;18:369–77.
104. Pogue-Geile K, Yothers G, Taniyama Y, Tanaka N, Gavin P, Colangelo L, Blackmon N, Lipchik C, Kim SR, Sharif S, Allegra C, Petrelli N, O'Connell MJ, Wolmark N, Paik S. Defective mismatch repair and benefit from bevacizumab for colon cancer: findings from NSABP C-08. J Natl Cancer Inst. 2013;105(13):989–92. doi:10.1093/jnci/djt140. PubMed PMID: 23821759. PubMed Central PMCID: PMC3699439.
105. Shi C, Daniels JA, Hruban RH. Molecular characterization of pancreatic neoplasms. Adv Anat Pathol. 2008;15:185–95.
106. Wood LD, Hruban RH. Pathology and molecular genetics of pancreatic neoplasms. Cancer J. 2012; 18:492–501.
107. Cubilla AL, Fitzgerald PJ. In: Tumors of the exocrine pancreas. Vol. 2nd series. Washington, DC: Armed Forces Institute of Pathology; 1984. p.
108. Banville N, Geraghty R, Fox E, Leahy DT, Green A, Keegan D, et al. Medullary carcinoma of the pancreas in a man with hereditary nonpolyposis colorectal cancer due to a mutation of the MSH2 mismatch repair gene. Hum Pathol. 2006;37: 1498–502.
109. Calhoun ES, Jones JB, Ashfaq R, Adsay NV, Baker SJ, Valentine V, et al. BRAF and FBXW7 (CDC4, FBW7, AGO, SEL10) mutations in distinct subsets of pancreatic cancer: potential therapeutic targets. Am J Pathol. 2003;163: 1255–60.

110. Goggins M, Offerhaus GJ, Hilgers W, Griffin CA, Shekher M, Tang D, et al. Pancreatic adenocarcinomas with DNA replication errors (RER+) are associated with wild-type K-ras and characteristic histopathology. Poor differentiation, a syncytial growth pattern, and pushing borders suggest RER +. Am J Pathol. 1998;152:1501–7.
111. Wilentz RE, Goggins M, Redston M, Marcus VA, Adsay NV, Sohn TA, et al. Genetic, immunohistochemical, and clinical features of medullary carcinoma of the pancreas: a newly described and characterized entity. Am J Pathol. 2000; 156:1641–51.
112. Winter JM, Ting AH, Vilardell F, Gallmeier E, Baylin SB, Hruban RH, et al. Absence of E-cadherin expression distinguishes noncohesive from cohesive pancreatic cancer. Clin Cancer Res. 2008;14:412–8.
113. Wei CH, Gorgan TR, Elashoff DA, Hines OJ, Farrell JJ, Donahue TR. A meta-analysis of gemcitabine biomarkers in patients with pancreaticobiliary cancers. Pancreas. 2013;42:1303–10.
114. Lowery MA, Kelsen DP, Stadler ZK, Yu KH, Janjigian YY, Ludwig E, et al. An emerging entity: pancreatic adenocarcinoma associated with a known BRCA mutation: clinical descriptors, treatment implications, and future directions. Oncologist. 2011;16:1397–402.
115. Villarroel MC, Rajeshkumar NV, Garrido-Laguna I, Jesus-Acosta A, Jones S, Maitra A, et al. Personalizing cancer treatment in the age of global genomic analyses: PALB2 gene mutations and the response to DNA damaging agents in pancreatic cancer. Mol Cancer Ther. 2011; 10:3–8.
116. Infante MH, Sato N, Tonascia J, Klein AP, Riall TS, Yeo CJ, et al. Peritumoral fibroblast SPARC expression and patient outcome with resectable pancreatic adenocarcinoma. J Clin Oncol. 2007;25:319–25.
117. Von Hoff DD, Ervin T, Arena FP, Chiorean EG, Infante J, Moore M, et al. Increased survival in pancreatic cancer with nab-paclitaxel plus gemcitabine. N Engl J Med. 2013;369: 1691–703.
118. In: WHO classification of tumours of the digestive system. 4th ed. Lyon: IARC; 2010. Vol. p.
119. Heaphy CM, de Wilde RF, Jiao Y, Klein AP, Edil BH, Shi C, et al. Altered telomeres in tumors with ATRX and DAXX mutations. Science. 2011;333:425.
120. Jiao Y, Shi C, Edil BH, de Wilde RF, Klimstra DS, Maitra A, et al. DAXX/ATRX, MEN1, and mTOR pathway genes are frequently altered in pancreatic neuroendocrine tumors. Science. 2011;331:1199–203.
121. Yachida S, Vakiani E, White CM, Zhong Y, Saunders T, Morgan R, et al. Small cell and large cell neuroendocrine carcinomas of the pancreas are genetically similar and distinct from well-differentiated pancreatic neuroendocrine tumors. Am J Surg Pathol. 2012;36:173–84.
122. Krueger DA, Care MM, Holland K, Agricola K, Tudor C, Mangeshkar P, et al. Everolimus for subependymal giant-cell astrocytomas in tuberous sclerosis. N Engl J Med. 2010;363:1801–11.
123. Yao JC, Shah MH, Ito T, Bohas CL, Wolin EM, Van Cutsem E, et al. Everolimus for advanced pancreatic neuroendocrine tumors. N Engl J Med. 2011;364:514–23.
124. Abraham SC, Klimstra DS, Wilentz RE, Wu TT, Cameron JL, Yeo CJ, et al. Solid-pseudopapillary tumors of the pancreas are genetically distinct from pancreatic ductal adenocarcinomas and almost always harbor beta-catenin mutations. Am J Pathol. 2002;160:1361–9.
125. Wu J, Jiao Y, Dal Molin M, Maitra A, de Wilde RF, Wood LD, et al. Whole-exome sequencing of neoplastic cysts of the pancreas reveals recurrent mutations in components of ubiquitin-dependent pathways. Proc Natl Acad Sci U S A. 2011;108:21188–93.
126. Wu J, Matthaei H, Maitra A, Dal Molin M, Wood LD, Eshleman JR, et al. Recurrent GNAS mutations define an unexpected pathway for pancreatic cyst development. Sci Transl Med. 2011; 3:92ra66.
127. Kanda M, Knight S, Topazian MD, Syngal S, Farrell J, Lee JH, et al. Mutant GNAS detected in duodenal collections of secretin-stimulated pancreatic juice indicates the presence or emergence of pancreatic cysts. Gut. 2012;62:1024–33.
128. Sawada T, Yamamoto E, Suzuki H, Nojima M, Maruyama R, Shioi Y, et al. Association between genomic alterations and metastatic behavior of colorectal cancer identified by array-based comparative genomic hybridization. Genes Chromosomes Cancer. 2013;52:140–9.
129. Bohanes P, LaBonte MJ, Winder T, Lenz HJ. Predictive molecular classifiers in colorectal cancer. Semin Oncol. 2011;38:576–87.
130. Sadanandam A, Lyssiotis CA, Homicsko K, Collisson EA, Gibb WJ, Wullschleger S, et al. A colorectal cancer classification system that associates cellular phenotype and responses to therapy. Nat Med. 2013;19:619–25.
131. Harris TJ, Peifer M. Decisions, decisions: beta-catenin chooses between adhesion and transcription. Trends Cell Biol. 2005;15:234–7.
132. Nieto MA. Epithelial plasticity: a common theme in embryonic and cancer cells. Science. 2013;342:1234850.

CHAPTER 26

Molecular Pathology of Genitourinary Cancers: Translating the Cancer Genome to the Clinic

JOSHUA I. WARRICK, SCOTT A. TOMLINS

Introduction 435
The Molecular Pathology of Prostate Cancer 436
The Molecular Pathology of Bladder Cancer 446
Conclusion 455
Acknowledgements 455

Introduction

Genitourinary malignancies, including cancers of the prostate, urinary bladder, kidney, testis, and penis, are major causes of cancer morbidity and mortality in the USA. Interrogation of the cancer genome and transcriptome, through single gene assays (including assessment of gene products by immunohistochemistry [IHC]), multiplexed panels, and targeted or full sequencing, has led to major advances in our understanding of the molecular underpinnings of numerous cancers. High-throughput technologies, such as DNA microarrays and next-generation sequencing (NGS), combined with large international efforts to comprehensively interrogate cancer genomes and transcriptomes such as The Cancer Genome Atlas (TCGA), will likely lead to a complete cataloguing of the aberrations present in genitourinary cancers. The shift toward selecting the right therapy for the molecular alteration(s) driving a patient's particular cancer in the era of precision medicine will increase the clinical demand for routine cancer genome/transcriptome assessment. Importantly, pathologists are ideally suited to be leading the efforts to understand the range and diversity of these aberrations, how they can be assessed in routine specimens, and which assay(s) can be used best to answer important clinical questions (e.g., "which is the best therapy for my patient with bladder cancer?"). In this chapter, we aim to provide an overview of the range of driving genome or transcriptome alterations in common genitourinary cancers. We have focused on important single genes, multigene panels, and findings from exome/genome-wide interrogation. We have attempted to place these lesions and related assays into a clinical context, particularly regarding current and future translation in relation to areas of clinical need. Lastly, given the explosion in reports and assays for interrogating the cancer genome from the single gene level (through techniques such as IHC, fluorescence in situ hybridization [FISH], capillary sequencing) to full genome scale sequencing, we have sought to point out salient points to be considered by the pathologist when thinking about implementing novel biomarkers or assays.

G.J. Netto and I. Schrijver (eds.), *Genomic Applications in Pathology*,
DOI 10.1007/978-1-4939-0727-4_26, © Springer Science+Business Media New York 2015

The Molecular Pathology of Prostate Cancer

Prostate cancer is the leading type of cancer and the second most common cause of cancer death affecting American men [1]. Current prognostic models rely heavily on pathologic grade and stage. Tumor grade is determined by the Gleason grading system, which assigns numeric values (range 1–5) to tumor architecture, and in simplified terms sums the two most prevalent patterns to achieve an overall Gleason score (range 2–10). Cancers with Gleason score 2–5 are uncommon. Tumors with Gleason score 3+3=6 have indolent behavior, with an extremely low chance of causing patient death. In contrast, those with a higher Gleason score have greater potential for metastasis and causing death, which increases with the score [2–4]. Pathologic stage also strongly correlates with prognosis, with higher rates of metastasis and death being associated with cancer extending outside the prostate.

Single Genes in Prostate Cancer

ETS GENE FUSIONS

Fusion genes resulting from rearrangements involving members of the ETS transcription factor family are the most common known molecular abnormality in prostate cancer, seen in ~50 % of cases detected by serum PSA screening [5–10]. The most common rearrangement involves either chromosomal deletion or insertion of chromosome 21, resulting in fusion of the 5′ untranslated region of *TMPRSS2*, an androgen regulated gene, with the ETS family member *ERG*. This fusion results in androgen-driven expression of full length (or minimally N-terminally truncated) ERG protein product (Fig. 26.1). The vast majority (>90 %) of ETS rearrangements involves *ERG*, while the remaining ETS fusions include *ETV1* (chromosome 7), *ETV5* (chromosome 3), or *ETV4* (chromosome 17) as common 3′ partners. Similarly, although *TMPRSS2* is the most common 5′ partner for ERG, other 5′ partners, including *SLC45A3* and *NDRG1*, have been identified. Non-*ERG* ETS gene rearrangements commonly have been identified with a variety of 5′ partners.

ETS rearrangements can readily be detected by FISH [6, 10, 11]. In the most commonly used method, fluorescently labeled probes (typically red and green) flanking the regions just 5′ and 3′ to the ETS gene of interest are used. If no ETS rearrangement is present, two fused signals (typically yellow) will be identified per cell, as the probes are close to one another with resulting color addition of red and green signals. Loss of the region 5′ to the ETS gene, as is seen in fusion through deletion, will result in loss of one probe, consequentially showing one yellow signal and one signal the color of the 3′ probe. Similarly, if the material 5′ to the ETS gene is lost to a separate chromosome through insertion, a single yellow signal will be seen in addition to separate red and green signals (Fig. 26.1). Alternative approaches, including three-color FISH, with probes located 5′ to *TMPRSS2*, as well as 5′ and 3′ to *ERG*, have also been described [12].

The *TMPRSS2-ERG* rearrangement results in overexpression of the fusion gene protein product, which is nearly full length ERG protein with no contribution from *TMPRSS2*. Monoclonal antibodies have therefore been developed against this target, and IHC utilizing these antibodies has been shown to strongly correlate with *TMPRSS2-ERG* fusion by FISH (>95 % sensitivity and specificity for detection of translocation) [5, 7, 8, 13–15]. ERG IHC has also been shown to be >99 % specific for prostate cancer and high grade prostatic intraepithelial neoplasia (HGPIN), which is nearly invariably adjacent to ERG^+ carcinoma in prostatectomy specimens [16]. Examples of *ERG* FISH and IHC are shown in Fig. 26.1.

Although FISH for *ERG* rearrangements and IHC for ERG expression are highly concordant in the great majority of cases of prostatic adenocarcinoma, important exceptions are neuroendocrine prostatic carcinoma (typically small-cell carcinoma) and poorly differentiated prostatic adenocarcinomas (e.g., acinar adenocarcinoma, Gleason score 5+5=10). Although this occurs infrequently at presentation, in the setting of prolonged androgen deprivation therapy, prostate cancers develop resistance to hormonal therapy and exhibit loss of androgen receptor (AR) signaling, which may be accompanied by development of a neuroendocrine/small-cell

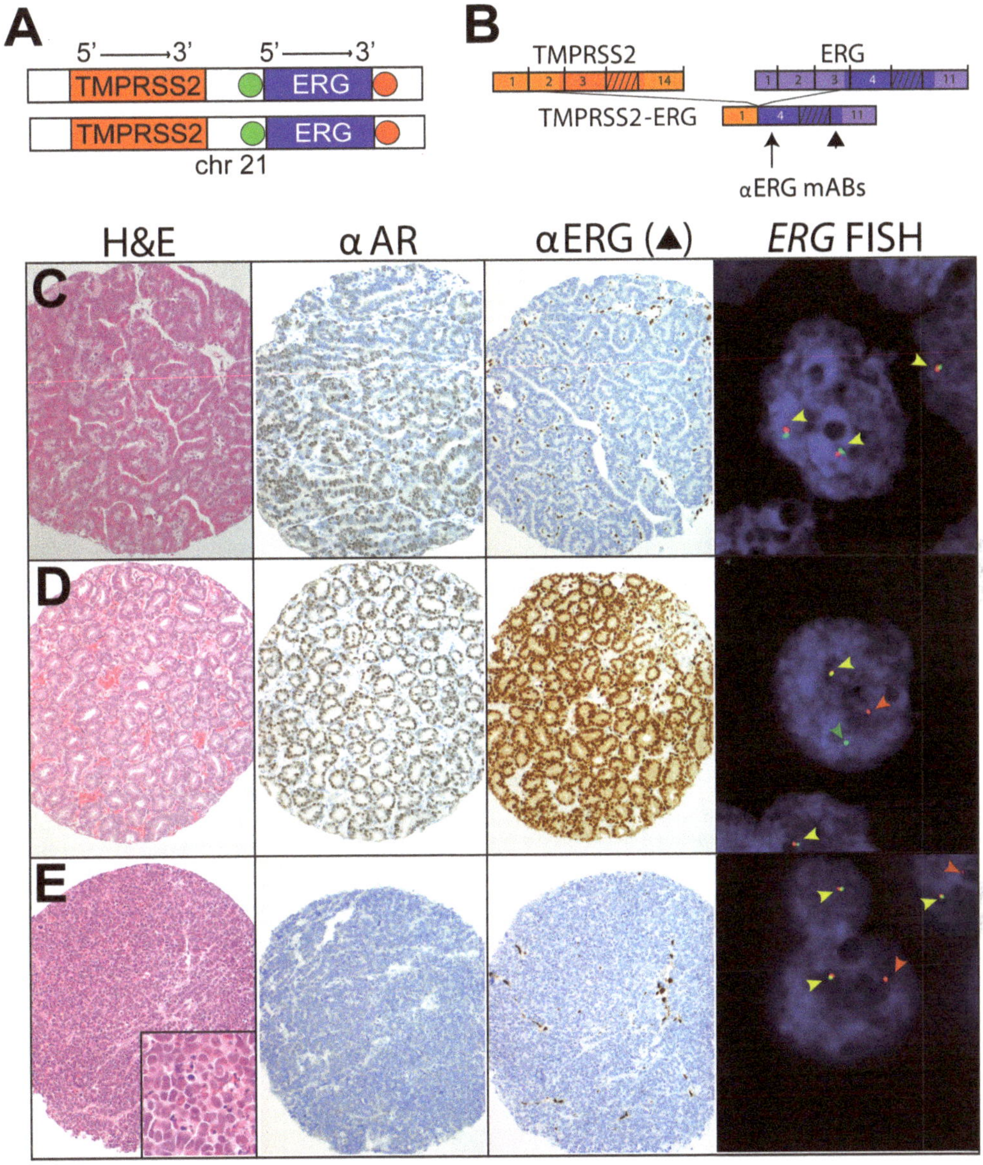

Figure 26-1 Fluorescence in situ hybridization (FISH) and immunohistochemistry (IHC) for detecting *ERG* gene fusions in prostate cancer. (**a**) FISH for *ERG* rearrangements is commonly performed using dual-color split probes flanking *ERG*. (**b**) The transcript structure of *TMPRSS2* and *ERG*, with *boxes* indicating exons, and coding regions in *darker colors*. The structure of the *TMPRSS2-ERG* fusion transcript, which encodes a slightly truncated ERG protein product, is indicated. The advent of monoclonal antibodies (mABs) specific for ERG (clone 9FY CPDR, raised against the N terminus of ERG, *arrow*) and ERG/FLI1 (EPR3864, raised against the C-terminus of ERG, *arrowhead*) has enabled IHC based detection of the *TMPRSS2-ERG* gene fusion product. (**c–e**) ERG staining is reflective of ERG rearrangement and androgen receptor (AR) signaling status. Tissue microarray cores from prostate cancer xenografts were characterized by H and E (*left most panel*), IHC for AR and ERG (using EPR3864), and FISH for *ERG* rearrangements (*right most panel*, as in (**a**)). (**c**) Positive AR staining and negative ERG staining in a xenograft with intact *AR* signal and wild-type *ERG* by FISH (*right panel* with all fused [*yellow*] signals). (**d**) Positive AR and ERG staining in a xenograft with intact *AR* signal and *ERG* rearrangement through insertion (*right panel* with separation of one pair of *red* and *green* signals). (**e**) Negative AR and ERG staining in a xenograft with neuroendocrine/small-cell morphology, loss of the *AR* signal and *ERG* rearrangement through deletion (*right panel* with loss of one 5′ [*green*] signal). Original magnification 10× (H and E and IHC), 60× (FISH and inset of (**e**) [*left panel*])

carcinoma phenotype. The *TMPRSS2:ERG* transcripts encode a slightly truncated ERG protein product, which is driven by the androgen response elements upstream of *TMPRSS2*. Hence, prostate cancers that have lost AR signaling (e.g., small-cell carcinomas in the setting of androgen deprivation therapy) will not express ERG by IHC, although the *TMPRSS2-ERG* rearrangement is still detectable by FISH (Fig. 26.1). This caveat is important in the setting of determining the site of origin of a cancer of unknown primary. That is, if other AR regulated products, such as tissue PSA (IHC), are negative in a cancer of unknown primary, there is little value in assessing ERG protein expression. On the other hand, given that *ERG* rearrangements are present at the DNA level in ~50 % of all prostate cancers (regardless of AR signaling status), and *ERG* rearrangements are maintained in prostate cancers that dedifferentiate, FISH for *ERG* rearrangement can be helpful in tumors that do not express PSA, but when clinical suspicion for a prostatic origin remains.

The clinical utility of ERG assessment is beginning to emerge, most commonly by IHC, given the ease of incorporation it into existing pathology workflows. Because ERG expression is highly specific for prostate cancer (orders or magnitude more specific than alpha-methylacyl-CoA racemase (AMACR), a protein preferentially expressed by prostate cancer), ERG immunoexpression has shown promise in classifying diagnostically challenging, small acinar foci identified on prostate needle core biopsy [17]. The majority of prostate cancers histologically are composed of crowded small acinar glands. Benign processes, including partial atrophy and adenosis, may mimic this histologic appearance, creating a diagnostic challenge. Because ERG is highly specific for cancer, positive ERG immunostaining in these cases is strong evidence that a focus of crowded glands without basal cells represents cancer and is not a benign mimic (Fig. 26.2).

A small subset of HGPIN lesions on needle biopsy are also ERG$^+$ by IHC (10–30 %), which must be considered when using ERG immunostains to support a diagnosis of cancer. Importantly, in prostatectomy sections, ERG$^+$ HGPIN is nearly always located immediately adjacent to ERG$^+$ cancer, while isolated ERG$^-$ HGPIN is frequently observed, supporting potential utility of ERG for risk stratifying isolated HGPIN identified on prostate biopsy [16]. Recently, Park et al. have assessed this question in a unique cohort of patients who were diagnosed with isolated HGPIN and enrolled on a phase III trial of toremifene (a selective estrogen receptor modulator) or placebo [18]. Importantly, all patients had central review of their initial biopsy containing HGPIN, and subsequently were biopsied at 1, 2, and 3 years. ERG status by IHC was assessed in 461 of the 1,590 initially randomized patients. ERG was expressed in 11 % of HGPIN, and by 3 years 53 % vs. 35 % of patients with initial ERG$^+$ vs. ERG$^-$ HGPIN were diagnosed with cancer, a statistically significant difference. Thus, we anticipate that ERG assessment may be incorporated into future efforts to risk stratify isolated HGPIN.

Specific Alterations in ETS Gene Fusion Negative Prostate Cancers

ETS gene fusions are clonal in nearly all cancer foci as demonstrated by FISH and IHC. Combined with their frequency, this provides a basis for basic molecular subtyping of prostate cancer. Importantly, multiple alterations have been found nearly exclusively in ETS fusion negative (ETS$^-$) cancers, supporting this molecular subtyping approach. Here, we briefly highlight specific alterations in ETS$^-$ cancers, as well as potentially relevant genes with alterations in both ETS$^+$ and ETS$^-$ cancers.

SPINK1/SPOP/CHD1

SPINK1 is a trypsin inhibitor first described in the pancreas, and has been shown to be overexpressed in ~5–10 % of prostate cancers [19, 20]. Across more than 10,000 samples, SPINK1 expression is seen nearly exclusively in ETS$^-$ cancer foci, assessed in part through dual ERG/SPINK1 staining [19–23]. Importantly, this simple dual IHC approach provides a simple assessment for assessing clonality on limited tissue specimens which may be useful in the future. Exome sequencing of prostate cancers identified mutations in

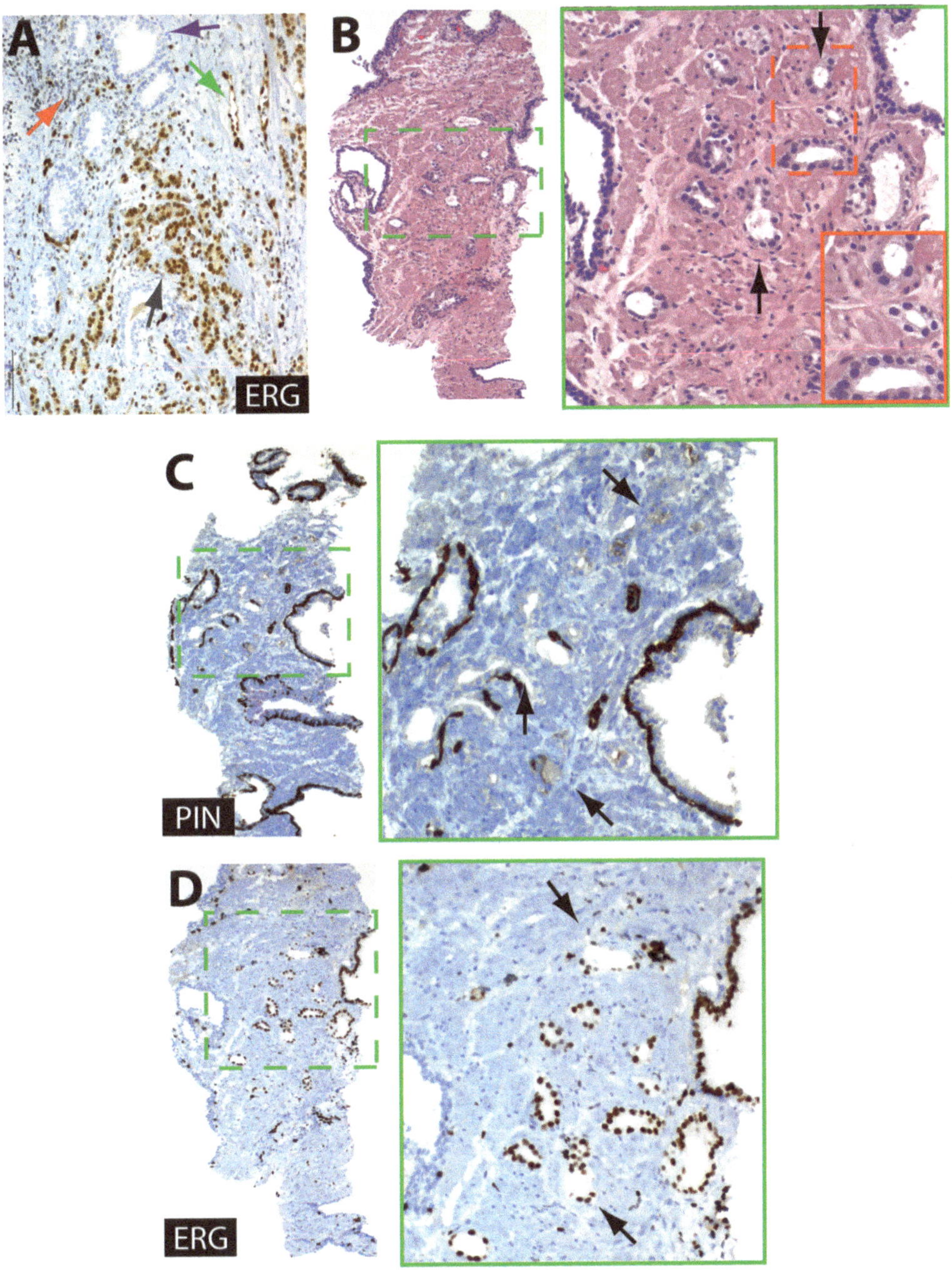

Figure 26-2 ERG immunohistochemistry (IHC) in prostate cancer. (**a**) Typical ERG staining in a focus of prostate cancer using the EPR3864 antibody. ERG shows strong, diffuse nuclear staining in cancerous glands (*grey arrow*) harboring *ERG* rearrangement. Staining is not present in adjacent benign glands (*purple arrow*). ERG antibodies used for IHC also detect wild-type ERG (and may cross-react with the related ETS protein FLI1), which results in diffuse strong nuclear staining in blood vessels (*green arrow*) and weak staining in tissue lymphocytes (only seen with EPR3864, *red arrow*). Original magnification 20×. (**b–d**) Utility of ERG IHC in the diagnostic workup of challenging cases. (**b**) A 12 core needle biopsy had two cores each with a small focus of architecturally and cytologically suspicious glands (*black arrows*, one focus shown). Original magnification 10×, 20× (*green box*) and 60× (*red box*). (**c–d**) The core was assessed by IHC for (**c**) basal cell markers (p63 and high molecular weight cytokeratin, brown chromogen) and AMACR (red chromogen) in a cocktail (PIN), and (**d**) ERG (brown chromogen). Original magnification 10×, and 20× (*green boxes*). In our opinion, as high grade PIN was not in the differential, the presence of ERG staining in the atypical glands (most of which showed only artefactual basal cell staining, see *black arrows* in (**c**)) is consistent with a diagnosis of carcinoma. The other focus on the separate core showed similar staining

the cullin ligase *SPOP* in approximately 6–15 % of prostate cancers [24], making it one of the most frequently mutated genes in prostate cancer. Missense mutations are clustered in the substrate biding cleft region, supporting a functional role in prostate cancer. *SPOP* mutations have been identified exclusively in ETS⁻ prostate cancers, and overlap with *SPINK1*⁺ cancers, as well as those harboring alterations in *CHD1*, a chromatin remodeling enzyme, which is deleted or mutated in 5–15 % of prostate cancers [24]. The clinical significance of the *SPINK1*⁺/*SPOP*^{mut}/*CHD1*^{del} subtype of ETS⁻ prostate cancers is not well defined, and no clear association with differential outcome after prostatectomy has been reported.

RAS/RAF/FGFR Family Fusions

Additionally, RNA-seq based studies have identified a number of potentially targetable gene fusions involving members of the *RAS*, *RAF*, and *FGFR* family fused typically to androgen regulated genes, in a total of ~2–5 % of prostate cancers, which are exclusively ETS⁻ [25–30]. Although point mutations in these gene families are very infrequent in Caucasian cohorts, they may be more frequent in other populations, which correspondingly also have lower ETS⁺ prevalence [31–36]. Given the development of inhibitors that target these alterations in other cancers, we expect that identification of cancers with these alterations may be important for directing therapy in the future.

PTEN

Alterations in classic tumor suppressors and oncogenes, such as *PTEN*, *TP53*, and *MYC* are relatively frequent in prostate cancer. *TP53* was identified as the most frequently mutated gene through point mutation or indels in CRPC [30], and *MYC* is frequently overexpressed in prostate cancer through broad amplifications of 8q, which occur early in prostate cancer development [37]. At present, the clinical utility of these markers is not clear. *PTEN*, a tumor suppressor located on chromosome 10q23, has received intense study in prostate cancer, in large part due to its association with aggressive disease [38–42]. We highlight it here as it provides an important study on the incorporation of prognostic biomarkers into clinical models. PTEN is the most commonly deleted gene in prostate cancer, being deleted in 15–50 % of cases [38–42]. *PTEN* deletions are up to three times more common in *ETS*⁺ prostate cancers than *ETS*⁻ prostate cancers [38–42]. Non-deletion mutations in *PTEN* are also present and observed in up to 21 % of prostate cancers [24, 43–46].

PTEN deletions have been associated with several clinical and pathologic features of aggressive tumors [39, 40, 42, 45, 47–52]. Presently, the gold standard for detections of *PTEN* deletions is FISH, which directly detects the abnormality. IHC directed against the protein product of *PTEN* has been developed, and has shown good sensitivity for deletion. However, 37–45 % of prostate cancers exhibiting PTEN loss by IHC have shown no detectable loss by FISH or SNP microarray, which may be explained by chromosomal rearrangements disrupting the function of PTEN. These are not identified by standard FISH probes used to detect *PTEN* deletions. Which type of assay best correlates with prostate cancer outcome remains unknown, and is an important question for future studies. In contrast to *ETS* status and *SPINK1* expression, which tend to exhibit diffuse expression within a prostate cancer focus, *PTEN* deletions may show considerable heterogeneity, and loss may be seen in only a focal portion of the tumor or nonuniformly in circulating tumor cells [45, 48, 49]. Despite numerous studies reporting association of *PTEN* deletion with clinicopathologic parameters of aggressive disease behavior, definite clinical value in assessing *PTEN* has not been shown to date, primarily because studies have not demonstrated that it adds prognostic information when combined with standard clinicopathologic data. For example, a recent large study with good follow-up data showed that, though statistically significant, *PTEN* deletion status showed a small hazard ratio (1.2) for the prediction of biochemical recurrence in a Cox multivariate regression incorporating *PTEN* deletion status, pathologic tumor stage, serum PSA level, and Gleason score. In contrast, Gleason score and pathologic stage showed maximum hazard ratios of 6.1 and 5.9, respectively [41]. Similarly, a recent retrospective study demonstrated PTEN loss by IHC was predictive of biochemical recurrence after prostatectomy in a cohort of men matched for age, Gleason score, and stage. Although the hazard ratio for recurrence in this study was higher (2.2), it remains unclear if this is of sufficient magnitude to demonstrate clinical utility.

These findings highlight a crucial consideration in incorporating molecular assays into clinical practice. That is, does the marker add to the best available model in a clinically meaningful manner? [53] For example, in deciding if *PTEN* status should be used to predict outcome (e.g., biochemical recurrence [BCR] after prostatectomy), evidence of mere independence from other easily assessed parameters at prostatectomy does not necessarily indicate added clinical value, as these parameters can all be easily obtained as a group. Hence, in addition to independence, the new marker must show added predictive value of sufficient magnitude to be clinically actionable when incorporated into the best currently available model (using markers that can be assessed easily). The most relevant method of comparing performance utilizes an area under the curve (AUC) comparison of receiver operator characteristic (ROC) curves or concordance indexes, comparing the current best model to the same model including the new biomarker (*PTEN* in this example).

Assays to assess the above mentioned single genes are beginning to reveal robust, reproducible molecular subtypes of prostate cancer, with specific clinicopathologic associations. For example, *ETS*$^{+}$/*SPINK1*$^{-}$/*SPOP*$^{(wt)}$/*PTEN*$^{(homozygous\ del)}$/*MYC*$^{(amp)}$ prostate cancer would likely represent an aggressive, high grade tumor, while an *ETS*$^{+}$/*SPINK1*$^{-}$/*SPOP*$^{(wt)}$/*PTEN*$^{(wt)}$/*MYC*$^{(wt)}$ tumor may represent an indolent, low grade tumor. Similarly, an *ETS*$^{-}$/*SPINK1*$^{+}$/*SPOP*$^{(mut)}$/*PTEN*$^{(wt)}$/*MYC*$^{(wt)}$ tumor may have pathologic and prognostic features intermediate to these two examples. Perhaps more importantly, the ability to reproducibly identify multiple subtypes will facilitate the stratification of prostate cancer cohorts for biologic behavior, response to therapy and will help towards the prioritization of subtypes for investigation of novel targeted therapies (even if not directed at these specific biomarkers). The approximate distribution of lesions defining major molecular subtypes, demonstrated by mutual exclusivity, is shown in Fig. 26.3.

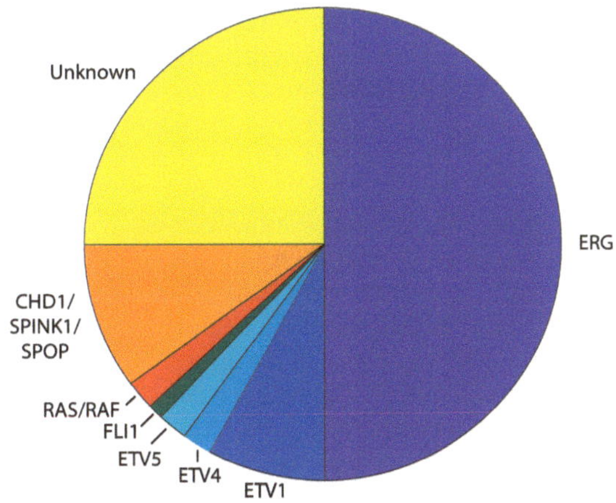

Figure 26-3 Molecular subtypes of prostate cancer. The approximate distribution of driving molecular lesions in prostate cancers amongst PSA-screened Caucasians are presented. *ETS* gene fusions (including those involving *ERG*, *ETV1*, *ETV4*, *ETV5*, and *FLI1*), are mutually exclusive with tumors harboring activating gene fusions or mutations in RAS and RAF family members or FGFR2 fusions. A subset of *ETS* and *RAS/RAF* wild-type tumors have *SPINK1* outlier expression, disruption of *CHD1*, and/or *SPOP* mutations. Approximately 25 % of prostate cancers have private or unclear drivers

Applications of the Hereditary Genome in Prostate Cancer: HOXB13

Although in this chapter we have focused on somatic genomic alterations in genitourinary neoplasms (alterations in the cancer genome), hereditary (germ line) mutations are known to predispose to multiple genitourinary neoplasms (e.g., *BRCA2* and prostate cancer or other entities discussed in the kidney cancer section later on). As with somatic mutations, the ease of interrogating the genome has led to a large number of studies directed at identifying relatively common hereditary variants that influence the overall risk of developing cancer or the risk of developing aggressive cancer forms. Such hereditary variants, including SNPs or copy number variants (CNVs), are commonly interrogated in genome-wide association studies (GWAS), typically through array based technology.

A critical point for the practicing pathologist to consider is that although studies may identify variants that show impressive statistically significant associations (often assessing tens of thousands of cases and controls), such highly significant results do not equate to clinical significance. For example, in an effort to personalize serum PSA cutpoints for prostate cancer detection, Gudmundsson et al. identified six SNPs that were associated with serum PSA levels, each at a significance of $p<3.0E-10$ [54]. However, using a model incorporating serum PSA together with these identified SNPs compared to PSA alone only increased the AUC by 0.5 % and 1.4 %, respectively, in two test cohorts. The limited increase in AUC in part reflects the modest effects of the identified SNPs on PSA levels. Similar findings of limited clinical impact have been frequently identified in several other GWAS studies [55].

An alternative strategy made feasible by the decreased cost of sequencing, which is allowing the interrogation of thousands of genomes from "case" and "control" patients, is to identify rare variants that may be more strongly associated with cancer development. As an example, through sequencing the 17q21-22 region which had previously been linked to prostate cancer through pedigree analysis of families with hereditary prostate cancer, Ewing et al. identified the same non-synonymous mutation in *HOXB13* (p.G84E), which co-segregated with prostate cancer in each family [56]. This rare variant (carrier frequency estimate 0.1–1.5 %) has consistently shown higher odds ratios (~3–10×) for associations with prostate cancer (including early onset and familial cancer) compared to variants identified by GWAS studies (odds ratios typically 1.1–1.5). Similar findings in other contexts will likely be enabled by WES/WGS efforts [57] and will likely define new "high-risk" criteria that then impact which patients undergo screening and subsequent biopsy for the pathologist to interpret.

Multigene Panels in Prostate Cancer

Using gene expression profiling using DNA microarrays, much has been learned about common gene expression alterations in various cancer types. However, this technique is difficult to perform on routinely processed formalin fixed paraffin embedded (FFPE) tissue, and is better suited for use with fresh or frozen tissue, which severely limits its clinical utility. As a research tool, DNA microarrays provide unique insight into cancer classification, and have been used in the breast, for example, to develop consistent gene classifiers validated across several studies. Although expression clusters have been investigated and identified in prostate cancer [58, 59], robust and reproducible classifiers based on the clustering of tens to thousands of genes have not been consistently identified. Despite that limitation, studies using DNA microarrays have identified genes and pathways of interest and potential prognostic signatures [60, 61]. Myriad diagnostics has developed a 40-gene qPCR panel (Polaris™), which may modestly improve on the ability of standard clinicopathologic data to predict death from prostate cancer [62, 63]. This panel, which includes 40 cell cycle genes, has been shown to have potential utility in both radical prostatectomy and needle core biopsy specimens. Genomic Health has recently announced the Oncotype DX® Prostate Cancer Assay, which measures expression of several genes using reverse transcription polymerase chain reaction (RT-PCR) from FFPE needle biopsy tissue [64], similar to the currently available Oncotype DX® Breast Cancer Assay. Likewise, GenomeDx has reported a genomic-clinical classifier model (Decipher™)using expression profiling data from Affymetrix exon microarray analysis of FFPE isolated RNA that reportedly showed improved performance compared to a clinical model for predicting clinical recurrence [65, 66].

For such expression-signature based classifiers to be routinely used in clinical practice, a number of conditions should be met. First, assays should be able to use standard, clinically relevant samples (FFPE tissues). Second, they need to be validated in prospective, independent tissue cohorts. If the assay is intended to be used to stratify risk among men under consideration of active surveillance (delaying definitive therapy), it should be applicable to prostate biopsy tissues and must account for potentially unsampled higher grade and/or multifocal prostate

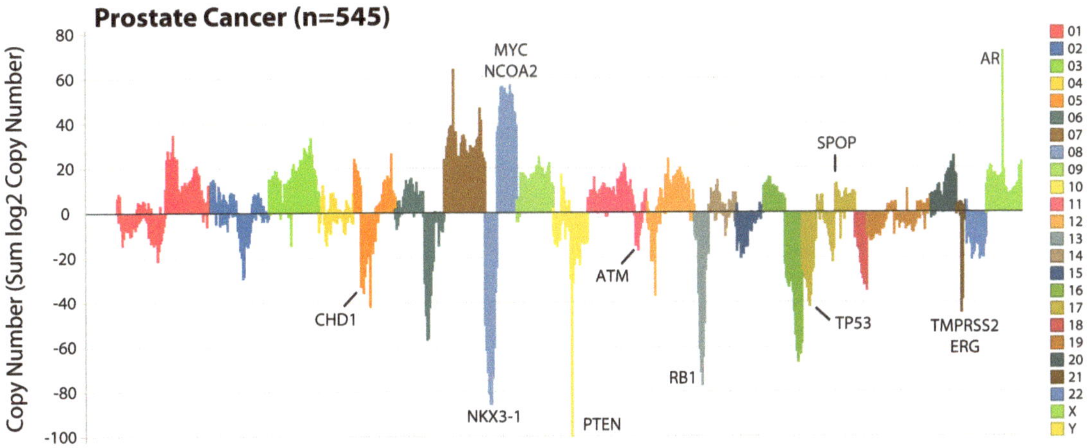

Figure 26-4 Copy number changes in prostate cancer. Genome-wide copy number profiles from 545 prostate cancers from four studies were visualized using the Oncomine Powertools DNA Copy Number Browser (powertools.oncomine.com). The sum of the log2 copy number for each segmented sample is plotted in genomic order (see legend for chromosomes). The locations of genes harboring recurrent copy number gains/losses or mutations are indicated

cancer, which may not be represented in the diagnostic tissue. Similarly, if assays are to be used to predict "aggressiveness" after prostatectomy (commonly determined by biochemical recurrence), it would be highly desirable that the data generated from the assay would actually influence treatment, such as addition or dose modification of radiation therapy, or addition or modification of the length of anti-androgen therapy. Lastly, it is imperative that the assay adds to the best available clinicopathologic model presently used for predicting the outcome of interest in a clinically meaningful way, rather than only being statistically independent of known clinicopathologic parameters.

Whole-Genome and Exome Sequencing in Prostate Cancer

As of fall 2013, WGS of over 60 prostate cancers has been reported [29, 67]. These studies and several exome-sequencing studies have helped further elucidate the genomic landscape of prostate cancer. Importantly, WGS has illuminated that prostate cancers harbor large numbers of rearrangements, many involving known cancer-associated genes. A distinctive "closed chain" pattern of rearrangement has also been identified, characterized by complex exchanges within and between chromosomes with no net loss of genetic material, resulting in unique, balanced combinations of chimeric chromosomes [29]. These features were commonly found in *ETS*$^+$ prostate cancers, with *ETS* members being involved in the closed chains, further supporting *ETS* gene fusions as defining a distinct molecular subtype.

By combining previous array CGH (aCGH) and SNP copy number profiling studies with copy number data from sequencing based studies, the emerging portrait of the prostate cancer genome suggests that focal gains or losses and point mutations are relatively infrequent (*PTEN* loss and *SPOP* and *TP53* mutations being the most frequent), with rearrangements and broad gains/losses playing important roles (Fig. 26.4).

Sequencing of Advanced Prostate Cancer for Precision Medicine

Patients with high grade or advanced prostate cancer are frequently treated with androgen deprivation therapy (medical castration), using drugs such as leuprolide and flutamide. These treatments frequently lead to a clinically meaningful, long term response, with many patients achieving more than 5 years of disease control. However, all patients will eventually become resistant to standard androgen deprivation therapies, a state known as castration resistant prostate cancer (CRPC). This aggressive state of the disease is associated with high mortality, and was previously thought to be androgen independent. However, studies evaluating the mechanisms of castrate resistance have demonstrated intense selection for maintenance of androgen signaling, using varying mechanisms including increased *AR* expression, upregulation of androgen synthesizing enzymes within cancer cells, development of splice variants of *AR*, and the development of truncated, formerly androgen-dependent proteins that act independently of AR binding [68]. Such observations subsequently led to the hypothesis that many CRPCs are still dependent on androgen signaling. The latter has been borne correct as demonstrated through the efficacy of next-generation anti-androgen signaling drugs, including abiraterone and enzalutamide, in post-chemotherapy CRPC patients.

In other solid tumors, such as lung cancer, molecular testing is often utilized for selection of optimal medical therapy in patients with advanced disease [69]. Given that CRPC is an aggressive disease state amenable to treatment with agents targeting novel aspects of prostate cancer biology, it is likely that molecular testing will play an increasing role in the management of advanced prostate cancer in the near future. In keeping with this, in an effort to understand driving genetic events in CRPC, Grasso et al. performed an integrative analysis, including exome sequencing, of 50 lethal CRPCs obtained during rapid autopsy, which allows for procurement of fresh frozen tissues [30]. This study confirmed alterations in genes known to harbor aberrations in CRPC, including amplification and mutation of *AR*, and loss and deletion of *TP53* and *PTEN*. Mutations were also identified frequently in chromatin/histone remodeling genes, including several members of the *MLL* complex. Perhaps more importantly, a substantial number of patients, despite having clinical resistance to androgen deprivation, showed strong expression of the *TMPRSS2-ERG* gene fusion, indicating active *AR* signaling. Such a profile of androgen signaling assessment including expression of *AR* and target genes, *ETS* gene fusion genomic and transcript status and presence of *AR* copy number alterations and mutations may be used in the future to select and predict responses to novel anti-androgen drugs.

Furthermore, several patients with CRPC analyzed by Grasso et al. harbored mutations in potentially targetable genes [30]. For example, one individual had homozygous *BRCA2* deletions, in addition to a statistically significant increase in mutations consistent with disruption of the DNA repair mechanism, suggesting potential response to poly(ADP-ribose) polymerase family (PARP) inhibitors [70]. Another case showed high level copy gain of cyclin-dependent kinase 4 (*CDK4*), suggesting possible benefit from a CDK4 inhibitor. Given the lack of prevalent targetable alterations in CRPC, other than those affecting *AR*, and the presence instead of rare targetable driving mutations (e.g., *RAF* alterations), therapy may need to be individualized for small subsets of patients with CRPC ("precision" or "personalized" medicine). An example of such an effort is the MI-ONCOSEQ study launched at the University of Michigan. In this ongoing study, comprehensive exome and transcriptome sequencing is performed in patients with advanced cancers (including CRPC patients), for whom available treatment options have been exhausted. Patients meet with genetic counselors, and sequencing results are discussed at a multidisciplinary tumor board in an effort to discuss potentially actionable alterations that could be targeted with established or investigational agents. Most recently, this program has been used to identify an androgen driven *FGFR2* fusion in a patient with CRPC, which could be targeted by both FGFR family inhibitors as well as anti-androgens [25].

The majority of genomic studies performed on prostate cancer have used fresh or frozen tissue, which is a suboptimal specimen for widespread clinical use. Sequencing based assays that are compatible with FFPE tissues are more applicable for routine use. For example, the company Foundation Medicine has developed a multiplexed sequencing/copy number assay that can be performed from FFPE [71]. Beltran et al. have recently used this assay to sequence 3,320 exons from 182 cancer-associated genes, as well as 37 introns in 14 commonly rearranged genes in a series of 45 prostate cancers including a subset of CRPCs [28]. Their results were consistent with previous studies, showing amplifications of *PTEN* and *MYC*, *TMPRSS2-ERG* gene fusions, *TP53* mutations, and *AR* amplifications in CRPC in similar frequencies as previously described. Such an assay is desirable, because it could provide nearly all useful ancillary genomic data on a cancer specimen from FFPE in a single assay. An important caveat is that optimal therapy selection may require assessment of both the genome and the transcriptome, in an effort to understand current driving aberrations. For example, perhaps the most important information to assess in CRPC is *AR* signaling status. Genomic identification of *AR* amplification or mutation, or evidence of *TMPRSS2-ERG* rearrangements, do not necessarily inform on *AR* signaling status. That is, CRPCs with evidence of active androgen signaling (retained expression of *AR* and of *AR* regulated genes, including *ERG*) may be candidates for more aggressive androgen deprivation therapy, irrespective of genomic status of *AR* and *TMPRSS2-ERG*, while those with androgen signaling loss (and no expression of *AR* or *TMPRSS2-ERG*) may not.

Genomic/Transcriptomic Alterations for Early Detection of Prostate Cancer

Screening for prostate cancer with serum PSA is currently controversial, and several professional organizations in the USA have recommended against general screening. PSA has both sensitivity and specificity limitations and leads to the detection of a significant proportion of small indolent prostate cancers that would not otherwise have become clinically relevant during a man's lifetime. Hence, intense efforts and resources are being spent on biomarkers for the early detection of prostate cancer in pursuit of a better screening approach.

Prostate Health Index (PHI), a combination of serum PSA, free PSA, and [-2] pro-PSA, has been FDA approved for risk stratification in the pre-biopsy setting and outperforms serum PSA or free PSA alone [72]. An inherent limitation in this test is that [-2] pro-PSA, the most cancer specific of the three markers, shows at least weak expression in >70 % of benign prostate epithelium, consistent with PSA and its derivatives being more prostate specific than prostate cancer specific [73].

The two most advanced urine based biomarkers for prostate cancer are *PCA3* and *TMPRSS2:ERG*. *PCA3* is a noncoding RNA expressed in prostate cancer and HGPIN [74–78], that is detectable in the urine of patients with this cancer [79]. Progensa® is a currently available commercial clinical assay for urine *PCA3* measurement, which calculates a score based on the ratio of *PCA3* RNA to PSA mRNA in urine samples collected following "attentive" digital rectal exam. Progensa® was cleared by the Food and Drug Administration (FDA) for use in determining whether men with a negative prostate biopsy (triggered most commonly by PSA screening) should obtain a repeat biopsy. Using the urine *PCA3* assay score at a cutoff of 25, a negative result has a 90 % negative predictive value (NPV) for prostate cancer [80]. Repeat biopsy in men with a negative urine *PCA3* study can thus be avoided. *PCA3* has also been shown to outperform serum PSA in predicting the presence of cancer in first time prostate biopsies, and incorporation into multivariate models such as the Prostate Cancer Prevention Trial Prostate Cancer Risk Calculator (PCPTRC) improves performance [81].

TMPRSS2-ERG gene fusion transcripts have also been detectable in the urine of patients with *ERG*+ prostate cancer [79, 82]. Like *PCA3*, models including urine *TMPRSS2-ERG* measurement have shown improvement over traditional predictors. For example, in a recent report urine *TMPRSS2-ERG* and *PCA3* combined with the variables in the PCPTRC had improved performance for the detection of prostate cancer (AUC 0.75–0.79) over the PCPTRC derived risk alone (AUC 0.64–0.66) [83]. Similar performance of urine *TMPRSS2-ERG* has been observed in multiple cohorts using different assays [84–86], including an assay analogous to the Progensa *PCA3* assay, which can be performed on the same urine specimen sent for *PCA3* testing. Importantly, analysis of ERG by IHC in prostatectomy samples paired with the testing of urine samples demonstrates that the urine *TMPRSS2-ERG* score (calculated similarly to the Progensa *PCA3* score) is strongly correlated to the total *ERG*+

tumor dimension [87], and at present there is no known biologic explanation for continued elevation of urine *TMPRSS2-ERG* following prostatectomy except for residual prostate cancer.

The use of urine as a biocompartment for *TMPRSS2-ERG* and *PCA3* detection highlights the utility of other sources besides tissues and serum for interrogating the cancer genome. For example, the quantity of circulating tumor cells (CTCs) detectable in the blood of patients with metastatic prostate cancer, and their change in number in response to therapy, can be prognostic [88, 89]. Perhaps more importantly, detection of important genetic driving events, such as *AR* amplification and signaling status, *PTEN* loss and *TMPRSS2-ERG* rearrangement status, have all been demonstrated in CTCs [48, 90–93], suggesting that such specimens may be utilized for genetic interrogation when tissue is not readily available. Similarly, the analysis of circulating free DNA in plasma or urine also allows for interrogation of the cancer genome given the depth of sequencing coverage enabled by second and third generation NGS platforms, and preliminary results suggest potential utility in prostate cancer [94–96]. Lastly, Olmos et al. and Ross et al. have each shown that two separate expression signatures derived from whole blood collected from patients with CRPC are prognostic, with such signatures likely reflecting host response rather than tumor characteristics directly [97, 98]. Given the advances made in both CTC and sequencing technologies, we anticipate that interrogation of non-tissue based biospecimens will play an increasingly relevant role in the future.

The Molecular Pathology of Bladder Cancer

The majority of bladder cancers are urothelial carcinomas, which have widely diverse histopathologic features and clinical behavior. Noninvasive low-grade papillary urothelial carcinoma progresses to invasive cancer in only a minority of cases, while flat urothelial carcinoma in situ has a much higher risk of progression to invasive cancer. Similarly, while patients with invasive urothelial carcinoma confined to the lamina propria (pathologic stage pT1) have a relatively good prognosis and are candidates for conservative therapy, patients with invasive disease involving the muscularis propria or beyond (pathologic stage T2 or greater, known as locally advanced disease) have high mortality rates and require aggressive therapy.

Current oncogenetic models in urothelial carcinomas suggest two separate molecular routes for superficial and muscle invasive disease [99, 100]. The first route is defined by early, activating point mutations in *FGFR3*, RAS family genes, *PIK3CA* or other oncogenes. These early changes are correlated with papillary noninvasive tumors and superficially (i.e., lamina propria only) invasive disease. A small fraction (~15 %) in this group subsequently develop loss of function of *TP53*, *RB1*, *PTEN*, or other tumor suppressor genes, resulting in progression to muscle-invasive carcinoma, which may ultimately lead to metastasis and death. In the second route, loss of function of tumors suppressor genes occurs early, without preceding mutations in *FGFR3*, RAS family genes, or *PIK3CA*. This route is associated with "flat" urothelial carcinoma in situ as a precursor lesion and a higher risk of progression to invasive carcinoma. Evidently, these associations between initiating events and subsequent behavior are not absolute.

Single Genes in Bladder Cancer

FGFR3 AND TP53

Despite the above delineated pathways of cancer progression, robust molecular classifications for urothelial carcinoma have not been fully defined. For example, although *FGFR3*$^{(mut)}$ and *TP53*$^{(mut)}$ tumors appear to define different pathways of tumor progression, many *FGFR3*$^{(mut)}$ tumors develop mutations in *TP53*. In fact, *TP53* mutations are thought to be a major molecular route by which many *FGFR3*$^{(mut)}$ tumors progress to invasive disease. As a single biomarker, *FGFR3* mutational status has good performance for predicting which patients will develop locally advanced disease. Burger et al. have recently described that *FGFR3* mutation is protective against progression in patients with high grade pTa/pT1 urothelial carcinoma. Progression free survival at 5 years was 100 % for patients with high grade disease and *FGFR3* mutation, but approximately 45 % for those with high grade disease without it [101]. Mutational status was not as useful in evaluating urothelial carcinomas as a whole, however, as multivariate analysis including tumor grade showed that *FGFR3* mutational status was not a statistically significant predictor of progression.

Although the association of *TP53* mutation with aggressive urothelial cancer supports an important functional role, studies of *TP53* as a biomarker have been somewhat disappointing. *TP53* mutational status only modestly improves on standard clinicopathologic data (pathological stage, presence of angiolymphatic invasion, concomitant flat urothelial carcinoma in situ (CIS), patient age, patient gender, and chemotherapy status) for the prediction of disease recurrence and cancer specific mortality in a cohort of patients with invasive bladder urothelial carcinoma [102]. Similarly, *TP53* mutational status, as a single biomarker, only modestly improves the ability to predict time to progression from pT1 to muscle-invasive disease in patients with high grade urothelial carcinoma (e.g., HR 1.47), and no such predictive value seems to be present in pTa tumors [103]. Despite this lack of demonstrated clinical utility, mutations in *TP53* are diverse in urothelial carcinoma, and more sophisticated *TP53* analysis may yield better results in the future. For example, George et al. have shown both clinical outcome and p53 status by IHC are associated with specific (i.e., location of exon, and single vs. multiple mutations) *TP53* gene mutations. Specifically, single mutations in exon 5 were shown to be associated with wild-type p53 expression by IHC and less aggressive behavior. In contrast, tumors with multiple *TP53* mutations showed p53 overexpression by IHC in most cases and aggressive tumor behavior. Tumors with single mutations in exon 8 of *TP53* also showed p53 overexpression in most cases but intermediately aggressive tumor behavior [104].

KI-67

Immunoexpression of proliferation marker Ki-67 has shown modest prognostic value. A large recent trial indicated that Ki-67 independently improved prediction of disease recurrence and cancer specific survival in multivariate model including pathologic stage, grade, presence of angiolymphatic invasion, concurrent flat CIS, age, and gender [105]. This study also reported that addition of Ki-67 status to a standard multivariable model enhanced its predictive accuracy by 2.9 % for disease recurrence and 2.4 % for cancer-specific survival.

OTHER SINGLE GENES IN BLADDER CANCER

Additional novel single gene biomarkers also have promise. For example, increased expression of the Cancer-Testis antigens MAGE-A3, NY-ESO-1, and LAGE-1, determined by quantitative RT-PCR (qRT-PCR), has been shown to correlate with shorter progression free survival [106]. Amplification of HER2/neu, a member of the EGFR family, has been shown to correlate with increased risk of progression from Ta/T1 to locally advanced disease [107].

Molecular Subtyping of Bladder Cancer

Lindgren et al. were among the first to highlight the considerable molecular diversity in urothelial carcinomas in two consecutive studies evaluating gene expression profiles by DNA microarray, chromosomal analysis by aCGH and mutational status of targeted genes in a large cohort of urothelial carcinoma. The first study unveiled the presence of three groups of urothelial carcinomas (Clusters I–III) based on unsupervised gene expression profile hierarchical clustering [108]. Cluster I was characterized by low expression of cell cycle genes, frequent *FGFR3* mutations, lack of *TP53* mutations, and no losses of chromosome 9. The majority (73 %) of cases in this group showed low grade histology. Cluster III showed overexpression of cell cycle genes, no *FGFR3* mutations, frequent *TP53* mutations, and frequent losses of chromosome 9. No cases in this cluster showed low grade histology. Cluster II was the largest cluster, and exhibited features intermediate between Clusters I and

III. In their subsequent study involving a subset of the original cohort, the authors were able to group tumors into two groups (MS1 and MS2), based on mutational status of genes in several pathways, including cell cycle related genes [109]. As is expected from reducing a three group classifier to one with only two groups, MS1 tumors carried frequent *FGFR3* mutations, but *TP53* mutations were present in several cases. MS2 tumors were likewise enriched in *TP53* mutations, although some had *FGFR3* mutations. MS2 showed much greater chromosomal instability, seen by increased focal genomic amplifications. *HRAS* and *KRAS* mutations were seen equally in MS1 and MS2. More importantly, this study used genes in its clustering to create multigene prognostic models, which were useful in predicting behavior in subgroups of patients. Specifically, a gene signature based on these clusters was able to predict time to metastasis in a group of patients with high grade disease undergoing cystectomy (cumulative proportion metastasis free 0.90 vs. 0.45 at 60 months).

Multigene Panels in Bladder Cancer

Because urothelial carcinomas harbor substantial molecular diversity, single gene assays may not capture molecular subsets large enough to be clinically meaningful. Therefore, multigene assays may provide greater insight into molecular abnormalities in a given tumor, and appear to have greater ability to predict disease behavior than assays evaluating any single gene. Several gene panels have been described. For example, Dyrskjøt et al. describe a 12-gene qRT-PCR expression panel, based on multiple incarnations and distillations of previous, larger panels described by the same group. The 12-gene panel had strong performance for the ability to predict progression from non-muscle-invasive cancer to locally advanced disease (HR 7.4, multivariate model including stage, grade, and treatment) [110]. Importantly, the full model incorporating the 12-gene panel with clinical variables showed improved predictive power over the 12-gene score alone and over clinical variables alone (Harrell's C was 82 % for the full model, 75 % for the 12-gene panel alone, and 72 % for clinical variables alone). However, this represents optimum prediction and was not independently validated. Similarly, this assay was performed on fresh frozen tissue, which limits direct clinical translation.

Gene expression panels have also been used to predict recurrence. For example, Shariat et al. report a three antibody IHC panel (p53, p27, and Ki-67) associated with disease recurrence and disease specific mortality in patients undergoing cystectomy [111]. Dobosq et al., using a 110-gene qRT-PCR panel, found three genes (*PPARG*, STATHMIN, and CAVEOLIN-2) to be associated with early recurrence in patients with low-grade, noninvasive urothelial carcinoma [112]. Panels predicting disease stage and lymph node status have also been described [113], including a 20-gene panel associated with lymph node positivity in cystectomy patients [114]. In a recent meta-analysis, Riester et al. evaluated numerous gene signatures from prior studies, in an attempt to define a gene signature that can improve upon established clinicopathologic nomograms for predicting survival in bladder cancer patients undergoing cystectomy [115]. A 20-gene panel was proposed among genes consistently showing correlation with survival across studies.

Markers predicting response to neoadjuvant chemotherapy in urothelial bladder carcinoma are also actively sought. Takata et al. reported a 14-gene qRT-PCR panel discovered using expression profiling of transurethral resection (TUR) specimens prior to neoadjuvant M-VAC (methotrexate, vinblastine, doxorubicin, and cisplatin) treatment [116]. The same group identified a 12-gene qRT-PCR panel using genome-wide expression profiling that is associated with response to combined gemcitabine and carboplatin (GC) neoadjuvant chemotherapy [117]. Interestingly, although the M-VAC panel and GC panel contain no overlapping genes, both panels contain genes in the soluble carrier group (*SLC16A3* and *SLC22A18* in M-VAC and GC panels, respectively.).

Unfortunately, the vast majority of these predictors suffer from the same limitations as those described for prostate, including

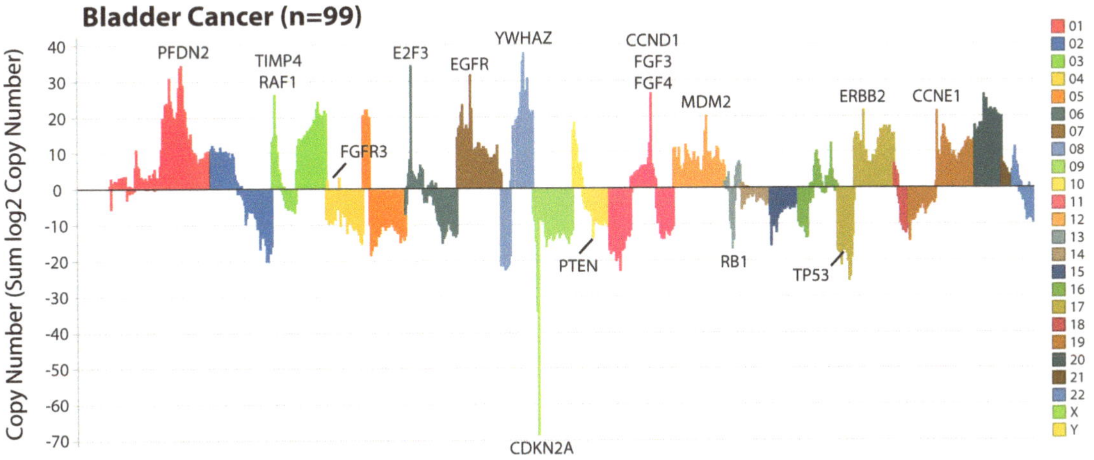

Figure 26-5 Copy number changes in bladder cancer. Genome-wide copy number profiles from 99 bladder cancers from a single study (TCGA) were visualized using the Oncomine Powertools DNA Copy Number Browser. The sum of the log2 copy number for each segmented sample is plotted in genomic order (see legend for chromosomes). The location of genes harboring recurrent copy number gains/losses or mutations are indicated

lack of validation in independent sample sets, and they have not been shown to add to the best available clinical models. As large efforts like the TCGA identify additional molecular subtypes and the ability to incorporate DNA based lesions, we anticipate that the predictive ability of such assays will likely improve.

Whole-Genome and Exome Sequencing in Bladder Cancer

Fewer WGS studies have been performed on urothelial carcinoma compared to prostate cancer, although TCGA data is expected soon. In the largest study to date, WGS was performed in nine urothelial carcinomas, followed by selective mutational analysis of WGS identified outlier mutations in a separate series of 88 tumors [118]. Consistent with previous studies, frequent mutations were seen in *TP53* (21 %), *RB1* (11 %), *HRAS* (10 %), *FGFR3* (9 %) and *KRAS* (6 %). None of the analyzed tumors contained simultaneous mutations in *TP53* and *FGFR3*, or *TP53* and *HRAS*. More importantly, a high (59 %) mutation rate was identified in chromatin remodeling genes, a finding not previously described. Mutations of chromatin and histone modifiers is likely to be a near uniform finding in cancer, as such genes are commonly mutated in cancers of diverse histologic origin, including prostate and kidney.

Recently, WGS has been performed on a patient with metastatic urothelial carcinoma demonstrating a durable response to mTOR inhibitor agent everolimus [119]. Somatic loss-of-function mutations were identified in *TSC1*, a known regulator of *mTOR* pathway activation. Subsequent targeted mutational analysis in a larger cohort (109 bladder cancers) revealed an 8 % *TSC1* mutation rate that also correlated with response to everolimus. This study demonstrates a potential method for identifying genes associated with therapy response, which could function as a future.

Preliminary analysis of copy number alterations in bladder cancer points to a more focal copy number gain and loss compared to other genitourologic malignancies (Fig. 26.5). *CDKN2A* is the most frequently deleted gene in bladder cancer. Loss of this gene locus on chromosome 9p21 has been exploited in urine multitarget FISH assays such as UroVysion. Unlike prostate and kidney cancers, bladder cancer shows focal high level amplifications of receptor tyrosine kinases such as *ERBB2* and *EGFR* in a mutually exclusive manner (Fig. 26.6). The amplification levels of the latter are similar to

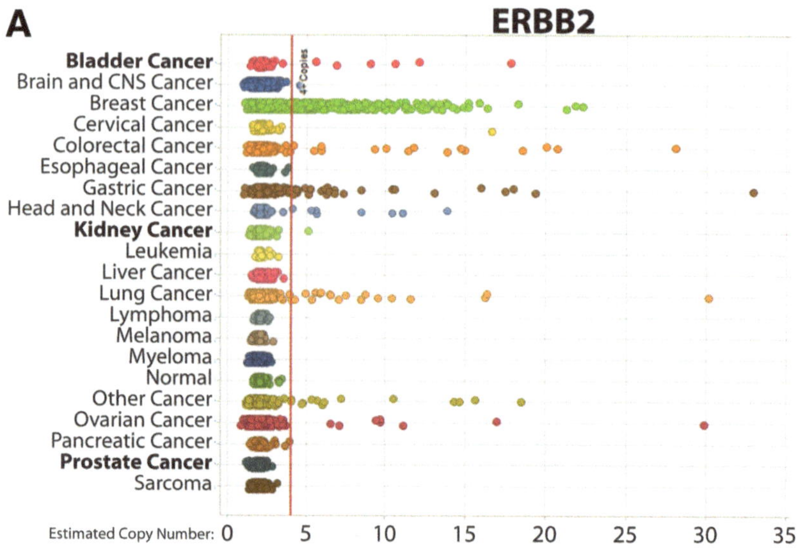

Cancer Type	>4 Copies (%)	Total Cases
Breast	10.1%	3470
Gastric	7.2%	417
Bladder	6.1%	99
Other	2.9%	488
Colorectal	1.8%	849
Head & Neck	1.1%	760
Ovarian	1.1%	857
Lung	1.0%	1837
Cervical	0.5%	186
Kidney	0.2%	635
Brain & CNS	0.1%	1454
Esophageal	0.0%	61
Leukemia	0.0%	240
Liver	0.0%	301
Lymphoma	0.0%	332
Melanoma	0.0%	124
Myeloma	0.0%	575
Normal	0.0%	5671
Pancreatic	0.0%	58
Prostate	0.0%	566
Sarcoma	0.0%	227

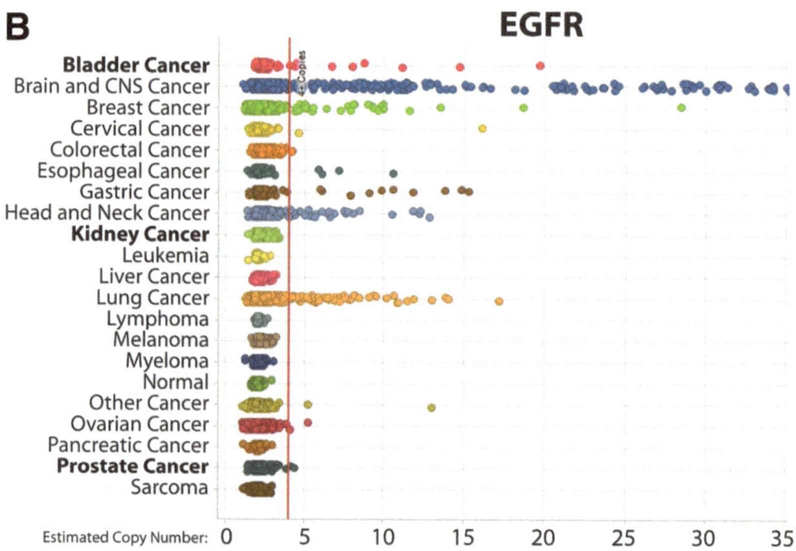

Cancer Type	>4 Copies (%)	Total Cases
Brain & CNS	22.1%	1455
Bladder	9.1%	99
Esophageal	6.6%	61
Head & Neck	3.6%	760
Lung	3.2%	1840
Gastric	2.4%	417
Cervical	1.1%	186
Breast	0.8%	3478
Other	0.4%	488
Prostate	0.4%	566
Colorectal	0.2%	849
Ovarian	0.2%	859
Kidney	0.0%	635
Leukemia	0.0%	240
Liver	0.0%	301
Lymphoma	0.0%	332
Melanoma	0.0%	122
Myeloma	0.0%	575
Normal	0.0%	5671
Pancreatic	0.0%	58
Sarcoma	0.0%	227

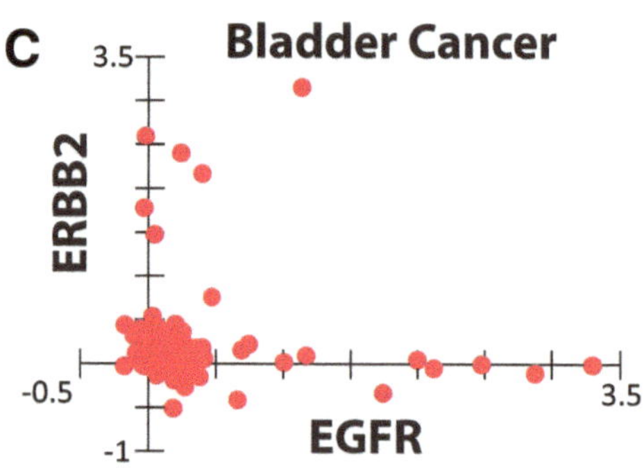

Figure 26-6 High level copy gains in *EGFR* and *ERBB2* in bladder cancer. (**a**) Estimated copy number for *ERBB2* across profiled cancer types, visualized using the Oncomine Powertools Integrated Gene Viewer (powertools.oncomine.com). The *red line* indicates four copies. The table shows a frequency of >4 copies across all profiled cases present in the Oncomine database. Genitourinary cancers are bolded. (**b**) As in (**a**), except for *EGFR*. (**c**) *ERBB2* and *EGFR* show mutually exclusive high level copy number gains in bladder cancer. *ERBB2* and *EGFR* copy number data (log2) for all 99 profiled TCGA bladder cancers downloaded from the Oncomine database (www.oncomine.com) are plotted

those encountered breast and lung cancers, where such aberrations play a driving role and are targetable [120]. It is expected that further interrogation of the bladder cancer genome will identify additional mutations and gene fusions that may drive cancer behavior or serve as prognostic or predictive markers.

Multigene Panels in Bladder Cancer for Early Detection/ Recurrence Monitoring

Urothelial carcinomas have been shown to harbor frequent gains of chromosome 3q, 7p, and 17q, and, as indicated above, frequent deletions of the 9p21 locus harboring *CDKN2A*. A multicolor FISH based urine assay, UroVysion, has been developed based on these abnormalities. Several studies have demonstrated the utility of multitarget FISH assays in urine samples with the goal of predicting recurrence of urothelial carcinoma in the surveillance setting with an overall sensitivity of 81–87 % and a specificity of 92–97 % [121, 122]. These sensitivities are superior to standard urine cytology, which has a sensitivity of ~30–70 % [123, 124]. The specificity of such multitarget FISH assays has also been consistently shown to be comparable to standard urine cytology [125–128]. The UroVysion assay has also been approved by the US FDA for primary screening for urothelial carcinoma in patients with hematuria [129]. Urine microsatellite, methylation, and targeted mutational analyses have also shown potential in monitoring patients with urothelial carcinoma [130]. The potential of NGS and digital PCR based urine assays promises further advancement in early bladder cancer detection.

Molecular Pathology of Kidney Cancer

Renal cell carcinomas (RCC) account for the majority of renal malignancies. Clear-cell RCC is the most common histologic type followed by papillary and chromophobe RCC. We will focus primarily on clear-cell RCC.

Single Genes in Kidney Cancer

VHL

The majority of clear-cell RCCs harbor abnormalities in the *VHL* gene located on chromosome 3p25, and inactivation of *VHL* is the single most common abnormality in clear-cell RCCs. The frequency of somatic mutations in *VHL* is as high as 80 %, as reported in a recent study employing sensitive high-throughput methods [131]. The majority (80 %) of mutations in this study were deletions, insertions, and missense mutations. Loss of 3p is also seen in ~80 % of clear-cell RCCs [132–134], (see Fig. 26.7). Homozygous inactivation of *VHL* involving a 3p deletion of one allele in association with inactivating mutation of the second *VHL* allele is common [133]. *VHL* encodes a member of the ubiquitin ligase complex, which participates in the controlled degradation of numerous cellular proteins, including HIF1α and HIF2α. HIF1 and HIF2 are a transcription factors (dimers composed of α and β subunits) that mediate the cellular response to tissue hypoxia. When activated, HIF1 and HIF2 stimulate angiogenesis and cellular proliferation via increased production of VEGF, erythropoietin, PDGF, TGF-α, and other mediators [135].

Deletions of 3p can be detected reliably with FISH. Importantly, 3p deletions are not seen in other variants of RCC, such as papillary RCC and chromophobe RCC. Therefore, 3p loss detectable by FISH is useful as a specific diagnostic biomarker for clear-cell RCC in cases that pose diagnostic difficulty [136, 137].

PBRM1

A recent exome sequencing study by Varela et al. identified *PBRM1*, which encodes a chromatin remodeling enzyme, as another commonly mutated gene in clear-cell RCCs, with truncating mutations seen in 41 % (92/227) of cases [138]. The majority of mutations were either indels or nonsense mutations. *PBRM1* is a member of the *SWI/SNF* chromatin remodeling complex involved in the maintenance of DNA stability and the regulation of gene transcription.

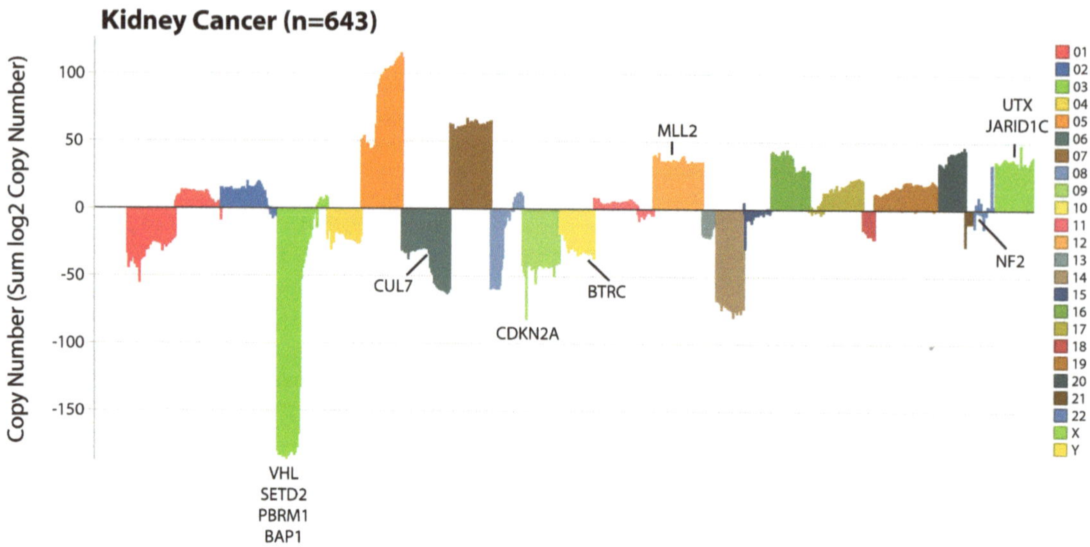

Figure 26-7 Copy number changes in kidney cancer. Genome-wide copy number profiles from 643 kidney cancers (predominantly clear-cell carcinoma) from two studies were visualized using the Oncomine Powertools DNA Copy Number Browser. The sum of the log2 copy number for each segmented sample is plotted in genomic order (see legend for chromosomes). The location of genes harboring recurrent copy number gains/losses or mutations are indicated

PBRM1 resides on chromosome 3p, centromeric to *VHL*. Hence, in addition to *VHL*, losses of 3p usually result in loss of one copy of *PBRM1*. Similar to *VHL*, homozygous inactivation of *PBRM1* via concomitant 3p loss and *PBRM1* mutation in the second allele is a common occurrence in clear-cell RCC. Also similar to *VHL*, the majority of *PBRM1*$^{(mut)}$ cases (95 % in the above Varela et al. study) express the hypoxia phenotype [138].

INHERITED SYNDROMES AND SINGLE GENE MUTATIONS IN NON-CLEAR-CELL RENAL CELL CARCINOMA

Although we have focused predominantly on somatic alterations in this chapter, assessment of the germ line for variants predisposing or causally associated with specific cancers will likely become more commonplace with the advent of routine NGS of the genome. For example, hereditary leiomyomatosis and renal cell carcinoma (HLRCC) syndrome is caused by a germ-line mutation in the fumarate dehydrogenase gene (*FH*), and is associated with an increased incidence of RCCs and cutaneous leiomyomata [139]. Although the syndrome predisposes to several RCC histologic types, a peculiar type of RCC with papillary architecture, prominent "cherry like" nucleoli, and perinucleolar clearing has been associated with this germ-line defect. Other germ-line mutation syndromes associated with increased familial risk of RCC include von Hippel–Lindau syndrome as a result of germ-line mutations in tumor suppressor *VHL* [140]; hereditary papillary renal cell carcinoma syndrome (HPRCC) due to germ-line activating mutations in the *MET* proto-oncogene on chromosome 7 [141], and Birt–Hogg–Dubé syndrome associated with germ-line mutations in the *BHD* gene which encodes the folliculin protein [142].

Single gene mutations have also been described in sporadic non-clear-cell RCCs. For example, sporadic papillary RCCs have been shown to harbor activating *MET* mutations in a small fraction of cases, similar to their hereditary counterparts [143, 144]. Recurrent gains of chromosomes 7 and 17 are also frequently identified in sporadic papillary

RCC [136, 145], as is overexpression of the protein product of *MET* [146].

Finally, a special type of RCC is defined by recurrent chromosomal translocations involving members of the *MiTF-TFE* genes family [147, 148]. The *MiTF-TFE* family of renal carcinomas contain balanced translocations of one member of the basic helix-loop-helix zipper transcription factors, most commonly *TFE3* on chromosome Xp11 and *TFEB* on chromosome 6p21. The most common translocation partners are *ASPL* (chromosome 17q25) and *PRCC* (chromosome 1q21). Tumors of this type frequently have a unique papillary architecture with abundant clear cytoplasm, are high grade, disproportionately affect younger patients, and are associated with high rates of metastasis. Diagnosis is usually confirmed using well-established *TFE3* and *TFEB* break-apart FISH assays [149–152]. Immunohistochemical assessment of TFE3, TFEB, and CATHEPSIN K proteins have also been used, albeit with less accuracy.

WHOLE-GENOME AND EXOME SEQUENCING IN RENAL CELL CARCINOMA

In addition to identification of mutation in *PBRM1*, exome sequencing studies of clear-cell RCC have shown frequent mutations in other chromatin/histone modifying genes. Among the most frequently identified are the histone methylases *MLL2* and *SETD2*, and the histone demethylases *UTX* and *JARID1C* (*KDM5C*) [138, 153, 154]. Collectively, mutations in these genes are seen in nearly 15 % of clear-cell RCC [138]. Mutations are commonly missense, splice site, or indels, consistent with a tumor suppressive function. Histone modification is a primary method of gene expression regulation and as with urothelial carcinoma and prostate cancer, aberrant modification of histone complexes appears to contribute to abnormal gene expression in a sizable fraction of clear-cell RCCs. A recently published molecular characterization of clear-cell RCC by TCGA Research Network has corroborated the importance of chromatin remodeling genes in clear-cell RCC, and has further demonstrated mutations in the SWI/SNF chromatin remodeling complex (which includes PBRM1) may have profound effects on numerous other pathways [155]. The TCGA study also found recurrent mutations in the PI(3)K/AKT pathway, as well as frequent mutations in genes involved in cellular metabolism.

A smaller group of clear-cell RCCs harboring truncating mutations in *NF2* has also been identified by exome sequencing [153]. *NF2* is a tumor suppressor. *NF2* germ-line mutations are associated with neurofibromatosis type 2, a syndrome characterized by predisposition to benign and malignant peripheral nerve sheath tumors, meningiomas, and gliomas. While the majority of tumors with *SETD2* or *JARID1C* mutations in this study showed either *VHL* mutations and/or the hypoxia phenotype, none of the clear-cell RCC harboring *NF2* mutations appear to contain *VHL* mutations or show the hypoxia phenotype [153]. These findings suggest that *NF2* mutated clear-cell RCC may represent a distinct molecular subtype.

As mentioned above, the ubiquitin-mediated proteolysis pathway (UMPP) includes *VHL*, and functions in the controlled degradation of many cellular proteins, including the HIFs. Recurrent mutations in several members of the UMPP pathway have been identified in clear-cell RCCs. Guo et al. sequenced all 135 genes in the UMPP pathway in a set of 98 clear-cell RCC, and found mutations in a member of that pathway in 50 % of tumors [156]. Comprehensive analysis of the clear-cell RCC genome demonstrates few focal aberrations, with nearly universal broad loss of chromosome 3p, as well as broad gains and losses of other chromosomes (Fig. 26.7).

Molecular Prediction of Treatment Response in Kidney Cancer

Traditionally, chemotherapy has not proven to be effective in clear-cell RCC. Immunotherapy with interleukin-2 (IL-2) and interferon-α (INF-α) is effective in a small subset (~10 %) of metastatic clear-cell RCCs but is limited by its high toxicity profile [157]. More recently, immunotherapy has been largely replaced by newer treatment strategies targeting VEGF/VEGFR including small molecule inhibitors of VEGFR and other tyrosine kinases (e.g., sorafenib and sunitinib) and monoclonal antibodies directed against circulating VEGF (e.g., bevacizumab). Targeted therapy response rates in clear-cell RCC are higher than seen with immunotherapy (~40 % vs. 10 %, respectively), although toxicity is still an issue [158].

Predictors of response to VEGF targeted therapy are beginning to emerge. Choueiri et al. have recently showed that loss-of-function mutations in *VHL* were associated with improved response to VEGF targeted therapy (51 % vs. 31 % response rate) [159]. In contrast, *VHL* inactivation by other means was not associated with response, as all pooled patients with *VHL* inactivation did not show a statistically significant difference in response rate compared to those with wild-type *VHL*. CA-IX (a zinc metalloenzyme upregulated by *HIF*) expression in clear-cell RCCs has been shown to correlate with response to INF-α [160], however CA-IX expression has not been reported to have a predictive response to sorafenib or sunitinib [161].

A subset of clear-cell RCCs demonstrate overexpression of members of the mTOR pathway which induces cellular proliferation and represses apoptosis. Phosphorylated Akt and S6 (phos-Akt and phos-S6) may be used as markers of pathway activation. *PTEN* is an upstream suppressor of the mTOR pathway. Small molecular inhibitors of the mTOR pathway have been developed (e.g., temsiroliumus, everolimus). These have shown superior overall survival and progression free survival compared to immunotherapy in clear-cell RCC [162], and improved progression free survival in a group of patients undergoing progression with VEGF inhibitor therapy [163]. Elevated phos-S6 and phos-Akt expression, as measured by IHC was associated with improved response to temsirolimus in one recent study. Patients with higher phos-S6 and/or phos-Akt expression demonstrated greater objective response rates. Higher phos-S6 was also associated with greater median overall survival [164]. In contrast, expression of *CA-IX* and *VHL* mutational status showed no association with response to mTOR inhibitors [164]. *PTEN* and *HIF1α* expression have similarly shown no association with response [165].

Molecular Pathology of Testis and Penile Cancers

Tumors of the testis and penis have received considerably less attention regarding genomic analysis compared to those of the prostate, bladder, and kidney. However, molecular perturbations described in both offer unique insights into tumor behavior and treatment response. The following is a brief discussion of some salient molecular alterations in these two organs.

Germ cell malignancies account for the great majority of testicular tumors, with seminoma being the predominant histologic type. As expected in germ cell lineage, seminomas express markers of totipotency, including *OCT 3/4* and *NANOG* [166, 167]. They also frequently show 12p gains, most commonly as isochromosome 12p [168]. In contrast to other solid tumors, *TP53* mutations are rare [169, 170]. Seminomas, as well as the majority of other testicular germ cell tumors, are extraordinarily responsive to chemotherapy and radiation with DNA damaging agents such as cisplatin, etoposide, and bleomycin with cure rates well exceeding 90 %. This is thought to be related to a strong propensity for germ cells to undergo apoptosis as a result of DNA damage [170]. Although relatively uncommon, chemotherapy resistance is well-known and is associated with microsatellite instability (MSI) and *BRAF* mutations [171, 172]. The mechanism of the resistance to chemotherapy is unclear, but may be related to diminished ability of tumors harboring these alterations to successfully complete programmed cell death.

Squamous cell carcinoma (SCC) represents the majority of malignant penile neoplasms. Like their cervicovaginal and oropharyngeal counterparts, a subset of penile SCC is associated with high-risk human papilloma virus (HPV) infection. As in other sites, high-risk HPV DNA becomes integrated into the host cell genome, leading to production of the viral proteins E6 and E7. These induce cellular proliferation by inactivation of *TP53* and *RB1*, which are associated with increased expression of p16 (*CDKN2A*). The immunophenotype of many HPV associated tumors is, therefore, *TP53*$^{-}$/RB^{-}/p16^{+}, and p16 may be used as a surrogate marker for HPV infection. As in the oropharynx [173], p16^{+} penile SCC has been shown to have better prognosis than p16^{-} SCC, demonstrating improved cancer specific survival [174]. Interestingly, penile SCCs with basaloid morphology are p16^{+}, but have a worse prognosis than conventional penile SCC [175], which stands in contrast to the improved prognosis seen in SCC with this morphology in the oropharynx. Identification of the molecular underpinnings of SCC in the penis is in its earliest stages. Further work will likely reveal important molecular classifiers, which will almost certainly include p16/HPV status.

Conclusion

Our increased understanding of the cancer genome will likely impact patients with all types of malignancies of the genitourinary tract. This includes identification of novel groups of "high-risk" patients who will then undergo biopsy after abnormal screening or imaging, to patients with advanced cancer who may undergo comprehensive whole-genome/transcriptome sequencing to optimize therapy. Similarly, although immunohistochemistry will remain an important methodology to rapidly interrogate expression of markers that may identify specific molecular subtypes in existing workflows (for diagnosis, prognosis, or prediction), we anticipate that genitourinary pathologists of the future will need to be familiar with diverse assays, including FISH, qRT-PCR, and NGS, in order to meet the demands of patients and referring physicians. An overview of the biospecimens, assays, and clinical indications discussed above is illustrated in Fig. 26.8. In this chapter, we provide an overview of the biomarkers and techniques that will likely have the most immediate clinical impact in genitourinary pathology. Nevertheless, the pace of genomic discovery almost guarantees that the most important biomarkers and techniques to impact clinical practice may be yet to be discovered.

Acknowledgements

S.A.T. was supported by a Career Development Award from the University of Michigan Prostate Cancer S.P.O.R.E. and has been supported by the Prostate Cancer Foundation. S.A.T. is a co-inventor on a patent issued to the University of Michigan on ETS fusions in prostate cancer. The diagnostic field of use has been licensed to Gen-Probe, Inc., who has sublicensed certain rights to Ventana Medical Systems. S.A.T. are co-inventors on a patent filed by the University of Michigan on SPINK1 in prostate cancer. The diagnostic field of use has been licensed to Gen-Probe, Inc., who has sublicensed certain rights to Ventana Medical Systems. S.A.T. is a consultant and has received honoraria from Ventana Medical Systems.

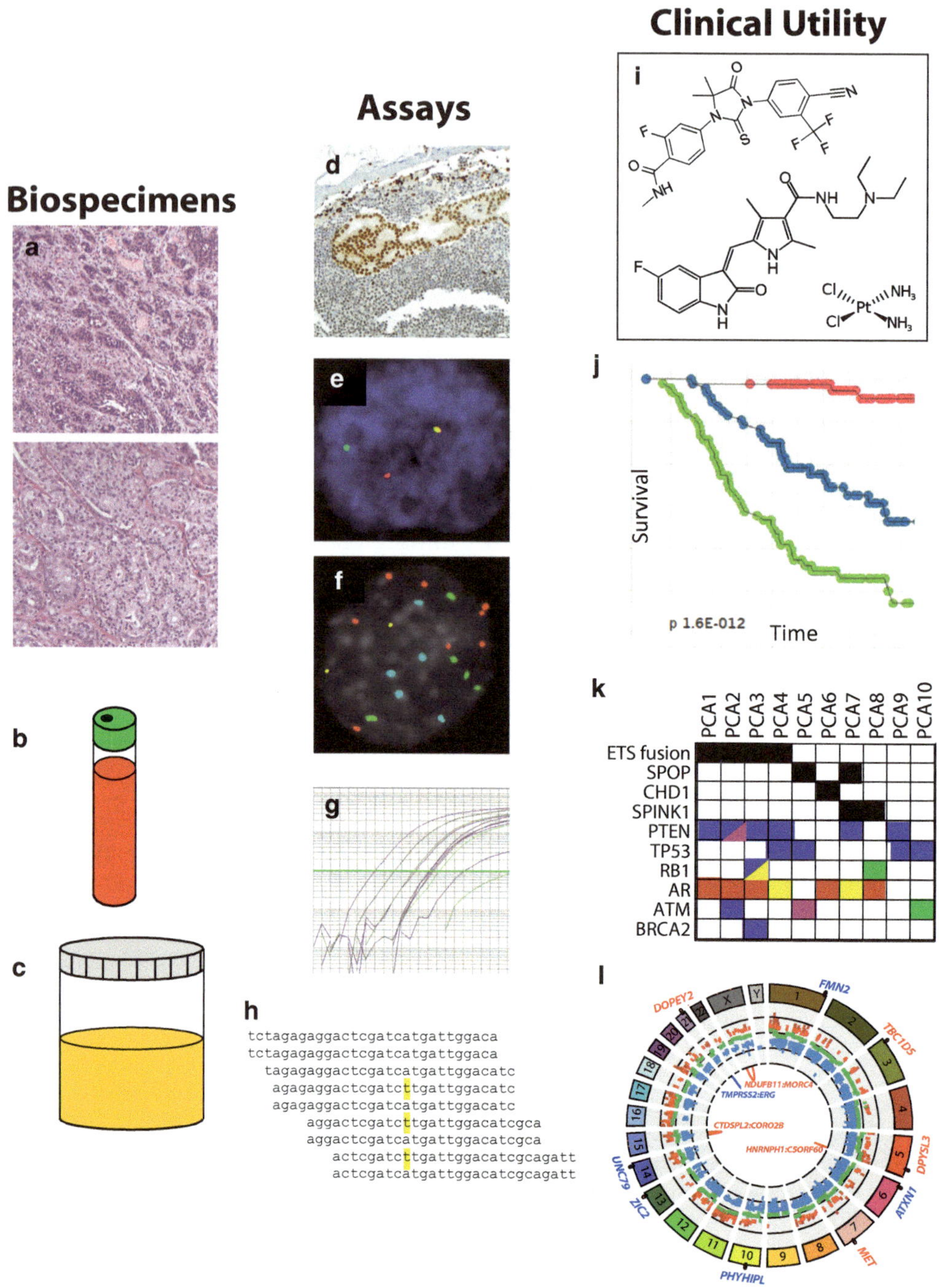

Figure 26-8 Current and future genomic applications in genitourinary pathology. Examples of biospecimens, assays and the clinical utility of genomic applications in genitourinary pathology are shown. A variety of specimen types can be utilized for interrogating the cancer genome, including (**a**) routine tissue specimens, (**b**) blood (as a source of protein, circulating tumor cells, or free nucleic acids), and (**c**) urine (as a source of protein, tumor cells, or free nucleic acids). These samples can be used for a variety of analyses, including (**d**) IHC (an ERG positive prostate cancer lymph node metastasis is shown), (**e**) FISH for rearrangements (Split probes showing a *BRAF* rearranged prostate cancer cell), or (**f**) copy number (Urovision FISH in a bladder cancer cell), gene expression or copy number profiling by microarrays or (**g**) qRT-PCR, or (**h**) NGS of the cancer genome/transcriptome. Such assays are applicable in numerous clinical scenarios, including (**i**) predicting response to therapy, (**j**) outcome, (**k**) basic molecular subtyping (driving alterations in prostate cancer are shown), or (**l**) comprehensive interrogation of genitourinary cancer genomes and transcriptome (circus plot for visualizing the cancer genome, including point mutations/indels, copy number alterations, and gene fusion)

REFERENCES

1. Siegel R, Naishadham D, Jemal A. Cancer statistics, 2012. CA Cancer J Clin. 2012;62(1):10–29.
2. Gleason DF. Classification of prostatic carcinomas. Cancer Chemother Rep. 1966;50(3):125–8.
3. Mellinger GT, Gleason D, Bailar 3rd J. The histology and prognosis of prostatic cancer. J Urol. 1967;97(2):331–7.
4. Uemura H, Hoshino K, Sasaki T, Miyoshi Y, Ishiguro H, Inayama Y, et al. Usefulness of the 2005 International Society of Urologic Pathology Gleason grading system in prostate biopsy and radical prostatectomy specimens. BJU Int. 2009;103(9):1190–4.
5. Park K, Tomlins SA, Mudaliar KM, Chiu Y-L, Esgueva R, Mehra R, et al. Antibody-based detection of ERG rearrangement-positive prostate cancer. Neoplasia. 2010;12(7):590–8.
6. Rubin MA, Maher CA, Chinnaiyan AM. Common gene rearrangements in prostate cancer. J Clin Oncol. 2011;29(27):3659–68.
7. Furusato B, Tan S-H, Young D, Dobi A, Sun C, Mohamed AA, et al. ERG oncoprotein expression in prostate cancer: clonal progression of ERG-positive tumor cells and potential for ERG-based stratification. Prostate Cancer Prostatic Dis. 2010;13(3):228–37.
8. Braun M, Goltz D, Shaikhibrahim Z, Vogel W, Böhm D, Scheble V, et al. ERG protein expression and genomic rearrangement status in primary and metastatic prostate cancer: a comparative study of two monoclonal antibodies. Prostate Cancer Prostatic Dis. 2012;15(2):165–9.
9. Tomlins SA, Rhodes DR, Perner S, Dhanasekaran SM, Mehra R, Sun X-W, et al. Recurrent fusion of TMPRSS2 and ETS transcription factor genes in prostate cancer. Science. 2005;310(5748):644–8.
10. Tomlins SA, Bjartell A, Chinnaiyan AM, Jenster G, Nam RK, Rubin MA, et al. ETS gene fusions in prostate cancer: from discovery to daily clinical practice. Eur Urol. 2009;56(2):275–86.
11. Kumar-Sinha C, Tomlins SA, Chinnaiyan AM. Recurrent gene fusions in prostate cancer. Nat Rev Cancer. 2008;8(7):497–511.
12. Yoshimoto M, Joshua AM, Chilton-Macneill S, Bayani J, Selvarajah S, Evans AJ, et al. Three-color FISH analysis of TMPRSS2/ERG fusions in prostate cancer indicates that genomic microdeletion of chromosome 21 is associated with rearrangement. Neoplasia. 2006;8(6):465–9.
13. Rosen P, Sesterhenn IA, Brassell SA, McLeod DG, Srivastava S, Dobi A. Clinical potential of the ERG oncoprotein in prostate cancer. Nat Rev Urol. 2012;9(3):131–7.
14. Van Leenders GJLH, Boormans JL, Vissers CJ, Hoogland AM, Bressers AAJWM, Furusato B, et al. Antibody EPR3864 is specific for ERG genomic fusions in prostate cancer: implications for pathological practice. Mod Pathol. 2011;24(8):1128–38.
15. Falzarano SM, Zhou M, Carver P, Tsuzuki T, Simmerman K, He H, et al. ERG gene rearrangement status in prostate cancer detected by immunohistochemistry. Virchows Arch. 2011;459(4):441–7.
16. Furusato B, Tan S-H, Young D, Dobi A, Sun C, Mohamed AA, et al. ERG oncoprotein expression in prostate cancer: clonal progression of ERG-positive tumor cells and potential for ERG-based stratification. Prostate Cancer Prostatic Dis. 2010;13(3):228–37.
17. Shah RB, Tadros Y, Brummell B, Zhou M. The diagnostic use of ERG in resolving an "atypical glands suspicious for cancer" diagnosis in prostate biopsies beyond that provided by basal cell and α-methylacyl-CoA-racemase markers. Hum Pathol. 2012;44:786–94.
18. Park K, Dalton J, Narayanan R, Barbieri CE, Hancock M, Bostwick DG, et al. TMPRSS2:ERG gene fusion predicts subsequent detection of prostate cancer in patients with high-grade prostatic intraepithelial neoplasia. J Clin Oncol. 2014;32:206–11.
19. Tomlins SA, Rhodes DR, Yu J, Varambally S, Mehra R, Perner S, et al. The role of SPINK1 in ETS rearrangement-negative prostate cancers. Cancer Cell. 2008;13(6):519–28.
20. Leinonen KA, Tolonen TT, Bracken H, Stenman U-H, Tammela TLJ, Saramäki OR, et al. Association of SPINK1 expression and TMPRSS2:ERG fusion with prognosis in endocrine-treated prostate cancer. Clin Cancer Res. 2010;16(10):2845–51.
21. Grupp K, Diebel F, Sirma H, Simon R, Breitmeyer K, Steurer S, et al. SPINK1 expression is tightly linked to 6q15- and 5q21-deleted ERG-fusion negative prostate cancers but unrelated to PSA recurrence. Prostate. 2013;73(15):1690–8.
22. Bismar TA, Yoshimoto M, Duan Q, Liu S, Sircar K, Squire JA. Interactions and relationships of PTEN, ERG, SPINK1 and AR in castration-resistant prostate cancer. Histopathology. 2012;60(4):645–52.
23. Lippolis G, Edsjö A, Stenman U-H, Bjartell A. A high-density tissue microarray from patients with clinically localized prostate cancer reveals ERG and TATI exclusivity in tumor cells. Prostate Cancer Prostatic Dis. 2013;16(2):145–50.
24. Barbieri CE, Baca SC, Lawrence MS, Demichelis F, Blattner M, Theurillat J-P, et al. Exome sequencing identifies recurrent SPOP, FOXA1

and MED12 mutations in prostate cancer. Nat Genet. 2012;44(6):685–9.
25. Wu Y-M, Su F, Kalyana-Sundaram S, Khazanov N, Ateeq B, Cao X, et al. Identification of targetable FGFR gene fusions in diverse cancers. Cancer Discov. 2013;3(6):636–47.
26. Palanisamy N, Ateeq B, Kalyana-Sundaram S, Pflueger D, Ramnarayanan K, Shankar S, et al. Rearrangements of the RAF kinase pathway in prostate cancer, gastric cancer and melanoma. Nat Med. 2010;16(7):793–8.
27. Wang X-S, Shankar S, Dhanasekaran SM, Ateeq B, Sasaki AT, Jing X, et al. Characterization of KRAS rearrangements in metastatic prostate cancer. Cancer Discov. 2011;1(1):35–43.
28. Beltran H, Yelensky R, Frampton GM, Park K, Downing SR, Macdonald TY, et al. Targeted next-generation sequencing of advanced prostate cancer identifies potential therapeutic targets and disease heterogeneity. Eur Urol. 2012; 63:920–6.
29. Berger MF, Lawrence MS, Demichelis F, Drier Y, Cibulskis K, Sivachenko AY, et al. The genomic complexity of primary human prostate cancer. Nature. 2011;470(7333):214–20.
30. Grasso CS, Wu Y-M, Robinson DR, Cao X, Dhanasekaran SM, Khan AP, et al. The mutational landscape of lethal castration-resistant prostate cancer. Nature. 2012;487(7406): 239–43.
31. Cho N-Y, Choi M, Kim B-H, Cho Y-M, Moon KC, Kang GH. BRAF and KRAS mutations in prostatic adenocarcinoma. Int J Cancer. 2006; 119(8):1858–62.
32. Shen Y, Lu Y, Yin X, Zhu G, Zhu J. KRAS and BRAF mutations in prostate carcinomas of Chinese patients. Cancer Genet Cytogenet. 2010;198(1):35–9.
33. Ren G, Liu X, Mao X, Zhang Y, Stankiewicz E, Hylands L, et al. Identification of frequent BRAF copy number gain and alterations of RAF genes in Chinese prostate cancer. Genes Chromosomes Cancer. 2012;51(11):1014–23.
34. Mao X, Yu Y, Boyd LK, Ren G, Lin D, Chaplin T, et al. Distinct genomic alterations in prostate cancers in Chinese and Western populations suggest alternative pathways of prostate carcinogenesis. Cancer Res. 2010;70(13):5207–12.
35. Miyagi Y, Sasaki T, Fujinami K, Sano J, Senga Y, Miura T, et al. ETS family-associated gene fusions in Japanese prostate cancer: analysis of 194 radical prostatectomy samples. Mod Pathol. 2010;23(11):1492–8.
36. Magi-Galluzzi C, Tsusuki T, Elson P, Simmerman K, LaFargue C, Esgueva R, et al. TMPRSS2-ERG gene fusion prevalence and class are significantly different in prostate cancer of Caucasian, African-American and Japanese patients. Prostate. 2011;71(5):489–97.
37. Jenkins RB, Qian J, Lieber MM, Bostwick DG. Detection of c-myc oncogene amplification and chromosomal anomalies in metastatic prostatic carcinoma by fluorescence in situ hybridization. Cancer Res. 1997;57(3):524–31.
38. Li Y, Su J, DingZhang X, Zhang J, Yoshimoto M, Liu S, et al. PTEN deletion and heme oxygenase-1 overexpression cooperate in prostate cancer progression and are associated with adverse clinical outcome. J Pathol. 2011;224(1):90–100.
39. Lotan TL, Gurel B, Sutcliffe S, Esopi D, Liu W, Xu J, et al. PTEN protein loss by immunostaining: analytic validation and prognostic indicator for a high risk surgical cohort of prostate cancer patients. Clin Cancer Res. 2011;17(20): 6563–73.
40. Antonarakis ES, Keizman D, Zhang Z, Gurel B, Lotan TL, Hicks JL, et al. An immunohistochemical signature comprising PTEN, MYC, and Ki67 predicts progression in prostate cancer patients receiving adjuvant docetaxel after prostatectomy. Cancer. 2012;118:6063–71.
41. Krohn A, Diedler T, Burkhardt L, Mayer P-S, De Silva C, Meyer-Kornblum M, et al. Genomic deletion of PTEN is associated with tumor progression and early PSA recurrence in ERG fusion-positive and fusion-negative prostate cancer. Am J Pathol. 2012;181(2):401–12.
42. Reid AHM, Attard G, Brewer D, Miranda S, Riisnaes R, Clark J, et al. Novel, gross chromosomal alterations involving PTEN cooperate with allelic loss in prostate cancer. Mod Pathol. 2012;25(6):902–10.
43. Cairns P, Okami K, Halachmi S, Halachmi N, Esteller M, Herman JG, et al. Frequent inactivation of PTEN/MMAC1 in primary prostate cancer. Cancer Res. 1997;57(22):4997–5000.
44. Dong JT, Sipe TW, Hyytinen ER, Li CL, Heise C, McClintock DE, et al. PTEN/MMAC1 is infrequently mutated in pT2 and pT3 carcinomas of the prostate. Oncogene. 1998;17(15):1979–82.
45. Suzuki H, Freije D, Nusskern DR, Okami K, Cairns P, Sidransky D, et al. Interfocal heterogeneity of PTEN/MMAC1 gene alterations in multiple metastatic prostate cancer tissues. Cancer Res. 1998;58(2):204–9.
46. Pourmand G, Ziaee A-A, Abedi AR, Mehrsai A, Alavi HA, Ahmadi A, et al. Role of PTEN gene in progression of prostate cancer. Urol J. 2007; 4(2):95–100.
47. Choucair K, Ejdelman J, Brimo F, Aprikian A, Chevalier S, Lapointe J. PTEN genomic deletion predicts prostate cancer recurrence and is associated with low AR expression and transcriptional activity. BMC Cancer. 2012;12(1):543.
48. Attard G, Swennenhuis JF, Olmos D, Reid AHM, Vickers E, A'Hern R, et al. Characterization of ERG, AR and PTEN gene status in circulating tumor cells from patients with castration-resistant

prostate cancer. Cancer Res. 2009; 69(7):2912–8.
49. Sowalsky A, Ye H, Bubley GJ, Balk SP. Clonal progression of prostate cancers from Gleason grade 3 to grade 4. Cancer Res. 2012;73: 1050–5.
50. Yoshimoto M, Ding K, Sweet JM, Ludkovski O, Trottier G, Song KS, et al. PTEN losses exhibit heterogeneity in multifocal prostatic adenocarcinoma and are associated with higher Gleason grade. Mod Pathol. 2012;26:435–47.
51. Chaux A, Peskoe SB, Gonzalez-Roibon N, Schultz L, Albadine R, Hicks J, et al. Loss of PTEN expression is associated with increased risk of recurrence after prostatectomy for clinically localized prostate cancer. Mod Pathol. 2012;25(11):1543–9.
52. Yoshimoto M, Cunha IW, Coudry RA, Fonseca FP, Torres CH, Soares FA, et al. FISH analysis of 107 prostate cancers shows that PTEN genomic deletion is associated with poor clinical outcome. Br J Cancer. 2007;97(5):678–85.
53. Kattan MW. Judging new markers by their ability to improve predictive accuracy. J Natl Cancer Inst. 2003;95(9):634–5.
54. Gudmundsson J, Besenbacher S, Sulem P, Gudbjartsson DF, Olafsson I, Arinbjarnarson S, et al. Genetic correction of PSA values using sequence variants associated with PSA levels. Sci Transl Med. 2010;2(62):62ra92.
55. Witte JS. Personalized prostate cancer screening: improving PSA tests with genomic information. Sci Transl Med. 2010;2(62):62ps55.
56. Ewing CM, Ray AM, Lange EM, Zuhlke KA, Robbins CM, Tembe WD, et al. Germline mutations in HOXB13 and prostate-cancer risk. N Engl J Med. 2012;366(2):141–9.
57. Gudmundsson J, Sulem P, Gudbjartsson DF, Masson G, Agnarsson BA, Benediktsdottir KR, et al. A study based on whole-genome sequencing yields a rare variant at 8q24 associated with prostate cancer. Nat Genet. 2012;44(12): 1326–9.
58. Markert EK, Mizuno H, Vazquez A, Levine AJ. Molecular classification of prostate cancer using curated expression signatures. Proc Natl Acad Sci U S A. 2011;108(52):21276–81.
59. Lapointe J, Li C, Higgins JP, van de Rijn M, Bair E, Montgomery K, et al. Gene expression profiling identifies clinically relevant subtypes of prostate cancer. Proc Natl Acad Sci U S A. 2004; 101(3):811–6.
60. Penney KL, Sinnott JA, Fall K, Pawitan Y, Hoshida Y, Kraft P, et al. mRNA expression signature of Gleason grade predicts lethal prostate cancer. J Clin Oncol. 2011;29(17):2391–6.
61. True L, Coleman I, Hawley S, Huang C-Y, Gifford D, Coleman R, et al. A molecular correlate to the Gleason grading system for prostate adenocarcinoma. Proc Natl Acad Sci U S A. 2006;103(29):10991–6.
62. Cuzick J, Berney DM, Fisher G, Mesher D, Møller H, Reid JE, et al. Prognostic value of a cell cycle progression signature for prostate cancer death in a conservatively managed needle biopsy cohort. Br J Cancer. 2012;106(6):1095–9.
63. Cuzick J, Swanson GP, Fisher G, Brothman AR, Berney DM, Reid JE, et al. Prognostic value of an RNA expression signature derived from cell cycle proliferation genes in patients with prostate cancer: a retrospective study. Lancet Oncol. 2011;12(3):245–55.
64. Klein E, Maddala T, Millward C, Cherbavaz D, Falzarano SM, Knezevic D, et al. Development of needle biopsy-based genomic test to improve discrimination of clinically aggressive from indolent prostate cancer. J Clin Oncol. 2012;30 (Suppl; abstr 4560).
65. Jenkins R, Bergstralh E, Davicioni E, Karnes J, Ballman K, Fink S, et al. Validation of a genomic-clinical classifier model for predicting clinical recurrence of patients with localized prostate cancer in a high-risk population. J Clin Oncol. 2012;30 (Suppl 5;abstr 175).
66. Karnes RJ, Ghadessi M, Bergstralh E, Erho N, Crisan A, Davicioni E, et al. Validation of a genomic classifier that predicts metastatic disease progression in men with high risk pathological features post-prostatectomy. J Urol. 189 (Suppl 4; abstr 2130).
67. Baca SC, Prandi D, Lawrence MS, Mosquera JM, Romanel A, Drier Y, et al. Punctuated evolution of prostate cancer genomes. Cell. 2013;153(3): 666–77.
68. Loriot Y, Zoubeidi A, Gleave ME. Targeted therapies in metastatic castration-resistant prostate cancer: beyond the androgen receptor. Urol Clin North Am. 2012;39(4):517–31.
69. Aisner DL, Marshall CB. Molecular pathology of non-small cell lung cancer: a practical guide. Am J Clin Pathol. 2012;138(3):332–46.
70. Fong PC, Boss DS, Yap TA, Tutt A, Wu P, Mergui-Roelvink M, et al. Inhibition of poly(ADP-ribose) polymerase in tumors from BRCA mutation carriers. N Engl J Med. 2009;361(2): 123–34.
71. Wagle N, Berger MF, Davis MJ, Blumenstiel B, Defelice M, Pochanard P, et al. High-throughput detection of actionable genomic alterations in clinical tumor samples by targeted, massively parallel sequencing. Cancer Discov. 2012; 2(1):82–93.
72. Ito K, Miyakubo M, Sekine Y, Koike H, Matsui H, Shibata Y, et al. Diagnostic significance of [−2]pro-PSA and prostate dimension-adjusted PSA-related indices in men with total PSA in the 2.0–10.0 ng/mL range. World J Urol. 2013; 31(2):305–11.

73. Chan TY, Mikolajczyk SD, Lecksell K, Shue MJ, Rittenhouse HG, Partin AW, et al. Immunohistochemical staining of prostate cancer with monoclonal antibodies to the precursor of prostate-specific antigen. Urology. 2003; 62(1):177–81.
74. Bussemakers MJ, van Bokhoven A, Verhaegh GW, Smit FP, Karthaus HF, Schalken JA, et al. DD3: a new prostate-specific gene, highly overexpressed in prostate cancer. Cancer Res. 1999;59(23):5975–9.
75. De Kok JB, Verhaegh GW, Roelofs RW, Hessels D, Kiemeney LA, Aalders TW, et al. DD3(PCA3), a very sensitive and specific marker to detect prostate tumors. Cancer Res. 2002; 62(9):2695–8.
76. Schmidt U, Fuessel S, Koch R, Baretton GB, Lohse A, Tomasetti S, et al. Quantitative multigene expression profiling of primary prostate cancer. Prostate. 2006;66(14):1521–34.
77. Popa I, Fradet Y, Beaudry G, Hovington H, Beaudry G, Têtu B. Identification of PCA3 (DD3) in prostatic carcinoma by in situ hybridization. Mod Pathol. 2007;20(11):1121–7.
78. Warrick JI, Tomlins SA, Carskadon SL, Young AM, Siddiqui J, Wei JT, et al. Evaluation of tissue PCA3 expression in prostate cancer by RNA in situ hybridization-a correlative study with urine PCA3 and TMPRSS2-ERG. Mod Pathol. [Epub ahead of print].
79. Salagierski M, Schalken JA. Molecular diagnosis of prostate cancer: PCA3 and TMPRSS2:ERG gene fusion. J Urol. 2012;187(3):795–801.
80. Hessels D, Klein Gunnewiek JMT, van Oort I, Karthaus HFM, van Leenders GJL, van Balken B, et al. DD3(PCA3)-based molecular urine analysis for the diagnosis of prostate cancer. Eur Urol. 2003;44(1):8–15. Discussion 15–16.
81. Ankerst DP, Groskopf J, Day JR, Blase A, Rittenhouse H, Pollock BH, et al. Predicting prostate cancer risk through incorporation of prostate cancer gene 3. J Urol. 2008;180(4):1303–8. Discussion 1308.
82. Laxman B, Tomlins SA, Mehra R, Morris DS, Wang L, Helgeson BE, et al. Noninvasive detection of TMPRSS2:ERG fusion transcripts in the urine of men with prostate cancer. Neoplasia. 2006;8(10):885–8.
83. Tomlins SA, Aubin SMJ, Siddiqui J, Lonigro RJ, Sefton-Miller L, Miick S, et al. Urine TMPRSS2:ERG fusion transcript stratifies prostate cancer risk in men with elevated serum PSA. Sci Transl Med. 2011;3(94):94ra72.
84. Leyten GHJM, Hessels D, Jannink SA, Smit FP, de Jong H, Cornel EB, et al. Prospective multicentre evaluation of PCA3 and TMPRSS2-ERG gene fusions as diagnostic and prognostic urinary biomarkers for prostate cancer. Eur Urol. 2012;65:534–42.
85. Cornu J-N, Cancel-Tassin G, Egrot C, Gaffory C, Haab F, Cussenot O. Urine TMPRSS2:ERG fusion transcript integrated with PCA3 score, genotyping, and biological features are correlated to the results of prostatic biopsies in men at risk of prostate cancer. Prostate. 2013;73(3): 242–9.
86. Stephan C, Jung K, Semjonow A, Schulze-Forster K, Cammann H, Hu X, et al. Comparative assessment of urinary prostate cancer antigen 3 and TMPRSS2:ERG gene fusion with the serum [−2] proprostate-specific antigen-based prostate health index for detection of prostate cancer. Clin Chem. 2013;59(1):280–8.
87. Young A, Palanisamy N, Siddiqui J, Wood DP, Wei JT, Chinnaiyan AM, et al. Correlation of urine TMPRSS2:ERG and PCA3 to ERG+ and total prostate cancer burden. Am J Clin Pathol. 2012;138(5):685–96.
88. Scher HI, Jia X, de Bono JS, Fleisher M, Pienta KJ, Raghavan D, et al. Circulating tumour cells as prognostic markers in progressive, castration-resistant prostate cancer: a reanalysis of IMMC38 trial data. Lancet Oncol. 2009;10(3):233–9.
89. Danila DC, Fleisher M, Scher HI. Circulating tumor cells as biomarkers in prostate cancer. Clin Cancer Res. 2011;17(12):3903–12.
90. Stott SL, Hsu C-H, Tsukrov DI, Yu M, Miyamoto DT, Waltman BA, et al. Isolation of circulating tumor cells using a microvortex-generating herringbone-chip. Proc Natl Acad Sci U S A. 2010;107(43):18392–7.
91. Stott SL, Lee RJ, Nagrath S, Yu M, Miyamoto DT, Ulkus L, et al. Isolation and characterization of circulating tumor cells from patients with localized and metastatic prostate cancer. Sci Transl Med. 2010;2(25):25ra23.
92. Danila DC, Anand A, Sung CC, Heller G, Leversha MA, Cao L, et al. TMPRSS2-ERG status in circulating tumor cells as a predictive biomarker of sensitivity in castration-resistant prostate cancer patients treated with abiraterone acetate. Eur Urol. 2011;60(5):897–904.
93. Miyamoto DT, Lee RJ, Stott SL, Ting DT, Wittner BS, Ulman M, et al. Androgen receptor signaling in circulating tumor cells as a marker of hormonally responsive prostate cancer. Cancer Discov. 2012;2(11):995–1003.
94. Leary RJ, Kinde I, Diehl F, Schmidt K, Clouser C, Duncan C, et al. Development of personalized tumor biomarkers using massively parallel sequencing. Sci Transl Med. 2010;2(20):20ra14.
95. Leary RJ, Sausen M, Kinde I, Papadopoulos N, Carpten JD, Craig D, et al. Detection of chromosomal alterations in the circulation of cancer patients with whole-genome sequencing. Sci Transl Med. 2012;4(162):162ra154.
96. Heitzer E, Ulz P, Belic J, Gutschi S, Quehenberger F, Fischereder K, et al. Tumor-associated copy

number changes in the circulation of patients with prostate cancer identified through whole-genome sequencing. Genome Med. 2013;5(4):30.

97. Olmos D, Brewer D, Clark J, Danila DC, Parker C, Attard G, et al. Prognostic value of blood mRNA expression signatures in castration-resistant prostate cancer: a prospective, two-stage study. Lancet Oncol. 2012;13(11): 1114–24.
98. Ross RW, Galsky MD, Scher HI, Magidson J, Wassmann K, Lee G-SM, et al. A whole-blood RNA transcript-based prognostic model in men with castration-resistant prostate cancer: a prospective study. Lancet Oncol. 2012;13(11): 1105–13.
99. Castillo-Martin M, Domingo-Domenech J, Karni-Schmidt O, Matos T, Cordon-Cardo C. Molecular pathways of urothelial development and bladder tumorigenesis. Urol Oncol. 2010;28(4):401–8.
100. Netto GJ. Molecular biomarkers in urothelial carcinoma of the bladder: are we there yet? Nat Rev Urol. 2012;9(1):41–51.
101. Burger M, van der Aa MNM, van Oers JMM, Brinkmann A, van der Kwast TH, Steyerberg EC, et al. Prediction of progression of non-muscle-invasive bladder cancer by WHO 1973 and 2004 grading and by FGFR3 mutation status: a prospective study. Eur Urol. 2008;54(4):835–43.
102. Goebell PJ, Groshen SG, Schmitz-Dräger BJ. p53 immunohistochemistry in bladder cancer: a new approach to an old question. Urol Oncol. 2010;28(4):377–88.
103. Van Rhijn BWG, Zuiverloon TCM, Vis AN, Radvanyi F, van Leenders GJLH, Ooms BCM, et al. Molecular grade (FGFR3/MIB-1) and EORTC risk scores are predictive in primary non-muscle-invasive bladder cancer. Eur Urol. 2010;58(3):433–41.
104. George B, Datar RH, Wu L, Cai J, Patten N, Beil SJ, et al. p53 gene and protein status: the role of p53 alterations in predicting outcome in patients with bladder cancer. J Clin Oncol. 2007;25(34): 5352–8.
105. Margulis V, Lotan Y, Karakiewicz PI, Fradet Y, Ashfaq R, Capitanio U, et al. Multi-institutional validation of the predictive value of Ki-67 labeling index in patients with urinary bladder cancer. J Natl Cancer Inst. 2009;101(2):114–9.
106. Dyrskjøt L, Zieger K, Kissow Lildal T, Reinert T, Gruselle O, Coche T, et al. Expression of MAGE-A3, NY-ESO-1, LAGE-1 and PRAME in urothelial carcinoma. Br J Cancer. 2012; 107(1):116–22.
107. Chen PC-H, Yu H-J, Chang Y-H, Pan C-C. Her2 amplification distinguishes a subset of non-muscle-invasive bladder cancers with a high risk of progression. J Clin Pathol. 2012;66:113–9.
108. Lindgren D, Liedberg F, Andersson A, Chebil G, Gudjonsson S, Borg A, et al. Molecular characterization of early-stage bladder carcinomas by expression profiles, FGFR3 mutation status, and loss of 9q. Oncogene. 2006;25(18): 2685–96.
109. Lindgren D, Frigyesi A, Gudjonsson S, Sjödahl G, Hallden C, Chebil G, et al. Combined gene expression and genomic profiling define two intrinsic molecular subtypes of urothelial carcinoma and gene signatures for molecular grading and outcome. Cancer Res. 2010;70(9):3463–72.
110. Dyrskjøt L, Reinert T, Novoradovsky A, Zuiverloon TCM, Beukers W, Zwarthoff E, et al. Analysis of molecular intra-patient variation and delineation of a prognostic 12-gene signature in non-muscle invasive bladder cancer; technology transfer from microarrays to PCR. Br J Cancer. 2012;107(8):1392–8.
111. Shariat SF, Bolenz C, Godoy G, Fradet Y, Ashfaq R, Karakiewicz PI, et al. Predictive value of combined immunohistochemical markers in patients with pT1 urothelial carcinoma at radical cystectomy. J Urol. 2009;182(1):78–84. Discussion 84.
112. Dubosq F, Ploussard G, Soliman H, Turpin E, Latil A, Desgrandchamps F, et al. Identification of a three-gene expression signature of early recurrence in non-muscle-invasive urothelial cell carcinoma of the bladder. Urol Oncol. 2011;30: 833–40.
113. Dyrskjøt L, Thykjaer T, Kruhøffer M, Jensen JL, Marcussen N, Hamilton-Dutoit S, et al. Identifying distinct classes of bladder carcinoma using microarrays. Nat Genet. 2003;33(1):90–6.
114. Smith SC, Baras AS, Dancik G, Ru Y, Ding K-F, Moskaluk CA, et al. A 20-gene model for molecular nodal staging of bladder cancer: development and prospective assessment. Lancet Oncol. 2011;12(2):137–43.
115. Riester M, Taylor JM, Feifer A, Koppie T, Rosenberg JE, Downey RJ, et al. Combination of a novel gene expression signature with a clinical nomogram improves the prediction of survival in high-risk bladder cancer. Clin Cancer Res. 2012;18(5):1323–33.
116. Takata R, Katagiri T, Kanehira M, Tsunoda T, Shuin T, Miki T, et al. Predicting response to methotrexate, vinblastine, doxorubicin, and cisplatin neoadjuvant chemotherapy for bladder cancers through genome-wide gene expression profiling. Clin Cancer Res. 2005;11(7):2625–36.
117. Kato Y, Zembutsu H, Takata R, Miya F, Tsunoda T, Obara W, et al. Predicting response of bladder cancers to gemcitabine and carboplatin neoadjuvant chemotherapy through genome-wide gene expression profiling. Exp Ther Med. 2011; 2(1):47–56.
118. Gui Y, Guo G, Huang Y, Hu X, Tang A, Gao S, et al. Frequent mutations of chromatin remodeling

genes in transitional cell carcinoma of the bladder. Nat Genet. 2011;43(9):875–8.
119. Iyer G, Hanrahan AJ, Milowsky MI, Al-Ahmadie H, Scott SN, Janakiraman M, et al. Genome sequencing identifies a basis for everolimus sensitivity. Science. 2012;338(6104):221.
120. Iyer G, Al-Ahmadie H, Schultz N, Hanrahan AJ, Ostrovnaya I, Balar AV, et al. Prevalence and co-occurrence of actionable genomic alterations in high-grade bladder cancer. J Clin Oncol. 2013; 31(25):3133–40.
121. Halling KC, King W, Sokolova IA, Meyer RG, Burkhardt HM, Halling AC, et al. A comparison of cytology and fluorescence in situ hybridization for the detection of urothelial carcinoma. J Urol. 2000;164(5):1768–75.
122. Sokolova IA, Halling KC, Jenkins RB, Burkhardt HM, Meyer RG, Seelig SA, et al. The development of a multitarget, multicolor fluorescence in situ hybridization assay for the detection of urothelial carcinoma in urine. J Mol Diagn. 2000; 2(3):116–23.
123. Rife CC, Farrow GM, Utz DC. Urine cytology of transitional cell neoplasms. Urol Clin North Am. 1979;6(3):599–612.
124. Kern WH. The grade and pathologic stage of bladder cancer. Cancer. 1984;53(5):1185–9.
125. Bubendorf L, Grilli B, Sauter G, Mihatsch MJ, Gasser TC, Dalquen P. Multiprobe FISH for enhanced detection of bladder cancer in voided urine specimens and bladder washings. Am J Clin Pathol. 2001;116(1):79–86.
126. Sarosdy MF, Schellhammer P, Bokinsky G, Kahn P, Chao R, Yore L, et al. Clinical evaluation of a multi-target fluorescent in situ hybridization assay for detection of bladder cancer. J Urol. 2002;168(5):1950–4.
127. Skacel M, Fahmy M, Brainard JA, Pettay JD, Biscotti CV, Liou LS, et al. Multitarget fluorescence in situ hybridization assay detects transitional cell carcinoma in the majority of patients with bladder cancer and atypical or negative urine cytology. J Urol. 2003;169(6):2101–5.
128. Veeramachaneni R, Nordberg ML, Shi R, Herrera GA, Turbat-Herrera EA. Evaluation of fluorescence in situ hybridization as an ancillary tool to urine cytology in diagnosing urothelial carcinoma. Diagn Cytopathol. 2003;28(6):301–7.
129. Sarosdy MF, Kahn PR, Ziffer MD, Love WR, Barkin J, Abara EO, et al. Use of a multitarget fluorescence in situ hybridization assay to diagnose bladder cancer in patients with hematuria. J Urol. 2006;176(1):44–7.
130. Zuiverloon TCM, Beukers W, van der Keur KA, Nieuweboer AJM, Reinert T, Dyrskjot L, et al. Combinations of urinary biomarkers for surveillance of patients with incident nonmuscle invasive bladder cancer: the European FP7 UROMOL project. J Urol. 2013;189(5): 1945–51.
131. Nickerson ML, Jaeger E, Shi Y, Durocher JA, Mahurkar S, Zaridze D, et al. Improved identification of von Hippel-Lindau gene alterations in clear cell renal tumors. Clin Cancer Res. 2008; 14(15):4726–34.
132. Kovacs G, Erlandsson R, Boldog F, Ingvarsson S, Müller-Brechlin R, Klein G, et al. Consistent chromosome 3p deletion and loss of heterozygosity in renal cell carcinoma. Proc Natl Acad Sci U S A. 1988;85(5):1571–5.
133. Furge KA, Tan MH, Dykema K, Kort E, Stadler W, Yao X, et al. Identification of deregulated oncogenic pathways in renal cell carcinoma: an integrated oncogenomic approach based on gene expression profiling. Oncogene. 2007;26(9): 1346–50.
134. Zbar B, Brauch H, Talmadge C, Linehan M. Loss of alleles of loci on the short arm of chromosome 3 in renal cell carcinoma. Nature. 1987; 327(6124):721–4.
135. Pfeifer J. Molecular genetic testing in surgical pathology. Philadelphia, PA: Lippincott Williams & Wilkins; 2006. Chapter 17, Urinary Tract.
136. Hughson MD, Dickman K, Bigler SA, Meloni AM, Sandberg AA. Clear-cell and papillary carcinoma of the kidney: an analysis of chromosome 3, 7, and 17 abnormalities by microsatellite amplification, cytogenetics, and fluorescence in situ hybridization. Cancer Genet Cytogenet. 1998;106(2):93–104.
137. Barocas DA, Mathew S, DelPizzo JJ, Vaughan Jr ED, Sosa RE, Fine RG, et al. Renal cell carcinoma sub-typing by histopathology and fluorescence in situ hybridization on a needle-biopsy specimen. BJU Int. 2007;99(2):290–5.
138. Varela I, Tarpey P, Raine K, Huang D, Ong CK, Stephens P, et al. Exome sequencing identifies frequent mutation of the SWI/SNF complex gene PBRM1 in renal carcinoma. Nature. 2011;469(7331):539–42.
139. Merino MJ, Torres-Cabala C, Pinto P, Linehan WM. The morphologic spectrum of kidney tumors in hereditary leiomyomatosis and renal cell carcinoma (HLRCC) syndrome. Am J Surg Pathol. 2007;31(10):1578–85.
140. Solomon D, Schwartz A. Renal pathology in von Hippel-Lindau disease. Hum Pathol. 1988;19(9): 1072–9.
141. Dharmawardana PG, Giubellino A, Bottaro DP. Hereditary papillary renal carcinoma type I. Curr Mol Med. 2004;4(8):855–68.
142. Toro JR, Wei M-H, Glenn GM, Weinreich M, Toure O, Vocke C, et al. BHD mutations, clinical and molecular genetic investigations of Birt-Hogg-Dubé syndrome: a new series of 50 families and a review of published reports. J Med Genet. 2008;45(6):321–31.
143. Schmidt L, Duh FM, Chen F, Kishida T, Glenn G, Choyke P, et al. Germline and somatic mutations in the tyrosine kinase domain of the MET

proto-oncogene in papillary renal carcinomas. Nat Genet. 1997;16(1):68–73.
144. Schmidt L, Junker K, Nakaigawa N, Kinjerski T, Weirich G, Miller M, et al. Novel mutations of the MET proto-oncogene in papillary renal carcinomas. Oncogene. 1999;18(14):2343–50.
145. Klatte T, Pantuck AJ, Said JW, Seligson DB, Rao NP, LaRochelle JC, et al. Cytogenetic and molecular tumor profiling for type 1 and type 2 papillary renal cell carcinoma. Clin Cancer Res. 2009;15(4):1162–9.
146. Sweeney P, El-Naggar AK, Lin S-H, Pisters LL. Biological significance of c-met over expression in papillary renal cell carcinoma. J Urol. 2002;168(1):51–5.
147. Argani P, Olgac S, Tickoo SK, Goldfischer M, Moch H, Chan DY, et al. Xp11 translocation renal cell carcinoma in adults: expanded clinical, pathologic, and genetic spectrum. Am J Surg Pathol. 2007;31(8):1149–60.
148. Camparo P, Vasiliu V, Molinie V, Couturier J, Dykema KJ, Petillo D, et al. Renal translocation carcinomas: clinicopathologic, immunohistochemical, and gene expression profiling analysis of 31 cases with a review of the literature. Am J Surg Pathol. 2008;32(5):656–70.
149. Green WM, Yonescu R, Morsberger L, Morris K, Netto GJ, Epstein JI, et al. Utilization of a TFE3 break-apart FISH assay in a renal tumor consultation service. Am J Surg Pathol. 2013; 37(8):1150–63.
150. Rao Q, Williamson SR, Zhang S, Eble JN, Grignon DJ, Wang M, et al. TFE3 break-apart FISH has a higher sensitivity for Xp11.2 translocation-associated renal cell carcinoma compared with TFE3 or cathepsin K immunohistochemical staining alone: expanding the morphologic spectrum. Am J Surg Pathol. 2013;37(6):804–15.
151. Mosquera J-M, Dal Cin P, Mertz KD, Perner S, Davis IJ, Fisher DE, et al. Validation of a TFE3 break-apart FISH assay for Xp11.2 translocation renal cell carcinomas. Diagn Mol Pathol. 2011;20(3):129–37.
152. Zhong M, De Angelo P, Osborne L, Keane-Tarchichi M, Goldfischer M, Edelmann L, et al. Dual-color, break-apart FISH assay on paraffin-embedded tissues as an adjunct to diagnosis of Xp11 translocation renal cell carcinoma and alveolar soft part sarcoma. Am J Surg Pathol. 2010;34(6):757–66.
153. Dalgliesh GL, Furge K, Greenman C, Chen L, Bignell G, Butler A, et al. Systematic sequencing of renal carcinoma reveals inactivation of histone modifying genes. Nature. 2010;463(7279): 360–3.
154. Van Haaften G, Dalgliesh GL, Davies H, Chen L, Bignell G, Greenman C, et al. Somatic mutations of the histone H3K27 demethylase gene UTX in human cancer. Nat Genet. 2009; 41(5):521–3.
155. Cancer Genome Atlas Research Network. Comprehensive molecular characterization of clear cell renal cell carcinoma. Nature. 2013;499(7456):43–9.
156. Guo G, Gui Y, Gao S, Tang A, Hu X, Huang Y, et al. Frequent mutations of genes encoding ubiquitin-mediated proteolysis pathway components in clear cell renal cell carcinoma. Nat Genet. 2012;44(1):17–9.
157. Sleijfer S, Bannink M, Van Gool AR, Kruit WHJ, Stoter G. Side effects of interferon-alpha therapy. Pharm World Sci. 2005;27(6):423–31.
158. Méndez-Vidal MJ, Martínez Ortega E, Montesa Pino A, Pérez Valderrama B, Viciana R. Management of adverse events of targeted therapies in normal and special patients with metastatic renal cell carcinoma. Cancer Metastasis Rev. 2012;31 Suppl 1:S19–27.
159. Choueiri TK, Plantade A, Elson P, Negrier S, Ravaud A, Oudard S, et al. Efficacy of sunitinib and sorafenib in metastatic papillary and chromophobe renal cell carcinoma. J Clin Oncol. 2008;26(1):127–31.
160. Atkins M, Regan M, McDermott D, Mier J, Stanbridge E, Youmans A, et al. Carbonic anhydrase IX expression predicts outcome of interleukin 2 therapy for renal cancer. Clin Cancer Res. 2005;11(10):3714–21.
161. Choueiri TK, Regan MM, Rosenberg JE, Oh WK, Clement J, Amato AM, et al. Carbonic anhydrase IX and pathological features as predictors of outcome in patients with metastatic clear-cell renal cell carcinoma receiving vascular endothelial growth factor-targeted therapy. BJU Int. 2010;106(6):772–8.
162. Atkins MB, Hidalgo M, Stadler WM, Logan TF, Dutcher JP, Hudes GR, et al. Randomized phase II study of multiple dose levels of CCI-779, a novel mammalian target of rapamycin kinase inhibitor, in patients with advanced refractory renal cell carcinoma. J Clin Oncol. 2004; 22(5):909–18.
163. Motzer RJ, Escudier B, Oudard S, Hutson TE, Porta C, Bracarda S, et al. Efficacy of everolimus in advanced renal cell carcinoma: a double-blind, randomised, placebo-controlled phase III trial. Lancet. 2008;372(9637):449–56.
164. Cho D, Signoretti S, Dabora S, Regan M, Seeley A, Mariotti M, et al. Potential histologic and molecular predictors of response to temsirolimus in patients with advanced renal cell carcinoma. Clin Genitourin Cancer. 2007; 5(6):379–85.
165. Figlin RA, de Souza P, McDermott D, Dutcher JP, Berkenblit A, Thiele A, et al. Analysis of PTEN and HIF-1alpha and correlation with efficacy in patients with advanced renal cell carcinoma

treated with temsirolimus versus interferon-alpha. Cancer. 2009;115(16):3651–60.

166. Looijenga LHJ, Stoop H, de Leeuw HPJC, de Gouveia Brazao CA, Gillis AJM, van Roozendaal KEP, et al. POU5F1 (OCT3/4) identifies cells with pluripotent potential in human germ cell tumors. Cancer Res. 2003;63(9):2244–50.
167. Cheng L, Sung M-T, Cossu-Rocca P, Jones TD, MacLennan GT, De Jong J, et al. OCT4: biological functions and clinical applications as a marker of germ cell neoplasia. J Pathol. 2007;211(1):1–9.
168. Bosl GJ, Dmitrovsky E, Reuter VE, Samaniego F, Rodriguez E, Geller NL, et al. Isochromosome of the short arm of chromosome 12: clinically useful markers for male germ cell tumors. J Natl Cancer Inst. 1989;81(24):1874–8.
169. Lutzker SG, Mathew R, Taller DR. A p53 dose–response relationship for sensitivity to DNA damage in isogenic teratocarcinoma cells. Oncogene. 2001;20(23):2982–6.
170. di Pietro A, de Vries EGE, Gietema JA, Spierings DCJ, de Jong S. Testicular germ cell tumours: the paradigm of chemo-sensitive solid tumours. Int J Biochem Cell Biol. 2005;37(12):2437–56.
171. Mayer F, Gillis AJM, Dinjens W, Oosterhuis JW, Bokemeyer C, Looijenga LHJ. Microsatellite instability of germ cell tumors is associated with resistance to systemic treatment. Cancer Res. 2002;62(10):2758–60.
172. Honecker F, Wermann H, Mayer F, Gillis AJM, Stoop H, van Gurp RJLM, et al. Microsatellite instability, mismatch repair deficiency, and BRAF mutation in treatment-resistant germ cell tumors. J Clin Oncol. 2009;27(13):2129–36.
173. El-Mofty SK. Human papillomavirus (HPV) related carcinomas of the upper aerodigestive tract. Head Neck Pathol. 2007;1(2):181–5.
174. Bethune G, Campbell J, Rocker A, Bell D, Rendon R, Merrimen J. Clinical and pathologic factors of prognostic significance in penile squamous cell carcinoma in a North American population. Urology. 2012;79(5):1092–7.
175. Chaux A, Reuter V, Lezcano C, Velazquez EF, Torres J, Cubilla AL. Comparison of morphologic features and outcome of resected recurrent and nonrecurrent squamous cell carcinoma of the penis: a study of 81 cases. Am J Surg Pathol. 2009;33(9):1299–306.

CHAPTER 27

GENOMIC APPLICATIONS IN GYNECOLOGIC MALIGNANCIES

SARAH CHIANG, LUCIANO G. MARTELOTTO, BRITTA WEIGELT

Introduction 465
Endometrial Carcinoma 465
Uterine Mesenchymal Tumors 472
Conclusions 476

Introduction

The uterine corpus represents the most common site for gynecologic malignancies in North America and Europe with an estimated 49,560 new cases and 8,190 deaths in 2013 in the United States [1]. The incidence of endometrial cancer, in particular endometrioid adenocarcinoma, is steadily increasing [2–4] and is likely to be reflective of the aging population, the use of hormone-replacement therapy, and the increased prevalence of overweight and obesity among women.

Cancer of the uterine corpus is a heterogeneous disease comprising multiple entities with distinct risk factors, histopathological features, and outcomes. Approximately 75–80 % of patients with endometrial carcinomas are diagnosed at an early stage (stage I/II; disease confined to the uterus) and are managed by surgery with or without adjuvant radiotherapy resulting in a 5-year overall survival rate of 74–91 % [5]. Of these early stage carcinomas, approximately 15–20 % recur, and despite advances in adjuvant chemotherapy and radiation strategies, the outcome of women diagnosed with advanced or recurrent disease remains poor with a median overall survival of 5–15 months [6–8]. Malignant mesenchymal uterine neoplasms, albeit rare, generally have an aggressive clinical behavior, and represent a challenge in regards to diagnosis and disease management. Thus, there is a need to improve current treatment strategies and incorporate targeted therapies in standard regimens, and to identify those patients with a high risk of recurrence and to select the optimal systemic treatment for a given patient with uterine cancer. Genomic studies aiming to identify specific molecular markers for classification, risk-stratification, and therapy decision-making are continuing to unveil the repertoire of alterations in uterine cancer. The molecular profiling efforts have focused on the most common histological subtypes of endometrial carcinomas, the endometrioid and serous adenocarcinomas, which we discuss in this chapter. In addition, we provide an overview of the recent developments in our understanding of the molecular basis of the most common uterine mesenchymal neoplasms, the endometrial stromal and smooth muscle tumors.

Endometrial Carcinoma

Classification and Biomarkers of Endometrial Carcinoma

For optimal disease management of gynecological cancers, the extent of disease is determined on the basis of the International

G.J. Netto and I. Schrijver (eds.), *Genomic Applications in Pathology*,
DOI 10.1007/978-1-4939-0727-4_27, © Springer Science+Business Media New York 2015

Federation of Gynecology and Obstetrics (FIGO) system. To improve the prognostic information provided by the FIGO staging, the guidelines for endometrial carcinoma underwent revision in 2009 [9] based on the knowledge gathered since the first surgical-pathological FIGO staging system described in 1988.

Traditionally, endometrial carcinomas are classified based on clinical, epidemiological, and "endocrine-metabolic" features into two types [10]: type I tumors, which are of low grade and associated with unopposed estrogen stimulation and endometrial hyperplasia, and type II tumors, which are typically of high grade and traditionally thought to be unrelated to hormonal factors or hyperplasia. A recent pooled analysis of endometrial cancer risk factors in 14,069 endometrial cancer cases and 35,312 controls, however, revealed that type I and type II endometrial cancers share many common etiologic factors, and it has been suggested that the etiology of type II tumors may not be completely estrogen-independent [11]. For histologic subtyping of endometrial cancer, modified versions of the World Health Organization (WHO) [12] and International Society of Gynecological Pathologists classifications [13, 14] have been used (Table 27.1). Noteworthy, although malignant carcinosarcomas (also called malignant mixed Mullerian tumors) are classified as mixed epithelial and mesenchymal tumors according to the WHO, recent studies have provided evidence to suggest that these neoplasms derive from a transformed epithelial cell and are now considered by many in the pathology community as high-grade carcinomas undergoing sarcomatous differentiation [15], and to some extent, being the uterine counterpart of metaplastic carcinoma of the breast [16].

A three-tiered FIGO grading system based on architectural pattern and nuclear features is used to grade some types of endometrial carcinomas, namely endometrioid and mucinous adenocarcinomas [17]. According to the percentage of solid non-glandular and non-squamous growth, tumors are assigned grade 1 (≤5 % solid growth), grade 2 (6–50 % of solid growth), or grade 3 (>50 % of solid growth). The overall grade of grade 1 or 2 tumors is raised by one if marked nuclear atypia is present [18]. FIGO grading has generally not been recommended for non-endometrioid (i.e., serous and clear cell), mixed epithelial, or morphologically heterogeneous tumors, due to potential lack of correlation with clinical outcome. Alternative grading schemes, such as a binary system irrespective of tumor subtype, have been proposed and may be more reproducible in practice [19, 20].

In addition to histological type and grade, several surgical and pathological parameters, including FIGO stage, depth of myometrial invasion, lymphovascular invasion, cervical involvement, lymph node status, and DNA ploidy, have been shown to be predictors of prognosis in patients with endometrial carcinoma [21], and are used to guide treatment. There is no definitive consensus, however, as to the prognostic and predictive factors to be used, and therefore the definition of risk groups in endometrial cancer is variable [22].

Whilst there are specific, established histologic criteria for the various subtypes of endometrial carcinoma, tumors that demonstrate overlapping morphologic features or are poorly differentiated continue to pose significant diagnostic challenges. Interobserver reproducibility in the diagnosis of high-grade endometrial carcinoma is limited, even among expert gynecologic pathologists [23]. The differential diagnosis of serous versus clear cell or serous versus FIGO grade 3 endometrioid carcinomas represents the most frequent areas of disagreement [23, 24]. In such cases, ancillary immunohistochemical studies in conjunction with morphologic interpretation have proven helpful in tumor classification [24–26] (Table 27.2). Endometrioid, serous, and clear cell carcinomas all express pan-cytokeratins (CKs), epithelial membrane antigen (EMA), and glycoprotein-associated markers CA125, Ber-EP4, and B72.3. All three subtypes are also usually CK7-positive and CK20-negative. Endometrioid adenocarcinomas, particularly those that are low-grade, are typically estrogen receptor (ER) and progesterone receptor (PR)-positive, display patchy and weak p53 and p16 expression, have a low proliferation index, and are negative for PTEN. It should be noted that a subset of grade 3 endometrioid carcinomas, however, shows diffuse and strong or completely absent p53 staining, indicating aberrant protein expression. Serous carcinomas typically lack diffuse ER and PR expression, have a high proliferation index, and show

Table 27-1 Histological Classification of Endometrial Adenocarcinoma

World Health Organization [12]	Blaustein [14]	Clement and Young [13]
Endometrioid Squamous differentiation Villoglandular Secretory	Endometrioid Squamous differentiation Villoglandular Secretory Ciliated cell	Endometrioid Secretory Typical With papillae Villoglandular Small non-villous papillae Microglandular Sertoliform Tumors with cords and hyalinization With metaplastic changes Squamous differentiation Clear cell change, not otherwise specified Surface changes resembling syncytial metaplasia or microglandular hyperplasia Oxyphilic (or oncocytic) cells With spindled epithelial cells (sarcomatoid)
Mucinous	Mucinous	Mucinous
Serous and serous endometrial intraepithelial carcinoma	Serous	Serous
Clear cell	Clear cell	Clear cell
Squamous	Squamous	Squamous
		Transitional
Neuroendocrine tumors		Neuroendocrine
Undifferentiated and dedifferentiated	Undifferentiated	Undifferentiated
Mixed (minor component accounts for at least 5 % of the tumor)	Mixed	Mixed (minor component accounts for at least 10 % of the tumor)
		Lymphoepithelioma-like Hepatoid Giant cell Glassy cell With trophoblastic differentiation

diffuse and strong p16 and aberrant p53 immunoexpression. Clear cell carcinoma is often ER- and PR-negative and shows p53, p16, and MIB-1 expression profiles that are intermediate between endometrioid and serous carcinomas (Table 27.2). Recently, a variant of clear cell carcinoma overexpressing p53 has been reported [27]. HNF-1b is expressed in clear cell carcinoma, but not in other endometrial adenocarcinomas [28]. Undifferentiated and the undifferentiated component of dedifferentiated carcinomas show only weak or focal keratin expression, but most cases demonstrate intense staining of rare cells with EMA and cytokeratin 18 [29].

Molecular Genetic Classification of Endometrial Cancer

The limited reproducibility of the classification of a subset of high-grade morphologically ambiguous endometrial carcinomas coupled with burgeoning molecular data on the genetic

Table 27-2 **Immunophenotype of Endometrioid, Serous, and Clear Cell Carcinomas of the Uterus**

Antibody	Endometrioid	Serous	Clear cell
ER	Strong, diffuse +	Weak, focal + or –	Weak, focal + or –
PR	Strong, diffuse +	Weak, focal + or –	Weak, focal + or –
MIB1 (Ki-67)	Low	High	Moderate
p53	Focal, weak + (grades 1 and 2) Strong, diffuse + or completely absent (grade 3)	Strong, diffuse +, or completely absent	Moderate +
p16	Patchy, variable +	Strong, diffuse +	Moderate +
Keratins	CK7+, CK20–	CK7+, CK20–	CK7+, CK20–
EMA	+	+	+
CA125, Ber-EP4, B72.3	+	+	+
PTEN	–	+	+
HNF-1b	–	–	+

CK cytokeratin, *EMA* epithelial membrane antigen, *ER* estrogen receptor, *PR* progesterone receptor, – negative, + positive

aberrations found in subsets of endometrial cancer has fueled efforts aiming to redefine the histologic classification of these tumors. Endometrioid endometrial cancers have been shown to be characterized by a high mutational frequency, mainly targeting *PTEN*, *PIK3CA*, *KRAS*, *FGFR2*, and *CTNNB1*, and by microsatellite instability (MSI) due to *MLH1* promoter hypermethylation [30] in up to 45 % cases (reviewed in [31]). More recently, recurrent mutations in *PIK3R1* and *ARID1A* have been reported [32–34]. In contrast, serous carcinomas harbor a high frequency of *TP53* and *PPP2R1A* mutations [34, 35], as well as overexpression and amplification of *HER2* in a subset of cases [36, 37]. Given that (1) frequencies of some mutations vary according to histological grade [33, 38], (2) mutational profiles of endometrioid and serous endometrial carcinomas partially overlap, and (3) there is genetic heterogeneity within each of these entities, single gene mutations or small gene panels, although not sufficient to allow for a purely mutation-based classification, may serve as an aid for morphological classification. In particular, given the binary output of mutational analyses, it has been suggested that these molecular markers may be easier to interpret than immunohistochemical results [38]. As some molecular alterations are preferentially but not exclusively found in serous versus endometrioid cancers, such as *HER2* amplification or aberrant p53 expression, it may not be entirely surprising that these markers are associated with prognosis when all types of endometrial carcinoma are considered together [39].

In breast cancer, gene expression profiling has led to a molecular classification of the disease [40], and to commercially available prognostic gene signatures as predictors of outcome and guides for treatment (reviewed in [41]). In endometrial cancer, the evidence from microarray-based expression profiling studies suggests that different histological subtypes harbor distinct transcriptomic [42–44] and distinct microRNA profiles [45], and that several genes are overexpressed and amplified in specific subtypes, such as STK15 in serous and clear cell carcinomas [46]. Furthermore, it has been observed that at the transcriptomic level, stage I serous cancers are similar to stage I, grade 3 endometrioid cancers [47], and that high-grade endometrial cancers can be classified into two subgroups with distinct molecular alterations using a panel of 22 genes involved in the PI3K-AKT pathway [48]. Gene expression sets associated with prognosis have also been reported, including a risk score stratifying

Table 27-3 **Genomic Subtypes of Endometrioid Type and Serous Endometrial Carcinoma**

	POLE **(ultramutated)**	***MSI*** **(hypermutated)**	***Copy-number low*** **(endometrioid)**	***Copy-number high*** **(serous-like)**
Mutation rate/Mb	232×10^{-6}	18×10^{-6}	2.9×10^{-6}	2.3×10^{-6}
Copy number aberrations	Few	Few	Few	High
Microsatellite instability	Mixed (high, low, stable)	High; *MLH1* promoter methylation	Stable	Stable
Characteristic genes mutated	*POLE*	*PTEN, RPL22, KRAS, ARID5B*	*PTEN, CTNNB1*	*TP53, FBXW7, PPP2R1A*
Histology	Endometrioid	Endometrioid	Endometrioid	Serous Endometrioid
Tumor grade	Mixed (1, 2, and 3)	Mixed (1, 2, and 3)	1 and 2	3
Progression-free survival	Best	Intermediate	Intermediate	Worst

Based on whole exome sequencing, gene copy number, and microsatellite instability (MSI) analysis of 248 endometrioid and serous endometrial cancers by The Cancer Genome Atlas Network [56]

clinically/pathologically intermediate-risk endometrial cancer patients into high- and low-risk recurrence groups with significant differences in time to recurrence [49]. Gene expression profiling has also yielded gene expression clusters that are predictive of recurrence in endometrioid and serous cancers [50]. Analysis of endometrial cancers at the genomic level using array-based methods has revealed that, in contrast to serous carcinomas and carcinosarcomas, endometrioid endometrial cancers harbor only few gene copy number aberrations [51]. These studies have also shown that within the group of endometrioid cancers, high levels of chromosomal instability are associated with poor prognosis [52]. Unlike breast cancer, however, due to the small number of cases analyzed so far and the lack of robust validation of gene sets or signatures in independent datasets, microarray-based studies are yet to yield clinically utilized assays for patients with endometrial cancer.

Recent advances in high-throughput sequencing technologies have allowed for the characterization of complete genomes at base pair resolution in a time and cost-effective manner. Using next generation sequencing (NGS) based technologies, whole exome sequencing of serous endometrial cancers has revealed frequent occurrence of somatic mutations in chromatin-remodeling genes (e.g., *CHD4*) and ubiquitin ligase complex genes such as *FBXW7* [53–55], and amplification of *CCNE1* [54, 55], a target of FBXW7-mediated ubiquitination, in addition to previously recognized mutations in *TP53*, *PIK3CA*, and *PPP2R1A*. Recently, The Cancer Genome Atlas (TCGA) [56] group performed a comprehensive integrative genomic analysis of endometrioid and serous endometrial cancers, which identified four groups (Table 27.3):

1. Cancers characterized by microsatellite instability (MSI) and associated hypermutation with a background mutation rate approximately tenfold greater than non-MSI tumors. These tumors were exclusively of the endometrioid type, and have few DNA copy number alterations and frequent frameshift deletions in *RPL22* and *KRAS* mutations.
2. Microsatellite stable endometrioid tumors with lower mutation frequency than MSI tumors, low copy number changes, and high frequency of *CTNNB1* mutations (52 %).
3. A group comprising serous endometrial carcinomas and a subset (25 %) of the grade 3 endometrioid cases characterized by extensive DNA copy number

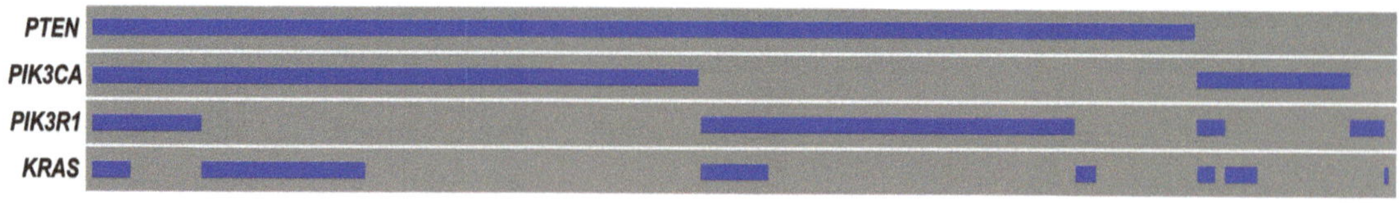

Figure 27-1 Coexisting PI3K pathway mutations in endometrioid endometrial cancer. Endometrial carcinomas of endometrioid histology with mutations in *PTEN, PIK3CA, PIK3R1,* and/or *KRAS* (*blue bars*) were selected from The Cancer Genome Atlas study through cBioPortal [56, 185] ($n = 187$), and the patterns of co-occurring mutations in these cases are shown. *Blue bars* indicate presence of a mutation

alterations and genomic instability, similar to those seen in high-grade serous ovarian carcinoma (see separate chapter). Most of these tumors were shown to harbor a *TP53* mutation and a high frequency of *FBXW7* (22 %) and *PP2R1A* (22 %) mutations.

4. A small subset of endometrioid endometrial carcinomas (7 %) with very high mutation rates ("ultra-mutated" group) characterized by hotspot mutations in the exonuclease domain of *POLE* [57], the catalytic subunit of DNA polymerase epsilon, and a favorable outcome.

This integrative genomic analysis allowed for the identification of mutations preferentially associated with specific subgroups of endometrial cancers, such as *ARID5B* mutations in 23.1 % of MSI endometrioid cancers versus 5.6 % and 0 % in microsatellite stable endometrioid and serous carcinomas, respectively [56]. However, an overlap in the mutational repertoire between these genomic groups was observed as was the genetic heterogeneity within a given group (i.e., not all cases in a genomically defined group harbored a specific mutation). Furthermore, a subset of high-grade endometrioid tumors was shown to harbor copy number and mutational profiles similar to those of serous carcinomas, providing evidence to suggest that these may share a similar biology and be driven by similar genetic aberrations, providing molecular evidence to warrant studies investigating whether these tumors could be treated similarly.

The PI3K Pathway in Endometrial Cancer

The PI3K/AKT/mTOR pathway is altered in the vast majority of endometrioid endometrial cancers. The most frequently mutated member genes are *PTEN* (78 %), the main negative regulator of the PI3K pathway, *PIK3CA* (53 %) and *PIK3R1* (37 %), the catalytic and regulatory subunits of PI3K, respectively, and *KRAS* (25 %) [56], which interacts with p110alpha (i.e., the catalytic subunit of PI3K) [58]. Furthermore, 42 % of serous carcinomas were reported to harbor a *PIK3CA* alteration [32, 56]. Unlike in some other cancer types, mutations in the PI3K pathway are not mutually exclusive and coexistence of *PTEN* and *PIK3CA* mutations and/or *KRAS* or *PTEN* and *PIK3R1* and/or *KRAS* mutations in endometrioid endometrial cancers is common [32, 33, 56, 59] (Fig. 27.1). The functional consequences of harboring multiple co-occurring mutations in a tumor and their epistatic interactions are not yet entirely understood. It has been suggested that DNA mismatch repair (MMR, see below) deficiency may partly contribute to the high frequency of mutations affecting different components of the PI3K pathway in endometrioid cancers [33]; however this phenomenon seems to be equally frequent in MMR proficient cases [56]. A significant proportion of *PIK3R1* alterations (p85; regulatory subunit of PI3K) have been found to be located within the iSH2 domain [32, 33], which mediates the binding of p85 to p110α and may constitute a mutational hotspot [60]. Given that not all *PIK3R1* mutations demonstrate gain of function in in vitro models [32, 33], it is unclear whether they are functionally equivalent to activating mutations in *PIK3CA*. In addition, the spectrum of somatic *PIK3CA* mutations within endometrial carcinomas has been found to be more varied than that of colorectal and breast cancers [61]. In contrast to colorectal, brain, gastric, or breast cancers, where >75 % of alterations have been shown to occur in two hotspots in the helical and kinase domains

[62], in endometrial cancer, *PIK3CA* mutations are distributed throughout the gene and somatic, activating mutations in the adapter binding domain (ABD), ABD-linker region, and C2 domains of p110a are also frequent [33, 61].

Given the high prevalence of PI3K pathway aberration in endometrial cancer, clinical trials are currently underway to assess the efficacy of inhibitors of this pathway in patients with endometrial cancer (reviewed in [31, 39, 63]). First results from early clinical trials revealed that a subset of chemotherapy-naïve patients with advanced endometrial cancer are responsive to single agent rapamycin analogs (i.e., allosteric mTOR inhibitors), and that therapeutic responses or stabilization of disease can be seen across histological types [8, 64–66]. These trials also suggest that not only endometrioid but also a subset of serous carcinomas may be dependent on the PI3K pathway. Predictive markers associated with response to mTORC1 inhibitors have yet to be identified. The latter could be due to the small number and heterogeneous groups of endometrial cancer patients included in the trials to date. In vitro analyses of endometrioid endometrial cancer cell lines have suggested that inhibitors targeting different components of the PI3K pathway may be associated with distinct genomic predictors [67]. However, results from ongoing clinical trials are eagerly awaited.

Lynch Syndrome

Although the majority of endometrial cancers occur sporadically, approximately 2 % of cases arise in the setting of the hereditary Lynch syndrome (or also referred to as hereditary non-polyposis colorectal carcinoma syndrome, HNPCC) [68, 69]. This autosomal dominant disease is associated with germline mutations in DNA mismatch repair (MMR) genes, *MLH1*, *MSH2*, *MSH6*, or *PMS2*, as well as *EPCAM* deletion resulting in *MSH2* promoter methylation. The syndrome is associated with increased risk of endometrial, colorectal, gastric, and other cancers. The risk for endometrial cancer is estimated to be up to 60 % [70–72]. Genetic alterations in DNA MMR genes lead to the accumulation of unstable microsatellite sequences throughout the genome (i.e., MSI). Women with Lynch syndrome who are diagnosed with endometrial carcinoma also have an increased risk of developing colorectal cancer [73] and several other cancers including breast cancer [74]; thus their recognition is important for family screening, early detection or interventions to reduce the risk of additional cancers.

Identification of individuals with Lynch syndrome is based on the Amsterdam and Bethesda criteria [72, 75], which, in contrast to colorectal cancer, have proven ineffective for endometrial cancer patients [68]. In fact, guidelines for the identification of endometrial cancer patients who have Lynch syndrome have yet to be developed, and different screening algorithms have been suggested [76, 77].

A definite diagnosis of Lynch syndrome is established by germline mutational analysis of DNA MMR genes *MLH1*, *MSH2*, *MSH6*, and/or *PMS2*. Also, MSI analysis using polymerase chain reaction of two mononucleotide (BAT25, BAT26) and three dinucleotide (D2S123, D5S346, D17S250) microsatellite markers has been recommended by the National Cancer Institute (Bethesda Guidelines) for the identification of HNPCC patients [78]. It has been shown, however, that a panel of five mononucleotide repeat markers rather than mono- and dinucleotide markers as used in the Bethesda Guidelines may provide a more accurate evaluation of tumor MSI in colorectal [75, 79] and endometrial cancers (BAT-25, BAT-26, NR-21, NR-24, and NR-27) [80]. When using the five Bethesda markers, high-frequency MSI (MSI-H) is present if two or more markers show instability, and low-frequency MSI (MSI-L) if one marker shows instability. In contrast, when using five mononucleotide repeat markers, three or more mutant alleles are typically required to indicate MSI-H [79]. It should be noted that endometrial cancer represents the most common clinical manifestation in female *MSH6* germline mutation carriers [81, 82]. However, carcinomas with *MSH6* mutations may be microsatellite stable (MSS) or MSI-L [76, 83], and, therefore, potentially may be missed by MSI analysis. In addition, as mentioned above, a subset of sporadic endometrial carcinomas show MSI due to *MLH1* promoter methylation, which cannot be differentiated by MSI analysis.

The role of the pathologist in the identification of patients with Lynch syndrome-associated endometrial cancer should not be underestimated. Given the cost of conventional sequencing of MMR genes and the low prevalence of Lynch syndrome in patients with endometrial cancer, immunohistochemical analysis of the four MMR proteins, which has been shown to be sensitive and specific for the detection of germline MMR abnormalities and of tumors with MSI [68, 84], has been advocated in the context of endometrial cancers arising (1) in patients with personal or family history of Lynch syndrome or Lynch syndrome-associated tumors, (2) in patients younger than 50 years of age, and (3) in patients of any age, whose tumors arise in the lower uterine segment, synchronously with ovarian clear cell carcinoma, are of de-differentiated/undifferentiated type, and/or show tumor heterogeneity, peritumoral lymphocytes, or tumor-infiltrating lymphocytes [76, 77]. Loss of MLH1/PMS2 expression in the absence of *MLH1* promoter methylation, loss of MSH2 and/or MSH6 expression, and potentially of PMS2 alone indicates the presence of an MMR germline mutation, and identifies high-risk Lynch syndrome patients to be considered for germline testing.

With the introduction of cost-effective multiplex NGS-based test panels evaluating cancer susceptibility genes, including MMR genes [85], one may speculate that in the future MMR mutational screening could be implemented as a primary screen rather than a confirmatory test in individuals with high risk of Lynch syndrome.

Uterine Mesenchymal Tumors

Uterine mesenchymal tumors comprise a heterogeneous group of neoplasms, the overwhelming majority of which are benign smooth muscle neoplasms. Malignant mesenchymal tumors are rare, accounting for 2–5 % of uterine cancers. Differentiation of benign mesenchymal lesions from their malignant counterparts is crucial due to their distinct clinical outcome. Some mesenchymal tumors pose diagnostic challenges in surgical pathology practice, in particular endometrial stromal tumors and the biologic spectrum of smooth muscle tumors, as these may display variable and overlapping morphologic features. There have been several important developments in our understanding of the molecular genetics of these tumors over the recent years that have improved classification, diagnosis, and prediction of outcome.

Smooth Muscle Tumors

Smooth muscle tumors represent the most prevalent mesenchymal neoplasms in the uterus, and are classified into leiomyoma, smooth muscle tumor of uncertain malignant potential, and leiomyosarcoma, epithelioid and myxoid leiomyosarcoma variants [12]. Leiomyoma represents the vast majority of uterine smooth muscle tumors, which frequently affects women of reproductive age and is the leading indication for hysterectomy in the United States [86, 87]. In contrast, leiomyosarcoma and smooth muscle tumor of uncertain malignant potential represent only a small percentage of these neoplasms (40 % of all uterine sarcomas; 1–3 % of all uterine malignancies [88]). Leiomyosarcomas are thought to arise independently from leiomyomas [89]. They are highly aggressive tumors with a 5-year survival of 15–60 % [90]. Diagnostic criteria for leiomyosarcomas vary slightly with histologic subtype, but are based on a combination of histologic features including the presence of moderate to severe nuclear atypia, high mitotic rate, and tumor cell necrosis [91]. Neoplasms that do not meet histologic criteria for the diagnosis of leiomyosarcoma are classified as smooth muscle tumors of uncertain malignant potential. There is, however, significant interobserver variability in the assessment of these morphologic features as well as of the mitotic index [92, 93].

Immunohistochemical makers are of limited help in distinguishing the different groups of smooth muscle tumors. Hormone receptors are positive in the vast majority of leiomyomas, and up to 60 % of leiomyosarcomas can express ER and PR [94–100]. p16 and p53 expression is usually present in leiomyosarcomas; however these markers can also be detected in the leiomyoma variant with bizarre nuclei and less frequently in smooth muscle tumors of uncertain malignant potential [101–111]. MIB1 (Ki-67) expression, a marker of proliferation, can be elevated in leiomyosarcoma, leiomyoma with bizarre nuclei, and mitotically active leiomyoma [104, 108–110]. Therefore, immunohistochemical markers are of limited utility in reliably discriminating between benign and malignant smooth muscle tumors.

Over the past years, various genetic aberrations have been discovered in uterine smooth muscle tumors. These include recurrent mutations in mediator complex subunit 12 (*MED12*), which have been reported in up to 80 % of conventional uterine leiomyomas [88, 112–118]. *MED12* is a component of the mediator complex involved in the transcription of RNA polymerase II-dependent genes [119], and also plays a role in the Wnt/β-catenin and Hedgehog signaling pathways [120, 121], and in the regulation of *Nanog* and *Nanog* target genes [122]. *MED12* mutations have also been found in a subset of leiomyoma variants, including cellular leiomyomas [123], leiomyomas with bizarre nuclei [117], and mitotically active leiomyomas [123]. In contrast, mutations in *MED12* are less frequently found in smooth muscle tumors of uncertain malignant potential [88, 123] and leiomyosarcomas [88, 114, 117, 118]. However, the number of samples analyzed to date remains relatively small. As mentioned above, it is generally thought that leiomyosarcomas arise independently from leiomyomas [89]; however the presence of *MED12* mutations in a subset of leiomyosarcomas suggests that a small subgroup of leiomyosarcomas may, in fact, originate from leiomyomas.

Somatic mutations in fumarate hydratase (*FH*), which encodes an enzyme of the tricarboxylic acid cycle, are found in a small subset of sporadic (i.e., nonhereditary) leiomyomas [124]. In addition, leiomyomas have been shown to harbor the recurrent t(12;14)(q15;q23–24) translocation in approximately 10 % of cases or other 12q14–15 chromosomal rearrangements involving the *HMGA2* and *RAD51B* loci, rearrangements involving 6p21 and Xq22 (affecting *COL4A5* and *COL4A6*), deletions of 7q, and trisomy 12, among other genetic aberrations [125, 126] (Table 27.4). Interestingly, complex chromosomal rearrangements resembling chromothripsis [127] have recently been documented in these benign tumors, and a subset of physically distinct uterine leiomyomas from the same patient have been shown to be clonally related [126]. These observations suggest that even hallmarks of catastrophic genetic events, such as chromothripsis, cannot be employed to differentiate benign from malignant smooth muscle tumors.

Leiomyosarcomas often display high levels of genetic instability, so much so that no identical karyotypes were found in 68 cases published in the literature between 1994 and 2004 (reviewed in [89]). Although chromosomal alterations involving 1p translocations and deletions [128] and *TP53* mutations [129, 130] have been reported in a subset of leiomyosarcomas, no single highly recurrent genetic aberration has been identified to date. Furthermore, it is currently unclear whether the genetic alterations recently described in leiomyomas, such as the chromosomal rearrangements affecting *COL4A5* and *COL4A6* [126], are unique to benign lesions.

In summary, no reliable immunohistochemical or molecular markers have been identified for the discrimination between benign and malignant smooth muscle tumors. The role of these ancillary markers is limited. If used at all, the above markers should be used in conjunction with thorough histologic analysis to establish a definitive diagnosis.

Endometrial Stromal Tumors

Endometrial stromal tumors are rare neoplasms that are classified into endometrial stromal nodule (ESN), low-grade and high-grade endometrial stromal sarcoma (ESS), and undifferentiated uterine sarcoma (US) according to the latest WHO classification [12]. Both ESNs and low-grade ESSs are composed of bland cells resembling proliferative endometrial stromal cells and can exhibit a number of variant histologic features. ESNs are benign and have well-circumscribed borders without invasion of surrounding tissue in contrast to ESS that demonstrates infiltrative growth and vascular invasion. Low-grade ESSs are relatively indolent with a 10-year

Table 27-4 **The Most Recurrent Genomic Aberrations in Sporadic Uterine Mesenchymal Tumors**

	Chromosomal rearrangement	Gene copy number aberration	Mutated gene
Leiomyoma	t(12;14)(q15;q23–24) and other 12q14–15 (involving *HMGA2* and *RAD51B*) 6p21 Chromothripsis	del(7)(q22q32) Trisomy chromosome 12	*MED12* *FH*
Leiomyosarcoma	Complex and frequent 1p	Complex and frequent 1p deletions	*TP53*
Endometrial stromal nodule	t(7;17)(p15;q21) (*JAZF1-SUZ12*)		
Endometrial stromal sarcoma	t(7;17)(p15;q21) (*JAZF1-SUZ12*) t(6;7)(p21;p15) (*JAZF1-PHF1*) t(6;10;10)(p21;q22;p11) (*PHF1-EPC1*) t(1;6)(p34;p21) (*MEAF6-PHF1*) der(22)t(X;22)(p11;q13) (*ZC3H7B-BCOR*) t(10;17)(q22;p13) (*YWHAE–FAM22A/B*)		
Undifferentiated uterine sarcoma	t(7;17)(p15;q21) (*JAZF1-SUZ12*) t(10;17)(q22;p13) (*YWHAE–FAM22A/B*)[a]	Complex karyotype Multiple numerical and structural aberrations	*TP53*

[a]*YWHAE-FAM22A/B* rearranged tumors are now recognized as high-grade ESSs in the new 2014 WHO classification [12]

overall survival rate ranging from 65 to 76 % [131]. High-grade ESSs, a newly recognized subset of ESS, may exhibit a high-grade round cell component in addition to a low-grade ESS and appear to have a prognosis intermediate between low-grade ESS and UUS. UUSs, previously known as uterine endometrial sarcomas by the WHO, demonstrate significant cytologic atypia, bear no morphologic similarity to endometrial stroma, often demonstrate destructive myometrial invasion, and have a poor prognosis (median overall survival 1–3 years) [131–134]. Given the heterogeneity observed within UUS tumors, some authors have suggested further subtyping of UUS into uniform (UUS-U) and pleomorphic types (UUS-P) to stratify morphologically endometrial stromal tumors into those that demonstrate uniform cytologic atypia with enlarged, hyperchromatic nuclei and prominent nucleoli and those that exhibit marked nuclear pleomorphism, respectively [135]. The clinical and biological significance of UUS-U and UUS-P remains to be fully determined.

Ancillary immunohistochemical studies may be useful in the diagnosis of endometrial stromal tumors. A panel of markers including CD10 and smooth muscle markers such as desmin, h-caldesmon, and HDAC8 has been recommended to discriminate ESNs and ESSs from smooth muscle tumors [136–138]. It should be noted that the expression of these markers should be interpreted in conjunction with tumor morphology given that areas of smooth muscle differentiation may express any of the aforementioned smooth muscle markers. Markers that are usually expressed in ovarian sex cord stromal tumors can also be found in “sex cord-like” foci of endometrial stromal tumors. The latter include inhibin, calretinin, CD99, melan A, and WT-1 [139–145]. In addition, low-grade ESSs frequently express ER and PR [135, 146–150], nuclear β-catenin [135, 151–154], EGFR [155–157], and c-kit [156, 158], and less frequently androgen receptor [159], aromatase [160], PDGF-α, PDGF-β, or VEGF [158, 161]. It should be noted, however, that mutations in *KIT*, *PDGFR-α*, *PDGFR-β*, or *EGFR* have not been detected in ESS [156, 157]; hence, the proteins they encode are unlikely to constitute optimal therapeutic targets for ESS. Cyclin D1 is diffusely expressed in high-grade ESSs that exhibit high-grade round cell morphology and harbor specific genetic alterations (i.e., *YWHAE-FAM22* fusion; see below). ER and PR expression is present in the low-grade component of these tumors, but absent in round cell areas.

Compared to ESS, the immunohistochemical profile of UUS is less well established (reviewed in [162]). ER and PR are expressed in up to 50 % of UUS-U, but not in UUS-P [135, 150, 163]. Most UUS-U and approximately one-third of UUS-P express nuclear β-catenin, whereas overexpression of p53 is more often associated with UUS-P than UUS-U [135, 150, 163]. Cyclin D1 immunoexpression has been found only rarely in UUS-U [164].

Endometrial stromal tumors are characterized by recurrent chromosomal translocations (Table 27.4), including the t(7;17)(p15;q21), t(6;7)(p21;p15), t(6;10;10)(p21;q22;p11), t(10;17)(q22;p13), t(1;6)(p34;p21), and der(22)t(X;22)(p11;q13) which result in the formation of the chimeric transcripts *JAZF1-SUZ12*, *PHF1-JAZF1*, *EPC1-PHF1*, *YWHAE-FAM22*, *MEAF6-PHF1*, and *ZC3H7B-BCOR*, respectively (reviewed in [162]). When present, gene fusions appear to be mutually exclusive and have been observed in both conventional endometrial stromal tumors and less frequently those that display variant morphology. The fusion of *JAZF1* and *SUZ12* (also known as *JJAZ1*) is the most common genetic alteration in endometrial stromal tumors and has been reported in 68 % of ESNs, 43 % of ESSs, and 9 % of UUSs [135, 162, 165–171]. *PHF1-JAZF1* and *EPC1-PHF1* rearrangements have been observed in 10 % and 6 % of ESSs, respectively [162, 170, 172]. *MEAF6-PHF1* and *ZC3H7B-BCOR* gene fusions have recently been characterized in rare low-grade ESSs [173, 174], and also *PHF1* and *JAZF1* rearrangements without known partners have also been observed [170, 171]. *YWHAE-FAM22* fusions have been recently described in ESSs harboring high-grade areas as well as in UUS-Us [149, 163, 175, 176], although its frequency among endometrial stromal tumors is not yet known.

The functions of the chimeric proteins resulting from the various gene rearrangements identified in endometrial stromal tumors are not resolved at this time. The most common fusion joins the first three exons of *JAZF1*, the function of which is relatively unknown, to the last 15 exons of *SUZ12* (*JJAZ1*), a component of the polycomb repression complex 2 [177]. Low levels of *JAZF1-SUZ12* mRNA, which are thought to arise from trans-splicing of precursor mRNAs for *JAZF1* and *SUZ12* genes, and the JAZF1-SUZ12 protein have been detected in normal endometrial stromal cells lines and in normal late-secretory and early-proliferative phase endometrium [177, 178]. Based on these findings, it has been suggested that acquisition of the *JAZF1-SUZ12* fusion via chromosomal translocation may be an early event in the pathogenesis of endometrial stromal tumors. Expression of *JAZF1-SUZ12* has been shown to promote cell proliferation in vitro, however only when accompanied by suppression of endogenous wild-type SUZ12 (from the unarranged allele) [177]. Interestingly, the non-rearranged wild-type *JAZF1* allele is active in ESNs but silenced in ESSs harboring the *JAZF1-SUZ12* fusion, providing evidence to suggest that ESSs arise from ESNs via genetic or epigenetic silencing of the non-rearranged allele [177]. The presence of *JAZF1-SUZ12* fusion in rare UUSs also raises the possibility that a small subset of UUSs originate from ESNs and ESSs via dedifferentiation [135, 165]; it should be noted, however, that the vast majority of UUSs are unlikely to stem from ESNs, given that these tumors appear to be genetically distinct.

Several studies have sought to correlate specific variant histologic features and gene fusions in endometrial stromal tumors. Only the recently characterized *YWHAE-FAM22* fusion transcripts, derived from the t(10;17) rearrangement, appear to be associated with specific clinicopathologic features. *YWHAE-FAM22* fusions have been shown to lead to nuclear accumulation of the functionally intact 14-3-3ε (*YWHAE*) protein-interaction domain, thereby likely redirecting known cytoplasmic 14-3-3ε protein interactions with phosphoserine-containing proteins to the nuclear compartment [175]. The putative mechanisms of action of chimeric products involving *MEAF6, EPC1, PHF1, ZC3H7B*, and *BCOR* remain to be determined. The *YWHAE-FAM22* fusion gene has been reported in ESSs and UUS-Us with high-grade round cell areas with increased mitotic activity, tumor cell necrosis, and adjacent low-grade fibroblastic or fibromyxoid areas [149, 163, 175]. Rare tumors harboring this novel gene fusion may also demonstrate rosettes, pseudoglandular/pseudopapillary patterns, or sex cord differentiation [149, 179]. It should be noted that whilst the *YWHAE-FAM22* fusion has also been described in a subset of clear cell sarcomas of the kidney [180, 181], it has not been found in conventional low-grade ESS or other uterine and non-gynecological mesenchymal tumors, adenosarcomas, or carcinosarcomas [175]. The fibroblastic component of tumors harboring *YWHAE-FAM22* fusion shares a similar immunoprofile with conventional low-grade ESS and expresses CD10, ER, and PR. Importantly, however, CD10 is typically negative and hormone receptors are often absent or only focally weakly-to-moderately expressed in the high-grade round cell areas. By gene expression profiling, *CCND1* (cyclin D1) has also been reported to be differentially expressed between *YWHAE-FAM22* rearranged tumors and conventional low-grade ESSs, with cyclin D1 being a sensitive and specific diagnostic marker for the identification of stromal sarcomas harboring the t(10;17) translocation [164]. The mechanism leading to cyclin D1 overexpression in *YWHAE-FAM22* rearranged tumors remains to be elucidated.

The characterization of the genetic underpinning of uterine mesenchymal neoplasms is providing fertile ground for the development of molecular genetic tests for their accurate diagnosis, opportunities for the development of immunohistochemical surrogate markers for the presence of these genetic aberrations, and novel targeted therapies. With the efforts of the TCGA, the International Cancer Genome Consortium (ICGC), and individual investigators, a wealth of data on the genetic aberrations found in rare types of uterine cancers will be available. These data will be instrumental to find answers for questions that have puzzled pathologists for decades, including the actual nature and molecular drivers of carcinosarcomas (i.e., mixed epithelial and mesenchymal tumors), as well as the similarities and differences between uterine tumors and their ovarian counterparts, including sex cord stromal tumors and adenosarcomas.

Conclusions

In the past, endometrial cancer was perceived as a "benign disease" given the large proportion of women with low-risk disease and high cure rates [182]. Advances in our understanding of uterine cancer revealed the complexity of the disease, in particular at the molecular level. This information may result in a refinement of the taxonomy of uterine neoplasms and in the identification of genetic biomarkers required for the introduction of molecular target-based therapy for the treatment of women with this disease. The ongoing efforts to characterize the genetic landscapes of cancers are expected to provide further insights into cancers of the uterine corpus and lead to the introduction of genomic applications for endometrial cancer in the near future. The insights from the integrated genomic characterization of endometrial carcinoma by the TCGA [56] have called the distinct nature of type I and type II endometrial cancers into question. In fact, these two subsets of endometrial cancers are highly heterogeneous, and similarities between entities classified as type I or type II have been documented.

It is probable that with the burgeoning information on the genetic makeup of different types of endometrial cancer, a combined morphological and molecular classification will emerge, which may result in more accurate diagnosis, in particular, of high-grade lesions. Based on the results of the integrative analysis carried out by TCGA, one may envision that *POLE* sequencing analysis, or potentially an immunohistochemistry assay sufficiently accurate to identify *POLE* mutant lesions, may be used to identify patients with a favorable expected outcome, who would be candidates for more conservative systemic treatment approaches. In contrast, the presence of *TP53* mutations in endometrioid endometrial cancers may define the subset of endometrioid carcinomas with a serous-like aggressive behavior. In addition, a proof-of-principle study has demonstrated that endometrial and ovarian cancer cells can be detected through massively parallel sequencing analysis of DNA extracted from routine liquid-based cervical cytological specimens, which may constitute a first step towards a new generation of cancer screening tests [183].

The recent identification of recurrent mutations and fusion genes in uterine mesenchymal tumors is expected to lead to the development of molecular tests, based on in situ hybridization, reverse transcriptase (RT)-PCR, and RNA-sequencing, or even immunohistochemical tests, such as cyclin D1 expression for the detection of high-grade ESS harboring *YWHAE-FAM22* rearrangements [164].

We anticipate that diagnostic and predictive biomarkers based on the genetic features of endometrioid cancers and subsets of mesenchymal tumors are likely to be incorporated into the armamentarium of diagnostic gynecological pathologists in the near future. Furthermore, immunohistochemical markers to identify prognostically and therapeutically distinct subsets of uterine disease are likely to be developed and incorporated to pathology practice.

Finally, the integration of biomarker assessment and/or routine tissue collection in clinical trials testing novel targeted agents should be encouraged, and this source of biological material is absolutely essential for the identification of predictive markers. To date, in particular PI3K pathway and FGFR2 inhibitors are thought to have great potential as targeted therapies in subsets of patients with endometrial carcinoma (reviewed in [31, 39, 63, 184]). It should be noted, however, that from a clinical perspective, the patients most in need for more effective therapies are those with advanced, recurrent, or metastatic disease. Genomic studies, however, have so far focused on the analysis of primary tumors. Conversely, early clinical trials are usually performed in pretreated patients with advanced disease. Therefore, germane to the realization of the potential of precision medicine is that studies ascertaining whether the genetic landscapes of metastatic endometrial cancer are similar or distinct from those of early lesions are carried out.

REFERENCES

1. American Cancer Society. Cancer facts & figures 2013. Atlanta: American Cancer Society; 2013.
2. Evans T, Sany O, Pearmain P, Ganesan R, Blann A, Sundar S. Differential trends in the rising incidence of endometrial cancer by type: data from a UK population-based registry from 1994 to 2006. Br J Cancer. 2011;104(9):1505–10. doi:10.1038/bjc.2011.68.
3. Wartko P, Sherman ME, Yang HP, Felix AS, Brinton LA, Trabert B. Recent changes in endometrial cancer trends among menopausal-age US women. Cancer Epidemiol. 2013;37:374–7. doi:10.1016/j.canep.2013.03.008.
4. Boll D, Karim-Kos HE, Verhoeven RHA, Burger CW, Coebergh JW, van de Poll-Franse LV, van Doorn HC. Increased incidence and improved survival in endometrioid endometrial cancer diagnosed since 1989 in The Netherlands: a population based study. Eur J Obstet Gynecol Reprod Biol. 2013;166(2):209–14. doi:10.1016/j.ejogrb.2012.10.028.
5. Creasman WT, Odicino F, Maisonneuve P, Quinn MA, Beller U, Benedet JL, Heintz AP, Ngan HY, Pecorelli S. Carcinoma of the corpus uteri. FIGO 26th Annual Report on the Results of Treatment in Gynecological Cancer. Int J Gynaecol Obstet. 2006;95 Suppl 1:S105–43. doi:10.1016/S0020-7292(06)60031-3.
6. Makker V, Hensley ML, Zhou Q, Iasonos A, Aghajanian CA. Treatment of advanced or recurrent endometrial carcinoma with doxorubicin in patients progressing after paclitaxel/carboplatin: memorial Sloan-Kettering Cancer Center experience from 1995 to 2009. Int J Gynecol Cancer. 2013;23:929–34. doi:10.1097/IGC.0b013e3182915c20.
7. Carey MS, Gawlik C, Fung-Kee-Fung M, Chambers A, Oliver T. Systematic review of systemic therapy for advanced or recurrent endometrial cancer. Gynecol Oncol. 2006;101(1):158–67. doi:10.1016/j.ygyno.2005.11.019.
8. Oza AM, Elit L, Tsao MS, Kamel-Reid S, Biagi J, Provencher DM, Gotlieb WH, Hoskins PJ, Ghatage P, Tonkin KS, Mackay HJ, Mazurka J,

Sederias J, Ivy P, Dancey JE, Eisenhauer EA. Phase II study of temsirolimus in women with recurrent or metastatic endometrial cancer: a trial of the NCIC Clinical Trials Group. J Clin Oncol. 2011;29(24):3278–85. doi:10.1200/JCO.2010.34.1578.
9. Pecorelli S. Revised FIGO staging for carcinoma of the vulva, cervix, and endometrium. Int J Gynaecol Obstet. 2009;105(2):103–4.
10. Bokhman JV. Two pathogenetic types of endometrial carcinoma. Gynecol Oncol. 1983;15(1):10–7.
11. Setiawan VW, Yang HP, Pike MC, McCann SE, Yu H, Xiang YB, Wolk A, Wentzensen N, Weiss NS, Webb PM, van den Brandt PA, van de Vijver K, Thompson PJ, Strom BL, Spurdle AB, Soslow RA, Shu XO, Schairer C, Sacerdote C, Rohan TE, Robien K, Risch HA, Ricceri F, Rebbeck TR, Rastogi R, Prescott J, Polidoro S, Park Y, Olson SH, Moysich KB, Miller AB, McCullough ML, Matsuno RK, Magliocco AM, Lurie G, Lu L, Lissowska J, Liang X, Lacey Jr JV, Kolonel LN, Henderson BE, Hankinson SE, Hakansson N, Goodman MT, Gaudet MM, Garcia-Closas M, Friedenreich CM, Freudenheim JL, Doherty J, De Vivo I, Courneya KS, Cook LS, Chen C, Cerhan JR, Cai H, Brinton LA, Bernstein L, Anderson KE, Anton-Culver H, Schouten LJ, Horn-Ross PL. Type I and II endometrial cancers: have they different risk factors? J Clin Oncol. 2013;31:2607–18. doi:10.1200/JCO.2012.48.2596.
12. Kurman RJ, Carcangiu M, Herrington CS, Young RH. World Health Organization classification of tumors of female reproductive organs. 4th ed. Lyon: IARC; 2014.
13. Clement PB, Young RH. Atlas of gynecologic surgical pathology. 3rd ed. Philadelphia: Saunders/Elsevier; 2014.
14. Kurman RJ, Ellenson LH, Ronnett BM, editors. Blaustein's pathology of the female genital tract. 6th ed. New York: Springer; 2011.
15. McCluggage WG. Uterine carcinosarcomas (malignant mixed Mullerian tumors) are metaplastic carcinomas. Int J Gynecol Cancer. 2002; 12(6):687–90.
16. McCluggage WG. Malignant biphasic uterine tumours: carcinosarcomas or metaplastic carcinomas? J Clin Pathol. 2002;55(5):321–5.
17. Announcement. FIGO stages-1988 revision. Gynecol Oncol. 1989;35:125–6.
18. Zaino RJ, Kurman RJ, Diana KL, Morrow CP. The utility of the revised International Federation of Gynecology and Obstetrics histologic grading of endometrial adenocarcinoma using a defined nuclear grading system. A Gynecologic Oncology Group study. Cancer. 1995;75(1):81–6.
19. Alkushi A, Abdul-Rahman ZH, Lim P, Schulzer M, Coldman A, Kalloger SE, Miller D, Gilks CB. Description of a novel system for grading of endometrial carcinoma and comparison with existing grading systems. Am J Surg Pathol. 2005;29(3):295–304.
20. Lax SF, Kurman RJ, Pizer ES, Wu L, Ronnett BM. A binary architectural grading system for uterine endometrial endometrioid carcinoma has superior reproducibility compared with FIGO grading and identifies subsets of advance-stage tumors with favorable and unfavorable prognosis. Am J Surg Pathol. 2000;24(9):1201–8.
21. Prat J. Prognostic parameters of endometrial carcinoma. Hum Pathol. 2004;35(6):649–62.
22. Sorbe B. Predictive and prognostic factors in definition of risk groups in endometrial carcinoma. ISRN Obstet Gynecol. 2012;2012:325790. doi:10.5402/2012/325790.
23. Gilks CB, Oliva E, Soslow RA. Poor interobserver reproducibility in the diagnosis of high-grade endometrial carcinoma. Am J Surg Pathol. 2013;37(6):874–81. doi:10.1097/PAS.0b013e31827f576a.
24. Soslow RA. High-grade endometrial carcinomas – strategies for typing. Histopathology. 2013;62(1):89–110. doi:10.1111/his.12029.
25. Clarke BA, Gilks CB. Endometrial carcinoma: controversies in histopathological assessment of grade and tumour cell type. J Clin Pathol. 2010;63(5):410–5.doi:10.1136/jcp.2009.071225.
26. Mittal K, Soslow R, McCluggage WG. Application of immunohistochemistry to gynecologic pathology. Arch Pathol Lab Med. 2008;132(3):402–23. doi:10.1043/1543-2165(2008)132[402:AOITGP]2.0.CO;2.
27. Delair D, Soslow RA (2012) Endometrial clear cell carcinomas with and without aberrant p53 expression: a study of 16 cases. Laboratory Investigation Meeting abstract; 92 (supp. 1) (101st Annual Meeting of the United States and Canadian Academy of Pathology):265A.
28. Fadare O, Liang SX. Diagnostic utility of hepatocyte nuclear factor 1-Beta immunoreactivity in endometrial carcinomas: lack of specificity for endometrial clear cell carcinoma. Appl Immunohistochem Mol Morphol. 2012;20(6): 580–7. doi:10.1097/PAI.0b013e31824973d1.
29. Tafe LJ, Garg K, Chew I, Tornos C, Soslow RA. Endometrial and ovarian carcinomas with undifferentiated components: clinically aggressive and frequently underrecognized neoplasms. Mod Pathol. 2010;23(6):781–9. doi:10.1038/modpathol.2010.41.
30. Simpkins SB, Bocker T, Swisher EM, Mutch DG, Gersell DJ, Kovatich AJ, Palazzo JP, Fishel R, Goodfellow PJ. MLH1 promoter methylation and gene silencing is the primary cause of microsatellite instability in sporadic endometrial cancers. Hum Mol Genet. 1999;8(4):661–6.
31. Dedes KJ, Wetterskog D, Mendes-Pereira AM, Natrajan R, Lambros MB, Geyer FC, Vatcheva R,

Savage K, Mackay A, Lord CJ, Ashworth A, Reis-Filho JS. PTEN deficiency in endometrioid endometrial adenocarcinomas predicts sensitivity to PARP inhibitors. Sci Transl Med. 2010; 2(53):53ra75. doi:10.1126/scitranslmed.3001538.
32. Urick ME, Rudd ML, Godwin AK, Sgroi D, Merino M, Bell DW. PIK3R1 (p85alpha) is somatically mutated at high frequency in primary endometrial cancer. Cancer Res. 2011;71(12):4061–7. doi:10.1158/0008-5472.CAN-11-0549.
33. Cheung LW, Hennessy BT, Li J, Yu S, Myers AP, Djordjevic B, Lu Y, Stemke-Hale K, Dyer MD, Zhang F, Ju Z, Cantley LC, Scherer SE, Liang H, Lu KH, Broaddus RR, Mills GB. High frequency of PIK3R1 and PIK3R2 mutations in endometrial cancer elucidates a novel mechanism for regulation of PTEN protein stability. Cancer Discov. 2011;1(2):170–85. doi:10.1158/2159-8290.CD-11-0039.
34. Wiegand KC, Lee AF, Al-Agha OM, Chow C, Kalloger SE, Scott DW, Steidl C, Wiseman SM, Gascoyne RD, Gilks B, Huntsman DG. Loss of BAF250a (ARID1A) is frequent in high-grade endometrial carcinomas. J Pathol. 2011;224(3):328–33. doi:10.1002/path.2911.
35. McConechy MK, Anglesio MS, Kalloger SE, Yang W, Senz J, Chow C, Heravi-Moussavi A, Morin GB, Mes-Masson AM, Carey MS, McAlpine JN, Kwon JS, Prentice LM, Boyd N, Shah SP, Gilks CB, Huntsman DG. Subtype-specific mutation of PPP2R1A in endometrial and ovarian carcinomas. J Pathol. 2011;223(5):567–73. doi:10.1002/path.2848.
36. Santin AD, Bellone S, Van Stedum S, Bushen W, Palmieri M, Siegel ER, De Las Casas LE, Roman JJ, Burnett A, Pecorelli S. Amplification of c-erbB2 oncogene: a major prognostic indicator in uterine serous papillary carcinoma. Cancer. 2005;104(7):1391–7. doi:10.1002/cncr.21308.
37. Hetzel DJ, Wilson TO, Keeney GL, Roche PC, Cha SS, Podratz KC. HER-2/neu expression: a major prognostic factor in endometrial cancer. Gynecol Oncol. 1992;47(2):179–85.
38. McConechy MK, Ding J, Cheang MC, Wiegand KC, Senz J, Tone AA, Yang W, Prentice LM, Tse K, Zeng T, McDonald H, Schmidt AP, Mutch DG, McAlpine JN, Hirst M, Shah SP, Lee CH, Goodfellow PJ, Gilks CB, Huntsman DG. Use of mutation profiles to refine the classification of endometrial carcinomas. J Pathol. 2012;228(1):20–30. doi:10.1002/path.4056.
39. Salvesen HB, Haldorsen IS, Trovik J. Markers for individualised therapy in endometrial carcinoma. Lancet Oncol. 2012;13(8):e353–61. doi:10.1016/S1470-2045(12)70213-9.
40. Perou CM, Sorlie T, Eisen MB, van de Rijn M, Jeffrey SS, Rees CA, Pollack JR, Ross DT, Johnsen H, Akslen LA, Fluge O, Pergamenschikov A, Williams C, Zhu SX, Lonning PE, Borresen-Dale AL, Brown PO, Botstein D. Molecular portraits of human breast tumours. Nature. 2000;406(6797):747–52.doi:10.1038/35021093.
41. Weigelt B, Baehner FL, Reis-Filho JS. The contribution of gene expression profiling to breast cancer classification, prognostication and prediction: a retrospective of the last decade. J Pathol. 2010;220(2):263–80. doi:10.1002/path.2648.
42. Risinger JI, Maxwell GL, Chandramouli GV, Jazaeri A, Aprelikova O, Patterson T, Berchuck A, Barrett JC. Microarray analysis reveals distinct gene expression profiles among different histologic types of endometrial cancer. Cancer Res. 2003;63(1):6–11.
43. Maxwell GL, Chandramouli GV, Dainty L, Litzi TJ, Berchuck A, Barrett JC, Risinger JI. Microarray analysis of endometrial carcinomas and mixed mullerian tumors reveals distinct gene expression profiles associated with different histologic types of uterine cancer. Clin Cancer Res. 2005;11(11):4056–66. doi:10.1158/1078-0432.CCR-04-2001.
44. Chen Y, Yao Y, Zhang L, Li X, Wang Y, Zhao L, Wang J, Wang G, Shen D, Wei L, Zhao J. cDNA microarray analysis and immunohistochemistry reveal a distinct molecular phenotype in serous endometrial cancer compared to endometrioid endometrial cancer. Exp Mol Pathol. 2011; 91(1):373–84.doi:10.1016/j.yexmp.2011.04.005.
45. Ratner ES, Tuck D, Richter C, Nallur S, Patel RM, Schultz V, Hui P, Schwartz PE, Rutherford TJ, Weidhaas JB. MicroRNA signatures differentiate uterine cancer tumor subtypes. Gynecol Oncol. 2010;118(3):251–7. doi:10.1016/j.ygyno.2010.05.010.
46. Moreno-Bueno G, Sanchez-Estevez C, Cassia R, Rodriguez-Perales S, Diaz-Uriarte R, Dominguez O, Hardisson D, Andujar M, Prat J, Matias-Guiu X, Cigudosa JC, Palacios J. Differential gene expression profile in endometrioid and nonendometrioid endometrial carcinoma: STK15 is frequently overexpressed and amplified in nonendometrioid carcinomas. Cancer Res. 2003;63(18):5697–702.
47. Mhawech-Fauceglia P, Wang D, Kesterson J, Syriac S, Clark K, Frederick PJ, Lele S, Liu S. Gene expression profiles in stage I uterine serous carcinoma in comparison to grade 3 and grade 1 stage I endometrioid adenocarcinoma. PLoS One. 2011;6(3):e18066. doi:10.1371/journal.pone.0018066.
48. Catasus L, D'Angelo E, Pons C, Espinosa I, Prat J. Expression profiling of 22 genes involved in the PI3K-AKT pathway identifies two subgroups of high-grade endometrial carcinomas with different molecular alterations. Mod Pathol. 2010;23(5):694–702. doi:10.1038/modpathol.2010.44.

49. Ferguson SE, Olshen AB, Viale A, Barakat RR, Boyd J. Stratification of intermediate-risk endometrial cancer patients into groups at high risk or low risk for recurrence based on tumor gene expression profiles. Clin Cancer Res. 2005;11(6):2252–7. doi:10.1158/1078-0432. CCR-04-1353.
50. Salvesen HB, Carter SL, Mannelqvist M, Dutt A, Getz G, Stefansson IM, Raeder MB, Sos ML, Engelsen IB, Trovik J, Wik E, Greulich H, Bo TH, Jonassen I, Thomas RK, Zander T, Garraway LA, Oyan AM, Sellers WR, Kalland KH, Meyerson M, Akslen LA, Beroukhim R. Integrated genomic profiling of endometrial carcinoma associates aggressive tumors with indicators of PI3 kinase activation. Proc Natl Acad Sci U S A. 2009;106(12):4834–9. doi:10.1073/pnas.0806514106.
51. Fles R, Hoogendoorn WE, Platteel I, Scheerman CE, de Leeuw-Mantel G, Mourits MJ, Hollema H, van Leeuwen FE, van Boven HH, Nederlof PM. Genomic profile of endometrial tumors depends on morphological subtype, not on tamoxifen exposure. Genes Chromosomes Cancer. 2010;49(8):699–710. doi:10.1002/gcc.20781.
52. Murayama-Hosokawa S, Oda K, Nakagawa S, Ishikawa S, Yamamoto S, Shoji K, Ikeda Y, Uehara Y, Fukayama M, McCormick F, Yano T, Taketani Y, Aburatani H. Genome-wide single-nucleotide polymorphism arrays in endometrial carcinomas associate extensive chromosomal instability with poor prognosis and unveil frequent chromosomal imbalances involved in the PI3-kinase pathway. Oncogene. 2010;29(13):1897–908. doi:10.1038/onc.2009.474.
53. Le Gallo M, O'Hara AJ, Rudd ML, Urick ME, Hansen NF, O'Neil NJ, Price JC, Zhang S, England BM, Godwin AK, Sgroi DC, Hieter P, Mullikin JC, Merino MJ, Bell DW. Exome sequencing of serous endometrial tumors identifies recurrent somatic mutations in chromatin-remodeling and ubiquitin ligase complex genes. Nat Genet. 2012;44(12):1310–5. doi:10.1038/ng.2455.
54. Kuhn E, Wu RC, Guan B, Wu G, Zhang J, Wang Y, Song L, Yuan X, Wei L, Roden RB, Kuo KT, Nakayama K, Clarke B, Shaw P, Olvera N, Kurman RJ, Levine DA, Wang TL, Shih Ie M. Identification of molecular pathway aberrations in uterine serous carcinoma by genome-wide analyses. J Natl Cancer Inst. 2012;104(19):1503–13. doi:10.1093/jnci/djs345.
55. Zhao S, Choi M, Overton JD, Bellone S, Roque DM, Cocco E, Guzzo F, English DP, Varughese J, Gasparrini S, Bortolomai I, Buza N, Hui P, Abu-Khalaf M, Ravaggi A, Bignotti E, Bandiera E, Romani C, Todeschini P, Tassi R, Zanotti L, Carrara L, Pecorelli S, Silasi DA, Ratner E, Azodi M, Schwartz PE, Rutherford TJ, Stiegler AL, Mane S, Boggon TJ, Schlessinger J, Lifton RP, Santin AD. Landscape of somatic single-nucleotide and copy-number mutations in uterine serous carcinoma. Proc Natl Acad Sci U S A. 2013;110(8):2916–21. doi:10.1073/pnas.1222577110.
56. Kandoth C, Schultz N, Cherniack AD, Akbani R, Liu Y, Shen H, Robertson AG, Pashtan I, Shen R, Benz CC, Yau C, Laird PW, Ding L, Zhang W, Mills GB, Kucherlapati R, Mardis ER, Levine DA. Integrated genomic characterization of endometrial carcinoma. Nature. 2013;497(7447):67–73. doi:10.1038/nature12113.
57. Church DN, Briggs SE, Palles C, Domingo E, Kearsey SJ, Grimes JM, Gorman M, Martin L, Howarth KM, Hodgson SV, Kaur K, Taylor J, Tomlinson IP. DNA polymerase varepsilon and delta exonuclease domain mutations in endometrial cancer. Hum Mol Genet. 2013;22:2820–8. doi:10.1093/hmg/ddt131.
58. Gupta S, Ramjaun AR, Haiko P, Wang Y, Warne PH, Nicke B, Nye E, Stamp G, Alitalo K, Downward J. Binding of ras to phosphoinositide 3-kinase p110alpha is required for ras-driven tumorigenesis in mice. Cell. 2007;129(5):957–68. doi:10.1016/j.cell.2007.03.051.
59. Oda K, Stokoe D, Taketani Y, McCormick F. High frequency of coexistent mutations of PIK3CA and PTEN genes in endometrial carcinoma. Cancer Res. 2005;65(23):10669–73. doi:10.1158/0008-5472.CAN-05-2620.
60. Jaiswal BS, Janakiraman V, Kljavin NM, Chaudhuri S, Stern HM, Wang W, Kan Z, Dbouk HA, Peters BA, Waring P, Dela Vega T, Kenski DM, Bowman KK, Lorenzo M, Li H, Wu J, Modrusan Z, Stinson J, Eby M, Yue P, Kaminker JS, de Sauvage FJ, Backer JM, Seshagiri S. Somatic mutations in p85alpha promote tumorigenesis through class IA PI3K activation. Cancer Cell. 2009;16(6):463–74. doi:10.1016/j.ccr.2009.10.016.
61. Rudd ML, Price JC, Fogoros S, Godwin AK, Sgroi DC, Merino MJ, Bell DW. A unique spectrum of somatic PIK3CA (p110alpha) mutations within primary endometrial carcinomas. Clin Cancer Res. 2011;17(6):1331–40. doi:10.1158/1078-0432.CCR-10-0540.
62. Samuels Y, Wang Z, Bardelli A, Silliman N, Ptak J, Szabo S, Yan H, Gazdar A, Powell SM, Riggins GJ, Willson JK, Markowitz S, Kinzler KW, Vogelstein B, Velculescu VE. High frequency of mutations of the PIK3CA gene in human cancers. Science. 2004;304(5670):554. doi:10.1126/science.1096502.
63. Slomovitz BM, Coleman RL. The PI3K/AKT/mTOR pathway as a therapeutic target in endometrial cancer. Clin Cancer Res. 2012;18(21):5856–64. doi:10.1158/1078-0432.CCR-12-0662.

64. Tredan O, Treilleux I, Wang Q, Gane N, Pissaloux D, Bonnin N, Petit T, Cretin J, Bonichon-Lamichhane N, Priou F, Lavau-Denes S, Mari V, Freyer G, Lebrun D, Alexandre J, Ray-Coquard I. Predicting everolimus treatment efficacy in patients with advanced endometrial carcinoma: a GINECO group study. Target Oncol. 2012;8:243–51. doi:10.1007/s11523-012-0242-9.
65. Ray-Coquard I, Favier L, Weber B, Roemer-Becuwe C, Bougnoux P, Fabbro M, Floquet A, Joly F, Plantade A, Paraiso D, Pujade-Lauraine E. Everolimus as second- or third-line treatment of advanced endometrial cancer: ENDORAD, a phase II trial of GINECO. Br J Cancer. 2013;108(9):1771–7. doi:10.1038/bjc.2013.183.
66. Colombo N, McMeekin DS, Schwartz PE, Sessa C, Gehrig PA, Holloway R, Braly P, Matei D, Morosky A, Dodion PF, Einstein MH, Haluska F. Ridaforolimus as a single agent in advanced endometrial cancer: results of a single-arm, phase 2 trial. Br J Cancer. 2013;108(5):1021–6. doi:10.1038/bjc.2013.59.
67. Weigelt B, Warne PH, Lambros MB, Reis-Filho JS, Downward J. PI3K pathway dependencies in endometrioid endometrial cancer cell lines. Clin Cancer Res. 2013;19:3533–44. doi:10.1158/1078-0432.CCR-12-3815.
68. Hampel H, Frankel W, Panescu J, Lockman J, Sotamaa K, Fix D, Comeras I, La Jeunesse J, Nakagawa H, Westman JA, Prior TW, Clendenning M, Penzone P, Lombardi J, Dunn P, Cohn DE, Copeland L, Eaton L, Fowler J, Lewandowski G, Vaccarello L, Bell J, Reid G, de la Chapelle A. Screening for Lynch syndrome (hereditary nonpolyposis colorectal cancer) among endometrial cancer patients. Cancer Res. 2006;66(15):7810–7. doi:10.1158/0008-5472.CAN-06-1114.
69. Ollikainen M, Abdel-Rahman WM, Moisio AL, Lindroos A, Kariola R, Jarvela I, Poyhonen M, Butzow R, Peltomaki P. Molecular analysis of familial endometrial carcinoma: a manifestation of hereditary nonpolyposis colorectal cancer or a separate syndrome? J Clin Oncol. 2005;23(21):4609–16. doi:10.1200/JCO.2005.06.055.
70. Aarnio M, Sankila R, Pukkala E, Salovaara R, Aaltonen LA, de la Chapelle A, Peltomaki P, Mecklin JP, Jarvinen HJ. Cancer risk in mutation carriers of DNA-mismatch-repair genes. Int J Cancer. 1999;81(2):214–8.
71. Meyer LA, Broaddus RR, Lu KH. Endometrial cancer and Lynch syndrome: clinical and pathologic considerations. Cancer Control. 2009;16(1):14–22.
72. Vasen HF, Watson P, Mecklin JP, Lynch HT. New clinical criteria for hereditary nonpolyposis colorectal cancer (HNPCC, Lynch syndrome) proposed by the International Collaborative group on HNPCC. Gastroenterology. 1999;116(6):1453–6.
73. Lu KH, Dinh M, Kohlmann W, Watson P, Green J, Syngal S, Bandipalliam P, Chen LM, Allen B, Conrad P, Terdiman J, Sun C, Daniels M, Burke T, Gershenson DM, Lynch H, Lynch P, Broaddus RR. Gynecologic cancer as a "sentinel cancer" for women with hereditary nonpolyposis colorectal cancer syndrome. Obstet Gynecol. 2005;105(3):569–74. doi:10.1097/01.AOG.0000154885.44002.ae.
74. Win AK, Lindor NM, Winship I, Tucker KM, Buchanan DD, Young JP, Rosty C, Leggett B, Giles GG, Goldblatt J, Macrae FA, Parry S, Kalady MF, Baron JA, Ahnen DJ, Marchand LL, Gallinger S, Haile RW, Newcomb PA, Hopper JL, Jenkins MA. Risks of colorectal and other cancers after endometrial cancer for women with Lynch syndrome. J Natl Cancer Inst. 2013;105(4):274–9. doi:10.1093/jnci/djs525.
75. Umar A, Boland CR, Terdiman JP, Syngal S, de la Chapelle A, Ruschoff J, Fishel R, Lindor NM, Burgart LJ, Hamelin R, Hamilton SR, Hiatt RA, Jass J, Lindblom A, Lynch HT, Peltomaki P, Ramsey SD, Rodriguez-Bigas MA, Vasen HF, Hawk ET, Barrett JC, Freedman AN, Srivastava S. Revised Bethesda guidelines for hereditary nonpolyposis colorectal cancer (lynch syndrome) and microsatellite instability. J Natl Cancer Inst. 2004;96(4):261–8.
76. Folkins AK, Longacre TA. Hereditary gynaecological malignancies: advances in screening and treatment. Histopathology. 2013;62(1):2–30. doi:10.1111/his.12028.
77. Garg K, Soslow RA. Lynch syndrome (hereditary non-polyposis colorectal cancer) and endometrial carcinoma. J Clin Pathol. 2009;62(8):679–84. doi:10.1136/jcp.2009.064949.
78. Boland CR, Thibodeau SN, Hamilton SR, Sidransky D, Eshleman JR, Burt RW, Meltzer SJ, Rodriguez-Bigas MA, Fodde R, Ranzani GN, Srivastava S. A National Cancer Institute Workshop on microsatellite instability for cancer detection and familial predisposition: development of international criteria for the determination of microsatellite instability in colorectal cancer. Cancer Res. 1998;58(22):5248–57.
79. Suraweera N, Duval A, Reperant M, Vaury C, Furlan D, Leroy K, Seruca R, Iacopetta B, Hamelin R. Evaluation of tumor microsatellite instability using five quasimonomorphic mononucleotide repeats and pentaplex PCR. Gastroenterology. 2002;123(6):1804–11. doi:10.1053/gast.2002.37070.
80. Wong YF, Cheung TH, Lo KW, Yim SF, Chan LK, Buhard O, Duval A, Chung TK, Hamelin R. Detection of microsatellite instability in endometrial cancer: advantages of a panel of five mononucleotide repeats over the National

Cancer Institute panel of markers. Carcinogenesis. 2006;27(5):951–5. doi:10.1093/carcin/bgi333.
81. Wijnen J, de Leeuw W, Vasen H, van der Klift H, Moller P, Stormorken A, Meijers-Heijboer H, Lindhout D, Menko F, Vossen S, Moslein G, Tops C, Brocker-Vriends A, Wu Y, Hofstra R, Sijmons R, Cornelisse C, Morreau H, Fodde R. Familial endometrial cancer in female carriers of MSH6 germline mutations. Nat Genet. 1999;23(2):142–4. doi:10.1038/13773.
82. Hendriks YM, Wagner A, Morreau H, Menko F, Stormorken A, Quehenberger F, Sandkuijl L, Moller P, Genuardi M, Van Houwelingen H, Tops C, Van Puijenbroek M, Verkuijlen P, Kenter G, Van Mil A, Meijers-Heijboer H, Tan GB, Breuning MH, Fodde R, Wijnen JT, Brocker-Vriends AH, Vasen H. Cancer risk in hereditary nonpolyposis colorectal cancer due to MSH6 mutations: impact on counseling and surveillance. Gastroenterology. 2004;127(1):17–25.
83. Wu Y, Berends MJ, Mensink RG, Kempinga C, Sijmons RH, van Der Zee AG, Hollema H, Kleibeuker JH, Buys CH, Hofstra RM. Association of hereditary nonpolyposis colorectal cancer-related tumors displaying low microsatellite instability with MSH6 germline mutations. Am J Hum Genet. 1999;65(5):1291–8. doi:10.1086/302612.
84. Modica I, Soslow RA, Black D, Tornos C, Kauff N, Shia J. Utility of immunohistochemistry in predicting microsatellite instability in endometrial carcinoma. Am J Surg Pathol. 2007;31(5):744–51. doi:10.1097/01.pas.0000213428.61374.06.
85. Domchek SM, Bradbury A, Garber JE, Offit K, Robson ME. Multiplex genetic testing for cancer susceptibility: out on the high wire without a net? J Clin Oncol. 2013;31(10):1267–70. doi:10.1200/JCO.2012.46.9403.
86. Zimmermann A, Bernuit D, Gerlinger C, Schaefers M, Geppert K. Prevalence, symptoms and management of uterine fibroids: an international internet-based survey of 21,746 women. BMC Womens Health. 2012;12:6. doi:10.1186/1472-6874-12-6.
87. Cramer SF, Patel A. The frequency of uterine leiomyomas. Am J Clin Pathol. 1990;94(4):435–8.
88. Perot G, Croce S, Ribeiro A, Lagarde P, Velasco V, Neuville A, Coindre JM, Stoeckle E, Floquet A, MacGrogan G, Chibon F. MED12 alterations in both human benign and malignant uterine soft tissue tumors. PLoS One. 2012;7(6):e40015. doi:10.1371/journal.pone.0040015.
89. Sandberg AA. Updates on the cytogenetics and molecular genetics of bone and soft tissue tumors: leiomyosarcoma. Cancer Genet Cytogenet. 2005;161(1):1–19.
90. Howlader N, Noone AM, Krapcho M, Garshell J, Neyman N, Altekruse SF, Kosary CL, Yu M, Ruhl J, Tatalovich Z, Cho H, Mariotto A, Lewis DR, Chen HS, Feuer EJ, Cronin KA, editors. SEER cancer statistics review, 1975–2010. Bethesda, MD: National Cancer Institute; 2013.
91. Bell SW, Kempson RL, Hendrickson MR. Problematic uterine smooth muscle neoplasms. A clinicopathologic study of 213 cases. Am J Surg Pathol. 1994;18(6):535–58.
92. Veras E, Malpica A, Deavers MT, Silva EG. Mitosis-specific marker phospho-histone H3 in the assessment of mitotic index in uterine smooth muscle tumors: a pilot study. Int J Gynecol Pathol. 2009;28(4):316–21. doi:10.1097/PGP.0b013e318193df97.
93. Lim D, Alvarez T, Nucci MR, Gilks B, Longacre T, Soslow RA, Oliva E. Interobserver variability in the interpretation of tumor cell necrosis in uterine leiomyosarcoma. Am J Surg Pathol. 2013;37(5):650–8. doi:10.1097/PAS.0b013e3182851162.
94. Soper JT, McCarty Jr KS, Hinshaw W, Creasman WT, McCarty Sr KS, Clarke-Pearson DL. Cytoplasmic estrogen and progesterone receptor content of uterine sarcomas. Am J Obstet Gynecol. 1984;150(4):342–8.
95. Sutton GP, Stehman FB, Michael H, Young PC, Ehrlich CE. Estrogen and progesterone receptors in uterine sarcomas. Obstet Gynecol. 1986;68(5):709–14.
96. Wade K, Quinn MA, Hammond I, Williams K, Cauchi M. Uterine sarcoma: steroid receptors and response to hormonal therapy. Gynecol Oncol. 1990;39(3):364–7.
97. Zhai YL, Kobayashi Y, Mori A, Orii A, Nikaido T, Konishi I, Fujii S. Expression of steroid receptors, Ki-67, and p53 in uterine leiomyosarcomas. Int J Gynecol Pathol. 1999;18(1):20–8.
98. Mittal K, Demopoulos RI. MIB-1 (Ki-67), p53, estrogen receptor, and progesterone receptor expression in uterine smooth muscle tumors. Hum Pathol. 2001;32(9):984–7. doi:10.1053/hupa.2001.27113.
99. Bodner K, Bodner-Adler B, Kimberger O, Czerwenka K, Leodolter S, Mayerhofer K. Estrogen and progesterone receptor expression in patients with uterine leiomyosarcoma and correlation with different clinicopathological parameters. Anticancer Res. 2003;23(1B):729–32.
100. Leitao MM, Soslow RA, Nonaka D, Olshen AB, Aghajanian C, Sabbatini P, Dupont J, Hensley M, Sonoda Y, Barakat RR, Anderson S. Tissue microarray immunohistochemical expression of estrogen, progesterone, and androgen receptors in uterine leiomyomata and leiomyosarcoma. Cancer. 2004;101(6):1455–62. doi:10.1002/cncr.20521.
101. Jeffers MD, Farquharson MA, Richmond JA, McNicol AM. p53 immunoreactivity and mutation of the p53 gene in smooth muscle tumours

of the uterine corpus. J Pathol. 1995;177(1): 65–70. doi:10.1002/path.1711770111.
102. Bodner-Adler B, Bodner K, Czerwenka K, Kimberger O, Leodolter S, Mayerhofer K. Expression of p16 protein in patients with uterine smooth muscle tumors: an immunohistochemical analysis. Gynecol Oncol. 2005;96(1):62–6. doi:10.1016/j.ygyno.2004.09.026.
103. Akhan SE, Yavuz E, Tecer A, Iyibozkurt CA, Topuz S, Tuzlali S, Bengisu E, Berkman S. The expression of Ki-67, p53, estrogen and progesterone receptors affecting survival in uterine leiomyosarcomas. A clinicopathologic study. Gynecol Oncol. 2005;99(1):36–42. doi:10.1016/j.ygyno.2005.05.019.
104. O'Neill CJ, McBride HA, Connolly LE, McCluggage WG. Uterine leiomyosarcomas are characterized by high p16, p53 and MIB1 expression in comparison with usual leiomyomas, leiomyoma variants and smooth muscle tumours of uncertain malignant potential. Histopathology. 2007;50(7):851–8. doi:10.1111/j.1365-2559.2007.02699.x.
105. Atkins KA, Arronte N, Darus CJ, Rice LW. The Use of p16 in enhancing the histologic classification of uterine smooth muscle tumors. Am J Surg Pathol. 2008;32(1):98–102. doi:10.1097/PAS.0b013e3181574d1e.
106. Gannon BR, Manduch M, Childs TJ. Differential Immunoreactivity of p16 in leiomyosarcomas and leiomyoma variants. Int J Gynecol Pathol. 2008;27(1):68–73. doi:10.1097/pgp.0b013e3180ca954f.
107. Chen L, Yang B. Immunohistochemical analysis of p16, p53, and Ki-67 expression in uterine smooth muscle tumors. Int J Gynecol Pathol. 2008;27(3):326–32. doi:10.1097/PGP.0b013e31815ea7f5.
108. Lee CH, Turbin DA, Sung YC, Espinosa I, Montgomery K, van de Rijn M, Gilks CB. A panel of antibodies to determine site of origin and malignancy in smooth muscle tumors. Mod Pathol. 2009;22(12):1519–31. doi:10.1038/modpathol.2009.122.
109. Ip PP, Cheung AN, Clement PB. Uterine smooth muscle tumors of uncertain malignant potential (STUMP): a clinicopathologic analysis of 16 cases. Am J Surg Pathol. 2009;33(7):992–1005. doi:10.1097/PAS.0b013e3181a02d1c.
110. D'Angelo E, Espinosa I, Ali R, Gilks CB, Rijn M, Lee CH, Prat J. Uterine leiomyosarcomas: tumor size, mitotic index, and biomarkers Ki67, and Bcl-2 identify two groups with different prognosis. Gynecol Oncol. 2011;121(2):328–33. doi:10.1016/j.ygyno.2011.01.022.
111. Hakverdi S, Gungoren A, Yaldiz M, Hakverdi AU, Toprak S. Immunohistochemical analysis of p16 expression in uterine smooth muscle tumors. Eur J Gynaecol Oncol. 2011;32(5):513–5.
112. Makinen N, Mehine M, Tolvanen J, Kaasinen E, Li Y, Lehtonen HJ, Gentile M, Yan J, Enge M, Taipale M, Aavikko M, Katainen R, Virolainen E, Bohling T, Koski TA, Launonen V, Sjoberg J, Taipale J, Vahteristo P, Aaltonen LA. MED12, the mediator complex subunit 12 gene, is mutated at high frequency in uterine leiomyomas. Science. 2011;334(6053):252–5. doi:10.1126/science.1208930.
113. Makinen N, Heinonen HR, Moore S, Tomlinson IP, van der Spuy ZM, Aaltonen LA. MED12 exon 2 mutations are common in uterine leiomyomas from South African patients. Oncotarget. 2011;2(12):966–9.
114. Je EM, Kim MR, Min KO, Yoo NJ, Lee SH. Mutational analysis of MED12 exon 2 in uterine leiomyoma and other common tumors. Int J Cancer. 2012;131(6):E1044–7. doi:10.1002/ijc.27610.
115. McGuire MM, Yatsenko A, Hoffner L, Jones M, Surti U, Rajkovic A. Whole exome sequencing in a random sample of North American women with leiomyomas identifies MED12 mutations in majority of uterine leiomyomas. PLoS One. 2012;7(3):e33251. doi:10.1371/journal.pone.0033251.
116. Markowski DN, Huhle S, Nimzyk R, Stenman G, Loning T, Bullerdiek J. MED12 mutations occurring in benign and malignant mammalian smooth muscle tumors. Genes Chromosomes Cancer. 2013;52(3):297–304. doi:10.1002/gcc.22029.
117. Matsubara A, Sekine S, Yoshida M, Yoshida A, Taniguchi H, Kushima R, Tsuda H, Kanai Y. Prevalence of MED12 mutations in uterine and extrauterine smooth muscle tumours. Histopathology. 2013;62(4):657–61. doi:10.1111/his.12039.
118. de Graaff MA, Cleton-Jansen AM, Szuhai K, Bovee JV. Mediator complex subunit 12 exon 2 mutation analysis in different subtypes of smooth muscle tumors confirms genetic heterogeneity. Hum Pathol. 2013;44(8):1597–604. doi:10.1016/j.humpath.2013.01.006.
119. Conaway RC, Conaway JW. Function and regulation of the mediator complex. Curr Opin Genet Dev. 2011;21(2):225–30. doi:10.1016/j.gde.2011.01.013.
120. Kim S, Xu X, Hecht A, Boyer TG. Mediator is a transducer of Wnt/beta-catenin signaling. J Biol Chem. 2006;281(20):14066–75. doi:10.1074/jbc.M602696200.
121. Zhou H, Kim S, Ishii S, Boyer TG. Mediator modulates Gli3-dependent Sonic hedgehog signaling. Mol Cell Biol. 2006;26(23):8667–82. doi:10.1128/MCB.00443-06.
122. Tutter AV, Kowalski MP, Baltus GA, Iourgenko V, Labow M, Li E, Kadam S. Role for Med12 in regulation of Nanog and Nanog target genes. J Biol Chem. 2009;284(6):3709–18. doi:10.1074/jbc.M805677200.

123. Makinen N, Vahteristo P, Kampjarvi K, Arola J, Butzow R, Aaltonen LA. MED12 exon 2 mutations in histopathological uterine leiomyoma variants. Eur J Hum Genet. 2013;21:1300–3. doi:10.1038/ejhg.2013.33.
124. Lehtonen R, Kiuru M, Vanharanta S, Sjoberg J, Aaltonen LM, Aittomaki K, Arola J, Butzow R, Eng C, Husgafvel-Pursiainen K, Isola J, Jarvinen H, Koivisto P, Mecklin JP, Peltomaki P, Salovaara R, Wasenius VM, Karhu A, Launonen V, Nupponen NN, Aaltonen LA. Biallelic inactivation of fumarate hydratase (FH) occurs in nonsyndromic uterine leiomyomas but is rare in other tumors. Am J Pathol. 2004;164(1):17–22. doi:10.1016/S0002-9440(10)63091-X.
125. Sandberg AA. Updates on the cytogenetics and molecular genetics of bone and soft tissue tumors: leiomyoma. Cancer Genet Cytogenet. 2005;158(1):1–26. doi:10.1016/j.cancergencyto.2004.08.025.
126. Mehine M, Kaasinen E, Makinen N, Katainen R, Kampjarvi K, Pitkanen E, Heinonen HR, Butzow R, Kilpivaara O, Kuosmanen A, Ristolainen H, Gentile M, Sjoberg J, Vahteristo P, Aaltonen LA. Characterization of uterine leiomyomas by whole-genome sequencing. N Engl J Med. 2013;369:43–53. doi:10.1056/NEJMoa1302736.
127. Stephens PJ, Greenman CD, Fu B, Yang F, Bignell GR, Mudie LJ, Pleasance ED, Lau KW, Beare D, Stebbings LA, McLaren S, Lin ML, McBride DJ, Varela I, Nik-Zainal S, Leroy C, Jia M, Menzies A, Butler AP, Teague JW, Quail MA, Burton J, Swerdlow H, Carter NP, Morsberger LA, Iacobuzio-Donahue C, Follows GA, Green AR, Flanagan AM, Stratton MR, Futreal PA, Campbell PJ. Massive genomic rearrangement acquired in a single catastrophic event during cancer development. Cell. 2011;144(1):27–40. doi:10.1016/j.cell.2010.11.055.
128. Hodge JC, Morton CC (2007) Genetic heterogeneity among uterine leiomyomata: insights into malignant progression. Hum Mol Genet. 16 Spec No 1:R7-13. doi:10.1093/hmg/ddm043.
129. Kuhn E, Yemelyanova A, Wang TL, Kurman R, Shih IeM. Abstract 5536: TP53 and MED12 mutations in uterine smooth muscle tumors. In: AACR 103rd Annual Meeting, Chicago, IL, USA, 2012. vol 8, Supplement 1. Cancer Res.
130. de Vos S, Wilczynski SP, Fleischhacker M, Koeffler P. p53 alterations in uterine leiomyosarcomas versus leiomyomas. Gynecol Oncol. 1994;54(2):205–8. doi:10.1006/gyno.1994.1194.
131. Chew I, Oliva E. Endometrial stromal sarcomas: a review of potential prognostic factors. Adv Anat Pathol. 2010;17(2):113–21. doi:10.1097/PAP.0b013e3181cfb7c2.
132. Tanner EJ, Garg K, Leitao Jr MM, Soslow RA, Hensley ML. High grade undifferentiated uterine sarcoma: surgery, treatment, and survival outcomes. Gynecol Oncol. 2012;127(1):27–31. doi:10.1016/j.ygyno.2012.06.030.
133. Malouf GG, Lhomme C, Duvillard P, Morice P, Haie-Meder C, Pautier P. Prognostic factors and outcome of undifferentiated endometrial sarcoma treated by multimodal therapy. Int J Gynaecol Obstet. 2013;122(1):57–61. doi:10.1016/j.ijgo.2013.01.025.
134. Jin Y, Pan L, Wang X, Dai Z, Huang H, Guo L, Shen K, Lian L. Clinical characteristics of endometrial stromal sarcoma from an academic medical hospital in China. Int J Gynecol Cancer. 2010;20(9):1535–9.
135. Kurihara S, Oda Y, Ohishi Y, Iwasa A, Takahira T, Kaneki E, Kobayashi H, Wake N, Tsuneyoshi M. Endometrial stromal sarcomas and related high-grade sarcomas: immunohistochemical and molecular genetic study of 31 cases. Am J Surg Pathol. 2008;32(8):1228–38. doi:10.1097/PAS.0b013e31816a3b42.
136. Nucci MR, O'Connell JT, Huettner PC, Cviko A, Sun D, Quade BJ. h-Caldesmon expression effectively distinguishes endometrial stromal tumors from uterine smooth muscle tumors. Am J Surg Pathol. 2001;25(4):455–63.
137. Oliva E, Young RH, Amin MB, Clement PB. An immunohistochemical analysis of endometrial stromal and smooth muscle tumors of the uterus: a study of 54 cases emphasizing the importance of using a panel because of overlap in immunoreactivity for individual antibodies. Am J Surg Pathol. 2002;26(4):403–12.
138. de Leval L, Waltregny D, Boniver J, Young RH, Castronovo V, Oliva E. Use of histone deacetylase 8 (HDAC8), a new marker of smooth muscle differentiation, in the classification of mesenchymal tumors of the uterus. Am J Surg Pathol. 2006;30(3):319–27. doi:10.1097/01.pas.0000188029.63706.31.
139. McCluggage WG, Date A, Bharucha H, Toner PG. Endometrial stromal sarcoma with sex cord-like areas and focal rhabdoid differentiation. Histopathology. 1996;29(4):369–74.
140. Fukunaga M, Miyazawa Y, Ushigome S. Endometrial low-grade stromal sarcoma with ovarian sex cord-like differentiation: report of two cases with an immunohistochemical and flow cytometric study. Pathol Int. 1997;47(6):412–5.
141. Zamecnik M, Michal M. Endometrial stromal nodule with retiform sex-cord-like differentiation. Pathol Res Pract. 1998;194(6):449–53.
142. Baker RJ, Hildebrandt RH, Rouse RV, Hendrickson MR, Longacre TA. Inhibin and CD99 (MIC2) expression in uterine stromal neoplasms with sex-cord-like elements. Hum Pathol. 1999;30(6):671–9.
143. Ohta Y, Suzuki T, Kojima M, Shiokawa A, Mitsuya T. Low-grade endometrial stromal

sarcoma with an extensive epithelial-like element. Pathol Int. 2003;53(4):246–51.
144. Sumathi VP, Al-Hussaini M, Connolly LE, Fullerton L, McCluggage WG. Endometrial stromal neoplasms are immunoreactive with WT-1 antibody. Int J Gynecol Pathol. 2004;23(3): 241–7.
145. Irving JA, Carinelli S, Prat J. Uterine tumors resembling ovarian sex cord tumors are polyphenotypic neoplasms with true sex cord differentiation. Mod Pathol. 2006;19(1):17–24. doi:10.1038/modpathol.3800475.
146. Reich O, Regauer S, Urdl W, Lahousen M, Winter R. Expression of oestrogen and progesterone receptors in low-grade endometrial stromal sarcomas. Br J Cancer. 2000;82(5):1030–4. doi:10.1054/bjoc.1999.1038.
147. Chu MC, Mor G, Lim C, Zheng W, Parkash V, Schwartz PE. Low-grade endometrial stromal sarcoma: hormonal aspects. Gynecol Oncol. 2003;90(1):170–6.
148. Balleine RL, Earls PJ, Webster LR, Mote PA, deFazio A, Harnett PR, Clarke CL. Expression of progesterone receptor A and B isoforms in low-grade endometrial stromal sarcoma. Int J Gynecol Pathol. 2004;23(2):138–44.
149. Lee CH, Marino-Enriquez A, Ou W, Zhu M, Ali RH, Chiang S, Amant F, Gilks CB, van de Rijn M, Oliva E, Debiec-Rychter M, Dal Cin P, Fletcher JA, Nucci MR. The clinicopathologic features of YWHAE-FAM22 endometrial stromal sarcomas: a histologically high-grade and clinically aggressive tumor. Am J Surg Pathol. 2012;36(5):641–53. doi:10.1097/PAS.0b013e31824a7b1a.
150. Jakate K, Azimi F, Ali RH, Lee CH, Clarke BA, Rasty G, Shaw PA, Melnyk N, Huntsman DG, Laframboise S, Rouzbahman M. Endometrial sarcomas: an immunohistochemical and JAZF1 re-arrangement study in low-grade and undifferentiated tumors. Mod Pathol. 2013;26(1):95–105. doi:10.1038/modpathol.2012.136.
151. Ng TL, Gown AM, Barry TS, Cheang MC, Chan AK, Turbin DA, Hsu FD, West RB, Nielsen TO. Nuclear beta-catenin in mesenchymal tumors. Mod Pathol. 2005;18(1):68–74. doi:10.1038/modpathol.3800272.
152. Jung CK, Jung JH, Lee A, Lee YS, Choi YJ, Yoon SK, Lee KY. Diagnostic use of nuclear beta-catenin expression for the assessment of endometrial stromal tumors. Mod Pathol. 2008;21(6):756–63. doi:10.1038/modpathol.2008.53.
153. Kildal W, Pradhan M, Abeler VM, Kristensen GB, Danielsen HE. Beta-catenin expression in uterine sarcomas and its relation to clinicopathological parameters. Eur J Cancer. 2009;45(13):2412–7. doi:10.1016/j.ejca.2009.06.017.
154. Kurihara S, Oda Y, Ohishi Y, Kaneki E, Kobayashi H, Wake N, Tsuneyoshi M. Coincident expression of beta-catenin and cyclin D1 in endometrial stromal tumors and related high-grade sarcomas. Mod Pathol. 2010;23(2):225–34. doi:10.1038/modpathol.2009.162.
155. Moinfar F, Gogg-Kamerer M, Sommersacher A, Regitnig P, Man YG, Zatloukal K, Denk H, Tavassoli FA. Endometrial stromal sarcomas frequently express epidermal growth factor receptor (EGFR, HER-1): potential basis for a new therapeutic approach. Am J Surg Pathol. 2005;29(4):485–9.
156. Cossu-Rocca P, Contini M, Uras MG, Muroni MR, Pili F, Carru C, Bosincu L, Massarelli G, Nogales FF, De Miglio MR. Tyrosine kinase receptor status in endometrial stromal sarcoma: an immunohistochemical and genetic-molecular analysis. Int J Gynecol Pathol. 2012;31(6):570–9. doi:10.1097/PGP.0b013e31824fe289.
157. Sardinha R, Hernandez T, Fraile S, Tresserra F, Vidal A, Gomez MC, Astudillo A, Hernandez N, Saenz de Santamaria J, Ordi J, Goncalves L, Ramos R, Balana C, de Alava E. Endometrial stromal tumors: immunohistochemical and molecular analysis of potential targets of tyrosine kinase inhibitors. Clin Sarcoma Res. 2013;3(1):3. doi:10.1186/2045-3329-3-3.
158. Cheng X, Yang G, Schmeler KM, Coleman RL, Tu X, Liu J, Kavanagh JJ. Recurrence patterns and prognosis of endometrial stromal sarcoma and the potential of tyrosine kinase-inhibiting therapy. Gynecol Oncol. 2011;121(2):323–7. doi:10.1016/j.ygyno.2010.12.360.
159. Moinfar F, Regitnig P, Tabrizi AD, Denk H, Tavassoli FA. Expression of androgen receptors in benign and malignant endometrial stromal neoplasms. Virchows Arch. 2004;444(5):410–4. doi:10.1007/s00428-004-0981-9.
160. Reich O, Regauer S. Aromatase expression in low-grade endometrial stromal sarcomas: an immunohistochemical study. Mod Pathol. 2004;17(1): 104–8. doi:10.1038/sj.modpathol.3800031.
161. Liegl B, Reich O, Nogales FF, Regauer S. PDGF-alpha and PDGF-beta are expressed in endometrial stromal sarcoma: a potential therapeutic target for tyrosine kinase inhibitors? Histopathology. 2006;49(5):545–6. doi:10.1111/j.1365-2559.2006.02529.x.
162. Chiang S, Oliva E. Recent developments in uterine mesenchymal neoplasms. Histopathology. 2013;62(1):124–37. doi:10.1111/his.12048.
163. Croce S, Hostein I, Ribeiro A, Garbay D, Velasco V, Stoeckle E, Guyon F, Floquet A, Neuville A, Coindre JM, Macgrogan G, Chibon F. YWHAE rearrangement identified by FISH and RT-PCR in endometrial stromal sarcomas: genetic and pathological correlations. Mod

Pathol. 2013;26:1390–400. doi:10.1038/modpathol.2013.69.
164. Lee CH, Ali RH, Rouzbahman M, Marino-Enriquez A, Zhu M, Guo X, Brunner AL, Chiang S, Leung S, Nelnyk N, Huntsman DG, Blake Gilks C, Nielsen TO, Dal Cin P, van de Rijn M, Oliva E, Fletcher JA, Nucci MR. Cyclin D1 as a diagnostic immunomarker for endometrial stromal sarcoma with YWHAE-FAM22 rearrangement. Am J Surg Pathol. 2012;36(10):1562–70. doi:10.1097/PAS.0b013e31825fa931.
165. Koontz JI, Soreng AL, Nucci M, Kuo FC, Pauwels P, van Den Berghe H, Dal Cin P, Fletcher JA, Sklar J. Frequent fusion of the JAZF1 and JJAZ1 genes in endometrial stromal tumors. Proc Natl Acad Sci U S A. 2001;98(11):6348–53. doi:10.1073/pnas.101132598.
166. Huang HY, Ladanyi M, Soslow RA. Molecular detection of JAZF1-JJAZ1 gene fusion in endometrial stromal neoplasms with classic and variant histology: evidence for genetic heterogeneity. Am J Surg Pathol. 2004;28(2):224–32.
167. Hrzenjak A, Moinfar F, Tavassoli FA, Strohmeier B, Kremser ML, Zatloukal K, Denk H. JAZF1/JJAZ1 gene fusion in endometrial stromal sarcomas: molecular analysis by reverse transcriptase-polymerase chain reaction optimized for paraffin-embedded tissue. J Mol Diagn. 2005;7(3):388–95. doi:10.1016/S1525-1578(10)60568-5.
168. Oliva E, de Leval L, Soslow RA, Herens C. High frequency of JAZF1-JJAZ1 gene fusion in endometrial stromal tumors with smooth muscle differentiation by interphase FISH detection. Am J Surg Pathol. 2007;31(8):1277–84. doi:10.1097/PAS.0b013e318031f012.
169. Nucci MR, Harburger D, Koontz J, Dal Cin P, Sklar J. Molecular analysis of the JAZF1-JJAZ1 gene fusion by RT-PCR and fluorescence in situ hybridization in endometrial stromal neoplasms. Am J Surg Pathol. 2007;31(1):65–70. doi:10.1097/01.pas.0000213327.86992.d1.
170. Chiang S, Ali R, Melnyk N, McAlpine JN, Huntsman DG, Gilks CB, Lee CH, Oliva E. Frequency of known gene rearrangements in endometrial stromal tumors. Am J Surg Pathol. 2011;35(9):1364–72. doi:10.1097/PAS.0b013e3182262743.
171. D'Angelo E, Ali RH, Espinosa I, Lee CH, Huntsman DG, Gilks B, Prat J. Endometrial stromal sarcomas with sex cord differentiation are associated with PHF1 rearrangement. Am J Surg Pathol. 2013;37(4):514–21. doi:10.1097/PAS.0b013e318272c612.
172. Micci F, Walter CU, Teixeira MR, Panagopoulos I, Bjerkehagen B, Saeter G, Heim S. Cytogenetic and molecular genetic analyses of endometrial stromal sarcoma: nonrandom involvement of chromosome arms 6p and 7p and confirmation of JAZF1/JJAZ1 gene fusion in t(7;17). Cancer Genet Cytogenet. 2003;144(2):119–24.
173. Panagopoulos I, Micci F, Thorsen J, Gorunova L, Eibak AM, Bjerkehagen B, Davidson B, Heim S. Novel fusion of MYST/Esa1-associated factor 6 and PHF1 in endometrial stromal sarcoma. PLoS One. 2012;7(6):e39354. doi:10.1371/journal.pone.0039354.
174. Panagopoulos I, Thorsen J, Gorunova L, Haugom L, Bjerkehagen B, Davidson B, Heim S, Micci F. Fusion of the ZC3H7B and BCOR genes in endometrial stromal sarcomas carrying an X;22-translocation. Genes Chromosomes Cancer. 2013;52(7):610–8. doi:10.1002/gcc.22057.
175. Lee CH, Ou WB, Marino-Enriquez A, Zhu M, Mayeda M, Wang Y, Guo X, Brunner AL, Amant F, French CA, West RB, McAlpine JN, Gilks CB, Yaffe MB, Prentice LM, McPherson A, Jones SJ, Marra MA, Shah SP, van de Rijn M, Huntsman DG, Dal Cin P, Debiec-Rychter M, Nucci MR, Fletcher JA. 14-3-3 fusion oncogenes in high-grade endometrial stromal sarcoma. Proc Natl Acad Sci U S A. 2012;109(3):929–34. doi:10.1073/pnas.1115528109.
176. Isphording A, Ali RH, Irving J, Goytain A, Nelnyk N, Hoang LN, Gilks CB, Huntsman DG, Nielsen TO, Nucci MR, Lee CH. YWHAE-FAM22 endometrial stromal sarcoma: diagnosis by reverse transcription-polymerase chain reaction in formalin-fixed, paraffin-embedded tumor. Hum Pathol. 2013;44(5):837–43. doi:10.1016/j.humpath.2012.08.007.
177. Li H, Ma X, Wang J, Koontz J, Nucci M, Sklar J. Effects of rearrangement and allelic exclusion of JJAZ1/SUZ12 on cell proliferation and survival. Proc Natl Acad Sci U S A. 2007;104(50):20001–6. doi:10.1073/pnas.0709986104.
178. Li H, Wang J, Mor G, Sklar J. A neoplastic gene fusion mimics trans-splicing of RNAs in normal human cells. Science. 2008;321(5894):1357–61. doi:10.1126/science.1156725.
179. Amant F, Tousseyn T, Coenegrachts L, Decloedt J, Moerman P, Debiec-Rychter M. Case report of a poorly differentiated uterine tumour with t(10;17) translocation and neuroectodermal phenotype. Anticancer Res. 2011;31(6):2367–71.
180. Fehr A, Hansson MC, Kindblom LG, Stenman G. YWHAE-FAM22 gene fusion in clear cell sarcoma of the kidney. J Pathol. 2012;227(4):e5–7. doi:10.1002/path.4040.
181. O'Meara E, Stack D, Lee CH, Garvin AJ, Morris T, Argani P, Han JS, Karlsson J, Gisselson D, Leuschner I, Gessler M, Graf N, Fletcher JA, O'Sullivan MJ. Characterization of the chromosomal translocation t(10;17)(q22;p13) in clear cell sarcoma of kidney. J Pathol. 2012;227(1):72–80. doi:10.1002/path.3985.

182. Boronow RC. Endometrial cancer: not a benign disease. Obstet Gynecol. 1976;47(5):630–4.
183. Kinde I, Bettegowda C, Wang Y, Wu J, Agrawal N, Shih IM, Kurman R, Dao F, Levine DA, Giuntoli R, Roden R, Eshleman JR, Carvalho JP, Marie SK, Papadopoulos N, Kinzler KW, Vogelstein B, Diaz Jr LA. Evaluation of DNA from the Papanicolaou test to detect ovarian and endometrial cancers. Sci Transl Med. 2013;5(167):167ra4. doi:10.1126/scitranslmed.3004952.
184. Byron SA, Pollock PM. FGFR2 as a molecular target in endometrial cancer. Future Oncol. 2009;5(1):27–32.doi:10.2217/14796694.5.1.27.
185. Cerami E, Gao J, Dogrusoz U, Gross BE, Sumer SO, Aksoy BA, Jacobsen A, Byrne CJ, Heuer ML, Larsson E, Antipin Y, Reva B, Goldberg AP, Sander C, Schultz N. The cBio cancer genomics portal: an open platform for exploring multidimensional cancer genomics data. Cancer Discov. 2012;2(5): 401–4. doi:10.1158/2159-8290.CD-12-0095.

CHAPTER 28

GENOMIC APPLICATIONS IN EPITHELIAL OVARIAN MALIGNANCIES

JAMES D. BRENTON

Introduction 489
Cell of Origin of EOC 489
Molecular Classification of EOC 490
Homologous Recombination and High-Grade Carcinoma 493
Intratumoral Heterogeneity in HGSOC May Contribute to Platinum Resistance 494
Conclusions 497

Introduction

Overall survival rates for women with advanced epithelial ovarian cancer (EOC) have remained unchanged over the past three decades, and fewer than 40 % of patients remain alive at 5 years after diagnosis [1]. This outcome has not been altered by the use of complex cytotoxic chemotherapy combinations [2–4]. The reasons for this may be explained by the unique biology and genomic characteristics of EOC. Here we review key approaches to diagnosis and stratification that are now needed to advance the field. New classifications have been developed following recent insights into the cell of origin of the different subtypes of EOC and their most critical driver events. The most important clinical questions for the pathologist remain how to unequivocally classify EOC and which additional genomic data may identify individuals with high chance of response or risk of relapse. This chapter concentrates on recent molecular insights that are likely to be highly relevant to clinical care over the next 5 years.

Cell of Origin of EOC

Genomic classifiers for cancer are strongly determined by the cell of origin of carcinoma. Arguably, longstanding confusion about the cellular origins of EOC has hindered the development of strong morphological and molecular classifications. Previously, the prevailing model was that EOC arose from metaplastic changes in the normal ovarian epithelium (NOE) and could arise from structural epithelial abnormalities on the surface of the ovary, such as clefts and inclusion cysts (reviewed in [5]). Transdifferentiation of the NOE into high-grade serous ovarian cancer (HGSOC), endometrioid, and mucinous subtypes was hypothesized to follow developmental pathways of Müllerian differentiation in the fallopian tube, endocervix, and endometrium, respectively. Dubeau [5] first suggested that "ovarian" cancers might arise from the secondary Müllerian system, including endosalpingiosis, endometriosis, and para-ovarian and paratubal cysts. Evidence that pre-invasive lesions of HGSOC were present in the distal fallopian tube was discovered by meticulous pathological studies on the fimbria of women with *BRCA1* or *BRCA2* mutations collected

G.J. Netto and I. Schrijver (eds.), *Genomic Applications in Pathology*,
DOI 10.1007/978-1-4939-0727-4_28, © Springer Science+Business Media New York 2015

after risk-reducing surgery [6–12]. Similar studies on sporadic HGSOC cases showed that tubal intraepithelial carcinomas (TICs) could be identified in 75 % of cases and that these fallopian tube lesions were underdiagnosed by standard histopathological examination [13]. The same *TP53* mutations were also shown in TIC and ovarian metastases which provided strong evidence for a single clonal origin for HGSOC arising in the tubal epithelium [10, 13]. These studies now strongly implicate TIC in the fallopian tube (and potentially in endosalpingiosis tissues) as the cells of origin for HGSOC.

The epidemiological risk factors for endometrioid and ovarian clear-cell adenocarcinoma (OCCA) are distinctly different from HGSOC and, in particular, are strongly linked to endometriosis [14, 15]. *ARID1A* was independently identified by two groups as a key driver in clear-cell and endometrioid EOC. Next-generation sequencing (NGS) of RNA and DNA revealed inactivating mutations in 46–57 % and 30 % of cases, respectively [16, 17]. Loss of *ARID1A* expression and *ARID1A* or *PIK3CA* mutation had also been demonstrated in endometriosis tissue contiguous with OCCA [16, 18]. This further supports the notion that endometriosis tissue is the precursor lesion. *ARID1A* encodes BAF250a which is a component of the SWI/SNF nucleosome remodeling complex (reviewed in [19]).

Molecular Classification of EOC

The Importance of Morphology and Immunohistochemical Markers

The pathological classification of EOC was last revised by the World Health Organization (WHO) in 2003 [20] and has not yet been revised in line with the newer data for cell of origin described above. New recommendations will be published in 2014 that will simplify the classification into five major types: HGSOC, OCCA, endometrioid, mucinous, and low-grade serous carcinoma. The majority of endometrioid EOC will be regarded as low-grade neoplasms. A major motivation for these changes has been the need to improve reproducibility of diagnosis [21]. The 2003 WHO classification (largely unchanged from 1973) is descriptive and importantly lacks specific criteria to distinguish between disease entities, leading to poor reproducibility, particularly between serous and endometrioid carcinoma, mixed versus pure cell types, and poorly differentiated tumors. These morphological challenges are compounded when the starting material is limited, such as from small radiologically guided biopsy specimens. Three main deficiencies that exist in previous classifications have now been addressed:

Firstly, diagnostic criteria for the five major types have been refined, based on morphology and specific immunohistochemical markers. The impetus for this change can be traced to a seminal study, which tested for variation between subtypes in the expression of 21 biomarkers and correlation with stage and prognosis across a population-based cohort of 500 ovarian carcinomas [22]. The expression of 20 biomarkers was significantly different *between* subtypes, but invariant across different stages of disease *within* each subtype. This suggested that subtypes are reproducible and distinct disease entities. The data also demonstrated that the subtype of EOC had the largest effect on prognosis. For example, 9 of the 21 biomarkers had prognostic value when applied across the entire cohort, but only three remained informative when analyzed by subtype-specific expression. The proliferation marker Ki-67 had previously been reported to be a poor prognostic marker in EOC, but was not prognostic *within* any of the subgroups. Thus, Ki-67 is a surrogate marker for HGSOC, and its apparent poor prognostic effect across all cases was due to the worse prognosis of HGSOC compared to other subtypes. A similar confounding effect was seen for the HGSOC-specific marker WT1, which was a poor prognostic marker overall, but was favorable within the HGSOC cases. These findings provided strong evidence for biological and survival differences between high-grade serous neoplasms and the lower-grade endometrioid and clear-cell subtypes. More importantly, they also showed that correct analysis of biomarker data in EOC requires stratification by subtype.

Secondly, the requirement for grading of EOC (as well, moderate and poorly differentiated) has been removed, owing to its poor reproducibility between pathologists [23–25] and because grade does not provide additional prognostic information independent of histological type. For example, in a study of 605 stage 1 or 2 EOC cases, tumor subtype and substage (stage IA or IB) had stronger prognostic value than grade [26] and stratifying HGSOC into grade 2 or 3 in cases from a randomized trial of different duration of adjuvant chemotherapy did not provide significant prognostic information [27]. Histological grade and subtype are covariates, but differences in grade do not reproduce the variability in outcome between subtypes—especially given the differences in stage at presentation. A comprehensive pathology review of 1,009 EOC cases showed that HGSOC accounted for 88 % of stage III/IV cancers and 36 % of stage I/II cancers [21, 24]. By contrast, low-stage EOCs were significantly enriched for endometrial and clear-cell subtypes (27 % versus 3 % and 26 % versus 5 %, respectively) compared to cases with stage III/IV disease. Therefore, the major differences are determined by the underlying biology of the EOC subtype and the only important distinction is between high-grade and low-grade serous carcinoma.

Thirdly, a minimal panel of immunohistochemical markers for classifying ovarian cancer subtype has been identified using nominal logistic regression analysis from tissue microarray data sets [28]. Expression of nine markers (CDKN2A, DKK1, HNF1B, MDM2, PGR, TFF3, TP53, VIM, and WT1) was the most predictive of EOC subtype when tested on a training set of 744 cases and 81 independent cases (κ=0.85 and 0.78, respectively) [28]. Use of these markers in the Calculator for Ovarian Subtype Probability (COSP) score results in a probability score for each subtype and it has been used to evaluate large research cohorts [29]. The proposed 2014 WHO classification for EOC will recommend use of WT1 as the main positive marker for serous carcinoma (low and high grade). Although the COSP is not part of the 2014 WHO recommendations, it represents the gold standard for EOC classification and should be used in cases of diagnostic difficulty instead of less discriminant markers such as CK7, CK20, and CA125. This is particularly important when only a small sample is available from an image-guided core biopsy and a non-EOC diagnosis needs to be excluded. This clinical situation is now more common owing to increased used of neoadjuvant chemotherapy which can cause difficulties in accurate tumor typing [30].

Taken together, these data show that subtypes of EOC differ in cell of origin, epidemiology, natural history and biology. It is therefore logical to view them as distinct and different diseases, and the term "subtype" should be deprecated as it implies a common tissue of origin. As genomically targeted therapies for endometrioid and clear-cell ovarian cancer are not currently in the clinic as standard of care treatments, the remainder of this chapter will concentrate on potential biomarkers for personalized treatment approaches in HGSOC and their relationship with platinum and PARP inhibitor sensitivity and resistance.

Expression Analysis

Initial attempts to define prognostic and predictive signatures for chemoresistance in EOC focused on the use of expression microarray studies [31–41]. These studies have had little success in identifying useful biomarkers [42–44]. This is in marked contrast to similar efforts in breast cancer, where expression profiling of a relatively small number of cases provided new molecular classifications with prognostic value, that have been quickly applied in research and in the clinic [45, 46]. These expression signatures are dominated by the transcriptional effects of estrogen receptor (ER), progesterone receptor (PR) and ERBB2 mediated pathways. Although it is assumed that HGSOC most closely resembles triple negative breast cancer (see [47] and below), a large analysis of 2,933 EOCs showed that strong PR staining was independently associated with improved disease-specific survival in HGSOC [48]. This provides strong evidence for the possible utility of measuring ER and PR in HGSOC, but trials are needed to determine if these markers are predictive for hormonal therapy.

Expression profiling has defined four main molecular subtypes in HGSOC, described as C1 (high stromal response), C2 (high immune signature), C4 (low stromal response), and

C5 (mesenchymal, low immune signature) [41, 47]. Although reproducible, the prognostic effects of these signatures are weaker than that of the ER/PR expression discussed above and have not had clinical impact. Substantial expansion of the classifications identified by Tothill has recently been performed by the cancer genome atlas (TCGA) [49]. Importantly, this has shown that individual HGSOC samples can show multiple subtype signatures, which questions the prognostic utility of classifying cases into mutually exclusive groups. The expression signatures are strongly modulated by stroma and infiltrating immune cells, and future work will need to study purified HGSOC cells to understand if classifiers exist.

By contrast, parallel proteomic analysis on 412 HGSOC samples with reverse phase protein arrays provided improved prognostic information as compared to expression signatures [50]. Refinement of the protein signature showed that five proteins (AR, BID, phosphorylated TAZ, phosphorylated EGFR, and HSP70) were associated with longer progression free survival (PFS) and increased expression in the low risk group. Four proteins (STAT5α, phosphorylated PKCα, phosphorylated MEK1 and EEF2) were associated with shorter PFS and increased expression in the high-risk group. AR may be a compelling marker and should now be tested across large sample sets using tissue microarrays.

Mutational Spectrum of High-Grade Serous Ovarian Carcinoma

The Cancer Genome Anatomy project has provided the most comprehensive mutation survey of fresh-frozen samples from HGSOC [47]. Whole exome sequencing from 489 HGSOC cases has confirmed profound mutational heterogeneity between patients and confirmed relatively few recurrent gene mutations, particularly in the tumor suppressor genes *TP53*, *BRCA1*, *BRCA2*, and *NF1*. Of note, oncogenic mutations in *EGFR*, *PIK3CA*, *BRAF*, and *KRAS* are extremely infrequent and may account for less than 1 % of mutations seen in HGSOC cases. These results differ from those obtained by Sequenom sequencing of 203 advanced stage HGSOC cases from FFPE specimens [51] which suggested that 56 % (113/203) of tumor samples harbored candidate mutations in 112 oncogenes. The most common somatic oncogene mutations were found in *EGFR* (9.4 %), *KRAS* (4.5 %), *PDGFRα* (4.5 %), *KIT* (3 %), and *PIK3CA* (3 %). At present, the reasons for this discrepancy are unclear. The depth of sequencing coverage for the somatic samples from TCGA may be limiting, but it is notable that no other groups have reported such a high mutation rate for these oncogenes. Where sequencing is being routinely carried out on EOC specimens, there is utility for screening *BRAF* and *KRAS* (along with *TP53*, see below) as this can provide diagnostically relevant information for low-grade serous carcinomas.

Mutations in *TP53* are Ubiquitous in High-Grade Serous Ovarian Carcinoma

The *TP53* gene encodes the p53 tumor suppressor protein and is among the most frequently mutated genes in human cancer [52, 53]. The frequency of *TP53* mutation is different between EOC subtypes [54–57]. Demonstration that *TP53* mutation is ubiquitous in HGSOC was first provided by Sanger sequencing results from 145 women with serous neoplasms [58]. Mutations were identified in 119 of 123 HGSOC cases (96.7 %) and 19 % of the mutations were present in exons 2–4 and 9–11, explaining the higher percentage of mutations seen in comparison to earlier studies. Whole exome sequencing by the Cancer Genome Atlas project also confirmed near-100 % mutation rate. These studies underscore the essential role that *TP53* mutation has as an early driver mutation for HGSOC and are consistent with the findings of p53 accumulation and mutation in fallopian tube p53 signature foci and TICs [12]. However, the fact that signature foci in the fallopian tube may have *TP53* mutation, but do not develop into TICs, suggests that loss of p53 function is required but not sufficient for the development of HGSOC.

Determining *TP53* mutational status should now be considered for all women with HGSOC although, at present, this result is not predictive of outcome. Firstly, presence of

TP53 mutation provides diagnostic information and excludes low-grade serous carcinoma. Secondly, the type of mutation may give prognostic information. As in other cancers, approximately 50 % of *TP53* mutations in HGSOC are missense and null mutations (including nonsense, frameshift, and splice site mutations) account for about 30 % [47, 58]. Loss of p53 protein by immunohistochemistry in HGSOC (which suggests a nonsense mutation) appears to be a poor prognostic factor, as compared with cases with strong nuclear staining (which suggests a missense mutation) which have a reduced risk of recurrence (HR 0.71, 95 % CI 0.51–0.99) [59]. This finding is supported by an earlier publication that showed with Sanger sequencing that nonsense mutations had worse outcome, but larger cohorts are needed to confirm a definite adverse prognostic effect [59, 60]. Thirdly, *TP53* mutation-specific therapy is now available or will soon be placed into early phase trials [61–64] and precise knowledge of the type of mutation will be important for eligibility or stratification.

High-Grade Serous Carcinoma is Defined by Profound Structural DNA Aberrations

In contrast to the relative lack of recurrent oncogenic mutation, the TCGA analysis showed profound and recurrent copy number aberrations (CNAs). In particular, previously described large regional aberrations were confirmed (8 gains; 22 losses) and the majority of these were present in >50 % of cases. Recurrent focal CNAs were identified in 63 regions including *CCNE1*, *MYC*, and *MECOM* genes. These data demonstrate that HGSOC is distinct from other epithelial cancers, such as breast cancer, in that oncogenic drivers result from complex copy number changes and not from classical mutations in oncogenes.

Amplification of *CCNE1*, which encodes the cell cycle checkpoint protein cyclin E, may be a therapeutic target. A previous study of 118 HGSOC cases showed that amplification of chromosome 19q12 was a strong negative prognostic factor and correlated with platinum resistance [65–67]. As well as *CCNE1*, this region also includes the anti-apoptotic oncogene *C19orf2* (also known as *URI*). Knockdown of *CCNE1* in EOC cell lines with 19q12 amplification paradoxically increased cisplatin resistance in short-term assays, although it did result in reduced clonogenic survival, suggesting an oncogenic effect [68]. Studies of the effects of amplification or overexpression of *C19orf2* in vitro and in vivo showed increased cisplatin resistance, mediated by increased S6K1-BAD survival signaling [69]. Mechanistic insights into platinum resistance in *CCNE1*-amplified cases have come from siRNA screens showing that *BRCA1* and members of the ubiquitin pathway are required for survival in cancers that have *CCNE1* amplification [70]. Use of CDK2 inhibitors in these cases may therefore be a relevant therapeutic approach [71].

Given the strong prognostic effect of 19q12 amplification, it is important that patients are stratified for this marker using assays to test for amplification of the region or overexpression of Cyclin E. However, determining which is the most important driver oncogene (and therefore therapeutic target) will require highly specific pharmacological inhibitors and biomarker-driven clinical trials.

Homologous Recombination and High-Grade Carcinoma

BRCA1/2 Mutation Has a Strong Survival Effect

In comparison to other epithelial cancers, HGSOC shows the highest sensitivity to platinum-based chemotherapy and initial response rates are 70–80 % when surgery is combined with chemotherapy. Uniquely, a substantial proportion of patients with relapsed disease will respond to re-treatment with platinum chemotherapy. The time interval between diagnosis and development of progressive disease is the strongest predictor of response rates to re-treatment, and is used clinically to define “platinum-resistant” and “platinum-sensitive” relapsed disease [72]. Despite the critical importance of platinum therapy, primary and acquired resistance is still poorly understood [73].

The relative hypersensitivity of HGSOC to treatment may be explained by high rates of intrinsic homologous recombination deficiency (HRD). Carboplatin induces inter- and intra-strand cross-linking that results in both single- and double-strand DNA breaks. In normal cells, double strand break damage can be repaired by either error-free homologous recombination (HR) or by the error-prone non-homologous end-joining (NHEJ) pathway. Cancer cells that lack BRCA1 or BRCA2 function cannot carry out HR repair and are therefore very sensitive to platinum-induced DNA damage that induces apoptotic death.

BRCA1 and *BRCA2* mutation carriers predominantly develop HGSOC, and previous data obtained from all EOC subtypes may underestimate the prevalence of these mutations in this subgroup. Recent studies of *BRCA1* and *BRCA2* in HGSOC patients have shown combined germ-line mutation rates of 3–23 %, and additional pathogenic somatic mutations can also be found in non-carrier cases [47, 74–76]. This is correlated with significantly longer survival in HGSOC patients with *BRCA1* or *BRCA2* germ-line mutations as compared with non-carriers [74, 77–79]. A large collaborative study has recently shown that 5-year overall survival was 36 % for non-carriers (n=2,666), 44 % for *BRCA1* carriers (n=909), and 52 % for *BRCA2* carriers (n=304). Highly significant survival differences remained after additional adjustment for major prognostic factors (*BRCA1*: HR, 0.73; *BRCA2*: HR, 0.49; $P<.001$ for both) [80].

These findings have important implications for the development of predictive biomarkers for women with HGSOC as they suggest that cases with *BRCA1/2* mutations will have longer progression-free survival and be over-represented in the clinically defined platinum-sensitive relapsed group (progression more than 6 months from primary treatment). Support for this hypothesis is provided by the Australian Ovarian Cancer Study (AOCS) of 1,001 cases, in which *BRCA1* or *BRCA2* carriers had improved rates of progression-free as well as overall survival [76]. Carriers were less frequent in patients who progressed in less than 6 months from primary treatment (platinum-resistant group). Mutation-negative patients who responded to multiple courses of platinum-based treatment were more likely to carry somatic *BRCA1/2* mutations. Therefore, biomarkers based on *BRCA1* and *BRCA2* sequencing may be highly predictive for outcome and now should be included for stratification in clinical trials. It is likely that this testing will be provided by the medical genetics service in most hospitals, but somatic sequencing of *BRCA1* and *BRCA2* may be coordinated by the molecular pathologist.

Functional Tests for Homologous Recombination May be a Strong Predictive Biomarker for HGSOC

Recent publications have suggested that HRD may be a common mutator phenotype in HGSOC patients. As discussed above, more than 10 % of HGSOC cases may have germ-line or somatic *BRCA1/2* mutations that will cause HRD. In addition, familial non-*BRCA1/2* cases are enriched for mutations in HR pathway genes, including *RAD51C* [81, 82], *RAD51D* [83] and *BRIP1* [84], suggesting that there is common involvement of DNA repair enzymes in the pathogenesis of HGSOC. Also, using a functional assay of HR status based on RAD51 focus formation after in vitro DNA damage, direct testing was performed on 24 primary cultures of EOC and showed that 16 (64 %) were HR deficient and this was highly correlated with in vitro response to PARPi [85]. A similar rate was predicted from the analysis carried out by the TCGA using survival data and mutation status in candidate HR-related genes [47].

Intratumoral Heterogeneity in HGSOC May Contribute to Platinum Resistance

Although platinum resistance has been intensively studied using in vitro models, there is only limited evidence that mechanisms commonly seen in cell lines, involving altered apoptosis pathways, increased drug excretion or tolerance to DNA adducts, occur in clinical samples [73]. The importance of HRD in determining platinum sensitivity is underscored by studies showing that in a small proportion of *BRCA1* and *BRCA2* carriers, resistance is caused by revertant or secondary mutations that restore somatic BRCA1 or BRCA2 protein function [86–90]. Although it is unknown if these mutations predate platinum treatment, their existence strongly argues that the acquisition of a resistant phenotype involves selection effects.

It has been proposed that genetic heterogeneity could explain the development of drug resistance in HGSOC [91] based on the demonstration of divergent evolution between sensitive and resistant subclones from three cases of HGSOC [92]. The genetic changes in the cell lines derived before and after clinical resistance developed were incompatible with a simple linear model, and the most parsimonious explanation was that resistant lineages were present as a minor subpopulation of the tumor mass at the time of first therapy [91, 92]. The possibility that significant genetic heterogeneity existed within many cancers was originally proposed by Nowell in 1976 and demonstrated using cytogenetic methods in 1978 [93, 94]. However, it is only recently that the degree and types of genetic variation present within an individual's cancer could be accurately characterized through the advances of NGS and high accuracy single nucleotide polymorphism (SNP) comparative genomic hybridization (CGH) arrays [91, 92, 95–102]. Loss of heterozygosity data have provided evidence for genetic heterogeneity in HGSOC [95, 96] and more detailed studies have shown that subclonal populations preexist in epithelial tumors [97, 98, 103, 104] and undergo treatment-related selection in leukemias [105–107], breast cancer [99], and renal cancer [108]. Therefore, characterizing which HGSOC cases may have intratumoral genetic heterogeneity may be critical to predict the risk of treatment failure, particularly when there is clonal diversity for the cells containing the "actionable" mutation [86, 107–109].

There is strong evidence that selective effects can explain drug resistance in hematological cancers [110, 111]. Point mutations conferring resistance to imatinib have been shown to be present at low frequency before treatment in both acute lymphocytic leukemia and chronic myeloid leukemia. At relapse, these mutations are present in high frequency in the leukemic blasts [105–107]. In acute lymphocytic leukemia, a higher frequency of resistant mutations in the initial presenting disease is directly correlated with shorter remission [105]. Similar changes have been shown in breast cancer using NGS of primary and relapsed disease in a single case of lobular breast cancer. Relapse-specific mutations could be detected at a frequency of 1–13 % in the primary tumor 7 years before disease recurrence [99].

Intratumoral genetic heterogeneity in HGSOC has been demonstrated both within a region of tumor and between metastatic sites [95, 96, 112]. These genetic differences could be expected to alter chemosensitivity. Consistent with this, variable in vitro responses to a variety of chemotherapeutic agents were observed in primary EOC cells obtained from different metastatic sites from the same individual, suggesting the existence of genetically or epigenetically diverse subpopulations [113]. Similar differential effects on response have also been shown during chemoradiation of advanced cervical cancer [91]. In three out of ten cases there were distinct genetic subpopulations before treatment and these regions showed differential responses to chemo-radiotherapy, leading to mixed response and selection of resistant disease. Studies of cell lines derived before and after relapse in three cases of HGSOC have found that presentation and resistant disease are not linearly genetically related, showing that the relapsed genotype cannot have arisen by direct descent from a dominant clone at disease presentation [92]. These studies suggest a hypothesis for platinum-resistant disease in HGSOC where strong selection for a minor resistant clonal population occurs, rather than genetic progression from presentation disease. These models may also be relevant in platinum-sensitive disease as it will be important to determine whether revertant mutations in *BRCA1* and *BRCA2* exist before treatment, as is seen for *ABL* mutations in chronic myeloid leukemia.

Understanding Tumor Heterogeneity and Platinum Resistance in HGSOC Requires Sequential Biopsy and Plasma Studies

In order to understand how platinum resistance evolves in HGSOC it is now essential to undertake genomic comparisons of sequential tissues and plasma circulating tumor DNA (ctDNA) from diagnosis to relapsed/progressive disease. Image-guided core biopsy (IGCB) of ovarian cancer, either under ultrasound or CT guidance, is safe and feasible and is the standard of care for diagnosis of patients with suspected ovarian cancer, particularly prior to neoadjuvant chemotherapy [114, 115]. However, re-biopsy of relapsed HGSOC tissue is not a standard of care and the cost for each biopsy may be significant. Although studies on malignant cells from abdominal ascites can offer a less morbid approach to assaying relapsed disease, only a small proportion of patients with relapsed disease will have ascites that can be aspirated. In addition, malignant cells in ascites do not allow examination of tumor stroma and tumor vasculature, which can both profoundly affect tumor biology.

Circulating Tumor DNA Can Be Used to Identify Mutations and Track Tumor Evolution

Although CT and serum CA125 are the standard of care for estimating disease burden and response in HGSOC, there is an urgent need for cheap and sensitive blood-based markers to provide molecular measures of response and to identify minimal residual disease. Circulating DNA in plasma and serum contains tumor-specific sequences that have recently been exploited in small numbers of patients as personalized biomarkers. Evidence that cancer patients had higher levels of circulating free DNA in serum was first shown in 1977 [116], which prompted exploratory studies using total DNA as a potential biomarker in several cancers, including EOC [117]. However, these assays were not specific and had confounding effects that prevented reliable clinical use.

Mutation-specific assays of ctDNA can be developed by genotyping or sequencing the tumor followed by the design of allele- or mutation-specific PCR-based assays to detect tumor DNA. The feasibility of this approach has been demonstrated by accurate quantitation of dynamic changes in ctDNA from colorectal cancer, non-small-cell lung cancer, breast cancer, and osteosarcoma, using assays developed for tumor specific mutations and rearrangements [104, 118–121]. These early data suggest that ctDNA dynamics compare favorably with other diagnostic modalities, including serum tumor markers and CT imaging. Use of ctDNA is potentially very important for studies in HGSOC because the detection of circulating tumor cells (CTC) has been very difficult. Using the Veridex CellSearch system only 14 % of patients with advanced EOC had more than two CTCs (median 0) indicating that CTC counts are too insensitive to be used as a measure of response [122].

Tagged-amplicon deep sequencing (TAm-Seq) was developed as a method that can amplify and directly sequence large genomic regions from low counts of fragmented DNA and has shown ability to identify mutations in the plasma of patients with HGSOC [123]. Using primers to amplify 5,995 bases covering *TP53*, *EGFR*, *BRAF*, and *KRAS* to screen plasma obtained from 38 HGSOC patients with high levels of ctDNA it was possible to identify mutations in *TP53* at allelic frequencies of 2–65 %. In one patient with relapsed HGSOC, a de novo mutation in *EGFR* was identified that was present at very low frequency in the tumor mass 15 months prior to disease relapse. TAm-Seq was also used to monitor the dynamics of ten mutations in plasma DNA of a single patient with metastatic breast cancer. Previous studies have followed one to two mutations in an individual patient [104, 118] but tracking multiple mutations could provide critical insights into clonal evolution and may increase the robustness for tumor monitoring by mitigating the effects of sampling noise or mutational drift. This work demonstrates that TAm-Seq of ctDNA offers the potential for a "liquid biopsy" that can be used for personalized genomic profiling and to explore clonal evolution and the potential of differential response to treatment. This has recently been demonstrated by the use of ctDNA to detect the emergence of subclonal *KRAS* mutations as the cause for secondary resistance to cetuximab [124, 125].

Conclusions

The emerging molecular data indicate that EOC is a group of distinct diseases, each with different clinical and epidemiological characteristics. Previous work to identify predictive and prognostic biomarkers has been greatly weakened by the inclusion of different ovarian subtypes. Platinum resistance in HGSOC may represent a distinct molecular subtype, but further work is required to define the driver mutations in this disease and the mechanisms of intrinsic resistance. The use of PARP inhibitors in platinum-sensitive HGSOC heralds the real possibility of curative strategies based on maintenance treatment. However, understanding how platinum sensitivity may be lost in HGSOC will require genomic comparison of sequential tissues taken at diagnosis and at time of relapse. Stratified medicine for EOC, particularly for platinum-sensitive disease, is now feasible but will require much wider access to *BRCA1* and *BRCA2* testing for women with HGSOC and genomic tests of HRD in tumor tissue.

REFERENCES

1. Coleman MP, Forman D, Bryant H, Butler J, Rachet B, Maringe C, et al. Cancer survival in Australia, Canada, Denmark, Norway, Sweden, and the UK, 1995-2007 (the International Cancer Benchmarking Partnership): an analysis of population-based cancer registry data. Lancet. 2011;377(9760):127–38. Epub 2010/12/25.
2. Bookman MA, Brady MF, McGuire WP, Harper PG, Alberts DS, Friedlander M, et al. Evaluation of new platinum-based treatment regimens in advanced-stage ovarian cancer: a phase III trial of the gynecologic cancer Intergroup. J Clin Oncol. 2009;27(9):1419–25.
3. du Bois A, Weber B, Rochon J, Meier W, Goupil A, Olbricht S, et al. Addition of epirubicin as a third drug to carboplatin-paclitaxel in first-line treatment of advanced ovarian cancer: a prospectively randomized gynecologic cancer intergroup trial by the Arbeitsgemeinschaft Gynaekologische Onkologie Ovarian Cancer Study Group and the Groupe d'Investigateurs Nationaux pour l'Etude des Cancers Ovariens. J Clin Oncol. 2006;24(7): 1127–35. Epub 2006/03/01.
4. du Bois A, Herrstedt J, Hardy-Bessard AC, Muller HH, Harter P, Kristensen G, et al. Phase III trial of carboplatin plus paclitaxel with or without gemcitabine in first-line treatment of epithelial ovarian cancer. J Clin Oncol. 2010;28(27):4162–9. Epub 2010/08/25.
5. Dubeau L. The cell of origin of ovarian epithelial tumors and the ovarian surface epithelium dogma: does the emperor have no clothes? Gynecol Oncol. 1999;72(3):437–42.
6. Piek JM, van Diest PJ, Zweemer RP, Kenemans P, Verheijen RH. Tubal ligation and risk of ovarian cancer. Lancet. 2001;358(9284):844.
7. Piek JM, van Diest PJ, Zweemer RP, Jansen JW, Poort-Keesom RJ, Menko FH, et al. Dysplastic changes in prophylactically removed Fallopian tubes of women predisposed to developing ovarian cancer. J Pathol. 2001;195(4):451–6.
8. Leeper K, Garcia R, Swisher E, Goff B, Greer B, Paley P. Pathologic findings in prophylactic oophorectomy specimens in high-risk women. Gynecol Oncol. 2002;87(1):52–6.
9. Piek JMJ, Verheijen Re HM, Kenemans P, Massuger LF, Bulten H, van Diest PJ. BRCA1/2-related ovarian cancers are of tubal origin: a hypothesis. Gynecol Oncol. 2003;90(2):491.
10. Lee Y, Miron A, Drapkin R, Nucci MR, Medeiros F, Saleemuddin A, et al. A candidate precursor to serous carcinoma that originates in the distal fallopian tube. J Pathol. 2007;211(1):26–35.
11. Crum CP, Drapkin R, Kindelberger D, Medeiros F, Miron A, Lee Y. Lessons from BRCA: the tubal fimbria emerges as an origin for pelvic serous cancer. Clin Med Res. 2007;5(1):35–44.
12. Callahan MJ, Crum CP, Medeiros F, Kindelberger DW, Elvin JA, Garber JE, et al. Primary fallopian tube malignancies in BRCA-positive women undergoing surgery for ovarian cancer risk reduction. J Clin Oncol. 2007;25(25):3985–90.
13. Kindelberger DW, Lee Y, Miron A, Hirsch MS, Feltmate C, Medeiros F, et al. Intraepithelial carcinoma of the fimbria and pelvic serous carcinoma: evidence for a causal relationship. Am J Surg Pathol. 2007;31(2):161–9.
14. Viganó P, Somigliana E, Chiodo I, Abbiati A, Vercellini P. Molecular mechanisms and biological plausibility underlying the malignant transformation of endometriosis: a critical analysis. Hum Reprod Update. 2006;12(1):77–89.
15. Ness RB. Endometriosis and ovarian cancer: thoughts on shared pathophysiology. Am J Obstet Gynecol. 2003;189(1):280–94.
16. Wiegand KC, Shah SP, Al-Agha OM, Zhao Y, Tse K, Zeng T, et al. ARID1A mutations in

endometriosis-associated ovarian carcinomas. N Engl J Med. 2010;363(16):1532–43.
17. Jones S, Wang T-L, Shih I-M, Mao T-L, Nakayama K, Roden R, et al. Frequent mutations of chromatin remodeling gene ARID1A in ovarian clear cell carcinoma. Science. 2010;330(6001): 228–31.
18. Yamamoto S, Tsuda H, Takano M, Iwaya K, Tamai S, Matsubara O. PIK3CA mutation is an early event in the development of endometriosis-associated ovarian clear cell adenocarcinoma. J Pathol. 2011;225(2):189–94. Epub 2011/07/08.
19. Wilson BG, Roberts CW. SWI/SNF nucleosome remodellers and cancer. Nat Rev Cancer. 2011;11(7):481–92. Epub 2011/06/10.
20. Tavassoli FA, Devilee P. Pathology and genetics of tumours of the breast and female genital tract. Lyon, PA: IARC; 2003. 432 p.
21. Köbel M, Kalloger SE, Baker PM, Ewanowich CA, Arseneau J, Zherebitskiy V, et al. Diagnosis of ovarian carcinoma cell type is highly reproducible: a transcanadian study. Am J Surg Pathol. 2010;34(7):984–93.
22. Köbel M, Kalloger SE, Boyd N, McKinney S, Mehl E, Palmer C, et al. Ovarian carcinoma subtypes are different diseases: implications for biomarker studies. PLoS Med. 2008;5(12):e232.
23. Singh N, Ayhan A, Menon U, Chin Aleong JA, Faruqi AZ, Gayther SA, et al. Grading of serous ovarian carcinoma: further evidence of a lack of agreement between conventional grading systems. Histopathology. 2008;52(3):393–5.
24. Gilks CB, Ionescu DN, Kalloger SE, Köbel M, Irving J, Clarke B, et al. Tumor cell type can be reproducibly diagnosed and is of independent prognostic significance in patients with maximally debulked ovarian carcinoma. Hum Pathol. 2008;39(8):1239–51.
25. Kommoss S, Schmidt D, Kommoss F, Hedderich J, Harter P, Pfisterer J, et al. Histological grading in a large series of advanced stage ovarian carcinomas by three widely used grading systems: consistent lack of prognostic significance. A translational research subprotocol of a prospective randomized phase III study (AGO-OVAR 3 protocol). Virchows Arch. 2009;454(3):249–56.
26. Köbel M, Kalloger SE, Santos JL, Huntsman DG, Gilks CB, Swenerton KD. Tumor type and substage predict survival in stage I and II ovarian carcinoma: insights and implications. Gynecol Oncol. 2010;116(1):50–6.
27. Bell J, Brady MF, Young RC, Lage J, Walker JL, Look KY, et al. Randomized phase III trial of three versus six cycles of adjuvant carboplatin and paclitaxel in early stage epithelial ovarian carcinoma: a Gynecologic Oncology Group study. Gynecol Oncol. 2006;102(3):432–9.
28. Kalloger SE, Kobel M, Leung S, Mehl E, Gao D, Marcon KM, et al. Calculator for ovarian carcinoma subtype prediction. Mod Pathol. 2011;24(4):512–21. Epub 2010/12/07.
29. Kobel M, Kalloger SE, Lee S, Duggan MA, Kelemen LE, Prentice L, et al. Biomarker-based ovarian carcinoma typing: a histologic investigation in the ovarian tumor tissue analysis consortium. Cancer Epidemiol Biomarkers Prev. 2013;22(10):1677–86. Epub 2013/07/25.
30. McCluggage WG, Lyness RW, Atkinson RJ, Dobbs SP, Harley I, McClelland HR, et al. Morphological effects of chemotherapy on ovarian carcinoma. J Clin Pathol. 2002;55(1):27–31.
31. Spentzos D, Levine DA, Ramoni MF, Joseph M, Gu X, Boyd J, et al. Gene expression signature with independent prognostic significance in epithelial ovarian cancer. J Clin Oncol. 2004;22(23): 4700–10.
32. Selvanayagam ZE, Cheung TH, Wei N, Vittal R, Lo KWK, Yeo W, et al. Prediction of chemotherapeutic response in ovarian cancer with DNA microarray expression profiling. Cancer Genet Cytogenet. 2004;154(1):63–6.
33. Spentzos D, Levine DA, Kolia S, Otu H, Boyd J, Libermann TA, et al. Unique gene expression profile based on pathologic response in epithelial ovarian cancer. J Clin Oncol. 2005;23(31): 7911–8.
34. Berchuck A, Iversen ES, Lancaster JM, Pittman J, Luo J, Lee P, et al. Patterns of gene expression that characterize long-term survival in advanced stage serous ovarian cancers. Clin Cancer Res. 2005;11(10):3686–96.
35. Hartmann LC, Lu KH, Linette GP, Cliby WA, Kalli KR, Gershenson D, et al. Gene expression profiles predict early relapse in ovarian cancer after platinum-paclitaxel chemotherapy. Clin Cancer Res. 2005;11(6):2149–55.
36. Bernardini M, Lee C-H, Beheshti B, Prasad M, Albert M, Marrano P, et al. High-resolution mapping of genomic imbalance and identification of gene expression profiles associated with differential chemotherapy response in serous epithelial ovarian cancer. Neoplasia. 2005; 7(6):603–13.
37. Jazaeri AA, Awtrey CS, Chandramouli GVR, Chuang YE, Khan J, Sotiriou C, et al. Gene expression profiles associated with response to chemotherapy in epithelial ovarian cancers. Clin Cancer Res. 2005;11(17):6300–10.
38. Helleman J, Jansen MPHM, Span PN, van Staveren IL, Massuger LFAG, Meijer-van Gelder ME, et al. Molecular profiling of platinum resistant ovarian cancer. Int J Cancer. 2006;118(8): 1963–71.
39. Dressman HK, Berchuck A, Chan G, Zhai J, Bild A, Sayer R, et al. An integrated genomic-based approach to individualized treatment of patients with advanced-stage ovarian cancer. J Clin Oncol. 2007;25(5):517–25.

40. Crijns APG, Fehrmann RSN, de Jong S, Gerbens F, Meersma GJ, Klip HG, et al. Survival-related profile, pathways, and transcription factors in ovarian cancer. PLoS Med. 2009;6(2):e24.
41. Tothill RW, Tinker AV, George J, Brown R, Fox SB, Lade S, et al. Novel molecular subtypes of serous and endometrioid ovarian cancer linked to clinical outcome. Clin Cancer Res. 2008; 14(16):5198–208.
42. Fehrmann RSN, Li X-Y, van der Zee AGJ, de Jong S, Te Meerman GJ, de Vries EGE, et al. Profiling studies in ovarian cancer: a review. Oncologist. 2007;12(8):960–6.
43. Agarwal R, Kaye SB. Expression profiling and individualisation of treatment for ovarian cancer. Curr Opin Pharmacol. 2006;6(4):345–9.
44. Na Y-J, Farley J, Zeh A, del Carmen M, Penson R, Birrer MJ. Ovarian cancer: markers of response. Int J Gynecol Cancer. 2009;19 Suppl 2:S21–9.
45. Sorlie T, Perou CM, Tibshirani R, Aas T, Geisler S, Johnsen H, et al. Gene expression patterns of breast carcinomas distinguish tumor subclasses with clinical implications. Proc Natl Acad Sci U S A. 2001;98(19):10869–74.
46. Sotiriou C, Neo S-Y, McShane LM, Korn EL, Long PM, Jazaeri A, et al. Breast cancer classification and prognosis based on gene expression profiles from a population-based study. Proc Natl Acad Sci U S A. 2003;100(18):10393–8.
47. The Cancer Genome Atlas Research Network. Integrated genomic analyses of ovarian carcinoma. Nature. 2011;474(7353):609–15. Epub 2011/07/02.
48. Sieh W, Kobel M, Longacre TA, Bowtell DD, deFazio A, Goodman MT, et al. Hormone-receptor expression and ovarian cancer survival: an ovarian tumor tissue analysis consortium study. Lancet Oncol. 2013;14(9):853–62. Epub 2013/07/13.
49. Verhaak RG, Tamayo P, Yang JY, Hubbard D, Zhang H, Creighton CJ, et al. Prognostically relevant gene signatures of high-grade serous ovarian carcinoma. J Clin Invest. 2013;123(1):517–25. Epub 2012/12/22.
50. Yang JY, Yoshihara K, Tanaka K, Hatae M, Masuzaki H, Itamochi H, et al. Predicting time to ovarian carcinoma recurrence using protein markers. J Clin Invest. 2013;123(9):3740–50. Epub 2013/08/16.
51. Matulonis UA, Hirsch M, Palescandolo E, Kim E, Liu J, van Hummelen P, et al. High throughput interrogation of somatic mutations in high grade serous cancer of the ovary. PLoS One. 2011;6(9):e24433. Epub 2011/09/21.
52. Nigro JM, Baker SJ, Preisinger AC, Jessup JM, Hostetter R, Cleary K, et al. Mutations in the p53 gene occur in diverse human tumour types. Nature. 1989;342(6250):705–8.
53. Vogelstein B, Kinzler KW. Cancer genes and the pathways they control. Nat Med. 2004;10(8): 789–99.
54. de Graeff P, Crijns APG, de Jong S, Boezen M, Post WJ, de Vries EGE, et al. Modest effect of p53, EGFR and HER-2/neu on prognosis in epithelial ovarian cancer: a meta-analysis. Br J Cancer. 2009;101(1):149–59.
55. Stewart RL, Royds JA, Burton JL, Heatley MK, Wells M. Direct sequencing of the p53 gene shows absence of mutations in endometrioid endometrial adenocarcinomas expressing p53 protein. Histopathology. 1998;33(5):440–5.
56. Singer G, Stöhr R, Cope L, Dehari R, Hartmann A, Cao DF, et al. Patterns of p53 mutations separate ovarian serous borderline tumors and low- and high-grade carcinomas and provide support for a new model of ovarian carcinogenesis: a mutational analysis with immunohistochemical correlation. Am J Surg Pathol. 2005;29(2):218–24.
57. Kuo K-T, Mao T-L, Jones S, Veras E, Ayhan A, et al. Frequent activating mutations of PIK3CA in ovarian clear cell carcinoma. Am J Pathol. 2009;174(5):1597–601.
58. Ahmed AA, Etemadmoghadam D, Temple J, Lynch AG, Riad M, Sharma R, et al. Driver mutations in TP53 are ubiquitous in high grade serous carcinoma of the ovary. J Pathol. 2010; 221(1):49–56.
59. Köbel M, Reuss A, Bois AD, Kommoss S, Kommoss F, Gao D, et al. The biological and clinical value of p53 expression in pelvic high-grade serous carcinomas. J Pathol. 2010;222(2): 191–8.
60. Shahin MS, Hughes JH, Sood AK, Buller RE. The prognostic significance of p53 tumor suppressor gene alterations in ovarian carcinoma. Cancer. 2000;89(9):2006–17.
61. Brown CJ, Lain S, Verma CS, Fersht AR, Lane DP. Awakening guardian angels: drugging the p53 pathway. Nat Rev Cancer. 2009;9(12): 862–73.
62. Cheok CF, Verma CS, Je B, Lane DP. Translating p53 into the clinic. Nat Rev Clin Oncol. 2011; 8(1):25–37.
63. Carrassa L, Chila R, Lupi M, Ricci F, Celenza C, Mazzoletti M, et al. Combined inhibition of Chk1 and Wee1: in vitro synergistic effect translates to tumor growth inhibition in vivo. Cell Cycle. 2012;11(13):2507–17. Epub 2012/06/21.
64. Emerling BM, Hurov JB, Poulogiannis G, Tsukazawa KS, Choo-Wing R, Wulf GM, et al. Depletion of a putatively druggable class of phosphatidylinositol kinases inhibits growth of p53-null tumors. Cell. 2013;155(4):844–57. Epub 2013/11/12.
65. Nakayama K, Nakayama N, Jinawath N, Salani R, Kurman RJ, Shih I-M, et al. Amplicon profiles

in ovarian serous carcinomas. Int J Cancer. 2007; 120(12):2613–7.
66. Farley J, Smith LM, Darcy KM, Sobel E, O'Connor D, Henderson B, et al. Cyclin E expression is a significant predictor of survival in advanced, suboptimally debulked ovarian epithelial cancers: a Gynecologic Oncology Group study. Cancer Res. 2003;63(6):1235–41.
67. Etemadmoghadam D, deFazio A, Beroukhim R, Mermel C, George J, Getz G, et al. Integrated genome-wide DNA copy number and expression analysis identifies distinct mechanisms of primary chemoresistance in ovarian carcinomas. Clin Cancer Res. 2009;15(4):1417–27.
68. Etemadmoghadam D, George J, Cowin PA, Cullinane C, Kansara M, Group AOCS, et al. Amplicon-dependent CCNE1 expression is critical for clonogenic survival after cisplatin treatment and is correlated with 20q11 gain in ovarian cancer. PLoS One. 2010;5(11):e15498.
69. Theurillat JP, Metzler SC, Henzi N, Djouder N, Helbling M, Zimmermann AK, et al. URI is an oncogene amplified in ovarian cancer cells and is required for their survival. Cancer Cell. 2011;19(3):317–32. Epub 2011/03/15.
70. Etemadmoghadam D, Weir BA, Au-Yeung G, Alsop K, Mitchell G, George J, et al. Synthetic lethality between CCNE1 amplification and loss of BRCA1. Proc Natl Acad Sci U S A. 2013;110:19489–94. Epub 2013/11/13.
71. Etemadmoghadam D, Au-Yeung G, Wall M, Mitchell C, Kansara M, Loehrer E, et al. Resistance to CDK2 inhibitors is associated with selection of polyploid cells in CCNE1-amplified ovarian cancer. Clin Cancer Res. 2013;19(21): 5960–71. Epub 2013/09/06.
72. Markman M, Rothman R, Hakes T, Reichman B, Hoskins W, Rubin S, et al. Second-line platinum therapy in patients with ovarian cancer previously treated with cisplatin. J Clin Oncol. 1991;9(3):389–93.
73. Borst P, Rottenberg S, Jonkers J. How do real tumors become resistant to cisplatin? Cell Cycle. 2008;7(10):1353–9.
74. Trainer AH, Meiser B, Watts K, Mitchell G, Tucker K, Friedlander M. Moving toward personalized medicine: treatment-focused genetic testing of women newly diagnosed with ovarian cancer. Int J Gynecol Cancer. 2010;20(5): 704–16.
75. Hennessy BTJ, Timms KM, Carey MS, Gutin A, Meyer LA, Flake N, Darl D, et al. Somatic mutations in BRCA1 and BRCA2 could expand the number of patients that benefit from poly (ADP ribose) polymerase inhibitors in ovarian cancer. J Clin Oncol. 2010;28(22):3570–6.
76. Alsop K, Fereday S, Meldrum C, deFazio A, Emmanuel C, George J, et al. BRCA mutation frequency and patterns of treatment response in BRCA mutation-positive women with ovarian cancer: a report from the Australian Ovarian Cancer Study Group. J Clin Oncol. 2012;30(21):2654–63. Epub 2012/06/20.
77. Chetrit A, Hirsh-Yechezkel G, Ben-David Y, Lubin F, Friedman E, Sadetzki S. Effect of BRCA1/2 mutations on long-term survival of patients with invasive ovarian cancer: the national Israeli study of ovarian cancer. J Clin Oncol. 2008;26(1):20–5.
78. Tan DSP, Rothermundt C, Thomas K, Bancroft E, Eeles R, Shanley S, et al. "BRCAness" syndrome in ovarian cancer: a case–control study describing the clinical features and outcome of patients with epithelial ovarian cancer associated with BRCA1 and BRCA2 mutations. J Clin Oncol. 2008;26(34):5530–6.
79. Yang D, Khan S, Sun Y, Hess K, Shmulevich I, Sood AK, et al. Association of BRCA1 and BRCA2 mutations with survival, chemotherapy sensitivity, and gene mutator phenotype in patients with ovarian cancer. JAMA. 2011;306(14):1557–65. Epub 2011/10/13.
80. Bolton KL, Chenevix-Trench G, Goh C, Sadetzki S, Ramus SJ, Karlan BY, et al. Association between BRCA1 and BRCA2 mutations and survival in women with invasive epithelial ovarian cancer. JAMA. 2012;307(4):382–90. Epub 2012/01/26.
81. Meindl A, Hellebrand H, Wiek C, Erven V, Wappenschmidt B, Niederacher D, et al. Germline mutations in breast and ovarian cancer pedigrees establish RAD51C as a human cancer susceptibility gene. Nat Genet. 2010;42(5):410–4. Epub 2010/04/20.
82. Pelttari LM, Heikkinen T, Thompson D, Kallioniemi A, Schleutker J, Holli K, et al. RAD51C is a susceptibility gene for ovarian cancer. Hum Mol Genet. 2011;20(16):3278–88. Epub 2011/05/28.
83. Loveday C, Turnbull C, Ramsay E, Hughes D, Ruark E, Frankum JR, et al. Germline mutations in RAD51D confer susceptibility to ovarian cancer. Nat Genet. 2011;43(9):879–82. Epub 2011/08/09.
84. Rafnar T, Gudbjartsson DF, Sulem P, Jonasdottir A, Sigurdsson A, Jonasdottir A, et al. Mutations in BRIP1 confer high risk of ovarian cancer. Nat Genet. 2011;43(11):1104–7. Epub 2011/10/04.
85. Mukhopadhyay A, Elattar A, Cerbinskaite A, Wilkinson SJ, Drew Y, Kyle S, et al. Development of a functional assay for homologous recombination status in primary cultures of epithelial ovarian tumor and correlation with sensitivity to poly(ADP-ribose) polymerase inhibitors. Clin Cancer Res. 2010;16(8):2344–51.
86. Norquist B, Wurz KA, Pennil CC, Garcia R, Gross J, Sakai W, et al. Secondary somatic mutations restoring BRCA1/2 predict chemotherapy

resistance in hereditary ovarian carcinomas. J Clin Oncol. 2011;29(22):3008–15. Epub 2011/06/29.
87. Sakai W, Swisher EM, Jacquemont C, Chandramohan KV, Couch FJ, Langdon SP, et al. Functional restoration of BRCA2 protein by secondary BRCA2 mutations in BRCA2-mutated ovarian carcinoma. Cancer Res. 2009;69(16):6381–6. Epub 2009/08/06.
88. Swisher EM, Sakai W, Karlan BY, Wurz K, Urban N, Taniguchi T. Secondary BRCA1 mutations in BRCA1-mutated ovarian carcinomas with platinum resistance. Cancer Res. 2008;68(8):2581–6.
89. Sakai W, Swisher EM, Karlan BY, Agarwal MK, Higgins J, Friedman C, et al. Secondary mutations as a mechanism of cisplatin resistance in BRCA2-mutated cancers. Nature. 2008;451(7182):1116–20.
90. Edwards SL, Brough R, Lord CJ, Natrajan R, Vatcheva R, Levine DA, et al. Resistance to therapy caused by intragenic deletion in BRCA2. Nature. 2008;451(7182):1111–5. Epub 2008/02/12.
91. Cooke SL, Temple J, Macarthur S, Zahra MA, Tan LT, Crawford RAF, et al. Intra-tumour genetic heterogeneity and poor chemoradiotherapy response in cervical cancer. Br J Cancer. 2011;104(2):361–8.
92. Cooke SL, Ng CKY, Melnyk N, Garcia MJ, Hardcastle T, Temple J, et al. Genomic analysis of genetic heterogeneity and evolution in high-grade serous ovarian carcinoma. Oncogene. 2010;29(35):4905–13.
93. Nowell PC. The clonal evolution of tumor cell populations. Science. 1976;194(4260):23–8.
94. Dexter DL, Kowalski HM, Blazar BA, Fligiel Z, Vogel R, Heppner GH. Heterogeneity of tumor cells from a single mouse mammary tumor. Cancer Res. 1978;38(10):3174–81.
95. Khalique L, Ayhan A, Weale ME, Jacobs IJ, Ramus SJ, Gayther SA. Genetic intra-tumour heterogeneity in epithelial ovarian cancer and its implications for molecular diagnosis of tumours. J Pathol. 2007;211(3):286–95.
96. Khalique L, Ayhan A, Whittaker JC, Singh N, Jacobs IJ, Gayther SA, et al. The clonal evolution of metastases from primary serous epithelial ovarian cancers. Int J Cancer. 2009;124(7):1579–86.
97. Navin N, Kendall J, Troge J, Andrews P, Rodgers L, McIndoo J, et al. Tumour evolution inferred by single-cell sequencing. Nature. 2011;472(7341):90–4. Epub 2011/03/15.
98. Navin N, Krasnitz A, Rodgers L, Cook K, Meth J, Kendall J, et al. Inferring tumor progression from genomic heterogeneity. Genome Res. 2010;20(1):68–80.
99. Shah SP, Morin RD, Khattra J, Prentice L, Pugh T, Burleigh A, et al. Mutational evolution in a lobular breast tumour profiled at single nucleotide resolution. Nature. 2009;461(7265):809–13.
100. Wu X, Northcott PA, Dubuc A, Dupuy AJ, Shih DJ, Witt H, et al. Clonal selection drives genetic divergence of metastatic medulloblastoma. Nature. 2012;482(7386):529–33. Epub 2012/02/22.
101. Vermaat JS, Nijman IJ, Koudijs MJ, Gerritse FL, Scherer SJ, Mokry M, et al. Primary colorectal cancers and their subsequent hepatic metastases are genetically different: implications for selection of patients for targeted treatment. Clin Cancer Res. 2012;18(3):688–99. Epub 2011/12/17.
102. Carter SL, Cibulskis K, Helman E, McKenna A, Shen H, Zack T, et al. Absolute quantification of somatic DNA alterations in human cancer. Nat Biotechnol. 2012;30(5):413–21. Epub 2012/05/01.
103. Shah SP, Roth A, Goya R, Oloumi A, Ha G, Zhao Y, et al. The clonal and mutational evolution spectrum of primary triple-negative breast cancers. Nature. 2012;486(7403):395–9. Epub 2012/04/13.
104. Campbell PJ, Yachida S, Mudie LJ, Stephens PJ, Pleasance ED, Stebbings LA, et al. The patterns and dynamics of genomic instability in metastatic pancreatic cancer. Nature. 2010;467(7319):1109–13. Epub 2010/10/29.
105. Choi S, Henderson MJ, Kwan E, Beesley AH, Sutton R, Bahar AY, et al. Relapse in children with acute lymphoblastic leukemia involving selection of a preexisting drug-resistant subclone. Blood. 2007;110(2):632–9.
106. Shah NP, Nicoll JM, Nagar B, Gorre ME, Paquette RL, Kuriyan J, et al. Multiple BCR-ABL kinase domain mutations confer polyclonal resistance to the tyrosine kinase inhibitor imatinib (STI571) in chronic phase and blast crisis chronic myeloid leukemia. Cancer Cell. 2002;2(2):117–25.
107. Roche-Lestienne C, Soenen-Cornu V, Grardel-Duflos N, Laï JL, Grardel-Duflos N, Philippe N, Facon T, Fenaux P, Preudhomme C. Several types of mutations of the Abl gene can be found in chronic myeloid leukemia patients resistant to STI571, and they can pre-exist to the onset of treatment. Blood. 2002;100(3):1014–8.
108. Gerlinger M, Rowan AJ, Horswell S, Larkin J, Endesfelder D, Gronroos E, et al. Intratumor heterogeneity and branched evolution revealed by multiregion sequencing. N Engl J Med. 2012;366(10):883–92. Epub 2012/03/09.
109. Sordella R, Bell DW, Haber DA, Settleman J. Gefitinib-sensitizing EGFR mutations in lung cancer activate anti-apoptotic pathways. Science. 2004;305(5687):1163–7. Epub 2004/07/31.

110. Shipitsin M, Campbell LL, Argani P, Weremowicz S, Bloushtain-Qimron N, Yao J, et al. Molecular definition of breast tumor heterogeneity. Cancer Cell. 2007;11(3):259–73.
111. Mullighan CG, Phillips LA, Su X, Ma J, Miller CB, Shurtleff SA, et al. Genomic analysis of the clonal origins of relapsed acute lymphoblastic leukemia. Science. 2008;322(5906):1377–80.
112. Micci F, Haugom L, Ahlquist T, Abeler VM, Trope CG, Lothe RA, et al. Tumor spreading to the contralateral ovary in bilateral ovarian carcinoma is a late event in clonal evolution. J Oncol. 2010;2010:646340.
113. McAlpine JN, Eisenkop SM, Spirtos NM. Tumor heterogeneity in ovarian cancer as demonstrated by in vitro chemoresistance assays. Gynecol Oncol. 2008;110(3):360–4.
114. Griffin N, Grant LA, Freeman SJ, Jimenez-Linan M, Berman LH, Earl H, et al. Image-guided biopsy in patients with suspected ovarian carcinoma: a safe and effective technique? Eur Radiol. 2009;19(1):230–5. Epub 2008/08/16.
115. Spencer JA, Weston MJ, Saidi SA, Wilkinson N, Hall GD. Clinical utility of image-guided peritoneal and omental biopsy. Nat Rev Clin Oncol. 2010;7(11):623–31.
116. Leon SA, Shapiro B, Sklaroff DM, Yaros MJ. Free DNA in the serum of cancer patients and the effect of therapy. Cancer Res. 1977;37(3): 646–50. Epub 1977/03/01.
117. Kamat AA, Baldwin M, Urbauer D, Dang D, Han LY, Godwin A, et al. Plasma cell-free DNA in ovarian cancer: an independent prognostic biomarker. Cancer. 2010;116(8):1918–25. Epub 2010/02/19.
118. Diehl F, Schmidt K, Choti MA, Romans K, Goodman S, Li M, et al. Circulating mutant DNA to assess tumor dynamics. Nat Med. 2008;14(9):985–90. Epub 2008/08/02.
119. Leary RJ, Kinde I, Diehl F, Schmidt K, Clouser C, Duncan C, et al. Development of personalized tumor biomarkers using massively parallel sequencing. Sci Transl Med. 2010;2(20):20. Epub 2010/04/08.
120. Yung TK, Chan KC, Mok TS, Tong J, To KF, Lo YM. Single-molecule detection of epidermal growth factor receptor mutations in plasma by microfluidics digital PCR in non-small cell lung cancer patients. Clin Cancer Res. 2009;15(6): 2076–84. Epub 2009/03/12.
121. Chen Z, Feng J, Buzin CH, Liu Q, Weiss L, Kernstine K, et al. Analysis of cancer mutation signatures in blood by a novel ultra-sensitive assay: monitoring of therapy or recurrence in non-metastatic breast cancer. PLoS One. 2009;4(9):e7220. Epub 2009/10/01.
122. Poveda A, Kaye SB, McCormack R, Wang S, Parekh T, Ricci D, et al. Circulating tumor cells predict progression free survival and overall survival in patients with relapsed/recurrent advanced ovarian cancer. Gynecol Oncol. 2011;122(3):567–72. Epub 2011/06/15.
123. Forshew T, Murtaza M, Parkinson C, Gale D, Tsui DW, Kaper F, et al. Noninvasive identification and monitoring of cancer mutations by targeted deep sequencing of plasma DNA. Sci Transl Med. 2012;4(136):136ra68. Epub 2012/06/01.
124. Misale S, Yaeger R, Hobor S, Scala E, Janakiraman M, Liska D, et al. Emergence of KRAS mutations and acquired resistance to anti-EGFR therapy in colorectal cancer. Nature. 2012;486(7404):532–6. Epub 2012/06/23.
125. Diaz Jr LA, Williams RT, Wu J, Kinde I, Hecht JR, Berlin J, et al. The molecular evolution of acquired resistance to targeted EGFR blockade in colorectal cancers. Nature. 2012; 486(7404):537–40. Epub 2012/06/23.

CHAPTER 29

Genomic Applications in Soft Tissue Sarcomas

EVA WARDELMANN, WOLFGANG HARTMANN

Introduction 503
Adipocytic Tumors 504
Fibroblastic/Myofibroblastic Tumors 507
Fibrohistiocytic Tumors 512
Skeletal Muscle Tumors 513
Vascular Tumors 514
Tumors of Uncertain Differentiation 516
Conclusions 523

Introduction

Sarcomas account for approximately 1 % of all malignancies. The 2013 WHO classification [49] recognizes over 70 types of soft tissue neoplasms. During the last few years, genomic alterations which are of diagnostic, prognostic, and/or predictive value have been detected in sarcomas. As a result, the development of molecular methods to subclassify sarcoma subtypes became warranted because molecular signatures may pinpoint potential areas of interest for diagnostic tools, prediction of clinical outcomes, and potential response to therapeutic targets.

In general, sarcomas can be subdivided into two different morphological subgroups: tumors with a non-pleomorphic morphology and those with a pleomorphic phenotype. The non-pleomorphic sarcomas more often carry specific molecular aberrations whereas pleomorphic sarcomas frequently have a complex karyotype. Sarcomas with such complex karyotypes account for approximately 50 % of all soft tissue sarcomas.

Three major types of genomic alterations occur in sarcomas: reciprocal translocations (~15 %), specific mutations (~25 %), and amplifications (~10 %). The chromosomal translocations lead to the formation of chimeric fusion genes the protein products of which function either as transcription factors, autocrine growth factors, or tyrosine kinases. Specific mutations are found preferentially in genes encoding tyrosine kinases. Amplifications mainly affect genes which encode important players in cell cycle control.

This chapter presents an overview of the genomic applications that currently play an increasingly important role in the diagnostic and treatment decision algorithms of soft tissue sarcomas. Soft tissue tumors are classified in Table 29.1. Table 29.2 depicts the various genomic aberrations that have been well recognized in sarcoma subtypes. Subtypes for which such conclusive molecular data are still missing (e.g., smooth muscle, pericytic, chondro-osseous, and nerve sheath tumors) are not listed. The vast majority of benign soft tissue tumors are also excluded whereas some neoplasms with intermediate biologic behavior are included.

G.J. Netto and I. Schrijver (eds.), *Genomic Applications in Pathology*,
DOI 10.1007/978-1-4939-0727-4_29, © Springer Science+Business Media New York 2015

Table 29-1 Chapter Content with Soft Tissue Tumor Classification

Adipocytic tumors
Atypical lipomatous tumor/well-differentiated liposarcoma; dedifferentiated liposarcoma
Myxoid liposarcoma
Pleomorphic liposarcoma
Fibroblastic/myofibroblastic tumors
Desmoid-type fibromatosis
Giant cell fibroblastoma
Dermatofibrosarcoma protuberans
Solitary fibrous tumor
Inflammatory myofibroblastic tumor
Infantile fibrosarcoma
Low-grade fibromyxoid sarcoma
Sclerosing epithelioid fibrosarcoma
So-called fibrohistiocytic tumors
Tenosynovial giant cell tumor, localized/diffuse type
Skeletal-muscle tumors
Embryonal rhabdomyosarcoma
Alveolar rhabdomyosarcoma
Vascular tumors
Haemangioendothelioma
Angiosarcoma
Tumors of uncertain differentiation
Synovial sarcoma
Epithelioid sarcoma
Alveolar soft part sarcoma
Clear cell sarcoma of soft tissue
Extraskeletal myxoid chondrosarcoma
Desmoplastic small round cell tumor
Extrarenal rhabdoid tumor
Gastrointestinal stromal tumors (GIST)

Adipocytic Tumors

Atypical Lipomatous Tumor/ Well-Differentiated Liposarcoma and Dedifferentiated Liposarcoma

CLINICOPATHOLOGICAL FEATURES

Atypical lipomatous tumors (ALT)/well-differentiated liposarcomas (WDLS) constitute approximately 40–45 % of all liposarcomas. They arise mainly in middle-aged adults [23, 42] and occur predominantly in the deep soft tissues of the limbs followed in frequency by the retroperitoneum, the paratesticular region, and the mediastinum [23]. ALT/WDLS are locally aggressive but do not carry a potential for metastasis. The distinction between ALT and WDLS is clinical where lesions arising in surgically accessible sites are referred to as ALT and those arising in deeper surgically less amenable sites (and therefore enduring more frequent local relapses) are termed WDLS. Dedifferentiated liposarcomas (DDLS) arise in the same group of patients at comparable sites with a significant predominance in the retroperitoneum. Approximately 90 % of DDLS arise de novo, whereas 10 % develop in recurrences of ALT/WDLS [23]. DDLS develop metastases in up to 20 % of the cases, thus contributing in part to the aggressive behavior of ALT/WDLS following recurrence [68, 107]. Consistent with their genetic hallmark (see below), most ALT/WDLS and DDLS immunohistochemically show positive nuclear staining with antibodies against MDM2 and CDK4 [23, 158].

GENOMIC ALTERATIONS

ALT/WDLS and DDLS are genetically characterized by the presence of supernumerary rings and giant marker chromosomes containing amplified sequences originating from the chromosomal region 12q14–15. The amplicon displays considerable heterogeneity, containing numerous oncogenes [79, 80]. *MDM2* is consistently amplified, acting as an antagonist to p53 by targeting the protein for degradation via its ubiquitin ligase function and through inhibition of its transcriptional activation function [23]. Almost 90 % of tumors display co-amplification of *CDK4*, leading to cell cycle progression via RB phosphorylation. Tumors with *MDM2* amplification lacking co-amplification of *CDK4* have been shown to be associated with a more mature histological phenotype and a better prognosis [78]. DDLS have been reported to be genomically more complex than ALT/WDLS [170].

PROGNOSIS AND TREATMENT

Completely excised ALT/WDLS arising in surgically amenable sites only rarely recur

Table 29-2 Genomic Aberrations in Soft Tissue Tumors

Tumor entity	Genomic aberration	Fusion gene, mutated gene
Alveolar rhabdomyosarcoma (ARMS)	t(2;13)(q35;q14) t(1;13)(p36;q14) t(2;2)(p23;q36) t(X;2)(q13;q36)	*PAX3-FOXO1A* *PAX7-FOXO1A* *PAX3-NCOA1* *PAX3-FOXO4*
Alveolar soft part sarcoma (ASPS)	t(X;17)(p11;q25)	*ASPSCR1-TFE3*
Angiomatoid fibrous histiocytoma (AFH)	t(12;16)(q13;p11) t(2;22)(q33;q12) t(12;22)(q13;q12)	*TLS-ATF1* *EWS1R-CREB1* *EWS1R-ATF 1*
Angiosarcoma (ASA)	Missense mutation Amplification	*KDR, FLT4* *c-MYC*
Clear cell sarcoma (CCS)	t(12;22)(q13;q12) t(2;22)(q33;q12)	*EWS1R-ATF1* *EWS1R-CREB1*
Congenital fibrosarcoma (CGFS)	t(12;15)(p13;q25)	*ETV6-NTRK3*
Dermatofibrosarcoma protuberans (DFSP)	t(17;22)(q22;q13) der(22)t(17;22) Ring chromosome	*COL1A1-PDGFB*
Desmoplastic small round cell tumor (DSRCT)	t(11;22)(p13;q12)	*EWS1R-WT1*
Endometrial stromal sarcoma (ESS)	t(7;17)(p15;q21) t(10;17)(q22;p13)	*JAZF1-JJAZ1* *YWHAE-FAM22A/B*
Epithelioid hemangioendothelioma (EHE)	t(1;3(p36.3;q25) t(X;11)	*WWTR1-CAMTA1* *YAP1-TFE3*
Epithelioid sarcoma (EWS)	Intragenic deletions	*SMARCB1/INI1*
Ewing sarcoma (ES)	t(11;22)(q24;q12) t(21;22)(q22;q12) t(7;22)(p22;q12) t(17;22)(q12;q12) t(2;22)(q33;q12) t(16;21)(p11;q22	*EWS1R-FLI1* *EWS1R-ERG* *EWS1R-ETV1* *EWS1R-E1AF* *EWS1R-FEV* *FUS-ERG*
Extrarenal rhabdoid tumor (ERT)	Homozygous inactivation by deletion	*hSNF/INI1/SMARCB1/BAF47* or *SMARCA4 (BRG1)* loss
Extraskelettal myxoid chondrosarcoma (EMCS)	t(9;22)(q22;q12) t(9;17)(q22;q11) t(3;9)(q12;q22) t(9;17)(q22;q11)	*EWS1R-CHN* *TAF2N-CHN* *TFG-NR4A3* *TCF12-NR4A3*
Fibromatosis (desmoid type)	CTNNB1 mutations, APC mutations	missense mutations
Gastrointestinal stromal tumor (GIST)	mutations	*KIT, PDGFRA, SDH, NF1, BRAF* or other genes
Inflammatory myofibroblastic tumor (IMFT)	t(2p23)	div. *ALK* fusion partners
Low-grade fibromyxoid sarcoma (LGFS) Sclerosing epithelioid fibrosarcoma (SEF)	t(7;16)(q33–34;p11) t(11;16)(p11;p11)	*FUS-CREB3L2* *FUS-CREB3L1* *EWS1R-CREB3L1*

(continued)

Table 29-2 **(continued)**

Tumor entity	Genomic aberration	Fusion gene, mutated gene
Myxoinflammatory fibroblastic sarcoma (MIFS)	t(1;10)(p22;q24) Ring chromosome	Deregulation of *FGF8+NPM3* Amplification of *VGLL3*
Solitary fibrous tumor/ hemangiopericytoma (SFT)	der(12)(q13–15)	*NAB2-STAT6*
Synovial sarcoma (SS)	t(X;18)(p11;q11) t(X;18)(p11;q11) t(X;18)(p11;q11) t(X;20)(p11;q13)	*SS18-SSX1* *SS18-SSX2* *SS18-SSX4* *SS181-SSX1*
Tenosynovial giant cell tumor (TGCT)	t(1;2)(p13;q37)	*CSF-COL6A3*
Well-differentiated liposarcoma (WDLS)/atypical lipomatous tumor (ALT)	Ring chromosome/giant marker	Amplification of *MDM2, CDK4, HMGA2, GLI1*

after complete excision, whereas retroperitoneal, mediastinal, or paratesticular lesions have a higher frequency of local recurrence [23]. ALT/WDLS are associated with a variable risk of dedifferentiation that is related to site of origin. The risk is estimated to be less than 5 % in lesions arising in the limbs and higher than 20 % in those arising in the retroperitoneum [23]. DDLS recur locally in more than 40 % of the cases and lead to distant metastases in up to 20 % [68, 107]. Overall mortality is estimated to be 30–40 % at 5 years [23]. Whereas complete surgical excision with wide margins represents the treatment of choice, recent trials have documented a favorable progression-free rate in patients with *CDK4*-amplified WDLS/DDLS upon treatment with a small molecule CDK4/CDK6 inhibitor [37] and efforts to target MDM2, thereby activating the p53 pathway, are also ongoing [142].

Myxoid Liposarcoma

CLINICOPATHOLOGICAL FEATURES

Myxoid liposarcoma (MLS) represents 15–20 % of liposarcomas and arises mainly in the deep soft tissues of the extremities, particularly the thigh [89] during the fourth and fifth decades of life. Thirty to forty percent of the patients develop distant metastases, frequently involving other soft tissue sites. Presence of more than 5 % tumor cells with round cell differentiation has been used to define high histological grade and is associated with an unfavorable outcome [7, 89].

GENOMIC ALTERATIONS

MLS is characterized by reciprocal translocations t(12;16)(q13;p11) that result in *FUS-DDIT3* (*CHOP*) gene fusions, which are present in over 95 % of cases [24, 138]. In rare instances, alternative t(12;22)(q13;q12) translocations are found resulting in *EWS-DDIT3* fusion oncogenes [5, 25, 127]. *FUS* and *EWS* encode RNA-binding proteins involved in transcriptional control; DDIT3 binds C/EBP transcription factors through their highly conserved leucine zipper domain and inhibits their function in adipocytic differentiation. *FUS-DDIT3* functions by inhibiting adipogenesis and maintaining immature adipocytes in a continuous cycle of proliferation without differentiation. There is strong evidence to suggest that these translocations are the primary oncogenic event in MLS [40]. Recent large-scale genomic approaches documented activation of the PIK3/AKT signaling cascade in MLS with mutations in the *PIK3CA* gene found in 18 % of cases in addition to rare inactivating mutations in the *PTEN* tumor suppressor gene. Importantly, patients whose tumors harbored mutations in PIK3CA had a shorter disease-specific survival [10, 35]. Alternative mechanisms leading to PIK3/AKT pathway activation include HGF/MET and RET signaling [118].

PROGNOSIS AND TREATMENT

Increased round cell histologic content, presence of necrosis, and alterations of the p53 tumor suppressor gene are associated with unfavorable outcome. MLS is associated with an overall favorable 5-year disease-free survival (85 %). Variability of the *DDIT3* translocation does not affect prognosis [7].

Complete excision with wide tumor-free margins is the treatment of choice. MLS without round cell differentiation is particularly radiosensitive and patients treated with adjuvant or neoadjuvant radiotherapy achieve 98 % five-year local control. Potential novel treatment agents include Trabectedin, the cytotoxic activity of which is ascribed to binding the minor groove of DNA. Trabectedin efficacy in leiomyosarcomas and liposarcomas is currently under investigation in a phase III trial (NCT01343277) [146].

Pleomorphic Liposarcoma

CLINICOPATHOLOGICAL FEATURES

Pleomorphic liposarcomas (PLS) are rare tumors accounting for 5 % of sarcomas with adipocytic differentiation. PLS affect patients older than 50 years [8]. Most cases arise in the deep soft tissues of the extremities, the lower extremity being involved more frequently. Thirty to fifty percent of patients develop metastases with lung and pleura representing the preferred sites of metastatic spread.

GENOMIC ALTERATIONS

PLS display complex genomic rearrangements with some recurrent losses reported concerning the chromosomal regions 13q14.2, 17q11.2, and 17p13.1, where the *RB*, *NF1*, and *TP53* tumor suppressor genes are located [10, 152, 169].

PROGNOSIS AND TREATMENT

Larger tumor size, central and deep locations, and high mitotic activity are associated with a worse prognosis. Overall, 40–50 % tumor-associated mortality is reported and a 5-year survival rate of 60–65 % is achieved [56, 74]. Treatment modalities include complete excision with wide tumor-free margins, chemotherapy, and radiotherapy.

Fibroblastic/Myofibroblastic Tumors

Desmoid-Type (Deep) Fibromatosis

CLINICOPATHOLOGICAL FEATURES

Fibromatoses are myofibroblastic proliferations with infiltrative growth pattern and high recurrence rate that lack metastatic potential. Cases can occur in the context of familial adenomatous polyposis (FAP) coli syndrome where they more often behave aggressively compared to sporadic desmoids, occasionally leading to death. Their overall incidence is 2–4 new cases per 100,000/year [87]. Immunohistochemically, the spindle cells may express smooth muscle actin but not desmin. They are negative for KIT receptor (CD117), DOG1, and CD34 which is essential to differentiate fibromatosis from gastrointestinal stromal tumors (GIST). The most important diagnostic marker is β-catenin which is typically expressed not only in the cytoplasm but also in the nucleus, the latter observation being crucial [171].

GENOMIC ALTERATIONS

At the molecular level, a sporadic mutation in the *CTNNB1* gene which encodes β-catenin is frequently detectable by selective sequencing of exon 3 [75, 76]. Apart from constituting a subunit of the cadherin protein complex β-catenin acts as an intracellular signal transducer in the Wnt signaling pathway. In the case of a mutation in exon 3, β-catenin cannot be degraded and is translocated into the nucleus where it accumulates and acts as a transcription factor [171] (Fig. 29.1). Alternatively, germline mutations in the *APC* (adenomatous polyposis coli) gene can lead to a nuclear accumulation of β-catenin [101, 121] and thus to fibromatosis. The risk for patients with FAP to develop a fibromatosis is 2.56/1,000 persons per year and increases with repeated surgical procedures.

PROGNOSIS AND TREATMENT

The prognosis of fibromatosis is not predictable by morphology or molecular markers.

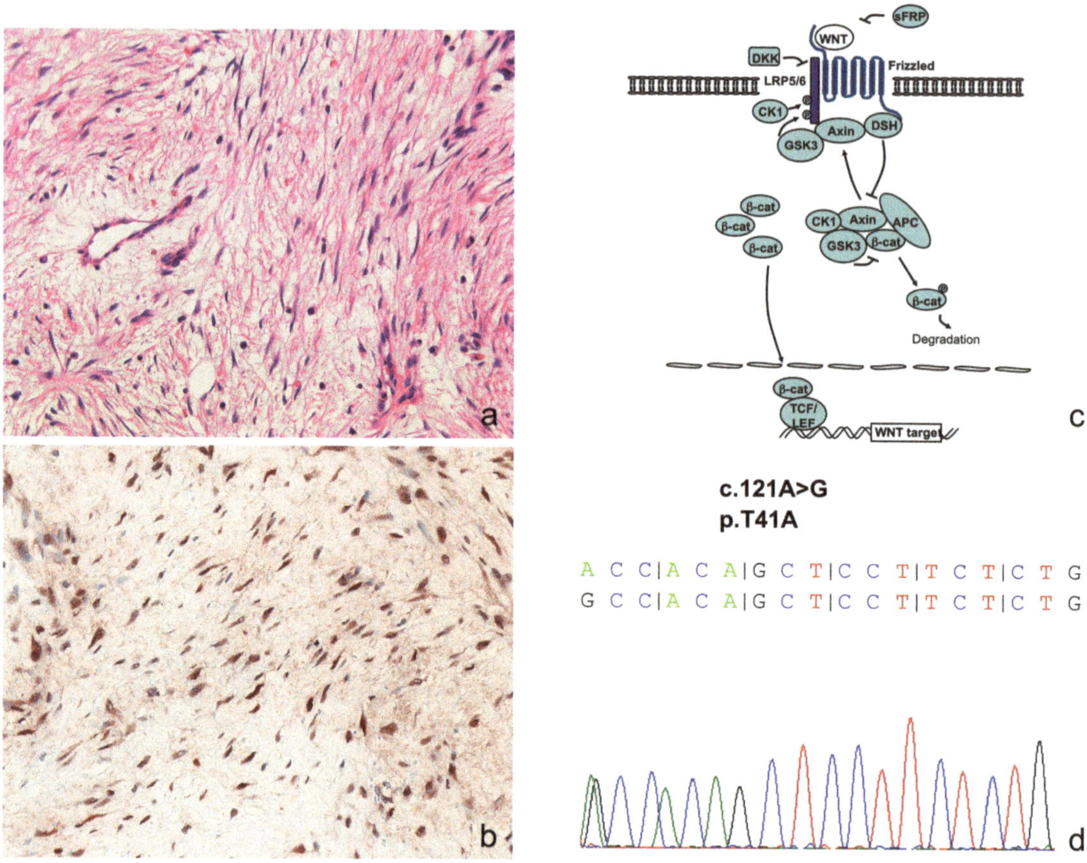

Figure 29-1 Sporadic mutation in the *CTNNB1* gene. (**a**) Characteristic aspect of a case of desmoid-type fibromatosis composed of a relatively monomorphic spindle cell proliferation of variable density displaying. (**b**) Nuclear accumulation of β-catenin detectable by immunohistochemistry. (**c**) Schematic view of the WNT signaling pathway with the central effector β-catenin being subject to degradation in the OFF state of the pathway and nuclear transfer and transcriptional activation in the ON state due to WNT signals, activating mutations of the *CTNNB1* gene or inactivating mutations of components of the degradation complex. (**d**) Heterozygous point mutation affecting codon 41 of the *CTNNB1* gene encoding β-catenin leading to nuclear accumulation (adapted from [77])

As yet, there is no evidence that the molecular subtype influences the outcome [75, 76, 87]. Spontaneous regression is observed in a subgroup of patients whereas occasional tumors may culminate in death following multiple recurrences. Treatment options in fibromatoses include simple surgical resection, chemotherapy, antihormonal treatment, tyrosine kinase inhibition, and/or radiation [87]. Currently, no reliable molecular biomarker has been shown to have a role in guiding treatment strategy.

Giant Cell Fibroblastoma/ Dermatofibrosarcoma Protuberans

CLINICOPATHOLOGICAL FINDINGS

Giant cell fibroblastoma (GCF) is now regarded as the juvenile form of dermatofibrosarcoma protuberans (DFSP) because both dermal sarcomas carry the same translocation [59, 84, 172]. GCF occurs mainly in childhood in the age group below 10 years and

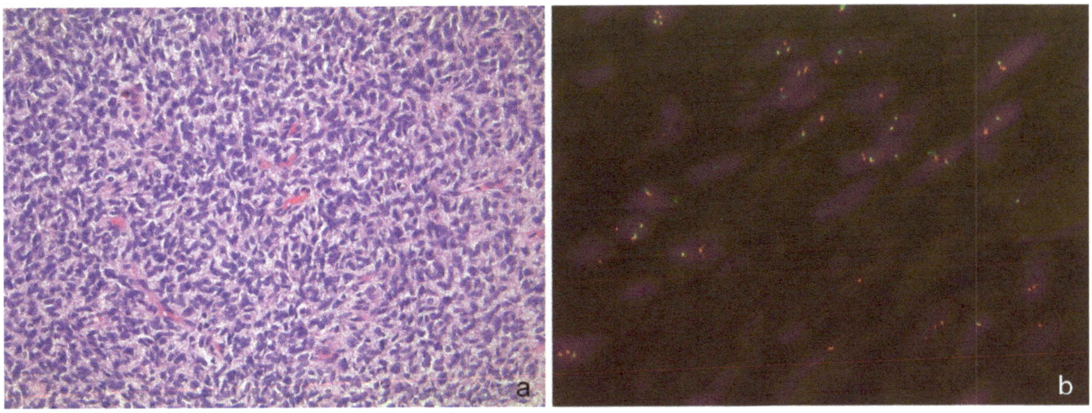

Figure 29-2 Dermatofibrosarcoma protuberans. (**a**) Characteristic aspect of a case of dermatofibrosarcoma protuberans composed of monomorphic spindle cells growing in a storiform pattern. (**b**) Fluorescence in situ hybridization of the tumor using a PDGFB dual color break apart probe showing several tumor cells with one red–green signal indicating a normal PDGFB locus and 1–3 extra copies of red signals representing indicating a break in the PDGFB locus

belongs to the group of fibroblastic tumors with intermediate malignant potential. The majority of patients are male. Typically, these tumors occur in the dermis and subcutis of the trunk or extremities and less often in the head and neck region. DFSP occurs in the third or fourth decades of life with slight male preponderance. It is characterized by its locally infiltrative nature and high tendency to recur (in up to 30 % of cases).

Immunohistochemically, CD34 is the most reliable marker, although not specific. Typically, it is strongly and diffusely positive. In the rare cases of fibrosarcomatous transformation in DFSP, CD34 expression may be lost. GCF and DFSP are negative for factor XIIIa, in contrast to histiocytic markers such as CD68, lysozyme, and CD10.

GENOMIC ALTERATIONS

Both GCF and DFSP are characterized by a specific reciprocal translocation t(17;22)(q22;q13) or more often as a supernumerary ring chromosome involving sequences of both chromosomes 17 and 22. These rearrangements lead to the fusion of the collagen 1A1 gene (*COL1A1*; alpha chain type 1 of collagen gene) and the platelet-derived growth factor B gene (*PDGFB*) (Fig. 29.2). As a result tumor cells produce high amounts of PDGFB leading to the constitutive activation of the PDGFB receptor, a type III receptor tyrosine kinase. Assays used for the analysis of the translocation include multiplex reverse transcription polymerase chain reaction and FISH assays, with COL1A1-PDGFB dual color dual fusion or PDGFB dual color break apart probes showing reliable results [133, 179].

PROGNOSIS AND TREATMENT

Both GCF and DFSP have a high risk for recurrence. Risk for metastasis is limited to those DFSP cases exhibiting fibrosarcomatous transformation. Treatment of advanced, inoperable or recurrent DFSP lesions is now based on targeting tyrosine kinase inhibition through PDGFR inhibition as reviewed in Llombart et al. [102]. The majority of patients show partial or even complete response to imatinib treatment with minimal toxicity [106]. As a result, imatinib is now considered the gold standard for patients with locally advanced or metastatic DFSP [145]. This therapy is also used in the neoadjuvant setting in order to reduce tumor size and to decrease morbidity prior to a surgical excision.

Solitary Fibrous Tumor

CLINICOPATHOLOGICAL FINDINGS

Solitary fibrous tumor (SFT) is a fibroblastic neoplasm characterized by a typical vascular pattern which is designated as hemangiopericytic. SFT was initially described in the pleura but is now recognized to occur in nearly any location. In less than 5 % of patients, SFT manifests with hypoglycemia that is thought to be mediated by insulin-like growth factors (IGFs) produced by the tumor [57]. Hypoglycemia disappears following tumor resection. SFT occurs in adults of all age groups and has equal incidence in both sexes. Rarely, an abrupt transition from SFT to high-grade sarcoma can be seen.

GENOMIC ALTERATIONS

Very recently, a recurrent translocation has been identified in SFT. It is the fusion of two neighboring, partly overlapping genes in chromosome band 12q13: *NAB2* and *STAT6* [21, 114]. *NAB2* is transcribed from telomere to centromere and *STAT6* vice versa. In both genes, several break points have been identified leading to diverse fusion products. Over 90 % of SFT display *NAB2-STAT6* fusions. The detection rate depends on primer combinations and perhaps also on sampling. Whether deregulation of *STAT6* or of *NAB2* is the driving force for tumor development is still under debate. *NAB2-STAT6* fusion results in a chimeric protein in which the carboxy-terminal repressor domain of NAB2 (repressing *EGR1*: early growth response gene 1) is replaced with a highly variable portion of STAT6 which seems to play a pivotal role in the development of SFT. However, functional assays will be needed to fully understand the role of *NAB2* and *STAT6* in the development of SFT. Interestingly, both *NAB2* and *STAT6* expression can be detected by immunohistochemistry in SFT [155]. Most if not all SFT show a strong nuclear NAB2 expression as well as nuclear expression of STAT6. The latter is in contrast to the cytoplasmic expression of STAT6 in nonneoplastic tissue. Thus, nuclear localization of STAT6 is possibly a surrogate immunohistochemical marker for *NAB2-STAT6* fusion [155]. Additional studies to elucidate the sensitivity and specificity of this observation are needed.

PROGNOSIS AND TREATMENT

SFT is yet another soft tissue neoplasm of intermediate biologic behavior with only 15 % of tumors showing aggressive biologic features that include metastatic potential. Prognostic histomorphological parameters of aggressive behavior include tumor size larger than 10 cm, presence of ≥ 4 mitoses/10 HPFs (high power fields) necrosis, and a strong or intermediate p53 expression in more than 5 % of tumor cells [28, 149]. The treatment of choice is complete surgical resection. In advanced SFT, small series of cases have been successfully treated with tyrosine kinase inhibitors such as sunitinib, figitumumab, or pazopanib with partial response obtained in half of the cases [98, 161].

Inflammatory Myofibroblastic Tumor

CLINICOPATHOLOGICAL FEATURES

Inflammatory myofibroblastic tumors (IMT) are soft tissue neoplasms of intermediate biologic behavior which frequently recur but only rarely metastasize. They present preferentially in the omentum and the mesentery. Systemic B-symptoms such as fever, anemia, and weight loss may be present at time of diagnosis. IMT more frequently affect children and young adults but older patients may also develop this neoplasm [32]. Immunohistochemistry shows ALK positivity in more than half of the cases in correlation with the presence of an underlying *ALK* rearrangement. Tumor cells also exhibit smooth muscle actin expression but usually lack other myogenic markers [20].

GENOMIC ALTERATIONS

In up to 60 % of the cases, rearrangements of the anaplastic lymphoma kinase (*ALK*) gene (chr. 2p23) have been identified, leading to aberrant constitutive activation of the ALK kinase. The fusion partners include genes encoding cytoplasmic proteins like *TPM3*, *TPM4*, *CARS*, *CLTC*, *ATIC*, and *SEC31L1*, as well as the *RANBP2* gene coding for a nuclear protein. Fusion with the former group of proteins leads to cytoplasmic ALK expression; rearrangements with *RANBP2* lead to nuclear ALK expression and a characteristic round cell histomorphological phenotype. Immunohistochemical detection of ALK expression reliably predicts ALK rearrangements in IMT, and RT-PCR assays as well as ALK dual color break apart FISH assays may be employed to prove the genomic aberration (reviewed in [58]).

PROGNOSIS AND TREATMENT

IMT are associated with a low risk of aggressive behavior and metastasis. Surgery is the mainstay of treatment. Treatment options in cases with advanced unresectable disease were limited until the detection of *ALK* fusions in IMT. The first successful ALK-directed therapy was performed with the ALK/MET-inhibitor crizotinib [17]. However, as in other tumor entities with genomic *ALK* alterations, the development of secondary resistance has emerged in single cases highlighting the need of more specific and diverse *ALK* inhibitors [174].

Infantile Fibrosarcoma

CLINICOPATHOLOGICAL FEATURES

Infantile fibrosarcomas are low-grade malignant neoplasms with a favorable prognosis that can be present at birth or develop in the first 2 years of life. Synonyms are congenital or juvenile fibrosarcoma and aggressive infantile fibromatosis. Most often, infantile fibrosarcomas occur in the extremities, usually presenting as a rapidly enlarging mass. Rarely, infantile fibrosarcomas involve the trunk or head and neck region. Infantile fibrosarcomas are negative for β-catenin and myogenic markers.

GENOMIC ALTERATIONS

The vast majority of cases carry a recurrent translocation t(12;15)(p13;q26) leading to the fusion of *NTRK3* and *ETV6*, which results in an oncogenic activation of NTRK3 tyrosine kinase [48]. Routine analysis may be performed by FISH employing *ETV6* dual color break apart probes or by RT-PCR. Additionally, trisomies of chromosomes 8, 11, 17, and 20 are characteristic. There is a close genomic relationship to cellular congenital mesoblastic nephroma, since they share the same translocation and are very similar in morphology leading to the consensus that cellular congenital mesoblastic nephroma actually represents infantile fibrosarcoma arising in the kidney [147].

PROGNOSIS AND TREATMENT

Infantile fibrosarcomas have a high recurrence rate but only rarely metastasize. The mortality rate ranges from 5 to 25 %. Single cases with spontaneous regression have been reported.

Complete surgical excision remains the treatment of choice. Adjuvant chemotherapy has been proven effective. To date, there are no molecularly based therapeutic approaches [126].

Low-Grade Fibromyxoid Sarcoma

CLINICOPATHOLOGICAL FEATURES

Low-grade fibromyxoid sarcomas (LGFMS) are deep-seated tumors that are usually located in extremities (especially proximal lower limbs) or trunk. LGFMS occur in adults (typically fourth decade) with a male predominance. Immunohistochemically, LGFMS frequently express CD34, EMA, and claudin-1. A helpful marker for the differential diagnosis is MUC-4 which is typically strongly positive [38]. No expression of myogenic markers, cytokeratins, or S100 protein is expected in LGFMS.

GENOMIC ALTERATIONS

LGFMS typically show a balanced translocation t(7;16)(q32–34;p11) or t(11;16)(p11;p11), leading to the fusion of *FUS* and *CREB3L2* or *CREB3L1*. A FUS dual color break apart FISH assay is suitable for the detection of the aberration. Rare cases have been reported to carry an *EWSR1*-*CREB3L1* translocation; those cases may be detected by an *EWSR1* dual color break apart assay in cases without a translocation of *FUS*. There is no evidence of a correlation between the presence and type of the translocation and clinical outcome or morphologic characteristics [38].

PROGNOSIS AND TREATMENT

Recurrences occur in up to 20 % of deeply located lesions. Metastases occur in about 30 % of cases [46]. In recurrent cases, LGFMS can progress to frank high-grade spindle cell sarcoma.

Surgical excision with wide margins is the treatment of choice. As late occurrence of metastases is frequent, long-term follow-up is recommended [46].

Sclerosing Epithelioid Fibrosarcoma

CLINICOPATHOLOGICAL FEATURES

As implied in the nomenclature, sclerosing epithelioid fibrosarcoma (SEF) is a variant of fibrosarcoma which is characterized, apart from a multinodular growth pattern, by at least focal epithelioid morphology and regions with dense fibrosis. SEF occurs preferentially on limbs, limb girdles, and the trunk and rarely in visceral sites. Often, there is a close connection to periosteum or fascia. Adjacent bone can be involved. SEF can mimic epithelial neoplasms. Furthermore, areas resembling LGFMS or adult-type fibrosarcoma may be encountered in SEF. Immunohistochemistry can help in the differential with metastatic epithelial neoplasms because SEF are negative for cytokeratins. The fact that both SEF and LGFMS strongly express MUC4 in the majority of cases has led to the speculation of whether both lesions belong to a spectrum of one tumor entity. In a study by Doyle et al. [38] all tumors displaying hybrid LGFMS and SEF zones showed a strong MUC4 expression. MUC4 expression was also found in 20 of 29 pure SEF.

GENOMIC ALTERATIONS

Like LGFMS, SEF may harbor translocations leading to the fusion of *FUS* and *CREB3L1* or *CREB3L2* [38, 60]. Furthermore, hybrid tumors exhibiting both phenotypes may carry *EWS1R* and *CREB3L1* rearrangements [180]. Wang et al. [180] found pure SEF to often lack *FUS* rearrangements, especially in the absence of MUC4 expression. Therefore, it appears that different genomic subgroups exist among pure LGFMS, pure SEF, and hybrid tumors. The prognostic and predictive value of the above molecular observations remains to be determined. Detection of *FUS* rearrangement can be a strong aid to achieving the correct diagnosis in SEF. Detection of the translocation may be performed by *FUS* or *EWSR1* dual color break apart probes.

PROGNOSIS AND TREATMENT

SEFs are aggressive sarcomas with a higher than 50 % local recurrence rate. Metastases have been reported in 43–86 % of cases. A 43–75 % range of 5-year survival rates is documented [108].

Complete surgical resection is the treatment of choice. In tumors involving bone, amputation may be required. There is no established role for adjuvant chemotherapy or radiation at this time.

Fibrohistiocytic Tumors

Tenosynovial Giant Cell Tumor

CLINICOPATHOLOGICAL FEATURES

Tenosynovial giant cell tumors (TGCT) of the tendon sheath arise from the synovium of joints, bursae, and tendon sheaths and are subdivided according to their growth pattern into localized and diffuse subtypes. The former is the more frequent subtype and predominantly occurs in the hand, particularly the fingers. Diffuse TGCT primarily affect the knee, hip, and foot [154]. Both subtypes of TGCT can occur at any age, with a peak incidence in the fourth decade. Diffuse-type TGCT form larger villous or nodular masses; localized TGCT mostly are well-circumscribed nodules with fibrous septae. Histologically, both tumors are composed of a mixture of stromal cells, macrophages, and osteoclast-like giant cells [176].

GENOMIC ALTERATIONS

A balanced t(1;2)(p13;q37) is present in most tumors. However, the translocation is present in only a minor proportion of lesional cells. Most often, it leads to the fusion of *CSF* (Colony stimulating factor) with *COL6A3* resulting in high level of CSF expression which in turn leads to tumor accumulation of macrophages expressing the CSF receptor [189].

PROGNOSIS AND TREATMENT

Less than one third of localized-type TGCT recur locally and these are usually cured by surgical excision. Diffuse-type TGCT are more likely to display locally aggressive behavior with a recurrence rate of up to 50 %. Few cases of malignant "sarcomatous" diffuse-type TGCT have been described; these tumors often show a significant increase in mitotic activity and have been shown to express increased levels of cyclin A and (wildtype) p53, and to carry chromosomal losses of the region 15q22–24 [12, 115, 120, 141, 160].

Complete surgical excision is the treatment of choice. Adjuvant radiotherapy has been proposed in cases of recurrent diffuse-type TGCT. In surgically inoperable tumors and in the setting of metastatic disease, tyrosine kinase inhibitors (e.g., imatinib) have been considered [15].

Skeletal Muscle Tumors

Embryonal Rhabdomyosarcoma

CLINICOPATHOLOGICAL FEATURES

Comprising 60–70 % of rhabdomyosarcomas, embryonal rhabdomyosarcomas (ERMS) represent the most frequent subtype of malignant soft tissue tumors with skeletal muscle differentiation. ERMS mainly affect children up to 10 years of age with those below 5 years making up about 36 % of patients [124]. The majority of these tumors arise in the head and neck and the genitourinary regions. Recognized subtypes include a botryoid subtype that occurs in visceral organs close to the mucosal surface and an anaplastic subtype, defined by the presence of bizarre atypical cells. Immunohistochemically, rhabdomyoblasts express myogenic markers including desmin, myogenin, and MyoD1 [116].

GENOMIC ALTERATIONS

ERMS frequently show numerical chromosomal aberrations. The imprinted chromosomal region 11p15.5 that harbors several growth-related genes including the *IGF2* (insulin-like growth factor 2) and *p57KIP2* genes is affected by preferential maternal allelic losses in most cases of ERMS [156, 192]. Deletion of *CDKN2A/B*, a key regulator of the p53 and Rb pathways, is found in the majority of ERMS tumors whereas inactivating mutations in *TP53* occur in approximately 30 % of the tumors [135, 167]. *NF1* deletions leading to activation of RAS signaling occur in 15 %. Alternatively, activating *RAS* mutations may be present. These are seen in an additional 40 % of cases [135]. Activation of the *FGFR4* tyrosine kinase by amplification of mutant alleles has also been observed in 20 % of ERMS [135]. Gene expression profiles indicating activation of the Hedgehog pathway, partially associated with *GLI1* amplification, have been reported to confer a poor prognosis in ERMS as well as translocation-negative alveolar rhabdomyosarcoma (ARMS) [196]. Recently, copy number gains and mutations in the *ALK* kinase have been found in ERMS [177].

PROGNOSIS AND TREATMENT

Established prognostic factors in rhabdomyosarcomas include patient age, histological classification, stage, and site of origin. Five-year survival in patients with conventional type ERMS is 66 % [13]. Lower patient age, embryonal (versus alveolar) histologic type, and the botryoid variant are associated with improved outcome. On the other hand, the presence of histologically anaplastic features in ERMS and involvement of the extremities or parameningeal sites are linked with a worse prognosis [85, 139]. Recently, *ALK* copy number gains were reported to be associated with metastatic disease and poor survival [177].

Treatment is usually multimodal and "risk-adapted." It includes surgery, chemotherapy, and usually radiotherapy. Molecularly targeted approaches are currently under investigation, including substances directed against the IGF-IR, mTOR, and VEGF/PDGF [65].

Alveolar Rhabdomyosarcoma

CLINICOPATHOLOGICAL FEATURES

Alveolar rhabdomyosarcomas (ARMS) account for 20–30 % of rhabdomyosarcomas. ARMS more frequently affect adolescents and younger adults and most commonly arise in the extremities followed by the head and neck and trunk regions [63, 131]. The term "alveolar" refers to their typical composition of monomorphous round cells situated in small nests, which are separated by fibrovascular septae. A solid variant also exists. Rhabdomyoblasts are less frequently encountered in ARMS compared to ERMS. Immunohistochemically, desmin, myogenin, and MyoD1 expression serve as markers of ARMS skeletal muscle differentiation [157].

GENOMIC ALTERATIONS

Approximately 75 % of ARMS are characterized by the presence of a reciprocal translocation involving the *FOXO1* gene and a partner member of the PAX gene family of transcription factors. t(2;13)(q35;q14) translocation occurs in 60 % of tumors and leads to the juxtaposition of the *PAX3* and the *FOXO1* genes, whereas t(1;13)(p36;q14) is present in an additional 10–15 % of ARMS, linking the *PAX7* and *FOXO1* genes [9, 27, 55]. For routine diagnostic purposes a *FKHR (FOXO)* dual color break apart FISH assay is well established, RT-PCR is comparable in terms of sensitivity but allows the detection of the translocation partner. *PAX3* and *PAX7* are members of the paired box transcription factor family and both are involved in skeletal muscle development. *FOXO1* represents a member of the forkhead transcription factor family. The resulting chimeric proteins activate transcription at *PAX3* and *PAX7* binding sites, respectively, but are 10- to 100-fold more potent than wild-type *PAX3* and *PAX7* [192]. While *PAX3-FOXO1* expression is driven by a transcriptional mechanism, *PAX7-FOXO1* gene expression is enhanced by an amplification of the fusion gene [26]. The oncogenic nature of the *PAX-FOXO1* fusion gene has been documented in animal studies [88]. Fusion-positive ARMS frequently carry further genomic amplifications [187] including co-amplification of the *MYCN* gene on chromosome 2p24 as well as a circumscribed region on chromosome 2q13–14 that includes the *CDK4* gene. Finally, *ALK* copy number alterations have been shown to be associated with strong *ALK* expression and the presence of metastatic disease at the time of diagnosis.

PROGNOSIS AND TREATMENT

ARMS are generally more aggressive than ERMS with a 5-year survival of 53 %, compared to the 66 % average cited in ERMS [140]. Importantly, presence of *PAX3-FOXO1* appears to be associated with a worse outcome [65]. Fusion-negative ARMS have a similar prognosis to ERMS. Like ERMS, treatment is risk-adapted and includes surgery, chemotherapy, and usually radiotherapy. Molecularly targeted approaches are also being pursued. The use of CDK4 inhibitors has been proposed in cases with *CDK4* amplification. Two phase I clinical trials are underway evaluating different CDK4 inhibitors [54].

Vascular Tumors

Haemangioendothelioma

Hemangioendotheliomas are vascular neoplasms occupying a spectrum of biological potential ranging from tumors with intermediate to aggressive malignant potential. Kaposiform, retiform, and composite hemangioendotheliomas are among the vascular tumors of intermediate biology whereas epithelioid hemangioendothelioma (EHE) is a malignant tumor that carries metastatic potential and will be discussed below.

CLINICOPATHOLOGICAL FEATURES

EHE is a rare tumor that usually occurs in the superficial or deep soft tissue, preferentially in the extremities and head and neck region. Visceral organs can be affected. In a subgroup of patients, EHE is characterized by multicentric growth. In these cases, it has been demonstrated that the different tumor foci are monoclonal in nature [43] and therefore represent metastatic implants of the same neoplastic clone rather than synchronous neoplasms. EHE occurs in all age groups with no gender preference. Immunohistochemically EHE tumor cells are positive for vascular markers including CD34, CD31, Fli-1, and ERG.

GENOMIC ALTERATIONS

A recurrent translocation t(1;3)(p36.3;q25) was initially described, in two cases, by

Mendlick et al. [109]. The diagnostic relevance of this translocation was subsequently confirmed [44, 191] and it was shown that this translocation, leading to *WWTR1-CAMTA1* gene fusion, is present in a high percentage of EHE of different anatomic sites but absent in benign epithelioid hemangiomas and epithelioid angiosarcomas that may enter the differential diagnosis of EHE. CAMTA1 belongs to the calmodulin-binding transcription activator family of proteins and is thought to be involved in cell cycle regulation. WWTR1 is a transcriptional co-activator with a PDZ-binding motif but without known DNA-binding domain. Multiple interaction partners of WWTR1 have been identified, and WWTR1 is a downstream effector of the Hippo pathway. Whether different subtypes of this translocation result in diverse biological behavior in EHE is yet to be determined [44]. Recently, Antonescu et al. described a subgroup of EHE that lacks *WWTR1-CAMTA1* translocation but displays nuclear expression of TFE3 due to an underlying *TFE3* rearrangement [6].

PROGNOSIS AND TREATMENT

The metastatic rate of EHE is 20–30 % leading to 10–20 % mortality. Adverse prognostic factors include the presence of high mitotic rate (>3/50 HPF) and tumor size larger than 3 cm.

The treatment of choice is complete surgical excision with wide margins.

Angiosarcoma

CLINICOPATHOLOGICAL FEATURES

Angiosarcomas (ASA) are rare tumors that represent only 1 % of all sarcomas. Males are more frequently affected. Although ASA can occur in any age group, they are far more frequently found in the elderly and are extremely rare in children. Frequent ASA locations include the soft tissue of extremities (lower more often than upper) and the trunk. A subset of ASA are "secondary" neoplasms that develop following radiotherapy (especially in the setting of adjuvant radiation for breast cancer) [110], chronic lymphedema, or due to exposure to carcinogenic agents such as thorotrast or vinylchloride (visceral ASA). Secondary ASA may also arise in a background of a preexisting tumor such as schwannoma, *NF1* associated malignant peripheral nerve sheath tumors, dedifferentiated liposarcoma, or germ cell tumors (for review see [136]).

Immunohistochemically, ASA strongly express endothelial markers such as CD34, CD31, Fli-1, and ERG. The potential for lack of expression of one of these markers in a given ASA exists; thus the use of a marker panel approach is favored. It is also important to remember that CD34 is not endothelial lineage specific and can be expressed in other types of sarcoma. Smooth muscle actin can be used to establish the absence of myopericytes in ASA, a feature of potential diagnostic value in difficult cases [110].

GENOMIC ALTERATIONS

Approximately 10 % of primary as well as secondary ASA reveal *KDR* (*VEGFR2*, *FLK-1*) mutations. KDR (kinase insert domain receptor) is a type III receptor tyrosine kinase, the encoding gene of which is located on chromosome 4q11–12. In secondary ASA (following radiotherapy or chronic lymphedema), a high level amplification of *MYC* is detected in the vast majority of cases [62, 103], leading to upregulation of the miRNA cluster 17–92 (13q31.3) [81]. The miR-cluster 17–92 is responsible for the pro-angiogenetic effect of *MYC* amplification by downregulating thrombospondin and connective tissue growth factor [36]. It is suggested that the detection of *MYC* amplification can help differentiate well-differentiated ASA from "atypical vascular lesions" which may also occur in association with prior radiotherapy. Very rare cases of primary ASA and non-ASA soft tissue sarcomas have also been shown to harbor *MYC* amplification.

FLT4 amplication is yet another genomic alteration and is encountered in 25 % of secondary ASA. Intriguingly, *FLT4* amplication can co-occur with *MYC* amplification but not with *KDR* mutations.

A rare subset of ASA develops in association with genetic syndromes such as Klippel-Trenaunay syndrome or Mafucci syndrome. The underlying genetic aberration in this setting remains unknown.

PROGNOSIS AND TREATMENT

ASA are very aggressive malignant neoplasms with poor prognosis independent of tumor grade. A dismal 5-year survival rate of 20–30 % is expected.

Radical surgical resection with wide tumor-free margins is the first choice of treatment. Adjuvant chemotherapy has been utilized. More recently, the role of targeted therapy with inhibitors of angiogenesis has been explored. In vitro studies have demonstrated that ASA harboring *KDR* mutations may respond to KDR inhibitors such as sorafenib or sunitinib. In secondary ASA with *FLT4* amplification, the role of tyrosine kinase inhibitors could be promising and should be further evaluated. Another promising approach is the pharmacologic blockade of Angiopoeitin 2 with the peptibody AMG386 (Amgen, Thousand Oaks, CA) which is currently under investigation in a phase I trial [54].

Tumors of Uncertain Differentiation

Synovial Sarcoma

CLINICOPATHOLOGICAL FEATURES

Synovial sarcomas (SS) represent 5–10 % of all malignant soft tissue tumors. They occur more frequently in adolescents and young adults and most commonly arise in the deep soft tissue of the lower and upper extremities followed by the trunk and the head and neck regions. Monophasic and biphasic subtypes of SS are recognized. The latter displays epithelial differentiation in addition to the uniform spindle mesenchymal component of monophasic SS. Immunohistochemically, the majority of SS, regardless of subtype, at least focally express epithelial markers such as EMA and keratins as well as CD99 and TLE1 [112, 125, 173].

GENOMIC ALTERATIONS

At the molecular level, SS are characterized by the presence of a reciprocal translocation t(X;18)(p11;q11), linking the *SS18* (*SYT*) gene and the *SSX1*, *SSX2*, or *SSX4* gene, in order of frequency. The *SYT-SSX1* fusion occurs in approximately two thirds of SS, whereas the *SYT-SSX2* fusion is found in almost one third of the cases. Detection of the translocation is well established by RT-PCR and *SS18 (SYT)* dual color break apart FISH assays, with the latter showing a higher sensitivity [166]. Neither *SS18* nor the *SSX* genes contain a DNA-binding domain [93]; hence the *SS18-SSX* chimeric protein exerts its oncogenic function as a part of a multiprotein complex, in which it associates with the transcription factor ATF2 and the repressor TLE1. The multiprotein complex acts by repressing the transcription of ATF2 target genes [164]. Recently, *SS18-SSX* fusion proteins have also been shown to disrupt the repressive action of SWI/SNF complexes on SOX2 expression. The latter protein expression is crucial for the proliferation control of SS cells [86].

Expression of the insulin-like growth factor receptor 1 (*IGF1R*) has been shown to be associated with an aggressive SS phenotype [193]. IGF1R has been proposed as a therapeutic target in SS [52].

PROGNOSIS AND TREATMENT

Prognostic factors include tumor stage, tumor size, and tumor grade. Ten-year disease-specific survival rates of 75 % are obtained in children and adolescents and 52 % in adults [165]. Presence of *SS18-SSX2* fusion appears to be associated with a more favorable prognosis and a lower rate of metastatic disease at diagnosis [94]. Treatment is multimodal and includes surgery, chemotherapy, and radiotherapy. Novel treatment strategies using bcl2 inhibitors such as ABT-263/Navitoclax (AbbVie Chicago, IL) have been proposed given the strong expression of bcl2 in SS [54].

Epithelioid Sarcoma

CLINICOPATHOLOGICAL FEATURES

Epithelioid sarcoma (ES) is a rare type of sarcoma, most frequently arising in distal extremities, especially the hand and the forearm and rarely the head and neck, penile, and vulvar regions. A proximal type of ES occurs in the proximal limb girdle; in axial locations such as perineum, pelvis, or mediastinum and on the chest wall. ES can involve subcutaneous tissue (typically presenting as a non-healing ulcer) or deep soft tissue.

Immunohistochemically, ES co-expresses vimentin, cytokeratins, EMA, and CD34 (in 50 % of cases). CD31, ERG, and S100 are not expressed in ES [159]. Unlike rhabdoid tumors, loss of INI1 expression in ES is usually not associated with *INI1* gene mutations and is thought to be due to epigenetic downregulation through promoter methylation. INI1 loss can be of utility in the differential diagnosis between ES and carcinomas [72].

GENOMIC ALTERATIONS

Cytogenetically, ES shows deletions of chromosome 22. In classical ES subtype, t(8;22)(q22;q11) is found, albeit inconsistently. Single case reports have illustrated the presence of a t(10;22) in proximal type ES, and single cases with intragenic *INI1 (SMARCB1)* deletions leading to loss of INI1 expression have been reported [50]. The usual absence of *INI1* mutations may help in distinguishing ES from malignant rhabdoid tumors that consistently show *INI1* genetic alterations.

PROGNOSIS AND TREATMENT

Both classic and proximal ES have a high rate of recurrence and can metastasize. However, the proximal type is associated with a higher mortality rate. Metastasis to regional lymph nodes is encountered in up to one third of cases, an occurrence that is rather unusual for a sarcoma. Hematogenous spread to lung, bones, brain, and secondary soft tissue locations also occurs in ES. Favorable prognostic parameters include young patient age at diagnosis, tumor size below 2 cm, and female sex. Adverse prognostic factors include proximal tumor location, presence of tumor necrosis, vascular invasion, and incomplete surgical excision.

An aggressive surgical approach is usually undertaken in ES due to the expected high recurrence rate. Amputation has to be considered in multinodular tumors of extremities. Targeted treatment approaches have not been pursued.

Alveolar Soft Part Sarcoma

CLINICOPATHOLOGICAL FEATURES

Alveolar soft part sarcoma (ASPS) mainly affects young adults and children. It owes its designation to its alveolar-like clusters of large tumor cell morphology. Whereas the classic alveolar subtype most often occurs in the buttocks and thigh, a solid variant predominates in the tongue and the eye. The majority of ASPS are intramuscular in location. Associated distant metastasis to the lung and brain is present in up to 25 % of cases [51]. Immunohistochemically, ASPS is uniquely negative for vimentin unlike most other sarcoma types. Desmin may be focally positive whereas myogenin expression is lacking. CD34, S100, and keratins are not expressed in ASPS. Strong expression of TFE3 in ASPS reflects the underlying *TFE3-ASPL* gene fusion [144].

GENOMIC ALTERATIONS

ASPS carries a specific unbalanced translocation der(17)t(X;17)(p11;p25) leading to the fusion of the *TFE3* gene located on Xp11.2 encoding a transcription factor and the *ASPL* gene on chromosome 17q25, which may be detected by a *TFE3* dual color break apart FISH [190, 195].

PROGNOSIS AND TREATMENT

Late recurrences and metastases are common in ASPS. The 5-, 10-, and 20-year survival rates are 60, 38, and 15 %, respectively; this is a reflection of the rather frequent occurrence of late metastases [51]. Complete surgical resection is the treatment of choice. Recently, antiangiogenic-targeted treatment has been shown to be effective. Furthermore, given the evidence for activation of the AKT/mTOR pathway and MET activation in ASPS, treatment with mTOR inhibitors (e.g., rapamycin) and MET inhibition have been suggested [143]. ASPS has been shown to be targetable by sunitinib (Pfizer, New York, NY) [54, 162].

Clear Cell Sarcoma of Soft Tissue

CLINICOPATHOLOGICAL FEATURES

Described first by Enzinger in 1965 [41], clear cell sarcoma of soft tissue (CCS) was subsequently demonstrated to be of melanocytic differentiation by electron microscopy [70], and distinguished from other sarcomas arising in the tenosynovial soft tissue. CCS occurs in young and middle-aged adults without sex predilection. The majority of cases arises in the ankle or foot whereas other parts of the extremities are rarely involved. Very rare cases may also occur in the head and neck region, trunk, penis, retroperitoneum, kidney, and gastrointestinal tract. CCS is located in the deep soft tissue and exhibits a relation to tendons or aponeuroses. Immunohistochemistry shows constant S100 positivity. Nearly all CCSs are additionally positive for HMB45 and often for MITF and Melan-A. Neuroendocrine markers can also be coexpressed as well as other non lineage-specific markers such as CD57 and bcl2 [69]. Myogenic markers, CD117 and CD34, are not expressed in CCS.

GENOMIC ALTERATIONS

CCS often exhibits a complex karyotype. The most relevant alteration is a reciprocal translocation t(12;22)(q13;q12) leading to the fusion of the *EWS1R* and the *ATF1* (activating transcription factor 1 gene) genes [128]. Both genes encode transcription factors and the translocation leads to the fusion of the N-terminal end of EWS1R with the bZIP domain of ATF1. The resulting chimeric protein can activate itself in a cAMP-independent manner. Alternatively, *EWS1R* can be fused to *CREB1* (cAMP-responsive element-binding protein 1) leading to the activation of MITF and thus to a melanocytic phenotype. The latter type of translocation is primarily encountered in CCS of the gastrointestinal tract [3] but can also be occasionally seen in CCS of other locations. The detection of either translocation can be very helpful in differentiating CCS from metastatic melanoma.

CCS of the gastrointestinal tract represents a specific subtype that shares some features with its soft tissue counterpart but differs in its higher biologic aggressiveness, behaving like high-grade sarcomas [92]. In contrast to the classical type, CCS of the gastrointestinal tract (CCSLGT) expresses S100 protein but not other melanocytic markers such as HMB45, Melan A, or MITF. CD57 and/or NSE expression can be encountered [53, 194]. At the molecular level both translocation subtypes (i.e., *EWS1R* and either *ATF1* or *CREB1*) can be observed.

PROGNOSIS AND TREATMENT

Local recurrence is common, especially following incomplete resection. Metastasis occurs in 30 % of cases, often late. A 47–67 % five-year survival rate is observed. The survival rate drops to 33 and 10 % at 10 and 20 years, respectively. Fifteen percent of cases develop metastases to lymph nodes. Common sites of distant metastases include lung and bone. A tumor size larger than 5 cm is an adverse prognostic factor. As indicated above, CCSLGT has an unfavorable prognosis. Local excision is eventually followed by adjuvant chemotherapy.

Extraskeletal Myxoid Chondrosarcoma

CLINICOPATHOLOGICAL FEATURES

Extraskeletal myxoid chondrosarcomas (ESMC) most frequently arise in men in their fourth–sixth decade. Common intramuscular locations include the lower limb girdle and buttock as well as the distal upper extremities. Less often, ESMC arise in the retroperitoneal and the head and neck regions.

GENOMIC ALTERATIONS

The majority of ESMC harbor translocations involving the *NR4A3* gene (previously designated as *CHN*). *NR4A3* encodes the nuclear receptor subfamily 4 group A type III belonging to the family of steroid and thyroid hormone receptors. In 75 % of ESMC, the *NR4A3* translocation involves *EWS1R* [t(9;22]. This fusion can be demonstrated using a break apart probe [129]. The resulting NR4A3-EWS1R fusion protein contains a transcriptional activation domain and a DNA-binding domain and functions as a transcription factor. Other translocation partners of *NR4A3* include *TAF15*, *TFG*, and *TCF12*.

PROGNOSIS AND TREATMENT

Typically, ESMC exhibit local recurrences and metastases, often many years following diagnosis. Cellular and high-grade tumors have aggressive behavior. Complete surgical resection with wide margins is the treatment of choice. The response to radio- or chemotherapy is poor.

Desmoplastic Small Round Cell Tumor

CLINICOPATHOLOGICAL FEATURES

Desmoplastic small round cell tumors (DSRCT) typically arise in young males presenting as an abdominal or pelvic mass. Other locations include the pleura, the paratesticular region, the brain, ovaries, the pancreas, soft tissue, and bones. Because of their frequent topographic relation to serosal surfaces, a derivation from mesothelial or submesothelial stem cells has been proposed. DSRCT are histologically composed of cords and nests of small round tumor cells separated by desmoplastic stroma, earning them their descriptive designation. Immunohistochemically, these unique tumors display multiphenotypic evidence of differentiation as indicated by their expression of epithelial markers (e.g., cytokeratins and EMA), mesenchymal markers such as desmin, and neuroectodermal markers such as NSE. WT1 expression is a consistent feature of DSRCT. CD99 is coexpressed in 20 % of cases [137, 144].

GENOMIC ALTERATIONS

DSCRT exhibit a typical reciprocal translocation t(11;22)(p13;q12) leading to the fusion of *EWSR1* and *WT1* (Wilms tumor 1) genes, which is easily detectable in an *EWSR1* dual color break apart FISH approach which may be more suitable than RT-PCR assays due to some variability of the breakpoints. The resulting fusion protein acts as a transcription factor the target of which may include *PDGF-A* that could be responsible for the associated prominent desmoplasia.

PROGNOSIS AND TREATMENT

Overall prognosis is poor. Like other aggressive sarcomas, DSRCT treatment is multimodal in approach. To date, no effective targeted therapy has been developed.

Extrarenal Rhabdoid Tumor

CLINICOPATHOLOGICAL FEATURES

Extrarenal rhabdoid tumors (ERT) are rare highly aggressive soft tissue tumors occurring predominantly in infancy and childhood [123]. The designation originates from the presence of morphologically and genetically identical tumors that arise in the kidney and the brain. ERT most frequently arise in deep axial locations such as the neck and paraspinal regions. Visceral manifestations (mainly in the liver) are also on record. The "rhabdoid" phenotype is due to the presence of juxtanuclear eosinophilic cytoplasmic inclusions. Immunohistochemically, ERT are positive for vimentin, EMA and keratins, CD99, as well as neuroectodermal markers (synaptophysin and NSE) [47, 91, 175]. ERT characteristically show loss of INI1 nuclear expression [73].

GENOMIC ALTERATIONS

Like their renal and cerebral counterparts, ERT demonstrate homozygous inactivation of the *SMARCB1* (*INI1/HSNF/BAF47*) genes, with a particularly high incidence of smaller deletions of 22q11.22–22q11.23 in soft tissue ERT [82]. The *SMARCB1* gene is a component of the mammalian SWI/SNF complex, which functions in an ATP-dependent manner to remodel chromatin. *SMARCB1* loss is associated with functional disruption of the p16INK4-CyclinD/CDK4-pRb-E2F mitotic checkpoint [178] and an activation of the Hedgehog pathway [83].

Very few cases of rhabdoid tumors retain *SMARCB1*, but alternatively display a loss of another SWI/SNF member, *SMARCA4* (*BRG1*) [64, 153]. Large-scale sequencing analyses have revealed only very few additional mutations in rhabdoid tumors further supporting the crucial oncogenic role of *SMARCB1* inactivation [97].

PROGNOSIS AND TREATMENT

Prognosis is poor. The 5-year overall survival has been reported to be <15 % [16]. Treatment options include surgery, chemotherapy, and radiotherapy.

Table 29-3 **Metastatic Risk of Gastrointestinal Stromal Tumors (Modified According to [111])**

Group	Size (cm)	Mitotic count (HPFs)	Metastatic risk Stomach	Jejunum/ileum	Duodenum	Rectum
1	≤2	≤5/50	ø	ø	ø	ø
2	>2–5	≤5/50	Very low	Low	Low	Low
3a	>5–10	≤5/50	Low	Moderate	High	High
3b	>10	≤5/50	Moderate	High	High	High
4	≤2	>5/50	ø[a]	High[a]	–	High
5	>2–5	>5/50	Moderate	High	High	High
6a	>5–10	>5/50	High	High	High	High
6b	>10	>5/50	High	High	High	High

Adapted from Miettinen and Lasota [111]
Based on previously published long-term follow-up studies on 1939 GISTs
– no cases available, HPFs high powered fields
[a]Tumor categories with very small numbers of cases

Gastrointestinal Stromal Tumors

CLINICOPATHOLOGICAL FEATURES

With an estimated yearly incidence of 10–15 cases per million inhabitants [122], gastrointestinal stromal tumors (GIST) are the most common mesenchymal tumors of the gastrointestinal tract. Two thirds are located in the stomach with the small intestine being the second most frequent tumor location. A small proportion of GIST (5 %) occurs in the rectum, and rare examples affecting the esophagus and the peritoneum without clear connection to the tubular gastrointestinal tract have been reported as well. The biologic behavior is highly correlated with its primary location with a less aggressive behavior expected in the gastric location compared to extragastric sites.

GIST are thought to be derived from a precursor of the interstitial cells of Cajal (ICCs) with which they share their characteristic expression of stem cell proteins CD34 and the KIT receptor. Both markers are of great diagnostic utility as immunohistochemical markers of GIST. More recently, DOG1 ("discovered on GIST1") was identified as the most sensitive and specific marker for GIST. DOG1 was identified by gene expression profiling and encodes an ion channel protein with eight transmembrane domains [45, 188]. DOG1 is especially helpful in KIT-negative GISTs [99, 113]. The diagnosis in these often epithelioid gastric GIST can be further supported by the detection of mutations in the *PDGFRα* gene [134, 183].

Aggressive behavior is encountered in approximately 50 % of GIST. At least three highly significant prognostic parameters are recognized. These include (a) primary location, (b) tumor size, and (c) mitotic count/5 mm^2. These three parameters are the basis for the most commonly used risk assessment system put forward by Miettinen and Lasota in 2006 [111]; see Table 29.3.

A clinical parameter which is not included in the Miettinen classification but is highly relevant for prognosis is tumor rupture (pre- or intraoperatively). A documented tumor rupture increases the risk of recurrence several fold to a greater than 90 % recurrence rate [71].

GENOMIC ALTERATIONS

GIST are among the best examples in oncology regarding how a single somatic mutation can influence prognosis and predict treatment response. Treatment of GIST is regarded as the paradigm for molecular targeted therapy. Genomic characterization is now well accepted as a central part of the diagnostic process and a prerequisite for treatment planning. Furthermore, the discovery of secondary mutations as a main cause for treatment resistance has pinpointed the most common mechanism for resistance to tyrosine kinase inhibition [181, 186].

Up to 90 % of GIST carry primary activating mutations in the *KIT* gene or the *PDGFRα* gene. Both genes are located on chromosome 4 and encode type III receptor tyrosine kinases which display homology in 30 % of their amino acids. The prognostic and therapeutically predictive relevance of the mutational status of these genes is now well accepted [31, 67, 96]. The reported frequencies of mutational subtypes differ considerably between anatomic locations of GIST and among different studies, probably due to case selection biases. Primary activating mutations can occur either in the extracellular domain of the receptor protein (i.e., *KIT* exon 9), in the juxtamembraneous domain (*KIT* exon 11, *PDGFRα* exon 12), in the first tyrosine kinase domain (*KIT* exon 13 and 14, *PDGFRα* exon 14), or in the second tyrosine kinase domain (*KIT* exon 17, *PDGFRα* exon 18). Mutations have been reported in *KIT* exon 8 as well, but these seem to be rare [75, 76]. Approximately 65 % of all GIST carry *KIT* exon 11 mutations, whereas *KIT* exon 9 and *PDGFRα* exon 18 mutations account for about 10 % of primary mutations each. Thus, at least 85 % of all GIST carry a mutation at one of these three sites. Another 5 % may carry mutations in exons 13, 14, or 17 of *KIT* or in exons 12 or 14 of *PDGFRα*, leading to a frequency of about 1 % in each of these regions [75, 76]. As a result, the number of cases with the latter locations of mutations is low in most trials, making it difficult to draw strong conclusions concerning their prognostic and predictive value at this time. The histological phenotype and location of GIST correlate with their *KIT* or the *PDGFRα* mutation status. The vast majority of gastric GIST carry *KIT* exon 11 or *PDGFRA* exon 18 mutations whereas *KIT* exon 11 and 9 mutations predominate in intestinal GIST [182].

The remaining 10–15 % of GIST seem to lack *KIT* or *PDGFRA* mutations and are termed "wild-type GIST." Recently, several small genomic subgroups have been identified among the wild-type GIST. One subgroup of sporadic wild-type GIST occurs as part of the Carney's triad in association with pulmonary hamartomas/chondromas and paragangliomas. The vast majority of patients are young females with multiple gastric GIST which can metastasize to regional lymph nodes [18]. Interestingly, with exception of GIST occurring as part of Carney's triad, lymphatic spread is exceedingly rare in sporadic GIST. The genomic background of Carney's triad remains to be fully characterized. Other subgroups among the so-called wild-type GIST rarely carry *BRAF* mutations as an underlying alteration [2]. Finally, mutations in the *SDHA*, *SDHB*, *SDHC*, and *SDHD* genes encoding the subunits of the succinate dehydrogenase complex have been identified, beyond the hereditary Carney–Stratakis syndrome, in sporadic "wild-type" GISTs [130, 132, 163].

MUTATIONAL STATUS IN FAMILIAL GIST

One third of neurofibromatosis type I (NF I) patients will develop one or more GIST during their lifetime. The majority of these lesions occur in the small bowel and show low aggressive behavior [90, 117, 168]. Another familial setting of GIST is that of a rare familial disorder resulting from a germline mutation in the *KIT* gene (exons 8, 11, 13, 17; for a review see [11, 119]). The latter leads to the development of multiple GIST, in some cases in combination with systemic mastocytosis and ICC hyperplasia throughout the GI tract, and associated dysphagia. Less than 30 kindreds of this *KIT* germline mutations have been described in the literature.

Finally, kindreds with multiple GIST carrying a *PDGFRA* mutation have been described in two reports [22, 29]. Whether these gastrointestinal mesenchymal tumors are genuine GIST or rather represent inflammatory fibroid polyps (IFP) would have to be further explored by novel more specific immunhistochemical markers such as DOG1. IFP carry identical types of *PDGFRA* mutations in the same hot spots and can occur anywhere throughout the gastrointestinal tract [77, 148].

OTHER GENETIC AND EPIGENETIC MECHANISMS IN GIST PATHOGENESIS

Compared to other sarcoma subtypes, the majority of GIST have a low cytogenetic complexity. The most frequent alterations are losses of the long arms of chromosomes 14 and/or 22 which are found both in benign and in malignant GIST. With tumor progression the number of chromosomal losses increases with additional losses in 1p, 9p, 9q, 11p, and 13q and amplifications in 5p, 8q, 17q, and 20q. Different types of chromosomal aberrations can be correlated with primary tumor location and are of prognostic value [61].

Inactivating alterations in the tumor suppressor gene *CDKN2A* may also be at play in GIST development. The *CDKN2A* gene is located on chromosome 9p21. The encoded $p16^{INK4}$protein inhibits cyclin-dependent kinases and leads to increased cell proliferation. It has been demonstrated that different types of *CDKN2A* alterations such as promoter methylation, point mutations, or homozygous deletions leading to loss of function are associated with aggressive biological behavior of GIST [150, 151].

PROGNOSTIC RELEVANCE OF SPORADIC *KIT/PDGFRα (ALPHA)* MUTATIONS

KIT/PDGFRα mutations are detected in a high proportion of the so-called micro-GISTs that are incidentally detected and measure less than 1 cm [1]. This observation demonstrates that *KIT/PDGFRα* mutations are early oncogenic events in GIST and that other additional genomic or epigenetic events modulate biologic behavior. However, several independent studies have demonstrated a strong correlation between GIST mutational status and their risk for metastasis. Our own group and others have found such a correlation between a specific 6 bp deletion in *KIT* exon 11 (on the protein level p.W557_K558del) and a high metastatic risk [104, 105, 184]. In contrast, the vast majority of GIST with *PDGFRα* mutation show a low level of aggressiveness [95, 183]. The relevance of *KIT* exon 9 mutations for biological behavior remains controversial [4] because these mutations are found almost always in non-gastric GIST, which usually behave more aggressively than gastric GIST. Several groups have proposed the inclusion of tumor mutational status as an additional prognostic parameter in a novel risk classification system.

PREDICTIVE VALUE OF *KIT/PDGFRA* MUTATIONS FOR TREATMENT RESPONSE

The relevance of mutational status for treatment response in metastatic GIST has become clear through multiple trials [30, 66]. In summary, GIST with *KIT* exon 11 mutation have the highest response rates (of 80–90 %) to the standard daily dose of 400 mg imatinib. Tumors with *KIT* exon 9 mutation have a lower response rate of about 45 %. The response rate in GIST with *PDGFRα* mutations also strongly depends on the mutational subtype [19]. A specific point mutation in *PDGFRα* exon 18 that leads to a substitution of aspartate with valine (p. D842V) results in primary imatinib resistance whereas tumors with other mutational subtypes in the same exon respond to imatinib. "Wild-type GIST" lacking activating mutations in either *KIT* or *PDGFRα* genes behave in a heterogeneous fashion but overall seem to have a low rate of treatment response. However, their low incidence makes it difficult to draw final conclusions at present.

Adjuvant imatinib treatment for at least 3 years following complete primary resection is associated with improvement in recurrence free and overall survival rates. Furthermore, imatinib can be used in the neoadjuvant setting in primary inoperable GIST. The subsequent reduction of tumor size allows for secondary resection with lower morbidity (reviewed in [39]). As a result, both the European Society for Medical Oncology and the National Comprehensive Cancer Network guidelines strongly recommend molecular typing in GIST [14, 33] to identify patients with primary imitinib resistance (e.g., *PDGFRα* exon 18 mutation p.D842V) and to appropriately adjust imatinib dosage (in case of a *KIT* exon 9 mutation).

MECHANISMS OF RESISTANCE IN GIST

The majority of patients with metastatic GIST develop secondary resistance to imatinib.

The frequency of such occurrence is estimated to be at least 80 %. It is thought to be due to the development of secondary mutations in the *KIT* gene or rarely in the *PDGFRα* gene that can be demonstrated in a large percentage of progressing tumor samples. The secondary mutations are preferentially located in the kinase domain (*KIT* exons 13, 14, or 17) [185] leading to the inhibition of imatinib binding. It has recently been shown that allele-specific PCR (AS-PCR) approaches are more sensitive than denaturing high-performance liquid chromatography (DHPLC) with regard to the detection rate of resistance mutations [100]. Depending on the type of secondary mutation, alternative second-line treatment may be successfully pursued. However, effective therapy is lacking as yet for a subgroup of cases with secondary *KIT* exon 17 mutations. In selected cases, resection of single progressing lesions can be an option. The role of pharmacokinetics in secondary imatinib resistance remains to be further clarified [34].

Conclusions

The extensive elucidation of genomic alterations in soft tissue tumors has allowed for a refinement of the morphology-based classification of these neoplasms and provided additional prognostic parameters and new targets of therapy with predictive markers of response.

REFERENCES

1. Agaimy A, Wünsch P, Hofstaedter F, Blaszyk H, Rümmele P, Gaumann A, Dietmaier W, Hartmann A. Minute gastric sclerosing stromal tumors (GIST tumorlets) are common in adults and frequently show c-KIT mutations. Am J Surg Pathol. 2007;31:113–20.
2. Agaram N, Wong G, Guo T, Maki R, Singer S, Dematteo R, Besmer P, Antonescu C. Novel V600E BRAF mutations in imatinib-naive and imatinib-resistant gastrointestinal stromal tumors. Genes Chromosomes Cancer. 2008;47:853–9.
3. Antonescu C, Nafa K, Segal N, Dal Cin P, Ladanyi M. EWS-CREB1: a recurrent variant fusion in clear cell sarcoma – association with gastrointestinal location and absence of melanocytic differentiation. Clin Cancer Res. 2006;12:5356–62.
4. Antonescu C, Sommer G, Sarran L, Tschernyavsky S, Riedel E, Woodruff J, Robson M, Maki R, Brennan M, Ladanyi M, et al. Association of KIT exon 9 mutations with nongastric primary site and aggressive behavior: KIT mutation analysis and clinical correlates of 120 gastrointestinal stromal tumors. Clin Cancer Res. 2003;9:3329–37.
5. Antonescu CR, Elahi A, Healey JH, Brennan MF, Lui MY, Lewis J, Jhanwar SC, Woodruff JM, Ladanyi M. Monoclonality of multifocal myxoid liposarcoma: confirmation by analysis of TLS-CHOP or EWS-CHOP rearrangements. Clin Cancer Res. 2000;6:2788–93.
6. Antonescu CR, Le Loarer F, Mosquera JM, Sboner A, Zhang L, Chen CL, Chen HW, Pathan N, Krausz T, Dickson BC, et al. Novel YAP1-TFE3 fusion defines a distinct subset of epithelioid hemangioendothelioma. Genes Chromosomes Cancer. 2013;52:775–84.
7. Antonescu CR, Tschernyavsky SJ, Decuseara R, Leung DH, Woodruff JM, Brennan MF, Bridge JA, Neff JR, Goldblum JR, Ladanyi M. Prognostic impact of P53 status, TLS-CHOP fusion transcript structure, and histological grade in myxoid liposarcoma: a molecular and clinicopathologic study of 82 cases. Clin Cancer Res. 2001;7:3977–87.
8. Azumi N, Curtis J, Kempson RL, Hendrickson MR. Atypical and malignant neoplasms showing lipomatous differentiation. A study of 111 cases. Am J Surg Pathol. 1987;11:161–83.
9. Barr FG. Gene fusions involving PAX and FOX family members in alveolar rhabdomyosarcoma. Oncogene. 2001;20:5736–46.
10. Barretina J, Taylor BS, Banerji S, Ramos AH, Lagos-Quintana M, Decarolis PL, Shah K, Socci ND, Weir BA, Ho A, et al. Subtype-specific genomic alterations define new targets for soft-tissue sarcoma therapy. Nat Genet. 2010;42:715–21.
11. Benesch M, Wardelmann E, Ferrari A, Brennan B, Verschuur A. Gastrointestinal stromal tumors (GIST) in children and adolescents: a comprehensive review of the current literature. Pediatr Blood Cancer. 2009;53:1171–9.
12. Bertoni F, Unni KK, Beabout JW, Sim FH. Malignant giant cell tumor of the tendon sheaths and joints (malignant pigmented villonodular synovitis). Am J Surg Pathol. 1997;21:153–63.
13. Betires-Alj M, Guillermo Paez J, David F, Keilhack H, Halmos B, Naoki K, Maris J, Richardson A, Bardelli A, Sugarbaker D, et al. Activating mutations of the Noonan syndrome-associated SHP2/PTPN11 gene in human solid

tumors and adult acute myelogenous leukemia. Cancer Res. 2004;64:8816–20.
14. Blay JY, Bonvalot S, Boukovinas I, Casali PG, De Alava E, Dei Tos AP, Dirksen U, Duffaud F, Eriksson M, Fedenko A, Ferrari A, Ferrari S, del Muro XG, Gelderblom H, Grimer R, Gronchi A, Hall KS, Hassan B, Hogendoorn P, Hohenberger P, Issels R, Joensuu H, Jost L, Jurgens H, Kager L, Le Cesne A, Leyvraz S, Martin J, Merimsky O, Nishida T, Picci P, Reichardt P, Rutkowski P, Schlemmer M, Sleijfer S, Stacchiotti S, Taminiau A, Wardelmann E. Gastrointestinal stromal tumors: ESMO clinical practice guidelines for diagnosis, treatment and follow-up. ESMO/ European Sarcoma Network Working Group. Ann Oncol. 2012;23:Vii49–55.
15. Blay JY, El Sayadi H, Thiesse P, Garret J, Ray-Coquard I. Complete response to imatinib in relapsing pigmented villonodular synovitis/tenosynovial giant cell tumor (PVNS/TGCT). Ann Oncol. 2008;19:821–2.
16. Bourdeaut F, Freneaux P, Thuille B, Bergeron C, Laurence V, Brugieres L, Verite C, Michon J, Delattre O, Orbach D. Extra-renal non-cerebral rhabdoid tumours. Pediatr Blood Cancer. 2008; 51:363–8.
17. Butrynski J, D'Adamo D, Hornick J, Dal Cin P, Antonescu C, Jhanwar S, Ladanyi M, Capelletti M, Rodig S, Ramaiya N, et al. Crizotinib in ALK-rearranged inflammatory myofibroblastic tumor. N Engl J Med. 2010;363:1727–33.
18. Carney J, Stratakis C. Familial paraganglioma and gastric stromal sarcoma: a new syndrome distinct from the Carney triad. Am J Med Gen. 2002;108:132–9.
19. Cassier P, Fumagalli E, Rutkowski P, Schöffski P, Van Glabbeke M, Debiec-Rychter M, Emile J, Duffaud F, Martin-Broto J, Landi B, et al. Outcome of patients with platelet-derived growth factor receptor alpha-mutated gastrointestinal stromal tumors in the tyrosine kinase inhibitor era. Clin Cancer Res. 2012;18: 4458–64.
20. Cessna M, Zhou H, Sanger W, Perkins S, Tripp S, Pickering D, Daines C, Coffin C. Expression of ALK1 and p80 in inflammatory myofibroblastic tumor and its mesenchymal mimics: a study of 135 cases. Mod Pathol. 2002;15:931–8.
21. Chmielecki J, Crago AM, Rosenberg M, O'Connor R, Walker SR, Ambrogio L, Auclair D, McKenna A, Heinrich MC, Frank DA, Meyerson M. Whole-exome sequencing identifies a recurrent NAB2-STAT6 fusion in solitary fibrous tumors. Nat Genet. 2013;45:131–2.
22. Chompret A, Kannengiesser C, Barrois M, Terrier P, Dahan P, Tursz T, Lenoir G, Bressac-de Paillerets B. PDGFRA germline mutation in a family with multiple cases of gastrointestinal stromal tumor. Gastroenterology. 2004;126: 318–21.
23. Coindre JM, Pedeutour F, Aurias A. Well-differentiated and dedifferentiated liposarcomas. Virchows Arch. 2010;456:167–79.
24. Crozat A, Aman P, Mandahl N, Ron D. Fusion of CHOP to a novel RNA-binding protein in human myxoid liposarcoma. Nature. 1993;363: 640–4.
25. Dal Cin P, Sciot R, Panagopoulos I, Aman P, Samson I, Mandahl N, Mitelman F, Van den Berghe H, Fletcher CD. Additional evidence of a variant translocation t(12;22) with EWS/CHOP fusion in myxoid liposarcoma: clinicopathological features. J Pathol. 1997;182:437–41.
26. Davis RJ, Barr FG. Fusion genes resulting from alternative chromosomal translocations are overexpressed by gene-specific mechanisms in alveolar rhabdomyosarcoma. Proc Natl Acad Sci U S A. 1997;94:8047–51.
27. Davis RJ, D'Cruz CM, Lovell MA, Biegel JA, Barr FG. Fusion of PAX7 to FKHR by the variant t(1;13)(p36;q14) translocation in alveolar rhabdomyosarcoma. Cancer Res. 1994;54:2869–72.
28. de Perrot M, Fischer S, Bründler M, Sekine Y, Keshavjee S. Solitary fibrous tumors of the pleura. Ann Thorac Surg. 2002;74:285–93.
29. De Raedt T, Cools J, Debiec-Rychter M, Brems H, Mentens N, Sciot R, Himpens J, De Wever I, Schöffski P, Marynen P, Legius E. Intestinal neurofibromatosis is a subtype of familial GIST and results from a dominant activating mutation in *PDGFRA*. Gastroenterology. 2006;131: 1907–12.
30. Debiec-Rychter M, Dumez H, Judson I, Wasag B, Verweij J, Brown M, Dimitrijevic S, Sciot R, Stul M, Vranck H, et al. Use of c-kit/PDGFRA mutational analysis to predict the clinical response to imatinib in patients with advanced gastrointestinal stromal tumours entered on phase I and II studies of the EORTC Soft Tissue and Bone Sarcoma Group. Eur J Cancer. 2004; 40:689–95.
31. Debiec-Rychter M, Sciot R, Le Cesne A, Schlemmer M, Hohenberger P, van Oosterom A, Blay J, Leyvraz S, Stul M, Casali P, et al. KIT mutations and dose selection for imatinib in patients with advanced gastrointestinal stromal tumours. Eur J Cancer. 2006;42:1093–103.
32. Dehner L. Inflammatory myofibroblastic tumor. The continued definition of one type of so-called inflammatory pseudotumor. Am J Surg Pathol. 2004;28:1652–4.
33. Demetri G, von Mehren M, Antonescu C, DeMatteo R, Ganjoo K, Maki R, Pisters P, Raut C, Riedel R, Schuetze S, et al. NCCN Task Force report: update on the management of patients with gastrointestinal stromal tumors. J Natl Compr Canc Netw. 2010;8:S1–41.
34. Demetri G, Wang Y, Wehrle E, Racine A, Nikolova Z, Blanke C, Joensuu H, von Mehren M. Imatinib plasma levels are correlated with

clinical benefit in patients with unresectable/metastatic gastrointestinal stromal tumors. J Clin Oncol. 2009;27:3141–7.
35. Demicco EG, Torres KE, Ghadimi MP, Colombo C, Bolshakov S, Hoffman A, Peng T, Bovee JV, Wang WL, Lev D, Lazar AJ. Involvement of the PI3K/Akt pathway in myxoid/round cell liposarcoma. Mod Pathol. 2012;25:212–21.
36. Dews M, Homayouni A, Yu D, Murphy D, Sevignani C, Wentzel E, Furth E, Lee W, Enders G, Mendell J, Thomas-Tikhonenko A. Augmentation of tumor angiogenesis by a Myc-activated microRNA cluster. Nat Genet. 2006;38:1060–5.
37. Dickson MA, Tap WD, Keohan ML, D'Angelo SP, Gounder MM, Antonescu CR, Landa J, Qin LX, Rathbone DD, Condy MM, et al. Phase II trial of the CDK4 inhibitor PD0332991 in patients with advanced CDK4-amplified well-differentiated or dedifferentiated liposarcoma. J Clin Oncol. 2013;31:2024–8.
38. Doyle L, Wang W, Dal Cin P, Lopez-Terrada D, Mertens F, Lazar A, Fletcher C, Hornick J. MUC4 is a sensitive and extremely useful marker for sclerosing epithelioid fibrosarcoma: association with FUS gene rearrangement. Am J Surg Pathol. 2012;36:1444–51.
39. Eisenberg B, Trent J. Adjuvant and neoadjuvant imatinib therapy: current role in the management of gastrointestinal stromal tumors. Int J Cancer. 2011;129:2533–42.
40. Engstrom K, Willen H, Kabjorn-Gustafsson C, Andersson C, Olsson M, Goransson M, Jarnum S, Olofsson A, Warnhammar E, Aman P. The myxoid/round cell liposarcoma fusion oncogene FUS-DDIT3 and the normal DDIT3 induce a liposarcoma phenotype in transfected human fibrosarcoma cells. Am J Pathol. 2006;168:1642–53.
41. Enzinger FM. Clear-cell sarcoma of tendons and aponeuroses. An analysis of 21 cases. Cancer. 1965;18:1163–74.
42. Enzinger FM, Winslow DJ. Liposarcoma. A study of 103 cases. Virchows Arch Pathol Anat Physiol Klin Med. 1962;335:367–88.
43. Errani C, Sung Y, Zhang L, Healey J, Antonescu C. Monoclonality of multifocal epithelioid hemangioendothelioma of the liver by analysis of WWTR1-CAMTA1 breakpoints. Cancer Genet. 2012;205:12–7.
44. Errani C, Zhang L, Sung Y, Hajdu M, Singer S, Maki R, Healey J, Antonescu C. A novel WWTR1-CAMTA1 gene fusion is a consistent abnormality in epithelioid hemangioendothelioma of different anatomic sites. Genes Chromosomes Cancer. 2011;50:644–53.
45. Espinosa I, Lee C, Kim M, Rouse B, Subramanian S, Montgomery K, Varma S, Corless C, Heinrich M, Smith K, et al. A novel monoclonal antibody against DOG1 is a sensitive and specific marker for gastrointestinal stromal tumors. Am J Surg Pathol. 2008;32:210–8.
46. Evans HL. Low-grade fibromyxoid sarcoma: a clinicopathologic study of 33 cases with long-term follow-up. Am J Surg Pathol. 2011;35:1450–62.
47. Fanburg-Smith J, Meis-Kindblom J, Fante R, Kindblom L. Malignant granular cell tumor of soft tissue: diagnostic criteria and clinicopathologic correlation. Am J Surg Pathol. 1998;22:779–94.
48. Fisher C. Soft tissue sarcomas with non-EWS translocations: molecular genetic features and pathologic and clinical correlations. Virchows Arch. 2010;456:153–66.
49. Fletcher C, Bridge J, Hogendoorn P, Mertens F, editors. WHO classification of tumours of soft tissue and bone. Lyon: IARC Press; 2013.
50. Flucke U, Slootweg P, Mentzel T, Pauwels P, Hulsebos T. Re: Infrequent SMARCB1/INI1 gene alteration in epithelioid sarcoma: a useful tool in distinguishing epithelioid sarcoma from malignant rhabdoid tumor: direct evidence of mutational inactivation of SMARCB1/INI1 in epithelioid sarcoma. Hum Pathol. 2009;40:1361–2.
51. Folpe AL, Deyrup AT. Alveolar soft-part sarcoma: a review and update. J Clin Pathol. 2006;59:1127–32.
52. Friedrichs N, Küchler J, Endl E, Koch A, Czerwitzki J, Wurst P, Metzger D, Schulte J, Holst M, Heukamp L, et al. Insulin-like growth factor-1 receptor acts as a growth regulator in synovial sarcoma. J Pathol. 2008;216:428–39.
53. Friedrichs N, Testi M, Moiraghi L, Modena P, Paggen E, Plötner A, Wiechmann V, Mantovani-Löffler L, Merkelbach-Bruse S, Buettner R, Wardelmann E. Clear cell sarcoma-like tumour with giant cells in the small bowel – further evidence for a new tumor entity? Int J Surg Pathol. 2005;13:313–8.
54. Frith AE, Hirbe AC, Van Tine BA. Novel pathways and molecular targets for the treatment of sarcoma. Curr Oncol Rep. 2013;15:378–85.
55. Galili N, Davis RJ, Fredericks WJ, Mukhopadhyay S, Rauscher 3rd FJ, Emanuel BS, Rovera G, Barr FG. Fusion of a fork head domain gene to PAX3 in the solid tumour alveolar rhabdomyosarcoma. Nat Genet. 1993;5:230–5.
56. Gebhard S, Coindre JM, Michels JJ, Terrier P, Bertrand G, Trassard M, Taylor S, Chateau MC, Marques B, Picot V, Guillou L. Pleomorphic liposarcoma: clinicopathologic, immunohistochemical, and follow-up analysis of 63 cases: a study from the French Federation of Cancer Centers Sarcoma Group. Am J Surg Pathol. 2002;26:601–16.

57. Gengler C, Guillou L. Solitary fibrous tumour and haemangiopericytoma: evolution of a concept. Histopathology. 2006;28:67–74.
58. Gleason BC, Hornick JL. Inflammatory myofibroblastic tumours: where are we now? J Clin Pathol. 2008;61:428–37.
59. Goldblum J. Giant cell fibroblastoma: a report of three cases with histologic and immunohistochemical evidence of a relationship to dermatofibrosarcoma protuberans. Arch Pathol Lab Med. 1996;120:1052–5.
60. Guillou L, Benhattar J, Gengler C, Gallagher G, Ranchère-Vince D, Collin F, Terrier P, Terrier-Lacombe M, Leroux A, Marquès B, et al. Translocation-positive low-grade fibromyxoid sarcoma: clinicopathologic and molecular analysis of a series expanding the morphologic spectrum and suggesting potential relationship to sclerosing epithelioid fibrosarcoma: a study from the French Sarcoma Group. Am J Surg Pathol. 2007;31:1387–402.
61. Gunawan B, von Heydebreck A, Sander B, Schulten H, Haller F, Langer C, Armbrust T, Bollmann M, Gasparov S, Kovac D, Füzesi F. An oncogenetic tree model in gastrointestinal stromal tumours (GISTs) identifies different pathways of cytogenetic evolution with prognostic implications. J Pathol. 2007;211:463–70.
62. Guo T, Zhang L, Chang N, Singer S, Maki R, Antonescu C. Consistent MYC and FLT4 gene amplification in radiation-induced angiosarcoma but not in other radiation-associated atypical vascular lesions. Genes Chromosomes Cancer. 2011;50:25–33.
63. Harms D. Alveolar rhabdomyosarcoma: a prognostically unfavorable rhabdomyosarcoma type and its necessary distinction from embryonal rhabdomyosarcoma. Curr Top Pathol. 1995;89:273–96.
64. Hasselblatt M, Isken S, Linge A, Eikmeier K, Jeibmann A, Oyen F, Nagel I, Richter J, Bartelheim K, Kordes U, et al. High-resolution genomic analysis suggests the absence of recurrent genomic alterations other than SMARCB1 aberrations in atypical teratoid/rhabdoid tumors. Genes Chromosomes Cancer. 2013;52:185–90.
65. Hawkins DS, Spunt SL, Skapek SX. Children's Oncology Group's 2013 blueprint for research: soft tissue sarcomas. Pediatr Blood Cancer. 2013;60:1001–8.
66. Heinrich M, Corless C, Blanke CD, Demetri G, Joensuu H, Roberts P, Eisenberg B, Von Mehren M, Fletcher C, Sandau K, et al. Molecular correlates of imatinib resistance in gastrointestinal stromal tumors. JCO. 2006;24:4764–74.
67. Heinrich M, Owzar K, Corless C, Hollis D, Borden E, Fletcher C, Ryan C, von Mehren M, Blanke C, Rankin C, et al. Correlation of kinase genotype and clinical outcome in the North American Intergroup Phase III Trial of imatinib mesylate for treatment of advanced gastrointestinal stromal tumor: CALGB 150105 Study by Cancer and Leukemia Group B and Southwest Oncology Group. J Clin Oncol. 2008;26:5360–7.
68. Henricks WH, Chu YC, Goldblum JR, Weiss SW. Dedifferentiated liposarcoma: a clinicopathological analysis of 155 cases with a proposal for an expanded definition of dedifferentiation. Am J Surg Pathol. 1997;21:271–81.
69. Hisaoka M, Ishida T, Kuo T, Matsuyama A, Imamura T, Nishida K, Kuroda H, Inayama Y, Oshiro H, Kobayashi H, et al. Clear cell sarcoma of soft tissue: a clinicopathologic, immunohistochemical, and molecular analysis of 33 cases. Am J Surg Pathol. 2008;32:452–60.
70. Hoffman GJ, Carter D. Clear cell sarcoma of tendons and aponeuroses with melanin. Arch Pathol. 1973;95:22–5.
71. Hohenberger P, Ronellenfitsch U, Oladeji O, Pink D, Ströbel P, Wardelmann E, Reichardt P. Pattern of recurrence in patients with ruptured primary gastrointestinal stromal tumour. Br J Surg. 2010;97:1854–9.
72. Hollmann TJ, Hornick JL. INI1-deficient tumors: diagnostic features and molecular genetics. Am J Surg Pathol. 2011;35(10):e47–63. doi:101097/PAS0b013e31822b325b.
73. Hoot AC, Russo P, Judkins AR, Perlman EJ, Biegel JA. Immunohistochemical analysis of hSNF5/INI1 distinguishes renal and extra-renal malignant rhabdoid tumors from other pediatric soft tissue tumors. Am J Surg Pathol. 2004;28:1485–91.
74. Hornick JL, Bosenberg MW, Mentzel T, McMenamin ME, Oliveira AM, Fletcher CD. Pleomorphic liposarcoma: clinicopathologic analysis of 57 cases. Am J Surg Pathol. 2004;28:1257–67.
75. Huss S, Künstlinger H, Wardelmann E, Kleine M, Binot E, Merkelbach-Bruse S, Rüdiger T, Mittler J, Hartmann W, Büttner R, Schildhaus H. A subset of gastrointestinal stromal tumors previously regarded as wild-type tumors carries somatic activating mutations in KIT exon 8 (p.D419del). Mod Pathol. 2013. doi:10.1038/modpathol.2013.47.
76. Huss S, Nehles J, Binot E, Wardelmann E, Mittler J, Kleine MA, Künstlinger H, Hartmann W, Hohenberger P, Merkelbach-Bruse S, Buettner R, Schildhaus HU. β-Catenin (CTNNB1) mutations and clinicopathological features of mesenteric desmoid-type fibromatosis. Histopathology. 2013;62:294–304.
77. Huss S, Wardelmann E, Goltz D, Binot E, Hartmann W, Merkelbach-Bruse S, Büttner R, Schildhaus H. Activating PDGFRA mutations in inflammatory fibroid polyps occur in exons 12,

14 and 18 and are associated with tumour localization. Histopathology. 2012;61:59–68.
78. Italiano A, Bianchini L, Gjernes E, Keslair F, Ranchere-Vince D, Dumollard JM, Haudebourg J, Leroux A, Mainguene C, Terrier P, et al. Clinical and biological significance of CDK4 amplification in well-differentiated and dedifferentiated liposarcomas. Clin Cancer Res. 2009;15:5696–703.
79. Italiano A, Bianchini L, Keslair F, Bonnafous S, Cardot-Leccia N, Coindre JM, Dumollard JM, Hofman P, Leroux A, Mainguene C, et al. HMGA2 is the partner of MDM2 in well-differentiated and dedifferentiated liposarcomas whereas CDK4 belongs to a distinct inconsistent amplicon. Int J Cancer. 2008;122:2233–41.
80. Italiano A, Cardot N, Dupre F, Monticelli I, Keslair F, Piche M, Mainguene C, Coindre JM, Pedeutour F. Gains and complex rearrangements of the 12q13-15 chromosomal region in ordinary lipomas: the "missing link" between lipomas and liposarcomas? Int J Cancer. 2007;121:308–15.
81. Italiano A, Thomas R, Breen M, Zhang L, Crago A, Singer S, Khanin R, Maki R, Mihailovic A, Hafner M, et al. The miR-17-92 cluster and its target THBS1 are differentially expressed in angiosarcomas dependent on MYC amplification. Genes Chromosomes Cancer. 2012;51:569–78.
82. Jackson EM, Sievert AJ, Gai X, Hakonarson H, Judkins AR, Tooke L, Perin JC, Xie H, Shaikh TH, Biegel JA. Genomic analysis using high-density single nucleotide polymorphism-based oligonucleotide arrays and multiplex ligation-dependent probe amplification provides a comprehensive analysis of INI1/SMARCB1 in malignant rhabdoid tumors. Clin Cancer Res. 2009;15:1923–30.
83. Jagani Z, Mora-Blanco EL, Sansam CG, McKenna ES, Wilson B, Chen D, Klekota J, Tamayo P, Nguyen PT, Tolstorukov M, et al. Loss of the tumor suppressor Snf5 leads to aberrant activation of the Hedgehog-Gli pathway. Nat Med. 2010;16:1429–33.
84. Jha P, Moosavi C, Fanburg-Smith J. Giant cell fibroblastoma: an update and addition of 86 new cases from the Armed Forces Institute of Pathology, in honor of Dr. Franz M. Enzinger. Ann Diagn Pathol. 2007;11:81–8.
85. Joshi D, Anderson JR, Paidas C, Breneman J, Parham DM, Crist W. Age is an independent prognostic factor in rhabdomyosarcoma: a report from the Soft Tissue Sarcoma Committee of the Children's Oncology Group. Pediatr Blood Cancer. 2004;42:64–73.
86. Kadoch C, Crabtree GR. Reversible disruption of mSWI/SNF (BAF) complexes by the SS18-SSX oncogenic fusion in synovial sarcoma. Cell. 2013;153:71–85.
87. Kasper B, Ströbel P, Hohenberger P. Desmoid tumors: clinical features and treatment options for advanced disease. Oncologist. 2011;16:682–93.
88. Keller C, Arenkiel BR, Coffin CM, El-Bardeesy N, DePinho RA, Capecchi MR. Alveolar rhabdomyosarcomas in conditional Pax3:Fkhr mice: cooperativity of Ink4a/ARF and Trp53 loss of function. Genes Dev. 2004;18:2614–26.
89. Kilpatrick SE, Doyon J, Choong PF, Sim FH, Nascimento AG. The clinicopathologic spectrum of myxoid and round cell liposarcoma. A study of 95 cases. Cancer. 1996;77:1450–8.
90. Kinoshita K, Hirota S, Isozaki K, Ohashi A, Nishida T, Kitamura Y, Shinomura Y, Matsuzawa Y. Absence of c-kit gene mutations in gastrointestinal stromal tumours from neurofibromatosis type I patients. J Pathol. 2004;202:80–5.
91. Kodet R, Newton Jr WA, Sachs N, Hamoudi AB, Raney RB, Asmar L, Gehan EA. Rhabdoid tumors of soft tissues: a clinicopathologic study of 26 cases enrolled on the Intergroup Rhabdomyosarcoma Study. Hum Pathol. 1991;22:674–84.
92. Kosemehmetoglu K, Folpe A. Clear cell sarcoma of tendons and aponeuroses, and osteoclast-rich tumour of the gastrointestinal tract with features resembling clear cell sarcoma of soft parts: a review and update. J Clin Pathol. 2010;63:416–23.
93. Ladanyi M. Fusions of the SYT and SSX genes in synovial sarcoma. Oncogene. 2001;20:5755–62.
94. Ladanyi M, Antonescu CR, Leung DH, Woodruff JM, Kawai A, Healey JH, Brennan MF, Bridge JA, Neff JR, Barr FG, et al. Impact of SYT-SSX fusion type on the clinical behavior of synovial sarcoma: a multi-institutional retrospective study of 243 patients. Cancer Res. 2002;62:135–40.
95. Lasota J, Dansonka-Mieszkowska A, Sobin L, Miettinen M. A great majority of GISTs with PDGFRA mutations represent gastric tumors of low or no malignant potential. Lab Invest. 2004;84:874–83.
96. Lasota J, Miettinen M. Clinical significance of oncogenic KIT and PDGFRA mutations in gastrointestinal stromal tumours. Histopathology. 2008;53:245–66.
97. Lee RS, Stewart C, Carter SL, Ambrogio L, Cibulskis K, Sougnez C, Lawrence MS, Auclair D, Mora J, Golub TR, et al. A remarkably simple genome underlies highly malignant pediatric rhabdoid cancers. J Clin Invest. 2012;122:2983–8.
98. Levard A, Derbel O, Méeus P, Ranchère D, Ray-Coquard I, Blay J, Cassier P. Outcome of patients with advanced solitary fibrous tumors: the Centre Léon Bérard experience. BMC Cancer. 2013;13:109.
99. Liegl B, Hornick J, Corless C, Fletcher C. Monoclonal antibody DOG1.1 shows higher

sensitivity than KIT in the diagnosis of gastrointestinal stromal tumors, including unusual subtypes. Am J Surg Pathol. 2009;33:437–46.
100. Liegl B, Kepten I, Le C, Zhu M, Demetri G, Heinrich M, Fletcher C, Corless C, Fletcher J. Heterogeneity of kinase inhibitor resistance mechanisms in GIST. J Pathol. 2008;216:64–74.
101. Lips DJ, Barker N, Clevers H, Hennipman A. The role of APC and beta-catenin in the aetiology of aggressive fibromatosis (desmoid tumors). Eur J Surg Oncol. 2009;35:3–10.
102. Llombart B, Serra-Guillén C, Monteagudo C, López Guerrero J, Sanmartín O. Dermatofibrosarcoma protuberans: a comprehensive review and update on diagnosis and management. Semin Diagn Pathol. 2013;30:13–28.
103. Manner J, Radlwimmer B, Hohenberger P, Mössinger K, Küffer S, Sauer C, Belharazem D, Zettl A, Coindre J, Hallermann C, et al. MYC high level gene amplification is a distinctive feature of angiosarcomas after irradiation or chronic lymphedema. Am J Pathol. 2010;176:34–9.
104. Martin J, Poveda A, Llombart-Bosch A, Ramos R, López-Guerrero J, Garcia del Muro J, Maurel J, Calabuig S, Gutierrez A, González de Sande J, et al. Deletions affecting codons 557–558 of the c-KIT gene indicate a poor prognosis in patients with completely resected gastrointestinal stromal tumors: a study by the Spanish Group for Sarcoma Research (GEIS). J Clin Oncol. 2005;23:6190–8.
105. Martin-Broto J, Gutierrez A, Garcia-del-Muro X, Lopez-Guerrero J, Martinez-Trufero J, de Sande L, Lainez N, Maurel J, De Juan A, Losa F, et al. Prognostic time dependence of deletions affecting codons 557 and/or 558 of KIT gene for relapse-free survival (RFS) in localized GIST: a Spanish Group for Sarcoma Research (GEIS) Study. Ann Oncol. 2010;21:1552–7.
106. McArthur G, Demetri G, van Oosterom A, Heinrich M, Debiec-Rychter M, Corless C, Nikolova Z, Dimitrijevic S, Fletcher F. Molecular and clinical analysis of locally advanced dermatofibrosarcoma protuberans treated with imatinib: Imatinib Target Exploration Consortium Study B2225. J Clin Oncol. 2005;23:866–73.
107. McCormick D, Mentzel T, Beham A, Fletcher CD. Dedifferentiated liposarcoma. Clinicopathologic analysis of 32 cases suggesting a better prognostic subgroup among pleomorphic sarcomas. Am J Surg Pathol. 1994;18:1213–23.
108. Meis-Kindblom JM, Kindblom LG, Enzinger FM. Sclerosing epithelioid fibrosarcoma. A variant of fibrosarcoma simulating carcinoma. Am J Surg Pathol. 1995;19:979–93.
109. Mendlick M, Nelson M, Pickering D, Johansson S, Seemayer T, Neff J, Vergara G, Rosenthal H, Bridge J. Translocation t(1;3)(p36.3;q25) is a nonrandom aberration in epithelioid hemangioendothelioma. Am J Surg Pathol. 2001;25:684–7.
110. Mentzel T, Schildhaus HU, Palmedo G, Buttner R, Kutzner H. Postradiation cutaneous angiosarcoma after treatment of breast carcinoma is characterized by MYC amplification in contrast to atypical vascular lesions after radiotherapy and control cases: clinicopathological, immunohistochemical and molecular analysis of 66 cases. Mod Pathol. 2012;25:75–85.
111. Miettinen M, Lasota J. Gastrointestinal stromal tumors: pathology and prognosis at different sites. Semin Diagn Pathol. 2006;23:70–83.
112. Miettinen M, Limon J, Niezabitowski A, Lasota J. Patterns of keratin polypeptides in 110 biphasic, monophasic, and poorly differentiated synovial sarcomas. Virchows Arch. 2000;437:275–83.
113. Miettinen M, Wang Z, Lasota J. DOG1 antibody in the differential diagnosis of gastrointestinal stromal tumors: a study of 1840 cases. Am J Surg Pathol. 2009;33:1401–8.
114. Mohajeri A, Tayebwa J, Collin A, Nilsson J, Magnusson L, von Steyern FV, Brosjo O, Domanski HA, Larsson O, Sciot R, et al. Comprehensive genetic analysis identifies a pathognomonic NAB2/STAT6 fusion gene, nonrandom secondary genomic imbalances, and a characteristic gene expression profile in solitary fibrous tumor. Genes Chromosomes Cancer. 2013. doi:101002/gcc22083.
115. Moore JR, Weiland AJ, Curtis RM. Localized nodular tenosynovitis: experience with 115 cases. J Hand Surg Am. 1984;9:412–7.
116. Morotti RA, Nicol KK, Parham DM, Teot LA, Moore J, Hayes J, Meyer W, Qualman SJ. An immunohistochemical algorithm to facilitate diagnosis and subtyping of rhabdomyosarcoma: the Children's Oncology Group experience. Am J Surg Pathol. 2006;30:962–8.
117. Mussi C, Schildhaus H, Gronchi A, Wardelmann E, Hohenberger P. Therapeutic consequences from molecular biology for GIST patients affected by neurofibromatosis type I. Clin Cancer Res. 2008;14:4550–5.
118. Negri T, Virdis E, Brich S, Bozzi F, Tamborini E, Tarantino E, Jocolle G, Cassinelli G, Grosso F, Sanfilippo R, et al. Functional mapping of receptor tyrosine kinases in myxoid liposarcoma. Clin Cancer Res. 2010;16:3581–93.
119. Neuhann T, Mansmann V, Merkelbach-Bruse S, Klink B, Hellinger A, Höffkes H, Wardelmann E, Schildhaus H, Tinschert S. A novel germline KIT mutation (p.L576P) in a family presenting with

juvenile onset of multiple gastrointestinal stromal tumors, skin hyperpigmentations, and esophageal stenosis. Am J Surg Pathol. 2013;37: 898–905.
120. Nielsen AL, Kiaer T. Malignant giant cell tumor of synovium and locally destructive pigmented villonodular synovitis: ultrastructural and immunohistochemical study and review of the literature. Hum Pathol. 1989;20:765–71.
121. Nieuwenhuis M, Lefevre J, Bülow S, Järvinen H, Bertario L, Kernéis S, Parc Y, Vasen H. Family history, surgery, and APC mutation are risk factors for desmoid tumors in familial adenomatous polyposis: an international cohort study. Dis Colon Rectum. 2011;54:1229–34.
122. Nilsson B, Bümming P, Meis-Kindblom JM, Odén A, Dortok A, Gustavsson B, Sablinska K, Kindblom LG. Gastrointestinal stromal tumors: the incidence, prevalence, clinical course, and prognostication in the preimatinib mesylate era – a population-based study in western Sweden. Cancer. 2005;103:821–9.
123. Oda Y, Tsuneyoshi M. Extrarenal rhabdoid tumors of soft tissue: clinicopathological and molecular genetic review and distinction from other soft-tissue sarcomas with rhabdoid features. Pathol Int. 2006;56:287–95.
124. Ognjanovic S, Linabery AM, Charbonneau B, Ross JA. Trends in childhood rhabdomyosarcoma incidence and survival in the United States, 1975–2005. Cancer. 2009;115:4218–26.
125. Olsen SH, Thomas DG, Lucas DR. Cluster analysis of immunohistochemical profiles in synovial sarcoma, malignant peripheral nerve sheath tumor, and ewing sarcoma. Mod Pathol. 2006; 19:659–68.
126. Orbach D, Rey A, Cecchetto G, Oberlin O, Casanova M, Thebaud E, Scopinaro M, Bisogno G, Carli M, Ferrari A. Infantile fibrosarcoma: management based on the European experience. J Clin Oncol. 2010;28:318–23.
127. Panagopoulos I, Hoglund M, Mertens F, Mandahl N, Mitelman F, Aman P. Fusion of the EWS and CHOP genes in myxoid liposarcoma. Oncogene. 1996;12:489–94.
128. Panagopoulos I, Mertens F, Debiec-Rychter M, Isaksson M, Limon J, Kardas I, Domanski HA, Sciot R, Perek D, Crnalic S, et al. Molecular genetic characterization of the EWS/ATF1 fusion gene in clear cell sarcoma of tendons and aponeuroses. Int J Cancer. 2002;99:560–7.
129. Panagopoulos I, Mertens F, Isaksson M, Domanski HA, Brosjo O, Heim S, Bjerkehagen B, Sciot R, Dal Cin P, Fletcher JA, et al. Molecular genetic characterization of the EWS/CHN and RBP56/CHN fusion genes in extraskeletal myxoid chondrosarcoma. Genes Chromosomes Cancer. 2002; 35(4):340–52.
130. Pantaleo M, Astolfi A, Indio V, Moore R, Thiessen N, Heinrich M, Gnocchi C, Santini D, Catena F, Formica S, et al. SDHA loss-of-function mutations in KIT-PDGFRA wild-type gastrointestinal stromal tumors identified by massively parallel sequencing. J Natl Cancer Inst. 2011;103: 983–7.
131. Parham DM, Alaggio R, Coffin CM. Myogenic tumors in children and adolescents. Pediatr Dev Pathol. 2012;15:211–38.
132. Pasini B, McWhinney S, Bei T, Matyakhina L, Stergiopoulos S, Muchow M, Boikos S, Ferrando B, Pacak K, Assie G, et al. Clinical and molecular genetics of patients with the Carney-Stratakis syndrome and germline mutations of the genes coding for the succinate dehydrogenase subunits SDHB, SDHC, and SDHD. Eur J Hum Genet. 2008;16:79–88.
133. Patel KU, Szabo SS, Hernandez VS, Prieto VG, Abruzzo LV, Lazar AJ, Lopez-Terrada D. Dermatofibrosarcoma protuberans COL1A1-PDGFB fusion is identified in virtually all dermatofibrosarcoma protuberans cases when investigated by newly developed multiplex reverse transcription polymerase chain reaction and fluorescence in situ hybridization assays. Hum Pathol. 2008;39:184–93.
134. Pauls K, Merkelbach-Bruse S, Thal D, Büttner R, Wardelmann E. PDGFR- and c-kit mutated gastrointestinal stromal tumors (GISTs) are characterized by distinctive histological and immunohistochemical features. Histopathology. 2004;46:166–75.
135. Paulson V, Chandler G, Rakheja D, Galindo RL, Wilson K, Amatruda JF, Cameron S. High-resolution array CGH identifies common mechanisms that drive embryonal rhabdomyosarcoma pathogenesis. Genes Chromosomes Cancer. 2011;50:397–408.
136. Penel N, Marreaud S, Robin YM, Hohenberger P. Angiosarcoma: state of the art and perspectives. Crit Rev Oncol Hematol. 2011;80: 257–63.
137. Philippe-Chomette P, Kabbara N, Andre N, Pierron G, Coulomb A, Laurence V, Blay JY, Delattre O, Schleiermacher G, Orbach D. Desmoplastic small round cell tumors with EWS-WT1 fusion transcript in children and young adults. Pediatr Blood Cancer. 2012;58: 891–7.
138. Rabbitts TH, Forster A, Larson R, Nathan P. Fusion of the dominant negative transcription regulator CHOP with a novel gene FUS by translocation t(12;16) in malignant liposarcoma. Nat Genet. 1993;4:175–80.
139. Raney RB, Anderson JR, Barr FG, Donaldson SS, Pappo AS, Qualman SJ, Wiener ES, Maurer HM, Crist WM. Rhabdomyosarcoma and

undifferentiated sarcoma in the first two decades of life: a selective review of intergroup rhabdomyosarcoma study group experience and rationale for Intergroup Rhabdomyosarcoma Study V. J Pediatr Hematol Oncol. 2001;23:215–20.
140. Raney RB, Anderson JR, Brown KL, Huh WW, Maurer HM, Meyer WH, Parham DM, Rodeberg DA, Wolden SL, Donaldson SS. Treatment results for patients with localized, completely resected (group I) alveolar rhabdomyosarcoma on Intergroup Rhabdomyosarcoma Study Group (IRSG) protocols III and IV, 1984–1997: a report from the Children's Oncology Group. Pediatr Blood Cancer. 2010;55:612–6.
141. Rao AS, Vigorita VJ. Pigmented villonodular synovitis (giant-cell tumor of the tendon sheath and synovial membrane). A review of eighty-one cases. J Bone Joint Surg Am. 1984;66:76–94.
142. Ray-Coquard I, Blay JY, Italiano A, Le Cesne A, Penel N, Zhi J, Heil F, Rueger R, Graves B, Ding M, et al. Effect of the MDM2 antagonist RG7112 on the P53 pathway in patients with MDM2-amplified, well-differentiated or dedifferentiated liposarcoma: an exploratory proof-of-mechanism study. Lancet Oncol. 2012;13:1133–40.
143. Reis H, Hager T, Wohlschlaeger J, Bauer S, Katenkamp K, Katenkamp D, Bubba H. Mammalian target of rapamycin pathway activity in alveolar soft part sarcoma. Hum Pathol. 2013;44(10):2266–74.
144. Rekhi B, Ingle A, Agarwal M, Puri A, Laskar S, Jambhekar N. Alveolar soft part sarcoma 'revisited': clinicopathological review of 47 cases from a tertiary cancer referral centre, including immunohistochemical expression of TFE3 in 22 cases and 21 other tumours. Pathology. 2012;44: 11–7.
145. Rutkowski P, Van Glabbeke M, Rankin C, Ruka W, Rubin B, Debiec-Rychter M, Lazar A, Gelderblom H, Sciot R, Lopez-Terrada D, et al. Imatinib mesylate in advanced dermatofibrosarcoma protuberans: pooled analysis of two phase II clinical trials. J Clin Oncol. 2010;28:1772–9.
146. Ryan CW, Desai J. The past, present, and future of cytotoxic chemotherapy and pathway-directed targeted agents for soft tissue sarcoma. Am Soc Clin Oncol Educ Book. 2013;2013: 386–93.
147. Sandberg A, Bridge J. Updates in the cytogenetics and molecular genetics of bone and soft tissue tumors. gastrointestinal stromal tumors. Cancer Genet Cytogenet. 2002;135:1–22.
148. Schildhaus H, Cavlar T, Binot E, Buettner R, Wardelmann E, Merkelbach-Bruse S. Inflammatory fibroid polyps harbour mutations in the platelet-derived growth factor receptor alpha (PDGFRA) gene. J Pathol. 2008;216:176–82.
149. Schirosi L, Lantuejoul S, Cavazza A, Murer B, Yves Brichon P, Migaldi M, Sartori G, Sgambato A, Rossi G. Pleuro-pulmonary solitary fibrous tumors: a clinicopathologic, immunohistochemical, and molecular study of 88 cases confirming the prognostic value of de Perrot staging system and p53 expression, and evaluating the role of c-kit, BRAF, PDGFRs (alpha/beta), c-met, and EGFR. Am J Surg Pathol. 2008;32:1627–42.
150. Schneider-Stock R, Boltze C, Lasota J, Miettinen M, Peters B, Pross M, Roessner A, Günther T. High prognostic value of p16INK4 alterations in gastrointestinal stromal tumors. J Clin Oncol. 2003;21:1688–97.
151. Schneider-Stock R, Boltze C, Lasota J, Peters B, Corless C, Ruemmele P, Terracciano L, Pross M, Insabato L, Di Vizio D, et al. Loss of p16 protein defines high-risk patients with gastrointestinal stromal tumors: a tissue microarray study. Clin Cancer Res. 2005;11:638–45.
152. Schneider-Stock R, Walter H, Radig K, Rys J, Bosse A, Kuhnen C, Hoang-Vu C, Roessner A. MDM2 amplification and loss of heterozygosity at Rb and p53 genes: no simultaneous alterations in the oncogenesis of liposarcomas. J Cancer Res Clin Oncol. 1998;124:532–40.
153. Schneppenheim R, Fruhwald MC, Gesk S, Hasselblatt M, Jeibmann A, Kordes U, Kreuz M, Leuschner I, Martin Subero JI, Obser T, et al. Germline nonsense mutation and somatic inactivation of SMARCA4/BRG1 in a family with rhabdoid tumor predisposition syndrome. Am J Hum Genet. 2010;86:279–84.
154. Schwartz HS, Unni KK, Pritchard DJ. Pigmented villonodular synovitis. A retrospective review of affected large joints. Clin Orthop Relat Res. 1989;243–255.
155. Schweizer L, Koelsche C, Sahm F, Piro R, Capper D, Reuss D, Pusch S, Habel A, Meyer J, Göck T, et al. Meningeal hemangiopericytoma and solitary fibrous tumors carry the NAB2-STAT6 fusion and can be diagnosed by nuclear expression of STAT6 protein. Acta Neuropathol. 2013;125:651–8.
156. Scrable H, Witte D, Shimada H, Seemayer T, Sheng WW, Soukup S, Koufos A, Houghton P, Lampkin B, Cavenee W. Molecular differential pathology of rhabdomyosarcoma. Genes Chromosomes Cancer. 1989;1:23–35.
157. Sebire NJ, Malone M. Myogenin and MyoD1 expression in paediatric rhabdomyosarcomas. J Clin Pathol. 2003;56:412–6.
158. Sirvent N, Coindre JM, Maire G, Hostein I, Keslair F, Guillou L, Ranchere-Vince D, Terrier P, Pedeutour F. Detection of MDM2-CDK4 amplification by fluorescence in situ hybridization in 200 paraffin-embedded tumor samples: utility in diagnosing adipocytic lesions and comparison

with immunohistochemistry and real-time PCR. Am J Surg Pathol. 2007;31:1476–89.

159. Sobanko JF, Meijer L, Nigra TP. Epithelioid sarcoma: a review and update. J Clin Aesthet Dermatol. 2009;2:49–54.

160. Somerhausen NS, Fletcher CD. Diffuse-type giant cell tumor: clinicopathologic and immunohistochemical analysis of 50 cases with extraarticular disease. Am J Surg Pathol. 2000;24: 479–92.

161. Stacchiotti S, Negri T, Palassini E, Conca E, Gronchi A, Morosi C, Messina A, Pastorino U, Pierotti M, Casali P, Pilotti S. Sunitinib malate and figitumumab in solitary fibrous tumor: patterns and molecular bases of tumor response. Mol Cancer Ther. 2010;9:1286–97.

162. Stacchiotti S, Tamborini E, Marrari A, Brich S, Rota SA, Orsenigo M, Crippa F, Morosi C, Gronchi A, Pierotti MA, et al. Response to sunitinib malate in advanced alveolar soft part sarcoma. Clin Cancer Res. 2009;15:1096–104.

163. Stratakis C, Carney J. The triad of paragangliomas, gastric stromal tumours and pulmonary chondromas (Carney triad), and the dyad of paragangliomas and gastric stromal sarcomas (Carney-Stratakis syndrome): molecular genetics and clinical implications. J Intern Med. 2009;266:43–52.

164. Su L, Sampaio AV, Jones KB, Pacheco M, Goytain A, Lin S, Poulin N, Yi L, Rossi FM, Kast J, et al. Deconstruction of the SS18-SSX fusion oncoprotein complex: insights into disease etiology and therapeutics. Cancer Cell. 2012;21:333–47.

165. Sultan I, Rodriguez-Galindo C, Saab R, Yasir S, Casanova M, Ferrari A. Comparing children and adults with synovial sarcoma in the surveillance, epidemiology, and end results program, 1983 to 2005: an analysis of 1268 patients. Cancer. 2009;115:3537–47.

166. Sun B, Sun Y, Wang J, Zhao X, Zhang S, Liu Y, Li X, Feng Y, Zhou H, Hao X. The diagnostic value of SYT-SSX detected by reverse transcriptase-polymerase chain reaction (RT-PCR) and fluorescence in situ hybridization (FISH) for synovial sarcoma: a review and prospective study of 255 cases. Cancer Sci. 2008;99:1355–61.

167. Takahashi Y, Oda Y, Kawaguchi K, Tamiya S, Yamamoto H, Suita S, Tsuneyoshi M. Altered expression and molecular abnormalities of cell-cycle-regulatory proteins in rhabdomyosarcoma. Mod Pathol. 2004;17:660–9.

168. Takazawa Y, Sakurai S, Sakuma Y, Ikeda T, Yamaguchi J, Hashizume Y, Yokoyama S, Motegi A, Fukayama M. Gastrointestinal stromal tumors of neurofibromatosis type I (von Recklingshausen's disease). Am J Surg Pathol. 2005;29:755–63.

169. Taylor BS, Barretina J, Socci ND, Decarolis P, Ladanyi M, Meyerson M, Singer S, Sander C. Functional copy-number alterations in cancer. PLoS One. 2008;3:e3179.

170. Taylor BS, DeCarolis PL, Angeles CV, Brenet F, Schultz N, Antonescu CR, Scandura JM, Sander C, Viale AJ, Socci ND, Singer S. Frequent alterations and epigenetic silencing of differentiation pathway genes in structurally rearranged liposarcomas. Cancer Discov. 2011;1:587–97.

171. Tejpar S, Nollet F, Li C, Wunder J, Michils G, dal Cin P, Van Cutsem E, Bapat B, van Roy F, Cassiman J, Alman B. Predominance of beta-catenin mutations and beta-catenin dysregulation in sporadic aggressive fibromatosis (desmoid tumor). Oncogene. 1999;18:6615–20.

172. Terrier-Lacombe M, Guillou L, Maire G, Terrier P, Vince D, de Saint Aubain Somerhausen N, Collin F, Pedeutour F, Coindre J. Dermatofibrosarcoma protuberans, giant cell fibroblastoma, and hybrid lesions in children: clinicopathologic comparative analysis of 28 cases with molecular data – a study from the French Federation of Cancer Centers Sarcoma Group. Am J Surg Pathol. 2003;27:27–39.

173. Terry J, Saito T, Subramanian S, Ruttan C, Antonescu CR, Goldblum JR, Downs-Kelly E, Corless CL, Rubin BP, van de Rijn M, et al. TLE1 as a diagnostic immunohistochemical marker for synovial sarcoma emerging from gene expression profiling studies. Am J Surg Pathol. 2007; 31:240–6.

174. Tothova Z, Wagner A. Anaplastic lymphoma kinase-directed therapy in inflammatory myofibroblastic tumors. Curr Opin Oncol. 2012;24: 409–13.

175. Tsuneyoshi M, Daimaru Y, Hashimoto H, Enjoji M. Malignant soft tissue neoplasms with the histologic features of renal rhabdoid tumors: an ultrastructural and immunohistochemical study. Hum Pathol. 1985;16:1235–42.

176. Ushijima M, Hashimoto H, Tsuneyoshi M, Enjoji M. Giant cell tumor of the tendon sheath (nodular tenosynovitis). A study of 207 cases to compare the large joint group with the common digit group. Cancer. 1986;57:875–84.

177. van Gaal JC, Flucke UE, Roeffen MH, de Bont ES, Sleijfer S, Mavinkurve-Groothuis AM, Suurmeijer AJ, van der Graaf WT, Versleijen-Jonkers YM. Anaplastic lymphoma kinase aberrations in rhabdomyosarcoma: clinical and prognostic implications. J Clin Oncol. 2012;30: 308–15.

178. Vries RG, Bezrookove V, Zuijderduijn LM, Kia SK, Houweling A, Oruetxebarria I, Raap AK, Verrijzer CP. Cancer-associated mutations in chromatin remodeler hSNF5 promote chromosomal instability by compromising the mitotic checkpoint. Genes Dev. 2005;19:665–70.

179. Walluks K, Chen Y, Woelfel C, Yang L, Cui T, Seliger C, Geier C, Knosel T, Hauke S, Petersen I.

Molecular and clinicopathological analysis of dermatofibrosarcoma protuberans. Pathol Res Pract. 2013;209:30–5.
180. Wang W, Evans H, Meis J, Liegl-Atzwanger B, Bovee J, Goldblum J, Billings S, Rubin B, López-Terrada D, Lazar A. FUS rearrangements are rare in 'pure' sclerosing epithelioid fibrosarcoma. Mod Pathol. 2012;25:846–53.
181. Wardelmann E, Biermann K, Merkelbach-Bruse S, Schildhaus H, Thomas N, Buettner R, Pietsch T, Heinicke T, Speidel N, Pink D, et al. Polyclonal resistance in gastrointestinal stromal tumor treated with sequential kinase inhibitors. Clin Cancer Res. 2006;12:6206.
182. Wardelmann E, Hohenberger P, Reichardt P, Merkelbach-Bruse S, Schildhaus H, Büttner R. Gastrointestinal stromal tumors of the stomach. Updates and differences compared to other locations. Pathologie. 2010;31:195–8.
183. Wardelmann E, Hrychyk A, Merkelbach-Bruse S, Pauls K, Goldstein J, Hohenberger P, Losen I, Manegold C, Büttner R, Pietsch T. Association of platelet-derived growth factor receptor alpha mutations with gastric primary site and epithelioid or mixed cell morphology in gastrointestinal stromal tumors. J Mol Diagn. 2004;6:197–204.
184. Wardelmann E, Losen I, Hans V, Neidt I, Speidel N, Bierhoff E, Heinicke T, Pietsch T, Büttner R, Merkelbach-Bruse S. Deletion of Trp-557 and Lys-558 in the juxtamembrane domain of the c-kit protooncogene is associated with metastatic behavior of gastrointestinal stromal tumors. Int J Cancer. 2003;106:887–95.
185. Wardelmann E, Merkelbach-Bruse S, Pauls K, Thomas N, Schildhaus H, Heinicke T, Speidel N, Pietsch T, Buettner R, Pink D, et al. Polyclonal evolution of multiple secondary *KIT* mutations in gastrointestinal stromal tumors under treatment with imatinib mesylate. Clin Cancer Res. 2006;12:1743–9.
186. Wardelmann E, Thomas N, Merkelbach-Bruse S, Pauls K, Speidel N, Büttner R, Bihl H, Leutner C, Heinicke T, Hohenberger P. Acquired resistance to imatinib in gastrointestinal stromal tumors caused by multiple *KIT* mutations. Lancet Oncol. 2005;6:249–51.
187. Weber-Hall S, Anderson J, McManus A, Abe S, Nojima T, Pinkerton R, Pritchard-Jones K, Shipley J. Gains, losses, and amplification of genomic material in rhabdomyosarcoma analyzed by comparative genomic hybridization. Cancer Res. 1996;56:3220–4.
188. West R, Corless C, Chen X, Rubin B, Subramanian S, Montgomery K, Zhu S, Ball C, Nielsen T, Patel R, et al. The novel marker, DOG1, is expressed ubiquitously in gastrointestinal stromal tumors irrespective of KIT or PDGFRA mutation status. Am J Pathol. 2004;165:107–13.
189. West RB, Rubin BP, Miller MA, Subramanian S, Kaygusuz G, Montgomery K, Zhu S, Marinelli RJ, De Luca A, Downs-Kelly E, et al. A landscape effect in tenosynovial giant-cell tumor from activation of CSF1 expression by a translocation in a minority of tumor cells. Proc Natl Acad Sci U S A. 2006;103:690–5.
190. Williams A, Bartle G, Sumathi V, Meis J, Mangham D, Grimer R, Kindblom L. Detection of ASPL/TFE3 fusion transcripts and the TFE3 antigen in formalin-fixed, paraffin-embedded tissue in a series of 18 cases of alveolar soft part sarcoma: useful diagnostic tools in cases with unusual histological features. Virchows Arch. 2011;458:291–300.
191. Woelfel C, Liehr T, Weise A, Langrehr J, Kotb W, Pacyna-Gengelbach M, Katenkamp D, Petersen I. Molecular cytogenetic characterization of epithelioid hemangioendothelioma. Cancer Genet. 2011;204:671–6.
192. Xia SJ, Pressey JG, Barr FG. Molecular pathogenesis of rhabdomyosarcoma. Cancer Biol Ther. 2002;1:97–104.
193. Xie Y, Skytting B, Nilsson G, Brodin B, Larsson O. Expression of insulin-like growth factor-1 receptor in synovial sarcoma: association with an aggressive phenotype. Cancer Res. 1999;59:3588–91.
194. Zambrano E, Reyes-Mugica M, Franchi A, Rosai J. An osteoclast-rich tumor of the gastrointestinal tract with features resembling clear cell sarcoma of soft parts: reports of 6 cases of a GIST simulator. Int J Surg Pathol. 2003;11:75–81.
195. Zhong M, De Angelo P, Osborne L, Keane-Tarchichi M, Goldfischer M, Edelmann L, Yang Y, Linehan WM, Merino MJ, Aisner S, Hameed M. Dual-color, break-apart FISH assay on paraffin-embedded tissues as an adjunct to diagnosis of Xp11 translocation renal cell carcinoma and alveolar soft part sarcoma. Am J Surg Pathol. 2010;34:757–66.
196. Zibat A, Missiaglia E, Rosenberger A, Pritchard-Jones K, Shipley J, Hahn H, Fulda S. Activation of the hedgehog pathway confers a poor prognosis in embryonal and fusion gene-negative alveolar rhabdomyosarcoma. Oncogene. 2010;29:6323–30.

PART V

Genomic Applications of Inherited and Infectious Diseases

CHAPTER 30

Genomic Applications in Inherited Genetic Disorders

BRYAN L. KROCK, RONG MAO, D. HUNTER BEST, ELAINE LYON

Introduction 535
Prenatal Applications 539
Fetal Applications 540
Newborn/Infant Applications 542
Adult or Young Adult Onset Applications 545
Exome and Genome 546
Conclusions 548

Introduction

The enhanced capabilities of next-generation sequencing (NGS) have changed how clinical laboratory geneticists approach test design and application. For instance, some disorders display extreme locus heterogeneity, and can be caused by mutations in one of many genes. Retinitis pigmentosa (RP) is an example of locus heterogeneity, and is caused by mutations in at least 60 different genes. Traditionally, molecular genetic testing for this disorder would use a tiered approach where initial testing would prioritize the most frequently mutated gene, *RHO* (rhodopsin), which accounts for 30–40 % of autosomal dominant retinitis pigmentosa [18]. Patients without a *RHO* gene mutation would be tested for the next common gene, and so on until either a causative mutation was found or the costs associated with additional testing outweighed the likelihood of success. The latter scenario is quickly realized for many patients, leaving them without a genetic diagnosis. This phenomenon has commonly been referred to as the "diagnostic odyssey," which can be frustrating for patients and families as well as complicate genetic counseling for family members. Additionally, genetic testing often is not available for very rare causes of a disorder, because these extremely low volume tests are not financially viable for most clinical laboratories. NGS based tests may represent the only diagnostic option for some patients, particularly those who are clearly affected with a genetic disorder but display an atypical presentation or have a rare, poorly studied genetic disorder. In these cases, the shotgun approach to genetic testing may prove to be an indispensable tool for clinical diagnosis.

Although clinical NGS is still in its infancy, numerous academic and commercial laboratories have adopted the technology. This is due, in part, to the ability of NGS to simultaneously sequence all genes associated with a disorder at prices and turnaround times that are competitive with traditional Sanger sequencing. In fact, for larger gene panels, NGS is already far more cost-effective while improving time to results over a sequential approach. As the technology, bioinformatics, and laboratory experience with NGS improves, NGS based assays will compare even more favorably to Sanger based methods, leading to widespread use for clinical NGS testing. The current applications of NGS genetic tests demonstrate its broad

G.J. Netto and I. Schrijver (eds.), *Genomic Applications in Pathology*,
DOI 10.1007/978-1-4939-0727-4_30, © Springer Science+Business Media New York 2015

Table 30-1 Examples of Clinically Available NGS Tests for Inherited or Germ-Line Disorders

NGS test	Age of application Fetal	Infant/child	Adult
Amyotrophic lateral sclerosis			X
Aortopathies		X	X
Autism spectrum disorder		X	
Brain malformations/holoprosencephaly	X	X	
Cardiomyopathy		X	X
Chromosomal aneuploidies from cff DNA	X		
Ciliopathy	X	X	
Congenital orders of glycosylation		X	
Epilepsy		X	X
Eye disorders		X	X
Immunodeficiency		X	X
Inherited cancer			X[a]
Mitochondrial disorders		X	
Neuromuscular		X	X
Noonan syndrome and related disorders	X	X	X
X-linked intellectual disability		X	
Skeletal dysplasia	X	X	
Whole-exome sequencing	X	X	X
Whole-genome sequencing	X	X	X

Cff cell free fetal
[a]For adult cancers. Children may be tested for pediatric or early-onset cancers

utility, as they can be appropriate to every stage of life – from preconception or fetal screening to newborn, childhood, and adult onset disorders.

Clinical NGS based tests for inherited diseases can focus on a subset of genes, all coding regions (exome), or even the whole genome. Each of these approaches has advantages and disadvantages that should be considered with regard to both test development and clinical application. Multi-gene panels generally focus on a specific disorder that displays genetic heterogeneity or a group of disorders that have overlapping genetics and/or phenotypes (see Table 30.1). NGS gene panels commonly have been developed for genetically heterogeneous disorders, as they are able to sequence hundreds of genes at very high coverage, imparting high sensitivity and specificity to the assay. This high depth of coverage, along with the digital nature of NGS allows identification of somatic mosaicism with greater sensitivity than Sanger sequencing. This advantage has clear utility for disorders commonly associated with somatic mosaicism, such as tuberous sclerosis or Proteus syndrome [31, 66]. In addition, recent reports suggest somatic mosaicism may be more common than previously appreciated [4, 50]. For instance, a patient with Cowden syndrome, a disorder not commonly associated with mosaicism, had tested negative for *PTEN* mutations by Sanger sequencing DNA from peripheral blood. Application of a NGS panel containing *PTEN* to the same sample revealed a low frequency frameshift *PTEN* mutation that was subsequently found in the heterozygous state in skin fibroblasts, confirming somatic mosaicism [50]. Continued application of NGS panels will likely improve our understanding of the frequency of mosaicism in many inherited genetic disorders. The high depth of sequencing obtained with NGS panels can also be exploited to assess copy number alterations with the appropriate bioinformatic tools [42]. While not commonly employed in clinical assays for inherited disorders, quantitative analyses may become more widely adopted because they provide clinically important data for no additional reagent costs.

In addition to genetically heterogeneous disorders, diseases that display overlapping phenotypic spectra are also good candidates for NGS panels, as sequencing of multiple genes may be required to clarify a clinical diagnosis. The aortopathies are an example and will be discussed in greater detail later in this chapter. While an NGS panel still costs more than the average single gene sequencing, gene panels will likely replace cascading genetic testing algorithms, especially as NGS reagent costs decrease. In some cases, an NGS panel will become part of a testing algorithm and complement existing Sanger sequencing tests. For example, mutations in *GJB2* and deletions of *GJB6* account for up to 20 % of hearing loss cases [12]. Accordingly, sequential *GJB2* mutation studies and *GJB6* targeted deletion analysis prior to an NGS panel would prove cost-effective. In addition to their advantages over Sanger based methods, NGS panels also have specific benefits over exome and whole-genome sequencing (WGS). First, bioinformatic processing and test interpretation for NGS panels is less challenging, as fewer genes are analyzed and fewer variants are recovered. Additionally, NGS panels will seldom yield incidental findings, making clinical reports less challenging. Next, whole-exome sequencing (WES)/WGS tests have gaps in coverage, even in known disease causing genes. NGS panels can fill in these gaps with Sanger sequencing to achieve complete coverage of targeted genes. Also, depth of sequencing is greater for NGS panels, which are therefore more sensitive for detecting somatic mosaicism and potentially copy number variation. Finally, the ability to multiplex more samples and use smaller capacity platforms makes NGS panels cheaper and faster than clinical exome and whole-genome tests.

WES involves selective enrichment for and sequencing of the entire coding region of the genome. While it represents only approximately 1.5 % of the total genome, mutations in the exome account for about 85 % of all known disease causing mutations [57]. By selectively sequencing the coding region of the genome, clinical molecular geneticists can interrogate all known genes at a reasonable cost. WES has been applied in a research setting for gene discovery with great success, often enabling the identification of a disease causing gene by studying a single individual or family. In this way, clinical WES can blur the lines between a clinical test and research assay and novel disease causing genes have likely been identified in patients referred for clinical WES. Although there are disadvantages of WES compared to NGS panels, its principal advantage is that it provides an unbiased view of a patient's mutational landscape. Along these lines, WES does not require a priori suspicion of a specific genetic disorder, and accordingly it represents the only molecular diagnostic option for some patients. Clinical WES may also yield unexpected genetic diagnoses that shape the clinical diagnosis of patients. For instance, WES may identify two or more disorders contributing to the overall phenotype. In an early demonstration of genomic analysis, a pair of siblings studied for the genetic cause of Miller syndrome was shown to have two recessive disorders, Miller syndrome with the concurrent identification of the responsible gene, and primary ciliary dyskinesia [46].

WES has been more widely adopted for clinical use than WGS, which involves sequencing the entire genome. This is mainly because genetic alterations in the exome are easier to interpret than intragenic and intergenic regions, particularly when they cause direct changes to the protein sequence or alter canonical splice donor/acceptor sites. Nonetheless, WGS is clinically available. In contrast to WES and NGS panels, WGS does not involve target enrichment, and accordingly, may represent a less biased representation of a patient's variant landscape. Additionally, large gene rearrangements can potentially be detected by WGS. While dramatic technological advances have enabled clinical geneticists to probe the entire genome, the value added over WES is currently unclear. The vast majority of noncoding regions are uninterpretable, and the large amounts of extra data take up tremendous storage space and computational time. Given the difficulty in interpreting variants outside of coding and intron/exon boundary regions, WGS is analyzed similar to an exome. If the causative mutation(s) is/are not found, exploring variants in other genomic regions could be attempted. Direct comparison of clinical WES and WGS has not yet been reported, but would be useful for evaluating the utility of WGS. Due to the similar nature of clinical WES and WGS applications, they will be discussed together for the remainder of this chapter.

Similar to other clinical genetic tests, selecting the appropriate tests and patients for NGS-based studies is critical to maximize diagnostic yield and minimize unproductive genetic testing. As NGS panels are targeted to a phenotype or group of disorders, patients with a clear clinical presentation consistent with a particular disorder are good candidates for a disease-specific NGS panel. In this way, patients for NGS panels are evaluated in a similar manner to traditional genetic tests. Some clinicians may prefer WES instead of gene panels because of the broader coverage and rationalize that WES will cover all of the genes in an NGS panel. However, most NGS panels will achieve complete coverage of the target genes (although filling in with Sanger sequencing may be needed), and should detect known mutations (provided that the type of mutation is detectable by NGS), giving a higher negative predictive value. In contrast, WES may have low coverage in clinically relevant genes and miss important mutations as a result. Moreover, poorly covered genes may not be clearly conveyed in a clinical report, potentially giving a false confidence in a test result. The American College of Medical Genetics and Genomics (ACMG) has outlined indications for WES and WGS diagnostic testing [1]. WES/WGS sequencing should be considered for patients in whom the phenotype or family history strongly implicates a disorder with a genetic etiology, but the phenotype does not correspond with a specific disorder for which a genetic test targeting a specific gene is available on a clinical basis. WES/WGS is also indicated for defined genetic disorders that demonstrate a high degree of genetic heterogeneity, making analysis of multiple genes simultaneously a more practical approach. Additionally, patients with an apparent genetic disorder, but who have failed to obtain a genetic diagnosis with available genetic tests are candidates for WES/WGS. Lastly, a fetus with a likely genetic disorder that has not obtained a diagnosis with specific genetic tests available for the phenotype is also a candidate for WES/WGS. Given the costs and technical/analytical limitations of WES/WGS, patients should be tested for copy number variations by a cytogenomic microarray prior to WES, as this is typically a more cost-effective testing paradigm. These selection criteria will obviously enrich for patients with novel genetic disorders or mutations in genes not previously associated with a disease. Accordingly, initial estimates of the diagnostic yield of clinical WES/WGS were low. Recent reports suggest a much higher success rate [74].

Whereas NGS represents an extremely powerful technology for genetic diagnostics, its limitations must be considered before its clinical application. First, WES and NGS panels cannot reliably detect chromosomal translocations. Approaches to assess copy number variations, large deletions/insertions and gene fusions are being developed or are in the early stages of clinical application, but are not widely in use. NGS, similar to traditional Sanger based methods, often cannot amplify GC rich regions and can suffer from allele drop out. Accordingly, clinically relevant promoter or exon 1 mutations can potentially be missed. Long repetitive elements, such as those observed in disorders with triplet repeat expansions are particularly problematic for NGS, both for the sequencing chemistry and bioinformatic sequencing alignment. When testing for disorders that can be caused by repeat expansion or more traditional mutations, it is important to rule out the repeat expansion before an NGS panel is ordered. For instance, in those patients tested for X-linked intellectual disability, triplet repeat expansion in *FMR1* should be ruled out prior to NGS-based tests. NGS panels and exome sequencing also rely on the enrichment of targeted exonic regions, the process of which is not 100 % efficient, leaving areas with low or no coverage. Although Sanger sequencing can supplement deficient regions in NGS panels to achieve 100 % coverage, this approach is not feasible for WES. Clinical groups have sought to enhance the coverage of WES by augmenting the enrichment reagents for known clinically relevant genes. These approaches represent improvements over initial clinical exome sequencing, and similar advancements in commercially available reagents are likely. Interpretation of rare and novel sequence variants is already a significant challenge for traditional Sanger sequencing-based tests. As the number of genes sequenced by NGS panels and by WES/WGS are orders of magnitude greater than targeted mutation studies and Sanger sequencing, there is a proportional increase in novel variants in NGS-based assays. These variants are challenging to evaluate and take significant amounts of time for manual review; as a result they represent a significant interpretive burden that will be encountered with nearly all WES/WGS cases. The inevitable increase in variants of uncertain significance (VUS) in clinical reports represents another difficulty, as they may be understood differently by ordering physicians. Additionally, the medical literature is evaluating novel variants in both known and novel genes at a record pace, which will likely alter the classification of previously reported variants. Integrating these new findings into previously reported cases is and will remain challenging, as standards of best practice have not been established and the tools to do so are unavailable. One approach to minimize reporting of VUS is to perform WES/WGS on trios to enable the bioinformatics sorting of variants based on the mode of inheritance. However, in our experience with WES, testing is often ordered only for the proband. This results in an increased number of variants that have to be manually evaluated, increasing the time required for analysis. Additionally, follow-up testing of family members to confirm *de novo* or biparental inheritance can be challenging for proband-only WES. While not insurmountable, these challenges increase the turnaround times and costs associated with these diagnostic tests. As clinical NGS is still a young field and has relatively high reagent costs, studies demonstrating the clinical utility and defining the performance characteristics of NGS tests are small, proof of concept studies. Although the initial consensus is that this platform performs very well for clinical purposes, large scale studies will be essential to fully characterize the diagnostic yield, sensitivity, specificity and cost-effectiveness of NGS tests compared to current gold standards.

Prenatal Applications

Carrier testing for inherited disorders has traditionally been performed for monogenetic, autosomal recessive disorders with very high allele frequencies, such as cystic fibrosis, or in populations with elevated risk for genetic disease, based on either family history or ethnicity, most notably the Ashkenazi Jewish (AJ) population [17]. Appropriate application of preconception screening can be a profoundly effective public health tool, as demonstrated by the >90 % reduction in the incidence of Tay Sachs disease in the American AJ population[23, 27, 44]. However, carrier screening outside of the few aforementioned situations has been impractical for technical and logistical reasons—until recently. NGS has enabled clinicians and laboratory professionals to consider expanded carrier screening, where prospective parents could be screened for the vast majority of known deleterious recessive mutations in a time-effective and cost-effective manner. This new approach could reduce the overall prevalence of severe genetic disorders, which collectively account for ~20 % of infant mortality [8], and facilitate the prenatal diagnosis of genetic disorders for at-risk pregnancies, leading to early intervention and improved outcomes for affected infants. Proof of concept studies were first reported in 2011, where a targeted NGS panel was designed to screen 7,717 genomic regions for known disease causing mutations in 437 genes that cause 448 severe recessive childhood diseases [5]. Preconception screening was performed in 104 unrelated individuals and the authors found the average carrier burden for severe pediatric recessive mutations was 2.8 per individual, consistent with previous estimates [28]. Importantly, the authors presented cost analyses that projected an overall analytical cost of $378, equating to less than $1 per condition tested. However, this figure does not include costs associated with test interpretation, reporting and other ancillary costs associated with a complex genetic test.

Given that large scale preconception screening has the potential to dramatically reduce the incidence of mortality and morbidity due to genetic disease, significant hurdles to the broad application of this screening paradigm remain. Variant classification is a significant challenge for all NGS based tests, even when clear phenotypes and modes of inheritance are known. Interpretation of novel variants in the absence of a phenotype adds another layer of complexity to this already difficult process. One study found that 27 % of mutations cited in the literature were found to be common polymorphisms or misannotated [5], a critical problem for all NGS based testing that necessitates careful examination of all variants reported. This will add to the already significant time and cost of data analysis. When screening asymptomatic carriers, rare nonconservative and nonsynonymous variants will likely be identified in many patients, some of which would likely be scored as VUS. The increased psychological stress for parents and perhaps increased reliance upon pre-implantation genetic diagnosis due to VUS reports need to be carefully considered for expanded carrier screening. Moreover, the concept of variable penetrance needs to be clearly conveyed to patients concerning appropriate genetic disorders. One approach to minimize many of the aforementioned concerns is to restrict analysis to known pathogenic mutations, using NGS as a multi-gene, multi-mutation, but targeted panel. However, rare, novel pathogenic mutations may be missed with this method. If a mutation is identified in a targeted panel, full gene analysis should be considered for the reproductive partner, but with the same analytic and interpretive issues previously described. Although a comprehensive single test, this technology in its current form will not be able to identify some of the most frequent disease causing mutations, of which triplet repeat expansion fragile X syndrome is an example. This limitation of NGS needs to be clearly communicated to both patients and ordering physicians. Ultimately, expanded carrier screening is an area likely to undergo dramatic changes due to NGS. In the face of drastic change, clinical guidelines need to be established to ensure appropriate application, reporting and counseling of NGS based expanded carrier screening.

Fetal Applications

Cell Free Fetal DNA in Maternal Plasma

Testing of fetal genotypes has traditionally relied upon invasive sampling of fetal cellular material through amniocentesis or chorionic villus sampling, both of which carry a small but significant risk for fetal loss. Therefore, a noninvasive means of fetal DNA sampling for genetic evaluation has long been pursued. Fetal lymphocytes are present in very small numbers in maternal blood, but their extreme rarity, challenges to their purification and concerns about persistence after birth have precluded their use in clinical testing. Cell free fetal DNA (cffDNA) was later identified in maternal plasma and serum, suggesting that this could be an easily accessible, noninvasive source of fetal DNA for genetic testing [34]. This source of cffDNA is the result of normal placental cell apoptosis, which releases highly fragmented DNA representative of the fetal genotype into the maternal circulation. Importantly, cffDNA can be detected from 4 weeks gestation until birth [22], making it amenable to genetic testing in at-risk pregnancies. The entire genome is represented in cffDNA [33], suggesting that this platform is appropriate for molecular testing for the vast majority of inherited disorders. Moreover, cffDNA is highly unstable in the maternal circulation and is cleared soon after birth, meaning that a sample will not be contaminated with fetal DNA from prior pregnancies[36]. Fetal DNA represents 5–10 % of total plasma DNA, with the remainder maternal in origin [40]. While the fraction of cffDNA increases with fetal age, pure fetal DNA cannot be extracted from maternal serum, and the maternal DNA background has been the major hurdle to the use of cffDNA for molecular diagnostics. Accordingly, the first application of cffDNA testing was to detect fetal Y chromosome sequences in maternal plasma, circumventing the issue of contaminating maternal DNA [34]. Other early clinical applications of cffDNA also reflect this constraint, as they include determining fetal Rh D status in Rh D negative mothers and detecting paternally inherited autosomal dominant and recessive mutations [11]. Initial efforts to broaden the applicability of cffDNA relied on allelic heterozygosity between the fetus and mother to determine fetal chromosomal dosage in testing for fetal aneuploidy, primarily trisomies 21, 18, and 13 [35]. Traditional methods, however, require heterozygosity at many loci on each chromosome, making clinical assay design challenging. Moreover, locus specific approaches require large amounts of DNA for multiple PCR reactions to achieve analytical precision [7], making them unrealistic for clinical application. NGS can alleviate some of the constraints encountered by traditional technologies for cffDNA testing. The massively parallel nature of NGS can intrinsically sequence hundreds of thousands of sites with great depth in a quantitative manner. This enables very small changes in chromosome DNA frequency to be detected, as would be predicted for a trisomy 21 fetus contributing 5 % of the total plasma DNA content. Thus, by counting NGS sequencing reads from maternal plasma mapping to chromosomes 21, 18, and 13 and assessing the over- or under-representation of these chromosomes in maternal plasma, two initial groups were able to accurately identify trisomy 21 fetuses [7, 15]. Fan et al. were also able to identify trisomy 13 and trisomy 18, demonstrating this approach is applicable to other common chromosomal aneuploidies [15]. Several genetic diagnostic companies now offer noninvasive prenatal testing of the common fetal aneuploidies by NGS analysis of cffDNA, as well as known microdeletion syndromes. While some of the most common genetic disorders are now accessible by testing cffDNA, recent advances suggest that dramatic expansion in this field is likely.

Remarkably, the entire genome of an 18.5 week gestation fetus has been sequenced from cffDNA isolated from maternal plasma [24]. This approach required extensive specialized bioinformatic processing and integration of maternal and paternal genomic data, and is not likely to become clinically available in the current form. However, it demonstrates that with the appropriate techniques, any region of the fetal genome can be queried by NGS. Accordingly, NGS based testing of cffDNA is likely to expand to gene sequencing and copy number variant detection in the future.

Skeletal Dysplasia

Skeletal dysplasias are a heterogeneous group of disorders characterized by abnormal bone or cartilage growth. Over 300 types of skeletal dysplasias have been described with causal mutations in over 200 genes [69]. Although rare disorders individually, they have an overall prevalence of approximately 3 per 10,000 births and 20.0 per 10,000 stillbirths [61]. They vary greatly in severity, with severe forms such as thanatophoric dysplasia being lethal in the prenatal or neonatal period and mild forms such as hypochondroplasia not detected until childhood. Routine ultrasound monitoring can identify skeletal dysplasias during fetal development; however, a specific diagnosis is challenging due to the limitations of noninvasive imaging procedures. Although some dysplasias are diagnosed by ultrasound and confirmed by molecular analysis prenatally, many may remain undiagnosed. Yet a diagnosis is important for appropriate prenatal and postnatal management, as well as determining recurrence risk. Most importantly, an accurate diagnosis can differentiate between lethal and nonlethal conditions. NGS panels for skeletal dysplasias have the potential to provide molecular diagnosis for those who would not have traditionally received a prenatal diagnosis. A gene panel designed for skeletal dysplasias detected prenatally by ultrasound includes gene families that cover the most commonly observed disorders, such as fibroblast growth factor receptor 3 (*FGFR3*), mutations in which cause thanatophoric dysplasia, achondroplasia, hypochondroplasia, and the Crouzon and Muenke craniosynostosis syndromes. *FGFR2* mutations also cause Crouzon syndrome, as well as Apert, Pfeiffer, and Jackson–Weiss syndromes, and should also be included in a skeletal dysplasia panel. The collagen gene *COL2A1* is responsible for achondrogenesis, spondyloepiphyseal dysplasia congenital and Stickler syndrome, while *COL1A1* and *COL1A2* mutations (as well as mutations in other genes) are responsible for many types of osteogenesis imperfecta, another common skeletal dysplasia. An advantage of an NGS gene panel is that the numerous genetic causes of rare skeletal dysplasias can be simultaneously evaluated along with common causes. Accordingly, genes such as *SOX9* (campomelic dysplasia), *SLC26A2* (diastrophic dysplasia, achondrogenesis), and *ALPL* (hypophophatasia) can be tested in fetuses with abnormal skeletal findings. As with all prenatal (fetal) testing, a prenatal skeletal dysplasia panel ideally will have a quick turnaround time and be suited for amniotic fluid or chorionic villus samples. With this consideration, careful thought should be given to the appropriate NGS platform for fetal testing, with a preference given to faster sequencing systems.

Noonan Syndrome

Noonan syndrome (NS) is an autosomal dominant condition that is caused by hyperactivation of the RAS/MAPK signaling pathway [54]. It is a relatively common disorder, with an estimated incidence as high as 1:1,000 live births. NS displays both phenotypic and genetic overlap with several related disorders that are also caused by inappropriate RAS/MAPK activity, including Neurofibromatosis type 1, Cardio-facio-cutaneous syndrome, Costello Syndrome, Noonan Syndrome with Multiple Lentigines (NSML, formerly known as LEOPARD syndrome), and Legius syndrome. These disorders are collectively referred to as RASopathies, and are characterized by short stature, congenital heart defects, facial dysmorphia, developmental delay, cryptorchidism, variable skeletal abnormalities and variable tumor predisposition. NS displays significant variability in clinical presentation, in part due to genetic heterogeneity, as it can be caused by mutations in at least seven genes. This, along with the phenotypic overlap with other RASopathies, can make a definitive diagnosis challenging in some cases. Indeed, some individuals are only diagnosed after they have more severely affected children. In children with a RASopathy phenotype who do not display clear symptoms of one of the related syndromes, a genetic diagnosis can guide the clinical diagnosis. Prenatal studies suggest NS phenotypes can manifest in early fetal development, with features that include increased nuchal translucency, cystic hygroma, hydrops fetalis, distended jugular lymphatic sacs and congenital heart disease [47]. These findings have led some clinicians to consider prenatal molecular testing for NS and related disorders based on ultrasound abnormalities [9, 29]. In the prenatal setting, a genetic diagnosis of a RASopathy may be the most significant finding that results in a clinical diagnosis, and can guide patient management as a result. The efficacy of prenatal genetic testing for NS was examined in a study that performed prenatal testing for the most common genetic causes of NS, namely, mutations in *PTPN11*, *KRAS*, *SOS1*, and *RAF1*, in pregnancies with increased nuchal translucency and one other abnormality [9]. De novo mutations in these genes were detected in 17.3 % of cases, indicating that prenatal genetic testing for NS can greatly aid in an early diagnosis of this disease. Of note, the authors only assessed the most commonly mutated genes for NS, suggesting that some cases might be missed with their testing strategy. Accordingly, an NGS panel for NS would presumably have greater diagnostic yield than sequencing of a few genes. NGS panels for NS and related disorders are currently available clinically, although there have been no published reports of their prenatal application. Some parents may elect to use a prenatal diagnosis of NS to make decisions regarding termination of pregnancy. For those who do not, a prenatal NS diagnosis has clinical value, as there is a 50–80 % penetrance of congenital heart disease in NS patients [55]. These defects are often not detectable by fetal ultrasound, and a prenatal NS diagnosis would facilitate early monitoring of affected newborns. In summary, NGS panels for NS and related disorders are applicable to both fetal and postnatal diagnostics.

Newborn/Infant Applications

Hearing Loss

Hearing loss is the most common sensory impairment in humans, with an incidence of approximately 1 in 1,000 newborns [41, 49]. Newborn screening programs have been very successful at identifying hearing impaired infants, the early diagnosis of which is crucial for affected patients to receive maximal benefit from hearing aids or cochlear implants for language development. Although hearing loss can be caused by environmental factors, genetic mutations account for up to 75 % of cases [56]. Around 70 % of hearing loss is considered nonsyndromic hearing loss and is not accompanied by other recognizable phenotypes [63]. Conversely, 30 % of genetic hearing loss is syndromic, and the diagnosis of some of these disorders would predict more severe manifestations developing with age. For instance, Usher syndrome cannot be clinically distinguished from nonsyndromic hearing loss at a young age; however, affected individuals develop progressive retinitis pigmentosa in early adolescence. Early identification of the causative mutation in Usher syndrome enables life planning to cope with progressive vision loss and potentially therapeutic intervention through gene therapy, which is currently in clinical trials for Usher syndrome type 1b. Hereditary hearing loss is a genetically heterogeneous disorder, with over 100 associated genetic loci and 46 causally implicated genes [64], a property that makes this disorder appropriate for NGS panels. Diagnostic testing for the most common genetic cause of hearing loss, mutation of the *GJB2* gene and deletions of *GJB6*, which together account for approximately 20 % of nonsyndromic hearing loss, is widely available [51]. However, other genetic causes of hearing loss individually account for a small percentage of cases, and their frequency in many populations is unknown, making sequential analysis of the remaining candidate genes impractical. Because of this limitation, less than 20 % of patients referred for genetic testing receive a molecular diagnosis [21]. Accordingly, a proof of concept study demonstrated the clinical applicability of a NGS gene panel for hereditary hearing loss by designing a panel of 54 genes known to cause nonsyndromic hereditary hearing loss [59]. The authors identified causative mutations in five out of six *GJB2* and *SLC26A4* mutation negative patients, suggesting a NGS-based hearing loss panel would have good diagnostic yield and clinical utility. Therefore, it would be cost-effective to screen patients for the common genetic causes of hearing loss by conventional methods prior to NGS hearing loss panel.

Epilepsy

Epilepsy represents a group of complex neurological disorders that are united by the presence of recurring seizures. With a prevalence of three million individuals in the USA and 3 % lifetime incidence, it is among the most common neurological disorders in the developed world [19]. The age of onset is most common in infancy and in advanced age, and can be followed by a widely variable clinical course depending on the subtype [20]. For instance, benign familial neonatal-infantile seizures have a mean age of onset of 3 months, but patients usually undergo remission by 12 months and have a very low risk of seizures later in life [68]. In contrast, Dravet syndrome patients experience seizures beginning at 6 months of age and generally do not respond to therapy [20]. Other seizure types manifest between 1 and 4 years of age, and patients suffer frequent severe seizures which slow the child's development. The causes of epilepsy are as varied as their clinical course and can be the result of a range of environmental or genetic factors. The heritable forms of epilepsy can exhibit either single gene or polygenic etiology, with mutations affecting the function of ion channels representing the best characterized pathophysiologic mechanism [20]. In addition, numerous genetic syndromes exhibit an epileptic phenotype, including but not limited to mitochondrial disorders, neuronal migration disorders, holoprosencephaly, metabolic disorders, and storage disorders. The numerous epileptic subtypes exhibit extensive phenotypic and genetic overlap with each other and with the aforementioned syndromes, which can make a definitive diagnosis exceedingly challenging in some patients. Accordingly, a molecular diagnosis can have substantial diagnostic and prognostic value for the epilepsies. Progressive myoclonic epilepsies are a prime example, as a molecular finding can distinguish between several highly related disorders [58]. Finding a genetic cause of epilepsy can have a significant impact on patient care. For instance, a ketogenic diet has proven effective at reducing seizures in patients with GLUT1 deficiency syndrome [25]. For the aforementioned reasons, epilepsy represents an ideal disease candidate for the application of an NGS panel. To this end, Lemke et al. developed an NGS panel of 265 genes encompassing all genes known to cause epilepsy or a syndrome with epilepsy as a feature [30]. In a pilot study, they analyzed 33 patients presenting with either concise epilepsy phenotypes of known or unknown genetic etiology or a severe but unspecified seizure disorder. Impressively, disease-causing mutations were identified in 16 of the 33 patients, including all 8 patients who had a previous genetic diagnosis. They also identified mutations in five of ten patients who displayed phenotypes for which the genetic background is largely unknown. Interestingly, three patients with either Dravet syndrome or myoclonic epilepsy were found to have *SCN1A* mutations that had been missed by Sanger sequencing or high-resolution melting analysis. A good illustration of the power of this diagnostic approach is the finding that three patients in their cohort received diagnoses of extremely rare disorders that likely would have gone undiagnosed in the absence of a genetic diagnosis. This pilot study has demonstrated significant advantages of NGS over traditional molecular diagnostic methods for epilepsy. However, larger studies will be required to accurately define the clinical performance of this methodology.

Mitochondrial Disorders

Mitochondrial dysfunction underlies a group of disorders that have a reported incidence of 1 in 5,000 live births [14]. Mitochondria-related disorders have a broad range of clinical presentation, display locus and allelic heterogeneity, and can be autosomal recessive, dominant, sex-linked, and sporadic, making them challenging to diagnose and thereby obvious candidates for NGS panels [10]. Further complicating matters, mitochondria are uniquely composed of proteins and RNA encoded by the nuclear and mitochondrial genomes, mutations in both of which can lead to mitochondrial dysfunction and disease. The mitochondrial genome is 16.6 kb and contains 37 genes that encode mitochondrial enzymes, and transfer and ribosomal RNAs. The nuclear genome contributes over 1,000 genes to mitochondrial function, mutations in at least 100 of which are associated with human disease [26, 48]. Although there is clear clinical utility for a NGS mitochondrial disorder panel, sequencing these disease genes poses some technical challenges. Mitochondrial genes have numerous nuclear encoded pseudogenes, which can be challenging to distinguish from the targeted coding gene given the short reads currently employed by NGS platforms. Additionally, a cell or population of cells can have more than one mitochondrial genome, a phenomenon known as heteroplasmy. A recent NGS based study has shown that the incidence of heteroplasmy varies between 10 and 50 %, which is significantly higher than previously appreciated [60]. Disease-causing mitochondrial mutations are frequently heteroplasmic, and the proportion of the mutant allele directly impacts disease manifestation and severity. While NGS is uniquely suited to detect heteroplasmy, NGS panels need deep coverage of the mitochondrial genome to sensitively detect low level heteroplasmic mutations. Moreover, heteroplasmy can complicate data analysis, interpretation and reporting, as there is currently no consensus as to what threshold of mutation burden is clinically significant. This is confounded by the fact that the cell population analyzed, usually peripheral blood, may not accurately represent the mitochondrial DNA (mtDNA) profile of the diseased tissue. Moreover, Sanger sequencing is incapable of confirming low-level heteroplasmic mutations and more sensitive methods such as allele-specific PCR or pyrosequencing are needed to confirm true positives. For rare mutations, the need to confirm low level-heteroplasmic variants presents a significant burden for clinical laboratories. In light of these challenges, laboratories should consider validating samples from other tissues when developing NGS based panels for mitochondrial disorders. Additionally, developing and validating mitochondrial NGS panels on multiple NGS platforms would provide a robust method for confirmation of heteroplasmic mutations.

Multiple groups have reported the effective development and application of NGS based mitochondrial gene panels [6, 10, 65]. They have employed either a hybrid capture or a combination of long-range PCR for selective enrichment of the mitochondrial genome and emulsion PCR for nuclear genes. The first proof of concept study applied an NGS panel containing the mitochondrial genome and 1,000 nuclear encoded genes to 42 unrelated infants with clinical and biochemical evidence of mitochondrial disease [6]. Ten patients were found to have clear disease causing mutations in known genes, while 13 patients had mutations in nuclear-encoded mitochondrial genes that had not been previously linked to disease. A second group reported a validation study of a clinical NGS panel, where 13 clinical samples and a group of 16 validation samples with known mitochondrial and nuclear mutations were sequenced [10]. All known variants in the validation samples were observed, while five clinical samples were found to have disease causing mutations. These studies collectively demonstrate that NGS panels are effective for the molecular diagnosis of mitochondrial disorders and should prove to be a valuable tool for clinicians. A recent study reported that commercially available exome sequencing reagents can evaluate both the mitochondrial and nuclear genomes, suggesting that mitochondrial disorders can be effectively evaluated by WES [13]. As mitochondrial disorders have significant phenotypic overlap with numerous other disorders, including epilepsy, hearing loss and retinitis pigmentosa, evaluation of the mitochondrial genome with WES is an important consideration.

Adult or Young Adult Onset Applications

Aortopathy

Aneurysm and dissection of the aorta is a common cause of mortality, accounting for 1–2 % of deaths every year [32]. Diseases of the aorta are collectively known as the aortopathies and can involve dilation, aneurysm or malformation of the aorta. These aortic disorders result in dissection of the aorta, most commonly the thoracic aorta. There is a strong genetic contribution to thoracic aortic aneurysm, with up to 20 % of cases the result of an inherited disorder [67]. One of the most common inherited aortopathies is Marfan syndrome, a connective tissue disorder with a prevalence of approximately 1 in 5,000 [52]. Affected individuals display a pleiotropic phenotype that primarily involves the skeletal and cardiovascular systems, with cardiovascular phenotypes being the primary cause of mortality. Mutations in the fibrillin gene (*FBN1*) cause 90–95 % of Marfan cases, with over 1,500 different mutations in the large, 65 exon *FBN1* gene described [37]. Loeys–Dietz syndrome (LDS) displays a clinical presentation similar to Marfan syndrome, but may include hypertelorism, cleft palate or bifid uvula, and arteriole tortuosity [38]. LDS patients also experience more severe phenotypes with widespread aneurysms, and those of the aorta tend to be more aggressive and more likely to dissect at younger ages. Importantly, diagnostic criteria have not been established for LDS, and a definitive diagnosis often relies mainly on molecular testing [2]. Numerous other disorders that involve aneurysm of the aorta have been described and include congenital contractural arachnodactyly, Shprintzen–Goldberg syndrome, Ehlers–Danlos syndrome, autosomal recessive cutis laxa type 1B, arterial tortuosity syndrome, Williams–Beuren syndrome and familial thoracic aortic aneurysm and dissection. These disorders are all caused by mutations in genes involved in TGFβ signaling or cytoskeletal organization, and as a result, can be challenging to differentiate from one another. Early clinical and molecular diagnosis of these disorders is essential, as there is a trend toward gene tailored management strategies [16]. For instance, surgical intervention is recommended for LDS patients with mutations in *TGFBR1* or *TGFBR2* when their ascending aorta reaches a diameter of 40–42 mm, whereas patients with familial thoracic aortic aneurysm and dissection with *MYH11* mutations should undergo surgery when their ascending aorta reaches a diameter of 45–50 mm. Considering the overlapping phenotypic and genetic spectrum of the aortopathies, this group of disorders is an excellent candidate for targeted NGS.

The first published report of an NGS panel for the aortopathies focused on the diagnosis of Marfan syndrome and LDS [2]. As small deletions in *FBN1* are a rare cause of Marfan syndrome, the authors proposed a testing strategy that performed multiplex ligation dependent probe amplification (MLPA) for

FBN1 on mutation negative samples. Using this approach, they achieved a mutation identification rate of 92 % in a group of 87 Marfan syndrome patients. A second published study expanded the number of genes and genetic disorders associated with thoracic aortic aneurism on an NGS panel, including Ehlers–Danlos syndrome type IV, congenital contractural arachnodactyly and nonsyndromic familial forms of thoracic aortic aneurysms [72]. This represents a common trend in NGS panels, where existing tests are expanded to include additional disorders or new genetic causes of a disease. In summary, NGS panels for the aortopathies represent an excellent application of this technology, because NGS can improve upon the yield of traditional testing methodologies and the resultant molecular diagnoses have high clinical utility.

Cardiomyopathy

Inherited cardiac disorders are a relatively common group of diseases that collectively affects about 1 in 390 people [53]. In contrast to age-related cardiac disease, inherited cardiomyopathies occur much earlier in life, ranging from adolescence to early adulthood. Hypertrophic cardiomyopathy (HCM), dilated cardiomyopathy (DCM) and arrhythmogenic right ventricular cardiomyopathy (ARVC) are the most common forms of inherited cardiomyopathy and, together, represent a major cause of heart disease in all age groups [70]. These disorders predispose to sudden cardiac death, but also may progress with age, leading to heart failure. Accordingly, individuals with inherited cardiomyopathies represent critical cases for early diagnosis so that appropriate clinical management can be employed to reduce morbidity and mortality associated with these disorders. Although cardiomyopathies are diagnosed based on clinical presentation, a genetic diagnosis of the proband greatly facilitates the genetic counseling and risk assessment for family members [71]. This is of particular importance for this group of disorders that can go undiagnosed for years.

Inherited cardiomyopathies represent an excellent application of NGS panels, as they exhibit locus heterogeneity and both genetic and phenotypic overlap [62]. For instance, HCM is caused by mutations in approximately 8 genes, although over 20 genes have published clinical associations with HCM. Similarly, DCM is caused by mutations in approximately 10 genes, though around 40 genes have published association with DCM. Lastly, ARVC is less genetically heterogeneous, with 6 genes clearly causing disease and an additional 4 disease-associated genes. In total, these disorders have been associated with over 50 genes, which certainly precludes comprehensive Sanger sequencing. An NGS cardiomyopathy panel can also facilitate a clinical diagnosis in some cases; for instance, later-stage HCM is morphologically similar to dilated cardiomyopathy. Cardiomyopathies, moreover, are a feature of numerous other syndromic conditions, including metabolic disorders, Noonan syndrome and assorted myopathies. In these cases, a genetic diagnosis can greatly facilitate differentiation between these disorders. In a recent study, a 41 gene NGS cardiomyopathy panel was evaluated for 223 unrelated patients presenting with HCM [39]. Published disease-causing mutations were found in 33.6 % of patients, while an additional 23.8 % of patients had novel mutations that were predicted to be pathogenic. Of note, 219 rare variants in the Titin (*TTN*) gene were found in 142 study patients, with 209 of these being missense variants. *TTN* is known to be highly polymorphic gene and in our evaluation of WES of patients referred for unrelated conditions, we find novel *TTN* variants in the vast majority of cases. Thus, for the cardiomyopathies, *TTN* represents a challenging gene for variant analysis, as it is highly polymorphic, but is a genuine disease causing gene for familial DCM. Nonetheless, the high diagnostic yield observed in this study is a promising finding that is consistent with other evaluations of NGS panels.

Exome and Genome

Clinical WES/WGS tests are unique among sequencing based tests in that they can be applied to the diagnosis of essentially any genetic disorder. As with most new technologies, it was first applied on a research basis with great success in the realm of gene discovery. Because the majority of patients subjected to WES/WGS have been tested for common genetic disorders, the diagnostic yield of WES/WGS was initially unclear. In the absence of large scale studies, the first initial case reports focused on successful applications of WES/WGS. A notable success story of clinical WES was the report of a young male child with intractable inflammatory bowel disease [73]. The patient suffered from severe stunting and malnutrition, bacterial sepsis and other severe complications related to severe gastrointestinal inflammation. A WES study revealed a missense mutation in the *X-Linked Inhibitor of Apoptosis* gene (*XIAP*), which had not previously been associated with bowel disease, but had been implicated in the proinflammatory response and bacterial sensing through the nucleotide-binding oligomerization domain (NOD) signaling pathway. Functional studies revealed defects in the NOD signaling pathway and subsequently an allogeneic hematopoietic progenitor cell transplant was performed on the patient based on these findings. Remarkably, the patient was reported to be disease free 42 days post-transplant. This study represents a unique case where gene discovery and novel therapy were made possible by the application of WES. A second study examined fraternal twins who suffered from dopa (3,4-dihydroxyphenylalanine)-responsive dystonia (DRD) [3]. They had no identified deleterious mutations in the two genes known to cause DRD for which clinical tests were available, but were not tested for two other known causative genes for which clinical tests were unavailable. WGS identified compound heterozygous mutations in the *SPR* gene, a gene previously implicated in DRD. This genetic diagnosis led to the supplementation of their current therapy with a second agent that resulted in clinical improvements for both twins. This study represents a more likely positive outcome in clinical WES/WGS studies, where a mutation is found in a known gene for which no clinical tests are available.

Although the studies discussed represent the great potential of WES/WGS, they do not address what the yield of these genetic tests would be in real-life clinical practice. As WES/WGS is becoming more widely adopted, studies are emerging that begin to address this critical question. One study performed WES on 12 patients with unexplained and apparent genetic conditions who had previously been subjected to chromosome microarrays and targeted testing [45]. The authors found likely genetic diagnoses in 6 of the 12 probands, with 4 of the cases caused by mutations in known Mendelian disease genes. The two remaining patients had mutations in candidate disease genes. A larger study examined the results of 250 unselected clinical WES cases [74]. The majority of probands (approximately 80 %) were children referred for neurological phenotypes that included behavioral and developmental disorders. In all, 25 % of cases received a genetic diagnosis that met their established criteria for molecular diagnosis. As all patients had been subjected to previous genetic and biochemical studies, the high yield of clinical WES was quite surprising. Patients with a nonspecific neurological disorder had the highest rate of positive diagnosis (33 %), with patients exhibiting a specific neurological disorder yielding a similar success rate of 31 %. These results suggest these two patient groups represent excellent candidates for clinical WES/WGS testing. Of note, the overall success rate of WES/WGS observed in this study compares favorably to the diagnostic yield for chromosomal microarray in patients with developmental disabilities or congenital anomalies [43]. For the 62 % of positive cases, a striking 83 % of autosomal dominant and 40 % of X-linked disorders were the result of de novo mutations. These results support the growing appreciation for *de novo* mutations in genetic disorders, and indicate that this mode of inheritance should be carefully examined for all NGS-based tests. Moreover, this observation demonstrates the utility of WES/WGS on parent/child trios, given that hundreds of rare variants are found in every exome study. Filtering by inheritance can reduce this list of variants by an order of magnitude, allowing the reviewer to focus time and resources on variants with a higher probability of pathogenicity. Interestingly, four patients received two separate genetic diagnoses of nonoverlapping genetic disorders. These cases represent an outcome that is only realistically possible using WES/WGS methods, and serve as a point of consideration for clinical geneticists when evaluating patients. Whereas there is strong evidence for the effectiveness of both cytogenomic and WES/WGS testing independently, it is unclear how the combination of these two methodologies would perform in the clinic. Future studies that evaluate the yield of a combination of WES/WGS and cytogenomic arrays will be useful. In conclusion, WES/WGS has made a rapid transition from a research assay to clinical diagnostic test. Early success stories have incited great excitement, but must be tempered with realistic expectations, as the lack of knowledge of many genes is a significant obstacle to positive outcomes in many clinical WES/WGS cases. The most common clinical applications of WES/WGS have been for the diagnosis of rare disorders in the pediatric population. Whether WES/WGS sequencing will be applied clinically for common adult onset disorders will be an interesting aspect of the field in the future.

Conclusions

The clinical application of NGS tests has already led to an improved understanding of numerous inherited disorders, both in discovering new disease causing genes and for novel therapeutic strategies. Both rare and common inherited disorders have benefitted from the increased capabilities NGS has provided to clinical laboratories and the striking success stories reported in the literature have fanned the flames of interest in the medical community. NGS brings the promise of improved healthcare for people in all stages of life. From couples seeking preconception genetic screening to noninvasive prenatal diagnosis to disorders suffered by young and old, we may soon realize a time when every person is able to have a whole-genome test. Though we are still in the early days of clinical NGS testing, it is clear that this technology will enjoy explosive growth over the coming decade.

As clinical NGS matures, technical improvements in both sequencing capabilities and bioinformatic analysis will impact the future of this testing methodology. One could envision a single NGS assay that combines copy number analysis and sequence variant detection. As a WES or WGS test, this assay would integrate all the relevant genetic information in a single test. As such, it would reduce turnaround times, and could be consolidated into a single report that would be easier for clinical laboratories and clinicians alike. Longer NGS read lengths will lead to longer distance haplotyping, a capability that would facilitate NGS testing of cffDNA. As NGS sequencing improves in quality and declines in cost, WES/WGS may someday be performed at high depth without sequencing gaps. In this scenario, NGS panels may become obsolete and the field may move towards WES/WGS as a singular testing methodology that is analyzed in a targeted fashion based on clinical phenotype.

As discussed previously, variant classification is a challenging, time-consuming endeavor. Clinical laboratories are combining efforts and sharing variant information in centralized databases. Variants will ideally be linked to phenotypic data so that this large data set can be queried to assess rare variants in an informed context. Such an effort, if widely adopted, will be a tremendous tool for both researchers working towards gene discovery and clinical laboratories evaluating patient NGS tests. Additionally, many ethnicities are not well represented in the large-scale genome sequencing efforts such as 1,000 genomes and the Exome Variant Server. As a result, clinical WES tests of these patients yield many rare, novel variants that reflect their ancestry, but are unrelated to their clinical presentation. Expansion of these large scale sequencing projects to include these underrepresented ethnicities will greatly facilitate the evaluation of rare variants and also enhance the quality of clinical reports generated from WES/WGS testing.

REFERENCES

1. ACMG Board of Directors. Points to consider in the clinical application of genomic sequencing. Genet Med. 2012;14:759–61.
2. Baetens M, Van Laer L, De Leeneer K, Hellemans J, De Schrijver J, Van De Voorde H, Renard M, Dietz H, Lacro RV, Menten B, Van Criekinge W, De Backer J, De Paepe A, Loeys B, Coucke PJ. Applying massive parallel sequencing to molecular diagnosis of Marfan and Loeys-Dietz syndromes. Hum Mutat. 2011;32: 1053–62.
3. Bainbridge MN, Wiszniewski W, Murdock DR, Friedman J, Gonzaga-Jauregui C, Newsham I, Reid

JG, Fink JK, Morgan MB, Gingras MC, Muzny DM, Hoang LD, Yousaf S, Lupski JR, Gibbs RA. Whole-genome sequencing for optimized patient management. Sci Transl Med. 2011;3:87re3.
4. Beicht S, Strobl-Wildemann G, Rath S, Wachter O, Alberer M, Kaminsky E, Weber LT, Hinrichsen T, Klein HG, Hoefele J. Next generation sequencing as a useful tool in the diagnostics of mosaicism in Alport syndrome. Gene. 2013;526:474–7.
5. Bell CJ, Dinwiddie DL, Miller NA, Hateley SL, Ganusova EE, Mudge J, Langley RJ, Zhang L, Lee CC, Schilkey FD, Sheth V, Woodward JE, Peckham HE, Schroth GP, Kim RW, Kingsmore SF. Carrier testing for severe childhood recessive diseases by next-generation sequencing. Sci Transl Med. 2011;3:65ra4.
6. Calvo SE, Compton AG, Hershman SG, Lim SC, Lieber DS, Tucker EJ, Laskowski A, Garone C, Liu S, Jaffe DB, Christodoulou J, Fletcher JM, Bruno DL, Goldblatt J, Dimauro S, Thorburn DR, Mootha VK. Molecular diagnosis of infantile mitochondrial disease with targeted next-generation sequencing. Sci Transl Med. 2012;4:118ra10.
7. Chiu RW, Chan KC, Gao Y, Lau VY, Zheng W, Leung TY, Foo CH, Xie B, Tsui NB, Lun FM, Zee BC, Lau TK, Cantor CR, Lo YM. Noninvasive prenatal diagnosis of fetal chromosomal aneuploidy by massively parallel genomic sequencing of DNA in maternal plasma. Proc Natl Acad Sci U S A. 2008;105:20458–63.
8. Costa T, Scriver CR, Childs B. The effect of Mendelian disease on human health: a measurement. Am J Med Genet. 1985;21:231–42.
9. Croonen EA, Nillesen WM, Stuurman KE, Oudesluijs G, Van De Laar IM, Martens L, Ockeloen C, Mathijssen IB, Schepens M, Ruiterkamp-Versteeg M, Scheffer H, Faas BH, Van Der Burgt I, Yntema HG. Prenatal diagnostic testing of the Noonan syndrome genes in fetuses with abnormal ultrasound findings. Eur J Hum Genet. 2013;21:936–42.
10. Dames S, Chou LS, Xiao Y, Wayman T, Stocks J, Singleton M, Eilbeck K, Mao R. The development of next-generation sequencing assays for the mitochondrial genome and 108 nuclear genes associated with mitochondrial disorders. J Mol Diagn. 2013;15:526–34.
11. Daniels G, Finning K, Martin P, Massey E. Noninvasive prenatal diagnosis of fetal blood group phenotypes: current practice and future prospects. Prenat Diagn. 2009;29:101–7.
12. Del Castillo I, Moreno-Pelayo MA, Del Castillo FJ, Brownstein Z, Marlin S, Adina Q, Cockburn DJ, Pandya A, Siemering KR, Chamberlin GP, Ballana E, Wuyts W, Maciel-Guerra AT, Alvarez A, Villamar M, Shohat M, Abeliovich D, Dahl HH, Estivill X, Gasparini P, Hutchin T, Nance WE, Sartorato EL, Smith RJ, Van Camp G, Avraham KB, Petit C, Moreno F. Prevalence and evolutionary origins of the del(GJB6-D13S1830) mutation in the DFNB1 locus in hearing-impaired subjects: a multicenter study. Am J Hum Genet. 2003;73:1452–8.
13. Dinwiddie DL, Smith LD, Miller NA, Atherton AM, Farrow EG, Strenk ME, Soden SE, Saunders CJ, Kingsmore SF. Diagnosis of mitochondrial disorders by concomitant next-generation sequencing of the exome and mitochondrial genome. Genomics. 2013;102:148–56.
14. Distelmaier F, Koopman WJ, Van Den Heuvel LP, Rodenburg RJ, Mayatepek E, Willems PH, Smeitink JA. Mitochondrial complex I deficiency: from organelle dysfunction to clinical disease. Brain. 2009;132:833–42.
15. Fan HC, Blumenfeld YJ, Chitkara U, Hudgins L, Quake SR. Noninvasive diagnosis of fetal aneuploidy by shotgun sequencing DNA from maternal blood. Proc Natl Acad Sci U S A. 2008;105:16266–71.
16. Gillis E, Van Laer L, Loeys BL. Genetics of thoracic aortic aneurysm: at the crossroad of transforming growth factor-beta signaling and vascular smooth muscle cell contractility. Circ Res. 2013;113:327–40.
17. Grody WW, Thompson BH, Gregg AR, Bean LH, Monaghan KG, Schneider A, Lebo RV. ACMG position statement on prenatal/preconception expanded carrier screening. Genet Med. 2013;15:482–3.
18. Hartong DT, Berson EL, Dryja TP. Retinitis pigmentosa. Lancet. 2006;368:1795–809.
19. Hauser WA, Annegers JF, Kurland LT. Incidence of epilepsy and unprovoked seizures in Rochester, Minnesota: 1935–1984. Epilepsia. 1993;34:453–68.
20. Helbig I, Scheffer IE, Mulley JC, Berkovic SF. Navigating the channels and beyond: unravelling the genetics of the epilepsies. Lancet Neurol. 2008;7:231–45.
21. Hilgert N, Smith RJ, Van Camp G. Forty-six genes causing nonsyndromic hearing impairment: which ones should be analyzed in DNA diagnostics? Mutat Res. 2009;681:189–96.
22. Illanes S, Denbow M, Kailasam C, Finning K, Soothill PW. Early detection of cell-free fetal DNA in maternal plasma. Early Hum Dev. 2007;83:563–6.
23. Kaback MM. Population-based genetic screening for reproductive counseling: the Tay-Sachs disease model. Eur J Pediatr. 2000;159 Suppl 3:S192–5.
24. Kitzman JO, Snyder MW, Ventura M, Lewis AP, Qiu R, Simmons LE, Gammill HS, Rubens CE, Santillan DA, Murray JC, Tabor HK, Bamshad MJ, Eichler EE, Shendure J. Noninvasive whole-genome sequencing of a human fetus. Sci Transl Med. 2012;4:137.
25. Klepper J, Scheffer H, Leiendecker B, Gertsen E, Binder S, Leferink M, Hertzberg C, Nake A, Voit T, Willemsen MA. Seizure control and acceptance of the ketogenic diet in GLUT1 deficiency syndrome: a 2- to 5-year follow-up of 15 children enrolled prospectively. Neuropediatrics. 2005;36:302–8.
26. Koene S, Smeitink J. Mitochondrial medicine: entering the era of treatment. J Intern Med. 2009;265:193–209.
27. Kronn D, Jansen V, Ostrer H. Carrier screening for cystic fibrosis, Gaucher disease, and Tay-Sachs disease in the ashkenazi jewish population: the first 1000 cases at New York University Medical Center, New York, NY. Arch Intern Med. 1998;158:777–81.
28. Lazarin GA, Haque IS, Nazareth S, Iori K, Patterson AS, Jacobson JL, Marshall JR, Seltzer WK, Patrizio P, Evans EA, Srinivasan BS. An empirical estimate of carrier frequencies for 400+ causal Mendelian variants: results from an ethnically diverse clinical sample of 23,453 individuals. Genet Med. 2012;15:178–86.
29. Lee KA, Williams B, Roza K, Ferguson H, David K, Eddleman K, Stone J, Edelmann L, Richard G, Gelb BD, Kornreich R. PTPN11 analysis for the prenatal

diagnosis of Noonan syndrome in fetuses with abnormal ultrasound findings. Clin Genet. 2009; 75:190–4.
30. Lemke JR, Riesch E, Scheurenbrand T, Schubach M, Wilhelm C, Steiner I, Hansen J, Courage C, Gallati S, Burki S, Strozzi S, Simonetti BG, Grunt S, Steinlin M, Alber M, Wolff M, Klopstock T, Prott EC, Lorenz R, Spaich C, Rona S, Lakshminarasimhan M, Kroll J, Dorn T, Kramer G, Synofzik M, Becker F, Weber YG, Lerche H, Bohm D, Biskup S. Targeted next generation sequencing as a diagnostic tool in epileptic disorders. Epilepsia. 2012;53:1387–98.
31. Lindhurst MJ, Sapp JC, Teer JK, Johnston JJ, Finn EM, Peters K, Turner J, Cannons JL, Bick D, Blakemore L, Blumhorst C, Brockmann K, Calder P, Cherman N, Deardorff MA, Everman DB, Golas G, Greenstein RM, Kato BM, Keppler-Noreuil KM, Kuznetsov SA, Miyamoto RT, Newman K, Ng D, O'brien K, Rothenberg S, Schwartzentruber DJ, Singhal V, Tirabosco R, Upton J, Wientroub S, Zackai EH, Hoag K, Whitewood-Neal T, Robey PG, Schwartzberg PL, Darling TN, Tosi LL, Mullikin JC, Biesecker LG. A mosaic activating mutation in AKT1 associated with the Proteus syndrome. N Engl J Med. 2011;365:611–9.
32. Lindsay ME, Dietz HC. Lessons on the pathogenesis of aneurysm from heritable conditions. Nature. 2011;473:308–16.
33. Lo YM, Chan KC, Sun H, Chen EZ, Jiang P, Lun FM, Zheng YW, Leung TY, Lau TK, Cantor CR, Chiu RW. Maternal plasma DNA sequencing reveals the genome-wide genetic and mutational profile of the fetus. Sci Transl Med. 2010;2:61.
34. Lo YM, Corbetta N, Chamberlain PF, Rai V, Sargent IL, Redman CW, Wainscoat JS. Presence of fetal DNA in maternal plasma and serum. Lancet. 1997; 350:485–7.
35. Lo YM, Tsui NB, Chiu RW, Lau TK, Leung TN, Heung MM, Gerovassili A, Jin Y, Nicolaides KH, Cantor CR, Ding C. Plasma placental RNA allelic ratio permits noninvasive prenatal chromosomal aneuploidy detection. Nat Med. 2007;13:218–23.
36. Lo YM, Zhang J, Leung TN, Lau TK, Chang AM, Hjelm NM. Rapid clearance of fetal DNA from maternal plasma. Am J Hum Genet. 1999;64: 218–24.
37. Loeys B, De Backer J, Van Acker P, Wettinck K, Pals G, Nuytinck L, Coucke P, De Paepe A. Comprehensive molecular screening of the FBN1 gene favors locus homogeneity of classical Marfan syndrome. Hum Mutat. 2004;24:140–6.
38. Loeys BL, Chen J, Neptune ER, Judge DP, Podowski M, Holm T, Meyers J, Leitch CC, Katsanis N, Sharifi N, Xu FL, Myers LA, Spevak PJ, Cameron DE, De Backer J, Hellemans J, Chen Y, Davis EC, Webb CL, Kress W, Coucke P, Rifkin DB, De Paepe AM, Dietz HC. A syndrome of altered cardiovascular, craniofacial, neurocognitive and skeletal development caused by mutations in TGFBR1 or TGFBR2. Nat Genet. 2005;37:275–81.
39. Lopes LR, Zekavati A, Syrris P, Hubank M, Giambartolomei C, Dalageorgou C, Jenkins S, Mckenna W, UK10K Consortium, Plagnol V, Elliott PM. Genetic complexity in hypertrophic cardiomyopathy revealed by high-throughput sequencing. J Med Genet. 2013;50:228–39.
40. Lun FM, Chiu RW, Allen Chan KC, Yeung Leung T, Kin Lau T, Dennis Lo YM. Microfluidics digital PCR reveals a higher than expected fraction of fetal DNA in maternal plasma. Clin Chem. 2008;54:1664–72.
41. Mason JA, Herrmann KR. Universal infant hearing screening by automated auditory brainstem response measurement. Pediatrics. 1998;101:221–8.
42. Medvedev P, Stanciu M, Brudno M. Computational methods for discovering structural variation with next-generation sequencing. Nat Methods. 2009;6: S13–20.
43. Miller DT, Adam MP, Aradhya S, Biesecker LG, Brothman AR, Carter NP, Church DM, Crolla JA, Eichler EE, Epstein CJ, Faucett WA, Feuk L, Friedman JM, Hamosh A, Jackson L, Kaminsky EB, Kok K, Krantz ID, Kuhn RM, Lee C, Ostell JM, Rosenberg C, Scherer SW, Spinner NB, Stavropoulos DJ, Tepperberg JH, Thorland EC, Vermeesch JR, Waggoner DJ, Watson MS, Martin CL, Ledbetter DH. Consensus statement: chromosomal microarray is a first-tier clinical diagnostic test for individuals with developmental disabilities or congenital anomalies. Am J Hum Genet. 2010;86:749–64.
44. Mitchell JJ, Capua A, Clow C, Scriver CR. Twenty-year outcome analysis of genetic screening programs for Tay-Sachs and beta-thalassemia disease carriers in high schools. Am J Hum Genet. 1996;59:793–8.
45. Need AC, Shashi V, Hitomi Y, Schoch K, Shianna KV, Mcdonald MT, Meisler MH, Goldstein DB. Clinical application of exome sequencing in undiagnosed genetic conditions. J Med Genet. 2012;49:353–61.
46. Ng SB, Buckingham KJ, Lee C, Bigham AW, Tabor HK, Dent KM, Huff CD, Shannon PT, Jabs EW, Nickerson DA, Shendure J, Bamshad MJ. Exome sequencing identifies the cause of a mendelian disorder. Nat Genet. 2009;42:30–5.
47. Nisbet DL, Griffin DR, Chitty LS. Prenatal features of Noonan syndrome. Prenat Diagn. 1999;19:642–7.
48. Pagliarini DJ, Calvo SE, Chang B, Sheth SA, Vafai SB, Ong SE, Walford GA, Sugiana C, Boneh A, Chen WK, Hill DE, Vidal M, Evans JG, Thorburn DR, Carr SA, Mootha VK. A mitochondrial protein compendium elucidates complex I disease biology. Cell. 2008;134:112–23.
49. Parving A. The need for universal neonatal hearing screening – some aspects of epidemiology and identification. Acta Paediatr Suppl. 1999;88:69–72.
50. Pritchard CC, Smith C, Marushchak T, Koehler K, Holmes H, Raskind W, Walsh T, Bennett RL. A mosaic PTEN mutation causing Cowden syndrome identified by deep sequencing. Genet Med. 2013; 15:1004–7.
51. Putcha GV, Bejjani BA, Bleoo S, Booker JK, Carey JC, Carson N, Das S, Dempsey MA, Gastier-Foster JM, Greinwald Jr JH, Hoffmann ML, Jeng LJ, Kenna MA, Khababa I, Lilley M, Mao R, Muralidharan K, Otani IM, Rehm HL, Schaefer F, Seltzer WK, Spector EB, Springer MA, Weck KE, Wenstrup RJ, Withrow S, Wu BL, Zariwala MA, Schrijver I. A multicenter study of the frequency and distribution of GJB2 and GJB6 mutations in a large North American cohort. Genet Med. 2007;9:413–26.
52. Pyeritz RE. Marfan syndrome: current and future clinical and genetic management of cardiovascular manifestations. Semin Thorac Cardiovasc Surg. 1993; 5:11–6.

53. Raju H, Alberg C, Sagoo GS, Burton H, Behr ER. Inherited cardiomyopathies. BMJ. 2011;343:d6966.
54. Rauen KA. The RASopathies. Annu Rev Genomics Hum Genet. 2013;14:355–69.
55. Sarkozy A, Conti E, Seripa D, Digilio MC, Grifone N, Tandoi C, Fazio VM, Di Ciommo V, Marino B, Pizzuti A, Dallapiccola B. Correlation between PTPN11 gene mutations and congenital heart defects in Noonan and LEOPARD syndromes. J Med Genet. 2003;40:704–8.
56. Schrijver I. Hereditary non-syndromic sensorineural hearing loss: transforming silence to sound. J Mol Diagn. 2004;6:275–84.
57. Scriver C, Beaud et al, Sly WS, Valle D, editors. The metabolic and molecular bases of inherited disease. New York: McGraw-Hill; 1995.
58. Shahwan A, Farrell M, Delanty N. Progressive myoclonic epilepsies: a review of genetic and therapeutic aspects. Lancet Neurol. 2005;4:239–48.
59. Shearer AE, Deluca AP, Hildebrand MS, Taylor KR, Gurrola 2nd J, Scherer S, Scheetz TE, Smith RJ. Comprehensive genetic testing for hereditary hearing loss using massively parallel sequencing. Proc Natl Acad Sci U S A. 2010;107:21104–9.
60. Sosa MX, Sivakumar IK, Maragh S, Veeramachaneni V, Hariharan R, Parulekar M, Fredrikson KM, Harkins TT, Lin J, Feldman AB, Tata P, Ehret GB, Chakravarti A. Next-generation sequencing of human mitochondrial reference genomes uncovers high heteroplasmy frequency. PLoS Comput Biol. 2012;8:e1002737.
61. Stevenson DA, Carey JC, Byrne JL, Srisukhumbowornchai S, Feldkamp ML. Analysis of skeletal dysplasias in the Utah population. Am J Med Genet A. 2012;158A:1046–54.
62. Teekakirikul P, Kelly MA, Rehm HL, Lakdawala NK, Funke BH. Inherited cardiomyopathies: molecular genetics and clinical genetic testing in the postgenomic era. J Mol Diagn. 2013;15:158–70.
63. Van Camp G, Willems PJ, Smith RJ. Nonsyndromic hearing impairment: unparalleled heterogeneity. Am J Hum Genet. 1997;60:758–64.
64. Van Camp GASR. 2013. www.hereditaryhearingloss.org [Online]. Accessed 7 Dec 2013.
65. Vasta V, Ng SB, Turner EH, Shendure J, Hahn SH. Next generation sequence analysis for mitochondrial disorders. Genome Med. 2009;1:100.
66. Verhoef S, Bakker L, Tempelaars AM, Hesseling-Janssen AL, Mazurczak T, Jozwiak S, Fois A, Bartalini G, Zonnenberg BA, Van Essen AJ, Lindhout D, Halley DJ, Van Den Ouweland AM. High rate of mosaicism in tuberous sclerosis complex. Am J Hum Genet. 1999;64:1632–7.
67. Verloes A, Sakalihasan N, Koulischer L, Limet R. Aneurysms of the abdominal aorta: familial and genetic aspects in three hundred thirteen pedigrees. J Vasc Surg. 1995;21:646–55.
68. Vigevano F. Benign familial infantile seizures. Brain Dev. 2005;27:172–7.
69. Warman ML, Cormier-Daire V, Hall C, Krakow D, Lachman R, Lemerrer M, Mortier G, Mundlos S, Nishimura G, Rimoin DL, Robertson S, Savarirayan R, Sillence D, Spranger J, Unger S, Zabel B, Superti-Furga A. Nosology and classification of genetic skeletal disorders: 2010 revision. Am J Med Genet A. 2011;155A:943–68.
70. Watkins H, Ashrafian H, Redwood C. Inherited cardiomyopathies. N Engl J Med. 2011;364:1643–56.
71. Wilde AA, Behr ER. Genetic testing for inherited cardiac disease. Nat Rev Cardiol. 2013;10:571–83.
72. Wooderchak-Donahue WL, O'fallon B, Furtado LV, Durtschi JD, Plant P, Ridge PG, Rope AF, Yetman AT, Bayrak-Toydemir P. A direct comparison of next generation sequencing enrichment methods using an aortopathy gene panel-clinical diagnostics perspective. BMC Med Genomics. 2012;5:50.
73. Worthey EA, Mayer AN, Syverson GD, Helbling D, Bonacci BB, Decker B, Serpe JM, Dasu T, Tschannen MR, Veith RL, Basehore MJ, Broeckel U, Tomita-Mitchell A, Arca MJ, Casper JT, Margolis DA, Bick DP, Hessner MJ, Routes JM, Verbsky JW, Jacob HJ, Dimmock DP. Making a definitive diagnosis: successful clinical application of whole exome sequencing in a child with intractable inflammatory bowel disease. Genet Med. 2011;13:255–62.
74. Yang Y, Muzny DM, Reid JG, Bainbridge MN, Willis A, Ward PA, Braxton A, Beuten J, Xia F, Niu Z, Hardison M, Person R, Bekheirnia MR, Leduc MS, Kirby A, Pham P, Scull J, Wang M, Ding Y, Plon SE, Lupski JR, Beaud et al, Gibbs RA, Eng CM. Clinical whole-exome sequencing for the diagnosis of mendelian disorders. N Engl J Med. 2013;369:1502–11.

CHAPTER 31

GENOMIC APPLICATIONS IN PHARMACOGENOMICS

REHAN QAYYUM

Introduction 553

Use of Genomics in Designing Pharmacogenomic Studies 554

Application of Genomics in Determining Drug Efficacy 557

Application of Genomics in Minimizing Adverse Drug Reactions 558

Application of Genomics in Vaccinomics 558

Genomic Applications in Research 558

Conclusions 559

Introduction

The main goal of pharmacogenomics is to understand the influence of genetic variations between individuals on drug efficacy, metabolism, and toxicity. For some, the terms pharmacogenomics and pharmacogenetics are interchangeable but for others both terms carry different meanings with the term pharmacogenetics having a more limited definition, being reserved for the study of inherited differences in drug response [1, 2]. The study of variations in drug response between individuals can be dated back to Pythagoras at around 510 B.C.E. [2]. More recently, English physiologist, Archibald Garrod, has been the first to propose in 1923 that genetic variants that affect metabolism of endogenous molecules may affect drug metabolism [3]. In 1932, Snyder reported the first inherited trait associated with an exogenous chemical compound (phenylthiourea nontaster trait) in a cohort of 800 families [4, 5]. The pace of progress in the field of pharmacogenomics has accelerated with the completion of the human genome project [6] and at the time of this writing, the Table of Pharmacogenomic Biomarkers in Drug Labels contained 118 entries [7]. With the current advances in the fields of -omics, the anticipated benefits of pharmacogenomics are closer to realization than ever before. A better understanding of the interaction between drugs and genetic variants will lead to discovery of drugs that are more powerful, efficacious, and safer. Physicians will be able to prescribe not just the right drug but also the correct dose for a patient, thus maximizing the efficacy while minimizing the adverse effects. Vaccines created making use of genetic information will be able to activate the immune system to a large number of pathogens without exposing an individual to the risk of an infection. With all these and other benefits, the promise is that overall cost of health care will decrease.

G.J. Netto and I. Schrijver (eds.), *Genomic Applications in Pathology*,
DOI 10.1007/978-1-4939-0727-4_31, © Springer Science+Business Media New York 2015

Use of Genomics in Designing Pharmacogenomic Studies

Study Design

The choice between a candidate gene study, genome-wide association study (GWAS), exome sequencing (ES), and whole-genome sequencing (WGS) study is usually dictated by the hypothesis, approach (discovery versus targeted), and available resources. In an ideal setting where resources are not a constraint, a WGS study design is comprehensive and provides more data than are obtainable from the other three study designs. However, obtaining WGS in a sizeable number of patients is prohibitively expensive (although the cost of WGS is coming down rapidly) and analyzing the vast amounts of data generated by WGS requires extensive bioinformatic support.

Candidate Gene Study Designs

Based on the prior knowledge of drug target molecules, metabolism, and excretion, investigators may hypothesize that a certain gene or a group of genes determines the observed effect of a particular drug. Although each drug is likely to have a unique set of genes that determine its response in any individual, certain genes are more likely to be involved in the absorption, distribution, metabolism, and excretion (ADME) of a wide variety of drugs. A group of 32 genes has been designated as the core ADME genes (Table 31.1) by PharmaADME Working Group, a panel of industry and academic experts, and includes genes for several enzymes in the cytochrome P450 system and genes for several proteins that belong to solute carrier family [8]. An extended ADME gene list contains 267 additional genes for proteins responsible for the modification of functional groups of drugs, conjugation of drugs with endogenous moieties, the uptake and excretion of drugs in and out of cells, and those that can either alter the expression of other ADME genes or affect the biochemistry of ADME enzymes [8]. Commercially available gene chips include variants within the core ADME genes such as the DMET™ chip by Affymetrix (containing 1,936 variants across 231 genes) and the VeraCode® ADME Core Panel by Illumina (184 variants in 34 genes). In addition, microarray chips that contain variants from specific gene list can be custom-built.

Table 31-1 **List of ADME (Absorption, Distribution, Metabolism, and Excretion) Genes [8]**

Phase I	Phase II	Transporter
CYP1A1	GSTM1	ABCB1
CYP1A2	GSTP1	ABCC2
CYP2A6	GSTT1	ABCG2
CYP2B6	NAT1	SLC15A2
CYP2C19	NAT2	SLC22A1
CYP2C8	SULT1A1	SLC22A2
CYP2C9	TPMT	SLC22A6
CYP2D6	UGT1A1	SLCO1B1
CYP2E1	UGT2B15	SLCO1B3
CYP3A4	UGT2B17	
CYP3A5	UGT2B7	
DPYD		

Phase I is a type of drug metabolism in which drug chemical structure is modified by addition of polar or reactive groups such as hydroxyl (OH) groups. Phase II is a type of metabolism in which drug chemical structure is conjugated with other charged small molecules such as glycine or glucuronic acid

Whereas this type of assay design can be used for a study with a small number of patients and limited resources, a more likely use is in the drug development process where early identification of drug safety issues may save lives and cost. The biggest challenge with this type of study design is that often our knowledge of a drug's pharmacodynamics and pharmacokinetic pathways is incomplete, making selection of all relevant genes very difficult, if not impossible. Therefore, an association of drug responses with variants that are not included on the chip cannot be discovered.

Genome-Wide Association Study Designs

Whereas in a candidate gene study different sources of information are incorporated to develop a list of genes potentially involved in a drug's metabolic pathways, in GWAS a comprehensive and unbiased search throughout the whole genome is performed, with the goal of identifying relatively frequent genetic variants that may be associated with drug response [9, 10]. Thus, GWAS allows the discovery of novel genetic variants that are not in the known pathways. As the cost of performing microarrays to conduct a genome-wide scan has come down recently and the currently available microarray chips have good coverage across much of the genome, GWAS designs should typically be used for identifying genomic regions of interest. The identification of a genetic variant responsible for variability in drug response depends on several factors such as the effect size and allele frequency of the genetic variant, as well as the sample size of the study [11]. Genetic variants associated with drug response tend to have larger effect sizes and hence are easier to discover than variants associated with disease phenotypes. However, the often relatively small sample size of pharmacogenomic studies makes it difficult to discover genetic variants. Because of the relatively large effect sizes of genetic variants that are identified in pharmacogenomic GWAS, it is much easier to translate these findings into clinical practice [12]. GWAS can be used to examine the role of different biological pathways and can thus provide important insights into the mechanisms underlying drug response. Several commercially available genotyping arrays provide a wide range of options to choose from, based on one's experimental needs and budget.

GWAS provides an unbiased approach to scanning the whole genome for genetic variants associated with drug response, but there are important limitations. One limitation of GWAS is the penalty for performing a large number of statistical tests. GWAS test the association of drug response with about a million single nucleotide polymorphisms (SNPs) spread across the genome and, due to the large number of tests, result in a very high type I error with a conventional statistical significance threshold. To decrease the risk of inflated type I error, a much lower threshold is often employed, thus requiring a large effect size and/or sample size [13]. Moreover, the effect of variants with a low minor allele frequency (MAF) cannot be studied with currently available sample sizes. Thus, only the effect of relatively common variants (with a minor allele frequency greater than 1 %) on drug response is studied [14]. This limitation is especially evident in studying the cytochrome P450 family of genes that play an important role in drug metabolism. The genes in this family have several isoforms, are polymorphic and have a wide range of allele frequencies including variants with very low MAFs [15]. As a result, there is limited coverage with the currently available GWAS platforms [16]. To summarize, an interesting aspect of pharmacogenomic studies is that an interaction between drug, disease, and genetic variants is potentially possible but a complete examination of this interaction requires very large sample sizes, well beyond those used today.

WES Study Designs

Not only are GWAS limited to examining alleles with relatively higher MAFs, but the association between a variant and a phenotype (such as drug response) is only an *association* and often the identified variants only "tag" the causal variant, requiring further examination of the region around the identified variant. Moreover, whereas heritability estimates of drug response are relatively large, the amount of variability explained by the variants discovered through GWAS is quite small, sending researchers to search for "missing heritability" [17–20]. Some have argued that the drug response may not be determined by common variants (or variants tagged to common variants) but by rare variants. Because GWAS have poor coverage of rare variants, a different method is needed to discover these.

Ideally, sequencing the whole genome should reveal all variants that contribute to drug response. However, despite recent advances in sequencing technology, WGS remains too expensive for large studies. An alternative, in between GWAS and WGS, is to sequence only the exonic regions of the genome, or roughly the 2–3 % of the genome that encodes proteins. By capturing and sequencing only exons, the cost of sequencing decreases significantly, yet our ability to identify rare variants increases markedly [21].

When designing a WES study, a researcher should pay attention to several important aspects of the study design. Sample selection for WES usually is limited to enrolling unrelated individuals because it is not possible to expose unaffected individuals in a family to a drug. Once a DNA sample is obtained from participants, one of several different methods (such as hybridization, circularization, or PCR) can be used to capture the exonic regions of the genome. Few commercial kits (such as those from Agilent, Illumina, and Nimblegen) are available that employ one of these methods for capture. Of note, these kits differ in their definition of "exome" and cover slightly different parts of the genome. Deciding the depth of coverage is the next step and depends on several factors; for most experiments, coverage between 20× and 50× at each nucleotide will suffice. Once data are available, the first step is alignment to the genome followed by variant calling. Several quality control measures are considered throughout the alignment and variant calling process. Data on called variants are usually saved in a variant call format (VCF), which is then annotated. The annotations may include genomic coordinates, population frequencies, conservation throughout evolution, effect on protein structure, expected severity due to protein change, and any known clinical associations. This is followed by several heuristic filtering algorithms to narrow down the list of variants of interest [21, 22]. Non-synonymous variants are of particular interest for obvious reasons and various statistical and computation methods have been developed to assess the functional impact on proteins [23–32]. Because the majority of the identified variants are rare, statistical methods used for GWAS studies have very low power to detect an effect. An alternative is to analyze a group of rare variants within a defined region, usually a gene, and several statistical methods have been proposed to use this approach [22, 33–40]. Although we are still waiting to see a report of a pharmacogenomic study utilizing WES, it is likely that such studies will provide new and important insights into drug response.

WGS Study Design

Exome sequencing focuses on those parts of the genome about which we have better knowledge, the exons from the known coding regions, but leaves out regulatory components that may control whether a gene will be expressed and the extent of its level of expression. Furthermore, it is very likely that our current knowledge of the protein-coding regions of the genome is incomplete and, therefore, undiscovered but important genes would not be captured by commercial exome capture kits. In contrast to WES, WGS provides information about all identified variants irrespective of the location in the genome and provides the opportunity to discover a large number of genetic variants important to drug response. The study design issues as well as analytical issues are similar to WES except that the volume of data is much bigger and the ultimate number of variants obtained per sample much larger. If GWAS are any guide, WGS is likely to identify a large number of rare variants in the noncoding regions or in pseudogenes, forcing us to further understand how these variants control gene expression.

Other Study Designs

Other genomic methods, such as RNA sequencing (RNA-seq), DNA methylation studies and other epigenetic methods, can be used to study the effect of genomics on drug response, as well. For most drugs, our current knowledge of molecular targets is limited at best. This is true even for drugs that have been in common use for many decades and have been studied extensively, such as aspirin. For example, RNA-seq may identify genes that are upregulated or downregulated in those individuals who respond to a drug as compared to those who do not [41].

Choice of Study Population

Perhaps the most economical way of conducting a pharmacogenomic study to identify genetic variants is in the setting of a clinical trial. In clinical trials, patients are already enrolled to receive a particular drug at (a) certain dose(s), or an alternative drug or placebo, and are being followed for outcomes. Genotyping all or a subset of patients may provide an opportunity to study the effect of genetic variants on drug response phenotypes. Patients for genotyping can be selected at the end of the study when it is known which patients did or did not have a particular outcome during the clinical trial. Clinical trials are a good setting for studying candidate genes, but a single clinical trial may not provide a large enough number of participants to conduct a pharmacogenomic GWAS. Therefore, results from several clinical trials may need to be combined to get to the needed sample size. While combining clinical trials, usually in a meta-analysis framework, certain issues may arise, however. These include the use of different doses of the drugs, use of concomitant drugs, or underlying differences in study populations. In these instances, decisions may need to be made as to which studies should or can be included.

An example of the use of a clinical trial cohort for conducting a pharmacogenomic study is the Trial to Assess Improvement in Therapeutic Outcomes by Optimizing Platelet Inhibition with Prasugrel–Thrombolysis in Myocardial Infarction 38 (TRITON-TIMI 38) trial. This trial compared two antiplatelet agents (clopidogrel and prasugrel) in patients with acute coronary syndromes who were scheduled to have percutaneous coronary interventions [42]. The drug response phenotype was a composite of clinical outcomes (cardiovascular deaths, myocardial infarction, or stroke). While the overall study enrolled 13,608 patients, 1 candidate-gene pharmacogenomic study examined 1,477 subjects who were in the clopidogrel arm and found that carriers of a reduced function allele in the *CYP2C19* gene were associated with a 53 % increased risk of the composite clinical outcome [43]. Another candidate-gene pharmacogenomic study from the same clinical trial included 2,932 patients (1,471 from the clopidogrel arm and 1,461 from the prasugrel arm) and found that the TT genotype of the c.3435C>T variant in the *ABCB1* gene was associated with a 72 % increased risk of a composite clinical outcome in individuals treated with clopidogrel but not with prasugrel [44].

An example of the use of GWAS in a clinical trial setting is a study by Ramsey et al. This study was conducted to identify genetic variants determining methotrexate clearance in children with acute lymphoblastic leukemia and used data from two studies (P9904 and P9905). Investigators found that methotrexate clearance was associated with polymorphisms in the organic anion transporter gene *SLCO1B1* and further confirmed their findings in independent cohorts [45]. Phase I/II clinical studies also provide an opportunity to perform candidate-gene pharmacogenomic studies (for examples see [46–48]).

Another important source for pharmacogenomic studies are clinical cohorts in which information is collected from electronic health records (EHR) [49]. In these studies, data are collected from hospital or other clinical records and DNA is collected for research at the time of contact with the patient. This model has several advantages. The data are collected on patients who are receiving regular medical care and are neither self-selected nor selected on the basis of some criteria. Thus, these subjects provide a "real world" opportunity for pharmacogenomic studies. As the data are extracted from already existing medical records, the cost of acquiring data is minimal although the cost of DNA isolation genotyping remains the same. However, the genotype information obtained for one study can be used for additional studies on the same patient cohort (especially if genotyping is performed using a genome-wide scan) and by doing so can potentially reduce costs further. The disadvantages include that data are not collected for the purposes of research and as a result data may not be of high quality. However, certain types of data are likely to be of reasonably good quality. This would include parameters such as vital signs, laboratory data, medication records, and billing information. Perhaps the most prominent example of the use of EHR is in the Electronic Medical Records and Genomics (eMERGE) network [50].

Application of Genomics in Determining Drug Efficacy

Finding the Right Drug

Some patients respond poorly to typically very effective medicines and often the underlying reason for this is based on differences in our genome. Poor responsiveness may stem from the effect of genomic variations on drug pharmacodynamics (such as changes in drug receptors) or on pharmacokinetics (such as drug metabolism). For example, β2 agonists, such as albuterol and salmeterol, are quite effective for the treatment of asthma but may not be as effective in a small group of individuals with a certain SNP in the *ADRB2* gene [51]. Identification of such asthmatic patients and treatment with alternative therapies may improve treatment response and decrease morbidity. Similarly, while clopidogrel is very effective in reducing adverse cardiac events in patients who undergo percutaneous coronary intervention with stent implantation, individuals with the *CYP2C19**2 allele remain at an increased risk of future events even with clopidogrel therapy [52]. Identifying individuals with this allele may result in the prescription of alternative anti-platelet agents for an adequate inhibition of platelet function.

Finding the Right Dose

As currently practiced, physicians look at certain features in determining the right dose of drugs for their patients. These include a patient's age, sex, and body weight. This becomes especially an issue when the therapeutic window of a drug is small, such as is the case with warfarin. Warfarin is the most commonly prescribed anticoagulant drug for the prevention and treatment of venous thromboembolism and for the prevention of stroke in atrial fibrillation and with mechanical heart valves. Variants in the *CYP2C9* and *VKORC1* genes have been shown to determine warfarin metabolism and a warfarin dosing regimen based on the genotype has been developed and validated [53, 54].

Application of Genomics in Minimizing Adverse Drug Reactions

Use of almost all drugs is associated with some adverse reactions ranging from very mild ones to very serious ones resulting in severe illness or even death. Identification of genetic variants that may predict these adverse events and choosing alternative therapies for patients with these variants may result in more optimal drug responses.

The earliest examples of the use of pharmacogenomics are studies of severe adverse drug reactions such as seen with the use of mercaptopurine, succinylcholine, and perhexiline [55]. More recent examples include studies of statins, which are a group of lipid-lowering agents that have been consistently shown to reduce cardiovascular morbidity and mortality in patients with coronary artery disease or in patients at risk of developing coronary artery disease [56]. However, some patients develop myopathy while taking statins. A GWAS identified rs4363657, a SNP in the *SLCO1B1* gene (gene product responsible for hepatic uptake of statins), linked to the development of myopathy. Using an alternative lipid-lowering agent or using lower doses of statins in patients with this SNP may avert the development of statin-induced myopathy [57, 58]. Similarly, polymorphisms in *CYP2D6* and *CYP2C19* genes affect the efficacy and safety of tricyclic antidepressants and presence of these variants may require either reduced dosing or alternative therapies [59].

Application of Genomics in Vaccinomics

The application of pharmacogenomics to vaccine design has been labeled as "vaccinomics" and this new field is using the tools of genomics and bioinformatics to develop novel vaccines. Some of the recent approaches in vaccinomics include the use of epitope determination and prediction algorithms for exploring the use of peptide epitopes as immunogens [60]. In addition to developing new vaccines, genomic applications can help to identify individuals who are likely to develop an adequate immune response with a particular vaccine and who will develop adverse effects [61].

Genomic Applications in Research

Drug Discovery

Genomic applications can help the pharmaceutical industry from the very beginning of the drug-discovery process [62]. Genomic approaches can identify suitable gene targets and may identify potential molecules that can be evaluated as drugs [63]. Furthermore, knowledge of genetic variants may allow for more appropriate and ultimately safer inclusion and exclusion criteria, resulting in a more successful passage of drugs through the pharmaceutical pipeline. Lastly, the interaction of drugs that are currently in clinical use with newly discovered gene targets can be examined and we may discover new uses of previously approved drugs whose safety has already been shown [64].

Discovery of Biological Mechanisms

The use of genomic applications in studying drugs is improving our understanding of biological mechanisms in two main ways. First, whereas drugs are exogenous molecules that are introduced from outside, these molecules may have certain similarities with endogenous molecules of the body. These similarities may include aspects of drug/ligand receptors and metabolic pathways. Identification of receptors and enzymes involved in the metabolism of a drug may provide insights into the metabolism of endogenous molecules. For example, the cytochrome P450 family of genes was initially discovered as encoding detoxifying enzymes, but subsequent studies have highlighted the importance of these enzymes in the metabolism of endogenous molecules [65]. Second, because our understanding of the pathophysiology underlying most diseases is incomplete, historically the presence or absence of signs and symptoms has been used to classify diseases. It is quite possible that different pathophysiological mechanisms may culminate in a similar set of signs and symptoms and hence become defined into one disease process. The use of genomics tools to increase our understanding of drug responses and biological mechanisms will deepen our understanding of the pathophysiological basis of disease and is likely to result in better classification of diseases and more appropriate and targeted therapy for patients.

Conclusions

Whereas there are several drugs with pharmacogenomic warnings from the Food and Drug Administration (FDA), challenges abound regarding the identification of relevant genetic variants, and in every step on the road to clinical implementation (Table 31.2). The full benefit of genomics in clinical practice can only be realized when drug therapy for each individual can be personalized to his/her lifestyle and genome. Advances in several fields are needed before this dream of personalized medicine can be realized [66]. At the same time, application of genomics principles in pharmacogenomics holds promise for not only personalized medicine but also new drug discovery and development and novel insights into the biological mechanisms.

Table 31-2 **Challenges in Pharmacogenomics Research**

Identification of genetic variant
– Issues with current technology
– Need for large sample sizes
– Requirements for validation studies
– Cost of WGS
– Variable definitions of drug response
– Large number of potential hits
– Need for better statistical tests and algorithms
– Bioinformatics support
Demonstration of Efficacy and Effectiveness of Pharmacogenomic Approaches
– Defining drug response
– Need for large sample size especially studies with hard clinical outcomes, such as death or myocardial infarction
– Need for large validation studies
– Need for technology that can be applied at the point of use
– Statistical methods
– Genotype-guided studies
– Concomitant use of other drugs
Implementation in Clinical Practice
– Patient and physician education
– Privacy and ethical concerns
– Regulatory and legal issues
– Bioinformatics support

REFERENCES

1. http://www.ncbi.nlm.nih.gov/About/primer/pharm.html. Accessed 30 Apr 2013.
2. Nebert DW. Pharmacogenetics and pharmacogenomics: why is this relevant to the clinical geneticist? Clin Genet. 1999;56(4):247–58.
3. Roden DM, Wilke RA, Kroemer HK, Stein CM. Pharmacogenomics: the genetics of variable drug responses. Circulation. 2011;123(15):1661–70. doi:10.1161/CIRCULATIONAHA.109.914820.
4. Snyder LH. Inherited taste deficiency. Science. 1931;74(1910):151–2. doi:10.1126/science.74.1910.151.
5. Snyder L. Studies in human inheritance. IX, the inheritance of taste deficiency in man. Ohio J Sci. 1932;32(5):436–40.
6. Mundy C. The human genome project: a historical perspective. Pharmacogenomics. 2001;2(1):37–49. doi:10.1517/14622416.2.1.37.
7. http://www.fda.gov/drugs/scienceresearch/researchareas/pharmacogenetics/ucm083378.htm. Accessed 30 Apr 2013.
8. http://www.pharmaadme.org/joomla/. Accessed 4 Sept 2013.
9. Daly AK. Genome-wide association studies in pharmacogenomics. Nat Rev Genet. 2010;11(4):241–6. doi:10.1038/nrg2751.
10. Daly AK. Using genome-wide association studies to identify genes important in serious adverse drug reactions. Annu Rev Pharmacol Toxicol. 2012;52:21–35. doi:10.1146/annurev-pharmtox-010611-134743.
11. Hirschhorn JN, Daly MJ. Genome-wide association studies for common diseases and complex traits. Nat Rev Genet. 2005;6(2):95–108. doi:10.1038/nrg1521.
12. Motsinger-Reif AA, Jorgenson E, Relling MV, Kroetz DL, Weinshilboum R, Cox NJ, Roden DM. Genome-wide association studies in pharmacogenomics: successes and lessons. Pharmacogenet Genomics. 2013;23(8):383–94. doi:10.1097/FPC.0b013e32833d7b45.
13. Moskvina V, Schmidt KM. On multiple-testing correction in genome-wide association studies. Genet Epidemiol. 2008;32(6):567–73. doi:10.1002/gepi.20331.
14. Wagner MJ. Rare-variant genome-wide association studies: a new frontier in genetic analysis of complex traits. Pharmacogenomics. 2013;14(4):413–24. doi:10.2217/pgs.13.36.
15. Nelson MR, Wegmann D, Ehm MG, Kessner D, St Jean P, Verzilli C, Shen J, Tang Z, Bacanu SA, Fraser D, Warren L, Aponte J, Zawistowski M, Liu X, Zhang H, Zhang Y, Li J, Li Y, Li L, Woollard P, Topp S, Hall MD, Nangle K, Wang J, Abecasis G, Cardon LR, Zollner S, Whittaker JC, Chissoe SL, Novembre J, Mooser V. An abundance of rare functional variants in 202 drug target genes sequenced in 14,002 people. Science. 2012;337(6090):100–4. doi:10.1126/science.1217876.
16. Hiratsuka M. In vitro assessment of the allelic variants of cytochrome P450. Drug Metab Pharmacokinet. 2012;27(1):68–84.
17. McGeachie MJ, Stahl EA, Himes BE, Pendergrass SA, Lima JJ, Irvin CG, Peters SP, Ritchie MD, Plenge RM, Tantisira KG. Polygenic heritability estimates in pharmacogenetics: focus on asthma and related phenotypes. Pharmacogenet Genomics. 2013;23(6):324–8. doi:10.1097/FPC.0b013e3283607acf.
18. Sadee W. The relevance of "missing heritability" in pharmacogenomics. Clin Pharmacol Ther. 2012;92(4):428–30. doi:10.1038/clpt.2012.116.
19. Faraday N, Yanek LR, Mathias R, Herrera-Galeano JE, Vaidya D, Moy TF, Fallin MD, Wilson AF, Bray PF, Becker LC, Becker DM. Heritability of platelet responsiveness to aspirin in activation pathways directly and indirectly related to cyclooxygenase-1. Circulation. 2007;115(19):2490–6. doi:10.1161/CIRCULATIONAHA.106.667584.
20. Shuldiner AR, O'Connell JR, Bliden KP, Gandhi A, Ryan K, Horenstein RB, Damcott CM, Pakyz R, Tantry US, Gibson Q, Pollin TI, Post W, Parsa A, Mitchell BD, Faraday N, Herzog W, Gurbel PA. Association of cytochrome P450 2C19 genotype with the antiplatelet effect and clinical efficacy of clopidogrel therapy. JAMA. 2009;302(8):849–57. doi:10.1001/jama.2009.1232.
21. Isakov O, Perrone M, Shomron N. Exome sequencing analysis: a guide to disease variant detection. Methods Mol Biol. 2013;1038:137–58. doi:10.1007/978-1-62703-514-9_8.
22. Stitziel NO, Kiezun A, Sunyaev S. Computational and statistical approaches to analyzing variants identified by exome sequencing. Genome Biol. 2011;12(9):227. doi:10.1186/gb-2011-12-9-227.
23. Kumar P, Henikoff S, Ng PC. Predicting the effects of coding non-synonymous variants on protein function using the SIFT algorithm. Nat Protoc. 2009;4(7):1073–81. doi:10.1038/nprot.2009.86.
24. Reva B, Antipin Y, Sander C. Predicting the functional impact of protein mutations: application to cancer genomics. Nucleic Acids Res. 2011; 39(17):e118. doi:10.1093/nar/gkr407.
25. Stone EA, Sidow A. Physicochemical constraint violation by missense substitutions mediates impairment of protein function and disease severity. Genome Res. 2005;15(7):978–86. doi:10.1101/gr.3804205.
26. Thomas PD, Campbell MJ, Kejariwal A, Mi H, Karlak B, Daverman R, Diemer K, Muruganujan A, Narechania A. PANTHER: a library of protein families and subfamilies indexed by function. Genome Res. 2003;13(9):2129–41. doi:10.1101/gr.772403.
27. Adzhubei IA, Schmidt S, Peshkin L, Ramensky VE, Gerasimova A, Bork P, Kondrashov AS, Sunyaev SR. A method and server for predicting damaging missense mutations. Nat Methods. 2010;7(4):248–9. doi:10.1038/nmeth0410-248.
28. Bromberg Y, Rost B. SNAP: predict effect of non-synonymous polymorphisms on function. Nucleic Acids Res. 2007;35(11):3823–35. doi:10.1093/nar/gkm238.
29. Yue P, Moult J. Identification and analysis of deleterious human SNPs. J Mol Biol. 2006;356(5):1263–74. doi:10.1016/j.jmb.2005.12.025.
30. Carter H, Douville C, Stenson PD, Cooper DN, Karchin R. Identifying Mendelian disease genes with the variant effect scoring tool. BMC Genomics. 2013;14 Suppl 3:S3. doi:10.1186/1471-2164-14-S3-S3.
31. Li B, Krishnan VG, Mort ME, Xin F, Kamati KK, Cooper DN, Mooney SD, Radivojac P. Automated

inference of molecular mechanisms of disease from amino acid substitutions. Bioinformatics. 2009; 25(21):2744–50. doi:10.1093/bioinformatics/btp528.
32. Schwarz JM, Rodelsperger C, Schuelke M, Seelow D. MutationTaster evaluates disease-causing potential of sequence alterations. Nat Methods. 2010;7(8):575–6. doi:10.1038/nmeth0810-575.
33. Morgenthaler S, Thilly WG. A strategy to discover genes that carry multi-allelic or mono-allelic risk for common diseases: a cohort allelic sums test (CAST). Mutat Res. 2007;615(1–2):28–56. doi:10.1016/j.mrfmmm.2006.09.003.
34. Li B, Leal SM. Methods for detecting associations with rare variants for common diseases: application to analysis of sequence data. Am J Hum Genet. 2008;83(3):311–21. doi:10.1016/j.ajhg.2008.06.024.
35. Price AL, Kryukov GV, de Bakker PI, Purcell SM, Staples J, Wei LJ, Sunyaev SR. Pooled association tests for rare variants in exon-resequencing studies. Am J Hum Genet. 2010;86(6):832–8. doi:10.1016/j.ajhg.2010.04.005.
36. Bhatia G, Bansal V, Harismendy O, Schork NJ, Topol EJ, Frazer K, Bafna V. A covering method for detecting genetic associations between rare variants and common phenotypes. PLoS Comput Biol. 2010;6(10):e1000954. doi:10.1371/journal.pcbi.1000954.
37. Luo L, Boerwinkle E, Xiong M. Association studies for next-generation sequencing. Genome Res. 2011;21(7):1099–108. doi:10.1101/gr.115998.110.
38. Madsen BE, Browning SR. A groupwise association test for rare mutations using a weighted sum statistic. PLoS Genet. 2009;5(2):e1000384. doi:10.1371/journal.pgen.1000384.
39. Liu DJ, Leal SM. A novel adaptive method for the analysis of next-generation sequencing data to detect complex trait associations with rare variants due to gene main effects and interactions. PLoS Genet. 2010; 6(10):e1001156. doi:10.1371/journal.pgen.1001156.
40. Sul JH, Han B, He D, Eskin E. An optimal weighted aggregated association test for identification of rare variants involved in common diseases. Genetics. 2011; 188(1):181–8. doi:10.1534/genetics.110.125070.
41. Yick CY, Zwinderman AH, Kunst PW, Grunberg K, Mauad T, Fluiter K, Bel EH, Lutter R, Baas F, Sterk PJ. Glucocorticoid-induced changes in gene expression of airway smooth muscle in patients with asthma. Am J Respir Crit Care Med. 2013;187(10): 1076–84. doi:10.1164/rccm.201210-1886OC.
42. Wiviott SD, Braunwald E, McCabe CH, Montalescot G, Ruzyllo W, Gottlieb S, Neumann FJ, Ardissino D, De Servi S, Murphy SA, Riesmeyer J, Weerakkody G, Gibson CM, Antman EM, Investigators T-T. Prasugrel versus clopidogrel in patients with acute coronary syndromes. N Engl J Med. 2007;357(20):2001–15. doi:10.1056/NEJMoa0706482.
43. Mega JL, Close SL, Wiviott SD, Shen L, Hockett RD, Brandt JT, Walker JR, Antman EM, Macias W, Braunwald E, Sabatine MS. Cytochrome p-450 polymorphisms and response to clopidogrel. N Engl J Med. 2009;360(4):354–62. doi:10.1056/NEJMoa0809171.
44. Mega JL, Close SL, Wiviott SD, Shen L, Walker JR, Simon T, Antman EM, Braunwald E, Sabatine MS. Genetic variants in ABCB1 and CYP2C19 and cardiovascular outcomes after treatment with clopidogrel and prasugrel in the TRITON-TIMI 38 trial: a pharmacogenetic analysis. Lancet. 2010;376(9749): 1312–9. doi:10.1016/S0140-6736(10)61273-1.
45. Ramsey LB, Panetta JC, Smith C, Yang W, Fan Y, Winick NJ, Martin PL, Cheng C, Devidas M, Pui CH, Evans WE, Hunger SP, Loh M, Relling MV. Genome-wide study of methotrexate clearance replicates SLCO1B1. Blood. 2013;121(6):898–904. doi:10.1182/blood-2012-08-452839.
46. Ramnath N, Daignault-Newton S, Dy GK, Muindi JR, Adjei A, Elingrod VL, Kalemkerian GP, Cease KB, Stella PJ, Brenner DE, Troeschel S, Johnson CS, Trump DL. A phase I/II pharmacokinetic and pharmacogenomic study of calcitriol in combination with cisplatin and docetaxel in advanced non-small-cell lung cancer. Cancer Chemother Pharmacol. 2013;71(5):1173–82. doi:10.1007/s00280-013-2109-x.
47. Anthoney DA, Naik J, Macpherson IR, Crawford D, Hartley JM, Hartley JA, Saito T, Abe M, Jones K, Miwa M, Twelves C, Evans TR. Phase I study of TP300 in patients with advanced solid tumors with pharmacokinetic, pharmacogenetic and pharmacodynamic analyses. BMC Cancer. 2012;12:536. doi:10.1186/1471-2407-12-536.
48. Hayes DN, Lucas AS, Tanvetyanon T, Krzyzanowska MK, Chung CH, Murphy BA, Gilbert J, Mehra R, Moore DT, Sheikh A, Hoskins J, Hayward MC, Zhao N, O'Connor W, Weck KE, Cohen RB, Cohen EE. Phase II efficacy and pharmacogenomic study of Selumetinib (AZD6244; ARRY-142886) in iodine-131 refractory papillary thyroid carcinoma with or without follicular elements. Clin Cancer Res. 2012;18(7):2056–65. doi:10.1158/1078-0432.CCR-11-0563.
49. Denny JC. Chapter 13: mining electronic health records in the genomics era. PLoS Comput Biol. 2012;8(12):e1002823. doi:10.1371/journal.pcbi.1002823.
50. Manolio TA. Collaborative genome-wide association studies of diverse diseases: programs of the NHGRI's office of population genomics. Pharmacogenomics. 2009;10(2):235–41. doi:10.2217/14622416.10.2.235.
51. Yu IW, Bukaveckas BL. Pharmacogenetic tests in asthma therapy. Clin Lab Med. 2008;28(4):645–65. doi:10.1016/j.cll.2008.05.001.
52. Roberts JD, Wells GA, Le May MR, Labinaz M, Glover C, Froeschl M, Dick A, Marquis JF, O'Brien E, Goncalves S, Druce I, Stewart A, Gollob MH, So DY. Point-of-care genetic testing for personalisation of antiplatelet treatment (RAPID GENE): a prospective, randomised, proof-of-concept trial. Lancet. 2012;379(9827):1705–11. doi:10.1016/S0140-6736(12)60161-5.
53. Sconce EA, Khan TI, Wynne HA, Avery P, Monkhouse L, King BP, Wood P, Kesteven P, Daly AK, Kamali F. The impact of CYP2C9 and VKORC1 genetic polymorphism and patient characteristics upon warfarin dose requirements: proposal for a new dosing regimen. Blood. 2005;106(7):2329–33. doi:10.1182/blood-2005-03-1108.
54. Anderson JL, Horne BD, Stevens SM, Woller SC, Samuelson KM, Mansfield JW, Robinson M, Barton S, Brunisholz K, Mower CP, Huntinghouse JA, Rollo JS, Siler D, Bair TL, Knight S, Muhlestein JB,

Carlquist JF. A randomized and clinical effectiveness trial comparing two pharmacogenetic algorithms and standard care for individualizing warfarin dosing (CoumaGen-II). Circulation. 2012;125(16):1997–2005. doi:10.1161/CIRCULATIONAHA.111.070920.
55. Karczewski KJ, Daneshjou R, Altman RB. Chapter 7: pharmacogenomics. PLoS Comput Biol. 2012;8(12):e1002817. doi:10.1371/journal.pcbi.1002817.
56. Baigent C, Keech A, Kearney PM, Blackwell L, Buck G, Pollicino C, Kirby A, Sourjina T, Peto R, Collins R, Simes R, Cholesterol Treatment Trialists C. Efficacy and safety of cholesterol-lowering treatment: prospective meta-analysis of data from 90,056 participants in 14 randomised trials of statins. Lancet. 2005;366(9493):1267–78. doi:10.1016/S0140-6736(05)67394-1.
57. Search E, Link E, Parish S, Armitage J, Bowman L, Heath S, Matsuda F, Gut I, Lathrop M, Collins R. SLCO1B1 variants and statin-induced myopathy—a genomewide study. N Engl J Med. 2008;359(8):789–99.doi:10.1056/NEJMoa0801936.
58. Danik JS, Chasman DI, MacFadyen JG, Nyberg F, Barratt BJ, Ridker PM. Lack of association between SLCO1B1 polymorphisms and clinical myalgia following rosuvastatin therapy. Am Heart J. 2013;165(6):1008–14. doi:10.1016/j.ahj.2013.01.025.
59. Hicks JK, Swen JJ, Thorn CF, Sangkuhl K, Kharasch ED, Ellingrod VL, Skaar TC, Muller DJ, Gaedigk A, Stingl JC, Clinical Pharmacogenetics Implementation. Clinical pharmacogenetics implementation consortium guideline for CYP2D6 and CYP2C19 genotypes and dosing of tricyclic antidepressants. Clin Pharmacol Ther. 2013;93(5):402–8. doi:10.1038/clpt.2013.2.
60. Sirskyj D, Diaz-Mitoma F, Golshani A, Kumar A, Azizi A. Innovative bioinformatic approaches for developing peptide-based vaccines against hypervariable viruses. Immunol Cell Biol. 2011;89(1):81–9. doi:10.1038/icb.2010.65.
61. Thomas C, Moridani M. Interindividual variations in the efficacy and toxicity of vaccines. Toxicology. 2010;278(2):204–10.doi:10.1016/j.tox.2009.10.008.
62. Russell C, Rahman A, Mohammed AR. Application of genomics, proteomics and metabolomics in drug discovery, development and clinic. Ther Deliv. 2013;4(3):395–413. doi:10.4155/tde.13.4.
63. Scheffler RJ, Colmer S, Tynan H, Demain AL, Gullo VP. Antimicrobials, drug discovery, and genome mining. Appl Microbiol Biotechnol. 2013;97(3):969–78. doi:10.1007/s00253-012-4609-8.
64. Liou SY, Stringer F, Hirayama M. The impact of pharmacogenomics research on drug development. Drug Metab Pharmacokinet. 2012;27(1):2–8.
65. Nebert DW, Dalton TP. The role of cytochrome P450 enzymes in endogenous signalling pathways and environmental carcinogenesis. Nat Rev Cancer. 2006;6(12):947–60. doi:10.1038/nrc2015.
66. Swen JJ, Huizinga TW, Gelderblom H, de Vries EG, Assendelft WJ, Kirchheiner J, Guchelaar HJ. Translating pharmacogenomics: challenges on the road to the clinic. PLoS Med. 2007;4(8):e209. doi:10.1371/journal.pmed.0040209.

CHAPTER 32

Sequencing Cell Free DNA in the Maternal Circulation to Screen for Down Syndrome and Other Common Aneuploidies

GLENN E. PALOMAKI

Introduction 563
Discovery, Identification and Initial Uses of Circulating Cell Free (ccf) DNA 565
Sequencing of ccfDNA to Identify Down Syndrome: Preliminary Studies 566
Preliminary Academic Studies 566
Down Syndrome Screening: Clinical Validation Studies in the USA 569
Other Disorders Currently Detected by Maternal Plasma DNA Testing 570
Testing in Specific Subgroups 574
Future Enhancements/ Improvements 575
Outstanding Implementation Issues 576
Conclusions 578

Introduction

Screening is different from most forms of medical testing in that the individuals opting for screening are generally healthy. Those offering or recommending screening tests need to ensure that the screening test performance is well characterized and acceptable, usually by determining the detection rate or sensitivity (the proportion of affected individuals with a positive test) and false positive rate (proportion of unaffected individuals with a positive test, or 1-specificity). A working definition of screening has been proposed by Wald [58], and will be adhered to in this chapter.

> Screening is the systematic application of a test or enquiry to identify individuals at sufficient risk of a specific disorder to warrant further investigation or direct preventive action, amongst persons who have not sought medical attention on account of symptoms of that disor der.

Although Down syndrome was first described in 1862 by Langdon Down [22], the cause of this disorder (an extra chromosome 21) was not determined until 1959 [33]. This led directly to the ability to prenatally diagnose Down syndrome via amniocentesis and karyotyping. Given the known association between increasing birth prevalence and advancing maternal age, the first

G.J. Netto and I. Schrijver (eds.), *Genomic Applications in Pathology*,
DOI 10.1007/978-1-4939-0727-4_32, © Springer Science+Business Media New York 2015

screening test for Down syndrome was the question "How old will you be at delivery?" Until 2007 [1], this question still formed the basis of most prenatal screening for Down syndrome, although laboratory tests apart from karyotyping were, by then, available.

In the early 1980s, women routinely had second trimester maternal serum alpha-fetoprotein (AFP) measured as a screening test for open neural tube defects. In 1984, a woman who had a very low AFP level asked her physician whether the low measurement may be related to the diagnosis of trisomy 18 in her newborn. This query prompted a line of investigation confirming that second trimester AFP is a marker not only for trisomy 18, but also for Down syndrome [37]. The reason for reduced AFP levels in Down syndrome and trisomy 18 fetuses is still unknown, but the discovery of this association was quickly verified [19] and a screening algorithm devised [44]. Further research identified additional early second trimester serum markers [8, 11, 31] leading to the development of the "triple test," which includes maternal age in combination with AFP, unconjugated estriol (uE3), and human chorionic gonadotropin (hCG) measurements. The current best second trimester combination (the triple test with dimeric inhibin-A measurements) is called the "quadruple" test and has a sensitivity of about 80 % at a false positive rate of about 5 % [28, 31]. During the same time period, additional markers were identified that could be measured in the late first trimester. These included not only serum markers (PAPP-A [57] and free beta hCG [53]), but also an ultrasound measurement of the translucent space between the spine and skin of the fetus [56]. Together with maternal age, these markers formed the basis of the "combined" test [63]. Sensitivity and specificity are similar to, or slightly better than, the quadruple test.

Throughout the 1990s, the first and second trimester tests were offered independently. Many screening programs elected to offer testing in the first or second trimester, but not both. In 1999, the concept of the "integrated" test was first reported [60], in which the best first and second trimester markers were combined into a single risk estimate. This "integrated" test has the best performance of any widely offered prenatal screening test based on biochemical and ultrasound measurements, with a sensitivity of about 90 % at a false positive rate of 2 % [59].

Table 32-1 Types and Numbers of Down Syndrome Screening Tests Performed in US Laboratories in 2012

Type of test	Laboratories	Median (*N*)	Number (%)
First trimester[a]	34	3,000	565,692 (19)
Second trimester	122	2,538	1,770,024 (60)
AFP only	85	720	235,492
Triple test	44	402	90,132
Quadruple test[b]	118	2,400	1,443,900
Integrated	30	4,176	583,416 (21)
Fully integrated	22	2,136	102,972
Serum integrated	21	888	119,760
Sequential	24	2,436	405,144
All	123[c]	3,660	2,963,592 (100)

[a]Includes all first trimester tests, including those using serum measurements of total/intact hCG, free beta hCG and dimeric inhibin-A
[b]Includes tests with five second trimester serum markers
[c]Does not add up, as some laboratories are counted in multiple "types of tests"

As part of an external proficiency testing program administered by the College of American Pathologists (CAP) and the American College of Medical Genetics and Genomics (www.cap.org), a survey of participating laboratories has been conducted within the first of three distributions each year, beginning in 2011. This survey asks participants what tests they offer and how many are performed each year. Results from the 2011 and 2012 survey have been compiled [46] and are summarized in Table 32.1. Overall, the survey documents that about 72 % of the 4.24 million pregnancies in the USA are screened for Down syndrome by a total of 123 laboratories. Testing most commonly occurs in the second trimester (60 % of tested women), followed by various forms of the integrated test (21 %) and combined testing (19 %). These numbers do not include women who opt directly for invasive testing (amniocentesis or chorionic villus sampling) and karyotyping.

Serum markers are readily available throughout the country with measurements made using FDA cleared instruments and reagents. The results are subject to external proficiency testing that shows that they can be implemented reliably in a variety of high-complexity laboratory settings. The costs of reagents are relatively low with test charges ranging from <$100 to $200 or more. Ultrasound measurements are operator dependent, but several organizations provide training and oversight (http://www.fetalmedicineusa.com/FMF/US, https://www.ntqr.org/). The turn-around time for combined or integrated testing is a few days. One implementation in the UK is the One Stop Clinic for Assessment of fetal Risk, or OSCAR clinic [7] where the ultrasound measurement, biochemistry testing, reporting of results, counseling, and the offer of a diagnostic test, if indicated, can be made within one hour.

Although the sensitivity and specificity of screening tests for Down syndrome, and to some extent trisomy 18 and trisomy 13, are adequate, the prenatal care community has two main objectives: (1) prenatal screening and diagnosis should occur early in pregnancy, preferably being completed by the end of the first trimester, and (2) the number of invasive procedures should be kept to a minimum to reduce the possibility of procedure-related loss. Optimally, a non-invasive prenatal diagnostic (NIPD) test might be developed that could identify not only Down syndrome and the other common trisomies but also other prenatally diagnosable conditions, many of which can now only be identified after amniocentesis or chorionic villus sampling (CVS) and using diagnostic techniques such as arrayed comparative genomic hybridization (aCGH). Testing cell free nucleic acids in maternal circulation is the next, and perhaps final, step in this process.

Discovery, Identification and Initial Uses of Circulating Cell Free (ccf) DNA

Identification of ccfDNA from the Fetus in Maternal Plasma

In 1997, a landmark publication [35] reported the presence of fetal DNA in maternal plasma and serum. This finding was based on earlier work showing circulating cell free cancer DNA in the plasma of affected patients [14, 38, 55]. In that 1997 study, pregnant and nonpregnant women provided serum, plasma and whole blood samples for testing. The fetal sex was determined at birth, or by karyotype after amniocentesis. After extraction, DNA was tested for presence of a Y-chromosome sequence (DYS14). Of the 30 pregnancies with a male fetus, plasma testing identified 24 (80 %); none of the 13 female fetuses were misclassified. Fewer males were identified using serum and whole blood, but again, none of samples from the 13 female fetuses were positive for the Y-probe. The 80 % detection rate for males was, in hindsight, likely due to the small plasma sample size (10 μL), as most of the errors occurred earlier in gestation.

Using ccfDNA in Maternal Plasma to Determine Fetal Sex

Soon after the report by Lo and colleagues [35], these findings were confirmed and expanded by other groups. Early fetal sexing can be useful to resolve ambiguous genitalia, manage X-linked conditions, and help identify some single-gene disorders (e.g., congenital adrenal hyperplasia). Noninvasive testing could thus be used in place of invasive cytogenetic testing. Because early fetal sex determination via ultrasound is not reliable in the first trimester [42], ccfDNA testing could be reliable alternative. A summary of 57 published studies on the use of ccfDNA to identify fetal sex was published in 2011 [20].

Overall, 3,524 male and 3,017 female pregnancy samples were included. The performance was better after 7 weeks of gestation, when plasma rather than serum was used, and when real time quantitative PCR (RQ-PCR) was employed. At 7–20 weeks gestation, the use of RQ-PCR results in an estimated sensitivity for determining fetal sex of 98 %, with a corresponding specificity of 99.1 %.

Use of ccfDNA in Maternal Plasma to Determine Fetal Rh Status

Among RhD negative women, RhD positive fetal cells crossing the placenta can cause the mother to make anti-RhD antibodies. This Rh incompatibility can lead to fetal complications. Rh incompatibility is preventable by providing injection of Rh immunoglobulin to RhD negative women in the second trimester. However, treatment is unnecessary if the fetus is also RhD negative. Testing ccfDNA can identify these pregnancies and avoid unnecessary treatment. In a recent nationwide study in Denmark [18], 2,312 RhD negative women were tested both by routine genotyping and by testing of ccfDNA for RhD status at 25 weeks gestation. Overall, the ccfDNA test had a 99.9 % sensitivity with 96.5 % accuracy. A total of 862 of these women avoided unnecessary treatment; 39 women with an RhD negative fetus still received treatment (unnecessary treatment), and two women with RhD positive fetuses were not detected and the mothers were not treated (false negatives).

Sequencing of ccfDNA to Identify Down Syndrome: Preliminary Studies

This chapter focuses on the development and validation of the four laboratory developed ccfDNA tests currently offered in the USA. Given the competitive commercial nature of ccfDNA test application in the USA, no academic laboratories appear willing, at this time, to validate or offer maternal plasma DNA testing for aneuploidy. Therefore, this section is devoted to preliminary academic proof-of-concept and in-house validation studies performed by commercial companies. Later sections will describe the results from collaborations with academic sites that performed external validation studies. Three laboratories located outside of the USA (Berry Genomics, Beijing, China; Beijing Genomics Institute (BGI) Health Institute, Beijing, China, and LifeCodexx, Konstanz, Germany) all use a variant of the shotgun sequencing described in the following sections.

Preliminary Academic Studies

In late 2008, two studies [16, 24] demonstrated the potential for shotgun sequencing of ccfDNA in maternal plasma to identify common autosomal trisomies. Both studies used similar counting methodologies and neither was blinded. After sequencing, the DNA fragments were mapped to the human genome to identify the chromosome of origin. A normalizing function was then applied to account for the varying number of total matched reads per sample. These normalized values were then compared with values from known euploid and trisomic samples.

In the study by Fan and colleagues [24], nine samples from Down syndrome pregnancies showed "normalized sequence tag densities" for chromosome 21 that were 4–18 % higher than the corresponding densities in samples from euploid pregnancies. Two trisomy 18 samples were also detected, but one trisomy 13 sample was only slightly elevated. On average, about ten million DNA fragments were sequenced per patient and about five million were matched and used for analysis. This study was limited in that most trisomic samples were collected after amniocentesis (which could influence the results), some samples were collected in the third trimester, no trisomic samples were collected prior to 14 weeks gestation, and the number of control samples (six) was very limited.

In the study by Chiu et al. [16], all 14 samples from Down syndrome pregnancies had chromosome 21 *z*-scores of 5 or higher, while none of the 14 samples from euploid pregnancies had values above 3. The z-score measures the difference between the chromosome 21 % in the patient minus the average percentage in euploid pregnancies, divided by the standard deviation of chromosome 21 percentages in known euploid pregnancies. All samples were collected at 20 weeks gestation or earlier. On average, about 11 million DNA fragments were sequenced and about 2.5 million were matched and used for analysis. Limitations of this study include the collection of samples after an invasive procedure for 11 of 14 cases, and no trisomy 18 or trisomy 13 samples were tested.

In a subsequent large-scale collaborative study [15], samples from three sites were available for testing. In order to scale testing for larger throughput, this group multiplexed eight samples per flow cell lane (Illumnia, San Diego, CA), allowing for up to 64 patients per run. Seven hundred and fifty-three samples were run, including 86 samples from Down syndrome pregnancies. Unfortunately, there was an average of only 300,000 matched reads per patient, resulting in a detection rate of 79 % with a false positive rate of 1.1 %. A subset of the samples was available to rerun at a 2-plex, with an average of 2.3 million matched reads per patient. The Down syndrome detection rate improved to 100 %, with a 2.1 % false positive rate. For the first time, a study reported that it was not possible to obtain results on a subset of samples (1.4 %) that passed specimen quality requirements, but not sequencing quality metrics. This "failure" rate is an important consideration when examining the performance of this type of testing. This was also the first report that showed the clear relationship between the fetal fraction (measured by Y-probes among male fetuses) and the chromosome 21 z-score. The median fetal fraction was 15 %. Limitations of the study include the un-blinded running of the 2-plex samples and the lack of a training set of known euploid samples prior to test interpretation.

Preliminary Studies from Sequenom, Inc.

A publication [23] from a commercial company (Sequenom, Inc., San Diego, CA) provided results from three small preliminary studies as well as a larger clinical validation of a laboratory developed test that could be suitable for introduction into practice. In the larger study, a total of 40 Down syndrome cases were matched with 440 euploid control pregnancies. Thirteen samples were not suitable for testing, and another 18 failed quality control, including one case of Down syndrome. The classification system relied on a single z-score cutoff of 3.0. The detection rate was 100 % (39/39) with no false positives (0/410). The failure rate was 4 %. At least 12 million matched reads were available for each patient. The median gestational age was 15 weeks, but some late second and third trimester samples were included, along with some cases that had been collected after diagnostic testing.

Preliminary Studies from Verinata, Inc.

Another study [50] from a privately held company described a laboratory developed test that was aimed at detecting Down syndrome, trisomy 18, and trisomy 13, as well as selected sex aneuploidies. Their method relies on comparing target chromosome results with (a) specific comparison chromosome(s). For example, the chromosome 21 counts are compared against counts from chromosome 9. The comparator chromosome(s) was/were chosen based on minimizing the variances of the ratio over runs. This approach may reduce variability, as the comparator chromosome may be similar in GC content to the chromosome of interest. The mean and standard deviation of these ratios in known euploid pregnancies can then be used to assign a "normalized chromosome value" which can be interpreted as a z-score. After training, a set of 48 samples were blindly tested. On average, ten million matched reads per patient were available for analysis. For Down syndrome, the detection rate was 100 % (13/13). The classification method used by this group employs two cutoffs: results below 2.5 are considered euploid while those above 4.0 are considered trisomic. Those in between the cutoffs are classified as "no-calls." No false positive results occurred.

Alternatives to Shotgun Sequencing

The previous sections focused on shotgun sequencing methodologies. If the testing is to remain focused on chromosomes 21, 18, 13, X and Y, then one might consider preferentially amplifying targets located on these chromosomes as a way to reduce "unnecessary" sequencing time and expense. These next two methodologies apply targeted sequencing to maternal plasma DNA sequencing for common aneuploidies.

Preliminary Studies from Ariosa, Inc.

Two preliminary papers described targeted sequencing for common aneuploidies. One [52] uses the sequencing methodology with a simple interpretive method, while another [51] replaces that method with the "FORTE" algorithm that allows for patient-specific risks. According to this second publication [51] several hundred non-polymorphic loci were identified on each chromosome of interest (chromosomes 21 and 18), and a subset of the most predictive loci was then identified using the training set. After normalizing the counts for assay and patient biases, the chromosome 21 results were compared to the chromosome 18 results (under the assumption that no fetus will be trisomic for both chromosomes). The result was expressed as a normalized *z*-score. The *z*-score was then transformed into a risk by multiplying a prior risk for Down syndrome (based on mother's age) and a likelihood ratio derived from the overlapping distributions of *z*-scores in affected and unaffected pregnancies, after accounting for fetal fraction. The test set consisted of samples from 123 euploid and 36 Down syndrome pregnancies. Some cases were collected after invasive testing and the control pregnancies were women at low risk of aneuploidy. Some samples were collected in the third trimester. All 36 cases were identified (100 % detection rate) and no false positive results occurred (0 % false positive rate). There were no failures. Similar results were reported using only the *z*-score [52].

Preliminary Proof-of-Concept Study from Natera, Inc.

The methodology behind this publication [62] relies on a highly multiplexed PCR reaction that amplifies thousands of single nucleotide polymorphisms on chromosomes 13, 18, 21, X, and Y. Testing of the plasma is supplemented by maternal genotyping of the same SNPs using the sample's buffycoat. A complex matrix of potential genotypes based on crossover frequencies, source of the extra chromosome, single nucleotide polymorphism (SNP) copies (mono, di, and tri), and fetal fraction are modeled and the paired genotypes fitted using a Bayesian maximum likelihood method. The most likely model is the sum of the likelihoods and is reported as an accuracy score (from 0 to 100 %). The algorithm is capable of incorporating the paternal genotype as well, if available. A total of 161 samples were collected, most between 9 and 25 weeks of gestation. Most aneuploidies were collected after invasive testing. Some euploid samples were collected prior to confirmatory invasive testing, but many were simply assumed to be euploid. Both maternal and paternal genotypes were available for analysis. Overall, 20 samples (12 %) failed quality parameters, many with low fetal fractions. Among the remaining samples, 11 Down syndrome cases were correctly identified (100 % detection rate) and all of the control samples were negative (0 % false positive rate).

Summary of Proof-of-Concept and Preliminary Examination of Laboratory Developed Tests

These preliminary studies all had important weaknesses including small sample sizes, samples taken at gestational ages not relevant for screening, samples taken after (rather than before) an invasive procedure, a limited range of abnormalities and the need for training sets of known euploid pregnancies. However, they all provide important information regarding required elements for such testing. Clearly, samples with low fetal fraction will be more difficult to interpret correctly, the total number of matched reads will be important, and there will be variability in how results are interpreted (two or three categories, or a risk assessment).

Down Syndrome Screening: Clinical Validation Studies in the USA

This section focuses on clinical validity studies which require that a locked down assay be blindly tested by an external group. The focus is on Down syndrome test performance. Other aneuploidies that may have been part of the reports will be discussed in a later section.

Shotgun Sequencing/Counting: Sequenom, Inc.

A large external validation study funded by Sequenom, Inc. relied on sample collection at 27 prenatal diagnostic centers around the world [45]. The study was administered by an academic center at Women & Infants Hospital, Providence, RI. Although this first report focused on Down syndrome, all testing relating to this sample set (including trisomy 18, trisomy 13, sex aneuploidies, and twins) was completed during a single 10 week time period. In total, 4,664 "high risk" pregnancies were sampled and outcomes from invasive testing sought. The median maternal age was 37 years, with all samples collected between nine and 21 completed weeks' gestation. A total of 212 cases (105 first trimester and 107 second trimester) were tested, matched 7:1 (controls:cases) with known euploid samples (1,484 controls). The standard operating protocol had been published earlier [23]. A positive test was defined as having a *z*-score at or above 3 while those falling below were negative. When no *z*-score was assigned, the test was considered to have failed. Among the 212 cases tested, 209 were positive (detection rate 98.6 %, 95 % confidence interval (CI) 95.9–99.7 %). Among the 1,471 controls with successful testing, three were positive (false positive rate 0.2 %, 95 % CI <0.1–0.6 %). There were 13 test failures (0.9 %, 95 % CI 0.5–1.5 %), all of which occurred among the euploid samples. Further examination indicated that routine adjustment for GC content of chromosome 21 would have improved performance with only two false negatives (detection rate 99.1 %) and one false positive (false positive rate 0.1 %). This is the only study to date that validated test results against an academic laboratory and provided information regarding throughput.

Shotgun Sequencing/Selective Comparison: Verinata

An external validation study funded by Verinata, Inc. [6], utilized the same basic methodology, but there were important changes from the earlier report [50]. New reference data for both the comparator chromosome and the referent ratio distribution parameters were derived from 110 known euploid samples. Data handling, blinding and interpretation were handled by two clinical research organizations. Among a cohort of 2,882 enrolled pregnancies from 60 prenatal diagnostic sites in the USA, 532 samples from singleton pregnancies were selected for testing, including 90 non-mosaic Down syndrome cases and 311 euploid controls (an additional 131 samples had trisomy 18, trisomy 13, and Turner syndrome and have been excluded from the present analysis). Among the 532 samples, 16 (3.0 %) had insufficient fetal DNA for sequencing, resulting in a test failure. One case and six control samples (7/401 or 1.7 %) were "no-calls." Among the remaining samples, all (89/89) Down syndrome cases were detected (100 %, 95 % CI 96.4–100 %). There were no false positives among the 404 control samples (0 %, 95 % CI 0.0–0.9 %).

Targeted Sequencing/Counting: Ariosa Diagnostics

An external validation study funded by Ariosa Diagnostics [41] recruited patients from three countries. Test results and clinical data were merged by a clinical research organization. Among a cohort of 4,002 women, 3,228 samples from singleton pregnancies were eligible for testing. This included 94 samples from

Down syndrome pregnancies and 2,888 samples from euploid pregnancies. Overall, 148 samples (4.6 %) were excluded due to low fetal fraction and/or assay failure. Among the remaining samples, 81 of 81 Down syndrome cases were detected (100 %, 95 % CI 95.6–100 %). There was one false positive (<0.1 %, 95 % CI <0.1–0.2 %). This study also reported performance in trisomy 18 pregnancies.

A second external validation study[3] was performed using stored samples collected prior to CVS. A total of 300 euploid and 50 Down syndrome samples were tested. Three euploid samples failed (1 %). All Down syndrome cases were positive and all euploid pregnancies were negative. Trisomy 18 samples were also tested.

Targeted Sequencing/ Genotyping: Natera

The Natera laboratory-developed test continues to evolve. The external validation study [40] included more SNPs (19,488) compared to the previous publication [62], which used 10,000 SNPs. In addition, the algorithm for interpretation changed. Based on this, some argue that this should also be considered a preliminary study, but there was a clear external independent challenge of this updated version. A total of 242 singleton pregnancies provided samples prior to having CVS. Both the buffycoat and plasma were tested, with 13 test failures (5.4 %), including two Down syndrome and 11 euploid pregnancies. Among the remaining pregnancies, all 25 Down syndrome cases were detected (100 %, 95 % CI 86–100 %), with no false positives among the remaining 192 euploid pregnancies (0 %, 95 % CI 0–1.9 %). The study also reported performance in a small number of trisomy 18, trisomy 13, and Turner syndrome pregnancies. It also included a single case of triploidy.

Summary of Commercial Laboratory-Developed and Externally Validated Tests for Down Syndrome

The four commercial laboratory-developed and externally validated tests [3, 6, 40, 41, 45] all had high Down syndrome detection (Table 32.2). The largest study, of 212 cases, reported a detection rate of 98.6 %, and all cases had a reported result. The detection rate for the four additional studies had 100 % detection rates, but three of these four did not report results for one or more cases. This highlights a potential philosophical difference among these approaches. Is it better to fail to make a call than to fail the call you make? Considering that these are screening tests, it is well understood that they are not expected to always be correct. It is for this reason that conventional serum/ultrasound based screening tests for Down syndrome report results in risk, rather than Yes/No or Positive/Negative. Usually, screening tests have very low failure rates. When testing maternal plasma DNA, it will not be possible to assign a result to every sample, especially when the fetal fraction is very low (e.g., <2 or <3 %). In contrast, diagnostic tests are designed to minimize incorrect calls by identifying borderline results and reporting them as being "in the grey zone" or "no-call." Diagnostic tests also aim to minimize failures, as they are intended to be the "gold standard" result. In the long run, some of these laboratory-developed tests will improve their detection and false positive rates, while others will aim at reducing their failure rates. Together, they demonstrate that the various available methods performed very well, but each had strengths and weaknesses.

Other Disorders Currently Detected by Maternal Plasma DNA Testing

Trisomy 18 and Trisomy 13

Trisomy 18 and 13 are more difficult to detect than Down syndrome. The GC content of these chromosomes makes the counting techniques more challenging [16, 24]. This problem can be addressed by improvements in the sequencing chemistry [16] and through bioinformatics [13]. In addition, it now appears that the fetal fractions may also be lower in trisomy 18 pregnancies due to a smaller placenta, resulting in more test failures and reduced average separation between euploid and trisomy 18 scores.

Table 32-2 **Comparison of Published External Clinical Validation Studies for Down Syndrome Screening**

Characteristic	Palomaki et al. [45]	Ashoor et al. [3]	Bianchi et al. [6]	Norton et al. [41]	Nicolaides et al. [40]
Published	11/2011	4/2012	5/2012	8/2012	6/2013
Testing performed by	Sequenom, Inc.	Ariosa Diagnostics	Verinata	Ariosa Diagnostics	Natera, Inc.
Study design	Nested case/control	Nested case/control	Nested case/control	Cohort high risk	Cohort high risk
Multiplexing	4	96	2	96	96
Down syndrome	212	50	90	81	27
Euploid/other	1,484	300	311	2,888	215
Illumina platform	HiSeq 2000	NR	HiSeq 2000	NR	NR
Matched counts ($\times 10^6$)	16	~1	14	~1	9
CLIA certification	Yes	NR	NR	NR	NR
Throughput (samples/week)	235	NR	NR	NR	NR
Average maternal age (years)	37	35	35	34	36
Average GA (range)	15 (9–21)	12 (11–13)	15 (8–22)	17 (10–39)	13 (11–13)
Trimester % (first/second/third)	50/50/0	100/0/0	31/69/0	25/58/7	100/0/0
Testing failures (%)	13/1,697 (0.8)	1/400 (0.7)	16/532 (3.0)	148/3,228 (4.6)	13/242 (5.4)
Classified as "No-call" (%)	0	0	7/532 (1.3)	0	0
Detection rate (%)	209/212(98.6)	50/50 (100)	89/89 (100)	81/81 (100)	25/25 (100)
Cases not called[a]	0/212 (0.0)	0/50 (0.0)	1/90 (1.1)	3/84 (3.6)	2/27 (7.4)
False positive rate (%)	3/1,471 (0.2)	0/300 (0.0)	0/311(0.0)	1/2,887 (<0.1)	0/204 (0.0)
Collection tube	EDTA (purple)	EDTA (purple)	ACD (yellow)	BCT (mottled)	BCT (mottled)
Plasma volume	>3.5 mL × 2	0.5 mL × 4	>7 mL	8–10 mL	8–10 mL
Median fetal fraction (%)	13	NR	NR[b]	11	NR

GA gestational age

[a]Not detected due to test failure, or result classified as "no-call"

[b]Test does not routinely include a fetal fraction estimate

Among the external validation studies summarized in Table 32.2 [3, 4, 6, 40, 41, 45] all now have publications regarding the detection of trisomy 18 and trisomy 13 (Table 32.3). Overall, the trisomy 18 detection rate is 97.3 % (95 % CI 93.9–99.1 %) with a false positive rate of 0.1 % (95 % CI <0.1–0.3 %). However, nine additional trisomy 18 samples in these five studies failed testing or were considered a no-call, resulting in a failure rate of 5.6 %.

Most of those studies also included trisomy 13. One publication has been added. Overall, the trisomy 13 detection rate is 84 % (95 % CI 68.0–93.8 %) with a false positive rate of 0.5 % (95 % CI 0.3–0.8 %). This is likely to be an overestimate. One study [45] is responsible for 16 of the 21 false positive

Table 32-3 **Detection and False Positive Rates for Trisomy 18 and Trisomy 13 in External Validation S tudies**

Study	Test	T18 DR (%)	T18 FPR (%)	T18 failed (%)	T13 DR (%)	T13 FPR (%)	T13 failed (%)
Palomaki et al. [45]	Sequenom	59/59 (100)	5/1,688 (0.3)	3/62 (4.8)	11/12 (92)	16/1,688 (0.9)	0/12 (0)
Ashoor et al. [3]	Ariosa	48/50 (96.0)	0/300 (0.0)	0/50 (0.0)	–	–	–
Bianchi et al. [6]	Verinata	35/37 (94.6)	1/461 (0.2)	2/39 (5.1)	11/14 (79)	3/488 (0.6)	2/16 (13)
Norton et al. [41]	Ariosa	37/38 (97.4)	2/2,888 (0.1)	4/42 (9.6)	–	–	–
Ashoor et al. [4]	Ariosa	–	–	–	8/10 (80)	2/1,939 (0.1)	0/10 (0)
Nicolaides et al. [40]	Natera	3/3 (100)	0/192 (0.0)	0/3 (0.0)	1/1 (100)	0/192 (0.0)	0/1 (0)
All	Any	182/187 (97.3)	8/5,509 (0.1)	9/196 (5.6)	31/37 (84)	21/4,307 (0.5)	2/39 (5)

T18 trisomy 18, *T13* trisomy 13, *DR* detection rate, *FPR* false positive rate

Table 32-4 **Detection and False Positive Rates for Sex Chromosome Aneuploidies in External Validation Studies**

	Turner syndrome (45,X)		Triple X syndrome (47,XXX)		Klinefelter syndrome 47,XXY		47,XYY	
Study	DR (%)	Fail	DR (%)	Fail	DR (%)	Fail	DR (%)	Fail
Bianchi et al. [6]	15/16 (94)	4	3/4 (75)	0	2/2 (100)	0	3/3 (100)	0
Nicolaides et al. [40]	2/2 (100)	0	–	–	–	–	–	–
Mazloom et al. [36]	17/18 (94)	3	1/1 (100)	0	5/5 (100)	0	2/2 (100)	1
All	34/36 (94)	7	4/5 (90)	0	7/7 (100)	0	5/5 (100)	1

DR detection rate

results. In that study, a *z*-score cutoff of 3.0 was used, but the lowest detected case had a *z*-score of 7, while all but three controls had a *z*-score below 5. That laboratory now uses a higher *z*-score cutoff for both trisomy 18 and trisomy 13. There were two more trisomy 13 cases that failed testing or were considered a no-call, for a total failure rate of 5 %.

Sex Aneuploidies and Fetal Sex

In order to identify the common sex aneuploidies, it is necessary to first identify fetal sex. Many of the same cohorts used in these external validation studies examined the potential to identify the more common sex aneuploidies as well as to identify fetal sex. Table 32.4 provides data from the three studies reporting sex aneuploidies [6, 36, 40]. By far the most common sex aneuploidy is Turner syndrome (45X). Turner syndrome is over-represented because it is far more common in the first trimester than at term, due to spontaneous loss, and many of these studies focus on first trimester samples. More importantly, many enrolled women were identified as "high risk" because of increased nuchal

Table 32-5 **Identification of Fetal Sex in External Validation Studies**

	Males			Females		
Study	**Called (%)**	**No-call (%)**	**FP (%)**	**Called (%)**	**No-call (%)**	**FP (%)**
Bianchi et al. [6]	184/184 (100)	39 (17)	1/233 (0.4)	232/233 (99.6)	7 (2.9)	0/184 (0.0)
Nicolaides et al. [40]	103/103 (100)	7[a] (6.4)	0/89 (0.0)	89/89 (100)	6[a] (6.3)	0/103 (0.0)
Mazloom et al. [36]	191/192 (99.5)	13 (6.3)	4/172 (2.3 %)	167/172 (97.1)	4 (2.3)	1/192 (0.5)
All	478/479 (99.8)	59 (11.0)	5/494 (1.0 %)	488/494 (98.8)	17 (3.3)	1/479 (0.3)

[a]There were 13 no-calls, but the fetal sex was not provided. *FP* false positives

translucency. In early pregnancy, a large nuchal translucency can also be an indication of non-immune fetal hydrops, which is strongly associated with Turner syndrome. Of the 36 cases of Turner syndrome with a DNA interpretation, 34 were detected. However, seven additional cases of Turner syndrome were tested, but either the test failed, or the result was inconclusive. There were far fewer cases of triple X syndrome (47,XXX), Klinefelter syndrome (XXY), and XYY syndrome, but most were detected.

Table 32.5 shows the ability of those same tests to correctly identify fetal sex. Usually, Y chromosome counts are used to identify males, with quantification of both X and Y counts to determine aneuploidy, once fetal sex is determined. The SNP methodology used in one study [40] relies on matching expected SNP genotypes with models defining what a sex aneuploidy might look like. Overall, the sex aneuploidy detection rate for males is higher than for females. All methods have an important proportion of samples that fail testing or for which no clinical call is made (range 6.3–17 % for males and 2.3–6.3 % for females).

Mosaicism

Identifying fetuses with mosaicism for fetal aneuploidies will be more difficult than for complete aneuploidy. An additional complication is that the ccfDNA methods rely on fetal DNA derived from the placenta and, therefore, confined placental mosaicism could be identified even if the fetal chromosomes are normal. This issue also complicates the interpretation of CVS test results. To help understand screening for mosaicism, consider the effective fetal fraction when a mosaicism for chromosome 21 is 50 % and the fetal fraction is 10 %. The mosaicism would indicate that about half of the fetal DNA would be derived from cells with three chromosomes 21 and half from euploid cells. Thus, the effective fetal fraction is only 5 %. This computation is complicated by the potential heterogeneity of the placenta, which would be difficult to quantify. In many instances, a low rate of mosaicism combined with an average or below average fetal fraction would result in a negative test result.

One study [6] reported on three mosaic karyotypes involving the common trisomies. Two were mosaic for chromosome 21 (29 %, 44 % mosaicism) and one for trisomy 18 (89 % mosaicism). No test results or fetal fractions were reported, but all three were reported as being detected. A second study [12] identified five relevant mosaic karyotypes involving chromosomes 21, 18 and 13. The median fetal fraction was 13 % (range 6–23 %), and mosaicism ranged from 10 to 50 %. The effective fetal fractions ranged from 0.6 to 10 %. As expected, the two mosaics with very low effective fetal fractions were identified as euploid, while one of the two with effective fetal fractions around 5 % was detected. The lone sample with a high effective fetal fraction of 10 % (45 % mosaicism × 23 % fetal fraction) was a trisomy 18 mosaic superimposed on a full trisomy 21 karyotype. The result was positive for Down syndrome, but the *z*-score for chromosome 18 was unremarkable.

In summary, maternal plasma DNA testing in its current implementation will likely identify most high level mosaicism when the fetal fraction is average or above. However, it will likely not identify most of the lower level mosaicism, especially when the fetal fraction is below average. In the future, deeper sequencing would likely allow for a higher proportion of mosaic pregnancies to be identified. As with all positive ccfDNA testing for aneuploidy, any woman with a positive test result should be offered an invasive procedure to allow for a definitive diagnosis.

Testing in Specific Subgroups

IVF Pregnancies

Many studies explicitly exclude pregnancies achieved by in vitro fertilization (IVF), whereas others did not document IVF status during enrollment. One study [6] initially reported that 17 of 38 IVF pregnancies had chromosomal abnormalities (all correctly identified) and that 21 were euploid (all correctly identified). No mention was made of the fetal fraction in this population or whether the normalized chromosome value (similar to a *z*-score) differed from non-IVF pregnancies.

A more comprehensive study by our group [31] identified a subset of 632 tested women from seven enrollment sites that had information available about the use of assisted reproductive technologies (ART). Among euploid pregnancies, the 33 ART pregnancies did not differ from the 599 naturally conceived pregnancies in gestational age, maternal weight, maternal age, total ccfDNA, fetal ccfDNA or fetal fraction. As expected, the mean chromosome 21 *z*-score was close to 0 (–0.13) in the naturally conceived pregnancies, but was significantly ($p=0.048$) higher among the ART pregnancies. Both the chromosome 18 and chromosome 13 average *z*-scores were also elevated, but only the chromosome 18 *z*-score was statistically significant ($p=0.0032$). No differences in demographic or fetal fraction measurements occurred among the 10 ART and 63 naturally conceived Down syndrome pregnancies. However, the average chromosome 21 *z*-score for the ART cases was lower (8.7 versus 11.4, $p=0.14$). This is an intriguing finding, given that these women seek alternatives to invasive testing that maternal plasma DNA testing provides, but these findings need to be confirmed.

Twin Pregnancies

Multiple gestations are now more common due to ART, with a current twin pregnancy rate of 1:30 in the USA (http://www.dd.gov/nchs/fastats/multiple.htm). Twin pregnancies discordant for a trisomy would likely be more difficult to classify correctly, as perhaps half of the fetal fragments would be derived from the placenta associated with the euploid fetus. Thus, at a fetal fraction of 10 %, a discordant twin pregnancy may have an effective fetal fraction of only 5 %. When the twins are concordant for a trisomy, it is likely that testing will have similar performance to that in singleton pregnancies.

The largest group of affected multiple gestations was analyzed by shotgun sequencing [10]. In that study, 17 euploid twin pregnancies, five discordant and two concordant for Down syndrome, and one discordant for trisomy 13 were tested. The first finding was that the fetal fraction for the pregnancy (estimated by using multiple differentially methylated markers) was significantly higher than for pregnancies with a single fetus (geometric mean 18.1 and 13.4 %, respectively). This may be due to a higher placental mass in twin pregnancies. All affected twin pregnancies were well above the *z*-score cutoff of 3.0 and all euploid twin pregnancies were under 3.0.

A second study [54] of 84 multiple gestations (75 twin pregnancies) correctly identified the two affected pregnancies (one discordant for Down syndrome and one concordant for trisomy 18). No false positives occurred. This group then indirectly estimated the fetal fraction per male fetus and found an estimated average of 12.6 % per euploid twin pregnancy compared to 12.2 % in singleton euploid pregnancies. One group [50] tested five twins (four in the training set and one in the test set). All were correctly classified, including one concordant, and one discordant, for Down syndrome. No information on fetal fraction was reported.

A related study [49] used deep sequencing (350- to 800-fold) of SNPs located in targeted regions in both maternal DNA and plasma DNA. The authors then compared the ratio of minor to major allele, for SNPs at which the mother was homozygous. In this way, zygosity of the twin pregnancy was identified. Using this model, they confirmed in dizygotic twins that roughly 50 % of the fetal fraction relates to each fetus, but that fraction varied considerably (40, 62, 67, and 56 % of the total fetal fraction for the four samples). The authors suggested that determining the zygosity and, when the fetuses are dizygotic, determining the contribution of each fetus to the fetal fraction would be helpful in determining the reliability of test results.

Future Enhancements/ Improvements

Will ccfDNA Testing Ever Be Diagnostic for Common Aneuploidies?

ccfDNA testing is still in its infancy, and much will be learned over the next few years as more samples are tested. Both sensitivity and specificity are expected to improve, and the range of fetal disorders identified will grow. However, all of these ccfDNA tests are using DNA of placental origin. Thus, these tests will "karyotype" the placenta but not the fetus, and will therefore not be as accurate as a second trimester amniocentesis and diagnostic testing via karyotype or aCGH array. In addition, there will likely always be infrequent causes of false positive test results. Recently, a false positive test result has been reported in association with a maternal cancer [43]. Because this result was positive for both chromosomes 13 and 18, it was viewed with suspicion from the outset. However, it demonstrates that false positive test results will continue to occur. Another example shows the problems with interpreting other autosomal chromosomes, a common practice for laboratories located in China. In that study [17], a trisomy 22 was identified in a second trimester ccfDNA test. Since trisomy 22 is uncommon this late in pregnancy, the potential for a confined placenta mosaicism must be considered. A subsequent test confirmed the results. The woman did not want an invasive procedure and delivered a healthy baby. Three placenta samples were karyotype and each confirmed the trisomy 22 finding. Other rare reasons for false positive fetal results have also been reported [47]. Thus, for the foreseeable future, amniocentesis and diagnostic testing should be offered to all women with a positive test result.

Detection of Smaller Deletions/ Duplications

Several groups have already shown the ability to identify deletions and duplications using ccfDNA [29, 48, 54]. All of these required deeper sequencing and/or adding targeted areas for sequencing. Ongoing research regarding aCGH [61] is helping to define which deletions/duplications are pathogenic and which are benign. If testing is not targeted to specific known duplications/deletions, the issue of variants of unknown clinical significance will be faced. Routine prenatal screening should not include strategies for identifying variants of unknown significance. Validation of laboratory-developed tests for these disorders will be difficult, given the relatively rare nature of each. One can anticipate that in the very near future, a ccfDNA-based clinical screening test offers the identification of select deletion/duplication syndromes.

Exome/Genome Sequencing of the Fetus

Research groups have applied multiple methods allowing for whole-exome/whole-genome sequencing of the fetus [25, 30, 34]. At this time, the resources needed to undertake such an effort far outweigh any direct benefit of this technology. However, if such testing becomes more common place among adults, and the technology becomes less expensive, there will likely be pressure and interest to perform such testing in the prenatal setting.

Outstanding Implementation Issues

Testing in the General Population

All of the studies with complete ascertainment of fetal status have occurred in a "high risk" population. This results from the availability of a karyotype (or in some instances a FISH or aCGH study) after an invasive procedure, such as an amniocentesis or CVS. Such procedures carry a small, but important risk of miscarriage related to the procedure and would not be suitable for use in a general population. Because of this, several professional organizations [2, 5, 21, 27, 32] have suggested that ccfDNA testing cannot be applied in a general pregnancy population due to lack of sufficient information.

Several groups have addressed the issue of whether there are any known factors that might indicate that the sensitivity/specificity of ccfDNA testing of maternal plasma might differ in a general population versus a high risk population. Table 32.6 lists potential differences in these two populations, along with studies that provided relevant information.

There are, however, certainly other important differences between offering the test in a high risk setting versus a general population setting. Women classified as being at high risk have already been screened positive (e.g., by a question like "How old are you," an abnormal ultrasound study, or a positive family history). They are aware of their risk status and are likely to be referred to trained counselors. In this setting, they can be provided additional information and have their questions answered. In addition, most screen positive women have Down syndrome risks in the range of 1:250–1:10. Only a very small proportion might have risks exceeding 1:1 (50 %). This allows for more objective decision-making by focusing on the fact that a 1:100 risk means that 99 of 100 outcomes are not Down syndrome. Thus, the high risk group has more access to information at a "teachable moment" without extreme risks that may complicate decision-making. Contrast this with a woman in the general population being offered ccfDNA testing. This is one of many options and she is unlikely to be focused on the potential impact of the testing results. If the woman were to be screen

Table 32-6 **Potential Differences Between a "High Risk" Population and a General Population that might Impact ccfDNA Test Performance**

Factor	Potential impact	Findings
Prevalence	A high risk population will have a higher prevalence of the disorder	Although true, the higher prevalence, by itself, will not impact the sensitivity and specific of the test. However, the predictive values (both positive and negative) will change depending on prevalence. However, this change is well described and screening programs deal with varying prevalence in current practice
Fetal fraction	A high risk population may have a higher fetal fraction leading to higher sensitivity	After review of many factors (e.g., maternal age, abnormal ultrasound findings, serum marker levels), only maternal weight and some serum markers may differ. The impact is likely to be very small and is sometimes in the wrong direction (low PAPP-A associated with low fetal fraction would reduce performance in high risk, but not low risk settings) [4, 9, 39, 40, 45]
Maternal age	A high risk population is older and this could impact test performance	The only known predictor of improved test performance is increasing the fetal fraction. Maternal age alone does not directly impact the test's final measure (e.g., *z*-score) [4, 9, 45]
Other test indications	Those with a positive family history or abnormal US might be easier to detect	All studies stratifying results by indication find no differences in detection or false positive rate or, when examined, the test statistics such as *z*-score [4, 9, 45]

US ultrasound studies

Table 32-7 Recommended Content for Patient Educational Material as Suggested by Professional Organizations

ACOG	ISPD	SOGC	NSGC	ACMG	Recommended content
				X	Provides an introductory statement about the purpose of testing
			X		Contains accurate and up-to-date information about the possible test results
X			X		Contains accurate and up-to-date information for available follow-up testing
X	X	X	X	X	Stresses the implications of a positive DNA test result
X	X	X	X	X	Explains that false positive results can occur and that there is a need for confirmatory testing
X	X				States that high Down syndrome risks will occur with positive DNA test results
	X			X	Explains the potential stress associated with the extended wait for test results
	X	X		X	Contains information that the test results may not be informative for some patients
X	X	X	X	X	Provides information that amniocentesis/chorionic villus sampling would still be indicated in order to diagnose other disorders that the DNA test is not designed to detect

ACOG American Congress of Obstetricians and Gynecologists, *ISPD* International Society of Prenatal Diagnosis, *SOGC* Society of Obstetricians and Gynecologists of Canada, *NSGC* National Society of Genetic Counselors, *ACMG* American College of Medical Genetics and Genomics

positive on this test, her risks will be very high compared to current group of "high risk" women. For women with a positive test for chromosome 21, the group risk may be 1:1 or even higher. This will almost certainly create tremendous anxiety that must be dealt with by the care providers quickly. It is also likely to put the couple in a difficult position to make decisions due to the very high assigned risk. It might also be confusing as to whether this is a diagnostic or screening test. Despite these concerns, this near diagnostic performance is no different than that provided by most newborn screening tests and many other genetic testing for inherited diseases. The keys to appropriate patient care are a well-informed cadre of health care providers, educational materials for patients, and a program in place to provide comprehensive care to the women with screen positive results.

Educational Materials

As part of the introduction of any new prenatal screening test, education of the providers and patients is of paramount importance. Each of the laboratories currently offers both provider and patient information on their Web sites (www.ariosadx.com, www.natera.com, www.sequenomcmm.com, and www.verinata.com). Several professional organizations have recommended specific content that should be covered in patient materials [2, 5, 21, 27, 32]. This content is covered in Table 32.7. None of the patient materials currently available from the commercial companies include all of the recommended content.

External Proficiency Testing

Developing and validating a laboratory developed next-generation sequencing test for ccfDNA to identify common aneuploidies is an expensive and time consuming undertaking requiring significant expertise and infrastructure. The field is also complicated by intellectual property issues. These issues are likely the factors limiting many high complexity laboratories from offering similar testing option. External proficiency testing of these tests is needed to help ensure reliability and quality, and to help harmonize the practice surrounding testing. It will be challenging to support an external proficiency testing program with only 5–10 laboratories worldwide. Creation and distribution of manufactured samples needs to be examined and piloted, and a formal proficiency testing program, coupled with laboratory inspections, are the best option for ensuring consistent and high quality testing.

Conclusions

The finding of ccfDNA in maternal plasma in 1997 [35] provided a potential path to the long-awaited noninvasive prenatal diagnostic test. However, the technology to allow for such a test was not readily available until a decade later. Once the technology was available, pilot trials began to show the feasibility of multiple methods to identify common aneuploidies. These were followed by larger external validation trials that confirmed very high sensitivity and specificity. Currently, four commercial companies in the USA have laboratory developed tests for clinical use. Within 1 year from the launching of the first commercial test, the American College of Obstetricians and Gynecologists recommended that ccfDNA testing be offered to high risk women as a secondary screening test prior to invasive diagnostic testing [2]. These tests are rapidly improving and expanding from common trisomies to fetal sex and common sex aneuploidies. It is expected that this expansion will continue into deletion/duplication syndromes and perhaps, in the farther future, whole fetal exome or genome testing. This is not to say that ccfDNA testing of maternal plasma is without problems. The test is expensive (list prices from $900 to $2,700) and turn-around time ranges from 5 to 14 days or more. Expansion of testing into the general population will be more challenging than offering testing to high risk women and the cost justification will be more difficult, as well. However, by any measure available, these new testing options have opened a new era in prenatal screening for an ever widening array of fetal disorders.

REFERENCES

1. ACOG. ACOG practice bulletin no. 77: screening for fetal chromosomal abnormalities. Obstet Gynecol. 2007;109:217–27.
2. ACOG. Committee opinion no. 545: noninvasive prenatal testing for fetal aneuploidy. Obstet Gynecol. 2012;120:1532–4.
3. Ashoor G, Syngelaki A, Wagner M, Birdir C, Nicolaides KH. Chromosome-selective sequencing of maternal plasma cell-free DNA for first-trimester detection of trisomy 21 and trisomy 18. Am J Obstet Gynecol. 2012;206(322):e1–5.
4. Ashoor G, Syngelaki A, Wang E, Struble C, Oliphant A, Song K, et al. Trisomy 13 detection in the first trimester of pregnancy using a chromosome-selective cell-free DNA analysis method. Ultrasound Obstet Gynecol. 2013;41:21–5.
5. Benn P, Borrell A, Crossley J, Cuckle H, Dugoff L, Gross S, et al. Aneuploidy screening: a position statement from a committee on behalf of the Board of the International Society for Prenatal Diagnosis, January 2011. Prenat Diagn. 2011;31:519–22.
6. Bianchi DW, Platt LD, Goldberg JD, Abuhamad AZ, Sehnert AJ, Rava RP, et al. Genome-wide fetal aneuploidy detection by maternal plasma DNA sequencing. Obstet Gynecol. 2012;119:890–901.
7. Bindra R, Heath V, Liao A, Spencer K, Nicolaides KH. One-stop clinic for assessment of risk for trisomy 21 at 11–14 weeks: a prospective study of 15,030 pregnancies. Ultrasound Obstet Gynecol. 2002;20:219–25.
8. Bogart MH, Pandian MR, Jones OW. Abnormal maternal serum chorionic gonadotropin levels in pregnancies with fetal chromosome abnormalities. Prenat Diagn. 1987;7:623–30.
9. Brar H, Wang E, Struble C, Musci TJ, Norton ME. The fetal fraction of cell-free DNA in maternal plasma is not affected by a priori risk of fetal trisomy. J Matern Fetal Neonatal Med. 2013;26:143–5.
10. Canick JA, Kloza EM, Lambert-Messerlian GM, Haddow JE, Ehrich M, van den Boom D, et al. DNA sequencing of maternal plasma to identify Down syndrome and other trisomies in multiple gestations. Prenat Diagn. 2012;32:730–4.
11. Canick JA, Knight GJ, Palomaki GE, Haddow JE, Cuckle HS, Wald NJ. Low second trimester maternal serum unconjugated oestriol in pregnancies with Down's syndrome. Br J Obstet Gynaecol. 1988;95: 330–3.
12. Canick JA, Palomaki GE, Kloza EM, Lambert-Messerlian GM, Haddow JE. The impact of maternal plasma DNA fetal fraction on next generation sequencing tests for common fetal aneuploidies. Prenat Diagn. 2013 Jul;33(7):667–74. doi: 10.1002/pd.4126. Epub 2013 May 31.
13. Chen EZ, Chiu RW, Sun H, Akolekar R, Chan KC, Leung TY, et al. Noninvasive prenatal diagnosis of fetal trisomy 18 and trisomy 13 by maternal plasma DNA sequencing. PLoS One. 2011;6:e21791.
14. Chen XQ, Stroun M, Magnenat JL, Nicod LP, Kurt AM, Lyautey J, et al. Microsatellite alterations in plasma DNA of small cell lung cancer patients. Nat Med. 1996;2:1033–5.
15. Chiu RW, Akolekar R, Zheng YW, Leung TY, Sun H, Chan KC, et al. Non-invasive prenatal assessment of

trisomy 21 by multiplexed maternal plasma DNA sequencing: large scale validity study. BMJ. 2011;342:c7401.
16. Chiu RW, Chan KC, Gao Y, Lau VY, Zheng W, Leung TY, et al. Noninvasive prenatal diagnosis of fetal chromosomal aneuploidy by massively parallel genomic sequencing of DNA in maternal plasma. Proc Natl Acad Sci U S A. 2008;A105:20458–63.
17. Choi H, Lau TK, Jiang FM, Chan MK, Zhang HY, Lo PS, et al. Fetal aneuploidy screening by maternal plasma DNA sequencing: 'false positive' due to confined placental mosaicism. Prenat Diagn. 2013;33: 198–200.
18. Clausen FB, Christiansen M, Steffensen R, Jorgensen S, Nielsen C, Jakobsen MA, et al. Report of the first nationally implemented clinical routine screening for fetal RHD in D-pregnant women to ascertain the requirement for antenatal RhD prophylaxis. Transfusion. 2012;52:752–8.
19. Cuckle HS, Wald NJ, Lindenbaum RH. Maternal serum alpha-fetoprotein measurement: a screening test for Down syndrome. Lancet. 1984;1:926–9.
20. Devaney SA, Palomaki GE, Scott JA, Bianchi DW. Noninvasive fetal sex determination using cell-free fetal DNA: a systematic review and meta-analysis. JAMA. 2011;306:627–36.
21. Devers PL, Cronister A, Ormond KE, Facio F, Brasington CK, Flodman P. Noninvasive prenatal testing/noninvasive prenatal diagnosis: the position of the national society of genetic counselors. J Genet Couns. 2013;22:291–5.
22. Down JLH. Observations on an ethnic classification of idiots. Lond Hosp Rep. 1866;3:259–62.
23. Ehrich M, Deciu C, Zwiefelhofer T, Tynan JA, Cagasan L, Tim R, et al. Noninvasive detection of fetal trisomy 21 by sequencing of DNA in maternal blood: a study in a clinical setting. Am J Obstet Gynecol. 2011;204(205):e1–11.
24. Fan HC, Blumenfeld YJ, Chitkara U, Hudgins L, Quake SR. Noninvasive diagnosis of fetal aneuploidy by shotgun sequencing DNA from maternal blood. Proc Natl Acad Sci U S A. 2008;A105:16266–71.
25. Fan HC, Gu W, Wang J, Blumenfeld YJ, El-Sayed YY, Quake SR. Non-invasive prenatal measurement of the fetal genome. Nature. 2012;487:320–4.
26. Lambert-Messerlian GM, Kloza EM, Williams JR III, Loucky J, O'Brien B, Wilkins-Haug L, Mahoney MJ, De Biasio P, Borrell A, Ehrich M, van den Boom D, Bombard AT, Deciu C, Palomaki GE. Maternal plasma DNA testing for aneuploidy in pregnancies achieved by assisted reproductive technologies. Submitted (Gen Med 2013)
27. Gregg AR, Gross SJ, Best RG, Monaghan KG, Bajaj K, Skotko BG, et al. ACMG statement on noninvasive prenatal screening for fetal aneuploidy. Genet Med. 2013;15:395–8.
28. Haddow JE, Palomaki GE, Knight GJ, Foster DL, Neveux LM. Second trimester screening for Down's syndrome using maternal serum dimeric inhibin A. J Med Screen. 1998;5:115–9.
29. Jensen TJ, Dzakula Z, Deciu C, van den Boom D, Ehrich M. Detection of microdeletion 22q11.2 in a fetus by next-generation sequencing of maternal plasma. Clin Chem. 2012;58:1148–51.
30. Kitzman JO, Snyder MW, Ventura M, Lewis AP, Qiu R, Simmons LE, et al. Noninvasive whole-genome sequencing of a human fetus. Sci Transl Med. 2012;4:137ra76.
31. Lambert-Messerlian G, Kloza EM, Williams J 3rd, Loucky J, O'Brien B, Wilkins-Haug L, Mahoney MJ, De Biasio P, Borrell A, Ehrich M, van den Boom D, Bombard AT, Deciu C, Palomaki GE. Maternal plasma DNA testing for aneuploidy in pregnancies achieved by assisted reproductive technologies. Genet Med. 2013 Oct 3. doi: 10.1038/gim.2013.149. [Epub ahead of print].
32. Langlois S, Brock JA, Genetics C, Wilson RD, Audibert F, Brock JA, et al. Current status in non-invasive prenatal detection of Down syndrome, trisomy 18, and trisomy 13 using cell-free DNA in maternal plasma. J Obstet Gynaecol Can. 2013;35:177–81.
33. Lejeune J, Gautier M, Turpin R. Study of somatic chromosomes from 9 mongoloid children. C R Hebd Seances Acad Sci. 1959;248:1721–2.
34. Lo YM, Chan KC, Sun H, Chen EZ, Jiang P, Lun FM, et al. Maternal plasma DNA sequencing reveals the genome-wide genetic and mutational profile of the fetus. Sci Transl Med. 2010;2:61ra91.
35. Lo YM, Corbetta N, Chamberlain PF, Rai V, Sargent IL, Redman CW, et al. Presence of fetal DNA in maternal plasma and serum. Lancet. 1997;350:485–7.
36. Mazloom AR, Dzakula Z, Oeth P, Wang H, Jensen T, Tynan J, et al. Noninvasive prenatal detection of sex chromosomal aneuploidies by sequencing circulating cell-free DNA from maternal plasma. Prenat Diagn. 2013;33:591–7.
37. Merkatz IR, Nitowsky HM, Macri JN, Johnson WE. An association between low maternal serum alpha-fetoprotein and fetal chromosomal abnormalities. Am J Obstet Gynecol. 1984;148:886–94.
38. Nawroz H, Koch W, Anker P, Stroun M, Sidransky D. Microsatellite alterations in serum DNA of head and neck cancer patients. Nat Med. 1996;2:1035–7.
39. Nicolaides KH, Syngelaki A, Ashoor G, Birdir C, Touzet G. Noninvasive prenatal testing for fetal trisomies in a routinely screened first-trimester population. Am J Obstet Gynecol. 2012;207(374):e1–6.
40. Nicolaides KH, Syngelaki A, Gil M, Atanasova V, Markova D. Validation of targeted sequencing of single-nucleotide polymorphisms for non-invasive prenatal detection of aneuploidy of chromosomes 13, 18, 21, X, and Y. Prenat Diagn. 2013;33:575–9.
41. Norton ME, Brar H, Weiss J, Karimi A, Laurent LC, Caughey AB, et al. Non-Invasive Chromosomal Evaluation (NICE) Study: results of a multicenter prospective cohort study for detection of fetal trisomy 21 and trisomy 18. Am J Obstet Gynecol. 2012;207(137):e1–8.
42. Odeh M, Granin V, Kais M, Ophir E, Bornstein J. Sonographic fetal sex determination. Obstet Gynecol Surv. 2009;64:50–7.
43. Osborne CM, Hardisty E, Devers P, Kaiser-Rogers K, Hayden MA, Goodnight W, et al. Discordant noninvasive prenatal testing results in a patient subsequently diagnosed with metastatic disease. Prenat Diagn. 2013;33:609–11.
44. Palomaki GE, Haddow JE. Maternal serum alpha-fetoprotein, age, and Down syndrome risk. Am J Obstet Gynecol. 1987;156:460–3.
45. Palomaki GE, Kloza EM, Lambert-Messerlian GM, Haddow JE, Neveux LM, Ehrich M, et al. DNA sequencing of maternal plasma to detect Down

syndrome: an international clinical validation study. Genet Med. 2011;13:913–20.
46. Palomaki GE, Knight, GJ, Ashwood ER, Best RG, Haddow JE: Screening for Down syndrome in the United States: Results of Surveys in 2011 and 2012. Arch Pathol Lab Med., 137:921-926, 2013. PMID: 23808464.
47. Pan M, Li FT, Li Y, Jiang FM, Li DZ, Lau TK, et al. Discordant results between fetal karyotyping and non-invasive prenatal testing by maternal plasma sequencing in a case of uniparental disomy 21 due to trisomic rescue. Prenat Diagn. 2013;33:598–601.
48. Peters D, Chu T, Yatsenko SA, Hendrix N, Hogge WA, Surti U, et al. Noninvasive prenatal diagnosis of a fetal microdeletion syndrome. N Engl J Med. 2011;365:1847–8.
49. Qu JZ, Leung TY, Jiang P, Liao GJ, Cheng YK, Sun H, et al. Noninvasive prenatal determination of twin zygosity by maternal plasma DNA analysis. Clin Chem. 2013;59:427–35.
50. Sehnert AJ, Rhees B, Comstock D, de Feo E, Heilek G, Burke J, et al. Optimal detection of fetal chromosomal abnormalities by massively parallel DNA sequencing of cell-free fetal DNA from maternal blood. Clin Chem. 2011;57:1042–9.
51. Sparks AB, Struble CA, Wang ET, Song K, Oliphant A. Noninvasive prenatal detection and selective analysis of cell-free DNA obtained from maternal blood: evaluation for trisomy 21 and trisomy 18. Am J Obstet Gynecol. 2012;206(319):e1–9.
52. Sparks AB, Wang ET, Struble CA, Barrett W, Stokowski R, McBride C, et al. Selective analysis of cell-free DNA in maternal blood for evaluation of fetal trisomy. Prenat Diagn. 2012;32:3–9.
53. Spencer K, Macri JN, Aitken DA, Connor JM. Free beta-hCG as first-trimester marker for fetal trisomy. Lancet. 1992;339:1480.
54. Srinivasan A, Bianchi D, Liao W, Sehnert A, Rava R. 52: Maternal plasma DNA sequencing: effects of multiple gestation on aneuploidy detection and the relative cell-free fetal DNA (cffDNA) per fetus. Am J Obstet Gynecol. 2013;208:S31.
55. Stroun M, Anker P, Maurice P, Lyautey J, Lederrey C, Beljanski M. Neoplastic characteristics of the DNA found in the plasma of cancer patients. Oncology. 1989;46:318–22.
56. Szabo J, Gellen J. Nuchal fluid accumulation in trisomy-21 detected by vaginosonography in first trimester. Lancet. 1990;336:1133.
57. Wald N, Stone R, Cuckle HS, Grudzinskas JG, Barkai G, Brambati B, et al. First trimester concentrations of pregnancy associated plasma protein A and placental protein 14 in Down's syndrome. BMJ. 1992;305:28.
58. Wald NJ. The definition of screening. J Med Screen. 2001;8:1.
59. Wald NJ, Rodeck C, Hackshaw AK, Walters J, Chitty L, Mackinson AM. First and second trimester antenatal screening for Down's syndrome: the results of the Serum, Urine and Ultrasound Screening Study (SURUSS). J Med Screen. 2003;10: 56–104.
60. Wald NJ, Watt HC, Hackshaw AK. Integrated screening for Down's syndrome on the basis of tests performed during the first and second trimesters. N Engl J Med. 1999;341:461–7.
61. Wapner RJ, Martin CL, Levy B, Ballif BC, Eng CM, Zachary JM, et al. Chromosomal microarray versus karyotyping for prenatal diagnosis. N Engl J Med. 2012;367:2175–84.
62. Zimmermann B, Hill M, Gemelos G, Demko Z, Banjevic M, Baner J, et al. Noninvasive prenatal aneuploidy testing of chromosomes 13, 18, 21, X, and Y, using targeted sequencing of polymorphic loci. Prenat Diagn. 2012;32:1233–41.
63. Zimmermann R, Hucha A, Savoldelli G, Binkert F, Achermann J, Grudzinskas JG. Serum parameters and nuchal translucency in first trimester screening for fetal chromosomal abnormalities. Br J Obstet Gynaecol. 1996;103:1009–14.

CHAPTER 33

Genomic Applications in the Clinical Management of Infectious Diseases

MARTINA I. LEFTEROVA, NIAZ BANAEI, BENJAMIN A. PINSKY

Introduction 581
NGS in Diagnostic Virology 583
NGS in Clinical Bacteriology 592
NGS in Clinical Mycology 595
Conclusions 595

Introduction

High-throughput sequencing technology, frequently referred to as next-generation sequencing (NGS), has demonstrated enormous potential for many fields of pathology, including microbiology [1]. Various platforms, which differ in their sequencing chemistries, read lengths, and throughput capabilities, are currently available (reviewed in [2, 3]). The large amounts of sequencing data that such instruments generate (Table 33.1) have made it possible to catalogue reference genomes for numerous species of bacteria, viruses, and parasites. Additionally, this sequencing information is helping to address important questions about mechanisms of virulence, immune evasion, antimicrobial resistance, and the spread of infection. However, NGS is no longer confined to the realm of research, as technological advances have led to the development of benchtop NGS platforms that are more efficient, less expensive, and easier to operate [4]. These advances have made NGS accessible to diagnostic microbiology and virology laboratories that already rely on molecular methods, including multiplex polymerase chain reaction (PCR)-based assays and traditional bidirectional dideoxynucleoside chain-termination (i.e., Sanger) sequencing, for pathogen identification and characterization of antimicrobial susceptibility. NGS approaches are particularly attractive for infectious disease testing given their increased sensitivity compared to Sanger sequencing, their potential for very high throughput, and the ability to test multiple targets and multiple specimens in a single run [5–7].

NGS studies of microorganisms typically follow one of two general strategies: targeted amplicon sequencing or whole-genome sequencing (WGS) (Fig. 33.1) [8, 9]. The first approach uses target-specific primers for PCR-mediated amplification, so that the genomic regions of interest are selectively enriched and sequenced. This method is particularly helpful when a rare variant is sought, e.g., a drug-resistant mutant in a primarily wild-type population. Sequencing of whole genomes, on the other hand, relies on non-targeted library preparation, and is often performed when microorganisms are unknown or the goal is to define the genomic content and predict its function in the organism under investigation. The choice of sequencing strategy often drives the selection of sequencing platform [8]: for example, 454 pyrosequencing instruments offer longer read lengths

G.J. Netto and I. Schrijver (eds.), *Genomic Applications in Pathology*,
DOI 10.1007/978-1-4939-0727-4_33,

Table 33-1 **Characteristics of Current High-throughput and Bench-top NGS Platforms**

Platform (manufacturer)	Sequencing Chemistry	Avg read length (base pairs)	Throughput (per run)	Run time (hours)	Applications and references
454 GS FLX Titanium XL+ (Roche)	Pyrosequencing (Light emission)	700 (Max 1,000)	700 Mb	23 h	Amplicon sequencing: CMV genotyping, Görzer et al. (2011)[a] WGS: *Acinetobacter baumannii*, Smith et al. [152][a]
HiSeq 2500 (Illumina)	Fluorescently labeled nucleotides	Rapid run mode 2×150	150–180 Gb	40 h	Metagenomics: Human microbiome project[a] WGS: Influenza, Yongfeng et al. [96][a]
		High-throughput mode 2×100	540–600 Gb	11 days	
SOLiD 5500×1 System (Life Technologies)	Ligation of 8 bp fluorescent probes	75×35 (Paired end)	10–15 Gb/day	7 days	WGS: *Coccidioides immitis* typing, Engelthaler et al. [175]
PacBio RS (Pacific Biosciences)	Single molecule real time	3,000 (Max 15,000)		90 min	WGS: Haitian Vibrio outbreak, Chin et al. [155]
Bench-top instruments					
454 GS Junior (Roche)	Pyrosequencing (Light emission)	400	35 Mb	10 h	Amplicon sequencing: HIV drug resistance, Dudley et al. [45]
MiSeq (Illumina)	Fluorescently labeled nucleotides	2×250	7.5–8.5 Gb	39 h	WGS: MRSA outbreak, Koser et al. [157]
Ion Torrent PGM Sequencer	Electrical detection of proton ions	100	1 Gb	2 h	WGS: German *E. coli* O104 outbreak, Mellmann et al. [149]

The specifications of the instruments were obtained from the manufacturers' websites and represent up-to-date information as of December 2012. Illustrative applications and the corresponding references are also listed
[a]These studies used older versions of the platforms than those for which specifications are listed. Avg = average

(Table 33.1) allowing a large amplicon to be sequenced in a single read. On the other hand, platforms like Illumina and SOLiD have higher throughputs (Table 33.1) which makes them well suited for studies that require a greater sequencing depth, for example for detection of a microorganism or resistant populations whose DNA or RNA represents <1 % of the pool of nucleic acids in a clinical specimen [10, 11].

Examples of these approaches in infectious disease testing will be discussed below with particular attention to technical and bioinformatics challenges that specific diagnostic scenarios pose. The use of NGS in clinical microbiology and virology laboratories is currently infrequent, though its role in patient management is anticipated to grow as standardized operational protocols and data analysis pipelines emerge. Because most studies that have used NGS in clinically applicable contexts have focused on viruses, a greater part of this chapter will be devoted to discussion of genomic applications in diagnostic virology. However, the potential utility of NGS methods in clinical bacteriology and mycology will also be reviewed.

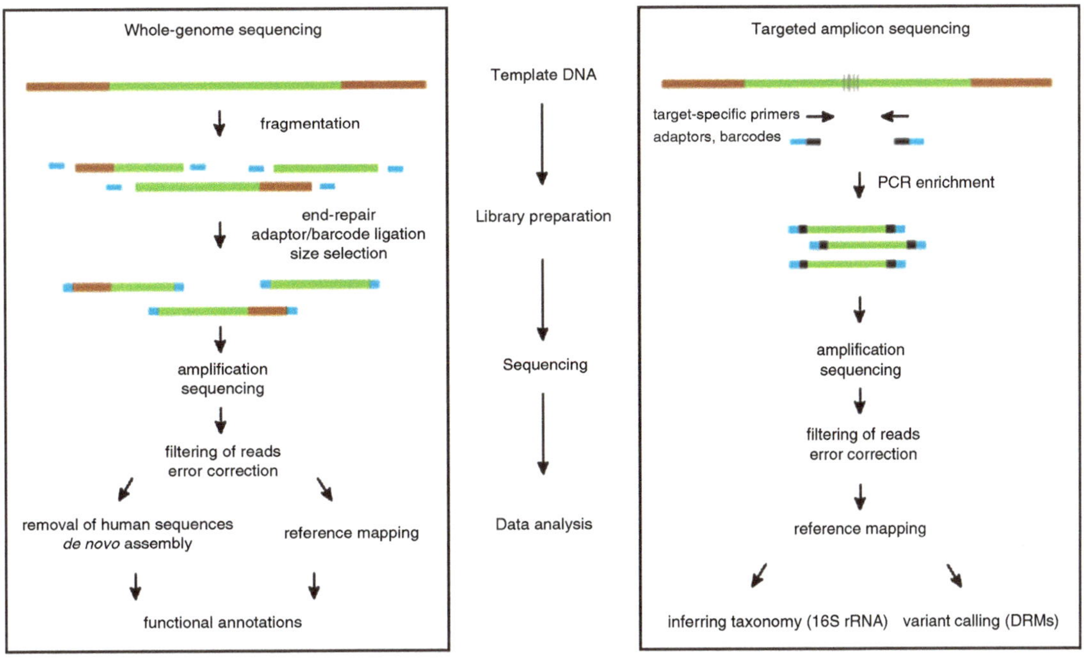

Figure 33-1 Illustration of whole-genome sequencing (WGS) and targeted amplicon sequencing approaches for microorganism investigation. Template enrichment in amplicon sequencing is accomplished by using target-specific primers, followed by primers that are partially complementary to the target-specific primers (*black bars*) and contain sequencing adaptors and bar codes (*blue bars*). Hatched marks on the template indicate drug resistance mutations (DRMs) in a gene of interest (*green*). In WGS, template enrichment is accomplished by nucleic acid fragmentation either with enzymatic or mechanical means, followed by enzymatic treatments (end-repair) to allow ligation of random primers that contain sequencing adaptors and barcodes (*blue bars*). Size selection allows only fragments of a predefined length to be used for sequencing. Applications of amplicon sequencing include 16S ribosomal RNA (rRNA) sequencing to interrogate bacterial taxonomic diversity as well as testing for viral drug resistant variants. WGS is suitable for de novo genome assembly of previously uncharacterized organisms as well as functional profiling by mapping to a reference sequence. Bioinformatics removal of human sequences is required since the nucleic acids of the organism of interest frequently constitute less than 1 % of the nucleic acid pool

NGS in Diagnostic Virology

Detection of Drug Resistance Mutations

HUMAN IMMUNODEFICIENCY VIRUS

The ability of human immunodeficiency virus (HIV) to mutate and escape antiretroviral drugs is largely due to the error-prone nature of the RNA reverse transcriptase and was recognized soon after highly active antiretroviral therapy (HAART) became available [12]. There are also important host factors that facilitate emergence of drug resistance such as therapy non-adherence, although resistant variants can also be acquired during initial infection [12]. The list of drug resistance mutations (DRMs) is continuously expanding and includes mutations that have been identified by correlation of genotypic variation with therapy failure, by phenotypic characterization of clinical isolates, or by in vitro passaging experiments [13]. Additionally, many epidemiologic studies have accumulated data on drug resistance rates in both treatment-experienced and treatment-naïve patients, and it has been shown that drug resistance testing can predict treatment outcome [14, 15]. In the context of such findings, expert panels currently recommend genotypic testing of therapy-naïve patients when they enter into care and therapy-experienced patients when they show evidence of virologic failure [12, 16].

Traditionally, genotypic HIV drug resistance testing has been performed using "population" or "bulk" sequencing, which involves Sanger sequencing of reverse transcription polymerase chain reaction (RT-PCR) amplicons of the HIV reverse transcriptase and protease genes. Several genotypic interpretation systems (GIS) have been developed that can facilitate downstream sequence analysis [17–19], which includes comparison of the subject sequence to a reference sequence, assignment of nucleotide and amino acid changes, query against a database of known DRMs, and drug resistance score calling. Such automated tools have been compared to manual sequence curation and are clearly superior in terms of improving sequence analysis as well as expediting and standardizing workflow [18]. Bioinformatics tools for therapy optimization based on genotypic resistance have also been developed [15, 19–21]. However, in their current state, they do not obviate the need for expert advice in the choice of salvage therapy [15].

Although substantial strides have been made in HIV drug resistance testing, Sanger sequencing has certain limitations that can be addressed through the use of NGS technologies. A major issue is the inability of Sanger sequencing to detect minority variants present at <20 % of the viral population [7]. Such variants are kept at low levels in the absence of therapy because they are less fit relative to wild-type virus. However, in the presence of selective pressure from antiretroviral therapy, the resistant variants become dominant [12]. Importantly, the presence of minority variants has implications for treatment outcome, as evidenced by studies that have demonstrated increased risk of first-line therapy failure in patients with baseline resistant variants present at <20 % of the viral population, i.e., below the limit of detection of Sanger sequencing [22–24].

Highly sensitive methods for DRM detection, such as allele-specific PCR [24] and real-time PCR [22] targeting DRMs with well-established roles in antiretroviral therapy response, require prior knowledge of the variants that will be targeted and are limited in the number of DRMs that can be interrogated at a given time. NGS methods, on the other hand, offer an unbiased approach and their superiority for capturing minority resistant variants relative to Sanger population sequencing has been demonstrated in several publications [25–30]. At least half of the DRMs identified by NGS are missed by Sanger sequencing [28, 29, 31] and represent additional mutations in patients in whom a drug resistant variant has already been identified or unsuspected DRMs in patients with presumed wild-type viral genotype [28]. NGS also allows for more sensitive monitoring of the dynamics of viral population changes, specifically the acquisition, emergence, disappearance, and "archiving" of drug resistance mutants on and off therapy [32]. Based on NGS data, the rate of transmitted drug resistance (TDR) to protease and reverse transcriptase is now estimated to be close to 30 % in treatment-naïve patients [29, 31] versus the previously estimated 7–15 % [12]. NGS approaches have also made it possible to assess genotypic drug resistance to latest generations of antiretroviral drugs such as integrase inhibitors and entry inhibitors. Whereas integrase inhibitors are typically not associated with baseline resistance [33, 34], entry inhibitors can be. These drugs target the interaction of the HIV envelope glycoprotein gp120 with the surface receptor CCR5 on host cells. Entry inhibitor resistance occurs mainly by viral utilization of a different cell surface receptor instead of CCR5, i.e., CXCR4 [13]. Several studies have applied NGS to identification of minority populations with CXCR4 tropism, demonstrating that the switch from an entry-inhibitor-sensitive CCR5-using population to one that is resistant and uses CXCR4 occurs from preexisting minority CXCR4-tropic variants [35–37].

Importantly, the clinical significance of low-abundance drug resistance variants detected by NGS remains to be fully characterized. Several studies have retrospectively evaluated the impact of low-abundance resistance variants detected by NGS in treatment-naïve patients [29, 31], as well as in treatment-experienced patients undergoing virologic failure [28, 38]. Although patients with low-abundance DRMs detected by NGS alone appear to have a modestly increased risk of failing therapy, in general, the risk of failure is substantially higher if a high-abundance mutant is present, which can be demonstrated both by NGS and Sanger sequencing. Thus, it is hoped that data accumulated through NGS studies will contribute to a better understanding of the role of specific minority resistant variants on therapy failure and will lead to more accurate therapy optimization tools.

The majority of the NGS studies discussed above have used an amplicon sequencing approach with 454 pyrosequencing technology (454 Life Sciences) [2, 7, 39]. This approach is well suited for diagnosis of viral drug resistance because the DRMs typically occur in the functional domains of the proteins that are targeted by the drug, so the regions can be easily amplified with primers that flank mutation hotspots. Another advantage of 454 pyrosequencing is the length of the sequencing reads (up to 1,000 base pairs (bp) with the GS FLX Titanium XL+ instrument and 400 bp with the 454 Junior Sequencer), which makes it possible to identify viruses carrying multiple DRMs.

A major challenge in NGS-based studies is distinguishing true viral variants from artifacts generated during PCR amplification or sequencing, such as mismatches, insertions/deletions, and PCR-mediated recombination products known as chimeric sequences [40–42]. In particular, 454 pyrosequencing is known to be error prone in homopolymer regions, i.e., repeats of three or more identical bases [43, 44]. Some investigators have taken the approach of estimating amplification and sequencing error rates by using plasmid clones [5, 25, 37, 45]. For instance, Wang et al. [25] subjected plasmids to 454 pyrosequencing and Sanger sequencing, with the assumption that all differences are due to library preparation and/or NGS errors [25]. Thus, empirical error rates were calculated for insertions, deletions, and mismatches in areas with homopolymers (average error frequency rate: 4.4×10^{-3}) and without homopolymers (average error frequency rate: 7.0×10^{-4}) [25]. Bioinformatics tools have also been developed for cleaning up sequencing data, which reduces error rates and allows for calling of authentic low-abundance viral variants in NGS studies [32, 46, 47]. Additionally, many studies validate NGS-identified minority variants from clinical specimens by conventional sequencing of plasmid subclones or dilutions of RT-PCR products [25, 46].

Another challenge is reproducible library preparation from clinical specimens with low viral counts because the number of viral copies that are used for library preparation will be small and a mixed viral population may not be accurately represented. Differential amplification of some variants can also skew the final PCR product mixture because of stochastic events in early PCR cycles or differences in the efficiency of primer annealing [7, 41]. This can be especially problematic for low abundance variants, as illustrated by Tsibris et al. [37] who performed amplification and sequencing of an identical mixed viral population in quadruplicate, and found the coefficient of variation for detecting a minor variant to be 22.8 % (mean ± SD, 2.426 ± 0.55 %). Jabara et al. modified the RT-PCR step with primers tagged with a random sequence, such that each template received a unique identifier that could be used to create a consensus sequence for each individual template, thus correcting for random errors during library preparation and sequencing [48]. Through these primer identifiers the authors demonstrated amplification skewing and a wide range of template representation, which was particularly pronounced for low-abundance variants. One approach that has been attempted for addressing PCR bias is performing multiple independent RT-PCRs from the same clinical specimen and pooling their products to serve as template for library preparation [49].

HEPATITIS B AND HEPATITIS C VIRUSES

Although drug resistance testing is not widely used in the management of hepatitis B virus (HBV) or hepatitis C virus (HCV) infections, resistance to current and experimental drugs has been documented [50–52]. This is consistent with the fact that the HCV RNA-dependent RNA polymerase and the HBV polymerase are error-prone and lack proofreading activity, leading to accumulation of large numbers of mutations during viral replication. As with HIV, genotypic resistance testing for HCV and HBV has primarily relied on Sanger sequencing, although some of the recent studies have applied NGS methods to characterizing HCV and HBV DRM incidence.

Drugs that are currently approved for treatment of HBV include the immunomodulatory agent interferon-α and nucleoside analogues (lamivudine, telbivudine, and entecavir) and two nucleotide analogues (adefovir and tenofovir). Nucleoside/nucleotide analogues (NAs) inhibit viral replication by interfering with the activity of the HBV polymerase. The prototype NA, lamivudine, has been highly effective for viral suppression and normalization of liver function. However, its long-term use is associated with frequent resistance mutations in the tyrosine (Y)–methionine (M)–aspartic acid (D)–aspartic acid (D) (YMDD) motif of the HBV polymerase [51]. DRMs in the YMDD motif are encountered in up to 75 % of treatment-experienced patients [53] and in 10–13 % of treatment-naive patients with chronic HBV infection [54]. At least one mutation conferring cross-resistance to lamivudine, adefovir, and tenofovir has been identified [55]. The remaining NAs have lower incidence of resistance, but the presence of lamivudine DRMs is a risk factor for development of resistance to other NA drugs through accumulation of additional mutations [51]. Databases for interpreting HBV genotypic resistance data are also emerging [56, 57]. Several recent studies have employed NGS approaches to interrogate the incidence of NA DRMs in treatment-experienced and treatment-naïve patients [58–62]. As with HIV, these studies have confirmed the improved sensitivity of NGS relative to Sanger sequencing for detection of low-abundance resistant mutants [59, 60, 62] and for investigating the evolution of DRMs before and during drug therapy [61]. Additionally, the length of 454 pyrosequencing reads has allowed linkage information to be derived from NGS data and has demonstrated that multiple DRMs occur in combination within individual viral genomes [58, 61]. In spite of these successful NGS applications, resistance testing is not frequently recommended clinically, because virologic breakthrough during single-agent therapy is typically managed empirically by addition of another agent with a different resistance profile [51]. Nevertheless, the improved ability to characterize the resistance profiles of current and investigational NAs will undoubtedly be of benefit as new combination and sequential therapies are attempted for the treatment of chronic HBV infections.

Drug resistance testing is also not regularly used for HCV management because, until recently, chronic HCV treatment relied primarily on the combination of pegylated interferon-α and the antiviral agent ribavirin. In the last few years, direct-acting antivirals (DAAs) targeting HCV have been developed and two protease inhibitors that affect HCV polyprotein processing, telaprevir and boceprevir, have been approved for treatment of chronic infection with HCV genotype 1. However, these drugs, as well as other DAAs in development, have been associated with high levels of resistance both in vitro and in clinical trials in which the drugs were used as monotherapy [52]. It has been hypothesized that such mutations preexist at low levels in the absence of therapy because of the high level and error-prone nature of HCV replication. NGS methods, utilizing 454 pyrosequencing and Illumina platforms, have been used to demonstrate that baseline resistant variants are present at <1 % of the pre-therapy viral population [63, 64]. Deep sequencing has also shown that HCV DRMs persist after cessation of a DAA, although this did not correlate with response to combination therapy with pegylated interferon-α, ribavirin, and the same DAA [65]. In fact, the outcomes of several recent clinical trials suggest that failure of combination therapy is largely dependent on lack of response to pegylated interferon-α and ribavirin [52]. Such findings justify the current practice not to include genotypic drug resistance testing in HCV management, although it is an important aspect of clinical trials investigating new DAAs. Conceivably, as such drugs become clinically available and new combination regimens are designed, resistance testing may become necessary, particularly for patients with prior virologic treatment failure.

CYTOMEGALOVIRUS

Another virus for which genotypic drug resistance testing is becoming routine is human cytomegalovirus (CMV), also known as human herpes virus 5 (HHV5) (reviewed in [66, 67]). Although CMV causes asymptomatic or mild infection in immune-competent hosts, it can lead to severe systemic and tissue-invasive disease in immune-compromised patients, such as solid organ transplant (SOT) and bone marrow transplant (BMT) or hematopoietic stem cell transplant (HSCT) recipients.

The number of antiviral drugs approved for CMV treatment is limited and includes ganciclovir (GCV) and its oral pro-drug valganciclovir (VCV), foscarnet (FOS), and cidofovir (CDV). Prophylaxis with GCV or VCV has been shown to substantially reduce rates of severe CMV disease in transplant recipients [68, 69], although regimens frequently last months or years. Importantly, duration of CMV antiviral therapy greater than 3 months is a risk factor for development of CMV drug resistance, in addition to intensive immunosuppression, multiple episodes of CMV reactivation, high viral counts, and suboptimal antiviral concentrations due to poor compliance or low absorption [66]. Rates of CMV drug resistance vary, based on patient populations: 5–12.5 % in SOT recipients, with CMV-negative lung transplant recipients at the highest risk [70–73]; 2–5 % in BMT/HSCT recipients [67, 73]; and 9 % in HIV-infected patients in the HAART era compared to >20 % before HAART [74–77].

Although no NGS assays for CMV drug resistance are currently available, this virus is an ideal target for drug resistance testing by NGS because resistance to current CMV drugs occurs in two well-defined, relatively small genomic regions. All currently available CMV antivirals target the CMV DNA polymerase UL54 and mutations in this gene can confer resistance. GCV and VCV require phosphorylation by the CMV phosphotransferase UL97 for antiviral activity and mutations in UL97 also result in resistance to these drugs. Numerous sequence variants have been identified in UL97 and UL54 from clinical specimens, although only some of those confer drug resistance in phenotypic assays [67]. Confirmed DRMs in UL97 cluster in a small number of codons, such that >80 % of clinical isolates with GCV resistance have one of seven frequently encountered DRMs: M460V/I, H520Q, C592G, A594V, L595S, and C603W [67]. In contrast, DRMs in CMV polymerase are spread throughout the UL54 gene, particularly the exonuclease and polymerase domains [67]. Many UL54 mutations are associated with cross-resistance, such that specific DRMs cause resistance to both GCV and CDV, and occasionally resistance to all three drugs [67, 73]. The majority of resistance mutations identified in patients are located in the UL97 gene rather than UL54, which may be related to the fact that GCV and VCV are the most commonly used drugs for prophylaxis and preemptive therapy [69, 78].

Timely detection of CMV drug resistance is critical because DRMs can accumulate with continued exposure to a drug [67], potentially leading to shortened graft survival and increased morbidity [71, 72]. On the other hand, rational change of therapy following identification of GCV resistance has been shown to lead to more rapid clearance of virus [70]. Although no standardized protocol for alternative therapy selection exists at this time, algorithms have been proposed by SOT and BMT expert panels [69, 78]. These algorithms advocate genotypic resistance testing when CMV viral load is rising or CMV-related disease develops in the context of CMV antiviral treatment lasting longer than 2 weeks. The current gold standard for genotypic detection of CMV DRMs is Sanger sequencing of PCR-amplified UL97 and UL54 gene segments, which requires viral loads >1,000 CMV copies/mL. Notably, numerous drug sensitive polymorphisms exist in the regions of UL97 and UL54 where DRMs have been identified. Thus, nucleotide changes in these genes are only considered DRMs if drug resistance has been demonstrated in phenotypic assays [67].

Given that resistance to current CMV drugs occurs in well-circumscribed regions of the CMV genome, this virus is well suited for an NGS-based DRM testing approach analogous to HIV protease and reverse transcriptase 454 pyrosequencing. Based on the HIV studies discussed above, conceivable advantages of NGS methods for CMV drug resistance testing over standard Sanger sequencing include increased sensitivity, the ability to multiplex samples through barcoding, and the potential to identify resistant variants that are present at <20 % of the viral population. Importantly, the impact of low-abundance drug resistance variants on CMV virologic failure has not been systematically investigated and NGS assays would make it possible to address this question. The rates of transmitted drug resistance are also unknown, although it is thought that baseline DRMs before anti-CMV therapy initiation do not occur commonly [76]. Data from NGS studies of CMV drug resistance may also contribute to DRM databases and genotypic interpretation systems, similar to those that exist for HIV [17, 18, 20]. In fact, a bioinformatics tool has recently been developed to link genotypic CMV data with resistance phenotypes [79].

Detection of Novel Viral Pathogens in Clinical Specimens

Another area of diagnostic virology where NGS is increasingly being applied is the discovery of novel viruses in clinical specimens of patients with diseases of unknown etiology. Importantly, many viruses cannot be cultured or identified by traditional molecular techniques, and although methods relying on cloning and Sanger sequencing have been used to identify novel viruses, those methods are generally laborious, time-consuming, and mainly applicable to sterile samples such as cerebrospinal fluid [8]. Microarrays targeting highly conserved regions within viral families are capable of detecting known viruses, but they cannot identify novel pathogens without sequence similarity to oligonucleotides on the array [11, 80]. In contrast, NGS offers an efficient, highly sensitive and unbiased methodology for detection of previously unknown viruses in specimens with mixed nucleic acids. The general approach in such studies is fundamentally different from that used in amplicon sequencing. Firstly, the virus of interest is usually not known and therefore cannot be selectively amplified with target-specific primers. Thus, various laboratory and bioinformatics techniques need to be used to enrich and separate viral RNA or DNA from the predominantly human nucleic acids. Secondly, a reference sequence may not be available for mapping of sequencing reads if the virus is truly novel or largely divergent from known related viruses. This necessitates de novo assembly of the viral genome, which requires use of different library preparation and sequencing approaches, as well as different bioinformatics tools.

Laboratory methods that have been used for viral particle purification and enrichment include viral culture, ultracentrifugation, density gradient centrifugation, and pretreatment of the sample with RNase or DNase in order to remove host nucleic acids, while preserving capsid-protected viral particles [8, 81]. PCR-based methods that selectively amplify viral genomes include rolling circle amplification for viruses with a circular genome [82]; use of restriction enzyme sites that are more frequently encountered in viral DNA than human, followed by ligation of adaptors and PCR amplification [83, 84]; as well as the use of oligonucleotides that specifically target host nucleic acids, e.g., ribosomal RNA, and interfere with their reverse transcription and amplification without impacting viral genome enrichment [83]. Newer methods have incorporated hybridization approaches to capture viral nucleic acids with antisense oligonucleotides as baits, although bait design requires at least some prior knowledge of the pathogen [85, 86]. Computational tools have also been developed for "subtracting" host sequences from the initial read pool containing mixed human and microbial sequences [11, 87, 88]. This filtering step is crucial because viral sequences may comprise less than 1 % of the initial aligned reads [10, 11].

Another challenge when attempting to identify an unknown pathogen is the selection of optimal amplification and library preparation protocols as well as the sequencing platform. Many clinical specimens have low viral concentrations and require an amplification step to generate sufficient DNA for library preparation, which is frequently performed with random primers. Because it is not always known whether the pathogen is a DNA or RNA virus, using an approach that amplifies both nucleic acids can be helpful. An example is the Phi29 bacteriophage DNA polymerase-based multiple displacement amplification with random primers [89, 90]. The choice of sequencing platform also requires significant consideration. Whereas 454 pyrosequencing produces long reads and therefore linkage information that facilitates assembly of contiguous sequences (contigs), it has relatively low throughput compared to Illumina and SOLiD (35–700 Mb vs. 7.5–600 Gb vs. 10–15 Gb, respectively). In fact, a recent study examined the performance of 454 pyrosequencing (GS FLX Titanium) and Illumina (Illumina GAII sequencer) for detection of viruses in cerebrospinal fluid and plasma samples artificially spiked with various concentrations of known viruses [90]. When reads were mapped to the reference genomes of the spiked-in viruses, the Illumina platform had higher sensitivity, which approximated the sensitivity of optimized real-time PCR assays. The Illumina platform also had an improved ability to detect unknown viruses by generating contigs for de novo genome assembly, although this could not be accomplished reliably at low viral loads [90].

Another critical aspect of successful viral discovery is the use of appropriate bioinformatics tools. When the reference genome is known, as in amplicon sequencing experiments, read mapping software typically applies stringent mismatch rules in order to minimize errors. In contrast, with unknown pathogens it may be impossible to map reads to publicly available viral databases, because the target virus is highly divergent. This challenge has been addressed by bioinformatics tools that assemble reads into contigs by identifying overlapping sequences between reads, followed by contig assembly into genomes [91]. Sequences assembled this way can be compared to public databases by using algorithms with relaxed stringency in order to identify related viruses. In some cases, comparison of amino acid sequences of predicted proteins from the novel virus to microbial protein databases is also necessary to identify phylogenetic links [92]. A major challenge in de novo assembly is the presence of repeat sequences, because those may present difficulties with PCR amplification or accurate assignment to a specific part of a genome. Computational and experimental strategies are being developed to address this issue [8, 91].

The approaches outlined above have been successfully used to identify viral pathogens in patients with infectious syndromes of unclear etiology. A specific example is a case in which three solid organ transplant recipients with the same organ donor developed a fatal febrile illness with sepsis and encephalopathy 4–6 weeks post-transplant [92]. Recipient specimens were subjected to 454 pyrosequencing, which identified a small number of sequences related to Old World arenaviruses, ultimately leading to assembly of the entire genome. Targeted approaches such as real-time PCR and immunostaining confirmed that the arenavirus was present in all three recipients, while the donor had evidence of specific IgG and IgM antibodies [92]. NGS was also used in the identification of a novel enterovirus (enterovirus 109) in a nose/throat swab sample from a Nicaraguan child with influenza-like illness [93]. In this case, an Illumina platform produced 119 reads (out of 4.66×10^6 nonhuman, nonbacterial reads) that tentatively could be assigned to an enterovirus-like virus, although the reads were not sufficient to assemble the novel genome. Virus-specific primers were designed based on the NGS-generated reads for amplification and targeted sequencing of the missing genomic regions. A similar targeted strategy was employed to assemble the genome of another arenavirus (Lujo virus), identified in a series of nosocomial hemorrhagic fever cases with high fatality rate in Southern Africa [94]. In this study, 454 pyrosequencing provided evidence of an arenavirus within 72 h of receiving the specimens from affected patients, which represents an acceptable turnaround time for public health efforts to control viral spread.

High-throughput sequencing was also used as proof of concept in a study of 17 patients with confirmed H1N1 influenza infection from the 2009 outbreak [10]. The authors showed that NGS with an Illumina GA IIx platform generated a sufficient number of reads to allow de novo assembly and up to 90 % coverage of the H1N1 genome when the sequencing reads of all 17 samples were pooled and used for assembly. This study and others have demonstrated the utility of NGS for characterizing new influenza reassortant viruses that may pose diagnostic challenges at the onset of epidemics if conventional molecular methods for influenza detection are unable to differentiate between the new virus and seasonal variants [95, 96].

The discovery of Merkel cell polyomavirus (MCV) illustrates NGS-based identification of a tumor-associated virus, specifically Merkel cell carcinoma (MCC), which is a rare but aggressive neuroectodermal tumor in immunecompromised patients [97]. Partial sequences from this small double-stranded DNA virus were detected by 454 pyrosequencing in cDNA libraries from four MCC tumors, and subsequently the complete viral genome was reconstructed using targeted molecular approaches, including rapid amplification of cDNA ends (RACE), PCR, and Sanger sequencing [97]. Another application of NGS in viral discovery is the investigation of arthropod-borne illnesses, including identification of novel pathogens such as Heartland virus, a new phlebovirus associated with cases of severe febrile illness in Missouri [98], and surveillance of mosquito vectors for the presence of known pathogens such as dengue virus [99].

An important caveat to these studies is that, although the association of the newly identified virus with disease is convincing, demonstrating causality is more difficult. Clearly, the presence of a novel virus in specimens from individuals with disease does not automatically imply pathogenicity. Traditionally, proving that a microorganism is the causative agent of disease has depended on fulfilling Koch's postulates, specifically the requirements that a putative etiologic agent be found in affected hosts but not healthy controls, that it is propagated in culture, and can reproduce the disease when a healthy host is inoculated. However, it is increasingly evident that many viruses cannot be cultured, which has necessitated revisions of Koch's postulates [100–102]. Such guidelines eliminate the requirement for microorganism isolation but expand on the rigor with which the association between microorganism and disease has to be established. For example, if a putative pathogen is identified by a sequence-based approach, the presence of the virus should be demonstrated in affected organs and tissues by immunostaining for viral antigens, by molecular methods for viral nucleic acids, or microscopy for virions. Phylogenetic analysis that shows that putative virulence factors in the novel virus are evolutionarily related to known pathogens, or a correlation between viral copy number and disease severity, can also be helpful in establishing causality.

Monitoring Viral Genome Diversity

NGS is also being applied to the study of viral population diversity, which has implications for understanding transmissibility, viral immune escape, and the response to antiviral therapies and vaccines. Rapid viral replication is associated with high mutation rates, especially for RNA viruses whose RNA polymerases are error prone and lack proofreading activity [103]. Thus, even if a virus is transmitted as a single copy, the mutations that accumulate with consecutive replication cycles within the host create a genetically diverse population that, theoretically, could contain every possible mutation across the viral genome [104]. This cloud-like swarm of variants has been termed "quasispecies" and is hypothesized to function cooperatively such that individual genotypes that are less fit are preserved because of their potential to become dominant in response to changing environmental conditions, thereby preserving the collective fitness of the population [104]. Understanding the dynamics of viral evolution, therefore, may be helpful for developing more successful antiviral therapies and vaccines.

As outlined in the preceding sections, NGS technologies are well suited for studying genotypic diversity because the high depth of coverage allows low-prevalence variants to be detected and faithfully represented by multiple sequencing reads. Investigation of viral genome variability has been performed on a genome-wide scale [105] and by targeting individual genomic regions known to be highly diverse [106–110]. Both 454 pyrosequencing and Illumina platforms have been used. Viral intrahost diversity has been interrogated for a number of viruses, including HIV [107–110], influenza A [95, 111–113], HCV [106], dengue virus [114], rhinovirus [115], and CMV [105, 116] to name a few. Major challenges in characterizing viral variants are sequencing/amplification error correction and haplotype reconstruction, because most platforms produce reads that are shorter than a complete viral genome. Bioinformatics tools have been developed specifically to quantify genetic diversity in NGS data derived from mixed viral populations [47, 117–119]. The general principle of such programs is to cluster overlapping reads within sliding windows of the reference genome and use probabilistic models to correct sequencing errors, assign a consensus sequence to each haplotype, concatenate adjacent windows to generate global haplotypes, and estimate variant prevalence.

NGS has been used to understand HIV viral population diversity in several clinically relevant contexts. For example, in clinical trials for transmission prevention within serodiscordant couples in which one partner is initially HIV-negative but subsequently becomes infected [109, 110], the intervention typically targets the partner who is infected at baseline, and efficacy is evaluated by the rate at which uninfected sexual partners remain infection-free. This means that in cases of seroconversion the ability to accurately link transmission to the enrolled infected partner is paramount. This can be done by sequencing HIV in both partners, followed by phylogenetic and Bayesian analyses to assess genetic distance between viral sequences from the two partners. Traditionally, this type of analysis has used Sanger sequencing which produces a single consensus viral sequence per infected individual, whereas NGS allows entire viral populations to be compared. In one study, NGS was able to resolve linkage status in several cases where traditional Sanger-based analysis could not [110]. A similar approach was employed to establish linked transmission of HCV between intravenous drug users [106]. Characterization of viral population diversity has also been applied in studies assessing rates of HIV superinfection [107], where the goal is to identify individuals in whom viral populations at baseline are phylogenetically unrelated to those in follow-up specimens.

Influenza A is another RNA virus with a propensity for rapid evolution, immune evasion, and high virulence. Consequently, assessing influenza virus population diversity by NGS is likely to play an important role in monitoring for emergence of new variants [120]. This virus has a single-stranded, negative-sense RNA genome organized in eight individual segments, and it is subject to mutation not only through the error-prone RNA polymerase (i.e., “antigenic drift”) but also through reassortments of the gene segments of different influenza subtypes (i.e., “antigenic shift”). The latter produces particularly virulent variants when the reassortment occurs between influenza subtypes that infect different host species. Given that hemagglutinin is the major target of the host immune system, there is partial immunity to influenza viruses that infect humans but not to those infecting birds or swine because their hemagglutinin epitopes are foreign to the human immune system. In fact, the three major pandemics of the twentieth century (H1N1 in 1918, H2N2 in 1957, and H3N2 in 1968) can be traced to reassortants of human, avian, and swine viruses, indicating that antigenic shift leads to more severe disease and a larger number of affected individuals [120]. Thus, the ability to track influenza virus evolution and to identify novel, potentially virulent variants is crucial for epidemic containment and vaccine development. Given the ability of influenza A to evolve rapidly, the unbiased approach of NGS can provide a diagnostic advantage when novel variants enter the human population. NGS can characterize the entire viral genome and identify reassortants early on, whereas existing diagnostic assays usually target previously seen influenza subtypes and may miss novel virulent variants, as occurred at the beginning of the 2009 H1N1 pandemic [121]. Genomics efforts to study influenza A have been centralized through the collaborative Influenza Genome Sequencing Project, which initially used Sanger sequencing to characterize 209 genomes [112, 122], but more recently has turned to NGS to sequence nearly 10,000 genomes from various outbreaks and epidemics and different host species (http://gsc.jcvi.org/projects/msc/influenza/). The availability of such longitudinal genomic data has made it possible to study the evolution of influenza A viruses within the human population and to guide surveillance efforts [123]. Influenza diversity has also been studied at the level of a single host, demonstrating evidence of mixed influenza infections and the emergence of drug resistance in an immune-compromised patient during the 2009 H1N1 pandemic [111].

Although DNA viruses have mutation rates that are orders of magnitude lower than RNA viruses [103], the question of intrahost viral genome variability has important implications for certain DNA viruses, such as CMV. CMV glycoproteins mediate viral entry and host immune response, suggesting that mutations in these genes may alter CMV biological properties and/or immune recognition of viral epitopes [124]. Because of this, sequence diversity in the CMV glycoprotein genes has been studied extensively in the context of vaccine development [124]. Discrete strains have been characterized based on glycoprotein sequence variation and it has been shown that immune-compromised patients frequently have co-infections with multiple CMV genotypes, which is associated with worse virologic outcomes compared to single-strain infections [125, 126]. Given the high sensitivity of NGS for low-abundance variants, it can be particularly useful for detection of multiple co-infecting genotypes. In fact, targeted 454 pyrosequencing of three glycoprotein genes in clinical specimens from lung transplant recipients [116] demonstrated that all specimens had evidence of mixed infections with one or two dominant strains at any given time. A different study, which performed WGS using an Illumina platform, found that the level of intrahost diversity at the glycoproteins was less than the diversity at open reading frames with other functions, including DNA replication [105]. Surprisingly, the intrahost CMV diversity on a genome-wide scale was of similar magnitude as that in RNA viruses such as HIV and dengue virus. Thus, it is possible that, in spite of the fidelity of the CMV DNA polymerase, mutations can accumulate to high levels as a result of high viral replication rates [105] or pressure from the host immune system [124]. Additional research is needed to replicate these findings and better characterize the source and significance of such high genomic diversity in DNA viral populations.

NGS in Clinical Bacteriology

Genomics approaches are likely to also gain a role in diagnosis and management of bacterial infections, although a variety of challenges remain to be overcome before NGS can become a routine part of clinical bacteriology. Diagnostic bacteriology has traditionally relied on isolation of a pathogen by culture, followed by biochemical tests that identify the organism to the genus or species level. Phenotypic antibiotic susceptibility testing is then performed to select optimal antimicrobial therapy. Virulence factor characterization is also important, because it can impact public health measures for vaccine development and outbreak containment, as well as management of individual patients, for example to inform the administration of antitoxin [127]. Given that the majority of bacterial organisms in the human body cannot be cultured [128], the utility of NGS for bacterial identification is easily imagined. However, the ability of NGS to replace or supplement conventional phenotypic methods for antimicrobial susceptibility testing remains a significant challenge. An important obstacle is that the genotypic determinants of antibiotic resistance as well as virulence are still largely unknown for many pathogens. Large-scale efforts, such as the Human Microbiome Project [128] and MetaHit [129], which aim to shed light on the genomic diversity of bacterial populations at various sites on the human body in health and disease, will be enormously helpful in elucidating genotype-to-phenotype relationships, especially when combined with proteomic and transcriptomic data. Such projects will also be instrumental for the development of standardized technical protocols and bioinformatics tools for sequence-based testing [130].

Studies that have employed NGS for identification of bacterial organisms in human specimens have used one of two general sequencing strategies: targeted amplicon sequencing of 16S ribosomal RNA (rRNA) genes or shotgun WGS. The first approach is particularly useful for characterizing taxonomic diversity of microbial communities and usually employs 454 pyrosequencing because of the longer read lengths (Table 33.1). It employs primers that anneal to 16S rRNA sequences that are conserved in bacteria, with intervening regions that are highly variable and typically provide sufficient sequence for taxonomic assignment. This strategy is at the core of the Human Microbiome Project, which has used 16S rRNA primers that cover hypervariable regions V3 to V5 (V35) and V1 to V3 (V13) to generate 5,177 bacterial taxonomic profiles from 242 healthy adults, sampled from 15 to 18 body sites at different time points [130]. Similarly, 16S rRNA hypervariable region pyrosequencing has been used to study diversity of bacterial communities in several diseases, including bacterial vaginosis, which is associated with increased bacterial heterogeneity compared to the healthy state [131]. On the other hand, examination of microbiota in the lower airway of cystic fibrosis patients [132, 133] and stool of patients with

Clostridium difficile-associated disease [134, 135] has shown that disease progression is marked by decreasing bacterial diversity, possibly related to escalating antibiotic exposures. Such results are beginning to show that certain infectious conditions may be driven by disturbances in the normal structure and diversity of a microbial community rather than the action of individual pathogens.

An important consideration in 16S rRNA-based NGS studies is the choice of primer targets because certain areas of 16S rRNA genes may allow amplification of a broader spectrum of bacteria, whereas other areas may lead to under-representation of some groups of bacteria [136]. Thus, knowledge of the diversity and specific genera present in a given specimen source may be helpful in selecting appropriate primer sets. Additionally, with 16S rRNA sequencing, classification may only be possible to the family or genus level because the amount of sequence variation is usually insufficient for species identification [137]. This limitation may be acceptable if the goal is to study an entire microbial community, but the approach would be suboptimal when the exact identity of a bacterial pathogen needs to be known for the purposes of patient management. One way to obtain species resolution is by sequencing multiple genomic regions, for example by the recently described ribosomal multilocus sequence typing (rMLST) method [138]. To illustrate the capabilities of this strategy, the authors used sequence data from 53 ribosomal subunit genes to achieve resolution to the strain level. However, the large number of loci that had to be sequenced indicated that this approach may necessitate a whole-genome, rather than targeted amplicon, sequencing approach.

Regardless of sequencing strategy, the accuracy of bacterial species identification will largely depend on the abundance and scope of reference sequences in the databases used for analysis. One of the advantages of 16S rRNA sequencing is that several extensive databases already exist. The SILVA database [139] contains >3 million small subunit and >250,000 large subunit bacterial rRNA gene sequences in its current version [140] and the Greengenes database calculates taxonomic relationships based on >400,000 16S rRNA sequences [141]. Bioinformatics tools that are specific for 16S rRNA data analysis include methods for error correction or de-noising [137, 142] and removal of amplification-derived chimeric sequences [143–145], which can artificially inflate the biodiversity estimate of a population [146]. De-noised data or raw sequences can be analyzed in dedicated pipelines such as QIIME [147] and mothur [144], that cluster similar sequences into operational taxonomic units (OTUs), followed by phylogenetic analyses. The purpose of OTUs is to group similar sequences into entities that represent genus or species level classifications depending on a threshold that is set for sequence similarity, typically 97 % for species assignment [137]. Importantly, the diagnostic utility of characterizing entire bacterial communities in the context of a clinical laboratory remains to be established, although genomics approaches are already deciphering the role of microbiota in health and disease.

While 16S rRNA-based NGS approaches are useful for studying diversity of entire bacterial communities, WGS approaches allow more detailed functional and taxonomic analyses of individual members of a community. For example, WGS can facilitate identification of individual bacterial pathogens in clinical situations in which the pathogen is not culturable or its DNA constitutes a small minority of the nucleic acids in a specimen. In such scenarios, WGS allows the comparison of multiple genomic regions of the pathogen to reference databases for more accurate classification. For instance, NGS was able to detect evidence of *Francisella tularensis* in an abscess sample in which bacterial DNA represented only 0.002 % of a total of 38,285,502 reads [148]. Such studies demonstrate the importance of sequencing depth and illustrate why shotgun WGS of specimens with diverse genomic composition has frequently relied on Illumina technology [130, 137]. Other platforms that have been used for WGS of bacteria include the Ion Torrent Personal Genome Machine (PGM) [149, 150], 454 pyrosequencing [151–153], SOLiD [154], and the PacBio RS sequencing system [155].

Whole-genome approaches have also been applied to better understand the mechanisms of virulence of pathogenic bacteria, including *Campylobacter jejuni* [151], *Staphylococcus aureus* [156], and *Acinetobacter baumannii* [152]. In addition, WGS approaches may provide important epidemiologic information in outbreaks of bacterial illness. For example, WGS was used to characterize and determine the genetic origins of the bacterial pathogens responsible for several recent outbreaks: the 2011 German Enterohemorrhagic *Escherichia coli* O104:H4 outbreak [149], the 2010 Haitian Vibrio cholera outbreak [155], and a cluster of methicillin-resistant *Staphylococcus aureus* (MRSA) cases in a neonatal intensive care unit [157].

Although phenotypic antimicrobial resistance testing is relatively well standardized, it is costly, available only for a limited number of organisms [158], and can take from a few days for fast-growing pathogens like *Escherichia coli* to several weeks for slow-growing organisms like *Mycobacterium tuberculosis* [127]. Molecular assays with improved sensitivity and turnaround times already exist for chromosomal or plasmid-encoded genes conferring antimicrobial resistance, including *mecA* in MRSA, and *vanA* or *vanB* in vancomycin-resistant enterococci [158]. Molecular tests have also been developed to detect several well-characterized mutations that confer resistance to first-line and second-line drugs in the *Mycobacterium tuberculosis* genome [159]. However, resistance to an antimicrobial class can be mediated by several molecular mechanisms. For example, penicillin resistance can be due to a beta-lactamase or a modified penicillin binding protein; high-level aminoglycoside resistance in enterococci can be due to a large number of aminoglycoside-modifying enzymes; and macrolide resistance can be either constitutive or inducible, with the inducible mechanisms frequently posing a diagnostic challenge [158]. NGS-based genotypic susceptibility testing, therefore, could simplify workflow and eliminate the need for individual PCR-based assays by simultaneously interrogating multiple resistance mechanisms. The application of NGS in a clinically relevant timeframe for detection of drug resistance patterns has been demonstrated in a proof-of-concept study utilizing 454 pyrosequencing of the bioterrorism agent *Bacillus anthracis*: starting with purified genomic DNA, draft bacterial genomes could be obtained within 24 h, and identification of genotypic antibiotic resistance was accomplished in the following 12 h [160]. Currently, NGS genotypic susceptibility testing is feasible only in a small number of situations in which the relationship between phenotypic and genotypic resistance has been established. However, NGS approaches are certain to be instrumental for the identification of novel clinically relevant mutations conferring resistance to existing antibacterial agents and newly developed antibiotics. For example, 454 pyrosequencing was used to explore the mode of action and potential resistance mechanisms of an experimental antituberculosis drug, R207910 [153]. Notably, a drawback of genotypic assays is that they do not provide a quantitative measure of antimicrobial susceptibility, and because some resistance genes may confer only low level of antibiotic resistance [158], sequence-based testing may not fully obviate the need for correlation with phenotypic susceptibility testing. In order to be useful as a screening or a stand-alone test, NGS must have a turnaround time that is shorter than, or at least equivalent to, phenotypic methods. Whether this is possible in a routine clinical laboratory remains to be demonstrated.

WGS strategies are also helpful for extracting functional information, both on the species and the population level: genomic sequences can be used to predict genes, as well as protein structure and function, metabolic pathways, and phenotypic correlations [161]. Thus, shotgun sequencing has become a driver of metagenomic studies, which examine the genomic content of entire bacterial communities within various habitats, and delineate not only taxonomic diversity but also the functional and metabolic pathways of the communities [137, 162]. The analysis of metagenomic data poses even more challenges than those discussed for 16S rRNA sequencing [137]. After sequencing data are filtered for human sequences and sequencing errors, the putative bacterial reads have to be aligned to reference genomes or subjected to de novo assembly of contigs in order for gene predictions to be made, as well as assignment of biological functions. Importantly, both taxonomic and functional annotations may be hindered by limited availability of reference genomes, although large endeavors exploring bacterial metagenomics in the human host, such as the Human Microbiome Project [128, 130] and MetaHit [129], are actively expanding bacterial genomic databases and are testing bioinformatics pipelines for metagenomic data analysis.

NGS in Clinical Mycology

Invasive fungal infections (IFI) can be life-threatening. However, microbiological culture is frequently unavailable or yields no growth, even when fungal elements are observed on microscopic examination. Culture-independent tests, such as serology and antigen assays, may be helpful, although many lack sensitivity and specificity [163, 164]. Because tailored antifungal therapy early in the course of disease is critical, particularly in immune-compromised patients [165], nucleic acid-based tests can be utilized for accurate and rapid diagnosis. In particular, broad-range PCR targeting ribosomal genes followed by Sanger sequencing of amplicons has allowed direct detection and identification of fungi from clinical specimens [166–169]. The fungal rRNA operon is present in multiple copies (≥100 copies per fungal genome) and therefore targeting this region improves sensitivity of fungal detection in clinical specimens. Specific areas of the fungal rRNA locus, the internal transcribed spacer regions (ITS1 and ITS2), have been the most frequently used targets for fungal identification by Sanger sequencing to date [170]. Importantly, primers for ITS1 and ITS2 amplification can be easily designed to prevent cross-priming with human or bacterial DNA.

Despite the many advantages of direct fungal identification with rRNA locus sequencing, for clinical purposes it also has limitations. Variables that are likely to affect the sensitivity and accuracy of the assay include DNA extraction methodology, primer selection, and phylogenetic breadth and taxonomic accuracy of reference sequence databases [166, 171]. Because fungal identification by sequence analysis is highly susceptible to contamination with fungal spores and commensal organisms, it is imperative that testing is strictly limited to samples obtained from sterile sources with visible fungal elements on the microscopic exam [172]. Additionally, not all fungi can be identified using the ITS region, including clinically important organisms such as *Aspergillus* section *Fumigati* and *Fusarium solani* species complex [172]. Accurate identification of these fungi requires sequence analysis of alternative genes or combination of multiple ribosomal targets [166]. Another problem is the paucity of publicly available genomic sequences for rare fungi. For example, the GenBank sequence database is of limited utility for identification of rare fungi, due to incomplete taxon sampling of disease-causing fungi, incorrect nomenclature assigned to sequence entries, and erroneous or truncated entries which can only be edited by the original depositor [173, 174].

NGS methods can offer solutions to several of these issues. As discussed for viruses and bacteria in previous sections, NGS can dramatically improve sensitivity by simultaneously sampling large numbers of genomic regions, thus increasing the probability of finding phylogenetic relationships, even in limited databases. The amount of sequence diversity that NGS datasets provide can also be useful for epidemiological tracing of outbreaks. Engelthaler et al. used the SOLiD platform to sequence *Coccidioides immitis* isolates from three patients who received solid organ transplants from a single donor. The sequence similarity of the isolates suggested origin from a single ancestor, most likely the common organ donor, who also had serologic evidence of coccidioidomycosis [175]. Because NGS approaches do not have to rely on target-specific primers, the danger of missing clinically important fungi due to lack of recognition by ITS primers is resolved, although contamination with environmental and commensal organisms remains a significant concern. Finally, NGS studies of clinically important fungi will undoubtedly expand the amount of publicly available genomic sequences and lead to development of improved reference databases.

Conclusions

In this chapter, we have discussed various areas of clinical microbiology in which genomic approaches have been used to identify or characterize medically important pathogens. The majority of these studies have been conducted as proof-of-concept experiments or research investigations. However, adopting high-throughput sequencing in diagnostic microbiology laboratories is certain to occur in the near future, as NGS procedures and data-analysis tools become more user friendly. Targeted NGS assays relying on amplicon sequencing, such as viral drug resistance testing, are likely to be introduced first given the sensitivity advantages over Sanger sequencing and the accumulating data supporting the clinical relevance of low-abundance resistance mutations, particularly in HIV. NGS-based amplicon sequencing of ribosomal RNA genes may also become routine for identification of pathogenic bacteria and fungi when there is high suspicion for infection and culture is negative or not available, for example in formalin-fixed, paraffin-embedded tissues. WGS strategies may also be useful for pathogen identification in sterile specimens if testing can be optimized to provide clinically actionable data faster than culture or currently available molecular methods. Importantly, the ability of NGS methods and bioinformatics pipelines to accurately identify and characterize pathogens will need to be rigorously validated and compared with traditional diagnostic techniques.

The greatest attraction of genomic approaches is that WGS could provide all relevant information about a pathogen in a single assay, including species identification, strain typing, virulence determination, and antimicrobial resistance. In practice, widespread implementation of NGS in clinical microbiology laboratories will require acquisition of costly new equipment and reagents, optimization of turnaround times, and personnel re-training from largely phenotypic testing approaches to methods that are increasingly reliant on bioinformatics expertise. Thus, NGS methods are expected to supplement, rather than replace, conventional diagnostic testing. Further, early clinical adoption of NGS methods is likely to take place in laboratories that already have significant molecular and Sanger sequencing experience. An important hurdle, even in the most sophisticated of clinical laboratories, is that genotype-phenotype correlations for many clinically relevant microorganisms are unknown, although large-scale metagenomic efforts like the Human Microbiome Project will undoubtedly define numerous new associations between sequence and function. It will also be crucial to construct standardized procedures for specimen handling, library preparation and sequencing, as well as data interpretation in order to ensure the accuracy and reproducibility of NGS-derived genotypic results. Ultimately, the tremendous promise of NGS methods for diagnostic infectious disease testing will also need the successful development of clinical microbiologists capable of interpreting and evaluating NGS data and placing these data in the appropriate clinical context.

REFERENCES

1. Boyd SD. Diagnostic applications of high-throughput DNA sequencing. Annu Rev Pathol. 2012.
2. Loman NJ, Constantinidou C, Chan JZ, Halachev M, Sergeant M, Penn CW, Robinson ER, Pallen MJ. High-throughput bacterial genome sequencing: an embarrassment of choice, a world of opportunity. Nat Rev Microbiol. 2012;10: 599–606.
3. Anderson MW, Schrijver I. Next generation DNA sequencing and the future of genomic medicine. Genes. 2010;1:38–69.
4. Loman NJ, Misra RV, Dallman TJ, Constantinidou C, Gharbia SE, Wain J, Pallen MJ. Performance comparison of benchtop high-throughput sequencing platforms. Nat Biotechnol. 2012; 30:434–9.
5. Hoffmann C, Minkah N, Leipzig J, Wang G, Arens MQ, Tebas P, Bushman FD. DNA bar coding and pyrosequencing to identify rare HIV drug resistance mutations. Nucleic Acids Res. 2007;35:e91.
6. Hamady M, Walker JJ, Harris JK, Gold NJ, Knight R. Error-correcting barcoded primers for pyrosequencing hundreds of samples in multiplex. Nat Methods. 2008;5:235–7.
7. Vrancken B, Lequime S, Theys K, Lemey P. Covering all bases in HIV research: unveiling a hidden world of viral evolution. AIDS Rev. 2010;12:89–102.
8. Radford AD, Chapman D, Dixon L, Chantrey J, Darby AC, Hall N. Application of next-generation sequencing technologies in virology. J Gen Virol. 2012;93:1853–68.
9. Weinstock GM. Genomic approaches to studying the human microbiota. Nature. 2012;489: 250–6.
10. Greninger AL, Chen EC, Sittler T, Scheinerman A, Roubinian N, Yu G, Kim E, Pillai DR, Guyard C, Mazzulli T, et al. A metagenomic analysis of

pandemic influenza A (2009 H1N1) infection in patients from North America. PLoS One. 2010;5:e13381.
11. Yozwiak NL, Skewes-Cox P, Stenglein MD, Balmaseda A, Harris E, DeRisi JL. Virus identification in unknown tropical febrile illness cases using deep sequencing. PLoS Negl Trop Dis. 2012;6:e1485.
12. Cortez KJ, Maldarelli F. Clinical management of HIV drug resistance. Viruses. 2011;3:347–78.
13. Johnson VA, Calvez V, Gunthard HF, Paredes R, Pillay D, Shafer R, Wensing AM, Richman DD. 2011 update of the drug resistance mutations in HIV-1. Top Antivir Med. 2011;19:156–64.
14. Grant PM, Zolopa AR. The use of resistance testing in the management of HIV-1-infected patients. Curr Opin HIV AIDS. 2009;4:474–80.
15. Dunn DT, Coughlin K, Cane PA. Genotypic resistance testing in routine clinical care. Curr Opin HIV AIDS. 2011;6:251–7.
16. Vandamme AM, Camacho RJ, Ceccherini-Silberstein F, de Luca A, Palmisano L, Paraskevis D, Paredes R, Poljak M, Schmit JC, Soriano V, et al. European recommendations for the clinical use of HIV drug resistance testing: 2011 update. AIDS Rev. 2011;13:77–108.
17. Tang MW, Liu TF, Shafer RW. The HIVdb system for HIV-1 genotypic resistance interpretation. Intervirology. 2012;55:98–101.
18. Woods CK, Brumme CJ, Liu TF, Chui CK, Chu AL, Wynhoven B, Hall TA, Trevino C, Shafer RW, Harrigan PR. Automating HIV drug resistance genotyping with RECall, a freely accessible sequence analysis tool. J Clin Microbiol. 2012;50:1936–42.
19. Lengauer T, Sing T. Bioinformatics-assisted anti-HIV therapy. Nat Rev Microbiol. 2006;4:790–7.
20. Zazzi M, Kaiser R, Sonnerborg A, Struck D, Altmann A, Prosperi M, Rosen-Zvi M, Petroczi A, Peres Y, Schulter E, et al. Prediction of response to antiretroviral therapy by human experts and by the EuResist data-driven expert system (the EVE study). HIV Med. 2011;12:211–8.
21. Beerenwinkel N, Daumer M, Oette M, Korn K, Hoffmann D, Kaiser R, Lengauer T, Selbig J, Walter H. Geno2pheno: estimating phenotypic drug resistance from HIV-1 genotypes. Nucleic Acids Res. 2003;31:3850–5.
22. Johnson JA, Li JF, Wei X, Lipscomb J, Irlbeck D, Craig C, Smith A, Bennett DE, Monsour M, Sandstrom P, et al. Minority HIV-1 drug resistance mutations are present in antiretroviral treatment-naive populations and associate with reduced treatment efficacy. PLoS Med. 2008;5:e158.
23. Li JZ, Paredes R, Ribaudo HJ, Svarovskaia ES, Metzner KJ, Kozal MJ, Hullsiek KH, Balduin M, Jakobsen MR, Geretti AM, et al. Low-frequency HIV-1 drug resistance mutations and risk of NNRTI-based antiretroviral treatment failure: a systematic review and pooled analysis. JAMA. 2011;305:1327–35.
24. Metzner KJ, Giulieri SG, Knoepfel SA, Rauch P, Burgisser P, Yerly S, Gunthard HF, Cavassini M. Minority quasispecies of drug-resistant HIV-1 that lead to early therapy failure in treatment-naive and -adherent patients. Clin Infect Dis. 2009;48:239–47.
25. Wang C, Mitsuya Y, Gharizadeh B, Ronaghi M, Shafer RW. Characterization of mutation spectra with ultra-deep pyrosequencing: application to HIV-1 drug resistance. Genome Res. 2007;17:1195–201.
26. Mitsuya Y, Varghese V, Wang C, Liu TF, Holmes SP, Jayakumar P, Gharizadeh B, Ronaghi M, Klein D, Fessel WJ, et al. Minority human immunodeficiency virus type 1 variants in antiretroviral-naive persons with reverse transcriptase codon 215 revertant mutations. J Virol. 2008;82:10747–55.
27. Fisher R, van Zyl GU, Travers SA, Kosakovsky Pond SL, Engelbrech S, Murrell B, Scheffler K, Smith D. Deep sequencing reveals minor protease resistance mutations in patients failing a protease inhibitor regimen. J Virol. 2012;86:6231–7.
28. Le T, Chiarella J, Simen BB, Hanczaruk B, Egholm M, Landry ML, Dieckhaus K, Rosen MI, Kozal MJ. Low-abundance HIV drug-resistant viral variants in treatment-experienced persons correlate with historical antiretroviral use. PLoS One. 2009;4:e6079.
29. Simen BB, Simons JF, Hullsiek KH, Novak RM, Macarthur RD, Baxter JD, Huang C, Lubeski C, Turenchalk GS, Braverman MS, et al. Low-abundance drug-resistant viral variants in chronically HIV-infected, antiretroviral treatment-naive patients significantly impact treatment outcomes. J Infect Dis. 2009;199:693–701.
30. Varghese V, Shahriar R, Rhee SY, Liu T, Simen BB, Egholm M, Hanczaruk B, Blake LA, Gharizadeh B, Babrzadeh F, et al. Minority variants associated with transmitted and acquired HIV-1 nonnucleoside reverse transcriptase inhibitor resistance: implications for the use of second-generation nonnucleoside reverse transcriptase inhibitors. J Acquir Immune Defic Syndr. 2009;52:309–15.
31. Lataillade M, Chiarella J, Yang R, Schnittman S, Wirtz V, Uy J, Seekins D, Krystal M, Mancini M, McGrath D, et al. Prevalence and clinical significance of HIV drug resistance mutations by ultra-deep sequencing in antiretroviral-naive subjects in the CASTLE study. PLoS One. 2010;5:e10952.
32. Hedskog C, Mild M, Jernberg J, Sherwood E, Bratt G, Leitner T, Lundeberg J, Andersson B, Albert J. Dynamics of HIV-1 quasispecies during antiviral treatment dissected using ultra-deep pyrosequencing. PLoS One. 2010;5:e11345.

33. Mukherjee R, Jensen ST, Male F, Bittinger K, Hodinka RL, Miller MD, Bushman FD. Switching between raltegravir resistance pathways analyzed by deep sequencing. AIDS. 2011;25: 1951–9.
34. Armenia D, Vandenbroucke I, Fabeni L, Van Marck H, Cento V, D'Arrigo R, Van Wesenbeeck L, Scopelliti F, Micheli V, Bruzzone B, et al. Study of genotypic and phenotypic HIV-1 dynamics of integrase mutations during raltegravir treatment: a refined analysis by ultra-deep 454 pyrosequencing. J Infect Dis. 2012;205: 557–67.
35. Archer J, Braverman MS, Taillon BE, Desany B, James I, Harrigan PR, Lewis M, Robertson DL. Detection of low-frequency pretherapy chemokine (CXC motif) receptor 4 (CXCR4)-using HIV-1 with ultra-deep pyrosequencing. AIDS. 2009;23:1209–18.
36. Bunnik EM, Swenson LC, Edo-Matas D, Huang W, Dong W, Frantzell A, Petropoulos CJ, Coakley E, Schuitemaker H, Harrigan PR, et al. Detection of inferred CCR5- and CXCR4-using HIV-1 variants and evolutionary intermediates using ultra-deep pyrosequencing. PLoS Pathog. 2011;7:e1002106.
37. Tsibris AM, Korber B, Arnaout R, Russ C, Lo CC, Leitner T, Gaschen B, Theiler J, Paredes R, Su Z, et al. Quantitative deep sequencing reveals dynamic HIV-1 escape and large population shifts during CCR5 antagonist therapy in vivo. PLoS One. 2009;4:e5683.
38. Codoner FM, Pou C, Thielen A, Garcia F, Delgado R, Dalmau D, Alvarez-Tejado M, Ruiz L, Clotet B, Paredes R. Added value of deep sequencing relative to population sequencing in heavily pre-treated HIV-1-infected subjects. PLoS One. 2011;6:e19461.
39. Rothberg JM, Leamon JH. The development and impact of 454 sequencing. Nat Biotechnol. 2008;26:1117–24.
40. Gorzer I, Guelly C, Trajanoski S, Puchhammer-Stockl E. The impact of PCR-generated recombination on diversity estimation of mixed viral populations by deep sequencing. J Virol Methods. 2010;169:248–52.
41. Mild M, Hedskog C, Jernberg J, Albert J. Performance of ultra-deep pyrosequencing in analysis of HIV-1 pol gene variation. PLoS One. 2011;6:e22741.
42. Harismendy O, Ng PC, Strausberg RL, Wang X, Stockwell TB, Beeson KY, Schork NJ, Murray SS, Topol EJ, Levy S, et al. Evaluation of next generation sequencing platforms for population targeted sequencing studies. Genome Biol. 2009;10:R32.
43. Becker EA, Burns CM, Leon EJ, Rajabojan S, Friedman R, Friedrich TC, O'Connor SL, Hughes AL. Experimental analysis of sources of error in evolutionary studies based on Roche/454 pyrosequencing of viral genomes. Genome Biol Evol. 2012;4:457–65.
44. Skums P, Dimitrova Z, Campo DS, Vaughan G, Rossi L, Forbi JC, Yokosawa J, Zelikovsky A, Khudyakov Y. Efficient error correction for next-generation sequencing of viral amplicons. BMC Bioinformatics. 2012;13 Suppl 10:S6.
45. Dudley DM, Chin EN, Bimber BN, Sanabani SS, Tarosso LF, Costa PR, Sauer MM, Kallas EG, O'Connor DH. Low-cost ultra-wide genotyping using Roche/454 pyrosequencing for surveillance of HIV drug resistance. PLoS One. 2012;7:e36494.
46. Eriksson N, Pachter L, Mitsuya Y, Rhee SY, Wang C, Gharizadeh B, Ronaghi M, Shafer RW, Beerenwinkel N. Viral population estimation using pyrosequencing. PLoS Comput Biol. 2008;4:e1000074.
47. Zagordi O, Geyrhofer L, Roth V, Beerenwinkel N. Deep sequencing of a genetically heterogeneous sample: local haplotype reconstruction and read error correction. J Comput Biol. 2010;17:417–28.
48. Jabara CB, Jones CD, Roach J, Anderson JA, Swanstrom R. Accurate sampling and deep sequencing of the HIV-1 protease gene using a Primer ID. Proc Natl Acad Sci U S A. 2011; 108:20166–71.
49. Vandenbroucke I, Van Marck H, Mostmans W, Van Eygen V, Rondelez E, Thys K, Van Baelen K, Fransen K, Vaira D, Kabeya K, et al. HIV-1V3 envelope deep sequencing for clinical plasma specimens failing in phenotypic tropism assays. AIDS Res Ther. 2010;7:4.
50. Chevaliez S, Rodriguez C, Pawlotsky JM. New virologic tools for management of chronic hepatitis B and C. Gastroenterology. 2012;142:1303–1313 e1301
51. Fung J, Lai CL, Seto WK, Yuen MF. Nucleoside/nucleotide analogues in the treatment of chronic hepatitis B. J Antimicrob Chemother. 2011;66: 2715–25.
52. Pawlotsky JM. Treatment failure and resistance with direct-acting antiviral drugs against hepatitis C virus. Hepatology. 2011;53:1742–51.
53. Yuen MF, Seto WK, Chow DH, Tsui K, Wong DK, Ngai VW, Wong BC, Fung J, Yuen JC, Lai CL. Long-term lamivudine therapy reduces the risk of long-term complications of chronic hepatitis B infection even in patients without advanced disease. Antivir Ther. 2007;12: 1295–303.
54. Tan Y, Ding K, Su J, Trinh X, Peng Z, Gong Y, Chen L, Cui Q, Lei N, Chen X, et al. The naturally occurring YMDD mutation among patients chronically infected HBV and untreated with lamivudine: a systematic review and meta-analysis. PLoS One. 2012;7:e32789.

55. Villet S, Pichoud C, Billioud G, Barraud L, Durantel S, Trepo C, Zoulim F. Impact of hepatitis B virus rtA181V/T mutants on hepatitis B treatment failure. J Hepatol. 2008;48:747–55.
56. Hayer J, Jadeau F, Deleage G, Kay A, Zoulim F, Combet C. HBVdb: a knowledge database for hepatitis B virus. Nucleic Acids Res. 2012.
57. Yuen LK, Ayres A, Littlejohn M, Colledge D, Edgely A, Maskill WJ, Locarnini SA, Bartholomeusz A. SeqHepB: a sequence analysis program and relational database system for chronic hepatitis B. Antiviral Res. 2007;75:64–74.
58. Ko SY, Oh HB, Park CW, Lee HC, Lee JE. Analysis of hepatitis B virus drug-resistant mutant haplotypes by ultra-deep pyrosequencing. Clin Microbiol Infect. 2012;18:E404–11.
59. Margeridon-Thermet S, Shulman NS, Ahmed A, Shahriar R, Liu T, Wang C, Holmes SP, Babrzadeh F, Gharizadeh B, Hanczaruk B, et al. Ultra-deep pyrosequencing of hepatitis B virus quasispecies from nucleoside and nucleotide reverse-transcriptase inhibitor (NRTI)-treated patients and NRTI-naive patients. J Infect Dis. 2009;199: 1275–85.
60. Margeridon-Thermet S, Svarovskaia ES, Babrzadeh F, Martin R, Liu TF, Pacold M, Reuman EC, Holmes SP, Borroto-Esoda K, Shafer RW. Low-level persistence of drug-resistance mutations in hepatitis B virus-infected subjects with a past history of Lamivudine treatment. Antimicrob Agents Chemother. 2012.
61. Rodriguez-Frias F, Tabernero D, Quer J, Esteban JI, Ortega I, Domingo E, Cubero M, Camos S, Ferrer-Costa C, Sanchez A, et al. Ultra-deep pyrosequencing detects conserved genomic sites and quantifies linkage of drug-resistant amino acid changes in the hepatitis B virus genome. PLoS One. 2012;7:e37874.
62. Solmone M, Vincenti D, Prosperi MC, Bruselles A, Ippolito G, Capobianchi MR. Use of massively parallel ultradeep pyrosequencing to characterize the genetic diversity of hepatitis B virus in drug-resistant and drug-naive patients and to detect minor variants in reverse transcriptase and hepatitis B S antigen. J Virol. 2009;83: 1718–26.
63. Nasu A, Marusawa H, Ueda Y, Nishijima N, Takahashi K, Osaki Y, Yamashita Y, Inokuma T, Tamada T, Fujiwara T, et al. Genetic heterogeneity of hepatitis C virus in association with antiviral therapy determined by ultra-deep sequencing. PLoS One. 2011;6:e24907.
64. Verbinnen T, Van Marck H, Vandenbroucke I, Vijgen L, Claes M, Lin TI, Simmen K, Neyts J, Fanning G, Lenz O. Tracking the evolution of multiple in vitro hepatitis C virus replicon variants under protease inhibitor selection pressure by 454 deep sequencing. J Virol. 2010; 84:11124–33.
65. Lenz O, de Bruijne J, Vijgen L, Verbinnen T, Weegink C, Van Marck H, Vandenbroucke I, Peeters M, Simmen K, Fanning G., et al. Efficacy of re-treatment with TMC435 as combination therapy in hepatitis c virus-infected patients following TMC435 monotherapy. Gastroenterology. 2012;143:1176–1178 e1176.
66. Drew WL. Cytomegalovirus resistance testing: pitfalls and problems for the clinician. Clin Infect Dis. 2010;50:733–6.
67. Lurain NS, Chou S. Antiviral drug resistance of human cytomegalovirus. Clin Microbiol Rev. 2010;23:689–712.
68. Baroco AL, Oldfield EC. Gastrointestinal cytomegalovirus disease in the immunocompromised patient. Curr Gastroenterol Rep. 2008; 10:409–16.
69. Ljungman P, Reusser P, de la Camara R, Einsele H, Engelhard D, Ribaud P, Ward K. Management of CMV infections: recommendations from the infectious diseases working party of the EBMT. Bone Marrow Transplant. 2004;33:1075–81.
70. Myhre HA, Haug Dorenberg D, Kristiansen KI, Rollag H, Leivestad T, Asberg A, Hartmann A. Incidence and outcomes of ganciclovir-resistant cytomegalovirus infections in 1244 kidney transplant recipients. Transplantation. 2011;92: 217–23.
71. Limaye AP, Corey L, Koelle DM, Davis CL, Boeckh M. Emergence of ganciclovir-resistant cytomegalovirus disease among recipients of solid-organ transplants. Lancet. 2000;356: 645–9.
72. Li F, Kenyon KW, Kirby KA, Fishbein DP, Boeckh M, Limaye AP. Incidence and clinical features of ganciclovir-resistant cytomegalovirus disease in heart transplant recipients. Clin Infect Dis. 2007;45:439–47.
73. Hantz S, Garnier-Geoffroy F, Mazeron MC, Garrigue I, Merville P, Mengelle C, Rostaing L, Saint Marcoux F, Essig M, Rerolle JP, et al. Drug-resistant cytomegalovirus in transplant recipients: a French cohort study. J Antimicrob Chemother. 2010;65:2628–40.
74. Boivin G, Gilbert C, Gaudreau A, Greenfield I, Sudlow R, Roberts NA. Rate of emergence of cytomegalovirus (CMV) mutations in leukocytes of patients with acquired immunodeficiency syndrome who are receiving valganciclovir as induction and maintenance therapy for CMV retinitis. J Infect Dis. 2001;184:1598–602.
75. Gilbert C, Handfield J, Toma E, Lalonde R, Bergeron MG, Boivin G. Emergence and prevalence of cytomegalovirus UL97 mutations associated with ganciclovir resistance in AIDS patients. AIDS. 1998;12:125–9.
76. Jabs DA, Dunn JP, Enger C, Forman M, Bressler N, Charache P. Cytomegalovirus retinitis and

viral resistance. Prevalence of resistance at diagnosis, 1994. Cytomegalovirus Retinitis and Viral Resistance Study Group. Arch Ophthalmol. 1996;114:809–14.
77. Martin BK, Ricks MO, Forman MS, Jabs DA. Change over time in incidence of ganciclovir resistance in patients with cytomegalovirus retinitis. Clin Infect Dis. 2007;44:1001–8.
78. Kotton CN, Kumar D, Caliendo AM, Asberg A, Chou S, Snydman DR, Allen U, Humar A. International consensus guidelines on the management of cytomegalovirus in solid organ transplantation. Transplantation. 2010;89:779–95.
79. Chevillotte M, von Einem J, Meier BM, Lin FM, Kestler HA, Mertens T. A new tool linking human cytomegalovirus drug resistance mutations to resistance phenotypes. Antiviral Res. 2010;85:318–27.
80. Wang D, Coscoy L, Zylberberg M, Avila PC, Boushey HA, Ganem D, DeRisi JL. Microarray-based detection and genotyping of viral pathogens. Proc Natl Acad Sci U S A. 2002;99: 15687–92.
81. Barzon L, Lavezzo E, Militello V, Toppo S, Palu G. Applications of next-generation sequencing technologies to diagnostic virology. Int J Mol Sci. 2011;12:7861–84.
82. Rector A, Tachezy R, Van Ranst M. A sequence-independent strategy for detection and cloning of circular DNA virus genomes by using multiply primed rolling-circle amplification. J Virol. 2004;78:4993–8.
83. de Vries M, Deijs M, Canuti M, van Schaik BD, Faria NR, van de Garde MD, Jachimowski LC, Jebbink MF, Jakobs M, Luyf AC, et al. A sensitive assay for virus discovery in respiratory clinical samples. PLoS One. 2011;6:e16118.
84. Pyrc K, Jebbink MF, Berkhout B, van der Hoek L. Detection of new viruses by VIDISCA. Virus discovery based on cDNA-amplified fragment length polymorphism. Methods Mol Biol. 2008;454:73–89.
85. Depledge DP, Palser AL, Watson SJ, Lai IY, Gray ER, Grant P, Kanda RK, Leproust E, Kellam P, Breuer J. Specific capture and whole-genome sequencing of viruses from clinical samples. PLoS One. 2011;6:e27805.
86. Duncavage EJ, Magrini V, Becker N, Armstrong JR, Demeter RT, Wylie T, Abel HJ, Pfeifer JD. Hybrid capture and next-generation sequencing identify viral integration sites from formalin-fixed, paraffin-embedded tissue. J Mol Diagn. 2011;13:325–33.
87. Bhaduri A, Qu K, Lee CS, Ungewickell A, Khavari PA. Rapid identification of non-human sequences in high-throughput sequencing datasets. Bioinformatics. 2012;28:1174–5.
88. Kostic AD, Ojesina AI, Pedamallu CS, Jung J, Verhaak RG, Getz G, Meyerson M. PathSeq: software to identify or discover microbes by deep sequencing of human tissue. Nat Biotechnol. 2011;29:393–6.
89. Berthet N, Reinhardt AK, Leclercq I, van Ooyen S, Batejat C, Dickinson P, Stamboliyska R, Old IG, Kong KA, Dacheux L, et al. Phi29 polymerase based random amplification of viral RNA as an alternative to random RT-PCR. BMC Mol Biol. 2008;9:77.
90. Cheval J, Sauvage V, Frangeul L, Dacheux L, Guigon G, Dumey N, Pariente K, Rousseaux C, Dorange F, Berthet N, et al. Evaluation of high-throughput sequencing for identifying known and unknown viruses in biological samples. J Clin Microbiol. 2011;49:3268–75.
91. Miller JR, Koren S, Sutton G. Assembly algorithms for next-generation sequencing data. Genomics. 2010;95:315–27.
92. Palacios G, Druce J, Du L, Tran T, Birch C, Briese T, Conlan S, Quan PL, Hui J, Marshall J, et al. A new arenavirus in a cluster of fatal transplant-associated diseases. N Engl J Med. 2008;358: 991–8.
93. Yozwiak NL, Skewes-Cox P, Gordon A, Saborio S, Kuan G, Balmaseda A, Ganem D, Harris E, DeRisi JL. Human enterovirus 109: a novel interspecies recombinant enterovirus isolated from a case of acute pediatric respiratory illness in Nicaragua. J Virol. 2010;84:9047–58.
94. Briese T, Paweska JT, McMullan LK, Hutchison SK, Street C, Palacios G, Khristova ML, Weyer J, Swanepoel R, Egholm M, et al. Genetic detection and characterization of Lujo virus, a new hemorrhagic fever-associated arenavirus from Southern Africa. PLoS Pathog. 2009;5: e1000455.
95. Hoper D, Hoffmann B, Beer M. A comprehensive deep sequencing strategy for full-length genomes of influenza A. PLoS One. 2011; 6:e19075.
96. Yongfeng H, Fan Y, Jie D, Jian Y, Ting Z, Lilian S, Jin Q. Direct pathogen detection from swab samples using a new high-throughput sequencing technology. Clin Microbiol Infect. 2011;17: 241–4.
97. Feng H, Shuda M, Chang Y, Moore PS. Clonal integration of a polyomavirus in human Merkel cell carcinoma. Science. 2008;319:1096–100.
98. McMullan LK, Folk SM, Kelly AJ, MacNeil A, Goldsmith CS, Metcalfe MG, Batten BC, Albarino CG, Zaki SR, Rollin PE, et al. A new phlebovirus associated with severe febrile illness in Missouri. N Engl J Med. 2012;367:834–41.
99. Bishop-Lilly KA, Turell MJ, Willner KM, Butani A, Nolan NM, Lentz SM, Akmal A, Mateczun A, Brahmbhatt TN, Sozhamannan S, et al. Arbovirus detection in insect vectors by rapid, high-throughput pyrosequencing. PLoS Negl Trop Dis. 2010;4:e878.

100. Falkow S. Molecular Koch's postulates applied to bacterial pathogenicity–a personal recollection 15 years later. Nat Rev Microbiol. 2004;2: 67–72.
101. Fredericks DN, Relman DA. Sequence-based identification of microbial pathogens: a reconsideration of Koch's postulates. Clin Microbiol Rev. 1996;9:18–33.
102. Mokili JL, Rohwer F, Dutilh BE. Metagenomics and future perspectives in virus discovery. Curr Opin Virol. 2012;2:63–77.
103. Sanjuan R, Nebot MR, Chirico N, Mansky LM, Belshaw R. Viral mutation rates. J Virol. 2010;84:9733–48.
104. Lauring AS, Andino R. Quasispecies theory and the behavior of RNA viruses. PLoS Pathog. 2010;6:e1001005.
105. Renzette N, Bhattacharjee B, Jensen JD, Gibson L, Kowalik TF. Extensive genome-wide variability of human cytomegalovirus in congenitally infected infants. PLoS Pathog. 2011;7:e1001344.
106. Escobar-Gutierrez A, Vazquez-Pichardo M, Cruz-Rivera M, Rivera-Osorio P, Carpio-Pedroza JC, Ruiz-Pacheco JA, Ruiz-Tovar K, Vaughan G. Identification of hepatitis C virus transmission using a next-generation sequencing approach. J Clin Microbiol. 2012;50:1461–3.
107. Redd AD, Mullis CE, Serwadda D, Kong X, Martens C, Ricklefs SM, Tobian AA, Xiao C, Grabowski MK, Nalugoda F, et al. The rates of HIV superinfection and primary HIV incidence in a general population in Rakai, Uganda. J Infect Dis. 2012;206:267–74.
108. Buzon MJ, Codoner FM, Frost SD, Pou C, Puertas MC, Massanella M, Dalmau J, Llibre JM, Stevenson M, Blanco J, et al. Deep molecular characterization of HIV-1 dynamics under suppressive HAART. PLoS Pathog. 2011;7:e1002314.
109. Campbell MS, Mullins JI, Hughes JP, Celum C, Wong KG, Raugi DN, Sorensen S, Stoddard JN, Zhao H, Deng W, et al. Viral linkage in HIV-1 seroconverters and their partners in an HIV-1 prevention clinical trial. PLoS One. 2011;6:e16986.
110. Eshleman SH, Hudelson SE, Redd AD, Wang L, Debes R, Chen YQ, Martens CA, Ricklefs SM, Selig EJ, Porcella SF, et al. Analysis of genetic linkage of HIV from couples enrolled in the HIV Prevention Trials Network 052 trial. J Infect Dis. 2011;204:1918–26.
111. Ghedin E, Laplante J, DePasse J, Wentworth DE, Santos RP, Lepow ML, Porter J, Stellrecht K, Lin X, Operario D, et al. Deep sequencing reveals mixed infection with 2009 pandemic influenza A (H1N1) virus strains and the emergence of oseltamivir resistance. J Infect Dis. 2011;203: 168–74.
112. Holmes EC, Ghedin E, Miller N, Taylor J, Bao Y, St George K, Grenfell BT, Salzberg SL, Fraser CM, Lipman DJ, et al. Whole-genome analysis of human influenza A virus reveals multiple persistent lineages and reassortment among recent H3N2 viruses. PLoS Biol. 2005;3:e300.
113. Roedig JV, Rapp E, Hoper D, Genzel Y, Reichl U. Impact of host cell line adaptation on quasispecies composition and glycosylation of influenza A virus hemagglutinin. PLoS One. 2011;6: e27989.
114. Parameswaran P, Charlebois P, Tellez Y, Nunez A, Ryan EM, Malboeuf CM, Levin JZ, Lennon NJ, Balmaseda A, Harris E, et al. Genome-wide patterns of intrahuman dengue virus diversity reveal associations with viral phylogenetic clade and interhost diversity. J Virol. 2012;86:8546–58.
115. Tapparel C, Cordey S, Junier T, Farinelli L, Van Belle S, Soccal PM, Aubert JD, Zdobnov E, Kaiser L. Rhinovirus genome variation during chronic upper and lower respiratory tract infections. PLoS One. 2011;6:e21163.
116. Gorzer I, Guelly C, Trajanoski S, Puchhammer-Stockl E. Deep sequencing reveals highly complex dynamics of human cytomegalovirus genotypes in transplant patients over time. J Virol. 2010;84:7195–203.
117. Astrovskaya I, Tork B, Mangul S, Westbrooks K, Mandoiu I, Balfe P, Zelikovsky A. Inferring viral quasispecies spectra from 454 pyrosequencing reads. BMC Bioinformatics. 2011;12 Suppl 6:S1.
118. Zagordi O, Bhattacharya A, Eriksson N, Beerenwinkel N. ShoRAH: estimating the genetic diversity of a mixed sample from next-generation sequencing data. BMC Bioinformatics. 2011;12:119.
119. Prosperi MC, Salemi M. QuRe: software for viral quasispecies reconstruction from next-generation sequencing data. Bioinformatics. 2012; 28:132–3.
120. McHardy AC, Adams B. The role of genomics in tracking the evolution of influenza A virus. PLoS Pathog. 2009;5:e1000566.
121. Wang R, Taubenberger JK. Methods for molecular surveillance of influenza. Expert Rev Anti Infect Ther. 2010;8:517–27.
122. Ghedin E, Sengamalay NA, Shumway M, Zaborsky J, Feldblyum T, Subbu V, Spiro DJ, Sitz J, Koo H, Bolotov P, et al. Large-scale sequencing of human influenza reveals the dynamic nature of viral genome evolution. Nature. 2005;437: 1162–6.
123. Rambaut A, Pybus OG, Nelson MI, Viboud C, Taubenberger JK, Holmes EC. The genomic and epidemiological dynamics of human influenza A virus. Nature. 2008;453:615–9.
124. Jackson SE, Mason GM, Wills MR. Human cytomegalovirus immunity and immune evasion. Virus Res. 2011;157:151–60.
125. Lisboa LF, Tong Y, Kumar D, Pang XL, Asberg A, Hartmann A, Rollag H, Jardine AG, Pescovitz MD, Humar A. Analysis and clinical correlation

of genetic variation in cytomegalovirus. Transpl Infect Dis. 2012;14:132–40.
126. Manuel O, Asberg A, Pang X, Rollag H, Emery VC, Preiksaitis JK, Kumar D, Pescovitz MD, Bignamini AA, Hartmann A, et al. Impact of genetic polymorphisms in cytomegalovirus glycoprotein B on outcomes in solid-organ transplant recipients with cytomegalovirus disease. Clin Infect Dis. 2009;49:1160–6.
127. Didelot X, Bowden R, Wilson DJ, Peto TE, Crook DW. Transforming clinical microbiology with bacterial genome sequencing. Nat Rev Genet. 2012;13:601–12.
128. Peterson J, Garges S, Giovanni M, McInnes P, Wang L, Schloss JA, Bonazzi V, McEwen JE, Wetterstrand KA, Deal C, et al. The NIH Human Microbiome Project. Genome Res. 2009;19: 2317–23.
129. Qin J, Li R, Raes J, Arumugam M, Burgdorf KS, Manichanh C, Nielsen T, Pons N, Levenez F, Yamada T, et al. A human gut microbial gene catalogue established by metagenomic sequencing. Nature. 2010;464:59–65.
130. A framework for human microbiome research. Nature. 2012;486:215–21.
131. Srinivasan S, Hoffman NG, Morgan MT, Matsen FA, Fiedler TL, Hall RW, Ross FJ, McCoy CO, Bumgarner R, Marrazzo JM, et al. Bacterial communities in women with bacterial vaginosis: high resolution phylogenetic analyses reveal relationships of microbiota to clinical criteria. PLoS One. 2012;7:e37818.
132. VanDevanter DR, LiPuma JJ. Microbial diversity in the cystic fibrosis airways: where is thy sting? Future Microbiol. 2012;7:801–3.
133. Zhao J, Schloss PD, Kalikin LM, Carmody LA, Foster BK, Petrosino JF, Cavalcoli JD, VanDevanter DR, Murray S, Li JZ, et al. Decade-long bacterial community dynamics in cystic fibrosis airways. Proc Natl Acad Sci U S A. 2012;109:5809–14.
134. Shahinas D, Silverman M, Sittler T, Chiu C, Kim P, Allen-Vercoe E, Weese S, Wong A, Low DE, Pillai DR. Toward an understanding of changes in diversity associated with fecal microbiome transplantation based on 16S rRNA gene deep sequencing. mBio. 2012;3.
135. Dethlefsen L, Huse S, Sogin ML, Relman DA. The pervasive effects of an antibiotic on the human gut microbiota, as revealed by deep 16S rRNA sequencing. PLoS Biol. 2008;6:e280.
136. Klindworth A, Pruesse E, Schweer T, Peplies J, Quast C, Horn M, Glockner FO. Evaluation of general 16S ribosomal RNA gene PCR primers for classical and next-generation sequencing-based diversity studies. Nucleic Acids Res. 2012.
137. Kuczynski J, Lauber CL, Walters WA, Parfrey LW, Clemente JC, Gevers D, Knight R. Experimental and analytical tools for studying the human microbiome. Nat Rev Genet. 2012;13:47–58.
138. Jolley KA, Bliss CM, Bennett JS, Bratcher HB, Brehony C, Colles FM, Wimalarathna H, Harrison OB, Sheppard SK, Cody AJ, et al. Ribosomal multilocus sequence typing: universal characterization of bacteria from domain to strain. Microbiology. 2012;158:1005–15.
139. Pruesse E, Quast C, Knittel K, Fuchs BM, Ludwig W, Peplies J, Glockner FO. SILVA: a comprehensive online resource for quality checked and aligned ribosomal RNA sequence data compatible with ARB. Nucleic Acids Res. 2007;35:7188–96.
140. Quast C, Pruesse E, Yilmaz P, Gerken J, Schweer T, Yarza P, Peplies J, Glockner FO. The SILVA ribosomal RNA gene database project: improved data processing and web-based tools. Nucleic Acids Res. 2012.
141. McDonald D, Price MN, Goodrich J, Nawrocki EP, DeSantis TZ, Probst A, Andersen GL, Knight R, Hugenholtz P. An improved Greengenes taxonomy with explicit ranks for ecological and evolutionary analyses of bacteria and archaea. ISME J. 2012;6:610–8.
142. Schloss PD, Gevers D, Westcott SL. Reducing the effects of PCR amplification and sequencing artifacts on 16S rRNA-based studies. PLoS One. 2011;6:e27310.
143. Edgar RC, Haas BJ, Clemente JC, Quince C, Knight R. UCHIME improves sensitivity and speed of chimera detection. Bioinformatics. 2011;27:2194–200.
144. Schloss PD, Westcott SL, Ryabin T, Hall JR, Hartmann M, Hollister EB, Lesniewski RA, Oakley BB, Parks DH, Robinson CJ, et al. Introducing mothur: open-source, platform-independent, community-supported software for describing and comparing microbial communities. Appl Environ Microbiol. 2009;75: 7537–41.
145. Haas BJ, Gevers D, Earl AM, Feldgarden M, Ward DV, Giannoukos G, Ciulla D, Tabbaa D, Highlander SK, Sodergren E, et al. Chimeric 16S rRNA sequence formation and detection in Sanger and 454-pyrosequenced PCR amplicons. Genome Res. 2011;21:494–504.
146. Reeder J, Knight R. Rapidly denoising pyrosequencing amplicon reads by exploiting rank-abundance distributions. Nat Methods. 2010; 7:668–9.
147. Caporaso JG, Kuczynski J, Stombaugh J, Bittinger K, Bushman FD, Costello EK, Fierer N, Pena AG, Goodrich JK, Gordon JI, et al. QIIME allows analysis of high-throughput community sequencing data. Nat Methods. 2010;7:335–6.
148. Kuroda M, Sekizuka T, Shinya F, Takeuchi F, Kanno T, Sata T, Asano S. Detection of a possible bioterrorism agent, Francisella sp., in a clinical

specimen by use of next-generation direct DNA sequencing. J Clin Microbiol. 2012; 50:1810–2.
149. Mellmann A, Harmsen D, Cummings CA, Zentz EB, Leopold SR, Rico A, Prior K, Szczepanowski R, Ji Y, Zhang W, et al. Prospective genomic characterization of the German enterohemorrhagic Escherichia coli O104:H4 outbreak by rapid next generation sequencing technology. PLoS One. 2011;6:e22751.
150. Vogel U, Szczepanowski R, Claus H, Junemann S, Prior K, Harmsen D. Ion torrent personal genome machine sequencing for genomic typing of Neisseria meningitidis for rapid determination of multiple layers of typing information. J Clin Microbiol. 2012;50:1889–94.
151. Hofreuter D, Tsai J, Watson RO, Novik V, Altman B, Benitez M, Clark C, Perbost C, Jarvie T, Du L, et al. Unique features of a highly pathogenic Campylobacter jejuni strain. Infect Immun. 2006;74:4694–707.
152. Smith MG, Gianoulis TA, Pukatzki S, Mekalanos JJ, Ornston LN, Gerstein M, Snyder M. New insights into Acinetobacter baumannii pathogenesis revealed by high-density pyrosequencing and transposon mutagenesis. Genes Dev. 2007;21:601–14.
153. Andries K, Verhasselt P, Guillemont J, Gohlmann HW, Neefs JM, Winkler H, Van Gestel J, Timmerman P, Zhu M, Lee E, et al. A diarylquinoline drug active on the ATP synthase of Mycobacterium tuberculosis. Science. 2005; 307:223–7.
154. Perez Chaparro PJ, McCulloch JA, Cerdeira LT, Al-Dilaimi A, Canto de Sa LL, de Oliveira R, Tauch A, de Carvalho Azevedo VA, Cruz Schneider MP, da Silva AL. Whole genome sequencing of environmental Vibrio cholerae O1 from 10 nanograms of DNA using short reads. J Microbiol Methods. 2011;87:208–12.
155. Chin CS, Sorenson J, Harris JB, Robins WP, Charles RC, Jean-Charles RR, Bullard J, Webster DR, Kasarskis A, Peluso P, et al. The origin of the Haitian cholera outbreak strain. N Engl J Med. 2011;364:33–42.
156. McAdam PR, Holmes A, Templeton KE, Fitzgerald JR. Adaptive evolution of Staphylococcus aureus during chronic endobronchial infection of a cystic fibrosis patient. PLoS One. 2011;6:e24301.
157. Koser CU, Holden MT, Ellington MJ, Cartwright EJ, Brown NM, Ogilvy-Stuart AL, Hsu LY, Chewapreecha C, Croucher NJ, Harris SR, et al. Rapid whole-genome sequencing for investigation of a neonatal MRSA outbreak. N Engl J Med. 2012;366:2267–75.
158. Jenkins SG, Schuetz AN. Current concepts in laboratory testing to guide antimicrobial therapy. Mayo Clin Proc. 2012;87:290–308.
159. Heysell SK, Houpt ER. The future of molecular diagnostics for drug-resistant tuberculosis. Expert Rev Mol Diagn. 2012;12:395–405.
160. Chen PE, Willner KM, Butani A, Dorsey S, George M, Stewart A, Lentz SM, Cook CE, Akmal A, Price LB, et al. Rapid identification of genetic modifications in Bacillus anthracis using whole genome draft sequences generated by 454 pyrosequencing. PLoS One. 2010;5:e12397.
161. Structure, function and diversity of the healthy human microbiome. Nature. 2012;486:207–14.
162. Williamson SJ, Yooseph S. From bacterial to microbial ecosystems (metagenomics). Methods Mol Biol. 2012;804:35–55.
163. Blair JE, Coakley B, Santelli AC, Hentz JG, Wengenack NL. Serologic testing for symptomatic coccidioidomycosis in immunocompetent and immunosuppressed hosts. Mycopathologia. 2006;162:317–24.
164. Hage CA, Ribes JA, Wengenack NL, Baddour LM, Assi M, McKinsey DS, Hammoud K, Alapat D, Babady NE, Parker M, et al. A multicenter evaluation of tests for diagnosis of histoplasmosis. Clin Infect Dis. 2011;53:448–54.
165. Pagano L, Caira M, Offidani M, Martino B, Candoni A, Valentini CG, Specchia G, Nosari A, Tosti ME, Leone G, et al. Adherence to international guidelines for the treatment of invasive aspergillosis in acute myeloid leukaemia: feasibility and utility (SEIFEM-2008B study). J Antimicrob Chemother. 2010;65:2013–8.
166. Balajee SA, Sigler L, Brandt ME. DNA and the classical way: identification of medically important molds in the 21st century. Med Mycol. 2007;45:475–90.
167. Ciardo DE, Schar G, Bottger EC, Altwegg M, Bosshard PP. Internal transcribed spacer sequencing versus biochemical profiling for identification of medically important yeasts. J Clin Microbiol. 2006;44:77–84.
168. Ciardo DE, Lucke K, Imhof A, Bloemberg GV, Bottger EC. Systematic internal transcribed spacer sequence analysis for identification of clinical mold isolates in diagnostic mycology: a 5-year study. J Clin Microbiol. 2010;48:2809–13.
169. Pounder JI, Simmon KE, Barton CA, Hohmann SL, Brandt ME, Petti CA. Discovering potential pathogens among fungi identified as nonsporulating molds. J Clin Microbiol. 2007;45:568–71.
170. Schoch CL, Seifert KA, Huhndorf S, Robert V, Spouge JL, Levesque CA, Chen W. Nuclear ribosomal internal transcribed spacer (ITS) region as a universal DNA barcode marker for Fungi. Proc Natl Acad Sci U S A. 2012;109:6241–6.
171. Munoz-Cadavid C, Rudd S, Zaki SR, Patel M, Moser SA, Brandt ME, Gomez BL. Improving molecular detection of fungal DNA in formalin-fixed paraffin-embedded tissues: comparison of five tissue DNA extraction methods using

panfungal PCR. J Clin Microbiol. 2010;48:2147–53.

172. Lau A, Chen S, Sorrell T, Carter D, Malik R, Martin P, Halliday C. Development and clinical application of a panfungal PCR assay to detect and identify fungal DNA in tissue specimens. J Clin Microbiol. 2007;45:380–5.

173. Romanelli AM, Sutton DA, Thompson EH, Rinaldi MG, Wickes BL. Sequence-based identification of filamentous basidiomycetous fungi from clinical specimens: a cautionary note. J Clin Microbiol. 2010;48:741–52.

174. Bidartondo MI. Preserving accuracy in GenBank. Science. 2008;319:1616.

175. Engelthaler DM, Chiller T, Schupp JA, Colvin J, Beckstrom-Sternberg SM, Driebe EM, Moses T, Tembe W, Sinari S, Beckstrom-Sternberg JS, et al. Next-generation sequencing of Coccidioides immitis isolated during cluster investigation. Emerg Infect Dis. 2011;17:227–32.

INDEX

A

Absorption, distribution, metabolism, and excretion (ADME) genes, 554
Accreditation Council for Graduate Medical Education (ACGME), 105, 108
Accuracy
  academic laboratories, testing and interpretation, 150
  Association for Molecular Pathology v. Myriad Genetics, Inc., 150
  CLIA, 143
  clinical side, 150
  DTC testing, 141
  ethics, 149
  Federal Food, Drug and Cosmetics Act, 143
  "Free the Data", 150
  "Global Alliance", 150–151
  HTT gene, 150
  human genome, 149
  LDTs, 143
  regulatory schemes, 143
  sequencing techniques, 149
  traditional genetic testing, 150
  WES, 150
  WGS, 149
  whole-genome sequences, 150
aCGH. *See* Array comparative genomic hybridization (aCGH)
ACGME. *See* Accreditation Council for Graduate Medical Education (ACGME)
ACMG. *See* American College of Medical Genetics and Genomics (ACMG)
Acute myeloid leukemias (AMLs)
  aCGH, 309
  "escape clone monitoring", 315
  FlT3 mutation, 300
  hematologic malignancies, 308
  pathogenesis, 315
  patients, 300
  T-cell gene and IGH rearrangements, 299–300
Acute promyelocytic leukemia (APML), 102
Adenylation master mix components, 216
Adipocytic tumors, 504–507
ADME genes. *See* Absorption, distribution, metabolism, and excretion (ADME) genes
Adult/young adult onset applications
  aortopathy, 545–546
  cardiomyopathy, 546
Affinity-based microchips, 72, 73
AFP. *See* Alpha-fetoprotein (AFP)
Aldehyde dehydrogenase (ALDH), 74, 75
Alpha-fetoprotein (AFP), 564
ALT. *See* Atypical lipomatous tumors (ALT)
Alveolar rhabdomyosarcoma, 513–514
Alveolar soft part sarcoma (ASPS), 517
AMA. *See* American Medical Association (AMA)
America Invents Act of 2011, 130
American College of Medical Genetics and Genomics (ACMG), 13, 115, 151, 153, 264
  genome/exome sequencing, 253NGS, 241
  WGS, 253
American Medical Association (AMA), 124
American Society for Clinical Pathology (ASCP), 107, 113
Amgen v. Chugai Pharmaceutical Co., 131
AMLs. *See* Acute myeloid leukemias (AMLs)
AMP. *See* Association for Molecular Pathology (AMP)
Amplicon sequencing
  genome, 589
  microorganisms, 582, 583
  NGS studies, 581
  rRNA, 592
AMP v. Myriad, 137
Aneuploidy, 575
Angiosarcoma, 515–516
Annotation, 12, 184, 189, 226, 264, 279
Aortopathy
  fibrillin gene *(FBN1)* mutation, 545
  Marfan syndrome, 545
  multiplex ligation dependent probe amplification (MLPA), 545–546
  TGFβ signaling genes mutation, 545
  thoracic aortic aneurism, 546
APML. *See* Acute promyelocytic leukemia (APML)
Array comparative genomic hybridization (aCGH), 35, 146, 147, 164, 309–311, 322, 325, 337, 443, 575
ASCP. *See* American Society for Clinical Pathology (ASCP)
Ashkenazi Jewish (AJ) population, 539
Association for Molecular Pathology (AMP), 107, 113, 125, 126, 216, 277, 425
Association for Molecular Pathology v. Myriad Genetics
  BRCA1 and *BRCA2* genes, 135, 137
  breast and ovarian cancer, 136–137
  CAFC decision, 137
  dissenting judge, 137
  the District Court, 135
  DNA sequence, 137
  molecular biology, 135–136
  Myriad's patent claims, 137–138

G.J. Netto and I. Schrijver (eds.), *Genomic Applications in Pathology*,
DOI 10.1007/978-1-4939-0727-4, © Springer Science+Business Media New York 2015

Association for Molecular Pathology v. Myriad Genetics (cont.)
Patent Act of 1952, 135
pertinent sections, 136
USPTO, 137
Atypical lipomatous tumors (ALT), 504–506
Atypical teratoid/rhabdoid tumor (AT/RT), 328
Australian Ovarian Cancer Study (AOCS), 494

B
Bacteriology, NGS
antimicrobial testing, 592
Clostridium difficile, 592–593
metagenomic studies, 594
MRSA, 594 *Mycobacterium tuberculosis*, 594OTUs, 593
phenotype and antibiotic testing, 592
rMLST, 593
rRNA, 592
virulence factor, 592
WGS, 593
BAM file format. *See* Binary alignment map (BAM) file format
Bayh-Dole Act, 130
BEAMing, ccfDNA
beads, 90
flow cytometry, 90, 91
genetic molecules, 90
human genomic DNA, 90
microemulsions, 90
oligonucleotides, 90
PCRs, 90
plasma tumor DNA (ptDNA), 90, 91
Beth Israel Deaconess Medical Center (BIDMC), 106
BIDMC. *See* Beth Israel Deaconess Medical Center (BIDMC)
Bilski v. Kappos, 132–133
Binary alignment map (BAM) file format, 180, 186, 189, 245
Bioinformatics
data storage, 116
definition, 177
exome and genome Sequencing, 190–193
financial investment, 116
genomics revolution., 193
illumina technology (*see* Illumina technology)
ion torrent, 184–190
laboratory, 113
NGS, 178
Biological samples, miRNA
expression and localization, 50
FFPE samples, 49
ISH, 49
LCM, 49
plasma and serum, 50
tumor cells, 50
Biomarker development, 92
analytic validity, 172
clinical validity and utility, 170, 172
delivery process, 171
ELSI, 172
OPHG, 172
REMARK, 171
STARD, 171
TMUGS, 171
USPSTF, 172, 173
Bladder cancer
Cancer-Testis antigens MAGE-A3, 447
copy number changes, 449
DNA based lesions, 449
EGFR and ERBB2, high level copy gains, 449, 450
FGFR3 and TP53, 446–447
gemcitabine and carboplatin (GC) neoadjuvant chemotherapy, 448
gene expression panels, 448
12-gene qRT-PCR expression panel, 448
KI-67, 447
LAGE-1, 447
molecular subtyping, 447–448
multigene panels, early detection/recurrence monitoring, 451
NY-ESO-1, 447
whole-genome and exome sequencing, 449–451
BRAF. *See* B-type RAS associated factor *(BRAF)*
Brain tumors
BRAF, 326–327
CDKN2A, 328–329
EGFR and receptor tyrosine kinases, 325–326
gliomas (*see* Gliomas)
IDH1 mutations, 324–325
INI1, 328
MGMT gene, 327–328
oligodendroglioma, 322–324
1p/19q, 322–324
PTEN, tumor suppressor, 328
whole-genome evaluation (*see* Medulloblastomas)
whole-genome sequencing, 321
Breast Cancer Index^{SM} (BCI), 369
Breast carcinoma, 36
adjuvant systemic therapy, 360
first-generation prognostic gene signatures (*see* First-generation prognostic gene signatures)
gene expression (*see* Gene expression)
gene expression profiling, 468–469
HER2 status, 359
"intrinsic" genes, 360
lymph node involvement, 359
management, 359
massively parallel sequencing and intra-tumor genetic heterogeneity, 371–373
microarray-based gene expression profiling, 360
molecular classification (*see* Molecular classification, breast cancer)
second-generation prognostic gene signatures (*see* Second-generation prognostic gene signatures)
Broadband genetic and genomic testing, 146–148
aCGH, 146
ethical and legal issues, 148
"partially broadband" technology, 147

SNP chips, 147
tandem mass spectrometry, 146–147
WGS, 147–148
B-type RAS associated factor *(BRAF)*, 87, 426
brain tumors, 326–327
mutation, activation and inactivation, 386
pulmonary malignancies, 386
Burkitt lymphoma, 306–308

C

CAFC. *See* The Court of Appeals for the Federal Circuit (CAFC)
Calculator for Ovarian Subtype Probability (COSP), 491
Cancer
APC mutations, 87
BC management, 36
bioinformatics and genomic datasets in public domain, 163–164
biomarker discovery, 162–163
BRAF mutation, 87
clinically applicable gene-based assays, 164–168
CLL, 46
CTCs, 36, 88
DNA integrity, 89
drug resistance, 47
EGFR, 88
genomics, 89, 162–168, 199, 232, 343
integration approaches, silico datasets, 164
liquid biopsy, 87
loss of heterozygosity (LOH), 89
lung cancer, 37
metastatic disease, 37, 87
miR-16-1, 46
miR-21, 46–47
MiR-129-2, 47
miR-155, 47
miR-15a, 46
MM, 47
NGS, 87
passenger mutations, 86
PIK3CA mutations, 88
proto-oncogenes, 87
pseudogene expression, 36
ptDNA, 86
radioimmunoassays, 86
recurrent gene fusions, 37
Sunitinib, 37
TCL-1, 46
tumor suppressor genes, 87
Cancer stem cell (CSC)
adjuvant chemotherapy, 76
bone marrow, 76
breast cancer studies, 76
DTC, 76
flow cytometry, 76
lung cancer and brain tumors, 76
"Cancer stemness" colorectal cancer, 416–418
CAP. *See* College of American Pathologists (CAP)
Cardiomyopathy, 112, 225, 277, 546
Castration resistant prostate cancers (CRPC), 167, 440, 444–445
Catalogue of Somatic Mutations in Cancer (COSMIC), 268, 281, 285
Causal and candidate gene
bioinformatics strategies, 190
heuristic and probabilistic approaches, 190–191
identification, 190
CBO. *See* Clinical Bioinformatic Ontology (CBO)
ccfDNA. *See* Circulating cell-free DNA (ccfDNA)
CCS. *See* Clear cell sarcoma (CCS)
CDKN2A
brain tumors, 328–329
P16, HNSCC, 345
CDR. *See* Complementary determining regions (CDR)
CDSS. *See* Clinical decision support systems (CDSS)
Cell free fetal DNA (cffDNA), in maternal plasma
fetal genotypes testing, 540
fetal Y chromosome sequences detection, 540
microdeletion syndromes, 541
NGS analysis, 541
Centers for Medicare and Medicaid Services (CMS), 124
Centromere enumeration probes (CEP), 306
CEP. *See* Centromere enumeration probes (CEP)
Cerebral spinal fluid (CSF), 267, 512–513
CGH. *See* Genomic hybridization (CGH)
Children, 146, 335
autosomal recessive genetic disorder, 142
BRCA1 mutations, 146
clinical genetics, 146
family member's genome, 146
gliomas, LGGs and DIPG, 335
pediatricians/geneticists, 142
WGS, 153
Chromogenic in situ hybridization (CISH), 166–167
Chromosomal inversions and translocations
ALK rearrangements, 388
EML4-ALK fusions, 388
evaluation, 402–403
HER family proteins, 388–389
KIF5B gene, 390
lung adenocarcinoma cell line, 390
ROS1 gene, 390
Chronic lymphocytic leukemias (CLL), 46
Chronic myeloid leukemia (CML), 88, 92, 162, 167, 303, 306
Circulating cell-free DNA (ccfDNA)
aneuploidy, 566
BEAMing, 90–91
cancer diagnostics, 86–89
cellular debris, 85
droplet digital PCR, 91–92
human fluids, 85
maternal-fetal medicine, 86
maternal plasma, 565–566
PARE, 92
shotgun sequencing, 566

Circulating tumor cells (CTC), 36, 88, 496
  ALDH, 74, 75
  cancer, 71
  CSC, 75–76
  EMT, 74, 77
  enrichment-based microchips, 71–73
  genomic analysis, 72
  genomic profile, 77–79
  hematopoietic cells, 74
  IHC/IF detection, 74
  intratumoral heterogeneity, 75
  mutation analyses, 79–80
  peripheral blood, 71
  sentinel nodes, 71
  tumor cell heterogeneity, 75
CISH. *See* Chromogenic in situ hybridization (CISH)
Classical GBM, 334
Clear cell sarcoma (CCS), 517–518
CLIA. *See* Clinical Laboratory Improvement Act (CLIA); Clinical Laboratory Improvement Amendments (CLIA)
Clinical Bioinformatic Ontology (CBO), 274
Clinical decision support systems (CDSS), 262
  Clinical Pharmacogenomics Implementation Consortium, 274
  genomic medicine information, 272, 273
  Health eDecision, 274
  informatics requirements, 274
Clinical genetics, 146, 153, 234, 538
Clinical genomics, 8, 14, 126, 259, 261, 445
  ACMG, 277
  cancer test, 280–281
  communication, 281–290
  high-throughput sequencing, 277
  pathologist, 290–291
  single gene/gene panel, 278–279
  whole-exome and whole-genome sequencing, 279
Clinical Laboratory Improvement Act (CLIA)
  clinical testing, 122
  entrepreneurial laboratories, 121
  LDTs, 120
Clinical Laboratory Improvement Amendments (CLIA), 104, 143, 165, 166, 264, 271
  CAP, 112
  challenges, 259
  clinical genomics, 259
  general laboratory, 112
  LIS *vs.* LIMS, 261–263
  molecular pathology, 112
  QA and QI activities, 271
  technical and IT staff, 266
Clinical sequencing. *See* Genome sequencing (GS) assay
Clinical trials design, 160, 168, 338
  genetic biomarkers, 168–169
  HPV-positive women, 161
  preclinical research, 170
CLL. *See* Chronic lymphocytic leukemias (CLL)
Clonality, 267, 438
  B-cell, 298–299
  diagnostic, 299–300
  T-cell, 299
CMOS. *See* Complementary metal-oxide-semiconductor (CMOS)
CMS. *See* Centers for Medicare and Medicaid Services (CMS)
CMV. *See* Cytomegalovirus (CMV)
CN-LOH. *See* Copy neutral loss of heterozygosity (CN-LOH)
CNVs. *See* Copy number variants (CNVs)
College of American Pathologists (CAP), 112, 115, 117, 123, 277
  ACMG, 247
  ASCO, 359, 360
  CLIA-certified, 112
  FDA, 123
  NGS, 9
  proficiency test, 256, 564
  USCAP, 107
Colorectal cancer
  "cancer stemness" colorectal cancer, 416–418
  core cellular processes and hallmarks, 421
  expression data based classification, 422
  genome-wide analysis, 422
  genotype-based classification, 420–422
  mismatch repair defective, evaluation, 423–424
  molecular classification, 418
  MUTYH/MYH colon cancer, 420
  oncogene-induced senescence (OIS), 418–420
  "passenger mutations", 420
  predictive markers, targeted therapeutics, 424–427
  translocations, 421
  types, 420
Companion diagnostics
  FDA, 122
  genomic sequencing assays, 119
  IVDs, 122
Complementary determining regions (CDR), 298
Complementary metal-oxide-semiconductor (CMOS), 563
Consent, 265
  laboratory, 243, 290
  LIS, 268
  medical appropriateness, 142–143, 148–149
  requisition and clinical information, 113–114
  WGS, 149
Copy neutral loss of heterozygosity (CN-LOH), 309, 311, 312
Copy number aberrations (CNAs), 493
Copy number variants (CNVs), 210, 229, 244, 264, 312
COSMIC. *See* Catalogue of Somatic Mutations in Cancer (COSMIC)
COSP. *See* Calculator for Ovarian Subtype Probability (COSP)
The Court of Appeals for the Federal Circuit (CAFC), 131, 133–135
CPT. *See* Current procedural terminology (CPT)
Crouzon syndrome, 541

CRPC. *See* Castration resistant prostate cancers (CRPC)
CRT. *See* Cyclic reversible termination (CRT)
CSC. *See* Cancer stem cell (CSC)
CSF. *See* Cerebral spinal fluid (CSF)
CTC. *See* Circulating tumor cells (CTC)
Current procedural terminology (CPT), 125, 269
  AMA, 124
  molecular pathology, 124
  novel medical services, 126
Cyclic reversible termination (CRT), 203
Cytomegalovirus (CMV)
  bioinformatics tool, 588
  GCV and VCV, 587
  HHV5, 586
  phenotypic assays, 587
  prophylaxis, 587

D
DAAs. *See* Direct-acting antivirals (DAAs)
Database, 11, 14, 153, 247, 274, 312, 586
Data warehousing, 260, 266, 270–272
DCDF. *See* Dual color dual fusion (DCDF)
DDLS. *See* Dedifferentiated liposarcomas (DDLS)
Dedifferentiated liposarcomas (DDLS), 504–506
Desmoid-type (deep) fibromatosis, 507–508
Desmoplastic small round cell tumor (DSRCT), 519
Detoxification enzymes, 349
Diamond v. Chakrabarty, 131
Diffuse intrinsic pontine gliomas (DIPG), 335
Digital PCR, 92
DIPG. *See* Diffuse intrinsic pontine gliomas (DIPG)
Direct-acting antivirals (DAAs), 586
Direct RNA sequencing (DRS), 35
Discrimination, genetic, 145, 153
Disease-targeted panels, 209, 210
Disseminated tumor cells (DTC), 76
DNA-based single gene assays
  B-cell immunoglobulin gene rearrangement, 297–299
  clonality testing, 299–300
  FlT3 mutation, 300–303
  JAK2 mutation, 300
  PCR, 297
  T-cell receptor gene rearrangement, 299
DNA isolation
  control reactions, 402
  FFPE tissue, 401
  HRM PCR setting, 402
  NGS to molecular lung cancer pathology, 393–394
  precautions, 401–402
DNA sequencing, 3, 4, 9, 197
DNETs. *See* Dysembryoplastic neuroepithelial tumors (DNETs)
Down syndrome
  AFP, 564
  Ariosa diagnostics, 569–570
  autosomal trisomies, 566
  ccf DNA, 565–566
  cell lane, 567
  clinical validation studies, 570, 571
  euploid pregnancies, 566, 568
  exome/genome sequencing, 575
  external proficiency testing, 577–578
  high risk population, 576
  IVF pregnancies, 574
  maternal plasma, 570–574
  Natera, 568, 570
  NIPD, 565
  patient educational material, 577
  prenatal screening, 564
  quadruple test, 564
  screening, 569–570
  sensitivity, 564
  Sequenom, 567, 569
  serum markers, 565
  serum/ultrasound, 570
  shotgun sequencing, 568
  twin pregnancies, 574–575
  ultrasound measurements, 565
  Verinata, 567, 569
Dravet syndrome, 544
DRM. *See* Drug resistance mutations (DRM)
DRS. *See* Direct RNA sequencing (DRS)
Drug efficacy determination
  individuals identification, 558
  responsiveness, 557–558
  right dose, 558
Drug resistance mutations (DRM)
  CMV, 586, 587
  HBV and HCV,585, 586
  HIV, 583–585
DSRCT. *See* Desmoplastic small round cell tumor (DSRCT)
DTC. *See* Disseminated tumor cells (DTC)
Dual color dual fusion (DCDF), 308
Dysembryoplastic neuroepithelial tumors (DNETs), 322, 325

E
Early Detection Research Network (EDRN), 171
EGFR. *See* Epidermal growth factor receptor (EGFR)
EHR. *See* Electronic health records (EHR)
Electronic health records (EHR), 260, 262, 265, 266, 268, 272, 274
Electronic Medical Records and Genomics (eMERGE), 274, 557
ELSI. *See* Ethical, Legal and Social Implications (ELSI)
Embryonal rhabdomyosarcoma, 513
eMERGE. *See* Electronic Medical Records and Genomics (eMERGE)
EMT. *See* Epithelial-mesenchymal transition (EMT)
Endometrial carcinoma
  epithelial membrane antigen (EMA), 466
  FIGO staging system, 466
  glycoprotein-associated markers, 466

Endometrial carcinoma (*cont.*)
histological classification, 467
HNF-1b expression, 467
immunophenotype, endometrioid, serous and clear cell carcinomas, 468
Lynch syndrome, 471–472
malignant mixed Mullerian tumors, 466
molecular genetic classification, 467–470
mTORC1 inhibitors, 471
pan-cytokeratins (CKs), 466
p53 and p16 expression, 466–467
PI3K pathway, 470–471
Endometrial stromal tumors
ancillary immunohistochemical studies, 474–475
chimeric proteins, 475
endometrial stromal nodule (ESN), 473–476
immunohistochemical surrogate markers, 476
International Cancer Genome Consortium (ICGC), 476
low-grade and high-grade endometrial stromal sarcoma (ESS), 473–476
recurrent chromosomal translocations, 474, 475
undifferentiated uterine sarcoma (USS), 473–476
YWHAE-FAM22 fusion gene, 476
EndoPredict test, 369
Enrichment-based microchips
affinity-based microchips, 72, 73
hydrodynamic flow, 73
immunomagnetic enrichment, 72
non-epithelial cancers, 72
size-based microchips, 72–73
tumor cells, 72
Ependymoma, 337
Epidermal growth factor receptor (EGFR), 79, 88, 101, 162, 171
biomarker evaluation, 384
composition, 383
exon, 385
gene copy number, 384
HNSCC, malignant transformation, 345
inhibitors, 351
inhibitory monoclonal antibodies, 384
MET amplification, 385
optical imaging, 348
receptor activation, 383–384
therapeutic applications, 351
tyrosine kinase inhibitors (TKIs), 384
Epilepsy, 543–544
Epithelial-mesenchymal transition (EMT), 74
Epithelial ovarian malignancies (EOM)
ARID1A, 490
expression analysis, 491–492
HGSOC, mutational spectrum, 492
high-grade serous ovarian cancer (HGSOC), 489
homologous recombination and high-grade carcinoma (*see* Homologous recombination and high-grade carcinoma)
morphology and immunohistochemical markers, 490–491
normal ovarian epithelium (NOE), 489
ovarian clear-cell adenocarcinoma (OCCA), 490
pre-invasive lesions, 489
resistance (*see* Intratumoral heterogeneity, in HGSOC)
structural DNA aberrations, 493
TP53 mutations, 490, 492–493
trans-differentiation, 489
tubal intraepithelial carcinomas (TICs), 490
Epithelioid sarcoma (ES), 516–517
eQTL. *See* Expression quantitative trait loci (eQTL)
ERT. *See* Extrarenal rhabdoid tumor (ERT)
ESHG. *See* European Society for Human Genetics (ESHG)
ESMC. *See* Extraskeletal myxoid chondrosarcoma (ESMC)
Ethical, Legal and Social Implications (ELSI), 172
Ethics, 13, 106, 125, 172, 235, 290
accuracy (*see* Accuracy)
adults with symptoms, 142
communicating results, 143
contemporary genetic testing, 141
Down syndrome, 142
DTC testing, 141, 143–144
genetic/chromosomal variants, 141
medical appropriateness and informed consent, 142–143
"targeted" genetic testing, 141
ETS gene fusions, 436–438
European Society for Human Genetics (ESHG), 151
Exome and genome
clinical WES/WGS tests, 546–548
congenital anomalies, 547
dopa (3,4-dihydroxyphenylalanine)-responsive dystonia (DRD), 547
inflammatory bowel disease, 547
nucleotide binding oligomerization domain (NOD) signaling pathway, 547
X-Linked Inhibitor of Apoptosis gene (XIAP) mutation, 547
Exome sequencing assay
Alzheimer's disease, 231
causative mutations, 234
clinical care, 232–233
gene panels, 235
GWAS, 231
heterogeneous disorders, 225–226
implemention, 233–234
and Mendelian disorders, 230–231
NGS, 225
novel disease, 231
Parkinson's disease, 231
traditional molecular diagnostic approaches, 226
WES (*see* Whole-exome sequencing (WES))
X-chromosome-linked intellectual disability, 231
Expression profiling, 491–492
biological processes, 53
ISH, 54, 55
LCM, 53
microarrays, 31, 54–58
NGS, 58–61

northern blotting, 53
qRT-PCR, 54
RNA-Seq, 35
signal intensity, 31
Expression quantitative trait loci (eQTL), 36
Extrarenal rhabdoid tumor (ERT), 519
Extraskeletal myxoid chondrosarcoma (ESMC), 518–519

F
FDA. *See* Food and Drug Administration (FDA)
Federation of Gynecology and Obstetrics (FIGO) system, 466
FFPE. *See* Formalin-fixed paraffin-embedded (FFPE)
FGFR1 and *DDR2*, in squamous cell carcinoma, 390, 403
Fibroblastic/myofibroblastic tumors, 507–512
Fibrohistiocytic tumors, 512–513
Fine needle aspirate (FNA), 123
First-generation prognostic gene signatures
Breast Cancer Index^{SM} (BCI), 369
characteristics, 365
EndoPredict test, 369
ER-positive disease, 364
MammaPrint®, 366–367
MapQuant Dx™, 367
Oncotype Dx®, 368
PAM50, 368–369
schematic representation, 366
tumors size and lymph node status, 364
Veridex 76-gene signatures, 367–368
FISH. *See* Fluorescence *in situ* hybridization (FISH)
Fluorescence in situ hybridization (FISH), 306–309, 397–398
Fluorescence resonance energy transfer (FRET), 20
FMS-like tyrosine kinase 3 (FLT3) mutation, 300–303, 305
FNA. *See* Fine needle aspirate (FNA)
Food and Drug Administration (FDA), 35
AMP, 124
biochemical assay, 123
CAP, 123
colorectal cancer, 123
enforcement discretion, 123
FNA, 123
metastatic malignant melanoma, 123
Formalin-fixed paraffin-embedded (FFPE), 49, 114, 166, 306
Fragment length analysis, 397
FRET. *See* Fluorescence resonance energy transfer (FRET)

G
Ganciclovir (GCV), 587
Gastrointestinal stromal tumors (GIST), 520–523
GATK. *See* Genome Analysis Tool Kit (GATK)
GBM. *See* Glioblastoma (GBM)
Gene discovery, 129, 131, 190–191, 229, 547
Gene expression
analysis, 360
chemotherapy resistance, 371
intra-tumor genetic heterogeneity, 371
prognostic signatures, 364
SET index, 371
Gene expression profile (GEP), 165, 468–469
Gene panels. *See also* Next-generation sequencing (NGS)
clinical genomics, 278–279
disease-targeted multi-gene panels, 209, 222
exome sequencing assay, 235
prostate cancer, 442–443
WES, 235
Gene patents, 14, 15, 129–131
Amgen v. Chugai Pharmaceutical Co., 131
Bayh-Dole Act, 130
CAFC, 131
chemical law, 131
Diamond v. Chakrabarty, 131
economic value, 130
Federal Circuit decisions, 131
legitimization, 130
proponents, 129
US biotech industry, 131
Genetic counseling, 152
Genetic Information Nondiscrimination Act (GINA), 103, 145
Genetic mosaicism and intratumoral heterogeneity, 335, 336
Genetics, 35, 39
Genetic testing, 13, 91, 103, 131, 136, 148, 153, 535, 538, 540, 543
ACCE Model Project, 172
adult-onset disorders, 13
Americans with private health coverage, 145
authorized disclosures, 145
BTK, 145
cDNA, 138
clinical genetics, 146
13 "CODIS" markers, 145
counseling, 253
ethical and legal issues, 141–144
families share genetic variants, 146
genetic discrimination, 145
GINA, 145
HIPAA, 145
implication, court decisions, 138
IRD, 200
medical process, 145
nontrivial percentage of men, 146
pathogenic/clinically insignificant, 146
and reports, 265, 268
specimen identification and tracking, 267–268
survey-based study, 105
"targeted" version, 144–145
traditional, 145
Genetic Test Report (GTR), 265
Gene variants, 189, 192, 193

Genitourinary cancers
bladder cancer (*see* Bladder cancer)
current and future genomic applications, 456
kidney cancer, 451–454
prostate cancer (*see* Prostate cancer)
testis and penile cancers, 454–455
Genome Analysis Tool Kit (GATK), 180–181, 184, 222
Genome sequencing (GS) assay
ACMG, 115
AMP, 113
analytical, 244–246
ASCP, 113
bioinformatics analysis, 116
CLIA, 112
clinical application, 235–236
clinical laboratory testing, 111
communication and support, 253–254
diagnostic tests, 114
FDA, 116
FFPE, 114
germline analysis, 114
incidental findings, 115
infrastructure, 254–255
IRB, 114
LDTs, 116
Mendelian disorders, 112
NGS, 111
post-analytical, 246–253
pre-analytical, 243–244
process and workflow, 241, 242
proficiency testing, 116–117
protein-coding regions, 112
quality assessment and control, 255–256
SNVs, 115
Genome wide association studies (GWAS), 36, 104, 163, 231, 442, 554–555, 557, 558
Genomic hybridization (CGH), 78
Genomic pathology
ACGME, 105
APC gene variant, 103
APML, 102
ASCP, 107
BIDMC, 106
chromosomal microarrays, 102
donor-recipient, 102
EGFR, 101
gene assay, 102
GINA, 103
GWAS, 104
horizontal strand, 103
molecular biology, 104
molecular pathology, 101
neoplastic disease, 104
oncology, 101
polymerase chain reaction, 106
salivary duct carcinoma, 102
SNPs, 101, 104
TRIG, 107
whole-genome and transcriptome sequencing, 102
XIAP, 102
Genomic profile, CTC
anticancer therapies, 79
CGH, 78
DTC, 77
HER-2, 78
IHC/IF labelling, 78
immunomagnetic enrichment, 78
MCF7 cells, 78
metastasis, 77
parylene filter, 78, 79
whole-genome, 77
The Genomic Research and Accessibility Act, 129
Genomic testing
cases, 260
CLIA, 120
CLIA testing, 259
clinical infrastructure, 260
CPT, 119
data in health care, 259–260
FDA, 122–124
gap analysis, 260–261
IT requirements, 261, 266
LDTs, 121–122
LIS *vs.* LIMS, 261–264, 266
maintenance and function, 120
novel technologies, 119
order entry systems, 266–267
regulatory framework, 119
reimbursement, 124–126
requirements, 260
sequencing assays, 121
standard file formats, 264–265
timeline and plan, 261
Genotypic interpretation systems (GIS), 584
Genotypic viral drug resistance, 584, 586
GEP. *See* Gene expression profile (GEP)
Germ cell malignancies, 454–455
Giant cell fibroblastoma/dermatofibrosarcoma protuberans, 508–509
GINA. *See* Genetic Information Nondiscrimination Act (GINA)
GIS. *See* Genotypic interpretation systems (GIS)
GIST. *See* Gastrointestinal stromal tumors (GIST)
Glioblastoma (GBM), 333
Gliomas, 321, 324
classical GBM, 334
CNS neoplasms, 321
ependymoma, 337
genetic mosaicism and intratumoral heterogeneity, 335, 336
glioblastoma (GBM), 333
LGGs and DIPG, in children, 335
meningioma, 335–337
mesenchymal GBM, 334
neural GBM, 334
proneural GBM, 334
secondary GBM (sGBM), 333
Graham v. John Deere Co., 134
GS assay. *See* Genome sequencing (GS) assay

GTR. *See* Genetic Test Report (GTR)
GWAS. *See* Genome wide association studies (GWAS)
Gynecologic malignancies
  endometrial carcinoma, 465–472
  uterine mesenchymal tumors, 472–476

H
HAART. *See* Highly active antiretroviral therapy (HAART)
Haemangioendothelioma, 514–515
HBV. *See* Hepatitis B virus (HBV)
Head and neck squamous cell carcinoma (HNSCC)
  adenovirus, 350–351
  CDKN2A/P16, 345
  EGFR (*see* Epidermal growth factor receptor (EGFR))
  HPV (*see* Human papillomavirus (HPV), HNSCC)
  NOTCH, 343–345
  PIK3CA gene, 346
  prognostic applications, 348–349
  RAS, 345–346
  therapeutic applications, 349–352
  tobacco and alcohol, 342
  TP53, tumor-suppressor gene, 343
Health Insurance Portability and Accountability Act (HIPAA), 145, 152, 316
Hearing loss, 542–543
Helicos sequencing chemistry, 20–21
Hematologic oncology
  aCGH, 309–311
  B-and T-cells clonality, 298–300
  BCR-ABl1 mRNA transcript, 303–305
  chromosomal abnormalities, 306
  chromosome assays, 305
  CN-LOH, 309
  conventional cytogenetics, 306
  dilemma, 300
  FISH, 306–309
  IGH and IGκ gene, 298
  NGS (*see* Next-generation sequencing (NGS))
  3' reverse primer binding site, 298
  RQ-PCR assays, 300
  single gene assays (*see* Single gene analysis)
  SNP arrays, 311–314
  whole genome assays, 314
α-Hemolysin, 25
Hepatitis B virus (HBV)
  DAAs, 586
  interferon-α and ribavirin, 586
  NAs, 586
  454 pyrosequencing, 587
  RNA polymerase, 585
  YMDD, 586–586
Hereditary genome applications, prostate cancer, 441–442
Hereditary leiomyomatosis and renal cell carcinoma (HLRCC) syndrome, 452
Hereditary papillary renal cell carcinoma syndrome (HPRCC), 452
HGMD. *See* Human Gene Mutation Database (HGMD)
HGVS. *See* Human Genome Variation Society (HGVS)
HHV5. *See* Human herpes virus 5 (HHV5)
Highly active antiretroviral therapy (HAART), 269, 583
High-resolution melting (HRM), 395–397
High throughput sequencing, 20, 23, 58, 279, 285, 290, 395–397, 581, 589
  cancer, 281
  clinical implementation, 281
  description, 277
  disruptive technology, 290
  germline mutations, 281, 282
  pathologists, 290–291
HIPAA. *See* Health Insurance Portability and Accountability Act (HIPAA)
Hippel–Lindau syndrome, 452
Hodgkin lymphoma, 299
Homologous recombination and high-grade carcinoma
  BRCA1/2 mutation, 493–494
  functional tests, 494
HRM. *See* High-resolution melting (HRM)
Human disease, miRNA
  adipogenesis, 48
  autoimmune diseases, 48
  B-cell lymphomas, 46
  cancer, 46–48
  cardiovascular diseases, 48
  embryogenesis, 46
  p53 pathways, 46
Human Gene Mutation Database (HGMD), 11
Human Genome Variation Society (HGVS), 281
Human herpes virus 5 (HHV5), 586
Human immunodeficiency virus (HIV)
  amplicon sequencing, 584
  antiretroviral drugs, 583
  chimeric sequences, 585
  epidemiologic studies, 583
  genotypic resistance, 584
  GIS, 584
  HAART, 583
  inhibitor resistance, 584
  low-abundance resistance, 584
  plasmid subclones, 585
  pyrosequencing, 585
  RT-PCR, 584
  substantial strides, 584
  TDR, 584
Human papillomavirus (HPV), HNSCC
  anal cancer, 342–343
  cervical cancer in women, 342–343, 346
  de-intensification, 349–350
  diagnostic applications, 347–348
  features, 346
  penile cancer in men, 342–343
  p16 immunohistochemistry (IHC), surrogate marker, 347

Human papillomavirus (HPV), HNSCC (*cont.*)
  prognostic applications, 348–349
  in regional/distant metastatic foci, 347
  risk factor, 342–343
  in situ hybridization (ISH), 347
  vaccines and immunotherapy, 350
Hybridization buffer components, 219

I
Illumina/solexa Genome analyzer, 202
Illumina technology
  algorithms, 180
  ANNOVAR, GATK and snpEff, 184
  chastity filter ratio, 178, 179
  cluster failure, 179
  FASTQ file, 180
  GATK, 180–181
  germline variant detection, 183–184
  local realignment, 181, 182
  mapping and aligning, 180
  quality (Q) score, 179–180
  and SAMtools, 180
  sequence reads and alignment, 181–183
  sequencing clusters, 178, 179
  software programs, 184
  VCF, 181
Image-guided core biopsy (IGCB), 496
Immune-related signatures, 370–371
Immunohistochemical/immunofluorescent (IHC/IF), 74
Index-specific hybridization blockers (ISHB), 214
Infantile fibrosarcoma, 511
Infectious diseases management
  amplicon sequencing, 583, 585
  antimicrobial susceptibility, 581
  epidemiologic studies, 583
  germ-line disorders, 581, 582
  microbiology and virology, 582
  microorganisms, 581
  NGS (*see* Next-generation sequencing (NGS))
  novel viral pathogens, 588–590
  nucleic acid fragmentation, 583, 583
  viral genome diversity, 590–592
  WGS, 583, 583
Inflammatory myofibroblastic tumor, 510–511
Informatics system requirements, 262, 274
Inherited genetic disorders
  adult/young adult onset applications, 545–546
  aortopathies, 537
  cell free fetal DNA, in maternal plasma, 540–541
  "diagnostic odyssey", 535
  exome and genome, 546–548
  Miller syndrome, 537
  multi-gene panels, 536
  newborn/infant applications, 542–545
  next-generation sequencing (NGS) tests, 535–536
  Noonan syndrome (NS), 542
  prenatal applications, 539–540
  Proteus syndrome, 536
  retinitis pigmentosa (RP), 535
  RHO (rhodopsin) gene mutation, 535
  Sanger sequencing, 536–537
  skeletal dysplasia, 541–542
  variants of uncertain significance (VUS), 539
  WES tests, 537–539
  WGS, 537–539
In Re Kubin, 133–134
In silico predictors
  computational scoring metric, 193, 194
  gene variants, 193
  genotype-phenotype outcomes, 193
  monogenic diseases, gene mutations, 192
  MutPred, 192
  PMut, 192
  PolyPhen and SIFT, 192
In situ hybridization (ISH), 49
  expression profiling, 54, 55
  HPV, HNSCC, 347
In-solution sequence capture, 210, 211
Institutional Review Board (IRB), 114
Internal tandem duplication (ITD) mutation, 300–302
International Cancer Genome Consortium (ICGC), 476
Interpretation and reporting, 283, 285, 287–289
  ACMG, 278
  analytical and clinical, 278
Intratumoral heterogeneity, in HGSOC
  "actionable" mutation, 495
  comparative genomic hybridization (CGH) array, 495
  epithelial tumors, 495
  genetic heterogeneity, 495
  in hematological cancers, 495
  leukemic blasts, 495
  metastatic sites, 495
  platinum sensitivity, 495
  sensitive and resistant subclones, 495
  sequential biopsy and plasma studies, 496
  single nucleotide polymorphism (SNP), 495
  tumor DNA circulation, 496–497
In vitro diagnostic (IVDs), 122
In vitro fertilization (IVF), 574
Ion-sensitive field-effect transistor (ISFET), 5, 6
Ion Torrent Personal Genome Machine (PGM), 7, 187, 199, 202, 314, 593
Ion Torrent™ technology
  base calling module, 184–185
  Browser's reports, 186
  Data Acquisition (DAT) data files, 184
  Ion Reporter^{T} software, 188–189
  nucleotide incorporation detection, 184
  pipeline processing, 184, 185
  realignment, 187
  sequencing chip, 184
  software approach, 187
  third-party software, 189–190
  TMAP, 185, 186
  Torrent Suite Software v3., 188
  TVC, 188

IRB. *See* Institutional Review Board (IRB)
ISFET. *See* Ion-sensitive field-effect transistor (ISFET)
ISH. *See* In situ hybridization (ISH)
ISHB. *See* Index-specific hybridization blockers (ISHB)
ITD mutation. *See* Internal tandem duplication (ITD) mutation
IVDs. *See* In vitro diagnostic (IVDs)
IVF. *See* In vitro fertilization (IVF)

J
Janus kinase 2 (JAK2) mutations, 300, 301

K
Kidney cancer
  copy number changes, 452
  inherited syndromes and single gene mutations, 452–453
  PBRM1, 451–452
  treatment response, molecular prediction, 453–454
  VHL gene, 451
  whole-genome and exome sequencing, 453
KRAS, TKI treatment and therapeutic options, 385–386
KSR Int'l Co. v. Teleflex Inc., 133
Kyoto Encyclopedia of Genes and Genomes (KEGG), 265–266

L
Laboratory developed tests (LDTs), 116, 143
  CLIA, 121
  diagnostic tests, 122
  FDA, 122
  IVDs, 122
  medical advancement, 121
Laboratory information management systems (LIMS)
  CLIA laboratory, 261–262, 264
  clinical and genomic LIS, 262, 263
  commercial clinical, 261
  institutions, 261
  NGS section, 264
  non-CLIA, 261
  testing pathways and informatics system requirements, 261, 262
  workflow process, 254
Laboratory information systems (LIS)
  CDSS, 274
  CLIA laboratory, 261–262, 264
  complex molecular tests, 270
  consents, 268
  data integration and reporting, 270, 271
  genomic testing, 266
  vs. LIMS (*see* Laboratory information management systems (LIMS))
Laser capture microdissection (LCM), 49
Law
  American case law, 151
  enforcement, 145
  genetic discrimination, 145
  GINA, 145
  HIPAA, 145
LCM. *See* Laser capture microdissection (LCM)
LDTs. *See* Laboratory developed tests (LDTs)
Leukemia, 37, 39, 310, 495
  AMLs, 300
  chronic lymphocytic, 309, 310
  diagnostic chromosomal abnormalities, 306
  lymphomas, 298–299
  pediatric acute, 315
  pre-B-acute lymphoblastic, 299–300
  pre-T-ALL, 299
  T-cell, 299
Ligation master mix components, 216
LIMS. *See* Laboratory information management systems (LIMS)
LINEs. *See* Long interspersed elements (LINEs)
LIS. *See* Laboratory information systems (LIS)
Locus specific indicators (LSI), 306
Logical Observation Identifiers Names and Codes (LOINC), 265, 269
LOINC. *See* Logical Observation Identifiers Names and Codes (LOINC)
Long interspersed elements (LINEs), 211
Low-grade fibromyxoid sarcoma, 511–512
LSI. *See* Locus specific indicators (LSI)
Lung cancer pathology
  amplicon clones, 391
  development, 400–401
  DNA isolation, 393–394
  fluorescence in situ hybridization (FISH), 397–398
  fluorochrome-labelled nucleotides, 391
  fragment length analysis, 397
  high-resolution melting (HRM), 395–397
  ion semiconductor sequencing system, 391, 392
  methods, 391–393
  MOSFET (metal–oxide–semiconductor field-effect transistor) flow cell, 391
  mutation analyses protocols, 394–395
  operation workflow, diagnostics, 398–399
  Sanger sequencing, 395
  selected protocols, 400
Lymphoma, 47, 167, 388
  Burkitt lymphoma, 306–308
  Hodgkin lymphoma, 299
  leukemia, 306
  T-cell, 299, 300
Lynch syndrome, 423
  endometrial carcinoma, 471–472
  mismatch repair defective colorectal cancer, 424
  oncogene-induced senescence (OIS) colorectal cancers, 418–419

M
MAF. *See* Minor allele frequency (MAF)
MammaPrint®, 366–367
MapQuant Dx™, 367

Massively parallel sequencing and intra-tumor genetic heterogeneity
ER-alpha and HER2, 371–372
genetic aberrations collection, 371
genetic diversity, 373
human cancers, genomic analyses, 372–373
tumor heterogeneity, 372
Maternal plasma, DNA
DYS14, 565
fetal Rh status, 566
fetal sex, 565–566
mosaicism, 573–574
serum, 565
sex aneuploidies, 572–573
trisomy 18 and 15, 570–572
Mayo v. Prometheus, 134–135
MCC. *See* Merkel cell carcinoma (MCC)
MDS. *See* Myelodysplastic syndrome (MDS)
Medically important pathogens, 595
Medulloblastomas
expression profiling, 330
genomic features, germ-line mutation and somatic loss, 331
Group 563, 332
Group 66, 332–333
molecular classification, 329
Shh group, 331–332
subtypes, 329
Turcot and Gorlin syndromes, 329
Wnt group, 330–331
Mendelian disorders, 230–231
Meningioma, 335–337
Merkel cell carcinoma (MCC), 589
Mesenchymal GBM, 334
Metagenomics
bioinformatics, 595
human host, 594
shotgun sequencing, 594
Methicillin-resistant *Staphylococcus* aureus (MRSA), 594
MGMT gene, brain tumors, 327–328
Microarrays
affymetrix, 58
hybridization reaction, 56
intraplatform and interplatform, 55
reproducibility, 55
target DNA and RNA, 54
Microbiology
bacterial gene expression, 38
pathogenomics, 38
rRNA gene sequencing, 38
small RNAs (sRNAs), 38
UTRs, 38
MicroRNAs (miRNAs)
antitumoral effect, 63
bioinformatics analysis, 64
biological samples, 48–50
cardiac hypertrophy, 63
cell/tissue lysis, 50
data processing and normalization, 58, 62
expression profiles, 53–58
extraction protocols, 50–52
gene interaction, 62–63
GenePix Pro software program, 62
human disease, 46–48
lung cancer cells, 64
magnetic beads, 51
mimics and antagomirs, 64
noncoding RNAs (ncRNAs), 45
oligonucleotides, 63
physiological and pathological processes, 45
qRT-PCR, 58
quality and quantity control, 52–53
SPE, 50, 51
statistical tools, 62
TRIzol/TRIReagent, 50, 53
Microsatellite instability (MSI), 469
Minor allele frequency (MAF), 555
miRNAs expression assays. *See* MicroRNAs (miRNAs)
Mismatch repair defective colorectal cancer
characteristics, 424, 425
hereditary non-polyposis colon cancer (HNPCC), 423
immunohistochemistry (IHC)-based screening strategies, 423
Lynch syndrome, 424
MMR-deficient (dMMR), 423
multitest screening algorithm strategy, 424
sporadic dMMR colorectal adenocarcinoma, 423
Mismatch repair defective, evaluation, 423–424
Mitochondrial disorders, 544–545
MLS. *See* Myxoid liposarcoma (MLS)
MM. *See* Multiple myeloma (MM)
Molecular classification, breast cancer
androgen receptor (AR) and AR-related genes, 361
claudin-low subtype, 361
ER-positive and ER-negative, 360
"intrinsic" classification, 361
"intrinsic" molecular subtypes, 360
luminal tumors, 360
METABRIC study, 363
molecular subtypes, 362
multiple microarray-based classification, 363
PAM50 assay, 361
PI3K and mTOR inhibitor, 363
single nucleotide polymorphism (SNP) arrays, 363–364
single sample predictors, 361
TNBC classification system, 363
triple-negative breast cancers (TNBCs), 361, 363
MRSA. *See* Methicillin-resistant *Staphylococcus* aureus (MRSA)
Multigene panels, prostate cancer, 442–443
Multiple myeloma (MM), 47
Multiplex ligation dependent probe amplification (MLPA), 322
Mutation, 101, 78, 92, 225
APC, 87
BRAF, 87
cancer DNA, 86

PIK3CA, 87
proto-oncogenes, 87
TP53, 87
Mutation analyses protocols
androgen receptor, 80
EGFR mutations, 79
epigenomic and miRNA, 80–81
flowchart, 394
KRAS mutations, 79
PCR amplification, 394
pyrosequencing, 394
shifted termination assays (STA), 394
transcription profiles, 80
tumor cells, 79
web-based resources, 394–395
MUTYH/MYH colon cancer, 420
Myelodysplastic syndrome (MDS), 300, 306, 309, 311
Myxoid liposarcoma (MLS), 506–507

N

NAIL. *See* Natural Killer Cell Activation Inducing Ligand (NAIL)
NAs. *See* Nucleotide analogues (NAs)
National Comprehensive Cancer Network (NCCN) guidelines, 425
National Human Genome Research Institute (NHGRI), 4
Natural Killer Cell Activation Inducing Ligand (NAIL), 133–134
Neoplastic diseases, 231–232
Neural GBM, 334
Neuroendocrine neoplasms, 428
Newborn/infant applications, inherited genetic disorders
epilepsy, 543–544
hearing loss, 542–543
mitochondrial disorders, 544–545
newborn screening programs, 542–543
Next-generation sequencing (NGS), 58–61, 87, 111, 125, 163, 314–316, 577
aCGH, 35
ACMG clinical laboratory, 243
adenylate end-repaired DNA, 216–217
Agilent SureSelect, 210
alignment algorithms, 180
amplification, 220–221
amplify pre-capture library, 217
assay design and considerations, 210–212
bacteriology, 592–594
bead preparation, 219–220
benchtop instruments, 7
bioinformatics, 9, 178
capture hybrid library, 220
characteristics, 314
colorectal cancer (*see* Colorectal cancer)
computing and bioinformatics infra structure, 257
cost, 203–204
CRT, 203
data analysis, 222
diagnostic assays, 35
diagnostic virology (*see* Drug resistance mutations (DRM))
DNA sequencing, 209
equipment, 212, 214
exome and genome sequencing, 209–210, 226
facility, 254
fetal trisomy, 102
first generation sequencing, 197
gene-by-gene diagnostic odysseys, 209
gene panels, 209, 210
genomic DNA, 314
hematolymphoid malignancy, 314–316
hybridization, 218–219
illumina sequencing, 222
illumina/solexa genome analyzer, 202
information technology, 209
in-solution hybridization-based technologies, 210
Ion Torrent PGM, 202
laboratories, 4–5, 210
library quantification by qPCR, 221–222
mapping, 245
materials, 212–214
microbiology, 38, 103
molecular diagnostic laboratories, 209
mutation detection, 9
mycology, 595
PCR, 210
platform chemistry, 5, 6
post-hybridization library cleanup, 220
pre-capture library assessment, 218
processes, quality metrics, 244–245
purify amplified post-capture libraries, 221
purify amplified pre-capture libraries, 217–218
pyrosequencing, 203
quality control analysis, 221
reagent and oligonucleotide preparation, 214
RNA-Seq, 32
Roche GS FLX Genome Sequencer, 202
sequencing by ligation, 203
shear DNA, 215
signals transformation, 245–246
single-gene testing, 198–202, 222–223
software, 189
SOLiD, 4
soLiD genetic analyzer, 202
staff, 253–254
stringency washes, 220
T4 polynucleotide kinase and DNA polymerase, 215–216
workflow process, 257
NGS. *See* Next-generation sequencing (NGS)
NHGRI. *See* National Human Genome Research Institute (NHGRI)
NIPD. *See* Non-invasive prenatal diagnostic (NIPD)
Non-invasive prenatal diagnostic (NIPD), 565
Non-small-cell lung cancers (NSCLC), 167, 171
Noonan syndrome (NS), 542
Noonan Syndrome with Multiple Lentigines (NSML), 542

NOTCH
HNSCC, 343–345
pathway inhibitors, 351–352
Novel viral pathogens
amplicon sequencing, 589
arenavirus, 589
contigs, 589
enterovirus, 589
Koch's postulates, 590
MCC, 589
microarrays targeting, 588
oligonucleotides, 588
PCR, 589
phylogenetic analysis, 591
454 pyrosequencing, 588
NS. *See* Noonan syndrome (NS)
NSCLC. *See* Non-small-cell lung cancers (NSCLC)
Nucleotide analogues (NAs), 586

O
Office of Public Health Genomics (OPHG), 172
Oligonucleotides, 4, 8, 20, 21, 25, 26, 31, 33, 63, 92, 200, 201, 214, 250, 311–313, 348, 580
OMIM. *See* Online Mendelian Inheritance in Man database (OMIM)
Oncogene-induced senescence (OIS) colorectal cancers
CpG island methylator phenotype (CIMP), 419–420
Lynch syndrome, 418–419
microsatellite defects, 419
microsatellite instability-high (MSI-H), 419
MLH1 gene, 419–420
Oncology, 37, 38
ONCOMINE platform, 163, 164
Oncotype Dx®, 368
Online Mendelian Inheritance in Man database (OMIM), 265
Operational taxonomic units (OTUs), 593
OPHG. *See* Office of Public Health Genomics (OPHG)
Order entry, genomic tests, 266–267
OTUs. *See* Operational taxonomic units (OTUs)

P
PacBio RS system, 23
PAM50, 368–369
Pancreatic cancer
cystic lesions, 428
neuroendocrine neoplasms, 428
pancreatic cystic lesions, 428
undifferentiated carcinomas, 427–428
Parallel sequencing technologies
bait hybridization, 7
benign polymorphisms, 12
bioinformatics, 9
Celera project, 3
cystic fibrosis screening, 12
gene patents, 14
genomic assay, 11
Haloplex system, 7
haploid genome, 3
HGMD, 123
incidental findings, 13–14
microfluidic platform, 88
mutation detection, 9, 14
NGS analysis, 10–11
platform chemistry, 4–5
polymorphisms, 14
QA, 9
rapid sequencing revolution, 5–7
shotgun sequencing, 3
SNPs, 123
VCF, 123
VUS, 12
WES, 8
wet bench process, 9
WGS, 8
PARE. *See* Personalized analysis of rearranged ends (PARE)
Pathology. *See* Genomic pathology
PCR. *See* Polymerase chain reaction (PCR)
Personal Genome Machine (PGM), 7
Personalized analysis of rearranged ends (PARE)
bioinformatics tools, 93
cancer DNA, 92, 93
CML, 92
genomic rearrangements, 93
ptDNA measurements, 92, 93
qPCR assays, 92
sepsis, 94
somatic rearrangements, 94
WES, 92
PFS. *See* Progression free survival (PFS)
PGM. *See* Personal Genome Machine (PGM)
Pharmacogenomics
adverse drug reactions, minimizing, 558
biological mechanisms, 559
candidate gene study designs, 554
challenges, 559
DNA methylation studies, 556
in drug discovery, 558–559
drug efficacy determination, 557–558
genome-wide association study (GWAS) designs, 554–555
goal, 553
RNA sequencing (RNA-seq) studies, 556
study population choice, 556–557
vaccinomics application, 558
WES study designs, 555–556
WGS study designs, 556
PIK3CA gene
alterations and signalling, 387–386
HNSCC, 346
Platform chemistry, sequencing
beads, 5, 6
clonal clusters, 5, 6
CMOS, 5

DNA sequencing, 4
flow cell, 5, 6
fluorescence emittance, 5
ISFET sensor, 5, 6
NGS platforms, 4
NHGRI, 4
nucleotides, 5, 6
pyrosequencing, 4
SOLiD, 4
"Platinum-sensitive" relapsed disease, 493
Pleomorphic liposarcoma, 507
Polymerase chain reaction (PCR), 20, 86, 90, 102, 167–168, 230, 347, 426, 442, 556
Polynucleotides, 25, 215
Post-capture library amplification thermal cycling potocol, 221
Pre-analytical considerations, genome sequencing
ACMG clinical laboratory, 243
CLSI publications, 243
WGS, 243, 246
Pre-capture library amplification
components, 217
thermal cycling protocol, 220
Predictive markers, targeted therapeutics
BRAF (B-type RAS associated factor), 426
laboratory developed tests (LDTs), 426
Sanger sequencing, 426
targeted anti-EGFR treatment, 425–426
"targeted" therapeutic strategies, 426–427
Prenatal screening
biochemical and ultrasound measurements, 564
and diagnosis, 565
Down syndrome, 564
fetal disorders, 578
Proficiency, 116–117
Progression free survival (PFS), 492
Proneural GBM, 334
Prostate cancer
alpha-methylacyl-CoA racemase (AMACR), 438
copy number changes, 443
ERG gene fusions, FISH and IHC, 436, 437, 439
ETS rearrangements, 436
genomic/transcriptomic alterations, early detection, 445–446
Gleason grading system, 436
hereditary genome applications, 441–442
HGPIN lesions, 438
molecular subtypes, 441
multigene panels, 442–443
RAS/RAF/FGFR family fusions, 440–441
sequencing, advanced, 443–445
SPINK1/SPOP/CHD1, 438–440
TMPRSS2-ERG rearrangement, 436–438
tumor grade, 436
whole-genome and exome sequencing, 443
Protein nanopores, 24
α-hemolysin, 25
oligonucleotide, 26
polynucleotides, 25
Staphylococcus aureus, 25
Pulmonary malignancies
ALK, *RET*, and *ROS1*, chromosomal inversions and translocations, 388–390
BRAF mutation, activation and inactivation, 386
EGFR, targeted therapy response and resistance, 383–385
FGFR1 and *DDR2*, in squamous cell carcinoma, 390
HER2, amplification and mutation, 386–387
KRAS, TKI treatment and therapeutic options, 385–386
MET activation, 388
NGS to molecular lung cancer pathology (*see* Next-generation sequencing (NGS))
PIK3CA, alterations and signalling, 387–386

Q

QA. *See* Quality assurance (QA)
qRT-PCR. *See* Quantitative reverse transcription-polymerase chain reaction (qRT-PCR)
Quality and quantity control, miRNA
NanoDrop/Bioanalyzer, 51, 52
nucleic acids, 52
ribosomal RNA, 52
RNA integrity, 52
Quality assurance (QA), 101, 126
Quantitative reverse transcription-polymerase chain reaction (qRT-PCR), 58

R

Rapid sequencing revolution
Benchtop sequencers, 5
DNA sequencing, 4, 5
HiSeq, 7
MiSeq platform, 7
NGS, 5
NHGRI, 4, 5
PGM, 7
RAS gene, HNSCC, 345–346
Real-time quantitative PCR (RQ-PCR), 300, 303
Receptor tyrosine kinase (RTK) genes, 325–326
Reimbursement, genomic testing
AMA, 124
analyte specific codes, 125
CMS, 124
CPT, 124
molecular diagnostic tests, 124
NGS, 125
QA, 126
sequencing assays, 126
Renal cell carcinomas (RCC). *See* Kidney cancer
Reporting
Browser's Reports, 186
Ion Reporter™, 188–189
TMAP, 187
Reporting recommendations for tumor MARKer prognostic studies (REMARK), 171

Research
clinical genetic testing, 146
"Global Alliance", 150–151
SNP chips, 147
WGS, 154
Reverse transcription polymerase chain reaction (RT-PCR), 584
Ribosomal multilocus sequence typing (rMLST), 593
Ribosomal RNA (rRNA) sequencing, 38
amplicon, 592, 596
Human Microbiome Project, 592
metagenomic data, 594
NGS, 593
rMLST. *See* Ribosomal multilocus sequence typing (rMLST)
RNase block dilution, 219
RNA sequencing (RNA-seq)
aCGH, 35
cDNA, 33
deoxyadenine, 33
DRS, 35
endothelial progenitor cells, 36
eQTL, 36
FDA, 35
gene expression profile, 31, 35
GWAS, 36
human genome, 35
hybridization, 32
leukemias and lymphomas, 32
microarray, 31
NGS, 32
nucleotide sequence, 32
signal intensity, 31
Roche GS FLX Genome Sequencer, 202
RQ-PCR. *See* Real-time quantitative PCR (RQ-PCR)
RT-PCR. *See* Reverse transcription polymerase chain reaction (RT-PCR)

S
SAGE. *See* Serial analysis of gene expression (SAGE)
Sanger sequencing, 32, 124, 197, 203, 204, 226, 395, 426, 536, 544, 588
SBS. *See* Sequencing by synthesis (SBS)
Scanning transmission electron microscopy (STEM), 23, 24
Sclerosing epithelioid fibrosarcoma, 512
Second-generation prognostic gene signatures
immune-related signatures, 370–371
stromal gene signatures, 370
Sequence variants, 291
HGMD, 279
interpretation, 278
Sequencing. *See also* Genome sequencing (GS) assay
bioinformatics analysis, 58
exome and genome, 190–191
human chromosome 13q31, 47
illumina, 180
Ion Torrent, 184–190
NGS, 178
northern blotting, 53
qRT-PCR, 54
Sequencing by Oligonucleotide Ligation and Detection (SOLiD), 4, 32, 202
Sequencing by synthesis (SBS), 32
Sequencing technologies, 5, 19–21, 58, 108, 165, 210, 225, 234, 360, 422, 446, 469, 555, 581
protein nanopores, 24–26
single-molecule, 19–21
SMRT®, 21–23
solid-state nanopore, 24–26
Serial analysis of gene expression (SAGE), 164
Sex aneuploidies
fetal sex, 572, 573
nuchal translucency, 573
SNP methodology, 573
Turner syndrome, 572
Short interspersed elements (SINEs), 211
SIFT. *See* Sorting intolerant from tolerant (SIFT)
Silver-enhanced in situ hybridization (SISH), 167
SINEs. *See* Short interspersed elements (SINEs)
Single gene analysis
archeological and mitochondrial studies, 201
clinical applications, 201–202
DNA-based single gene assays (*see* DNA-based single gene assays)
human leukocyte antigen (HLA), 200–201
infectious diseases, 199–200
inherited diseases, 200
oncologic testing, 198–199
residual disease detection, 198
RNA-based single gene assays, 303
Single-molecule real-time (SMRT®) sequencing, 22, 23, 33, 234
FRET, 23
phosphate chain, 22
ZMW, 21
Single nucleotide polymorphism (SNP), 123, 19, 101, 104, 145, 147, 149, 305, 309, 312, 313, 316, 573
abnormalities in EBF1 and PAX5 genes, 311
affymetrix technology, 311
bead-based, 311, 312
buccal mucosal epithelial cell, 312–313
chip-based, 313
CNVs, 312
genomic DNA, 311
genomic loci, 311
hematologic malignancies, 311, 314
MAX pathway, 311
medicine, 311
single allele, 311
Single-nucleotide variants (SNVs), 115, 278, 285
SISH. *See* Silver-enhanced in situ hybridization (SISH)
Skeletal dysplasia, 541–542
Skeletal muscle tumors, 513–514
Smooth muscle tumors
fumarate hydratase (FH), somatic mutations, 473
genomic aberrations, 473, 474

immunohistochemical makers, 472–473
leiomyoma, 472
leiomyosarcomas, 472, 473
MED12 mutations, 473
MIB1 (Ki-67) expression, 473
SMRT®. *See* Single-molecule real-time (SMRT®) sequencing
SNOMED-CT. *See* Systematized Nomenclature of Medicine-Clinical Terms (SNOMED-CT)
SNP. *See* Single nucleotide polymorphism (SNP)
SNVs. *See* Single-nucleotide variants (SNVs)
Soft tissue sarcomas
adipocytic tumors, 504–507
alveolar soft part sarcoma, 517
clear cell sarcoma, 517–518
desmoplastic small round cell tumor, 519
epithelioid sarcoma, 516–517
extrarenal rhabdoid tumor, 519
extraskeletal myxoid chondrosarcoma, 518–519
fibroblastic/myofibroblastic tumors, 507–512
fibrohistiocytic tumors, 512–513
gastrointestinal stromal tumors, 520–523
skeletal muscle tumors, 513–514
synovial sarcoma, 516
vascular tumors, 514–516
Software
alignment, 186
ANNOVAR, 184
calling variants, 187
chastity filter, 179
DNASTAR (SeqMan) and SoftGenetics (NextGENe), 189
GATK, 184
GATK and SAMtools, 180
heuristic and probabilistic approaches, 190–191
Ion Reporter™, 188–189
Partek and Avadis NGS, 189
SIFT, Polyphen2 and Mutation Taster, 184
snpEff, 184
TMAP, 185–186
Torrent Suite Software v3.4, 188
TVC, 187–189
SOLiD. *See* Sequencing by Oligonucleotide Ligation and Detection (SOLiD)
Solid-phase extraction (SPE), 50, 51
Solid-state nanopore, 26–27
carbon nanotubes, 26
DNA transistor, 27
Solitary fibrous tumor, 509–510
Sorting intolerant from tolerant (SIFT), 184, 190, 192
SPE. *See* Solid-phase extraction (SPE)
Squamous cell carcinoma (SCC), 454–455
Staphylococcus aureus, 25, 38, 583
STEM. *See* Scanning transmission electron microscopy (STEM)
Stromal gene signatures, 370
SureSelect adapter block mix, 219
Synovial sarcoma, 516
Systematized Nomenclature of Medicine-Clinical Terms (SNOMED-CT), 269

T
Tagged-amplicon deep sequencing (TAm-Seq), 496
TAm-Seq. *See* Tagged-amplicon deep sequencing (TAm-Seq)
Target enrichment, 210, 211, 215, 219
TBC. *See* The Biomarkers Consortium (TBC)
TDR. *See* Transmitted drug resistance (TDR)
Tenosynovial giant cell tumor, 512–513
The Biomarkers Consortium (TBC), 171
The Cancer Genome Atlas (TCGA), 469–470
TMAP. *See* Torrent Mapping Alignment Program (TMAP)
TMUGS. *See* Tumor Marker Utility Grading System (TMUGS)
Torrent Mapping Alignment Program (TMAP), 185–187
Torrent Variant Caller (TVC), 187–188, 190
TP53, HNSCC
prognostic applications, 348
tumor-suppressor gene, 343
Training Residents in Genomics (TRIG), 107
Transcriptome sequencing
benign and malignant, 31
cancer, 36–37
clinical microbiology, 38
cloud computing, 35
de novo assembly, 33
genome, 33, 34
Illumina platform, 32
Ion Torrent technology, 33
nucleotides, 32
protein coding RNA, 31
RNA-seq (*see* RNA sequencing (RNA-seq))
SBS, 32
SOLiD, 32
Translational research, 169–170
Transmitted drug resistance (TDR), 584
TRIG. *See* Training Residents in Genomics (TRIG)
Trisomy, Down syndrome
detection rates, 571
false positive rates, 571, 572
fetal fractions, 570
Tumor Marker Utility Grading System (TMUGS), 171
TVC. *See* Torrent Variant Caller (TVC)
Tyrosine (Y)-methionine (M)-aspartic acid (D)-aspartic acid (D) (YMDD), 585–586

U
UMLS. *See* Unified Medical Language System (UMLS)
Undergraduate medical education, 108
Undifferentiated carcinomas, pancreatic cancer, 427–428
Unified Medical Language System (UMLS), 269
The United States Patent and Trademark Office (USPTO), 130
Untranslated regions (UTRs), 38
The US Preventive Services Task Force (USPSTF), 172, 173

USPSTF. *See* The US Preventive Services Task Force (USPSTF)
USPTO. *See* The United States Patent and Trademark Office (USPTO)
Uterine mesenchymal tumors
  endometrial stromal tumors, 473–476
  smooth muscle tumors, 472–473
Utility patent, 129
UTRs. *See* Untranslated regions (UTRs)

V
Vaccinomics, 558
Valganciclovir (VCV), 587
Variant Call Format (VCF), 123, 181, 264, 265
Variants of uncertain clinical significance (VUS), 12
Variants of unknown significance (VUS), 278
Vascular tumors, 514–516
VCF. *See* Variant Call Format (VCF)
Veridex 76-gene signatures, 367–368
Verification, 116
Viral discovery, 589
Viral genome diversity
  CMV glycoproteins, 591
  HIV, 591
  influenza A, 591
  intrahost diversity, 590
  phylogenetic and Bayesian analyses, 591
  quasispecies, 590
  rapid viral replication, 590
  WGS, 592
VUS. *See* Variants of uncertain clinical significance (VUS); Variants of unknown significance (VUS)

W
Well-differentiated liposarcomas (WDLS), 504–506
WES. *See* Whole-exome sequencing (WES)
WGS. *See* Whole genome sequencing (WGS)
Whole-exome sequencing (WES), 279, 280, 285, 420, 537–539, 547, 548
  analysis and detection, 226, 227
  challenges, 233–234
  clinical care, 232–233
  clinical diagnostic, 232
  comprehensive gene panels, 235
  diagnostic assay-proven potential, 228–229
  and exome capture, 226, 227
  facilitates gene discovery, 229
  gene panels, 235
  heterozygous mutation, *CPSI* gene, 233
  homozygous mutations, *SMOC2* gene, 231
  implementation, 234
  incidental findings, 13
  medical genetics to medical genomics, 230
  molecular diagnosis, 227–229
  multifactorial and multigenic complex disease, 231
  neoplastic diseases, 231–232
  and NGS, 226, 227
  panel assays, 8
  protein-coding genes, 8
  study designs, 555–556
  third-generation sequencing technologies, 234
Whole genome sequencing (WGS), 13, 141, 154, 279, 280, 285, 593
  clinical laboratory, 246
  considerations, 243
  cost and time investment, 250
  microarray testing, 244
  process and workflow, 242
  prohibitive cost and analysis, 232
  resources and infrastructure, 254–255
  screening tests, 251
  stage processes, quality metrics, 244–245
  test definition, 243, 243–244
  tumors molecular profile, 250

X
XIAP. *See* X-linked inhibitor of apoptosis gene (XIAP)
X-linked inhibitor of apoptosis gene (XIAP), 102

Z
Zero-mode waveguide (ZMW), 23
  DNA sequencing applications, 21–22
  fluorescent signals, 22
  laser energy, 22

FSC
www.fsc.org
MIX
Papier | Fördert gute Waldnutzung
FSC® C083411

Kolibri 360 GmbH
Ferdinand-Jühlke-Straße 7
99095 Erfurt, Deutschland
produktsicherheit@kolibri360.de